MW00754247

Harper's Illustrated Biochemistry

Twenty-Eighth Edition

Robert K. Murray, MD, PhD

Professor (Emeritus) of Biochemistry
University of Toronto
Toronto, Ontario

David A. Bender, PhD

Sub-Dean
University College Medical School
Senior Lecturer in Biochemistry
Department of Structural and Molecular
Biology and Division of Medical Education
University College London

Kathleen M. Botham, PhD, DSc

Professor of Biochemistry
Royal Veterinary College
University of London

Peter J. Kennelly, PhD

Professor and Head
Department of Biochemistry
Virginia Polytechnic Institute and State
University
Blacksburg, Virginia

Victor W. Rodwell, PhD

Emeritus Professor of Biochemistry
Purdue University
West Lafayette, Indiana

P. Anthony Weil, PhD

Professor of Molecular Physiology and
Biophysics
Vanderbilt University School of Medicine
Nashville, Tennessee

Medical

New York Chicago San Francisco Lisbon London
Madrid Mexico City Milan New Delhi San Juan
Seoul Singapore Sydney Toronto

The McGraw·Hill Companies

Harper's Illustrated Biochemistry, Twenty-Eighth Edition

1 2 3 4 5 6 7 8 9 10 CTP/CTP 12 11 10 9

ISBN 978-0-07-162591-3
MHID 0-07-162591-7
ISSN 1043-9811

Notice

This book was set in Minion Pro by Newgen Publishing and Imaging Services.
The editors were Michael Weitz and Kim J. Davis.
The production supervisor was Sherri Souffrance.
Project management was provided by Newgen Publishing and Imaging Services.
The text designer was Elise Lansdon.
The cover art director was Margaret Webster-Shapiro.
China Translation & Printing Services Ltd. was the printer and binder.

This book is printed on acid-free paper.

Cover: Green fluorescent protein molecule. Computer model showing the secondary structure of a molecule of green fluorescent protein (GFP). Some central atoms are represented as rods. The molecule has a cylindrical structure formed from beta sheets (ribbons). GFP, which fluoresces green in blue light, is widely used as a research tool in biology and medicine. The gene that encodes GFP can be fused to genes that encode a previously invisible target protein to facilitate study of its movement inside intact cells, and to tag cancer cells to track their spread through the body. *Credit:* Laguna Design/Photo Researchers, Inc.

International Edition ISBN 978-0-07-163827-2; MHID 0-07-163827-X

Co-Authors

Daryl K. Granner, MD

Emcritus Professor of Molecular Physiology and Biophysics and Medicine, Vanderbilt University, Nashville, Tennessee

Peter L. Gross, MD, MSc, FRCP(C)

Associate Professor, Department of Medicine, McMaster University, Hamilton, Ontario

Frederick W. Keeley, PhD

Associate Director and Senior Scientist, Research Institute, Hospital for Sick Children, Toronto, and Professor, Department of Biochemistry, University of Toronto, Toronto, Ontario

Peter A. Mayes, PhD, DSc

Emeritus Professor of Veterinary Biochemistry, Royal Veterinary College, University of London, London

Margaret L. Rand, PhD

Associate Senior Scientist, Hospital for Sick Children, Toronto, and Professor, Departments of Laboratory Medicine & Pathobiology and Biochemistry, University of Toronto, Toronto, Ontario

New Key Features in This Edition!

- Influence of the Human Genome Project on various biomedical fields
- Revision of the use of enzymes in medical diagnosis
- New material on computer-aided drug discovery
- Compilation of some conformational diseases
- New material on advanced glycation end-products and their importance in diabetes mellitus
- New material on the attachment of influenza virus to human cells
- Some major challenges facing medicine

Important biochemical updates include:

- Expanded coverage of mass spectrometry, a key analytical method in contemporary biochemistry
- New figures revealing various aspects of protein structure
- Expanded coverage of active sites of enzymes and transition states
- New information on methods of assaying enzymes
- Expanded coverage of aspects of enzyme kinetics
- New information on micro and silencing RNAs
- New information on eukaryotic transcription mechanisms, including the biogenesis of micro RNA and the role of nucleosomes
- Description of activities of micro RNAs
- New material on next generation sequencing (NGS) platforms
- New material on chromatin immunoprecipitation (CHIP) technology and its uses
- New information on subcellular localization of key signaling enzymes (kinases, phosphatases)
- New information on how hormones affect gene transcription

Full-color presentation, including 600+ illustrations

Easy-to-read tables provide great content in a succinct manner

Every chapter opens with a Summary of Biomedical Importance

Chapters conclude with chapter summaries

Two all-new chapters: "Free Radicals and Antioxidant Nutrients" and "Biochemical Case Histories" which offers an extensive presentation of 16 pathophysical conditions

Contents

Preface   ix

1. **Biochemistry & Medicine**
   Robert K. Murray, MD, PhD   1

2. **Water & pH**
   Peter J. Kennelly, PhD & Victor W. Rodwell, PhD   6

SECTION **I**

STRUCTURES & FUNCTIONS OF PROTEINS & ENZYMES   14

3. **Amino Acids & Peptides**
   Peter J. Kennelly, PhD & Victor W. Rodwell, PhD   14

4. **Proteins: Determination of Primary Structure**
   Peter J. Kennelly, PhD & Victor W. Rodwell, PhD   21

5. **Proteins: Higher Orders of Structure**
   Peter J. Kennelly, PhD & Victor W. Rodwell, PhD   31

6. **Proteins: Myoglobin & Hemoglobin**
   Peter J. Kennelly, PhD & Victor W. Rodwell, PhD   43

7. **Enzymes: Mechanism of Action**
   Peter J. Kennelly, PhD & Victor W. Rodwell, PhD   51

8. **Enzymes: Kinetics**
   Peter J. Kennelly, PhD & Victor W. Rodwell, PhD   62

9. **Enzymes: Regulation of Activities**
   Peter J. Kennelly, PhD & Victor W. Rodwell, PhD   75

10. **Bioinformatics & Computational Biology**
    Peter J. Kennelly, PhD & Victor W. Rodwell, PhD   84

SECTION **II**

BIOENERGETICS & THE METABOLISM OF CARBOHYDRATES & LIPIDS   92

11. **Bioenergetics: The Role of ATP**
    Kathleen M. Botham, PhD, DSc & Peter A. Mayes, PhD, DSc   92

12. **Biologic Oxidation**
    Kathleen M. Botham, PhD, DSc & Peter A. Mayes, PhD, DSc   98

13. **The Respiratory Chain & Oxidative Phosphorylation**
    Kathleen M. Botham, PhD, DSc & Peter A. Mayes, PhD, DSc   103

14. **Carbohydrates of Physiologic Significance**
    David A. Bender, PhD   113

15. **Lipids of Physiologic Significance**
    Kathleen M. Botham, PhD, DSc & Peter A. Mayes, PhD, DSc   121

16. **Overview of Metabolism & the Provision of Metabolic Fuels**
    David A. Bender, PhD   131

17. **The Citric Acid Cycle: The Catabolism of Acetyl-CoA**
    David A. Bender, PhD   143

18. **Glycolysis & the Oxidation of Pyruvate**
    David A. Bender, PhD   149

19. **Metabolism of Glycogen**
    David A. Bender, PhD   157

20. **Gluconeogenesis & the Control of Blood Glucose**
    David A. Bender, PhD **165**

21. **The Pentose Phosphate Pathway & Other Pathways of Hexose Metabolism**
    David A. Bender, PhD **174**

22. **Oxidation of Fatty Acids: Ketogenesis**
    Kathleen M. Botham, PhD, DSc & Peter A. Mayes, PhD, DSc **184**

23. **Biosynthesis of Fatty Acids & Eicosanoids**
    Kathleen M. Botham, PhD, DSc & Peter A. Mayes, PhD, DSc **193**

24. **Metabolism of Acylglycerols & Sphingolipids**
    Kathleen M. Botham, PhD, DSc & Peter A. Mayes, PhD, DSc **205**

25. **Lipid Transport & Storage**
    Kathleen M. Botham, PhD, DSc & Peter A. Mayes, PhD, DSc **212**

26. **Cholesterol Synthesis, Transport, & Excretion**
    Kathleen M. Botham, PhD, DSc & Peter A. Mayes, PhD, DSc **224**

SECTION III
METABOLISM OF PROTEINS & AMINO ACIDS   234

27. **Biosynthesis of the Nutritionally Nonessential Amino Acids**
    Victor W. Rodwell, PhD **234**

28. **Catabolism of Proteins & of Amino Acid Nitrogen**
    Victor W. Rodwell, PhD **239**

29. **Catabolism of the Carbon Skeletons of Amino Acids**
    Victor W. Rodwell, PhD **248**

30. **Conversion of Amino Acids to Specialized Products**
    Victor W. Rodwell, PhD **262**

31. **Porphyrins & Bile Pigments**
    Robert K. Murray, MD, PhD **271**

SECTION IV
STRUCTURE, FUNCTION, & REPLICATION OF INFORMATIONAL MACROMOLECULES   285

32. **Nucleotides**
    Victor W. Rodwell, PhD **285**

33. **Metabolism of Purine & Pyrimidine Nucleotides**
    Victor W. Rodwell, PhD **292**

34. **Nucleic Acid Structure & Function**
    P. Anthony Weil, PhD **302**

35. **DNA Organization, Replication, & Repair**
    P. Anthony Weil, PhD **312**

36. **RNA Synthesis, Processing, & Modification**
    P. Anthony Weil, PhD **335**

37. **Protein Synthesis & the Genetic Code**
    P. Anthony Weil, PhD **353**

38. **Regulation of Gene Expression**
    P. Anthony Weil, PhD **369**

39. **Molecular Genetics, Recombinant DNA, & Genomic Technology**
    P. Anthony Weil, PhD **388**

SECTION V
BIOCHEMISTRY OF EXTRACELLULAR & INTRACELLULAR COMMUNICATION   406

40. **Membranes: Structure & Function**
    Robert K. Murray, MD, PhD & Daryl K. Granner, MD **406**

41. **The Diversity of the Endocrine System**
    P. Anthony Weil, PhD **425**

42. **Hormone Action & Signal Transduction**
    P. Anthony Weil, PhD **444**

SECTION **VI**

SPECIAL TOPICS   459

43. **Nutrition, Digestion, & Absorption**
    David A. Bender, PhD   459

44. **Micronutrients: Vitamins & Minerals**
    David A. Bender, PhD   467

45. **Free Radicals and Antioxidant Nutrients**
    David A. Bender, PhD   482

46. **Intracellular Traffic & Sorting of Proteins**
    Robert K. Murray, MD, PhD   487

47. **Glycoproteins**
    Robert K. Murray, MD, PhD   506

48. **The Extracellular Matrix**
    Robert K. Murray, MD, PhD &
    Frederick W. Keeley, PhD   527

49. **Muscle & the Cytoskeleton**
    Robert K. Murray, MD, PhD   545

50. **Plasma Proteins & Immunoglobulins**
    Robert K. Murray, MD, PhD   566

51. **Hemostasis & Thrombosis**
    Peter L. Gross, MD, Robert K. Murray, MD, PhD &
    Margaret L. Rand, PhD   583

52. **Red & White Blood Cells**
    Robert K. Murray, MD, PhD   593

53. **Metabolism of Xenobiotics**
    Robert K. Murray, MD, PhD   609

54. **Biochemical Case Histories**
    Robert K. Murray, MD, PhD &
    Peter L. Gross, MD   616

Appendix I   **647**

Appendix II   **648**

Index   **651**

Preface

The authors and publisher are pleased to present the twenty-eighth edition of *Harper's Illustrated Biochemistry*. This edition features for the first time multiple color images, many entirely new, that vividly emphasize the ever-increasing complexity of biochemical knowledge. The cover picture of green fluorescent protein (GFP), which recognizes the award of the 2008 Nobel Prize in Chemistry to Martin Chalfie, Roger Y. Tsien, and Osamu Shimomura, reflects the book's emphasis on new developments. Together with its derivatives, GFP fulfills an ever-widening role in tracking protein movement in intact cells and tissues, and has multiple applications to cell biology, biochemistry and medicine.

In this edition, we bid a regretful farewell to long-time author and editor, Daryl Granner. In 1983, in preparation for the 20th edition, Daryl was asked to write new chapters on the endocrine system and the molecular mechanism of hormones, which he did with great success. He assumed responsibility for the chapters on membranes, protein synthesis and molecular biology in the 21st edition, and wrote a highly informative new chapter on the then emerging field of recombinant DNA technology. Over the ensuing 25 years, through the 27th edition, Daryl continuously revised his chapters to provide concise, instructive descriptions of these rapidly changing, complex fields. Daryl's editorial colleagues express their gratitude for his many invaluable contributions as an author, editor and a friend, and wish him all the best in his future endeavors.

David Bender, Kathleen Botham, Peter Kennelly, and Anthony Weil, formerly co-authors, are now full authors. Rob Murray gratefully acknowledges the major contributions of Peter Gross, Fred Keeley, and Margaret Rand to specific chapters, and thanks Reinhart Reithmeier, Alan Volchuk, and David B. Williams for reviewing and making invaluable suggestions for the revision of Chapters 40 and 46. In addition, he is grateful to Kasra Haghighat and Mohammad Rassouli-Rashti for reading and suggesting improvements to Chapter 54.

Changes in the Twenty-Eighth Edition

Consistent with our goal of providing students with a text that describes and illustrates biochemistry in a comprehensive, concise, and readily accessible manner, the authors have incorporated substantial new material in this edition. Many new figures and tables have been added. Every chapter has been revised, updated and in several instances substantially rewritten to incorporate the latest advances in both knowledge and technology of importance to the understanding and practice of medicine.

Two new chapters have been added. Chapter 45, entitled "Free Radicals and Antioxidant Nutrients," describes the sources of free radicals; their damaging effects on DNA, proteins, and lipids; and their roles in causing diseases such as cancer and atherosclerosis. The role of antioxidants in counteracting their deleterious effects is assessed.

Chapter 54, entitled "Biochemical Case Histories," provides extensive presentations of 16 pathophysiologic conditions: adenosine deaminase deficiency, Alzheimer disease, cholera, colorectal cancer, cystic fibrosis, diabetic ketoacidosis, Duchenne muscular dystrophy, ethanol intoxication, gout, hereditary hemochromatosis, hypothyroidism, kwashiorkor (and protein-energy malnutrition), myocardial infarction, obesity, osteoporosis, and xeroderma pigmentosum.

Important new features of medical interest include:

- Influence of the Human Genome Project on various biomedical fields.
- Re-write of the use of enzymes in medical diagnosis.
- New material on computer-aided drug discovery.
- Compilation of some conformational diseases.
- New material on advanced glycation end-products and their importance in diabetes mellitus.
- New material on the attachment of influenza virus to human cells.
- Some major challenges facing medicine.

The following topics that have been added to various chapters are of basic biochemical interest:

- Expanded coverage of mass spectrometry, a key analytical method in contemporary biochemistry.
- New figures revealing various aspects of protein structure.
- Expanded coverage of active sites of enzymes and transition states.
- New information on methods of assaying enzymes.
- Expanded coverage of aspects of enzyme kinetics.

- New information on micro- and silencing RNAs.
- New information on eukaryotic transcription mechanisms, including the biogenesis of mRNA and the role of nucleosomes.
- Description of activities of miRNAs.
- New material on Next Generation Sequencing (NGS) platforms.
- New material on the Chromatin Immunoprecipitation (CHIP) technology and its uses.
- New information on subcellular localization of key signaling enzymes (kinases, phosphatases).
- New information on how hormones affect gene transcription.

Every chapter begins with a summary of the biomedical importance of its contents and concludes with a summary reviewing the major topics covered.

Organization of the Book

Following two introductory chapters ("Biochemistry and Medicine" and "Water and pH"), the text is divided into six main sections. All sections and chapters emphasize the medical relevance of biochemistry.

Section I addresses the structures and functions of proteins and enzymes. Because almost all of the reactions in cells are catalyzed by enzymes, it is vital to understand the properties of enzymes before considering other topics. This section also contains a chapter on bioinformatics and computational biology, reflecting the increasing importance of these topics in modern biochemistry, biology and medicine.

Section II explains how various cellular reactions either utilize or release energy, and traces the pathways by which carbohydrates and lipids are synthesized and degraded. Also described are the many functions of these two classes of molecules.

Section III deals with the amino acids, their many metabolic fates, certain key features of protein catabolism, and the biochemistry of the porphyrins and bile pigments.

Section IV describes the structures and functions of the nucleotides and nucleic acids, and includes topics such as DNA replication and repair, RNA synthesis and modification, protein synthesis, the principles of recombinant DNA and genomic technology, and new understanding of how gene expression is regulated.

Section V deals with aspects of extracellular and intracellular communication. Topics include membrane structure and function, the molecular bases of the actions of hormones, and the key field of signal transduction.

Section VI discusses twelve special topics: nutrition, digestion and absorption; vitamins and minerals; free radicals and antioxidants; intracellular trafficking and sorting of proteins; glycoproteins; the extracellular matrix; muscle and the cytoskeleton; plasma proteins and immunoglobulins; hemostasis and thrombosis; red and white blood cells; the metabolism of xenobiotics; and 16 biochemically oriented case histories. The latter chapter concludes with a brief Epilog indicating some major challenges for medicine in whose solution biochemistry and related disciplines will play key roles.

Appendix I contains a list of laboratory results relevant to the cases discussed in Chapter 54.

Appendix II contains a list of useful web sites and a list of biochemical journals or journals with considerable biochemical content.

Acknowledgments

The authors thank Michael Weitz for his vital role in the planning and actualization of this edition. It has been a pleasure to work with him. We are also very grateful to Kim Davis for her highly professional supervising of the editing of the text, to Sherri Souffrance for supervising its production, to Elise Langdon for its design, and to Margaret Webster-Shapiro for her work on the cover art. We warmly acknowledge the work of the artists, typesetters, and other individuals not known to us who participated in the production of the twenty-eighth edition of *Harper's Illustrated Biochemistry*. In particular, we are very grateful to Joanne Jay of Newgen North America for her central role in the management of the entire project and to Joseph Varghese of Thomson Digital for his skilled supervision of the large amount of art work that was necessary for this edition.

Suggestions from students and colleagues around the world have been most helpful in the formulation of this edition. We look forward to receiving similar input in the future.

Robert K. Murray, Toronto, Ontario, Canada
David A. Bender, London, UK
Kathleen M. Botham, London, UK
Peter J. Kennelly, Blacksburg, Virginia, USA
Victor W. Rodwell, West Lafayette, Indiana, USA
P. Anthony Weil, Nashville, Tennessee, USA

Biochemistry & Medicine

Robert K. Murray, MD, PhD

INTRODUCTION

Biochemistry can be defined as *the science of the chemical basis of life* (Gk *bios* "life"). The **cell** is the structural unit of living systems. Thus, biochemistry can also be described as *the science of the chemical constituents of living cells and of the reactions and processes they undergo.* By this definition, biochemistry encompasses large areas of **cell biology, molecular biology,** and **molecular genetics.**

The Aim of Biochemistry Is to Describe & Explain, in Molecular Terms, All Chemical Processes of Living Cells

The major objective of biochemistry is the complete understanding, at the molecular level, of all of the chemical processes associated with living cells. To achieve this objective, biochemists have sought to isolate the numerous molecules found in cells, determine their structures, and analyze how they function. Many techniques have been used for these purposes; some of them are summarized in Table 1–1.

A Knowledge of Biochemistry Is Essential to All Life Sciences

The biochemistry of the nucleic acids lies at the heart of **genetics;** in turn, the use of genetic approaches has been critical for elucidating many areas of biochemistry. **Physiology,** the study of body function, overlaps with biochemistry almost completely. **Immunology** employs numerous biochemical techniques, and many immunologic approaches have found wide use by biochemists. **Pharmacology** and **pharmacy** rest on a sound knowledge of biochemistry and physiology; in particular, most drugs are metabolized by enzyme-catalyzed reactions. Poisons act on biochemical reactions or processes; this is the subject matter of **toxicology.** Biochemical approaches are being used increasingly to study basic aspects of **pathology** (the study of disease), such as inflammation, cell injury, and cancer. Many workers in **microbiology, zoology,** and **botany** employ biochemical approaches almost exclusively. These relationships are not surprising, because life as we know it depends on biochemical reactions and processes. In fact, the old barriers among the life sciences are breaking down, and biochemistry is increasingly becoming their common language.

A Reciprocal Relationship Between Biochemistry & Medicine Has Stimulated Mutual Advances

The two major concerns for workers in the health sciences—and particularly physicians—are the understanding and maintenance of **health** and the understanding and effective treatment of **diseases.** Biochemistry impacts enormously on both of these fundamental concerns of medicine. In fact, the interrelationship of biochemistry and medicine is a wide, two-way street. Biochemical studies have illuminated many aspects of health and disease, and conversely, the study of various aspects of health and disease has opened up new areas of biochemistry. Some examples of this two-way street are shown in Figure 1–1. For instance, knowledge of protein structure and function was necessary to elucidate the single biochemical difference between normal hemoglobin and sickle cell hemoglobin. On the other hand, analysis of sickle cell hemoglobin has contributed significantly to our understanding of the structure and function of both normal hemoglobin and other proteins. Analogous examples of reciprocal benefit between biochemistry and medicine could be cited for the other paired items shown in Figure 1–1. Another example is the pioneering work of Archibald Garrod, a physician in England during the early 1900s. He studied patients with a number of relatively rare disorders (alkaptonuria, albinism, cystinuria, and pentosuria; these are described in later chapters) and established that these conditions were genetically determined. Garrod designated these conditions as **inborn errors of metabolism.** His insights provided a major foundation for the development of the field of human biochemical genetics. More recent efforts to understand the basis of the genetic disease known as **familial hypercholesterolemia,** which results in severe atherosclerosis at an early age, have led to dramatic progress in understanding of cell receptors and of mechanisms of uptake of cholesterol into cells. Studies of **oncogenes** in cancer cells have directed attention to the molecular mechanisms involved in the control of normal

TABLE 1–1   The Principal Methods and Preparations Used in Biochemical Laboratories

Methods for Separating and Purifying Biomolecules[1]
Salt fractionation (eg, precipitation of proteins with ammonium sulfate)
Chromatography: Paper, ion exchange, affinity, thin-layer, gas–liquid, high-pressure liquid, gel filtration
Electrophoresis: Paper, high-voltage, agarose, cellulose acetate, starch gel, polyacrylamide gel, SDS-polyacrylamide gel
Ultracentrifugation

Methods for Determining Biomolecular Structures
Elemental analysis
UV, visible, infrared, and NMR spectroscopy
Use of acid or alkaline hydrolysis to degrade the biomolecule under study into its basic constituents
Use of a battery of enzymes of known specificity to degrade the biomolecule under study (eg, proteases, nucleases, glycosidases)
Mass spectrometry
Specific sequencing methods (eg, for proteins and nucleic acids)
X-ray crystallography

Preparations for Studying Biochemical Processes
Whole animal (includes transgenic animals and animals with gene knockouts)
Isolated perfused organ
Tissue slice
Whole cells
Homogenate
Isolated cell organelles
Subfractionation of organelles
Purified metabolites and enzymes
Isolated genes (including polymerase chain reaction and site-directed mutagenesis)

[1]Most of these methods are suitable for analyzing the components present in cell homogenates and other biochemical preparations. The sequential use of several techniques will generally permit purification of most biomolecules. The reader is referred to texts on methods of biochemical research for details.

cell growth. These and many other examples emphasize how the study of disease can open up areas of cell function for basic biochemical research.

The relationship between medicine and biochemistry has important implications for the former. As long as medical treatment is firmly grounded in the knowledge of biochemistry and other basic sciences, the practice of medicine will have a rational basis that can be adapted to accommodate new knowledge. This contrasts with unorthodox health cults and at least some "alternative medicine" practices that are often founded on little more than myth and wishful thinking and generally lack any intellectual basis.

NORMAL BIOCHEMICAL PROCESSES ARE THE BASIS OF HEALTH

The World Health Organization (WHO) defines **health** as a state of "complete physical, mental and social well-being and not merely the absence of disease and infirmity." From a strictly biochemical viewpoint, health may be considered that situation in which all of the many thousands of intra- and extracellular reactions that occur in the body are proceeding at rates commensurate with the organism's maximal survival in the physiologic state. However, this is an extremely reductionist view, and it should be apparent that caring for the health of patients requires not only a wide knowledge of biologic principles but also of psychologic and social principles.

Biochemical Research Has Impact on Nutrition & Preventive Medicine

One major prerequisite for the maintenance of health is that there be optimal dietary intake of a number of chemicals; the chief of these are **vitamins,** certain **amino acids,** certain **fatty acids,** various **minerals,** and **water.** Because much of the subject matter of both biochemistry and nutrition is concerned with the study of various aspects of these chemicals, there is a close relationship between these two sciences. Moreover, more emphasis is being placed on systematic attempts to maintain health and forestall disease, that is, on **preventive medicine.** Thus, nutritional approaches to—for example—the prevention of atherosclerosis and cancer are receiving increased emphasis. Understanding nutrition depends to a great extent on knowledge of biochemistry.

Most & Perhaps All Diseases Have a Biochemical Basis

We believe that most if not all diseases are manifestations of abnormalities of molecules, chemical reactions, or biochemical processes. The **major factors responsible for causing diseases** in animals and humans are listed in Table 1–2. All of them affect one or more critical chemical reactions or molecules in the body. Numerous examples of the biochemical bases of diseases will be encountered in this text. In most of these conditions, biochemical studies contribute to both the diagnosis and treatment. Some **major uses of biochemical investigations and of laboratory tests in relation to diseases** are summarized in Table 1–3. Chapter 54 of this text further helps to illustrate the relationship of biochemistry to disease by discussing in some detail biochemical aspects of 16 different medical cases.

Some of the **major challenges that medicine and related health sciences** face are also outlined very briefly at the end of Chapter 54. In addressing these challenges, biochemical studies are already and will continue to be interwoven with studies in various other disciplines, such as genetics, immunology, nutrition, pathology and pharmacology.

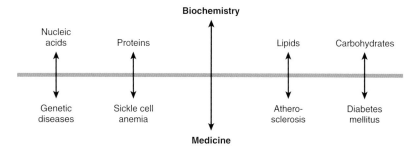

FIGURE 1–1 Examples of the two-way street connecting biochemistry and medicine. Knowledge of the biochemical molecules shown in the top part of the diagram has clarified our understanding of the diseases shown on the bottom half—and conversely, analyses of the diseases shown below have cast light on many areas of biochemistry. Note that sickle cell anemia is a genetic disease and that both atherosclerosis and diabetes mellitus have genetic components.

Impact of the Human Genome Project (HGP) on Biochemistry, Biology, & Medicine

Remarkable progress was made in the late 1990s in sequencing the human genome by the HGP. This culminated in July 2000, when leaders of the two groups involved in this effort (the International Human Genome Sequencing Consortium and Celera Genomics, a private company) announced that over 90% of the genome had been sequenced. Draft versions of the sequence were published in early 2001. With the exception of a few gaps, the sequence of the entire human genome was completed in 2003, 50 years after the description of the double-helical nature of DNA by Watson and Crick.

The **implications** of the HGP for **biochemistry**, all of **biology**, and for **medicine** and **related health sciences** are tremendous, and only a few points are mentioned here. It is **now possible to isolate any gene and usually determine its structure and function** (eg, by sequencing and knockout experiments). Many **previously unknown genes** have been re-

vealed; their products have already been established, or are under study. New light has been thrown on **human evolution**, and procedures for **tracking disease genes** have been greatly refined. Reference to the human genome will be made in various sections of this text.

Figure 1–2 shows **areas of great current interest** that have developed either directly as a result of the progress made in the HGP, or have been spurred on by it. As an outgrowth of the HGP, many so-called -**omics** fields have sprung up, involving comprehensive studies of the structures and functions of the molecules with which each is concerned. Definitions of the fields listed below are given in the Glossary of this

TABLE 1–2 The Major Causes of Diseases[1]

1. *Physical agents:* Mechanical trauma, extremes of temperature, sudden changes in atmospheric pressure, radiation, electric shock.

2. *Chemical agents, including drugs:* Certain toxic compounds, therapeutic drugs, etc.

3. *Biologic agents:* Viruses, bacteria, fungi, higher forms of parasites.

4. *Oxygen lack:* Loss of blood supply, depletion of the oxygen-carrying capacity of the blood, poisoning of the oxidative enzymes.

5. *Genetic disorders:* Congenital, molecular.

6. *Immunologic reactions:* Anaphylaxis, autoimmune disease.

7. *Nutritional imbalances:* Deficiencies, excesses.

8. *Endocrine imbalances:* Hormonal deficiencies, excesses.

[1]**Note:** All of the causes listed act by influencing the various biochemical mechanisms in the cell or in the body.
(Adapted, with permission, from Robbins SL, Cotram RS, Kumar V: *The Pathologic Basis of Disease,* 3rd ed. Saunders, 1984. Copyright © 1984 Elsevier Inc. with permission from Elsevier.)

TABLE 1–3 Some Uses of Biochemical Investigations and Laboratory Tests in Relation to Diseases

Use	Example
1. To reveal the fundamental causes and mechanisms of diseases	Demonstration of the nature of the genetic defects in cystic fibrosis.
2. To suggest rational treatments of diseases based on item 1 above	A diet low in phenylalanine for treatment of phenylketonuria.
3. To assist in the diagnosis of specific diseases	Use of the plasma levels of troponin I or T in the diagnosis of myocardial infarction.
4. To act as screening tests for the early diagnosis of certain diseases	Use of measurement of blood thyroxine or thyroid-stimulating hormone (TSH) in the neonatal diagnosis of congenital hypothyroidism.
5. To assist in monitoring the progress (ie, recovery, worsening, remission, or relapse) of certain diseases	Use of the plasma enzyme alanine aminotransferase (ALT) in monitoring the progress of infectious hepatitis.
6. To assist in assessing the response of diseases to therapy	Use of measurement of blood carcinoembryonic antigen (CEA) in certain patients who have been treated for cancer of the colon.

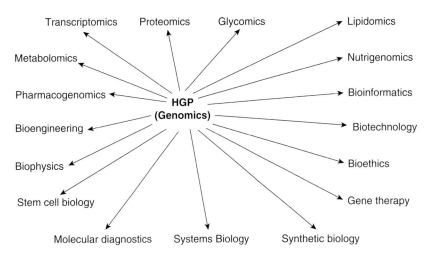

FIGURE 1–2 The Human Genome Project (HGP) has influenced many disciplines and areas of research.

chapter. The products of genes (RNA molecules and proteins) are being studied using the technics of **transcriptomics** and **proteomics**. One spectacular example of the speed of progress in transcriptomics is the explosion of knowledge about small RNA molecules as regulators of gene activity. Other -omics fields include **glycomics, lipidomics, metabolomics, nutrigenomics,** and **pharmacogenomics**. To keep pace with the amount of information being generated, **bioinformatics** has received much attention. Other related fields to which the impetus from the HGP has carried over are **biotechnology, bioengineering**, **biophysics,** and **bioethics**. **Stem cell biology** is at the center of much current research. **Gene therapy** has yet to deliver the promise that it contains, but it seems probable that will occur sooner or later. Many new **molecular diagnostic tests** have developed in areas such as genetic, microbiologic, and immunologic testing and diagnosis. **Systems biology** is also burgeoning. **Synthetic biology** is perhaps the most intriguing of all. This has the potential for creating living organisms (eg, initially small bacteria) from genetic material in vitro. These could perhaps be designed to carry out specific tasks (eg, to mop up petroleum spills). As in the case of stem cells, this area will attract much attention from bioethicists and others. Many of the above topics are referred to later in this text.

All of the above have made the present time a very exciting one for studying or to be directly involved in biology and medicine. The outcomes of research in the various areas mentioned above will impact tremendously on the future of biology, medicine and the health sciences.

SUMMARY

- Biochemistry is the science concerned with studying the various molecules that occur in living cells and organisms and with their chemical reactions. Because life depends on biochemical reactions, biochemistry has become the basic language of all biologic sciences.

- Biochemistry is concerned with the entire spectrum of life forms, from relatively simple viruses and bacteria to complex human beings.

- Biochemistry and medicine are intimately related. Health depends on a harmonious balance of biochemical reactions occurring in the body, and disease reflects abnormalities in biomolecules, biochemical reactions, or biochemical processes.

- Advances in biochemical knowledge have illuminated many areas of medicine. Conversely, the study of diseases has often revealed previously unsuspected aspects of biochemistry. Biochemical approaches are often fundamental in illuminating the causes of diseases and in designing appropriate therapies.

- The judicious use of various biochemical laboratory tests is an integral component of diagnosis and monitoring of treatment.

- A sound knowledge of biochemistry and of other related basic disciplines is essential for the rational practice of medicine and related health sciences.

- Results of the HGP and of research in related areas will have a profound influence on the future of biology, medicine and other health sciences.

REFERENCES

Burtis CA, Ashwood ER, Bruns DE: *Tietz Textbook of Clinical Chemistry and Molecular Diagnostics,* 4th ed. Elsevier Inc, 2006.

Encyclopedia of Life Sciences. John Wiley, 2001. (Contains some 3000 comprehensive articles on various aspects of the life sciences. Accessible online at www.els.net via libraries with a subscription.)

Fruton JS: *Proteins, Enzymes, Genes: The Interplay of Chemistry and Biology.* Yale University Press, 1999. (Provides the historical background for much of today's biochemical research.)

Garrod AE: Inborn errors of metabolism. (Croonian Lectures.) Lancet 1908;2:1, 73, 142, 214.

Guttmacher AE, Collins FS: Genomic medicine—A primer. N Engl J Med 2002;347:1512. (This article was the first of a series of 11 monthly articles published in the *New England Journal of Medicine* describing various aspects of genomic medicine.)

Guttmacher AE, Collins FS: Realizing the promise of genomics in biomedical research. JAMA 2005;294(11):1399.

Kornberg A: Basic research: The lifeline of medicine. FASEB J 1992;6:3143.

Kornberg A: Centenary of the birth of modern biochemistry. FASEB J 1997;11:1209.

Manolio TA, Collins FS: Genes, environment, health, and disease: Facing up to complexity. Hum Hered 2007;63:63.

McKusick VA: *Mendelian Inheritance in Man. Catalogs of Human Genes and Genetic Disorders,* 12th ed. Johns Hopkins University Press, 1998. [Abbreviated MIM]

Online Mendelian Inheritance in Man (OMIM): Center for Medical Genetics, Johns Hopkins University and National Center for Biotechnology Information, National Library of Medicine, 1997. http://www.ncbi.nlm.nih.gov/omim/ (The numbers assigned to the entries in OMIM will be cited in selected chapters of this work. Consulting this extensive collection of diseases and other relevant entries—specific proteins, enzymes, etc—will greatly expand the reader's knowledge and understanding of various topics referred to and discussed in this text. The online version is updated almost daily.)

Oxford Dictionary of Biochemistry and Molecular Biology, rev. ed. Oxford University Press, 2000.

Scriver CR et al (editors): *The Metabolic and Molecular Bases of Inherited Disease,* 8th ed. McGraw-Hill, 2001 (This text is now available online and updated as *The Online Metabolic & Molecular Bases of Inherited Disease* at www.ommbid.com. Subscription is required, although access may be available via university and hospital libraries and other sources).

Scherer S: *A Short Guide to the Human Genome.* CSHL Press, 2008.

GLOSSARY

Bioengineering: The application of engineering to biology and medicine.

Bioethics: The area of ethics that is concerned with the application of moral and ethical principles to biology and medicine.

Bioinformatics: The discipline concerned with the collection, storage and analysis of biologic data, mainly DNA and protein sequences (see Chapter 10).

Biophysics: The application of physics and its technics to biology and medicine.

Biotechnology: The field in which biochemical, engineering, and other approaches are combined to develop biological products of use in medicine and industry.

Gene Therapy: Applies to the use of genetically engineered genes to treat various diseases (see Chapter 39).

Genomics: The genome is the complete set of genes of an organism (eg, the human genome) and genomics is the in depth study of the structures and functions of genomes (see Chapter 10 and other chapters).

Glycomics: The glycome is the total complement of simple and complex carbohydrates in an organism. Glycomics is the systematic study of the structures and functions of glycomes (eg, the human glycome; see Chapter 47).

Lipidomics: The lipidome is the complete complement of lipids found in an organism. Lipidomics is the in depth study of the structures and functions of all members of the lipidome and of their interactions, in both health and disease.

Metabolomics: The metabolome is the complete complement of metabolites (small molecules involved in metabolism) found in an organism. Metabolomics is the in depth study of their structures, functions, and changes in various metabolic states.

Molecular Diagnostics: The use of molecular approaches (eg, DNA probes) to assist in the diagnosis of various biochemical, genetic, immunologic, microbiologic, and other medical conditions.

Nutrigenomics: The systematic study of the effects of nutrients on genetic expression and also of the effects of genetic variations on the handling of nutrients.

Pharmacogenomics: The use of genomic information and technologies to optimize the discovery and development of drug targets and drugs (see Chapter 54).

Proteomics: The proteome is the complete complement of proteins of an organism. Proteomics is the systematic study of the structures and functions of proteomes, including variations in health and disease (see Chapter 4).

Stem Cell Biology: A stem cell is an undifferentiated cell that has the potential to renew itself and to differentiate into any of the adult cells found in the organism. Stem cell biology is concerned with the biology of stem cells and their uses in various diseases.

Synthetic Biology: The field that combines biomolecular technics with engineering approaches to build new biological functions and systems.

Systems Biology: The field of science in which complex biologic systems are studied as integrated wholes (as opposed to the reductionist approach of, for example, classic biochemistry).

Transcriptomics: The transcriptome is the complete set of RNA transcripts produced by the genome at a fixed period in time. Transcriptomics is the comprehensive study of gene expression at the RNA level (see Chapter 36 and other chapters).

2

Water & pH

Peter J. Kennelly, PhD & Victor W. Rodwell, PhD

BIOMEDICAL IMPORTANCE

Water is the predominant chemical component of living organisms. Its unique physical properties, which include the ability to solvate a wide range of organic and inorganic molecules, derive from water's dipolar structure and exceptional capacity for forming hydrogen bonds. The manner in which water interacts with a solvated biomolecule influences the structure of each. An excellent nucleophile, water is a reactant or product in many metabolic reactions. Water has a slight propensity to dissociate into hydroxide ions and protons. The acidity of aqueous solutions is generally reported using the logarithmic pH scale. Bicarbonate and other buffers normally maintain the pH of extracellular fluid between 7.35 and 7.45. Suspected disturbances of acid–base balance are verified by measuring the pH of arterial blood and the CO_2 content of venous blood. Causes of acidosis (blood pH <7.35) include diabetic ketosis and lactic acidosis. Alkalosis (pH >7.45) may follow vomiting of acidic gastric contents. Regulation of water balance depends upon hypothalamic mechanisms that control thirst, on antidiuretic hormone (ADH), on retention or excretion of water by the kidneys, and on evaporative loss. Nephrogenic diabetes insipidus, which involves the inability to concentrate urine or adjust to subtle changes in extracellular fluid osmolarity, results from the unresponsiveness of renal tubular osmoreceptors to ADH.

WATER IS AN IDEAL BIOLOGIC SOLVENT

Water Molecules Form Dipoles

A water molecule is an irregular, slightly skewed tetrahedron with oxygen at its center (Figure 2–1). The two hydrogens and the unshared electrons of the remaining two sp^3-hybridized orbitals occupy the corners of the tetrahedron. The 105-degree angle between the hydrogens differs slightly from the ideal tetrahedral angle, 109.5 degrees. Ammonia is also tetrahedral, with a 107-degree angle between its hydrogens. Water is a **dipole,** a molecule with electrical charge distributed asymmetrically about its structure. The strongly electronegative oxygen atom pulls electrons away from the hydrogen nuclei, leaving them with a partial positive charge, while its two unshared electron pairs constitute a region of local negative charge.

Water, a strong dipole, has a high **dielectric constant.** As described quantitatively by Coulomb's law, the strength of interaction F between oppositely charged particles is inversely proportionate to the dielectric constant ε of the surrounding medium. The dielectric constant for a vacuum is unity; for hexane it is 1.9; for ethanol, 24.3; and for water, 78.5. Water therefore greatly decreases the force of attraction between charged and polar species relative to water-free environments with lower dielectric constants. Its strong dipole and high dielectric constant enable water to dissolve large quantities of charged compounds such as salts.

Water Molecules Form Hydrogen Bonds

A partially unshielded hydrogen nucleus covalently bound to an electron-withdrawing oxygen or nitrogen atom can interact with an unshared electron pair on another oxygen or nitrogen atom to form a **hydrogen bond.** Since water molecules contain both of these features, hydrogen bonding favors the self-association of water molecules into ordered arrays (Figure 2–2). Hydrogen bonding profoundly influences the physical properties of water and accounts for its exceptionally high viscosity, surface tension, and boiling point. On average, each molecule in liquid water associates through hydrogen bonds with 3.5 others. These bonds are both relatively weak and transient, with a half-life of one microsecond or less. Rupture of a hydrogen bond in liquid water requires only about 4.5 kcal/mol, less than 5% of the energy required to rupture a covalent O—H bond.

Hydrogen bonding enables water to dissolve many organic biomolecules that contain functional groups which can participate in hydrogen bonding. The oxygen atoms of aldehydes, ketones, and amides, for example, provide lone pairs of electrons that can serve as hydrogen acceptors. Alcohols and amines can serve both as hydrogen acceptors and as donors of unshielded hydrogen atoms for formation of hydrogen bonds (Figure 2–3).

FIGURE 2–1 The water molecule has tetrahedral geometry.

FIGURE 2–2 **Left:** Association of two dipolar water molecules by a hydrogen bond (dotted line). **Right:** Hydrogen-bonded cluster of four water molecules. Note that water can serve simultaneously both as a hydrogen donor and as a hydrogen acceptor.

FIGURE 2–3 Additional polar groups participate in hydrogen bonding. Shown are hydrogen bonds formed between alcohol and water, between two molecules of ethanol, and between the peptide carbonyl oxygen and the peptide nitrogen hydrogen of an adjacent amino acid.

INTERACTION WITH WATER INFLUENCES THE STRUCTURE OF BIOMOLECULES

Covalent & Noncovalent Bonds Stabilize Biologic Molecules

The covalent bond is the strongest force that holds molecules together (Table 2–1). Noncovalent forces, while of lesser magnitude, make significant contributions to the structure, stability, and functional competence of macromolecules in living cells. These forces, which can be either attractive or repulsive, involve interactions both within the biomolecule and between it and the water that forms the principal component of the surrounding environment.

TABLE 2–1 Bond Energies for Atoms of Biologic Significance

Bond Type	Energy (kcal/mol)	Bond Type	Energy (kcal/mol)
O—O	34	O=O	96
S—S	51	C—H	99
C—N	70	C=S	108
S—H	81	O—H	110
C—C	82	C=C	147
C—O	84	C=N	147
N—H	94	C=O	164

Biomolecules Fold to Position Polar & Charged Groups on Their Surfaces

Most biomolecules are **amphipathic;** that is, they possess regions rich in charged or polar functional groups as well as regions with hydrophobic character. Proteins tend to fold with the R-groups of amino acids with hydrophobic side chains in the interior. Amino acids with charged or polar amino acid side chains (eg, arginine, glutamate, serine) generally are present on the surface in contact with water. A similar pattern prevails in a phospholipid bilayer, where the charged head groups of phosphatidyl serine or phosphatidyl ethanolamine contact water while their hydrophobic fatty acyl side chains cluster together, excluding water. This pattern maximizes the opportunities for the formation of energetically favorable charge–dipole, dipole–dipole, and hydrogen bonding interactions between polar groups on the biomolecule and water. It also minimizes energetically unfavorable contacts between water and hydrophobic groups.

Hydrophobic Interactions

Hydrophobic interaction refers to the tendency of nonpolar compounds to self-associate in an aqueous environment. This self-association is driven neither by mutual attraction nor by what are sometimes incorrectly referred to as "hydrophobic bonds." Self-association minimizes energetically unfavorable interactions between nonpolar groups and water.

While the hydrogens of nonpolar groups such as the methylene groups of hydrocarbons do not form hydrogen bonds, they do affect the structure of the water that surrounds them. Water molecules adjacent to a hydrophobic group are restricted in the number of orientations (degrees of freedom) that permit them to participate in the maximum number of energetically favorable hydrogen bonds. Maximal formation of multiple hydrogen bonds can be maintained only by increasing the order of the adjacent water molecules, with an accompanying decrease in entropy.

It follows from the second law of thermodynamics that the optimal free energy of a hydrocarbon–water mixture is a function of both maximal enthalpy (from hydrogen bonding) and minimum entropy (maximum degrees of freedom). Thus, nonpolar molecules tend to form droplets in order to minimize exposed surface area and reduce the number of water molecules affected. Similarly, in the aqueous environment of the living cell the hydrophobic portions of biopolymers tend to be buried inside the structure of the molecule, or within a lipid bilayer, minimizing contact with water.

Electrostatic Interactions

Interactions between charged groups help shape biomolecular structure. Electrostatic interactions between oppositely charged groups within or between biomolecules are termed **salt bridges.** Salt bridges are comparable in strength to hydrogen bonds but act over larger distances. They therefore often facilitate the binding of charged molecules and ions to proteins and nucleic acids.

van der Waals Forces

van der Waals forces arise from attractions between transient dipoles generated by the rapid movement of electrons of all neutral atoms. Significantly weaker than hydrogen bonds but potentially extremely numerous, van der Waals forces decrease as the sixth power of the distance separating atoms. Thus, they act over very short distances, typically 2–4 Å.

Multiple Forces Stabilize Biomolecules

The DNA double helix illustrates the contribution of multiple forces to the structure of biomolecules. While each individual DNA strand is held together by covalent bonds, the two strands of the helix are held together exclusively by noncovalent interactions. These noncovalent interactions include hydrogen bonds between nucleotide bases (Watson–Crick base pairing) and van der Waals interactions between the stacked purine and pyrimidine bases. The helix presents the charged phosphate groups and polar ribose sugars of the backbone to water while burying the relatively hydrophobic nucleotide bases inside. The extended backbone maximizes the distance between negatively charged phosphates, minimizing unfavorable electrostatic interactions.

WATER IS AN EXCELLENT NUCLEOPHILE

Metabolic reactions often involve the attack by lone pairs of electrons residing on electron-rich molecules termed **nucleophiles** upon electron-poor atoms called **electrophiles.** Nucleophiles and electrophiles do not necessarily possess a formal negative or positive charge. Water, whose two lone pairs of sp^3 electrons bear a partial negative charge, is an excellent nucleophile. Other nucleophiles of biologic importance include the oxygen atoms of phosphates, alcohols, and carboxylic acids; the sulfur of thiols; the nitrogen of amines; and the imidazole ring of histidine. Common electrophiles include the carbonyl carbons in amides, esters, aldehydes, and ketones and the phosphorus atoms of phosphoesters.

Nucleophilic attack by water generally results in the cleavage of the amide, glycoside, or ester bonds that hold biopolymers together. This process is termed **hydrolysis.** Conversely, when monomer units are joined together to form biopolymers such as proteins or glycogen, water is a product, for example, during the formation of a peptide bond between two amino acids:

While hydrolysis is a thermodynamically favored reaction, the amide and phosphoester bonds of polypeptides and oligonucleotides are stable in the aqueous environment of the cell. This seemingly paradoxic behavior reflects the fact that the thermodynamics governing the equilibrium of a reaction do not determine the rate at which it will proceed. In the cell, protein catalysts called **enzymes** accelerate the rate of hydrolytic reactions when needed. **Proteases** catalyze the hydrolysis of proteins into their component amino acids, while **nucleases** catalyze the hydrolysis of the phosphoester bonds in DNA and RNA. Careful control of the activities of these enzymes is required to ensure that they act only on appropriate target molecules at appropriate times.

Many Metabolic Reactions Involve Group Transfer

Many of the enzymic reactions responsible for synthesis and breakdown of biomolecules involve the transfer of a chemical group G from a donor D to an acceptor A to form an acceptor group complex, A–G:

$$D–G + A \rightleftarrows A–G + D$$

The hydrolysis and phosphorolysis of glycogen, for example, involve the transfer of glucosyl groups to water or to orthophosphate. The equilibrium constant for the hydrolysis of covalent bonds strongly favors the formation of split products. Conversely, in many cases the group transfer reactions responsible for the biosynthesis of macromolecules involve the thermodynamically unfavored formation of covalent bonds. Enzymes surmount this barrier by coupling these group transfer reactions to other, favored reactions so that the overall change in free energy favors biopolymer synthesis. Given the nucleophilic character of water and its high concentration in cells, why are biopolymers such as proteins and DNA relatively stable? And how can synthesis of biopolymers occur in an aqueous environment? Central to both questions are the properties of enzymes. In the absence of enzymic catalysis, even reactions that are highly favored thermodynamically do not necessarily take place rapidly. Precise and differential control of enzyme activity and the sequestration of enzymes in specific organelles determine under what physiologic conditions a given biopolymer will be synthesized or degraded. Newly synthesized biopolymers are not immediately hydrolyzed, in part because the active sites of biosynthetic enzymes sequester substrates in an environment from which water can be excluded.

Water Molecules Exhibit a Slight but Important Tendency to Dissociate

The ability of water to ionize, while slight, is of central importance for life. Since water can act both as an acid and as a base, its ionization may be represented as an intermolecular proton transfer that forms a hydronium ion (H_3O^+) and a hydroxide ion (OH^-):

$$H_2O + H_2O \rightleftharpoons H_3O^+ + OH^-$$

The transferred proton is actually associated with a cluster of water molecules. Protons exist in solution not only as H_3O^+, but also as multimers such as $H_5O_2^+$ and $H_7O_3^+$. The proton is nevertheless routinely represented as H^+, even though it is in fact highly hydrated.

Since hydronium and hydroxide ions continuously recombine to form water molecules, an *individual* hydrogen or oxygen cannot be stated to be present as an ion or as part of a water molecule. At one instant it is an ion; an instant later it is part of a water molecule. Individual ions or molecules are therefore not considered. We refer instead to the *probability* that at any instant in time a hydrogen will be present as an ion or as part of a water molecule. Since 1 g of water contains 3.46×10^{22} molecules, the ionization of water can be described statistically. To state that the probability that a hydrogen exists as an ion is 0.01 means that at any given moment in time, a hydrogen atom has 1 chance in 100 of being an ion and 99 chances out of 100 of being part of a water molecule. The actual probability of a hydrogen atom in pure water existing as a hydrogen ion is approximately 1.8×10^{-9}. The probability of its being part of a water molecule thus is almost unity. Stated another way, for every hydrogen ion and hydroxide ion in pure water, there are 1.8 billion or 1.8×10^9 water molecules. Hydrogen ions and hydroxide ions nevertheless contribute significantly to the properties of water.

For dissociation of water,

$$K = \frac{\left[H^+\right]\left[OH^-\right]}{\left[H_2O\right]}$$

where the brackets represent molar concentrations (strictly speaking, molar activities) and K is the **dissociation constant.** Since 1 mole (mol) of water weighs 18 g, 1 liter (L) (1000 g) of water contains $1000 \div 18 = 55.56$ mol. Pure water thus is 55.56 molar. Since the probability that a hydrogen in pure water will exist as a hydrogen ion is 1.8×10^{-9}, the molar concentration

of H^+ ions (or of OH^- ions) in pure water is the product of the probability, 1.8×10^{-9}, times the molar concentration of water, 55.56 mol/L. The result is 1.0×10^{-7} mol/L.

We can now calculate K for pure water:

$$K = \frac{\left[H^+\right]\left[OH^-\right]}{\left[H_2O\right]} = \frac{\left[10^{-7}\right]\left[10^{-7}\right]}{\left[55.56\right]}$$
$$= 0.018 \times 10^{-14} = 1.8 \times 10^{-16} \text{mol/L}$$

The molar concentration of water, 55.56 mol/L, is too great to be significantly affected by dissociation. It therefore is considered to be essentially constant. This constant may therefore be incorporated into the dissociation constant K to provide a useful new constant K_w termed the **ion product** for water. The relationship between K_w and K is shown below:

$$K = \frac{\left[H^+\right]\left[OH^-\right]}{\left[H_2O\right]} = 1.8 \times 10^{-16} \text{mol/L}$$
$$K_w = (K)\left[H_2O\right] = \left[H^+\right]\left[OH^-\right]$$
$$= \left(1.8 \times 10^{-16} \text{mol/L}\right)(55.56 \text{ mol/L})$$
$$= 1.00 \times 10^{-14} \left(\text{mol/L}\right)^2$$

Note that the dimensions of K are moles per liter and those of K_w are moles² per liter². As its name suggests, the ion product K_w is numerically equal to the product of the molar concentrations of H^+ and OH^-:

$$K_w = \left[H^+\right]\left[OH^-\right]$$

At 25°C, $K_w = (10^{-7})^2$, or 10^{-14} (mol/L)². At temperatures below 25°C, K_w is somewhat less than 10^{-14}, and at temperatures above 25°C it is somewhat greater than 10^{-14}. Within the stated limitations of the effect of temperature, K_w equals 10^{-14} (mol/L)² for all aqueous solutions, even solutions of acids or bases. We use K_w to calculate the pH of acidic and basic solutions.

pH IS THE NEGATIVE LOG OF THE HYDROGEN ION CONCENTRATION

The term **pH** was introduced in 1909 by Sörensen, who defined pH as the negative log of the hydrogen ion concentration:

$$pH = -\log\left[H^+\right]$$

This definition, while not rigorous, suffices for many biochemical purposes. To calculate the pH of a solution:

1. Calculate the hydrogen ion concentration [H^+].
2. Calculate the base 10 logarithm of [H^+].
3. pH is the negative of the value found in step 2.

For example, for pure water at 25°C,

$$pH = -\log\left[H^+\right] = -\log 10^{-7} = -(-7) = 7.0$$

This value is also known as the *power* (English), *puissant* (French), or *potennz* (German) of the exponent, hence the use of the term "p."

Low pH values correspond to high concentrations of H^+ and high pH values correspond to low concentrations of H^+.

Acids are **proton donors** and bases are **proton acceptors.** **Strong acids** (eg, HCl, H_2SO_4) completely dissociate into anions and cations even in strongly acidic solutions (low pH). **Weak acids** dissociate only partially in acidic solutions. Similarly, **strong bases** (eg, KOH, NaOH)—but not **weak bases** (eg, $Ca[OH]_2$)—are completely dissociated at high pH. Many biochemicals are weak acids. Exceptions include phosphorylated intermediates, whose phosphoryl group contains two dissociable protons, the first of which is strongly acidic.

The following examples illustrate how to calculate the pH of acidic and basic solutions.

Example 1: What is the pH of a solution whose hydrogen ion concentration is 3.2×10^{-4} mol/L?

$$pH = -\log\left[H^+\right]$$
$$= -\log\left(3.2 \times 10^{-4}\right)$$
$$= -\log\left(3.2\right) - \log\left(10^{-4}\right)$$
$$= -0.5 + 4.0$$
$$= 3.5$$

Example 2: What is the pH of a solution whose hydroxide ion concentration is 4.0×10^{-4} mol/L? We first define a quantity **pOH** that is equal to $-\log[OH^-]$ and that may be derived from the definition of K_w:

$$K_w = \left[H^+\right]\left[OH^-\right] = 10^{-14}$$

Therefore

$$\log\left[H^+\right] + \log\left[OH^-\right] = \log 10^{-14}$$

or

$$pH + pOH = 14$$

To solve the problem by this approach:

$$\left[OH^-\right] = 4.0 \times 10^{-4}$$
$$pOH = -\log\left[OH^-\right]$$
$$= -\log\left(4.0 \times 10^{-4}\right)$$
$$= -\log\left(4.0\right) - \log\left(10^{-4}\right)$$
$$= -0.60 + 4.0$$
$$= 3.4$$

Now:

$$pH = 14 - pOH = 14 - 3.4$$
$$= 10.6$$

The examples above illustrate how the logarithmic pH scale facilitates reporting and comparing hydrogen ion concentrations that differ by orders of magnitude from one another, ie, 0.00032 M (pH 3.5) and 0.000000000025 M (pH 10.6).

Example 3: What are the pH values of (a) 2.0×10^{-2} mol/L KOH and of (b) 2.0×10^{-6} mol/L KOH? The OH^- arises from

two sources, KOH and water. Since pH is determined by the total $[H^+]$ (and pOH by the total $[OH^-]$), both sources must be considered. In the first case (a), the contribution of water to the total $[OH^-]$ is negligible. The same cannot be said for the second case (b):

	Concentration (mol/L)	
	(a)	**(b)**
Molarity of KOH	2.0×10^{-2}	2.0×10^{-6}
$[OH^-]$ from KOH	2.0×10^{-2}	2.0×10^{-6}
$[OH^-]$ from water	1.0×10^{-7}	1.0×10^{-7}
Total $[OH^-]$	2.00001×10^{-2}	2.1×10^{-6}

Once a decision has been reached about the significance of the contribution by water, pH may be calculated as above.

The above examples assume that the strong base KOH is completely dissociated in solution and that the concentration of OH^- ions was thus equal to that of the KOH plus that present initially in the water. This assumption is valid for dilute solutions of strong bases or acids but not for weak bases or acids. Since weak electrolytes dissociate only slightly in solution, we must use the **dissociation constant** to calculate the concentration of $[H^+]$ (or $[OH^-]$) produced by a given molarity of a weak acid (or base) before calculating total $[H^+]$ (or total $[OH^-]$) and subsequently pH.

Functional Groups That Are Weak Acids Have Great Physiologic Significance

Many biochemicals possess functional groups that are weak acids or bases. Carboxyl groups, amino groups, and phosphate esters, whose second dissociation falls within the physiologic range, are present in proteins and nucleic acids, most coenzymes, and most intermediary metabolites. Knowledge of the dissociation of weak acids and bases thus is basic to understanding the influence of intracellular pH on structure and biologic activity. Charge-based separations such as electrophoresis and ion exchange chromatography also are best understood in terms of the dissociation behavior of functional groups.

We term the protonated species (eg, HA or $R—NH_3^+$) the **acid** and the unprotonated species (eg, A^- or $R—NH_2$) its **conjugate base.** Similarly, we may refer to a **base** (eg, A^- or $R—NH_2$) and its **conjugate acid** (eg, HA or $R—NH_3^+$). Representative weak acids (left), their conjugate bases (center), and pK_a values (right) include the following:

$R—CH_2—COOH$	$R—CH_2—COO^-$	$pK_a = 4-5$
$R—CH_2—NH_3^+$	$R—CH_2—NH_2$	$pK_a = 9-10$
H_2CO_3	HCO_3^-	$pK_a = 6.4$
$H_2PO_4^-$	HPO_4^{-2}	$pK_a = 7.2$

We express the relative strengths of weak acids and bases in terms of their dissociation constants. Shown below are the expressions for the dissociation constant (K_a) for two representative weak acids, R—COOH and R—NH$_3^+$.

$$R\text{—COOH} \rightleftharpoons R\text{—COO}^- + H^+$$

$$K_a = \frac{\left[R\text{—COO}^-\right]\left[H^+\right]}{[R\text{—COOH}]}$$

$$R\text{—NH}_3^+ \rightleftharpoons R\text{—NH}_2 + H^+$$

$$K_a = \frac{\left[R\text{—NH}_2\right]\left[H^+\right]}{\left[R\text{—NH}_3^+\right]}$$

Since the numeric values of K_a for weak acids are negative exponential numbers, we express K_a as pK_a, where

$$pK_a = -\log K_a$$

Note that pK_a is related to K_a as pH is to [H$^+$]. The stronger the acid, the lower is its pK_a value.

pK_a is used to express the relative strengths of both acids and bases. For any weak acid, its conjugate is a strong base. Similarly, the conjugate of a strong base is a weak acid. The relative strengths of bases are expressed in terms of the pK_a of their conjugate acids. For polyprotic compounds containing more than one dissociable proton, a numerical subscript is assigned to each dissociation in order of relative acidity. For a dissociation of the type

$$R\text{—NH}_3^+ \rightarrow R\text{—NH}_2 + H^+$$

the pK_a is the pH at which the concentration of the acid R—NH$_3^+$ equals that of the base R—NH$_2$.

From the above equations that relate K_a to [H$^+$] and to the concentrations of undissociated acid and its conjugate base, when

$$\left[R\text{—COO}^-\right] = [R\text{—COOH}]$$

or when

$$\left[R\text{—NH}_2\right] = \left[R\text{—NH}_3^+\right]$$

then

$$K_a = \left[H^+\right]$$

Thus, when the associated (protonated) and dissociated (conjugate base) species are present at equal concentrations, the prevailing hydrogen ion concentration [H$^+$] is numerically equal to the dissociation constant, K_a. If the logarithms of both sides of the above equation are taken and both sides are multiplied by –1, the expressions would be as follows:

$$K_a = \left[H^+\right]$$

$$-\log K_a = -\log\left[H^+\right]$$

Since $-\log K_a$ is defined as pK_a, and $-\log$ [H$^+$] defines pH, the equation may be rewritten as

$$pK_a = pH$$

ie, the pK_a of an acid group is the pH at which the protonated and unprotonated species are present at equal concentrations.

The pK_a for an acid may be determined by adding 0.5 equivalent of alkali per equivalent of acid. The resulting pH will equal the pK_a of the acid.

The Henderson–Hasselbalch Equation Describes the Behavior of Weak Acids & Buffers

The Henderson–Hasselbalch equation is derived below.
A weak acid, HA, ionizes as follows:

$$HA \rightleftharpoons H^+ + A^-$$

The equilibrium constant for this dissociation is

$$K_a = \frac{\left[H^+\right]\left[A^-\right]}{[HA]}$$

Cross-multiplication gives

$$\left[H^+\right]\left[A^-\right] = K_a[HA]$$

Divide both sides by [A$^-$]:

$$\left[H^+\right] = K_a\frac{[HA]}{\left[A^-\right]}$$

Take the log of both sides:

$$\log\left[H^+\right] = \log\left(K_a\frac{[HA]}{\left[A^-\right]}\right)$$

$$= \log K_a + \log\frac{[HA]}{\left[A^-\right]}$$

Multiply through by –1:

$$-\log\left[H^+\right] = -\log K_a - \log\frac{[HA]}{\left[A^-\right]}$$

Substitute pH and pK_a for $-\log$ [H$^+$] and $-\log$ K_a, respectively; then:

$$pH = pK_a - \log\frac{[HA]}{\left[A^-\right]}$$

Inversion of the last term removes the minus sign and gives the Henderson–Hasselbalch equation:

$$\mathbf{pH = pK_a + \log\frac{\left[A^-\right]}{[HA]}}$$

The Henderson–Hasselbalch equation has great predictive value in protonic equilibria. For example,

1. When an acid is exactly half-neutralized, [A$^-$] = [HA]. Under these conditions,

$$pH = pK_a + \log\frac{\left[A^-\right]}{[HA]} = pK_a + \log\frac{1}{1} = pK_a + 0$$

Therefore, at half-neutralization, pH = pK_a.

2. When the ratio $[A^-]/[HA] = 100:1$,

$$pH = pK_a + \log \frac{[A^-]}{[HA]}$$

$$pH = pK_a + \log 100/1 = pK_a + 2$$

3. When the ratio $[A^-]/[HA] = 1:10$,

$$pH = pK_a + \log 1/1\,0 = pK_a + (-1)$$

If the equation is evaluated at ratios of $[A^-]/[HA]$ ranging from 10^3 to 10^{-3} and the calculated pH values are plotted, the resulting graph describes the titration curve for a weak acid (Figure 2–4).

Solutions of Weak Acids & Their Salts Buffer Changes in pH

Solutions of weak acids or bases and their conjugates exhibit buffering, the ability to resist a change in pH following addition of strong acid or base. Since many metabolic reactions are accompanied by the release or uptake of protons, most intracellular reactions are buffered. Oxidative metabolism produces CO_2, the anhydride of carbonic acid, which if not buffered would produce severe acidosis. Maintenance of a constant pH involves buffering by phosphate, bicarbonate, and proteins, which accept or release protons to resist a change in pH. For experiments using tissue extracts or enzymes, constant pH is maintained by the addition of buffers such as MES ([2-N-morpholino]ethanesulfonic acid, pK_a 6.1), inorganic orthophosphate (pK_{a2} 7.2), HEPES (N-hydroxyethylpiperazine-N'-2-ethanesulfonic acid, pK_a 6.8), or Tris (tris[hydroxymethyl] aminomethane, pK_a 8.3). The value of pK_a relative to the desired pH is the major determinant of which buffer is selected.

Buffering can be observed by using a pH meter while titrating a weak acid or base (Figure 2–4). We can also calculate the pH shift that accompanies addition of acid or base to a buffered solution. In the example, the buffered solution (a

weak acid, $pK_a = 5.0$, and its conjugate base) is initially at one of four pH values. We will calculate the pH shift that results when 0.1 meq of KOH is added to 1 meq of each solution:

Initial pH	5.00	5.37	5.60	5.86
$[A^-]_{initial}$	0.50	0.70	0.80	0.88
$[HA]_{initial}$	0.50	0.30	0.20	0.12
$([A^-]/[HA])_{initial}$	1.00	2.33	4.00	7.33
Addition of 0.1 meq of KOH produces				
$[A^-]_{final}$	0.60	0.80	0.90	0.98
$[HA]_{final}$	0.40	0.20	0.10	0.02
$([A^-]/[HA])_{final}$	1.50	4.00	9.00	49.0
$\log ([A^-]/[HA])_{final}$	0.18	0.60	0.95	1.69
Final pH	5.18	5.60	5.95	6.69
ΔpH	**0.18**	**0.60**	**0.95**	**1.69**

Notice that the change in pH per milliequivalent of OH^- added depends on the initial pH. The solution resists changes in pH most effectively at pH values close to the pK_a. A solution of a weak acid and its conjugate base buffers most effectively in the pH range $pK_a \pm 1.0$ pH unit.

Figure 2–4 also illustrates the net charge on one molecule of the acid as a function of pH. A fractional charge of –0.5 does not mean that an individual molecule bears a fractional charge but that the *probability* is 0.5 that a given molecule has a unit negative charge at any given moment in time. Consideration of the net charge on macromolecules as a function of pH provides the basis for separatory techniques such as ion exchange chromatography and electrophoresis.

Acid Strength Depends on Molecular Structure

Many acids of biologic interest possess more than one dissociating group. The presence of adjacent negative charge hinders the release of a proton from a nearby group, raising its pK_a. This is apparent from the pK_a values for the three dissociating groups of phosphoric acid and citric acid (Table 2–2). The effect of adjacent charge decreases with distance. The second pK_a for succinic acid, which has two methylene groups between its carboxyl groups, is 5.6, whereas the second pK_a for glutaric acid, which has one additional methylene group, is 5.4.

pK_a Values Depend on the Properties of the Medium

The pK_a of a functional group is also profoundly influenced by the surrounding medium. The medium may either raise or lower the pK_a depending on whether the undissociated acid

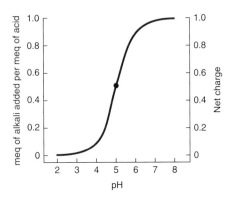

FIGURE 2–4   Titration curve for an acid of the type HA. The heavy dot in the center of the curve indicates the pK_a 5.0.

TABLE 2–2 Relative Strengths of Selected Acids of Biologic Significance[1]

Monoprotic Acids		
Formic	pK	3.75
Lactic	pK	3.86
Acetic	pK	4.76
Ammonium ion	pK	9.25
Diprotic Acids		
Carbonic	pK_1	6.37
	pK_2	10.25
Succinic	pK_1	4.21
	pK_2	5.64
Glutaric	pK_1	4.34
	pK_2	5.41
Triprotic Acids		
Phosphoric	pK_1	2.15
	pK_2	6.82
	pK_3	12.38
Citric	pK_1	3.08
	pK_2	4.74
	pK_3	5.40

[1]**Note:** Tabulated values are the pK_a values (−log of the dissociation constant) of selected monoprotic, diprotic, and triprotic acids.

or its conjugate base is the charged species. The effect of dielectric constant on pK_a may be observed by adding ethanol to water. The pK_a of a carboxylic acid *increases,* whereas that of an amine *decreases* because ethanol decreases the ability of water to solvate a charged species. The pK_a values of dissociating groups in the interiors of proteins thus are profoundly affected by their local environment, including the presence or absence of water.

SUMMARY

- Water forms hydrogen-bonded clusters with itself and with other proton donors or acceptors. Hydrogen bonds account for the surface tension, viscosity, liquid state at room temperature, and solvent power of water.
- Compounds that contain O, N, or S can serve as hydrogen bond donors or acceptors.
- Macromolecules exchange internal surface hydrogen bonds for hydrogen bonds to water. Entropic forces dictate that macromolecules expose polar regions to an aqueous interface and bury nonpolar regions.
- Salt bridges, hydrophobic interactions, and van der Waals forces participate in maintaining molecular structure.
- pH is the negative log of [H+]. A low pH characterizes an acidic solution, and a high pH denotes a basic solution.
- The strength of weak acids is expressed by pK_a, the negative log of the acid dissociation constant. Strong acids have low pK_a values and weak acids have high pK_a values.
- Buffers resist a change in pH when protons are produced or consumed. Maximum buffering capacity occurs ± 1 pH unit on either side of pK_a. Physiologic buffers include bicarbonate, orthophosphate, and proteins.

REFERENCES

Reese KM: Whence came the symbol pH. Chem & Eng News 2004;82:64.
Segel IM: *Biochemical Calculations*. Wiley, 1968.
Stillinger FH: Water revisited. Science 1980;209:451.
Suresh SJ, Naik VM: Hydrogen bond thermodynamic properties of water from dielectric constant data. J Chem Phys 2000;113:9727.
Wiggins PM: Role of water in some biological processes. Microbiol Rev 1990;54:432.

C H A P T E R

3

Amino Acids & Peptides

Peter J. Kennelly, PhD & Victor W. Rodwell, PhD

BIOMEDICAL IMPORTANCE

In addition to providing the monomer units from which the long polypeptide chains of proteins are synthesized, the L-α-amino acids and their derivatives participate in cellular functions as diverse as nerve transmission and the biosynthesis of porphyrins, purines, pyrimidines, and urea. Short polymers of amino acids called *peptides* perform prominent roles in the neuroendocrine system as hormones, hormone-releasing factors, neuromodulators, or neurotransmitters. While proteins contain only L-α-amino acids, microorganisms elaborate peptides that contain both D- and L-α-amino acids. Several of these peptides are of therapeutic value, including the antibiotics bacitracin and gramicidin A and the antitumor agent bleomycin. Certain other microbial peptides are toxic. The cyanobacterial peptides microcystin and nodularin are lethal in large doses, while small quantities promote the formation of hepatic tumors. Humans and other higher animals lack the capability to synthesize 10 of the 20 common L-α-amino acids in amounts adequate to support infant growth or to maintain health in adults. Consequently, the human diet must contain adequate quantities of these nutritionally essential amino acids.

PROPERTIES OF AMINO ACIDS

The Genetic Code Specifies 20 L-α-Amino Acids

Of the over 300 naturally occurring amino acids, 20 constitute the monomer units of proteins. While a nonredundant three-letter genetic code could accommodate more than 20 amino

acids, its redundancy limits the available codons to the 20 L-α-amino acids listed in Table 3–1, classified according to the polarity of their R groups. Both one- and three-letter abbreviations for each amino acid can be used to represent the amino acids in peptides and proteins (Table 3–1). Some proteins contain additional amino acids that arise by modification of an amino acid already present in a peptide. Examples include conversion of peptidyl proline and lysine to 4-hydroxyproline and 5-hydroxylysine; the conversion of peptidyl glutamate to γ-carboxyglutamate; and the methylation, formylation, acetylation, prenylation, and phosphorylation of certain aminoacyl residues. These modifications extend the biologic diversity of proteins by altering their solubility, stability, and interaction with other proteins.

Selenocysteine, the 21st L-α-Amino Acid?

Selenocysteine is an L-α-amino acid found in a handful of proteins, including certain peroxidases and reductases where it participates in the catalysis of electron transfer reactions. As its name implies, a selenium atom replaces the sulfur of its structural analog, cysteine. The pK_3 of selenocysteine, 5.2, is 3 units lower than that of cysteine. Since selenocysteine is inserted into polypeptides during translation, it is commonly referred to as the "21st amino acid." However, unlike the other 20 genetically encoded amino acids, selenocysteine is not specified by a simple three-letter codon (see Chapter 27).

Only L-α-Amino Acids Occur in Proteins

With the sole exception of glycine, the α-carbon of amino acids is chiral. Although some protein amino acids are dextrorotato-

TABLE 3–1   ʟ-α-Amino Acids Present in Proteins

Name	Symbol	Structural Formula	pK_1	pK_2	pK_3
With Aliphatic Side Chains			α-COOH	α-NH₃⁺	R Group
Glycine	Gly [G]	H—CH—COO⁻ | NH₃⁺	2.4	9.8	
Alanine	Ala [A]	CH₃—CH—COO⁻ | NH₃⁺	2.4	9.9	
Valine	Val [V]	H₃C \ CH—CH—COO⁻ / | H₃C NH₃⁺	2.2	9.7	
Leucine	Leu [L]	H₃C \ CH—CH₂—CH—COO⁻ / | H₃C NH₃⁺	2.3	9.7	
Isoleucine	Ile [I]	CH₃ | CH₂ | CH—CH—COO⁻ | | CH₃ NH₃⁺	2.3	9.8	
With Side Chains Containing Hydroxylic (OH) Groups					
Serine	Ser [S]	CH₂—CH—COO⁻ | | OH NH₃⁺	2.2	9.2	about 13
Threonine	Thr [T]	CH₃—CH—CH—COO⁻ | | OH NH₃⁺	2.1	9.1	about 13
Tyrosine	Tyr [Y]	See below.			
With Side Chains Containing Sulfur Atoms					
Cysteine	Cys [C]	CH₂—CH—COO⁻ | | SH NH₃⁺	1.9	10.8	8.3
Methionine	Met [M]	CH₂—CH₂—CH—COO⁻ | | S—CH₃ NH₃⁺	2.1	9.3	
With Side Chains Containing Acidic Groups or Their Amides					
Aspartic acid	Asp [D]	⁻OOC—CH₂—CH—COO⁻ | NH₃⁺	2.1	9.9	3.9
Asparagine	Asn [N]	H₂N—C—CH₂—CH—COO⁻ || | O NH₃⁺	2.1	8.8	
Glutamic acid	Glu [E]	⁻OOC—CH₂—CH₂—CH—COO⁻ | NH₃⁺	2.1	9.5	4.1
Glutamine	Gln [Q]	H₂N—C—CH₂—CH₂—CH—COO⁻ || | O NH₃⁺	2.2	9.1	

(continued)

TABLE 3–1 L-α-Amino Acids Present in Proteins (continued)

Name	Symbol	Structural Formula	pK$_1$	pK$_2$	pK$_3$
With Side Chains Containing Basic Groups			α-COOH	α-NH$_3$⁺	R Group
Arginine	Arg [R]		1.8	9.0	12.5
Lysine	Lys [K]		2.2	9.2	10.8
Histidine	His [H]		1.8	9.3	6.0
Containing Aromatic Rings					
Histidine	His [H]	See above.			
Phenylalanine	Phe [F]		2.2	9.2	
Tyrosine	Tyr [Y]		2.2	9.1	10.1
Tryptophan	Trp [W]		2.4	9.4	
Imino Acid					
Proline	Pro [P]		2.0	10.6	

ry and some levorotatory, all share the absolute configuration of L-glyceraldehyde and thus are L-α-amino acids. Several free L-α-amino acids fulfill important roles in metabolic processes. Examples include ornithine, citrulline, and argininosuccinate that participate in urea synthesis; tyrosine in formation of thyroid hormones; and glutamate in neurotransmitter biosynthesis. D-Amino acids that occur naturally include free D-serine and D-aspartate in brain tissue, D-alanine and D-glutamate in the cell walls of gram-positive bacteria, and D-amino acids in certain peptides and antibiotics produced by bacteria, fungi, reptiles, and other nonmammalian species.

Amino Acids May Have Positive, Negative, or Zero Net Charge

Charged and uncharged forms of the ionizable —COOH and —NH$_3$⁺ weak acid groups exist in solution in protonic equilibrium:

$$R-COOH \rightleftarrows R-COO^- + H^+$$
$$R-NH_3^+ \rightleftarrows R-NH_2 + H^+$$

While both R—COOH and R—NH$_3$⁺ are weak acids, R—COOH is a far stronger acid than R—NH$_3$⁺. At physiologic pH (pH 7.4), carboxyl groups exist almost entirely as R—COO⁻ and amino groups predominantly as R—NH$_3$⁺. Figure 3–1 illustrates the effect of pH on the charged state of aspartic acid.

Molecules that contain an equal number of ionizable groups of opposite charge and that therefore bear no net charge are termed **zwitterions.** Amino acids in blood and most tissues thus should be represented as in **A,** below.

FIGURE 3–1 Protonic equilibria of aspartic acid.

Structure **B** cannot exist in aqueous solution because at any pH low enough to protonate the carboxyl group, the amino group would also be protonated. Similarly, at any pH sufficiently high for an uncharged amino group to predominate, a carboxyl group will be present as R—COO⁻. The uncharged representation **B** is, however, often used for reactions that do not involve protonic equilibria.

pKₐ Values Express the Strengths of Weak Acids

The acid strengths of weak acids are expressed as their **pKₐ**. For molecules with multiple dissociable protons, the pKₐ for each acidic group is designated by replacing the subscript "a" with a number (Table 3–1). The imidazole group of histidine and the guanidino group of arginine exist as resonance hybrids with positive charge distributed between both nitrogens (histidine) or all three nitrogens (arginine) (Figure 3–2). The net charge on an amino acid—the algebraic sum of all the positively and negatively charged groups present—depends upon the pKₐ values of its functional groups and on the pH of the surrounding medium. Altering the charge on amino acids and their derivatives by varying the pH facilitates the physical separation of amino acids, peptides, and proteins (see Chapter 4).

At Its Isoelectric pH (pI), an Amino Acid Bears No Net Charge

Zwitterions are one example of an **isoelectric** species—the form of a molecule that has an equal number of positive and

FIGURE 3–2 Resonance hybrids of the protonated forms of the R groups of histidine and arginine.

negative charges and thus is electrically neutral. The isoelectric pH, also called the pI, is the pH midway between pKₐ values on either side of the isoelectric species. For an amino acid such as alanine that has only two dissociating groups, there is no ambiguity. The first pKₐ (R—COOH) is 2.35 and the second pKₐ (R—NH₃⁺) is 9.69. The isoelectric pH (pI) of alanine thus is

$$pI = \frac{pK_1 + pK_2}{2} = \frac{2.35 + 9.69}{2} = 6.02$$

For polyfunctional acids, pI is also the pH midway between the pKₐ values on either side of the isoionic species. For example, the pI for aspartic acid is

$$pI = \frac{pK_1 + pK_2}{2} = \frac{2.09 + 3.96}{2} = 3.02$$

For lysine, pI is calculated from:

$$pI = \frac{pK_2 + pK_3}{2}$$

Similar considerations apply to all polyprotic acids (eg, proteins), regardless of the number of dissociating groups present. In the clinical laboratory, knowledge of the pI guides selection of conditions for electrophoretic separations. For example, electrophoresis at pH 7.0 will separate two molecules with pI values of 6.0 and 8.0, because at pH 7.0 the molecule with a pI of 6.0 will have a net positive charge, and that with a pI of 8.0, a net negative charge. Similar considerations apply to understanding chromatographic separations on ionic supports such as diethylaminoethyl (DEAE) cellulose (see Chapter 4).

pKₐ Values Vary With the Environment

The environment of a dissociable group affects its pKₐ. The pKₐ values of the R groups of free amino acids in aqueous solution (Table 3–1) thus provide only an approximate guide to the pKₐ values of the same amino acids when present in proteins. A polar environment favors the charged form (R—COO⁻ or R—NH₃⁺), and a nonpolar environment favors the uncharged form (R—COOH or R—NH₂). A nonpolar environment thus *raises* the pKₐ of a carboxyl group (making it a weaker acid) but

TABLE 3–2   Typical Range of pK_a Values for Ionizable Groups in Proteins

Dissociating Group	pK_a Range
α-Carboxyl	3.5–4.0
Non-α COOH of Asp or Glu	4.0–4.8
Imidazole of His	6.5–7.4
SH of Cys	8.5–9.0
OH of Tyr	9.5–10.5
α-Amino	8.0–9.0
ε-Amino of Lys	9.8–10.4
Guanidinium of Arg	~12.0

lowers that of an amino group (making it a stronger acid). The presence of adjacent charged groups can reinforce or counteract solvent effects. The pK_a of a functional group thus will depend upon its location within a given protein. Variations in pK_a can encompass whole pH units (Table 3–2). pK_a values that diverge from those listed by as much as 3 pH units are common at the active sites of enzymes. An extreme example, a buried aspartic acid of thioredoxin, has a pK_a above 9—a shift of more than 6 pH units!

The Solubility of Amino Acids Reflects Their Ionic Character

The charged functional groups of amino acids ensure that they are readily solvated by—and thus soluble in—polar solvents such as water and ethanol but insoluble in nonpolar solvents such as benzene, hexane, or ether.

Amino acids do not absorb visible light and thus are colorless. However, tyrosine, phenylalanine, and especially tryptophan absorb high-wavelength (250–290 nm) ultraviolet light. Because it absorbs ultraviolet light about ten times more efficiently than either phenylalanine or tyrosine, tryptophan makes the major contribution to the ability of most proteins to absorb light in the region of 280 nm.

THE α-R GROUPS DETERMINE THE PROPERTIES OF AMINO ACIDS

Since glycine, the smallest amino acid, can be accommodated in places inaccessible to other amino acids, it often occurs where peptides bend sharply. The hydrophobic R groups of alanine, valine, leucine, and isoleucine and the aromatic R groups of phenylalanine, tyrosine, and tryptophan typically occur primarily in the interior of cytosolic proteins. The charged R groups of basic and acidic amino acids stabilize specific protein conformations via ionic interactions, or salt bridges. These interactions also function in "charge relay" systems dur-

ing enzymatic catalysis and electron transport in respiring mitochondria. Histidine plays unique roles in enzymatic catalysis. The pK_a of its imidazole proton permits it to function at neutral pH as either a base or an acid catalyst. The primary alcohol group of serine and the primary thioalcohol (—SH) group of cysteine are excellent nucleophiles and can function as such during enzymatic catalysis. However, the secondary alcohol group of threonine, while a good nucleophile, does not fulfill an analogous role in catalysis. The —OH groups of serine, tyrosine, and threonine also participate in regulation of the activity of enzymes whose catalytic activity depends on the phosphorylation state of these residues.

FUNCTIONAL GROUPS DICTATE THE CHEMICAL REACTIONS OF AMINO ACIDS

Each functional group of an amino acid exhibits all of its characteristic chemical reactions. For carboxylic acid groups, these reactions include the formation of esters, amides, and acid anhydrides; for amino groups, acylation, amidation, and esterification; and for —OH and —SH groups, oxidation and esterification. The most important reaction of amino acids is the formation of a peptide bond (shaded).

Alanyl   Cysteinyl   Valine

Amino Acid Sequence Determines Primary Structure

The number and order of all of the amino acid residues in a polypeptide constitute its primary structure. Amino acids present in peptides are called aminoacyl residues and are named by replacing the *-ate* or *-ine* suffixes of free amino acids with *-yl* (eg, alan*yl*, aspart*yl*, tyros*yl*). Peptides are then named as derivatives of the carboxyl terminal aminoacyl residue. For example, Lys-Leu-Tyr-Gln is called lys*yl*-leuc*yl*-tyros*yl*-glutam*ine*. The *-ine* ending on glutamine indicates that its α-carboxyl group is *not* involved in peptide bond formation.

Peptide Structures Are Easy to Draw

Prefixes like *tri-* or *octa-* denote peptides with three or eight **residues,** respectively. By convention, peptides are written with the residue that bears the free α-amino group at the left. To draw a peptide, use a zigzag to represent the main chain or backbone. Add the main chain atoms, which occur in the repeating order: α-nitrogen, α-carbon, carbonyl carbon. Now add a hydrogen atom to each α-carbon and to each peptide nitrogen, and an oxygen to the carbonyl carbon. Finally, add the appropriate R groups (shaded) to each α-carbon atom.

Three-letter abbreviations linked by straight lines represent an unambiguous primary structure. Lines are omitted for single-letter abbreviations.

Glu - Ala - Lys - Gly - Tyr - Ala
E   A   K   G   Y   A

Some Peptides Contain Unusual Amino Acids

In mammals, peptide hormones typically contain only the α-amino acids of proteins linked by standard peptide bonds. Other peptides may, however, contain nonprotein amino acids, derivatives of the protein amino acids, or amino acids linked by an atypical peptide bond. For example, the amino terminal glutamate of glutathione, which participates in protein folding and in the metabolism of xenobiotics (Chapter 53), is linked to cysteine by a non-α peptide bond (Figure 3–3). The amino terminal glutamate of thyrotropin-releasing hormone (TRH) is cyclized to pyroglutamic acid, and the carboxyl group of the carboxyl terminal prolyl residue is amidated. The nonprotein amino acids D-phenylalanine and ornithine are present in the cyclic peptide antibiotics tyrocidin and gramicidin S, while the heptapeptide opioids dermorphin and deltophorin in the skin of South American tree frogs contain D-tyrosine and D-alanine.

Peptides Are Polyelectrolytes

The peptide bond is uncharged at any pH of physiologic interest. Formation of peptides from amino acids is therefore accompanied by a net loss of one positive and one negative

charge per peptide bond formed. Peptides nevertheless are charged at physiologic pH owing to their carboxyl and amino terminal groups and, where present, their acidic or basic R groups. As for amino acids, the net charge on a peptide depends on the pH of its environment and on the pK_a values of its dissociating groups.

The Peptide Bond Has Partial Double-Bond Character

Although peptides are written as if a single bond linked the α-carboxyl and α-nitrogen atoms, this bond in fact exhibits partial double-bond character:

There thus is no freedom of rotation about the bond that connects the carbonyl carbon and the nitrogen of a peptide bond. Consequently, the O, C, N and H atoms of a peptide bond are coplanar. The imposed semi-rigidity of the peptide bond has important consequences for the manner in which peptides and proteins fold to generate higher orders of structure. Encircling brown arrows (Figure 3–4) indicate free rotation about the remaining bonds of the polypeptide backbone.

Noncovalent Forces Constrain Peptide Conformations

Folding of a peptide probably occurs coincident with its biosynthesis (see Chapter 37). The physiologically active confor-

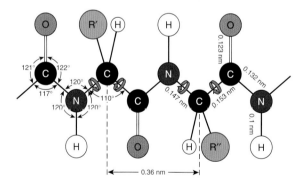

FIGURE 3–4 Dimensions of a fully extended polypeptide chain. The four atoms of the peptide bond are coplanar. Free rotation can occur about the bonds that connect the α-carbon with the α-nitrogen and with the α-carbonyl carbon (brown arrows). The extended polypeptide chain is thus a semirigid structure with two-thirds of the atoms of the backbone held in a fixed planar relationship one to another. The distance between adjacent α-carbon atoms is 0.36 nm (3.6 Å). The interatomic distances and bond angles, which are not equivalent, are also shown. (Redrawn and reproduced, with permission, from Pauling L, Corey LP, Branson HR: The structure of proteins: Two hydrogen-bonded helical configurations of the polypeptide chain. Proc Natl Acad Sci U S A 1951;37:205.)

FIGURE 3–3 Glutathione (γ-glutamyl-cysteinyl-glycine). Note the non-α peptide bond that links Glu to Cys.

mation reflects the collective contributions of the amino acid sequence, steric hindrance, and noncovalent interactions (eg, hydrogen bonding, hydrophobic interactions) between residues. Common conformations include α-helices and β-pleated sheets (see Chapter 5).

ANALYSIS OF THE AMINO ACID CONTENT OF BIOLOGIC MATERIALS

In order to determine the identity and quantity of each amino acid in a sample of biologic material, it is first necessary to hydrolyze the peptide bonds that link the amino acids together by treatment with hot HCl. The resulting mixture of free amino acids is then treated with 6-amino-*N*-hydroxysuccinimidyl carbamate, which reacts with their α-amino groups to form fluorescent derivatives that are separated and identified using high-pressure liquid chromatography (see Chapter 4). Ninhydrin, also widely used for detecting amino acids, forms a purple product with α-amino acids and a yellow adduct with the imine groups of proline and hydroxyproline.

SUMMARY

- Both D-amino acids and non-α-amino acids occur in nature, but only L-α-amino acids are present in proteins.
- All amino acids possess at least two weakly acidic functional groups, $R-NH_3^+$ and $R-COOH$. Many also possess additional weakly acidic functional groups such as —OH, —SH, guanidino, or imidazole moieties.
- The pK_a values of all functional groups of an amino acid dictate its net charge at a given pH. pI is the pH at which an amino

acid bears no net charge and thus does not move in a direct current electrical field.

- Of the biochemical reactions of amino acids, the most important is the formation of peptide bonds.
- The R groups of amino acids determine their unique biochemical functions. Amino acids are classified as basic, acidic, aromatic, aliphatic, or sulfur-containing based on the properties of their R groups.
- Peptides are named for the number of amino acid residues present, and as derivatives of the carboxyl terminal residue. The primary structure of a peptide is its amino acid sequence, starting from the amino-terminal residue.
- The partial double-bond character of the bond that links the carbonyl carbon and the nitrogen of a peptide renders four atoms of the peptide bond coplanar and restricts the number of possible peptide conformations.

REFERENCES

Doolittle RF: Reconstructing history with amino acid sequences. Protein Sci 1992;1:191.

Gladyschev VN, Hatfield DL: Selenocysteine-containing proteins in mammals. J Biomed Sci 1999;6:151.

Kreil G: D-Amino acids in animal peptides. Annu Rev Biochem 1997;66:337.

Nokihara K, Gerhardt J: Development of an improved automated gas-chromatographic chiral analysis system: application to nonnatural amino acids and natural protein hydrolysates. Chirality 2001;13:431.

Sanger F: Sequences, sequences, and sequences. Annu Rev Biochem 1988;57:1.

Stadtman TC: Selenocysteine. Annu Rev Biochem 1996;65:83.

Wilson NA et al: Aspartic acid 26 in reduced *Escherichia coli* thioredoxin has a pK_a greater than 9. Biochemistry 1995;34:8931.

Proteins: Determination of Primary Structure

Peter J. Kennelly, PhD & Victor W. Rodwell, PhD

BIOMEDICAL IMPORTANCE

Proteins are physically and functionally complex macromolecules that perform multiple critically important roles. An internal protein network, the cytoskeleton (Chapter 49), maintains cellular shape and physical integrity. Actin and myosin filaments form the contractile machinery of muscle (Chapter 49). Hemoglobin transports oxygen (Chapter 6), while circulating antibodies search out foreign invaders (Chapter 50). Enzymes catalyze reactions that generate energy, synthesize and degrade biomolecules, replicate and transcribe genes, process mRNAs, etc (Chapter 7). Receptors enable cells to sense and respond to hormones and other environmental cues (Chapters 41 & 42). Proteins are subject to physical and functional changes that mirror the life cycle of the organisms in which they reside. A typical protein is born at translation (Chapter 37), matures through posttranslational processing events such as partial proteolysis (Chapters 9 & 37), alternates between working and resting states through the intervention of regulatory factors (Chapter 9), ages through oxidation, deamidation, etc (Chapter 52), and dies when it is degraded to its component amino acids (Chapter 29). An important goal of molecular medicine is the identification of proteins and those events in their life cycle whose presence, absence, or deficiency is associated with specific physiologic states or diseases (Figure 4–1). The primary sequence of a protein provides both a molecular fingerprint for its identification and information that can be used to identify and clone the gene or genes that encode it.

PROTEINS & PEPTIDES MUST BE PURIFIED PRIOR TO ANALYSIS

Highly purified protein is essential for the detailed examination of its physical and functional properties. Cells contain thousands of different proteins, each in widely varying amounts. The isolation of a specific protein in quantities sufficient for analysis thus presents a formidable challenge that may require multiple successive purification techniques. Classic approaches exploit differences in relative solubility of individual proteins as a function of pH (isoelectric precipitation), polarity (precipitation with ethanol or acetone), or salt concentration (salting out with ammonium sulfate). Chromato-

graphic separations partition molecules between two phases, one mobile and the other stationary. For separation of amino acids or sugars, the stationary phase, or matrix, may be a sheet of filter paper (paper chromatography) or a thin layer of cellulose, silica, or alumina (thin-layer chromatography [TLC]).

Column Chromatography

Column chromatography of proteins employs as the stationary phase small spherical beads of modified cellulose, acrylamide, or silica whose surface typically has been coated with chemical functional groups. The beads are packed in a cylindrical container, or column, comprised of glass, plastic, or metal. These stationary phase matrices interact with proteins based on their charge, hydrophobicity, and ligand-binding properties. A protein mixture is applied to the column and the liquid mobile phase is percolated through it. Small portions of the mobile phase or eluant are collected as they emerge (Figure 4–2).

Partition Chromatography

Column chromatographic separations depend on the relative affinity of different proteins for a given stationary phase and for the mobile phase. In partition chromatography, association between each protein and the matrix is weak and transient. Proteins that interact more strongly with the stationary phase are retained longer. The length of time that a protein is associated with the stationary phase is a function of the composition of both the stationary and mobile phases. Optimal separation of the protein of interest from other proteins thus can be achieved by careful manipulation of the composition of the two phases.

Size Exclusion Chromatography

Size exclusion—or gel filtration—chromatography separates proteins based on their **Stokes radius**, the radius of the sphere they occupy as they tumble in solution. The Stokes radius is a function of molecular mass and shape. A tumbling elongated protein occupies a larger volume than a spherical protein of the same mass. Size-exclusion chromatography employs porous beads (Figure 4–3). The pores are analogous to indentations in a river bank. As objects move downstream, those that enter an indentation are retarded until they drift back into the

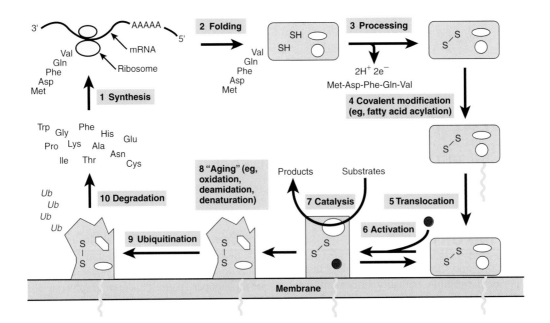

FIGURE 4–1 Diagrammatic representation of the life cycle of a hypothetical protein. **(1)** The life cycle begins with the synthesis on a ribosome of a polypeptide chain, whose primary structure is dictated by an mRNA. **(2)** As synthesis proceeds, the polypeptide begins to fold into its native conformation (blue). **(3)** Folding may be accompanied by processing events such as proteolytic cleavage of an *N*-terminal leader sequence (Met-Asp-Phe-Gln-Val)) or the formation of disulfide bonds (S—S). **(4)** Subsequent covalent modification may, for example, attach a fatty acid molecule (yellow) for **(5)** translocation of the modified protein to a membrane. **(6)** Binding an allosteric effector (red) may trigger the adoption of a catalytically active conformation. **(7)** Over time, proteins become damaged by chemical attack, deamidation, or denaturation, and **(8)** may be "labeled" by the covalent attachment of several ubiquitin molecules (Ub). **(9)** The ubiquitinated protein is subsequently degraded to its component amino acids, which become available for the synthesis of new proteins.

main current. Similarly, proteins with Stokes radii too large to enter the pores (excluded proteins) remain in the flowing mobile phase and emerge *before* proteins that can enter the pores (included proteins). Proteins thus emerge from a gel filtration column in descending order of their Stokes radii.

Absorption Chromatography

For absorption chromatography, the protein mixture is applied to a column under conditions where the protein of interest associates with the stationary phase so tightly that its partition coefficient is essentially unity. Nonadhering molecules are first eluted and discarded. Proteins are then sequentially released by disrupting the forces that stabilize the protein-stationary phase complex, most often by using a gradient of increasing salt concentration. The composition of the mobile phase is altered gradually so that molecules are selectively released in descending order of their affinity for the stationary phase.

Ion Exchange Chromatography

In ion exchange chromatography, proteins interact with the stationary phase by charge-charge interactions. Proteins with a net positive charge at a given pH adhere to beads with negatively charged functional groups such as carboxylates or sulfates (cat-

ion exchangers). Similarly, proteins with a net negative charge adhere to beads with positively charged functional groups, typically tertiary or quaternary amines (anion exchangers). Proteins, which are polyanions, compete against monovalent ions for binding to the support—thus the term "ion exchange." For example, proteins bind to diethylaminoethyl (DEAE) cellulose by replacing the counter-ions (generally Cl^- or CH_3COO^-) that neutralize the protonated amine. Bound proteins are selectively displaced by gradually raising the concentration of monovalent ions in the mobile phase. Proteins elute in inverse order of the strength of their interactions with the stationary phase.

Since the net charge on a protein is determined by the pH (see Chapter 3), sequential elution of proteins may be achieved by changing the pH of the mobile phase. Alternatively, a protein can be subjected to consecutive rounds of ion exchange chromatography, each at a different pH, such that proteins that coelute at one pH elute at different salt concentrations at another pH.

Hydrophobic Interaction Chromatography

Hydrophobic interaction chromatography separates proteins based on their tendency to associate with a stationary phase matrix coated with hydrophobic groups (eg, phenyl Sepharose, octyl Sepharose). Proteins with exposed hydrophobic

FIGURE 4–2 Components of a typical liquid chromatography apparatus. **R1** and **R2:** Reservoirs of mobile phase liquid. **P:** Programable pumping system containing two pumps, 1 and 2, and a mixing chamber, **M**. The system can be set to pump liquid from only one reservoir, to switch reservoirs at some predetermined point to generate a step gradient, or to mix liquids from the to reservoirs in proportions that vary over time to create a continuous gradient. **C:** Glass, metal, or plastic column containing stationary phase. **F:** Fraction collector for collecting portions, called *fractions,* of the eluant liquid in separate test tubes.

surfaces adhere to the matrix via hydrophobic interactions that are enhanced by a mobile phase of high ionic strength. Nonadherent proteins are first washed away. The polarity of the mobile phase is then decreased by gradually lowering the salt concentration. If the interaction between protein and stationary phase is particularly strong, ethanol or glycerol may be added to the mobile phase to decrease its polarity and further weaken hydrophobic interactions.

Affinity Chromatography

Affinity chromatography exploits the high selectivity of most proteins for their ligands. Enzymes may be purified by affinity chromatography using immobilized substrates, products, coenzymes, or inhibitors. In theory, only proteins that interact with the immobilized ligand adhere. Bound proteins are then eluted either by competition with soluble ligand or, less selectively, by disrupting protein-ligand interactions using urea, guanidine hydrochloride, mildly acidic pH, or high salt concentrations. Stationary phase matrices available commercially contain ligands such as NAD^+ or ATP analogs. Among the most powerful and widely applicable affinity matrices are those used for the purification of suitably modified recombinant proteins. These include a Ni^{2+} matrix that binds proteins with an attached poly-

histidine "tag," and a glutathione matrix that binds a recombinant protein linked to glutathione *S*-transferase.

Peptides Are Purified by Reversed-Phase High-Pressure Chromatography

The stationary phase matrices used in classic column chromatography are spongy materials whose compressibility limits flow of the mobile phase. High-pressure liquid chromatography (HPLC) employs incompressible silica or alumina microbeads as the stationary phase and pressures of up to a few thousand psi. Incompressible matrices permit both high flow rates and enhanced resolution. HPLC can resolve complex mixtures of lipids or peptides whose properties differ only slightly. Reversed-phase HPLC exploits a hydrophobic stationary phase of aliphatic polymers 3–18 carbon atoms in length. Peptide mixtures are eluted using a gradient of a water-miscible organic solvent such as acetonitrile or methanol.

Protein Purity Is Assessed by Polyacrylamide Gel Electrophoresis (PAGE)

The most widely used method for determining the purity of a protein is SDS-PAGE—polyacrylamide gel electrophoresis (PAGE)

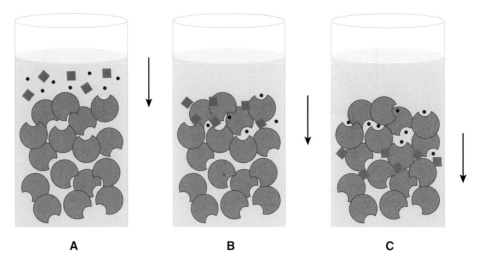

FIGURE 4–3 Size-exclusion chromatography. **A:** A mixture of large molecules (brown) and small molecules (red) are applied to the top of a gel filtration column. **B:** Upon entering the column, the small molecules enter pores in the stationary phase matrix (gray) from which the large molecules are excluded. **C:** As the mobile phase (blue) flows down the column, the large, excluded molecules flow with it, while the small molecules, which are temporarily sheltered from the flow when inside the pores, lag farther and farther behind.

in the presence of the anionic detergent sodium dodecyl sulfate (SDS). Electrophoresis separates charged biomolecules based on the rates at which they migrate in an applied electrical field. For SDS-PAGE, acrylamide is polymerized and cross-linked to form a porous matrix. SDS denatures and binds to proteins at a ratio of one molecule of SDS per two peptide bonds. When used in conjunction with 2-mercaptoethanol or dithiothreitol to reduce and break disulfide bonds (Figure 4–4), SDS-PAGE separates the component polypeptides of multimeric proteins. The large number of anionic SDS molecules, each bearing a charge of −1, overwhelms the charge contributions of the amino acid functional groups endogenous to the polypeptides. Since the charge-to-mass ratio of each SDS-polypeptide complex is approximately equal, the physical resistance each peptide encounters as it moves through the acrylamide matrix determines the rate of migration. Since large complexes encounter greater resistance, polypeptides separate based on their relative molecular mass (M_r). Individual polypeptides trapped in the acrylamide gel are visualized by staining with dyes such as Coomassie blue (Figure 4–5).

Isoelectric Focusing (IEF)

Ionic buffers called ampholytes and an applied electric field are used to generate a pH gradient within a polyacrylamide matrix. Applied proteins migrate until they reach the region of the matrix where the pH matches their isoelectric point (pI), the pH at which a molecule's net charge is zero. IEF is used in conjunction with SDS-PAGE for two-dimensional electrophoresis, which separates polypeptides based on pI in one dimension and based on M_r in the second (Figure 4–6). Two-dimensional electrophoresis is particularly well suited for separating the components of complex mixtures of proteins.

FIGURE 4–4 Oxidative cleavage of adjacent polypeptide chains linked by disulfide bonds (highlighted in blue) by performic acid **(left)** or reductive cleavage by β-mercaptoethanol **(right)** forms two peptides that contain cysteic acid residues or cysteinyl residues, respectively.

SANGER WAS THE FIRST TO DETERMINE THE SEQUENCE OF A POLYPEPTIDE

Mature insulin consists of the 21-residue A chain and the 30-residue B chain linked by disulfide bonds. Frederick Sanger

FIGURE 4–5 Use of SDS-PAGE to observe successive purification of a recombinant protein. The gel was stained with Coomassie blue. Shown are protein standards (lane **S**) of the indicated M_r, in kDa, crude cell extract **(E),** cytosol **(C),** high-speed supernatant liquid **(H),** and the DEAE-Sepharose fraction **(D).** The recombinant protein has a mass of about 45 kDa.

FIGURE 4–6 Two-dimensional IEF-SDS-PAGE. The gel was stained with Coomassie blue. A crude bacterial extract was first subjected to isoelectric focusing (IEF) in a pH 3–10 gradient. The IEF gel was then placed horizontally on the top of an SDS gel, and the proteins then further resolved by SDS-PAGE. Notice the greatly improved resolution of distinct polypeptides relative to ordinary SDS-PAGE gel (Figure 4–5).

reduced the disulfide bonds (Figure 4–4), separated the A and B chains, and cleaved each chain into smaller peptides using trypsin, chymotrypsin, and pepsin. The resulting peptides were then isolated and treated with acid to hydrolyze peptide bonds and generate peptides with as few as two or three amino acids. Each peptide was reacted with 1-fluoro-2,4-dini-

trobenzene (Sanger's reagent), which derivatizes the exposed α-amino groups of the amino-terminal residues. The amino acid content of each peptide was then determined and the amino-terminal amino acid identified. The ε-amino group of lysine also reacts with Sanger's reagent, but since an amino-terminal lysine reacts with 2 mol of Sanger's reagent, it is readily distinguished from a lysine in the interior of a peptide. Working from di- and tri-peptides up through progressively larger fragments, Sanger was able to reconstruct the complete sequence of insulin, an accomplishment for which he received a Nobel Prize in 1958.

THE EDMAN REACTION ENABLES PEPTIDES & PROTEINS TO BE SEQUENCED

Pehr Edman introduced phenylisothiocyanate (Edman's reagent) to selectively label the amino-terminal residue of a peptide. In contrast to Sanger's reagent, the phenylthiohydantoin (PTH) derivative can be removed under mild conditions to generate a new amino-terminal residue (Figure 4–7). Successive rounds of derivatization with Edman's reagent can therefore be used to sequence many residues of a single sample of peptide. While the first 20–30 residues of a peptide can readily be determined by the Edman method, most polypeptides contain several hundred amino acids. Consequently, most polypeptides must first be cleaved into smaller peptides prior to Edman sequencing. Cleavage also may be necessary to circumvent posttranslational modifications that render a protein's α-amino group "blocked," or unreactive with the Edman reagent.

It usually is necessary to generate several peptides using more than one method of cleavage. This reflects both inconsistency in the spacing of chemically or enzymatically susceptible cleavage sites and the need for sets of peptides whose sequences overlap so one can infer the sequence of the polypeptide from which they derive. Following cleavage, the resulting peptides are purified by reversed-phase HPLC and sequenced.

MOLECULAR BIOLOGY REVOLUTIONIZED THE DETERMINATION OF PRIMARY STRUCTURE

Knowledge of DNA sequences permits deduction of the primary structures of polypeptides. DNA sequencing requires only minute amounts of DNA and can readily yield the sequence of hundreds of nucleotides. To clone and sequence the DNA that encodes a particular protein, some means of identifying the correct clone—eg, knowledge of a portion of its nucleotide sequence—is essential. A hybrid approach thus has emerged. Edman sequencing is used to provide a partial amino acid sequence. Oligonucleotide primers modeled on

FIGURE 4–7 The Edman reaction. Phenylisothiocyanate derivatizes the amino-terminal residue of a peptide as a phenylthiohydantoic acid. Treatment with acid in a nonhydroxylic solvent releases a phenylthiohydantoin, which is subsequently identified by its chromatographic mobility, and a peptide one residue shorter. The process is then repeated.

posttranslational modifications such as proteolytic processing, methylation, glycosylation, phosphorylation, hydroxylation of proline and lysine, and disulfide bond formation that accompany maturation. While Edman sequencing can detect the presence of most posttranslational events, technical limitations often prevent identification of a specific modification.

MASS SPECTROMETRY DETECTS COVALENT MODIFICATIONS

On account of its superior sensitivity, speed, and versatility, mass spectrometry (MS) has replaced the Edman technique as the principal method for determining the sequences of peptides and proteins. Similarly, the posttranslational modification of proteins by the addition or deletion of carbohydrate moieties, phosphoryl, hydroxyl, or other groups adds or subtracts specific and readily identified increments of mass (Table 4–1). Mass spectrometry, which discriminates molecules based solely on their mass, thus can detect the comparatively subtle physical changes in proteins that occur during the life cycle of a cell or organism. In a simple, single quadrapole, mass spectrometer a sample in a vacuum is vaporized under conditions where protonation can occur, imparting positive charge. An electrical field then propels the cations through a magnetic field, which deflects them at a right angle to their original direction of flight and focuses them onto a detector (Figure 4–8). The magnetic force required to deflect the path of each ionic species onto the detector, measured as the current applied to the electromagnet, is recorded. For ions of identical net charge, this force is proportionate to their mass. In a time-of-flight mass spectrometer, a briefly applied electric field accelerates the ions towards a detector that records the time at which each ion arrives. For molecules of identical charge, the velocity to which they are accelerated—and hence the time required to reach the detector—will be inversely proportionate to their mass.

Conventional mass spectrometers generally are used to determine the masses of molecules of 4000 Da or less, whereas

this partial sequence can then be used to identify clones or to amplify the appropriate gene by the polymerase chain reaction (PCR) (see Chapter 39). Once an authentic DNA clone is obtained, its oligonucleotide sequence can be determined and the genetic code used to infer the primary structure of the encoded polypeptide.

The hybrid approach enhances the speed and efficiency of primary structure analysis and the range of proteins that can be sequenced. It also circumvents obstacles such as the presence of an amino-terminal blocking group or the lack of a key overlap peptide. Only a few segments of primary structure must be determined by Edman analysis.

DNA sequencing reveals the order in which amino acids are added to the nascent polypeptide chain as it is synthesized on the ribosome. However, it provides no information about

TABLE 4–1   Mass Increases Resulting from Common Posttranslational Modifications

Modification	Mass Increase (Da)
Phosphorylation	80
Hydroxylation	16
Methylation	14
Acetylation	42
Myristylation	210
Palmitoylation	238
Glycosylation	162

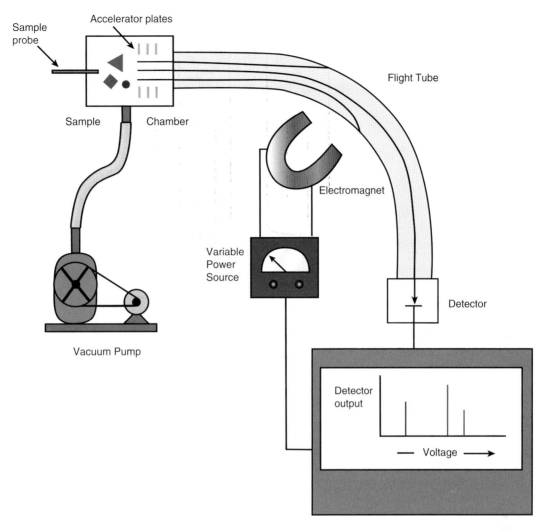

FIGURE 4–8 Basic components of a simple mass spectrometer. A mixture of molecules, represented by a red circle, green triangle, and blue diamond, is vaporized in an ionized state in the sample chamber. These molecules are then accelerated down the flight tube by an electrical potential applied to accelerator grid (yellow). An adjustable electromagnet applies a magnetic field that deflects the flight of the individual ions until they strike the detector. The greater the mass of the ion, the higher the magnetic field required to focus it onto the detector.

time-of-flight mass spectrometers are suited for determining the large masses of proteins. The analysis of peptides and proteins by mass spectometry initially was hindered by difficulties in volatilizing large organic molecules. While small organic molecules could be readily vaporized by heating in a vacuum (Figure 4–9), proteins, oligonucleotides, etc., were destroyed under these conditions. The two most commonly employed methods for dispersing peptides, proteins, and other large biomolecules into the vapor phase for mass spectrometric analysis are **electrospray ionization** and **matrix-assisted laser desorption and ionization,** or **MALDI**. In electrospray ionization, the molecules to be analyzed are dissolved in a volatile solvent and introduced into the sample chamber in a minute stream through a capillary (Figure 4–9). As the droplet of liquid emerges into the sample chamber, the solvent rapidly disperses leaving the macromolecule suspended in the

gaseous phase. The charged probe serves to ionize the sample. Electrospray ionization is frequently used to analyze peptides and proteins as they elute from an HPLC or other chromatography column. In MALDI, the sample is mixed with a liquid matrix containing a light-absorbing dye and a source of protons. In the sample chamber, the mixture is excited using a laser, causing the surrounding matrix to disperse into the vapor phase so rapidly as to avoid heating embedded peptides or proteins (Figure 4–9).

Peptides inside the mass spectrometer can be broken down into smaller units by collisions with neutral helium or argon atoms (collision-induced dissociation), and the masses of the individual fragments determined. Since peptide bonds are much more labile than carbon-carbon bonds, the most abundant fragments will differ from one another by units equivalent to one or two amino acids. Since—with the exceptions of

FIGURE 4–9 Three common methods for vaporizing molecules in the sample chamber of a mass spectrometer.

(1) leucine and isoleucine and (2) glutamine and lysine—the molecular mass of each amino acid is unique, the sequence of the peptide can be reconstructed from the masses of its fragments.

Tandem Mass Spectrometry

Complex peptide mixtures can now be analyzed, without prior purification, by tandem mass spectrometry, which employs the equivalent of two mass spectrometers linked in series. The first spectrometer separates individual peptides based upon their differences in mass. By adjusting the field strength of the first magnet, a single peptide can be directed into the second mass spectrometer, where fragments are generated and their masses determined.

Tandem Mass Spectrometry Can Detect Metabolic Abnormalities

Tandem mass spectrometry can be used to screen blood samples from newborns for the presence and concentrations of amino acids, fatty acids, and other metabolites. Abnormalities in metabolite levels can serve as diagnostic indicators for a variety of genetic disorders, such as phenylketonuria, ethylmalonic encephalopathy, and glutaric acidemia type 1.

GENOMICS ENABLES PROTEINS TO BE IDENTIFIED FROM SMALL AMOUNTS OF SEQUENCE DATA

Primary structure analysis has been revolutionized by genomics, the application of automated oligonucleotide sequencing and computerized data retrieval and analysis to sequence an organism's entire genetic complement. Since the determina-tion in 1995 of the complete genome sequence of *Haemophilus influenzae,* the genomes of hundreds of organisms have been deciphered. Where genome sequence is known, the task of determining a protein's DNA-derived primary sequence is materially simplified. In essence, the second half of the hybrid approach has already been completed. All that remains is to acquire sufficient information to permit the open reading frame (ORF) that encodes the protein to be retrieved from an Internet-accessible genome database and identified. In some cases, a segment of amino acid sequence only four or five residues in length may be sufficient to identify the correct ORF.

Computerized search algorithms assist the identification of the gene encoding a given protein. In peptide mass profiling, for example, a peptide digest is introduced into the mass spectrometer and the sizes of the peptides are determined. A computer is then used to find an ORF whose predicted protein product would, if broken down into peptides by the cleavage method selected, produce a set of peptides whose masses match those observed by MS.

PROTEOMICS & THE PROTEOME

The Goal of Proteomics Is to Identify the Entire Complement of Proteins Elaborated by a Cell Under Diverse Conditions

While the sequence of the human genome is known, the picture provided by genomics alone is both static and incomplete. Proteomics aims to identify the entire complement of proteins elaborated by a cell under diverse conditions. As genes are switched on and off, proteins are synthesized in particular cell types at specific times of growth or differentiation and in response to external stimuli. Muscle cells express proteins not expressed by neural cells, and the type of subunits present in the hemoglobin tetramer undergo change pre- and postpartum.

Many proteins undergo posttranslational modifications during maturation into functionally competent forms or as a means of regulating their properties. Knowledge of the human genome therefore represents only the beginning of the task of describing living organisms in molecular detail and understanding the dynamics of processes such as growth, aging, and disease. As the human body contains thousands of cell types, each containing thousands of proteins, the proteome—the set of all the proteins expressed by an individual cell at a particular time—represents a moving target of formidable dimensions.

Two-Dimensional Electrophoresis & Gene Array Chips Are Used to Survey Protein Expression

One goal of proteomics is the identification of proteins whose levels of expression correlate with medically significant events. The presumption is that proteins whose appearance or disappearance is associated with a specific physiologic condition or disease will provide insights into root causes and mechanisms. Determination of the proteomes characteristic of each cell type requires the utmost efficiency in the isolation and identification of individual proteins. The contemporary approach utilizes robotic automation to speed sample preparation and large two-dimensional gels to resolve cellular proteins. Individual polypeptides are then extracted and analyzed by Edman sequencing or mass spectroscopy. While only about 1000 proteins can be resolved on a single gel, two-dimensional electrophoresis has a major advantage in that it examines the proteins themselves. An alternative approach, called Multidimensional Protein Identification Technology, or MudPIT, employs successive rounds of chromatography to resolve the peptides produced from the digestion of a complex biologic sample into several, simpler fractions that can be analyzed separately by MS. Gene arrays, sometimes called DNA chips, in which the expression of the mRNAs that encode proteins is detected, offer a complementary approach to proteomics. While changes in the expression of the mRNA encoding a protein do not necessarily reflect comparable changes in the level of the corresponding protein, gene arrays are more sensitive than two-dimensional gels, particularly with respect to low abundance proteins, and thus can examine a wider range of gene products.

Bioinformatics Assists Identification of Protein Functions

The functions of a large proportion of the proteins encoded by the human genome are presently unknown. The development of protein arrays or chips for directly testing the potential functions of proteins on a mass scale remains in its infancy. However, recent advances in bioinformatics permit researchers to compare amino acid sequences to discover clues to potential properties, physiologic roles, and mechanisms of action of proteins. Algorithms exploit the tendency of nature to employ variations of a structural theme to per-

form similar functions in several proteins [eg, the Rossmann nucleotide binding fold to bind NAD(P)H, nuclear targeting sequences, and EF hands to bind Ca^{2+}]. These domains generally are detected in the primary structure by conservation of particular amino acids at key positions. Insights into the properties and physiologic role of a newly discovered protein thus may be inferred by comparing its primary structure with that of known proteins.

SUMMARY

- Long amino acid polymers or polypeptides constitute the basic structural unit of proteins, and the structure of a protein provides insight into how it fulfills its functions.

- Proteins undergo posttransitional alterations during their lifetime that influence their function and determine their fate.

- The Edman method has been largely replaced by MS, a sensitive and versatile tool for determining primary structure, for identifying posttranslational modifications, and for detecting metabolic abnormalities.

- DNA cloning and molecular biology coupled with protein chemistry provide a hybrid approach that greatly increases the speed and efficiency for determination of primary structures of proteins.

- Genomics—the analysis of the entire oligonucleotide sequence of an organism's complete genetic material—has provided further enhancements.

- Computer algorithms facilitate identification of the ORFs that encode a given protein by using partial sequences and peptide mass profiling to search sequence databases.

- Scientists are now trying to determine the primary sequence and functional role of every protein expressed in a living cell, known as its proteome.

- A major goal is the identification of proteins and of their posttranslational modifications whose appearance or disappearance correlates with physiologic phenomena, aging, or specific diseases.

REFERENCES

Arnaud CH: Mass spec tackles proteins. Chem Eng News 2006;84:17.

Austin CP: The impact of the completed human genome sequence on the development of novel therapeutics for human disease. Annu Rev Med 2004;55:1.

Cutler P: Protein arrays: the current state-of-the-art. Proteomics 2003;3:3.

Deutscher MP (editor): *Guide to Protein Purification.* Methods Enzymol 1990;182. (Entire volume.)

Elofsson A, von Heijne G: Membrane protein structure: predictions versus reality. Annu Rev Biochem 2007;76:125.

Geveart K, Vandekerckhove J: Protein identification methods in proteomics. Electrophoresis 2000;21:1145.

Gwynne P, Heebner G: Mass spectrometry in drug discovery and development: from physics to pharma. Science 2006;313:1315.

Khan J et al: DNA microarray technology: the anticipated impact on the study of human disease. Biochim Biophys Acta 1999; 1423:M17.

Kislinger T et al: Multidimensional protein identification technology (MudPIT): technical overview of a profiling method optimized for the comprehensive proteomic investigation of normal and diseased heart tissue. J Am Soc Mass Spectrom 2005; 16:1207.

Patnaik SK, Blumenfeld OO: Use of on-line tools and databases for routine sequence analyses. Anal Biochem 2001;289:1.

Rinaldo P, Tortorelli S, Matern D: Recent developments and new applications of tandem mass spectrometry in newborn screening. Curr Opin Pediatrics 2004;16:427.

Rodland KD: Proteomics and cancer diagnosis: the potential of mass spectrometry. Clin Biochem 2004;37:579.

Schena M et al: Quantitative monitoring of gene expression patterns with a complementary DNA microarray. Science 1995;270:467.

Scopes RK: *Protein Purification. Principles and Practice* (3rd ed.). Springer, 1994.

Semsarian C, Seidman CE: Molecular medicine in the 21st century. Intern Med J 2001;31:53.

Sharon M, Robinson CV: The role of mass spectrometry in structure elucidation of dynamic protein complexes. Annu Rev Biochem 2007;76:167.

Temple LK et al: Essays on science and society: defining disease in the genomics era. Science 2001;293:807.

Woodage T, Broder S: The human genome and comparative genomics: understanding human evolution, biology, and medicine. J Gastroenterol 2003;15:68.

Proteins: Higher Orders of Structure

Peter J. Kennelly, PhD & Victor W. Rodwell, PhD

BIOMEDICAL IMPORTANCE

In nature, form follows function. In order for a newly synthesized polypeptide to mature into a biologically functional protein capable of catalyzing a metabolic reaction, powering cellular motion, or forming the macromolecular rods and cables that provide structural integrity to hair, bones, tendons, and teeth, it must fold into a specific three-dimensional arrangement, or **conformation.** In addition, during maturation **posttranslational modifications** may add new chemical groups or remove transiently needed peptide segments. Genetic or nutritional deficiencies that impede protein maturation are deleterious to health. Examples of the former include Creutzfeldt–Jakob disease, scrapie, Alzheimer's disease, and bovine spongiform encephalopathy ("mad cow disease"). Scurvy represents a nutritional deficiency that impairs protein maturation.

CONFORMATION VERSUS CONFIGURATION

The terms configuration and conformation are often confused. **Configuration** refers to the geometric relationship between a given set of atoms, for example, those that distinguish L- from D-amino acids. Interconversion of *configurational* alternatives requires breaking covalent bonds. **Conformation** refers to the spatial relationship of every atom in a molecule. Interconversion between conformers occurs without covalent bond rupture, with retention of configuration, and typically via rotation about single bonds.

PROTEINS WERE INITIALLY CLASSIFIED BY THEIR GROSS CHARACTERISTICS

Scientists initially approached structure–function relationships in proteins by separating them into classes based upon properties such as solubility, shape, or the presence of nonprotein groups. For example, the proteins that can be extracted from cells using aqueous solutions at physiologic pH and ionic strength are classified as **soluble.** Extraction of **integral membrane proteins** requires dissolution of the membrane with detergents. **Globular proteins** are compact, roughly spherical molecules that have **axial ratios** (the ratio of their shortest to longest dimensions) of not over 3. Most enzymes are globular proteins. By contrast, many structural proteins adopt highly extended conformations. These **fibrous proteins** possess axial ratios of 10 or more.

Lipoproteins and **glycoproteins** contain covalently bound lipid and carbohydrate, respectively. Myoglobin, hemoglobin, cytochromes, and many other **metalloproteins** contain tightly associated metal ions. While more precise classification schemes have emerged based upon similarity, or **homology,** in amino acid sequence and three-dimensional structure, many early classification terms remain in use.

PROTEINS ARE CONSTRUCTED USING MODULAR PRINCIPLES

Proteins perform complex physical and catalytic functions by positioning specific chemical groups in a precise three-dimensional arrangement. The polypeptide scaffold containing these groups must adopt a conformation that is both functionally efficient and physically strong. At first glance, the biosynthesis of polypeptides comprised of tens of thousands of individual atoms would appear to be extremely challenging. When one considers that a typical polypeptide can adopt $\geq 10^{50}$ distinct conformations, folding into the conformation appropriate to their biologic function would appear to be even more difficult. As described in Chapters 3 & 4, synthesis of the polypeptide backbones of proteins employs a small set of common building blocks or modules, the amino acids, joined by a common linkage, the peptide bond. A stepwise modular pathway simplifies the folding and processing of newly synthesized polypeptides into mature proteins.

THE FOUR ORDERS OF PROTEIN STRUCTURE

The modular nature of protein synthesis and folding are embodied in the concept of orders of protein structure: **primary structure,** the sequence of the amino acids in a polypeptide chain; **secondary structure,** the folding of short (3- to 30-residue), contiguous segments of polypeptide into geometrically ordered units; **tertiary structure,** the assembly of secondary structural units into larger functional units such as the mature polypeptide and its component domains; and **quaternary structure,** the number and types of polypeptide units of oligomeric proteins and their spatial arrangement.

SECONDARY STRUCTURE

Peptide Bonds Restrict Possible Secondary Conformations

Free rotation is possible about only two of the three covalent bonds of the polypeptide backbone: the α-carbon (Cα) to the carbonyl carbon (Co) bond, and the Cα to nitrogen bond (Figure 3–4). The partial double-bond character of the peptide bond that links Co to the α-nitrogen requires that the carbonyl carbon, carbonyl oxygen, and α-nitrogen remain coplanar, thus preventing rotation. The angle about the Cα—N bond is termed the phi (Φ) angle, and that about the Co—Cα bond the psi (ψ) angle. For amino acids other than glycine, most combinations of phi and psi angles are disallowed because of steric hindrance (Figure 5–1). The conformations of proline are even more restricted due to the absence of free rotation of the N—Cα bond.

Regions of ordered secondary structure arise when a series of aminoacyl residues adopt similar phi and psi angles. Extended segments of polypeptide (eg, loops) can possess a variety of such angles. The angles that define the two most common types of secondary structure, the **α helix** and the **β sheet,** fall within the lower and upper left-hand quadrants of a Ramachandran plot, respectively (Figure 5–1).

The Alpha Helix

The polypeptide backbone of an α helix is twisted by an equal amount about each α-carbon with a phi angle of approximately −57 degrees and a psi angle of approximately −47 degrees. A complete turn of the helix contains an average of 3.6 aminoacyl residues, and the distance it rises per turn (its *pitch*) is 0.54 nm (Figure 5–2). The R groups of each aminoacyl residue in an α helix face outward (Figure 5–3). Proteins contain only L-amino acids, for which a right-handed α helix is by far the more stable, and only right-handed α helices are present in proteins. Schematic diagrams of proteins represent α helices as coils or cylinders.

The stability of an α helix arises primarily from hydrogen bonds formed between the oxygen of the peptide bond carbonyl and the hydrogen atom of the peptide bond nitrogen of the fourth residue down the polypeptide chain (Figure 5–4). The ability to form the maximum number of hydrogen bonds, supplemented by van der Waals interactions in the core of this tightly packed structure, provides the thermodynamic driving force for the formation of an α helix. Since the peptide bond nitrogen of proline lacks a hydrogen atom to contribute to a hydrogen bond, proline can only be stably accommodated within

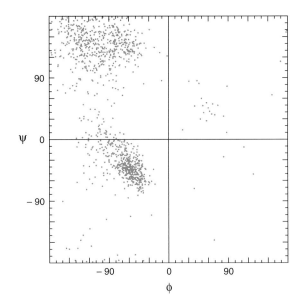

FIGURE 5–1 Ramachandran plot of the main chain phi (Φ) and psi (ψ) angles for approximately 1000 non-glycine residues in eight proteins whose structures were solved at high resolution. The dots represent allowable combinations, and the spaces prohibited combinations, of phi and psi angles. (Reproduced, with permission, from Richardson JS: The anatomy and taxonomy of protein structures. Adv Protein Chem 1981;34:167. Copyright © 1981. Reprinted with permission from Elsevier.)

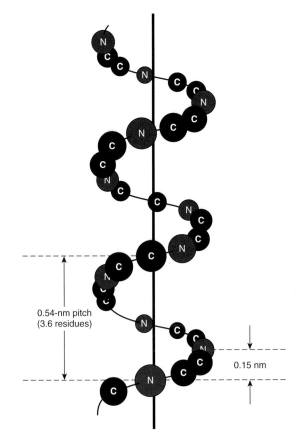

FIGURE 5–2 Orientation of the main chain atoms of a peptide about the axis of an α helix.

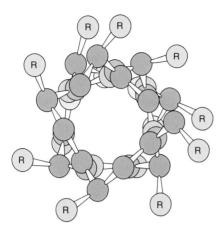

FIGURE 5–3 View down the axis of an α helix. The side chains (R) are on the outside of the helix. The van der Waals radii of the atoms are larger than shown here; hence, there is almost no free space inside the helix. (Slightly modified and reproduced, with permission, from Stryer L: *Biochemistry,* 3rd ed. Freeman, 1995. Copyright © 1995 W.H. Freeman and Company.)

the first turn of an α helix. When present elsewhere, proline disrupts the conformation of the helix, producing a bend. Because of its small size, glycine also often induces bends in α helices.

Many α helices have predominantly hydrophobic R groups on one side of the axis of the helix and predominantly hydrophilic ones on the other. These **amphipathic helices** are well adapted to the formation of interfaces between polar and nonpolar regions such as the hydrophobic interior of a protein and its aqueous environment. Clusters of amphipathic helices can create a channel, or pore, that permits specific polar molecules to pass through hydrophobic cell membranes.

The Beta Sheet

The second (hence "beta") recognizable regular secondary structure in proteins is the β sheet. The amino acid residues of a β sheet, when viewed edge-on, form a zigzag or pleated pattern in which the R groups of adjacent residues point in opposite directions. Unlike the compact backbone of the α helix, the peptide backbone of the β sheet is highly extended. But like the α helix, β sheets derive much of their stability from hydrogen bonds between the carbonyl oxygens and amide hydrogens of peptide bonds. However, in contrast to the α helix, these bonds are formed with adjacent segments of β sheet (Figure 5–5).

FIGURE 5–4 Hydrogen bonds (dotted lines) formed between H and O atoms stabilize a polypeptide in an α-helical conformation. (Reprinted, with permission, from Haggis GH et al: *Introduction to Molecular Biology.* Wiley, 1964, with permission of Pearson Education Limited.)

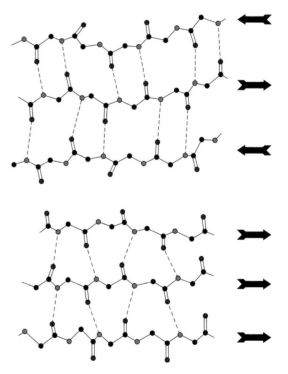

FIGURE 5–5 Spacing and bond angles of the hydrogen bonds of antiparallel and parallel pleated β sheets. Arrows indicate the direction of each strand. Hydrogen bonds are indicated by dotted lines with the participating α-nitrogen atoms (hydrogen donors) and oxygen atoms (hydrogen acceptors) shown in blue and red, respectively. Backbone carbon atoms are shown in black. For clarity in presentation, R groups and hydrogen atoms are omitted. **Top:** Antiparallel β sheet. Pairs of hydrogen bonds alternate between being close together and wide apart and are oriented approximately perpendicular to the polypeptide backbone. **Bottom:** Parallel β sheet. The hydrogen bonds are evenly spaced but slant in alternate directions.

FIGURE 5–6 Examples of tertiary structure of proteins. **Top:** The enzyme triose phosphate isomerase complexed with the substrate analog 2-phosphoglycerate (red). Note the elegant and symmetrical arrangement of alternating β sheets (light blue) and α helices (green), with the β sheets forming a β-barrel core surrounded by the helices. (Adapted from Protein Data Bank ID no. 1o5x.) **Bottom:** Lysozyme complexed with the substrate analog penta-N-acetyl chitopentaose (red). The color of the polypeptide chain is graded along the visible spectrum from purple (N-terminal) to tan (C-terminal). Notice how the concave shape of the domain forms a binding pocket for the pentasaccharide, the lack of β sheet, and the high proportion of loops and bends. (Adapted from Protein Data Bank ID no. 1sfb.)

Interacting β sheets can be arranged either to form a **parallel** β sheet, in which the adjacent segments of the polypeptide chain proceed in the same direction amino to carboxyl, or an **antiparallel** sheet, in which they proceed in opposite directions (Figure 5–5). Either configuration permits the maximum number of hydrogen bonds between segments, or strands, of the sheet. Most β sheets are not perfectly flat but tend to have a right-handed twist. Clusters of twisted strands of β sheet form the core of many globular proteins (Figure 5–6). Schematic diagrams represent β sheets as arrows that point in the amino to carboxyl terminal direction.

Loops & Bends

Roughly half of the residues in a "typical" globular protein reside in α helices and β sheets and half in loops, turns, bends, and other extended conformational features. Turns and bends refer to short segments of amino acids that join two units of secondary structure, such as two adjacent strands of an antiparallel β sheet. A β turn involves four aminoacyl residues, in which the first residue is hydrogen-bonded to the fourth, resulting in a tight 180-degree turn (Figure 5–7). Proline and glycine often are present in β turns.

FIGURE 5–7 A β turn that links two segments of anti-parallel β sheet. The dotted line indicates the hydrogen bond between the first and fourth amino acids of the four-residue segment Ala-Gly-Asp-Ser.

Loops are regions that contain residues beyond the minimum number necessary to connect adjacent regions of secondary structure. Irregular in conformation, loops nevertheless serve key biologic roles. For many enzymes, the loops that bridge domains responsible for binding substrates often contain aminoacyl residues that participate in catalysis. **Helix-loop-helix motifs** provide the oligonucleotide-binding portion of DNA-binding proteins such as repressors and transcription factors. Structural motifs such as the helix-loop-helix motif that are intermediate between secondary and tertiary structures are often termed **supersecondary structures.** Since many loops and bends reside on the surface of proteins and are thus exposed to solvent, they constitute readily accessible sites, or **epitopes,** for recognition and binding of antibodies.

While loops lack apparent structural regularity, they exist in a specific conformation stabilized through hydrogen bonding, salt bridges, and hydrophobic interactions with other portions of the protein. However, not all portions of proteins are necessarily ordered. Proteins may contain "disordered" regions, often at the extreme amino or carboxyl terminal, characterized by high conformational flexibility. In many instances, these disordered regions assume an ordered conformation upon binding of a ligand. This structural flexibility enables such regions to act as ligand-controlled switches that affect protein structure and function.

Tertiary & Quaternary Structure

The term "tertiary structure" refers to the entire three-dimensional conformation of a polypeptide. It indicates, in three-dimensional space, how secondary structural features—helices, sheets, bends, turns, and loops—assemble to form domains and how these domains relate spatially to one another. A **domain** is a section of protein structure sufficient to perform a particular chemical or physical task such as binding of a substrate or other ligand. Most domains are modular in nature,

contiguous in both primary sequence and three-dimensional space (Figure 5–8). Simple proteins, particularly those that interact with a single substrate, such as lysozyme or triose phosphate isomerase (Figure 5–6) and the oxygen storage protein myoglobin (Chapter 6), often consist of a single domain. By contrast, lactate dehydrogenase is comprised of two domains, an N-terminal NAD$^+$-binding domain and a C-terminal-binding domain for the second substrate, pyruvate (Figure 5–8). Lactate dehydrogenase is one a family of oxidoreductases which share a common N-terminal NAD(P)$^+$-binding domain known as the **Rossmann fold.** By fusing the Rossmann fold module to a variety of C-terminal domains, a large family of oxidoreductases have evolved that utilize NAD(P)$^+$/NAD(P)H for the oxidation and reduction of a wide range of metabolites. Examples include alcohol dehydrogenase, glyceraldehyde-3-phosphate dehydrogenase, malate dehydrogenase, quinone oxidoreductase, 6-phosphogluconate dehydrogenase, D-glycerate dehydrogenase, formate dehydrogenase, and 3α, 20β-hydroxysteroid dehydrogenase.

Not all domains bind substrates. Hydrophobic membrane domains anchor proteins to membranes or enable them to span membranes. Localization sequences target proteins to specific subcellular or extracellular locations such as the nucleus, mitochondria, secretory vesicles, etc. Regulatory domains trigger changes in protein function in response to the binding of allosteric effectors or covalent modifications (Chapter 9). Combining domain modules provides a facile route for generating proteins of great structural complexity and functional sophistication (Figure 5–9).

Proteins containing multiple domains also can be assembled through the association of multiple polypeptides, or protomers. Quaternary structure defines the polypeptide composition of a protein and, for an oligomeric protein, the spatial relationships between its protomers or subunits. **Monomeric** proteins consist of a single polypeptide chain. **Dimeric** proteins contain two polypeptide chains. **Homodimers** contain two copies of the same polypeptide chain, while in a **heterodimer** the polypeptides differ. Greek letters (α, β, γ, etc.) are used to distinguish different subunits of a heterooligomeric protein, and subscripts indicate the number of each subunit type. For example, α_4 designates a homotetrameric protein, and $\alpha_2\beta_2\gamma$ a protein with five subunits of three different types.

Since even small proteins contain many thousands of atoms, depictions of protein structure that indicate the position of every atom are generally too complex to be readily interpreted. Simplified schematic diagrams thus are used to depict key features of a protein's tertiary and quaternary structure. Ribbon diagrams (Figures 5–6 & 5–8) trace the conformation of the polypeptide backbone, with cylinders and arrows indicating regions of α helix and β sheet, respectively. In an even simpler representation, line segments that link the α carbons indicate the path of the polypeptide backbone. These schematic diagrams often include the side chains of selected amino acids that emphasize specific structure-function relationships.

FIGURE 5–8 Polypeptides containing two domains. **Top:** Shown is the three dimensional structure of lactate dehydrogenase with the substrates NADH (red) and pyruvate (blue) bound. Not all bonds in NADH are shown. The color of the polypeptide chain is graded along the visible spectrum from blue (N-terminal) to orange (C-terminal). Note how the N-terminal portion of the polypeptide forms a contiguous domain, encompassing the left portion of the enzyme, responsible for binding NADH. Similarly, the C-terminal portion forms a contiguous domain responsible for binding pyruvate. (Adapted from Protein Data Bank ID no. 3ldh.) **Bottom:** Shown is the three dimensional structure of the catalytic subunit of the cAMP-dependent protein kinase (Chapter 42) with the substrate analogs ADP (red) and peptide (purple) bound . The color of the polypeptide chain is graded along the visible spectrum from blue (N-terminal) to orange (C-terminal). Protein kinases transfer the γ-phosphate group of ATP to protein and peptide substrates (Chapter 9). Note how the N-terminal portion of the polypeptide forms a contiguous domain rich in β sheet that binds ADP. Similarly, the C-terminal portion forms a contiguous, α helix-rich domain responsible for binding the peptide substrate. (Adapted from Protein Data Bank ID no. 1jbp.)

MULTIPLE FACTORS STABILIZE TERTIARY & QUATERNARY STRUCTURE

Higher orders of protein structure are stabilized primarily—and often exclusively—by noncovalent interactions. Principal among these are hydrophobic interactions that drive most hydrophobic amino acid side chains into the interior of the protein, shielding them from water. Other significant contributors include hydrogen bonds and salt bridges between the carboxylates of aspartic and glutamic acid and the oppositely charged side chains of protonated lysyl, argininyl, and histidyl residues. While individually weak relative to a typical covalent bond of 80–120 kcal/mol, collectively these numerous interactions confer a high degree of stability to the biologically functional

FIGURE 5–9 Some multidomain proteins. The rectangles represent the polypeptides sequences of a forkhead transcription factor; 6-phosphofructo-2-kinase/fructose-2,6-bisphosphatase, a bifunctional enzyme whose activities are controlled in a reciprocal fashion by allosteric effectors and covalent modification (Chapter 20); phenylalanine hydroxylase (Chapters 27 & 29), whose activity is stimulated by phosphorylation of its regulatory domain; and the epidermal growth factor (EGF) receptor (Chapter 41), a transmembrane protein whose intracellular protein kinase domain is regulated via the binding of the peptide hormone EGF to its extracellular domain. Regulatory domains are colored green, catalytic domains dark blue and light blue, protein-protein interaction domains light orange, DNA binding domains dark orange, nuclear localization sequences medium blue, and transmembrane domains yellow. The kinase and bisphosphatase activities of 6-phosphofructo-2-kinase/fructose-2,6-bisphosphatase are catalyzed by the N- and C-terminal proximate catalytic domains, respectively.

conformation of a protein, just as a Velcro fastener harnesses the cumulative strength of multiple plastic loops and hooks.

Some proteins contain covalent disulfide (S—S) bonds that link the sulfhydryl groups of cysteinyl residues. Formation of disulfide bonds involves oxidation of the cysteinyl sulfhydryl groups and requires oxygen. Intrapolypeptide disulfide bonds further enhance the stability of the folded conformation of a peptide, while interpolypeptide disulfide bonds stabilize the quaternary structure of certain oligomeric proteins.

THREE-DIMENSIONAL STRUCTURE IS DETERMINED BY X-RAY CRYSTALLOGRAPHY OR BY NMR SPECTROSCOPY

X-Ray Crystallography

Following the solution in 1960 by John Kendrew of the three-dimensional structure of myoglobin, x-ray crystallography revealed the structures of thousands of proteins and of many oligonucleotides and viruses. For solution of its structure by x-ray crystallography, a protein is first precipitated under conditions that form large, well-ordered crystals. To establish appropriate

conditions, crystallization trials use a few microliters of protein solution and a matrix of variables (temperature, pH, presence of salts or organic solutes such as polyethylene glycol) to establish optimal conditions for crystal formation. Crystals mounted in quartz capillaries are first irradiated with monochromatic x-rays of approximate wavelength 0.15 nm to confirm that they are protein, not salt. Protein crystals may then be frozen in liquid nitrogen for subsequent collection of a high-resolution data set. The patterns formed by the x-rays that are diffracted by the atoms in their path are recorded on a photographic plate or its computer equivalent as a circular pattern of spots of varying intensity. The data inherent in these spots are then analyzed using a mathematical approach termed a *Fourier synthesis,* which summates wave functions. The wave amplitudes are related to spot intensity, but since the waves are not in phase, the relationship between their phases must next be determined. The traditional approach to solution of the "phase problem" employs **isomorphous displacement.** Prior to irradiation, an atom with a distinctive x-ray "signature" is introduced into a crystal at known positions in the primary structure of the protein. Heavy atom isomorphous displacement generally uses mercury or uranium, which bind to cysteine residues. An alternative approach uses the expression of plasmid-encoded recombinant proteins in which selenium replaces the sulfur of methionine.

Expression uses a bacterial host auxotrophic for methionine biosynthesis and a defined medium in which selenomethionine replaces methionine. Alternatively, if the unknown structure is similar to one that has already been solved, **molecular replacement** on an existing model provides an attractive way to phase the data without the use of heavy atoms. Finally, the results from the phasing and Fourier summations provide an electron density profile or three-dimensional map of how the atoms are connected or related to one another.

Laue X-Ray Crystallography

The ability of some crystallized enzymes to catalyze chemical reactions strongly suggests that structures determined by crystallography are indeed representative of the structures present in free solution. Classic crystallography provides, however, an essentially static picture of a protein that may undergo significant structural changes such as those that accompany enzymic catalysis. The Laue approach uses diffraction of polychromatic x-rays, and many crystals. The time-consuming process of rotating the crystal in the x-ray beam is avoided, which permits the use of extremely short exposure times. Detection of the motions of residues or domains of an enzyme during catalysis uses crystals that contain an inactive or "caged" substrate analog. An intense flash of visible light cleaves the caged precursor to release free substrate and initiate catalysis in a precisely controlled manner. Using this approach, data can be collected over time periods as short as a few nanoseconds.

Nuclear Magnetic Resonance Spectroscopy

Nuclear magnetic resonance (NMR) spectroscopy, a powerful complement to x-ray crystallography, measures the absorbance of radio frequency electromagnetic energy by certain atomic nuclei. "NMR-active" isotopes of biologically relevant elements include ^1H, ^{13}C, ^{15}N, and ^{31}P. The frequency, or chemical shift, at which a particular nucleus absorbs energy is a function of both the functional group within which it resides and the proximity of other NMR-active nuclei. Once limited to metabolites and relatively small macromolecules, $\leq 30\,kDa$, today proteins and protein complexes of $>100\,kDa$ can be analyzed by NMR. Two-dimensional NMR spectroscopy permits a three-dimensional representation of a protein to be constructed by determining the proximity of these nuclei to one another. NMR spectroscopy analyzes proteins in aqueous solution. Not only does this obviate the need to form crystals (a particular advantage when dealing with difficult to crystallize membrane proteins), it renders real-time observation of the changes in conformation that accompany ligand binding or catalysis possible. It also offers the possibility of perhaps one day being able to observe the structure and dynamics of proteins (and metabolites) within living cells.

Molecular Modeling

An increasingly useful adjunct to the empirical determination of the three-dimensional structure of proteins is the use of computer technology for molecular modeling. When the three-dimensional structure is known, **molecular dynamics** programs can be used to simulate the conformational dynamics of a protein and the manner in which factors such as temperature, pH, ionic strength, or amino acid substitutions influence these motions. **Molecular docking** programs simulate the interactions that take place when a protein encounters a substrate, inhibitor, or other ligand. Virtual screening for molecules likely to interact with key sites on a protein of biomedical interest is extensively used to facilitate the discovery of new drugs.

Molecular modeling is also employed to infer the structure of proteins for which x-ray crystallographic or NMR structures are not yet available. Secondary structure algorithms weigh the propensity of specific residues to become incorporated into α helices or β sheets in previously-studied proteins to predict the secondary structure of other polypeptides. In **homology modeling,** the known three-dimensional structure of a protein is used as a template upon which to erect a model of the probable structure of a related protein. Scientists are working to devise computer programs that will reliably predict the three-dimensional conformation of a protein directly from its primary sequence, thereby permitting the structures of the many unknown proteins for which templates are currently lacking to be determined.

PROTEIN FOLDING

Proteins are conformationally dynamic molecules that can fold into their functionally competent conformation in a time frame of milliseconds, and oftentimes can refold if their conformation becomes disrupted, or denatured. How is this remarkable process of folding achieved? Folding into the native state does not involve a haphazard search of all possible structures. Denatured proteins are not just random coils. Native contacts are favored, and regions of native structure persist even in the denatured state. Discussed below are factors that facilitate folding and refolding, and the current concepts and proposed mechanisms based on more than 40 years of largely in vitro experimentation.

The Native Conformation of a Protein Is Thermodynamically Favored

The number of distinct combinations of phi and psi angles specifying potential conformations of even a relatively small—15-kDa—polypeptide is unbelievably vast. Proteins are guided through this vast labyrinth of possibilities by thermodynamics. Since the biologically relevant—or native—conformation of a protein generally is that which is most energetically favored, knowledge of the native conformation is specified in the primary sequence. However, if one were to wait for a polypeptide to find its native conformation by random exploration of all possible conformations, the process would require billions of years to complete. Clearly, protein folding in cells takes place in a more orderly and guided fashion.

Folding Is Modular

Protein folding generally occurs via a stepwise process. In the first stage, as the newly synthesized polypeptide emerges from the ribosome, short segments fold into secondary structural units that provide local regions of organized structure. Folding is now reduced to the selection of an appropriate arrangement of this relatively small number of secondary structural elements. In the second stage, the hydrophobic regions segregate into the interior of the protein away from solvent, forming a "molten globule," a partially folded polypeptide in which the modules of secondary structure rearrange until the mature conformation of the protein is attained. This process is orderly, but not rigid. Considerable flexibility exists in the ways and in the order in which elements of secondary structure can be rearranged. In general, each element of secondary or super-secondary structure facilitates proper folding by directing the folding process toward the native conformation and away from unproductive alternatives. For oligomeric proteins, individual protomers tend to fold before they associate with other subunits.

Auxiliary Proteins Assist Folding

Under appropriate laboratory conditions, many proteins will spontaneously refold after being **denatured** (ie, unfolded) by treatment with acid or base, chaotropic agents, or detergents. However, refolding under these conditions is slow—minutes to hours. Moreover, some proteins fail to spontaneously refold in vitro, often forming insoluble **aggregates,** disordered complexes of unfolded or partially folded polypeptides held together by hydrophobic interactions. Aggregates represent unproductive dead ends in the folding process. Cells employ auxiliary proteins to speed the process of folding and to guide it toward a productive conclusion.

Chaperones

Chaperone proteins participate in the folding of over half of mammalian proteins. The hsp70 (70-kDa heat shock protein) family of chaperones binds short sequences of hydrophobic amino acids in newly synthesized polypeptides, shielding them from solvent. Chaperones prevent aggregation, thus providing an opportunity for the formation of appropriate secondary structural elements and their subsequent coalescence into a molten globule. The hsp60 family of chaperones, sometimes called **chaperonins,** differ in sequence and structure from hsp70 and its homologs. Hsp60 acts later in the folding process, often together with an hsp70 chaperone. The central cavity of the donut-shaped hsp60 chaperone provides a sheltered environment in which a polypeptide can fold until all hydrophobic regions are buried in its interior, eliminating aggregation.

Protein Disulfide Isomerase

Disulfide bonds between and within polypeptides stabilize tertiary and quaternary structure. However, disulfide bond formation is nonspecific. Under oxidizing conditions, a given

FIGURE 5–10 Isomerization of the N-α_1 prolyl peptide bond from a *cis* to a *trans* configuration relative to the backbone of the polypeptide.

cysteine can form a disulfide bond with the—SH of any accessible cysteinyl residue. By catalyzing disulfide exchange, the rupture of an S—S bond and its reformation with a different partner cysteine, protein disulfide isomerase facilitates the formation of disulfide bonds that stabilize a protein's native conformation.

Proline-*cis, trans*-Isomerase

All X-Pro peptide bonds—where X represents any residue—are synthesized in the *trans* configuration. However, of the X-Pro bonds of mature proteins, approximately 6% are *cis*. The *cis* configuration is particularly common in β turns. Isomerization from *trans* to *cis* is catalyzed by the enzyme proline-*cis, trans*-isomerase (Figure 5–10).

Folding Is a Dynamic Process

Proteins are conformationally dynamic molecules that can fold and unfold hundreds or thousands of times in their lifetime. How do proteins, once unfolded, refold and restore their functional conformation? First, unfolding rarely leads to the complete randomization of the polypeptide chain inside the cell. Unfolded proteins generally retain a number of contacts and regions of secondary structure that facilitate the refolding process. Second, chaperone proteins can "rescue" unfolded proteins that have become thermodynamically trapped in a misfolded dead end by unfolding hydrophobic regions and providing a second chance to fold productively. Glutathione can reduce inappropriate disulfide bonds that may be formed upon exposure to oxidizing agents such as O_2, hydrogen peroxide, or superoxide (Chapter 52).

PERTURBATION OF PROTEIN CONFORMATION MAY HAVE PATHOLOGIC CONSEQUENCES

Prions

The transmissible spongiform encephalopathies, or **prion diseases,** are fatal neurodegenerative diseases characterized by spongiform changes, astrocytic gliomas, and neuronal loss resulting from the deposition of insoluble protein aggregates in neural cells. They include Creutzfeldt–Jakob disease in humans, scrapie in sheep, and bovine spongiform encephalopathy (mad cow disease) in cattle. vCJD, a variant form of Creutzfeldt–Jakob disease that afflicts younger patients, is associated with early-onset psychiatric and behavioral disorders. Prion diseases may manifest themselves as infectious, genetic, or sporadic disorders. Because no viral or bacterial gene encoding the pathologic prion protein could be identified, the source and mechanism of transmission of prion disease long remained elusive. Today it is recognized that prion diseases are protein conformation diseases transmitted by altering the conformation, and hence the physical properties, of proteins endogenous to the host. Human prion-related protein, PrP, a glycoprotein encoded on the short arm of chromosome 20, normally is monomeric and rich in α helix. Pathologic prion proteins serve as the templates for the conformational transformation of normal PrP, known as PrPc, into PrPsc. PrPsc is rich in β sheet with many hydrophobic aminoacyl side chains exposed to solvent. As each new PrPsc molecule is formed, it triggers the production of yet more pathologic variants in a conformational chain reaction. Because PrPsc molecules associate strongly with one other through their exposed hydrophobic regions, the accumulating PrPsc units coalesce to form insoluble protease-resistant aggregates. Since one pathologic prion or prion-related protein can serve as template for the conformational transformation of many times its number of PrPc molecules, prion diseases can be transmitted by the protein alone without involvement of DNA or RNA.

Alzheimer's Disease

Refolding or misfolding of another protein endogenous to human brain tissue, β-amyloid, is a prominent feature of Alzheimer's disease. While the main cause of Alzheimer's disease remains elusive, the characteristic senile plaques and neurofibrillary bundles contain aggregates of the protein β-amyloid, a 4.3-kDa polypeptide produced by proteolytic cleavage of a larger protein known as amyloid precursor protein. In Alzheimer's disease patients, levels of β-amyloid become elevated, and this protein undergoes a conformational transformation from a soluble α helix–rich state to a state rich in β sheet and prone to self-aggregation. Apolipoprotein E has been implicated as a potential mediator of this conformational transformation.

Beta-Thalassemias

Thalassemias are caused by genetic defects that impair the synthesis of one of the polypeptide subunits of hemoglobin (Chapter 6). During the burst of hemoglobin synthesis that occurs during red cell development, a specific chaperone called α-hemoglobin-stabilizing protein (AHSP) binds to free hemoglobin α-subunits awaiting incorporation into the hemoglobin multimer. In the absence of this chaperone, free α-hemoglobin subunits aggregate, and the resulting precipitate has cytotoxic effects on the developing erythrocyte. Investigations using genetically modified mice suggest a role for AHSP in modulating the severity of β-thalassemia in human subjects.

COLLAGEN ILLUSTRATES THE ROLE OF POSTTRANSLATIONAL PROCESSING IN PROTEIN MATURATION

Protein Maturation Often Involves Making & Breaking Covalent Bonds

The maturation of proteins into their final structural state often involves the cleavage or formation (or both) of covalent bonds, a process of **posttranslational modification.** Many polypeptides are initially synthesized as larger precursors called **proproteins.** The "extra" polypeptide segments in these proproteins often serve as leader sequences that target a polypeptide to a particular organelle or facilitate its passage through a membrane. Other segments ensure that the potentially harmful activity of a protein such as the proteases trypsin and chymotrypsin remains inhibited until these proteins reach their final destination. However, once these transient requirements are fulfilled, the now superfluous peptide regions are removed by selective proteolysis. Other covalent modifications may take place that add new chemical functionalities to a protein. The maturation of collagen illustrates both of these processes.

Collagen Is a Fibrous Protein

Collagen is the most abundant of the fibrous proteins that constitute more than 25% of the protein mass in the human body. Other prominent fibrous proteins include keratin and myosin. These fibrous proteins represent a primary source of structural strength for cells (ie, the cytoskeleton) and tissues. Skin derives its strength and flexibility from an intertwined mesh of collagen and keratin fibers, while bones and teeth are buttressed by an underlying network of collagen fibers analogous to the steel strands in reinforced concrete. Collagen also is present in connective tissues such as ligaments and tendons. The high degree of tensile strength required to fulfill these structural roles requires elongated proteins characterized by repetitive amino acid sequences and a regular secondary structure.

Collagen Forms a Unique Triple Helix

Tropocollagen consists of three fibers, each containing about 1000 amino acids, bundled together in a unique conformation, the collagen triple helix (Figure 5–11). A mature collagen fiber forms an elongated rod with an axial ratio of about 200. Three intertwined polypeptide strands, which twist to the left, wrap around one another in a right-handed fashion to form the collagen triple helix. The opposing handedness of this superhelix and its component polypeptides makes the collagen triple helix highly resistant to unwinding—the same principle used in the steel cables of suspension bridges. A collagen triple helix has 3.3 residues per turn and a rise per residue nearly twice that of an α helix. The R groups of each polypeptide strand of the triple helix pack so closely that, in order to fit, one must be glycine. Thus, every third amino acid residue in collagen

FIGURE 5–11 Primary, secondary, and tertiary structures of collagen.

is a glycine residue. Staggering of the three strands provides appropriate positioning of the requisite glycines throughout the helix. Collagen is also rich in proline and hydroxyproline, yielding a repetitive Gly-X-Y pattern (Figure 5–11) in which Y generally is proline or hydroxyproline.

Collagen triple helices are stabilized by hydrogen bonds between residues in *different* polypeptide chains, a process aided by the hydroxyl groups of hydroxyprolyl residues. Additional stability is provided by covalent cross-links formed between modified lysyl residues both within and between polypeptide chains.

Collagen Is Synthesized as a Larger Precursor

Collagen is initially synthesized as a larger precursor polypeptide, procollagen. Numerous prolyl and lysyl residues of procollagen are hydroxylated by prolyl hydroxylase and lysyl hydroxylase, enzymes that require ascorbic acid (vitamin C; see Chapters 27 & 44). Hydroxyprolyl and hydroxylysyl residues provide additional hydrogen bonding capability that stabilizes the mature protein. In addition, glucosyl and galactosyl transferases attach glucosyl or galactosyl residues to the hydroxyl groups of specific hydroxylysyl residues.

The central portion of the precursor polypeptide then associates with other molecules to form the characteristic triple helix. This process is accompanied by the removal of the globular amino terminal and carboxyl terminal extensions of the precursor polypeptide by selective proteolysis. Certain lysyl residues are modified by lysyl oxidase, a copper-containing protein that converts ε-amino groups to aldehydes. The aldehydes can either undergo an aldol condensation to form a C=C double bond or to form a Schiff base (eneimine) with the ε-amino group of an unmodified lysyl residue, which is subsequently reduced to form a C—N single bond. These covalent bonds cross-link the individual polypeptides and imbue the fiber with exceptional strength and rigidity.

Nutritional & Genetic Disorders Can Impair Collagen Maturation

The complex series of events in collagen maturation provide a model that illustrates the biologic consequences of incomplete polypeptide maturation. The best-known defect in collagen biosynthesis is scurvy, a result of a dietary deficiency of vitamin C required by prolyl and lysyl hydroxylases. The resulting deficit in the number of hydroxyproline and hydroxylysine residues undermines the conformational stability of collagen fibers, leading to bleeding gums, swelling joints, poor wound healing, and ultimately death. Menkes' syndrome, characterized by kinky hair and growth retardation, reflects a dietary deficiency of the copper required by lysyl oxidase, which catalyzes a key step in formation of the covalent cross-links that strengthen collagen fibers.

Genetic disorders of collagen biosynthesis include several forms of osteogenesis imperfecta, characterized by fragile bones. In Ehlers-Danlos syndrome, a group of connective tissue disorders that involve impaired integrity of supporting structures, defects in the genes that encode α collagen-1, procollagen N-peptidase, or lysyl hydroxylase result in mobile joints and skin abnormalities (see also Chapter 48).

SUMMARY

- Proteins may be classified based on their solubility, shape, or function or on the presence of a prosthetic group, such as heme.

- The gene-encoded primary structure of a polypeptide is the sequence of its amino acids. Its secondary structure results from folding of polypeptides into hydrogen-bonded motifs such as the α helix, the β pleated sheet, β bends, and loops. Combinations of these motifs can form supersecondary motifs.

- Tertiary structure concerns the relationships between secondary structural domains. Quaternary structure of proteins with two or more polypeptides (oligomeric proteins) concerns the spatial relationships between various types of polypeptides.

- Primary structures are stabilized by covalent peptide bonds. Higher orders of structure are stabilized by weak forces—multiple hydrogen bonds, salt (electrostatic) bonds, and association of hydrophobic R groups.

- The phi (Φ) angle of a polypeptide is the angle about the C_a—N bond; the psi (ψ) angle is that about the C_a—C_o bond. Most combinations of phi-psi angles are disallowed due to steric hindrance. The phi-psi angles that form the α helix and the β sheet fall within the lower and upper left-hand quadrants of a Ramachandran plot, respectively.

- Protein folding is a poorly understood process. Broadly speaking, short segments of newly synthesized polypeptide fold into secondary structural units. Forces that bury hydrophobic regions from solvent then drive the partially folded polypeptide into a "molten globule" in which the modules of secondary structure are rearranged to give the native conformation of the protein.

- Proteins that assist folding include protein disulfide isomerase, proline-*cis, trans*-isomerase, and the chaperones that participate in the folding of over half of mammalian proteins. Chaperones shield newly synthesized polypeptides from solvent and provide an environment for elements of secondary structure to emerge and coalesce into molten globules.

- Techniques for study of higher orders of protein structure include x-ray crystallography, NMR spectroscopy, analytical ultracentrifugation, gel filtration, and gel electrophoresis.

- Collagen illustrates the close linkage between protein structure and biologic function. Diseases of collagen maturation include

Ehlers-Danlos syndrome and the vitamin C deficiency disease scurvy.

■ Prions—protein particles that lack nucleic acid—cause fatal transmissible spongiform encephalopathies such as Creutzfeldt–Jakob disease, scrapie, and bovine spongiform encephalopathy. Prion diseases involve an altered secondary-tertiary structure of a naturally occurring protein, PrPc. When PrPc interacts with its pathologic isoform PrPSc, its conformation is transformed from a predominantly α-helical structure to the β-sheet structure characteristic of PrPSc.

REFERENCES

Branden C, Tooze J: *Introduction to Protein Structure.* Garland, 1991.

Chiti F, Dobson CM: Protein misfolding, functional amyloid, and human disease. Annu Rev Biochem 2006;75:519.

Collinge J: Prion diseases of humans and animals: their causes and molecular basis. Annu Rev Neurosci 2001;24:519.

Foster MP et al: Solution NMR of large molecules and assemblies. Biochemistry 2007;46:331.

Frydman J: Folding of newly translated proteins in vivo: the role of molecular chaperones. Annu Rev Biochem 2001; 70:603.

Gothel SF, Marahiel MA: Peptidyl-prolyl *cis-trans* isomerases, a superfamily of ubiquitous folding catalysts. Cell Mol Life Sci 1999;55:423.

Hajdu J et al: Analyzing protein functions in four dimensions. Nat Struct Biol 2000;7:1006.

Hardy J: Toward Alzheimer therapies based on genetic knowledge. Annu Rev Med 2004;55:15.

Ho BK, Thomas A, Brasseur R: Revisiting the Ramachandran plot: hard-sphere repulsion, electrostatics, and H-bonding in the α-helix. Protein Sci 2003;12:2508.

Irani DN, Johnson RT: Diagnosis and prevention of bovine spongiform encephalopathy and variant Creutzfeldt-Jakob disease. Annu Rev Med 2003;54:305.

Jorgensen WL: The many roles of computation in drug discovery. Science 2004;303:1813.

Khare SD, Dokholyan NV: Molecular mechanisms of polypeptide aggregation in human disease. Curr Protein Pept Sci 2007;8:573.

Kong Y et al: Loss of alpha-hemoglobin-stabilizing protein impairs erythropoiesis and exacerbates beta-thalassemia. J Clin Invest 2004;114:1457.

Myers JK, Oas TG: Mechanism of fast protein folding. Annu Rev Biochem 2002;71:783.

Myllyharju J: Prolyl 4-hydroxylases, the key enzymes of collagen biosynthesis. Matrix Biol 2003;22:15.

Radord S: Protein folding: progress made and promises ahead. Trends Biochem Sci 2000;25:611.

Rider MH et al: 6-Phosphofructo-2-kinase/fructose-2,6-bisphosphatase: head-to-head with a bifunctional enzyme that controls glycolysis. Biochem J 2004;381:561.

Segrest MP et al: The amphipathic alpha-helix: a multifunctional structural motif in plasma lipoproteins. Adv Protein Chem 1995;45:1.

Stoddard BL et al: Millisecond Laue structures of an enzyme-product complex using photocaged substrate analogs. Nat Struct Biol 1998;5:891.

Wegrzyn RD, Deuerling E: Molecular guardians for newborn proteins: ribosome-associated chaperones and their role in protein folding. Cellular Mol Life Sci 2005;62:2727.

Young JC, Moarefi I, Hartl FU: *Hsp*90: a specialized but essential protein-folding tool. J Cell Biol 2001;154:267.

Proteins: Myoglobin & Hemoglobin

Peter J. Kennelly, PhD & Victor W. Rodwell, PhD

BIOMEDICAL IMPORTANCE

The heme proteins myoglobin and hemoglobin maintain a supply of oxygen essential for oxidative metabolism. Myoglobin, a monomeric protein of red muscle, stores oxygen as a reserve against oxygen deprivation. Hemoglobin, a tetrameric protein of erythrocytes, transports O_2 to the tissues and returns CO_2 and protons to the lungs. Cyanide and carbon monoxide kill because they disrupt the physiologic function of the heme proteins cytochrome oxidase and hemoglobin, respectively. The secondary-tertiary structure of the subunits of hemoglobin resembles myoglobin. However, the tetrameric structure of hemoglobin permits cooperative interactions that are central to its function. For example, 2,3-bisphosphoglycerate (BPG) promotes the efficient release of O_2 by stabilizing the quaternary structure of deoxyhemoglobin. Hemoglobin and myoglobin illustrate both protein structure-function relationships and the molecular basis of genetic diseases such as sickle cell disease and the thalassemias.

HEME & FERROUS IRON CONFER THE ABILITY TO STORE & TO TRANSPORT OXYGEN

Myoglobin and hemoglobin contain **heme,** a cyclic tetrapyrrole consisting of four molecules of pyrrole linked by α-methylene bridges. This planar network of conjugated double bonds absorbs visible light and colors heme deep red. The substituents at the β-positions of heme are methyl (M), vinyl (V), and propionate (Pr) groups arranged in the order M, V, M, V, M, Pr, Pr, M (Figure 6–1). One atom of ferrous iron (Fe^{2+}) resides at the center of the planar tetrapyrrole. Other proteins with metal-containing tetrapyrrole prosthetic groups include the cytochromes (Fe and Cu) and chlorophyll (Mg) (see Chapter 31). Oxidation and reduction of the Fe and Cu atoms of cytochromes are essential to their biologic function as carriers of electrons. By contrast, oxidation of the Fe^{2+} of myoglobin or hemoglobin to Fe^{3+} destroys their biologic activity.

Myoglobin Is Rich in α Helix

Oxygen stored in red muscle myoglobin is released during O_2 deprivation (eg, severe exercise) for use in muscle mitochondria for aerobic synthesis of ATP (see Chapter 13). A 153-aminoacyl residue polypeptide (MW 17,000), myoglobin folds into a compact shape that measures $4.5 \times 3.5 \times 2.5$ nm (Figure 6–2). Unusually high proportions, about 75%, of the residues are present in eight right-handed, 7–20 residue α helices. Starting at the amino terminal, these are termed helices A–H. Typical of globular proteins, the surface of myoglobin is polar, while—with only two exceptions—the interior contains only nonpolar residues such as Leu, Val, Phe, and Met. The exceptions are His E7 and His F8, the seventh and eighth residues in helices E and F, which lie close to the heme iron, where they function in O_2 binding.

Histidines F8 & E7 Perform Unique Roles in Oxygen Binding

The heme of myoglobin lies in a crevice between helices E and F oriented with its polar propionate groups facing the surface of the globin (Figure 6–2). The remainder resides in the nonpolar interior. The fifth coordination position of the iron is linked to a ring nitrogen of the **proximal histidine,** His F8. The **distal histidine,** His E7, lies on the side of the heme ring opposite to His F8.

The Iron Moves Toward the Plane of the Heme When Oxygen Is Bound

The iron of unoxygenated myoglobin lies 0.03 nm (0.3 Å) outside the plane of the heme ring, toward His F8. The heme therefore "puckers" slightly. When O_2 occupies the sixth coordination position, the iron moves to within 0.01 nm (0.1 Å) of the plane of the heme ring. Oxygenation of myoglobin, thus, is accompanied by motion of the iron, of His F8, and of residues linked to His F8.

Apomyoglobin Provides a Hindered Environment for Heme Iron

When O_2 binds to myoglobin, the bond between the first oxygen atom and the Fe^{2+} is perpendicular to the plane of the heme ring. The bond linking the first and second oxygen atoms lies at an angle of 121 degrees to the plane of the heme,

FIGURE 6–1 Heme. The pyrrole rings and methylene bridge carbons are coplanar, and the iron atom (Fe^{2+}) resides in almost the same plane. The fifth and sixth coordination positions of Fe^{2+} are directed perpendicular to—and directly above and below—the plane of the heme ring. Observe the nature of the methyl (blue), vinyl (green), and propionate (orange) substituent groups on the β carbons of the pyrrole rings, the central iron atom (red), and the location of the polar side of the heme ring (at about 7 o'clock) that faces the surface of the myoglobin molecule.

FIGURE 6–2 Three-dimensional structure of myoglobin. Shown is a ribbon diagram tracing the polypeptide backbone of myoglobin. The color of the polypeptide chain is graded along the visible spectrum from blue (N-terminal) to tan (C-terminal). The heme prosthetic group is red. The α-helical regions are designated A through H. The distal (E7) and proximal (F8) histidine residues are highlighted in blue and orange, respectively. Note how the polar propionate substituents (Pr) project out of the heme toward solvent. (Adapted from Protein Data Bank ID no. 1a6n.)

orienting the second oxygen away from the distal histidine (Figure 6–3, left). Isolated heme binds carbon monoxide (CO) 25,000 times more strongly than oxygen. Since CO is present in small quantities in the atmosphere and arises in cells from the catabolism of heme, why is it that CO does not completely displace O_2 from heme iron? The accepted explanation is that

FIGURE 6–3 Angles for bonding of oxygen and carbon monoxide (CO) to the heme iron of myoglobin. The distal E7 histidine hinders bonding of CO at the preferred (90-degree) angle to the plane of the heme ring.

the apoproteins of myoglobin and hemoglobin create a **hindered environment.** When CO binds to isolated heme, all three atoms (Fe, C, and O) lie perpendicular to the plane of the heme. However, in myoglobin and hemoglobin the distal histidine sterically precludes this preferred, high-affinity orientation. Binding at a less favored angle reduces the strength of the heme-CO bond to about 200 times that of the heme-O_2 bond (Figure 6–3, right) at which level the great excess of O_2 over CO normally present dominates. Nevertheless, about 1% of myoglobin typically is present combined with CO.

THE OXYGEN DISSOCIATION CURVES FOR MYOGLOBIN & HEMOGLOBIN SUIT THEIR PHYSIOLOGIC ROLES

Why is myoglobin unsuitable as an O_2 transport protein but well suited for O_2 storage? The relationship between the concentration, or partial pressure, of O_2 (Po_2) and the quantity of O_2 bound is expressed as an O_2 saturation isotherm (Figure 6–4). The oxygen-binding curve for myoglobin is hyperbolic. Myoglobin therefore loads O_2 readily at the Po_2 of the lung capillary bed (100 mm Hg). However, since myoglobin releases only a small fraction of its bound O_2 at the Po_2 values typically encountered in active muscle (20 mm Hg) or other tissues (40 mm Hg), it represents an ineffective vehicle for delivery of O_2. When strenuous exercise lowers the Po_2 of muscle tissue to about 5 mm Hg, myoglobin releases O_2 for mitochondrial synthesis of ATP, permitting continued muscular activity.

THE ALLOSTERIC PROPERTIES OF HEMOGLOBINS RESULT FROM THEIR QUATERNARY STRUCTURES

The properties of individual hemoglobins are consequences of their quaternary as well as of their secondary and tertiary structures. The quaternary structure of hemoglobin confers

FIGURE 6–4 Oxygen-binding curves of both hemoglobin and myoglobin. Arterial oxygen tension is about 100 mm Hg; mixed venous oxygen tension is about 40 mm Hg; capillary (active muscle) oxygen tension is about 20 mm Hg; and the minimum oxygen tension required for cytochrome oxidase is about 5 mm Hg. Association of chains into a tetrameric structure (hemoglobin) results in much greater oxygen delivery than would be possible with single chains. (Modified, with permission, from Scriver CR et al (editors): *The Molecular and Metabolic Bases of Inherited Disease,* 7th ed. McGraw-Hill, 1995.)

striking additional properties, absent from monomeric myoglobin, which adapts it to its unique biologic roles. The **allosteric** (Gk *allos* "other," *steros* "space") properties of hemo-

globin provide, in addition, a model for understanding other allosteric proteins (see Chapter 18).

Hemoglobin Is Tetrameric

Hemoglobins are tetramers composed of pairs of two different polypeptide subunits (Figure 6–5). Greek letters are used to designate each subunit type. The subunit composition of the principal hemoglobins are $\alpha_2\beta_2$ (HbA; normal adult hemoglobin), $\alpha_2\gamma_2$ (HbF; fetal hemoglobin), α_2S_2 (HbS; sickle cell hemoglobin), and $\alpha_2\delta_2$ (HbA$_2$; a minor adult hemoglobin). The primary structures of the β, γ, and δ chains of human hemoglobin are highly conserved.

Myoglobin & the β Subunits of Hemoglobin Share Almost Identical Secondary and Tertiary Structures

Despite differences in the kind and number of amino acids present, myoglobin and the β polypeptide of hemoglobin A have almost identical secondary and tertiary structures. Similarities include the location of the heme and the helical regions, and the presence of amino acids with similar properties at comparable locations. Although it possesses seven rather

FIGURE 6–5 Hemoglobin. Shown is the three-dimensional structure of deoxyhemoglobin with a molecule of 2,3-bisphosphoglycerate (dark blue) bound. The two α subunits are colored in the darker shades of green and blue, the two β subunits in the lighter shades of green and blue, and the heme prosthetic groups in red. (Adapted from Protein Data Bank ID no. 1b86.)

than eight helical regions, the α polypeptide of hemoglobin also closely resembles myoglobin.

Oxygenation of Hemoglobin Triggers Conformational Changes in the Apoprotein

Hemoglobins bind four molecules of O_2 per tetramer, one per heme. A molecule of O_2 binds to a hemoglobin tetramer more readily if other O_2 molecules are already bound (Figure 6–4). Termed **cooperative binding,** this phenomenon permits hemoglobin to maximize both the quantity of O_2 loaded at the Po_2 of the lungs and the quantity of O_2 released at the Po_2 of the peripheral tissues. Cooperative interactions, an exclusive property of multimeric proteins, are critically important to aerobic life.

P_{50} Expresses the Relative Affinities of Different Hemoglobins for Oxygen

The quantity P_{50}, a measure of O_2 concentration, is the partial pressure of O_2 that half-saturates a given hemoglobin. Depending on the organism, P_{50} can vary widely, but in all instances it will exceed the Po_2 of the peripheral tissues. For example, values of P_{50} for HbA and HbF are 26 and 20 mm Hg, respectively. In the placenta, this difference enables HbF to extract oxygen from the HbA in the mother's blood. However, HbF is suboptimal postpartum since its high affinity for O_2 limits the quantity of O_2 delivered to the tissues.

The subunit composition of hemoglobin tetramers undergoes complex changes during development. The human fetus initially synthesizes a $\zeta_2\epsilon_2$ tetramer. By the end of the first trimester, ζ and ϵ subunits have been replaced by α and γ subunits, forming HbF ($\alpha_2\gamma_2$), the hemoglobin of late fetal life. While synthesis of β subunits begins in the third trimester, β subunits do not completely replace γ subunits to yield adult HbA ($\alpha_2\beta_2$) until some weeks postpartum (Figure 6–6).

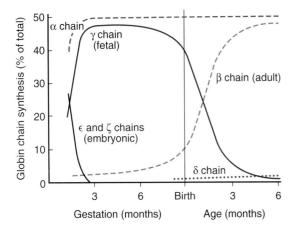

FIGURE 6–6 Developmental pattern of the quaternary structure of fetal and newborn hemoglobins. (Reproduced, with permission, from Ganong WF: *Review of Medical Physiology,* 20th ed. McGraw-Hill, 2001.)

Oxygenation of Hemoglobin Is Accompanied by Large Conformational Changes

The binding of the first O_2 molecule to deoxyHb shifts the heme iron toward the plane of the heme ring from a position about 0.04 nm beyond it (Figure 6–7). This motion is transmitted to the proximal (F8) histidine and to the residues attached thereto, which in turn causes the rupture of salt bridges between the carboxyl terminal residues of all four subunits. As a result, one pair of α/β subunits rotates 15 degrees with re-

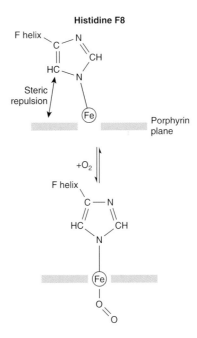

FIGURE 6–7 The iron atom moves into the plane of the heme on oxygenation. Histidine F8 and its associated residues are pulled along with the iron atom. (Slightly modified and reproduced, with permission, from Stryer L: *Biochemistry,* 4th ed. Freeman, 1995. Copyright © 1995 W. H. Freeman and Company.)

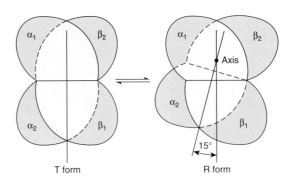

FIGURE 6–8 During transition of the T form to the R form of hemoglobin, the $\alpha_2\beta_2$ pair of subunits (green) rotates through 15 degrees relative to the pair of $\alpha_1\beta_1$ subunits (yellow). The axis of rotation is eccentric, and the $\alpha_2\beta_2$ pair also shifts toward the axis somewhat. In the representation, the tan $\alpha_1\beta_1$ pair is shown fixed while the green $\alpha_2\beta_2$ pair of subunits both shifts and rotates.

spect to the other, compacting the tetramer (Figure 6–8). Profound changes in secondary, tertiary, and quaternary structure accompany the high-affinity O_2-induced transition of hemoglobin from the low-affinity **T (taut) state** to the high-affinity **R (relaxed) state.** These changes significantly increase the affinity of the remaining unoxygenated hemes for O_2, as subsequent binding events require the rupture of fewer salt bridges (Figure 6–9). The terms T and R also are used to refer to the low-affinity and high-affinity conformations of allosteric enzymes, respectively.

After Releasing O_2 at the Tissues, Hemoglobin Transports CO_2 & Protons to the Lungs

In addition to transporting O_2 from the lungs to peripheral tissues, hemoglobin transports CO_2, the byproduct of respiration, and protons from peripheral tissues to the lungs. Hemoglobin carries CO_2 as carbamates formed with the amino terminal nitrogens of the polypeptide chains.

$$CO_2 + Hb-NH_3^+ \rightleftharpoons 2H^+ + Hb-\overset{H}{\underset{}{N}}-\overset{\overset{O}{\|}}{C}-O^-$$

Carbamates change the charge on amino terminals from positive to negative, favoring salt bridge formation between the α and β chains.

Hemoglobin carbamates account for about 15% of the CO_2 in venous blood. Much of the remaining CO_2 is carried as bicarbonate, which is formed in erythrocytes by the hydra-

tion of CO_2 to carbonic acid (H_2CO_3), a process catalyzed by carbonic anhydrase. At the pH of venous blood, H_2CO_3 dissociates into bicarbonate and a proton.

$$CO_2 + H_2O \underset{\text{CARBONIC ANHYDRASE}}{\rightleftharpoons} \underset{\text{Carbonic acid}}{H_2CO_3} \overset{\text{(Spontaneous)}}{\rightleftharpoons} HCO_3^- + H^+$$

Deoxyhemoglobin binds one proton for every two O_2 molecules released, contributing significantly to the buffering capacity of blood. The somewhat lower pH of peripheral tissues, aided by carbamation, stabilizes the T state and thus enhances the delivery of O_2. In the lungs, the process reverses. As O_2 binds to deoxyhemoglobin, protons are released and combine with bicarbonate to form carbonic acid. Dehydration of H_2CO_3 catalyzed by carbonic anhydrase forms CO_2, which is exhaled. Binding of oxygen thus drives the exhalation of CO_2 (Figure 6–10). This reciprocal coupling of proton and O_2 binding is termed the **Bohr effect.** The Bohr effect is dependent upon **cooperative interactions between the hemes of the hemoglobin tetramer.** Myoglobin, a monomer, exhibits no Bohr effect.

Protons Arise from Rupture of Salt Bridges When O_2 Binds

Protons responsible for the Bohr effect arise from rupture of salt bridges during the binding of O_2 to T state hemoglobin. Conversion to the oxygenated R state breaks salt bridges involving β chain residue His 146. The subsequent dissociation

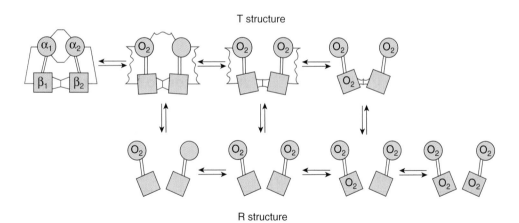

T structure

R structure

FIGURE 6–9 Transition from the T structure to the R structure. In this model, salt bridges (red lines) linking the subunits in the T structure break progressively as oxygen is added, and even those salt bridges that have not yet ruptured are progressively weakened (wavy red lines). The transition from T to R does not take place after a fixed number of oxygen molecules have been bound but becomes more probable as each successive oxygen binds. The transition between the two structures is influenced by protons, carbon dioxide, chloride, and BPG; the higher their concentration, the more oxygen must be bound to trigger the transition. Fully oxygenated molecules in the T structure and fully deoxygenated molecules in the R structure are not shown because they are unstable. (Modified and redrawn, with permission, from Perutz MF: Hemoglobin structure and respiratory transport. Sci Am [Dec] 1978; 239:92.)

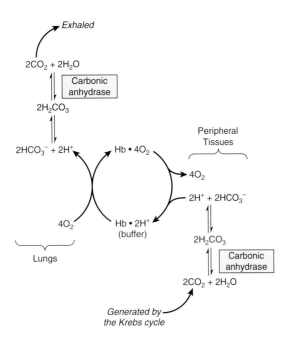

FIGURE 6–10 The Bohr effect. Carbon dioxide generated in peripheral tissues combines with water to form carbonic acid, which dissociates into protons and bicarbonate ions. Deoxyhemoglobin acts as a buffer by binding protons and delivering them to the lungs. In the lungs, the uptake of oxygen by hemoglobin releases protons that combine with bicarbonate ion, forming carbonic acid, which when dehydrated by carbonic anhydrase becomes carbon dioxide, which then is exhaled.

of protons from His 146 drives the conversion of bicarbonate to carbonic acid (Figure 6–10). Upon the release of O_2, the T structure and its salt bridges re-form. This conformational change increases the pK_a of the β chain His 146 residues, which bind protons. By facilitating the re-formation of salt bridges, an increase in proton concentration enhances the release of O_2 from oxygenated (R state) hemoglobin. Conversely, an increase in Po_2 promotes proton release.

2,3-Bisphosphoglycerate (BPG) Stabilizes the T Structure of Hemoglobin

A low Po_2 in peripheral tissues promotes the synthesis of 2,3-BPG in erythrocytes from the glycolytic intermediate 1,3-BPG.

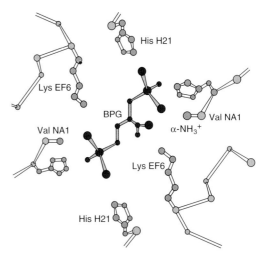

The hemoglobin tetramer binds one molecule of BPG in the central cavity formed by its four subunits (Figure 6–4).

FIGURE 6–11 Mode of binding of 2,3-bisphosphoglycerate (BPG) to human deoxyhemoglobin. BPG interacts with three positively charged groups on each β chain. (Based on Arnone A: X-ray diffraction study of binding of 2,3-diphosphoglycerate to human deoxyhemoglobin. Nature 1972; 237:146. Copyright © 1972. Adapted by permission from Macmillan Publishers Ltd.)

However, the space between the H helices of the β chains lining the cavity is sufficiently wide to accommodate BPG only when hemoglobin is in the T state. BPG forms salt bridges with the terminal amino groups of both β chains via Val NA1 and with Lys EF6 and His H21 (Figure 6–11). BPG therefore stabilizes deoxygenated (T state) hemoglobin by forming additional salt bridges that must be broken prior to conversion to the R state.

Residue H21 of the γ subunit of HbF is Ser rather than His. Since Ser cannot form a salt bridge, BPG binds more weakly to HbF than to HbA. The lower stabilization afforded to the T state by BPG accounts for HbF having a higher affinity for O_2 than HbA.

Adaptation to High Altitude

Physiologic changes that accompany prolonged exposure to high altitude include an increase in the number of erythrocytes and in their concentrations of hemoglobin and of BPG. Elevated BPG lowers the affinity of HbA for O_2 (increases P_{50}), which enhances release of O_2 at the tissues.

NUMEROUS MUTATIONS AFFECTING HUMAN HEMOGLOBINS HAVE BEEN IDENTIFIED

Mutations in the genes that encode the α or β subunits of hemoglobin potentially can affect its biologic function. However, almost all of the over 900 known genetic mutations affecting human hemoglobins are both extremely rare and benign, presenting no clinical abnormalities. When a mutation does compromise biologic function, the condition is termed a

hemoglobinopathy. It is estimated that more than 7% of the globe's population are carriers for hemoglobin disorders. The URL http://globin.cse.psu.edu/ (Globin Gene Server) provides information about—and links for—normal and mutant hemoglobins. Selected examples are described below.

Methemoglobin & Hemoglobin M

In methemoglobinemia, the heme iron is ferric rather than ferrous. Methemoglobin thus can neither bind nor transport O_2. Normally, the enzyme methemoglobin reductase reduces the Fe^{3+} of methemoglobin to Fe^{2+}. Methemoglobin can arise by oxidation of Fe^{2+} to Fe^{3+} as a side effect of agents such as sulfonamides, from hereditary hemoglobin M, or consequent to reduced activity of the enzyme methemoglobin reductase.

In hemoglobin M, histidine F8 (His F8) has been replaced by tyrosine. The iron of HbM forms a tight ionic complex with the phenolate anion of tyrosine that stabilizes the Fe^{3+} form. In α-chain hemoglobin M variants, the R-T equilibrium favors the T state. Oxygen affinity is reduced, and the Bohr effect is absent. β-Chain hemoglobin M variants exhibit R-T switching, and the Bohr effect is therefore present.

Mutations that favor the R state (eg, hemoglobin Chesapeake) increase O_2 affinity. These hemoglobins therefore fail to deliver adequate O_2 to peripheral tissues. The resulting tissue hypoxia leads to **polycythemia,** an increased concentration of erythrocytes.

Hemoglobin S

In HbS, the nonpolar amino acid valine has replaced the polar surface residue Glu6 of the β subunit, generating a hydrophobic **"sticky patch"** on the surface of the β subunit of both oxyHbS and deoxyHbS. Both HbA and HbS contain a complementary sticky patch on their surfaces that is exposed only in the deoxygenated T state. Thus, at low Po_2, deoxyHbS can polymerize to form long, insoluble fibers. Binding of deoxy-HbA terminates fiber polymerization, since HbA lacks the second sticky patch necessary to bind another Hb molecule (Figure 6–12). These twisted helical fibers distort the erythrocyte into a characteristic sickle shape, rendering it vulnerable to lysis in the interstices of the splenic sinusoids. They also cause multiple secondary clinical effects. A low Po_2 such as that at high altitudes exacerbates the tendency to polymerize. Emerging treatments for sickle cell disease include inducing HbF expression to inhibit the polymerization of HbS, stem cell transplantation, and, in the future, gene therapy.

BIOMEDICAL IMPLICATIONS

Myoglobinuria

Following massive crush injury, myoglobin released from damaged muscle fibers colors the urine dark red. Myoglobin can be detected in plasma following a myocardial infarction, but assay of serum enzymes (see Chapter 7) provides a more sensitive index of myocardial injury.

Anemias

Anemias, reductions in the number of red blood cells or of hemoglobin in the blood, can reflect impaired synthesis of hemoglobin (eg, in iron deficiency; Chapter 50) or impaired production of erythrocytes (eg, in folic acid or vitamin B_{12} deficiency; Chapter 44). Diagnosis of anemias begins with spectroscopic measurement of blood hemoglobin levels.

Thalassemias

The genetic defects known as thalassemias result from the partial or total absence of one or more α or β chains of hemoglobin. Over 750 different mutations have been identified, but only three are common. Either the α chain (alpha thalas-

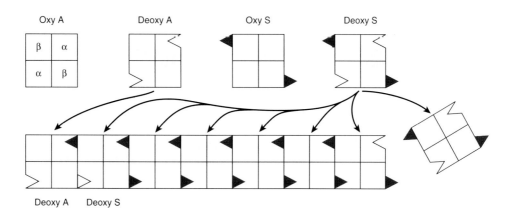

FIGURE 6–12 Representation of the sticky patch (▲) on hemoglobin S and its "receptor" (△) on deoxyhemoglobin A and deoxyhemoglobin S. The complementary surfaces allow deoxyhemoglobin S to polymerize into a fibrous structure, but the presence of deoxyhemoglobin A will terminate the polymerization by failing to provide sticky patches. (Modified and reproduced, with permission, from Stryer L: *Biochemistry,* 4th ed. Freeman, 1995. Copyright © 1995 W. H. Freeman and Company.)

semias) or β chain (beta thalassemias) can be affected. A superscript indicates whether a subunit is completely absent (α^0 or β^0) or whether its synthesis is reduced (α^+ or β^+). Apart from marrow transplantation, treatment is symptomatic.

Certain mutant hemoglobins are common in many populations, and a patient may inherit more than one type. Hemoglobin disorders thus present a complex pattern of clinical phenotypes. The use of DNA probes for their diagnosis is considered in Chapter 39.

Glycated Hemoglobin (HbA$_{1c}$)

When blood glucose enters the erythrocytes, it glycates the ε-amino group of lysine residues and the amino terminals of hemoglobin. The fraction of hemoglobin glycated, normally about 5%, is proportionate to blood glucose concentration. Since the half-life of an erythrocyte is typically 60 days, the level of glycated hemoglobin (HbA$_{1c}$) reflects the mean blood glucose concentration over the preceding 6–8 weeks. Measurement of HbA$_{1c}$ therefore provides valuable information for management of diabetes mellitus.

SUMMARY

- Myoglobin is monomeric; hemoglobin is a tetramer of two subunit types ($\alpha_2\beta_2$ in HbA). Despite having different primary structures, myoglobin and the subunits of hemoglobin have nearly identical secondary and tertiary structures.

- Heme, an essentially planar, slightly puckered, cyclic tetrapyrrole, has a central Fe^{2+} linked to all four nitrogen atoms of the heme, to histidine F8, and, in oxyMb and oxyHb, also to O_2.

- The O_2-binding curve for myoglobin is hyperbolic, but for hemoglobin it is sigmoidal, a consequence of cooperative interactions in the tetramer. Cooperativity maximizes the ability of hemoglobin both to load O_2 at the Po$_2$ of the lungs and to deliver O_2 at the Po$_2$ of the tissues.

- Relative affinities of different hemoglobins for oxygen are expressed as P$_{50}$, the Po$_2$ that half-saturates them with O_2. Hemoglobins saturate at the partial pressures of their respective respiratory organ, eg, the lung or placenta.

- On oxygenation of hemoglobin, the iron, histidine F8, and linked residues move toward the heme ring. Conformational changes that accompany oxygenation include rupture of salt bonds and loosening of quaternary structure, facilitating binding of additional O_2.

- 2,3-Bisphosphoglycerate (BPG) in the central cavity of deoxyHb forms salt bonds with the β subunits that stabilize deoxyHb. On oxygenation, the central cavity contracts, BPG is extruded, and the quaternary structure loosens.

- Hemoglobin also functions in CO_2 and proton transport from tissues to lungs. Release of O_2 from oxyHb at the tissues is accompanied by uptake of protons due to lowering of the pK_a of histidine residues.

- In sickle cell hemoglobin (HbS), Val replaces the β6 Glu of HbA, creating a "sticky patch" that has a complement on deoxyHb (but not on oxyHb). DeoxyHbS polymerizes at low O_2 concentrations, forming fibers that distort erythrocytes into sickle shapes.

- Alpha and beta thalassemias are anemias that result from reduced production of α and β subunits of HbA, respectively.

REFERENCES

Bettati S et al: Allosteric mechanism of haemoglobin: Rupture of salt-bridges raises the oxygen affinity of the T-structure. J Mol Biol 1998;281:581.

Frauenfelder H, McMahon BH, Fenimore PW: Myoglobin: The hydrogen atom of biology and a paradigm of complexity. Proc Natl Acad Sci USA 2003;100:8615.

Hardison RC et al: Databases of human hemoglobin variants and other resources at the globin gene server. Hemoglobin 2001;25:183.

Lukin JA, Ho C: The structure-function relationship of hemoglobin in solution at atomic resolution. Chem Rev 2004;104:1219.

Ordway GA, Garry DJ: Myoglobin: An essential hemoprotein in striated muscle. J Exp Biol 2004;207:3441.

Persons DA: Update on gene therapy for hemoglobin disorders. Curr Opin Mol Ther 2003;5:508.

Schrier SL, Angelucci E: New strategies in the treatment of the thalassemias. Annu Rev Med 2005;56:157.

Steinberg MH, Brugnara C: Pathophysiological-based approaches to treatment of sickle-cell disease. Annu Rev Med 2003;54:89.

Weatherall DJ, Clegg JD: *The Thalassemia Syndromes.* Blackwell Science. 2001.

Weatherall DJ et al: The hemoglobinopathies. In: *The Metabolic and Molecular Bases of Inherited Disease,* 8th ed., Scriver CR et al (editors). McGraw-Hill, 2000;4571–4635.

Weatherall DJ et al: Inherited disorders of hemoglobin. In: *Disease Control Priorities in Developing Countries,* Jamison D et al (editors). Oxford University Press and the World Bank, 2006;663–680.

Enzymes: Mechanism of Action

Peter J. Kennelly, PhD & Victor W. Rodwell, PhD

BIOMEDICAL IMPORTANCE

Enzymes are biologic polymers that catalyze the chemical reactions that make life as we know it possible. The presence and maintenance of a complete and balanced set of enzymes is essential for the breakdown of nutrients to supply energy and chemical building blocks; the assembly of those building blocks into proteins, DNA, membranes, cells, and tissues; and the harnessing of energy to power cell motility, neural function, and muscle contraction. With the exception of catalytic RNA molecules, or ribozymes, enzymes are proteins. The ability to assay the activity of specific enzymes in blood, other tissue fluids, or cell extracts aids in the diagnosis and prognosis of disease. Deficiencies in the quantity or catalytic activity of key enzymes can result from genetic defects, nutritional deficits, or toxins. Defective enzymes can result from genetic mutations or infection by viral or bacterial pathogens (eg, *Vibrio cholerae*). Medical scientists address imbalances in enzyme activity by using pharmacologic agents to inhibit specific enzymes and are investigating gene therapy as a means to remedy deficits in enzyme level or function.

In addition to serving as the catalysts for all metabolic processes, their impressive catalytic activity, substrate specificity, and stereospecificity enable enzymes to fulfill key roles in other processes related to human health and well-being. The absolute stereospecificity of enzymes is of particular value for use as soluble or immobilized catalysts for specific reactions in the synthesis of a drug or antibiotic. Proteolytic enzymes augment the capacity of detergents to remove dirt and stains. Enzymes play an important role in producing or enhancing the nutrient value of food products for both humans and animals. The protease rennin, for example, is utilized in the production of cheeses while lactase is employed to remove lactose from milk for the benefit of persons who suffer from lactose intolerance as a consequence of a deficiency in this hydrolytic enzyme (Chapter 43).

ENZYMES ARE EFFECTIVE & HIGHLY SPECIFIC CATALYSTS

The enzymes that catalyze the conversion of one or more compounds (**substrates**) into one or more different compounds

(**products**) enhance the rates of the corresponding non-catalyzed reaction by factors of at least 10^6. Like all catalysts, enzymes are neither consumed nor permanently altered as a consequence of their participation in a reaction.

In addition to being highly efficient, enzymes are also extremely selective catalysts. Unlike most catalysts used in synthetic chemistry, enzymes are specific both for the type of reaction catalyzed and for a single substrate or a small set of closely related substrates. Enzymes are also stereospecific catalysts and typically catalyze reactions of only one stereoisomer of a given compound—for example, D- but not L-sugars, L- but not D-amino acids. Since they bind substrates through at least "three points of attachment," enzymes can even convert nonchiral substrates to chiral products. Figure 7–1 illustrates why the enzyme-catalyzed reduction of the nonchiral substrate pyruvate produces L-lactate—rather than a racemic mixture of D- and L-lactate. The exquisite specificity of enzyme catalysts imbues living cells with the ability to simultaneously conduct and independently control a broad spectrum of chemical processes.

ENZYMES ARE CLASSIFIED BY REACTION TYPE

The commonly used names for most enzymes describe the type of reaction catalyzed, followed by the suffix *-ase*. For example, dehydrogen*ases* remove hydrogen atoms, prote*ases* hydrolyze proteins, and isomer*ases* catalyze rearrangements in configuration. Modifiers may precede the name to indicate the substrate (*xanthine* oxidase), the source of the enzyme (*pancreatic* ribonuclease), its regulation (*hormone-sensitive* lipase), or a feature of its mechanism of action (*cysteine* protease). Where needed, alphanumeric designators are added to identify multiple forms of an enzyme (eg, RNA polymerase *III*; protein kinase *Cβ*).

To address ambiguities, the International Union of Biochemists (IUB) developed an unambiguous system of enzyme nomenclature in which each enzyme has a unique name and code number that identify the type of reaction catalyzed and the substrates involved. Enzymes are grouped into six classes:

1. **Oxidoreductases** (catalyze oxidations and reductions)
2. **Transferases** (catalyze transfer of moieties such as glycosyl, methyl, or phosphoryl groups)
3. **Hydrolases** (catalyze *hydrolytic* cleavage of C—C, C—O, C—N, and other bonds)
4. **Lyases** (catalyze cleavage of C—C, C—O, C—N, and other bonds by *atom elimination,* leaving double bonds)
5. **Isomerases** (catalyze geometric or structural changes within a molecule)
6. **Ligases** (catalyze the joining together of two molecules coupled to the hydrolysis of ATP)

Despite the clarity of the IUB system, the names are lengthy and relatively cumbersome, so we generally continue to refer to enzymes by their traditional, albeit sometimes ambiguous names. The IUB name for hexokinase illustrates both the clarity of the IUB system and its complexities. The IUB name of hexokinase is ATP:D-hexose 6-phosphotransferase E.C. 2.7.1.1. This name identifies hexokinase as a member of class 2 (transferases), subclass 7 (transfer of a phosphoryl group), sub-subclass 1 (alcohol is the phosphoryl acceptor), and "hexose-6" indicates that the alcohol phosphorylated is on carbon six of a hexose. However, we continue to call it hexokinase.

PROSTHETIC GROUPS, COFACTORS, & COENZYMES PLAY IMPORTANT ROLES IN CATALYSIS

Many enzymes contain small nonprotein molecules and metal ions that participate directly in substrate binding or catalysis. Termed **prosthetic groups, cofactors,** and **coenzymes,** these extend the repertoire of catalytic capabilities beyond those afforded by the limited number of functional groups present on the aminoacyl side chains of peptides.

Prosthetic Groups Are Tightly Integrated Into an Enzyme's Structure

Prosthetic groups are distinguished by their tight, stable incorporation into a protein's structure by covalent or noncovalent

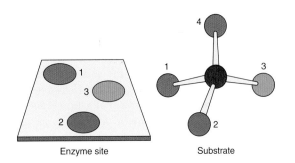

FIGURE 7–1 Planar representation of the "three-point attachment" of a substrate to the active site of an enzyme. Although atoms 1 and 4 are identical, once atoms 2 and 3 are bound to their complementary sites on the enzyme, only atom 1 can bind. Once bound to an enzyme, apparently identical atoms thus may be distinguishable, permitting a stereospecific chemical change.

forces. Examples include pyridoxal phosphate, flavin mononucleotide (FMN), flavin adenine dinucleotide (FAD), thiamin pyrophosphate, biotin, and the metal ions of Co, Cu, Mg, Mn, and Zn. Metals are the most common prosthetic groups. The roughly one-third of all enzymes that contain tightly bound metal ions are termed **metalloenzymes.** Metal ions that participate in redox reactions generally are complexed to prosthetic groups such as heme (Chapter 6) or iron-sulfur clusters (Chapter 12). Metals also may facilitate the binding and orientation of substrates, the formation of covalent bonds with reaction intermediates (Co^{2+} in coenzyme B_{12}), or interact with substrates to render them more **electrophilic** (electron-poor) or **nucleophilic** (electron-rich).

Cofactors Associate Reversibly With Enzymes or Substrates

Cofactors serve functions similar to those of prosthetic groups but bind in a transient, dissociable manner either to the enzyme or to a substrate such as ATP. Unlike the stably associated prosthetic groups, cofactors therefore must be present in the medium surrounding the enzyme for catalysis to occur. The most common cofactors also are metal ions. Enzymes that require a metal ion cofactor are termed **metal-activated enzymes** to distinguish them from the **metalloenzymes** for which metal ions serve as prosthetic groups.

Coenzymes Serve as Substrate Shuttles

Coenzymes serve as recyclable shuttles—or group transfer agents—that transport many substrates from their point of generation to their point of utilization. Association with the coenzyme also stabilizes substrates such as hydrogen atoms or hydride ions that are unstable in the aqueous environment of the cell. Other chemical moieties transported by coenzymes include methyl groups (folates), acyl groups (coenzyme A), and oligosaccharides (dolichol).

Many Coenzymes, Cofactors & Prosthetic Groups Are Derivatives of B Vitamins

The water-soluble B vitamins supply important components of numerous coenzymes. Several coenzymes contain, in addition, the adenine, ribose, and phosphoryl moieties of AMP or ADP (Figure 7–2). Nicotinamide is a component of the redox coenzymes NAD and NADP, whereas riboflavin is a component of the redox coenzymes FMN and FAD. **Pantothenic acid** is a component of the acyl group carrier coenzyme A. As its pyrophosphate, **thiamin** participates in decarboxylation of α-keto acids, and **folic acid** and **cobamide** coenzymes function in one-carbon metabolism.

CATALYSIS OCCURS AT THE ACTIVE SITE

An important early 20th-century insight into enzymic catalysis sprang from the observation that the presence of substrates

FIGURE 7–2 Structure of NAD⁺ and NADP⁺. For NAD⁺, R = H. For NADP⁺, R = PO₃²⁻.

FIGURE 7–3 Two-dimensional representation of a dipeptide substrate, glycyl-tyrosine, bound within the active site of carboxypeptidase A.

renders enzymes more resistant to the denaturing effects of elevated temperatures. This observation led Emil Fischer to propose that enzymes and their substrates interact to form an enzyme-substrate (ES) complex whose thermal stability was greater than that of the enzyme itself. This insight profoundly shaped our understanding of both the chemical nature and kinetic behavior (Chapter 8) of enzymic catalysis.

Fischer reasoned that the exquisitely high specificity with which enzymes recognize their substrates when forming an ES complex was analogous to the manner in which a mechanical lock discriminates the proper key. This enzymatic "lock" is referred to as the **active site**. In most enzymes, the active site takes the form of a cleft or pocket on the enzyme's surface (Figures 5–6 & 5–8) or, for some multimeric enzymes, at the interface between subunits. As its name implies, the active site is much more than simply a recognition site for binding substrates. It provides a three-dimensional environment that both shields substrates from solvent and facilitates catalysis. It also binds any cofactors and prosthetic groups that may be required for catalysis. Within the active site, substrate molecules are aligned in close proximity and optimal orientation to the functional groups of peptidyl aminoacyl residues, cofactors, and prosthetic groups responsible for catalyzing their chemical transformation into products (Figure 7–3).

ENZYMES EMPLOY MULTIPLE MECHANISMS TO FACILITATE CATALYSIS

Enzymes use various combinations of four general mechanisms to achieve dramatic catalytic enhancement of the rates of chemical reactions.

Catalysis by Proximity

For molecules to react, they must come within bond-forming distance of one another. The higher their concentration, the more frequently they will encounter one another, and the greater will be the rate of their reaction. When an enzyme binds substrate molecules at its active site, it creates a region of high local substrate concentration. This environment also orients the substrate molecules spatially in a position ideal for them to interact, resulting in rate enhancements of at least a thousand-fold.

Acid-Base Catalysis

The ionizable functional groups of aminoacyl side chains and (where present) of prosthetic groups contribute to catalysis by acting as acids or bases. Acid-base catalysis can be either specific or general. By "specific" we mean only protons (H₃O⁺) or OH⁻ ions. In **specific acid** or **specific base catalysis,** the rate of reaction is sensitive to changes in the concentration of protons but independent of the concentrations of other acids (proton donors) or bases (proton acceptors) present in solution or at the active site. Reactions whose rates are responsive to *all* the acids or bases present are said to be subject to **general acid** or **general base catalysis.**

Catalysis by Strain

Enzymes that catalyze -*lytic* reactions that involve breaking a covalent bond typically bind their substrates in a conformation slightly unfavorable for the bond that will undergo cleavage. This conformation mimics that of the **transition state intermediate**, a transient species representing the transition or half-way point in the transformation of substrates to products. The resulting strain stretches or distorts the targeted bond, weakening it and making it more vulnerable to cleavage. Nobel Laureate Linus Pauling was the first to suggest a role for transition state stabilization as a general mechanism by which

$$E-CHO \xrightarrow[\text{Ala}]{\text{Ala}} E \overset{\text{CHO}}{\underset{\text{Ala}}{}} \longrightarrow E \overset{\text{CH}_2\text{NH}_2}{\underset{\text{Pyr}}{}} \xrightarrow{\text{Pyr}} E-CH_2NH_2 \xrightarrow[\text{KG}]{\text{KG}} E \overset{\text{CH}_2\text{NH}_2}{\underset{\text{KG}}{}} \longrightarrow E \overset{\text{CHO}}{\underset{\text{Glu}}{}} \xrightarrow{\text{Glu}} E-CHO$$

FIGURE 7–4 "Ping-pong" mechanism for transamination. E—CHO and E—CH$_2$NH$_2$ represent the enzyme-pyridoxal phosphate and enzyme-pyridoxamine complexes, respectively. (Ala, alanine; Pyr, pyruvate; KG, α-ketoglutarate; Glu, glutamate.)

enzymes accelerate the rates of chemical reactions. Knowledge of the transition state of an enzyme-catalyzed reaction is frequently exploited by chemists to design and create more effective enzyme inhibitors, called **transition state analogs**, as potential pharmacophores.

Covalent Catalysis

The process of **covalent catalysis** involves the formation of a covalent bond between the enzyme and one or more substrates. The modified enzyme then becomes a reactant. Covalent catalysis introduces a new reaction pathway whose activation energy is lower—and therefore is faster—than the reaction pathway in homogeneous solution. The chemical modification of the enzyme is, however, transient. On completion of the reaction, the enzyme returns to its original unmodified state. Its role thus remains catalytic. Covalent catalysis is particularly common among enzymes that catalyze group transfer reactions. Residues on the enzyme that participate in covalent catalysis generally are cysteine or serine and occasionally histidine. Covalent catalysis often follows a "ping-pong" mechanism—one in which the first substrate is bound and its product released prior to the binding of the second substrate (Figure 7–4).

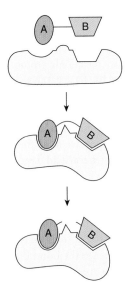

FIGURE 7–5 Two-dimensional representation of Koshland's induced fit model of the active site of a lyase. Binding of the substrate A—B induces conformational changes in the enzyme that aligns catalytic residues which participate in catalysis and strains the bond between A and B, facilitating its cleavage.

SUBSTRATES INDUCE CONFORMATIONAL CHANGES IN ENZYMES

While Fischer's "lock and key model" accounted for the exquisite specificity of enzyme-substrate interactions, the implied rigidity of the enzyme's active site failed to account for the dynamic changes that accompany catalysis. This drawback was addressed by Daniel Koshland's **induced fit** model, which states that when substrates approach and bind to an enzyme they induce a conformational change, a change analogous to placing a hand (substrate) into a glove (enzyme) (Figure 7–5). A corollary is that the enzyme induces reciprocal changes in its substrates, harnessing the energy of binding to facilitate the transformation of substrates into products. The induced fit model has been amply confirmed by biophysical studies of enzyme motion during substrate binding.

HIV PROTEASE ILLUSTRATES ACID-BASE CATALYSIS

Enzymes of the **aspartic protease family,** which includes the digestive enzyme pepsin, the lysosomal cathepsins, and the protease produced by the human immunodeficiency virus (HIV), share a common catalytic mechanism. Catalysis involves two conserved aspartyl residues, which act as acid-base catalysts. In the first stage of the reaction, an aspartate functioning as a general base (Asp X, Figure 7–6) extracts a proton from a water molecule, making it more nucleophilic. The resulting nucleophile then attacks the electrophilic carbonyl carbon of the peptide bond targeted for hydrolysis, forming a **tetrahedral transition state intermediate.** A second aspartate (Asp Y, Figure 7–6) then facilitates the decomposition of this tetrahedral intermediate by donating a proton to the amino group produced by rupture of the peptide bond. The two different active site aspartates can act simultaneously as a general base or as a general acid because their immediate environment favors ionization of one, but not the other.

CHYMOTRYPSIN & FRUCTOSE-2,6-BISPHOSPHATASE ILLUSTRATE COVALENT CATALYSIS

Chymotrypsin

While catalysis by aspartic proteases involves the direct hydrolytic attack of water on a peptide bond, catalysis by the **serine**

FIGURE 7–6 Mechanism for catalysis by an aspartic protease such as HIV protease. Curved arrows indicate directions of electron movement. ① Aspartate X acts as a base to activate a water molecule by abstracting a proton. ② The activated water molecule attacks the peptide bond, forming a transient tetrahedral intermediate. ③ Aspartate Y acts as an acid to facilitate breakdown of the tetrahedral intermediate and release of the split products by donating a proton to the newly formed amino group. Subsequent shuttling of the proton on Asp X to Asp Y restores the protease to its initial state.

FIGURE 7–7 Catalysis by chymotrypsin. ① The charge-relay system removes a proton from Ser 195, making it a stronger nucleophile. ② Activated Ser 195 attacks the peptide bond, forming a transient tetrahedral intermediate. ③ Release of the amino terminal peptide is facilitated by donation of a proton to the newly formed amino group by His 57 of the charge-relay system, yielding an acyl-Ser 195 intermediate. ④ His 57 and Asp 102 collaborate to activate a water molecule, which attacks the acyl-Ser 195, forming a second tetrahedral intermediate. ⑤ The charge-relay system donates a proton to Ser 195, facilitating breakdown of tetrahedral intermediate to release the carboxyl terminal peptide ⑥.

protease chymotrypsin involves prior formation of a covalent acyl enzyme intermediate. A highly reactive seryl residue, serine 195, participates in a charge-relay network with histidine 57 and aspartate 102. Far apart in primary structure, in the active site these residues are within bond-forming distance of one another. Aligned in the order Asp 102-His 57-Ser 195, they constitute a "charge-relay network" that functions as a "proton shuttle."

Binding of substrate initiates proton shifts that in effect transfer the hydroxyl proton of Ser 195 to Asp 102 (Figure 7–7). The enhanced nucleophilicity of the seryl oxygen facilitates its attack on the carbonyl carbon of the peptide bond of the substrate, forming a covalent acyl-enzyme intermediate. The proton on Asp 102 then shuttles through His 57 to

the amino group liberated when the peptide bond is cleaved. The portion of the original peptide with a free amino group then leaves the active site and is replaced by a water molecule. The charge-relay network now activates the water molecule by withdrawing a proton through His 57 to Asp 102. The resulting hydroxide ion attacks the acyl-enzyme intermediate and a reverse proton shuttle returns a proton to Ser 195, restoring its original state. While modified during the process of catalysis, chymotrypsin emerges unchanged on completion of the reaction. Trypsin and elastase employ a similar catalytic mechanism, but the numbers of the residues in their Ser-His-Asp proton shuttles differ.

Fructose-2,6-Bisphosphatase

Fructose-2,6-bisphosphatase, a regulatory enzyme of gluconeogenesis (Chapter 20), catalyzes the hydrolytic release of the phosphate on carbon 2 of fructose 2,6-bisphosphate. Figure 7–8 illustrates the roles of seven active site residues. Catalysis involves a "catalytic triad" of one Glu and two His residues and a covalent phosphohistidyl intermediate.

CATALYTIC RESIDUES ARE HIGHLY CONSERVED

Members of an enzyme family such as the aspartic or serine proteases employ a similar mechanism to catalyze a common reaction type but act on different substrates. Most enzyme families arose through gene duplication events that create a second copy of the gene that encodes a particular enzyme. The proteins encoded by the two genes can then evolve independently to recognize different substrates—resulting, for example, in chymotrypsin, which cleaves peptide bonds on the carboxyl terminal side of large hydrophobic amino acids, and trypsin, which cleaves peptide bonds on the carboxyl terminal side of basic amino acids. Proteins that diverged from a common ancestor are said to be **homologous** to one another. The common ancestry of enzymes can be inferred from the presence of specific amino acids in the same position in each family member. These residues are said to be **conserved residues.** Table 7–1 illustrates the primary structural conservation of two components of the charge-relay network for several serine proteases. Among the most highly conserved residues are those that participate directly in catalysis.

ISOZYMES ARE DISTINCT ENZYME FORMS THAT CATALYZE THE SAME REACTION

Higher organisms often elaborate several physically distinct versions of a given enzyme, each of which catalyzes the same reaction. Like the members of other protein families, these protein catalysts or **isozymes** arise through gene duplication. Isozymes may exhibit subtle differences in properties such as sensitivity to particular regulatory factors (Chapter 9) or substrate affinity (eg, hexokinase and glucokinase) that adapt them to specific tissues or circumstances. Some isozymes may also enhance survival by providing a "backup" copy of an essential enzyme.

E • Fru-2,6-P$_2$ E-P • Fru-6-P

E-P • H$_2$O   E • P$_i$

FIGURE 7–8 Catalysis by fructose-2,6-bisphosphatase. **(1)** Lys 356 and Arg 257, 307, and 352 stabilize the quadruple negative charge of the substrate by charge-charge interactions. Glu 327 stabilizes the positive charge on His 392. **(2)** The nucleophile His 392 attacks the C-2 phosphoryl group and transfers it to His 258, forming a phosphoryl-enzyme intermediate. Fructose 6-phosphate leaves the enzyme. **(3)** Nucleophilic attack by a water molecule, possibly assisted by Glu 327 acting as a base, forms inorganic phosphate. **(4)** Inorganic orthophosphate is released from Arg 257 and Arg 307. (Reproduced, with permission, from Pilkis SJ, et al: 6-Phosphofructo-2-kinase/fructose-2,6-bisphosphatase: A metabolic signaling enzyme. Annu Rev Biochem 1995;64:799. © 1995 by Annual Reviews, www.annualreviews.org. Reprinted with permission.)

THE CATALYTIC ACTIVITY OF ENZYMES FACILITATES THEIR DETECTION

The relatively small quantities of enzymes present in cells complicate determination of their presence and concentration. However, the amplification conferred by their ability to rapidly transform thousands of molecules of a specific substrate into products imbues each enzyme with the ability to reveal its presence. Assays of the catalytic activity of enzymes are frequently used in research and clinical laboratories. Under appropriate conditions (see Chapter 8), the rate of the catalytic reaction being monitored is proportional to the amount of enzyme present, which allows its concentration to be inferred.

Single-Molecule Enzymology

The limited sensitivity of traditional enzyme assays necessitates the use of a large group, or ensemble, of enzyme molecules in order to produce measurable quantities of product. The data obtained thus reflect the average catalytic capability

TABLE 7–1   Amino Acid Sequences in the Neighborhood of the Catalytic Sites of Several Bovine Proteases

Enzyme	Sequence Around Serine Ⓢ														Sequence Around Histidine Ⓗ									
Trypsin	D	S	C	Q	D	G	Ⓢ	G	G	P	V	V	C	S	G	K	V	V	S	A	A	Ⓗ	C	Y	K	S	G
Chymotrypsin A	S	S	C	M	G	D	Ⓢ	G	G	P	L	V	C	K	K	N	V	V	T	A	A	Ⓗ	G	G	V	T	T
Chymotrypsin B	S	S	C	M	G	D	Ⓢ	G	G	P	L	V	C	Q	K	N	V	V	T	A	A	Ⓗ	C	G	V	T	T
Thrombin	D	A	C	E	G	D	Ⓢ	G	G	P	F	V	M	K	S	P	V	L	T	A	A	Ⓗ	C	L	L	Y	P

Note: Regions shown are those on either side of the catalytic site seryl (S) and histidyl (H) residues.

of individual molecules. Recent advances in **nanotechnology** have made it possible to observe, usually by fluorescence microscopy, catalysis by individual enzyme and substrate molecules. Consequently, scientists can now measure the rate of single catalytic events and sometimes the individual steps in catalysis by a process called **single-molecule enzymology** (Figure 7–9).

Drug Discovery Requires Enzyme Assays Suitable for "High-Throughput" Screening

Enzymes constitute one of the primary classes of biomolecules targeted for the development of drugs and other therapeutic agents. Many antibiotics, for example, inhibit enzymes that are unique to microbial pathogens. The discovery of new drugs is greatly facilitated when a large number of potential pharmacophores can be assayed in a rapid, automated fashion—a process referred to as **high-throughput screening.** High-throughput screening takes advantage of recent advances in robotics, optics, data processing, and microfluidics to conduct and analyze many thousands of simultaneous assays of the activity of a given enzyme. The most commonly used high-throughput screening devices employ 10–100 μL volumes in 96, 384, or 1536 well plastic plates and fully automated equipment capable of dispensing substrates, coenzymes, enzymes, and potential inhibitors in a multiplicity of combinations and concentrations. High-throughput screening is ideal for surveying the numerous products of **combinatorial chemistry,** the simultaneous synthesis of large libraries of chemical compounds that contain all possible combinations of a set of chemical precursors. Enzyme assays that produce a chromagenic or fluorescent product are ideal, since optical detectors are readily engineered to permit the rapid analysis of multiple samples. At present, the sophisticated equipment required for truly large numbers of assays is available only in pharmaceutical houses, government-sponsored laboratories, and research universities. As described in Chapter 8, its principal use is the analysis of inhibitory compounds with ultimate potential for use as drugs

Enzyme-Linked Immunoassays

The sensitivity of enzyme assays can be exploited to detect proteins that lack catalytic activity. **Enzyme-linked immunosorbent assays** (ELISAs) use antibodies covalently linked to a "reporter enzyme" such as alkaline phosphatase or horserad-

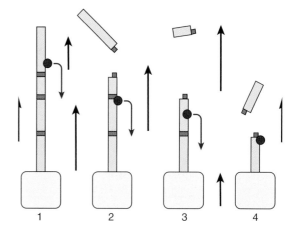

FIGURE 7–9 Direct observation of single DNA cleavage events catalyzed by a restriction endonuclease. DNA molecules immobilized to beads (pale yellow) are placed in a flowing stream of buffer (black arrows), which causes them to assume an extended conformation. Cleavage at one of the restriction sites (orange) by an endonuclease leads to a shortening of the DNA molecule, which can be observed directly in a microscope since the nucleotide bases in DNA are fluorescent. Although the endonuclease (red) does not fluoresce, and hence is invisible, the progressive manner in which the DNA molecule is shortened (1→4) reveals that the endonuclease binds to the free end of the DNA molecule and moves along it from site to site.

ish peroxidase whose products are readily detected, generally by the absorbance of light or by fluorescence. Serum or other biologic samples to be tested are placed in a plastic microtiter plate, where the proteins adhere to the plastic surface and are immobilized. Any remaining absorbing areas of the well are then "blocked" by adding a nonantigenic protein such as bovine serum albumin. A solution of antibody covalently linked to a reporter enzyme is then added. The antibodies adhere to the immobilized antigen and are themselves immobilized. Excess free antibody molecules are then removed by washing. The presence and quantity of bound antibody is then determined by adding the substrate for the reporter enzyme.

NAD(P)⁺-Dependent Dehydrogenases Are Assayed Spectrophotometrically

The physicochemical properties of the reactants in an enzyme-catalyzed reaction dictate the options for the assay of enzyme

FIGURE 7–10 Absorption spectra of NAD⁺ and NADH. Densities are for a 44 mg/L solution in a cell with a 1 cm light path. NADP⁺ and NADPH have spectrums analogous to NAD⁺ and NADH, respectively.

FIGURE 7–11 Coupled enzyme assay for hexokinase activity. The production of glucose 6-phosphate by hexokinase is coupled to the oxidation of this product by glucose-6-phosphate dehydrogenase in the presence of added enzyme and NADP⁺. When an excess of glucose-6-phosphate dehydrogenase is present, the rate of formation of NADPH, which can be measured at 340 nm, is governed by the rate of formation of glucose-6-phosphate by hexokinase.

activity. Spectrophotometric assays exploit the ability of a substrate or product to absorb light. The reduced coenzymes NADH and NADPH, written as NAD(P)H, absorb light at a wavelength of 340 nm, whereas their oxidized forms NAD(P)⁺ do not (Figure 7–10). When NAD(P)⁺ is reduced, the absorbance at 340 nm therefore increases in proportion to—and at a rate determined by—the quantity of NAD(P)H produced. Conversely, for a dehydrogenase that catalyzes the oxidation of NAD(P)H, a decrease in absorbance at 340 nm will be observed. In each case, the rate of change in optical density at 340 nm will be proportionate to the quantity of enzyme present.

Many Enzymes Are Assayed by Coupling to a Dehydrogenase

The assay of enzymes whose reactions are not accompanied by a change in absorbance or fluorescence is generally more difficult. In some instances, the product or remaining substrate can be transformed into a more readily detected compound. In other instances, the reaction product may have to be separated from unreacted substrate prior to measurement. An alternative strategy is to devise a synthetic substrate whose product absorbs light or fluoresces. For example, *p*-nitrophenyl phosphate is an artificial substrate for certain phosphatases and for chymotrypsin that does not absorb visible light. However, following hydrolysis, the resulting *p*-nitrophenylate anion absorbs light at 419 nm.

Another quite general approach is to employ a "coupled" assay (Figure 7–11). Typically, a dehydrogenase whose substrate is the product of the enzyme of interest is added in catalytic excess. The rate of appearance or disappearance of NAD(P)H then depends on the rate of the enzyme reaction to which the dehydrogenase has been coupled.

THE ANALYSIS OF CERTAIN ENZYMES AIDS DIAGNOSIS

The analysis of enzymes in blood plasma has played a central role in the diagnosis of several disease processes. Many enzymes are functional constituents of blood. Examples include pseudocholinesterase, lipoprotein lipase, and components of the cascade of events in blood clotting and clot dissolution. Other enzymes are released into plasma following cell death or injury. While these latter enzymes perform no physiologic function in plasma, their appearance or levels can assist in the diagnosis and prognosis of diseases and injuries affecting specific tissues. Following injury, the plasma concentration of a released enzyme may rise early or late, and may decline rapidly or slowly. Proteins from the cytoplasm tend to appear more rapidly than those from subcellular organelles. The speed with which enzymes and other proteins are removed from plasma varies with their susceptibility to proteolysis and permeability through renal glomeruli.

Quantitative analysis of the activity of released enzymes or other proteins, typically in plasma or serum but also in urine or various cells, provides information concerning diagnosis, prognosis, and response to treatment. Assays of enzyme *activity* typically employ standard kinetic assays of initial reaction rates. Alternatively, radioimmunoassays (RIA) provide quantitation of the absolute *amount* of an enzyme or of noncatalytic protein. Table 7–2 lists several enzymes of value in clinical diagnosis. These enzymes are, however, not absolutely specific for the indicated disease. For example, elevated blood levels of prostatic acid phosphatase are associated typically with prostate cancer, but also with certain other cancers and noncancerous conditions. Consequently, enzyme assay data must be considered together with other factors elicited through a comprehensive clinical examination. Factors to be considered in interpreting enzyme data include patient age, sex, prior history, possible drug use, and the sensitivity and the diagnostic specificity of the enzyme test.

TABLE 7–2 Principal Serum Enzymes Used in Clinical Diagnosis

Serum Enzyme	Major Diagnostic Use
Aminotransferases	
Aspartate aminotransferase (AST, or SGOT)	Myocardial infarction
Alanine aminotransferase (ALT, or SGPT)	Viral hepatitis
Amylase	Acute pancreatitis
Ceruloplasmin	Hepatolenticular degeneration (Wilson's disease)
Creatine kinase	Muscle disorders and myocardial infarction
γ-Glutamyl transferase	Various liver diseases
Lactate dehydrogenase isozyme 5	Liver diseases
Lipase	Acute pancreatitis
Phosphatase, acid	Metastatic carcinoma of the prostate
Phosphatase, alkaline (isozymes)	Various bone disorders, obstructive liver diseases

Note: Many of the above enzymes are not specific to the disease listed.

Enzymes Assist Diagnosis of Myocardial Infarction

An enzyme useful for diagnostic enzymology should be relatively specific for the tissue or organ under study, should appear in the plasma or other fluid at a time useful for diagnosis (the "diagnostic window"), and should be amenable to automated assay. The enzymes used to confirm a myocardial infarction (MI) illustrate the concept of a "diagnostic window," and provide a historical perspective on the use of different enzymes for this purpose.

Detection of an enzyme must be possible within a few hours of an MI to confirm a preliminary diagnosis and permit initiation of appropriate therapy. Enzymes that only appear in the plasma 12 h or more following injury are thus of limited utility. The first enzymes used to diagnose MI were aspartate aminotransferase (AST), alanine aminotransferase (ALT), and lactate dehydrogenase. AST and ALT proved less than ideal, however, as they appear in plasma relatively slowly and are not specific to heart muscle. While LDH also is released relatively slowly into plasma, it offered the advantage of tissue specificity as a consequence of its quaternary structure.

Lactate dehydrogenase (LDH) is a tetrameric enzyme consisting of two monomer types: H (for heart) and M (for muscle) that combine to yield five LDH isozymes: HHHH (I_1), HHHM (I_2), HHMM (I_3), HMMM (I_4), and MMMM (I_5). Tissue-specific expression of the H and M genes determines the relative proportions of each subunit in different tissues. Isozyme I_1 predominates in heart tissue, and isozyme I_5 in liver. Thus, tissue injury releases a characteristic pattern of LDH isozymes that can be separated by electrophoresis and detected using a coupled assay (Figure 7–12). Today, LDH has been superseded as a marker for MI by other proteins that appear more rapidly in plasma.

Creatine kinase (CK) has three isozymes: CK-MM (skeletal muscle), CK-BB (brain), and CK-MB (heart and skeletal muscle). CK-MB has a useful diagnostic window. It appears within 4–6 h of an MI, peaks at 24 h, and returns to baseline by 48–72 h. As for LDH, individual CK isozymes are separable by electrophoresis, thus facilitating detection.

Today, in most clinical laboratories for the diagnosis of MI CK has been supplemented by troponin. However, as CK-MB also is released upon skeletal muscle injury, assay of plasma CK levels continues in use to assess skeletal muscle disorders such as Duchenne muscular dystrophy.

Troponin is a complex of three proteins involved in muscle contraction in *skeletal* and *cardiac muscle* but not in *smooth mus-*

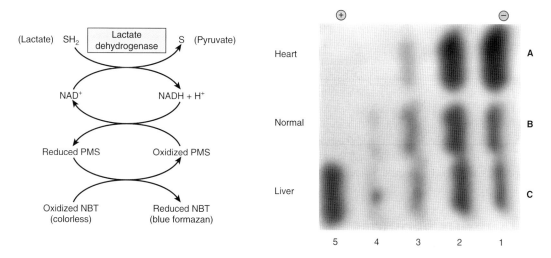

FIGURE 7–12 Normal and pathologic patterns of lactate dehydrogenase (LDH) isozymes in human serum. LDH isozymes of serum were separated by electrophoresis and visualized using the coupled reaction scheme shown on the left. (NBT, nitroblue tetrazolium; PMS, phenazine methylsulfate.) At right is shown the stained electropherogram. Pattern A is serum from a patient with a myocardial infarct; B is normal serum; and C is serum from a patient with liver disease. Arabic numerals denote specific LDH isozymes.

cle (see Chapter 49). Immunological measurement of plasma levels of cardiac troponins I and T provide sensitive and specific indicators of damage to heart muscle. Troponin levels rise 2–6 h after an MI and remain elevated for 4–10 days. In addition to MI, other heart muscle damage also elevates serum troponin levels. Cardiac troponins thus serves as a marker of all heart muscle damage. The search for additional markers for heart disease, such as ischemia modified albumin, and the simultaneous assessment of a spectrum of diagnostic markers via proteomics, continues to be an active area of clinical research.

Enzymes also can be employed in the clinical laboratory as tools for determining the concentration of critical metabolites. For example, glucose oxidase is frequently utilized to measure plasma glucose concentration. Enzymes are employed with increasing frequency as tools for the treatment of injury and disease. Tissue plasminogen activator (tPA) or streptokinase is used in the treatment of acute MI, while trypsin has been used in the treatment of cystic fibrosis (see Chapter 54).

ENZYMES FACILITATE DIAGNOSIS OF GENETIC AND INFECTIOUS DISEASES

Many diagnostic techniques take advantage of the specificity and efficiency of the enzymes that act on oligonucleotides such as DNA (Chapter 39). Enzymes known as **restriction endonucleases**, for example, cleave double-stranded DNA at sites specified by a sequence of four, six, or more base pairs called **restriction sites**. Cleavage of a sample of DNA with a restriction enzyme produces a characteristic set of smaller DNA fragments (see Chapter 39). Deviations in the normal product pattern, called **restriction fragment length polymorphisms (RFLPs)**, occur if a mutation renders a restriction site unrecognizable to its cognate restriction endonuclease or, alternatively, generates a new recognition site. RFLPs are currently utilized to facilitate prenatal detection of a number of hereditary disorders, including sickle cell trait, beta-thalassemia, infant phenylketonuria, and Huntington's disease.

The **polymerase chain reaction (PCR)** employs a thermostable DNA polymerase and appropriate oligonucleotide primers to produce thousands of copies of a defined segment of DNA from a minute quantity of starting material (see Chapter 39). PCR enables medical, biological, and forensic scientists to detect and characterize DNA present initially at levels too low for direct detection. In addition to screening for genetic mutations, PCR can be used to detect and identify pathogens and parasites such as *Trypanosoma cruzi*, the causative agent of Chagas' disease, and *Neisseria meningitides*, the causative agent of bacterial meningitis, through the selective amplification of their DNA.

RECOMBINANT DNA PROVIDES AN IMPORTANT TOOL FOR STUDYING ENZYMES

Recombinant DNA technology has emerged as an important asset in the study of enzymes. Highly purified samples

of enzymes are necessary for the study of their structure and function. The isolation of an individual enzyme, particularly one present in low concentration, from among the thousands of proteins present in a cell can be extremely difficult. If the gene for the enzyme of interest has been cloned, it generally is possible to produce large quantities of its encoded protein in *Escherichia coli* or yeast. However, not all animal proteins can be expressed in active form in microbial cells, nor do microbes perform certain posttranslational processing tasks. For these reasons, a gene may be expressed in cultured animal cell systems employing the baculovirus expression vector to transform cultured insect cells. For more details concerning recombinant DNA techniques, see Chapter 39.

Recombinant Fusion Proteins Are Purified by Affinity Chromatography

Recombinant DNA technology can also be used to create modified proteins that are readily purified by affinity chromatography. The gene of interest is linked to an oligonucleotide sequence that encodes a carboxyl or amino terminal extension to the encoded protein. The resulting modified protein, termed a **fusion protein**, contains a domain tailored to interact with a specific affinity support. One popular approach is to attach an oligonucleotide that encodes six consecutive histidine residues. The expressed "His tag" protein binds to chromatographic supports that contain an immobilized divalent metal ion such as Ni^{2+}. Alternatively, the substrate-binding do-

FIGURE 7–13 Use of glutathione S-transferase (GST) fusion proteins to purify recombinant proteins. (GSH, glutathione.)

main of glutathione S-transferase (GST) can serve as a "GST tag." Figure 7–13 illustrates the purification of a GST-fusion protein using an affinity support containing bound glutathione. Fusion proteins also often encode a cleavage site for a highly specific protease such as thrombin in the region that links the two portions of the protein. This permits removal of the added fusion domain following affinity purification.

Site-Directed Mutagenesis Provides Mechanistic Insights

Once the ability to express a protein from its cloned gene has been established, it is possible to employ **site-directed mutagenesis** to change specific aminoacyl residues by altering their codons. Used in combination with kinetic analyses and x-ray crystallography, this approach facilitates identification of the specific roles of given aminoacyl residues in substrate binding and catalysis. For example, the inference that a particular aminoacyl residue functions as a general acid can be tested by replacing it with an aminoacyl residue incapable of donating a proton.

SUMMARY

- Enzymes are highly effective and extremely specific catalysts.
- Organic and inorganic prosthetic groups, cofactors, and coenzymes play important roles in catalysis. Coenzymes, many of which are derivatives of B vitamins, serve as "shuttles."
- Catalytic mechanisms employed by enzymes include the introduction of strain, approximation of reactants, acid-base catalysis, and covalent catalysis.
- Aminoacyl residues that participate in catalysis are highly conserved among all classes of a given enzyme.
- Substrates and enzymes induce mutual conformational changes in one another that facilitate substrate recognition and catalysis.
- The catalytic activity of enzymes reveals their presence, facilitates their detection, and provides the basis for enzyme-linked immunoassays.
- Many enzymes can be assayed spectrophotometrically by coupling them to an NAD(P)⁺-dependent dehydrogenase.
- Combinatorial chemistry generates extensive libraries of potential enzyme activators and inhibitors that can be tested by high-throughput screening.
- Assay of plasma enzymes aids diagnosis and prognosis, for example, of myocardial infarction.
- Restriction endonucleases facilitate diagnosis of genetic diseases by revealing restriction fragment length polymorphisms.

- Site-directed mutagenesis, used to change residues suspected of being important in catalysis or substrate binding, provides insights into the mechanisms of enzyme action.
- Recombinant fusion proteins such as His-tagged or GST fusion enzymes are readily purified by affinity chromatography.
- Specific proteases can then remove affinity "tags" and generate the native enzyme.

REFERENCES

Brik A, Wong C-H: HIV-1 protease: Mechanism and drug discovery. Org Biomol Chem 2003;1:5.

Burtis CA et al. (editors): *Tietz Textbook of Clinical Chemistry and Molecular Diagnostics.* 4th ed. Elsevier, 2006.

Conyers GB et al: Metal requirements of a diadenosine pyrophosphatase from *Bartonella bacilliformis*. Magnetic resonance and kinetic studies of the role of Mn²⁺. Biochemistry 2000;39:2347.

Cornish PV, Ha T: A survey of single-molecule techniques in chemical biology. ACS Chem Biol 2007;2:53.

Fersht A: *Structure and Mechanism in Protein Science: A Guide to Enzyme Catalysis and Protein Folding.* Freeman, 1999.

Frank RAW, et al: A molecular switch and proton wire synchronize the active sites in thiamine enzymes. Science 2004;306:872.

Geysen HM, Schoenen F, Wagner D, Wagner R: Combinatorial compound libraries for drug discovery: An ongoing challenge. Nature Rev Drug Disc 2003;2:222.

Goddard J-P, Reymond J-L: Enzyme assays for high-throughput screening. Curr Opin Biotech 2004;15:314.

Gupta S, de Lemos JA: Use and misuse of cardiac troponins in clinical practice. Prog Cardiovasc Dis 2007;50:151.

Hedstrom L: Serine protease mechanism and specificity. Chem Rev 2002;102:4501.

Melanson SF, Tanasijevic MJ. Laboratory diagnosis of acute myocardial injury. Cardiovascular Pathol 2005;14:156.

Pereira DA, Williams JA: Origin and evolution of high throughput screening. Br J Pharmacol 2007;152:53.

Schafer B, Gemeinhardt H, Greulich KO: Direct microscopic observation of the time course of single-molecule DNA restriction reactions. Agnew Chem Int Ed 2001;40:4663.

Silverman RB: *The Organic Chemistry of Enzyme-Catalyzed Reactions.* Academic Press, 2002.

Suckling CJ: *Enzyme Chemistry.* Chapman & Hall, 1990.

Sundaresan V, Abrol R: Towards a general model for protein-substrate stereoselectivity. Protein Sci 2002;11:1330.

Todd AE, Orengo CA, Thornton JM: Plasticity of enzyme active sites. Trends Biochem Sci 2002;27:419.

Urich T et al: X-ray structure of a self-compartmentalizing sulfur cycle metalloenzyme. Science 2006;311:996.

Walsh CT: *Enzymatic Reaction Mechanisms.* Freeman, 1979.

Enzymes: Kinetics

Peter J. Kennelly, PhD & Victor W. Rodwell, PhD

BIOMEDICAL IMPORTANCE

Enzyme kinetics is the field of biochemistry concerned with the quantitative measurement of the rates of enzyme-catalyzed reactions and the systematic study of factors that affect these rates. A complete, balanced set of enzyme activities is of fundamental importance for maintaining homeostasis. An understanding of enzyme kinetics, thus, is important to understanding how physiologic stresses such as anoxia, metabolic acidosis or alkalosis, toxins, and pharmacologic agents affect that balance. Kinetic analysis can reveal the number and order of the individual steps by which enzymes transform substrates into products. Together with site-directed mutagenesis and other techniques that probe protein structure, kinetic analyses can reveal details of the catalytic mechanism of a given enzyme. The involvement of enzymes in virtually all physiologic processes makes them the targets of choice for drugs that cure or ameliorate human disease. Applied enzyme kinetics represents the principal tool by which scientists identify and characterize therapeutic agents that selectively inhibit the rates of specific enzyme-catalyzed processes. Enzyme kinetics thus plays a central and critical role in drug discovery and comparative pharmacodynamics, as well as in elucidating the mode of action of drugs.

CHEMICAL REACTIONS ARE DESCRIBED USING BALANCED EQUATIONS

A **balanced chemical equation** lists the initial chemical species (substrates) present and the new chemical species (products) formed for a particular chemical reaction, all in their correct proportions or **stoichiometry.** For example, balanced equation (1) below describes the reaction of one molecule each of substrates A and B to form one molecule each of products P and Q.

$$A + B \rightleftarrows P + Q \tag{1}$$

The double arrows indicate reversibility, an intrinsic property of all chemical reactions. Thus, for reaction (1), if A and B can form P and Q, then P and Q can also form A and B. Designation of a particular reactant as a "substrate" or "product" is

therefore somewhat arbitrary since the products for a reaction written in one direction are the substrates for the reverse reaction. The term "products" is, however, often used to designate the reactants whose formation is thermodynamically favored. Reactions for which thermodynamic factors strongly favor formation of the products to which the arrow points often are represented with a single arrow as if they were "irreversible":

$$A + B \rightarrow P + Q \tag{2}$$

Unidirectional arrows are also used to describe reactions in living cells where the products of reaction (2) are immediately consumed by a subsequent enzyme-catalyzed reaction. The rapid removal of product P or Q therefore effectively precludes occurrence of the reverse reaction, rendering equation (2) **functionally irreversible under physiologic conditions.**

CHANGES IN FREE ENERGY DETERMINE THE DIRECTION & EQUILIBRIUM STATE OF CHEMICAL REACTIONS

The Gibbs free energy change ΔG (also called either the free energy or Gibbs energy) describes both the *direction* in which a chemical reaction will tend to proceed and the concentrations of reactants and products that will be present at equilibrium. ΔG for a chemical reaction equals the sum of the free energies of formation of the reaction products ΔG_p minus the sum of the free energies of formation of the substrates ΔG_s. ΔG^0 denotes the change in free energy that accompanies transition from the standard state, one-molar concentrations of substrates and products, to equilibrium. A more useful biochemical term is $\Delta G^{0'}$, which defines ΔG^0 at a standard state of 10^{-7} M protons, pH 7.0 (Chapter 11). If the free energy of formation of the products is lower than that of the substrates, the signs of ΔG^0 and $\Delta G^{0'}$ will be negative, indicating that the reaction as written is favored in the direction left to right. Such reactions are referred to as **spontaneous.** The **sign** and the **magnitude** of the free energy change determine how far the reaction will proceed. Equation (3) illustrates the relationship between the equilibrium constant K_{eq} and ΔG^0,

$$\Delta G^0 = -RT \ln K_{eq} \qquad (3)$$

where R is the gas constant (1.98 cal/mol°K or 8.31 J/mol°K) and T is the absolute temperature in degrees Kelvin. K_{eq} is equal to the product of the concentrations of the reaction products, each raised to the power of their stoichiometry, divided by the product of the substrates, each raised to the power of their stoichiometry.

For the reaction $A + B \rightleftarrows P + Q$

$$K_{eq} = \frac{[P][Q]}{[A][B]} \qquad (4)$$

and for reaction (5)

$$A + A \rightleftarrows P \qquad (5)$$

$$K_{eq} = \frac{[P]}{[A]^2} \qquad (6)$$

ΔG^0 may be calculated from equation (3) if the molar concentrations of substrates and products present at equilibrium are known. If ΔG^0 is a negative number, K_{eq} will be greater than unity, and the concentration of products at equilibrium will exceed that of the substrates. If ΔG^0 is positive, K_{eq} will be less than unity, and the formation of substrates will be favored.

Notice that, since ΔG^0 is a function exclusively of the initial and final states of the reacting species, it can provide information only about the *direction* and *equilibrium state* of the reaction. ΔG^0 is independent of the mechanism of the reaction and therefore provides no information concerning **rates** of reactions. Consequently—and as explained below—although a reaction may have a large negative ΔG^0 or $\Delta G^{0'}$, it may nevertheless take place at a negligible rate.

THE RATES OF REACTIONS ARE DETERMINED BY THEIR ACTIVATION ENERGY

Reactions Proceed via Transition States

The concept of the **transition state** is fundamental to understanding the chemical and thermodynamic basis of catalysis. Equation (7) depicts a group transfer reaction in which an entering group E displaces a leaving group L, attached initially to R.

$$E + R - L \rightleftarrows E - R + L \qquad (7)$$

The net result of this process is to transfer group R from L to E. Midway through the displacement, the bond between R and L has weakened but has not yet been completely severed, and the new bond between E and R is as yet incompletely formed. This transient intermediate—in which neither free substrate nor product exists—is termed the **transition state, E···R···L.** Dotted lines represent the "partial" bonds that are undergoing

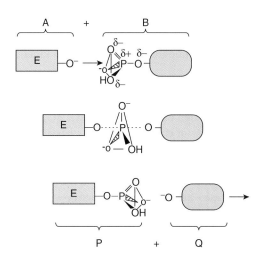

FIGURE 8–1 Formation of a transition state intermediate during a simple chemical reaction, $A + B \rightarrow P + Q$. Shown are three stages of a chemical reaction in which a phosphoryl group is transferred from leaving group L to entering group E. **Top:** Entering group E (A) approaches the other reactant, L-phosphate (B). Notice how the three oxygen atoms linked by the triangular lines and the phosphorus atom of the phosphoryl group form a pyramid. **Center:** As E approaches L-phosphate, the new bond between E and the phosphate group begins to form (dotted line) as that linking L to the phosphate group weakens. These partially-formed bonds are indicated by dotted lines. **Bottom:** Formation of the new product, E-phosphate (P), is now complete as the leaving group L (Q) exits. Notice how the geometry of the phosphoryl group differs between the transition state and the substrate or product. Notice how the phosphorus and three oxygen atoms that occupy the four corners of a pyramid in the substrate and product become coplanar, as emphasized by the triangle, in the transition state.

formation and rupture. Figure 8–1 provides a more detailed illustration of the transition state intermediate formed during the transfer of a phosphoryl group.

Reaction (7) can be thought of as consisting of two "partial reactions," the first corresponding to the formation (F) and the second to the subsequent decay (D) of the transition state intermediate. As for all reactions, characteristic changes in free energy, ΔG_F and ΔG_D are associated with each partial reaction.

$$E + R - L \rightleftarrows E \cdots R \cdots L \quad \Delta G_F \qquad (8)$$

$$E \cdots R \cdots L \rightleftarrows E - R + L \quad \Delta G_D \qquad (9)$$

$$E + R - L \rightleftarrows E - R + L \quad \Delta G = \Delta G_F + \Delta G_D \qquad (10)$$

For the overall reaction (10), ΔG is the sum of ΔG_F and ΔG_D. As for any equation of two terms, it is not possible to infer from ΔG either the sign or the magnitude of ΔG_F or ΔG_D.

Many reactions involve multiple transition states, each with an associated change in free energy. For these reactions, the overall ΔG represents the sum of *all* of the free energy changes associated with the formation and decay of *all* of the transition states. **Therefore, it is not possible to infer** from the overall ΔG **the number or type of transition states through which the reaction proceeds.** Stated another way, overall thermodynamics tells us nothing about kinetics.

ΔG_F Defines the Activation Energy

Regardless of the sign or magnitude of ΔG, ΔG_F for the overwhelming majority of chemical reactions has a positive sign. The formation of transition state intermediates therefore requires surmounting energy barriers. For this reason, ΔG_F for reaching a transition state is often termed the **activation energy, E_{act}**. The ease—and hence the frequency—with which this barrier is overcome is inversely related to E_{act}. The thermodynamic parameters that determine how *fast* a reaction proceeds thus are the ΔG_F values for formation of the transition states through which the reaction proceeds. For a simple reaction, where £ means "proportionate to,"

$$\text{Rate} \propto e^{-E_{act}/RT} \tag{11}$$

The activation energy for the reaction proceeding in the opposite direction to that drawn is equal to $-\Delta G_D$.

NUMEROUS FACTORS AFFECT THE REACTION RATE

The **kinetic theory**—also called the **collision theory**—of chemical kinetics states that for two molecules to react they (1) must approach within bond-forming distance of one another, or "collide"; and (2) must possess sufficient kinetic energy to overcome the energy barrier for reaching the transition state. It therefore follows that anything that increases the *frequency* or *energy* of collision between substrates will increase the rate of the reaction in which they participate.

Temperature

Raising the temperature increases the kinetic energy of molecules. As illustrated in Figure 8–2, the total number of molecules whose kinetic energy exceeds the energy barrier E_{act} (vertical bar) for formation of products increases from low (A) through intermediate (B) to high (C) temperatures. Increasing the kinetic energy of molecules also increases their rapidity of

FIGURE 8–2 The energy barrier for chemical reactions. (See text for discussion.)

motion and therefore the frequency with which they collide. This combination of more frequent and more highly energetic, and hence productive, collisions increases the reaction rate.

Reactant Concentration

The frequency with which molecules collide is directly proportionate to their concentrations. For two different molecules A and B, the frequency with which they collide will double if the concentration of either A or B is doubled. If the concentrations of both A and B are doubled, the probability of collision will increase 4-fold.

For a chemical reaction proceeding at constant temperature that involves one molecule each of A and B,

$$A + B \rightarrow P \tag{12}$$

the number of molecules that possess kinetic energy sufficient to overcome the activation energy barrier will be a constant. The number of collisions with sufficient energy to produce product P, therefore, will be directly proportionate to the number of collisions between A and B and, thus, to their molar concentrations, denoted by square brackets.

$$\text{Rate} \propto [A][B] \tag{13}$$

Similarly, for the reaction represented by

$$A + 2B \rightarrow P \tag{14}$$

which can also be written as

$$A + B + B \rightarrow P \tag{15}$$

the corresponding rate expression is

$$\text{Rate} \propto [A][B][B] \tag{16}$$

or

$$\text{Rate} \propto [A][B]^2 \tag{17}$$

For the general case, when n molecules of A react with m molecules of B,

$$nA + mB \rightarrow P \tag{18}$$

the rate expression is

$$\text{Rate} \propto [A]^n[B]^m \tag{19}$$

Replacing the proportionality sign with an equals sign by introducing a **rate constant k** characteristic of the reaction under study gives equations (20) and (21), in which the subscripts 1 and −1 refer to the forward and reverse reactions, respectively.

$$\text{Rate}_1 = k_1[A]^n[B]^m \tag{20}$$

$$\text{Rate}_{-1} = k_{-1}[P] \tag{21}$$

The sum of the molar ratios of the reactants defines the **kinetic order** of the reaction. Consider reaction (5). The stoichiometric coefficient for the sole reactant, A, is two. Therefore, the rate of production of P is proportional to the square of [A] and the reaction is said to be *second order* with respect to reactant A. In this instance, the overall reaction is also *second order*. Therefore, k_1 is referred to as a *second order rate constant*.

Reaction (12) describes a simple second order reaction between two different reactants, A and B. The stoichiometric coefficient for each reactant is one. Therefore, while the overall order of the reaction is two, it is said to be *first order* with respect to A and *first order* with respect to B. In the laboratory, the kinetic order of a reaction with respect to a particular reactant, referred to as the variable reactant or substrate, can be determined by maintaining the concentration of the other reactants at a constant, or fixed, concentration in large excess over the variable reactant. Under these *pseudo-first-order conditions*, the concentration of the fixed reactant(s) remains virtually constant. Thus, the rate of reaction will depend exclusively on the concentration of the variable reactant, sometimes also called the limiting reactant. The concepts of reaction order and pseudo-first-order conditions apply not only to simple chemical reactions but also to enzyme-catalyzed reactions.

K_{eq} Is a Ratio of Rate Constants

While all chemical reactions are to some extent reversible, at equilibrium the *overall* concentrations of reactants and products remain constant. At equilibrium, the rate of conversion of substrates to products therefore equals the rate at which products are converted to substrates.

$$Rate_1 = Rate_{-1} \qquad (22)$$

Therefore,

$$k_1[A]^n[B]^m = k_{-1}[P] \qquad (23)$$

and

$$\frac{k_1}{k_{-1}} = \frac{[P]}{[A]^n[B]^m} \qquad (24)$$

The ratio of k_1 to k_{-1} is termed the equilibrium constant, K_{eq}. The following important properties of a system at equilibrium must be kept in mind:

1. The equilibrium constant is a ratio of the reaction *rate constants* (not the reaction *rates*).
2. At equilibrium, the reaction *rates* (not the *rate constants*) of the forward and back reactions are equal.
3. Equilibrium is a *dynamic* state. Although there is no *net* change in the concentration of substrates or products, individual substrate and product molecules are continually being interconverted.

4. The numeric value of the equilibrium constant K_{eq} can be calculated either from the concentrations of substrates and products at equilibrium or from the ratio k_1/k_{-1}.

THE KINETICS OF ENZYMATIC CATALYSIS

Enzymes Lower the Activation Energy Barrier for a Reaction

All enzymes accelerate reaction rates by lowering ΔG_F for the formation of transition states. However, they may differ in the way this is achieved. Where the mechanism or the sequence of chemical steps at the active site is essentially equivalent to those for the same reaction proceeding in the absence of a catalyst, **the environment of the active site lowers ΔG_F by stabilizing the transition state intermediates.** To put it another way, the enzyme can be envisioned as binding to the transition state intermediate (Figure 8–1) more tightly than it does to either substrates or products. As discussed in Chapter 7, stabilization can involve (1) acid-base groups suitably positioned to transfer protons to or from the developing transition state intermediate, (2) suitably positioned charged groups or metal ions that stabilize developing charges, or (3) the imposition of steric strain on substrates so that their geometry approaches that of the transition state. HIV protease (see Figure 7–6) illustrates catalysis by an enzyme that lowers the activation barrier by stabilizing a transition state intermediate.

Catalysis by enzymes that proceeds via a *unique* reaction mechanism typically occurs when the transition state intermediate forms a covalent bond with the enzyme (**covalent catalysis**). The catalytic mechanism of the serine protease chymotrypsin (see Figure 7–7) illustrates how an enzyme utilizes covalent catalysis to provide a unique reaction pathway.

ENZYMES DO NOT AFFECT K_{eq}

While enzymes may undergo transient modifications during the process of catalysis, they always emerge unchanged at the completion of the reaction. **The presence of an enzyme therefore has no effect on ΔG^0 for the *overall* reaction,** which is a function solely of the *initial and final states* of the reactants. Equation (25) shows the relationship between the equilibrium constant for a reaction and the standard free energy change for that reaction:

$$\Delta G^0 = -RT \ln K_{eq} \qquad (25)$$

This principle is perhaps most readily illustrated by including the presence of the enzyme (Enz) in the calculation of the equilibrium constant for an enzyme-catalyzed reaction:

$$A + B + Enz \rightleftarrows P + Q + Enz \qquad (26)$$

Since the enzyme on both sides of the double arrows is present in equal quantity and identical form, the expression for the equilibrium constant,

$$K_{eq} = \frac{[P][Q][Enz]}{[A][B][Enz]} \qquad (27)$$

reduces to one identical to that for the reaction in the absence of the enzyme:

$$K_{eq} = \frac{[P][Q]}{[A][B]} \qquad (28)$$

Enzymes therefore have no effect on K_{eq}.

MULTIPLE FACTORS AFFECT THE RATES OF ENZYME-CATALYZED REACTIONS

Temperature

Raising the temperature increases the rate of both uncatalyzed and enzyme-catalyzed reactions by increasing the kinetic energy and the collision frequency of the reacting molecules. However, heat energy can also increase the kinetic energy of the enzyme to a point that exceeds the energy barrier for disrupting the noncovalent interactions that maintain its three-dimensional structure. The polypeptide chain then begins to unfold, or **denature,** with an accompanying loss of catalytic activity. The temperature range over which an enzyme maintains a stable, catalytically competent conformation depends upon—and typically moderately exceeds—the normal temperature of the cells in which it resides. Enzymes from humans generally exhibit stability at temperatures up to 45–55°C. By contrast, enzymes from the thermophilic microorganisms that reside in volcanic hot springs or undersea hydrothermal vents may be stable up to or even above 100°C.

The **Q_{10}, or temperature coefficient,** is the factor by which the rate of a biologic process increases for a 10°C increase in temperature. For the temperatures over which enzymes are stable, the rates of most biologic processes typically double for a 10°C rise in temperature ($Q_{10} = 2$). Changes in the rates of enzyme-catalyzed reactions that accompany a rise or fall in body temperature constitute a prominent survival feature for "cold-blooded" life forms such as lizards or fish, whose body temperatures are dictated by the external environment. However, for mammals and other homeothermic organisms, changes in enzyme reaction rates with temperature assume physiologic importance only in circumstances such as fever or hypothermia.

Hydrogen Ion Concentration

The rate of almost all enzyme-catalyzed reactions exhibits a significant dependence on hydrogen ion concentration. Most intracellular enzymes exhibit optimal activity at pH values between 5 and 9. The relationship of activity to hydrogen ion concentration (Figure 8–3) reflects the balance between

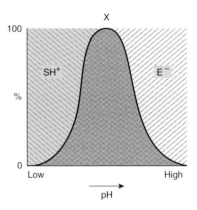

FIGURE 8–3 Effect of pH on enzyme activity. Consider, for example, a negatively charged enzyme (E^-) that binds a positively charged substrate (SH^+). Shown is the proportion (%) of SH^+ [\\\] and of E^- [///] as a function of pH. Only in the cross-hatched area do both the enzyme and the substrate bear an appropriate charge.

enzyme denaturation at high or low pH and effects on the charged state of the enzyme, the substrates, or both. For enzymes whose mechanism involves acid-base catalysis, the residues involved must be in the appropriate state of protonation for the reaction to proceed. The binding and recognition of substrate molecules with dissociable groups also typically involves the formation of salt bridges with the enzyme. The most common charged groups are carboxylate groups (negative) and protonated amines (positive). Gain or loss of critical charged groups adversely affects substrate binding and thus will retard or abolish catalysis.

ASSAYS OF ENZYME-CATALYZED REACTIONS TYPICALLY MEASURE THE INITIAL VELOCITY

Most measurements of the rates of enzyme-catalyzed reactions employ relatively short time periods, conditions that approximate **initial rate conditions.** Under these conditions, only traces of product accumulate, rendering the rate of the reverse reaction negligible. The **initial velocity (v_i)** of the reaction thus is essentially that of the rate of the forward reaction. Assays of enzyme activity almost always employ a large (10^3–10^7) molar excess of substrate over enzyme. Under these conditions, v_i is proportionate to the concentration of enzyme. Measuring the initial velocity therefore permits one to estimate the quantity of enzyme present in a biologic sample.

SUBSTRATE CONCENTRATION AFFECTS REACTION RATE

In what follows, enzyme reactions are treated as if they had only a single substrate and a single product. For enzymes with multiple substrates, the principles discussed below apply with equal validity. Moreover, by employing pseudo-first-order conditions (see above), scientists can study the dependence of reaction rate upon an individual reactant through the appropriate choice of fixed and variable substrates. In other words, under pseudo-first-order conditions the behavior of a multisubstrate enzyme will imitate one having a single substrate, In this instance, however, the observed rate constant will be a function of the rate constant k_1 for the reaction as well as the concentration of the fixed substrate(s).

For a typical enzyme, as substrate concentration is increased, v_i increases until it reaches a maximum value V_{max} (Figure 8–4). When further increases in substrate concentration do not further increase v_i, the enzyme is said to be "saturated" with substrate. Note that the shape of the curve that relates activity to substrate concentration (Figure 8–4) is hyperbolic. At any given instant, only substrate molecules that are combined with the enzyme as an ES complex can be transformed into product. Second, the equilibrium constant for the formation of the enzyme-substrate complex is not infinitely large. Therefore, even when the substrate is present in excess (points A and B of Figure 8–5), only a fraction of the enzyme may be present as an ES complex. At points A or B, increasing or decreasing [S] therefore will increase or decrease the number of ES complexes with a corresponding change in v_i. At point C (Figure 8–5), essentially all the enzyme is present as the ES complex. Since no free enzyme remains available for

forming ES, further increases in [S] cannot increase the rate of the reaction. Under these saturating conditions, v_i depends solely on—and thus is limited by—the rapidity with which product dissociates from the enzyme so that it may combine with more substrate.

THE MICHAELIS-MENTEN & HILL EQUATIONS MODEL THE EFFECTS OF SUBSTRATE CONCENTRATION

The Michaelis-Menten Equation

The Michaelis-Menten equation (29) illustrates in mathematical terms the relationship between initial reaction velocity v_i and substrate concentration [S], shown graphically in Figure 8–4.

$$v_i = \frac{V_{max}[S]}{K_m + [S]} \qquad (29)$$

The Michaelis constant K_m is the substrate concentration at which v_i is half the maximal velocity ($V_{max}/2$) attainable at a particular concentration of enzyme. K_m thus has the dimensions of substrate concentration. The dependence of initial reaction velocity on [S] and K_m may be illustrated by evaluating the Michaelis-Menten equation under three conditions.

1. When [S] is much less than K_m (point A in Figures 8–4 & 8–5), the term K_m + [S] is essentially equal to K_m. Replacing K_m + [S] with K_m reduces equation (29) to

$$v_i = \frac{V_{max}[S]}{K_m + [S]} \quad v_i \approx \frac{V_{max}[S]}{K_m} \approx \left(\frac{V_{max}}{K_m}\right)[S] \qquad (30)$$

where $\approx$ means "approximately equal to." Since V_{max} and K_m are both constants, their ratio is a constant. In other words, when [S] is considerably below K_m, v_i is proportionate to k[S]. The initial reaction velocity therefore is directly proportionate to [S].

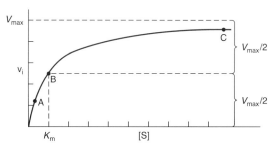

FIGURE 8–4 Effect of substrate concentration on the initial velocity of an enzyme-catalyzed reaction.

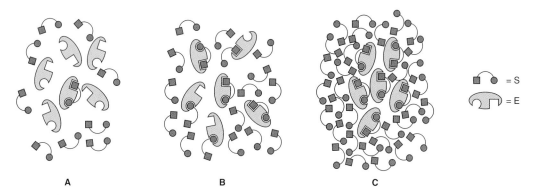

FIGURE 8–5 Representation of an enzyme in the presence of a concentration of substrate that is below K_m **(A)**, at a concentration equal to K_m **(B)**, and at a concentration well above K_m **(C)**. Points A, B, and C correspond to those points in Figure 8–4.

2. When [S] is much greater than K_m (point C in Figures 8–4 & 8–5), the term $K_m + [S]$ is essentially equal to [S]. Replacing $K_m + [S]$ with [S] reduces equation (29) to

$$v_i = \frac{V_{max}[S]}{K_m + [S]} \quad v_i \approx \frac{V_{max}[S]}{[S]} \approx V_{max} \quad (31)$$

Thus, when [S] greatly exceeds K_m, the reaction velocity is maximal (V_{max}) and unaffected by further increases in substrate concentration.

3. When $[S] = K_m$ (point B in Figures 8–4 & 8–5).

$$v_i = \frac{V_{max}[S]}{K_m + [S]} = \frac{V_{max}[S]}{2[S]} = \frac{V_{max}}{2} \quad (32)$$

Equation (32) states that when [S] equals K_m, the initial velocity is half-maximal. Equation (32) also reveals that K_m is—and may be determined experimentally from—the substrate concentration at which the initial velocity is half-maximal.

A Linear Form of the Michaelis-Menten Equation Is Used to Determine K_m & V_{max}

The direct measurement of the numeric value of V_{max}, and therefore the calculation of K_m, often requires impractically high concentrations of substrate to achieve saturating conditions. A linear form of the Michaelis-Menten equation circumvents this difficulty and permits V_{max} and K_m to be extrapolated from initial velocity data obtained at less than saturating concentrations of substrate. Start with equation (29),

$$v_i = \frac{V_{max}[S]}{K_m + [S]} \quad (29)$$

invert

$$\frac{1}{v_i} = \frac{K_m + [S]}{V_{max}[S]} \quad (33)$$

factor

$$\frac{1}{v_i} = \frac{K_m}{V_{max}[S]} + \frac{[S]}{V_{max}[S]} \quad (34)$$

and simplify

$$\frac{1}{v_i} = \left(\frac{K_m}{V_{max}}\right)\frac{1}{[S]} + \frac{1}{V_{max}} \quad (35)$$

Equation (35) is the equation for a straight line, $y = ax + b$, where $y = 1/v_i$ and $x = 1/[S]$. A plot of $1/v_i$ as y as a function of $1/[S]$ as x therefore gives a straight line whose y intercept is $1/V_{max}$ and whose slope is K_m/V_{max}. Such a plot is called a **double-reciprocal** or **Lineweaver-Burk plot** (Figure 8–6). Setting the y term of equation (36) equal to zero and solving for x reveals that the x intercept is $-1/K_m$.

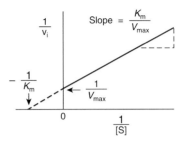

FIGURE 8–6 Double-reciprocal or Lineweaver-Burk plot of $1/v_i$ versus $1/[S]$ used to evaluate K_m and V_{max}.

$$0 = ax + b; \quad \text{therefore,} \quad x = \frac{-b}{a} = \frac{-1}{K_m} \quad (36)$$

K_m is thus most readily calculated from the negative x intercept.

The greatest virtue of the Lineweaver-Burk plot resides in the facility with which it can be used to determine the kinetic mechanisms of enzyme inhibitors (see below). However, in using a double-reciprocal plot to determine kinetic constants, it is important to avoid the introduction of bias through the clustering of data at low values of $1/[S]$. To achieve this, prepare a solution of substrate whose dilution into an assay will produce the maximum desired concentration of substrate. Now use the same volume of solutions prepared by diluting the stock solution by factors of 1:2, 1:3, 1:4, 1:5, etc. The data will then fall on the $1/[S]$ axis at intervals of 1, 2, 3, 4, 5, etc. Alternatively, a single-reciprocal plot such as the Eadie-Hofstee (v_i versus $v_i/[S]$) or Hanes-Woolf ($[S]/v_i$ versus [S]) plot can be used to minimize clustering.

The Catalytic Constant, k_{cat}

Several parameters may be used to compare the relative activity of different enzymes, or of different preparations of the same enzyme. The activity of impure enzyme preparations typically is expressed as *specific activity* (V_{max} divided by the protein concentration). For a homogeneous enzyme one may calculate its *turnover number* (V_{max} divided by the moles of enzyme present). But if the number of active sites present is known, the catalytic activity of a homogeneous enzyme is best expressed as its *catalytic constant*, k_{cat} (V_{max} divided by the number active sites, S_t).

$$k_{cat} = \frac{V_{max}}{S_t} \quad (37)$$

Since the units of concentration cancel out, the units of k_{cat} are reciprocal time.

Catalytic Efficiency, k_{cat}/K_m

By what measure should the efficiency of different enzymes, different substrates for a given enzyme, and the efficiency with which an enzyme catalyzes a reaction in the forward and reverse directions be quantified and compared? While the maximum capacity of a given enzyme to convert substrate to product is important, the benefits of a high k_{cat} can only be realized if K_m is sufficiently low. Thus, *catalytic efficiency* of enzymes is best expressed in terms of the ratio of these two kinetic constants, k_{cat}/K_m.

For certain enzymes, once substrate binds to the active site, it is converted to product and released so rapidly as to render these events effectively instantaneous. For these exceptionally efficient catalysts, the rate-limiting step is the formation of the ES complex. Such enzymes are said to be *diffusion-limited*, or catalytically perfect, since the fastest possible rate of catalysis is determined by the rate at which molecules move or diffuse through the solution. Examples of enzymes for which k_{cat}/K_m approaches the diffusion limit of 10^8–10^9 M^{-1}s^{-1} include triosephosphate isomerase, carbonic anhydrase, acetylcholinesterase, and adenosine deaminase.

Nature frequently circumvents the limitations on k_{cat}/K_m imposed by diffusion by assembling sets of related enzymes into multi-enzyme complexes. The geometric relationships of the enzymes in these complexes are such that the substrates and products are not permitted to diffuse into bulk solution until the last step in the sequence of catalytic steps is complete. Fatty acid synthetase extends this concept one step further by covalently attaching the substrate to a biotin tether that rotates from active site to active site within the complex until synthesis of a palmitic acid molecule is complete (Chapter 23).

K_m May Approximate a Binding Constant

The affinity of an enzyme for its substrate is the inverse of the dissociation constant K_d for dissociation of the enzyme-substrate complex ES.

$$E + S \underset{k_{-1}}{\overset{k_1}{\rightleftharpoons}} ES \tag{38}$$

$$K_d = \frac{k_{-1}}{k_1} \tag{39}$$

Stated another way, the smaller the tendency of the enzyme and its substrate to *dissociate*, the *greater* the affinity of the enzyme for its substrate. While the Michaelis constant K_m often approximates the dissociation constant K_d, this is by no means always the case. For a typical enzyme-catalyzed reaction,

$$E + S \underset{k_{-1}}{\overset{k_1}{\rightleftharpoons}} ES \overset{k_2}{\longrightarrow} E + P \tag{40}$$

the value of [S] that gives $v_i = V_{max}/2$ is

$$[S] = \frac{k_{-1} + k_2}{k_1} = K_m \tag{41}$$

When $k_{-1} \gg k_2$, then

$$k_{-1} + k_2 \approx k_{-1} \tag{42}$$

and

$$[S] \approx \frac{k_1}{k_{-1}} = K_d \tag{43}$$

Hence, $1/K_m$ only approximates $1/K_d$ under conditions where the association and dissociation of the ES complex are rapid relative to catalysis. For the many enzyme-catalyzed reactions for which $k_{-1} + k_2$ is *not* approximately equal to k_{-1}, $1/K_m$ will underestimate $1/K_d$.

The Hill Equation Describes the Behavior of Enzymes That Exhibit Cooperative Binding of Substrate

While most enzymes display the simple **saturation kinetics** depicted in Figure 8–4 and are adequately described by the Michaelis-Menten expression, some enzymes bind their substrates in a *cooperative* fashion analogous to the binding of oxygen by hemoglobin (Chapter 6). Cooperative behavior is an exclusive property of multimeric enzymes that bind substrate at multiple sites.

For enzymes that display positive cooperativity in binding substrate, the shape of the curve that relates changes in v_i to changes in [S] is sigmoidal (Figure 8–7). Neither the Michaelis-Menten expression nor its derived plots can be used to evaluate cooperative kinetics. Enzymologists therefore employ a graphic representation of the **Hill equation** originally derived to describe the cooperative binding of O_2 by hemoglobin. Equation (44) represents the Hill equation arranged in a form that predicts a straight line, where k' is a complex constant.

$$\frac{\log v_i}{V_{max} - v_i} = n \log [S] - \log k' \tag{44}$$

Equation (44) states that when [S] is low relative to k', the initial reaction velocity increases as the nth power of [S].

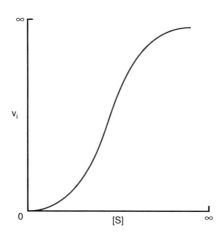

FIGURE 8–7 Representation of sigmoid substrate saturation kinetics.

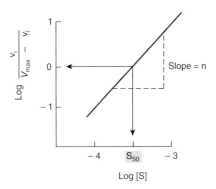

FIGURE 8–8 A graphic representation of a linear form of the Hill equation is used to evaluate S_{50}, the substrate concentration that produces half-maximal velocity, and the degree of cooperativity **n**.

A graph of $\log v_i/(V_{max} - v_i)$ versus $\log[S]$ gives a straight line (Figure 8–8), where the slope of the line **n** is the **Hill coefficient,** an empirical parameter whose value is a function of the number, kind, and strength of the interactions of the multiple substrate-binding sites on the enzyme. When n = 1, all binding sites behave independently, and simple Michaelis-Menten kinetic behavior is observed. If n is greater than 1, the enzyme is said to exhibit positive cooperativity. Binding of substrate to one site then enhances the affinity of the remaining sites to bind additional substrate. The greater the value for n, the higher the degree of cooperativity and the more markedly sigmoidal will be the plot of v_i versus [S]. A perpendicular dropped from the point where the *y* term $\log v_i/(V_{max} - v_i)$ is zero intersects the *x*-axis at a substrate concentration termed **S_{50}**, the substrate concentration that results in half-maximal velocity. S_{50} thus is analogous to the P_{50} for oxygen binding to hemoglobin (Chapter 6).

KINETIC ANALYSIS DISTINGUISHES COMPETITIVE FROM NONCOMPETITIVE INHIBITION

Inhibitors of the catalytic activities of enzymes provide both pharmacologic agents and research tools for study of the mechanism of enzyme action. The strength of the interaction between an inhibitor and an enzyme depends on the forces important in protein structure and ligand binding (hydrogen bonds, electrostatic interactions, hydrophobic interactions, and van der Waals forces; see Chapter 5). Inhibitors can be classified on the basis of their site of action on the enzyme, on whether they chemically modify the enzyme, or on the kinetic parameters they influence. Compounds that mimic the transition state of an enzyme-catalyzed reaction (transition state analogs) or that take advantage of the catalytic machinery of an enzyme (mechanism-based inhibitors) can be particularly potent inhibitors. Kinetically, we distinguish two classes of inhibitors based upon whether raising the substrate concentration does or does not overcome the inhibition.

Competitive Inhibitors Typically Resemble Substrates

The effects of competitive inhibitors can be overcome by raising the concentration of substrate. Most frequently, in competitive inhibition the inhibitor (**I**) binds to the substrate-binding portion of the active site—thereby blocking access by the substrate. The structures of most classic competitive inhibitors therefore tend to resemble the structures of a substrate, and thus are termed **substrate analogs.** Inhibition of the enzyme succinate dehydrogenase by malonate illustrates competitive inhibition by a substrate analog. Succinate dehydrogenase catalyzes the removal of one hydrogen atom from each of the two methylene carbons of succinate (Figure 8–9). Both succinate and its structural analog malonate ($^-OOC—CH_2—COO^-$) can bind to the active site of succinate dehydrogenase, forming an ES or an EI complex, respectively. However, since malonate contains only one methylene carbon, it cannot undergo dehydrogenation. The formation and dissociation of the EI complex is a dynamic process described by

$$E - I \underset{k_{-1}}{\overset{k_1}{\rightleftharpoons}} E + I \qquad (45)$$

for which the equilibrium constant K_i is

$$K_i = \frac{[E][I]}{[E - I]} = \frac{k_i}{k_{-i}} \qquad (46)$$

In effect, a **competitive inhibitor acts by decreasing the number of free enzyme molecules available to bind substrate, ie, to form ES, and thus eventually to form product,** as described below:

A competitive inhibitor and substrate exert reciprocal effects on the concentration of the EI and ES complexes. Since the formation of ES complexes removes free enzyme available to combine with the inhibitor, increasing [S] decreases the concentration of the EI complex and raises the reaction velocity. The extent to which [S] must be increased to completely overcome the inhibition depends upon the concentration of the inhibitor present, its affinity for the enzyme, K_i, and the affinity, K_m, of the enzyme for its substrate.

FIGURE 8–9 The succinate dehydrogenase reaction.

Double-Reciprocal Plots Facilitate the Evaluation of Inhibitors

Double-reciprocal plots distinguish between competitive and noncompetitive inhibitors and simplify evaluation of inhibition constants. v_i is determined at several substrate concentrations both in the presence and in the absence of inhibitor. For classic competitive inhibition, the lines that connect the experimental data points converge at the y-axis (Figure 8–10). Since the y intercept is equal to $1/V_{max}$, this pattern indicates that **when 1/[S] approaches 0, v_i is independent of the presence of inhibitor.** Note, however, that the intercept on the x-axis does vary with *inhibitor* concentration—and that since $-1/K'_m$ is smaller than $1/K_m$, K'_m (the "apparent K_m") becomes larger in the presence of increasing concentrations of the inhibitor. Thus, **a competitive inhibitor has no effect on V_{max} but raises K'_m, the apparent K_m for the substrate.** For simple competitive inhibition, the intercept on the x-axis is

$$x = \frac{-1}{K_m}\left(1 + \frac{[I]}{K_i}\right) \qquad (47)$$

Once K_m has been determined in the absence of inhibitor, K_i can be calculated from equation (47). K_i values are used to compare different inhibitors of the same enzyme. The lower the value for K_i, the more effective the inhibitor. For example, the statin drugs that act as competitive inhibitors of HMG-CoA reductase (Chapter 26) have K_i values several orders of magnitude lower than the K_m for the substrate HMG-CoA.

Simple Noncompetitive Inhibitors Lower V_{max} But Do Not Affect K_m

In noncompetitive inhibition, binding of the inhibitor does not affect binding of substrate. Formation of both EI and EIS complexes is therefore possible. However, while the enzyme-inhibitor complex can still bind substrate, its efficiency at transforming substrate to product, reflected by V_{max}, is decreased. Noncompetitive inhibitors bind enzymes at sites distinct from the substrate-binding site and generally bear little or no structural resemblance to the substrate.

For simple noncompetitive inhibition, E and EI possess identical affinity for substrate, and the EIS complex generates product at a negligible rate (Figure 8–11). More complex noncompetitive inhibition occurs when binding of the inhibitor *does* affect the apparent affinity of the enzyme for substrate, causing the lines to intercept in either the third or fourth quadrants of a double-reciprocal plot (not shown). While certain inhibitors exhibit characteristics of a mixture of competitive and noncompetitive inhibition, the evaluation of these inhibitors exceeds the scope of this chapter.

Dixon Plot

A Dixon plot is sometimes employed as an alternative to the Lineweaver-Burk plot for determining inhibition constants. The initial velocity (v_i) is measured at several concentrations of inhibitor, but at a fixed concentration of substrate (S). For a simple competitive or noncompetitive inhibitor, a plot of $1/v_i$ versus inhibitor concentration [I] yields a straight line. The experiment is repeated at different fixed concentrations of substrate. The resulting set of lines intersects to the left of the y-axis. For *competitive* inhibition, a perpendicular dropped to the negative x-axis from the point of intersection of the lines gives $-K_i$ (Figure 8–12, top). For *noncompetitive* inhibition the intercept on the negative x-axis is $-K_i$ (Figure 8–12, bottom). Pharmaceutical publications frequently employ Dixon plots to evaluate the comparative potency of competitive inhibitors.

IC_{50}

A less rigorous, but frequently used, alternative to K_i as a measure of inhibitory potency is the concentration of inhibitor that produces 50% inhibition, IC_{50}. Unlike the equilibrium dissociation constant K_i, the numeric value of IC_{50} varies as a function of the specific circumstances of substrate concentration, etc., under which it is determined.

Tightly Bound Inhibitors

Some inhibitors bind to enzymes with such high affinity, $K_i \leq 10^{-9}$ M, that the concentration of inhibitor required to measure K_i falls below the concentration of enzyme typically present in

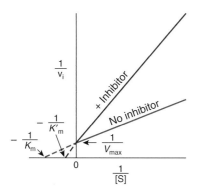

FIGURE 8–10 Lineweaver-Burk plot of competitive inhibition. Note the complete relief of inhibition at high [S] (ie, low 1/[S]).

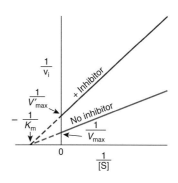

FIGURE 8–11 Lineweaver-Burk plot for simple noncompetitive inhibition.

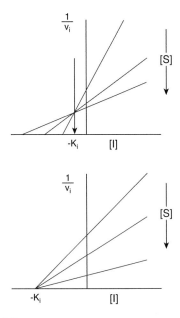

FIGURE 8–12 Applications of Dixon plots. **Top:** Competitive inhibition, estimation of K_i. **Bottom:** Noncompetitive inhibition, estimation of K_i.

an assay. Under these circumstances a significant fraction of the total inhibitor may be present as an EI complex. If so, this violates the assumption, implicit in classical steady-state kinetics, that the concentration of free inhibitor is independent of the concentration of enzyme. The kinetic analysis of these tightly bound inhibitors requires specialized kinetic equations that incorporate the concentration of enzyme to estimate K_i or IC_{50} and to distinguish competitive from noncompetitive tightly bound inhibitors.

Irreversible Inhibitors "Poison" Enzymes

In the above examples, the inhibitors form a dissociable, dynamic complex with the enzyme. Fully active enzyme can therefore be recovered simply by removing the inhibitor from the surrounding medium. However, a variety of other inhibitors act irreversibly by chemically modifying the enzyme. These modifications generally involve making or breaking covalent bonds with aminoacyl residues essential for substrate binding, catalysis, or maintenance of the enzyme's functional conformation. Since these covalent changes are relatively stable, an enzyme that has been "poisoned" by an irreversible inhibitor such as a heavy metal atom or an acylating reagent remains inhibited even after removal of the remaining inhibitor from the surrounding medium.

Mechanism-Based Inhibition

"Mechanism-based" or "suicide" inhibitors are specialized substrate analogs that contain a chemical group that can be transformed by the catalytic machinery of the target enzyme. After binding to the active site, catalysis by the enzyme gen-

erates a highly reactive group that forms a covalent bond to, and blocks function of, a catalytically essential residue. The specificity and persistence of suicide inhibitors, which are both enzyme specific and unreactive outside the confines of the enzyme active site, render them promising leads for the development of enzyme-specific drugs. The kinetic analysis of suicide inhibitors lies beyond the scope of this chapter. Neither the Lineweaver-Burk nor Dixon approach is applicable since suicide inhibitors violate a key boundary condition common to both approaches, namely that the activity of the enzyme does not decrease during the course of the assay.

MOST ENZYME-CATALYZED REACTIONS INVOLVE TWO OR MORE SUBSTRATES

While many enzymes have a single substrate, many others have two—and sometimes more—substrates and products. The fundamental principles discussed above, while illustrated for single-substrate enzymes, apply also to multisubstrate enzymes. The mathematical expressions used to evaluate multisubstrate reactions are, however, complex. While a detailed analysis of the full range of multisubstrate reactions exceeds the scope of this chapter, some common types of kinetic behavior for two-substrate, two-product reactions (termed "Bi-Bi" reactions) are considered below.

Sequential or Single-Displacement Reactions

In **sequential reactions,** both substrates must combine with the enzyme to form a ternary complex before catalysis can proceed (Figure 8–13, top). Sequential reactions are sometimes referred to as single-displacement reactions because the group undergoing transfer is usually passed directly, in a single step, from one substrate to the other. Sequential Bi-Bi reactions can be further distinguished on the basis of whether the two substrates add in a **random** or in a **compulsory** order. For random-order reactions, either substrate A or substrate B may combine first with the enzyme to form an EA or an EB complex (Figure 8–13, center). For compulsory-order reactions, A must first combine with E before B can combine with the EA complex. One explanation for a compulsory-order mechanism is that the addition of A induces a conformational change in the enzyme that aligns residues that recognize and bind B.

Ping-Pong Reactions

The term **"ping-pong"** applies to mechanisms in which one or more products are released from the enzyme before all the substrates have been added. Ping-pong reactions involve covalent catalysis and a transient, modified form of the enzyme (see Figure 7–4). Ping-pong Bi-Bi reactions are **double displacement reactions.** The group undergoing transfer is first displaced from substrate A by the enzyme to form product P and a modified form of the enzyme (F). The subsequent

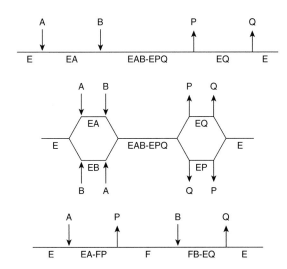

FIGURE 8–13 Representations of three classes of Bi-Bi reaction mechanisms. Horizontal lines represent the enzyme. Arrows indicate the addition of substrates and departure of products. **Top:** An ordered Bi-Bi reaction, characteristic of many NAD(P)H-dependent oxidoreductases. **Center:** A random Bi-Bi reaction, characteristic of many kinases and some dehydrogenases. **Bottom:** A ping-pong reaction, characteristic of aminotransferases and serine proteases.

group transfer from F to the second substrate B, forming product Q and regenerating E, constitutes the second displacement (Figure 8–13, bottom).

Most Bi-Bi Reactions Conform to Michaelis-Menten Kinetics

Most Bi-Bi reactions conform to a somewhat more complex form of Michaelis-Menten kinetics in which V_{max} refers to the reaction rate attained when both substrates are present at saturating levels. Each substrate has its own characteristic K_m value, which corresponds to the concentration that yields half-maximal velocity when the second substrate is present at saturating levels. As for single-substrate reactions, double-reciprocal plots can be used to determine V_{max} and K_m. v_i is measured as a function of the concentration of one substrate (the variable substrate) while the concentration of the other substrate (the fixed substrate) is maintained constant. If the lines obtained for several fixed-substrate concentrations are plotted on the same graph, it is possible to distinguish between a ping-pong enzyme, which yields parallel lines, and a sequential mechanism, which yields a pattern of intersecting lines (Figure 8–14).

Product inhibition studies are used to complement kinetic analyses and to distinguish between ordered and random Bi-Bi reactions. For example, in a random-order Bi-Bi reaction, each product will be a competitive inhibitor regardless of which substrate is designated the variable substrate. However, for a sequential mechanism (Figure 8–13, top), only product Q will give the pattern indicative of competitive inhibition when A is the variable substrate, while only product P will produce this pattern with B as the variable substrate. The other combi-

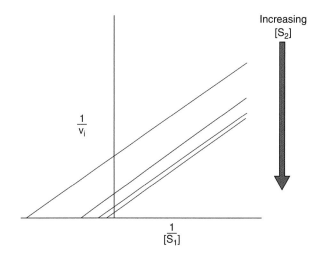

FIGURE 8–14 Lineweaver-Burk plot for a two-substrate ping-pong reaction. An increase in concentration of one substrate (S_1) while that of the other substrate (S_2) is maintained constant changes both the x and y intercepts, but not the slope.

nations of product inhibitor and variable substrate will produce forms of complex noncompetitive inhibition.

KNOWLEDGE OF ENZYME KINETICS, MECHANISM, AND INHIBITION AIDS DRUG DEVELOPMENT

Many Drugs Act as Enzyme Inhibitors

The goal of pharmacology is to identify agents that can

1. Destroy or impair the growth, invasiveness, or development of invading pathogens
2. Stimulate endogenous defense mechanisms
3. Halt or impede aberrant molecular processes triggered by genetic, environmental, or biologic stimuli with minimal perturbation of the host's normal cellular functions.

By virtue of their diverse physiologic roles and high degree of substrate selectivity, enzymes constitute natural targets for the development of pharmacologic agents that are both potent and specific. Statin drugs, for example, lower cholesterol production by inhibiting 3-hydroxy-3-methylglutaryl coenzyme A reductase (Chapter 26), while emtricitabine and tenofovir disoproxil fumarate block replication of the human immunodeficiency virus by inhibiting the viral reverse transcriptase (Chapter 34). Pharmacologic treatment of hypertension often includes the administration of an inhibitor of angiotensin-converting enzyme, thus lowering the level of angiotensin II, a vasoconstrictor (Chapter 42).

Enzyme Kinetics Defines Appropriate Screening Conditions

Enzyme kinetics plays a crucial role in drug discovery. Knowledge of the kinetic behavior of the enzyme of interest is necessary, first and foremost, to select appropriate assay conditions that readily detect the presence of an inhibitor. The concentration of substrate, for example, must be adjusted such that sufficient product is generated to permit facile detection of the enzyme's activity without being so high that it masks the presence of an inhibitor. Second, enzyme kinetics provides the means for quantifying and comparing the potency of different inhibitors and defining their mode of action. Noncompetitive inhibitors are particularly desirable, because—by contrast to competitive inhibitors—their effects can never be completely overcome by increases in substrate concentration.

Many Drugs Are Metabolized In Vivo

Drug development often involves more than the kinetic evaluation of the interaction of inhibitors with the target enzyme. Drugs are acted upon by enzymes present in the patient or pathogen, a process termed **drug metabolism.** For example, penicillin and other β-lactam antibiotics block cell wall synthesis in bacteria by irreversibly poisoning the enzyme alanyl alanine carboxypeptidase-transpeptidase. Many bacteria, however, produce β-lactamases that hydrolyze the critical β-lactam function in penicillin and related drugs. One strategy for overcoming the resulting antibiotic resistance is to simultaneously administer a β-lactamase inhibitor and a β-lactam antibiotic.

Metabolic transformation is also required to convert an inactive drug precursor, or **prodrug,** into its biologically active form (Chapter 53). 2′-Deoxy-5-fluorouridylic acid, a potent inhibitor of thymidylate synthase, a common target of cancer chemotherapy, is produced from 5-fluorouracil via a series of enzymatic transformations catalyzed by a phosphoribosyl transferase and the enzymes of the deoxyribonucleoside salvage pathway (Chapter 33). Effective design and administration of prodrugs requires knowledge of the kinetics and mechanisms of the enzymes responsible for transforming them into their biologically active forms.

SUMMARY

- The study of enzyme kinetics—the factors that affect the rates of enzyme-catalyzed reactions—reveals the individual steps by which enzymes transform substrates into products.
- ΔG, the overall change in free energy for a reaction, is independent of reaction mechanism and provides no information concerning *rates* of reactions.
- Enzymes do not affect K_{eq}. K_{eq}, a ratio of reaction *rate constants,* may be calculated from the concentrations of substrates and products at equilibrium or from the ratio k_1/k_{-1}.
- Reactions proceed via transition states in which ΔG_F is the activation energy. Temperature, hydrogen ion concentration, enzyme concentration, substrate concentration, and inhibitors all affect the rates of enzyme-catalyzed reactions.

- Measurement of the rate of an enzyme-catalyzed reaction generally employs initial rate conditions, for which the essential absence of product precludes the reverse reaction.
- Linear forms of the Michaelis-Menten equation simplify determination of K_m and V_{max}.
- A linear form of the Hill equation is used to evaluate the cooperative substrate-binding kinetics exhibited by some multimeric enzymes. The slope n, the Hill coefficient, reflects the number, nature, and strength of the interactions of the substrate-binding sites. A value of n greater than 1 indicates positive cooperativity.
- The effects of simple competitive inhibitors, which typically resemble substrates, are overcome by raising the concentration of the substrate. Simple noncompetitive inhibitors lower V_{max} but do not affect K_m.
- For simple competitive and noncompetitive inhibitors, the inhibitory constant K_i is equal to the equilibrium dissociation constant for the relevant enzyme-inhibitor complex. A simpler and less rigorous term for evaluating the effectiveness of an inhibitor is IC_{50}, the concentration of inhibitor that produces 50% inhibition under the particular circumstances of the experiment.
- Substrates may add in a random order (either substrate may combine first with the enzyme) or in a compulsory order (substrate A must bind before substrate B).
- In ping-pong reactions, one or more products are released from the enzyme before all the substrates have been added.
- Applied enzyme kinetics facilitates the identification and characterization of drugs that selectively inhibit specific enzymes. Enzyme kinetics thus plays a central and critical role in drug discovery, in comparative pharmacodynamics, and in determining the mode of action of drugs.

REFERENCES

Dixon M: The determination of enzyme inhibitor constants. Biochem J 1953;55:170.

Dixon M: The graphical determination of K_m and K_i. Biochem J 1972;129:197.

Fersht A: *Structure and Mechanism in Protein Science: A Guide to Enzyme Catalysis and Protein Folding.* Freeman, 1999.

Fraser CM, Rappuoli R: Application of microbial genomic science to advanced therapeutics. Annu Rev Med 2005;56:459.

Henderson PJF: A linear equation that describes the steady-state kinetics of enzymes and subcellular particles interacting with tightly bound inhibitors. Biochem J 1972;127:321.

Schramm, VL: Enzymatic transition-state theory and transition-state analogue design. J Biol Chem 2007;282:28297.

Schultz AR: *Enzyme Kinetics: From Diastase to Multi-enzyme Systems.* Cambridge University Press, 1994.

Segel IH: *Enzyme Kinetics.* Wiley Interscience, 1975.

Wlodawer A: Rational approach to AIDS drug design through structural biology. Annu Rev Med 2002;53:595.

Enzymes: Regulation of Activities

9

Peter J. Kennelly, PhD & Victor W. Rodwell, PhD

BIOMEDICAL IMPORTANCE

The 19th-century physiologist Claude Bernard enunciated the conceptual basis for metabolic regulation. He observed that living organisms respond in ways that are both quantitatively and temporally appropriate to permit them to survive the multiple challenges posed by changes in their external and internal environments. Walter Cannon subsequently coined the term "homeostasis" to describe the ability of animals to maintain a constant intracellular environment despite changes in their external environment. We now know that organisms respond to changes in their external and internal environment by balanced, coordinated adjustments in the rates of specific metabolic reactions. Perturbations of the sensor-response machinery responsible for maintaining homeostatic balance can be deleterious to human health. Cancer, diabetes, cystic fibrosis, and Alzheimer's disease, for example, are all characterized by regulatory dysfunctions triggered by pathogenic agents or genetic mutations. Many oncogenic viruses elaborate protein-tyrosine kinases that modify the regulatory events that control patterns of gene expression, contributing to the initiation and progression of cancer. The toxin from *Vibrio cholerae,* the causative agent of cholera, disables sensor-response pathways in intestinal epithelial cells by ADP-ribosylating the GTP-binding proteins (G-proteins) that link cell surface receptors to adenylyl cyclase. The consequent activation of the cyclase leads to the unrestricted flow of water into the intestines, resulting in massive diarrhea and dehydration. *Yersinia pestis,* the causative agent of plague, elaborates a protein-tyrosine phosphatase that hydrolyzes phosphoryl groups on key cytoskeletal proteins. Dysfunctions in the proteolytic systems responsible for the degradation of defective or abnormal proteins are believed to play a role in neurodegenerative diseases such as Alzheimer's and Parkinson's. Knowledge of factors that control the rates of enzyme-catalyzed reactions thus is essential to an understanding of the molecular basis of disease. This chapter outlines the patterns by which metabolic processes are controlled and provides illustrative examples. Subsequent chapters provide additional examples.

REGULATION OF METABOLITE FLOW CAN BE ACTIVE OR PASSIVE

Enzymes that operate at their maximal rate cannot respond to an increase in substrate concentration, and can respond only to a precipitous decrease in substrate concentration. The K_m values for most enzymes, therefore, tend to be close to the average intracellular concentration of their substrates, so that changes in substrate concentration generate corresponding changes in metabolite flux (Figure 9–1). Responses to changes in substrate level represent an important but *passive* means for coordinating metabolite flow and maintaining homeostasis in quiescent cells. However, they offer limited scope for responding to changes in environmental variables. The mechanisms that regulate enzyme efficiency in an *active* manner in response to internal and external signals are discussed below.

Metabolite Flow Tends to Be Unidirectional

Despite the existence of short-term oscillations in metabolite concentrations and enzyme levels, living cells exist in a dynamic steady state in which the mean concentrations of metabolic intermediates remain relatively constant over time (Figure 9–2). While all chemical reactions are to some extent reversible, in living cells the reaction products serve as substrates for—and are removed by—other enzyme-catalyzed reactions. Many nominally reversible reactions thus occur unidirectionally. This succession of coupled metabolic reactions is accompanied by an overall change in free energy that favors unidirectional metabolite flow (Chapter 11). The unidirectional flow of metabolites through a pathway with a large overall negative change in free energy is analogous to the flow of water through a pipe in which one end is lower than the other. Bends or kinks in the pipe simulate individual enzyme-catalyzed steps with a small negative or positive change in free energy. Flow of water through the pipe nevertheless remains unidirectional due to the overall change in height, which corresponds to the overall change in free energy in a pathway (Figure 9–3).

COMPARTMENTATION ENSURES METABOLIC EFFICIENCY & SIMPLIFIES REGULATION

In eukaryotes, anabolic and catabolic pathways that interconvert common products may take place in specific subcellular compartments. For example, many of the enzymes that degrade proteins and polysaccharides reside inside organelles called lysosomes.

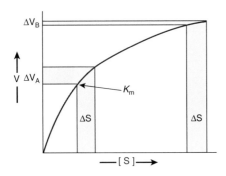

FIGURE 9–1 Differential response of the rate of an enzyme-catalyzed reaction, ΔV, to the same incremental change in substrate concentration at a substrate concentration of K_m (ΔV_A) or far above K_m (ΔV_B).

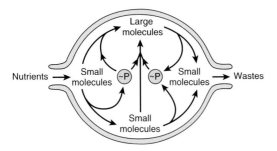

FIGURE 9–2 An idealized cell in steady state. Note that metabolite flow is unidirectional.

Similarly, fatty acid biosynthesis occurs in the cytosol, whereas fatty acid oxidation takes place within mitochondria (Chapters 22 & 23). Segregation of certain metabolic pathways within specialized cell types can provide further physical compartmentation. Alternatively, possession of one or more *unique intermediates* can permit apparently opposing pathways to coexist even in the absence of physical barriers. For example, despite many shared intermediates and enzymes, both glycolysis and gluconeogenesis are favored energetically. This could not be true if *all* the reactions were the same. If one pathway were favored energetically, the other would be accompanied by a change in free energy ΔG equal in magnitude but opposite in sign. Simultaneous spontaneity of both pathways results from substitution of one or more reactions by different reactions favored thermodynamically in the opposite direction. The glycolytic enzyme phosphofructokinase (Chapter 18) is replaced by the gluconeogenic enzyme fructose-1,6-bisphosphatase (Chapter 20). The ability of enzymes to discriminate between the structurally similar coenzymes NAD$^+$ and NADP$^+$ also results in a form of compartmentation, since it segregates the electrons of NADH that are destined for ATP generation from those of NADPH that participate in the reductive steps in many biosynthetic pathways.

Controlling an Enzyme That Catalyzes a Rate-Limiting Reaction Regulates an Entire Metabolic Pathway

While the flux of metabolites through metabolic pathways involves catalysis by numerous enzymes, active control of

FIGURE 9–3 Hydrostatic analogy for a pathway with a rate-limiting step (A) and a step with a ΔG value near zero (B).

homeostasis is achieved by regulation of only a select subset of these enzymes. The ideal enzyme for regulatory intervention is one whose quantity or catalytic efficiency dictates that the reaction it catalyzes is slow relative to all others in the pathway. Decreasing the catalytic efficiency or the quantity of the catalyst for the "bottleneck" or **rate-limiting reaction** therefore reduces metabolite flux through the entire pathway. Conversely, an increase in either its quantity or catalytic efficiency enhances flux through the pathway as a whole. For example, acetyl-CoA carboxylase catalyzes the synthesis of malonyl-CoA, the first committed reaction of fatty acid biosynthesis (Chapter 23). When synthesis of malonyl-CoA is inhibited, subsequent reactions of fatty acid synthesis cease for lack of substrates. As natural "governors" of metabolic flux, the enzymes that catalyze rate-limiting steps constitute efficient targets for regulatory intervention by drugs. For example, inhibition by "statin" drugs of HMG-CoA reductase, which catalyzes the rate-limiting reaction of cholesterogenesis, curtails synthesis of cholesterol.

REGULATION OF ENZYME QUANTITY

The catalytic capacity of the rate-limiting reaction in a metabolic pathway is the product of the concentration of enzyme molecules and their intrinsic catalytic efficiency. It therefore follows that catalytic capacity can be influenced both by changing the quantity of enzyme present and by altering its intrinsic catalytic efficiency.

Proteins Are Continually Synthesized and Degraded

By measuring the rates of incorporation of ^{15}N-labeled amino acids into protein and the rates of loss of ^{15}N from protein, Schoenheimer deduced that body proteins are in a state of "dynamic equilibrium" in which they are continuously synthesized and degraded—a process referred to as **protein turnover.** While the steady-state concentrations of some enzymes and other proteins remain essentially constant, or **constitutive,** over time, the concentrations of many enzymes are influenced by a wide range of physiologic, hormonal, or dietary factors.

The absolute quantity of an enzyme reflects the net balance between its rate of synthesis and its rate of degradation. In human subjects, alterations in the levels of specific enzymes can be affected by a change in the rate constant for the overall processes of synthesis (k_s), degradation (k_{deg}), or both.

Control of Enzyme Synthesis

The synthesis of certain enzymes depends upon the presence of **inducers,** typically substrates or structurally related compounds that initiate their synthesis. *Escherichia coli* grown on glucose will, for example, only catabolize lactose after addition of a β-galactoside, an inducer that initiates synthesis of a β-galactosidase and a galactoside permease (Figure 38–3). Inducible enzymes of humans include tryptophan pyrrolase, threonine dehydratase, tyrosine-α-ketoglutarate aminotransferase, enzymes of the urea cycle, HMG-CoA reductase, and cytochrome P450. Conversely, an excess of a metabolite may curtail synthesis of its cognate enzyme via **repression.** Both induction and repression involve *cis* elements, specific DNA sequences located upstream of regulated genes, and trans-acting regulatory proteins. The molecular mechanisms of induction and repression are discussed in Chapter 38. The synthesis of other enzymes can be stimulated by the binding to specific cell receptors of hormones and other extracellular signals. Detailed information on the control of protein synthesis in response to hormonal stimuli can be found in Chapter 42.

Control of Enzyme Degradation

In animals, many proteins are degraded by the ubiquitin-proteasome pathway, the discovery of which earned Aaron Ciechanover, Avram Hershko, and Irwin Rose a Nobel Prize. The 26S proteasome comprises more than 30 polypeptide subunits arranged in the form of a hollow cylinder. The active sites of its proteolytic subunits face the interior of the cylinder, thus preventing indiscriminate degradation of cellular proteins. Proteins are targeted to the proteasome by "ubiquitination," the covalent attachment of one or more ubiquitin molecules. Ubiquitin is a small, approximately 75 residue protein that is highly conserved among eukaryotes. Ubiquitination is catalyzed by a large family of enzymes called E3 ligases, which attach ubiquitin to the side-chain amino group of lysyl residues.

The ubiquitin-proteasome pathway is responsible both for the regulated degradation of selected cellular proteins (for example, cyclins, in response to specific intra- or extracellular signals) and for the removal of defective or aberrant protein species. The key to the versatility and selectivity of the ubiquitin-proteasome system resides in both the variety of intracellular E3 ligases and their ability to discriminate between different physical or conformational states of a target protein. Thus, the ubiquitin-proteasome pathway can selectively degrade proteins whose physical integrity and functional competency have been compromised by the loss of or damage to a prosthetic group, oxidation of cysteine or histidine residues, or deamidation of asparagine or glutamine residues. Recognition by proteolytic enzymes also can be regulated by covalent modifications such as phosphorylation; binding of substrates or allosteric effectors; or association with membranes, oligonucleotides, or other proteins. A growing body of evidence suggests that dysfunctions of the ubiquitin-proteasome pathway contribute to the accumulation of aberrantly folded protein species characteristic of several neurodegenerative diseases.

MULTIPLE OPTIONS ARE AVAILABLE FOR REGULATING CATALYTIC ACTIVITY

In humans, the induction of protein synthesis is a complex multistep process that typically requires hours to produce significant changes in overall enzyme level. By contrast, changes in intrinsic catalytic efficiency effected by binding of dissociable ligands (**allosteric regulation**) or by **covalent modification** achieve regulation of enzymic activity within seconds. Changes in protein level serve long-term adaptive requirements, whereas changes in catalytic efficiency are best suited for rapid and transient alterations in metabolite flux.

ALLOSTERIC EFFECTORS REGULATE CERTAIN ENZYMES

Feedback inhibition refers to inhibition of an enzyme in a biosynthetic pathway by an end product of that pathway. In the following example, for the biosynthesis of D from A catalyzed by enzymes Enz$_1$ through Enz$_3$,

$$\text{Enz}_1 \quad\quad \text{Enz}_2 \quad\quad \text{Enz}_3$$
$$A \;\;\rightarrow\;\; B \;\;\rightarrow\;\; C \;\;\rightarrow\;\; D$$

high concentrations of D inhibit the conversion of A to B. Inhibition results not from the "backing up" of intermediates but from the ability of D to bind to and inhibit Enz$_1$. Generally, D binds at an **allosteric site** spatially distinct from the catalytic site of the target enzyme. Feedback inhibitors thus typically bear little or no structural similarity to the substrates of the enzymes they inhibit. In this example, the feedback inhibitor D acts as a **negative allosteric effector** of Enz$_1$.

In a branched biosynthetic pathway, the initial reactions participate in the synthesis of multiple end products. Figure 9–4 shows a hypothetical branched biosynthetic pathway in which curved arrows lead from feedback inhibitors to the enzymes whose activity they inhibit. The sequences S$_3$ → A, S$_4$ → B, S$_4$ → C, and S$_3$ → → D each represent linear reaction sequences that are feedback-inhibited by their end products. The pathways of nucleotide biosynthesis (Chapter 33) provide specific examples.

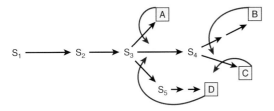

FIGURE 9–4 Sites of feedback inhibition in a branched biosynthetic pathway. S_1–S_5 are intermediates in the biosynthesis of end products A–D. Straight arrows represent enzymes catalyzing the indicated conversions. Curved red arrows represent feedback loops and indicate sites of feedback inhibition by specific end products.

The kinetics of feedback inhibition may be competitive, noncompetitive, partially competitive, or mixed. Feedback inhibitors, which frequently are the small molecule building blocks of macromolecules (eg, amino acids for proteins, nucleotides for nucleic acids), typically inhibit the first committed step in a particular biosynthetic sequence. A much-studied example is inhibition of bacterial aspartate transcarbamoylase by CTP (see below and Chapter 33).

Multiple feedback loops can provide additional fine control. For example, as shown in Figure 9–5, the presence of excess product B decreases the requirement for substrate S_2. However, S_2 is also required for synthesis of A, C, and D. So for this pathway, excess B curtails synthesis of all four end products, regardless of the need for the other three. To circumvent this potential difficulty, each end product may only partially inhibit catalytic activity. The effect of an excess of two or more end products may be strictly additive or, alternatively, greater than their individual effect (cooperative feedback inhibition).

Aspartate Transcarbamoylase Is a Model Allosteric Enzyme

Aspartate transcarbamoylase (ATCase), the catalyst for the first reaction unique to pyrimidine biosynthesis (Figure 33–9), is a target of feedback regulation by cytidine triphosphate

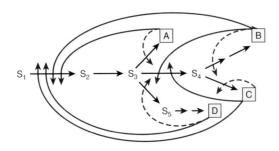

FIGURE 9–5 Multiple feedback inhibition in a branched biosynthetic pathway. Superimposed on simple feedback loops (dashed red arrows) are multiple feedback loops (solid red arrows) that regulate enzymes common to biosynthesis of several end products.

(CTP) and adenosine triphosphate. CTP, an end product of the pyrimidine biosynthetic pathway, inhibits ATCase, whereas ATP activates it. Moreover, high levels of ATP can overcome the inhibition by CTP, enabling synthesis of *pyrimidine* nucleotides to proceed when *purine* nucleotide levels are elevated.

Allosteric & Catalytic Sites Are Spatially Distinct

The lack of structural similarity between a feedback inhibitor and the substrate for the enzyme whose activity it regulates suggests that these effectors are not **isosteric** with a substrate but **allosteric** ("occupy another space"). Jacques Monod therefore proposed the existence of allosteric sites that are physically distinct from the catalytic site. **Allosteric enzymes** thus are those for which catalysis at the active site may be modulated by the presence of effectors at an allosteric site. This hypothesis has been confirmed by many lines of evidence, including x-ray crystallography and site-directed mutagenesis, demonstrating the existence of spatially distinct active and allosteric sites on a variety of enzymes. For example, the ATCase of *Escherichia coli* is a dodecamer consisting of six catalytic subunits and six regulatory subunits, the latter of which bind the nucleotide triphosphates that modulate activity. In general, binding of an allosteric regulator induces a conformational change in the enzyme that encompasses the active site.

Allosteric Effects May Be on K_m or on V_{max}

To refer to the kinetics of allosteric inhibition as "competitive" or "noncompetitive" with substrate carries misleading mechanistic implications. We refer instead to two classes of regulated enzymes: K-series and V-series enzymes. For K-series allosteric enzymes, the substrate saturation kinetics are competitive in the sense that K_m is raised without an effect on V_{max}. For V-series allosteric enzymes, the allosteric inhibitor lowers V_{max} without affecting the K_m. Alterations in K_m or V_{max} probably result from conformational changes at the catalytic site induced by binding of the allosteric effector at its site. For a K-series allosteric enzyme, this conformational change may weaken the bonds between substrate and substrate-binding residues. For a V-series allosteric enzyme, the primary effect may be to alter the orientation or charge of catalytic residues, lowering V_{max}. Intermediate effects on K_m and V_{max}, however, may be observed consequent to these conformational changes.

FEEDBACK REGULATION IS NOT SYNONYMOUS WITH FEEDBACK INHIBITION

In both mammalian and bacterial cells, end products "feed back" and control their own synthesis, in many instances by feedback inhibition of an early biosynthetic enzyme. We must, however, distinguish between **feedback regulation,** a phenomenologic term devoid of mechanistic implications, and feedback **inhibition,** a mechanism for regulation of enzyme activity. For example, while dietary cholesterol decreases hepatic synthesis of cholesterol, this feedback **regulation** does not involve feedback **inhibition.** HMG-CoA reductase, the rate-limiting enzyme of cholesterogenesis, is affected, but cholesterol does not feedback-inhibit its activity. Regulation in response to dietary cholesterol involves curtailment by cholesterol or a cholesterol metabolite of the expression of the gene that encodes HMG-CoA reductase (enzyme repression) (Chapter 26).

MANY HORMONES ACT THROUGH ALLOSTERIC SECOND MESSENGERS

Nerve impulses—and binding of hormones to cell surface receptors—elicit changes in the rate of enzyme-catalyzed reactions within target cells by inducing the release or synthesis of specialized allosteric effectors called **second messengers.** The primary, or "first," messenger is the hormone molecule or nerve impulse. Second messengers include $3',5'$-cAMP, synthesized from ATP by the enzyme adenylyl cyclase in response to the hormone epinephrine, and Ca^{2+}, which is stored inside the endoplasmic reticulum of most cells. Membrane depolarization resulting from a nerve impulse opens a membrane channel that releases calcium ion into the cytoplasm, where it binds to and activates enzymes involved in the regulation of contraction and the mobilization of stored glucose from glycogen. Glucose then supplies the increased energy demands of muscle contraction. Other second messengers include $3',5'$-cGMP and polyphosphoinositols, produced by the hydrolysis of inositol phospholipids by hormone-regulated phospholipases. Specific examples of the participation of second messengers in the regulation of cellular processes can be found in Chapters 19, 42, & 48.

REGULATORY COVALENT MODIFICATIONS CAN BE REVERSIBLE OR IRREVERSIBLE

In mammalian cells, the two most common forms of regulatory covalent modification are **partial proteolysis** and **phosphorylation.** Because organisms lack the ability to reunite the two portions of a protein produced by hydrolysis of a peptide bond, proteolysis constitutes an irreversible modification. By contrast, phosphorylation is a reversible modification process. The phosphorylation of proteins on seryl, threonyl, or tyrosyl residues, catalyzed by protein kinases, is thermodynamically favored. Equally favored is the hydrolytic removal of these phosphoryl groups by enzymes called protein phosphatases. The activities of protein kinases and protein phosphatases are themselves regulated, for if they were not, their concerted action would be both thermodynamically and biologically unproductive.

PROTEASES MAY BE SECRETED AS CATALYTICALLY INACTIVE PROENZYMES

Certain proteins are synthesized and secreted as inactive precursor proteins known as **proproteins.** The proproteins of enzymes are termed **proenzymes** or **zymogens.** Selective proteolysis converts a proprotein by one or more successive proteolytic "clips" to a form that exhibits the characteristic activity of the mature protein, for example, its enzymatic activity. Proteins synthesized as proproteins include the hormone insulin (proprotein = proinsulin), the digestive enzymes pepsin, trypsin, and chymotrypsin (proproteins = pepsinogen, trypsinogen, and chymotrypsinogen, respectively), several factors of the blood-clotting and blood clot dissolution cascades (see Chapter 51), and the connective tissue protein collagen (proprotein = procollagen).

Proenzymes Facilitate Rapid Mobilization of an Activity in Response to Physiologic Demand

The synthesis and secretion of proteases as catalytically inactive proenzymes protects the tissue of origin (eg, the pancreas) from autodigestion, such as can occur in pancreatitis. Certain physiologic processes such as digestion are intermittent but fairly regular and predictable. Others such as blood clot formation, clot dissolution, and tissue repair are brought "on line" only in response to pressing physiologic or pathophysiologic need. The processes of blood clot formation and dissolution clearly must be temporally coordinated to achieve homeostasis. Enzymes needed intermittently but rapidly often are secreted in an initially inactive form since the secretion process or new synthesis of the required proteins might be insufficiently rapid to respond to a pressing pathophysiologic demand such as the loss of blood (see Chapter 51).

Activation of Prochymotrypsin Requires Selective Proteolysis

Selective proteolysis involves one or more highly specific proteolytic clips that may or may not be accompanied by separation of the resulting peptides. Most importantly, selective proteolysis often results in conformational changes that "create" the catalytic site of an enzyme. Note that while His 57 and Asp 102 reside on the B peptide of α-chymotrypsin, Ser 195 resides on the C peptide (Figure 9–6). The conformational changes that accompany selective proteolysis of prochymotrypsin (chymotrypsinogen) align the three residues of the charge-relay network (see Figure 7–7), forming the catalytic site. Note also that contact and catalytic residues can be located on different peptide chains but still be within bond-forming distance of bound substrate.

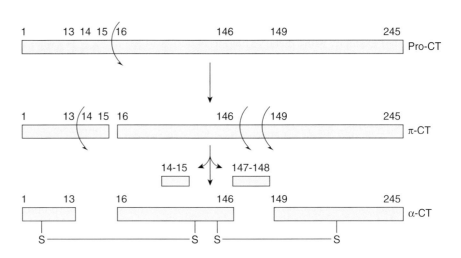

FIGURE 9–6 Two- dimensional representation of the sequence of proteolytic events that ultimately result in formation of the catalytic site of chymotrypsin, which includes the Asp 102-His57-Ser195 catalytic triad (see Figure 7–7). Successive proteolysis forms prochymotrypsin (pro-CT), π-chymotrypsin (π-Ct), and ultimately α-chymotrypsin (α-CT), an active protease whose three peptides remain associated by covalent inter-chain disulfide bonds.

REVERSIBLE COVALENT MODIFICATION REGULATES KEY MAMMALIAN ENZYMES

Mammalian proteins are the targets of a wide range of covalent modification processes. Modifications such as prenylation, glycosylation, hydroxylation, and fatty acid acylation introduce unique structural features into newly synthesized proteins that tend to persist for the lifetime of the protein. Among the covalent modifications that regulate protein function (eg, methylation, acetylation), the most common by far is phosphorylation-dephosphorylation. **Protein kinases** phosphorylate proteins by catalyzing transfer of the terminal phosphoryl group of ATP to the hydroxyl groups of seryl, threonyl, or tyrosyl residues, forming *O*-phosphoseryl, *O*-phosphothreonyl, or *O*-phosphotyrosyl residues, respectively (Figure 9–7). Some protein kinases target the side chains of histidyl, lysyl, arginyl, and aspartyl residues. The unmodified form of the protein can be regenerated by hydrolytic removal of phosphoryl groups, catalyzed by **protein phosphatases.**

A typical mammalian cell possesses thousands of phosphorylated proteins and several hundred protein kinases and protein phosphatases that catalyze their interconversion. The ease of interconversion of enzymes between their phospho-

and dephospho- forms accounts, in part, for the frequency with which phosphorylation-dephosphorylation is utilized as a mechanism for regulatory control. Phosphorylation-dephosphorylation permits the functional properties of the affected enzyme to be altered only for as long as it serves a specific need. Once the need has passed, the enzyme can be converted back to its original form, poised to respond to the next stimulatory event. A second factor underlying the widespread use of protein phosphorylation-dephosphorylation lies in the chemical properties of the phosphoryl group itself. In order to alter an enzyme's functional properties, any modification of its chemical structure must influence the protein's three-dimensional configuration. The high charge density of protein-bound phosphoryl groups—generally −2 at physiologic pH—and their propensity to form strong salt bridges with arginyl and lysyl residues renders them potent agents for modifying protein structure and function. Phosphorylation generally influences an enzyme's intrinsic catalytic efficiency or other properties by inducing conformational changes. Consequently, the amino acids targeted by phosphorylation can be and typically are relatively distant from the catalytic site itself.

Covalent Modification Regulates Metabolite Flow

In many respects, sites of protein phosphorylation and other covalent modifications can be considered another form of allosteric site. However, in this case, the "allosteric ligand" binds covalently to the protein. Both phosphorylation-dephosphorylation and feedback inhibition provide short-term, readily reversible regulation of metabolite flow in response to specific physiologic signals. Both act without altering gene expression. Both act on early enzymes of a protracted and often biosynthetic metabolic sequence, and both act at allosteric rather than catalytic sites. Feedback inhibition, however, involves a

FIGURE 9–7 Covalent modification of a regulated enzyme by phosphorylation-dephosphorylation of a seryl residue.

single protein and lacks hormonal and neural features. By contrast, regulation of mammalian enzymes by phosphorylation-dephosphorylation involves several proteins and ATP, and is under direct neural and hormonal control.

PROTEIN PHOSPHORYLATION IS EXTREMELY VERSATILE

Protein phosphorylation-dephosphorylation is a highly versatile and selective process. Not all proteins are subject to phosphorylation, and of the many hydroxyl groups on a protein's surface, only one or a small subset are targeted. While the most common enzyme function affected is the protein's catalytic efficiency, phosphorylation can also alter its location within the cell, susceptibility to proteolytic degradation, or responsiveness to regulation by allosteric ligands. Phosphorylation can increase an enzyme's catalytic efficiency, converting it to its active form in one protein, while phosphorylation of another protein converts it to an intrinsically inefficient, or inactive, form (Table 9–1).

Many proteins can be phosphorylated at multiple sites. Others are subject to regulation both by phosphorylation-dephosphorylation and by the binding of allosteric ligands, or by phosphorylation-dephosphorylation and another covalent modification. Phosphorylation-dephosphorylation at any one site can be catalyzed by multiple protein kinases or protein phosphatases. Many protein kinases and most protein phosphatases act on more than one protein and are themselves interconverted between active and inactive forms by the binding of second messengers or by covalent modification by phosphorylation-dephosphorylation.

The interplay between protein kinases and protein phosphatases, between the functional consequences of phosphorylation at different sites, between phosphorylation sites and allosteric sites, or between phosphorylation sites and other sites of covalent modification provides the basis for regulatory

networks that integrate multiple environmental input signals to evoke an appropriate coordinated cellular response. For example, modification of histones by a combination of acetylation and phosphorylation constitutes the basis for the "histone code," which modulates chromatin structure to enhance or silence the expression of genes (Chapter 38).

In these sophisticated regulatory networks, individual enzymes respond to different environmental signals. For example, if an enzyme can be phosphorylated at a single site by more than one protein kinase, it can be converted from a catalytically efficient to an inefficient (inactive) form, or vice versa, in response to any one of several signals. If the protein kinase is activated in response to a signal different from the signal that activates the protein phosphatase, the phosphoprotein becomes a decision node. The functional output, generally catalytic activity, reflects the phosphorylation state. This state or degree of phosphorylation is determined by the relative activities of the protein kinase and protein phosphatase, a reflection of the presence and relative strength of the environmental signals that act through each.

The ability of many protein kinases and protein phosphatases to target more than one protein provides a means for an environmental signal to coordinately regulate multiple metabolic processes. For example, the enzymes 3-hydroxy-3-methylglutaryl-CoA reductase and acetyl-CoA carboxylase—the rate-controlling enzymes for cholesterol and fatty acid biosynthesis, respectively—are phosphorylated and inactivated by the AMP-activated protein kinase. When this protein kinase is activated either through phosphorylation by yet another protein kinase or in response to the binding of its allosteric activator 5′-AMP, the two major pathways responsible for the synthesis of lipids from acetyl-CoA are both inhibited.

INDIVIDUAL REGULATORY EVENTS COMBINE TO FORM SOPHISTICATED CONTROL NETWORKS

Cells carry out a complex array of metabolic processes that must be regulated in response to a broad spectrum of environmental factors. Hence, interconvertible enzymes and the enzymes responsible for their interconvesion do not act as isolated "on" and "off" switches. In order to meet the demands of maintaining homeostasis, these building blocks are linked to form integrated regulatory networks.

One well-studied example of such a network is the eukaryotic cell cycle that controls cell division. Upon emergence from the quiescent, or G_0, state, the extremely complex process of cell division proceeds through a series of specific phases designated G_1, S, G_2, and M (Figure 9–8). Elaborate monitoring systems, called checkpoints, assess key indicators of progress to ensure that no phase of the cycle is initiated until the prior phase is complete. Figure 9–8 outlines, in simplified form, part of the checkpoint that controls the initiation of DNA rep-

TABLE 9–1 Examples of Mammalian Enzymes Whose Catalytic Activity Is Altered by Covalent Phosphorylation-Dephosphorylation

	Activity State	
Enzyme	Low	High
Acetyl-CoA carboxylase	EP	E
Glycogen synthase	EP	E
Pyruvate dehydrogenase	EP	E
HMG-CoA reductase	EP	E
Glycogen phosphorylase	E	EP
Citrate lyase	E	EP
Phosphorylase b kinase	E	EP
HMG-CoA reductase kinase	E	EP

Abbreviations: E, dephosphoenzyme; EP, phosphoenzyme.

FIGURE 9–8 A simplified representation of the G_1 to S checkpoint of the eukaryotic cell cycle. The circle shows the various stages in the eukaryotic cell cycle. The genome is replicated during S phase, while the two copies of the genome are segregated and cell division occurs during M phase. Each of these phases is separated by a *G*, or growth, phase characterized by an increase in cell size and the accumulation of the precursors required for the assembly of the large macromolecular complexes formed during S and M phases.

lication, called the S phase. A protein kinase called ATM is associated with the genome. If the DNA contains a double-stranded break, the resulting change in the conformation of the chromatin activates ATM. Upon activation, one subunit of the activated ATM dimer dissociates and initiates a series, or cascade, of protein phosphorylation-dephosphorylation events mediated by the CHK1 and CHK2 protein kinases, the Cdc25 protein phosphatase, and finally a complex between a cyclin and a cyclin-dependent protein kinase, or Cdk. Activation of the Cdk-cyclin complex blocks the G_1 to S transition, thus preventing the replication of damaged DNA. Failure at this checkpoint can lead to mutations in DNA that may lead to cancer or other diseases. Each step in the cascade provides a conduit for monitoring additional indicators of cell status prior to entering S phase.

SUMMARY

- Homeostasis involves maintaining a relatively constant intracellular and intra-organ environment despite wide fluctuations in the external environment. This is achieved via appropriate changes in the rates of biochemical reactions in response to physiologic need.

- The substrates for most enzymes are usually present at a concentration close to their K_m. This facilitates passive control of the rates of product formation in response to changes in levels of metabolic intermediates.

- Active control of metabolite flux involves changes in the concentration, catalytic activity, or both of an enzyme that catalyzes a committed, rate-limiting reaction.

- Selective proteolysis of catalytically inactive proenzymes initiates conformational changes that form the active site. Secretion as

an inactive proenzyme facilitates rapid mobilization of activity in response to injury or physiologic need and may protect the tissue of origin (eg, autodigestion by proteases).

- Binding of metabolites and second messengers to sites distinct from the catalytic site of enzymes triggers conformational changes that alter V_{max} or K_m.

- Phosphorylation by protein kinases of specific seryl, threonyl, or tyrosyl residues—and subsequent dephosphorylation by protein phosphatases—regulates the activity of many human enzymes. The protein kinases and phosphatases that participate in regulatory cascades that respond to hormonal or second messenger signals constitute regulatory networks that can process and integrate complex environmental information to produce an appropriate and comprehensive cellular response.

REFERENCES

Bray D: Protein molecules as computational elements in living cells. Nature 1995;376:307.

Ciechanover A, Schwartz AL: The ubiquitin system: Pathogenesis of human diseases and drug targeting. Biochim Biophys Acta 2004;1695:3.

Graves DJ, Martin BL, Wang JH: *Co- and Post-translational Modification of Proteins: Chemical Principles and Biological Effects.* Oxford Univ Press, 1994.

Johnson LN, Lewis RJ: Structural basis for control by phosphorylation. Chem Rev 2001;101:2209.

Muoio DM, Newgard CB: Obseity-related derangements in metabolic regulation. Anu Rev Biochem 2006;75:403.

Pilkis SJ et al: 6-Phosphofructo-2-kinase/fructose-2,6-bisphosphatase: a metabolic signaling enzyme. Annu Rev Biochem 1995;64:799.

Rovner SL: Hold that thought. Slowly but surely scientists are unveiling the complex chemical underpinnings of memory. Chem Eng News 2007;85,13.

Scriver CR et al (editors): *The Metabolic and Molecular Bases of Inherited Disease,* 8th ed. McGraw-Hill, 2000.

Sitaramayya A (editor): *Introduction to Cellular Signal Transduction.* Birkhauser, 1999.

Stadtman ER, Chock PB (editors): *Current Topics in Cellular Regulation.* Academic Press, 1969.

Stieglitz K et al: Monitoring the transition from the T to the R state in *E. coli* aspartate transcarbamoylase by x-ray crystallography: Crystal structures of the E50A mutant enzyme in four distinct allosteric states. J Mol Biol 2004;341:853.

Tu B et al: Logic of the yeast metabolic cycle: Temporal compartmentalization of cellular processes. Science 2005; 310:1152.

Weber G (editor): *Advances in Enzyme Regulation.* Pergamon Press, 1963.

Bioinformatics & Computational Biology

Peter J. Kennelly, PhD & Victor W. Rodwell, PhD

BIOMEDICAL IMPORTANCE

The first scientific models of pathogenesis, such as Louis Pasteur's seminal germ theory of disease, were binary in nature: each disease possessed a single, definable causal agent. Malaria was caused by the amoeba *Plasmodium falciparum*, tuberculosis by the bacterium *Mycobacterium tuberculosis*, sickle cell disease by a mutation in a gene encoding one of the subunits of hemoglobin, poliomyelitis by poliovirus, and scurvy by a deficiency in ascorbic acid. The strategy for treating or preventing disease thus could be reduced to a straightforward process of tracing the causal agent and then devising some means of eliminating it, neutralizing its effects, or blocking its route of transmission. This approach has been successfully employed to understand and treat a wide range of infectious and genetic diseases. However, it has become clear that the determinants of many pathologies—including cancer, coronary heart disease, type II diabetes, and Alzheimer's disease—are **multifactorial** in nature. Rather than having a specific causal agent or agents whose presence is both necessary and sufficient, the appearance and progression of the aforementioned diseases reflect the complex interplay between each individual's genetic makeup, diet, and lifestyle, as well as a range of environmental factors such as the presence of toxins, viruses, or bacteria.

The challenge posed by multifactorial diseases demands a quantum increase in the breadth and depth of our knowledge of living organisms capable of matching their sophistication and complexity. We must identify the many as yet unknown proteins encoded within the genomes of humans and of the organisms with which they interact, the functional relationships between these proteins, and the impact of dietary, genetic, and environmental factors thereupon. The sheer mass of information that must be processed to understand, as completely and comprehensively as possible, the molecular mechanisms that underlie the behavior of living organisms, as well as the manner in which perturbations can lead to disease or dysfunction, lies well beyond the ability of the human mind to review and analyze. Biomedical scientists therefore have turned to sophisticated computational tools to collect and evaluate biologic information on a mass scale.

GENOMICS: AN INFORMATION AVALANCHE

Physicians and scientists have long understood that the genome, the complete complement of genetic information of a living organism, represented a rich source of information concerning topics ranging from basic metabolism and mechanisms of evolution to longevity and aging. However, the massive size of the human genome, 3×10^9 nucleotide base pairs, required a paradigm shift in the manner in which scientists approached the determination of DNA sequences. Recent advances in bioinformatics and computational biology have, in turn, been fueled by the need to develop new approaches to "mine" the mass of genome sequence data that continues to emerge from the Human Genome Project (HGP) and its correlates.

The Human Genome Project

The successful completion of the HGP represents the culmination of more than six decades of achievements in molecular biology, genetics, and biochemistry. The chronology below lists several of the milestone events that led to the determination of the entire sequence of the human genome.

1944—DNA is shown to be the hereditary material
1953—Concept of the double helix is posited
1966—The genetic code is solved
1972—Recombinant DNA technology is developed
1977—Practical DNA sequencing technology emerges
1983—The gene for Huntington's disease is mapped
1985—The Polymerase Chain Reaction (PCR) is invented
1986—DNA sequencing becomes automated
1986—The gene for Duchenne muscular dystrophy is identified
1989—The gene for cystic fibrosis is identified
1990—The Human Genome Project is launched in the United States
1994—Human genetic mapping is completed
1996—The first human gene map is established
1999—The Single Nucleotide Polymorphism Initiative is started

1999—The first sequence of a human chromosome, number 22, is completed

2000—"First draft" of the human genome is completed

2003—Sequencing of the first human genome is completed

Coincident with these advances were the sequencing of the genomes of hundreds of other organisms, including *Haemophilus influenzae* (1995), yeast (1996), *Escherichia coli* (1997), *Caenorhabditis elegans* (1998), *Mycobacterium tuberculosis* (1998), rice (2000), *Listeria monocytogenes* (2001), the SARS coronavirus (2003), rat (2004), and chimpanzee (2005).

Two groups were responsible for sequencing the human genome. The Human Genome Sequencing Consortium employed hierarchical shotgun sequencing. For this approach, the entire genome was fragmented into approximately 100–200 kb pieces and inserted into bacterial artificial chromosomes (BACs). The BACs were then positioned on individual chromosomes by looking for marker sequences known as sequence-tagged sites (short unique genomic loci for which a PCR assay is available) whose locations were known. Clones of the BACs were then broken into small fragments (shotgunning), and sequenced. Computer algorithms were then used to identify matching sequence information from overlapping fragments to piece together the complete sequence. The team from Celera used the whole genome shotgun approach. Shotgun fragments were assembled into contigs (sets of DNA that possess overlapping sequences) that were arranged on scaffolds in the correct order, although not necessarily connected in one continuous sequence. The correct positions of these scaffolds were then determined by using sequence-tagged sites. High-throughput sequenators, powerful computer programs, and the element of competition each contributed to the rapid progress made by both groups that culminated in the speedy completion of the HGP. A detailed description of the two approaches can be found at www.genomenewsnetwork.org/articles/06_00/sequence_primer.shtml.

Genomes and Medicine

Ready access to genome sequences from organisms spanning all three phylogenetic domains, the *Archaea, Bacteria,* and *Eukarya,* coupled with access to powerful algorithms for manipulating and transforming data derived from these sequences, has already effected major transformations in biology and biochemistry. The early decades of the 21st century will witness the expansion of the "Genomics Revolution" into the practice of medicine as physicians and scientists exploit new knowledge of the human genome and of the genomes of the organisms that colonize, feed, and infect *Homo sapiens.*

Today, comparisons between the genomes of pathogenic and nonpathogenic strains of a microorganism can highlight likely determinants of virulence. Similarly, **comparative genomics** is being applied to pathogens and their hosts to identify lists of gene products unique to the former from which to select potential drug targets. In future, physicians will diagnose

and treat patients guided by insights provided by an individual's genome sequence. While genome-based "designer medicine" promises to be efficient and effective, significant technical and scientific challenges remain to be addressed before the promise of genomics can be completely fulfilled in both biology and medicine. These challenges, summarized with admirable clarity by FS Collins and associates in an article entitled "A vision for the future of genomic research. A blueprint for the genomic era" (*Nature* 2003; 422; 6934), are as follows:

- To comprehensively identify the structural and functional components encoded in a biologically diverse range of organisms.
- To elucidate the organization of genetic networks and protein pathways and establish how they contribute to cellular and organismal phenotypes.
- To develop a detailed understanding of the heritable variation in genomes.
- To understand evolutionary variation across species and the mechanisms that underlie this variation.
- To develop policy options that facilitate appropriate widespread use of genomics in both research and clinical settings.
- To develop robust strategies for identifying the genetic contributions to disease and drug response.
- To identify gene variants that contribute to good health and resistance to disease.
- To develop genome-based approaches for the prediction of disease susceptibility and drug response, the early detection of illness, and the delineation of the molecular taxonomy of disease states.
- To exploit new insights into genes and pathways for the development of more effective therapeutic approaches to disease.
- To determine how genetic risk information should be conveyed in clinical settings and how that information should guide health strategies and behavior.

The opportunities offered by the advancing genomic revolution will present society with profound challenges in the areas of ethics, law, and public policy. The first harbingers of these challenges can be glimpsed in the ongoing controversies regarding genetically modified foods, the cloning of whole animals, and the utilization of human embryonic stem cells in research. Forthcoming insights into the molecular and genetic contributions to human traits and behavior, as well as to physical health or to disease, will require the development of a new generation of national and international policies in the areas of law, medicine, agriculture, etc.

BIOINFORMATICS

Bioinformatics exploits the formidable information storage and processing capabilities of the computer to develop tools for the collection, collation, retrieval, and analysis of biologic data on a mass scale. Many bioinformatic resources (see below) can be accessed via the Internet, which provides them with global reach and impact. The central objective of a typical bioinformatics project is to assemble all of the available information relevant to a particular topic in a single location, often referred to as a **library** or **database,** in a uniform format that renders the data amenable to manipulation and analysis by computer algorithms.

The size and capabilities of bioinformatic databases can vary widely depending upon the scope and nature of their objectives. The PubMed database (http://www.ncbi.nlm.nih.gov/sites/entrez?db=pubmed) compiles citations for all articles published in thousands of journals devoted to biomedical and biological research. Currently, PubMed contains over 17 million citations. By contrast, the Cytochrome P450 Home-page (http://drnelson.utmem.edu/CytochromeP450.html) confines itself to the approximately 6000 members of the cytochrome P450 family of enzymes that are involved in the metabolism of many drugs and other xenobiotics (see Chapter 53).

The construction of a comprehensive and user-friendly database presents many challenges. First, biomedical information comes in a wide variety of forms. For example, the coding information in a genome, although voluminous, is composed of simple linear sequences of four nucleotide bases. While the number of amino acid residues that define a protein's primary structure is minute relative to the number of base pairs in a genome, a description of a protein's x-ray structure requires that the location of each atom be specified in three-dimensional space.

Second, anticipating the manner in which users may wish to search or analyze the information within a database, and devising algorithms for coping with these variables, can prove extremely challenging. For example, even the simple task of searching a gene database commonly employs, alone or in various combinations, criteria as diverse as the name of the gene, the name of the protein that it encodes, the biologic function of the gene product, a nucleotide sequence within the gene, a sequence of amino acids within the protein it encodes, the organism in which it is present, or the name of an investigator who works on that gene. Researchers wishing to determine whether the impact of a genetic polymorphism on longevity is influenced by the nature of the climate where a person resides may need to compare data from multiple databases. Similarly, a diverse range of criteria may apply when describing the subjects of a biomedical study: height; weight; age; gender; body mass index; diet; ethnicity; medical history; profession; use of drugs, alcohol, or tobacco products; exercise; blood pressure; habitat; marital status; blood type; serum cholesterol level; etc.

BIOINFORMATIC AND GENOMIC RESOURCES

The large collection of databases that have been developed for the assembly, annotation, analysis and distribution of biological and biomedical data reflects the breadth and variety of contemporary molecular, biochemical, epidemiological, and clinical research. Below are listed examples of prominent bioinformatics resources.

GenBank, UniProt, and the Protein Database

GenBank, UniProt, and the Protein Database (PDB) represent three of the oldest and most widely used bioinformatics databases. Each complements the other by focusing on a different aspect of macromolecular structure. UniProt Knowledgebase (http://www.pir.uniprot.org/) can trace its origins to the *Atlas of Protein Sequence and Structure,* a printed encyclopedia of protein sequences first published in 1968 by Margaret Dayhoff and the National Biomedical Research Foundation at Georgetown University. The aim of the atlas was to facilitate studies of protein evolution using the amino acid sequences being generated consequent to the development of the Edman method for protein sequencing (Chapter 4). In partnership with the Martinsreid Institute for Protein Sequences and the International Protein Information Database of Japan, the atlas made the transition to electronic form as the Protein Information Resource (PIR) Protein Sequence Database in 1984. In 2002, PIR integrated its database of protein sequence and function with the Swiss-Prot protein database, established by Amos Bairoch in 1986 under the auspices of the Swiss Institute of Bioinformatics and the European Bioinformatics Institute, to form the world's most comprehensive resource on protein structure and function, the UniProt Knowledgebase.

GenBank (http://www.ncbi.nlm.nih.gov/Genbank/), the National Institutes of Health's (NIH) genetic sequence database, is a collection of nucleotide sequences and their translations. GenBank, established in 1979 by Walter Goad of Los Alamos National Laboratory, currently is maintained by the National Center for Biotechnology Information at NIH. GenBank constitutes one of the cornerstones of the International Sequence Database Collaboration, a consortium that includes the DNA Database of Japan and the European Molecular Biology Laboratory.

The Protein Database (http://www.rcsb.org/pdb/home/home.do), a repository of the three-dimensional structures of proteins, polynucleotides, and other biological macromolecules, was established in 1971 by Edgar Meyer and Walter Hamilton of Brookhaven National Laboratories. In 1998, responsibility for the PDB was transferred to the Research Collaboration for Structural Bioinformatics formed by Rutgers University, the University of California at San Diego, and the University of Wisconsin.

The HapMap Database

While the genome sequence of any two individuals is 99.9% identical, human DNA contains approximately 10 million sites where individuals differ by a single-nucleotide base. These sites are called **single-nucleotide polymorphisms** or **SNPs.** When sets of SNPs localized to the same chromosome are inherited together in blocks, the pattern of SNPs in each block is termed a **haplotype.** By comparing the haplotype distributions in groups of individuals with or without a given disease or response, biomedical scientists can identify SNPs that are associated with specific phenotypic traits. This process can be facilitated by focusing on **Tag SNPs,** a subset of the SNPs in a given block sufficient to provide a unique marker for a given haplotype. Detailed study of each region should reveal variants in genes that contribute to a specific disease or response.

In 2002, scientists from the United States, Canada, China, Japan, Nigeria, and the United Kingdom launched the International HapMap Project (http://www.hapmap.org), a comprehensive effort to identify SNPs associated with common human diseases and differential responses to pharmaceuticals. The resulting **haplotype map (HapMap)** should lead to earlier and more accurate diagnosis, and hopefully also to improved prevention and patient management. Knowledge of an individual's genetic profile also will be used to guide the selection of safe and effective drugs or vaccines, a process termed **pharmacogenomics.** These genetic markers will also provide labels with which to identify and track specific genes as scientists seek to learn more about the critical processes of genetic inheritance and selection.

The ENCODE Project

Identification of all the *functional elements* of the genome will vastly expand our understanding of the molecular events that underlie human development, health, and disease. To address this goal, in late 2003, the National Human Genome Research Institute (NHGRI) initiated the ENCODE (*Encyclopedia Of DNA Elements*) Project. Based at the University of California at Santa Cruz, ENCODE (http://www.genome.gov/10005107) is a collaborative effort that combines laboratory and computational approaches to identify every functional element in the human genome. Consortium investigators with diverse backgrounds and expertise collaborate in the development and evaluation of new high-throughput techniques, technologies, and strategies to address current deficiencies in our ability to identify functional elements.

The pilot phase of ENCODE will target approximately 1% (30 Mb) of the human genome for rigorous computational and experimental analysis. The consortium faces many challenges. In addition to the sheer size of the human genome and the cryptic nature of much of its sequence, scientists must cope with the variations in genome function that characterize different cell types and developmental stages. Given the complexity of the issues, it is clear that no single experimental approach or cell type will suffice to provide a complete overview of the interrelationships between genome sequence, architecture, and function.

Entrez Gene and dbGAP

Entrez Gene (http://www.ncbi.nlm.nih.gov/sites/entrez?db=gene), a database maintained by the NCBI, provides a variety of information about individual human genes. The information includes the sequence of the genome in and around the gene, gene structure (exon-intron boundaries), the sequence of the mRNA(s) produced from the gene, and any phenotypes associated with a given mutation. *Entrez* Gene also lists, where known, the function of the encoded protein and the impact of known single-nucleotide polymorphisms in the coding region.

An NCBI database that complements *Entrez* Gene is db-GAP, the **Database of Genotype and Phenotype** (http://www.ncbi.nlm.nih.gov/sites/entrez?db=gap). dbGAP compiles the results of research into the links between specific genotypes and phenotypes. To protect the confidentiality of sensitive clinical data, the information contained in dbGAP is organized into open- and controlled-access sections. Access to sensitive data requires that the user apply for authorization to a data access committee.

Other databases dealing with human genetics and health include Online Mendelian Inheritance in Man (http://www.ncbi.nlm.nih.gov/sites/entrez?db=omim), the Human Gene Mutation Database (http;//www.hgmd.cf.ac.uk/ac/index.php), the Cancer Genome Atlas (http://cancergenome.nih.gov/), and GeneCards (http://www.genecards.org/), which tries to collect all relevant information on a given gene from databases worldwide to create a single, comprehensive "card" for each gene.

COMPUTATIONAL BIOLOGY

The primary objective of **computational biology** is to develop computer models that apply physical, chemical, and biological principles that mirror the behavior of biologic molecules and processes. Unlike bioinformatics, whose major focus is the collection and evaluation of existing data, computational biology is experimental and exploratory in nature. By performing virtual experiments and analyses "*in silico,*" computational biology offers the promise of greatly accelerating the pace and efficiency of scientific discovery.

Computational biologists are attempting to develop predictive models that will (1) permit the three-dimensional structure of a protein to be determined directly from its primary sequence, (2) determine the function of unknown proteins from their sequence and structure, (3) screen for potential inhibitors of a protein *in silico,* and (4) construct virtual cells that reproduce the behavior and predict the responses of their living counterparts to pathogens, toxins, diet, and drugs. The creation of computer algorithms that accurately mimic the behavior of proteins, enzymes, cells, etc will enhance the speed, efficiency, and the safety of biomedical research. Computational biology will also enable scientists to perform experiments *in silico* whose scope, hazard, or nature renders them inaccessible to or inappropriate for conventional laboratory or clinical approaches.

IDENTIFICATION OF PROTEINS BY HOMOLOGY

One important method for the identification, also called **annotation**, of novel proteins and gene products is via compar-

Language	Word	Alignment
English	PHYSIOLOGICAL	`P H Y S I O L O G I - C A L`
French	PHYSIOLOGIQUE	`P H Y S I O L O G I - Q U E`
German	PHYSIOLOGISCH	`P H Y S I O L O G I S C H`
Dutch	FYSIOLOGISCH	`F - Y S I O L O G I S C H`
Spanish	FYSIOLOGICO	`F - Y S I O L O G I - C O`
Polish	FIZJOLOGICZNY	`F - I Z J O L O G I - C Z N Y`

FIGURE 10–1 Representation of a multiple sequence alignment. Languages evolve in a fashion that mimics that of genes and proteins. Shown is the English word "physiological" in several languages. The alignment demonstrates their conserved features. Identities with the English word are shown in dark red; linguistic similarities in light red. Multiple sequence alignment algorithms identify conserved nucleotide and amino acid letters in DNA, RNA, and polypeptides in an analogous fashion.

ison with proteins of known sequence or structure. Simply put, homology searches and multiple sequence comparisons operate on the principle that proteins that perform similar functions will share conserved domains or other sequence features or **motifs,** and vice versa (Figure 10–1). Of the many algorithms developed for this purpose, the most widely used is **BLAST**.

BLAST (**Basic Local Alignment Search Tool)** and other sequence comparison/alignment algorithms trace their origins to the efforts of early molecular biologists to determine whether observed similarities in sequence among proteins that performed parallel metabolic functions were indicative of progressive change from a common origin. The major evolutionary question addressed was whether the similarities reflected (1) descent from a common ancestral protein (divergent evolution) or (2) the independent selection of a common mechanism for meeting some specific cellular need (convergent evolution), as would be anticipated if one particular solution was overwhelmingly superior to the alternatives. Calculation of the minimum number of nucleotide changes required to interconvert putative protein isoforms allows inferences to be drawn concerning whether or not the similarities and differences exhibit a pattern indicative of progressive change from a shared origin.

BLAST has evolved into a family of programs optimized to address specific needs and data sets. Thus, **blastp** compares an *amino acid* query sequence against a *protein* sequence database, **blastn** compares a *nucleotide* query sequence against a *nucleotide* sequence database, **blastx** compares a *nucleotide* query sequence translated in all reading frames against a *protein* sequence database to reveal potential translation products, **tblastn** compares a *protein* query sequence against a *nucleotide* sequence database dynamically translated in all six reading frames, and **tblastx** compares the six-frame translations of a *nucleotide* query sequence against the six-frame translations of a *nucleotide* sequence database. Unlike multiple sequence alignment programs that rely on *global* alignments, the **BLAST** algorithms emphasize regions of *local* alignment to detect relationships among sequences with only isolated regions of similarity. This approach provides speed and increased sensitivity for distant sequence relationships. Input

or "query" sequences are broken into "words" (default size 11 for nucleotides, 3 for amino acids). Word hits to databases are then extended in both directions.

IDENTIFICATION OF "UNKNOWN" PROTEINS

A substantial portion of the genes discovered by genome sequencing projects code for "unknown" or hypothetical proteins, proteins of unknown function or structure for which homologues of known function are lacking. Bioinformatics scientists are developing tools to enable scientists to deduce from the amino acid sequences of unknown proteins their three-dimensional structure and function. Currently, the list of unknown proteins uncovered by genomics contains thousands of entries, with new entries being added as more genome sequences are solved. The ability to generate structures and infer function *in silico* promises to significantly accelerate protein identification and provide insight into the mechanism by which proteins fold. This knowledge will aid in understanding the underlying mechanisms of various protein folding diseases, and will assist molecular engineers to design new proteins to perform novel functions.

Comparison of proteins whose three-dimensional structures have been determined by x-ray crystallography or NMR spectroscopy can reveal patterns that link specific primary sequence features to specific primary, secondary, and tertiary structures—sometimes called the folding code. The first algorithms used the frequency with which individual amino acids occurred in α-helices, β-sheets, turns, and loops to predict the secondary structure topography of a polypeptide. For example, a segment of protein sequence rich in amino acids frequently found in α-helices was predicted to adopt this conformation. By extending this process, for example, by weighing the impact of hydrophobic interactions in the formation of the protein core, algorithms of remarkable predictive reliability are being developed. However, while current programs perform well in generating the conformations of proteins formed of single domains, projecting the likely structure of membrane proteins and those composed of multiple domains remain problematic.

Scientists are also attempting to discern patterns between three-dimensional structure and physiologic function. Where a complete three-dimensional structure can be determined or predicted, the protein's surface can be scanned for the types of pockets and crevices indicative of likely binding sites for substrates, allosteric effectors, etc. The shape of the pocket and the distribution of hydrophobic, hydrophilic, and potentially charged amino acids within it can then be used to infer the structure of the biomolecule that binds or "docks" at that site. The space-filling representation of the enzyme HMG-CoA reductase and its complex with the drug lovastatin, shown in Figure 10–2, provides some perspective on the challenges inherent in identifying ligand-binding sites from scratch.

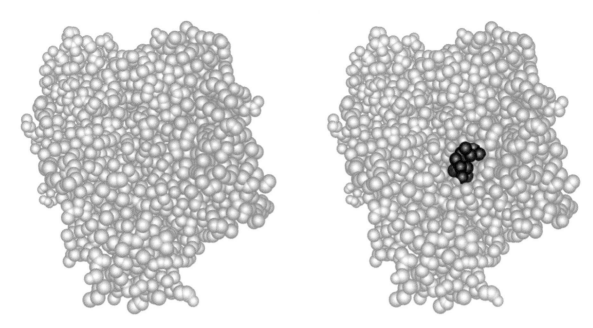

FIGURE 10–2 Shown are space-filling representations of the homodimeric HMG-CoA reductase from *Pseudomonas mevalonii* with (**right**) and without (**left**) the statin drug lovastatin bound. Each atom is represented by a sphere the size of its van der Waals' radius. The two polypeptide chains are colored gray and blue. The carbon atoms of lovastatin are colored black and the oxygen atoms red. Compare this model with the backbone representations of proteins shown in Chapters 5 & 6. (Adapted from Protein Data Bank ID no. 1t02.)

COMPUTER-AIDED DRUG DESIGN

Computer-Aided Drug Design (CADD) employs the same type of molecular-docking algorithms used to identify ligands for unknown proteins. However, in this case the set of potential ligands to be considered is not confined to those occurring in nature, and is aided by empirical knowledge of the structure or functional characteristics of the target protein.

For proteins of known three-dimensional structure, molecular-docking approaches employ programs that attempt to fit a series of potential ligand "pegs" into a designated binding site "hole" on a protein template. To identify optimum ligands, docking programs must account both for matching shapes and for complementary hydrophobic, hydrophilic, and charge attributes. The binding affinities of the inhibitors selected on the basis of early docking studies were disappointing, as the rigid models for proteins and ligands employed were incapable of replicating the conformational changes that occur in both ligand and protein as a consequence of binding, a phenomenon referred to as induced fit (Chapter 7).

Imbuing proteins and ligands with conformational flexibility requires massive computing power, however. Hybrid approaches have thus evolved that employ a set, or ensemble, of templates representing slightly different conformations of the protein (Figure 10–3) and either ensembles of ligand conformers (Figure 10–4) or ligands in which only a few select bonds are permitted to rotate freely. Once the set of potential ligands has been narrowed, more sophisticated docking analyses can be undertaken to identify high-affinity ligands able to

FIGURE 10–3 Two-dimensional representation of a set of conformers of a protein. Notice how the shape of the binding site changes.

interact with their protein target across the latter's spectrum of conformational states.

If no structural template is available for the protein of interest, computers can be used to assist the search for high-affinity inhibitors by calculating and projecting **structure-activity relationships (SARs)**. Binding affinities for several known inhibitors are compared and contrasted to determine the positive or negative thermodynamic contributions that specific chemical features contribute to ligand binding. This information is used to identify compounds offering the best combination of features.

SYSTEMS BIOLOGY & VIRTUAL CELLS

What if a scientist could detect, in a few moments, the effect of inhibiting a particular enzyme, of replacing a particular gene,

FIGURE 10–4 Conformers of a simple ligand. Shown are three of the many different conformations of glucose, commonly referred to as chair (**top**), twist boat (**middle**), and half chair (**bottom**). Note the differences not only in shape and compactness but in the position of the hydroxyl groups, potential participants in hydrogen bonds, as highlighted in red.

the response of a muscle cell to insulin, the proliferation of a cancer cell, or the production of beta amyloid by entering the appropriate query into a computer? The goal of **systems biology** is to construct computer models which faithfully reproduce and can predict the behavior of specific functional units that range from the enzymes and metabolites in a biosynthetic pathway to the network of proteins that controls the cell division cycle, to entire cells and organisms.

By constructing virtual molecular networks, scientists have been able to determine how cyanobacteria construct a reliable circadian clock using only four proteins. Models of the T cell receptor signaling pathway have revealed how its molecular circuitry has been arranged to produce switch-like responses upon stimulation by agonist peptide-major histocompatability complex (MHC) complexes on an antigen-presenting cell. Just as a person constructs a jigsaw puzzle, in part, by surveying the remaining pieces for matches to the gaps in the puzzle, so too scientists can use the gaps encountered in modeling molecular and cellular systems to guide the identification and annotation of the remaining protein pieces. This reverse engineering approach has been successfully used to define the function of type II glycerate 2-kinases in bacteria and to identify "missing" folate synthesis and transport genes in plants. Recently, scientists have been able to successfully create a sustainable metabolic network, composed of nearly two hundred proteins—an important step toward the creation of a **virtual cell.**

The "holy grail" of systems biologists is to replicate the behavior of living human cells *in silico*. The potential benefits of such virtual cells will be enormous. Not only will they permit

optimum sites for therapeutic intervention to be identified in a rapid and unbiased manner, but unintended side effects may be revealed prior to the decision to invest time and resources in the synthesis, analysis, and trials of a potential pharmacophore. The ability to conduct fast, economic toxicological screening of materials ranging from herbicides to cosmetics will benefit human health. Virtual cells can also aid in diagnosis. By manipulating a virtual cell to reproduce the metabolic profile of a patient, underlying genetic abnormalities may be revealed. The interplay of the various environmental, dietary, and genetic factors that contribute to multifactorial diseases such as cancer can be systematically analyzed. Preliminary trials of potential gene therapies can be assessed safely and rapidly *in silico*.

The duplication of a living cell *in silico* represents an extremely formidable undertaking. Not only must the virtual cell possess all of the proteins and metabolites for the type of cell to be modeled (eg, from brain, liver, nerve, muscle, or adipose), but these must be present in the appropriate concentration and subcellular location. The model must also account for the functional dynamics of its components, binding affinities, catalytic efficiency, covalent modifications, etc. To render a virtual cell capable of dividing or differentiating will entail a further quantum leap in complexity and sophistication.

CONCLUSION

The rapidly evolving fields of bioinformatics and computational biology hold unparalleled promise for the future both of medicine and of basic biology. Some promises are at present perceived clearly, others dimly, while yet others remain unimagined. A major objective of computational biologists is to develop computational tools that will enhance the efficiency, effectiveness, and speed of drug development. There seems little doubt that their impact on medical practice in the 21st century will equal or surpass that of the discovery of bacterial pathogenesis in the 19th century.

SUMMARY

- Genomics has yielded a massive quantity of information of great potential value to scientists and physicians.
- Bioinformatics involves the design of computer algorithms and construction of databases that enable biomedical scientists to access and analyze the growing avalanche of biomedical data.
- Major challenges in the construction of user-friendly databases include devising means for storing and organizing complex data that accommodate a wide range of potential search criteria.
- The goal of the Encode Project is to identify all the functional elements within the human genome.
- The HapMap, *Entrez* Gene, and dbGAP databases contain data concerning the relation of genetic mutations to pathological conditions.
- Computational biology uses computer algorithms to identify unknown proteins and conduct virtual experiments.
- BLAST is used to identify unknown proteins and genes by searching for sequence homologs of known function.

- Computational biologists are developing programs that will predict the three-dimensional structure of proteins directly from their primary sequence.

- Computer-aided drug design speeds drug discovery by trying to dock potential inhibitors to selected protein targets *in silico*.

- A major goal of systems biologists is to create faithful models of individual pathways and networks in order to elucidate functional principles and perform virtual experiments.

- The ultimate goal of systems biologists is to create virtual cells that can be used to more safely and efficiently diagnose and treat diseases, particularly those of a multifactorial nature.

REFERENCES

Altschul, SF et al: Basic local alignment search tool. J Mol Biol 1990;215:403.

Carroll SB: Genetics and the making of *Homo sapiens*. Nature 2003;422:849.

Collins FS, Barker AD: Mapping the human cancer genome. Sci Am 2007;296:50.

Collins FS et al: A vision for the future of genomics research. A blueprint for the genomic era. Nature 2003;422:835.

Couzin J: The HapMap gold rush: researchers mine a rich deposit. Science 2006;312:1131.

Cravat BF et al: Activity-based protein profiling: from enzyme chemistry to proteomic chemistry. Annu Rev Biochem 2008;77:383.

Debes JD, Urrutia R: Bioinformatics tools to understand human diseases. Surgery 2004;135:579.

Dill KA et al: The protein folding problem: when will it be solved? Curr Opin Struct Biol 2007;17:342.

Ekins S, Mestres J, Testa B: *In silico* pharmacology for drug discovery: applications to targets and beyond. Br J Pharmacol 2007;152:21.

Ekins S, Mestres J, Testa B: *In silico* pharmacology for drug discovery: methods for virtual ligand screening and profiling. Br J Pharmacol 2007;152:9.

Hotopp JCD et al: Comparative genomics of *Neisseria meningitides*: core genome, islands of horizontal transfer and pathogen-specific genes. Microbiology 2006;152:3733.

Kim JH: Bioinformatics and genomic medicine. Genet Med 2002;4:62S.

Koonin EV, Galperin MY: *Sequence—Evolution—Function. Computational Approaches to Comparative Genomics*. Kluwer Academic Publishers, 2003.

McInnes C: Virtual screening strategies in drug discovery. Curr Opin Cell Biol 2007;11:494.

Sansom C: Genes and disease. The Scientist 2008;30:34.

Slepchencko BM et al: Quantitative cell biology with the Virtual Cell. Trends Cell Biol 2003;13:570.

Villoutreix BO et al: Free resources to assist structure-based virtual ligand screening experiments. Curr Protein Pept Sci 2007;8:381.

C H A P T E R

11

Bioenergetics: The Role of ATP

Kathleen M. Botham, PhD, DSc & Peter A. Mayes, PhD, DSc

BIOMEDICAL IMPORTANCE

Bioenergetics, or biochemical thermodynamics, is the study of the energy changes accompanying biochemical reactions. Biologic systems are essentially **isothermic** and use chemical energy to power living processes. How an animal obtains suitable fuel from its food to provide this energy is basic to the understanding of normal nutrition and metabolism. Death from **starvation** occurs when available energy reserves are depleted, and certain forms of malnutrition are associated with energy imbalance (**marasmus**). Thyroid hormones control the rate of energy release (metabolic rate), and disease results when they malfunction. Excess storage of surplus energy causes **obesity,** an increasingly common disease of Western society, which predisposes to many diseases, including cardiovascular disease and diabetes mellitus type 2, and lowers life expectancy.

FREE ENERGY IS THE USEFUL ENERGY IN A SYSTEM

Gibbs change in free energy (ΔG) is that portion of the total energy change in a system that is available for doing work—ie, the useful energy, also known as the chemical potential.

Biologic Systems Conform to the General Laws of Thermodynamics

The first law of thermodynamics states that **the total energy of a system, including its surroundings, remains constant.** It implies that within the total system, energy is neither lost nor gained during any change. However, energy may be transferred from one part of the system to another or may be transformed into another form of energy. In living systems, chemical energy may be transformed into heat or into electrical, radiant, or mechanical energy.

The second law of thermodynamics states that **the total entropy of a system must increase if a process is to occur spontaneously. Entropy** is the extent of disorder or randomness of the system and becomes maximum as equilibrium is approached. Under conditions of constant temperature and pressure, the relationship between the free-energy change (ΔG) of a reacting system and the change in entropy (ΔS) is expressed by the following equation, which combines the two laws of thermodynamics:

$$\Delta G = \Delta H - T\Delta S$$

where ΔH is the change in **enthalpy** (heat) and T is the absolute temperature.

In biochemical reactions, because ΔH is approximately equal to ΔE, the total change in internal energy of the reaction, the above relationship may be expressed in the following way:

$$\Delta G = \Delta E - T\Delta S$$

If ΔG is negative, the reaction proceeds spontaneously with loss of free energy; ie, it is **exergonic.** If, in addition, ΔG is of great magnitude, the reaction goes virtually to completion and is essentially irreversible. On the other hand, if ΔG is positive, the reaction proceeds only if free energy can be gained; ie, it is **endergonic.** If, in addition, the magnitude of ΔG is great, the system is stable, with little or no tendency for a reaction to occur. If ΔG is zero, the system is at equilibrium and no net change takes place.

When the reactants are present in concentrations of 1.0 mol/L, ΔG^0 is the standard free-energy change. For biochemical reactions, a standard state is defined as having a pH of 7.0. The standard free-energy change at this standard state is denoted by $\Delta G^{0'}$.

The standard free-energy change can be calculated from the equilibrium constant K_{eq}.

$$\Delta G^{0'} = -RT \ln K'_{eq}$$

where R is the gas constant and T is the absolute temperature (Chapter 8). It is important to note that the actual ΔG may be larger or smaller than $\Delta G^{0'}$ depending on the concentrations of the various reactants, including the solvent, various ions, and proteins.

In a biochemical system, an enzyme only speeds up the attainment of equilibrium; it never alters the final concentrations of the reactants at equilibrium.

ENDERGONIC PROCESSES PROCEED BY COUPLING TO EXERGONIC PROCESSES

The vital processes—eg, synthetic reactions, muscular contraction, nerve impulse conduction, active transport—obtain energy by chemical linkage, or **coupling,** to oxidative reactions. In its simplest form, this type of coupling may be represented as shown in Figure 11–1. The conversion of metabolite A to metabolite B occurs with release of free energy and is coupled to another reaction in which free energy is required to convert metabolite C to metabolite D. The terms **exergonic** and **endergonic,** rather than the normal chemical terms "exothermic" and "endothermic," are used to indicate that a process is accompanied by loss or gain, respectively, of free energy in any form, not necessarily as heat. In practice, an endergonic process cannot exist independently, but must be a component of a coupled exergonic-endergonic system where the overall net change is exergonic. The exergonic reactions are termed **catabolism** (generally, the breakdown or oxidation of fuel

molecules), whereas the synthetic reactions that build up substances are termed **anabolism.** The combined catabolic and anabolic processes constitute **metabolism.**

If the reaction shown in Figure 11–1 is to go from left to right, then the overall process must be accompanied by loss of free energy as heat. One possible mechanism of coupling could be envisaged if a common obligatory intermediate (I) took part in both reactions, ie,

$$A + C \rightarrow I \rightarrow B + D$$

Some exergonic and endergonic reactions in biologic systems are coupled in this way. This type of system has a built-in mechanism for biologic control of the rate of oxidative processes, since the common obligatory intermediate allows the rate of utilization of the product of the synthetic path (D) to determine by mass action the rate at which A is oxidized. Indeed, these relationships supply a basis for the concept of **respiratory control**, the process that prevents an organism from burning out of control. An extension of the coupling concept is provided by dehydrogenation reactions, which are coupled to hydrogenations by an intermediate carrier (Figure 11–2).

An alternative method of coupling an exergonic to an endergonic process is to synthesize a compound of high-energy potential in the exergonic reaction and to incorporate this new compound into the endergonic reaction, thus effecting a transference of free energy from the exergonic to the endergonic pathway (Figure 11–3). The biologic advantage of this mechanism is that the compound of high potential energy, ~Ⓔ, unlike I in the previous system, need not be structurally

FIGURE 11–2 Coupling of dehydrogenation and hydrogenation reactions by an intermediate carrier.

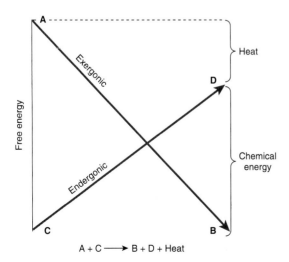

FIGURE 11–1 Coupling of an exergonic to an endergonic reaction.

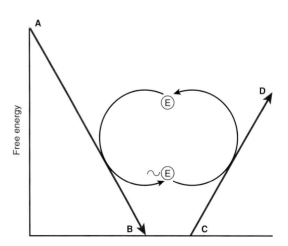

FIGURE 11–3 Transfer of free energy from an exergonic to an endergonic reaction via a high-energy intermediate compound (~Ⓔ).

related to A, B, C, or D, allowing Ⓔ to serve as a transducer of energy from a wide range of exergonic reactions to an equally wide range of endergonic reactions or processes, such as biosyntheses, muscular contraction, nervous excitation, and active transport. In the living cell, the principal high-energy intermediate or carrier compound (designated ~Ⓔ in Figure 11–3) is **adenosine triphosphate (ATP).**

HIGH-ENERGY PHOSPHATES PLAY A CENTRAL ROLE IN ENERGY CAPTURE AND TRANSFER

In order to maintain living processes, all organisms must obtain supplies of free energy from their environment. **Autotrophic** organisms utilize simple exergonic processes; eg, the energy of sunlight (green plants), the reaction $Fe^{2+} \rightarrow Fe^{3+}$ (some bacteria). On the other hand, **heterotrophic** organisms obtain free energy by coupling their metabolism to the breakdown of complex organic molecules in their environment. In all these organisms, ATP plays a central role in the transference of free energy from the exergonic to the endergonic processes (Figure 11–3). ATP is a nucleoside triphosphate containing adenine, ribose, and three phosphate groups. In its reactions in the cell, it functions as the Mg^{2+} complex (Figure 11–4).

The importance of phosphates in intermediary metabolism became evident with the discovery of the role of ATP, adenosine diphosphate (ADP), and inorganic phosphate (P_i) in glycolysis (Chapter 18).

The Intermediate Value for the Free Energy of Hydrolysis of ATP Has Important Bioenergetic Significance

The standard free energy of hydrolysis of a number of biochemically important phosphates is shown in Table 11–1. An estimate of the comparative tendency of each of the phosphate groups to transfer to a suitable acceptor may be obtained from the $\Delta G^{0'}$ of hydrolysis at 37°C. The value for the hydrolysis of the terminal phosphate of ATP divides the list into two groups. **Low-energy phosphates,** exemplified by the ester phosphates

TABLE 11–1   Standard Free Energy of Hydrolysis of Some Organophosphates of Biochemical Importance

Compound	$\Delta G^{0'}$ kJ/mol	kcal/mol
Phosphoenolpyruvate	−61.9	−14.8
Carbamoyl phosphate	−51.4	−12.3
1,3-Bisphosphoglycerate (to 3-phosphoglycerate)	−49.3	−11.8
Creatine phosphate	−43.1	−10.3
ATP → AMP + PP_i	−32.2	−7.7
ATP → ADP + P_i	−30.5	−7.3
Glucose 1-phosphate	−20.9	−5.0
PP_i	−19.2	−4.6
Fructose 6-phosphate	−15.9	−3.8
Glucose 6-phosphate	−13.8	−3.3
Glycerol 3-phosphate	−9.2	−2.2

Abbreviations: PP_i, pyrophosphate; P_i, inorganic orthophosphate.

Note: All values taken from Jencks (1976), except that for PP_i which is from Frey and Arabshahi (1995). Values differ between investigators, depending on the precise conditions under which the measurements were made.

found in the intermediates of glycolysis, have $G^{0'}$ values smaller than that of ATP, while in **high-energy phosphates** the value is higher than that of ATP. The components of this latter group, including ATP, are usually anhydrides (eg, the 1-phosphate of 1,3-bisphosphoglycerate), enolphosphates (eg, phosphoenolpyruvate), and phosphoguanidines (eg, creatine phosphate, arginine phosphate). The intermediate position of ATP allows it to play an important role in energy transfer. The high free-energy change on hydrolysis of ATP is due to relief of charge repulsion of adjacent negatively charged oxygen atoms and to stabilization of the reaction products, especially phosphate, as resonance hybrids. Other "high-energy compounds" are thiol esters involving coenzyme A (eg, acetyl-CoA), acyl carrier protein, amino acid esters involved in protein synthesis, S-adenosylmethionine (active methionine), UDPGlc (uridine diphosphate glucose), and PRPP (5-phosphoribosyl-1-pyrophosphate).

High-Energy Phosphates Are Designated by ~Ⓟ

The symbol ~Ⓟ indicates that the group attached to the bond, on transfer to an appropriate acceptor, results in transfer of the larger quantity of free energy. For this reason, the term **group transfer potential**, rather than "high-energy bond," is preferred by some. Thus, ATP contains two high-energy phosphate groups and ADP contains one, whereas the phosphate in AMP (adenosine monophosphate) is of the low-energy type, since it is a normal ester link (Figure 11–5).

FIGURE 11–4 Adenosine triphosphate (ATP) shown as the magnesium complex. ADP forms a similar complex with Mg^{2+}.

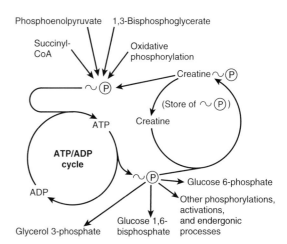

FIGURE 11–6 Role of ATP/ADP cycle in transfer of high-energy phosphate.

maintained, but when the ATP/ADP ratio is high, their concentration can increase to act as a store of high-energy phosphate (Figure 11–7).

When ATP acts as a phosphate donor to form those compounds of lower free energy of hydrolysis (Table 11–1), the phosphate group is invariably converted to one of low energy, eg,

$$\text{Glycerol} + \text{Adenosine} - \textcircled{P} \sim \textcircled{P} \sim \textcircled{P} \xrightarrow{\boxed{\text{GLYCEROL KINASE}}}$$

$$\text{Glycerol} - \textcircled{P} + \text{Adenosine} - \textcircled{P} \sim \textcircled{P}$$

ATP Allows the Coupling of Thermodynamically Unfavorable Reactions to Favorable Ones

The phosphorylation of glucose to glucose 6-phosphate, the first reaction of glycolysis (Figure 18–2), is highly endergonic and cannot proceed under physiologic conditions.

$$\text{Glucose} + P_i \rightarrow \text{Glucose 6-phosphate} + H_2O \qquad (1)$$

$$(\Delta G^{0'} = +13.8 \text{ kJ/mol})$$

FIGURE 11–7 Transfer of high-energy phosphate between ATP and creatine.

O⁻   O⁻   O⁻
| | |
Adenosine — O — P — O∿P — O∿P — O⁻
‖   ‖   ‖
O   O   O

or Adenosine — ⓟ∿ⓟ∿ⓟ

Adenosine triphosphate (ATP)

O⁻   O⁻
| |
Adenosine — O — P — O∿P — O⁻
‖   ‖
O   O

or Adenosine — ⓟ∿ⓟ

Adenosine diphosphate (ADP)

O⁻
|
Adenosine — O — P — O⁻
‖
O

or Adenosine — ⓟ

Adenosine monophosphate (AMP)

FIGURE 11–5 Structure of ATP, ADP, and AMP showing the position and the number of high-energy phosphates (~ⓟ).

HIGH-ENERGY PHOSPHATES ACT AS THE "ENERGY CURRENCY" OF THE CELL

ATP is able to act as a donor of high-energy phosphate to form those compounds below it in Table 11–1. Likewise, with the necessary enzymes, ADP can accept high-energy phosphate to form ATP from those compounds above ATP in the table. In effect, an **ATP/ADP cycle** connects those processes that generate ~ⓟ to those processes that utilize ~ⓟ (Figure 11–6), continuously consuming and regenerating ATP. This occurs at a very rapid rate, since the total ATP/ADP pool is extremely small and sufficient to maintain an active tissue for only a few seconds.

There are three major sources of ~ⓟ taking part in **energy conservation** or **energy capture**:

1. **Oxidative phosphorylation:** The greatest quantitative source of ~ⓟ in aerobic organisms. Free energy comes from respiratory chain oxidation using molecular O_2 within mitochondria (Chapter 12).
2. **Glycolysis:** A net formation of two ~ⓟ results from the formation of lactate from one molecule of glucose, generated in two reactions catalyzed by phosphoglycerate kinase and pyruvate kinase, respectively (Figure 18–2).
3. **The citric acid cycle:** One ~ⓟ is generated directly in the cycle at the succinate thiokinase step (Figure 17–3).

Phosphagens act as storage forms of high-energy phosphate and include creatine phosphate, which occurs in vertebrate skeletal muscle, heart, spermatozoa, and brain; and arginine phosphate, which occurs in invertebrate muscle. When ATP is rapidly being utilized as a source of energy for muscular contraction, phosphagens permit its concentrations to be

To take place, the reaction must be coupled with another—more exergonic—reaction such as the hydrolysis of the terminal phosphate of ATP.

$$ATP \rightarrow ADP + P_i \ (\Delta G^{0'} = -30.5 \ kJ/mol) \qquad (2)$$

When (1) and (2) are coupled in a reaction catalyzed by hexokinase, phosphorylation of glucose readily proceeds in a highly exergonic reaction that under physiologic conditions is irreversible. Many "activation" reactions follow this pattern.

Adenylyl Kinase (Myokinase) Interconverts Adenine Nucleotides

This enzyme is present in most cells. It catalyzes the following reaction:

$$ATP + AMP \xleftrightarrow{\text{ADENYLYL KINASE}} 2ADP$$

This allows:

1. High-energy phosphate in ADP to be used in the synthesis of ATP.
2. AMP, formed as a consequence of several activating reactions involving ATP, to be recovered by rephosphorylation to ADP.
3. AMP to increase in concentration when ATP becomes depleted and act as a metabolic (allosteric) signal to increase the rate of catabolic reactions, which in turn lead to the generation of more ATP (Chapter 20).

When ATP Forms AMP, Inorganic Pyrophosphate (PP$_i$) Is Produced

ATP can also be hydrolyzed directly to AMP, with the release of PP$_i$ (Table 11–1). This occurs, for example, in the activation of long-chain fatty acids (Chapter 22):

$$ATP + CoA \cdot SH + R \cdot COOH \xrightarrow{\text{ACYL-CoA SYNTHETASE}} AMP + PP_i + R \cdot CO - SCoA$$

This reaction is accompanied by loss of free energy as heat, which ensures that the activation reaction will go to the right; and is further aided by the hydrolytic splitting of PP$_i$, catalyzed by **inorganic pyrophosphatase,** a reaction that itself has a large $\Delta G^{0'}$ of −19.2 kJ/mol. Note that activations via the pyrophosphate pathway result in the loss of two ~P rather than one, as occurs when ADP and P$_i$ are formed.

$$PP_i + H_2O \xrightarrow{\text{INORGANIC PYROPHOSPHATASE}} 2P_i$$

A combination of the above reactions makes it possible for phosphate to be recycled and the adenine nucleotides to interchange (Figure 11–8).

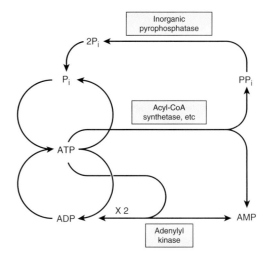

FIGURE 11–8 Phosphate cycles and interchange of adenine nucleotides.

Other Nucleoside Triphosphates Participate in the Transfer of High-Energy Phosphate

By means of the enzyme **nucleoside diphosphate kinase,** UTP, GTP, and CTP can be synthesized from their diphosphates, eg,

$$ATP + UDP \xleftrightarrow{\text{NUCLEOSIDE DIPHOSPHATE KINASE}} ADP + UTP \ \text{(uridine triphosphate)}$$

All of these triphosphates take part in phosphorylations in the cell. Similarly, specific nucleoside monophosphate kinases catalyze the formation of nucleoside diphosphates from the corresponding monophosphates.

Thus, adenylyl kinase is a specialized monophosphate kinase.

SUMMARY

- Biologic systems use chemical energy to power living processes.
- Exergonic reactions take place spontaneously with loss of free energy (ΔG is negative). Endergonic reactions require the gain of free energy (ΔG is positive) and occur only when coupled to exergonic reactions.
- ATP acts as the "energy currency" of the cell, transferring free energy derived from substances of higher energy potential to those of lower energy potential.

REFERENCES

de Meis L: The concept of energy-rich phosphate compounds: water, transport ATPases, and entropy energy. Arch Biochem Biophys 1993;306:287.

Frey PA, Arabshahi A: Standard free-energy change for the hydrolysis of the alpha, beta-phosphoanhydride bridge in ATP. Biochemistry 1995;34:11307.

Harold FM: *The Vital Force: A Study of Bioenergetics.* Freeman, 1986.

Harris DA: *Bioenergetics at a Glance: An Illustrated Introduction.* Blackwell Publishing, 1995.

Haynie D: *Biological Thermodynamics.* Cambridge University Press, 2008.

Jencks WP: Free energies of hydrolysis and decarboxylation. In: *Handbook of Biochemistry and Molecular Biology,* vol 1. *Physical and Chemical Data.* Fasman GD (editor). CRC Press, 1976:296–304.

Klotz IM: *Introduction to Biomolecular Energetics.* Academic Press, 1986.

Nicholls D, Ferguson F: *Bioenergetics.* Elsevier, 2003.

Biologic Oxidation

Kathleen M. Botham, PhD, DSc & Peter A. Mayes, PhD, DSc

BIOMEDICAL IMPORTANCE

Chemically, **oxidation** is defined as the removal of electrons and **reduction** as the gain of electrons. Thus, oxidation is always accompanied by reduction of an electron acceptor. This principle of oxidation-reduction applies equally to biochemical systems and is an important concept underlying understanding of the nature of biologic oxidation. Note that many biologic oxidations can take place without the participation of molecular oxygen, eg, dehydrogenations. The life of higher animals is absolutely dependent upon a supply of oxygen for **respiration,** the process by which cells derive energy in the form of ATP from the controlled reaction of hydrogen with oxygen to form water. In addition, molecular oxygen is incorporated into a variety of substrates by enzymes designated as **oxygenases;** many drugs, pollutants, and chemical carcinogens (xenobiotics) are metabolized by enzymes of this class, known as the **cytochrome P450 system.** Administration of oxygen can be lifesaving in the treatment of patients with respiratory or circulatory failure.

FREE ENERGY CHANGES CAN BE EXPRESSED IN TERMS OF REDOX POTENTIAL

In reactions involving oxidation and reduction, the free energy change is proportionate to the tendency of reactants to donate or accept electrons. Thus, in addition to expressing free energy change in terms of $\Delta G^{0'}$ (Chapter 11), it is possible, in an analogous manner, to express it numerically as an **oxidation-reduction** or **redox potential** (E'_0). The redox potential of a system (E_0) is usually compared with the potential of the hydrogen electrode (0.0 volts at pH 0.0). However, for biologic systems, the redox potential (E'_0) is normally expressed at pH 7.0, at which pH the electrode potential of the hydrogen electrode is −0.42 volts. The redox potentials of some redox systems of special interest in mammalian biochemistry are shown in Table 12–1. The relative positions of redox systems in the table allow prediction of the direction of flow of electrons from one redox couple to another.

Enzymes involved in oxidation and reduction are called **oxidoreductases** and are classified into four groups: **oxidases, dehydrogenases, hydroperoxidases,** and **oxygenases.**

OXIDASES USE OXYGEN AS A HYDROGEN ACCEPTOR

Oxidases catalyze the removal of hydrogen from a substrate using oxygen as a hydrogen acceptor.* They form water or hydrogen peroxide as a reaction product (Figure 12–1).

Some Oxidases Contain Copper

Cytochrome oxidase is a hemoprotein widely distributed in many tissues, having the typical heme prosthetic group present in myoglobin, hemoglobin, and other cytochromes (Chapter 6). It is the terminal component of the chain of respiratory carriers found in mitochondria (Chapter 13) and transfers electrons resulting from the oxidation of substrate molecules by dehydrogenases to their final acceptor, oxygen. The enzyme is poisoned by carbon monoxide, cyanide, and hydrogen sulfide. It has also been termed "cytochrome a_3." However, it is now known that the heme a_3 is combined with another heme, heme a, in a single protein to form the cytochrome oxidase enzyme complex and so it is more correctly termed **cytochrome aa_3.** It contains two molecules of heme, each having one Fe atom that oscillates between Fe^{3+} and Fe^{2+} during oxidation and reduction. Furthermore, two atoms of Cu are present, each associated with a heme unit.

Other Oxidases Are Flavoproteins

Flavoprotein enzymes contain **flavin mononucleotide (FMN)** or **flavin adenine dinucleotide (FAD)** as prosthetic groups. FMN and FAD are formed in the body from the vitamin **riboflavin** (Chapter 44). FMN and FAD are usually tightly—but not covalently—bound to their respective apoenzyme proteins. Metalloflavoproteins contain one or more metals

---

*The term "oxidase" is sometimes used collectively to denote all enzymes that catalyze reactions involving molecular oxygen.

TABLE 12–1   Some Redox Potentials of Special Interest in Mammalian Oxidation Systems

System	E'_0 Volts
H^+/H_2	−0.42
$NAD^+/NADH$	−0.32
Lipoate; ox/red	−0.29
Acetoacetate/3-hydroxybutyrate	−0.27
Pyruvate/lactate	−0.19
Oxaloacetate/malate	−0.17
Fumarate/succinate	+0.03
Cytochrome b; Fe^{3+}/Fe^{2+}	+0.08
Ubiquinone; ox/red	+0.10
Cytochrome c_1; Fe^{3+}/Fe^{2+}	+0.22
Cytochrome a; Fe^{3+}/Fe^{2+}	+0.29
Oxygen/water	+0.82

FIGURE 12–1 Oxidation of a metabolite catalyzed by an oxidase (**A**) forming H_2O, (**B**) forming H_2O_2.

as essential cofactors. Examples of flavoprotein enzymes include L-**amino acid oxidase,** an FMN-linked enzyme found in kidney with general specificity for the oxidative deamination of the naturally occurring L-amino acids; **xanthine oxidase,** which contains molybdenum and plays an important role in the conversion of purine bases to uric acid (Chapter 33), and is of particular significance in uricotelic animals (Chapter 28); and **aldehyde dehydrogenase,** an FAD-linked enzyme present in mammalian livers, which contains molybdenum and non-heme iron and acts upon aldehydes and N-heterocyclic substrates. The mechanisms of oxidation and reduction of these enzymes are complex. Evidence suggests a two-step reaction as shown in Figure 12–2.

DEHYDROGENASES CANNOT USE OXYGEN AS A HYDROGEN ACCEPTOR

There are a large number of enzymes in this class. They perform two main functions:

1. Transfer of hydrogen from one substrate to another in a coupled oxidation-reduction reaction (Figure 12–3). These dehydrogenases are specific for their substrates but often utilize common coenzymes or hydrogen carriers, eg, NAD^+. Since the reactions are reversible, these properties enable reducing equivalents to be freely transferred within the cell. This type of reaction, which enables one substrate to be oxidized at the expense of another, is particularly useful in enabling oxidative processes to occur in the absence of oxygen, such as during the anaerobic phase of glycolysis (Figure 18–2).

2. As components in the **respiratory chain** of electron transport from substrate to oxygen (Figure 13–3).

Many Dehydrogenases Depend on Nicotinamide Coenzymes

These dehydrogenases use **nicotinamide adenine dinucleotide (NAD⁺)** or **nicotinamide adenine dinucleotide phosphate (NADP⁺)**—or both—which are formed in the body from the vitamin **niacin** (Chapter 44). The coenzymes are reduced by the specific substrate of the dehydrogenase and reoxidized by a suitable electron acceptor (Figure 12–4). They may freely and reversibly dissociate from their respective apoenzymes.

Generally, **NAD-linked dehydrogenases** catalyze oxidoreduction reactions in the oxidative pathways of metabolism, particularly in glycolysis (Chapter 18), in the citric acid cycle (Chapter 17), and in the respiratory chain of mitochondria (Chapter 13). NADP-linked dehydrogenases are found characteristically in reductive syntheses, as in the extramitochondrial pathway of fatty acid synthesis (Chapter 23) and steroid synthesis Chapter 26)—and also in the pentose phosphate pathway (Chapter 21).

Other Dehydrogenases Depend on Riboflavin

The **flavin groups** associated with these dehydrogenases are similar to FMN and FAD occurring in oxidases. They are generally more tightly bound to their apoenzymes than are the nicotinamide coenzymes. Most of the **riboflavin-linked**

FIGURE 12–2 Oxidoreduction of isoalloxazine ring in flavin nucleotides via a semiquinone (free radical) intermediate (center).

FIGURE 12–3 Oxidation of a metabolite catalyzed by coupled dehydrogenases.

dehydrogenases are concerned with electron transport in (or to) the respiratory chain (Chapter 13). **NADH dehydrogenase** acts as a carrier of electrons between NADH and the components of higher redox potential (Figure 13–3). Other dehydrogenases such as **succinate dehydrogenase, acyl-CoA dehydrogenase**, and **mitochondrial glycerol-3-phosphate dehydrogenase** transfer reducing equivalents directly from the substrate to the respiratory chain (Figure 13–5). Another role of the flavin-dependent dehydrogenases is in the dehydrogenation (by **dihydrolipoyl dehydrogenase**) of reduced lipoate, an intermediate in the oxidative decarboxylation of pyruvate and α-ketoglutarate (Figures 13–5 & 18–5). The **electron-transferring flavoprotein (ETF)** is an intermediary carrier of electrons between acyl-CoA dehydrogenase and the respiratory chain (Figure 13–5).

Cytochromes May Also Be Regarded as Dehydrogenases

The **cytochromes** are iron-containing hemoproteins in which the iron atom oscillates between Fe^{3+} and Fe^{2+} during oxidation and reduction. Except for cytochrome oxidase (previously described), they are classified as dehydrogenases. In the respiratory chain, they are involved as carriers of electrons from flavoproteins on the one hand to cytochrome oxidase on the other (Figure 13–5). Several identifiable cytochromes occur in the respiratory chain, ie, cytochromes b, c_1, c, and cytochrome

oxidase. Cytochromes are also found in other locations, eg, the endoplasmic reticulum (cytochromes P450 and b_5), and in plant cells, bacteria, and yeasts.

HYDROPEROXIDASES USE HYDROGEN PEROXIDE OR AN ORGANIC PEROXIDE AS SUBSTRATE

Two type of enzymes found both in animals and plants fall into this category: **peroxidases** and **catalase.**

Hydroperoxidases protect the body against harmful peroxides. Accumulation of peroxides can lead to generation of free radicals, which in turn can disrupt membranes and perhaps cause diseases including cancer and atherosclerosis (see Chapters 15 & 44).

Peroxidases Reduce Peroxides Using Various Electron Acceptors

Peroxidases are found in milk and in leukocytes, platelets, and other tissues involved in eicosanoid metabolism (Chapter 23). The prosthetic group is protoheme. In the reaction catalyzed by peroxidase, hydrogen peroxide is reduced at the expense of several substances that will act as electron acceptors, such as ascorbate, quinones, and cytochrome c. The reaction catalyzed by peroxidase is complex, but the overall reaction is as follows:

$$H_2O_2 + AH_2 \xrightarrow{\text{PEROXIDASE}} 2H_2O + A$$

In erythrocytes and other tissues, the enzyme **glutathione peroxidase,** containing **selenium** as a prosthetic group, catalyzes the destruction of H_2O_2 and lipid hydroperoxides through the conversion of reduced glutathione to its oxidized form, protecting membrane lipids and hemoglobin against oxidation by peroxides (Chapter 21).

FIGURE 12–4 Mechanism of oxidation and reduction of nicotinamide coenzymes. There is stereospecificity about position 4 of nicotinamide when it is reduced by a substrate AH_2. One of the hydrogen atoms is removed from the substrate as a hydrogen nucleus with two electrons (hydride ion, H^-) and is transferred to the 4 position, where it may be attached in either the A or the B form according to the specificity determined by the particular dehydrogenase catalyzing the reaction. The remaining hydrogen of the hydrogen pair removed from the substrate remains free as a hydrogen ion.

$$NAD^+ + AH_2 \rightleftharpoons NADH + H^+ + A$$

Catalase Uses Hydrogen Peroxide as Electron Donor & Electron Acceptor

Catalase is a hemoprotein containing four heme groups. In addition to possessing peroxidase activity, it is able to use one molecule of H_2O_2 as a substrate electron donor and another molecule of H_2O_2 as an oxidant or electron acceptor.

$$2H_2O_2 \xrightarrow{\boxed{\text{CATALASE}}} 2H_2O_2 + O_2$$

Under most conditions in vivo, the peroxidase activity of catalase seems to be favored. Catalase is found in blood, bone marrow, mucous membranes, kidney, and liver. It functions to destroy of hydrogen peroxide formed by the action of oxidases. **Peroxisomes** are found in many tissues, including liver. They are rich in oxidases and in catalase. Thus, the enzymes that produce H_2O_2 are grouped with the enzyme that destroys it. However, mitochondrial and microsomal electron transport systems as well as xanthine oxidase must be considered as additional sources of H_2O_2.

OXYGENASES CATALYZE THE DIRECT TRANSFER & INCORPORATION OF OXYGEN INTO A SUBSTRATE MOLECULE

Oxygenases are concerned with the synthesis or degradation of many different types of metabolites. They catalyze the incorporation of oxygen into a substrate molecule in two steps: (1) oxygen is bound to the enzyme at the active site and (2) the bound oxygen is reduced or transferred to the substrate. Oxygenases may be divided into two subgroups, dioxygenases and monooxygenases.

Dioxygenases Incorporate Both Atoms of Molecular Oxygen into the Substrate

The basic reaction is shown below:

$$A + O_2 \rightarrow AO_2$$

Examples include the liver enzymes, **homogentisate dioxygenase** (oxidase) and **3-hydroxyanthranilate dioxygenase**

(oxidase), which contain iron; and L-**tryptophan dioxygenase** (tryptophan pyrolase) (Chapter 29), which utilizes heme.

Monooxygenases (Mixed-Function Oxidases, Hydroxylases) Incorporate Only One Atom of Molecular Oxygen into the Substrate

The other oxygen atom is reduced to water, an additional electron donor or cosubstrate (Z) being necessary for this purpose.

$$A-H + O_2 + ZH_2 \rightarrow A-OH + H_2O + Z$$

Cytochromes P450 Are Monooxygenases Important for the Detoxification of Many Drugs & for the Hydroxylation of Steroids

Cytochromes P450 are an important superfamily of heme-containing monooxgenases, and more than 50 such enzymes have been found in the human genome. These cytochromes are located mainly in the endoplasmic reticulum in the liver and intestine, but are also found in the mitochondria in some tissues. Both NADH and NADPH donate reducing equivalents for the reduction of these cytochromes (Figure 12–5), which in turn are oxidized by substrates in a series of enzymatic reactions collectively known as the **hydroxylase cycle** (Figure 12–6). In the endoplasmic reticulum of the liver, cytochromes P450 are found together with **cytochrome b_5** and have an important role in detoxification. The rate of detoxification of many medicinal drugs by cytochromes P450 determines the duration of their action. Benzpyrene, aminopyrine, aniline, morphine, and benzphetamine are hydroxylated, increasing their solubility and aiding their excretion. Many drugs such as phenobarbital have the ability to induce the synthesis of cytochromes P450.

Mitochondrial cytochrome P450 systems are found in steroidogenic tissues such as adrenal cortex, testis, ovary, and placenta and are concerned with the biosynthesis of steroid hormones from cholesterol (hydroxylation at C_{22} and C_{20} in side-chain cleavage and at the 11β and 18 positions). In addition, renal systems catalyzing 1α- and 24-hydroxylations of 25-hydroxycholecalciferol in vitamin D metabolism—and cholesterol 7α-hydroxylase and sterol 27-hydroxylase involved in bile acid biosynthesis in the liver (Chapter 26)—are P450 enzymes.

FIGURE 12–5 Electron transport chain in the endoplasmic reticulum. Cyanide (CN^-) inhibits the indicated step.

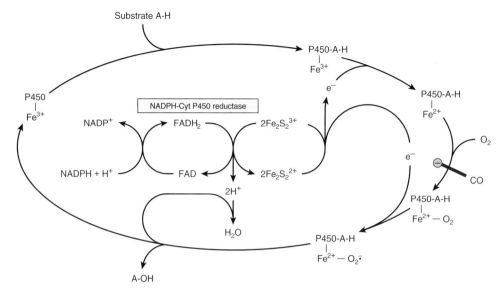

FIGURE 12–6 Cytochrome P450 hydroxylase cycle. The system shown is typical of steroid hydroxylases of the adrenal cortex. Liver microsomal cytochrome P450 hydroxylase does not require the iron-sulfur protein Fe$_2$S$_2$. Carbon monoxide (CO) inhibits the indicated step.

SUPEROXIDE DISMUTASE PROTECTS AEROBIC ORGANISMS AGAINST OXYGEN TOXICITY

Transfer of a single electron to O$_2$ generates the potentially damaging **superoxide anion free radical (O$_2^{\bar{\cdot}}$)**, the destructive effects of which are amplified by its giving rise to free-radical chain reactions (Chapter 15). The ease with which superoxide can be formed from oxygen in tissues and the occurrence of **superoxide dismutase,** the enzyme responsible for its removal in all aerobic organisms (although not in obligate anaerobes) indicate that the potential toxicity of oxygen is due to its conversion to superoxide.

Superoxide is formed when reduced flavins—present, for example, in xanthine oxidase—are reoxidized univalently by molecular oxygen.

$$Enz - Flavin - H_2 + O_2 \rightarrow Enz - Flavin - H + O_2^{\bar{\cdot}} + H^+$$

Superoxide can reduce oxidized cytochrome c

$$O_2^{\bar{\cdot}} + Cyt\,c\,(Fe^{3+}) \rightarrow O_2 + Cyt\,c\,(Fe^{2+})$$

or be removed by superoxide dismutase:

In this reaction, superoxide acts as both oxidant and reductant. Thus, superoxide dismutase protects aerobic organisms against the potential deleterious effects of superoxide. The enzyme occurs in all major aerobic tissues in the mitochondria and the cytosol. Although exposure of animals to an atmosphere of 100% oxygen causes an adaptive increase in superoxide dismutase, particularly in the lungs, prolonged exposure leads to lung damage and death. Antioxidants, eg,

α-tocopherol (vitamin E), act as scavengers of free radicals and reduce the toxicity of oxygen (Chapter 44).

SUMMARY

- In biologic systems, as in chemical systems, oxidation (loss of electrons) is always accompanied by reduction of an electron acceptor.

- Oxidoreductases have a variety of functions in metabolism; oxidases and dehydrogenases play major roles in respiration; hydroperoxidases protect the body against damage by free radicals; and oxygenases mediate the hydroxylation of drugs and steroids.

- Tissues are protected from oxygen toxicity caused by the superoxide free radical by the specific enzyme superoxide dismutase.

REFERENCES

Babcock GT, Wikstrom M: Oxygen activation and the conservation of energy in cell respiration. Nature 1992;356:301.

Coon MJ: Cytochrome P450: Nature's most versatile biological catalyst. Annu Rev Pharmacol Toxicol 2005;4:1.

Harris DA: *Bioenergetics at a Glance: An Illustrated Introduction.* Blackwell Publishing, 1995.

Johnson F, Giulivi C: Superoxide dismutases and their impact upon human health. Mol Aspects Med 2005;26.

Nicholls D, Ferguson F: *Bioenergetics.* Elsevier, 2003.

Raha S, Robinson BH: Mitochondria, oxygen free radicals, disease and aging. Trends Biochem Sci 2000;25:502.

Tyler DD: *The Mitochondrion in Health and Disease.* VCH Publishers, 1992.

The Respiratory Chain & Oxidative Phosphorylation

Kathleen M. Botham, PhD, DSc & Peter A. Mayes, PhD, DSc

BIOMEDICAL IMPORTANCE

Aerobic organisms are able to capture a far greater proportion of the available free energy of respiratory substrates than anaerobic organisms. Most of this takes place inside mitochondria, which have been termed the "powerhouses" of the cell. Respiration is coupled to the generation of the high-energy intermediate, ATP, by **oxidative phosphorylation**. A number of drugs (eg, **amobarbital**) and poisons (eg, **cyanide, carbon monoxide**) inhibit oxidative phosphorylation, usually with fatal consequences. Several inherited defects of mitochondria involving components of the respiratory chain and oxidative phosphorylation have been reported. Patients present with **myopathy** and **encephalopathy** and often have **lactic acidosis.**

SPECIFIC ENZYMES ACT AS MARKERS OF COMPARTMENTS SEPARATED BY THE MITOCHONDRIAL MEMBRANES

Mitochondria have an **outer membrane** that is permeable to most metabolites and an **inner membrane** that is selectively permeable, enclosing a **matrix** within (Figure 13–1). The outer membrane is characterized by the presence of various enzymes, including acyl-CoA synthetase and glycerolphosphate acyltransferase. Adenylyl kinase and creatine kinase are found in the intermembrane space. The phospholipid cardiolipin is concentrated in the inner membrane together with the enzymes of the respiratory chain, ATP synthase and various membrane transporters.

THE RESPIRATORY CHAIN OXIDIZES REDUCING EQUIVALENTS & ACTS AS A PROTON PUMP

Most of the energy liberated during the oxidation of carbohydrate, fatty acids, and amino acids is made available within mitochondria as reducing equivalents (—H or electrons) (Figure 13–2). Note that the enzymes of the citric acid cycle and β-oxidation (Chapters 22 & 17) are contained in mitochondria, together with the **respiratory chain,** which collects and transports reducing equivalents, directing them to their final reaction with oxygen to form water, and the machinery for oxidative phosphorylation, the process by which the liberated free energy is trapped as **high-energy phosphate**.

Components of the Respiratory Chain Are Contained in Four Large Protein Complexes Embedded in the Inner Mitochondrial Membrane

Electrons flow through the respiratory chain through a redox span of 1.1 V from $NAD^+/NADH$ to $O_2/2H_2O$ (Table 12–1), passing through three large protein complexes: **NADH-Q oxidoreductase (Complex I),** where electrons are transferred from NADH to coenzyme Q (Q) (also called **ubiquinone**); **Q-cytochrome c oxidoreductase (Complex III),** which passes the electrons on to cytochrome c; and **cytochrome c oxidase (Complex IV),** which completes the chain, passing the electrons to O_2 and causing it to be reduced to H_2O (Figure 13–3). Some substrates with more positive redox potentials than $NAD^+/NADH$ (eg, succinate) pass electrons to Q via a fourth complex, **succinate-Q reductase (Complex II),** rather than Complex I. The four complexes are embedded in the inner mitochondrial membrane, but Q and cytochrome c are mobile. Q diffuses rapidly within the membrane, while cytochrome c is a soluble protein. The flow of electrons through Complexes I, III, and IV results in the pumping of protons from the matrix across the inner mitochondrial membrane into the intermembrane space (Figure 13–7).

Flavoproteins and Iron-Sulfur Proteins (Fe-S) Are Components of the Respiratory Chain Complexes

Flavoproteins (Chapter 12) are important components of Complexes I and II. The oxidized flavin nucleotide (FMN or FAD) can be reduced in reactions involving the transfer of two electrons (to form $FMNH_2$ or $FADH_2$), but they can also accept one electron to form the semiquinone (Figure 12–2). **Iron-sulfur proteins (nonheme iron proteins, Fe-S)** are found in Complexes I, II, and III. These may contain one, two, or four Fe atoms linked to inorganic sulfur atoms and/or via cysteine-SH groups to the protein (Figure 13–4). The Fe-S take

FIGURE 13–1 Structure of the mitochondrial membranes. Note that the inner membrane contains many folds, or cristae.

part in single electron transfer reactions in which one Fe atom undergoes oxidoreduction between Fe^{2+} and Fe^{3+}.

Q Accepts Electrons via Complex I and Complex II

NADH-Q oxidoreductase or Complex I is a large L-shaped multi-subunit protein that catalyzes electron transfer from NADH to Q, coupled with the transfer of four H^+ across the membrane:

$$NADH + Q + 5H^+_{matrix} \rightarrow$$
$$NAD + QH_2 + 4H^+_{intermembrane space}$$

Electrons are transferred from NADH to FMN initially, then to a series of Fe-S centers, and finally to Q (Figure 13–5). In Complex II (succinate-Q reductase), $FADH_2$ is formed during

the conversion of succinate to fumarate in the citric acid cycle (Figure 17–3) and electrons are then passed via several Fe-S centers to Q (Figure 13–5). Glycerol-3-phosphate (generated in the breakdown of triacylglycerols or from glycolysis, Figure 18–2) and acyl-CoA also pass electrons to Q via different pathways involving flavoproteins (Figure 13–5).

The Q Cycle Couples Electron Transfer to Proton Transport in Complex III

Electrons are passed from QH_2 to cytochrome c via Complex III (Q-cytochrome *c* oxidoreductase):

$$QH_2 + 2Cyt\ c_{oxidized} + 2H^+_{matrix} \rightarrow$$
$$Q + 2Cyt\ c_{reduced} + 4H^+_{intermembrane space}$$

The process is believed to involve **cytochromes c_1, b_L, and b_H and a Rieske Fe-S** (an unusual Fe-S in which one of the Fe atoms is linked to two histidine-SH groups rather than two cysteine-SH groups) (Figure 13–5), and is known as the **Q cycle** (Figure 13–6). Q may exist in three forms, the oxidized quinone, the reduced quinol, or the semiquinone (Figure 13–6). The semiquinone is formed transiently during the cycle, one turn of which results in the oxidation of $2QH_2$ to Q, releasing $4H^+$ into the intermembrane space, and the reduction of one Q to QH_2, causing $2H^+$ to be taken up from the matrix (Figure 13–6). Note that while Q carries two electrons, the cytochromes carry only one, thus the oxidation of one QH_2 is coupled to the reduction of two molecules of cytochrome *c* via the Q cycle.

Molecular Oxygen Is Reduced to Water via Complex IV

Reduced cytochrome *c* is oxidized by Complex IV (cytochrome *c* oxidase), with the concomitant reduction of O_2 to two molecules of water:

$$4Cyt\ c_{reduced} + O_2 + 8H^+_{matrix} \rightarrow$$
$$4Cyt\ c_{oxidized} + 2H_2O + 4H^+_{intermembrane space}$$

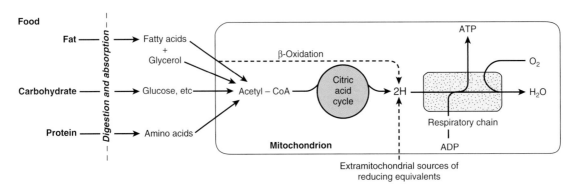

FIGURE 13–2 Role of the respiratory chain of mitochondria in the conversion of food energy to ATP. Oxidation of the major foodstuffs leads to the generation of reducing equivalents (2H) that are collected by the respiratory chain for oxidation and coupled generation of ATP.

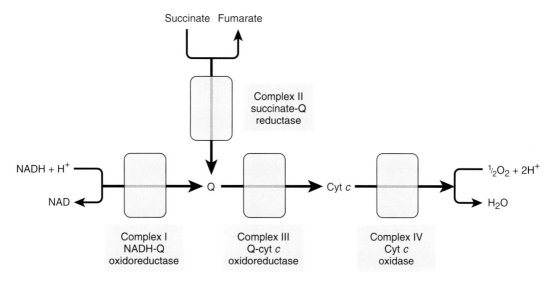

FIGURE 13–3 Overview of electron flow through the respiratory chain. (Q, coenzyme Q or ubiquinone; cyt, cytochrome.)

This transfer of four electrons from cytochrome c to O_2 involves two heme groups, a and a_3, and Cu (Figure 13–5). Electrons are passed initially to a Cu center (Cu_A), which contains 2Cu atoms linked to two protein cysteine-SH groups (resembling an Fe-S), then in sequence to heme a, heme a_3, a second Cu center, Cu_B, which is linked to heme a_3, and finally to O_2. Of the eight H⁺ removed from the matrix, four are used to form two water molecules and four are pumped into the intermembrane space. Thus, for every pair of electrons passing down the chain from NADH or $FADH_2$, 2H⁺ are pumped across the membrane by Complex IV. The O_2 remains tightly bound to Complex IV until it is fully reduced, and this minimizes the release of potentially damaging intermediates such as superoxide anions or peroxide which are formed when O_2 accepts one or two electrons, respectively (Chapter 12).

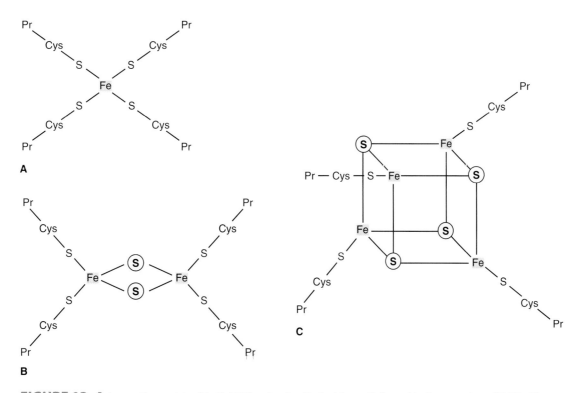

FIGURE 13–4 Iron-sulfur proteins (Fe-S). **(A)** The simplest Fe-S with one Fe bound by four cysteines. **(B)** 2Fe-2S center. **(C)** 4Fe-4S center. (Ⓢ, Inorganic sulfur; Pr, apoprotein; Cys, cysteine.)

FIGURE 13–5 Flow of electrons through the respiratory chain complexes, showing the entry points for reducing equivalents from important substrates. Q and cyt *c* are mobile components of the system as indicated by the dotted arrows. The flow through Complex III (the Q cycle) is shown in more detail in Figure 13–6. (Fe-S, iron-sulfur protein; ETF, electron transferring flavoprotein; Q, coenzyme Q or ubiquinone; cyt, cytochrome.)

ELECTRON TRANSPORT VIA THE RESPIRATORY CHAIN CREATES A PROTON GRADIENT WHICH DRIVES THE SYNTHESIS OF ATP

The flow of electrons through the respiratory chain generates ATP by the process of **oxidative phosphorylation. The chemiosmotic theory,** proposed by Peter Mitchell in 1961, postulates that the two processes are coupled by a proton gradient across the inner mitochondrial membrane so that **the proton motive force** caused by the electrochemical potential difference (negative on the matrix side) drives the mechanism of ATP synthesis. As we have seen, Complexes I, III, and IV act as **proton pumps.** Since the inner mitochondrial membrane is impermeable to ions in general and particularly to protons, these accumulate in the intermembrane space, creating the proton motive force predicted by the chemiosmotic theory.

A Membrane-Located ATP Synthase Functions as a Rotary Motor to Form ATP

The proton motive force drives a membrane-located **ATP synthase** that in the presence of P_i + ADP forms ATP. ATP synthase is embedded in the inner membrane, together with the respiratory chain complexes (Figure 13–7). Several subunits of the protein form a ball-like shape arranged around an axis known as F_1, which projects into the matrix and contains the phosphorylation mechanism (Figure 13–8). F_1 is attached to a membrane protein complex known as F_0, which also consists of several protein subunits. F_0 spans the membrane and forms a proton channel. The flow of protons through F_0 causes it to rotate, driving the production of ATP in the F_1 complex (Figures

13–7 & 13–8). This is thought to occur by a **binding change mechanism** in which the conformation of the β-subunits in F_1 is changed as the axis rotates from one that binds ATP tightly to one that releases ATP and binds ADP and P_i so that the next ATP can be formed. Estimates suggest that for each NADH oxidized, Complexes I and III translocate four protons each and Complex IV translocates two.

THE RESPIRATORY CHAIN PROVIDES MOST OF THE ENERGY CAPTURED DURING CATABOLISM

ADP captures, in the form of high-energy phosphate, a significant proportion of the free energy released by catabolic processes. The resulting ATP has been called the **energy "currency"** of the cell because it passes on this free energy to drive those processes requiring energy (Figure 11–6).

There is a net direct capture of two high-energy phosphate groups in the glycolytic reactions (Table 18–1). Two more high-energy phosphates per mole of glucose are captured in the citric acid cycle during the conversion of succinyl CoA to succinate. All of these phosphorylations occur at the **substrate level.** For each mol of substrate oxidized via Complexes I, III, and IV in the respiratory chain (ie, via NADH), 2.5 mol of ATP are formed per 0.5 mol of O_2 consumed; ie, the P:O ratio = 2.5 (Figure 13–7). On the other hand, when 1 mol of substrate (eg, succinate or 3-phophoglycerate) is oxidized via Complexes II, III, and IV, only 1.5 mol of ATP are formed; ie, P:O = 1.5. These reactions are known as **oxidative phosphorylation at the respiratory chain level.** Taking these values into account, it can be estimated that nearly 90% of the high-energy phosphates produced from the complete oxidation of

QH$_2$: Reduced (quinol) form (QH$_2$)   **Q:** Fully oxidized (quinone) form   **·Q$^-$:** Semiquinone (free radical) form

FIGURE 13–6 The Q cycle. During the oxidation of QH$_2$ to Q, one electron is donated to cyt *c* via a Rieske Fe-S and cyt *c$_1$* and the second to a Q to form the semiquinone via cyt *b$_L$* and cyt *b$_H$*, with 2H$^+$ being released into the intermembrane space. A similar process then occurs with a second QH$_2$, but in this case the second electron is donated to the semiquinone, reducing it to QH$_2$, and 2H$^+$ are taken up from the matrix. (Fe-S, iron-sulfur protein; Q, coenzyme Q or ubiquinone; cyt, cytochrome.)

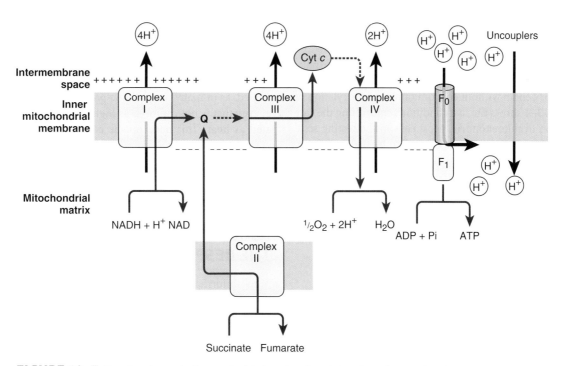

FIGURE 13–7 The chemiosmotic theory of oxidative phosphorylation. Complexes I, III, and IV act as proton pumps creating a proton gradient across the membrane, which is negative on the matrix side. The proton motive force generated drives the synthesis of ATP as the protons flow back into the matrix through the ATP synthase enzyme (see Figure 13–8). Uncouplers increase the permeability of the membrane to ions, collapsing the proton gradient by allowing the H$^+$ to pass across without going through the ATP synthase, and thus uncouple electron flow through the respiratory complexes from ATP synthesis. (Q, coenzyme Q or ubiquinone; cyt, cytochrome.)

FIGURE 13–8 Mechanism of ATP production by ATP synthase. The enzyme complex consists of an F_0 subcomplex which is a disk of "C" protein subunits. Attached is a γ subunit in the form of a "bent axle." Protons passing through the disk of "C" units cause it and the attached γ subunit to rotate. The γ subunit fits inside the F_1 subcomplex of three α and three β subunits, which are fixed to the membrane and do not rotate. ADP and P_i are taken up sequentially by the β subunits to form ATP, which is expelled as the rotating γ subunit squeezes each β subunit in turn and changes its conformation. Thus, three ATP molecules are generated per revolution. For clarity, not all the subunits that have been identified are shown—eg, the "axle" also contains an subunit.

1 mol glucose is obtained via oxidative phosphorylation coupled to the respiratory chain (Table 18–1).

Respiratory Control Ensures a Constant Supply of ATP

The rate of respiration of mitochondria can be controlled by the availability of ADP. This is because oxidation and phosphorylation are **tightly coupled**; ie, oxidation cannot proceed via the respiratory chain without concomitant phosphorylation of ADP. Table 13–1 shows the five conditions controlling the rate of respiration in mitochondria. Most cells in the resting state are in state 4, and respiration is controlled by the availability of ADP. When work is performed, ATP is converted to ADP, allowing more respiration to occur, which in turn replenishes the store of ATP. Under certain conditions, the concentration of inorganic phosphate can also affect the rate of functioning of the respiratory chain. As respiration increases (as in exercise), the cell approaches state 3 or state 5 when either the

capacity of the respiratory chain becomes saturated or the PO_2 decreases below the K_m for heme a_3. There is also the possibility that the ADP/ATP transporter, which facilitates entry of cytosolic ADP into and ATP out of the mitochondrion, becomes rate-limiting.

Thus, the manner in which biologic oxidative processes allow the free energy resulting from the oxidation of foodstuffs to become available and to be captured is stepwise, efficient, and controlled—rather than explosive, inefficient, and uncontrolled, as in many nonbiologic processes. The remaining free energy that is not captured as high-energy phosphate is liberated as **heat.** This need not be considered "wasted," since it ensures that the respiratory system as a whole is sufficiently exergonic to be removed from equilibrium, allowing continuous unidirectional flow and constant provision of ATP. It also contributes to maintenance of body temperature.

MANY POISONS INHIBIT THE RESPIRATORY CHAIN

Much information about the respiratory chain has been obtained by the use of inhibitors, and, conversely, this has provided knowledge about the mechanism of action of several poisons (Figure 13–9). They may be classified as inhibitors of the respiratory chain, inhibitors of oxidative phosphorylation, and uncouplers of oxidative phosphorylation.

Barbiturates such as amobarbital inhibit electron transport via Complex I by blocking the transfer from Fe-S to Q. At sufficient dosage, they are fatal in vivo. **Antimycin A** and **dimercaprol** inhibit the respiratory chain at Complex III. The classic poisons H_2S, **carbon monoxide** and **cyanide** inhibit

TABLE 13–1   States of Respiratory Control

Conditions Limiting the Rate of Respiration
State 1   Availability of ADP and substrate
State 2   Availability of substrate only
State 3   The capacity of the respiratory chain itself, when all substrates and components are present in saturating amounts
State 4   Availability of ADP only
State 5   Availability of oxygen only

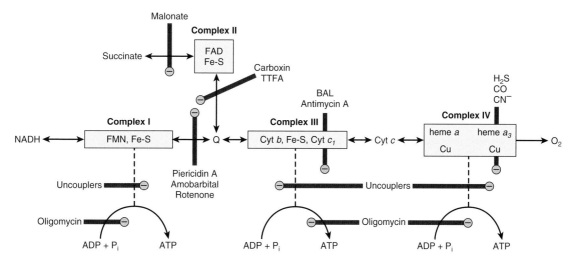

FIGURE 13–9 Sites of inhibition ($\ominus$) of the respiratory chain by specific drugs, chemicals, and antibiotics. (BAL, dimercaprol; TTFA, an Fe-chelating agent. Other abbreviations as in Figure 13–5.)

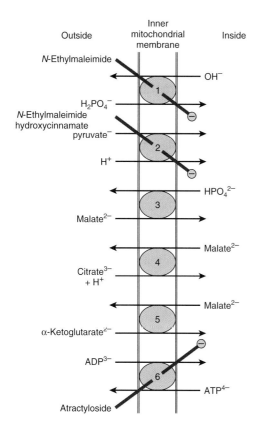

FIGURE 13–10 Transporter systems in the inner mitochondrial membrane. ① Phosphate transporter, ② pyruvate symport, ③ dicarboxylate transporter, ④ tricarboxylate transporter, ⑤ α-ketoglutarate transporter, ⑥ adenine nucleotide transporter. *N*-Ethylmaleimide, hydroxycinnamate, and atractyloside inhibit ($\ominus$) the indicated systems. Also present (but not shown) are transporter systems for glutamate/aspartate (Figure 13–13), glutamine, ornithine, neutral amino acids, and carnitine (Figure 22–1).

Complex IV and can therefore totally arrest respiration. **Malonate** is a competitive inhibitor of Complex II.

Atractyloside inhibits oxidative phosphorylation by inhibiting the transporter of ADP into and ATP out of the mitochondrion (Figure 13–10). The antibiotic **oligomycin** completely blocks oxidation and phosphorylation by blocking the flow of protons through ATP synthase (Figure 13–9).

Uncouplers dissociate oxidation in the respiratory chain from phosphorylation (Figure 13–7). These compounds are toxic in vivo, causing respiration to become uncontrolled, since the rate is no longer limited by the concentration of ADP or P_i. The uncoupler that has been used most frequently is **2,4-dinitrophenol,** but other compounds act in a similar manner. **Thermogenin (or the uncoupling protein)** is a physiological uncoupler found in brown adipose tissue that functions to generate body heat, particularly for the newborn and during hibernation in animals (Chapter 25).

THE CHEMIOSMOTIC THEORY CAN ACCOUNT FOR RESPIRATORY CONTROL AND THE ACTION OF UNCOUPLERS

The electrochemical potential difference across the membrane, once established as a result of proton translocation, inhibits further transport of reducing equivalents through the respiratory chain unless discharged by back-translocation of protons across the membrane through the ATP synthase. This in turn depends on availability of ADP and P_i.

Uncouplers (eg, dinitrophenol) are amphipathic (Chapter 15) and increase the permeability of the lipoid inner mitochondrial membrane to protons, thus reducing the electrochemical potential and short-circuiting the ATP synthase (Figure 13–7). In this way, oxidation can proceed without phosphorylation.

THE RELATIVE IMPERMEABILITY OF THE INNER MITOCHONDRIAL MEMBRANE NECESSITATES EXCHANGE TRANSPORTERS

Exchange diffusion systems involving transporter proteins that span the membrane are present in the membrane for exchange of anions against OH^- ions and cations against H^+ ions. Such systems are necessary for uptake and output of ionized metabolites while preserving electrical and osmotic equilibrium. The inner mitochondrial membrane is freely permeable to uncharged small molecules, such as oxygen, water, CO_2, NH_3, and to monocarboxylic acids, such as 3-hydroxybutyric, acetoacetic, and acetic. Long-chain fatty acids are transported into mitochondria via the carnitine system (Figure 22–1), and there is also a special carrier for pyruvate involving a symport that utilizes the H^+ gradient from outside to inside the mitochondrion (Figure 13–10). However, dicarboxylate and tricarboxylate anions and amino acids require specific transporter or carrier systems to facilitate their passage across the membrane. Monocarboxylic acids penetrate more readily in their undissociated and more lipid-soluble form.

The transport of di- and tricarboxylate anions is closely linked to that of inorganic phosphate, which penetrates readily as the $H_2PO_4^-$ ion in exchange for OH^-. The net uptake of malate by the dicarboxylate transporter requires inorganic phosphate for exchange in the opposite direction. The net uptake of citrate, isocitrate, or *cis*-aconitate by the tricarboxylate transporter requires malate in exchange. α-Ketoglutarate transport also requires an exchange with malate. The adenine nucleotide transporter allows the exchange of ATP and ADP but not AMP. It is vital in allowing ATP exit from mitochondria to the sites of extramitochondrial utilization and in allowing the return of ADP for ATP production within the mitochondrion (Figure 13–11). Since in this translocation four negative charges are removed from the matrix for every three taken in, the electrochemical gradient across the membrane (the proton motive force) favors the export of ATP. Na^+ can be exchanged for H^+, driven by the proton gradient. It is believed that active uptake of Ca^{2+} by mi-

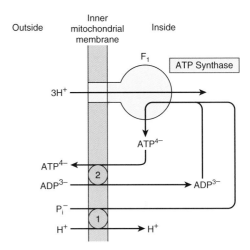

FIGURE 13–11 Combination of phosphate transporter ① with the adenine nucleotide transporter ② in ATP synthesis. The H^+/P_i symport shown is equivalent to the P_i/OH^- antiport shown in Figure 13–10.

tochondria occurs with a net charge transfer of 1 (Ca^+ uniport), possibly through a Ca^{2+}/H^+ antiport. Calcium release from mitochondria is facilitated by exchange with Na^+.

Ionophores Permit Specific Cations to Penetrate Membranes

Ionophores are lipophilic molecules that complex specific cations and facilitate their transport through biologic membranes, eg, **valinomycin** (K^+). The classic uncouplers such as dinitrophenol are, in fact, proton ionophores.

A Proton-Translocating Transhydrogenase Is a Source of Intramitochondrial NADPH

Energy-linked transhydrogenase, a protein in the inner mitochondrial membrane, couples the passage of protons down the electrochemical gradient from outside to inside the mitochondrion with the transfer of H from intramitochondrial NADH

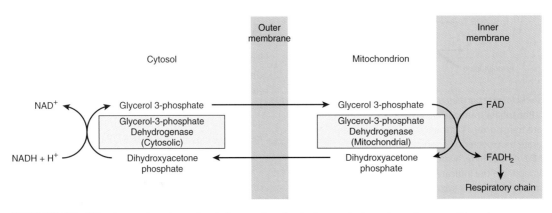

FIGURE 13–12 Glycerophosphate shuttle for transfer of reducing equivalents from the cytosol into the mitochondrion.

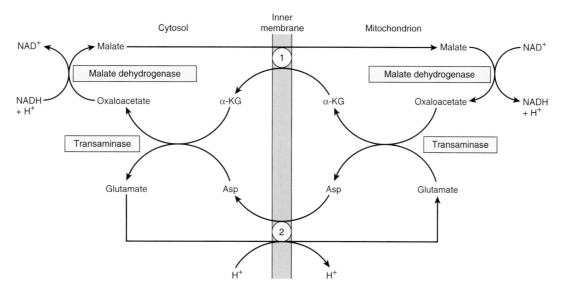

FIGURE 13–13 Malate shuttle for transfer of reducing equivalents from the cytosol into the mitochondrion. ① α-Ketoglutarate transporter, ② glutamate/aspartate transporter (note the proton symport with glutamate).

to NADPH for intramitochondrial enzymes such as glutamate dehydrogenase and hydroxylases involved in steroid synthesis.

Oxidation of Extramitochondrial NADH Is Mediated by Substrate Shuttles

NADH cannot penetrate the mitochondrial membrane, but it is produced continuously in the cytosol by 3-phosphoglyceraldehyde dehydrogenase, an enzyme in the glycolysis sequence (Figure 18–2). However, under aerobic conditions, extramitochondrial NADH does not accumulate and is presumed to be oxidized by the respiratory chain in mitochondria. The transfer of reducing equivalents through the mitochondrial membrane requires substrate pairs, linked by suitable dehydrogenases on each side of the mitochondrial membrane. The mechanism of transfer using the **glycerophosphate shuttle** is shown in Figure 13–12. Since the mitochondrial enzyme is linked to the respiratory chain via a flavoprotein rather than NAD, only 1.5 mol rather than 2.5 mol of ATP are formed per atom of oxygen consumed. Although this shuttle is present in some tissues (eg, brain, white muscle), in others (eg, heart muscle) it is deficient. It is therefore believed that the **malate shuttle** system (Figure 13–13) is of more universal utility. The complexity of this system is due to the impermeability of the mitochondrial membrane to oxaloacetate, which must react with glutamate to form aspartate and α-ketoglutarate by transamination before transport through the mitochondrial membrane and reconstitution to oxaloacetate in the cytosol.

Ion Transport in Mitochondria Is Energy Linked

Mitochondria maintain or accumulate cations such as K^+, Na^+, Ca^{2+}, and Mg^{2+}, and P_i. It is assumed that a primary proton pump drives cation exchange.

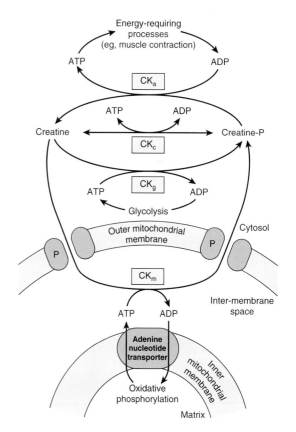

FIGURE 13–14 The creatine phosphate shuttle of heart and skeletal muscle. The shuttle allows rapid transport of high-energy phosphate from the mitochondrial matrix into the cytosol. (CK_a, creatine kinase concerned with large requirements for ATP, eg, muscular contraction; CK_c, creatine kinase for maintaining equilibrium between creatine and creatine phosphate and ATP/ADP; CK_g, creatine kinase coupling glycolysis to creatine phosphate synthesis; CK_m, mitochondrial creatine kinase mediating creatine phosphate production from ATP formed in oxidative phosphorylation; P, pore protein in outer mitochondrial membrane.)

The Creatine Phosphate Shuttle Facilitates Transport of High-Energy Phosphate from Mitochondria

This shuttle (Figure 13–14) augments the functions of **creatine phosphate** as an energy buffer by acting as a dynamic system for transfer of high-energy phosphate from mitochondria in active tissues such as heart and skeletal muscle. An isoenzyme of **creatine kinase** (CK_m) is found in the mitochondrial intermembrane space, catalyzing the transfer of high-energy phosphate to creatine from ATP emerging from the adenine nucleotide transporter. In turn, the creatine phosphate is transported into the cytosol via protein pores in the outer mitochondrial membrane, becoming available for generation of extramitochondrial ATP.

CLINICAL ASPECTS

The condition known as **fatal infantile mitochondrial myopathy and renal dysfunction** involves severe diminution or absence of most oxidoreductases of the respiratory chain. **MELAS** (mitochondrial encephalopathy, lactic acidosis, and stroke) is an inherited condition due to NADH-Q oxidoreductase (Complex I) or cytochrome oxidase (Complex IV) deficiency. It is caused by a mutation in mitochondrial DNA and may be involved in **Alzheimer's disease** and **diabetes mellitus**. A number of drugs and poisons act by inhibition of oxidative phosphorylation.

SUMMARY

- Virtually all energy released from the oxidation of carbohydrate, fat, and protein is made available in mitochondria as reducing equivalents (—H or e⁻). These are funneled into the respiratory chain, where they are passed down a redox gradient of carriers to their final reaction with oxygen to form water.

- The redox carriers are grouped into four respiratory chain complexes in the inner mitochondrial membrane. Three of the four complexes are able to use the energy released in the redox gradient to pump protons to the outside of the membrane, creating an electrochemical potential between the matrix and the inner membrane space.

- ATP synthase spans the membrane and acts like a rotary motor using the potential energy of the proton gradient or proton motive force to synthesize ATP from ADP and P_i. In this way, oxidation is closely coupled to phosphorylation to meet the energy needs of the cell.

- Because the inner mitochondrial membrane is impermeable to protons and other ions, special exchange transporters span the membrane to allow ions such as OH^-, ATP^{4-}, ADP^{3-}, and metabolites to pass through without discharging the electrochemical gradient across the membrane.

- Many well-known poisons such as cyanide arrest respiration by inhibition of the respiratory chain.

REFERENCES

Hinkle PC et al: P/O ratios of mitochondrial oxidative phosphorylation. Biochem Biophys Acta 2005;1706:1.

Mitchell P: Keilin's respiratory chain concept and its chemiosmotic consequences. Science 1979;206:1148.

Schultz BE, Chan SI: Structures and proton-pumping strategies of mitochondrial respiratory enzymes. Annu Rev Biophys Biomol Struct 2001;30:23.

Smeitink J et al: The genetics and pathology of oxidative phosphorylation. Nat Rev Genet 2001;2:342.

Tyler DD: *The Mitochondrion in Health and Disease*. VCH Publishers, 1992.

Wallace DC: Mitochondrial DNA in aging and disease. Sci Am 1997;277:22.

Yoshida M et al: ATP synthase—a marvellous rotary engine of the cell. Nat Rev Mol Cell Biol 2001;2:669.

Carbohydrates of Physiologic Significance

David A. Bender, PhD

BIOMEDICAL IMPORTANCE

Carbohydrates are widely distributed in plants and animals; they have important structural and metabolic roles. In plants, glucose is synthesized from carbon dioxide and water by photosynthesis and stored as starch or used to synthesize the cellulose of the plant cell walls. Animals can synthesize carbohydrates from amino acids, but most are derived ultimately from plants. **Glucose** is the most important carbohydrate; most dietary carbohydrate is absorbed into the bloodstream as glucose formed by hydrolysis of dietary starch and disaccharides, and other sugars are converted to glucose in the liver. Glucose is the major metabolic fuel of mammals (except ruminants) and a universal fuel of the fetus. It is the precursor for synthesis of all the other carbohydrates in the body, including **glycogen** for storage; **ribose** and **deoxyribose** in nucleic acids; **galactose** in lactose of milk, in glycolipids, and in combination with protein in glycoproteins and proteoglycans. Diseases associated with carbohydrate metabolism include **diabetes mellitus, galactosemia, glycogen storage diseases,** and **lactose intolerance.**

CARBOHYDRATES ARE ALDEHYDE OR KETONE DERIVATIVES OF POLYHYDRIC ALCOHOLS

Carbohydrates are classified as follows:

1. **Monosaccharides** are those sugars that cannot be hydrolyzed into simpler carbohydrates. They may be classified as **trioses, tetroses, pentoses, hexoses,** or **heptoses,** depending upon the number of carbon atoms, and as **aldoses** or **ketoses,** depending upon whether they have an aldehyde or ketone group. Examples are listed in Table 14–1. In addition to aldehydes and ketones, the polyhydric alcohols (sugar alcohols or **polyols**), in which the aldehyde or ketone group has been reduced to an alcohol group, also occur naturally in foods. They are synthesized by reduction of monosaccharides for use in the manufacture of foods for weight reduction and for diabetics. They are poorly absorbed, and have about half the energy yield of sugars.

2. **Disaccharides** are condensation products of two monosaccharide units; examples are maltose and sucrose.

3. **Oligosaccharides** are condensation products of three to ten monosaccharides. Most are not digested by human enzymes.

4. **Polysaccharides** are condensation products of more than ten monosaccharide units; examples are the starches and dextrins, which may be linear or branched polymers. Polysaccharides are sometimes classified as hexosans or pentosans, depending on the identity of the constituent monosaccharides (hexoses and pentoses, respectively). In addition to starches and dextrins, foods contain a wide variety of other polysaccharides that are collectively known as nonstarch polysaccharides; they are not digested by human enzymes, and are the major component of dietary fiber. Examples are cellulose from plant cell walls (a glucose polymer) and inulin, the storage carbohydrate in some plants (a fructose polymer).

BIOMEDICALLY, GLUCOSE IS THE MOST IMPORTANT MONOSACCHARIDE

The Structure of Glucose Can Be Represented in Three Ways

The straight-chain structural formula (aldohexose; Figure 14–1A) can account for some of the properties of glucose, but a cyclic structure (a **hemiacetal** formed by reaction between the aldehyde group and a hydroxyl group) is thermodynamically favored and accounts for other properties. The cyclic structure is normally drawn as shown in Figure 14–1B, the Haworth projection, in which the molecule is viewed from the side and above the plane of the ring; the bonds nearest to the viewer are bold and thickened, and the hydroxyl groups are above or below the plane of the ring. The six-membered ring containing one oxygen atom is actually in the form of a chair (Figure 14–1C).

Sugars Exhibit Various Forms of Isomerism

Glucose, with four asymmetric carbon atoms, can form 16 isomers. The more important types of isomerism found with glucose are as follows:

1. **D and L isomerism:** The designation of a sugar isomer as the D form or of its mirror image as the L form is determined

TABLE 14–1 Classification of Important Sugars

	Aldoses	Ketoses
Trioses ($C_3H_6O_3$)	Glycerose (glyceraldehyde)	Dihydroxyacetone
Tetroses($C_4H_8O_4$)	Erythrose	Erythrulose
Pentoses ($C_5H_{10}O_5$)	Ribose	Ribulose
Hexoses ($C_6H_{12}O_6$)	Glucose	Fructose
Heptoses ($C_7H_{14}O_7$)	—	Sedoheptulose

by its spatial relationship to the parent compound of the carbohydrates, the three-carbon sugar glycerose (glyceraldehyde).
The L and D forms of this sugar, and of glucose, are shown
in Figure 14–2. The orientation of the —H and —OH groups
around the carbon atom adjacent to the terminal primary alcohol carbon (carbon 5 in glucose) determines whether the
sugar belongs to the D or L series. When the —OH group on
this carbon is on the right (as seen in Figure 14–2), the sugar
is the D isomer; when it is on the left, it is the L isomer. Most
of the monosaccharides occurring in mammals are D sugars,
and the enzymes responsible for their metabolism are specific
for this configuration.

The presence of asymmetric carbon atoms also confers
optical activity on the compound. When a beam of plane-
polarized light is passed through a solution of an **optical isomer,** it rotates either to the right, dextrorotatory (+), or to the
left, levorotatory (–). The direction of rotation of polarized

FIGURE 14–2 D- and L-isomerism of glycerose and glucose.

light is independent of the stereochemistry of the sugar, so
it may be designated D(–), D(+), L(–), or L(+). For example,
the naturally occurring form of fructose is the D(–) isomer. In
solution, glucose is dextrorotatory, and glucose solutions are
sometimes known as **dextrose.**

2. Pyranose and furanose ring structures: The ring
structures of monosaccharides are similar to the ring structures of either pyran (a six-membered ring) or furan (a five-
membered ring) (Figures 14–3 & 14–4). For glucose in solution, more than 99% is in the pyranose form.

3. Alpha and beta anomers: The ring structure of an
aldose is a hemiacetal, since it is formed by combination of
an aldehyde and an alcohol group. Similarly, the ring structure of a ketose is a hemiketal. Crystalline glucose is α-D-
glucopyranose. The cyclic structure is retained in solution, but
isomerism occurs about position 1, the carbonyl or **anomeric**

FIGURE 14–1 D-Glucose. **(A)** Straight-chain form. **(B)** α-D-
glucose; Haworth projection. **(C)** α-D-glucose; chair form.

FIGURE 14–3 Pyranose and furanose forms of glucose.

FIGURE 14–4 Pyranose and furanose forms of fructose.

FIGURE 14–5 Epimers of glucose.

carbon atom, to give a mixture of α-glucopyranose (38%) and β-glucopyranose (62%). Less than 0.3% is represented by α and β anomers of glucofuranose.

4. **Epimers:** Isomers differing as a result of variations in configuration of the —OH and —H on carbon atoms 2, 3, and 4 of glucose are known as epimers. Biologically, the most important epimers of glucose are mannose (epimerized at carbon 2) and galactose (epimerized at carbon 4) (Figure 14–5).

5. **Aldose-ketose isomerism:** Fructose has the same molecular formula as glucose but differs in its structural formula, since there is a potential keto group in position 2, the anomeric carbon of fructose (Figures 14–4 & 14–6), whereas there is a potential aldehyde group in position 1, the anomeric carbon of glucose (Figures 14–2 & 14–7).

Many Monosaccharides Are Physiologically Important

Derivatives of trioses, tetroses, and pentoses and of a seven-carbon sugar (sedoheptulose) are formed as metabolic intermediates in glycolysis (Chapter 18) and the pentose phosphate pathway (Chapter 21). Pentoses are important in nucleotides, nucleic acids, and several coenzymes (Table 14–2). Glucose, galactose, fructose, and mannose are physiologically the most important hexoses (Table 14–3). The biochemically important ketoses are shown in Figure 14–6, and aldoses in Figure 14–7.

In addition, carboxylic acid derivatives of glucose are important, including D-glucuronate (for glucuronide formation and in glycosaminoglycans) and its metabolic derivative, L-iduronate (in glycosaminoglycans) (Figure 14–8) and L-gulonate (an intermediate in the uronic acid pathway; see Figure 21–4).

FIGURE 14–6 Examples of aldoses of physiologic significance.

FIGURE 14–7 Examples of ketoses of physiologic significance.

TABLE 14–2   Pentoses of Physiologic Importance

Sugar	Source	Biochemical and Clinical Importance
D-Ribose	Nucleic acids and metabolic intermediate	Structural component of nucleic acids and coenzymes, including ATP, NAD(P), and flavin coenzymes
D-Ribulose	Metabolic intermediate	Intermediate in the pentose phosphate pathway
D-Arabinose	Plant gums	Constituent of glycoproteins
D-Xylose	Plant gums, proteoglycans, glycosaminoglycans	Constituent of glycoproteins
L-Xylulose	Metabolic intermediate	Excreted in the urine in essential pentosuria

TABLE 14–3   Hexoses of Physiologic Importance

Sugar	Source	Biochemical Importance	Clinical Significance
D-Glucose	Fruit juices, hydrolysis of starch, cane or beet sugar, maltose and lactose	The main metabolic fuel for tissues; "blood sugar"	Excreted in the urine (glucosuria) in poorly controlled diabetes mellitus as a result of hyperglycemia
D-Fructose	Fruit juices, honey, hydrolysis of cane or beet sugar and inulin, enzymic isomerization of glucose syrups for food manufacture	Readily metabolized either via glucose or directly	Hereditary fructose intolerance leads to fructose accumulation and hypoglycemia
D-Galactose	Hydrolysis of lactose	Readily metabolized to glucose; synthesized in the mammary gland for synthesis of lactose in milk. A constituent of glycolipids and glycoproteins	Hereditary galactosemia as a result of failure to metabolize galactose leads to cataracts
D-Mannose	Hydrolysis of plant mannan gums	Constituent of glycoproteins	

FIGURE 14–8 α-D-Glucuronate (*left*) and β-L-iduronate (*right*).

Sugars Form Glycosides with Other Compounds & with Each Other

Glycosides are formed by condensation between the hydroxyl group of the anomeric carbon of a monosaccharide, and a second compound that may or may not (in the case of an **aglycone**) be another monosaccharide. If the second group is a hydroxyl, the O-glycosidic bond is an **acetal** link because it results from a reaction between a hemiacetal group (formed from an aldehyde and an —OH group) and another —OH group. If the hemiacetal portion is glucose, the resulting compound is a **glucoside;** if galactose, a **galactoside;** and so on. If the second group is an amine, an *N*-glycosidic bond is formed, eg, between adenine and ribose in nucleotides such as ATP (Figure 11–4).

Glycosides are widely distributed in nature; the aglycone may be methanol, glycerol, a sterol, a phenol, or a base such as adenine. The glycosides that are important in medicine because of their action on the heart (**cardiac glycosides**) all contain steroids as the aglycone. These include derivatives of digitalis and strophanthus such as **ouabain,** an inhibitor of the Na⁺-K⁺ ATPase of cell membranes. Other glycosides include antibiotics such as **streptomycin.**

Deoxy Sugars Lack an Oxygen Atom

Deoxy sugars are those in which one hydroxyl group has been replaced by hydrogen. An example is **deoxyribose** (Figure 14–9) in DNA. The deoxy sugar L-fucose (Figure 14–13) occurs in glycoproteins; 2-deoxyglucose is used experimentally as an inhibitor of glucose metabolism.

Amino Sugars (Hexosamines) Are Components of Glycoproteins, Gangliosides, & Glycosaminoglycans

The amino sugars include D-glucosamine, a constituent of hyaluronic acid (Figure 14–10), D-galactosamine (also known as chondrosamine), a constituent of chondroitin and D-mannosamine. Several **antibiotics** (eg, erythromycin) contain amino sugars, which are important for their antibiotic activity.

FIGURE 14–9 2-Deoxy-D-ribofuranose (β form).

FIGURE 14–10 Glucosamine (2-amino-D-glucopyranose) (α form). Galactosamine is 2-amino-D-galactopyranose. Both glucosamine and galactosamine occur as *N*-acetyl derivatives in more complex carbohydrates, eg, glycoproteins.

Maltose, Sucrose, & Lactose Are Important Disaccharides

The disaccharides are sugars composed of two monosaccharide residues linked by a glycoside bond (Figure 14–11). The physiologically important disaccharides are maltose, sucrose, and lactose (Table 14–4). Hydrolysis of sucrose yields a mixture of glucose and fructose called "invert sugar" because fructose is strongly levorotatory and changes (inverts) the weaker dextrorotatory action of sucrose.

POLYSACCHARIDES SERVE STORAGE & STRUCTURAL FUNCTIONS

Polysaccharides include the following physiologically important carbohydrates:

Starch is a homopolymer of glucose forming an α-glucosidic chain, called a **glucosan** or **glucan.** It is the most important dietary carbohydrate in cereals, potatoes, legumes, and other vegetables. The two main constituents are **amylose**

Maltose   Lactose   Sucrose

O-α-D-glucopyranosyl-(1 → 4)-α-D-glucopyranose   *O*-β-D-Galactopyranosyl-(1 → 4)-β-D-glucopyranose   *O*-α-D-Glucopyranosyl-(1 → 2)-β-D-fructofuranoside

FIGURE 14–11 Structures of important disaccharides. The α and β refer to the configuration at the anomeric carbon atom (*). When the anomeric carbon of the second residue takes part in the formation of the glycosidic bond, as in sucrose, the residue becomes a glycoside known as a furanoside or a pyranoside. As the disaccharide no longer has an anomeric carbon with a free potential aldehyde or ketone group, it no longer exhibits reducing properties. The configuration of the β-fructofuranose residue in sucrose results from turning the β-fructofuranose molecule depicted in Figure 14–4 through 180 degrees and inverting it.

TABLE 14–4 Disaccharides of Physiologic Importance

Sugar	Composition	Source	Clinical Significance
Sucrose	O-α-D-glucopyranosyl-(1→2)-β-D-fructofuranoside	Cane and beet sugar, sorghum and some fruits and vegetables	Rare genetic lack of sucrase leads to sucrose intolerance—diarrhea and flatulence
Lactose	O-α-D-galactopyranosyl-(1→4)-β-D-glucopyranose	Milk (and many pharmaceutical preparations as a filler)	Lack of lactase (alactasia) leads to lactose intolerance—diarrhea and flatulence; may be excreted in the urine in pregnancy
Maltose	O-α-D-glucopyranosyl-(1→4)-α-D-glucopyranose	Enzymic hydrolysis of starch (amylase); germinating cereals and malt	
Isomaltose	O-α-D-glucopyranosyl-(1→6)-α-D-glucopyranose	Enzymic hydrolysis of starch (the branch points in amylopectin)	
Lactulose	O-α-D-galactopyranosyl-(1→4)-β-D-fructofuranose	Heated milk (small amounts), mainly synthetic	Not hydrolyzed by intestinal enzymes, but fermented by intestinal bacteria; used as a mild osmotic laxative
Trehalose	O-α-D-glucopyranosyl-(1→1)-α-D-glucopyranoside	Yeasts and fungi; the main sugar of insect hemolymph	

FIGURE 14–12 Structure of starch. **(A)** Amylose, showing helical coil structure. **(B)** Amylopectin, showing 1 → 6 branch point.

(13–20%), which has a nonbranching helical structure, and **amylopectin** (80–85%), which consists of branched chains composed of 24–30 glucose residues united by α1 → 4 linkages in the chains and by α1 → 6 linkages at the branch points (Figure 14–12).

The extent to which starch in foods is hydrolyzed by amylase is determined by its structure, the degree of crystallization or hydration (the result of cooking), and whether it is enclosed in intact (and indigestible) plant cells walls. The **glycemic index** of a starchy food is a measure of its digestibility, based on the extent to which it raises the blood concentration of glucose compared with an equivalent amount of glucose or a reference food such as white bread or boiled rice.

Glycogen (Figure 14–13) is the storage polysaccharide in animals and is sometimes called animal starch. It is a more highly branched structure than amylopectin with chains of 12–14 α-ᴅ-glucopyranose residues (in α1 → 4 glucosidic linkage) with branching by means of α1 → 6 glucosidic bonds. **Inulin** is a polysaccharide of fructose (and hence a fructosan) found in tubers and roots of dahlias, artichokes, and dandelions. It is readily soluble in water and is used to determine the glomerular filtration rate, but it is not hydrolyzed by intestinal enzymes. **Dextrins** are intermediates in the hydrolysis of starch. **Cellulose** is the chief constituent of plant cell walls. It is insoluble and consists of β-ᴅ-glucopyranose units linked by β1 → 4 bonds to form long, straight chains strengthened

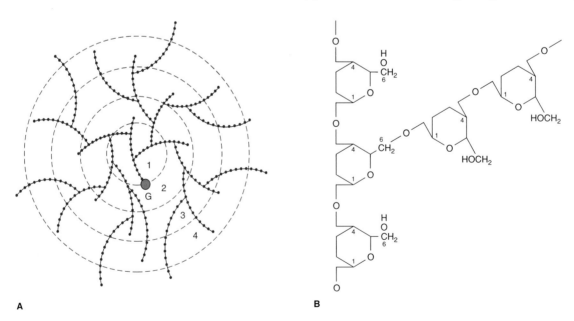

FIGURE 14–13 The glycogen molecule. **(A)** General structure. **(B)** Enlargement of structure at a branch point. The molecule is a sphere ~21 nm in diameter that can be seen in electron micrographs. It has a molecular mass of ~10^7 Da and consists of polysaccharide chains, each containing about 13 glucose residues. The chains are either branched or unbranched and are arranged in 12 concentric layers (only four are shown in the figure). The branched chains (each has two branches) are found in the inner layers and the unbranched chains in the outer layer. (G, glycogenin, the primer molecule for glycogen synthesis.)

by cross-linking hydrogen bonds. Mammals lack any enzyme that hydrolyzes the β1 → 4 bonds, and so cannot digest cellulose. It is an important source of "bulk" in the diet, and the major component of dietary fiber. Microorganisms in the gut of ruminants and other herbivores can hydrolyze the linkage and ferment the products to short-chain fatty acids as a major energy source. There is some bacterial metabolism of cellulose

TABLE 14–5 **Carbohydrates Found in Glycoproteins**

Hexoses	Mannose (Man), galactose (Gal)
Acetyl hexosamines	N-Acetylglucosamine (GlcNAc), N-acetyl-galactosamine (GalNAc)
Pentoses	Arabinose (Ara), Xylose (Xyl)
Methyl pentose	L-Fucose (Fuc, see Fig. 14–15)
Sialic acids	N-Acyl derivatives of neuraminic acid; the predominant sialic acid is N-acetyl-neuraminic acid (NeuAc, see Fig. 14–16)

FIGURE 14–15 β-L-Fucose (6-deoxy-β-L-galactose).

in the human colon. **Chitin** is a structural polysaccharide in the exoskeleton of crustaceans and insects, and also in mushrooms. It consists of N-acetyl-D-glucosamine units joined by β1 → 4 glycosidic bonds (Figure 14–14).

Glycosaminoglycans (mucopolysaccharides) are complex carbohydrates containing **amino sugars** and **uronic acids.** They may be attached to a protein molecule to form a **proteoglycan.** Proteoglycans provide the ground or packing substance of connective tissue. They hold large quantities of water and occupy space, thus cushioning or lubricating other structures, because of the large number of —OH groups and negative charges on the molecule which, by repulsion, keep the carbohydrate chains apart. Examples are **hyaluronic acid, chondroitin sulfate,** and **heparin** (Figure 14–14).

Glycoproteins (also known as mucoproteins) are proteins containing branched or unbranched oligosaccharide chains (Table 14–5, Figure 14–15); they occur in cell membranes (Chapters 40 & 47) and many other situations; serum albumin is a glycoprotein. The **sialic acids** are N- or O-acyl derivatives of neuraminic acid (Figure 14–16). **Neuraminic acid** is a nine-carbon sugar derived from mannosamine (an epimer of glucosamine) and pyruvate. Sialic acids are constituents of both **glycoproteins** and **gangliosides**.

FIGURE 14–14 Structure of some complex polysaccharides and glycosaminoglycans.

FIGURE 14–16 Structure of N-acetylneuraminic acid, a sialic acid (Ac = CH₃—CO—).

CARBOHYDRATES OCCUR IN CELL MEMBRANES & IN LIPOPROTEINS

Approximately 5% of the weight of cell membranes is carbohydrate in glycoproteins and glycolipids. Their presence on the outer surface of the plasma membrane (the **glycocalyx**) has been shown with the use of plant **lectins,** protein agglutinins that bind specific glycosyl residues. For example, **concanavalin A** binds α-glucosyl and α-mannosyl residues. **Glycophorin** is a major integral membrane glycoprotein of human erythrocytes. It has 130 amino acid residues and spans the lipid membrane, with polypeptide regions outside both the external and internal (cytoplasmic) surfaces. Carbohydrate chains are attached to the amino terminal portion outside the external surface. Carbohydrates are also present in apo-protein B of plasma lipoproteins.

SUMMARY

- Carbohydrates are major constituents of animal food and animal tissues. They are characterized by the type and number of monosaccharide residues in their molecules.
- Glucose is the most important carbohydrate in mammalian biochemistry because nearly all carbohydrate in food is converted to glucose for metabolism.
- Sugars have large numbers of stereoisomers because they contain several asymmetric carbon atoms.
- The physiologically important monosaccharides include glucose, the "blood sugar," and ribose, an important constituent of nucleotides and nucleic acids.
- The important disaccharides include maltose (glucosyl glucose), an intermediate in the digestion of starch; sucrose (glucosyl fructose), important as a dietary constituent containing fructose; and lactose (galactosyl glucose), in milk.
- Starch and glycogen are storage polymers of glucose in plants and animals, respectively. Starch is the major source of energy in the diet.
- Complex carbohydrates contain other sugar derivatives such as amino sugars, uronic acids, and sialic acids. They include proteoglycans and glycosaminoglycans, which are associated with structural elements of the tissues, and glycoproteins, which are proteins containing oligosaccharide chains; they are found in many situations including the cell membrane.

REFERENCES

Davis BG, Fairbanks AJ: *Carbohydrate Chemistry*. Oxford University Press, 2002.
Ernst B, Hart GW, Sinay P: *Carbohydrates in Chemistry and Biology*. Wiley-VCH, 2000.
Lindhorst TK, Thisbe K: *Essentials of Carbohydrate Chemistry and Biochemistry*, 3rd ed. Wiley-VCH, 2007.

Lipids of Physiologic Significance

Kathleen M. Botham, PhD, DSc & Peter A. Mayes, PhD, DSc

BIOMEDICAL IMPORTANCE

The lipids are a heterogeneous group of compounds, including fats, oils, steroids, waxes, and related compounds, that are related more by their physical than by their chemical properties. They have the common property of being (1) relatively **insoluble in water** and (2) **soluble in nonpolar solvents** such as ether and chloroform. They are important dietary constituents not only because of their high energy value, but also because of the fat-soluble vitamins and the essential fatty acids contained in the fat of natural foods. Fat is stored in **adipose tissue,** where it also serves as a thermal insulator in the subcutaneous tissues and around certain organs. Nonpolar lipids act as **electrical insulators,** allowing rapid propagation of depolarization waves along **myelinated nerves.** Combinations of lipid and protein (lipoproteins) serve as the means of **transporting lipids** in the blood. Knowledge of lipid biochemistry is necessary in understanding many important biomedical areas, eg, **obesity, diabetes mellitus, atherosclerosis,** and the role of various **polyunsaturated fatty acids** in nutrition and health.

LIPIDS ARE CLASSIFIED AS SIMPLE OR COMPLEX

1. **Simple lipids:** Esters of fatty acids with various alcohols.
   a. **Fats:** Esters of fatty acids with glycerol. **Oils** are fats in the liquid state.
   b. **Waxes:** Esters of fatty acids with higher molecular weight monohydric alcohols.
2. **Complex lipids:** Esters of fatty acids containing groups in addition to an alcohol and a fatty acid.
   a. **Phospholipids:** Lipids containing, in addition to fatty acids and an alcohol, a phosphoric acid residue. They frequently have nitrogen containing bases and other substituents, eg, in **glycerophospholipids** the alcohol is glycerol and in **sphingophospholipids** the alcohol is sphingosine.
   b. **Glycolipids (glycosphingolipids):** Lipids containing a fatty acid, sphingosine, and carbohydrate.
   c. **Other complex lipids:** Lipids such as sulfolipids and aminolipids. Lipoproteins may also be placed in this category.
3. **Precursor and derived lipids:** These include fatty acids, glycerol, steroids, other alcohols, fatty aldehydes, ketone bodies (Chapter 22), hydrocarbons, lipid-soluble vitamins, and hormones.

Because they are uncharged, acylglycerols (glycerides), cholesterol, and cholesteryl esters are termed **neutral lipids.**

FATTY ACIDS ARE ALIPHATIC CARBOXYLIC ACIDS

Fatty acids occur in the body mainly as esters in natural fats and oils, but are found in the unesterified form as **free fatty acids,** a transport form in the plasma. Fatty acids that occur in natural fats usually contain an even number of carbon atoms. The chain may be **saturated** (containing no double bonds) or **unsaturated** (containing one or more double bonds).

Fatty Acids Are Named after Corresponding Hydrocarbons

The most frequently used systematic nomenclature names the fatty acid after the hydrocarbon with the same number and arrangement of carbon atoms, with **-oic** being substituted for the final **-e** (Genevan system). Thus, saturated acids end in **-anoic,** eg, octanoic acid, and unsaturated acids with double bonds end in **-enoic,** eg, octadecenoic acid (oleic acid).

Carbon atoms are numbered from the carboxyl carbon (carbon No. 1). The carbon atoms adjacent to the carboxyl carbon (Nos. 2, 3, and 4) are also known as the α, β, and γ carbons, respectively, and the terminal methyl carbon is known as the ω- or n-carbon.

Various conventions use Δ for indicating the number and position of the double bonds (Figure 15–1); eg, Δ^9 indicates a double bond between carbons 9 and 10 of the fatty acid; ω9 indicates a double bond on the ninth carbon counting from the ω-carbon. In animals, additional double bonds are introduced only between the existing double bond (eg, ω9, ω6, or ω3) and the carboxyl carbon, leading to three series of fatty acids known as the ω9, ω6, and ω3 families, respectively.

18:1;9 *or* Δ^9 18:1

$$\underset{18}{CH_3}(CH_2)_7\underset{10}{CH} = \underset{9}{CH}(CH_2)_7\underset{1}{COOH}$$

or

ω9,C18:1 *or* n–9, 18:1

$$\underset{\substack{\omega \\ n \; 17}}{CH_3}\underset{2}{CH_2}\underset{3}{CH_2}\underset{4}{CH_2}\underset{5}{CH_2}\underset{6}{CH_2}\underset{7}{CH_2}\underset{8}{CH_2}\underset{\substack{9 \\ 10}}{CH} = \underset{\substack{10 \\ 9}}{CH}(CH_2)_7\underset{1}{COOH}$$

FIGURE 15–1 Oleic acid. n – 9 (n minus 9) is equivalent to ω9.

Saturated Fatty Acids Contain No Double Bonds

Saturated fatty acids may be envisaged as based on acetic acid (CH_3—COOH) as the first member of the series in which —CH2— is progressively added between the terminal CH_3— and —COOH groups. Examples are shown in Table 15–1. Other higher members of the series are known to occur, particularly in waxes. A few branched-chain fatty acids have also been isolated from both plant and animal sources.

Unsaturated Fatty Acids Contain One or More Double Bonds

Unsaturated fatty acids (Table 15–2) may be further subdivided as follows:

1. **Monounsaturated** (monoethenoid, monoenoic) acids, containing one double bond.
2. **Polyunsaturated** (polyethenoid, polyenoic) acids, containing two or more double bonds.
3. **Eicosanoids:** These compounds, derived from eicosa (20-carbon) polyenoic fatty acids, comprise the **prostanoids, leukotrienes (LTs),** and **lipoxins (LXs).** Prostanoids include **prostaglandins (PGs), prostacyclins (PGIs),** and **thromboxanes (TXs).**

TABLE 15–1   Saturated Fatty Acids

Common Name	Number of C Atoms	
Acetic	2	Major end product of carbohydrate fermentation by rumen organisms
Butyric	4	In certain fats in small amounts (especially butter). An end product of carbohydrate fermentation by rumen organisms[1]
Valeric	5	
Caproic	6	
Lauric	12	Spermaceti, cinnamon, palm kernel, coconut oils, laurels, butter
Myristic	14	Nutmeg, palm kernel, coconut oils, myrtles, butter
Palmitic	16	Common in all animal and plant fats
Stearic	18	

[1]Also formed in the cecum of herbivores and to a lesser extent in the colon of humans.

Prostaglandins exist in virtually every mammalian tissue, acting as local hormones; they have important physiologic and pharmacologic activities. They are synthesized in vivo by cyclization of the center of the carbon chain of 20-carbon (eicosanoic) polyunsaturated fatty acids (eg, arachidonic acid) to form a cyclopentane ring (Figure 15–2). A related series of compounds, the **thromboxanes,** have the cyclopentane ring interrupted with an oxygen atom (oxane ring) (Figure 15–3). Three different eicosanoic fatty acids give rise to three groups of eicosanoids characterized by the number of double bonds in the side chains, eg, PG_1, PG_2, PG_3. Different substituent groups attached to the rings give rise to series of prostaglandins and thromboxanes, labeled A, B, etc—eg, the "E" type of prostaglandin (as in PGE_2) has a keto group in position 9, whereas the "F" type has a hydroxyl group in this position. The **leukotrienes** and **lipoxins** (Figure 15–4) are a third group of eicosanoid derivatives formed via the lipoxygenase pathway (Figure 23–11). They are characterized by the presence of three or four conjugated double bonds, respectively. Leukotrienes cause bronchoconstriction as well as being potent proinflammatory agents, and play a part in **asthma.**

Most Naturally Occurring Unsaturated Fatty Acids Have *cis* Double Bonds

The carbon chains of saturated fatty acids form a zigzag pattern when extended at low temperatures. At higher temperatures, some bonds rotate, causing chain shortening, which explains why biomembranes become thinner with increases in temperature. A type of **geometric isomerism** occurs in unsaturated fatty acids, depending on the orientation of atoms or groups around the axes of double bonds, which do not allow rotation. If the acyl chains are on the same side of the bond, it is *cis-*, as in oleic acid; if on opposite sides, it is *trans-*, as in elaidic acid, the *trans* isomer of oleic acid (Figure 15–5). Double bonds in naturally occurring unsaturated long-chain fatty acids are nearly all in the *cis* configuration, the molecules being "bent" 120 degrees at the double bond. Thus, oleic acid has an L shape, whereas elaidic acid remains "straight." Increase in the number of *cis* double bonds in a fatty acid leads to a variety of possible spatial configurations of the molecule—eg, arachidonic acid, with four *cis* double bonds, is bent into a U shape. This has profound significance for molecular packing in cell membranes and on the positions occupied by fatty acids in more complex molecules such as phospholipids. *Trans* double bonds alter these spatial relationships. ***Trans* fatty acids** are present in certain foods, arising as a by-product of the saturation of fatty acids during hydrogenation, or "hardening," of natural oils in the manufacture of margarine. An additional small contribution comes from the ingestion of ruminant fat that contains *trans* fatty acids arising from the action of microorganisms in the rumen. Consumption of *trans* fatty acids is now known to be detrimental to health and is associated with increased risk of diseases including cardiovascular disease and diabetes mellitus. This has led to improved technology to produce soft margarine low in *trans* fatty acids or containing none at all.

TABLE 15–2 Unsaturated Fatty Acids of Physiologic and Nutritional Significance

Number of C Atoms and Number and Position of Common Double Bonds	Family	Common Name	Systematic Name	Occurrence
Monoenoic acids (one double bond)				
16:1;9	ω7	Palmitoleic	cis-9-Hexadecenoic	In nearly all fats.
18:1;9	ω9	Oleic	cis-9-Octadecenoic	Possibly the most common fatty acid in natural fats; particularly high in olive oil.
18:1;9	ω9	Elaidic	trans-9-Octadecenoic	Hydrogenated and ruminant fats.
Dienoic acids (two double bonds)				
18:2;9,12	ω6	Linoleic	all-cis-9,12-Octadecadienoic	Corn, peanut, cottonseed, soy bean, and many plant oils.
Trienoic acids (three double bonds)				
18:3;6,9,12	ω6	γ-Linolenic	all-cis-6,9,12-Octadecatrienoic	Some plants, eg, oil of evening primrose, borage oil; minor fatty acid in animals.
18:3;9,12,15	ω3	α-Linolenic	all-cis-9,12,15-Octadecatrienoic	Frequently found with linoleic acid but particularly in linseed oil.
Tetraenoic acids (four double bonds)				
20:4;5,8,11,14	ω6	Arachidonic	all-cis-5,8,11,14-Eicosatetraenoic	Found in animal fats; important component of phospholipids in animals.
Pentaenoic acids (five double bonds)				
20:5;5,8,11,14,17	ω3	Timnodonic	all-cis-5,8,11,14,17-Eicosapentaenoic	Important component of fish oils, eg, cod liver, mackerel, menhaden, salmon oils.
Hexaenoic acids (six double bonds)				
22:6;4,7,10,13,16,19	ω3	Cervonic	all-cis-4,7,10,13,16,19-Docosahexaenoic	Fish oils, phospholipids in brain.

FIGURE 15–2 Prostaglandin E$_2$ (PGE$_2$).

FIGURE 15–3 Thromboxane A$_2$ (TXA$_2$).

FIGURE 15–4 Leukotriene A$_4$ (LTA$_4$).

FIGURE 15–5 Geometric isomerism of Δ^9, 18:1 fatty acids (oleic and elaidic acids).

Physical and Physiologic Properties of Fatty Acids Reflect Chain Length and Degree of Unsaturation

The melting points of even-numbered carbon fatty acids increase with chain length and decrease according to unsaturation. A triacylglycerol containing three saturated fatty acids of 12 carbons or more is solid at body temperature, whereas if the fatty acid residues are 18:2, it is liquid to below 0°C. In practice, natural acylglycerols contain a mixture of fatty acids tailored to suit their functional roles. The membrane lipids, which must be fluid at all environmental temperatures, are more unsaturated than storage lipids. Lipids in tissues that are subject to cooling, eg, in hibernators or in the extremities of animals, are more unsaturated.

TRIACYLGLYCEROLS (TRIGLYCERIDES)* ARE THE MAIN STORAGE FORMS OF FATTY ACIDS

The triacylglycerols (Figure 15–6) are esters of the trihydric alcohol glycerol and fatty acids. Mono- and diacylglycerols, wherein one or two fatty acids are esterified with glycerol, are also found in the tissues. These are of particular significance in the synthesis and hydrolysis of triacylglycerols.

Carbons 1 & 3 of Glycerol Are Not Identical

To number the carbon atoms of glycerol unambiguously, the -sn (stereochemical numbering) system is used. It is important to realize that carbons 1 and 3 of glycerol are not identical when viewed in three dimensions (shown as a projection formula in Figure 15–7). Enzymes readily distinguish between them and are nearly always specific for one or the other carbon; eg, glycerol is always phosphorylated on sn-3 by glycerol kinase to give glycerol 3-phosphate and not glycerol 1-phosphate.

PHOSPHOLIPIDS ARE THE MAIN LIPID CONSTITUENTS OF MEMBRANES

Phospholipids may be regarded as derivatives of **phosphatidic acid** (Figure 15–8), in which the phosphate is esterified with the —OH of a suitable alcohol. Phosphatidic acid is important

FIGURE 15–6 Triacylglycerol.

FIGURE 15–7 Triacyl-sn-glycerol.

as an intermediate in the synthesis of triacylglycerols as well as phosphoglycerols but is not found in any great quantity in tissues.

Phosphatidylcholines (Lecithins) Occur in Cell Membranes

Phosphoacylglycerols containing choline (Figure 15–8) are the most abundant phospholipids of the cell membrane and represent a large proportion of the body's store of choline. Choline is important in nervous transmission, as acetylcholine, and as a store of labile methyl groups. **Dipalmitoyl lecithin** is a very effective surface-active agent and a major constituent of the **surfactant** preventing adherence, due to surface tension, of the inner surfaces of the lungs. Its absence from the lungs of premature infants causes **respiratory distress syndrome.** Most phospholipids have a saturated acyl radical in the sn-1 position but an unsaturated radical in the sn-2 position of glycerol.

Phosphatidylethanolamine (cephalin) and **phosphatidylserine** (found in most tissues) are also found in cell membranes and differ from phosphatidylcholine only in that ethanolamine or serine, respectively, replaces choline (Figure 15–8). Phosphatidylserine also plays a role in apoptosis (programmed cell death).

Phosphatidylinositol Is a Precursor of Second Messengers

The inositol is present in **phosphatidylinositol** as the stereoisomer, myoinositol (Figure 15–8). **Phosphatidylinositol 4,5-bisphosphate** is an important constituent of cell membrane phospholipids; upon stimulation by a suitable hormone agonist, it is cleaved into **diacylglycerol** and **inositol trisphosphate**, both of which act as internal signals or second messengers.

---

*According to the standardized terminology of the International Union of Pure and Applied Chemistry (IUPAC) and the International Union of Biochemistry (IUB), the monoglycerides, diglycerides, and triglycerides should be designated monoacylglycerols, diacylglycerols, and triacylglycerols, respectively. However, the older terminology is still widely used, particularly in clinical medicine.

Phosphatidic acid

FIGURE 15–8 Phosphatidic acid and its derivatives. The O⁻ shown shaded in phosphatidic acid is substituted by the substituents shown to form in **(A)** 3-phosphatidylcholine, **(B)** 3-phosphatidylethanolamine, **(C)** 3-phosphatidylserine, **(D)** 3-phosphatidylinositol, and **(E)** cardiolipin (diphosphatidylglycerol).

Cardiolipin Is a Major Lipid of Mitochondrial Membranes

Phosphatidic acid is a precursor of **phosphatidylglycerol** which, in turn, gives rise to **cardiolipin** (Figure 15–8). This phospholipid is found only in mitochondria and is essential for mitochondrial function. Decreased cardiolipin levels or alterations in its structure or metabolism cause mitochondrial dysfunction in aging and in pathological conditions including heart failure, hypothyroidism and Barth syndrome (cardioskeletal myopathy).

FIGURE 15–9 Lysophosphatidylcholine (lysolecithin).

FIGURE 15–10 Plasmalogen.

Lysophospholipids Are Intermediates in the Metabolism of Phosphoglycerols

These are phosphoacylglycerols containing only one acyl radical, eg, **lysophosphatidylcholine (lysolecithin)** (Figure 15–9), important in the metabolism and interconversion of phospholipids. It is also found in oxidized lipoproteins and has been implicated in some of their effects in promoting **atherosclerosis.**

Plasmalogens Occur in Brain & Muscle

These compounds constitute as much as 10% of the phospholipids of brain and muscle. Structurally, the plasmalogens resemble phosphatidylethanolamine but possess an ether link on the *sn*-1 carbon instead of the ester link found in acylglycerols. Typically, the alkyl radical is an unsaturated alcohol (Figure 15–10). In some instances, choline, serine, or inositol may be substituted for ethanolamine.

Sphingomyelins Are Found in the Nervous System

Sphingomyelins are found in large quantities in brain and nerve tissue. On hydrolysis, the sphingomyelins yield a fatty acid, phosphoric acid, choline, and a complex amino alcohol, **sphingosine** (Figure 15–11). No glycerol is present. The combination of sphingosine plus fatty acid is known as **ceramide,** a structure also found in the glycosphingolipids (see below).

GLYCOLIPIDS (GLYCOSPHINGOLIPIDS) ARE IMPORTANT IN NERVE TISSUES & IN THE CELL MEMBRANE

Glycolipids are widely distributed in every tissue of the body, particularly in nervous tissue such as brain. They occur par-

FIGURE 15–11 A sphingomyelin.

FIGURE 15–13 GM1 ganglioside, a monosialoganglioside, the receptor in human intestine for cholera toxin.

GM_3, is of considerable biologic interest, as it is known to be the receptor in human intestine for cholera toxin. Other gangliosides can contain anywhere from one to five molecules of sialic acid, giving rise to di-, trisialogangliosides, etc.

STEROIDS PLAY MANY PHYSIOLOGICALLY IMPORTANT ROLES

Cholesterol is probably the best known steroid because of its association with **atherosclerosis** and heart disease. However, biochemically it is also of significance because it is the precursor of a large number of equally important **steroids** that include the bile acids, adrenocortical hormones, sex hormones, D vitamins, cardiac glycosides, sitosterols of the plant kingdom, and some alkaloids.

All steroids have a similar cyclic nucleus resembling phenanthrene (rings A, B, and C) to which a cyclopentane ring (D) is attached. The carbon positions on the steroid nucleus are numbered as shown in Figure 15–14. It is important to

ticularly in the outer leaflet of the plasma membrane, where they contribute to **cell surface carbohydrates.**

The major glycolipids found in animal tissues are glycosphingolipids. They contain ceramide and one or more sugars. **Galactosylceramide** is a major glycosphingolipid of brain and other nervous tissue, found in relatively low amounts elsewhere. It contains a number of characteristic C24 fatty acids, eg, cerebronic acid.

Galactosylceramide (Figure 15–12) can be converted to sulfogalactosylceramide **(sulfatide),** present in high amounts in myelin. Glucosylceramide is the predominant simple glycosphingolipid of extraneural tissues, also occurring in the brain in small amounts. **Gangliosides** are complex glycosphingolipids derived from glucosylceramide that contain in addition one or more molecules of a **sialic acid.** Neuraminic acid (NeuAc; see Chapter 14) is the principal sialic acid found in human tissues. Gangliosides are also present in nervous tissues in high concentration. They appear to have receptor and other functions. The simplest ganglioside found in tissues is GM_3, which contains ceramide, one molecule of glucose, one molecule of galactose, and one molecule of NeuAc. In the shorthand nomenclature used, G represents ganglioside; M is a monosialo-containing species; and the subscript 3 is a number assigned on the basis of chromatographic migration. GM1 (Figure 15–13), a more complex ganglioside derived from

FIGURE 15–14 The steroid nucleus.

FIGURE 15–12 Structure of galactosylceramide (galactocerebroside, R = H), and sulfogalactosylceramide (a sulfatide, R = SO_4^{2-}).

FIGURE 15–15 Conformations of stereoisomers of the steroid nucleus.

realize that in structural formulas of steroids, a simple hexagonal ring denotes a completely saturated six-carbon ring with all valences satisfied by hydrogen bonds unless shown otherwise; ie, it is not a benzene ring. All double bonds are shown as such. Methyl side chains are shown as single bonds unattached at the farther (methyl) end. These occur typically at positions 10 and 13 (constituting C atoms 19 and 18). A side chain at position 17 is usual (as in cholesterol). If the compound has one or more hydroxyl groups and no carbonyl or carboxyl groups, it is a **sterol,** and the name terminates in -*ol.*

Because of Asymmetry in the Steroid Molecule, Many Stereoisomers Are Possible

Each of the six-carbon rings of the steroid nucleus is capable of existing in the three-dimensional conformation either of a "chair" or a "boat" (Figure 15–15). In naturally occurring steroids, virtually all the rings are in the "chair" form, which is the more stable conformation. With respect to each other, the rings can be either *cis* or *trans* (Figure 15–16). The junction between the A and B rings can be *cis* or *trans* in naturally occurring steroids. That between B and C is *trans,* as is usually the C/D junction. Bonds attaching substituent groups above the plane of the rings (β bonds) are shown with bold solid lines, whereas those bonds attaching groups below (α bonds) are indicated with broken lines. The A ring of a 5α steroid is always *trans* to the B ring, whereas it is *cis* in a 5β steroid. The methyl groups attached to C10 and C13 are invariably in the β configuration.

FIGURE 15–17 Cholesterol, 3-hydroxy-5,6-cholestene.

FIGURE 15–18 Ergosterol.

Cholesterol Is a Significant Constituent of Many Tissues

Cholesterol (Figure 15–17) is widely distributed in all cells of the body but particularly in nervous tissue. It is a major constituent of the plasma membrane and of plasma lipoproteins. It is often found as **cholesteryl ester,** where the hydroxyl group on position 3 is esterified with a long-chain fatty acid. It occurs in animals but not in plants or bacteria.

Ergosterol Is a Precursor of Vitamin D

Ergosterol occurs in plants and yeast and is important as a precursor of vitamin D (Figure 15–18). When irradiated with ultraviolet light, ring B is opened to form vitamin D_2 in a process similar to that which forms vitamin D_3 from 7-dehydrocholesterol in the skin (Figure 44–3).

FIGURE 15–16 Generalized steroid nucleus, showing **(A)** an all-*trans* configuration between adjacent rings and **(B)** a *cis* configuration between rings A and B.

$$-CH=\overset{\overset{\displaystyle CH_3}{|}}{C}-CH=CH-$$

FIGURE 15–19 Isoprene unit.

CH₂OH

FIGURE 15–20 Dolichol—a C95 alcohol.

Polyprenoids Share the Same Parent Compound as Cholesterol

Although not steroids, these compounds are related because they are synthesized, like cholesterol (Figure 26–2), from five-carbon isoprene units (Figure 15–19). They include **ubiquinone** (Chapter 13), which participates in the respiratory chain in mitochondria, and the long-chain alcohol **dolichol** (Figure 15–20), which takes part in glycoprotein synthesis by transferring carbohydrate residues to asparagine residues of the polypeptide (Chapter 47). Plant-derived isoprenoid compounds include rubber, camphor, the fat-soluble vitamins A, D, E, and K, and β-carotene (provitamin A).

LIPID PEROXIDATION IS A SOURCE OF FREE RADICALS

Peroxidation (**auto-oxidation**) of lipids exposed to oxygen is responsible not only for deterioration of foods (**rancidity**) but also for damage to tissues in vivo, where it may be a cause of cancer, inflammatory diseases, atherosclerosis, and aging. The deleterious effects are considered to be caused by **free radicals** (ROO•, RO•, OH•) produced during peroxide formation from fatty acids containing methylene-interrupted double bonds, ie,

those found in the naturally occurring polyunsaturated fatty acids (Figure 15–21). Lipid peroxidation is a chain reaction providing a continuous supply of free radicals that initiate further peroxidation and thus has potentially devastating effects. The whole process can be depicted as follows:

1. Initiation:

$$ROOH + Metal^{(n)+} \rightarrow ROO^{\bullet} + Metal^{(n-1)+} + H^+$$
$$X^{\bullet} + RH \rightarrow R^{\bullet} + XH$$

2. Propagation:

$$R^{\bullet} + O_2 \rightarrow ROO^{\bullet}$$
$$ROO^{\bullet} + RH \rightarrow ROOH + R^{\bullet}, etc$$

3. Termination:

$$ROO^{\bullet} + ROO^{\bullet} \rightarrow ROOR + O_2$$
$$ROO^{\bullet} + R^{\bullet} \rightarrow ROOR$$
$$R^{\bullet} + R^{\bullet} \rightarrow RR$$

To control and reduce lipid peroxidation, both humans in their activities and nature invoke the use of **antioxidants.** Propyl gallate, butylated hydroxyanisole (BHA), and butylated hydroxytoluene (BHT) are antioxidants used as food additives. Naturally occurring antioxidants include vitamin E (tocopherol), which is lipid-soluble, and urate and vitamin C, which are water-soluble. Beta-carotene is an antioxidant at low PO_2. Antioxidants fall into two classes: (1) preventive antioxidants, which reduce the rate of chain initiation and (2) chain-breaking antioxidants, which interfere with chain propagation. Preventive antioxidants include catalase and other peroxidases such as glutathione peroxidase (Figure 21–3) that react with ROOH; selenium, which is an essential component of glutathione peroxidase and regulates its activity, and chelators of metal ions such as EDTA (ethylenediaminetetraacetate) and DTPA (diethylenetriaminepentaacetate). In vivo, the principal chain-breaking antioxidants are superoxide dismutase, which

Malondialdehyde   **Endoperoxide**   **Hydroperoxide**
                                                        ROOH

FIGURE 15–21 Lipid peroxidation. The reaction is initiated by an existing free radical (X·), by light, or by metal ions. Malondialdehyde is only formed by fatty acids with three or more double bonds and is used as a measure of lipid peroxidation together with ethane from the terminal two carbons of ω3 fatty acids and pentane from the terminal five carbons of ω6 fatty acids.

FIGURE 15–22 Formation of lipid membranes, micelles, emulsions, and liposomes from amphipathic lipids, eg, phospholipids.

acts in the aqueous phase to trap superoxide free radicals (O_2^-) urate, and vitamin E, which acts in the lipid phase to trap ROO• radicals (Figure 44–6).

Peroxidation is also catalyzed in vivo by heme compounds and by **lipoxygenases** found in platelets and leukocytes. Other products of auto-oxidation or enzymic oxidation of physiologic significance include **oxysterols** (formed from cholesterol) and **isoprostanes** (formed from the peroxidation of polyunsaturated fatty acids such as arachidonic acid).

AMPHIPATHIC LIPIDS SELF-ORIENT AT OIL: WATER INTERFACES

They Form Membranes, Micelles, Liposomes, & Emulsions

In general, lipids are insoluble in water since they contain a predominance of nonpolar (hydrocarbon) groups. However, fatty acids, phospholipids, sphingolipids, bile salts, and, to a lesser extent, cholesterol contain polar groups. Therefore, part of the molecule is **hydrophobic,** or water-insoluble; and part is **hydrophilic,** or water-soluble. Such molecules are described as **amphipathic** (Figure 15–22). They become oriented at oil:water interfaces with the polar group in the water phase and the nonpolar group in the oil phase. A bilayer of such amphipathic lipids is the basic structure in biologic **membranes** (Chapter 40). When a critical concentration of these lipids is present in an aqueous medium, they form **micelles. Liposomes** may be formed by sonicating an amphipathic lipid in an aqueous medium. They consist of spheres of lipid bilayers that enclose part of the aqueous medium. Aggregations of bile salts into micelles and liposomes and the formation of mixed micelles with the products of fat digestion are important in facilitating absorption of lipids from the intestine. Liposomes are of potential clinical use—particularly when combined with tissue-specific antibodies—as carriers of drugs in the circulation, targeted to specific organs, eg, in cancer therapy. In addition, they are used for gene transfer into vascular cells and as carriers for topical and transdermal delivery of drugs and cosmetics. **Emulsions** are much larger particles, formed usually by nonpolar lipids in an aqueous medium. These are stabilized by emulsifying agents such as amphipathic lipids (eg, lecithin), which form a surface layer separating the main bulk of the nonpolar material from the aqueous phase (Figure 15–22).

SUMMARY

- Lipids have the common property of being relatively insoluble in water (hydrophobic) but soluble in nonpolar solvents. Amphipathic lipids also contain one or more polar groups, making them suitable as constituents of membranes at lipid-water interfaces.

- The lipids of major physiologic significance are fatty acids and their esters, together with cholesterol and other steroids.

- Long-chain fatty acids may be saturated, monounsaturated, or polyunsaturated, according to the number of double bonds present. Their fluidity decreases with chain length and increases according to degree of unsaturation.

- Eicosanoids are formed from 20-carbon polyunsaturated fatty acids and make up an important group of physiologically and pharmacologically active compounds known as prostaglandins, thromboxanes, leukotrienes, and lipoxins.

- The esters of glycerol are quantitatively the most significant lipids, represented by triacylglycerol ("fat"), a major constituent of lipoproteins and the storage form of lipid in adipose tissue. Phosphoacylglycerols are amphipathic lipids and have important roles—as major constituents of membranes and the outer layer of lipoproteins, as surfactant in the lung, as precursors of second messengers, and as constituents of nervous tissue.

- Glycolipids are also important constituents of nervous tissue such as brain and the outer leaflet of the cell membrane, where they contribute to the carbohydrates on the cell surface.

- Cholesterol, an amphipathic lipid, is an important component of membranes. It is the parent molecule from which all other steroids in the body, including major hormones such as the adrenocortical and sex hormones, D vitamins, and bile acids, are synthesized.

- Peroxidation of lipids containing polyunsaturated fatty acids leads to generation of free radicals that damage tissues and cause disease.

REFERENCES

Benzie IFF: Lipid peroxidation: a review of causes, consequences, measurement and dietary influences. Int J Food Sci Nutr 1996;47:233.

Christie WW: *Lipid Analysis,* 3rd ed. The Oily Press, 2003.

Dowhan W, Bodanov H: Functional roles of lipids in membranes. In: *Biochemistry of Lipids, Lipoproteins and Membranes,* 4th ed. Vance DE, Vance JE (editors). Elsevier, 2002.

Gurr MI: *Lipids in Nutrition and Health: A Reappraisal.* The Oily Press, 1999.

Gurr MI, Harwood JL, Frayn K: *Lipid Biochemistry.* Blackwell Publishing, 2002.

Overview of Metabolism & the Provision of Metabolic Fuels

David A. Bender, PhD

BIOMEDICAL IMPORTANCE

Metabolism is the term used to describe the interconversion of chemical compounds in the body, the pathways taken by individual molecules, their interrelationships, and the mechanisms that regulate the flow of metabolites through the pathways. Metabolic pathways fall into three categories: (1) **Anabolic pathways,** which are those involved in the synthesis of larger and more complex compounds from smaller precursors—eg, the synthesis of protein from amino acids and the synthesis of reserves of triacylglycerol and glycogen. Anabolic pathways are endothermic. (2) **Catabolic pathways,** which are involved in the breakdown of larger molecules, commonly involving oxidative reactions; they are exothermic, producing reducing equivalents, and, mainly via the respiratory chain, ATP. (3) **Amphibolic pathways,** which occur at the "crossroads" of metabolism, acting as links between the anabolic and catabolic pathways, eg, the citric acid cycle.

Knowledge of normal metabolism is essential for an understanding of abnormalities underlying disease. Normal metabolism includes adaptation to periods of starvation, exercise, pregnancy, and lactation. Abnormal metabolism may result from nutritional deficiency, enzyme deficiency, abnormal secretion of hormones, or the actions of drugs and toxins.

A 70-kg adult human being requires about 8–12 MJ (1920–2900 kcal) from metabolic fuels each day, depending on physical activity. Larger animals require less, and smaller animals more, per kg body weight, and growing children and animals have a proportionally higher requirement to allow for the energy cost of growth. For human beings this requirement is met from carbohydrates (40–60%), lipids (mainly triacylglycerol, 30–40%), and protein (10–15%), as well as alcohol. The mix of carbohydrate, lipid, and protein being oxidized varies, depending on whether the subject is in the fed or fasting state, and on the duration and intensity of physical work.

The requirement for metabolic fuels is relatively constant throughout the day, since average physical activity increases metabolic rate only by about 40–50% over the basal metabolic rate. However, most people consume their daily intake of metabolic fuels in two or three meals, so there is a need to form reserves of carbohydrate (glycogen in liver and muscle) and lipid (triacylglycerol in adipose tissue) in the period following a meal, for use during the intervening time when there is no intake of food.

If the intake of metabolic fuels is consistently greater than energy expenditure, the surplus is stored, largely as triacylglycerol in adipose tissue, leading to the development of **obesity** and its associated health hazards. By contrast, if the intake of metabolic fuels is consistently lower than energy expenditure, there are negligible reserves of fat and carbohydrate, and amino acids arising from protein turnover are used for energy-yielding metabolism rather than replacement protein synthesis, leading to **emaciation,** wasting, and, eventually, death (see Chapter 43).

In the fed state, after a meal, there is an ample supply of carbohydrate, and the metabolic fuel for most tissues is glucose. In the fasting state glucose must be spared for use by the central nervous system (which is largely dependent on glucose) and the red blood cells (which are wholly reliant on glucose). Therefore, tissues that can use fuels other than glucose do so; muscle and liver oxidize fatty acids and the liver synthesizes ketone bodies from fatty acids to export to muscle and other tissues. As glycogen reserves become depleted, amino acids arising from protein turnover are used for **gluconeogenesis.**

The formation and utilization of reserves of triacylglycerol and glycogen, and the extent to which tissues take up and oxidize glucose, are largely controlled by the hormones **insulin** and **glucagon.** In **diabetes mellitus**, there is either impaired synthesis and secretion of insulin (juvenile onset, or type I diabetes) or impaired sensitivity of tissues to insulin action (adult onset, or type II diabetes), leading to severe metabolic derangement. In cattle, the demands of heavy lactation can lead to ketosis, as can the demands of twin pregnancy in sheep.

PATHWAYS THAT PROCESS THE MAJOR PRODUCTS OF DIGESTION

The nature of the diet sets the basic pattern of metabolism. There is a need to process the products of digestion of dietary carbohydrate, lipid, and protein. These are mainly glucose, fatty acids and glycerol, and amino acids, respectively. In ruminants (and, to a lesser extent, other herbivores), dietary cellulose is fermented by symbiotic microorganisms to short-chain fatty acids (acetic, propionic, butyric), and metabolism

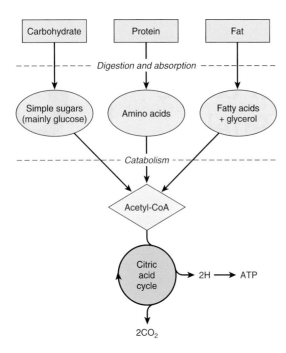

FIGURE 16–1 Outline of the pathways for the catabolism of dietary carbohydrate, protein, and fat. All the pathways lead to the production of acetyl-CoA, which is oxidized in the citric acid cycle, ultimately yielding ATP by the process of oxidative phosphorylation.

in these animals is adapted to use these fatty acids as major substrates. All the products of digestion are metabolized to a **common product, acetyl-CoA,** which is then oxidized by the **citric acid cycle** (Figure 16–1).

Carbohydrate Metabolism Is Centered on the Provision & Fate of Glucose

Glucose is the major fuel of most tissues (Figure 16–2). It is metabolized to pyruvate by the pathway of **glycolysis.** Aerobic tissues metabolize pyruvate to **acetyl-CoA,** which can enter the citric acid cycle for complete oxidation to CO_2 and H_2O, linked to the formation of ATP in the process of **oxidative phosphorylation** (Figure 13–2). Glycolysis can also occur anaerobically (in the absence of oxygen) when the end product is lactate.

Glucose and its metabolites also take part in other processes, eg: (1) Synthesis of the storage polymer **glycogen** in skeletal muscle and liver. (2) The **pentose phosphate pathway,** an alternative to part of the pathway of glycolysis. It is a source of reducing equivalents (NADPH) for fatty acid synthesis and the source of **ribose** for nucleotide and nucleic acid synthesis. (3) Triose phosphates gives rise to the **glycerol moiety** of triacylglycerols. (4) Pyruvate and intermediates of the citric acid cycle provide the carbon skeletons for the synthesis of **amino acids,** and acetyl-CoA is the precursor of **fatty acids** and **cholesterol** (and hence of all steroids synthesized in the body). **Gluconeogenesis** is the process of forming glucose from noncarbohydrate precursors, eg, lactate, amino acids, and glycerol.

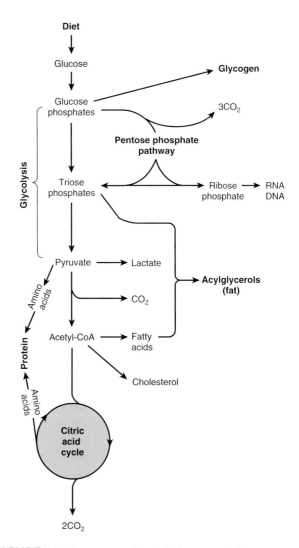

FIGURE 16–2 Overview of carbohydrate metabolism showing the major pathways and end products. Gluconeogenesis is not shown.

Lipid Metabolism Is Concerned Mainly with Fatty Acids & Cholesterol

The source of long-chain fatty acids is either dietary lipid or de novo synthesis from acetyl-CoA derived from carbohydrate or amino acids. Fatty acids may be oxidized to **acetyl-CoA (β-oxidation)** or esterified with glycerol, forming **triacylglycerol** (fat) as the body's main fuel reserve.

Acetyl-CoA formed by β-oxidation may undergo three fates (Figure 16–3).

1. As with acetyl-CoA arising from glycolysis, it is **oxidized** to $CO_2 + H_2O$ via the citric acid cycle.
2. It is the precursor for synthesis of **cholesterol** and other **steroids.**
3. In the liver, it is used to form **ketone bodies** (acetoacetate and 3-hydroxybutyrate) that are important fuels in prolonged fasting.

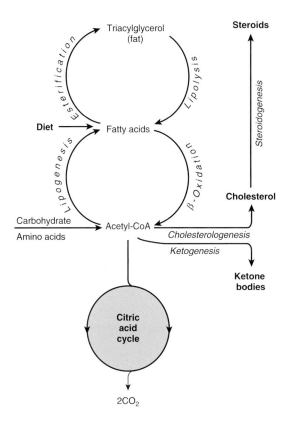

FIGURE 16–3 Overview of fatty acid metabolism showing the major pathways and end products. The ketone bodies are acetoacetate, 3-hydroxybutyrate, and acetone.

Much of Amino Acid Metabolism Involves Transamination

The amino acids are required for protein synthesis (Figure 16–4). Some must be supplied in the diet (the **essential amino acids**), since they cannot be synthesized in the body. The remainder are **nonessential amino acids,** which are supplied in the diet, but can also be formed from metabolic intermediates by **transamination** using the amino nitrogen from other amino acids. After **deamination,** amino nitrogen is excreted as **urea,** and the carbon skeletons that remain after transamination may (1) be oxidized to CO_2 via the citric acid cycle, (2) be used to synthesize glucose (gluconeogenesis), or (3) form ketone bodies, which may be oxidized or be used for synthesis of fatty acids.

Several amino acids are also the precursors of other compounds, eg, purines, pyrimidines, hormones such as epinephrine and thyroxine, and neurotransmitters.

METABOLIC PATHWAYS MAY BE STUDIED AT DIFFERENT LEVELS OF ORGANIZATION

In addition to studies in the whole organism, the location and integration of metabolic pathways is revealed by studies at several levels of organization. (1) At the **tissue and organ level** the nature of the substrates entering and metabolites leaving tissues and organs is defined. (2) At the **subcellular level**

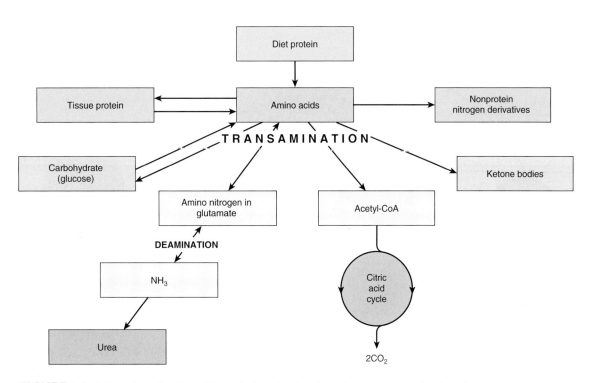

FIGURE 16–4 Overview of amino acid metabolism showing the major pathways and end products.

each cell organelle (eg, the mitochondrion) or compartment (eg, the cytosol) has specific roles that form part of a subcellular pattern of metabolic pathways.

At the Tissue & Organ Level, the Blood Circulation Integrates Metabolism

Amino acids resulting from the digestion of dietary protein and **glucose** resulting from the digestion of carbohydrate are absorbed via the hepatic portal vein. The liver has the role of regulating the blood concentration of water-soluble metabolites (Figure 16–5). In the case of glucose, this is achieved by taking up glucose in excess of immediate requirements and converting it to glycogen (**glycogenesis,** Chapter 19) or to fatty acids (**lipogenesis,** Chapter 23). Between meals, the liver acts to maintain the blood glucose concentration by breaking down glycogen (**glycogenolysis**, Chapter 19) and, together with the kidney, by converting noncarbohydrate metabolites such as lactate, glycerol, and amino acids to glucose (**gluconeogenesis**, Chapter 20). The maintenance of an adequate concentration of blood glucose is vital for those tissues in which it is the major fuel (the brain) or the only fuel (erythrocytes). The liver also **synthesizes the major plasma proteins** (eg, albumin) and **deaminates amino acids** that are in excess of requirements, forming urea, which is transported to the kidney and excreted (Chapter 28).

Skeletal muscle utilizes glucose as a fuel, both aerobically, forming CO_2, and anaerobically, forming lactate. It stores gly-

cogen as a fuel for use in muscle contraction and synthesizes muscle protein from plasma amino acids. Muscle accounts for approximately 50% of body mass and consequently represents a considerable store of protein that can be drawn upon to supply amino acids for gluconeogenesis in starvation (Chapter 20).

Lipids in the diet (Figure 16–6) are mainly triacylglycerol, and are hydrolyzed to monoacylglycerols and fatty acids in the gut, then re-esterified in the intestinal mucosa. Here they are packaged with protein and secreted into the lymphatic system and thence into the bloodstream as **chylomicrons,** the largest of the plasma **lipoproteins.** Chylomicrons also contain other lipid-soluble nutrients. Unlike glucose and amino acids, chylomicron triacylglycerol is not taken up directly by the liver. It is first metabolized by tissues that have **lipoprotein lipase,** which hydrolyzes the triacylglycerol, releasing fatty acids that are incorporated into tissue lipids or oxidized as fuel. The chylomicron remnants are cleared by the liver. The other major source of long-chain fatty acids is synthesis **(lipogenesis)** from carbohydrate, in adipose tissue and the liver.

Adipose tissue triacylglycerol is the main fuel reserve of the body. It is hydrolyzed **(lipolysis)** and glycerol and free fatty acids are released into the circulation. Glycerol is a substrate for gluconeogenesis. The fatty acids are transported bound to serum albumin; they are taken up by most tissues (but not brain or erythrocytes) and either esterified to triacylglycerols for storage or oxidized as a fuel. In the liver, triacylglycerol aris-

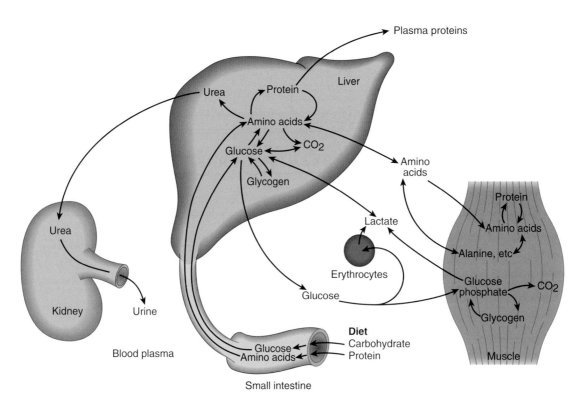

FIGURE 16–5 Transport and fate of major carbohydrate and amino acid substrates and metabolites. Note that there is little free glucose in muscle, since it is rapidly phosphorylated upon entry.

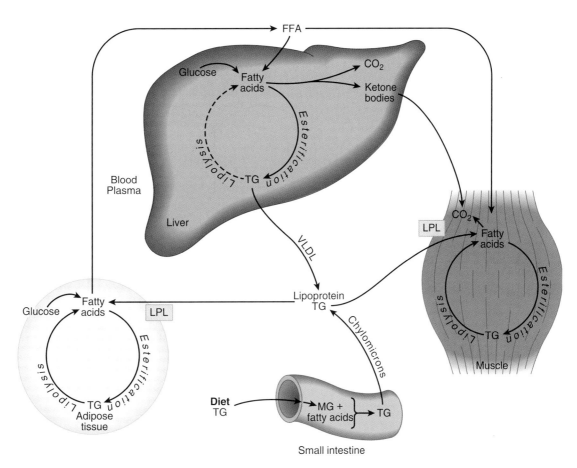

FIGURE 16–6 Transport and fate of major lipid substrates and metabolites. (FFA, free fatty acids; LPL, lipoprotein lipase; MG, monoacylglycerol; TG, triacylglycerol; VLDL, very low density lipoprotein.)

ing from lipogenesis, free fatty acids, and chylomicron remnants (see Figure 25–3) is secreted into the circulation in **very low density lipoprotein (VLDL).** This triacylglycerol undergoes a fate similar to that of chylomicrons. Partial oxidation of fatty acids in the liver leads to **ketone body** production (**ketogenesis, Chapter 22**). Ketone bodies are transported to extrahepatic tissues, where they act as a fuel in prolonged fasting and starvation.

At the Subcellular Level, Glycolysis Occurs in the Cytosol & the Citric Acid Cycle in the Mitochondria

Compartmentation of pathways in separate subcellular compartments or organelles permits integration and regulation of metabolism. Not all pathways are of equal importance in all cells. Figure 16–7 depicts the subcellular compartmentation of metabolic pathways in a liver parenchymal cell.

The central role of the **mitochondrion** is immediately apparent, since it acts as the focus of carbohydrate, lipid, and amino acid metabolism. It contains the enzymes of the citric acid cycle (Chapter 17), β-oxidation of fatty acids and ketogenesis (Chapter 22), as well as the respiratory chain and ATP synthase (Chapter 13).

Glycolysis (Chapter 18), the pentose phosphate pathway (Chapter 21), and fatty acid synthesis (Chapter 23) all occur in the cytosol. In gluconeogenesis (Chapter 20), substrates such as lactate and pyruvate, which are formed in the cytosol, enter the mitochondrion to yield **oxaloacetate** as a precursor for the synthesis of glucose in the cytosol.

The membranes of the **endoplasmic reticulum** contain the enzyme system for **triacylglycerol synthesis** (Chapter 24), and the **ribosomes** are responsible for **protein synthesis** (Chapter 37).

THE FLUX OF METABOLITES THROUGH METABOLIC PATHWAYS MUST BE REGULATED IN A CONCERTED MANNER

Regulation of the overall flux through a pathway is important to ensure an appropriate supply of the products of that pathway. It is achieved by control of one or more key reactions in the pathway, catalyzed by **regulatory enzymes.** The physicochemical factors that control the rate of an enzyme-catalyzed reaction, such as substrate concentration, are of primary importance in the control of the overall rate of a metabolic pathway (Chapter 9).

FIGURE 16–7 Intracellular location and overview of major metabolic pathways in a liver parenchymal cell. (AA →, metabolism of one or more essential amino acids; AA ↔, metabolism of one or more nonessential amino acids.)

Nonequilibrium Reactions Are Potential Control Points

In a reaction at equilibrium, the forward and reverse reactions occur at equal rates, and there is therefore no net flux in either direction.

$$A \leftrightarrow B \leftrightarrow C \leftrightarrow D$$

In vivo, under "steady-state" conditions, there is a net flux from left to right because there is a continuous supply of A and removal of D. In practice, there are invariably one or more **nonequilibrium** reactions in a metabolic pathway, where the reactants are present in concentrations that are far from equilibrium. In attempting to reach equilibrium, large losses of free energy occur, making this type of reaction essentially irreversible.

$$A \leftrightarrow B \xrightarrow{\text{Heat}} C \leftrightarrow D$$

Such a pathway has both flow and direction. The enzymes catalyzing nonequilibrium reactions are usually present in low concentration and are subject to a variety of regulatory mechanisms. However, most reactions in metabolic pathways cannot be classified as equilibrium or nonequilibrium, but fall somewhere between the two extremes.

The Flux-Generating Reaction Is the First Reaction in a Pathway That Is Saturated with Substrate

It may be identified as a nonequilibrium reaction in which the K_m of the enzyme is considerably lower than the normal substrate concentration. The first reaction in glycolysis, catalyzed by hexokinase (Figure 18–2), is such a flux-generating step because its K_m for glucose of 0.05 mmol/L is well below the normal blood glucose concentration of 5 mmol/L.

ALLOSTERIC & HORMONAL MECHANISMS ARE IMPORTANT IN THE METABOLIC CONTROL OF ENZYME-CATALYZED REACTIONS

A hypothetical metabolic pathway is shown in Figure 16–8, in which reactions A ↔ B and C ↔ D are equilibrium reactions and B ↔ C is a nonequilibrium reaction. The flux through such a pathway can be regulated by the availability of substrate A. This depends on its supply from the blood, which in turn depends on either food intake or key reactions that release substrates from tissue reserves into the bloodstream, eg, glycogen phosphorylase in liver (Figure 19–1) and hormone-sensitive lipase in adipose tissue (Figure 25–8). It also depends on the transport of substrate A into the cell. Flux is also determined by removal of the end product D and the availability of cosubstrates or cofactors represented by X and Y. Enzymes catalyzing nonequilibrium reactions are often allosteric proteins subject to the rapid actions of "feed-back" or "feed-forward" control by **allosteric modifiers,** in immediate response to the needs of the cell (Chapter 9). Frequently, the product of a biosynthetic pathway inhibits the enzyme catalyzing the first reaction in the pathway. Other control mechanisms depend on the action of **hormones** responding to the needs of the body as a whole; they may act rapidly by altering the activity of existing enzyme molecules, or slowly by altering the rate of enzyme synthesis (see Chapter 42).

MANY METABOLIC FUELS ARE INTERCONVERTIBLE

Carbohydrate in excess of requirements for immediate energy-yielding metabolism and formation of glycogen reserves in muscle and liver can readily be used for synthesis of fatty acids, and hence triacylglycerol in both adipose tissue and liver (whence it is exported in very low density lipoprotein). The importance of lipogenesis in humans is unclear; in Western countries dietary fat provides 35–45% of energy intake, while in less-developed countries, where carbohydrate may provide

60–75% of energy intake, the total intake of food is so low that there is little surplus for lipogenesis anyway. A high intake of fat inhibits lipogenesis in adipose tissue and liver.

Fatty acids (and ketone bodies formed from them) cannot be used for the synthesis of glucose. The reaction of pyruvate dehydrogenase, forming acetyl-CoA, is irreversible, and for every two-carbon unit from acetyl-CoA that enters the citric acid cycle, there is a loss of two carbon atoms as carbon dioxide before oxaloacetate is reformed. This means that acetyl-CoA (and hence any substrates that yield acetyl-CoA) can never be used for gluconeogenesis. The (relatively rare) fatty acids with an odd number of carbon atoms yield propionyl CoA as the product of the final cycle of β oxidation, and this can be a substrate for gluconeogenesis, as can the glycerol released by lipolysis of adipose tissue triacylglycerol reserves.

Most of the amino acids in excess of requirements for protein synthesis (arising from the diet or from tissue protein turnover) yield pyruvate, or four- and five-carbon intermediates of the citric acid cycle (Chapter 29). Pyruvate can be carboxylated to oxaloacetate, which is the primary substrate for gluconeogenesis, and the other intermediates of the cycle also result in a net increase in the formation of oxaloacetate, which is then available for gluconeogenesis. These amino acids are classified as **glucogenic.** Two amino acids (lysine and leucine) yield only acetyl-CoA on oxidation, and hence cannot be used for gluconeogenesis, and four others (ie, phenylalanine, tyrosine, tryptophan, and isoleucine) give rise to both acetyl-CoA and intermediates that can be used for gluconeogenesis. Those amino acids that give rise to acetyl-CoA are referred to as **ketogenic,** because in prolonged fasting and starvation much of the acetyl-CoA is used for synthesis of ketone bodies in the liver.

A SUPPLY OF METABOLIC FUELS IS PROVIDED IN BOTH THE FED & FASTING STATES

Glucose Is Always Required by the Central Nervous System and Erythrocytes

Erythrocytes lack mitochondria and hence are wholly reliant on (anaerobic) glycolysis and the pentose phosphate pathway at all times. The brain can metabolize ketone bodies to meet about 20% of its energy requirements; the remainder must be supplied by glucose. The metabolic changes that occur in the fasting state and starvation are the consequences of the need to preserve glucose and the limited reserves of glycogen in liver and muscle for use by the brain and red blood cells, and to ensure the provision of alternative metabolic fuels for other tissues. In pregnancy the fetus requires a significant amount of glucose, as does the synthesis of lactose in lactation (Figure 16–9).

In the Fed State, Metabolic Fuel Reserves Are Laid Down

For several hours after a meal, while the products of digestion are being absorbed, there is an abundant supply of meta-

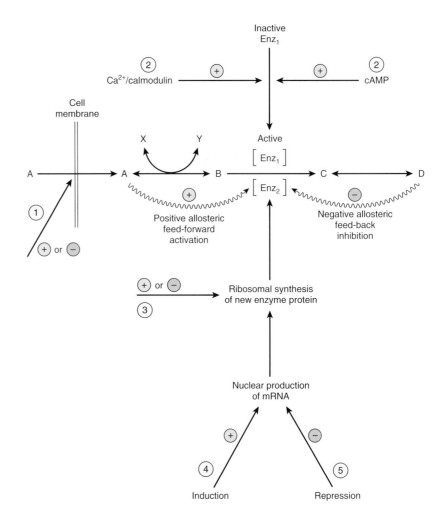

FIGURE 16–8 Mechanisms of control of an enzyme-catalyzed reaction. Circled numbers indicate possible sites of action of hormones: ① alteration of membrane permeability; ② conversion of an inactive to an active enzyme, usually involving phosphorylation/dephosphorylation reactions; ③ alteration of the rate translation of mRNA at the ribosomal level; ④ induction of new mRNA formation; and ⑤ repression of mRNA formation. ① and ② are rapid, whereas ③ through ⑤ are slower ways of regulating enzyme activity.

bolic fuels. Under these conditions, glucose is the major fuel for oxidation in most tissues; this is observed as an increase in the respiratory quotient (the ratio of carbon dioxide produced/oxygen consumed) from about 0.8 in the fasting state to near 1 (Table 16–1).

Glucose uptake into muscle and adipose tissue is controlled by **insulin,** which is secreted by the β-islet cells of the pancreas in response to an increased concentration of glucose in the portal blood. In the fasting state the glucose transporter of muscle and adipose tissue (GLUT-4) is in intracellular vesicles. An early response to insulin is the migration of these vesicles to the cell surface, where they fuse with the plasma membrane, exposing active glucose transporters. These insulin sensitive tissues only take up glucose from the bloodstream to any significant extent in the presence of the hormone. As insulin secretion falls in the fasting state, so the receptors are internalized again, reducing glucose uptake.

The uptake of glucose into the liver is independent of insulin, but liver has an isoenzyme of hexokinase (glucokinase) with a high K_m, so that as the concentration of glucose entering the liver increases, so does the rate of synthesis of glucose 6-phosphate. This is in excess of the liver's requirement for energy-yielding metabolism, and is used mainly for synthesis of **glycogen.** In both liver and skeletal muscle, insulin acts to stimulate glycogen synthetase and inhibit glycogen phospho-

rylase. Some of the additional glucose entering the liver may also be used for lipogenesis and hence triacylglycerol synthesis. In adipose tissue, insulin stimulates glucose uptake, its conversion to fatty acids and their esterification to triacylglycerol. It inhibits intracellular lipolysis and the release of free fatty acids.

The products of lipid digestion enter the circulation as **chylomicrons,** the largest of the plasma lipoproteins, especially rich in triacylglycerol (see Chapter 25). In adipose tissue and skeletal muscle, extracellular lipoprotein lipase is synthesized and activated in response to insulin; the resultant nonesterified fatty acids are largely taken up by the tissue and used for synthesis of triacylglycerol, while the glycerol remains in the bloodstream and is taken up by the liver and used for either gluconeogenesis and glycogen synthesis or lipogenesis. Fatty acids remaining in the bloodstream are taken up by the liver and reesterified. The lipid-depleted chylomicron remnants are cleared by the liver, and the remaining triacylglycerol is exported, together with that synthesized in the liver, in **very low density lipoprotein.**

Under normal conditions, the rate of tissue protein catabolism is more or less constant throughout the day; it is only in **cachexia** associated with advanced cancer and other diseases that there is an increased rate of protein catabolism. There is net protein catabolism in the fasting state, and net protein

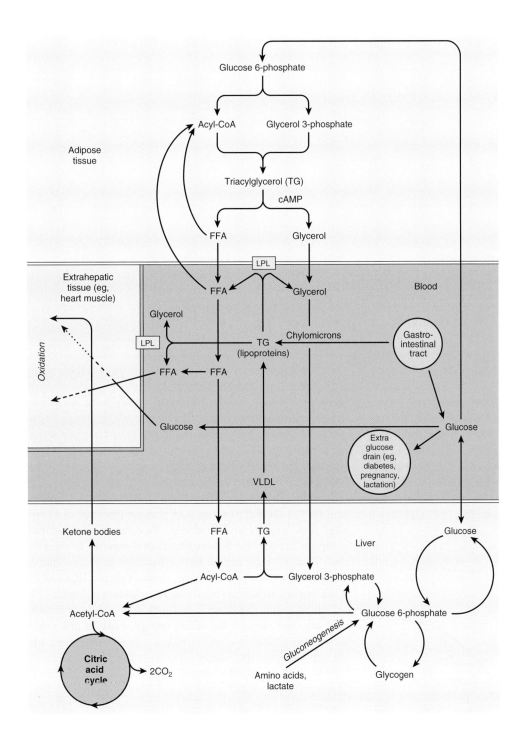

FIGURE 16–9 Metabolic interrelationships among adipose tissue, the liver, and extrahepatic tissues. In tissues such as heart, metabolic fuels are oxidized in the following order of preference: ketone bodies > fatty acids > glucose. (LPL, lipoprotein lipase; FFA, free fatty acids; VLDL, very low density lipoproteins.)

synthesis in the fed state, when the rate of synthesis increases by 20–25%. The increased rate of protein synthesis in response to increased availability of amino acids and metabolic fuel is again a response to insulin action. Protein synthesis is an energy expensive process; it may account for up to 20% of resting energy expenditure after a meal, but only 9% in the fasting state.

Metabolic Fuel Reserves Are Mobilized in the Fasting State

There is a small fall in plasma glucose in the fasting state, and then little change as fasting is prolonged into starvation.

Plasma free fatty acids increase in fasting, but then rise little more in starvation; as fasting is prolonged, so the plasma concentration of ketone bodies (acetoacetate and 3-hydroxybutyrate) increases markedly (Table 16–2, Figure 16–10).

In the fasting state, as the concentration of glucose in the portal blood falls, so insulin secretion decreases, and skeletal muscle and adipose tissue take up less glucose. The increase in secretion of **glucagon** by the α cells of the pancreas inhibits glycogen synthetase, and activates glycogen phosphorylase in the liver. The resulting glucose 6-phosphate is hydrolyzed by glucose 6-phosphatase, and glucose is released into the bloodstream for use by the brain and erythrocytes.

TABLE 16–1   Energy Yields, Oxygen Consumption, and Carbon Dioxide Production in the Oxidation of Metabolic Fuels

	Energy Yield (kJ/g)	O$_2$ Consumed (L/g)	CO$_2$ Produced (L/g)	RQ (CO$_2$ Produced/O$_2$ Consumed)	Energy (kJ)/L O$_2$
Carbohydrate	16	0.829	0.829	1.00	20
Protein	17	0.966	0.782	0.81	20
Fat	37	2.016	1.427	0.71	20
Alcohol	29	1.429	0.966	0.66	20

TABLE 16–2   Plasma Concentrations of Metabolic Fuels (mmol/L) in the Fed and Fasting States

	Fed	40 Hours Fasting	7 Days Starvation
Glucose	5.5	3.6	3.5
Free fatty acids	0.30	1.15	1.19
Ketone bodies	Negligible	2.9	4.5

Muscle glycogen cannot contribute directly to plasma glucose, since muscle lacks glucose 6-phosphatase, and the primary purpose of muscle glycogen is to provide a source of glucose 6-phosphate for energy-yielding metabolism in the muscle itself. However, acetyl-CoA formed by oxidation of fatty acids in muscle inhibits pyruvate dehydrogenase, leading to an accumulation of pyruvate. Most of this is transaminated to alanine, at the expense of amino acids arising from breakdown of "labile" protein reserves synthesized in the fed state. The alanine, and much of the keto acids resulting from this transamination are exported from muscle, and taken up by the liver, where the alanine is transaminated to yield pyruvate. The resultant amino acids are largely exported back to muscle, to provide amino groups for formation of more alanine, while the pyruvate is a major substrate for gluconeogenesis in the liver.

In adipose tissue the decrease in insulin and increase in glucagon results in inhibition of lipogenesis, inactivation of lipoprotein lipase, and activation of intracellular hormone-sensitive lipase (Chapter 25). This leads to release from adipose tissue of increased amounts of glycerol (which is a substrate for gluconeogenesis in the liver) and free fatty acids, which are used by liver, heart, and skeletal muscle as their preferred metabolic fuel, therefore sparing glucose.

Although muscle preferentially takes up and metabolizes free fatty acids in the fasting state, it cannot meet all of its energy requirements by β-oxidation. By contrast, the liver has a greater capacity for β-oxidation than it requires to meet its own energy needs, and as fasting becomes more prolonged, it forms more acetyl-CoA than can be oxidized. This acetyl-CoA is used to synthesize the **ketone bodies** (Chapter 22), which are major metabolic fuels for skeletal and heart muscle and can meet some of the brain's energy needs. In prolonged starvation, glucose may represent less than 10% of whole body energy-yielding metabolism.

Were there no other source of glucose, liver and muscle glycogen would be exhausted after about 18 h fasting. As fasting becomes more prolonged, so an increasing amount of the amino acids released as a result of protein catabolism is utilized in the liver and kidneys for gluconeogenesis (Table 16–3).

CLINICAL ASPECTS

In prolonged starvation, as adipose tissue reserves are depleted, there is a very considerable increase in the net rate of protein catabolism to provide amino acids, not only as substrates for gluconeogenesis, but also as the main metabolic fuel of all tissues. Death results when essential tissue proteins are catabolized and not replaced. In patients with **cachexia** as a result of release of **cytokines** in response to tumors and a number of other pathologic conditions, there is an increase in the rate

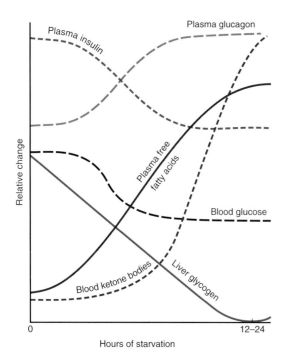

FIGURE 16–10 Relative changes in metabolic parameters during the onset of starvation.

TABLE 16–3   Summary of the Major Metabolic Features of the Principal Organs

Organ	Major Pathways	Main Substrates	Major Products Exported	Specialist Enzymes
Liver	Glycolysis, gluconeogenesis, lipogenesis, β-oxidation, citric acid cycle, ketogenesis, lipoprotein metabolism, drug metabolism, synthesis of bile salts, urea, uric acid, cholesterol, plasma proteins	Free fatty acids, glucose (in fed state), lactate, glycerol, fructose, amino acids, alcohol	Glucose, triacylglycerol in VLDL, ketone bodies, urea, uric acid, bile salts, cholesterol, plasma proteins	Glucokinase, glucose 6-phosphatase, glycerol kinase, phosphoenolpyruvate carboxykinase, fructokinase, arginase, HMG CoA synthase, HMG CoA lyase, alcohol dehydrogenase
Brain	Glycolysis, citric acid cycle, amino acid metabolism, neurotransmitter synthesis	Glucose, amino acids, ketone bodies in prolonged starvation	Lactate, end products of neurotransmitter metabolism	Those for synthesis and catabolism of neurotransmitters
Heart	β-Oxidation and citric acid cycle	Ketone bodies, free fatty acids, lactate, chylomicron and VLDL triacylglycerol, some glucose	—	Lipoprotein lipase, very active electron transport chain
Adipose tissue	Lipogenesis, esterification of fatty acids, lipolysis (in fasting)	Glucose, chylomicron and VLDL triacylglycerol	Free fatty acids, glycerol	Lipoprotein lipase, hormone-sensitive lipase, enzymes of pentose phosphate pathway
Fast twitch muscle	Glycolysis	Glucose, glycogen	Lactate, (alanine and ketoacids in fasting)	—
Slow twitch muscle	β-Oxidation and citric acid cycle	Ketone bodies, chylomicron and VLDL triacylglycerol	—	Lipoprotein lipase, very active electron transport chain
Kidney	Gluconeogenesis	Free fatty acids, lactate, glycerol, glucose	Glucose	Glycerol kinase, phosphoenolpyruvate carboxykinase
Erythrocytes	Anaerobic glycolysis, pentose phosphate pathway	Glucose	Lactate	Hemoglobin, enzymes of pentose phosphate pathway

Abbreviation: VLDL, very low density lipoproteins.

of tissue protein catabolism, as well as a considerably increased metabolic rate, so they are in a state of advanced starvation. Again, death results when essential tissue proteins are catabolized and not replaced.

The high demand for glucose by the fetus, and for lactose synthesis in lactation, can lead to ketosis. This may be seen as mild ketosis with hypoglycemia in human beings; in lactating cattle and in ewes carrying a twin pregnancy, there may be very pronounced ketoacidosis and profound hypoglycemia.

In poorly controlled type 1 **diabetes mellitus,** patients may become hyperglycemic, partly as a result of lack of insulin to stimulate uptake and utilization of glucose, and partly because in the absence of insulin there is increased gluconeogenesis from amino acids in the liver. At the same time, the lack of insulin results in increased lipolysis in adipose tissue, and the resultant free fatty acids are substrates for ketogenesis in the liver.

Utilization of these ketone bodies in muscle (and other tissues) may be impaired because of the lack of oxaloacetate (all tissues have a requirement for some glucose metabolism to maintain an adequate amount of oxaloacetate for citric acid cycle activity). In uncontrolled diabetes, the ketosis may be

severe enough to result in pronounced acidosis (**ketoacidosis**) since acetoacetate and 3-hydroxybutyrate are relatively strong acids. Coma results from both the acidosis and also the considerably increased osmolality of extracellular fluid (mainly as a result of the hyperglycemia).

SUMMARY

- The products of digestion provide the tissues with the building blocks for the biosynthesis of complex molecules and also with the fuel to power the living processes.
- Nearly all products of digestion of carbohydrate, fat, and protein are metabolized to a common metabolite, acetyl-CoA, before oxidation to CO_2 in the citric acid cycle.
- Acetyl-CoA is also the precursor for synthesis of long-chain fatty acids and steroids, including cholesterol and ketone bodies.
- Glucose provides carbon skeletons for the glycerol of triacylglycerols and nonessential amino acids.
- Water-soluble products of digestion are transported directly to the liver via the hepatic portal vein. The liver regulates the blood concentrations of glucose and amino acids.

- Pathways are compartmentalized within the cell. Glycolysis, glycogenesis, glycogenolysis, the pentose phosphate pathway, and lipogenesis occur in the cytosol. The mitochondria contain the enzymes of the citric acid cycle, β-oxidation of fatty acids, and the respiratory chain and ATP synthase. The membranes of the endoplasmic reticulum contain the enzymes for a number of other processes, including triacylglycerol synthesis and drug metabolism.

- Metabolic pathways are regulated by rapid mechanisms affecting the activity of existing enzymes, ie, allosteric and covalent modification (often in response to hormone action) and slow mechanisms affecting the synthesis of enzymes.

- Dietary carbohydrate and amino acids in excess of requirements can be used for fatty acid and hence triacylglycerol synthesis.

- In fasting and starvation, glucose must be provided for the brain and red blood cells; in the early fasting state, this is supplied from glycogen reserves. In order to spare glucose, muscle and other tissues do not take up glucose when insulin secretion is low; they utilize fatty acids (and later ketone bodies) as their preferred fuel.

- Adipose tissue releases free fatty acids in the fasting state. In prolonged fasting and starvation these are used by the liver for synthesis of ketone bodies, which are exported to provide the major fuel for muscle.

- Most amino acids, arising from the diet or from tissue protein turnover, can be used for gluconeogenesis, as can the glycerol from triacylglycerol.

- Neither fatty acids, arising from the diet or from lipolysis of adipose tissue triacylglycerol, nor ketone bodies, formed from fatty acids in the fasting state, can provide substrates for gluconeogenesis.

REFERENCES

Bender DA: *Introduction to Nutrition and Metabolism*, 4th ed. CRC Press, 2007.

Brosnan JT: Comments on the metabolic needs for glucose and the role of gluconeogenesis. European Journal of Clinical Nutrition 1999;53:Suppl 1,S107–S111.

Frayn KN: Integration of substrate flow in vivo: some insights into metabolic control. Clinical Nutrition 1997;16:277–282.

Frayn KN: *Metabolic Regulation: A Human Perspective*, 2nd ed. Blackwell Science, 2003.

Zierler K: Whole body metabolism of glucose. American Journal of Physiology 1999;276:E409–E426.

The Citric Acid Cycle: The Catabolism of Acetyl-CoA

David A. Bender, PhD

BIOMEDICAL IMPORTANCE

The citric acid cycle (Krebs cycle, tricarboxylic acid cycle) is a sequence of reactions in mitochondria that oxidizes the acetyl moiety of acetyl-CoA and reduces coenzymes that are reoxidized through the electron transport chain, linked to the formation of ATP.

The citric acid cycle is the final common pathway for the oxidation of carbohydrate, lipid, and protein because glucose, fatty acids, and most amino acids are metabolized to acetyl-CoA or intermediates of the cycle. It also has a central role in gluconeogenesis, lipogenesis, and interconversion of amino acids. Many of these processes occur in most tissues, but the liver is the only tissue in which all occur to a significant extent. The repercussions are therefore profound when, for example, large numbers of hepatic cells are damaged as in acute **hepatitis** or replaced by connective tissue (as in **cirrhosis**). The few genetic defects of citric acid cycle enzymes that have been reported are associated with severe neurological damage as a result of very considerably impaired ATP formation in the central nervous system.

THE CITRIC ACID CYCLE PROVIDES SUBSTRATE FOR THE RESPIRATORY CHAIN

The cycle starts with reaction between the acetyl moiety of acetyl-CoA and the four-carbon dicarboxylic acid oxaloacetate, forming a six-carbon tricarboxylic acid, citrate. In the subsequent reactions, two molecules of CO_2 are released and oxaloacetate is regenerated (Figure 17–1). Only a small quantity of oxaloacetate is needed for the oxidation of a large quantity of acetyl-CoA; it can be considered as playing a **catalytic role.**

The citric acid cycle is an integral part of the process by which much of the free energy liberated during the oxidation of fuels is made available. During the oxidation of acetyl-CoA, coenzymes are reduced and subsequently reoxidized in the respiratory chain, linked to the formation of ATP (oxidative phosphorylation, Figure 17–2; see also Chapter 13). This process is **aerobic,** requiring oxygen as the final oxidant of the reduced coenzymes. The enzymes of the citric acid cycle are located in the **mitochondrial matrix,** either free or attached to the inner mitochondrial membrane and the crista membrane, where the enzymes and coenzymes of the respiratory chain are also found (Chapter 13).

REACTIONS OF THE CITRIC ACID CYCLE LIBERATE REDUCING EQUIVALENTS & CO_2

The initial reaction between acetyl-CoA and oxaloacetate to form citrate is catalyzed by **citrate synthase,** which forms a carbon-carbon bond between the methyl carbon of acetyl-CoA and the carbonyl carbon of oxaloacetate (Figure 17–3). The thioester bond of the resultant citryl-CoA is hydrolyzed, releasing citrate and CoASH—an exothermic reaction.

Citrate is isomerized to isocitrate by the enzyme **aconitase** (aconitate hydratase); the reaction occurs in two steps: dehydration to *cis*-aconitate and rehydration to isocitrate. Although citrate is a symmetric molecule, aconitase reacts with citrate asymmetrically, so that the two carbon atoms that are lost in subsequent reactions of the cycle are not those that were added from acetyl-CoA. This asymmetric behavior is the result of **channeling**—transfer of the product of citrate synthase directly onto the active site of aconitase, without entering free solution. This provides integration of citric acid cycle activity and the provision of citrate in the cytosol as a source of acetyl-CoA for fatty acid synthesis. The poison **fluoroacetate** is toxic, because fluoroacetyl-CoA condenses with oxaloacetate to form fluorocitrate, which inhibits aconitase, causing citrate to accumulate.

Isocitrate undergoes dehydrogenation catalyzed by **isocitrate dehydrogenase** to form, initially, oxalosuccinate, which remains enzyme bound and undergoes decarboxylation to α-ketoglutarate. The decarboxylation requires Mg^{++} or Mn^{++} ions. There are three isoenzymes of isocitrate dehydrogenase. One, which uses NAD^+, is found only in mitochondria. The other two use $NADP^+$ and are found in mitochondria and the cytosol. Respiratory-chain-linked oxidation of isocitrate proceeds almost completely through the NAD^+-dependent enzyme.

α-Ketoglutarate undergoes **oxidative decarboxylation** in a reaction catalyzed by a multi-enzyme complex similar to that

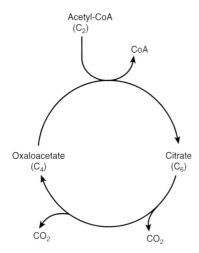

FIGURE 17–1 The citric acid cycle, illustrating the catalytic role of oxaloacetate.

involved in the oxidative decarboxylation of pyruvate (Figure 18–5). The **α-ketoglutarate dehydrogenase complex** requires the same cofactors as the pyruvate dehydrogenase complex—thiamin diphosphate, lipoate, NAD⁺, FAD, and CoA—and results in the formation of succinyl-CoA. The equilibrium of this reaction is so much in favor of succinyl-CoA formation that it must be considered to be physiologically unidirectional. As in the case of pyruvate oxidation (Chapter 18), arsenite inhibits the reaction, causing the substrate, **α-ketoglutarate,** to accumulate.

Succinyl-CoA is converted to succinate by the enzyme **succinate thiokinase (succinyl-CoA synthetase).** This is the only example in the citric acid cycle of substrate-level phosphorylation. Tissues in which gluconeogenesis occurs (the liver and kidney) contain two isoenzymes of succinate thiokinase, one specific for GDP and the other for ADP. The GTP formed is used for the decarboxylation of oxaloacetate to phosphoenolpyruvate in gluconeogenesis, and provides a regulatory link between citric acid cycle activity and the withdrawal of oxaloacetate for gluconeogenesis. Nongluconeogenic tissues have only the isoenzyme that uses ADP.

When ketone bodies are being metabolized in extrahepatic tissues there is an alternative reaction catalyzed by **succinyl-CoA-acetoacetate-CoA transferase (thiophorase),** involving transfer of CoA from succinyl-CoA to acetoacetate, forming acetoacetyl-CoA (Chapter 22).

The onward metabolism of succinate, leading to the regeneration of oxaloacetate, is the same sequence of chemical reactions as occurs in the β-oxidation of fatty acids: dehydrogenation to form a carbon-carbon double bond, addition of water to form a hydroxyl group, and a further dehydrogenation to yield the oxo-group of oxaloacetate.

The first dehydrogenation reaction, forming fumarate, is catalyzed by **succinate dehydrogenase,** which is bound to the inner surface of the inner mitochondrial membrane. The enzyme contains FAD and iron-sulfur (Fe:S) protein, and directly reduces ubiquinone in the electron transport chain. **Fumarase**

FIGURE 17–2 The citric acid cycle: the major catabolic pathway for acetyl-CoA in aerobic organisms. Acetyl-CoA, the product of carbohydrate, protein, and lipid catabolism, is taken into the cycle and oxidized to CO_2 with the release of reducing equivalents (2H). Subsequent oxidation of 2H in the respiratory chain leads to phosphorylation of ADP to ATP. For one turn of the cycle, nine ATP are generated via oxidative phosphorylation and one ATP (or GTP) arises at substrate level from the conversion of succinyl-CoA to succinate.

(fumarate hydratase) catalyzes the addition of water across the double bond of fumarate, yielding malate. Malate is converted to oxaloacetate by **malate dehydrogenase,** a reaction requiring NAD⁺. Although the equilibrium of this reaction strongly favors malate, the net flux is to oxaloacetate because of the continual removal of oxaloacetate (to form citrate, as a substrate for gluconeogenesis, or to undergo transamination to aspartate) and also the continual reoxidation of NADH.

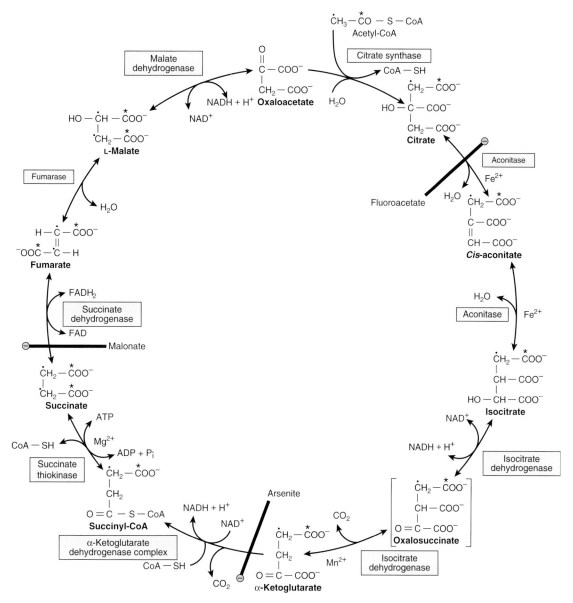

FIGURE 17–3 The citric acid (Krebs) cycle. Oxidation of NADH and FADH$_2$ in the respiratory chain leads to the formation of ATP via oxidative phosphorylation. In order to follow the passage of acetyl-CoA through the cycle, the two carbon atoms of the acetyl radical are shown labeled on the carboxyl carbon (*) and on the methyl carbon (•). Although two carbon atoms are lost as CO$_2$ in one turn of the cycle, these atoms are not derived from the acetyl-CoA that has immediately entered the cycle, but from that portion of the citrate molecule that was derived from oxaloacetate. However, on completion of a single turn of the cycle, the oxaloacetate that is regenerated is now labeled, which leads to labeled CO$_2$ being evolved during the second turn of the cycle. Because succinate is a symmetric compound, "randomization" of label occurs at this step so that all four carbon atoms of oxaloacetate appear to be labeled after one turn of the cycle. During gluconeogenesis, some of the label in oxaloacetate is incorporated into glucose and glycogen (Figure 20–1). The sites of inhibition (⊖) by fluoroacetate, malonate, and arsenite are indicated.

TEN ATP ARE FORMED PER TURN OF THE CITRIC ACID CYCLE

As a result of oxidations catalyzed by the dehydrogenases of the citric acid cycle, three molecules of NADH and one of FADH$_2$ are produced for each molecule of acetyl-CoA catabo-lized in one turn of the cycle. These reducing equivalents are transferred to the respiratory chain (see Figure 13–3), where reoxidation of each NADH results in formation of ~2.5 ATP, and of FADH$_2$, ~1.5 ATP. In addition, 1 ATP (or GTP) is formed by substrate-level phosphorylation catalyzed by suc-cinate thiokinase.

VITAMINS PLAY KEY ROLES IN THE CITRIC ACID CYCLE

Four of the B vitamins (Chapter 44) are essential in the citric acid cycle and hence energy-yielding metabolism: (1) **riboflavin,** in the form of flavin adenine dinucleotide (FAD), a cofactor for succinate dehydrogenase; (2) **niacin,** in the form of nicotinamide adenine dinucleotide (NAD), the electron acceptor for isocitrate dehydrogenase, α-ketoglutarate dehydrogenase, and malate dehydrogenase; (3) thiamin **(vitamin B$_1$),** as thiamin diphosphate, the coenzyme for decarboxylation in the α-ketoglutarate dehydrogenase reaction; and (4) **pantothenic acid,** as part of coenzyme A, the cofactor attached to "active" carboxylic acid residues such as acetyl-CoA and succinyl-CoA.

THE CITRIC ACID CYCLE PLAYS A PIVOTAL ROLE IN METABOLISM

The citric acid cycle is not only a pathway for oxidation of two-carbon units, but is also a major pathway for interconversion of metabolites arising from **transamination** and **deamination** of amino acids (Chapters 28 & 29), and providing the substrates for **amino acid synthesis** by transamination (Chapter 27), as well as for **gluconeogenesis** (Chapter 20) and **fatty acid**

synthesis (Chapter 23). Because it functions in both oxidative and synthetic processes, it is **amphibolic** (Figure 17–4).

The Citric Acid Cycle Takes Part in Gluconeogenesis, Transamination, & Deamination

All the intermediates of the cycle are potentially **glucogenic,** since they can give rise to oxaloacetate, and hence net production of glucose (in the liver and kidney, the organs that carry out gluconeogenesis; see Chapter 20). The key enzyme that catalyzes net transfer out of the cycle into gluconeogenesis is **phosphoenolpyruvate carboxykinase,** which catalyzes the decarboxylation of oxaloacetate to phosphoenolpyruvate, with GTP acting as the phosphate donor (see Figure 20–1).

Net transfer into the cycle occurs as a result of several reactions. Among the most important of such **anaplerotic** reactions is the formation of oxaloacetate by the carboxylation of pyruvate, catalyzed by **pyruvate carboxylase.** This reaction is important in maintaining an adequate concentration of oxaloacetate for the condensation reaction with acetyl-CoA. If acetyl-CoA accumulates, it acts as both an allosteric activator of pyruvate carboxylase and an inhibitor of pyruvate dehydrogenase, thereby ensuring a supply of oxaloacetate. Lactate, an important substrate for gluconeogenesis, enters the cycle via oxidation to pyruvate and then carboxylation to oxaloacetate.

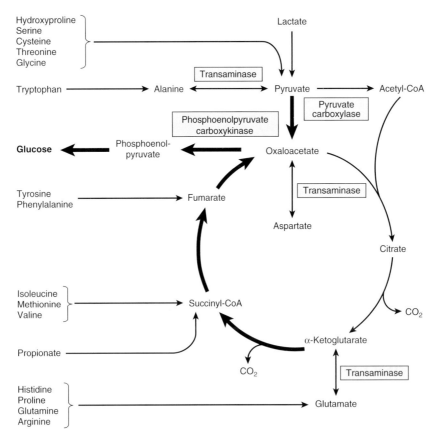

FIGURE 17–4 Involvement of the citric acid cycle in transamination and gluconeogenesis. The bold arrows indicate the main pathway of gluconeogenesis.

Aminotransferase (transaminase) reactions form pyruvate from alanine, oxaloacetate from aspartate, and α-ketoglutarate from glutamate. Because these reactions are reversible, the cycle also serves as a source of carbon skeletons for the synthesis of these amino acids. Other amino acids contribute to gluconeogenesis because their carbon skeletons give rise to citric acid cycle intermediates. Alanine, cysteine, glycine, hydroxyproline, serine, threonine, and tryptophan yield pyruvate; arginine, histidine, glutamine, and proline yield α-ketoglutarate; isoleucine, methionine, and valine yield succinyl-CoA; tyrosine and phenylalanine yield fumarate (see Figure 17–4).

In ruminants, whose main metabolic fuel is short-chain fatty acids formed by bacterial fermentation, the conversion of propionate, the major glucogenic product of rumen fermentation, to succinyl-CoA via the methylmalonyl-CoA pathway (Figure 20–2) is especially important.

The Citric Acid Cycle Takes Part in Fatty Acid Synthesis

Acetyl-CoA, formed from pyruvate by the action of pyruvate dehydrogenase, is the major substrate for long-chain fatty acid synthesis in nonruminants (Figure 17–5). (In ruminants, acetyl-CoA is derived directly from acetate.) Pyruvate dehydrogenase is a mitochondrial enzyme, and fatty acid synthesis is a cytosolic pathway; the mitochondrial membrane is impermeable to acetyl-CoA. Acetyl-CoA is made available in the cytosol from citrate synthesized in the mitochondrion, transported into the cytosol, and cleaved in a reaction catalyzed by **ATP-citrate lyase.** Citrate is only available for transport out of the mitochondrion when aconitase is saturated with its substrate, and citrate cannot be channeled directly from citrate synthase onto aconitase. This ensures that citrate is used for fatty acid synthesis only when there is an adequate amount to ensure continued activity of the cycle.

Regulation of the Citric Acid Cycle Depends Primarily on a Supply of Oxidized Cofactors

In most tissues, where the primary role of the citric acid cycle is in energy-yielding metabolism, **respiratory control** via the respiratory chain and oxidative phosphorylation regulates citric acid cycle activity (Chapter 13). Thus, activity is immediately dependent on the supply of NAD⁺, which in turn, because of the tight coupling between oxidation and phosphorylation, is dependent on the availability of ADP and hence, ultimately on the rate of utilization of ATP in chemical and physical work. In addition, individual enzymes of the cycle are regulated. The most likely sites for regulation are the nonequilibrium reactions catalyzed by pyruvate dehydrogenase, citrate synthase, isocitrate dehydrogenase, and α-ketoglutarate dehydrogenase. The dehydrogenases are activated by Ca^{2+}, which increases in concentration during muscular contraction and secretion, when there is increased energy demand. In a tissue such as brain, which is largely dependent on carbohydrate to supply acetyl-CoA, control of the citric acid cycle may occur at pyruvate dehydrogenase. Several enzymes are responsive to the energy status as shown by the [ATP]/[ADP] and [NADH]/[NAD⁺] ratios. Thus, there is allosteric inhibition of citrate synthase by ATP and long-chain fatty acyl-CoA. Allosteric activation of mitochondrial NAD-dependent isocitrate dehydrogenase by ADP is counteracted by ATP and NADH. The α-ketoglutarate dehydrogenase complex is regulated in the same way as is pyruvate dehydrogenase (Figure 18–6). Succinate dehydrogenase is inhibited by oxaloacetate, and the availability of oxaloacetate, as controlled by malate dehydrogenase, depends on the [NADH]/[NAD⁺] ratio. Since the K_m for oxaloacetate of citrate synthase is of the same order of magnitude as the intramitochondrial concentration, it is likely that the concentration of oxaloacetate controls the rate of citrate formation. Which of these mechanisms are important in vivo is still to be resolved.

SUMMARY

- The citric acid cycle is the final pathway for the oxidation of carbohydrate, lipid, and protein. Their common end-metabolite, acetyl-CoA, reacts with oxaloacetate to form citrate. By a series of dehydrogenations and decarboxylations, citrate is degraded, reducing coenzymes, releasing $2CO_2$, and regenerating oxaloacetate.

- The reduced coenzymes are oxidized by the respiratory chain linked to formation of ATP. Thus, the cycle is the major pathway for the formation of ATP and is located in the matrix of mitochondria adjacent to the enzymes of the respiratory chain and oxidative phosphorylation.

- The citric acid cycle is amphibolic, since in addition to oxidation it is important in the provision of carbon skeletons for gluconeogenesis, fatty acid synthesis, and interconversion of amino acids.

FIGURE 17–5 Participation of the citric acid cycle in fatty acid synthesis from glucose. See also Figure 23–5.

REFERENCES

Baldwin JE, Krebs HA: The evolution of metabolic cycles. Nature 1981;291:381.

Bowtell JL, Bruce M: Glutamine: an anaplerotic precursor. Nutrition 2002;18:222.

Briere JJ, Favier J, et al: Tricarboxylic acid cycle dysfunction as a cause of human diseases and tumor formation. Am J Physiol Cell Physiol 2006;291:C1114.

Brunengraber H & Roe CR: Anaplerotic molecules: current and future. J Inherit Metab Dis 2006;29:327.

De Meirleir L: Defects of pyruvate metabolism and the Krebs cycle. J Child Neurol 2002;Suppl 3:3S26.

Gibala MJ, Young ME: Anaplerosis of the citric acid cycle: role in energy metabolism of heart and skeletal muscle. Acta Physiol Scand 2000;168:657.

Jitrapakdee S, Vidal-Puig A, et al: Anaplerotic roles of pyruvate carboxylase in mammalian tissues. Cell Mol Life Sci 2006; 63:843.

Kay J, Weitzman PDJ (editors): Krebs' Citric Acid Cycle—Half a Century and Still Turning. Biochemical Society, 1987.

Kornberg H: Krebs and his trinity of cycles. Nat Rev Mol Cell Biol 2000;1:225.

Ott P, Clemmesen O, et al: Cerebral metabolic disturbances in the brain during acute liver failure: from hyperammonemia to energy failure and proteolysis. Neurochem Int 2003;47:13.

Owen OE, Kalhan SC: The key role of anaplerosis and cataplerosis for citric acid cycle function. J Biol Chem 2002;277:30409.

Pithukpakorn, M: Disorders of pyruvate metabolism and the tricarboxylic acid cycle. Mol Genet Metab 2005;85:243.

Sumegi B, Sherry AD: Is there tight channelling in the tricarboxylic acid cycle metabolon? Biochem Soc Trans 1991;19:1002.

Glycolysis & the Oxidation of Pyruvate

David A. Bender, PhD

BIOMEDICAL IMPORTANCE

Most tissues have at least some requirement for glucose. In the brain, the requirement is substantial. Glycolysis, the major pathway for glucose metabolism, occurs in the cytosol of all cells. It is unique, in that it can function either aerobically or anaerobically, depending on the availability of oxygen and the electron transport chain. Erythrocytes, which lack mitochondria, are completely reliant on glucose as their metabolic fuel, and metabolize it by anaerobic glycolysis. However, to oxidize glucose beyond pyruvate (the end product of glycolysis) requires both oxygen and mitochondrial enzyme systems: the pyruvate dehydrogenase complex, the citric acid cycle, and the respiratory chain.

Glycolysis is both the principal route for glucose metabolism and also the main pathway for the metabolism of fructose, galactose, and other carbohydrates derived from the diet. The ability of glycolysis to provide ATP in the absence of oxygen is especially important, because this allows skeletal muscle to perform at very high levels when oxygen supply is insufficient, and it allows tissues to survive anoxic episodes. However, heart muscle, which is adapted for aerobic performance, has relatively low glycolytic activity and poor survival under conditions of **ischemia.** Diseases in which enzymes of glycolysis (eg, pyruvate kinase) are deficient are mainly seen as **hemolytic anemias** or, if the defect affects skeletal muscle (eg, phosphofructokinase), as **fatigue.** In fast-growing cancer cells, glycolysis proceeds at a high rate, forming large amounts of pyruvate, which is reduced to lactate and exported. This produces a relatively acidic local environment in the tumor, which may have implications for cancer therapy. The lactate is used for gluconeogenesis in the liver, an energy-expensive process, which is responsible for much of the **hypermetabolism** seen in **cancer cachexia. Lactic acidosis** results from several causes, including impaired activity of pyruvate dehydrogenase.

GLYCOLYSIS CAN FUNCTION UNDER ANAEROBIC CONDITIONS

Early in the investigations of glycolysis it was realized that fermentation in yeast was similar to the breakdown of glycogen in muscle. It was noted that when a muscle contracts in an anaerobic medium, ie, one from which oxygen is excluded, **glycogen disappears** and **lactate appears.** When oxygen is admitted, aerobic recovery takes place and lactate is no longer produced. However, if contraction occurs under aerobic conditions, lactate does not accumulate and pyruvate is the major end product of glycolysis. Pyruvate is oxidized further to CO_2 and water (Figure 18–1). When oxygen is in short supply, mitochondrial reoxidation of NADH formed during glycolysis is impaired, and NADH is reoxidized by reducing pyruvate to lactate, so permitting glycolysis to proceed (Figure 18–1). While glycolysis can occur under anaerobic conditions, this has a price, for it limits the amount of ATP formed per mole of glucose oxidized, so that much more glucose must be metabolized under anaerobic than aerobic conditions. In yeast and some other microorganisms, pyruvate formed in anaerobic glycolysis is not reduced to lactate, but is decarboxylated and reduced to ethanol.

THE REACTIONS OF GLYCOLYSIS CONSTITUTE THE MAIN PATHWAY OF GLUCOSE UTILIZATION

The overall equation for glycolysis from glucose to lactate is as follows:

$$\text{Glucose} + 2\,\text{ADP} + 2\,\text{P}_i \rightarrow 2\,\text{Lactate} + 2\,\text{ATP} + 2\,\text{H}_2\text{O}$$

All of the enzymes of glycolysis (Figure 18–2) are found in the cytosol. Glucose enters glycolysis by phosphorylation to glucose 6-phosphate, catalyzed by **hexokinase,** using ATP as the phosphate donor. Under physiologic conditions, the phosphorylation of glucose to glucose 6-phosphate can be regarded as irreversible. Hexokinase is inhibited allosterically by its product, glucose 6-phosphate.

In tissues other than the liver (and pancreatic β-islet cells), the availability of glucose for glycolysis (or glycogen synthesis in muscle, Chapter 19, and lipogenesis in adipose tissue, Chapter 23) is controlled by transport into the cell, which in turn is regulated by **insulin.** Hexokinase has a high affinity (low K_m) for glucose, and in the liver it is saturated under normal

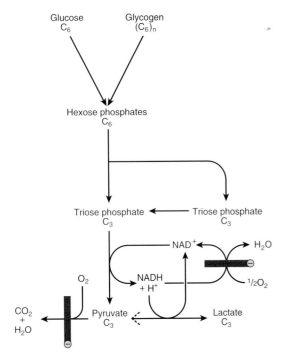

FIGURE 18–1 Summary of glycolysis. ⊖, blocked by anaerobic conditions or by absence of mitochondria containing key respiratory enzymes, as in erythrocytes.

conditions, and so acts at a constant rate to provide glucose 6-phosphate to meet the cell's need. Liver cells also contain an isoenzyme of hexokinase, **glucokinase,** which has a K_m very much higher than the normal intracellular concentration of glucose. The function of glucokinase in the liver is to remove glucose from the blood following a meal, providing glucose 6-phosphate in excess of requirements for glycolysis, which is used for glycogen synthesis and lipogenesis.

Glucose 6-phosphate is an important compound at the junction of several metabolic pathways: glycolysis, gluconeogenesis, the pentose phosphate pathway, glycogenesis, and glycogenolysis. In glycolysis it is converted to fructose 6-phosphate by **phosphohexose isomerase,** which involves an aldose-ketose isomerization. This reaction is followed by another phosphorylation catalyzed by the enzyme **phosphofructokinase** (phosphofructokinase-1) forming fructose 1,6-bisphosphate. The phosphofructokinase reaction may be considered to be functionally irreversible under physiologic conditions; it is both inducible and subject to allosteric regulation, and has a major role in regulating the rate of glycolysis. Fructose 1,6-bisphosphate is cleaved by **aldolase** (fructose 1,6-bisphosphate aldolase) into two triose phosphates, glyceraldehyde 3-phosphate and dihydroxyacetone phosphate. Glyceraldehyde 3-phosphate and dihydroxyacetone phosphate are interconverted by the enzyme **phosphotriose isomerase.**

Glycolysis continues with the oxidation of glyceraldehyde 3-phosphate to 1,3-bisphosphoglycerate. The enzyme catalyzing this oxidation, **glyceraldehyde 3-phosphate dehydrogenase,** is NAD-dependent. Structurally, it consists of four identical polypeptides (monomers) forming a tetramer. Four —SH

groups are present on each polypeptide, derived from cysteine residues within the polypeptide chain. One of the —SH groups is found at the active site of the enzyme (Figure 18–3). The substrate initially combines with this —SH group, forming a thiohemiacetal that is oxidized to a thiol ester; the hydrogens removed in this oxidation are transferred to NAD$^+$. The thiol ester then undergoes phosphorolysis; inorganic phosphate (P_i) is added, forming 1,3-bisphosphoglycerate, and the —SH group is reconstituted. In the next reaction, catalyzed by **phosphoglycerate kinase,** phosphate is transferred from 1,3-bisphosphoglycerate onto ADP, forming ATP (substrate-level phosphorylation) and 3-phosphoglycerate.

Since two molecules of triose phosphate are formed per molecule of glucose undergoing glycolysis, two molecules of ATP are formed at this stage per molecule of glucose undergoing glycolysis. The toxicity of arsenic is the result of competition of arsenate with inorganic phosphate (P_i) in this above reaction to give 1-arseno-3-phosphoglycerate, which hydrolyzes spontaneously to 3-phosphoglycerate without forming ATP. 3-Phosphoglycerate is isomerized to 2-phosphoglycerate by **phosphoglycerate mutase.** It is likely that 2,3-bisphosphoglycerate (diphosphoglycerate, DPG) is an intermediate in this reaction.

The subsequent step is catalyzed by **enolase** and involves a dehydration, forming phosphoenolpyruvate. Enolase is inhibited by **fluoride,** and when blood samples are taken for measurement of glucose, it is collected in tubes containing fluoride to inhibit glycolysis. The enzyme is also dependent on the presence of either Mg^{2+} or Mn^{2+}. The phosphate of phosphoenolpyruvate is transferred to ADP by **pyruvate kinase** to form two molecules of ATP per molecule of glucose oxidized.

The redox state of the tissue now determines which of two pathways is followed. Under **anaerobic** conditions, the NADH cannot be reoxidized through the respiratory chain to oxygen. Pyruvate is reduced by the NADH to lactate, catalyzed by **lactate dehydrogenase.** There are different tissue specific isoenzymes lactate dehydrogenase that have clinical significance (Chapter 7). The reoxidation of NADH via lactate formation allows glycolysis to proceed in the absence of oxygen by regenerating sufficient NAD$^+$ for another cycle of the reaction catalyzed by glyceraldehyde 3-phosphate dehydrogenase. Under **aerobic conditions,** pyruvate is taken up into mitochondria, and after oxidative decarboxylation to acetyl-CoA is oxidized to CO$_2$ by the citric acid cycle (Chapter 17). The reducing equivalents from the NADH formed in glycolysis are taken up into mitochondria for oxidation via one of the two shuttles described in Chapter 13.

Tissues That Function Under Hypoxic Conditions Produce Lactate

This is true of skeletal muscle, particularly the white fibers, where the rate of work output, and hence the need for ATP formation, may exceed the rate at which oxygen can be taken up and utilized (Figure 18–2). Glycolysis in erythrocytes always terminates in lactate, because the subsequent reactions of

FIGURE 18–2 The pathway of glycolysis. ($\textcircled{P}$, —PO_3^{2-}; P_i, $HOPO_3^{2-}$; $\ominus$, inhibition.) *Carbons 1–3 of fructose bisphosphate form dihydroxyacetone phosphate, and carbons 4–6 form glyceraldehyde 3-phosphate. The term "bis-," as in bisphosphate, indicates that the phosphate groups are separated, whereas the term "di-," as in adenosine diphosphate, indicates that they are joined.

pyruvate oxidation are mitochondrial, and erythrocytes lack mitochondria. Other tissues that normally derive much of their energy from glycolysis and produce lactate include brain, gastrointestinal tract, renal medulla, retina, and skin. Lactate production is also increased in septic shock, and many can-

cers also produce lactate. The liver, kidneys, and heart usually take up lactate and oxidize it, but produce it under hypoxic conditions.

When lactate production is high, as in vigorous exercise, septic shock, and cancer cachexia, much is used in the liver for

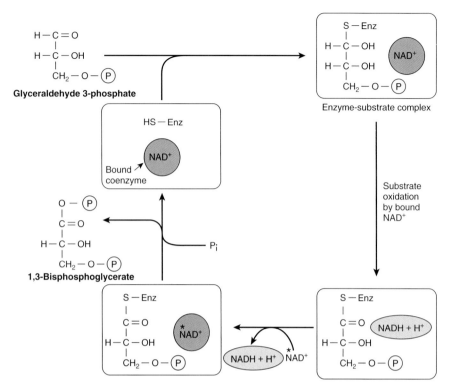

FIGURE 18–3 Mechanism of oxidation of glyceraldehyde 3-phosphate. (Enz, glyceraldehyde 3-phosphate dehydrogenase.) The enzyme is inhibited by the —SH poison iodoacetate, which is thus able to inhibit glycolysis. The NADH produced on the enzyme is not so firmly bound to the enzyme as is NAD⁺. Consequently, NADH is easily displaced by another molecule of NAD⁺.

gluconeogenesis (Chapter 20), leading to an increase in metabolic rate to provide the ATP and GTP needed. The resultant increase in oxygen consumption is seen as **oxygen debt** after vigorous exercise.

Under some conditions lactate may be formed in the cytosol, but then enter the mitochondrion to be oxidized to pyruvate for onward metabolism. This provides a pathway for the transfer of reducing equivalents from the cytotol into the mitochondrion for the electron transport chain in addition to the glycerophosphate (Figure 13–12) and malate (Figure 13–13) shuttles.

GLYCOLYSIS IS REGULATED AT THREE STEPS INVOLVING NONEQUILIBRIUM REACTIONS

Although most of the reactions of glycolysis are reversible, three are markedly exergonic and must therefore be considered physiologically irreversible. These reactions, catalyzed by **hexokinase** (and glucokinase), **phosphofructokinase,** and **pyruvate kinase,** are the major sites of regulation of glycolysis. Phosphofructokinase is significantly inhibited at normal intracellular concentrations of ATP; as discussed in Chapter 20, this inhibition can be rapidly relieved by 5′AMP that is formed as ADP begins to accumulate, signaling the need for an increased rate of glycolysis. Cells that are capable of **gluconeogenesis** (reversing the glycolytic pathway, Chapter 20) have different enzymes that catalyze reactions to reverse these irreversible

FIGURE 18–4 2,3-Bisphosphoglycerate pathway in erythrocytes.

steps; glucose 6-phosphatase, fructose 1,6-bisphosphatase and, to reverse the reaction of pyruvate kinase, pyruvate carboxylase and phosphoenolpyruvate carboxykinase. **Fructose** enters glycolysis by phosphorylation to fructose 1-phosphate, and bypasses the main regulatory steps, so resulting in formation of more pyruvate (and acetyl-CoA) than is required for ATP formation (Chapter 21). In the liver and adipose tissue, this leads to increased lipogenesis, and a high intake of fructose may be a factor in the development of obesity.

In Erythrocytes, the First Site of ATP Formation in Glycolysis May Be Bypassed

In erythrocytes, the reaction catalyzed by **phosphoglycerate kinase** may be bypassed to some extent by the reaction of **bisphosphoglycerate mutase,** which catalyzes the conversion of

1,3-bisphosphoglycerate to 2,3-bisphosphoglycerate, followed by hydrolysis to 3-phosphoglycerate and P_i, catalyzed by **2,3-bisphosphoglycerate phosphatase** (Figure 18–4). This alternative pathway involves no net yield of ATP from glycolysis. However, it does serve to provide 2,3-bisphosphoglycerate, which binds to hemoglobin, decreasing its affinity for oxygen, and so making oxygen more readily available to tissues (see Chapter 6).

THE OXIDATION OF PYRUVATE TO ACETYL-CoA IS THE IRREVERSIBLE ROUTE FROM GLYCOLYSIS TO THE CITRIC ACID CYCLE

Pyruvate, formed in the cytosol, is transported into the mitochondrion by a proton symporter (Figure 13–10). Inside the

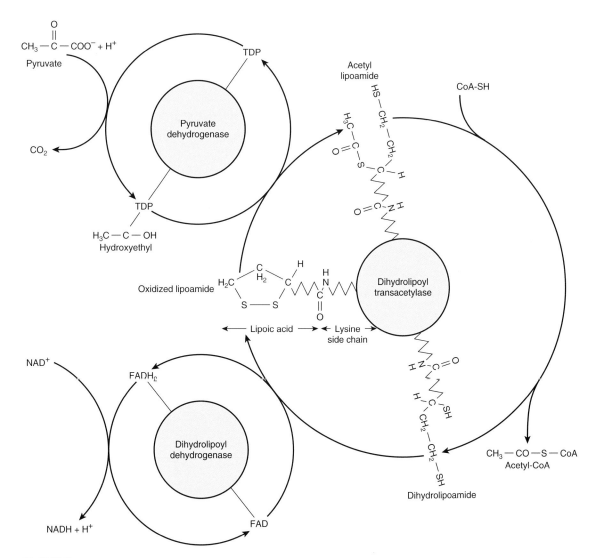

FIGURE 18–5 Oxidative decarboxylation of pyruvate by the pyruvate dehydrogenase complex. Lipoic acid is joined by an amide link to a lysine residue of the transacetylase component of the enzyme complex. It forms a long flexible arm, allowing the lipoic acid prosthetic group to rotate sequentially between the active sites of each of the enzymes of the complex. (NAD^+, nicotinamide adenine dinucleotide; FAD, flavin adenine dinucleotide; TDP, thiamin diphosphate.)

mitochondrion, it is oxidatively decarboxylated to acetyl-CoA by a multienzyme complex that is associated with the inner mitochondrial membrane. This **pyruvate dehydrogenase complex** is analogous to the α-ketoglutarate dehydrogenase complex of the citric acid cycle (Figure 17–3). Pyruvate is decarboxylated by the **pyruvate dehydrogenase** component of the enzyme complex to a hydroxyethyl derivative of the thiazole ring of enzyme-bound **thiamin diphosphate,** which in turn reacts with oxidized lipoamide, the prosthetic group of **dihydrolipoyl transacetylase,** to form acetyl lipoamide (Figure 18–5). Thiamin is vitamin B_1 (Chapter 44) and in deficiency, glucose metabolism is impaired, and there is significant (and potentially life-threatening) lactic and pyruvic acidosis. Acetyl lipoamide reacts with coenzyme A to form acetyl-CoA and reduced lipoamide. The reaction is completed when the reduced lipoamide is reoxidized by a flavoprotein, **dihydrolipoyl dehydrogenase,** containing FAD. Finally, the reduced flavoprotein is oxidized by NAD^+, which in turn transfers reducing equivalents to the respiratory chain.

$$Pyruvate + NAD^+ + CoA \rightarrow Acetyl\text{-}CoA + NADH + H^+ + CO_2$$

The pyruvate dehydrogenase complex consists of a number of polypeptide chains of each of the three component enzymes, and the intermediates do not dissociate, but remain bound to the enzymes. Such a complex of enzymes, in which the substrates are handed on from one enzyme to the next, increases the reaction rate and eliminates side reactions, increasing overall efficiency.

Pyruvate Dehydrogenase Is Regulated by End-Product Inhibition & Covalent Modification

Pyruvate dehydrogenase is inhibited by its products, acetyl-CoA and NADH (Figure 18–6). It is also regulated by phosphorylation by a kinase of three serine residues on the pyruvate dehydrogenase component of the multienzyme complex, resulting in decreased activity and by dephosphorylation by a phosphatase that causes an increase in activity. The kinase is activated by increases in the [ATP]/[ADP], [acetyl-CoA]/[CoA], and [NADH]/[NAD$^+$] ratios. Thus, pyruvate dehydrogenase, and therefore glycolysis, is inhibited both when there

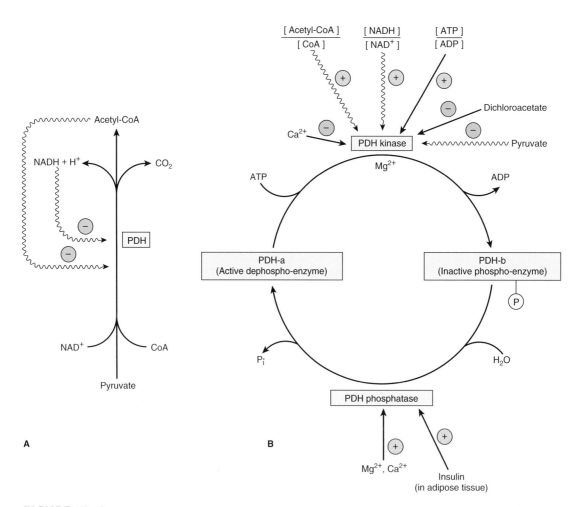

FIGURE 18–6 Regulation of pyruvate dehydrogenase (PDH). Arrows with wavy shafts indicate allosteric effects. **(A)** Regulation by end-product inhibition. **(B)** Regulation by interconversion of active and inactive forms.

TABLE 18–1 ATP Formation in the Catabolism of Glucose

Pathway	Reaction Catalyzed by	Method of ATP Formation	ATP per Mol of Glucose
Glycolysis	Glyceraldehyde 3-phosphate dehydrogenase	Respiratory chain oxidation of 2 NADH	5*
	Phosphoglycerate kinase	Substrate level phosphorylation	2
	Pyruvate kinase	Substrate level phosphorylation	2
			10
	Consumption of ATP for reactions of hexokinase and phosphofructokinase		−2
			Net 7
Citric acid cycle	Pyruvate dehydrogenase	Respiratory chain oxidation of 2 NADH	5
	Isocitrate dehydrogenase	Respiratory chain oxidation of 2 NADH	5
	α-Ketoglutarate dehydrogenase	Respiratory chain oxidation of 2 NADH	5
	Succinate thiokinase	Substrate level phosphorylation	2
	Succinate dehydrogenase	Respiratory chain oxidation of 2 $FADH_2$	3
	Malate dehydrogenase	Respiratory chain oxidation of 2 NADH	5
			Net 25
	Total per mol of glucose under aerobic conditions		32
	Total per mol of glucose under anaerobic conditions		2

*This assumes that NADH formed in glycolysis is transported into mitochondria by the malate shuttle (Figure 13–13). If the glycerophosphate shuttle is used, then only 1.5 ATP will be formed per mol of NADH. Note that there is a considerable advantage in using glycogen rather than glucose for anaerobic glycolysis in muscle, since the product of glycogen phosphorylase is glucose 1-phosphate (Figure 19–1), which is interconvertible with glucose 6-phosphate. This saves the ATP that would otherwise be used by hexokinase, increasing the net yield of ATP from 2 to 3 per glucose.

is adequate ATP (and reduced coenzymes for ATP formation) available, and also when fatty acids are being oxidized. In fasting, when free fatty acid concentrations increase, there is a decrease in the proportion of the enzyme in the active form, leading to a sparing of carbohydrate. In adipose tissue, where glucose provides acetyl-CoA for lipogenesis, the enzyme is activated in response to insulin.

Oxidation of Glucose Yields Up to 32 Mol of ATP Under Aerobic Conditions, But Only 2 Mol When O_2 Is Absent

When 1 mol of glucose is combusted in a calorimeter to CO_2 and water, approximately 2870 kJ are liberated as heat. When oxidation occurs in the tissues, approximately 32 mol of ATP are generated per molecule of glucose oxidized to CO_2 and water. In vivo, ΔG for the ATP synthase reaction has been calculated as approximately 51.6 kJ. It follows that the total energy captured in ATP per mole of glucose oxidized is 1651 kJ, or approximately 58% of the energy of combustion. Most of the ATP is formed by oxidative phosphorylation resulting from the reoxidation of reduced coenzymes by the respiratory chain. The remainder is formed by substrate-level phosphorylation (Table 18–1).

CLINICAL ASPECTS

Inhibition of Pyruvate Metabolism Leads to Lactic Acidosis

Arsenite and mercuric ions react with the —SH groups of lipoic acid and inhibit pyruvate dehydrogenase, as does a **dietary deficiency of thiamin,** allowing pyruvate to accumulate. Many alcoholics are thiamin-deficient (both because of a poor diet and also because alcohol inhibits thiamin absorption), and may develop potentially fatal pyruvic and lactic acidosis. Patients with **inherited pyruvate dehydrogenase deficiency,** which can be the result of defects in one or more of the components of the enzyme complex, also present with lactic acidosis, particularly after a glucose load. Because of the dependence of the brain on glucose as a fuel, these metabolic defects commonly cause neurologic disturbances.

Inherited aldolase A deficiency and pyruvate kinase deficiency in erythrocytes cause **hemolytic anemia.** The exercise capacity of patients with **muscle phosphofructokinase deficiency** is low, particularly if they are on high-carbohydrate diets. By providing lipid as an alternative fuel, eg, during starvation, when blood free fatty acid and ketone bodies are increased, work capacity is improved.

SUMMARY

- Glycolysis is the cytosolic pathway of all mammalian cells for the metabolism of glucose (or glycogen) to pyruvate and lactate.

- It can function anaerobically by regenerating oxidized NAD^+ (required in the glyceraldehyde-3-phosphate dehydrogenase reaction), by reducing pyruvate to lactate.

- Lactate is the end product of glycolysis under anaerobic conditions (eg, in exercising muscle) or when the metabolic machinery is absent for the further oxidation of pyruvate (eg, in erythrocytes).

- Glycolysis is regulated by three enzymes catalyzing nonequilibrium reactions: hexokinase, phosphofructokinase, and pyruvate kinase.

- In erythrocytes, the first site in glycolysis for generation of ATP may be bypassed, leading to the formation of 2,3-bisphosphoglycerate, which is important in decreasing the affinity of hemoglobin for O_2.

- Pyruvate is oxidized to acetyl-CoA by a multienzyme complex, pyruvate dehydrogenase, which is dependent on the vitamin-derived cofactor thiamin diphosphate.

- Conditions that involve an impairment of pyruvate metabolism frequently lead to lactic acidosis.

REFERENCES

Behal RH, Buxton DB, Robertson JG, Olson MS: Regulation of the pyruvate dehydrogenase multienzyme complex. Annu Rev Nutr 1993;13:497.

Boiteux A, Hess B: Design of glycolysis. Philos Trans R Soc Lond B Biol Sci 1981;293:5.

Fothergill-Gilmore LA: The evolution of the glycolytic pathway. Trends Biochem Sci 1986;11:47.

Gladden LB: Lactate metabolism: a new paradigm for the third millennium. J Physiol 2004;558:5.

Kim J-W, Dang CV: Multifaceted roles of glycolytic enzymes. Trends Biochem Sci 2005;30:142.

Levy B: Lactate and shock state: the metabolic view. Curr Opin Crit Care 2006;1:315.

Maj MC, Cameron JM, et al: Pyruvate dehydrogenase phosphatase deficiency: orphan disease or an under-diagnosed condition? Mol Cell Endocrinol 2006;249:1.

Martin E, Rosenthal RE, et al: Pyruvate dehydrogenase complex: metabolic link to ischemic brain injury and target of oxidative stress. J Neurosci Res 2005;79:240.

Patel MS, Korotchkina LG: Regulation of the pyruvate dehydrogenase complex. Biochem Soc Trans 2006;34:217.

Philp A, Macdonald AL, et al: Lactate—a signal coordinating cell and systemic function. J Exp Biol 2005;208:4561.

Pumain R, Laschet J: A key glycolytic enzyme plays a dual role in GABAergic neurotransmission and in human epilepsy. Crit Rev Neurobiol 2006;18:197.

Rider MH, Bertrand L, et al: 6-phosphofructo-2-kinase/fructose-2,6-bisphosphatase: head-to-head with a bifunctional enzyme that controls glycolysis. Biochem J 2004;381:561.

Robergs RA, Ghiasvand F, Parker D: Biochemistry of exercise-induced metabolic acidosis. Am J Physiol 2004; 287:R502.

Sugden MC, Holness MJ: Mechanisms underlying regulation of the expression and activities of the mammalian pyruvate dehydrogenase kinases. Arch Physiol Biochem 2006; 112:139.

Wasserman DH: Regulation of glucose fluxes during exercise in the postabsorptive state. Annu Rev Physiol 1995;57:191.

Metabolism of Glycogen

David A. Bender, PhD

BIOMEDICAL IMPORTANCE

Glycogen is the major storage carbohydrate in animals, corresponding to starch in plants; it is a branched polymer of α-D-glucose (Figure 14–13). It occurs mainly in liver and muscle. Although the liver content of glycogen is greater than that of muscle, because the muscle mass of the body is considerably greater than that of the liver, about three-quarters of total body glycogen is in muscle (Table 19–1).

Muscle glycogen provides a readily available source of glucose for glycolysis within the muscle itself. Liver glycogen functions to store and export glucose to maintain **blood glucose** between meals. The liver concentration of glycogen is about 450 mM after a meal, falling to about 200 mM after an overnight fast; after 12–18 h of fasting, liver glycogen is almost totally depleted. Although muscle glycogen does not directly yield free glucose (because muscle lacks glucose 6-phosphatase), pyruvate formed by glycolysis in muscle can undergo transamination to alanine, which is exported from muscle and used for gluconeogenesis in the liver (Figure 20–4). **Glycogen storage diseases** are a group of inherited disorders characterized by deficient mobilization of glycogen or deposition of abnormal forms of glycogen, leading to muscle weakness; some glycogen storage diseases result in early death.

The highly branched structure of glycogen provides a large number of sites for glycogenolysis, permitting rapid release of glucose 1-phosphate for muscle activity. Endurance athletes require a slower, more sustained release of glucose 1-phosphate. The formation of branch points in glycogen is slower than the addition of glucose units to a linear chain, and some endurance athletes practice carbohydrate loading—exercise to exhaustion (when muscle glycogen in largely depleted) followed by a high-carbohydrate meal, which results in rapid glycogen synthesis, with fewer branch points than normal.

GLYCOGENESIS OCCURS MAINLY IN MUSCLE & LIVER

The Pathway of Glycogen Biosynthesis Involves a Special Nucleotide of Glucose

As in glycolysis, glucose is phosphorylated to glucose 6-phosphate, catalyzed by **hexokinase** in muscle and **glucokinase** in liver (Figure 19–1). Glucose 6-phosphate is isomerized to glucose 1-phosphate by **phosphoglucomutase.** The enzyme itself is phosphorylated, and the phospho-group takes part in a reversible reaction in which glucose 1,6-bisphosphate is an intermediate. Next, glucose 1-phosphate reacts with uridine triphosphate (UTP) to form the active nucleotide **uridine diphosphate glucose (UDPGlc)** and pyrophosphate (Figure 19–2), catalyzed by **UDPGlc pyrophosphorylase.** The reaction proceeds in the direction of UDPGlc formation because **pyrophosphatase** catalyzes hydrolysis of pyrophosphate to $2 \times$ phosphate, so removing one of the reaction products.

Glycogen synthase catalyzes the formation of a glycoside bond between C-1 of the glucose of UDPGlc and C-4 of a terminal glucose residue of glycogen, liberating uridine diphosphate (UDP). A preexisting glycogen molecule, or "glycogen primer," must be present to initiate this reaction. The glycogen primer in turn is formed on a protein primer known as **glycogenin.** Glycogenin is a 37 kDa protein that is glucosylated on a specific tyrosine residue by UDPGlc. Further glucose residues are attached in the $1 \rightarrow 4$ position (catalyzed by glycogenin itself) to form a short chain that is a substrate for glycogen synthase. In skeletal muscle, glycogenin remains attached in the center of the glycogen molecule (Figure 14–13); in liver the number of glycogen molecules is greater than the number of glycogenin molecules.

Branching Involves Detachment of Existing Glycogen Chains

The addition of a glucose residue to a preexisting glycogen chain, or "primer," occurs at the nonreducing, outer end of the molecule, so that the branches of the glycogen molecule become elongated as successive $1 \rightarrow 4$ linkages are formed (Figure 19–3). When the chain is at least 11 glucose residues long, **branching enzyme** transfers a part of the $1 \rightarrow 4$-chain (at least six glucose residues) to a neighboring chain to form a $1 \rightarrow 6$ linkage, establishing a **branch point.** The branches grow by further additions of $1 \rightarrow 4$-glucosyl units and further branching.

GLYCOGENOLYSIS IS NOT THE REVERSE OF GLYCOGENESIS, BUT IS A SEPARATE PATHWAY

Glycogen phosphorylase catalyzes the rate-limiting step in glycogenolysis by catalyzing the phosphoroylytic cleavage (phos-

TABLE 19–1 Storage of Carbohydrate in a 70-kg Human Being

	Percentage of Tissue Weight	Tissue Weight	Body Content
Liver glycogen	5.0	1.8 kg	90 g
Muscle glycogen	0.7	35 kg	245 g
Extracellular glucose	0.1	10 L	10 g

phorolysis; of hydrolysis) of the 1 → 4 linkages of glycogen to yield glucose 1-phosphate (Figure 19–4). Glycogen phosphorylase requires pyridoxal phosphate (see Chapter 44) as its coenzyme. Unlike the reactions of amino acid metabolism (Chapter 29), in which the aldehyde is the reactive group, in phosphorylase it is the phosphate group that it catalytically active. The terminal glucosyl residues from the outermost chains of the glycogen molecule are removed sequentially until approximately

FIGURE 19–2 Uridine diphosphate glucose (UDPGlc).

four glucose residues remain on either side of a 1 → 6 branch (Figure 19–4). Another enzyme **(α-[1 → 4] → α-[1 → 4] glucan transferase)** transfers a trisaccharide unit from one branch to the other, exposing the 1 → 6 branch point. **Hydrolysis** of the

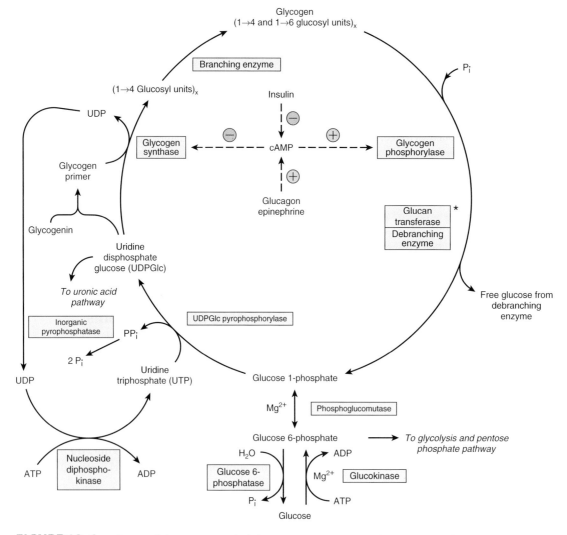

FIGURE 19–1 Pathways of glycogenesis and of glycogenolysis in the liver. (⊕, Stimulation; ⊖, inhibition.) Insulin decreases the level of cAMP only after it has been raised by glucagon or epinephrine; ie, it antagonizes their action. Glucagon is active in heart muscle but not in skeletal muscle. *Glucan transferase and debranching enzyme appear to be two separate activities of the same enzyme.

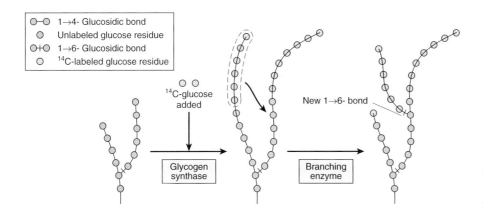

FIGURE 19–3 The biosynthesis of glycogen. The mechanism of branching as revealed by feeding ^{14}C-labeled glucose and examining liver glycogen at intervals.

$1 \rightarrow 6$ linkages requires the **debranching enzyme;** glucan transferase and the debranching enzyme are separate activities of a single protein with two catalytic sites. Further phosphorylase action can then proceed. The combined action of phosphorylase and these other enzymes leads to the complete breakdown of glycogen. The reaction catalyzed by phosphoglucomutase is reversible, so that glucose 6-phosphate can be formed from glucose 1-phosphate. In **liver** (and **kidney**), but not in muscle, **glucose 6-phosphatase** hydrolyzes glucose 6-phosphate, yielding glucose that is exported, leading to an increase in the blood glucose concentration. Glucose 6-phosphatase is in the lumen of the smooth endoplasmic reticulum, and genetic defects of the glucose 6-phosphate transporter can cause a variant of type I glycogen storage disease (see Table 19–2).

CYCLIC AMP INTEGRATES THE REGULATION OF GLYCOGENOLYSIS & GLYCOGENESIS

The principal enzymes controlling glycogen metabolism—glycogen phosphorylase and glycogen synthase—are regulated by

FIGURE 19–4 Steps in glycogenolysis.

allosteric mechanisms and covalent modification by reversible phosphorylation and dephosphorylation of enzyme protein in response to hormone action (Chapter 9).

Phosphorylation is increased in response to cyclic AMP (cAMP) (Figure 19–5) formed from ATP by **adenylyl cyclase** at the inner surface of cell membranes in response to hormones such as **epinephrine, norepinephrine,** and **glucagon.** cAMP is hydrolyzed by **phosphodiesterase,** so terminating hormone action; in liver insulin increases the activity of phosphodiesterase.

The Control of Phosphorylase Differs Between Liver & Muscle

In the liver the role of glycogen is to provide free glucose for export to maintain the blood concentration of glucose; in muscle the role of glycogen is to provide a source of glucose 6-phosphate for glycolysis in response to the need for ATP for muscle contraction. In both tissues, the enzyme is activated by phosphorylation catalyzed by phosphorylase kinase (to yield phosphorylase a) and inactivated by dephosphorylation catalyzed by phosphoprotein phosphatase (to yield phosphorylase b), in response to hormonal and other signals.

There is instantaneous overriding of this hormonal control. Active phosphorylase a in both tissues is allosterically inhibited by ATP and glucose 6-phosphate; in liver, but not muscle, free glucose is also an inhibitor. Muscle phosphorylase differs from the liver isoenzyme in having a binding site for 5′AMP, which acts as an allosteric activator of the (inactive) dephosphorylated b-form of the enzyme. 5′AMP acts as a potent signal of the energy state of the muscle cell; it is formed as the concentration of ADP begins to increase (indicating the need for increased substrate metabolism to permit ATP formation), as a result of the reaction of adenylate kinase: $2 \times \text{ADP} \leftrightarrow \text{ATP} + 5′\text{AMP}.$

cAMP Activates Phosphorylase

Phosphorylase kinase is activated in response to cAMP (Figure 19–6). Increasing the concentration of cAMP activates **cAMP-dependent protein kinase,** which catalyzes the phosphorylation by ATP of inactive **phosphorylase kinase b** to ac-

TABLE 19–2   Glycogen Storage Diseases

Type	Name	Enzyme Deficiency	Clinical Features
0	—	Glycogen synthase	Hypoglycemia; hyperketonemia; early death
Ia	Von Gierke's disease	Glucose 6-phosphatase	Glycogen accumulation in liver and renal tubule cells; hypoglycemia; lactic acidemia; ketosis; hyperlipemia
Ib	—	Endoplasmic reticulum glucose 6-phosphate transporter	As type Ia; neutropenia and impaired neutrophil function leading to recurrent infections
II	Pompe's disease	Lysosomal α1→4 and α1→6 glucosidase (acid maltase)	Accumulation of glycogen in lysosomes: juvenile onset variant, muscle hypotonia, death from heart failure by age 2; adult onset variant, muscle dystrophy
IIIa	Limit dextrinosis, Forbe's or Cori's disease	Liver and muscle debranching enzyme	Fasting hypoglycemia; hepatomegaly in infancy; accumulation of characteristic branched polysaccharide (limit dextrin); muscle weakness
IIIb	Limit dextrinosis	Liver debranching enzyme	As type IIIa, but no muscle weakness
IV	Amylopectinosis, Andersen's disease	Branching enzyme	Hepatosplenomegaly; accumulation of polysaccharide with few branch points; death from heart or liver failure before age 5
V	Myophosphorlylase deficiency, McArdle's syndrome	Muscle phosphorylase	Poor exercise tolerance; muscle glycogen abnormally high (2.5–4%); blood lactate very low after exercise
VI	Hers' disease	Liver phosphorylase	Hepatomegaly; accumulation of glycogen in liver; mild hypoglycemia; generally good prognosis
VII	Tarui's disease	Muscle and erythrocyte phosphofructokinase 1	Poor exercise tolerance; muscle glycogen abnormally high (2.5–4%); blood lactate very low after exercise; also hemolytic anemia
VIII		Liver phosphorylase kinase	Hepatomegaly; accumulation of glycogen in liver; mild hypoglycemia; generally good prognosis
IX		Liver and muscle phosphorylase kinase	Hepatomegaly; accumulation of glycogen in liver and muscle; mild hypoglycemia; generally good prognosis
X		cAMP-dependent protein kinase A	Hepatomegaly; accumulation of glycogen in liver

FIGURE 19–5   3′,5′-Adenylic acid (cyclic AMP; cAMP).

tive **phosphorylase kinase a,** which in turn, phosphorylates phosphorylase b to phosphorylase a. In the liver, cAMP is formed in response to glucagon, which is secreted in response to falling blood glucose. Muscle is insensitive to glucagon; in muscle, the signal for increased cAMP formation is the action of norepinephrine, which is secreted in response to fear or fright, when there is a need for increased glycogenolysis to permit rapid muscle activity.

Ca²⁺ Synchronizes the Activation of Phosphorylase With Muscle Contraction

Glycogenolysis in muscle increases several 100-fold at the onset of contraction; the same signal (increased cytosolic Ca^{2+} ion concentration) is responsible for initiation of both contraction and glycogenolysis. Muscle phosphorylase kinase, which activates glycogen phosphorylase, is a tetramer of four different subunits, α, β, γ, and δ. The α and β subunits contain serine residues that are phosphorylated by cAMP-dependent protein kinase. The δ subunit is identical to the Ca^{2+}-binding protein **calmodulin** (Chapter 42), and binds four Ca^{2+}. The binding of Ca^{2+} activates the catalytic site of the γ subunit even while the enzyme is in the dephosphorylated b state; the phosphorylated a form is only fully activated in the presence of high concentrations of Ca^{2+}.

Glycogenolysis in Liver Can Be cAMP-Independent

In the liver, there is cAMP-independent activation of glycogenolysis in response to stimulation of α₁ **adrenergic** receptors by epinephrine and norepinephrine. This involves mobilization of Ca^{2+} into the cytosol, followed by the stimulation of a

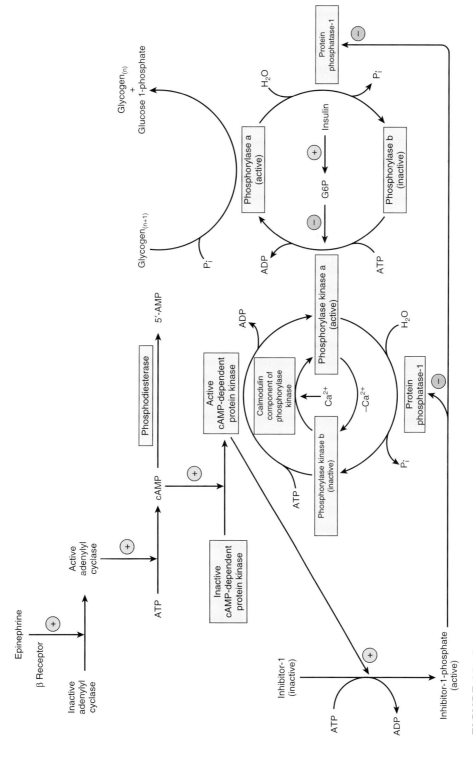

FIGURE 19–6 Control of phosphorylase in muscle. The sequence of reactions arranged as a cascade allows amplification of the hormonal signal at each step. (n, number of glucose residues; G6P, glucose 6-phosphate.)

Ca²⁺/calmodulin-sensitive phosphorylase kinase. cAMP-independent glycogenolysis is also activated by vasopressin, oxytocin, and angiotensin II acting either through calcium or the phosphatidylinositol bisphosphate pathway (Figure 42–10).

Protein Phosphatase-1 Inactivates Phosphorylase

Both phosphorylase a and phosphorylase kinase a are dephosphorylated and inactivated by **protein phosphatase-1.** Protein phosphatase-1 is inhibited by a protein, **inhibitor-1,** which is active only after it has been phosphorylated by cAMP-dependent protein kinase. Thus, cAMP controls both the activation and inactivation of phosphorylase (Figure 19–6). **Insulin** reinforces this effect by inhibiting the activation of phosphorylase b. It does this indirectly by increasing uptake of glucose, leading to increased formation of glucose 6-phosphate, which is an inhibitor of phosphorylase kinase.

Glycogen Synthase & Phosphorylase Activity Are Reciprocally Regulated

Like phosphorylase, glycogen synthase exists in both phosphorylated and nonphosphorylated states, and the effect of

phosphorylation is the reverse of that seen in phosphorylase (Figure 19–7). Active **glycogen synthase a** is dephosphorylated and inactive **glycogen synthase b** is phosphorylated.

Six different protein kinases act on glycogen synthase. Two are Ca²⁺/calmodulin-dependent (one of these is phosphorylase kinase). Another kinase is cAMP-dependent protein kinase, which allows cAMP-mediated hormonal action to inhibit glycogen synthesis synchronously with the activation of glycogenolysis. Insulin also promotes glycogenesis in muscle at the same time as inhibiting glycogenolysis by raising glucose 6-phosphate concentrations, which stimulates the dephosphorylation and activation of glycogen synthase. Dephosphorylation of glycogen synthase b is carried out by protein phosphatase-1, which is under the control of cAMP-dependent protein kinase.

REGULATION OF GLYCOGEN METABOLISM IS EFFECTED BY A BALANCE IN ACTIVITIES BETWEEN GLYCOGEN SYNTHASE & PHOSPHORYLASE

At the same time as phosphorylase is activated by a rise in concentration of cAMP (via phosphorylase kinase), glycogen syn-

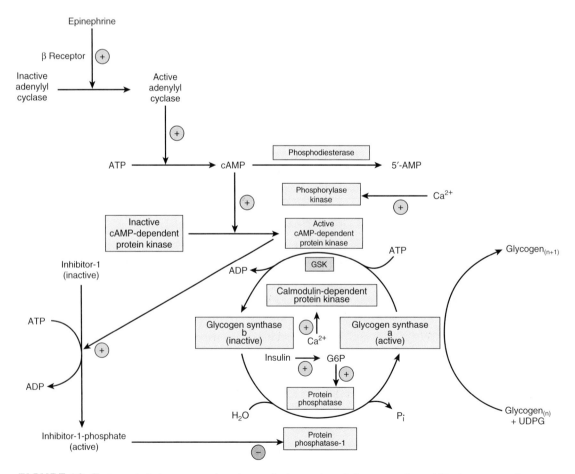

FIGURE 19–7 Control of glycogen synthase in muscle. (n, number of glucose residues; GSK, glycogen synthase kinase; G6P, glucose 6-phosphate.)

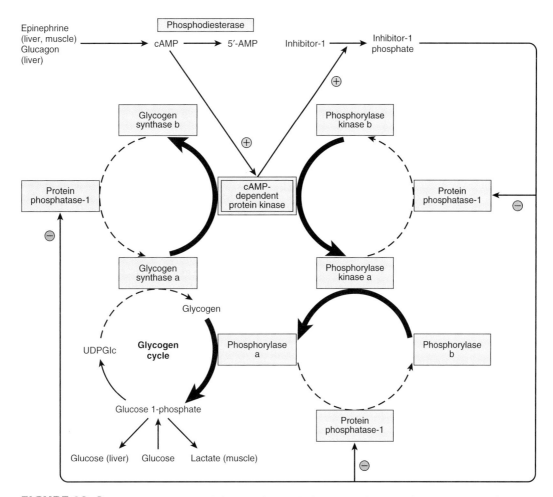

FIGURE 19–8 Coordinated control of glycogenolysis and glycogenesis by cAMP-dependent protein kinase. The reactions that lead to glycogenolysis as a result of an increase in cAMP concentrations are shown with bold arrows, and those that are inhibited by activation of protein phosphatase-1 are shown with dashed arrows. The reverse occurs when cAMP concentrations decrease as a result of phosphodiesterase activity, leading to glycogenesis.

thase is converted to the inactive form; both effects are mediated via **cAMP-dependent protein kinase** (Figure 19–8). Thus, inhibition of glycogenolysis enhances net glycogenesis, and inhibition of glycogenesis enhances net glycogenolysis. Also, the dephosphorylation of phosphorylase a, phosphorylase kinase, and glycogen synthase b is catalyzed by a single enzyme with broad specificity—**protein phosphatase-1.** In turn, protein phosphatase-1 is inhibited by cAMP-dependent protein kinase via inhibitor-1. Thus, glycogenolysis can be terminated and glycogenesis can be stimulated, or vice versa, synchronously, because both processes are dependent on the activity of cAMP-dependent protein kinase. Both phosphorylase kinase and glycogen synthase may be reversibly phosphorylated at more than one site by separate kinases and phosphatases. These secondary phosphorylations modify the sensitivity of the primary sites to phosphorylation and dephosphorylation **(multisite phosphorylation).** Also, they allow insulin, by way of increased glucose 6-phosphate, to have effects that act reciprocally to those of cAMP (see Figures 19–6 & 19–7).

CLINICAL ASPECTS

Glycogen Storage Diseases Are Inherited

"Glycogen storage disease" is a generic term to describe a group of inherited disorders characterized by deposition of an abnormal type or quantity of glycogen in tissues, or failure to mobilize glycogen. The principal diseases are summarized in Table 19–2.

SUMMARY

- Glycogen represents the principal storage carbohydrate in the body, mainly in the liver and muscle.
- In the liver, its major function is to provide glucose for extrahepatic tissues. In muscle, it serves mainly as a ready source of metabolic fuel for use in muscle. Muscle lacks glucose 6-phosphatase and cannot release free glucose from glycogen.
- Glycogen is synthesized from glucose by the pathway of glycogenesis. It is broken down by a separate pathway, glycogenolysis.

■ Cyclic AMP integrates the regulation of glycogenolysis and glycogenesis by promoting the simultaneous activation of phosphorylase and inhibition of glycogen synthase. Insulin acts reciprocally by inhibiting glycogenolysis and stimulating glycogenesis.

■ Inherited deficiencies of enzymes of glycogen metabolism in both liver and muscle cause glycogen storage diseases.

REFERENCES

Bollen M, Keppens S, Stalmans W: Specific features of glycogen metabolism in the liver. Biochem J 1998;336:19.

Ferrer JC, Favre C, Gomis RR, Fernandez-Novell JM, Garcia-Rocha M, de la Iglesia N, Cid E, Guinovart JJ: Control of glycogen deposition. FEBS Lett 2003;546:127.

Forde JE, Dale TC: Glycogen synthase kinase 3: a key regulator of cellular fate. Cell Mol Life Sci 2007;64:1930.

Greenberg CC, Jurczak MJ, et al: Glycogen branches out: new perspectives on the role of glycogen metabolism in the integration of metabolic pathways. Am J Physiol Endocrinol Metab 2006;291:E1.

Jentjens R, Jeukendrup A: Determinants of post-exercise glycogen synthesis during short-term recovery. Sports Med 2003;33:117.

McGarry JD, Kuwajima M, Newgard CB, Foster DW, Katz J: From dietary glucose to liver glycogen: the full circle round. Annu Rev Nutr 1987;7:51.

Meléndez-Hevia E, Waddell TG, Shelton ED: Optimization of molecular design in the evolution of metabolism: the glycogen molecule. Biochem J 1993;295:477.

Ozen H: Glycogen storage diseases: new perspectives. World J Gastroenterol 2007;13:2541.

Radziuk J, Pye S: Hepatic glucose uptake, gluconeogenesis and the regulation of glycogen synthesis. Diabetes Metab Res Rev 2001;17(4):250.

Roach PJ: Glycogen and its metabolism. Curr Mol Med 2002;2(2):101.

Roden M, Bernroider E: Hepatic glucose metabolism in humans—its role in health and disease. Best Pract Res Clin Endocrinol Metab 2003;17:365.

Shearer J, Graham TE: New perspectives on the storage and organization of muscle glycogen. Can J Appl Physiol 2002;27:179.

Shin YS: Glycogen storage disease: clinical, biochemical, and molecular heterogeneity. Semin Pediatr Neurol 2006;13:115.

Wolfsdorf JI, Holm IA: Glycogen storage diseases. Phenotypic, genetic, and biochemical characteristics, and therapy. Endocrinol Metab Clin North Am 1999;28:801.

Yeaman SJ, Armstrong JL, et al: Regulation of glycogen synthesis in human muscle cells. Biochem Soc Trans 2001;29:537.

Gluconeogenesis & the Control of Blood Glucose

David A. Bender, PhD

BIOMEDICAL IMPORTANCE

Gluconeogenesis is the process of synthesizing glucose or glycogen from noncarbohydrate precursors. The major substrates are the glucogenic amino acids (Chapter 29), lactate, glycerol, and propionate. Liver and kidney are the major gluconeogenic tissues, but the small intestine may also be a source of glucose in the fasting state.

Gluconeogenesis meets the needs of the body for glucose when insufficient carbohydrate is available from the diet or glycogen reserves. A supply of glucose is necessary especially for the nervous system and erythrocytes. Failure of gluconeogenesis is usually fatal. **Hypoglycemia** causes brain dysfunction, which can lead to coma and death. Glucose is also important in maintaining the level of intermediates of the citric acid cycle even when fatty acids are the main source of acetyl-CoA in the tissues. In addition, gluconeogenesis clears lactate produced by muscle and erythrocytes and glycerol produced by adipose tissue. In ruminants, propionate is a product of rumen metabolism of carbohydrates, and is a major substrate for gluconeogenesis.

GLUCONEOGENESIS INVOLVES GLYCOLYSIS, THE CITRIC ACID CYCLE, PLUS SOME SPECIAL REACTIONS

Thermodynamic Barriers Prevent a Simple Reversal of Glycolysis

Three nonequilibrium reactions in glycolysis (Chapter 18), catalyzed by hexokinase, phosphofructokinase and pyruvate kinase, prevent simple reversal of glycolysis for glucose synthesis (Figure 20–1). They are circumvented as follows.

Pyruvate & Phosphoenolpyruvate

Reversal of the reaction catalyzed by pyruvate kinase in glycolysis involves two endothermic reactions. Mitochondrial **pyruvate carboxylase** catalyzes the carboxylation of pyruvate to oxaloacetate, an ATP-requiring reaction in which the vitamin biotin is the coenzyme. Biotin binds CO_2 from bicarbonate as carboxybiotin prior to the addition of the CO_2 to pyruvate (Figure 44–17). A second enzyme, **phosphoenolpyruvate car-**

boxykinase, catalyzes the decarboxylation and phosphorylation of oxaloacetate to phosphoenolpyruvate using GTP as the phosphate donor. In liver and kidney, the reaction of succinate thiokinase in the citric acid cycle (Chapter 17) produces GTP (rather than ATP as in other tissues), and this GTP is used for the reaction of phosphoenolpyruvate carboxykinase, thus providing a link between citric acid cycle activity and gluconeogenesis, to prevent excessive removal of oxaloacetate for gluconeogenesis, which would impair citric acid cycle activity.

Fructose 1,6-Bisphosphate & Fructose 6-Phosphate

The conversion of fructose 1,6-bisphosphate to fructose 6-phosphate, for the reversal of glycolysis, is catalyzed by **fructose 1,6-bisphosphatase.** Its presence determines whether a tissue is capable of synthesizing glucose (or glycogen) not only from pyruvate, but also from triose phosphates. It is present in liver, kidney, and skeletal muscle, but is probably absent from heart and smooth muscle.

Glucose 6-Phosphate & Glucose

The conversion of glucose 6-phosphate to glucose is catalyzed by **glucose 6-phosphatase.** It is present in liver and kidney, but absent from muscle and adipose tissue, which, therefore, cannot export glucose into the bloodstream.

Glucose 1-Phosphate & Glycogen

The breakdown of glycogen to glucose 1-phosphate is catalyzed by phosphorylase. Glycogen synthesis involves a different pathway via uridine diphosphate glucose and **glycogen synthase** (Figure 19–1).

The relationships between gluconeogenesis and the glycolytic pathway are shown in Figure 20–1. After transamination or deamination, glucogenic amino acids yield either pyruvate or intermediates of the citric acid cycle. Therefore, the reactions described above can account for the conversion of both lactate and glucogenic amino acids to glucose or glycogen.

Propionate is a major precursor of glucose in ruminants; it enters gluconeogenesis via the citric acid cycle. After esterification with CoA, propionyl-CoA is carboxylated to D-methylmalonyl-CoA, catalyzed by **propionyl-CoA carbox-**

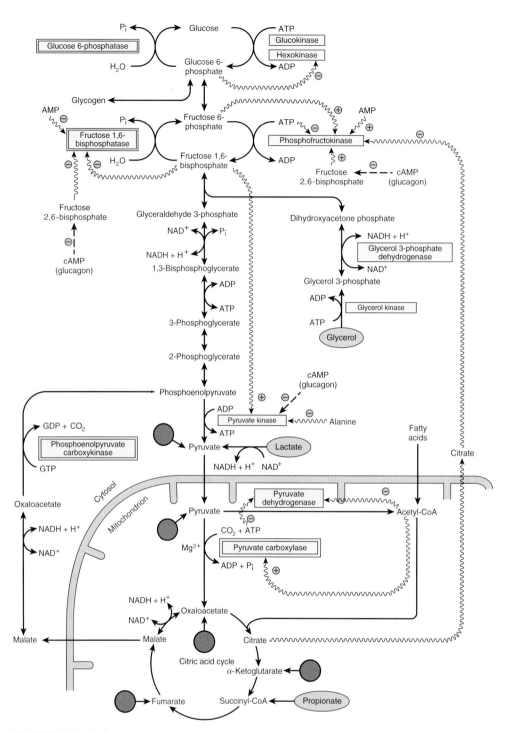

FIGURE 20–1 Major pathways and regulation of gluconeogenesis and glycolysis in the liver. Entry points of glucogenic amino acids after transamination are indicated by arrows extended from circles (see also Figure 17–4). The key gluconeogenic enzymes are enclosed in double-bordered boxes. The ATP required for gluconeogenesis is supplied by the oxidation of fatty acids. Propionate is of quantitative importance only in ruminants. Arrows with wavy shafts signify allosteric effects; dash-shafted arrows, covalent modification by reversible phosphorylation. High concentrations of alanine act as a "gluconeogenic signal" by inhibiting glycolysis at the pyruvate kinase step.

ylase, a biotin-dependent enzyme (Figure 20–2). **Methylmal-onyl-CoA racemase** catalyzes the conversion of D-methylma-lonyl-CoA to L-methylmalonyl-CoA, which then undergoes

isomerization to succinyl-CoA catalyzed by **methylmalonyl-CoA mutase.** In nonruminants, including humans, propi-onate arises from the β-oxidation of odd-chain fatty acids that occur in ruminant lipids (Chapter 22), as well as the oxidation of isoleucine and the side chain of cholesterol, and is a (relatively minor) substrate for gluconeogenesis. Methylmalonyl-CoA mutase is a vitamin B_{12}-dependent enzyme, and in deficiency methylmalonic acid is excreted in the urine (**methylmalonicaciduria**).

Glycerol is released from adipose tissue as a result of lipolysis of lipoprotein triacylglycerol in the fed state; it may be used for reesterification of free fatty acids to triacylglycerol in adipose tissue or liver, or may be a substrate for gluconeogenesis in the liver. In the fasting state glycerol released from lipolysis of adipose tissue triacylglycerol is used solely as a substrate for gluconeogenesis in the liver and kidneys.

SINCE GLYCOLYSIS & GLUCONEOGENESIS SHARE THE SAME PATHWAY BUT IN OPPOSITE DIRECTIONS, THEY MUST BE REGULATED RECIPROCALLY

Changes in the availability of substrates are responsible for most changes in metabolism either directly or indirectly acting via changes in hormone secretion. Three mechanisms are responsible for regulating the activity of enzymes concerned in carbohydrate metabolism: (1) changes in the rate of enzyme synthesis, (2) covalent modification by reversible phosphorylation, and (3) allosteric effects.

Induction & Repression of Key Enzymes Requires Several Hours

The changes in enzyme activity in the liver that occur under various metabolic conditions are listed in Table 20–1. The enzymes involved catalyze nonequilibrium (physiologically irreversible) reactions. The effects are generally reinforced because the activity of the enzymes catalyzing the reactions in

the opposite direction varies reciprocally (see Figure 20–1). The enzymes involved in the utilization of glucose (ie, those of glycolysis and lipogenesis) become more active when there is a superfluity of glucose, and under these conditions the enzymes of gluconeogenesis have low activity. Insulin, secreted in response to increased blood glucose, enhances the synthesis of the key enzymes in glycolysis. It also antagonizes the effect of the glucocorticoids and glucagon-stimulated cAMP, which induce synthesis of the key enzymes of gluconeogenesis.

Covalent Modification by Reversible Phosphorylation Is Rapid

Glucagon and **epinephrine,** hormones that are responsive to a decrease in blood glucose, inhibit glycolysis and stimulate gluconeogenesis in the liver by increasing the concentration of cAMP. This in turn activates cAMP-dependent protein kinase, leading to the phosphorylation and inactivation of **pyruvate kinase.** They also affect the concentration of fructose 2,6-bisphosphate and therefore glycolysis and gluconeogenesis, as described below.

Allosteric Modification Is Instantaneous

In gluconeogenesis, pyruvate carboxylase, which catalyzes the synthesis of oxaloacetate from pyruvate, requires acetyl-CoA as an **allosteric activator.** The addition of acetyl-CoA results in a change in the tertiary structure of the protein, lowering the K_m for bicarbonate. This means that as acetyl-CoA is formed from pyruvate, it automatically ensures the provision of oxaloacetate and, therefore, its further oxidation in the citric acid cycle, by activating pyruvate carboxylase. The activation of pyruvate carboxylase and the reciprocal inhibition of pyruvate dehydrogenase by acetyl-CoA derived from the oxidation of fatty acids explain the action of fatty acid oxidation in sparing the oxidation of pyruvate and in stimulating gluconeogenesis. The reciprocal relationship between these two enzymes

FIGURE 20–2 Metabolism of propionate.

TABLE 20–1   Regulatory and Adaptive Enzymes Associated with Carbohydrate Metabolism

	Activity in		Inducer	Repressor	Activator	Inhibitor
	Carbo-hydrate Feeding	Fasting and Diabetes				
Glycogenolysis, glycolysis, and pyruvate oxidation						
Glycogen synthase	↑	↓			Insulin, glucose 6-phosphate	Glucagon
Hexokinase						Glucose 6-phosphate
Glucokinase	↑	↓	Insulin	Glucagon		
Phosphofructokinase-1	↑	↓	Insulin	Glucagon	5′AMP, fructose 6-phosphate, fructose 2,6-bisphosphate, P_i	Citrate, ATP, glucagon
Pyruvate kinase	↑	↓	Insulin, fructose	Glucagon	Fructose 1,6-bisphosphate, insulin	ATP, alanine, glucagon, norepinephrine
Pyruvate dehydrogenase	↑	↓			CoA, NAD^+, insulin, ADP, pyruvate	Acetyl CoA, NADH, ATP (fatty acids, ketone bodies)
Gluconeogenesis						
Pyruvate carboxylase	↓	↑	Glucocorticoids, glucagon, epinephrine	Insulin	Acetyl CoA	ADP
Phosphoenolpyruvate carboxykinase	↓	↑	Glucocorticoids, glucagon, epinephrine	Insulin	Glucagon?	
Glucose 6-phosphatase	↓	↑	Glucocorticoids, glucagon, epinephrine	Insulin		

alters the metabolic fate of pyruvate as the tissue changes from carbohydrate oxidation (glycolysis) to gluconeogenesis during the transition from the fed to fasting state (see Figure 20–1). A major role of fatty acid oxidation in promoting gluconeogenesis is to supply the ATP required.

Phosphofructokinase (phosphofructokinase-1) occupies a key position in regulating glycolysis and is also subject to feedback control. It is inhibited by citrate and by normal intracellular concentrations of ATP and is activated by 5′AMP. 5′AMP acts as an indicator of the energy status of the cell. The presence of **adenylyl kinase** in liver and many other tissues allows rapid equilibration of the reaction

$$2ADP \leftrightarrow ATP + 5′AMP$$

Thus, when ATP is used in energy-requiring processes resulting in formation of ADP, [AMP] increases. A relatively small decrease in [ATP] causes a several-fold increase in [AMP], so that [AMP] acts as a metabolic amplifier of a small change in [ATP], and hence a sensitive signal of the energy state of the cell. The activity of phosphofructokinase-1 is thus regulated in response to the energy status of the cell to control the quantity of carbohydrate undergoing glycolysis prior to its entry into the citric acid cycle. Simultaneously, AMP activates phospho-

rylase, increasing glycogenolysis. A consequence of the inhibition of phosphofructokinase-1 is an accumulation of glucose 6-phosphate, which in turn inhibits further uptake of glucose in extrahepatic tissues by inhibition of μ hexokinase.

Fructose 2,6-Bisphosphate Plays a Unique Role in the Regulation of Glycolysis & Gluconeogenesis in Liver

The most potent positive allosteric activator of phosphofructokinase-1 and inhibitor of fructose 1,6-bisphosphatase in liver is **fructose 2,6-bisphosphate.** It relieves inhibition of phosphofructokinase-1 by ATP and increases the affinity for fructose 6-phosphate. It inhibits fructose 1,6-bisphosphatase by increasing the K_m for fructose 1,6-bisphosphate. Its concentration is under both substrate (allosteric) and hormonal control (covalent modification) (Figure 20–3).

Fructose 2,6-bisphosphate is formed by phosphorylation of fructose 6-phosphate by **phosphofructokinase-2.** The same enzyme protein is also responsible for its breakdown, since it has **fructose 2,6-bisphosphatase** activity. This **bifunctional enzyme** is under the allosteric control of fructose 6-phosphate, which stimulates the kinase and inhibits the phosphatase.

Hence, when there is an abundant supply of glucose, the concentration of fructose 2,6-bisphosphate increases, stimulating glycolysis by activating phosphofructokinase-1 and inhibiting fructose 1,6-bisphosphatase. In the fasting state, glucagon stimulates the production of cAMP, activating cAMP-dependent protein kinase, which in turn inactivates phosphofructokinase-2 and activates fructose 2,6-bisphosphatase by phosphorylation. Hence, gluconeogenesis is stimulated by a decrease in the concentration of fructose 2,6-bisphosphate, which inactivates phosphofructokinase-1 and relieves the inhibition of fructose 1,6-bisphosphatase.

Substrate (Futile) Cycles Allow Fine Tuning & Rapid Response

The control points in glycolysis and glycogen metabolism involve a cycle of phosphorylation and dephosphorylation catalyzed by glucokinase and glucose 6-phosphatase; phos-

FIGURE 20–3 Control of glycolysis and gluconeogenesis in the liver by fructose 2,6-bisphosphate and the bifunctional enzyme PFK-2/F-2,6-Pase (6-phosphofructo-2-kinase/fructose 2,6-bisphosphatase). (PFK-1, phosphofructokinase-1 [6-phosphofructo-1-kinase]; F-1,6-Pase, fructose 1,6-bisphosphatase.) Arrows with wavy shafts indicate allosteric effects.

phofructokinase-1 and fructose 1,6-bisphosphatase; pyruvate kinase, pyruvate carboxylase, and phosphoenolpyruvate carboxykinase; and glycogen synthase and phosphorylase. It would seem obvious that these opposing enzymes are regulated in such a way that when those involved in glycolysis are active, those involved in gluconeogenesis are inactive, since otherwise there would be cycling between phosphorylated and nonphosphorylated intermediates, with net hydrolysis of ATP. While this is so, in muscle both phosphofructokinase and fructose 1,6-bisphosphatase have some activity at all times, so that there is indeed some measure of (wasteful) substrate cycling. This permits the very rapid increase in the rate of glycolysis necessary for muscle contraction. At rest the rate of phosphofructokinase activity is some 10-fold higher than that of fructose 1,6-bisphosphatase; in anticipation of muscle contraction, the activity of both enzymes increases, fructose 1,6-bisphosphatase ten times more than phosphofructokinase, maintaining the same net rate of glycolysis. At the start of muscle contraction, the activity of phosphofructokinase increases further, and that of fructose 1,6-bisphosphatase falls, so increasing the net rate of glycolysis (and hence ATP formation) as much as a 1000-fold.

THE BLOOD CONCENTRATION OF GLUCOSE IS REGULATED WITHIN NARROW LIMITS

In the postabsorptive state, the concentration of blood glucose in most mammals is maintained between 4.5–5.5 mmol/L. After the ingestion of a carbohydrate meal, it may rise to 6.5–7.2 mmol/L, and in starvation, it may fall to 3.3–3.9 mmol/L. A sudden decrease in blood glucose (eg, in response to insulin overdose) causes convulsions, because of the dependence of the brain on a supply of glucose. However, much lower concentrations can be tolerated if hypoglycemia develops slowly enough for adaptation to occur. The blood glucose level in birds is considerably higher (14.0 mmol/L) and in ruminants considerably lower (approximately 2.2 mmol/L in sheep and 3.3 mmol/L in cattle). These lower normal levels appear to be associated with the fact that ruminants ferment virtually all dietary carbohydrate to short-chain fatty acids, and these largely replace glucose as the main metabolic fuel of the tissues in the fed state.

BLOOD GLUCOSE IS DERIVED FROM THE DIET, GLUCONEOGENESIS, & GLYCOGENOLYSIS

The digestible dietary carbohydrates yield glucose, galactose, and fructose that are transported to the liver via the **hepatic portal vein.** Galactose and fructose are readily converted to glucose in the liver (Chapter 21).

Glucose is formed from two groups of compounds that undergo gluconeogenesis (see Figures 17–4 & 20–1): (1) those which involve a direct net conversion to glucose, including most **amino acids** and **propionate;** and (2) those which are the products of the metabolism of glucose in tissues. Thus **lactate,** formed by glycolysis in skeletal muscle and erythrocytes, is transported to the liver and kidney where it reforms glucose, which again becomes available via the circulation for oxidation in the tissues. This process is known as the **Cori cycle,** or the **lactic acid cycle** (Figure 20–4).

In the fasting state, there is a considerable output of alanine from skeletal muscle, far in excess of its concentration in the muscle proteins that are being catabolized. It is formed by transamination of pyruvate produced by glycolysis of muscle glycogen, and is exported to the liver, where, after transamination back to pyruvate, it is a substrate for gluconeogenesis. This **glucose-alanine cycle** (see Figure 20–4) thus provides an indirect way of utilizing muscle glycogen to maintain blood glucose in the fasting state. The ATP required for the hepatic synthesis of glucose from pyruvate is derived from the oxidation of fatty acids.

Glucose is also formed from liver glycogen by glycogenolysis (Chapter 19).

Metabolic & Hormonal Mechanisms Regulate the Concentration of Blood Glucose

The maintenance of stable levels of glucose in the blood is one of the most finely regulated of all homeostatic mechanisms, involving the liver, extrahepatic tissues, and several hormones. Liver cells are freely permeable to glucose (via the GLUT 2 transporter), whereas cells of extrahepatic tissues (apart from pancreatic β-islets) are relatively impermeable, and their glucose transporters are regulated by insulin. As a result, uptake from the bloodstream is the rate-limiting step in the utiliza-

tion of glucose in extrahepatic tissues. The role of various glucose transporter proteins found in cell membranes is shown in Table 20–2.

Glucokinase Is Important in Regulating Blood Glucose After a Meal

Hexokinase has a low K_m for glucose, and in the liver is saturated and acting at a constant rate under all normal conditions. Glucokinase has a considerably higher K_m (lower affinity) for glucose, so that its activity increases with increases in the concentration of glucose in the hepatic portal vein (Figure 20–5). It promotes hepatic uptake of large amounts of glucose after a carbohydrate meal. It is absent from the liver of ruminants, which have little glucose entering the portal circulation from the intestines.

At normal systemic-blood glucose concentrations (4.5–5.5 mmol/L), the liver is a net producer of glucose. However, as the glucose level rises, the output of glucose ceases, and there is a net uptake.

Insulin Plays a Central Role in Regulating Blood Glucose

In addition to the direct effects of hyperglycemia in enhancing the uptake of glucose into the liver, the hormone insulin plays a central role in regulating blood glucose. It is produced by the β cells of the islets of Langerhans in the pancreas in response to hyperglycemia. The β-islet cells are freely permeable to glucose via the GLUT 2 transporter, and the glucose is phosphorylated by glucokinase. Therefore, increasing blood glucose increases metabolic flux through glycolysis, the citric acid cycle, and the generation of ATP. The increase in [ATP] inhibits ATP-sensitive K⁺ channels, causing depolarization of the cell membrane,

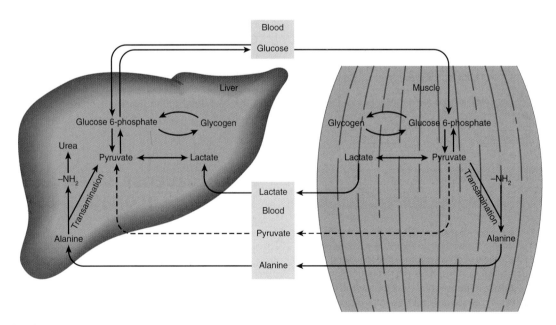

FIGURE 20–4 The lactic acid (Cori cycle) and glucose-alanine cycles.

TABLE 20–2 **Major Glucose Transporters**

	Tissue Location	Functions
Facilitative bidirectional transporters		
GLUT 1	Brain, kidney, colon, placenta, erythrocytes	Glucose uptake
GLUT 2	Liver, pancreatic β cell, small intestine, kidney	Rapid uptake or release of glucose
GLUT 3	Brain, kidney, placenta	Glucose uptake
GLUT 4	Heart and skeletal muscle, adipose tissue	Insulin-stimulated glucose uptake
GLUT 5	Small intestine	Absorption of glucose
Sodium-dependent unidirectional transporter		
SGLT 1	Small intestine and kidney	Active uptake of glucose against a concentration gradient

which increases Ca^{2+} influx via voltage-sensitive Ca^{2+} channels, stimulating exocytosis of insulin. Thus, the concentration of insulin in the blood parallels that of the blood glucose. Other substances causing release of insulin from the pancreas include amino acids, free fatty acids, ketone bodies, glucagon, secretin, and the sulfonylurea drugs tolbutamide and glyburide. These drugs are used to stimulate insulin secretion in type 2 diabetes mellitus (NIDDM, noninsulin-dependent diabetes mellitus); they act by inhibiting the ATP-sensitive K^+ channels. Epinephrine and norepinephrine block the release of insulin. Insulin lowers blood glucose immediately by enhancing glucose transport into adipose tissue and muscle by recruitment of glucose transporters (GLUT 4) from the interior of the cell to the plasma membrane. Although it does not affect glucose uptake into the liver directly, insulin does enhance long-term uptake as a result of its actions on the enzymes controlling glycolysis, glycogenesis, and gluconeogenesis (Chapter 19 & Table 20–1).

Glucagon Opposes the Actions of Insulin

Glucagon is the hormone produced by the α cells of the pancreatic islets. Its secretion is stimulated by hypoglycemia. In the liver it stimulates glycogenolysis by activating phosphorylase. Unlike epinephrine, glucagon does not have an effect on muscle phosphorylase. Glucagon also enhances gluconeogenesis from amino acids and lactate. In all these actions, glucagon acts via generation of cAMP (see Table 20–1). Both hepatic glycogenolysis and gluconeogenesis contribute to the **hyperglycemic effect** of glucagon, whose actions oppose those of insulin. Most of the endogenous glucagon (and insulin) is cleared from the circulation by the liver (Table 20–3).

Other Hormones Affect Blood Glucose

The **anterior pituitary gland** secretes hormones that tend to elevate the blood glucose and therefore antagonize the action of insulin. These are growth hormone, ACTH (corticotropin), and possibly other "diabetogenic" hormones. Growth hormone secretion is stimulated by hypoglycemia; it decreases glucose uptake in muscle. Some of this effect may be indirect, since it stimulates mobilization of free fatty acids from adipose tissue, which themselves inhibit glucose utilization. The **glucocorticoids** (11-oxysteroids) are secreted by the adrenal cortex, and are also synthesized in an unregulated manner in adipose tissue. They act to increase gluconeogenesis as a result of enhanced hepatic catabolism of amino acids, due to induction of aminotransferases (and other enzymes such as tryptophan dioxygenase) and key enzymes of gluconeogenesis. In addition, glucocorticoids inhibit the utilization of glucose in extrahepatic tissues. In all these actions, glucocorticoids act in a manner antagonistic to insulin. A number of **cytokines** secreted by macrophages infiltrating adipose tissue also have insulin antagonistic actions; together with glucocorticoids secreted by adipose tissue, this explains the insulin resistance that commonly occurs in obese people.

Epinephrine is secreted by the adrenal medulla as a result of stressful stimuli (fear, excitement, hemorrhage, hypoxia, hypoglycemia, etc.) and leads to glycogenolysis in liver and muscle owing to stimulation of phosphorylase via generation of cAMP. In muscle, glycogenolysis results in increased glycolysis, whereas in liver it results in the release of glucose into the bloodstream.

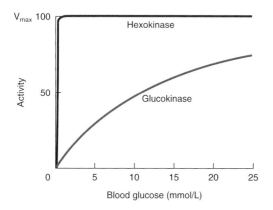

FIGURE 20–5 Variation in glucose phosphorylating activity of hexokinase and glucokinase with increasing blood glucose concentration. The K_m for glucose of hexokinase is 0.05 mmol/L and of glucokinase is 10 mmol/L.

TABLE 20–3   Tissue Responses to Insulin and Glucagon

	Liver	Adipose Tissue	Muscle
Increased by insulin	Fatty acid synthesis Glycogen synthesis Protein synthesis	Glucose uptake Fatty acid synthesis	Glucose uptake Glycogen synthesis Protein synthesis
Decreased by insulin	Ketogenesis Gluconeogenesis	Lipolysis	
Increased by glucagon	Glycogenolysis Gluconeogenesis Ketogenesis	Lipolysis	

FURTHER CLINICAL ASPECTS

Glucosuria Occurs When the Renal Threshold for Glucose Is Exceeded

When the blood glucose rises to relatively high levels, the kidney also exerts a regulatory effect. Glucose is continuously filtered by the glomeruli, but is normally completely reabsorbed in the renal tubules by active transport. The capacity of the tubular system to reabsorb glucose is limited to a rate of about 2 mmol/min, and in hyperglycemia (as occurs in poorly controlled diabetes mellitus), the glomerular filtrate may contain more glucose than can be reabsorbed, resulting in **glucosuria.** Glucosuria occurs when the venous blood glucose concentration exceeds about 10 mmol/L; this is termed the **renal threshold** for glucose.

Hypoglycemia May Occur During Pregnancy & in the Neonate

During pregnancy, fetal glucose consumption increases and there is a risk of maternal and possibly fetal hypoglycemia, particularly if there are long intervals between meals or at night. Furthermore, premature and low-birth-weight babies are more susceptible to hypoglycemia, since they have little adipose tissue to provide free fatty acids. The enzymes of gluconeogenesis may not be completely functional at this time, and gluconeogenesis is anyway dependent on a supply of free fatty acids for energy. Little glycerol, which would normally be released from adipose tissue, is available for gluconeogenesis.

The Body's Ability to Utilize Glucose May Be Ascertained by Measuring Glucose Tolerance

Glucose tolerance is the ability to regulate the blood glucose concentration after the administration of a test dose of glucose (normally 1 g/kg body weight) (Figure 20–6). **Diabetes mellitus** (type 1, or insulin-dependent diabetes mellitus; IDDM) is characterized by decreased glucose tolerance as a result of decreased secretion of insulin as a result of progressive destruction of pancreatic β-islet cells. Glucose tolerance is also impaired in type 2 diabetes mellitus (NIDDM) as a result of impaired sensitivity of tissues to insulin action. Insulin

resistance associated with obesity (and especially abdominal obesity) leading to the development of hyperlipidemia, then atherosclerosis and coronary heart disease, as well as overt diabetes, is known as the **metabolic syndrome.** Impaired glucose tolerance also occurs in conditions where the liver is damaged, in some infections, and in response to some drugs, as well as in conditions that lead to hyperactivity of the pituitary or adrenal cortex because of the antagonism of the hormones secreted by these glands to the action of insulin.

Administration of insulin (as in the treatment of diabetes mellitus) lowers the blood glucose concentration and increases its utilization and storage in the liver and muscle as glycogen. An excess of insulin may cause **hypoglycemia,** resulting in convulsions and even in death unless glucose is administered promptly. Increased tolerance to glucose is observed in pituitary or adrenocortical insufficiency, attributable to a decrease in the antagonism to insulin by the hormones normally secreted by these glands.

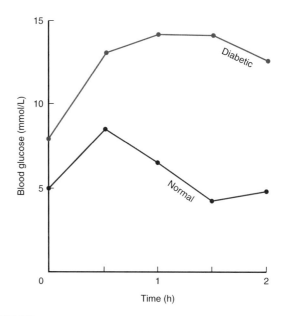

FIGURE 20–6 Glucose tolerance test. Blood glucose curves of a normal and a diabetic person after oral administration of 1 g of glucose/kg body weight. Note the initial raised concentration in the fasting diabetic. A criterion of normality is the return of the curve to the initial value within 2 h.

The Energy Cost of Gluconeogenesis Explains Why Very Low Carbohydrate Diets Promote Weight Loss

Very low carbohydrate diets, providing only 20 g per day of carbohydrate or less (compared with a desirable intake of 100–120 g/day), but permitting unlimited consumption of fat and protein, have been promoted as an effective regime for weight loss, although such diets are counter to all advice on a prudent diet for health. Since there is a continual demand for glucose, there will be a considerable amount of gluconeogenesis from amino acids; the associated high ATP cost must then be met by oxidation of fatty acids.

SUMMARY

■ Gluconeogenesis is the process of synthesizing glucose or glycogen from noncarbohydrate precursors. It is of particular importance when carbohydrate is not available from the diet. Significant substrates are amino acids, lactate, glycerol, and propionate.

■ The pathway of gluconeogenesis in the liver and kidney utilizes those reactions in glycolysis that are reversible plus four additional reactions that circumvent the irreversible nonequilibrium reactions.

■ Since glycolysis and gluconeogenesis share the same pathway but operate in opposite directions, their activities must be regulated reciprocally.

■ The liver regulates the blood glucose after a meal because it contains the high-K_m glucokinase that promotes increased hepatic utilization of glucose.

■ Insulin is secreted as a direct response to hyperglycemia; it stimulates the liver to store glucose as glycogen and facilitates uptake of glucose into extrahepatic tissues.

■ Glucagon is secreted as a response to hypoglycemia and activates both glycogenolysis and gluconeogenesis in the liver, causing release of glucose into the blood.

REFERENCES

Barthel A, Schmoll D: Novel concepts in insulin regulation of hepatic gluconeogenesis. Am J Physiol Endocrinol Metab 2003;285:E685.

Boden G: Gluconeogenesis and glycogenolysis in health and diabetes. J Investig Med 2004;52:375.

Dzugaj, A: Localization and regulation of muscle fructose 1,6-bisphosphatase, the key enzyme of glyconeogenesis. Adv Enzyme Regul 2006;46:51.

Jiang G, Zhang BB: Glucagon and regulation of glucose metabolism. Am J Physiol Endocrinol Metab 2003; 284:E671.

Jitrapakdee S, Vidal-Puig A, et al: Anaplerotic roles of pyruvate carboxylase in mammalian tissues. Cell Mol Life Sci 2006; 63:843.

Klover PJ, Mooney RA: Hepatocytes: critical for glucose homeostasis. Int J Biochem Cell Biol 2004;36:753.

McGuinness OP: Defective glucose homeostasis during infection. Ann Rev Nutr 2005;25:9.

Mlinar B, Marc J, et al: Molecular mechanisms of insulin resistance and associated diseases.Clin Chim Acta 2007;375:20.

Nordlie RC, Foster JD, Lange AJ: Regulation of glucose production by the liver. Ann Rev Nutr 1999;19:379.

Pilkis SJ, Claus TH: Hepatic gluconeogenesis/glycolysis: regulation and structure/function relationships of substrate cycle enzymes. Ann Rev Nutr 1991;11:465.

Pilkis SJ, Granner DK: Molecular physiology of the regulation of hepatic gluconeogenesis and glycolysis. Ann Rev Physiol 1992;54:885.

Postic C, Shiota M, Magnuson MA: Cell-specific roles of glucokinase in glucose homeostasis. Rec Prog Horm Res 2001;56:195.

Quinn PG, Yeagley D: Insulin regulation of PEPCK gene expression: a model for rapid and reversible modulation. Curr Drug Targets Immune Endocr Metabol Disord 2005;5:423.

Reaven GM: The insulin resistance syndrome: definition and dietary approaches to treatment. Ann Rev Nutr 2005;25:391.

Roden M, Bernroider E: Hepatic glucose metabolism in humans—its role in health and disease. Best Pract Res Clin Endocrinol Metab 2003;17:365.

Schuit FC, Huypens P, Heimberg H, Pipeleers DG: Glucose sensing in pancreatic beta-cells: a model for the study of other glucose-regulated cells in gut, pancreas, and hypothal amus. Diabetes 2001;50:1.

Suh SH, Paik IY, et al: Regulation of blood glucose homeostasis during prolonged exercise. Mol Cells 2007;23:272.

Wahren J, Ekberg K: Splanchnic regulation of glucose production. Ann Rev Nutr 2007;27:329.

Young, A: Inhibition of glucagon secretion. Adv Pharmacol 2005;52:151.

The Pentose Phosphate Pathway & Other Pathways of Hexose Metabolism

David A. Bender, PhD

BIOMEDICAL IMPORTANCE

The pentose phosphate pathway is an alternative route for the metabolism of glucose. It does not lead to formation of ATP but has two major functions: (1) the formation of **NADPH** for synthesis of fatty acids and steroids, and (2) the synthesis of **ribose** for nucleotide and nucleic acid formation. Glucose, fructose, and galactose are the main hexoses absorbed from the gastrointestinal tract, derived from dietary starch, sucrose, and lactose, respectively. Fructose and galactose can be converted to glucose, mainly in the liver.

Genetic deficiency of **glucose 6-phosphate dehydrogenase,** the first enzyme of the pentose phosphate pathway, is a major cause of hemolysis of red blood cells, resulting in **hemolytic anemia**. Glucuronic acid is synthesized from glucose via the **uronic acid pathway,** of minor quantitative importance, but of major significance for the excretion of metabolites and foreign chemicals (xenobiotics) as **glucuronides.** A deficiency in the pathway leads to the condition of **essential pentosuria.** The lack of one enzyme of the pathway (gulonolactone oxidase) in primates and some other animals explains why **ascorbic acid** (vitamin C) is a dietary requirement for humans but not most other mammals. Deficiencies in the enzymes of fructose and galactose metabolism lead to metabolic diseases such as **essential fructosuria, hereditary fructose intolerance,** and **galactosemia.**

THE PENTOSE PHOSPHATE PATHWAY FORMS NADPH & RIBOSE PHOSPHATE

The pentose phosphate pathway (hexose monophosphate shunt) is a more complex pathway than glycolysis (Figure 21–1). Three molecules of glucose 6-phosphate give rise to three molecules of CO_2 and three 5-carbon sugars. These are rearranged to regenerate two molecules of glucose 6-phosphate and one molecule of the glycolytic intermediate, glyceraldehyde 3-phosphate. Since two molecules of glyceraldehyde 3-phosphate can regenerate glucose 6-phosphate, the pathway can account for the complete oxidation of glucose.

REACTIONS OF THE PENTOSE PHOSPHATE PATHWAY OCCUR IN THE CYTOSOL

Like glycolysis, the enzymes of the pentose phosphate pathway are cytosolic. Unlike glycolysis, oxidation is achieved by dehydrogenation using **NADP⁺**, not **NAD⁺**, as the hydrogen acceptor. The sequence of reactions of the pathway may be divided into two phases: an **oxidative nonreversible phase** and a **nonoxidative reversible phase.** In the first phase, glucose 6-phosphate undergoes dehydrogenation and decarboxylation to yield a pentose, ribulose 5-phosphate. In the second phase, ribulose 5-phosphate is converted back to glucose 6-phosphate by a series of reactions involving mainly two enzymes: **transketolase** and **transaldolase** (see Figure 21–1).

The Oxidative Phase Generates NADPH

Dehydrogenation of glucose 6-phosphate to 6-phosphogluconate occurs via the formation of 6-phosphogluconolactone catalyzed by **glucose 6-phosphate dehydrogenase,** an NADP-dependent enzyme (Figures 21–1 & 21–2). The hydrolysis of 6-phosphogluconolactone is accomplished by the enzyme **gluconolactone hydrolase.** A second oxidative step is catalyzed by **6-phosphogluconate dehydrogenase,** which also requires NADP⁺ as hydrogen acceptor. Decarboxylation follows with the formation of the ketopentose ribulose 5-phosphate.

The Nonoxidative Phase Generates Ribose Precursors

Ribulose 5-phosphate is the substrate for two enzymes. **Ribulose 5-phosphate 3-epimerase** alters the configuration about carbon 3, forming the epimer xylulose 5-phosphate, also a ketopentose. **Ribose 5-phosphate ketoisomerase** converts ribulose 5-phosphate to the corresponding aldopentose, ribose 5-phosphate, which is used for nucleotide and nucleic acid synthesis. **Transketolase** transfers the two-carbon unit comprising carbons 1 and 2 of a ketose onto the aldehyde carbon of an aldose sugar. It therefore affects the conversion of a ketose

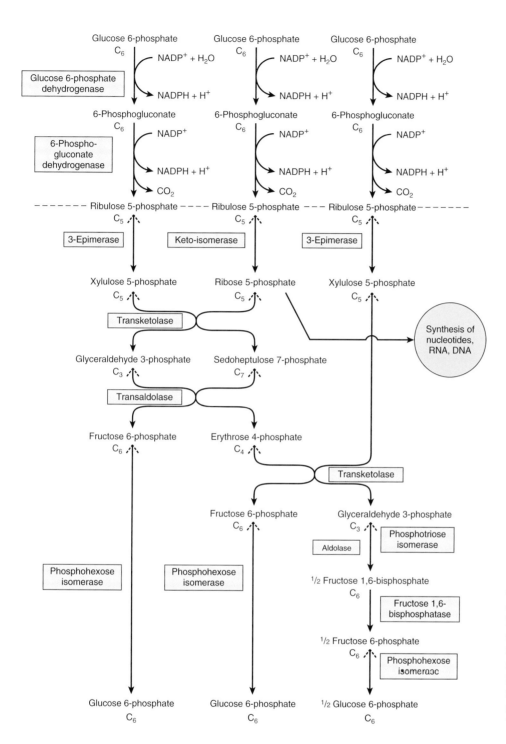

FIGURE 21–1 Flow chart of pentose phosphate pathway and its connections with the pathway of glycolysis. The full pathway, as indicated, consists of three interconnected cycles in which glucose 6-phosphate is both substrate and end-product. The reactions above the broken line are nonreversible, whereas all reactions under that line are freely reversible apart from that catalyzed by fructose 1,6-bisphosphatase.

sugar into an aldose with two carbons less and an aldose sugar into a ketose with two carbons more. The reaction requires Mg^{2+} and **thiamin diphosphate** (vitamin B_1) as coenzyme. The two-carbon moiety transferred is probably glycolaldehyde bound to thiamin diphosphate. Thus, transketolase catalyzes the transfer of the two-carbon unit from xylulose 5-phosphate to ribose 5-phosphate, producing the seven-carbon ketose sedoheptulose 7-phosphate and the aldose glyceraldehyde 3-phosphate. These two products then undergo transaldolation. **Transaldolase** catalyzes the transfer of a three-carbon

dihydroxyacetone moiety (carbons 1–3) from the ketose sedoheptulose 7-phosphate onto the aldose glyceraldehyde 3-phosphate to form the ketose fructose 6-phosphate and the four-carbon aldose erythrose 4-phosphate. In a further reaction catalyzed by **transketolase,** xylulose 5-phosphate serves as a donor of glycolaldehyde. In this case erythrose 4-phosphate is the acceptor, and the products of the reaction are fructose 6-phosphate and glyceraldehyde 3-phosphate.

In order to oxidize glucose completely to CO_2 via the pentose phosphate pathway, there must be enzymes present

FIGURE 21–2 The pentose phosphate pathway. (P, —PO_3^{2-}; PRPP, 5-phosphoribosyl 1-pyrophosphate.)

in the tissue to convert glyceraldehyde 3-phosphate to glucose 6-phosphate. This involves reversal of glycolysis and the gluconeogenic enzyme **fructose 1,6-bisphosphatase.** In tissues that lack this enzyme, glyceraldehyde 3-phosphate follows the normal pathway of glycolysis to pyruvate.

The Two Major Pathways for the Catabolism of Glucose Have Little in Common

Although glucose 6-phosphate is common to both pathways, the pentose phosphate pathway is markedly different from glycolysis. Oxidation utilizes NADP rather than NAD, and CO_2, which is not produced in glycolysis, is a characteristic product. No ATP is generated in the pentose phosphate pathway, whereas it is a major product of glycolysis.

Reducing Equivalents Are Generated in Those Tissues Specializing in Reductive Syntheses

The pentose phosphate pathway is active in liver, adipose tissue, adrenal cortex, thyroid, erythrocytes, testis, and lactating mammary gland. Its activity is low in nonlactating mammary gland and skeletal muscle. Those tissues in which the pathway is active use NADPH in reductive syntheses, eg, of fatty acids, steroids, amino acids via glutamate dehydrogenase, and reduced glutathione. The synthesis of glucose 6-phosphate dehydrogenase and 6-phosphogluconate dehydrogenase may also be induced by insulin in the fed state, when lipogenesis increases.

Ribose Can Be Synthesized in Virtually All Tissues

Little or no ribose circulates in the bloodstream, so tissues have to synthesize the ribose they require for nucleotide and nucleic acid synthesis using the pentose phosphate pathway (see Figure 21–2). It is not necessary to have a completely functioning pentose phosphate pathway for a tissue to synthesize ribose 5-phosphate. Muscle has only low activity of glucose 6-phosphate dehydrogenase and 6-phosphogluconate dehydrogenase, but, like most other tissues, it is capable of synthesizing ribose 5-phosphate by reversal of the nonoxidative phase of the pentose phosphate pathway utilizing fructose 6-phosphate.

THE PENTOSE PHOSPHATE PATHWAY & GLUTATHIONE PEROXIDASE PROTECT ERYTHROCYTES AGAINST HEMOLYSIS

In red blood cells the pentose phosphate pathway provides NADPH for the reduction of oxidized glutathione catalyzed by **glutathione reductase,** a flavoprotein containing FAD. Re-

duced glutathione removes H_2O_2 in a reaction catalyzed by **glutathione peroxidase,** an enzyme that contains the **selenium** analogue of cysteine (selenocysteine) at the active site (Figure 21–3). The reaction is important, since accumulation of H_2O_2 may decrease the life span of the erythrocyte by causing oxidative damage to the cell membrane, leading to hemolysis.

GLUCURONATE, A PRECURSOR OF PROTEOGLYCANS & CONJUGATED GLUCURONIDES, IS A PRODUCT OF THE URONIC ACID PATHWAY

In liver, the **uronic acid pathway** catalyzes the conversion of glucose to glucuronic acid, ascorbic acid (except in human beings and other species for which ascorbate is a vitamin), and pentoses (Figure 21–4). It is also an alternative oxidative pathway for glucose that, like the pentose phosphate pathway, does not lead to the formation of ATP. Glucose 6-phosphate is isomerized to glucose 1-phosphate, which then reacts with uridine triphosphate (UTP) to form uridine diphosphate glucose (UDPGlc) in a reaction catalyzed by **UDPGlc pyrophosphorylase,** as occurs in glycogen synthesis (Chapter 19). UDPGlc is oxidized at carbon 6 by NAD-dependent **UDPGlc dehydrogenase** in a two-step reaction to yield UDP-glucuronate.

UDP-glucuronate is the source of glucuronate for reactions involving its incorporation into proteoglycans or for reaction with substrates such as steroid hormones, bilirubin, and a number of drugs that are excreted in urine or bile as glucuronide conjugates (Figure 31–13).

Glucuronate is reduced to L-gulonate, the direct precursor of **ascorbate** in those animals capable of synthesizing this vitamin, in an NADPH-dependent reaction. In humans and other primates, as well as guinea pigs, bats, and some birds and fishes, ascorbic acid cannot be synthesized because of the absence of L-**gulonolactone oxidase.** L-Gulonate is oxidized to 3-keto-L-gulonate, which is then decarboxylated to L-xylulose. L-Xylulose is converted to the D isomer by an NADPH-dependent reduction to xylitol, followed by oxidation in an NAD-dependent reaction to D-xylulose. After conversion to D-xylulose 5-phosphate, it is metabolized via the pentose phosphate pathway.

FIGURE 21–3 Role of the pentose phosphate pathway in the glutathione peroxidase reaction of erythrocytes. (G-S-S-G, oxidized glutathione; G-SH, reduced glutathione; Se, selenium-containing enzyme.)

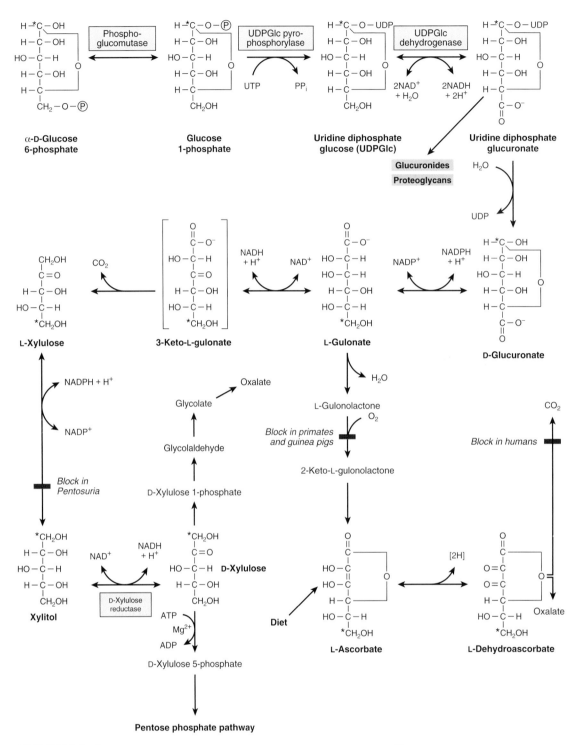

FIGURE 21–4 Uronic acid pathway. (*Indicates the fate of carbon 1 of glucose; ℗, $-PO_3^{2-}$.)

INGESTION OF LARGE QUANTITIES OF FRUCTOSE HAS PROFOUND METABOLIC CONSEQUENCES

Diets high in sucrose or in high-fructose syrups (HFS) used in manufactured foods and beverages lead to large amounts of fructose (and glucose) entering the hepatic portal vein.

Fructose undergoes more rapid glycolysis in the liver than does glucose, because it bypasses the regulatory step catalyzed by phosphofructokinase (Figure 21–5). This allows fructose to flood the pathways in the liver, leading to enhanced fatty acid synthesis, increased esterification of fatty acids, and increased VLDL secretion, which may raise serum triacylglycerols and ultimately raise LDL cholesterol concentrations. A specific

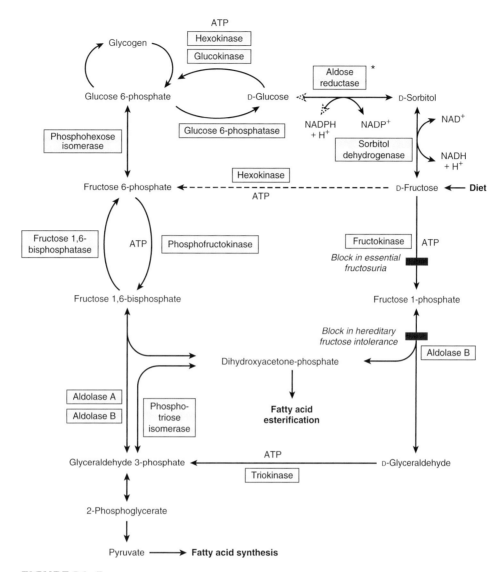

FIGURE 21–5 Metabolism of fructose. Aldolase A is found in all tissues, whereas aldolase B is the predominant form in liver. (*Not found in liver.)

kinase, **fructokinase,** in liver, kidney, and intestine, catalyzes the phosphorylation of fructose to fructose 1-phosphate. This enzyme does not act on glucose, and, unlike glucokinase, its activity is not affected by fasting or by insulin, which may explain why fructose is cleared from the blood of diabetic patients at a normal rate. Fructose 1-phosphate is cleaved to D-glyceraldehyde and dihydroxyacetone phosphate by **aldolase B,** an enzyme found in the liver, which also functions in glycolysis in the liver by cleaving fructose 1,6-bisphosphate. D-Glyceraldehyde enters glycolysis via phosphorylation to glyceraldehyde 3-phosphate catalyzed by **triokinase.** The two triose phosphates, dihydroxyacetone phosphate and glyceraldehyde 3-phosphate, may either be degraded by glycolysis or may be substrates for aldolase and hence gluconeogenesis, which is the fate of much of the fructose metabolized in the liver.

In extrahepatic tissues, hexokinase catalyzes the phosphorylation of most hexose sugars, including fructose, but glucose inhibits the phosphorylation of fructose, since it is a

better substrate for hexokinase. Nevertheless, some fructose can be metabolized in adipose tissue and muscle. Fructose is found in seminal plasma and in the fetal circulation of ungulates and whales. Aldose reductase is found in the placenta of the ewe and is responsible for the secretion of sorbitol into the fetal blood. The presence of sorbitol dehydrogenase in the liver, including the fetal liver, is responsible for the conversion of sorbitol into fructose. This pathway is also responsible for the occurrence of fructose in seminal fluid.

GALACTOSE IS NEEDED FOR THE SYNTHESIS OF LACTOSE, GLYCOLIPIDS, PROTEOGLYCANS, & GLYCOPROTEINS

Galactose is derived from intestinal hydrolysis of the disaccharide **lactose,** the sugar of milk. It is readily converted in the liver to glucose. **Galactokinase** catalyses the phosphorylation

FIGURE 21–6 Pathway of conversion of **(A)** galactose to glucose in the liver and **(B)** glucose to lactose in the lactating mammary gland.

{AQ}

of galactose, using ATP as phosphate donor (Figure 21–6). Galactose 1-phosphate reacts with uridine diphosphate glucose (UDPGlc) to form uridine diphosphate galactose (UDPGal) and glucose 1-phosphate, in a reaction catalyzed by **galactose 1-phosphate uridyl transferase.** The conversion of UDPGal to UDPGlc is catalyzed by **UDPGal 4-epimerase.** The reaction involves oxidation, then reduction, at carbon 4, with NAD+ as coenzyme. The UDPGlc is then incorporated into glycogen (Chapter 19).

Since the epimerase reaction is freely reversible, glucose can be converted to galactose, so that galactose is not a dietary essential. Galactose is required in the body not only in the formation of lactose but also as a constituent of glycolipids (cerebrosides), proteoglycans, and glycoproteins. In the synthesis of lactose in the mammary gland, UDPGal condenses with glucose to yield lactose, catalyzed by **lactose synthase** (see Figure 21–6).

Glucose Is the Precursor of Amino Sugars (Hexosamines)

Amino sugars are important components of **glycoproteins** (Chapter 47), of certain **glycosphingolipids** (eg, gangliosides; Chapter 15), and of **glycosaminoglycans** (Chapter 48). The major amino sugars are the hexosamines **glucosamine, ga-**lactosamine, and **mannosamine,** and the nine-carbon compound **sialic acid.** The principal sialic acid found in human tissues is *N*-acetylneuraminic acid (NeuAc). A summary of the metabolic interrelationships among the amino sugars is shown in Figure 21–7.

CLINICAL ASPECTS

Impairment of the Pentose Phosphate Pathway Leads to Erythrocyte Hemolysis

Genetic defects of glucose 6-phosphate dehydrogenase, with consequent impairment of the generation of NADPH, are common in populations of Mediterranean and Afro-Caribbean origin. The gene is on the X chromosome, so it is mainly males who are affected. Some 400 million people carry a mutated gene for glucose 6-phosphate dehydrogenase, making it the most common genetic defect, but most are asymptomatic. The distribution of mutant genes parallels that of malaria, suggesting that being heterozygous confers resistance against malaria. The defect is manifested as red cell hemolysis (**hemolytic anemia**) when susceptible individuals are subjected to oxidative stress (Chapter 52) from infection, drugs such as the antimalarial primaquine, and sulfonamides, or when they have eaten fava beans (*Vicia fava*—hence the

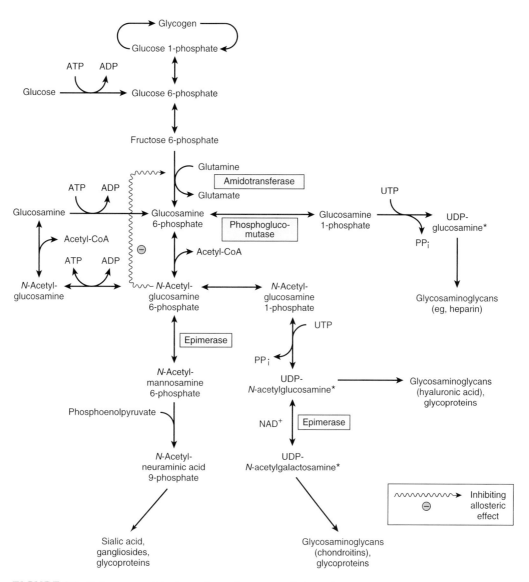

FIGURE 21–7 Summary of the interrelationships in metabolism of amino sugars. (*Analogous to UDPGlc.) Other purine or pyrimidine nucleotides may be similarly linked to sugars or amino sugars. Examples are thymidine diphosphate (TDP)-glucosamine and TDP-*N*-acetylglucosamine.

name of the disease, **favism**). There are two main variants of favism. In the Afro-Caribbean variant the enzyme is unstable, so that while average red cell activities are low, it is only the older erythrocytes that are affected by oxidative stress, and the hemolytic crises tend to be self-limiting. By contrast, in the Mediterranean variant the enzyme is stable, but has low activity in all erythrocytes. Hemolytic crises in these people are more severe and can be fatal. Glutathione peroxidase is dependent upon a supply of NADPH, which in erythrocytes can be formed only via the pentose phosphate pathway. It reduces organic peroxides and H_2O_2, as part of the body's defense against lipid peroxidation (Figure 15–21). Measurement of erythrocyte **transketolase,** and its activation by thiamin diphosphate is used to assess thiamin nutritional status (Chapter 44).

Disruption of the Uronic Acid Pathway Is Caused by Enzyme Defects & Some Drugs

In the rare benign hereditary condition **essential pentosuria,** considerable quantities of L-**xylulose** appear in the urine, because of absence of the enzyme necessary to reduce L-xylulose to xylitol. Various drugs increase the rate at which glucose enters the uronic acid pathway. For example, administration of barbital or chlorobutanol to rats results in a significant increase in the conversion of glucose to glucuronate, L-gulonate, and ascorbate. Aminopyrine and antipyrine increase the excretion of L-xylulose in pentosuric subjects. Pentosuria also occurs after consumption of relatively large amounts of fruits such as pears that are rich sources of pentoses (**alimentary pentosuria**).

Loading of the Liver with Fructose May Potentiate Hypertriacylglycerolemia, Hypercholesterolemia, & Hyperuricemia

In the liver, fructose increases fatty acid and triacylglycerol synthesis and VLDL secretion, leading to hypertriacylglycerolemia—and increased LDL cholesterol—which can be regarded as potentially atherogenic (Chapter 26). This is because fructose enters glycolysis via fructokinase, and the resulting fructose 1-phosphate bypasses the regulatory step catalyzed by phosphofructokinase (Chapter 18). In addition, acute loading of the liver with fructose, as can occur with intravenous infusion or following very high fructose intakes, causes sequestration of inorganic phosphate in fructose 1-phosphate and diminished ATP synthesis. As a result there is less inhibition of de novo purine synthesis by ATP, and uric acid formation is increased, causing hyperuricemia, which is a cause of gout (Chapter 33).

Defects in Fructose Metabolism Cause Disease

A lack of hepatic fructokinase causes **essential fructosuria,** which is a benign and asymptomatic condition. Absence of aldolase B, which cleaves fructose 1-phosphate, leads to **hereditary fructose intolerance**, which is characterized by profound hypoglycemia and vomiting after consumption of fructose (or sucrose, which yields fructose on digestion) (Figure 21–5). Diets low in fructose, sorbitol, and sucrose are beneficial for both conditions. One consequence of hereditary fructose intolerance and of a related condition as a result of **fructose 1,6-bisphosphatase deficiency** is fructose-induced **hypoglycemia** despite the presence of high glycogen reserves, because of fructose 1-phosphate and 1,6-bisphosphate allosterically inhibit liver phosphorylase. The sequestration of inorganic phosphate also leads to depletion of ATP and hyperuricemia.

Fructose & Sorbitol in the Lens Are Associated with Diabetic Cataract

Both fructose and sorbitol are found in the lens of the eye in increased concentrations in diabetes mellitus, and may be involved in the pathogenesis of **diabetic cataract.** The **sorbitol (polyol) pathway** (not found in liver) is responsible for fructose formation from glucose (see Figure 21–5) and increases in activity as the glucose concentration rises in those tissues that are not insulin-sensitive, ie, the lens, peripheral nerves, and renal glomeruli. Glucose is reduced to sorbitol by **aldose reductase,** followed by oxidation of sorbitol to fructose in the presence of NAD^+ and sorbitol dehydrogenase (polyol dehydrogenase). Sorbitol does not diffuse through cell membranes, but accumulates, causing osmotic damage. Simultaneously, myoinositol levels fall. Sorbitol accumulation and myoinositol depletion, as well as diabetic cataract, can be prevented by aldose reductase inhibitors in experimental animals, but to date there is no evidence that inhibitors are effective in preventing cataract or diabetic neuropathy in humans.

Enzyme Deficiencies in the Galactose Pathway Cause Galactosemia

Inability to metabolize galactose occurs in the **galactosemias,** which may be caused by inherited defects of galactokinase, uridyl transferase, or 4-epimerase (Figure 21–6A), though deficiency of **uridyl transferase** is the best known. Galactose is a substrate for aldose reductase, forming galactitol, which accumulates in the lens of the eye, causing cataract. The general condition is more severe if it is the result of a defect in the uridyl transferase, since galactose 1-phosphate accumulates and depletes the liver of inorganic phosphate. Ultimately, liver failure and mental deterioration result. In uridyl transferase deficiency, the epimerase is present in adequate amounts, so that the galactosemic individual can still form UDPGal from glucose. This explains how it is possible for normal growth and development of affected children to occur despite the galactose-free diets used to control the symptoms of the disease.

SUMMARY

- The pentose phosphate pathway, present in the cytosol, can account for the complete oxidation of glucose, producing NADPH and CO_2 but not ATP.

- The pathway has an oxidative phase, which is irreversible and generates NADPH, and a nonoxidative phase, which is reversible and provides ribose precursors for nucleotide synthesis. The complete pathway is present only in those tissues having a requirement for NADPH for reductive syntheses, eg, lipogenesis or steroidogenesis, whereas the nonoxidative phase is present in all cells requiring ribose.

- In erythrocytes, the pathway has a major function in preventing hemolysis by providing NADPH to maintain glutathione in the reduced state as the substrate for glutathione peroxidase.

- The uronic acid pathway is the source of glucuronic acid for conjugation of many endogenous and exogenous substances before excretion as glucuronides in urine and bile.

- Fructose bypasses the main regulatory step in glycolysis, catalyzed by phosphofructokinase, and stimulates fatty acid synthesis and hepatic triacylglycerol secretion.

- Galactose is synthesized from glucose in the lactating mammary gland and in other tissues where it is required for the synthesis of glycolipids, proteoglycans, and glycoproteins

REFERENCES

Ali M, Rellos P, Cox TM: Hereditary fructose intolerance. J Med Gen 1998;35:353.

Cappellini MD, Fiorelli G: Glucose 6-phosphate dehydrogenase deficiency. Lancet 2008;371:64.

Dunlop M: Aldose reductase and the role of the polyol pathway in diabetic nephropathy. Kidney Int 2000;77:S3.

Hers HG, Hue L: Gluconeogenesis and related aspects of glycolysis. Annu Rev Bioche 1983;52:617.

Horecker BL: The pentose phosphate pathway. J Biol Chem 2002;277:47965.

Le KA, Tappy L: Metabolic effects of fructose. Curr Opin Clin Nutr Metab Care 2006;9:469.

Leslie ND: Insights into the pathogenesis of galactosemia. Ann Rev Nutr 2003;23:59.

Mayes PA: Intermediary metabolism of fructose. Amer J Clin Nutr 1993;58:754.

Van den Berghe G: Inborn errors of fructose metabolism. Ann Rev Nutr 1994;14:41.

Wong D: Hereditary fructose intolerance. Mol Genet Metab 2005;85:165.

Oxidation of Fatty Acids: Ketogenesis

Kathleen M. Botham, PhD, DSc & Peter A. Mayes, PhD, DSc

BIOMEDICAL IMPORTANCE

Although fatty acids are both oxidized to acetyl-CoA and synthesized from acetyl-CoA, fatty acid oxidation is not the simple reverse of fatty acid biosynthesis but an entirely different process taking place in a separate compartment of the cell. The separation of fatty acid oxidation in mitochondria from biosynthesis in the cytosol allows each process to be individually controlled and integrated with tissue requirements. Each step in fatty acid oxidation involves acyl-CoA derivatives and is catalyzed by separate enzymes, utilizes NAD^+ and FAD as coenzymes, and generates ATP. It is an aerobic process, requiring the presence of oxygen.

Increased fatty acid oxidation is a characteristic of starvation and of diabetes mellitus, leading to **ketone body** production by the liver **(ketosis).** Ketone bodies are acidic and when produced in excess over long periods, as in diabetes, cause **ketoacidosis,** which is ultimately fatal. Because gluconeogenesis is dependent upon fatty acid oxidation, any impairment in fatty acid oxidation leads to **hypoglycemia.** This occurs in various states of **carnitine deficiency** or deficiency of essential enzymes in fatty acid oxidation, eg, **carnitine palmitoyltransferase,** or inhibition of fatty acid oxidation by poisons, eg, **hypoglycin.**

OXIDATION OF FATTY ACIDS OCCURS IN MITOCHONDRIA

Fatty Acids Are Transported in the Blood as Free Fatty Acids (FFA)

Free fatty acids—also called unesterified (UFA) or nonesterified (NEFA) fatty acids—are fatty acids that are in the **unesterified state.** In plasma, longer-chain FFA are combined with **albumin,** and in the cell they are attached to a **fatty acid-binding protein,** so that in fact they are never really "free." Shorter-chain fatty acids are more water-soluble and exist as the unionized acid or as a fatty acid anion.

Fatty Acids Are Activated Before Being Catabolized

Fatty acids must first be converted to an active intermediate before they can be catabolized. This is the only step in the complete degradation of a fatty acid that requires energy from ATP. In the presence of ATP and coenzyme A, the enzyme **acyl-CoA synthetase (thiokinase)** catalyzes the conversion of a fatty acid (or free fatty acid) to an "active fatty acid" or acyl-CoA, which uses one high-energy phosphate with the formation of AMP and PP_i (Figure 22–1). The PP_i is hydrolyzed by **inorganic pyrophosphatase** with the loss of a further high-energy phosphate, ensuring that the overall reaction goes to completion. Acyl-CoA synthetases are found in the endoplasmic reticulum, peroxisomes, and inside and on the outer membrane of mitochondria.

Long-Chain Fatty Acids Penetrate the Inner Mitochondrial Membrane as Carnitine Derivatives

Carnitine (β-hydroxy-γ-trimethylammonium butyrate), $(CH_3)_3N^+—CH_2—CH(OH)—CH_2—COO^-$, is widely distributed and is particularly abundant in muscle. Long-chain acyl-CoA (or FFA) cannot penetrate the inner membrane of mitochondria. In the presence of carnitine, however, **carnitine palmitoyltransferase-I,** located in the outer mitochondrial membrane, converts long-chain acyl-CoA to acylcarnitine, which is able to penetrate the inner membrane and gain access to the β-oxidation system of enzymes (Figure 22–1). **Carnitine-acylcarnitine translocase** acts as an inner membrane exchange transporter. Acylcarnitine is transported in, coupled with the transport out of one molecule of carnitine. The acylcarnitine then reacts with CoA, catalyzed by **carnitine palmitoyltransferase-II,** located on the inside of the inner membrane, reforming acyl-CoA in the mitochondrial matrix, and carnitine is liberated.

β-OXIDATION OF FATTY ACIDS INVOLVES SUCCESSIVE CLEAVAGE WITH RELEASE OF ACETYL-CoA

In β-oxidation (Figure 22–2), two carbons at a time are cleaved from acyl-CoA molecules, starting at the carboxyl end. The chain is broken between the α(2)- and β(3)-carbon atoms—

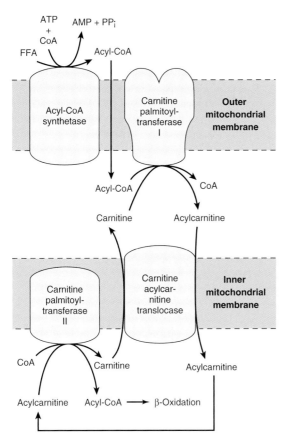

FIGURE 22–1 Role of carnitine in the transport of long-chain fatty acids through the inner mitochondrial membrane. Long-chain acyl-CoA cannot pass through the inner mitochondrial membrane, but its metabolic product, acylcarnitine, can.

FIGURE 22–2 Overview of β-oxidation of fatty acids.

hence the name β-oxidation. The two-carbon units formed are acetyl-CoA; thus, palmitoyl-CoA forms eight acetyl-CoA molecules.

The Cyclic Reaction Sequence Generates FADH₂ & NADH

Several enzymes, known collectively as "fatty acid oxidase," are found in the mitochondrial matrix or inner membrane adjacent to the respiratory chain. These catalyze the oxidation of acyl-CoA to acetyl-CoA, the system being coupled with the phosphorylation of ADP to ATP (Figure 22–3).

The first step is the removal of two hydrogen atoms from the 2(α)- and 3(β)-carbon atoms, catalyzed by **acyl-CoA dehydrogenase** and requiring FAD. This results in the formation of Δ^2-*trans*-enoyl-CoA and FADH₂. The reoxidation of FADH₂ by the respiratory chain requires the mediation of another flavoprotein, termed **electron-transferring flavoprotein** (Chapter 12). Water is added to saturate the double bond and form 3-hydroxyacyl-CoA, catalyzed by Δ^2-**enoyl-CoA hydratase.** The 3-hydroxy derivative undergoes further dehydrogenation on the 3-carbon catalyzed by **L(+)-3-hydroxyacyl-CoA dehydrogenase** to form the corresponding 3-ketoacyl-CoA compound. In this case, NAD⁺ is the coenzyme involved. Finally, 3-ketoacyl-CoA is split at the 2,3-position by **thiolase** (3-ketoacyl-CoA-thiolase), forming acetyl-CoA and a new acyl-CoA two carbons shorter than the original acyl-CoA molecule. The acyl-CoA formed in the cleavage reaction reenters the oxidative pathway at reaction 2 (Figure 22–3). In this way, a long-chain fatty acid may be degraded completely to acetyl-CoA (C_2 units). Since acetyl-CoA can be oxidized to CO_2 and water via the citric acid cycle (which is also found within the mitochondria), the complete oxidation of fatty acids is achieved.

Oxidation of a Fatty Acid with an Odd Number of Carbon Atoms Yields Acetyl-CoA Plus a Molecule of Propionyl-CoA

Fatty acids with an odd number of carbon atoms are oxidized by the pathway of β-oxidation, producing acetyl-CoA, until a three-carbon (propionyl-CoA) residue remains. This compound is converted to succinyl-CoA, a constituent of the citric acid cycle (Figure 20–2). Hence, **the propionyl residue from an odd-chain fatty acid is the only part of a fatty acid that is glucogenic.**

Oxidation of Fatty Acids Produces a Large Quantity of ATP

Transport in the respiratory chain of electrons from FADH₂ and NADH leads to the synthesis of four high-energy phosphates (Chapter 13) for each of the seven cycles needed for the breakdown of palmitate to acetyl-CoA ($7 \times 4 = 28$). A total of 8 mol of acetyl-CoA is formed, and each gives rise to 10 mol of ATP on oxidation in the citric acid cycle, making $8 \times 10 = 80$ mol. Two must be subtracted for the initial activation of the fatty acid, yielding a net gain of 106 mol of ATP per mole of

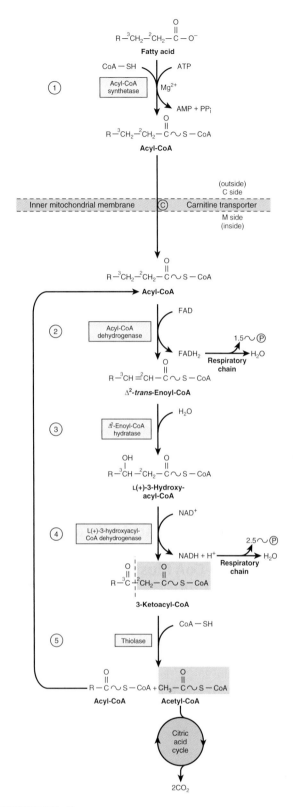

FIGURE 22–3 β-Oxidation of fatty acids. Long-chain acyl-CoA is cycled through reactions ②–⑤, acetyl-CoA being split off, each cycle, by thiolase (reaction ⑤). When the acyl radical is only four carbon atoms in length, two acetyl-CoA molecules are formed in reaction ⑤.

FIGURE 22–4 Sequence of reactions in the oxidation of unsaturated fatty acids, eg, linoleic acid. Δ^4-*cis*-fatty acids or fatty acids forming Δ^4-*cis*-enoyl-CoA enter the pathway at the position shown. NADPH for the dienoyl-CoA reductase step is supplied by intramitochondrial sources such as glutamate dehydrogenase, isocitrate dehydrogenase, and NAD(P)H transhydrogenase.

palmitate, or $106 \times 51.6^* = 5470$ kJ. This represents 68% of the free energy of combustion of palmitic acid.

Peroxisomes Oxidize Very Long Chain Fatty Acids

A modified form of β oxidation is found in **peroxisomes** and leads to the formation of acetyl-CoA and H_2O_2 (from the flavoprotein-linked dehydrogenase step), which is broken down by catalase (Chapter 12). Thus, this dehydrogenation in peroxisomes is not linked directly to phosphorylation and the generation of ATP. The system facilitates the oxidation of very long chain fatty acids (eg, C_{20}, C_{22}). These enzymes are induced by high-fat diets and in some species by hypolipidemic drugs such as clofibrate.

The enzymes in peroxisomes do not attack shorter-chain fatty acids; the β-oxidation sequence ends at octanoyl-CoA. Octanoyl and acetyl groups are both further oxidized in mitochondria. Another role of peroxisomal β oxidation is to shorten the side chain of cholesterol in bile acid formation (Chapter 26). Peroxisomes also take part in the synthesis of ether glycerolipids (Chapter 24), cholesterol, and dolichol (Figure 26–2).

OXIDATION OF UNSATURATED FATTY ACIDS OCCURS BY A MODIFIED β-OXIDATION PATHWAY

The CoA esters of these acids are degraded by the enzymes normally responsible for β oxidation until either a Δ^3-*cis*-acyl-CoA compound or a Δ^4-*cis*-acyl-CoA compound is formed, depending upon the position of the double bonds (Figure 22–4). The former compound is isomerized (Δ^3*cis* → Δ^2-***trans*-enoyl-CoA isomerase**) to the corresponding Δ^2-*trans*-CoA stage of β-oxidation for subsequent hydration and oxidation. Any Δ^4-*cis*-acyl-CoA either remaining, as in the case of linoleic acid, or entering the pathway at this point after conversion by acyl-CoA dehydrogenase to Δ^2 *trans* Δ^4-*cis*-dienoyl-CoA, is then metabolized as indicated in Figure 22–4.

KETOGENESIS OCCURS WHEN THERE IS A HIGH RATE OF FATTY ACID OXIDATION IN THE LIVER

Under metabolic conditions associated with a high rate of fatty acid oxidation, the liver produces considerable quantities of **acetoacetate** and **D(−)-3-hydroxybutyrate** (β-hydroxybutyrate). Acetoacetate continually undergoes spontaneous decarboxy-

FIGURE 22–5 Interrelationships of the ketone bodies. D(−)-3-hydroxybutyrate dehydrogenase is a mitochondrial enzyme.

lation to yield **acetone**. These three substances are collectively known as the **ketone bodies** (also called acetone bodies or [incorrectly*] "ketones") (Figure 22–5). Acetoacetate and 3-hydroxybutyrate are interconverted by the mitochondrial enzyme **D(−)-3-hydroxybutyrate dehydrogenase;** the equilibrium is controlled by the mitochondrial [NAD+]/[NADH] ratio, ie, the **redox state.** The concentration of total ketone bodies in the blood of well-fed mammals does not normally exceed 0.2 mmol/L except in ruminants, where 3-hydroxybutyrate is formed continuously from butyric acid (a product of ruminal fermentation) in the rumen wall. In vivo, the liver appears to be the only organ in nonruminants to add significant quantities of ketone bodies to the blood. Extrahepatic tissues utilize them as respiratory substrates. The net flow of ketone bodies from the liver to the extrahepatic tissues results from active hepatic synthesis coupled with very low utilization. The reverse situation occurs in extrahepatic tissues (Figure 22–6).

3-Hydroxy-3-Methylglutaryl-CoA (HMG-CoA) Is an Intermediate in the Pathway of Ketogenesis

Enzymes responsible for ketone body formation are associated mainly with the mitochondria. Two acetyl-CoA molecules formed in β oxidation condense with one another to form acetoacetyl-CoA by a reversal of the **thiolase** reaction. Acetoacetyl-CoA, which is the starting material for ketogen-

---

*ΔG for the ATP reaction, as explained in Chapter 18

*The term "ketones" should not be used because 3-hydroxybutyrate is not a ketone and there are ketones in blood that are not ketone bodies, eg, pyruvate and fructose.

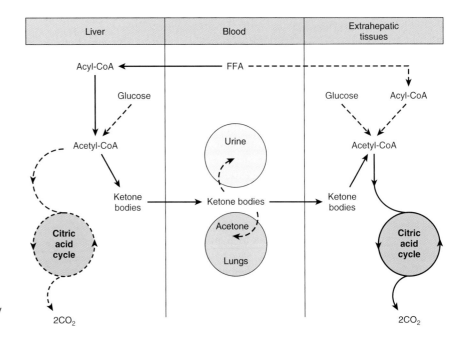

FIGURE 22–6 Formation, utilization, and excretion of ketone bodies. (The main pathway is indicated by the solid arrows.)

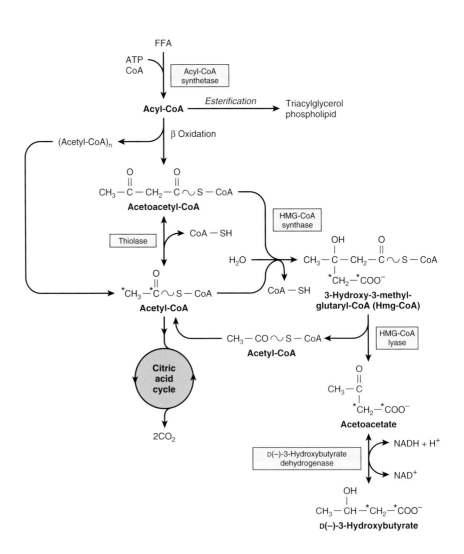

FIGURE 22–7 Pathways of ketogenesis in the liver. (FFA, free fatty acids.)

esis, also arises directly from the terminal four carbons of a fatty acid during β-oxidation (Figure 22–7). Condensation of acetoacetyl-CoA with another molecule of acetyl-CoA by **3-hydroxy-3-methylglutaryl-CoA synthase** forms 3-hydroxy-3-methylglutaryl-CoA **(HMG-CoA). 3-Hydroxy-3-methylglutaryl-CoA lyase** then causes acetyl-CoA to split off from the HMG-CoA, leaving free acetoacetate. The carbon atoms split off in the acetyl-CoA molecule are derived from the original acetoacetyl-CoA molecule. **Both enzymes must be present in mitochondria for ketogenesis to take place.** This occurs solely in liver and rumen epithelium. D(–)-3-Hydroxybutyrate is quantitatively the predominant ketone body present in the blood and urine in ketosis.

Ketone Bodies Serve as a Fuel for Extrahepatic Tissues

While an active enzymatic mechanism produces acetoacetate from acetoacetyl-CoA in the liver, acetoacetate once formed cannot be reactivated directly except in the cytosol, where it is used in a much less active pathway as a precursor in cholesterol synthesis. This accounts for the net production of ketone bodies by the liver.

In extrahepatic tissues, acetoacetate is activated to acetoacetyl-CoA by **succinyl-CoA-acetoacetate CoA transferase.** CoA is transferred from succinyl-CoA to form acetoacetyl-CoA (Figure 22–8). With the addition of a CoA, the acetoacetyl-CoA is split into two acetyl-CoAs by thiolase and oxidized in the citric acid cycle. If the blood level is raised, oxidation of ketone bodies increases until, at a concentration of approximately 12 mmol/L, they saturate the oxidative ma-

chinery. When this occurs, a large proportion of the oxygen consumption may be accounted for by the oxidation of ketone bodies.

In most cases, **ketonemia is due to increased production of ketone bodies** by the liver rather than to a deficiency in their utilization by extrahepatic tissues. While acetoacetate and D(–)-3-hydroxybutyrate are readily oxidized by extrahepatic tissues, acetone is difficult to oxidize in vivo and to a large extent is volatilized in the lungs.

In moderate ketonemia, the loss of ketone bodies via the urine is only a few percent of the total ketone body production and utilization. Since there are renal threshold-like effects (there is not a true threshold) that vary between species and individuals, measurement of the ketonemia, not the ketonuria, is the preferred method of assessing the severity of ketosis.

KETOGENESIS IS REGULATED AT THREE CRUCIAL STEPS

1. Ketosis does not occur in vivo unless there is an increase in the level of circulating free fatty acids that arise from lipolysis of triacylglycerol in adipose tissue. **Free fatty acids are the precursors of ketone bodies in the liver.** The liver, both in fed and in fasting conditions, extracts about 30% of the free fatty acids passing through it, so that at high concentrations the flux passing into the liver is substantial. **Therefore, the factors regulating mobilization of free fatty acids from adipose tissue are important in controlling ketogenesis** (Figures 22–9 & 25–8).

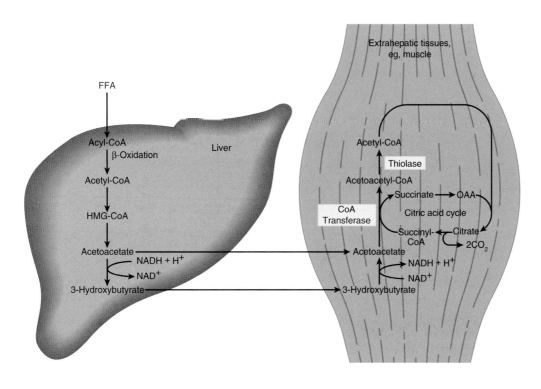

FIGURE 22–8 Transport of ketone bodies from the liver and pathways of utilization and oxidation in extrahepatic tissues.

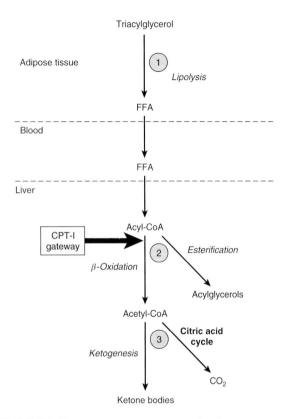

FIGURE 22–9 Regulation of ketogenesis. ①–③ show three crucial steps in the pathway of metabolism of free fatty acids (FFA) that determine the magnitude of ketogenesis. (CPT-I, carnitine palmitoyltransferase-I.)

2. After uptake by the liver, free fatty acids are either **β-oxidized** to CO_2 or ketone bodies or **esterified** to triacylglycerol and phospholipid. There is regulation of entry of fatty acids into the oxidative pathway by **carnitine palmitoyltransferase-I** (CPT-I), and the remainder of the fatty acid taken up is esterified. CPT-I activity is low in the fed state, leading to depression of fatty acid oxidation, and high in starvation, allowing fatty acid oxidation to increase. **Malonyl-CoA**, the initial intermediate in fatty acid biosynthesis (Figure 23–1) formed by acetyl-CoA carboxylase in the fed state, is a potent inhibitor of CPT-I (Figure 22–10). Under these conditions, free fatty acids enter the liver cell in low concentrations and are nearly all esterified to acylglycerols and transported out of the liver in very low density lipoproteins (VLDL). However, as the concentration of free fatty acids increases with the onset of starvation, acetyl-CoA carboxylase is inhibited directly by acyl-CoA, and [malonyl-CoA] decreases, releasing the inhibition of CPT-I and allowing more acyl-CoA to be β-oxidized. These events are reinforced in starvation by a decrease in the **[insulin]/[glucagon] ratio**. Thus, β-oxidation from free fatty acids is controlled by the CPT-I gateway into the mitochondria, and the balance of the free fatty acid uptake not oxidized is esterified.

3. In turn, the acetyl-CoA formed in β-oxidation is oxidized in the citric acid cycle, or it enters the pathway of ketogenesis to form ketone bodies. As the level of serum free fatty acids is raised, proportionately more free fatty acid is converted to ketone bodies and less is oxidized via the citric acid cycle to CO_2. The partition of acetyl-CoA between the ketogenic

{AQ}

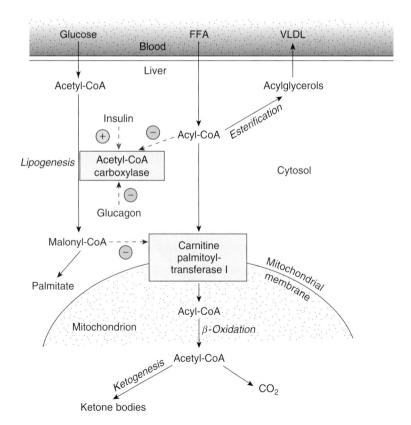

FIGURE 22–10 Regulation of long-chain fatty acid oxidation in the liver. (FFA, free fatty acids; VLDL, very low density lipoprotein.) Positive (⊕) and negative (⊖) regulatory effects are represented by broken arrows and substrate flow by solid arrows.

pathway and the pathway of oxidation to CO_2 is regulated so that the total free energy captured in ATP which results from the oxidation of free fatty acids remains constant as their concentration in the serum changes. This may be appreciated when it is realized that complete oxidation of 1 mol of palmitate involves a net production of 106 mol of ATP via β-oxidation and CO_2 production in the citric acid cycle (see above), whereas only 26 mol of ATP are produced when acetoacetate is the end product and only 21 mol when 3-hydroxybutyrate is the end product. Thus, ketogenesis may be regarded as a mechanism that allows the liver to oxidize increasing quantities of fatty acids within the constraints of a tightly coupled system of oxidative phosphorylation.

A fall in the concentration of oxaloacetate, particularly within the mitochondria, can impair the ability of the citric acid cycle to metabolize acetyl-CoA and divert fatty acid oxidation toward ketogenesis. Such a fall may occur because of an increase in the [NADH]/[NAD⁺] ratio caused by increased β-oxidation of fatty acids affecting the equilibrium between oxaloacetate and malate, leading to a decrease in the concentration of oxaloacetate, and when gluconeogenesis is elevated, which occurs when blood glucose levels are low. The activation of pyruvate carboxylase, which catalyzes the conversion of pyruvate to oxaloacetate, by acetyl-CoA partially alleviates this problem, but in conditions such as starvation and untreated diabetes mellitus, ketone bodies are overproduced causing ketosis.

CLINICAL ASPECTS

Impaired Oxidation of Fatty Acids Gives Rise to Diseases Often Associated with Hypoglycemia

Carnitine deficiency can occur particularly in the newborn—and especially in preterm infants—owing to inadequate biosynthesis or renal leakage. Losses can also occur in hemodialysis. This suggests a vitamin-like dietary requirement for carnitine in some individuals. Symptoms of deficiency include hypoglycemia, which is a consequence of impaired fatty acid oxidation and lipid accumulation with muscular weakness. Treatment is by oral supplementation with carnitine.

Inherited **CPT-I deficiency** affects only the liver, resulting in reduced fatty acid oxidation and ketogenesis, with hypoglycemia. **CPT-II deficiency** affects primarily skeletal muscle and, when severe, the liver. The sulfonylurea drugs (**glyburide** [**glibenclamide**] and **tolbutamide**), used in the treatment of type 2 diabetes mellitus, reduce fatty acid oxidation and, therefore, hyperglycemia by inhibiting CPT-I.

Inherited defects in the enzymes of β-oxidation and ketogenesis also lead to nonketotic hypoglycemia, coma, and fatty liver. Defects are known in long- and short-chain 3-hydroxyacyl-CoA dehydrogenase (deficiency of the long-chain enzyme may be a cause of **acute fatty liver of pregnancy**). **3-Ketoacyl-CoA thiolase** and **HMG-CoA lyase deficiency**

also affect the degradation of leucine, a ketogenic amino acid (Chapter 29).

Jamaican vomiting sickness is caused by eating the unripe fruit of the akee tree, which contains the toxin **hypoglycin.** This inactivates medium- and short-chain acyl-CoA dehydrogenase, inhibiting β-oxidation and causing hypoglycemia. **Dicarboxylic aciduria** is characterized by the excretion of C_6–C_{10} ω-dicarboxylic acids and by nonketotic hypoglycemia, and is caused by a lack of mitochondrial **medium-chain acyl-CoA dehydrogenase. Refsum's disease** is a rare neurologic disorder due to a metabolic defect that results in the accumulation of phytanic acid, which is found in dairy products and ruminant fat and meat. Phytanic acid is thought to have pathological effects on membrane function, protein prenylation, and gene expression. **Zellweger's (cerebrohepatorenal) syndrome** occurs in individuals with a rare inherited absence of peroxisomes in all tissues. They accumulate C_{26}–C_{38} polyenoic acids in brain tissue and also exhibit a generalized loss of peroxisomal functions. The disease causes severe neurological symptoms, and most patients die in the first year of life.

Ketoacidosis Results from Prolonged Ketosis

Higher than normal quantities of ketone bodies present in the blood or urine constitute **ketonemia** (hyperketonemia) or **ketonuria,** respectively. The overall condition is called **ketosis.** The basic form of ketosis occurs in **starvation** and involves depletion of available carbohydrate coupled with mobilization of free fatty acids. This general pattern of metabolism is exaggerated to produce the pathologic states found in **diabetes mellitus, the type 2 form of which is increasingly common in Western countries; twin lamb disease;** and **ketosis in lactating cattle.** Nonpathologic forms of ketosis are found under conditions of high-fat feeding and after severe exercise in the postabsorptive state.

Acetoacetic and 3-hydroxybutyric acids are both moderately strong acids and are buffered when present in blood or other tissues. However, their continual excretion in quantity progressively depletes the alkali reserve, causing **ketoacidosis.** This may be fatal in uncontrolled **diabetes mellitus.**

SUMMARY

- Fatty acid oxidation in mitochondria leads to the generation of large quantities of ATP by a process called β-oxidation that cleaves acetyl-CoA units sequentially from fatty acyl chains. The acetyl-CoA is oxidized in the citric acid cycle, generating further ATP.

- The ketone bodies (acetoacetate, 3-hydroxybutyrate, and acetone) are formed in hepatic mitochondria when there is a high rate of fatty acid oxidation. The pathway of ketogenesis involves synthesis and breakdown of 3-hydroxy-3-methylglutaryl-CoA (HMG-CoA) by two key enzymes, HMG-CoA synthase and HMG-CoA lyase.

- Ketone bodies are important fuels in extrahepatic tissues.
- Ketogenesis is regulated at three crucial steps: (1) control of free fatty acid mobilization from adipose tissue; (2) the activity of carnitine palmitoyltransferase-I in liver, which determines the proportion of the fatty acid flux that is oxidized rather than esterified; and (3) partition of acetyl-CoA between the pathway of ketogenesis and the citric acid cycle.
- Diseases associated with impairment of fatty acid oxidation lead to hypoglycemia, fatty infiltration of organs, and hypoketonemia.
- Ketosis is mild in starvation but severe in diabetes mellitus and ruminant ketosis.

REFERENCES

Eaton S, Bartlett K, Pourfarzam M: Mammalian mitochondrial β-oxidation. Biochem J 1996;320:345.

Fukao T, Lopaschuk GD, Mitchell GA: Pathways and control of ketone body metabolism: on the fringe of lipid metabolism. Prostaglandins Leukot Essent Fatty Acids 2004;70:243.

Gurr MI, Harwood JL, Frayn K: *Lipid Biochemistry.* Blackwell Publishing, 2002.

Reddy JK, Mannaerts GP: Peroxisomal lipid metabolism. Annu Rev Nutr 1994;14:343.

Scriver CR et al (editors): *The Metabolic and Molecular Bases of Inherited Disease,* 8th ed. McGraw-Hill, 2001.

Wood PA: Defects in mitochondrial beta-oxidation of fatty acids. Curr Opin Lipidol 1999;10:107.

Biosynthesis of Fatty Acids & Eicosanoids

Kathleen M. Botham, PhD, DSc & Peter A. Mayes, PhD, DSc

BIOMEDICAL IMPORTANCE

Fatty acids are synthesized by an **extramitochondrial system,** which is responsible for the complete synthesis of palmitate from acetyl-CoA in the cytosol. In most mammals, glucose is the primary substrate for lipogenesis, but in ruminants it is acetate, the main fuel molecule produced by the diet. Critical diseases of the pathway have not been reported in humans. However, inhibition of lipogenesis occurs in type 1 (insulin-dependent) **diabetes mellitus,** and variations in its activity affect the nature and extent of **obesity.**

Unsaturated fatty acids in phospholipids of the cell membrane are important in maintaining membrane fluidity. A high ratio of polyunsaturated fatty acids to saturated fatty acids (P:S ratio) in the diet is considered to be beneficial in preventing coronary heart disease. Animal tissues have limited capacity for desaturating fatty acids, and require certain dietary polyunsaturated fatty acids derived from plants. These **essential fatty acids** are used to form eicosanoic (C_{20}) fatty acids, which give rise to the **eicosanoids** prostaglandins, thromboxanes, leukotrienes, and lipoxins. Prostaglandins mediate **inflammation, pain,** and induce **sleep** and also regulate **blood coagulation** and **reproduction.** **Nonsteroidal anti-inflammatory drugs (NSAIDs)** such as **aspirin and ibuprofen** act by inhibiting prostaglandin synthesis. Leukotrienes have muscle contractant and chemotactic properties and are important in allergic reactions and inflammation.

THE MAIN PATHWAY FOR DE NOVO SYNTHESIS OF FATTY ACIDS (LIPOGENESIS) OCCURS IN THE CYTOSOL

This system is present in many tissues, including liver, kidney, brain, lung, mammary gland, and adipose tissue. Its cofactor requirements include NADPH, ATP, Mn^{2+}, biotin, and HCO_3^- (as a source of CO_2). **Acetyl-CoA** is the immediate substrate, and **free palmitate** is the end product.

Production of Malonyl-CoA Is the Initial & Controlling Step in Fatty Acid Synthesis

Bicarbonate as a source of CO_2 is required in the initial reaction for the carboxylation of acetyl-CoA to **malonyl-CoA** in the presence of ATP and **acetyl-CoA carboxylase.** Acetyl-CoA carboxylase has a requirement for the B vitamin **biotin** (Figure 23–1). The enzyme is a **multienzyme protein** containing a variable number of identical subunits, each containing biotin, biotin carboxylase, biotin carboxyl carrier protein, and transcarboxylase, as well as a regulatory allosteric site. The reaction takes place in two steps: (1) carboxylation of biotin involving ATP and (2) transfer of the carboxyl group to acetyl-CoA to form malonyl-CoA.

The Fatty Acid Synthase Complex Is a Polypeptide Containing Seven Enzyme Activities

In bacteria and plants, the individual enzymes of the **fatty acid synthase** system are separate, and the acyl radicals are found in combination with a protein called the **acyl carrier protein (ACP).** However, in vertebrates, the synthase system is a multienzyme polypeptide complex that incorporates ACP, which takes over the role of CoA. It contains the vitamin **pantothenic acid** in the form of 4′-phosphopantetheine (Figure 44–18). The use of one multienzyme functional unit has the advantages of achieving the effect of compartmentalization of the process within the cell without the erection of permeability barriers, and synthesis of all enzymes in the complex is coordinated since it is encoded by a single gene.

In mammals, the fatty acid synthase complex is a dimer comprising two identical monomers, each containing all seven enzyme activities of fatty acid synthase on one polypeptide chain (Figure 23–2). Initially, a priming molecule of acetyl-CoA combines with a cysteine —SH group catalyzed by **acetyl transacylase** (Figure 23–3, reaction 1a). Malonyl-CoA combines with the adjacent —SH on the 4′-phosphopantetheine of ACP of the other monomer, catalyzed by **malonyl transacylase** (reaction 1b), to form **acetyl (acyl)-malonyl enzyme.** The acetyl group attacks the methylene group of the malonyl residue, catalyzed by **3-ketoacyl synthase,** and liberates CO_2, forming 3-ketoacyl enzyme (acetoacetyl enzyme) (reaction 2), freeing the cysteine —SH group. Decarboxylation allows the reaction to go to completion, pulling the whole sequence of reactions in the forward direction. The 3-ketoacyl group is reduced, dehydrated, and reduced again (reactions 3, 4, 5) to

FIGURE 23–1 Biosynthesis of malonyl-CoA. (Enz, acetyl-CoA carboxylase.)

form the corresponding saturated acyl-S-enzyme. A new malonyl-CoA molecule combines with the —SH of 4′-phosphopantetheine, displacing the saturated acyl residue onto the free cysteine —SH group. The sequence of reactions is repeated six more times until a saturated 16-carbon acyl radical (palmityl) has been assembled. It is liberated from the enzyme complex by the activity of a seventh enzyme in the complex, **thioesterase** (deacylase). The free palmitate must be activated to acyl-CoA before it can proceed via any other metabolic pathway. Its usual fate is esterification into acylglycerols, chain elongation or desaturation, or esterification to cholesteryl ester. In mammary gland, there is a separate thioesterase specific for acyl residues of C_8, C_{10}, or C_{12}, which are subsequently found in milk lipids.

The equation for the overall synthesis of palmitate from acetyl-CoA and malonyl-CoA is

$$CH_3CO \cdot S \cdot CoA + 7HOOC \cdot CH\,CO \cdot S \cdot CoA + 14NADPH + 14H^+$$
$$\rightarrow CH_3\left(CH_2\right)_{14} COOH + 7CO_2 + 6H_2O + 8CoA \cdot SH + 14NADP^+$$

The acetyl-CoA used as a primer forms carbon atoms 15 and 16 of palmitate. The addition of all the subsequent C_2 units is via malonyl-CoA. Propionyl CoA acts as primer for the synthesis of long-chain fatty acids having an odd number of carbon atoms, found particularly in ruminant fat and milk.

The Main Source of NADPH for Lipogenesis Is the Pentose Phosphate Pathway

NADPH is involved as donor of reducing equivalents in both the reduction of the 3-ketoacyl and of the 2,3-unsaturated acyl derivatives (Figure 23–3, reactions 3 & 5). The oxidative reactions of the pentose phosphate pathway (see Chapter 21) are the chief source of the hydrogen required for the reductive synthesis of fatty acids. Significantly, tissues specializing in active lipogenesis—ie, liver, adipose tissue, and the lactating mammary gland—also possess an active pentose phosphate

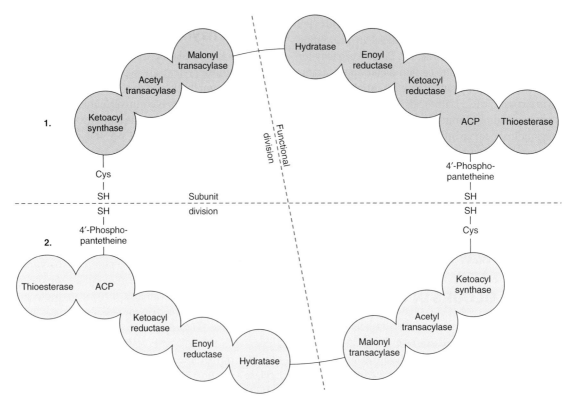

FIGURE 23–2 Fatty acid synthase multienzyme complex. The complex is a dimer of two identical polypeptide monomers, 1 and 2, each consisting of seven enzyme activities and the acyl carrier protein (ACP). (Cys—SH, cysteine thiol.) The —SH of the 4′-phosphopantetheine of one monomer is in close proximity to the —SH of the cysteine residue of the ketoacyl synthase of the other monomer, suggesting a "head-to-tail" arrangement of the two monomers. Though each monomer contains all the activities of the reaction sequence, the actual functional unit consists of one-half of one monomer interacting with the complementary half of the other. Thus, two acyl chains are produced simultaneously.

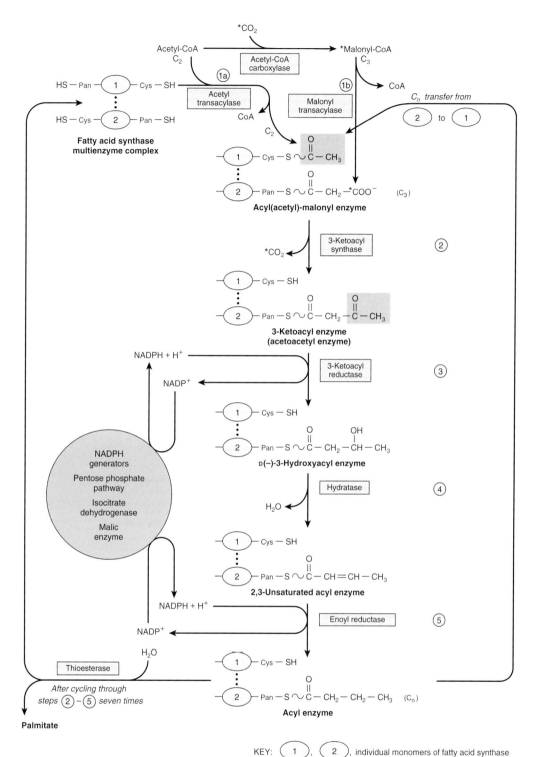

FIGURE 23–3

Biosynthesis of long-chain fatty acids. Details of how addition of a malonyl residue causes the acyl chain to grow by two carbon atoms. (Cys, cysteine residue; Pan, 4'-phosphopantetheine.) The blocks highlighted in blue contain initially a C_2 unit derived from acetyl-CoA (as illustrated) and subsequently the C_n unit formed in reaction 5.

pathway. Moreover, both metabolic pathways are found in the cytosol of the cell; so, there are no membranes or permeability barriers against the transfer of NADPH. Other sources of NADPH include the reaction that converts malate to pyruvate catalyzed by the **"malic enzyme"** (NADP malate dehydrogenase) (Figure 23–4) and the extramitochondrial **isocitrate dehydrogenase** reaction (probably not a substantial source, except in ruminants).

Acetyl-CoA Is the Principal Building Block of Fatty Acids

Acetyl-CoA is formed from glucose via the oxidation of pyruvate within the mitochondria. However, it does not diffuse readily into the extramitochondrial cytosol, the principal site of fatty acid synthesis. Citrate, formed after condensation of acetyl-CoA with oxaloacetate in the citric acid cycle within

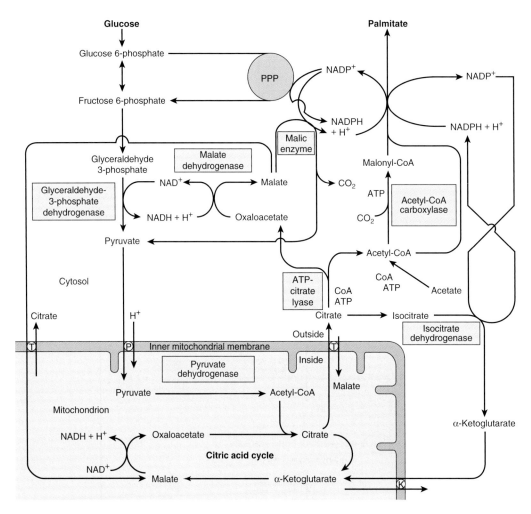

FIGURE 23–4 The provision of acetyl-CoA and NADPH for lipogenesis. (PPP, pentose phosphate pathway; T, tricarboxylate transporter; K, α-ketoglutarate transporter; P, pyruvate transporter.)

mitochondria, is translocated into the extramitochondrial compartment via the tricarboxylate transporter, where in the presence of CoA and ATP, it undergoes cleavage to acetyl-CoA and oxaloacetate catalyzed by **ATP-citrate lyase,** which increases in activity in the well-fed state. The acetyl-CoA is then available for malonyl-CoA formation and synthesis to palmitate (Figure 23–4). The resulting oxaloacetate can form malate via NADH-linked malate dehydrogenase, followed by the generation of NADPH via the malic enzyme. The NADPH becomes available for lipogenesis, and the pyruvate can be used to regenerate acetyl-CoA after transport into the mitochondrion. This pathway is a means of transferring reducing equivalents from extramitochondrial NADH to NADP. Alternatively, malate itself can be transported into the mitochondrion, where it is able to re-form oxaloacetate. Note that the citrate (tricarboxylate) transporter in the mitochondrial membrane requires malate to exchange with citrate (see Figure 13–10). There is little ATP-citrate lyase or malic enzyme in ruminants, probably because in these species acetate (derived from carbohydrate digestion in the rumen and activated to acetyl-CoA extramitochondrially) is the main source of acetyl-CoA.

Elongation of Fatty Acid Chains Occurs in the Endoplasmic Reticulum

This pathway (**the "microsomal system"**) elongates saturated and unsaturated fatty acyl-CoAs (from C_{10} upward) by two carbons, using malonyl-CoA as the acetyl donor and NADPH as the reductant, and is catalyzed by the microsomal **fatty acid elongase** system of enzymes (Figure 23–5). Elongation of stearyl-CoA in brain increases rapidly during myelination in order to provide C_{22} and C_{24} fatty acids for sphingolipids.

THE NUTRITIONAL STATE REGULATES LIPOGENESIS

Excess carbohydrate is stored as fat in many animals in anticipation of periods of caloric deficiency such as starvation, hibernation, etc, and to provide energy for use between meals in animals, including humans, that take their food at spaced intervals. Lipogenesis converts surplus glucose and intermediates such as pyruvate, lactate, and acetyl-CoA to fat, assisting the anabolic

FIGURE 23–5 Microsomal elongase system for fatty acid chain elongation. NADH is also used by the reductases, but NADPH is preferred.

phase of this feeding cycle. The nutritional state of the organism is the main factor regulating the rate of lipogenesis. Thus, the rate is high in the well-fed animal whose diet contains a high proportion of carbohydrate. It is depressed by restricted caloric intake, high-fat diet, or a deficiency of insulin, as in diabetes mellitus. These latter conditions are associated with increased concentrations of plasma free fatty acids, and an inverse relationship has been demonstrated between hepatic lipogenesis and the concentration of serum-free fatty acids. Lipogenesis is increased when sucrose is fed instead of glucose because fructose bypasses the phosphofructokinase control point in glycolysis and floods the lipogenic pathway (Figure 21–5).

SHORT- & LONG-TERM MECHANISMS REGULATE LIPOGENESIS

Long-chain fatty acid synthesis is controlled in the short term by allosteric and covalent modification of enzymes and in the long term by changes in gene expression governing rates of synthesis of enzymes.

Acetyl-CoA Carboxylase Is the Most Important Enzyme in the Regulation of Lipogenesis

Acetyl-CoA carboxylase is an allosteric enzyme and is activated by **citrate,** which increases in concentration in the well-fed state and is an indicator of a plentiful supply of acetyl-CoA. Citrate converts the enzyme from an inactive dimer to an active polymeric form, with a molecular mass of several million. Inactivation is promoted by phosphorylation of the enzyme and by long-chain acyl-CoA molecules, an example of negative feedback inhibition by a product of a reaction. Thus, if acyl-CoA accumulates because it is not esterified quickly enough or because of increased lipolysis or an influx of free fatty acids into the tissue, it will automatically reduce the synthesis of new fatty acid. Acyl-CoA also inhibits the mitochondrial **tricarboxylate transporter,** thus preventing activation of the enzyme by egress of citrate from the mitochondria into the cytosol.

Acetyl-CoA carboxylase is also regulated by hormones such as **glucagon, epinephrine,** and **insulin** via changes in its phosphorylation state (details in Figure 23–6).

Pyruvate Dehydrogenase Is Also Regulated by Acyl-CoA

Acyl-CoA causes an inhibition of pyruvate dehydrogenase by inhibiting the ATP-ADP exchange transporter of the inner mitochondrial membrane, which leads to increased intramitochondrial [ATP]/[ADP] ratios and therefore to conversion of active to inactive pyruvate dehydrogenase (see Figure 18–6), thus regulating the availability of acetyl-CoA for lipogenesis. Furthermore, oxidation of acyl-CoA due to increased levels of free fatty acids may increase the ratios of [acetyl-CoA]/[CoA] and [NADH]/[NAD$^+$] in mitochondria, inhibiting pyruvate dehydrogenase.

Insulin Also Regulates Lipogenesis by Other Mechanisms

Insulin stimulates lipogenesis by several other mechanisms as well as by increasing acetyl-CoA carboxylase activity. It increases the transport of glucose into the cell (eg, in adipose tissue), increasing the availability of both pyruvate for fatty acid synthesis and glycerol 3-phosphate for esterification of the newly formed fatty acids, and also converts the inactive form of pyruvate dehydrogenase to the active form in adipose tissue, but not in liver. Insulin also—by its ability to depress the level of intracellular cAMP—**inhibits lipolysis** in adipose tissue and reducing the concentration of plasma free fatty acids and, therefore, long-chain acyl-CoA, which are inhibitors of lipogenesis.

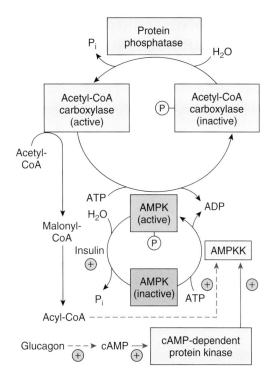

FIGURE 23–6 Regulation of acetyl-CoA carboxylase by phosphorylation/dephosphorylation. The enzyme is inactivated by phosphorylation by AMP-activated protein kinase (AMPK), which in turn is phosphorylated and activated by AMP-activated protein kinase kinase (AMPKK). Glucagon (and epinephrine) increase cAMP, and thus activate this latter enzyme via cAMP-dependent protein kinase. The kinase kinase enzyme is also believed to be activated by acyl-CoA. Insulin activates acetyl-CoA carboxylase via dephosphorylation of AMPK.

FIGURE 23–7 Structure of some unsaturated fatty acids. Although the carbon atoms in the molecules are conventionally numbered—ie, numbered from the carboxyl terminal—the ω numbers (eg, ω7 in palmitoleic acid) are calculated from the reverse end (the methyl terminal) of the molecules. The information in parentheses shows, for instance, that α-linolenic acid contains double bonds starting at the third carbon from the methyl terminal, has 18 carbons and 3 double bonds, and has these double bonds at the 9th, 12th, and 15th carbons from the carboxyl terminal. (*Classified as "essential fatty acids.")

The Fatty Acid Synthase Complex & Acetyl-CoA Carboxylase Are Adaptive Enzymes

These enzymes adapt to the body's physiologic needs by increasing in total amount in the fed state and by decreasing in during intake of a high-fat diet and in conditions such as starvation, and diabetes mellitus. **Insulin** is an important hormone causing gene expression and induction of enzyme biosynthesis, and **glucagon** (via cAMP) antagonizes this effect. Feeding fats containing polyunsaturated fatty acids coordinately regulates the inhibition of expression of key enzymes of glycolysis and lipogenesis. These mechanisms for longer-term regulation of lipogenesis take several days to become fully manifested and augment the direct and immediate effect of free fatty acids and hormones such as insulin and glucagon.

SOME POLYUNSATURATED FATTY ACIDS CANNOT BE SYNTHESIZED BY MAMMALS & ARE NUTRITIONALLY ESSENTIAL

Certain long-chain unsaturated fatty acids of metabolic significance in mammals are shown in Figure 23–7. Other C_{20},

C_{22}, and C_{24} polyenoic fatty acids may be derived from oleic, linoleic, and α-linolenic acids by chain elongation. Palmitoleic and oleic acids are not essential in the diet because the tissues can introduce a double bond at the Δ^9 position of a saturated fatty acid. **Linoleic and α-linolenic acids** are the only fatty acids known to be essential for the complete nutrition of many species of animals, including humans, and are known as the **nutritionally essential fatty acids.** In most mammals, **arachidonic acid** can be formed from linoleic acid (Figure 23–10). Double bonds can be introduced at the Δ^4, Δ^5, Δ^6, and Δ^9 positions (see Chapter 15) in most animals, but never beyond the Δ^9 position. In contrast, plants are able to synthesize the nutritionally essential fatty acids by introducing double bonds at the Δ^{12} and Δ^{15} positions.

MONOUNSATURATED FATTY ACIDS ARE SYNTHESIZED BY A Δ^9 DESATURASE SYSTEM

Several tissues including the liver are considered to be responsible for the formation of nonessential monounsaturated fatty acids from saturated fatty acids. The first double bond introduced into a saturated fatty acid is nearly always in the Δ^9 posi-

Stearoyl — CoA

O_2 + NADH + H^+

Δ^9 Desaturase   Cyt b_5

NAD^++ $2H_2O$

Oleoyl — CoA

FIGURE 23–8 Microsomal Δ^9 desaturase.

tion. An enzyme system—Δ^9 **desaturase** (Figure 23–8)—in the endoplasmic reticulum catalyzes the conversion of palmitoyl-CoA or stearoyl-CoA to palmitoleoyl-CoA or oleoyl-CoA, respectively. Oxygen and either NADH or NADPH are necessary for the reaction. The enzymes appear to be similar to a monooxygenase system involving cytochrome b_5 (Chapter 12).

SYNTHESIS OF POLYUNSATURATED FATTY ACIDS INVOLVES DESATURASE & ELONGASE ENZYME SYSTEMS

Additional double bonds introduced into existing monounsaturated fatty acids are always separated from each other by a methylene group (methylene interrupted) except in bacteria. Since animals have a Δ^9 desaturase, they are able to synthesize the ω9 (oleic acid) family of unsaturated fatty acids completely by a combination of chain elongation and desaturation (Figure 23–9). However, as indicated above, linoleic (ω6) or α-linolenic (ω3) acids required for the synthesis of the other members of the ω6 or ω3 families must be supplied in the diet. Linoleate may be converted to arachidonate via **γ-linolenate** by the pathway shown in Figure 23–10. The nutritional requirement for arachidonate may thus be dispensed with if there is adequate linoleate in the diet. The desaturation and chain elongation system is greatly diminished in the starving state, in response to glucagon and epinephrine administration, and in the absence of insulin as in type 1 diabetes mellitus.

Linoleoyl-CoA ($\Delta^{9,12}$-octadecadienoyl-CoA)

O_2 + NADH + H^+

Δ^6 Desaturase

$2H_2O$ + NAD^+

γ-Linolenoyl-CoA ($\Delta^{6,9,12}$-octadecatrienoyl-CoA)

C_2 (Malonyl-CoA, NADPH)

Microsomal chain elongation system (Elongase)

Dihomo-γ-linolenoyl-CoA ($\Delta^{8,11,14}$-eicosatrienoyl-CoA)

O_2 + NADH + H^+

Δ^5 Desaturase

$2H_2O$ + NAD^+

Arachidonoyl-CoA ($\Delta^{5,8,11,14}$-eicosatetraenoyl-CoA)

FIGURE 23–10 Conversion of linoleate to arachidonate. Cats cannot carry out this conversion owing to absence of Δ^6 desaturase and must obtain arachidonate in their diet.

ω9 Family — Oleic acid 18:1 $\xrightarrow{2}$ 18:2 $\xrightarrow{1}$ 20:2 $\xrightarrow{3}$ 20:3 $\xrightarrow{1}$ 22:3 $\xrightarrow{4}$ 22:4

24:1 $\xleftarrow{1}$ 22:1 $\xleftarrow{1}$ 20:1 $\xleftarrow{1}$ (Oleic acid)

Accumulates in essential fatty acid deficiency

ω6 Family — Linoleic acid 18:2 $\xrightarrow{2}$ 18:3 $\xrightarrow{1}$ 20:3 $\xrightarrow{3}$ 20:4 $\xrightarrow{1}$ 22:4 $\xrightarrow{4}$ 22:5

20:2

ω3 Family — α-Linolenic acid 18:3 $\xrightarrow{1}$ 20:3 $\xrightarrow{3}$ 20:4 $\xrightarrow{1}$ 22:4 $\xrightarrow{4}$ 22:5

FIGURE 23–9 Biosynthesis of the ω9, ω6, and ω3 families of polyunsaturated fatty acids. Each step is catalyzed by the microsomal chain elongation or desaturase system: 1, elongase; 2, Δ^6 desaturase; 3, Δ^5 desaturase; 4, Δ^4 desaturase. (⊖, Inhibition.)

DEFICIENCY SYMPTOMS ARE PRODUCED WHEN THE ESSENTIAL FATTY ACIDS (EFA) ARE ABSENT FROM THE DIET

Rats fed a purified nonlipid diet containing vitamins A and D exhibit a reduced growth rate and reproductive deficiency which may be cured by the addition of **linoleic, α-linolenic,** and **arachidonic acids** to the diet. These fatty acids are found in high concentrations in vegetable oils (Table 15–2) and in small amounts in animal carcasses. Essential fatty acids are required for prostaglandin, thromboxane, leukotriene, and lipoxin formation (see below), and they also have various other functions that are less well defined. They are found in the structural lipids of the cell, often in the 2 position of phospholipids, and are concerned with the structural integrity of the mitochondrial membrane.

Arachidonic acid is present in membranes and accounts for 5–15% of the fatty acids in phospholipids. Docosahexaenoic acid (DHA; ω3, 22:6), which is synthesized to a limited extent from α-linolenic acid or obtained directly from fish oils, is present in high concentrations in retina, cerebral cortex, testis, and sperm. DHA is particularly needed for development of the brain and retina and is supplied via the placenta and milk. Patients with **retinitis pigmentosa** are reported to have low blood levels of DHA. In **essential fatty acid deficiency,** nonessential polyenoic acids of the ω9 family, particularly $\Delta^{5,8,11}$-eicosatrienoic acid (ω9 20:3) (Figure 23–9), replace the essential fatty acids in phospholipids, other complex lipids, and membranes. The triene:tetraene ratio in plasma lipids can be used to diagnose the extent of essential fatty acid deficiency.

Trans Fatty Acids Are Implicated in Various Disorders

Small amounts of trans-unsaturated fatty acids are found in ruminant fat (eg, butter fat has 2–7%), where they arise from the action of microorganisms in the rumen, but the main source in the human diet is from partially hydrogenated vegetable oils (eg, margarine). Trans fatty acids compete with essential fatty acids and may exacerbate essential fatty acid deficiency. Moreover, they are structurally similar to saturated fatty acids (Chapter 15) and have comparable effects in the promotion of hypercholesterolemia and atherosclerosis (Chapter 26).

EICOSANOIDS ARE FORMED FROM C$_{20}$ POLYUNSATURATED FATTY ACIDS

Arachidonate and some other C_{20} polyunsaturated fatty acids give rise to **eicosanoids,** physiologically and pharmacologically active compounds known as **prostaglandins (PG), thromboxanes (TX), leukotrienes (LT),** and **lipoxins (LX)** (Chapter 15). Physiologically, they are considered to act as lo-cal hormones functioning through G-protein-linked receptors to elicit their biochemical effects.

There are three groups of eicosanoids that are synthesized from C_{20} eicosanoic acids derived from the essential fatty acids **linoleate** and **α-linolenate,** or directly from dietary arachidonate and eicosapentaenoate (Figure 23–11). Arachidonate, which may be obtained from the diet, but is usually derived from the 2 position of phospholipids in the plasma membrane by the action of phospholipase A$_2$ (Figure 24–6), is the substrate for the synthesis of the PG$_2$, TX$_2$ series (**prostanoids**) by the **cyclooxygenase pathway,** or the LT$_4$ and LX$_4$ series by the **lipoxygenase pathway,** with the two pathways competing for the arachidonate substrate (Figure 23–11).

THE CYCLOOXYGENASE PATHWAY IS RESPONSIBLE FOR PROSTANOID SYNTHESIS

Prostanoid synthesis (Figure 23–12) involves the consumption of two molecules of O$_2$ catalyzed by cyclooxygenase (COX) (also called **prostaglandin H synthase**), an enzyme that has two activities, a **cyclooxygenase** and **peroxidase.** COX is present as two isoenzymes, **COX-1 and COX-2.** The product, an endoperoxide (PGH), is converted to prostaglandins D and E as well as to a thromboxane (TXA$_2$) and prostacyclin (PGI$_2$). Each cell type produces only one type of prostanoid. The NSAID **aspirin** inhibits COX-1 and COX-2. Other NSAIDs include indomethacin and ibuprofen, and usually inhibit cyclooxygenases by competing with arachidonate. Since inhibition of COX-1 causes the stomach irritation often associated with taking NSAIDs, attempts have been made to develop drugs which selectively inhibit COX-2 (**coxibs**). Unfortunately, however, the success of this approach has been limited and some coxibs have been withdrawn or suspended from the market due to undesirable side effects and safety issues. Transcription of COX-2—but not of COX-1—is completely inhibited by **anti-inflammatory corticosteroids.**

Essential Fatty Acids Do Not Exert All Their Physiologic Effects Via Prostaglandin Synthesis

The role of essential fatty acids in membrane formation is unrelated to prostaglandin formation. Prostaglandins do not relieve symptoms of essential fatty acid deficiency, and an essential fatty acid deficiency is not caused by inhibition of prostaglandin synthesis.

Cyclooxygenase Is a "Suicide Enzyme"

"Switching off" of prostaglandin activity is partly achieved by a remarkable property of cyclooxygenase—that of self-catalyzed destruction; ie, it is a "suicide enzyme." Furthermore, the inactivation of prostaglandins by **15-hydroxyprostaglandin**

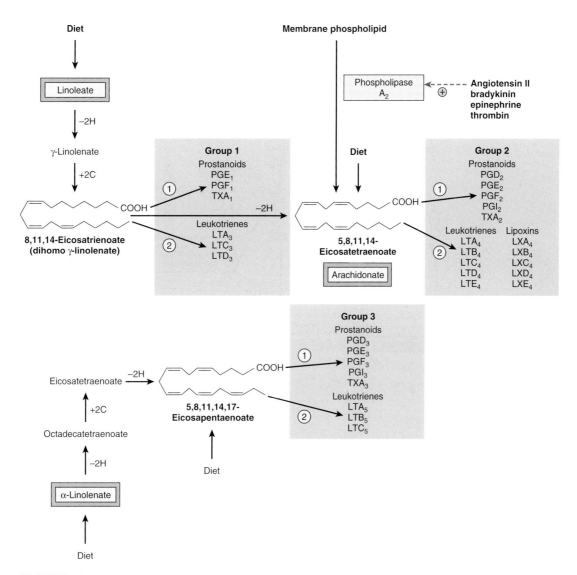

FIGURE 23–11 The three groups of eicosanoids and their biosynthetic origins. (PG, prostaglandin; PGI, prostacyclin; TX, thromboxane; LT, leukotriene; LX, lipoxin; ①, cyclooxygenase pathway; ②, lipoxygenase pathway.) The subscript denotes the total number of double bonds in the molecule and the series to which the compound belongs.

dehydrogenase is rapid. Blocking the action of this enzyme with sulfasalazine or indomethacin can prolong the half-life of prostaglandins in the body.

LEUKOTRIENES & LIPOXINS ARE FORMED BY THE LIPOXYGENASE PATHWAY

The **leukotrienes** are a family of conjugated trienes formed from eicosanoic acids in leukocytes, mastocytoma cells, platelets, and macrophages by the **lipoxygenase pathway** in response to both immunologic and nonimmunologic stimuli. Three different lipoxygenases (dioxygenases) insert oxygen into the 5, 12, and 15 positions of arachidonic acid, giving

rise to hydroperoxides (HPETE). Only **5 lipoxygenase** forms leukotrienes (details in Figure 23–13). **Lipoxins** are a family of conjugated tetraenes also arising in leukocytes. They are formed by the combined action of more than one lipoxygenase (Figure 23–13).

CLINICAL ASPECTS

Symptoms of Essential Fatty Acid Deficiency in Humans Include Skin Lesions & Impairment of Lipid Transport

In adults subsisting on ordinary diets, no signs of essential fatty acid deficiencies have been reported. However, infants receiving formula diets low in fat and patients maintained for long periods

FIGURE 23–12 Conversion of arachidonic acid to prostaglandins and thromboxanes of series 2. (PG, prostaglandin; TX, thromboxane; PGI, prostacyclin; HHT, hydroxyheptadecatrienoate.) (*Both of these starred activities are attributed to the cyclooxygenase enzyme [prostaglandin H synthase]. Similar conversions occur in prostaglandins and thromboxanes of series 1 and 3.)

exclusively by intravenous nutrition low in essential fatty acids show deficiency symptoms that can be prevented by an essential fatty acid intake of 1–2% of the total caloric requirement.

Abnormal Metabolism of Essential Fatty Acids Occurs in Several Diseases

Abnormal metabolism of essential fatty acids, which may be connected with dietary insufficiency, has been noted in cystic fibrosis, acrodermatitis enteropathica, hepatorenal syndrome, Sjögren-Larsson syndrome, multisystem neuronal degeneration, Crohn's disease, cirrhosis and alcoholism, and Reye's syndrome. Elevated levels of very-long-chain polyenoic acids have been found in the brains of patients with Zellweger's syndrome (Chapter 22). Diets with a high P:S (polyunsaturated:saturated fatty acid) ratio reduce serum cholesterol levels and are considered to be beneficial in terms of the risk of development of coronary heart disease.

Prostanoids Are Potent, Biologically Active Substances

Thromboxanes are synthesized in platelets and upon release cause vasoconstriction and platelet aggregation. Their synthesis is specifically inhibited by low-dose aspirin. **Prostacy-**

clins (PGI$_2$) are produced by blood vessel walls and are potent inhibitors of platelet aggregation. Thus, thromboxanes and prostacyclins are antagonistic. PG$_3$ and TX$_3$, formed from eicosapentaenoic acid (EPA), inhibit the release of arachidonate from phospholipids and the formation of PG$_2$ and TX$_2$. PGI$_3$ is as potent an antiaggregator of platelets as PGI$_2$, but TXA$_3$ is a weaker aggregator than TXA$_2$, changing the balance of activity and favoring longer clotting times. As little as 1 ng/mL of plasma prostaglandins causes contraction of smooth muscle in animals. Potential therapeutic uses include prevention of conception, induction of labor at term, termination of pregnancy, prevention or alleviation of gastric ulcers, control of inflammation and of blood pressure, and relief of asthma and nasal congestion. In addition, PGD$_2$ is a potent sleep-promoting substance. Prostaglandins increase cAMP in platelets, thyroid, corpus luteum, fetal bone, adenohypophysis, and lung but reduce cAMP in renal tubule cells and adipose tissue (Chapter 25).

Leukotrienes & Lipoxins Are Potent Regulators of Many Disease Processes

Slow-reacting substance of anaphylaxis (**SRS-A**) is a mixture of leukotrienes C$_4$, D$_4$, and E$_4$. This mixture of leukotrienes is a potent constrictor of the bronchial airway musculature. These leukotrienes together with **leukotriene B$_4$** also cause vascular

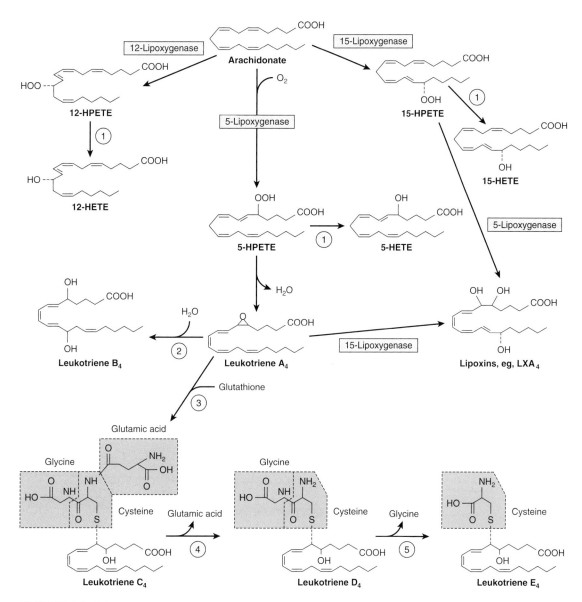

FIGURE 23–13 Conversion of arachidonic acid to leukotrienes and lipoxins of series 4 via the lipoxygenase pathway. Some similar conversions occur in series 3 and 5 leukotrienes. (HPETE, hydroperoxyeicosatetraenoate; HETE, hydroxyeicosatetraenoate; ①, peroxidase; ②, leukotriene A₄ epoxide hydrolase; ③, glutathione S-transferase; ④, γ-glutamyltranspeptidase; ⑤, cysteinyl-glycine dipeptidase.)

permeability and attraction and activation of leukocytes and are important regulators in many diseases involving inflammatory or immediate hypersensitivity reactions, such as asthma. Leukotrienes are vasoactive, and 5-lipoxygenase has been found in arterial walls. Evidence supports an anti-inflammatory role for lipoxins in vasoactive and immunoregulatory function, eg, as counterregulatory compounds (**chalones**) of the immune response.

SUMMARY

- The synthesis of long-chain fatty acids (lipogenesis) is carried out by two enzyme systems: acetyl-CoA carboxylase and fatty acid synthase.

- The pathway converts acetyl-CoA to palmitate and requires NADPH, ATP, Mn^{2+}, biotin, pantothenic acid, and as cofactors.

- Acetyl-CoA carboxylase converts acetyl-CoA to malonyl-CoA, then fatty acid synthase, a multienzyme complex of one polypeptide chain with seven separate enzymatic activities, catalyzes the formation of palmitate from one acetyl-CoA and seven malonyl-CoA molecules.

- Lipogenesis is regulated at the acetyl-CoA carboxylase step by allosteric modifiers, phosphorylation/dephosphorylation, and induction and repression of enzyme synthesis. The enzyme is allosterically activated by citrate and deactivated by long-chain acyl-CoA. Dephosphorylation (eg, by insulin) promotes its activity, while phosphorylation (eg, by glucagon or epinephrine) is inhibitory.

- Biosynthesis of unsaturated long-chain fatty acids is achieved by desaturase and elongase enzymes, which introduce double bonds and lengthen existing acyl chains, respectively.

- Higher animals have Δ^4, Δ^5, Δ^6, and Δ^9 desaturases but cannot insert new double bonds beyond the 9 position of fatty acids. Thus, the essential fatty acids linoleic ($\omega6$) and α-linolenic ($\omega3$) must be obtained from the diet.

- Eicosanoids are derived from C_{20} (eicosanoic) fatty acids synthesized from the essential fatty acids and make up important groups of physiologically and pharmacologically active compounds, including the prostaglandins, thromboxanes, leukotrienes, and lipoxins.

REFERENCES

Cook HW, McMaster CR: Fatty acid desaturation and chain elongation in eukaryotes. In: *Biochemistry of Lipids, Lipoproteins and Membranes,* 4th ed. Vance DE, Vance JE (editors). Elsevier, 2002;181–204.

Fischer S: Dietary polyunsaturated fatty acids and eicosanoid formation in humans. Adv Lipid Res 1989;23:169.

Fitzpatrick FA: Cyclooxygenase enzymes: regulation and function. Curr Pharm Des 2004;10:577.

Tong L: Acetyle-coenzyme A carboxylase: crucial metabolic enzyme and an attractive target for drug discovery. Cell Mol Life Sci 2005;62:1784.

McMahon B et al: Lipoxins: revelations on resolution. Trends Pharmacol Sci 2001;22:391.

Rangan VS, Smith S: Fatty acid synthesis in eukaryotes. In: *Biochemistry of Lipids, Lipoproteins and Membranes,* 4th ed. Vance DE, Vance JE (editors). Elsevier, 2002;151–180.

Sith S, Witkowski A, Joshi AK: Structural and functional organisation of the animal fatty acid synthase. Prog Lipid Res 2003;42:289.

Smith WL, Murphy RC: The eicosanoids: cyclooxygenase, lipoxygenase, and epoxygenase pathways. In: *Biochemistry of Lipids, Lipoproteins and Membranes,* 4th ed. Vance DE, Vance JE (editors). Elsevier, 2002;341–372.

Wijendran V, Hayes KC: Dietary n-6 and n-3 fatty acid balance and cardiovascular health. Annu Rev Nutr 2004;24:597.

Metabolism of Acylglycerols & Sphingolipids

Kathleen M. Botham, PhD, DSc & Peter A. Mayes, PhD, DSc

BIOMEDICAL IMPORTANCE

Acylglycerols constitute the majority of lipids in the body. Triacylglycerols are the major lipids in fat deposits and in food, and their roles in lipid transport and storage and in various diseases such as obesity, diabetes, and hyperlipoproteinemia will be described in subsequent chapters. The amphipathic nature of phospholipids and sphingolipids makes them ideally suitable as the main lipid component of cell membranes. Phospholipids also take part in the metabolism of many other lipids. Some phospholipids have specialized functions; eg, dipalmitoyl lecithin is a major component of **lung surfactant,** which is lacking in **respiratory distress syndrome** of the newborn. Inositol phospholipids in the cell membrane act as precursors of **hormone second messengers,** and **platelet-activating factor** is an alkylphospholipid. Glycosphingolipids, containing sphingosine and sugar residues as well as fatty acid that are found in the outer leaflet of the plasma membrane with their oligosaccharide chains facing outward, form part of the glycocalyx of the cell surface and are important (1) in cell adhesion and cell recognition, (2) as receptors for bacterial toxins (eg, the toxin that causes cholera), and (3) as ABO blood group substances. A dozen or so **glycolipid storage diseases** have been described (eg, Gaucher's disease, Tay-Sachs disease), each due to a genetic defect in the pathway for glycolipid degradation in the lysosomes.

HYDROLYSIS INITIATES CATABOLISM OF TRIACYLGLYCEROLS

Triacylglycerols must be hydrolyzed by a **lipase** to their constituent fatty acids and glycerol before further catabolism can proceed. Much of this hydrolysis (lipolysis) occurs in adipose tissue with release of free fatty acids into the plasma, where they are found combined with serum albumin. This is followed by free fatty acid uptake into tissues (including liver, heart, kidney, muscle, lung, testis, and adipose tissue, but not readily by brain), where they are oxidized or reesterified. The utilization of glycerol depends upon whether such tissues possess **glycerol kinase,** found in significant amounts in liver, kidney, intestine, brown adipose tissue, and lactating mammary gland.

TRIACYLGLYCEROLS & PHOSPHOGLYCEROLS ARE FORMED BY ACYLATION OF TRIOSE PHOSPHATES

The major pathways of triacylglycerol and phosphoglycerol biosynthesis are outlined in Figure 24–1. Important substances such as triacylglycerols, phosphatidylcholine, phosphatidylethanolamine, phosphatidylinositol, and cardiolipin, a constituent of mitochondrial membranes, are formed from **glycerol-3-phosphate.** Significant branch points in the pathway occur at the phosphatidate and diacylglycerol steps. From dihydroxyacetone phosphate are derived phosphoglycerols containing an ether link (—C—O—C—), the best known of which are plasmalogens and platelet-activating factor (PAF). Glycerol 3-phosphate and dihydroxyacetone phosphate are intermediates in glycolysis, making a very important connection between carbohydrate and lipid metabolism.

Phosphatidate Is the Common Precursor in the Biosynthesis of Triacylglycerols, Many Phosphoglycerols, & Cardiolipin

Both glycerol and fatty acids must be activated by ATP before they can be incorporated into acylglycerols. **Glycerol kinase** catalyzes the activation of glycerol to *sn*-glycerol 3-phosphate. If the activity of this enzyme is absent or low, as in muscle or

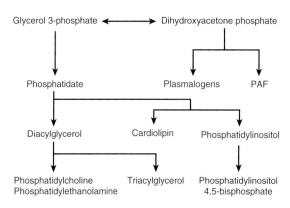

FIGURE 24–1 Overview of acylglycerol biosynthesis. (PAF, platelet-activating factor.)

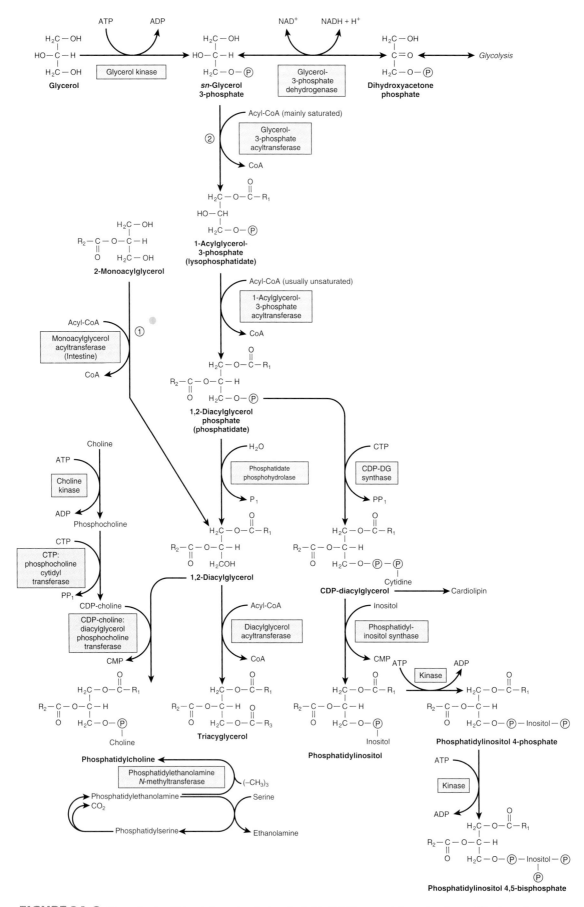

FIGURE 24–2 Biosynthesis of triacylglycerol and phospholipids. (①, Monoacylglycerol pathway; ②, glycerol phosphate pathway.) Phosphatidylethanolamine may be formed from ethanolamine by a pathway similar to that shown for the formation of phosphatidylcholine from choline.

adipose tissue, most of the glycerol 3-phosphate is formed from dihydroxyacetone phosphate by **glycerol-3-phosphate dehydrogenase** (Figure 24–2).

Biosynthesis of Triacylglycerols

Two molecules of acyl-CoA, formed by the activation of fatty acids by **acyl-CoA synthetase** (Chapter 22), combine with glycerol 3-phosphate to form **phosphatidate** (1,2-diacylglycerol phosphate). This takes place in two stages, catalyzed by **glycerol-3-phosphate acyltransferase** and **1-acylglycerol-3-phosphate acyltransferase**. Phosphatidate is converted by **phosphatidate phosphohydrolase** and **diacylglycerol acyltransferase (DGAT)** to 1,2-diacylglycerol and then triacylglycerol. DGAT catalyzes the only step specific for triacylglycerol synthesis and is thought to be rate limiting in most circumstances. In intestinal mucosa, **monoacylglycerol acyltransferase** converts **monoacylglycerol** to 1,2-diacylglycerol in the **monoacylglycerol pathway**. Most of the activity of these enzymes resides in the endoplasmic reticulum, but some is found in mitochondria. Phosphatidate phosphohydrolase is found mainly in the cytosol, but the active form of the enzyme is membrane-bound.

In the biosynthesis of phosphatidylcholine and phosphatidylethanolamine (Figure 24–2), choline or ethanolamine must first be activated by phosphorylation by ATP followed by linkage to CDP. The resulting CDP-choline or CDP-ethanolamine reacts with 1,2-diacylglycerol to form either phosphatidylcholine or phosphatidylethanolamine, respectively. Phosphatidylserine is formed from phosphatidylethanolamine directly by reaction with serine (Figure 24–2). Phosphatidylserine may re-form phosphatidylethanolamine by decarboxylation. An alternative pathway in liver enables phosphatidylethanolamine to give rise directly to phosphatidylcholine by progressive methylation of the ethanolamine residue. In spite of these sources of choline, it is considered to be an essential nutrient in many mammalian species, but this has not been established in humans.

The regulation of triacylglycerol, phosphatidylcholine, and phosphatidylethanolamine biosynthesis is driven by the availability of free fatty acids. Those that escape oxidation are preferentially converted to phospholipids, and when this requirement is satisfied they are used for triacylglycerol synthesis.

A phospholipid present in mitochondria is **cardiolipin** (diphosphatidylglycerol; Figure 15–8). It is formed from phosphatidylglycerol, which in turn is synthesized from CDP-diacylglycerol (Figure 24–2) and glycerol 3-phosphate according to the scheme shown in Figure 24–3. Cardiolipin, found in the inner membrane of mitochondria, has a key role in mitochondrial structure and function, and is also thought to be involved in programmed cell death (**apoptosis**).

Biosynthesis of Glycerol Ether Phospholipids

This pathway is located in peroxisomes. Dihydroxyacetone phosphate is the precursor of the glycerol moiety of glycerol ether phospholipids (Figure 24–4). This compound combines with acyl-CoA to give 1-acyldihydroxyacetone phosphate. The ether link is formed in the next reaction, producing 1-alkyldihydroxyacetone phosphate, which is then converted to 1-alkyl-

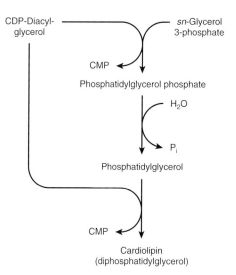

FIGURE 24–3 Biosynthesis of cardiolipin.

glycerol 3-phosphate. After further acylation in the 2 position, the resulting 1-alkyl-2-acylglycerol 3-phosphate (analogous to phosphatidate in Figure 24–2) is hydrolyzed to give the free glycerol derivative. **Plasmalogens,** which comprise much of the phospholipid in mitochondria, are formed by desaturation of the analogous 3-phosphoethanolamine derivative (Figure 24–4). **Platelet-activating factor (PAF)** (1-alkyl-2-acetyl-*sn*-glycerol-3-phosphocholine) is synthesized from the corresponding 3-phosphocholine derivative. It is formed by many blood cells and other tissues and aggregates platelets at concentrations as low as 10^{-11} mol/L. It also has hypotensive and ulcerogenic properties and is involved in a variety of biologic responses, including inflammation, chemotaxis, and protein phosphorylation.

Phospholipases Allow Degradation & Remodeling of Phosphoglycerols

Although phospholipids are actively degraded, each portion of the molecule turns over at a different rate—eg, the turnover time of the phosphate group is different from that of the 1-acyl group. This is due to the presence of enzymes that allow partial degradation followed by resynthesis (Figure 24–5). **Phospholipase A_2** catalyzes the hydrolysis of glycerophospholipids to form a free fatty acid and lysophospholipid, which in turn may be reacylated by acyl-CoA in the presence of an acyltransferase. Alternatively, lysophospholipid (eg, lysolecithin) is attacked by **lysophospholipase,** forming the corresponding glyceryl phosphoryl base, which may then be split by a hydrolase liberating glycerol 3-phosphate plus base. **Phospholipases A_1, A_2, B, C, and D** attack the bonds indicated in Figure 24–6. **Phospholipase A_2** is found in pancreatic fluid and snake venom as well as in many types of cells; **phospholipase C** is one of the major toxins secreted by bacteria; and **phospholipase D** is known to be involved in mammalian signal transduction.

Lysolecithin (lysophosphatidylcholine) may be formed by an alternative route that involves **lecithin:cholesterol acyltransferase (LCAT).** This enzyme, found in plasma, catalyzes

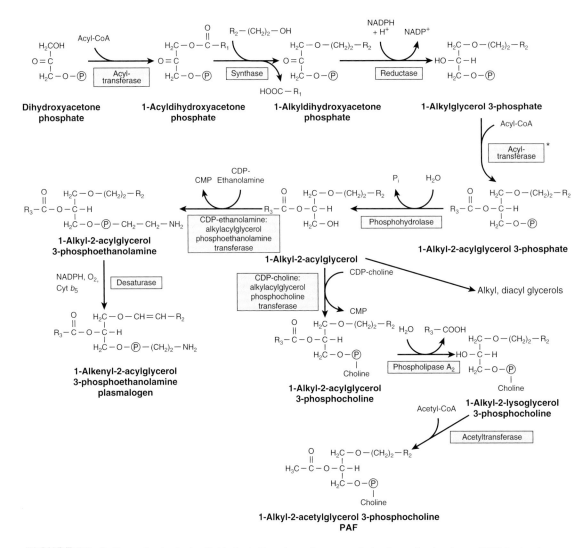

FIGURE 24–4 Biosynthesis of ether lipids, including plasmalogens, and platelet-activating factor (PAF). In the de novo pathway for PAF synthesis, acetyl-CoA is incorporated at stage*, avoiding the last two steps in the pathway shown here.

the transfer of a fatty acid residue from the 2 position of lecithin to cholesterol to form cholesteryl ester and lysolecithin and is considered to be responsible for much of the cholesteryl ester in plasma lipoproteins. Long-chain saturated fatty acids are found predominantly in the 1 position of phospholipids, whereas the polyunsaturated acids (eg, the precursors of prostaglandins) are incorporated more frequently into the 2 position. The incorporation of fatty acids into lecithin occurs by complete synthesis of the phospholipid, by transacylation between cholesteryl ester and lysolecithin, and by direct acylation of lysolecithin by acyl-CoA. Thus, a continuous exchange of the fatty acids is possible, particularly with regard to introducing essential fatty acids into phospholipid molecules.

ALL SPHINGOLIPIDS ARE FORMED FROM CERAMIDE

Ceramide is synthesized in the endoplasmic reticulum from the amino acid serine as shown in Figure 24–7. Ceramide is an

important signaling molecule (second messenger) regulating pathways including programmed cell death **(apoptosis),** the cell cycle and cell differentiation and senescence.

Sphingomyelins (Figure 15–11) are phospholipids and are formed when ceramide reacts with phosphatidylcholine to form sphingomyelin plus diacylglycerol (Figure 24–8A). This occurs mainly in the Golgi apparatus and to a lesser extent in the plasma membrane.

Glycosphingolipids Are a Combination of Ceramide with One or More Sugar Residues

The simplest glycosphingolipids **(cerebrosides)** are **galactosylceramide (GalCer)** and **glucosylceramide (GlcCer)**. GalCer is a major lipid of myelin, whereas GlcCer is the major glycosphingolipid of extraneural tissues and a precursor of most of the more complex glycosphingolipids. GalCer (Figure 24–8B) is formed in a reaction between ceramide and UDPGal (formed by epimerization from UDPGlc—Figure 21–6).

FIGURE 24–5 Metabolism of phosphatidylcholine (lecithin).

FIGURE 24–6 Sites of the hydrolytic activity of phospholipases on a phospholipid substrate.

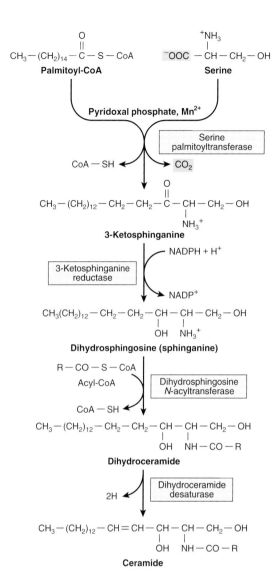

FIGURE 24–7 Biosynthesis of ceramide.

ber of gangliosides of increasing molecular weight may be formed. Most of the enzymes transferring sugars from nucleotide sugars (glycosyl transferases) are found in the Golgi apparatus.

Glycosphingolipids are constituents of the outer leaflet of plasma membranes and are important in **cell adhesion** and **cell recognition.** Some are antigens, eg, ABO blood group substances. Certain gangliosides function as receptors for bacterial toxins (eg, for **cholera toxin**, which subsequently activates adenylyl cyclase).

CLINICAL ASPECTS

Deficiency of Lung Surfactant Causes Respiratory Distress Syndrome

Lung surfactant is composed mainly of lipid with some proteins and carbohydrate and prevents the alveoli from collaps-

Sulfogalactosylceramide and other sulfolipids such as the **sulfo(galacto)-glycerolipids** and the **steroid sulfates** are formed after further reactions involving 3'-phosphoadenosine-5'-phosphosulfate (PAPS; "active sulfate"). **Gangliosides** are synthesized from ceramide by the stepwise addition of activated sugars (eg, UDPGlc and UDPGal) and a **sialic acid,** usually N-acetylneuraminic acid (Figure 24–9). A large num-

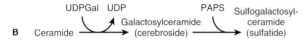

FIGURE 24–8 Biosynthesis of (**A**) sphingomyelin, (**B**) galactosylceramide and its sulfo derivative. (PAPS, "active sulfate," adenosine 3′-phosphate-5′-phosphosulfate.)

ing. The phospholipid **dipalmitoyl-phosphatidylcholine** decreases surface tension at the air–liquid interface and thus greatly reduces the work of breathing, but other surfactant lipid and protein components are also important in surfactant function. Deficiency of lung surfactant in the lungs of many preterm newborns gives rise to **infant respiratory distress syndrome (IRDS).** Administration of either natural or artificial surfactant is of therapeutic benefit.

Phospholipids & Sphingolipids Are Involved in Multiple Sclerosis and Lipidoses

Certain diseases are characterized by abnormal quantities of these lipids in the tissues, often in the nervous system. They may be classified into two groups: (1) true demyelinating diseases and (2) sphingolipidoses.

In **multiple sclerosis,** which is a demyelinating disease, there is loss of both phospholipids (particularly ethanolamine plasmalogen) and of sphingolipids from white matter. Thus, the lipid composition of white matter resembles that of gray matter. The cerebrospinal fluid shows raised phospholipid levels.

The **sphingolipidoses (lipid storage diseases)** are a group of inherited diseases that are caused by a genetic defect in the catabolism of lipids containing sphingosine. They are part of a larger group of lysosomal disorders and exhibit several constant features: (1) Complex lipids containing ceramide accumulate in cells, particularly neurons, causing neurodegeneration and

FIGURE 24–9 Biosynthesis of gangliosides. (NeuAc, N-acetylneuraminic acid.)

TABLE 24–1   Examples of Sphingolipidoses

Disease	Enzyme Deficiency	Lipid Accumulating	Clinical Symptoms
Tay-Sachs disease	Hexosaminidase A	Cer—Glc—Gal(NeuAc)÷GalNAc G_{M2} Ganglioside	Mental retardation, blindness, muscular weakness
Fabry's disease	α-Galactosidase	Cer—Glc—Gal÷Gal Globotriaosylceramide	Skin rash, kidney failure (full symptoms only in males; X-linked recessive)
Metachromatic leukodystrophy	Arylsulfatase A	Cer—Gal÷OSO₃ 3-Sulfogalactosylceramide	Mental retardation and psychologic disturbances in adults; demyelination
Krabbe's disease	β-Galactosidase	Cer÷Gal Galactosylceramide	Mental retardation; myelin almost absent
Gaucher's disease	β-Glucosidase	Cer÷Glc Glucosylceramide	Enlarged liver and spleen, erosion of long bones, mental retardation in infants
Niemann-Pick disease	Sphingomyelinase	Cer÷P—choline Sphingomyelin	Enlarged liver and spleen, mental retardation; fatal in early life
Farber's disease	Ceramidase	Acyl÷Sphingosine Ceramide	Hoarseness, dermatitis, skeletal deformation, mental retardation; fatal in early life

Abbreviations: NeuAc, N-acetylneuraminic acid; Cer, ceramide; Glc, glucose; Gal, galactose. ÷, site of deficient enzyme reaction.

shortening the life span. (2) The rate of **synthesis** of the stored lipid is normal. (3) The enzymatic defect is in the **lysosomal degradation pathway** of sphingolipids. (4) The extent to which the activity of the affected enzyme is decreased is similar in all tissues. There is no effective treatment for many of the diseases, although some success has been achieved with **enzyme replacement therapy** and **bone marrow transplantation** in the treatment of Gaucher's and Fabry's diseases. Other promising approaches are **substrate deprivation therapy** to inhibit the synthesis of sphingolipids and **chemical chaperone therapy**. **Gene therapy** for lysosomal disorders is also currently under investigation. Some examples of the more important lipid storage diseases are shown in Table 24–1.

Multiple sulfatase deficiency results in accumulation of sulfogalactosylceramide, steroid sulfates, and proteoglycans owing to a combined deficiency of arylsulfatases A, B, and C and steroid sulfatase.

SUMMARY

- Triacylglycerols are the major energy-storing lipids, whereas phosphoglycerols, sphingomyelin, and glycosphingolipids are amphipathic and have structural functions in cell membranes as well as other specialized roles.

- Triacylglycerols and some phosphoglycerols are synthesized by progressive acylation of glycerol 3-phosphate. The pathway bifurcates at phosphatidate, forming inositol phospholipids and cardiolipin on the one hand and triacylglycerol and choline and ethanolamine phospholipids on the other.

- Plasmalogens and platelet-activating factor (PAF) are ether phospholipids formed from dihydroxyacetone phosphate.

- Sphingolipids are formed from ceramide (N-acylsphingosine). Sphingomyelin is present in membranes of organelles involved in secretory processes (eg, Golgi apparatus). The simplest

glycosphingolipids are a combination of ceramide plus a sugar residue (eg, GalCer in myelin). Gangliosides are more complex glycosphingolipids containing more sugar residues plus sialic acid. They are present in the outer layer of the plasma membrane, where they contribute to the glycocalyx and are important as antigens and cell receptors.

- Phospholipids and sphingolipids are involved in several disease processes, including infant respiratory distress syndrome (lack of lung surfactant), multiple sclerosis (demyelination), and sphingolipidoses (inability to break down sphingolipids in lysosomes due to inherited defects in hydrolase enzymes).

REFERENCES

McPhail LC: Glycerolipid in signal transduction. In: *Biochemistry of Lipids, Lipoproteins and Membranes,* 4th ed. Vance DE, Vance JE (Eds.). Elsevier, 2002:315–340.

Merrill AH, Sandhoff K: Sphingolipids: metabolism and cell signaling. In: *Biochemistry of Lipids, Lipoproteins and Membranes,* 4th ed. Vance DE, Vance JE (Eds.). Elsevier, 2002:373–408.

Meyer KC, Zimmerman JJ: Inflammation and surfactant. Paediatr Respir Rev 2002;3:308.

Prescott SM et al: Platelet-activating factor and related lipid mediators. Annu Rev Biochem 2000;69:419.

Ruvolo PP: Intracellular signal transduction pathways activated by ceramide and its metabolites. Pharmacol Res 2003;47:383.

Scriver CR et al (editors): *The Metabolic and Molecular Bases of Inherited Disease,* 8th ed. McGraw-Hill, 2001.

Vance DE: Phospholipid biosynthesis in eukaryotes. In: *Biochemistry of Lipids, Lipoproteins and Membranes,* 4th ed. Vance DE, Vance JE (Eds.). Elsevier, 2002:205–232.

van Echten G, Sandhoff K: Ganglioside metabolism. Enzymology, topology, and regulation. J Biol Chem 1993;268:5341.

Lipid Transport & Storage

Kathleen M. Botham, PhD, DSc & Peter A. Mayes, PhD, DSc

BIOMEDICAL IMPORTANCE

Fat absorbed from the diet and lipids synthesized by the liver and adipose tissue must be transported between the various tissues and organs for utilization and storage. Since lipids are insoluble in water, the problem of how to transport them in the aqueous blood plasma is solved by associating nonpolar lipids (triacylglycerol and cholesteryl esters) with amphipathic lipids (phospholipids and cholesterol) and proteins to make water-miscible lipoproteins.

In a meal-eating omnivore such as the human, excess calories are ingested in the anabolic phase of the feeding cycle, followed by a period of negative caloric balance when the organism draws upon its carbohydrate and fat stores. Lipoproteins mediate this cycle by transporting lipids from the intestines as chylomicrons—and from the liver as very low density lipoproteins (VLDL)—to most tissues for oxidation and to adipose tissue for storage. Lipid is mobilized from adipose tissue as free fatty acids (FFA) bound to serum albumin. Abnormalities of lipoprotein metabolism cause various **hypo-** or **hyperlipoproteinemias** (Table 26–1). The most common of these is in **diabetes mellitus,** where insulin deficiency causes excessive mobilization of FFA and underutilization of chylomicrons and VLDL, leading to **hypertriacylglycerolemia.** Most other pathologic conditions affecting lipid transport are due primarily to inherited defects, some of which cause **hypercholesterolemia** and premature **atherosclerosis. Obesity**—particularly abdominal obesity—is a risk factor for increased mortality, hypertension, type 2 diabetes mellitus, hyperlipidemia, hyperglycemia, and various endocrine dysfunctions.

LIPIDS ARE TRANSPORTED IN THE PLASMA AS LIPOPROTEINS

Four Major Lipid Classes Are Present in Lipoproteins

Plasma lipids consist of **triacylglycerols** (16%), **phospholipids** (30%), **cholesterol** (14%), and **cholesteryl esters** (36%) and a much smaller fraction of unesterified long-chain fatty acids (4%). This latter fraction, the **free fatty acids (FFA),** is metabolically the most active of the plasma lipids.

Four Major Groups of Plasma Lipoproteins Have Been Identified

Because fat is less dense than water, the density of a lipoprotein decreases as the proportion of lipid to protein increases (Table 25–1). Four major groups of lipoproteins have been identified that are important physiologically and in clinical diagnosis. These are (1) **chylomicrons**, derived from intestinal absorption of triacylglycerol and other lipids; (2) **very low density lipoproteins** (VLDL, or pre-β-lipoproteins), derived from the liver for the export of triacylglycerol; (3) **low-density lipoproteins** (LDL, or β-lipoproteins), representing a final stage in the catabolism of VLDL; and (4) **high-density lipoproteins** (HDL, or β-lipoproteins), involved in cholesterol transport and also in VLDL and chylomicron metabolism. Triacylglycerol is the predominant lipid in chylomicrons and VLDL, whereas cholesterol and phospholipid are the predominant lipids in LDL and HDL, respectively (Table 25–1). Lipoproteins may be separated according to their electrophoretic properties into **α-, β-,** and **pre-β-lipoproteins.**

Lipoproteins Consist of a Nonpolar Core & a Single Surface Layer of Amphipathic Lipids

The **nonpolar lipid core** consists of mainly **triacylglycerol** and **cholesteryl ester** and is surrounded by a **single surface layer** of **amphipathic phospholipid** and **cholesterol** molecules (Figure 25–1). These are oriented so that their polar groups face outward to the aqueous medium, as in the cell membrane (Chapter 15). The protein moiety of a lipoprotein is known as an **apolipoprotein** or **apoprotein**, constituting nearly 70% of some HDL and as little as 1% of chylomicrons. Some apolipoproteins are integral and cannot be removed, whereas others are free to transfer to other lipoproteins.

The Distribution of Apolipoproteins Characterizes the Lipoprotein

One or more apolipoproteins (proteins or polypeptides) are present in each lipoprotein. The major apolipoproteins of HDL (β-lipoprotein) are designated A (Table 25–1). The main apolipoprotein of LDL (β-lipoprotein) is apolipoprotein B

TABLE 25–1   Composition of the Lipoproteins in Plasma of Humans

Lipoprotein	Source	Diameter (nm)	Density (g/mL)	Composition Protein (%)	Composition Lipid (%)	Main Lipid Components	Apolipoproteins
Chylomicrons	Intestine	90–1000	< 0.95	1–2	98–99	Triacylglycerol	A-I, A-II, A-IV,[1] B-48, C-I, C-II, C-III, E
Chylomicron remnants	Chylomicrons	45–150	< 1.006	6–8	92–94	Triacylglycerol, phospholipids, cholesterol	B-48, E
VLDL	Liver (intestine)	30–90	0.95–1.006	7–10	90–93	Triacylglycerol	B-100, C-I, C-II, C-III
IDL	VLDL	25–35	1.006–1.019	11	89	Triacylglycerol, cholesterol	B-100, E
LDL	VLDL	20–25	1.019–1.063	21	79	Cholesterol	B-100
HDL	Liver, intestine, VLDL, chylomicrons					Phospholipids, cholesterol	A-I, A-II, A-IV, C-I, C-II, C-III, D,[2] E
HDL₁		20–25	1.019–1.063	32	68		
HDL₂		10–20	1.063–1.125	33	67		
HDL₃		5–10	1.125–1.210	57	43		
Preβ-HDL[3]		< 5	> 1.210				A-I
Albumin/free fatty acids	Adipose tissue		> 1.281	99	1	Free fatty acids	

Abbreviations: HDL, high-density lipoproteins; IDL, intermediate-density lipoproteins; LDL, low-density lipoproteins; VLDL, very low density lipoproteins.
[1]Secreted with chylomicrons but transfers to HDL.
[2]Associated with HDL₂ and HDL₃ subfractions.
[3]Part of a minor fraction known as very high density lipoproteins (VHDL).

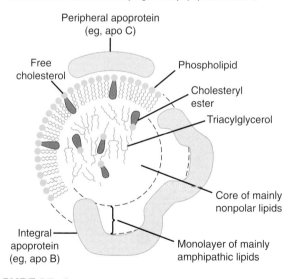

FIGURE 25–1 Generalized structure of a plasma lipoprotein. The similarities with the structure of the plasma membrane are to be noted. Small amounts of cholesteryl ester and triacylglycerol are found in the surface layer and a little free cholesterol in the core.

(B-100), which is found also in VLDL. Chylomicrons contain a truncated form of apo B (B-48) that is synthesized in the intestine, while B-100 is synthesized in the liver. Apo B-100 is one of the longest single polypeptide chains known, having 4536 amino acids and a molecular mass of 550,000 Da. Apo B-48 (48% of B-100) is formed from the same mRNA as apo B-100 after the introduction of a stop signal by an RNA edit-

ing enzyme. Apo C-I, C-II, and C-III are smaller polypeptides (molecular mass 7000–9000 Da) freely transferable between several different lipoproteins. Apo E is found in VLDL, HDL, chylomicrons, and chylomicron remnants; it accounts for 5–10% of total VLDL apolipoproteins in normal subjects.

Apolipoproteins carry out several roles: (1) they can form part of the structure of the lipoprotein, eg, apo B; (2) they are enzyme cofactors, eg, C-II for lipoprotein lipase, A-I for lecithin:cholesterol acyltransferase, or enzyme inhibitors, eg, apo A-II and apo C-III for lipoprotein lipase, apo C-I for cholesteryl ester transfer protein; and (3) they act as ligands for interaction with lipoprotein receptors in tissues, eg, apo B-100 and apo E for the LDL receptor, apo E for the LDL receptor-related protein (LRP), which has been identified as the remnant receptor, and apo A-I for the HDL receptor. The functions of apo A-IV and apo D, however, are not yet clearly defined, although apo D is believed to be an important factor in human neurodegenerative disorders.

FREE FATTY ACIDS ARE RAPIDLY METABOLIZED

The **free fatty acids** (FFA, nonesterified fatty acids, unesterified fatty acids) arise in the plasma from the breakdown of triacylglycerol in adipose tissue or as a result of the action of lipoprotein lipase on the plasma triacylglycerols. They are found **in combination with albumin,** a very effective solubilizer, in concentrations varying between 0.1 and 2.0 µeq/mL of plasma. Levels are low in the fully fed condition and rise to 0.7–0.8 μeq/mL in the starved state. In uncontrolled **diabetes mellitus,** the level may rise to as much as 2 μeq/mL.

Free fatty acids are removed from the blood extremely rapidly and oxidized (fulfilling 25–50% of energy requirements in starvation) or esterified to form triacylglycerol in the tissues. In starvation, esterified lipids from the circulation or in the tissues are oxidized as well, particularly in heart and skeletal muscle cells, where considerable stores of lipid are to be found.

The free fatty acid uptake by tissues is related directly to the plasma free fatty acid concentration, which in turn is determined by the rate of lipolysis in adipose tissue. After dissociation of the fatty acid-albumin complex at the plasma membrane, fatty acids bind to a **membrane fatty acid transport protein** that acts as a transmembrane cotransporter with Na⁺. On entering the cytosol, free fatty acids are bound by intracellular **fatty-acid-binding proteins.** The role of these proteins in intracellular transport is thought to be similar to that of serum albumin in extracellular transport of long-chain fatty acids.

TRIACYLGLYCEROL IS TRANSPORTED FROM THE INTESTINES IN CHYLOMICRONS & FROM THE LIVER IN VERY LOW DENSITY LIPOPROTEINS

By definition, **chylomicrons** are found in **chyle** formed only by the lymphatic system **draining the intestine.** They are responsible for the transport of all dietary lipids into the circulation. Small quantities of VLDL are also to be found in chyle; however, most **VLDL in the plasma** are of hepatic origin. **They are the vehicles of transport of triacylglycerol from the liver to the extrahepatic tissues.**

There are striking similarities in the mechanisms of formation of chylomicrons by intestinal cells and of VLDL by hepatic parenchymal cells (Figure 25–2), perhaps because— apart from the mammary gland—the intestine and liver are the only tissues from which particulate lipid is secreted. Newly secreted or "nascent" chylomicrons and VLDL contain only a small amount of apolipoproteins C and E, and the full complement is acquired from HDL in the circulation (Figures 25–3 & 25–4). Apo B is essential for chylomicron and VLDL formation. In **abetalipoproteinemia** (a rare disease), lipoproteins containing apo B are not formed and lipid droplets accumulate in the intestine and liver.

A more detailed account of the factors controlling hepatic VLDL secretion is given below.

CHYLOMICRONS & VERY LOW DENSITY LIPOPROTEINS ARE RAPIDLY CATABOLIZED

The clearance of chylomicrons from the blood is rapid, the half-time of disappearance being under 1 h in humans. Larger particles are catabolized more quickly than smaller ones. Fatty

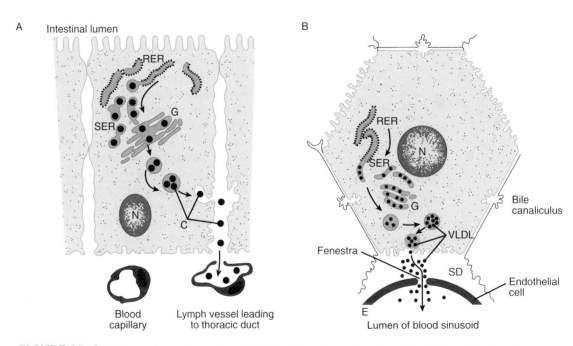

FIGURE 25–2 The formation and secretion of **(A)** chylomicrons by an intestinal cell and **(B)** very low density lipoproteins by a hepatic cell. (RER, rough endoplasmic reticulum; SER, smooth endoplasmic reticulum; G, Golgi apparatus; N, nucleus; C, chylomicrons; VLDL, very low density lipoproteins; E, endothelium; SD, space of Disse, containing blood plasma.) Apolipoprotein B, synthesized in the RER, is incorporated into lipoproteins in the SER, the main site of synthesis of triacylglycerol. After addition of carbohydrate residues in G, they are released from the cell by reverse pinocytosis. Chylomicrons pass into the lymphatic system. VLDL are secreted into the space of Disse and then into the hepatic sinusoids through fenestrae in the endothelial lining.

acids originating from chylomicron triacylglycerol are delivered mainly to adipose tissue, heart, and muscle (80%), while about 20% goes to the liver. However, **the liver does not metabolize native chylomicrons or VLDL significantly;** thus, the fatty acids in the liver must be secondary to their metabolism in extrahepatic tissues.

Triacylglycerols of Chylomicrons & VLDL Are Hydrolyzed by Lipoprotein Lipase

Lipoprotein lipase is located on the walls of blood capillaries, anchored to the endothelium by negatively charged proteoglycan chains of heparan sulfate. It has been found in heart, adipose tissue, spleen, lung, renal medulla, aorta, diaphragm, and lactating mammary gland, although it is not active in adult liver. It is not normally found in blood; however, following injection of **heparin**, lipoprotein lipase is released from its heparan sulfate binding sites into the circulation. **Hepatic lipase** is bound to the sinusoidal surface of liver cells and is also released by heparin. This enzyme, however, does not react readily with chylomicrons or VLDL but is involved in chylomicron remnant and HDL metabolism.

Both **phospholipids** and **apo C-II** are required as cofactors for lipoprotein lipase activity, while **apo A-II and apo C-III** act as inhibitors. Hydrolysis takes place while the lipoproteins are attached to the enzyme on the endothelium. Triacylglycerol is hydrolyzed progressively through a diacylglycerol to a monoacylglycerol and finally to free fatty acids plus glycerol. Some of the released free fatty acids return to the circulation, attached to albumin, but the bulk is transported into the tissue (Figures 25–3 & 25–4). Heart lipoprotein lipase has a low K_m for triacylglycerol, about one-tenth of that for the enzyme in adipose tissue. This enables the delivery of fatty acids from triacylglycerol to be **redirected from adipose tissue to the heart in the starved state** when the plasma triacylglycerol decreases. A similar redirection to the mammary gland occurs during lactation, allowing uptake of lipoprotein triacylglycerol fatty acid for **milk fat** synthesis. The **VLDL receptor** plays an important part in the delivery of fatty acids from VLDL triacylglycerol to adipocytes by binding VLDL and bringing it into close contact with lipoprotein lipase. In adipose tissue, **insulin** enhances lipoprotein lipase synthesis in adipocytes and its translocation to the luminal surface of the capillary endothelium.

The Action of Lipoprotein Lipase Forms Remnant Lipoproteins

Reaction with lipoprotein lipase results in the loss of 70–90% of the triacylglycerol of chylomicrons and in the loss of apo C (which returns to HDL) but not apo E, which is retained. The resulting **chylomicron remnant** is about half the diameter of the parent chylomicron and is relatively enriched in cholesterol and cholesteryl esters because of the loss of triacylglycerol (Figure 25–3). Similar changes occur to VLDL, with the formation of **VLDL remnants** (also called **intermediate-density lipoprotein [IDL]**) (Figure 25–4).

The Liver Is Responsible for the Uptake of Remnant Lipoproteins

Chylomicron remnants are taken up by the liver by receptor-mediated endocytosis, and the cholesteryl esters and triacyl-

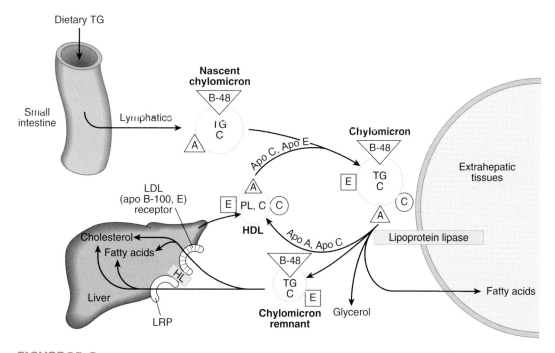

FIGURE 25–3 Metabolic fate of chylomicrons. (A, apolipoprotein A; B-48, apolipoprotein B-48; ©, apolipoprotein C; E, apolipoprotein E; HDL, high-density lipoprotein; TG, triacylglycerol; C, cholesterol and cholesteryl ester; PL, phospholipid; HL, hepatic lipase; LRP, LDL-receptor-related protein.) Only the predominant lipids are shown.

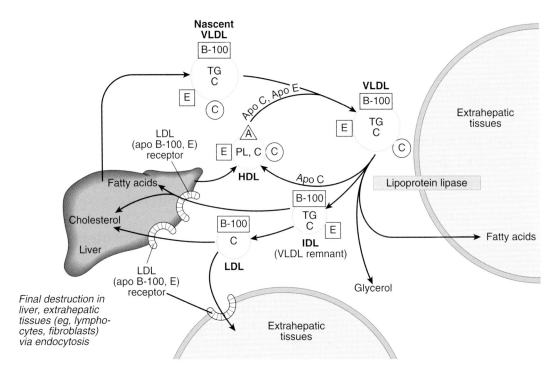

FIGURE 25–4 Metabolic fate of very low density lipoproteins (VLDL) and production of low-density lipoproteins (LDL). (A, apolipoprotein A; B-100, apolipoprotein B-100; ©, apolipoprotein C; E, apolipoprotein E; HDL, high-density lipoprotein; TG, triacylglycerol; IDL, intermediate-density lipoprotein; C, cholesterol and cholesteryl ester; PL, phospholipid.) Only the predominant lipids are shown. It is possible that some IDL is also metabolized via the LRP.

glycerols are hydrolyzed and metabolized. Uptake is mediated by **apo E** (Figure 25–3), via two apo E-dependent receptors, the **LDL (apo B-100, E) receptor** and the **LRP (LDL receptor-related protein)**. Hepatic lipase has a dual role: (1) it acts as a ligand to facilitate remnant uptake and (2) it hydrolyzes remnant triacylglycerol and phospholipid.

After metabolism to IDL, VLDL may be taken up by the liver directly via the LDL (apo B-100, E) receptor, or it may be converted to LDL. Only one molecule of apo B-100 is present in each of these lipoprotein particles, and this is conserved during the transformations. Thus, each LDL particle is derived from a single precursor VLDL particle (Figure 25–4). In humans, a relatively large proportion of IDL forms LDL, accounting for the increased concentrations of LDL in humans compared with many other mammals.

LDL IS METABOLIZED VIA THE LDL RECEPTOR

The liver and many extrahepatic tissues express the **LDL (apo B-100, E) receptor.** It is so designated because it is specific for apo B-100 but not B-48, which lacks the carboxyl terminal domain of B-100 containing the LDL receptor ligand, and it also takes up lipoproteins rich in apo E. Approximately 30% of LDL is degraded in extrahepatic tissues and 70% in the liver. A positive correlation exists between the incidence of **athero-sclerosis** and the plasma concentration of LDL cholesterol. The LDL (apoB-100, E) receptor is defective in **familial hyper-**

cholesterolemia, a genetic condition which increases blood LDL cholesterol levels and causes premature atherosclerosis. For further discussion of the regulation of the LDL receptor, see Chapter 26.

HDL TAKES PART IN BOTH LIPOPROTEIN TRIACYLGLYCEROL & CHOLESTEROL METABOLISM

HDL is synthesized and secreted from both liver and intestine (Figure 25–5). However, apo C and apo E are synthesized in the liver and transferred from liver HDL to intestinal HDL when the latter enters the plasma. A major function of HDL is to act as a repository for the apo C and apo E required in the metabolism of chylomicrons and VLDL. Nascent HDL consists of discoid phospholipid bilayers containing apo A and free cholesterol. These lipoproteins are similar to the particles found in the plasma of patients with a deficiency of the plasma enzyme **lecithin:cholesterol acyltransferase (LCAT)** and in the plasma of patients with **obstructive jaundice**. LCAT—and the LCAT activator apo A-I—bind to the discoidal particles, and the surface phospholipid and free cholesterol are converted into cholesteryl esters and lysolecithin (Chapter 24). The nonpolar cholesteryl esters move into the hydrophobic interior of the bilayer, whereas lysolecithin is transferred to plasma albumin. Thus, a nonpolar core is generated, forming a spherical, pseudomicellar HDL covered by a surface film of

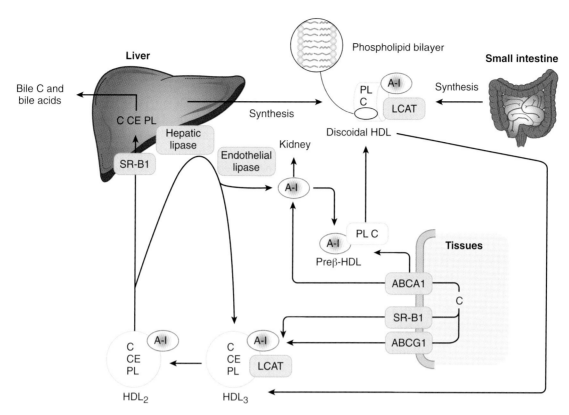

FIGURE 25–5 Metabolism of high-density lipoprotein (HDL) in reverse cholesterol transport. (LCAT, lecithin:cholesterol acyltransferase; C, cholesterol; CE, cholesteryl ester; PL, phospholipid; A-I, apolipoprotein A-I; SR-B1, scavenger receptor B1; ABCA 1, ATP-binding cassette transporter A1; ABCG1, ATP-binding cassette transporter G1.) Preβ-HDL, HDL$_2$, HDL$_3$—see Table 25–1. Surplus surface constituents from the action of lipoprotein lipase on chylomicrons and VLDL are another source of preβ-HDL. Hepatic lipase activity is increased by androgens and decreased by estrogens, which may account for higher concentrations of plasma HDL$_2$ in women.

polar lipids and apolipoproteins. This aids the removal of excess unesterified cholesterol from lipoproteins and tissues as described below. The **class B scavenger receptor B1 (SR-B1)** has been identified as an **HDL receptor with a dual role in HDL metabolism.** In the liver and in steroidogenic tissues, it binds HDL via apo A-I, and cholesteryl ester is selectively delivered to the cells, although the particle itself, including apo A-I, is not taken up. In the tissues, on the other hand, SR-B1 mediates the acceptance of cholesterol effluxed from the cells by HDL, which then transports it to the liver for excretion via the bile (either as cholesterol or after conversion to bile acids) in the process known as **reverse cholesterol transport** (Figure 25–5). HDL$_3$, generated from discoidal HDL by the action of LCAT, accepts cholesterol from the tissues via the **SR-B1** and the cholesterol is then esterified by LCAT, increasing the size of the particles to form the less dense HDL$_2$. HDL$_3$ is then reformed, either after selective delivery of cholesteryl ester to the liver via the SR-B1 or by hydrolysis of HDL$_2$ phospholipid and triacylglycerol by hepatic lipase and endothelial lipase. This interchange of HDL$_2$ and HDL$_3$ is called the **HDL cycle** (Figure 25–5). Free apo A-I is released by these processes and forms **preβ-HDL** after associating with a minimum amount of phospholipid and cholesterol. Surplus apo A-I is destroyed in the kidney. A second important mechanism for reverse cho-

lesterol transport involves the **ATP-binding cassette transporters A1 (ABCA1) and G1 (ABCG1).** These transporters are members of a family of transporter proteins that couple the hydrolysis of ATP to the binding of a substrate, enabling it to be transported across the membrane. ABCG1 mediates the transport of cholesterol from cells to HDL, while ABCA1 preferentially promotes efflux to poorly lipidated particles such as preβ-HDL or apo A-1, which are then converted to HDL$_3$ via discoidal HDL (Figure 25–5). Preβ-HDL is the most potent form of HDL inducing cholesterol efflux from the tissues.

HDL concentrations vary reciprocally with plasma triacylglycerol concentrations and directly with the activity of lipoprotein lipase. This may be due to surplus surface constituents, eg, phospholipid and apo A-I, being released during hydrolysis of chylomicrons and VLDL and contributing toward the formation of preβ-HDL and discoidal HDL. HDL$_2$ concentrations are **inversely related to the incidence of atherosclerosis,** possibly because they reflect the efficiency of reverse cholesterol transport. HDL$_c$ (HDL$_1$) is found in the blood of diet-induced hypercholesterolemic animals. It is rich in cholesterol, and its sole apolipoprotein is apo E. It appears that all plasma lipoproteins are interrelated components of one or more metabolic cycles that together are responsible for the complex process of plasma lipid transport.

THE LIVER PLAYS A CENTRAL ROLE IN LIPID TRANSPORT & METABOLISM

The liver carries out the following major functions in lipid metabolism:

1. It facilitates the digestion and absorption of lipids by the production of **bile**, which contains cholesterol and bile salts synthesized within the liver de novo or after uptake of lipoprotein cholesterol (Chapter 26).
2. It actively **synthesizes and oxidizes fatty acids** (Chapters 22 & 23) and also synthesizes triacylglycerols and phospholipids (Chapter 24).
3. It **converts fatty acids to ketone bodies (ketogenesis)** (Chapter 22).
4. It plays an integral part in the **synthesis and metabolism of plasma lipoproteins** (this chapter).

Hepatic VLDL Secretion Is Related to Dietary & Hormonal Status

The cellular events involved in VLDL formation and secretion have been described above (Figure 25–2) and are shown in Figure 25–6. Hepatic triacylglycerol synthesis provides the immediate stimulus for the formation and secretion of VLDL. The fatty acids used are derived from two possible sources: (1) synthesis within the liver from **acetyl-CoA** derived mainly from carbohydrate (perhaps not so important in humans) and (2) uptake of **free fatty acids** from the circulation. The first source is predominant in the well-fed condition, when fatty acid synthesis is high and the level of circulating free fatty acids is low. As triacylglycerol does not normally accumulate in the liver in these conditions, it must be inferred that it is transported from the liver in VLDL as rapidly as it is synthesized and that the synthesis of apo B-100 is not rate-limiting. Free fatty acids from the circulation are the main source during starvation, the feeding of high-fat diets, or in diabetes mellitus, when hepatic lipogenesis is inhibited. Factors that enhance both the synthesis of triacylglycerol and the secretion of VLDL by the liver include (1) the fed state rather than the starved state; (2) the feeding of diets high in carbohydrate (particularly if they contain sucrose or fructose), leading to high rates of lipogenesis and esterification of fatty acids; (3) high levels of circulating free fatty acids; (4) ingestion of ethanol; and (5) the presence of high concentrations of insulin and low concentrations of glucagon, which enhance fatty acid synthesis and esterification and inhibit their oxidation (Figure 25–6).

CLINICAL ASPECTS

Imbalance in the Rate of Triacylglycerol Formation & Export Causes Fatty Liver

For a variety of reasons, lipid—mainly as triacylglycerol—can accumulate in the liver (Figure 25–6). Extensive accumulation is regarded as a pathologic condition. **Nonalcoholic fatty liver disease (NAFLD)** is the most common liver disorder worldwide. When accumulation of lipid in the liver becomes chronic, inflammatory and fibrotic changes may develop leading to **nonalcoholic steatohepatitis (NASH),** which can progress to liver diseases including **cirrhosis, hepatocarcinoma, and liver failure**.

Fatty livers fall into two main categories. The first type is associated with **raised levels of plasma free fatty acids** resulting from mobilization of fat from adipose tissue or from the hydrolysis of lipoprotein triacylglycerol by lipoprotein lipase in extrahepatic tissues. The production of VLDL does not keep pace with the increasing influx and esterification of free fatty acids, allowing triacylglycerol to accumulate, which in turn causes a fatty liver. This occurs during **starvation** and the feeding of **high-fat diets.** The ability to secrete VLDL may also be impaired (eg, in starvation). In uncontrolled **diabetes mellitus, twin lamb disease,** and **ketosis in cattle,** fatty infiltration is sufficiently severe to cause visible pallor (fatty appearance) and enlargement of the liver with possible liver dysfunction.

The second type of fatty liver is usually due to a **metabolic block in the production of plasma lipoproteins,** thus allowing triacylglycerol to accumulate. Theoretically, the lesion may be due to (1) a block in apolipoprotein synthesis, (2) a block in the synthesis of the lipoprotein from lipid and apolipoprotein, (3) a failure in provision of phospholipids that are found in lipoproteins, or (4) a failure in the secretory mechanism itself.

One type of fatty liver that has been studied extensively in rats is caused by a deficiency of **choline,** which has therefore been called a **lipotropic factor.** The antibiotic puromycin, ethionine (α-amino-γ-mercaptobutyric acid), carbon tetrachloride, chloroform, phosphorus, lead, and arsenic all cause fatty liver and a marked reduction in concentration of VLDL in rats. Choline will not protect the organism against these agents but appears to aid in recovery. The action of carbon tetrachloride probably involves formation of free radicals causing lipid peroxidation. Some protection against this is provided by the antioxidant action of **vitamin E**-supplemented diets. The action of ethionine is thought to be caused by a reduction in availability of ATP due to its replacing methionine in S-adenosylmethionine, trapping available adenine and preventing synthesis of ATP. **Orotic acid** also causes fatty liver; it is believed to interfere with glycosylation of the lipoprotein, thus inhibiting release, and may also impair the recruitment of triacylglycerol to the particles. A deficiency of vitamin E enhances the hepatic necrosis of the choline deficiency type of fatty liver. Added vitamin E or a source of **selenium** has a protective effect by combating lipid peroxidation. In addition to protein deficiency, essential fatty acid and vitamin deficiencies (eg, linoleic acid, pyridoxine, and pantothenic acid) can cause fatty infiltration of the liver.

Ethanol Also Causes Fatty Liver

Alcoholic fatty liver is the first stage in **alcoholic liver disease (ALD)** which is caused by **alcoholism** and ultimately leads to **cirrhosis**. The fat accumulation in the liver is caused by a combination of impaired fatty acid oxidation and increased lipogenesis, which is thought to be due to changes in the [NADH]/[NAD$^+$] redox potential in the liver, and also to interference with

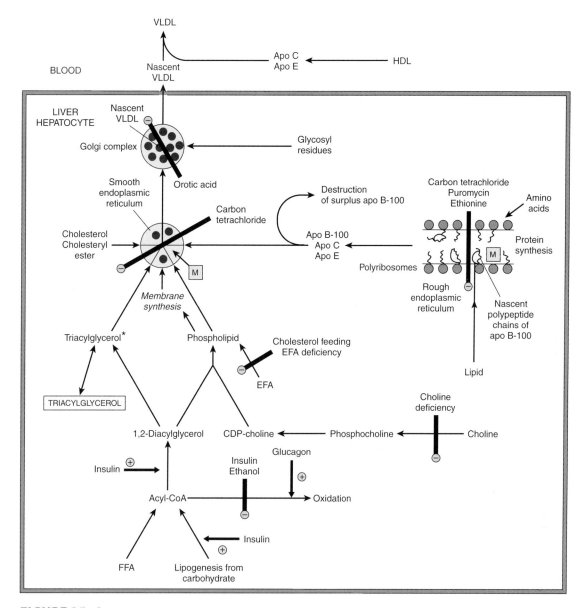

FIGURE 25–6 The synthesis of very low density lipoprotein (VLDL) in the liver and the possible loci of action of factors causing accumulation of triacylglycerol and fatty liver. (EFA, essential fatty acids; FFA, free fatty acids; HDL, high-density lipoproteins; Apo, apolipoprotein; M, microsomal triacylglycerol transfer protein.) The pathways indicated form a basis for events depicted in Figure 25–2. The main triacylglycerol pool (highlighted in orange square) in liver is not on the direct pathway of VLDL synthesis from acyl-CoA. Thus, FFA, insulin, and glucagon have immediate effects on VLDL secretion as their effects impinge directly on the small triacylglycerol precursor pool*. In the fully fed state, apo B-100 is synthesized in excess of requirements for VLDL secretion and the surplus is destroyed in the liver. During translation of apo B-100, microsomal transfer protein-mediated lipid transport enables lipid to become associated with the nascent polypeptide chain. After release from the ribosomes, these particles fuse with more lipids from the smooth endoplasmic reticulum, producing nascent VLDL.

the action of transcription factors regulating the expression of the enzymes involved in the pathways. Oxidation of ethanol by **alcohol dehydrogenase** leads to excess production of NADH.

$$CH_3-CH_2-OH \xrightarrow[\text{NAD}^+ \quad \text{NADH} + \text{H}^+]{\boxed{\text{ALCOHOL DEHYDROGENASE}}} CH_3-CHO$$

Ethanol   **Acetaldehyde**

The NADH generated competes with reducing equivalents from other substrates, including fatty acids, for the respiratory chain, inhibiting their oxidation and causing increased esterification of fatty acids to form triacylglycerol, resulting in the fatty liver. Oxidation of ethanol leads to the formation of acetaldehyde, which is oxidized by **aldehyde dehydrogenase,** producing acetate. The increased [NADH]/[NAD⁺] ratio also causes increased [lactate]/[pyruvate], resulting in **hyperlacticacidemia,** which decreases excretion of uric acid, aggravating **gout**. Some metabolism of ethanol takes place via a cytochrome P450-dependent microsomal ethanol oxidizing system (MEOS) involving NADPH and O_2. This system increases in activity in **chronic alcoholism** and may account for the increased metabolic clearance in this condition. Ethanol also inhibits the metabolism of some drugs, eg, barbiturates, by competing for cytochrome P450-dependent enzymes.

$$\underset{\textbf{Ethanol}}{CH_3-CH_2-OH} + NADPH + H^+ + O_2 \xrightarrow{\boxed{MEOS}} \underset{\textbf{Acetaldehyde}}{CH_3-CHO} + NADP^+ + 2H_2O$$

In some Asian populations and Native Americans, alcohol consumption results in increased adverse reactions to acetaldehyde owing to a genetic defect of mitochondrial aldehyde dehydrogenase.

ADIPOSE TISSUE IS THE MAIN STORE OF TRIACYLGLYCEROL IN THE BODY

Triacylglycerols are stored in adipose tissue in large **lipid droplets** and are continually undergoing lipolysis (hydrolysis) and re-esterification (Figure 25–7). These two processes are entirely different pathways involving different reactants and enzymes. This allows the processes of esterification or lipolysis to be regulated separately by many nutritional, metabolic, and hormonal factors. The balance between these two processes determines the magnitude of the free fatty acid pool in adipose tissue, which in turn determines the level of free fatty acids circulating in the plasma. Since the latter has most profound effects upon the metabolism of other tissues, particularly liver and muscle, the factors operating in adipose tissue that regulate the outflow of free fatty acids exert an influence far beyond the tissue itself.

The Provision of Glycerol 3-Phosphate Regulates Esterification: Lipolysis Is Controlled by Hormone-Sensitive Lipase

Triacylglycerol is synthesized from acyl-CoA and glycerol 3-phosphate (Figure 24–2). Because the enzyme **glycerol kinase** is not expressed in adipose tissue, glycerol cannot be utilized for the provision of glycerol 3-phosphate, which must be supplied from glucose via glycolysis.

Triacylglycerol undergoes hydrolysis by a **hormone-sensitive lipase** to form free fatty acids and glycerol. This lipase is distinct from lipoprotein lipase, which catalyzes lipoprotein triacylglycerol hydrolysis before its uptake into extrahepatic tissues (see above). Since the glycerol cannot be utilized, it enters the blood and is taken up and transported to tissues such as the liver and kidney, which possess an active glycerol kinase. The free fatty acids formed by lipolysis can be reconverted in adipose tissue to acyl-CoA by **acyl-CoA synthetase** and reesterified with glycerol 3-phosphate to form triacylglycerol. Thus, **there is a continuous cycle of lipolysis and re-esterification within the tissue.** However, when the rate of re-esterification is

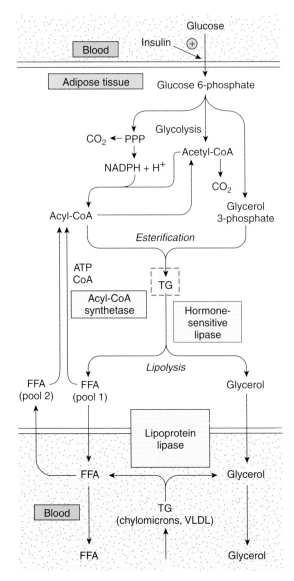

FIGURE 25–7 Triacylglycerol metabolism in adipose tissue. Hormone-sensitive lipase is activated by ACTH, TSH, glucagon, epinephrine, norepinephrine, and vasopressin and inhibited by insulin, prostaglandin E_1, and nicotinic acid. Details of the formation of glycerol 3-phosphate from intermediates of glycolysis are shown in Figure 24–2. (PPP, pentose phosphate pathway; TG, triacylglycerol; FFA, free fatty acids; VLDL, very low density lipoprotein.)

not sufficient to match the rate of lipolysis, free fatty acids accumulate and diffuse into the plasma, where they bind to albumin and raise the concentration of plasma free fatty acids.

Increased Glucose Metabolism Reduces the Output of Free Fatty Acids

When the utilization of glucose by adipose tissue is increased, the free fatty acid outflow decreases. However, the release of glycerol continues, demonstrating that the effect of glucose is not mediated by reducing the rate of lipolysis. The effect is due to the provision of glycerol 3-phosphate, which enhances esterification of free fatty acids. Glucose can take several pathways in adipose tissue, including oxidation to CO_2 via the citric acid cycle, oxidation in the pentose phosphate pathway, conversion to long-chain fatty acids, and formation of acylglycerol via glycerol 3-phosphate (Figure 25–7). When glucose utilization is high, a larger proportion of the uptake is oxidized to CO_2 and converted to fatty acids. However, as total glucose utilization decreases, the greater proportion of the glucose is directed to the formation of glycerol 3-phosphate for the esterification of acyl-CoA, which helps to minimize the efflux of free fatty acids.

HORMONES REGULATE FAT MOBILIZATION

Adipose Tissue Lipolysis Is Inhibited by Insulin

The rate of release of free fatty acids from adipose tissue is affected by many hormones that influence either the rate of esterification or the rate of lipolysis. **Insulin** inhibits the release of free fatty acids from adipose tissue, which is followed by a fall in circulating plasma free fatty acids. It enhances lipogenesis

and the synthesis of acylglycerol and increases the oxidation of glucose to CO_2 via the pentose phosphate pathway. All of these effects are dependent on the presence of glucose and can be explained, to a large extent, on the basis of the ability of insulin to enhance the uptake of glucose into adipose cells via the **GLUT 4 transporter.** Insulin also increases the activity of pyruvate dehydrogenase, acetyl-CoA carboxylase, and glycerol phosphate acyltransferase, reinforcing the effects of increased glucose uptake on the enhancement of fatty acid and acylglycerol synthesis. These three enzymes are regulated in a coordinate manner by phosphorylation-dephosphorylation mechanisms.

A principal action of insulin in adipose tissue is to inhibit the activity of **hormone-sensitive lipase,** reducing the release not only of free fatty acids but of glycerol as well. Adipose tissue is much more sensitive to insulin than are many other tissues, which points to adipose tissue as a major site of insulin action in vivo.

Several Hormones Promote Lipolysis

Other hormones accelerate the release of free fatty acids from adipose tissue and raise the plasma free fatty acid concentration by increasing the rate of lipolysis of the triacylglycerol

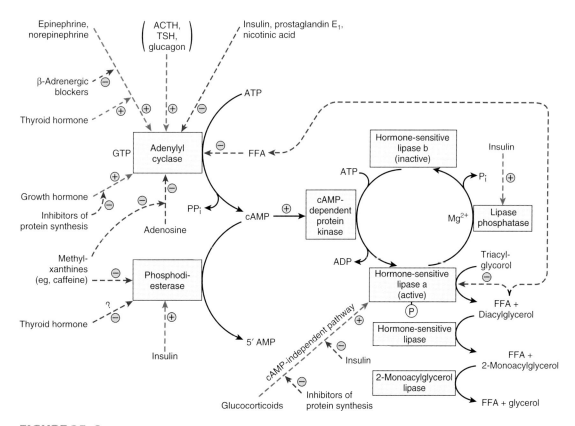

FIGURE 25–8 Control of adipose tissue lipolysis. (TSH, thyroid-stimulating hormone; FFA, free fatty acids.) Note the cascade sequence of reactions affording amplification at each step. The lipolytic stimulus is "switched off" by removal of the stimulating hormone; the action of lipase phosphatase; the inhibition of the lipase and adenylyl cyclase by high concentrations of FFA; the inhibition of adenylyl cyclase by adenosine; and the removal of cAMP by the action of phosphodiesterase. ACTH, TSH, and glucagon may not activate adenylyl cyclase in vivo, since the concentration of each hormone required in vitro is much higher than is found in the circulation. Positive (⊕) and negative (⊖) regulatory effects are represented by broken lines and substrate flow by solid lines.

stores (Figure 25–8). These include epinephrine, norepinephrine, glucagon, adrenocorticotropic hormone (ACTH), α- and β-melanocyte-stimulating hormones (MSH), thyroid-stimulating hormone (TSH), growth hormone (GH), and vasopressin. Many of these activate hormone-sensitive lipase. For an optimal effect, most of these lipolytic processes require the presence of **glucocorticoids** and **thyroid hormones.** These hormones act in a **facilitatory** or **permissive** capacity with respect to other lipolytic endocrine factors.

The hormones that act rapidly in promoting lipolysis, ie, catecholamines, do so by stimulating the activity of **adenylyl cyclase,** the enzyme that converts ATP to cAMP. The mechanism is analogous to that responsible for hormonal stimulation of glycogenolysis (Chapter 19). cAMP, by stimulating **cAMP-dependent protein kinase,** activates hormone-sensitive lipase. Thus, processes which destroy or preserve cAMP influence lipolysis. cAMP is degraded to 5′-AMP by the enzyme **cyclic 3′,5′-nucleotide phosphodiesterase.** This enzyme is inhibited by methylxanthines such as **caffeine** and **theophylline. Insulin** antagonizes the effect of the lipolytic hormones. Lipolysis appears to be more sensitive to changes in concentration of insulin than are glucose utilization and esterification. The antilipolytic effects of insulin, nicotinic acid, and prostaglandin E_1 are accounted for by inhibition of the synthesis of cAMP at the adenylyl cyclase site, acting through a G_i protein. Insulin also stimulates phosphodiesterase and the lipase phosphatase that inactivates hormone-sensitive lipase. The effect of growth hormone in promoting lipolysis is dependent on synthesis of proteins involved in the formation of cAMP. Glucocorticoids promote lipolysis via synthesis of new lipase protein by a cAMP-independent pathway, which may be inhibited by insulin, and also by promoting transcription of genes involved in the cAMP signal cascade. These findings help to explain the role of the pituitary gland and the adrenal cortex in enhancing fat mobilization. Adipose tissue secretes hormones such as **adiponectin,** which modulates glucose and lipid metabolism in muscle and liver, and **leptin,** which regulates energy homeostasis. Although leptin protects against obesity in rodents, current evidence suggests that its main role in humans is to act as a signal of energy sufficiency rather than energy excess.

The sympathetic nervous system, through liberation of norepinephrine in adipose tissue, plays a central role in the mobilization of free fatty acids. Thus, the increased lipolysis caused by many of the factors described above can be reduced or abolished by denervation of adipose tissue or by ganglionic blockade.

Perilipin Regulates the Balance Between Triacylglycerol Storage and Lipolysis in Adipocytes

Perilipin, a protein involved in the formation of lipid droplets in adipocytes, inhibits lipolysis in basal conditions by preventing access of the lipase enzymes to the stored triacylglycerols. On stimulation with hormones which promote triacylglycerol degradation, however, the protein targets hormone-sensitive

lipase to the lipid droplet surface and thus promotes lipolysis. Perilipin, therefore, enables the storage and breakdown of triacylglycerol to be coordinated according to the metabolic needs of the body.

Human Adipose Tissue May Not Be an Important Site of Lipogenesis

There is no significant incorporation of glucose or pyruvate into long-chain fatty acids; ATP-citrate lyase, a key enzyme in

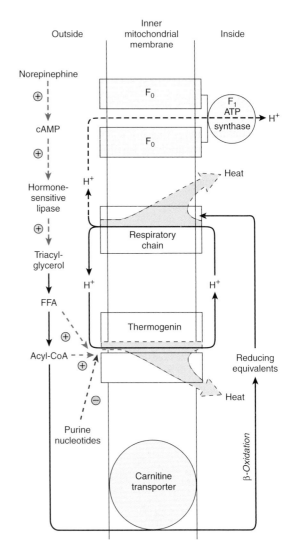

FIGURE 25–9 Thermogenesis in brown adipose tissue. Activity of the respiratory chain produces heat in addition to translocating protons (Chapter 13). These protons dissipate more heat when returned to the inner mitochondrial compartment via thermogenin instead of via the F_1 ATP synthase, the route that generates ATP (Figure 13–7). The passage of H^+ via thermogenin is inhibited by purine nucleotides when brown adipose tissue is unstimulated. Under the influence of norepinephrine, the inhibition is removed by the production of free fatty acids (FFA) and acyl-CoA. Note the dual role of acyl-CoA in both facilitating the action of thermogenin and supplying reducing equivalents for the respiratory chain. ⊕ and ⊖ signify positive or negative regulatory effects.

lipogenesis, does not appear to be present, and other lipogenic enzymes—eg, glucose-6-phosphate dehydrogenase and the malic enzyme—do not undergo adaptive changes. Indeed, it has been suggested that in humans there is a "carbohydrate excess syndrome" due to a unique limitation in ability to dispose of excess carbohydrate by lipogenesis. In birds, lipogenesis is confined to the liver, where it is particularly important in providing lipids for egg formation, stimulated by estrogens.

BROWN ADIPOSE TISSUE PROMOTES THERMOGENESIS

Brown adipose tissue is involved in metabolism, particularly at times when heat generation is necessary. Thus, the tissue is extremely active in some species in arousal from hibernation, in animals exposed to cold (nonshivering thermogenesis), and in heat production in the newborn animal. Though not a prominent tissue in humans, it is present in normal individuals, where it could be responsible for **"diet-induced thermogenesis."** It is noteworthy that brown adipose tissue is reduced or absent in obese persons. The tissue is characterized by a well-developed blood supply and a high content of mitochondria and cytochromes but low activity of ATP synthase. Metabolic emphasis is placed on oxidation of both glucose and fatty acids. **Norepinephrine** liberated from sympathetic nerve endings is important in increasing lipolysis in the tissue and increasing synthesis of lipoprotein lipase to enhance utilization of triacylglycerol-rich lipoproteins from the circulation. Oxidation and phosphorylation are not coupled in mitochondria of this tissue, and the phosphorylation that does occur is at the substrate level, eg, at the succinate thiokinase step and in glycolysis. Thus, **oxidation produces much heat, and little free energy is trapped in ATP.** A thermogenic uncoupling protein, **thermogenin,** acts as a proton conductance pathway dissipating the electrochemical potential across the mitochondrial membrane (Figure 25–9).

SUMMARY

- Since nonpolar lipids are insoluble in water, for transport between the tissues in the aqueous blood plasma they are combined with amphipathic lipids and proteins to make water-miscible lipoproteins.

- Four major groups of lipoproteins are recognized: Chylomicrons transport lipids resulting from digestion and absorption. Very low density lipoproteins (VLDL) transport triacylglycerol from the liver. Low-density lipoproteins (LDL) deliver cholesterol to the tissues, and high-density lipoproteins (HDL) remove cholesterol from the tissues and return it to the liver for excretion in the process known as reverse cholesterol transport.

- Chylomicrons and VLDL are metabolized by hydrolysis of their triacylglycerol, and lipoprotein remnants are left in the circulation. These are taken up by liver, but some of the remnants (IDL), resulting from VLDL form LDL, which is taken up by the liver and other tissues via the LDL receptor.

- Apolipoproteins constitute the protein moiety of lipoproteins. They act as enzyme activators (eg, apo C-II and apo A-I) or as ligands for cell receptors (eg, apo A-I, apo E, and apo B-100).

- Triacylglycerol is the main storage lipid in adipose tissue. Upon mobilization, free fatty acids and glycerol are released. Free fatty acids are an important fuel source.

- Brown adipose tissue is the site of "nonshivering thermogenesis." It is found in hibernating and newborn animals and is present in small quantity in humans. Thermogenesis results from the presence of an uncoupling protein, thermogenin, in the inner mitochondrial membrane.

REFERENCES

Arner P: Human fat cell lipolysis: biochemistry, regulation and clinical role. Best Pract Res Clin Endocrinol Metab 2005;19:471.

Brasaemle DL: Thematic review series: adipocyte biology. The perilipin family of structural lipid droplet proteins: stabilization of lipid droplets and control of lipolysis. J Lipid Res 48:2547.

Goldberg IJ, Merkel M: Lipoprotein lipase: physiology, biochemistry and molecular biology. Front Biosci 2001;6:D388.

Holm C et al: Molecular mechanisms regulating hormone sensitive lipase and lipolysis. Annu Rev Nutr 2000;20:365.

Kershaw EE, Flier JS: Adipose tissue as an endocrine organ. J Clin Endocrinol Metab 2004;89:2548.

Lenz A, Diamond FB: Obesity: the hormonal milieu. Curr Opin Endocrinol Diabetes Obes 2008;15:9.

Redgrave TG: Chylomicron metabolism. Biochem Soc Trans 2004;32:79.

Rye K-A et al: Overview of plasma lipid transport. In: *Plasma Lipids and Their Role in Disease.* Barter PJ, Rye K-A (editors). Harwood Academic Publishers, 1999:1–16

Schreuder TC et al: Nonalcoholic fatty liver disease: an overview of current insights in pathogenesis, diagnosis and treatment. World J Gastroenterol 2008;14:2474.

Sell H, Deshaies Y, Richard D: The brown adipocyte: update on its metabolic role. Int J Biochem Cell Biol 2004;36:2098.

Shelness GS, Sellers JA: Very-low-density lipoprotein assembly and secretion. Curr Opin Lipidol 2001;12:151.

Various authors: *Biochemistry of Lipids, Lipoproteins and Membranes,* 5th ed. Vance DE, Vance JE (editors). Elsevier, 2008.

Cholesterol Synthesis, Transport, & Excretion

Kathleen M. Botham, PhD, DSc & Peter A. Mayes, PhD, DSc

BIOMEDICAL IMPORTANCE

Cholesterol is present in tissues and in plasma either as free cholesterol or combined with a long-chain fatty acid as cholesteryl ester, the storage form. In plasma, both forms are transported in lipoproteins (Chapter 25). Cholesterol is an amphipathic lipid and as such is an essential structural component of membranes and of the outer layer of plasma lipoproteins. It is synthesized in many tissues from acetyl-CoA and is the precursor of all other steroids in the body, including corticosteroids, sex hormones, bile acids, and vitamin D. As a typical product of animal metabolism, cholesterol occurs in foods of animal origin such as egg yolk, meat, liver, and brain. Plasma low-density lipoprotein (LDL) is the vehicle of uptake of cholesterol and cholesteryl ester into many tissues. Free cholesterol is removed from tissues by plasma high-density lipoprotein (HDL) and transported to the liver, where it is eliminated from the body either unchanged or after conversion to bile acids in the process known as **reverse cholesterol transport** (Chapter 25). Cholesterol is a major constituent of **gallstones**. However, its chief role in pathologic processes is as a factor in the genesis of **atherosclerosis** of vital arteries, causing cerebrovascular, coronary, and peripheral vascular disease.

CHOLESTEROL IS DERIVED ABOUT EQUALLY FROM THE DIET & FROM BIOSYNTHESIS

A little more than half the cholesterol of the body arises by synthesis (about 700 mg/d), and the remainder is provided by the average diet. The liver and intestine account for approximately 10% each of total synthesis in humans. Virtually all tissues containing nucleated cells are capable of cholesterol synthesis, which occurs in the endoplasmic reticulum and the cytosol.

Acetyl-CoA Is the Source of All Carbon Atoms in Cholesterol

The biosynthesis of cholesterol may be divided into five steps: (1) Synthesis of mevalonate from acetyl-CoA (Figure 26–1).

(2) Formation of isoprenoid units from mevalonate by loss of CO_2 (Figure 26–2). (3) Condensation of six isoprenoid units form squalene (Figure 26–2). (4) Cyclization of squalene give rise to the parent steroid, lanosterol. (5) Formation of cholesterol from lanosterol (Figure 26–3).

Step 1—Biosynthesis of Mevalonate: HMG-CoA (3-hydroxy-3-methylglutaryl-CoA) is formed by the reactions used in mitochondria to synthesize ketone bodies (Figure 22–7). However, since cholesterol synthesis is extramitochondrial, the two pathways are distinct. Initially, two molecules of acetyl-CoA condense to form acetoacetyl-CoA catalyzed by cytosolic **thiolase.** Acetoacetyl-CoA condenses with a further molecule of acetyl-CoA catalyzed by **HMG-CoA synthase** to form HMG-CoA, which is reduced to **mevalonate** by NADPH catalyzed by **HMG-CoA reductase.** This is the principal regulatory step in the pathway of cholesterol synthesis and is the site of action of the most effective class of cholesterol-lowering drugs, the statins, which are HMG-CoA reductase inhibitors (Figure 26–1).

Step 2—Formation of Isoprenoid Units: Mevalonate is phosphorylated sequentially by ATP by three kinases, and after decarboxylation (Figure 26–2) the active isoprenoid unit, **isopentenyl diphosphate,** is formed.

Step 3—Six Isoprenoid Units Form Squalene: Isopentenyl diphosphate is isomerized by a shift of the double bond to form **dimethylallyl diphosphate,** then condensed with another molecule of isopentenyl diphosphate to form the ten-carbon intermediate **geranyl diphosphate** (Figure 26–2). A further condensation with isopentenyl diphosphate forms **farnesyl diphosphate.** Two molecules of farnesyl diphosphate condense at the diphosphate end to form **squalene.** Initially, inorganic pyrophosphate is eliminated, forming presqualene diphosphate, which is then reduced by NADPH with elimination of a further inorganic pyrophosphate molecule.

Step 4—Formation of Lanosterol: Squalene can fold into a structure that closely resembles the steroid nucleus (Figure 26–3). Before ring closure occurs, squalene is converted to squalene 2,3-epoxide by a mixed-function oxidase in the endoplasmic reticulum, **squalene epoxidase.** The methyl group on C_{14} is transferred to C_{13} and that on C_8–C_{14} as cyclization occurs, catalyzed by **oxidosqualene:lanosterol cyclase.**

FIGURE 26–1 Biosynthesis of mevalonate. HMG-CoA reductase is inhibited by statins. The open and solid circles indicate the fate of each of the carbons in the acetyl moiety of acetyl-CoA.

Step 5—Formation of Cholesterol: The formation of cholesterol from **lanosterol** takes place in the membranes of the endoplasmic reticulum and involves changes in the steroid nucleus and side chain (Figure 26–3). The methyl groups on C_{14} and C_4 are removed to form 14-desmethyl lanosterol and then zymosterol. The double bond at C_8–C_9 is subsequently moved to C_5–C_6 in two steps, forming **desmosterol**. Finally, the double bond of the side chain is reduced, producing cholesterol.

Farnesyl Diphosphate Gives Rise to Dolichol & Ubiquinone

The polyisoprenoids **dolichol** (Figure 15–20 & Chapter 47) and **ubiquinone** (Figure 13–5) are formed from farnesyl diphosphate by the further addition of up to 16 (dolichol) or 3–7 (ubiquinone) isopentenyl diphosphate residues (Figure 26–2). Some **GTP-binding proteins** in the cell membrane are prenylated with farnesyl or geranylgeranyl (20 carbon) residues. **Protein prenylation** is believed to facilitate the anchoring of proteins into lipoid membranes and may also be involved in protein-protein interactions and membrane-associated protein trafficking.

CHOLESTEROL SYNTHESIS IS CONTROLLED BY REGULATION OF HMG-CoA REDUCTASE

Regulation of cholesterol synthesis is exerted near the beginning of the pathway, at the HMG-CoA reductase step. The reduced synthesis of cholesterol in starving animals is accompanied by a decrease in the activity of the enzyme. However, it is only hepatic synthesis that is inhibited by dietary cholesterol. HMG-CoA reductase in liver is inhibited by mevalonate, the immediate product of the pathway, and by cholesterol, the main product. Cholesterol and metabolites repress transcription of the HMG-CoA reductase via activation of a **sterol regulatory element-binding protein (SREBP)** transcription factor. SREBPs are a family of proteins that regulate the transcription of a range of genes involved in the cellular uptake and metabolism of cholesterol and other lipids. A **diurnal variation** occurs in both cholesterol synthesis and reductase activity. In addition to these mechanisms regulating the rate of protein synthesis, the enzyme activity is also modulated more rapidly by post-translational modification (Figure 26–4). **Insulin or thyroid hormone** increases HMG-CoA reductase activity, whereas **glucagon or glucocorticoids** decrease it. Activity is reversibly modified by phosphorylation-dephosphorylation mechanisms, some of which may be cAMP-dependent and therefore immediately responsive to glucagon. Attempts to lower plasma cholesterol in humans by reducing the amount of cholesterol in the diet produce variable results. Generally, a decrease of 100 mg in dietary cholesterol causes a decrease of approximately 0.13 mmol/L of serum.

MANY FACTORS INFLUENCE THE CHOLESTEROL BALANCE IN TISSUES

In tissues, cholesterol balance is regulated as follows (Figure 26–5): Cell cholesterol increase is due to uptake of cholesterol-containing lipoproteins by receptors, eg, the LDL receptor or the scavenger receptor, uptake of free cholesterol from cholesterol-rich lipoproteins to the cell membrane, cholesterol synthesis, and hydrolysis of cholesteryl esters by the enzyme **cholesteryl ester hydrolase.** Decrease is due to efflux of cholesterol from the membrane to HDL via the ABCA1, ABCG1 or SR-B1 (Figure 25–5); esterification of cholesterol by **ACAT** (acyl-CoA:cholesterol acyltransferase); and utilization of cholesterol for synthesis of other steroids, such as hormones, or bile acids in the liver.

The LDL Receptor Is Highly Regulated

LDL (apo B-100, E) receptors occur on the cell surface in pits that are coated on the cytosolic side of the cell membrane with a protein called clathrin. The glycoprotein receptor spans the membrane, the B-100 binding region being at the exposed amino terminal end. After binding, LDL is taken up intact by endocytosis. The apoprotein and cholesteryl ester are then hydrolyzed in the lysosomes, and cholesterol is translocated

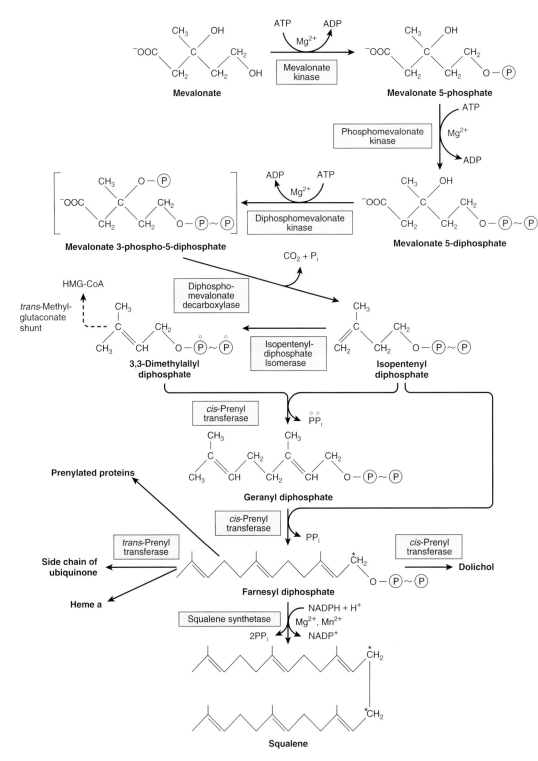

FIGURE 26–2 Biosynthesis of squalene, ubiquinone, dolichol, and other polyisoprene derivatives. (HMG, 3-hydroxy-3-methylglutaryl.) A farnesyl residue is present in heme a of cytochrome oxidase. The carbon marked with an asterisk becomes C_{11} or C_{12} in squalene. Squalene synthetase is a microsomal enzyme; all other enzymes indicated are soluble cytosolic proteins, and some are found in peroxisomes.

into the cell. The receptors are recycled to the cell surface. This influx of cholesterol inhibits the transcription of the genes encoding HMG-CoA synthase, HMG-CoA reductase and other enzymes involved in cholesterol synthesis, as well as the LDL receptor itself, via the SREBP pathway, and thus coordinately suppresses cholesterol synthesis and uptake. In addition, ACAT activity is stimulated, promoting cholesterol esterification. In this way, LDL receptor activity on the cell surface is regulated

FIGURE 26–3 Biosynthesis of cholesterol. The numbered positions are those of the steroid nucleus and the open and solid circles indicate the fate of each of the carbons in the acetyl moiety of acetyl-CoA. (*Refer to labeling of squalene in Figure 26–2.)

by the cholesterol requirement for membranes, steroid hormones, or bile acid synthesis (Figure 26–5).

CHOLESTEROL IS TRANSPORTED BETWEEN TISSUES IN PLASMA LIPOPROTEINS

Cholesterol is transported in plasma in lipoproteins, with the greater part in the form of cholesteryl ester (Figure 26–6), and

in humans the highest proportion is found in LDL. Dietary cholesterol equilibrates with plasma cholesterol in days and with tissue cholesterol in weeks. Cholesteryl ester in the diet is hydrolyzed to cholesterol, which is then absorbed by the intestine together with dietary unesterified cholesterol and other lipids. With cholesterol synthesized in the intestines, it is then incorporated into chylomicrons (Chapter 25). Of the cholesterol absorbed, 80–90% is esterified with long-chain fatty acids in the intestinal mucosa. Ninety-five percent of the chylomicron cholesterol is delivered to the liver in chylomicron

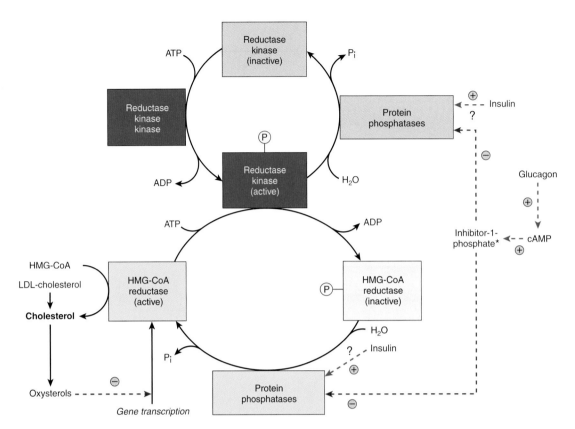

FIGURE 26–4 Possible mechanisms in the regulation of cholesterol synthesis by HMG-CoA reductase. Insulin has a dominant role compared with glucagon. (*See Figure 19–6.)

remnants, and most of the cholesterol secreted by the liver in VLDL is retained during the formation of IDL and ultimately LDL, which is taken up by the LDL receptor in liver and extrahepatic tissues (Chapter 25).

Plasma LCAT Is Responsible for Virtually All Plasma Cholesteryl Ester in Humans

Lecithin: cholesterol acyltransferase (LCAT) activity is associated with HDL containing apo A-I. As cholesterol in HDL becomes esterified, it creates a concentration gradient and draws in cholesterol from tissues and from other lipoproteins (Figures 26–5 & 26–6), thus enabling HDL to function in **reverse cholesterol transport** (Figure 25–5).

Cholesteryl Ester Transfer Protein Facilitates Transfer of Cholesteryl Ester from HDL to Other Lipoproteins

This protein, associated with HDL, is found in plasma of humans and many other species. It facilitates transfer of cholesteryl ester from HDL to VLDL, IDL, and LDL in exchange for triacylglycerol, relieving product inhibition of LCAT activity in HDL. Thus, in humans, much of the cholesteryl ester formed by LCAT finds its way to the liver via VLDL remnants (IDL) or LDL (Figure 26–6). The triacylglycerol-enriched HDL$_2$ delivers its cholesterol to the liver in the HDL cycle (Figure 25–5).

CHOLESTEROL IS EXCRETED FROM THE BODY IN THE BILE AS CHOLESTEROL OR BILE ACIDS (SALTS)

Cholesterol is excreted from the body via the bile either in the unesterified form or after conversion into bile acids in the liver. **Coprostanol** is the principal sterol in the feces; it is formed from cholesterol by the bacteria in the lower intestine.

Bile Acids Are Formed from Cholesterol

The **primary bile acids** are synthesized in the liver from cholesterol. These are **cholic acid** (found in the largest amount) and **chenodeoxycholic acid** (Figure 26–7). The 7α-hydroxylation of cholesterol is the first and principal regulatory step in the biosynthesis of bile acids and is catalyzed by **cholesterol 7α-hydroxylase,** a microsomal enzyme. A typical monooxygenase, it requires oxygen, NADPH, and cytochrome P450. Subsequent hydroxylation steps are also catalyzed by monooxygenases. The pathway of bile acid biosynthesis divides early into one sub-pathway leading to **cholyl-CoA,** characterized by an extra α-OH group on position 12, and another pathway leading to **chenodeoxycholyl-CoA** (Figure 26–7). A second pathway in mitochondria involving the 27-hydroxylation of cholesterol by **sterol 27-hydroxylase** as the first step is respon-

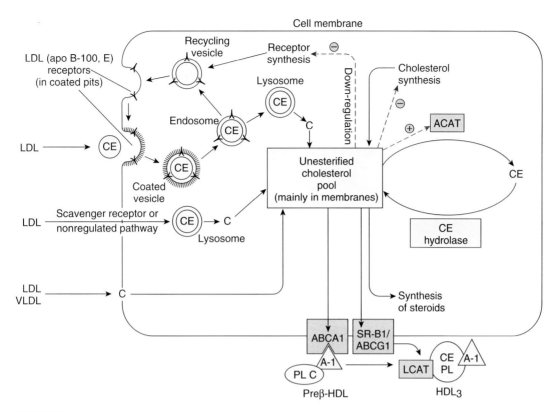

FIGURE 26–5 Factors affecting cholesterol balance at the cellular level. Reverse cholesterol transport may be mediated via the ABCA-1 transporter protein (with preβ-HDL as the exogenous acceptor) or the SR-B1 or ABCG-1 (with HDL$_3$ as the exogenous acceptor). (C, cholesterol; CE, cholesteryl ester; PL, phospholipid; ACAT, acyl-CoA:cholesterol acyltransferase; LCAT, lecithin:cholesterol acyltransferase; A-I, apolipoprotein A-I; LDL, low-density lipoprotein; VLDL, very low density lipoprotein.) LDL and HDL are not shown to scale.

sible for a significant proportion of the primary bile acids synthesized. The primary bile acids (Figure 26–7) enter the bile as glycine or taurine conjugates. Conjugation takes place in peroxisomes. In humans, the ratio of the glycine to the taurine conjugates is normally 3:1. In the alkaline bile (pH 7.6–8.4), the bile acids and their conjugates are assumed to be in a salt form—hence the term "bile salts."

A portion of the primary bile acids in the intestine is subjected to further changes by the activity of the intestinal bacteria. These include deconjugation and 7α-dehydroxylation, which produce the **secondary bile acids,** deoxycholic acid and lithocholic acid.

Most Bile Acids Return to the Liver in the Enterohepatic Circulation

Although products of fat digestion, including cholesterol, are absorbed in the first 100 cm of small intestine, the primary and secondary bile acids are absorbed almost exclusively in the ileum, and 98–99% is returned to the liver via the portal circulation. This is known as the **enterohepatic circulation** (Figure 26–6). However, lithocholic acid, because of its insolubility, is not reabsorbed to any significant extent. Only a small fraction

of the bile salts escapes absorption and is therefore eliminated in the feces. Nonetheless, this represents a major pathway for the elimination of cholesterol. Each day the small pool of bile acids (about 3–5 g) is cycled through the intestine six to ten times and an amount of bile acid equivalent to that lost in the feces is synthesized from cholesterol, so that a pool of bile acids of constant size is maintained. This is accomplished by a system of feedback controls.

Bile Acid Synthesis Is Regulated at the 7α-Hydroxylase Step

The principal rate-limiting step in the biosynthesis of bile acids is at the **cholesterol 7α-hydroxylase reaction** (Figure 26–7). The activity of the enzyme is feedback-regulated via the nuclear bile acid-binding receptor **farnesoid X receptor (FXR).** When the size of the bile acid pool in the enterohepatic circulation increases, FXR is activated and transcription of the cholesterol 7α-hydroxylase gene is suppressed. Chenodeoxycholic acid is particularly important in activating FXR. Cholesterol 7α-hydroxylase activity is also enhanced by cholesterol of endogenous and dietary origin and regulated by insulin, glucagon, glucocorticoids, and thyroid hormone.

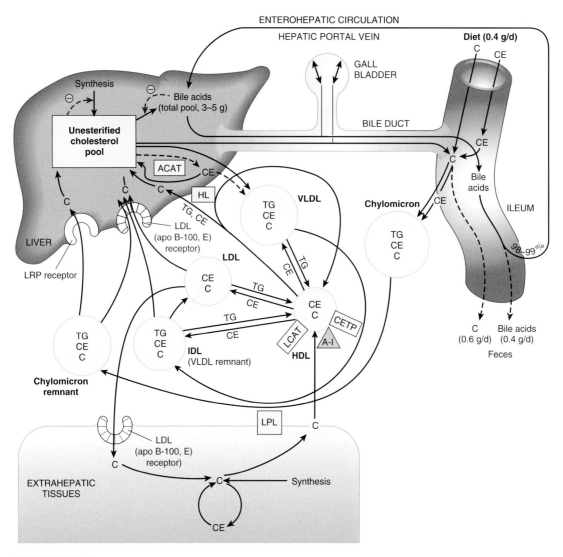

FIGURE 26–6 Transport of cholesterol between the tissues in humans. (C, unesterified cholesterol; CE, cholesteryl ester; TG, triacylglycerol; VLDL, very low density lipoprotein; IDL, intermediate-density lipoprotein; LDL, low-density lipoprotein; HDL, high-density lipoprotein; ACAT, acyl-CoA:cholesterol acyltransferase; LCAT, lecithin:cholesterol acyltransferase; A-I, apolipoprotein A-I; CETP, cholesteryl ester transfer protein; LPL, lipoprotein lipase; HL, hepatic lipase; LRP, LDL receptor-related protein.)

CLINICAL ASPECTS

The Serum Cholesterol Is Correlated with the Incidence of Atherosclerosis & Coronary Heart Disease

While elevated plasma cholesterol levels (>5.2 mmol/L) are believed to be a major factor in promoting atherosclerosis, it is now recognized that triacylglycerols are an independent risk factor. Atherosclerosis is characterized by the deposition of cholesterol and cholesteryl ester from the plasma lipoproteins into the artery wall. Diseases in which there is a prolonged elevation of levels of VLDL, IDL, chylomicron remnants, or LDL in the blood (eg, diabetes mellitus, lipid nephrosis, hypothyroidism, and other conditions of hyperlipidemia) are often accompanied by premature or more severe atherosclerosis. There is also an inverse

relationship between HDL (HDL$_2$) concentrations and coronary heart disease, making the **LDL:HDL cholesterol ratio a good predictive parameter.** This is consistent with the function of HDL in reverse cholesterol transport. Susceptibility to atherosclerosis varies widely among species, and humans are one of the few in which the disease can be induced by diets high in cholesterol.

Diet Can Play an Important Role in Reducing Serum Cholesterol

Hereditary factors play the greatest role in determining individual serum cholesterol concentrations; however, dietary and environmental factors also play a part, and the most beneficial of these is the substitution in the diet of **polyunsaturated and monounsaturated fatty acids** for saturated fatty acids. Plant oils such as corn oil and sunflower seed oil contain a high

FIGURE 26–7 Biosynthesis and degradation of bile acids. A second pathway in mitochondria involves hydroxylation of cholesterol by sterol 27-hydroxylase. *Catalyzed by microbial enzymes.

proportion of polyunsaturated fatty acids, while olive oil contains a high concentration of monounsaturated fatty acids. On the other hand, butter fat, beef fat, and palm oil contain a high proportion of saturated fatty acids. Sucrose and fructose have a greater effect in raising blood lipids, particularly triacylglycerols, than do other carbohydrates.

The reason for the cholesterol-lowering effect of polyunsaturated fatty acids is still not fully understood. It is clear, however, that one of the mechanisms involved is the upregulation of LDL receptors by poly- and monounsaturated as compared with saturated fatty acids, causing an increase in the catabolic rate of LDL, the main atherogenic lipoprotein. In addition, saturated fatty acids cause the formation of smaller VLDL particles that contain relatively more cholesterol, and they are utilized by extrahepatic tissues at a slower rate than are larger particles—tendencies that may be regarded as atherogenic.

Lifestyle Affects the Serum Cholesterol Level

Additional factors considered to play a part in coronary heart disease include **high blood pressure, smoking, male gender, obesity (particularly abdominal obesity), lack of exercise, and drinking soft as opposed to hard water**. Factors associated with elevation of plasma FFA followed by increased output of triacylglycerol and cholesterol into the circulation in VLDL include **emotional stress and coffee drinking**. Premenopausal women appear to be protected against many of these deleterious factors, and this is thought to be related to the beneficial effects of estrogen. There is an association between moderate alcohol consumption and a lower incidence of coronary heart disease. This may be due to elevation of HDL concentrations resulting from increased synthesis of apo A-I and changes in activity of cholesteryl ester transfer protein. It has been claimed that red wine is particularly beneficial, perhaps because of its content of antioxidants. Regular exercise lowers plasma LDL but raises HDL. Triacylglycerol concentrations are also reduced, due most likely to increased insulin sensitivity, which enhances the expression of lipoprotein lipase.

When Diet Changes Fail, Hypolipidemic Drugs Will Reduce Serum Cholesterol & Triacylglycerol

A family of drugs known as statins have proved highly efficacious in lowering plasma cholesterol and preventing heart disease. Statins act by inhibiting HMG-CoA reductase and up-regulating LDL receptor activity. Examples currently in use include **atorvastatin, simvastatin, fluvastatin,** and **pravastatin. Ezetimibe** reduces blood cholesterol levels by inhibit-

ing the absorption of cholesterol by the intestine by blocking uptake via the **Neimann-Pick C-like 1 protein**. Other drugs used include fibrates such as **clofibrate** and **gemfibrozil and nicotinic acid,** which act mainly to lower plasma triacylglycerols by decreasing the secretion of triacylglycerol and cholesterol-containing VLDL by the liver.

Primary Disorders of the Plasma Lipoproteins (Dyslipoproteinemias) Are Inherited

Inherited defects in lipoprotein metabolism lead to the primary condition of either **hypo-** or **hyperlipoproteinemia** (Table 26–1). In addition, diseases such as diabetes mellitus, hypothyroidism, kidney disease (nephrotic syndrome), and atherosclerosis are associated with secondary abnormal lipoprotein patterns that are very similar to one or another of the

TABLE 26–1   **Primary Disorders of Plasma Lipoproteins (Dyslipoproteinemias)**

Name	Defect	Remarks
Hypolipoproteinemias Abetalipoproteinemia	No chylomicrons, VLDL, or LDL are formed because of defect in the loading of apo B with lipid.	Rare; blood acylglycerols low; intestine and liver accumulate acylglycerols. Intestinal malabsorption. Early death avoidable by administration of large doses of fat-soluble vitamins, particularly vitamin E.
Familial alpha-lipoprotein deficiency Tangier disease Fish-eye disease Apo-A-I deficiencies	All have low or near absence of HDL.	Tendency toward hypertriacylglycerolemia as a result of absence of apo C-II, causing inactive LPL. Low LDL levels. Atherosclerosis in the elderly.
Hyperlipoproteinemias Familial lipoprotein lipase deficiency (type I)	Hypertriacylglycerolemia due to deficiency of LPL, abnormal LPL, or apo C-II deficiency causing inactive LPL.	Slow clearance of chylomicrons and VLDL. Low levels of LDL and HDL. No increased risk of coronary disease.
Familial hypercholesterolemia (type IIa)	Defective LDL receptors or mutation in ligand region of apo B-100.	Elevated LDL levels and hypercholesterolemia, resulting in atherosclerosis and coronary disease.
Familial type III hyperlipoproteinemia (broad beta disease, remnant removal disease, familial dysbetalipoproteinemia	Deficiency in remnant clearance by the liver is due to abnormality in apo E. Patients lack isoforms E3 and E4 and have only E2, which does not react with the E receptor.[1]	Increase in chylomicron and VLDL remnants of density < 1.019 (β-VLDL). Causes hypercholesterolemia, xanthomas, and atherosclerosis.
Familial hypertriacylglycerolemia (type IV)	Overproduction of VLDL often associated with glucose intolerance and hyperinsulinemia.	Cholesterol levels rise with the VLDL concentration. LDL and HDL tend to be subnormal. This type of pattern is commonly associated with coronary heart disease, type II diabetes mellitus, obesity, alcoholism, and administration of progestational hormones.
Familial hyperalphalipoproteinemia	Increased concentrations of HDL.	A rare condition apparently beneficial to health and longevity.
Hepatic lipase deficiency	Deficiency of the enzyme leads to accumulation of large triacylglycerolrich HDL and VLDL remnants.	Patients have xanthomas and coronary heart disease.
Familial lecithin:cholesterol acyltransferase (LCAT) deficiency	Absence of LCAT leads to block in reverse cholesterol transport. HDL remains as nascent disks incapable of taking up and esterifying cholesterol.	Plasma concentrations of cholesteryl esters and lysolecithin are low. Present is an abnormal LDL fraction, lipoprotein X, found also in patients with cholestasis. VLDL is abnormal (β-VLDL).
Familial lipoprotein(a) excess	Lp(a) consists of 1 mol of LDL attached to 1 mol of apo(a). Apo(a) shows structural homologies to plasminogen.	Premature coronary heart disease due to atherosclerosis, plus thrombosis due to inhibition of fibrinolysis.

[1]There is an association between patients possessing the apo E4 allele and the incidence of Alzheimer's disease. Apparently, apo E4 binds more avidly to β-amyloid found in neuritic plaques.

primary inherited conditions. Virtually all of the primary conditions are due to a defect at a stage in lipoprotein formation, transport, or destruction (see Figures 25–4, 26–5, & 26–6). Not all of the abnormalities are harmful.

SUMMARY

- Cholesterol is the precursor of all other steroids in the body, eg, corticosteroids, sex hormones, bile acids, and vitamin D. It also plays an important structural role in membranes and in the outer layer of lipoproteins.

- Cholesterol is synthesized in the body entirely from acetyl-CoA. Three molecules of acetyl-CoA form mevalonate via the important regulatory reaction for the pathway, catalyzed by HMG-CoA reductase. Next, a five-carbon isoprenoid unit is formed, and six of these condense to form squalene. Squalene undergoes cyclization to form the parent steroid lanosterol, which, after the loss of three methyl groups, forms cholesterol.

- Cholesterol synthesis in the liver is regulated partly by cholesterol in the diet. In tissues, cholesterol balance is maintained between the factors causing gain of cholesterol (eg, synthesis, uptake via the LDL or scavenger receptors) and the factors causing loss of cholesterol (eg, steroid synthesis, cholesteryl ester formation, excretion). The activity of the LDL receptor is modulated by cellular cholesterol levels to achieve this balance. In reverse cholesterol transport, HDL takes up cholesterol from the tissues and LCAT esterifies it and deposits it in the core of the particles. The cholesteryl ester in HDL is taken up by the liver, either directly or after transfer to VLDL, IDL, or LDL via the cholesteryl ester transfer protein.

- Excess cholesterol is excreted from the liver in the bile as cholesterol or bile salts. A large proportion of bile salts is absorbed into the portal circulation and returned to the liver as part of the enterohepatic circulation.

- Elevated levels of cholesterol present in VLDL, IDL, or LDL are associated with atherosclerosis, whereas high levels of HDL have a protective effect.

- Inherited defects in lipoprotein metabolism lead to a primary condition of hypo- or hyperlipoproteinemia. Conditions such as diabetes mellitus, hypothyroidism, kidney disease, and atherosclerosis exhibit secondary abnormal lipoprotein patterns that resemble certain primary conditions.

REFERENCES

Chiang JL: Regulation of bile acid synthesis: pathways, nuclear receptors and mechanisms. J Hepatol 2004;40:539.

Denke MA: Dietary fats, fatty acids and their effects on lipoproteins. Curr Atheroscler Rep 2006;8:466.

Jiang XC, Zhou HW: Plasma lipid transfer proteins. Curr Opin Lipidol 2006;17:302.

Ness GC, Chambers CM: Feedback and hormonal regulation of hepatic 3-hydroxy-3-methylglutaryl coenzyme A reductase: the concept of cholesterol buffering capacity. Proc Soc Exp Biol Med 2000;224:8.

Parks DJ et al: Bile acids: natural ligands for a nuclear orphan receptor. Science 1999;284:1365.

Russell DW: Cholesterol biosynthesis and metabolism. Cardiovascular Drugs Therap 1992;6:103.

Spady DK, Woollett LA, Dietschy JM: Regulation of plasma LDL cholesterol levels by dietary cholesterol and fatty acids. Annu Rev Nutr 1993;13:355.

Various authors: *Biochemistry of Lipids, Lipoproteins and Membranes,* 5th ed. Vance DE, Vance JE (editors). Elsevier, 2008.

Zhang FL, Casey PJ: Protein prenylation: molecular mechanisms and functional consequences. Annu Rev Biochem 1996;65:241.

Biosynthesis of the Nutritionally Nonessential Amino Acids

Victor W. Rodwell, PhD

BIOMEDICAL IMPORTANCE

Medical implications of the material in this chapter relate to the amino acid deficiency states that can result if nutritionally essential amino acids are absent from the diet, or are present in inadequate amounts. Amino acid deficiency states, while comparatively rare in the Western world, are endemic in certain regions of West Africa where diets rely heavily on grains that are poor sources of tryptophan and lysine. These nutritional disorders include **kwashiorkor**, which results when a child is weaned onto a starchy diet poor in protein; and **marasmus**, in which both caloric intake and specific amino acids are deficient. Patients with short bowel syndrome unable to absorb sufficient quantities of calories and nutrients suffer from significant nutritional and metabolic abnormalities. Both the nutritional disorder **scurvy,** a dietary deficiency of vitamin C, and specific genetic disorders are associated with an impaired ability of connective tissue to form hydroxyproline and hydroxylysine. The resulting conformational instability of collagen results in bleeding gums, swelling joints, poor wound healing, and ultimately in death. **Menkes' syndrome**, characterized by kinky hair and growth retardation, results from a dietary deficiency of copper, which is an essential cofactor for lysyl oxidase, an enzyme that functions in formation of the covalent cross-links that strengthen collagen fibers. Genetic disorders of collagen biosynthesis include several forms of **osteogenesis imperfecta**, characterized by fragile bones, and **Ehlers-Danlos syndrome**, a group of connective tissue disorders that result in mobile joints and skin abnormalities due to defects in the genes that encode enzymes that include lysyl hydroxylase (see Chapter 47).

NUTRITIONALLY ESSENTIAL & NUTRITIONALLY NONESSENTIAL AMINO ACIDS

As applied to amino acids, the terms "essential" and "nonessential" are misleading since all 20 common amino acids are essential to ensure health. Of these 20 amino acids, 8 *must* be present in the human diet, and thus are best termed "*nutritionally essential.*" The other 12 amino acids are "*nutritionally nonessential*" since they need not be present in the diet (Table 27–1). The distinction between these two classes of amino acids was established in the 1930s by feeding human subjects purified amino acids in place of protein. Subsequent biochemical investigations revealed the reactions and intermediates involved in the biosynthesis of all 20 amino acids.

Lengthy Metabolic Pathways Form the Nutritionally Essential Amino Acids

The existence of nutritional requirements suggests that dependence on an external supply of a given nutrient can be of greater survival value than the ability to biosynthesize it. Why? If a specific nutrient is present in the food, an organism that can synthesize it will transfer to its progeny genetic information of *negative* survival value. The survival value is negative rather than nil because ATP and nutrients are required to synthesize "unnecessary" DNA—even if specific encoded genes are no longer expressed. The number of enzymes required by prokaryotic cells to synthesize the nutritionally essential amino acids is large relative to the number of enzymes required to synthesize the nutritionally nonessential amino acids (Table 27–2). This suggests a survival advantage in retaining

TABLE 27–1   Amino Acid Requirements of Humans

Nutritionally Essential	Nutritionally Nonessential
Arginine[1]	Alanine
Histidine	Asparagine
Isoleucine	Aspartate
Leucine	Cysteine
Lysine	Glutamate
Methionine	Glutamine
Phenylalanine	Glycine
Threonine	Hydroxyproline[2]
Tryptophan	Hydroxylysine[2]
Valine	Proline
	Serine
	Tyrosine

[1]Nutritionally "semiessential." Synthesized at rates inadequate to support growth of children.
[2]Not necessary for protein synthesis, but is formed during post-translational processing of collagen.

TABLE 27–2   Enzymes Required for the Synthesis of Amino Acids from Amphibolic Intermediates

Number of Enzymes Required to Synthesize			
Nutritionally Essential		Nutritionally Nonessential	
Arg[1]	7	Ala	1
His	6	Asp	1
Thr	6	Asn[2]	1
Met	5 (4 shared)	Glu	1
Lys	8	Gln[1]	1
Ile	8 (6 shared)	Hyl[3]	1
Val	1 (7 shared)	Hyp[4]	1
Leu	3 (7 shared)	Pro[1]	3
Phe	10	Ser	3
Trp	5 (8 shared)	Gly[5]	1
	59	Cys[6]	2
		Tyr[7]	1
			17

[1]From Glu. [2]From Asp. [3]From Lys. [4]From Pro. [5]From Ser. [6]From Ser plus S[2−]. [7]From Phe.

the ability to manufacture "easy" amino acids while losing the ability to make "difficult" amino acids. This chapter addresses the reactions and intermediates involved in the biosynthesis by human tissues of the 12 nutritionally *nonessential* amino acids and selected nutritional and metabolic disorders associated with their metabolism.

Glutamate Dehydrogenase, Glutamine Synthetase, & Aminotransferases Play Central Roles in Amino Acid Biosynthesis

The combined action of the enzymes glutamate dehydrogenase, glutamine synthetase, and the aminotransferases converts ammonium ion into the α-amino nitrogen of amino acids. This is illustrated below by the short pathways that convert amphibolic intermediates to the nutritionally nonessential amino acids.

Glutamate

Reductive amidation of α-ketoglutarate is catalyzed by glutamate dehydrogenase (Figure 27–1). This reaction constitutes the first step in biosynthesis of the "glutamate family" of amino acids.

Glutamine

The amidation of glutamate to glutamine catalyzed by glutamine synthetase (Figure 27–2) involves the intermediate formation of γ-glutamyl phosphate. Following the ordered binding of glutamate and ATP, glutamate attacks the γ-phosphorus of ATP, forming γ-glutamyl phosphate and ADP. NH_4^+ then binds, and as NH_3, attacks γ-glutamyl phosphate to form a tetrahedral intermediate. Release of P_i and of a proton from the γ-amino group of the tetrahedral intermediate then facilitates release of the product, glutamine.

Alanine & Aspartate

Transamination of pyruvate forms alanine (Figure 27–3). Similarly, transamination of oxaloacetate forms aspartate.

Asparagine

The conversion of aspartate to asparagine, catalyzed by asparagine synthetase (Figure 27–4), resembles the glutamine

FIGURE 27–1 The glutamate dehydrogenase reaction.

FIGURE 27–2 The glutamine synthetase reaction.

FIGURE 27–3 Formation of alanine by transamination of pyruvate. The amino donor may be glutamate or aspartate. The other product thus is α-ketoglutarate or oxaloacetate.

FIGURE 27–4 The asparagine synthetase reaction. Note similarities to and differences from the glutamine synthetase reaction (Figure 27–2).

synthetase reaction (Figure 27–2), but glutamine rather than ammonium ion, provides the nitrogen. Bacterial asparagine synthetases can, however, also use ammonium ion. The reaction involves the intermediate formation of aspartyl phosphate. The coupled hydrolysis of PP_i to P_i by pyrophosphatase ensures that the reaction is strongly favored.

Serine

Oxidation of the α-hydroxyl group of the glycolytic intermediate 3-phosphoglycerate by 3-phosphoglycerate dehydrogenase converts it to 3-phosphohydroxypyruvate. Transamination and subsequent dephosphorylation then forms serine (Figure 27–5).

FIGURE 27–5 Serine biosynthesis. (α-AA, α-amino acids; α-KA, α-keto acids.)

Glycine

Glycine aminotransferases can catalyze the synthesis of glycine from glyoxylate and glutamate or alanine. Unlike most aminotransferase reactions, these strongly favor glycine synthesis. Additional important mammalian routes for glycine formation are from choline (Figure 27–6) and from serine (Figure 27–7).

Proline

Biosynthesis of proline from glutamate employs reactions similar to those of proline catabolism, but in which glutamate γ-phosphate is an intermediate (Figure 27–8).

Cysteine

While not nutritionally essential, cysteine is formed from methionine, which is nutritionally essential. Following conversion of methionine to homocysteine (see Figure 29–19), homocysteine and serine form cystathionine, whose hydrolysis forms cysteine and homoserine (Figure 27–9).

FIGURE 27–6 Formation of glycine from choline. The enzymes that catalyze the reactions shown are choline dehydrogenase, betaine dehydrogenase, betaine-homocysteine N-methyltransferase, sarcosine demethylase, and sarcosine oxidase, respectively.

FIGURE 27–7 The serine hydroxymethyltransferase reaction. The reaction is freely reversible. (H_4 folate, tetrahydrofolate.)

FIGURE 27–8 Biosynthesis of proline from glutamate. The catalysts for these reactions are glutamate 5-kinase, glutamate semialdehyde dehydrogenase, noncatalyzed ring closure, and pyrolline 5-carboxylate reductase.

FIGURE 27–9 Conversion of homocysteine and serine to homoserine and cysteine. The sulfur of cysteine derives from methionine and the carbon skeleton from serine.

FIGURE 27–10 The phenylalanine hydroxylase reaction. Two distinct enzymatic activities are involved. Activity II catalyzes reduction of dihydrobiopterin by NADPH, and activity I the reduction of O_2 to H_2O and of phenylalanine to tyrosine. This reaction is associated with several defects of phenylalanine metabolism discussed in Chapter 29.

Tyrosine

Phenylalanine hydroxylase converts phenylalanine to tyrosine (Figure 27–10). If the diet contains adequate quantities of the nutritionally essential amino acid phenylalanine, tyrosine is nutritionally nonessential. But since the phenylalanine hydroxylase reaction is irreversible, dietary tyrosine cannot replace phenylalanine. Catalysis by this mixed-function oxygenase incorporates one atom of O_2 into the *para* position of phenylalanine and reduces the other atom to water. Reducing power, provided as tetrahydrobiopterin (Figure 27–10), derives ultimately from NADPH.

Hydroxyproline & Hydroxylysine

Hydroxyproline and hydroxylysine occur principally in collagen. Since there is no tRNA for either hydroxylated amino acid, neither dietary hydroxyproline nor hydroxylysine is incorporated during protein synthesis. Peptidyl hydroxyproline and hydroxylysine arise from proline and lysine, but only af-

FIGURE 27–11 The prolyl hydroxylase reaction. The substrate is a proline-rich peptide. During the course of the reaction, molecular oxygen is incorporated into both succinate and proline. Lysyl hydroxylase catalyzes an analogous reaction.

$$H-Se-CH_2-\overset{\displaystyle H}{\underset{\displaystyle NH_3^+}{C}}-COO^-$$

$$Se + ATP \longrightarrow AMP + P_i + H-Se-\overset{\displaystyle O}{\underset{\displaystyle O^-}{P}}-O^-$$

FIGURE 27–12 Selenocysteine (**top**) and the reaction catalyzed by selenophosphate synthetase (**bottom**).

ter these amino acids have been incorporated into peptides. Hydroxylation of peptidyl prolyl and lysyl residues, catalyzed by **prolyl hydroxylase** and **lysyl hydroxylase** of skin, skeletal muscle, and granulating wounds requires, in addition to the substrate, molecular O_2, ascorbate, Fe^{2+}, and α-ketoglutarate (Figure 27–11). For every mole of proline or lysine hydroxylated, one mole of α-ketoglutarate is decarboxylated to succinate. The hydroxylases are mixed-function oxygenases. One atom of O_2 is incorporated into proline or lysine, the other into succinate (Figure 27–11). A deficiency of the vitamin C required for these hydroxylases results in **scurvy**.

Valine, Leucine, & Isoleucine

While leucine, valine, and isoleucine are all nutritionally essential amino acids, tissue aminotransferases reversibly interconvert all three amino acids and their corresponding α-keto acids. These α-keto acids thus can replace their amino acids in the diet.

Selenocysteine, the 21st Amino Acid

While its occurrence in proteins is uncommon, selenocysteine(Figure 27–12) is present at the active site of several human enzymes that catalyze redox reactions. Examples include thioredoxin reductase, glutathione peroxidase, and the deiodinase that converts thyroxine to triiodothyronine. Where present, selenocysteine participates in the catalytic mechanism of these enzymes. Significantly, replacement of selenocysteine by cysteine can significantly decrease catalytic activity. Impairments in human selenoproteins have been implicated in tumorigenesis and atherosclerosis, and are associated with selenium deficiency cardiomyopathy (Keshan disease).

Biosynthesis of selenocysteine requires cysteine, selenate (SeO_4^{2-}), ATP, a specific tRNA, and several enzymes. Serine provides the carbon skeleton of selenocysteine. Selenophosphate, formed from ATP and selenate (Figure 27–12),

serves as the selenium donor. Unlike hydroxyproline or hydroxylysine, selenocysteine arises co-translationally during its incorporation into peptides. The UGA anticodon of the unusual tRNA designated tRNA^{Sec} normally signals STOP. The ability of the protein synthetic apparatus to identify a selenocysteine-specific UGA codon involves the selenocysteine insertion element, a stem-loop structure in the untranslated region of the mRNA. Selenocysteine-tRNA^{Sec} is first charged with serine by the ligase that charges tRNA^{Ser}. Subsequent replacement of the serine oxygen by selenium involves selenophosphate formed by selenophosphate synthase (Figure 27–12). Successive enzyme-catalyzed reactions convert cysteyl-tRNA^{Sec} to aminoacryl-tRNA^{Sec} and then to selenocysteyl-tRNA^{Sec}. In the presence of a specific elongation factor that recognizes selenocysteyl-tRNA^{Sec}, selenocysteine can then be incorporated into proteins.

SUMMARY

- All vertebrates can form certain amino acids from amphibolic intermediates or from other dietary amino acids. The intermediates and the amino acids to which they give rise are α-ketoglutarate (Glu, Gln, Pro, Hyp), oxaloacetate (Asp, Asn), and 3-phosphoglycerate (Ser, Gly).

- Cysteine, tyrosine, and hydroxylysine are formed from nutritionally essential amino acids. Serine provides the carbon skeleton and homocysteine the sulfur for cysteine biosynthesis. Phenylalanine hydroxylase converts phenylalanine to tyrosine in an irreversible reaction.

- Neither dietary hydroxyproline nor hydroxylysine is incorporated into proteins because no codon or tRNA dictates their insertion into peptides.

- Peptidyl hydroxyproline and hydroxylysine are formed by hydroxylation of peptidyl proline or lysine in reactions catalyzed by mixed-function oxidases that require vitamin C as cofactor. The nutritional disease scurvy reflects impaired hydroxylation due to a deficiency of vitamin C.

- Selenocysteine, an essential active site residue in several mammalian enzymes, arises by co-translational insertion from a previously modified tRNA.

REFERENCES

Beckett GJ, Arthur JR: Selenium and endocrine systems. J Endocrinol 2005;184:455.

Brown KM, Arthur JR: Selenium, selenoproteins and human health: a review. Public Health Nutr 2001;4:593.

Kilberg MS: Asparagine synthetase chemotherapy. Annu Rev Biochem 2006;75:629.

Kohrl J et al: Selenium in biology: facts and medical perspectives. Biol Chem 2000;381:849.

Lobanov AV, Hatfield DL, Gladyshev VN: Reduced reliance on the trace element selenium during evolution of mammals. Genome Biol 2008;9:R62.

Scriver CR et al (editors): *The Metabolic and Molecular Bases of Inherited Disease,* 8th ed. McGraw-Hill, 2001.

Stickel F et al: Role of nutrition in liver transplantation for end-stage chronic liver disease. Nutr Rev 2008;66:47.

Catabolism of Proteins & of Amino Acid Nitrogen

Victor W. Rodwell, PhD

BIOMEDICAL IMPORTANCE

This chapter describes how the nitrogen of amino acids is converted to urea and the rare metabolic disorders that accompany defects in urea biosynthesis. In normal adults, nitrogen intake matches nitrogen excreted. Positive nitrogen balance, an excess of ingested over excreted nitrogen, accompanies growth and pregnancy. Negative nitrogen balance, where output exceeds intake, may follow surgery, advanced cancer, kwashiorkor, and marasmus. Ammonia, derived mainly from the α-amino nitrogen of amino acids, is highly toxic. Tissues therefore convert ammonia to the amide nitrogen of the nontoxic amino acid glutamine. Subsequent deamination of glutamine in the liver releases ammonia, which is then converted to urea, which is not toxic. If liver function is compromised, as in cirrhosis or hepatitis, elevated blood ammonia levels generate clinical signs and symptoms. Each enzyme of the urea cycle provides examples of metabolic defects and their physiologic consequences, and the cycle as a whole serves as a molecular model for the study of human metabolic defects.

PROTEIN TURNOVER OCCURS IN ALL FORMS OF LIFE

The continuous degradation and synthesis of cellular proteins occur in all forms of life. Each day, humans turn over 1–2% of their total body protein, principally muscle protein. High rates of protein degradation occur in tissues that are undergoing structural rearrangement, for example, uterine tissue during pregnancy, skeletal muscle in starvation, and tadpole tail tissue during metamorphosis. Of the liberated amino acids, approximately 75% are reutilized. Since excess free amino acids are not stored, those not immediately incorporated into new protein are rapidly degraded. The major portion of the carbon skeletons of the amino acids is converted to amphibolic intermediates, while the amino nitrogen is converted to urea and excreted in the urine.

PROTEASES & PEPTIDASES DEGRADE PROTEINS TO AMINO ACIDS

The relative susceptibility of a protein to degradation is expressed as its half-life ($t_{1/2}$), the time required to lower its concentration to half the initial value. Half-lives of liver proteins range from under 30 min to over 150 h. Typical "housekeeping" enzymes have $t_{1/2}$ values of over 100 h. By contrast, many key regulatory enzymes have $t_{1/2}$ values as low as 0.5–2 h. PEST sequences, regions rich in proline (P), glutamate (E), serine (S), and threonine (T), target some proteins for rapid degradation. Intracellular proteases hydrolyze internal peptide bonds. The resulting peptides are then degraded to amino acids by endopeptidases that cleave internal peptide bonds, and by aminopeptidases and carboxypeptidases that remove amino acids sequentially from the amino and carboxyl terminals, respectively. Degradation of circulating peptides such as hormones follows loss of a sialic acid moiety from the nonreducing ends of their oligosaccharide chains. Asialoglycoproteins are internalized by liver cell asialoglycoprotein receptors and degraded by lysosomal proteases termed cathepsins.

Extracellular, membrane-associated, and long-lived intracellular proteins are degraded in lysosomes by ATP-independent processes. By contrast, degradation of regulatory proteins with short half-lives and of abnormal or misfolded proteins occurs in the cytosol, and requires ATP and ubiquitin. **Ubiquitin**, so named because it is present in all eukaryotic cells, is a small (8.5 kDa, 76 residues) polypeptide that targets many intracellular proteins for degradation. The primary structure of ubiquitin is highly conserved. Only 3 of 76 residues differ between yeast and human ubiquitin. Ubiquitin molecules are attached by **non-α-peptide bonds** formed between the carboxyl terminal of ubiquitin and the α-amino groups of lysyl residues in the target protein (Figure 28–1). The residue present at its amino terminal affects whether a protein is ubiquitinated. Amino terminal Met or Ser retards, whereas Asp or Arg accelerates ubiquitination. Attachment of a single ubiquitin molecule to transmembrane proteins alters their subcellular localization and targets them for degradation. Soluble proteins undergo polyubiquitination, further ligase-catalyzed attachment of four or more additional ubiquitin molecules. Subsequent degradation of ubiquitin-tagged proteins takes place in the **proteasome**, a macromolecule with multiple different subunits that also is ubiquitous in eukaryotic cells. For the discovery of ubiquitin-mediated protein degradation, Aaron Ciechanover and Avram Hershko of Israel and Irwin Rose of the United States were awarded the 2004 Nobel Prize

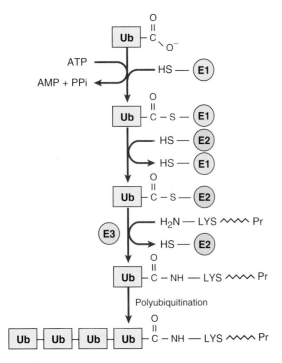

FIGURE 28–1 Reactions involved in the attachment of ubiquitin (Ub) to proteins. Three enzymes are involved. E1 is an activating enzyme, E2 is a ligase, and E3 is a transferase. While depicted as single entities, there are several types of E1, and over 500 types of E2. The terminal COOH of ubiquitin first forms a thioester. The coupled hydrolysis of PP$_i$ by pyrophosphatase ensures that the reaction will proceed readily. A thioester exchange reaction now transfers activated ubiquitin to E2. E3 then catalyzes the transfer of ubiquitin to the ε-amino group of a lysyl residue of the target protein. Additional rounds of ubiquitination result in subsequent polyubiquitination.

in Chemistry. Metabolic diseases associated with defects of ubiquitination include the Angelman syndrome and the von Hippel-Lindau syndrome in which there is a defect in the ubiquitin E3 ligase. For additional aspects of protein degradation and of ubiquitination, including its role in the cell cycle, see Chapter 4.

INTERORGAN EXCHANGE MAINTAINS CIRCULATING LEVELS OF AMINO ACIDS

The maintenance of steady-state concentrations of circulating plasma amino acids between meals depends on the net balance between release from endogenous protein stores and utilization by various tissues. Muscle generates over half of the total body pool of free amino acids, and liver is the site of the urea cycle enzymes necessary for disposal of excess nitrogen. Muscle and liver thus play major roles in maintaining circulating amino acid levels.

Figure 28–2 summarizes the post absorptive state. Free amino acids, particularly alanine and glutamine, are released

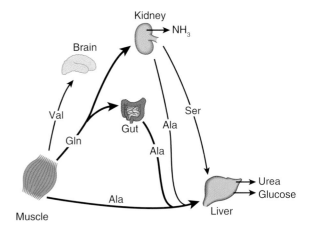

FIGURE 28–2 Interorgan amino acid exchange in normal post absorptive humans. The key role of alanine in amino acid output from muscle and gut and uptake by the liver is shown.

from muscle into the circulation. Alanine, which appears to be the vehicle of nitrogen transport in the plasma, is extracted primarily by the liver. Glutamine is extracted by the gut and the kidney, both of which convert a significant portion to alanine. Glutamine also serves as a source of ammonia for excretion by the kidney. The kidney provides a major source of serine for uptake by peripheral tissues, including liver and muscle. Branched-chain amino acids, particularly valine, are released by muscle and taken up predominantly by the brain.

Alanine is a key **gluconeogenic amino acid** (Figure 28–3). The rate of hepatic gluconeogenesis from alanine is far higher than from all other amino acids. The capacity of the liver for gluconeogenesis from alanine does not reach saturation until the alanine concentration reaches 20–30 times its physiologic level. Following a protein-rich meal, the splanchnic tissues release amino acids (Figure 28–4) while the peripheral muscles extract amino acids, in both instances predominantly branched-chain amino acids. Branched-chain amino acids thus serve a special role in nitrogen metabolism, both in the fasting state, when they provide the brain with an energy source, and after feeding, when they are extracted predominantly by muscles, having been spared by the liver.

ANIMALS CONVERT α-AMINO NITROGEN TO VARIED END PRODUCTS

Different animals excrete excess nitrogen as ammonia, uric acid, or urea. The aqueous environment of teleostean fish, which are **ammonotelic** (excrete ammonia), compels them to excrete water continuously to facilitate excretion of the highly toxic molecule ammonia. Birds, which must conserve water and maintain low weight, are **uricotelic** and excrete uric acid as semisolid guano. Many land animals, including humans, are **ureotelic** and excrete nontoxic, water-soluble urea. High blood

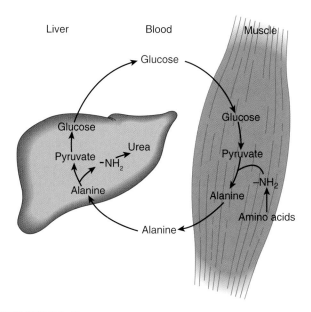

FIGURE 28–3 The glucose-alanine cycle. Alanine is synthesized in muscle by transamination of glucose-derived pyruvate, released into the bloodstream, and taken up by the liver. In the liver, the carbon skeleton of alanine is reconverted to glucose and released into the bloodstream, where it is available for uptake by muscle and resynthesis of alanine.

urea levels in renal disease are a consequence, not a cause, of impaired renal function.

BIOSYNTHESIS OF UREA

Urea biosynthesis occurs in four stages: (1) transamination, (2) oxidative deamination of glutamate, (3) ammonia transport, and (4) reactions of the urea cycle (Figure 28–5). The use of complementary DNA probes has shown that the expression in liver of the RNAs for all the enzymes of the urea cycle increases several fold in starvation.

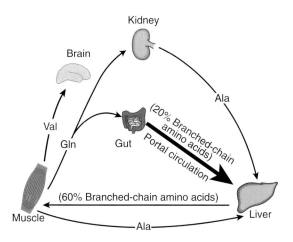

FIGURE 28–4 Summary of amino acid exchange between organs immediately after feeding.

Transamination Transfers α-Amino Nitrogen to α-Ketoglutarate, Forming Glutamate

Transamination reactions, which interconvert pairs of α-amino acids and α-keto acids (Figure 28–6), are readily reversible, and also function in amino acid biosynthesis. Of the protein amino acids, all except lysine, threonine, proline, and hydroxyproline participate in transamination. Transamination occurs via a "ping-pong" mechanism (see Figure 7–4). The coenzyme pyridoxal phosphate (PLP) is present at the catalytic site of all aminotransferases (and of many other enzymes that act on amino acids). A derivative of vitamin B_6, PLP forms an enzyme-bound Schiff base intermediate that can rearrange in various ways. During transamination, enzyme-bound PLP serves as a carrier of amino groups. Rearrangement forms an α-keto acid and enzyme-bound pyridoxamine phosphate, which then forms a Schiff base with a second keto acid. Following removal of its α-amino nitrogen by transamination, the remaining carbon "skeleton" of an amino acid is degraded by pathways discussed in Chapter 29.

Alanine-pyruvate aminotransferase (alanine aminotransferase) and glutamate-α-ketoglutarate aminotransferase (glutamate aminotransferase) catalyze the transfer of amino groups to pyruvate (forming alanine) or to α-ketoglutarate (forming glutamate) (Figure 28–7). Each aminotransferase

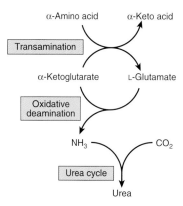

FIGURE 28–5 Overall flow of nitrogen in amino acid catabolism.

FIGURE 28–6 Transamination. The reaction is freely reversible with an equilibrium constant close to unity.

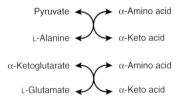

FIGURE 28–7 Alanine aminotransferase (**top**) and glutamate aminotransferase (**bottom**).

is specific for one pair of substrates, but nonspecific for the other pair. Since alanine is also a substrate for glutamate aminotransferase, all the amino nitrogen from amino acids that undergo transamination can be concentrated in glutamate. This is important because L-glutamate is the only amino acid that undergoes oxidative deamination at an appreciable rate in mammalian tissues. The formation of ammonia from α-amino groups thus occurs mainly via the α-amino nitrogen of L-glutamate. Transamination is not restricted to α-amino groups. The δ-amino group of ornithine (but not the ε-amino group of lysine) readily undergoes transamination. Elevated serum levels of aminotransferases characterize certain diseases (see Table 7–2).

L-GLUTAMATE DEHYDROGENASE OCCUPIES A CENTRAL POSITION IN NITROGEN METABOLISM

Transfer of amino nitrogen to α-ketoglutarate forms L-glutamate. Hepatic L-**glutamate dehydrogenase (GDH)**, which can use either NAD^+ or $NADP^+$, releases this nitrogen as ammonia (Figure 28–8). Conversion of α-amino nitrogen to ammonia by the concerted action of glutamate aminotransferase and GDH is often termed "transdeamination." Liver GDH activity is allosterically inhibited by ATP, GTP, and NADH, and is activated by ADP. The GDH reaction is freely reversible, and also functions in amino acid biosynthesis (see Figure 27–1).

Amino Acid Oxidases Also Remove Nitrogen as Ammonia

While their physiologic importance is uncertain, L-amino acid oxidases of liver and kidney convert an amino acid to an α-imino acid that decomposes to an α-keto acid with re-

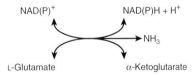

FIGURE 28–8 The L-glutamate dehydrogenase reaction. NAD(P)$^+$ means that either NAD^+ or $NADP^+$ can serve as the oxidoreductant. The reaction is reversible, but favors glutamate formation.

FIGURE 28–9 Oxidative deamination catalyzed by L-amino acid oxidase (L-α-amino acid:O$_2$ oxidoreductase). The α-imino acid, shown in brackets, is not a stable intermediate.

lease of ammonium ion (Figure 28–9). The reduced flavin is reoxidized by molecular oxygen, forming hydrogen peroxide (H_2O_2), which then is split to O_2 and H_2O by **catalase.**

Ammonia Intoxication Is Life-Threatening

The ammonia produced by enteric bacteria and absorbed into portal venous blood and the ammonia produced by tissues are rapidly removed from circulation by the liver and converted to urea. Thus, only traces (10–20 μg/dL) normally are present in peripheral blood. This is essential, since ammonia is toxic to the central nervous system. Should portal blood bypass the liver, systemic blood ammonia levels may attain toxic levels. This occurs in severely impaired hepatic function or the development of collateral links between the portal and systemic veins in cirrhosis. Symptoms of **ammonia intoxication** include tremor, slurred speech, blurred vision, coma, and ultimately death. Ammonia may be toxic to the brain in part because it reacts with α-ketoglutarate to form glutamate. The resulting depletion of levels of α-ketoglutarate then impairs function of the tricarboxylic acid (TCA) cycle in neurons.

Glutamine Synthetase Fixes Ammonia as Glutamine

Formation of glutamine is catalyzed by mitochondrial **glutamine synthetase** (Figure 28–10). Since amide bond synthesis is coupled to the hydrolysis of ATP to ADP and P$_i$, the reaction strongly favors glutamine synthesis. One key function of glutamine is to sequester ammonia in a nontoxic form.

Glutaminase & Asparaginase Deamidate Glutamine & Asparagine

Glutamine synthetase plays a major role in ammonia detoxification, interorgan nitrogen flux, and acid-base homeostasis.

FIGURE 28–10 The glutamine synthetase reaction strongly favors glutamine synthesis.

FIGURE 28–11 The glutaminase reaction proceeds essentially irreversibly in the direction of glutamate and NH_4^+ formation. Note that the *amide* nitrogen, not the α-amino nitrogen, is removed.

Hydrolytic release of the amide nitrogen of glutamine as ammonia, catalyzed by **glutaminase** (Figure 28–11), strongly favors glutamate formation. An analogous reaction is catalyzed by L-asparaginase. The concerted action of glutamine synthetase and glutaminase thus catalyzes the interconversion of free ammonium ion and glutamine. A rare deficiency in neonate glutamine synthetase results in severe brain damage, multi-organ failure, and death.

Formation & Secretion of Ammonia Maintains Acid-Base Balance

Excretion into urine of ammonia produced by renal tubular cells facilitates cation conservation and regulation of acid-base balance. Ammonia production from intracellular renal amino acids, especially glutamine, increases in **metabolic acidosis** and decreases in **metabolic alkalosis.**

UREA IS THE MAJOR END PRODUCT OF NITROGEN CATABOLISM IN HUMANS

Synthesis of 1 mol of urea requires 3 mol of ATP plus 1 mol each of ammonium ion and of the α-amino nitrogen of aspartate. Five enzymes catalyze the numbered reactions of Figure 28–12. Of the six participating amino acids, *N*-acetylglutamate functions solely as an enzyme activator. The others serve as carriers of the atoms that ultimately become urea. The major metabolic role of **ornithine, citrulline,** and **argininosuccinate** in mammals is urea synthesis. Urea synthesis is a cyclic process. The ornithine consumed in reaction 2 is regenerated in reaction 5, and so there is no net loss or gain of ornithine, citrulline, argininosuccinate, or arginine. Ammonium ion, CO_2, ATP, and aspartate are, however, consumed. Some reactions of urea synthesis occur in the matrix of the mitochondrion, and other reactions in the cytosol (Figure 28–12).

Carbamoyl Phosphate Synthetase I Initiates Urea Biosynthesis

Condensation of CO_2, ammonia, and ATP to form **carbamoyl phosphate** is catalyzed by mitochondrial **carbamoyl phosphate synthetase I**. A cytosolic form of this enzyme, carbamoyl phosphate synthetase II, uses glutamine rather than ammonia as the nitrogen donor and functions in pyrimidine biosynthesis (see Chapter 33). The concerted action of GDH and carbamoyl phosphate synthetase 1 thus shuttle amino nitrogen into carbamoyl phosphate, a compound with high group transfer potential.

Carbamoyl phosphate synthetase I, the rate-limiting enzyme of the urea cycle, is active only in the presence of *N*-**acetylglutamate**, an allosteric activator that enhances the affinity of the synthetase for ATP. Synthesis of one mol of carbamoyl phosphate requires 2 mol of ATP. One ATP serves as the phosphoryl donor for formation of the mixed acid anhydride bond of carbamoyl phosphate. The second ATP provides the driving force for synthesis of the amide bond of carbamoyl phosphate. The other products are two mol of ADP and one mol of P_i (reaction 1, Figure 28–12). The reaction proceeds stepwise. Reaction of bicarbonate with ATP forms carbonyl phosphate and ADP. Ammonia then displaces ADP, forming carbamate and orthophosphate. Phosphorylation of carbamate by the second ATP then forms carbamoyl phosphate.

Carbamoyl Phosphate Plus Ornithine Forms Citrulline

L-**Ornithine transcarbamoylase** catalyzes transfer of the carbamoyl group of carbamoyl phosphate to ornithine, forming citrulline and orthophosphate (reaction 2, Figure 28–12). While the reaction occurs in the mitochondrial matrix, both the formation of ornithine and the subsequent metabolism of citrulline take place in the cytosol. Entry of ornithine into mitochondria and exodus of citrulline from mitochondria therefore involve mitochondrial inner membrane transport systems (Figure 28–12).

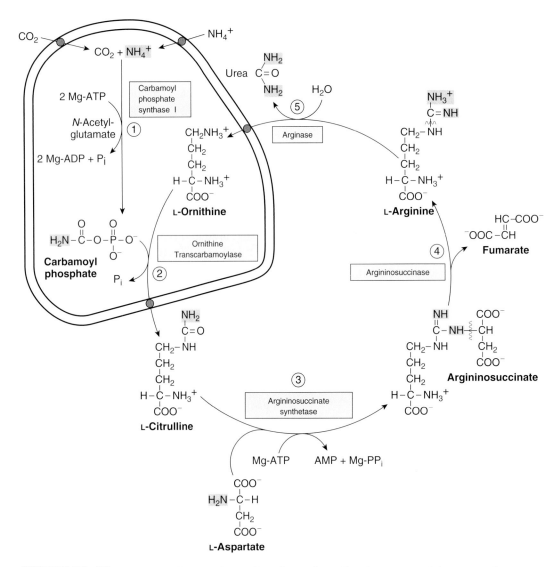

FIGURE 28–12 Reactions and intermediates of urea biosynthesis. The nitrogen-containing groups that contribute to the formation of urea are shaded. Reactions ① and ② occur in the matrix of liver mitochondria and reactions ③, ④, and ⑤ in liver cytosol. CO_2 (as bicarbonate), ammonium ion, ornithine, and citrulline enter the mitochondrial matrix via specific carriers (see red dots) present in the inner membrane of liver mitochondria.

Citrulline Plus Aspartate Forms Argininosuccinate

Argininosuccinate synthetase links aspartate and citrulline via the amino group of aspartate (reaction 3, Figure 28–12) and provides the second nitrogen of urea. The reaction requires ATP and involves intermediate formation of citrullyl-AMP. Subsequent displacement of AMP by aspartate then forms arginosuccinate.

Cleavage of Argininosuccinate Forms Arginine & Fumarate

Cleavage of argininosuccinate, catalyzed by **argininosuccinase,** proceeds with retention of nitrogen in arginine and release of the aspartate skeleton as fumarate (reaction 4, Figure

28–12). Addition of water to fumarate forms L-malate, which subsequent NAD⁺-dependent oxidation converts to oxaloacetate. These two reactions are analogous to reactions of the citric acid cycle (see Figure 17–3), but are catalyzed by *cytosolic* **fumarase and malate dehydrogenase**. Transamination of oxaloacetate by glutamate aminotransferase then re-forms aspartate. The carbon skeleton of aspartate-fumarate thus acts as a carrier of the nitrogen of glutamate into a precursor of urea.

Cleavage of Arginine Releases Urea & Re-forms Ornithine

Hydrolytic cleavage of the guanidino group of arginine, catalyzed by liver **arginase,** releases urea (reaction 5, Figure 28–12). The other product, ornithine, reenters liver mitochondria and participates in additional rounds of urea synthesis. Ornithine and lysine are potent inhibitors of arginase, and compete with arginine. Arginine also serves as the precursor of the potent muscle relaxant nitric oxide (NO) in a Ca^{2+}-dependent reaction catalyzed by NO synthase (see Figure 48–12).

Carbamoyl Phosphate Synthetase I Is the Pacemaker Enzyme of the Urea Cycle

The activity of carbamoyl phosphate synthetase I is determined by N-acetylglutamate, whose steady-state level is dictated by its rate of synthesis from acetyl-CoA and glutamate and its rate of hydrolysis to acetate and glutamate. These reactions are catalyzed by N-acetylglutamate synthase and N-acetylglutamate hydrolase, respectively. Major changes in diet can increase the concentrations of individual urea cycle enzymes 10- to 20-fold. For example, starvation elevates enzyme levels, presumably to cope with the increased production of ammonia that accompanies enhanced starvation-induced degradation of protein.

GENERAL FEATURES OF METABOLIC DISORDERS

The comparatively rare, but medically devastating, metabolic disorders associated with the enzymes of urea biosynthesis illustrate the following general principles of metabolic diseases:

1. Similar or identical clinical signs and symptoms can characterize any number of different molecular level defects in a given enzyme.
2. Rational therapy must be based on an understanding of the relevant biochemical enzyme-catalyzed reactions in both normal and impaired individuals.
3. The identification of intermediates and of ancillary products that accumulate prior to a metabolic block provides the basis for metabolic screening tests and can implicate the reaction that is impaired.
4. Precise diagnosis requires quantitative assay of the activity of the enzyme suspected to be defective.
5. The DNA sequence of the gene that encodes a given mutant enzyme is compared to that of the wild-type gene to identify the specific mutation(s) that cause the disease.

METABOLIC DISORDERS ARE ASSOCIATED WITH EACH REACTION OF THE UREA CYCLE

Defects in each enzyme of the urea cycle have been described. Many of the causative mutations have been mapped, and specific defects in the encoded enzymes have been identified. Five well-documented diseases represent defects in the biosynthesis of enzymes of the urea cycle. Molecular genetic analysis has pinpointed the loci of mutations associated with each de-

ficiency, each of which exhibits considerable genetic and phenotypic variability.

Urea cycle disorders are characterized by hyperammonemia, encephalopathy, and respiratory alkalosis. Four of the five metabolic diseases, deficiencies of carbamoyl phosphate synthetase, ornithine transcarbamylase, argininosuccinate synthetase, and argininosuccinate lyase, result in the accumulation of precursors of urea, principally ammonia and glutamine. Ammonia intoxication is most severe when the metabolic block occurs at reactions 1 or 2 (Figure 28–12), for if citrulline can be synthesized, some ammonia has already been removed by being covalently linked to an organic metabolite. Clinical symptoms common to all urea cycle disorders include vomiting, avoidance of high-protein foods, intermittent ataxia, irritability, lethargy, and severe mental retardation. The most dramatic clinical presentation occurs in full-term infants who initially appear normal, then exhibit progressive lethargy, hypothermia, and apnea due to high plasma ammonia levels. The clinical features and treatment of all five disorders are similar. Significant improvement and minimization of brain damage can accompany a low-protein diet ingested as frequent small meals to avoid sudden increases in blood ammonia levels. The goal of dietary therapy is to provide sufficient protein, arginine, and energy to promote growth and development while simultaneously minimizing the metabolic perturbations associated with these diseases.

Carbamoyl Phosphate Synthetase I

N-Acetylglutamate is essential for the activity of carbamoyl phosphate synthetase I (reaction 1, Figure 28–12). Defects in carbamoyl phosphate synthetase I are responsible for the relatively rare (estimated frequency 1:62,000) metabolic disease termed "hyperammonemia type 1."

N-Acetylglutamate Synthase

N-Acetylglutamate synthase (NAGS) catalyzes the formation from acetyl-CoA and glutamate of the N-acetylglutamate essential for carbamoyl phosphate synthetase I activity.

$$\text{L-Glutamate} + \text{Acetyl-CoA} \rightarrow N\text{-Acetyl-L-glutamate} + \text{CoASH}$$

While the clinical and biochemical features of NAGS deficiency are indistinguishable from those arising from a defect in phosphate synthetase I, a deficiency in NAGS may respond to administered N-acetylglutamate.

The Ornithine Transporter

Hyperornithinemia, hyperammonemia, and homocitrullinuria syndrome (**HHH syndrome**) results from mutation of the ORNT1 gene that encodes the mitochondrial membrane ornithine transporter. The failure to import cytosolic ornithine into the mitochondrial matrix renders the urea cycle inoperable, with consequent hyperammonemia, and hyperornithinemia due to the accompanying accumulation of cytosolic ornithine.

In the absence of its normal acceptor ornithine, mitochondrial carbamoyl phosphate carbamoylates lysine to homocitrulline, resulting in homocitrullinuria.

Ornithine Transcarbamoylase

The X-chromosome linked deficiency termed "hyperammonemia type 2" reflects a defect in ornithine transcarbamoylase (reaction 2, Figure 28–12). The mothers also exhibit hyperammonemia and an aversion to high-protein foods. Levels of glutamine are elevated in blood, cerebrospinal fluid, and urine, probably as a result of enhanced glutamine synthesis in response to elevated levels of tissue ammonia.

Argininosuccinate Synthetase

In addition to patients that lack detectable argininosuccinate synthetase activity (reaction 3, Figure 28–12), a 25-fold elevated K_m for citrulline has been reported. In the resulting citrullinemia, plasma and cerebrospinal fluid citrulline levels are elevated, and 1–2 g of citrulline are excreted daily.

Argininosuccinase (Argininosuccinate Lyase)

Argininosuccinicaciduria, accompanied by elevated levels of argininosuccinate in blood, cerebrospinal fluid, and urine, is associated with friable, tufted hair (trichorrhexis nodosa). Both early- and late-onset types are known. The metabolic defect is in argininosuccinase (argininosuccinate lyase; reaction 4, Figure 28–12). Diagnosis by measurement of erythrocyte argininosuccinase activity can be performed on umbilical cord blood or amniotic fluid cells.

Arginase

Hyperargininemia is an autosomal recessive defect in the gene for arginase (reaction 5, Figure 28–12). Unlike other urea cycle disorders, the first symptoms of hyperargininemia typically do not appear until age 2 to 4 years. Blood and cerebrospinal fluid levels of arginine are elevated. The urinary amino acid pattern, which resembles that of lysine-cystinuria, may reflect competition by arginine with lysine and cysteine for reabsorption in the renal tubule.

Analysis of Neonate Blood by Tandem Mass Spectrometry Can Detect Metabolic Diseases

Metabolic diseases caused by the absence or functional impairment of metabolic enzymes can be devastating. Early dietary intervention, however, can in many instances ameliorate the otherwise inevitable dire effects. The early detection of such metabolic diseases is thus is of primary importance. Since the initiation in the United States of newborn screening programs in the 1960s, all states now conduct metabolic screening of newborns, although the scope of screen employed varies among states. The powerful and sensitive technique of **tandem mass spectrometry** (see Chapter 4) can in a few minutes detect over 40 analytes of significance in the detection of metabolic disorders. Most states employ tandem MS to screen newborns to detect metabolic disorders such as organic acidemias, aminoacidemias, disorders of fatty acid oxidation, and defects in the enzymes of the urea cycle. However, at present there remain significant differences in analyte coverage between states. A recent article reviews the theory of tandem MS, its application to the detection of metabolic disorders, and situations that can yield false positives. It also includes a lengthy table of detectable analytes and the relevant metabolic diseases (see *Clinical Chemistry* 39, 315–332, 2006).

Can Gene Therapy Offer Promise for Correcting Defects in Urea Biosynthesis?

Gene therapy of defects in the enzymes of the urea cycle is an area of active investigation. Encouraging preliminary results in animal models, for example, using an adenoviral vector to treat citrullinemia, suggest potential, but at present gene therapy provides no effective solution for human subjects.

SUMMARY

- Human subjects degrade 1–2% of their body protein daily at rates that vary widely between proteins and with physiologic state. Key regulatory enzymes often have short half-lives.
- Proteins are degraded by both ATP-dependent and ATP-independent pathways. Ubiquitin targets many intracellular proteins for degradation. Liver cell surface receptors bind and internalize circulating asialoglycoproteins destined for lysosomal degradation.
- Fish excrete highly toxic NH_3 directly. Birds convert NH_3 to uric acid. Higher vertebrates convert NH_3 to urea.
- Transamination channels amino acid nitrogen into glutamate. L-Glutamate dehydrogenase (GDH) occupies a central position in nitrogen metabolism.
- Glutamine synthetase converts NH_3 to nontoxic glutamine. Glutaminase releases NH_3 for use in urea synthesis.
- NH_3, CO_2, and the amide nitrogen of aspartate provide the atoms of urea.
- Hepatic urea synthesis takes place in part in the mitochondrial matrix and in part in the cytosol.
- Changes in enzyme levels and allosteric regulation of carbamoyl phosphate synthetase by N-acetylglutamate regulate urea biosynthesis.
- Metabolic diseases are associated with defects in each enzyme of the urea cycle, of the membrane-associated ornithine transporter, and of N-acetylglutamate synthetase.

▪ Tandem mass spectrometry is the technique of choice for screening neonates for inherited metabolic diseases.

REFERENCES

Brooks P et al: Subcellular localization of proteasomes and their regulatory complexes in mammalian cells. Biochem J 2000;346:155.

Caldovic L et al: Mutations and polymorphisms in the human N-acetylglutamate synthase (NAGS) gene. Hum Mutat 2007;28:754.

Crombez EA, Cederbaum SD: Hyperargininemia due to liver arginase deficiency. Mol Genet Metab 2005;84:243.

Garg U, Dasouki M: Expanded newborn screening of inherited metabolic disorders by tandem mass spectrometry. Clinical and laboratory aspects. Clin Biochem 2006;39:315.

Elpeleg O et al: N-acetylglutamate synthase deficiency and the treatment of hyperammonemic encephalopathy. Ann Neurol 2002;52:845.

Gyato K et al: Metabolic and neuropsychological phenotype in women heterozygous for ornithine transcarbamylase deficiency. Ann Neurol 2004;55:80.

Haberle J et al: Diagnosis of N-acetylglutamate synthase deficiency by use of cultured fibroblasts and avoidance of nonsense-mediated mRNA decay. J Inherit Metab Dis 2003;26:601.

Haberle J et al: Mild citrullinemia in caucasians is an allelic variant of argininosuccinate synthetase deficiency (citrullinemia type 1). Mol Genet Metab 2003;80:302.

Haberle J et al: Congenital glutamine deficiency with glutamine synthetase mutations. N Engl J Med 2005;353:1926.

Iyer R et al: The human arginases and arginase deficiency. J Inherit Metab Dis 1998;21:86.

Pickart CM: Mechanisms underlying ubiquitination. Annu Rev Biochem 2001;70:503.

Scriver CR: Garrod's foresight; our hindsight. J Inherit Metab Dis 2001;24:93.

Scriver CR et al (editors): The Metabolic and Molecular Bases of Inherited Disease, 8th ed. McGraw-Hill, 2001.

Yi JJ, Ehlers MD: Emerging roles for ubiquitin and protein degradation in neuronal function. Pharmacol Rev 2007;59:206.

Catabolism of the Carbon Skeletons of Amino Acids

Victor W. Rodwell, PhD

BIOMEDICAL IMPORTANCE

This chapter discusses conversion of the carbon skeletons of the common L-α-amino acids to amphibolic intermediates, and the metabolic diseases or "inborn errors of metabolism" associated with these processes. Left untreated, these disorders can result in irreversible brain damage and early mortality. Prenatal or early postnatal detection and timely initiation of treatment thus are essential. Many of the enzymes concerned can be detected in cultured amniotic fluid cells, which facilitates prenatal diagnosis by amniocentesis. Almost all states conduct screening tests for up to as many as 30 metabolic diseases. These tests include, but are not limited to, disorders that result from defects in the catabolism of amino acids. The best screening tests use tandem mass spectrometry to detect, in a few drops of neonate blood, catabolites suggestive of a metabolic defect. While many changes in the primary structure of enzymes have no adverse effects, others modify the three-dimensional structure of catalytic or regulatory sites, lower catalytic efficiency (lower V_{max} or elevate K_m), or alter the affinity for an allosteric regulator of activity. A variety of mutations

thus may give rise to the same clinical signs and symptoms. Treatment consists primarily of feeding diets low in the amino acids whose catabolism is impaired.

TRANSAMINATION TYPICALLY INITIATES AMINO ACID CATABOLISM

Removal of α-amino nitrogen by transamination (see Figure 28–6) is the first catabolic reaction of amino acids except for proline, hydroxyproline, threonine, and lysine. The hydrocarbon skeleton that remains is then degraded to amphibolic intermediates as outlined in Figure 29–1.

Asparagine, Aspartate, Glutamine, and Glutamate

All four carbons of asparagine and aspartate form **oxaloacetate** (Figure 29–2, top). Analogous reactions convert glutamine and glutamate to **α-ketoglutarate** (Figure 29–2, bottom). No metabolic defects are associated with the catabolism of these four amino acids.

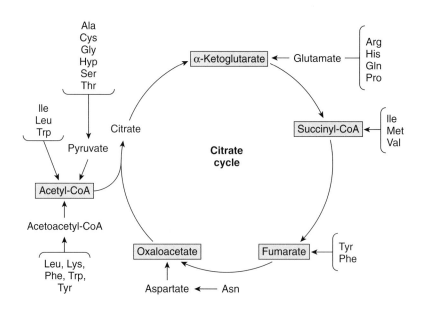

FIGURE 29–1 Amphibolic intermediates formed from the carbon skeletons of amino acids.

FIGURE 29–2 Catabolism of L-asparagine (**top**) and of L-glutamine (**bottom**) to amphibolic intermediates. (PYR, pyruvate; ALA, L-alanine.) In this and subsequent figures, color highlights portions of the molecules undergoing chemical change.

Proline

The catabolism of proline takes place in mitochodria. Since proline does not participate in transamination, the nitrogen of this imino acid is retained throughout its oxidation to Δ^1-pyrolline-5-carboxylate, ring opening to glutamate-γ-semialdehyde, and oxidation to glutamate, and is only removed during transamination of glutamate to α-ketoglutarate (Figure 29–3). There are two metabolic disorders of proline catabolism. Both types are inherited as autosomal recessive traits, and are consistent with a normal adult life. The metabolic block in **type I hyperprolinemia** is at **proline dehydrogenase.** There is no associated impairment of hydroxyproline catabolism. The metabolic block

FIGURE 29–3 Catabolism of proline. Numerals indicate sites of the metabolic defects in ① type I and ② type II hyperprolinemias.

FIGURE 29–4 Catabolism of arginine. Arginase-catalyzed cleavage of L-arginine forms urea and L-ornithine. This reaction represents the site of the metabolic defect in hyperargininemia. Subsequent transamination of L-ornithine to glutamate-γ-semialdehyde is followed by conversion to α-ketoglutarate (see Figure 29–3).

in **type II hyperprolinemia** is at **glutamate-γ-semialdehyde dehydrogenase,** an enzyme that also functions in hydroxyproline catabolism. Both proline and hydroxyproline catabolism thus are affected, and both Δ^1-pyroline-5-carboxylate and Δ^1-pyroline-3-hydroxy-5-carboxylate (see Figure 29–12) are excreted.

Arginine and Ornithine

Arginine is converted to ornithine then to glutamate–γ-semialdehyde (Figure 29–4). Subsequent catabolism to

FIGURE 29–5 Catabolism of L-histidine to α-ketoglutarate. (H₄ folate, tetrahydrofolate.) Histidase is the probable site of the metabolic defect in histidinemia.

α-ketoglutarate occurs as described above for proline (see Figure 29–3). Mutations in **ornithine δ-aminotransferase** elevate plasma and urinary ornithine and cause **gyrate atrophy of the retina.** Treatment involves restricting dietary arginine. In **hyperornithinemia-hyperammonemia syndrome,** a defective mitochondrial **ornithine-citrulline antiporter** (see Figure 28–12) impairs transport of ornithine into mitochondria for use in urea synthesis.

Histidine

Catabolism of histidine proceeds via urocanate, 4-imidazolone-5-propionate, and *N*-formiminoglutamate (Figlu). Formimino group transfer to tetrahydrofolate forms glutamate, then **α-ketoglutarate** (Figure 29–5). In folic acid deficiency, group transfer of the formimino group is impaired, and Figlu is excreted. Excretion of Figlu following a dose of histidine thus can be used to detect folic acid deficiency. Benign disorders of histidine catabolism include **histidinemia** and **urocanic aciduria** associated with impaired **histidase.**

CATABOLISM OF GLYCINE, SERINE, ALANINE, CYSTEINE, THREONINE AND 4-HYDROXYPROLINE

Glycine

The **glycine cleavage complex** of liver mitochondria splits glycine to CO_2 and NH_4^+ and forms N^5,N^{10}-methylene tetrahydrofolate.

$$\text{Glycine} + H_4\text{folate} + NAD^+ \rightarrow CO_2 + NH_3$$
$$+ 5,10\text{-}CH_2\text{-}H_4\text{folate} + NADH + H^+$$

FIGURE 29–6 The glycine cleavage system of liver mitochondria. The complex consists of three enzymes and an "H-protein" that has covalently attached dihyrolipoate. Catalysts for the numbered reactions are ① glycine dehydrogenase (decarboxylating), ② an ammonia-forming aminomethyltransferase, and ③ dihydrolipoamide dehydrogenase. (H₄folate, tetrahydrofolate).

The glycine cleavage system consists of three enzymes, and an "H-protein" that has a covalently attached dihyrolipoyl moiety. Figure 29–6 illustrates the individual reactions and intermediates in glycine cleavage. In **nonketotic hyperglycinemia**, a rare inborn error of glycine degradation presently known only in Finland, glycine accumulates in all body tissues including the central nervous system. The defect in **primary hyperoxaluria** is the failure to catabolize glyoxylate formed by the deamination of glycine. Subsequent oxidation of glyoxylate to oxalate results in urolithiasis, nephrocalcinosis, and early mortality from renal failure or hypertension. **Glycinuria** results from a defect in renal tubular reabsorption.

Serine

Following conversion to glycine, catalyzed by **serine hydroxymethyltransferase,** serine catabolism merges with that of glycine (Figure 29–7).

Alanine

Transamination of α-alanine forms pyruvate. Probably on account of its central role in metabolism there is no known metabolic defect of α-alanine catabolism.

FIGURE 29–7 Interconversion of serine and glycine by serine hydroxymethyltransferase. (H$_4$folate, tetrahydrofolate)

FIGURE 29–8 The cystine reductase reaction.

FIGURE 29–9 Catabolism of L-cysteine via the cysteine sulfinate pathway (**top**) and by the 3-mercaptopyruvate pathway (**bottom**).

FIGURE 29–10 Mixed disulfide of cysteine and homocysteine.

Cysteine

Cystine is first reduced to cysteine by **cystine reductase** (Figure 29–8). Two different pathways then convert cysteine to pyruvate (Figure 29–9). There are numerous abnormalities of cysteine metabolism. Cystine, lysine, arginine, and ornithine are excreted in **cystine-lysinuria (cystinuria),** a defect in renal reabsorption of these amino acids. Apart from cystine calculi, cystinuria is benign. The mixed disulfide of L-cysteine and L-homocysteine (Figure 29–10) excreted by cystinuric patients is more soluble than cystine and reduces formation of cystine calculi.

Several metabolic defects result in vitamin B$_6$-responsive or vitamin B$_6$-unresponsive **homocystinurias.** These include a deficiency in the reaction catalyzed by cystathionine β-synthase:

$$\text{Serine} + \text{Homocysteine} \rightarrow \text{Cystathionine} + H_2O$$

Consequences include osteoporosis and mental retardation. Defective carrier-mediated transport of cystine results in **cystinosis (cystine storage disease)** with deposition of cystine crystals in tissues and early mortality from acute

FIGURE 29–11 Conversion of threonine to glycine and acetyl-CoA.

FIGURE 29–12 Intermediates in L-hydroxyproline catabolism. (α-KA, α-keto acid; α-AA, α-amino acid.) Numerals identify sites of metabolic defects in ① hyperhydroxyprolinemia and ② type II hyperprolinemia.

FIGURE 29–13 Intermediates in tyrosine catabolism. Carbons are numbered to emphasize their ultimate fate. (α-KG, α-ketoglutarate; Glu, glutamate; PLP, pyridoxal phosphate.) Circled numerals represent the probable sites of the metabolic defects in ① type II tyrosinemia; ② neonatal tyrosinemia; ③ alkaptonuria; and ④ type I tyrosinemia, or tyrosinosis.

renal failure. Epidemiologic and other data link plasma homocysteine levels to cardiovascular risk, but the role of homocysteine as a causal cardiovascular risk factor remains at present controversial.

Threonine

Threonine aldolase cleaves threonine to acetaldehyde and glycine. Oxidation of acetaldehyde to acetate is followed by formation of acetyl-CoA (Figure 29–11). Catabolism of glycine is discussed above.

4-Hydroxyproline

Catabolism of 4-hydroxy-L-proline forms, successively, L-Δ^1-pyrroline-3-hydroxy-5-carboxylate, γ-hydroxy-L-glutamate-γ-semialdehyde, erythro-γ-hydroxy-L-glutamate, and α-keto-γ-hydroxyglutarate. An aldol-type cleavage then forms glyoxylate plus pyruvate (Figure 29–12). A defect in **4-hydroxyproline dehydrogenase** results in **hyperhydroxyprolinemia,** which is benign. There is no associated impairment of proline catabolism.

ADDITIONAL AMINO ACIDS THAT FORM ACETYL-CoA

Tyrosine

Figure 29–13 diagrams the conversion of tyrosine to amphibolic intermediates. Since ascorbate is the reductant for conversion of *p*-hydroxyphenylpyruvate to homogentisate, scorbutic patients excrete incompletely oxidized products of tyrosine catabolism. Subsequent reactions form maleylacetoacetate, fumarylacetoacetate, fumarate, acetoacetate, and ultimately acetyl-CoA.

The probable metabolic defect in **type I tyrosinemia (tyrosinosis)** is at **fumarylacetoacetate hydrolase** (Figure 29–13). Therapy employs a diet low in tyrosine and phenylalanine. Untreated acute and chronic tyrosinosis leads to death from liver failure. Alternate metabolites of tyrosine are also excreted in **type II tyrosinemia (Richner-Hanhart syndrome),** a defect in **tyrosine aminotransferase** (reaction 1, Figure 29–13), and in **neonatal tyrosinemia,** due to lowered *p*-hydroxyphenylpyruvate hydroxylase activity (reaction 2, Figure 29–13). Therapy employs a diet low in protein.

Alkaptonuria was first recognized and described in the 16th century. Characterized in 1859, it provided the basis for Garrod's classic ideas concerning heritable metabolic disorders. The defect is lack of **homogentisate oxidase** (reaction 3, Figure 29–13). The urine darkens on exposure to air due to oxidation of excreted homogentisate. Late in the disease, there is arthritis and connective tissue pigmentation (ochronosis) due to oxidation of homogentisate to benzoquinone acetate, which polymerizes and binds to connective tissue.

FIGURE 29–14 Alternative pathways of phenylalanine catabolism in phenylketonuria. The reactions also occur in normal liver tissue but are of minor significance.

Phenylalanine

Phenylalanine is first converted to tyrosine (see Figure 27–10). Subsequent reactions are those of tyrosine (Figure 29–13). **Hyperphenylalaninemias** arise from defects in phenylalanine hydroxylase itself (**type I, classic phenylketonuria or PKU,** frequency 1 in 10,000 births), in dihydrobiopterin reductase **(types II and III),** or in dihydrobiopterin biosynthesis **(types IV and V)** (see Figure 27–10). Alternative catabolites are excreted (Figure 29–14). A diet low in phenylalanine can prevent the mental retardation of PKU.

DNA probes facilitate prenatal diagnosis of defects in phenylalanine hydroxylase or dihydrobiopterin reductase. Elevated blood phenylalanine may not be detectable until 3–4 days postpartum. False-positives in premature infants may re-

FIGURE 29–15 Reactions and intermediates in the catabolism of L-lysine. (α-KG, α-ketoglutarate; Glu, L-glutamate.) Shown on the left are the reactions, and on the right the structures, of the intermediates. The numbered reactions and the metabolic defects associated with lysine catabolism are discussed in the accompanying text.

flect delayed maturation of enzymes of phenylalanine catabolism. An older and less reliable screening test employs $FeCl_3$ to detect urinary phenylpyruvate. $FeCl_3$ screening for PKU of the urine of newborn infants is compulsory in many countries, but in the United States has been largely supplanted by tandem mass spectrometry.

Lysine

The first six reactions of L-lysine catabolism in human liver form crotonyl-CoA, which is then degraded to acetyl-CoA and CO_2 by the reactions of fatty acid catabolism (see Figure 22–3). In what follows, circled numerals refer to the corresponding numbered reactions of Figure 29–15. Reactions 1 and 2 convert the Schiff base formed between α-ketoglutarate and the ε-amino group of lysine to L-α-aminoadipate-δ-semialdehyde. Both reactions are catalyzed by a single bifunctional enzyme, aminoadipate semialdehde synthase (also called lysine 2-oxo-

glutarate reductase-saccharopine dehydrogenase). Reduction of L-α-aminoadipate-δ-semialdehde to L-α-aminoadipate (reaction 3) is followed by transamination to α-ketoadipate (reaction 4). Conversion to the thioester glutaryl-CoA (reaction 5) is followed by the decarboxylation of glutaryl-CoA to crotonyl-CoA (reaction 6). The subsequent reactions are those of the catabolism of α-unsaturated fatty acids with an odd number of carbons.

Metabolic defects associated with reactions of the lysine catabolic pathway include hyperlysinemias. Hyperlysinemia can result from a defect in activity 1 or 2 of the bifunctional enzyme aminoadipate semialdehde synthase. Hyperlysinemia is accompanied by elevated levels of blood saccharopine only if the defect involves activity 2. A metabolic defect at reaction 6 results in an inherited metabolic disease that is associated with striatal and cortical degeneration, and characterized by elevated concentrations of glutarate and its metabolites,

FIGURE 29–16 Catabolism of L-tryptophan. (PLP, pyridoxal phosphate.)

3-Hydroxykynurenine

NH_4^+

Xanthurenate

FIGURE 29–17 Formation of xanthurenate in vitamin B_6 deficiency. Conversion of the tryptophan metabolite 3-hydroxykynurenine to 3-hydroxyanthranilate is impaired (see Figure 29–16). A large portion is therefore converted to xanthurenate.

glutaconate and 3-hydroxyglutarate. The challenge in management of these metabolic defects is to restrict dietary intake of L-lysine without accompanying malnutrition.

Tryptophan

Tryptophan is degraded to amphibolic intermediates via the kynurenine-anthranilate pathway (Figure 29–16) **Tryptophan oxygenase (tryptophan pyrolase)** opens the indole ring, incorporates molecular oxygen, and forms *N*-formylkynurenine. Tryptophan oxygenase, an iron porphyrin metalloprotein that

is inducible in liver by adrenal corticosteroids and by tryptophan, is feedback-inhibited by nicotinic acid derivatives, including NADPH. Hydrolytic removal of the formyl group of *N*-formylkynurenine, catalyzed by **kynurenine formylase,** produces kynurenine. Since **kynureninase** requires pyridoxal phosphate, excretion of xanthurenate (Figure 29–17) in response to a tryptophan load is diagnostic of vitamin B_6 deficiency. **Hartnup disease** reflects impaired intestinal and renal transport of tryptophan and other neutral amino acids. Indole derivatives of unabsorbed tryptophan formed by intestinal bacteria are excreted. The defect limits tryptophan availability for niacin biosynthesis and accounts for the pellagra-like signs and symptoms.

Methionine

Methionine reacts with ATP forming *S*-adenosylmethionine, "active methionine" (Figure 29–18). Subsequent reactions form propionyl-CoA (Figure 29–19) and ultimately succinyl-CoA (see Figure 20–2).

THE INITIAL REACTIONS ARE COMMON TO ALL THREE BRANCHED-CHAIN AMINO ACIDS

Reactions 1–3 of Figure 29–20 are analogous to those of fatty acid catabolism (see Figure 22–3). Following transamination, all three α-keto acids undergo oxidative decarboxylation catalyzed by mitochondrial **branched-chain α-keto acid dehydrogenase.** This multimeric enzyme complex of a decarboxylase, a transacylase, and a dihydrolipoyl dehydrogenase closely resembles pyruvate dehydrogenase (see Figure 18–5). Its regulation also parallels that of pyruvate dehydrogenase, being inactivated by phosphorylation and reactivated by dephosphorylation (see Figure 18–6). Reaction 3 is analogous to the dehydrogenation of fatty acyl-CoA thioesters (see Figure 22–3). In **isovaleric acidemia,** ingestion of protein-rich foods

FIGURE 29–18 Formation of *S*-adenosylmethionine. ~CH_3 represents the high group transfer potential of "active methionine."

FIGURE 29–19 Conversion of methionine to propionyl-CoA.

elevates isovalerate, the deacylation product of isovaleryl-CoA. Figures 29–21, 29–22, and 29–23 illustrate the subsequent reactions unique to each amino acid skeleton.

METABOLIC DISORDERS OF BRANCHED-CHAIN AMINO ACID CATABOLISM

As the name implies, the odor of urine in **maple syrup urine disease (branched-chain ketonuria)** suggests maple syrup or burnt sugar. The biochemical defect involves the **α-keto acid decarboxylase complex** (reaction 2, Figure 29–20). Plasma and urinary levels of leucine, isoleucine, valine, α-keto acids, and α-hydroxy acids (reduced α-keto acids) are elevated. The mechanism of toxicity is unknown. Early diagnosis, especially prior to 1 week of age, employs enzymatic analysis. Prompt replacement of dietary protein by an amino acid mixture that lacks leucine, isoleucine, and valine averts brain damage and early mortality.

Mutation of the dihydrolipoate reductase component impairs decarboxylation of branched-chain α-keto acids, of pyruvate, and of α-ketoglutarate. In **intermittent branched-chain ketonuria,** the α-keto acid decarboxylase retains some activity, and symptoms occur later in life. The impaired enzyme in **isovaleric acidemia** is **isovaleryl-CoA dehydrogenase** (reaction 3, Figure 29–20). Vomiting, acidosis, and coma follow ingestion of excess protein. Accumulated isovaleryl-CoA is hydrolyzed to isovalerate and excreted.

SUMMARY

- Excess amino acids are catabolized to amphibolic intermediates that serve as sources of energy or for the biosynthesis of carbohydrates and lipids.

- Transamination is the most common initial reaction of amino acid catabolism. Subsequent reactions remove any additional nitrogen and restructure hydrocarbon skeletons for conversion to oxaloacetate, α-ketoglutarate, pyruvate, and acetyl-CoA.

- Metabolic diseases associated with glycine catabolism include glycinuria and primary hyperoxaluria.

- Two distinct pathways convert cysteine to pyruvate. Metabolic disorders of cysteine catabolism include cystine-lysinuria, cystine storage disease, and the homocystinurias.

- Threonine catabolism merges with that of glycine after threonine aldolase cleaves threonine to glycine and acetaldehyde.

- Following transamination, the carbon skeleton of tyrosine is degraded to fumarate and acetoacetate. Metabolic diseases of tyrosine catabolism include tyrosinosis, Richner-Hanhart syndrome, neonatal tyrosinemia, and alkaptonuria.

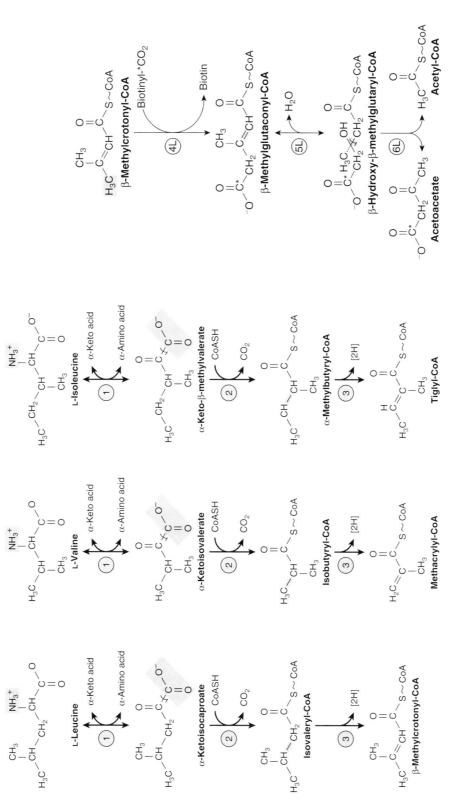

FIGURE 29–21 Catabolism of the β-methylcrotonyl-CoA formed from L-leucine. Asterisks indicate carbon atoms derived from CO_2.

FIGURE 29–20 The analogous first three reactions in the catabolism of leucine, valine, and isoleucine. Note also the analogy of reactions 2 and 3 to reactions of the catabolism of fatty acids (see Figure 22–3). The analogy to fatty acid catabolism continues, as shown in subsequent figures.

- Metabolic disorders of phenylalanine catabolism include phenylketonuria (PKU) and several hyperphenylalaninemias.

- Neither nitrogen of lysine undergoes direct transamination. The same effect is, however, achieved by the intermediate formation of saccharopine. Metabolic diseases of lysine catabolism include periodic and persistent forms of hyperlysinemia-ammonemia.

- The catabolism of leucine, valine, and isoleucine presents many analogies to fatty acid catabolism. Metabolic disorders of branched-chain amino acid catabolism include hypervalinemia, maple syrup urine disease, intermittent branched-chain ketonuria, isovaleric acidemia, and methylmalonic aciduria.

FIGURE 29–22 Subsequent catabolism of the tiglyl-CoA formed from L-isoleucine.

FIGURE 29–23 Subsequent catabolism of the methacrylyl-CoA formed from L-valine (see Figure 29–20). (α-KA, α-keto acid; α-AA, α-amino acid.)

REFERENCES

Blacher J, Safar ME: Homocysteine, folic acid, B vitamins and cardiovascular risk. J Nutr Health Aging 2001;5:196.

Bliksrud YT et al: Tyrosinemia type I, de novo mutation in liver tissue suppressing an inborn splicing defect. J Mol Med 2005;83:406.

Flusser H et al: Mild glycine encephalopathy (NKH) in a large kindred due to a silent exonic GLDC splice mutation. Neurology 2005;64:1426.

Garg U, Dasouki M: Expanded newborn screening of inherited metabolic disorders by tandem mass spectrometry. Clinical and laboratory aspects. Clin Biochem 2006;39:315.

Gerstner B et al: Glutaric acid and its metabolites cause apoptosis in immature oligodendrocytes: a novel mechanism of white matter degeneration in glutaryl-CoA dehydrogenase deficiency. Pediatr Res 2005;57;771.

Häussinger D, Schliess F: Glutamine metabolism and signaling in the liver. Front Biosci 2007;12:371.

Gjetting T et al: A phenylalanine hydroxylase amino acid polymorphism with implications for molecular diagnostics. Mol Genet Metab 2001;73:280.

Heldt K et al: Diagnosis of maple syrup urine disease by newborn screening allows early intervention without extraneous detoxification. Mol Genet Metab 2005;84:313.

Moshal K et al: Cardiac dys-synchronization and arrhythmia in hyperhomocysteinemia. Curr Neurovasc Res |2007;4:289.

Muller E, Kolker S: Reduction of lysine intake while avoiding malnutrition: major goals and major problems in dietary treatment of glutaryl-CoA dehydrogenase deficiency. J Inherit Metab Dis 2004;27:903.

Sacksteder KA et al: Identification of the alpha-aminoadipic semialdehyde synthase gene which is defective in familial hyperlysinemia. Am J Hum Genet 2000;66:1736.

Scriver CR et al (editors): The Metabolic and Molecular Bases of Inherited Disease, 8th ed. McGraw-Hill, 2001.

Waters PJ, Scriver CR, Parniak MA: Homomeric and heteromeric interactions between wild-type and mutant phenylalanine hydroxylase subunits: evaluation of two-hybrid approaches for functional analysis of mutations causing hyperphenylalaninemia. Mol Genet Metab 2001;73:230.

Conversion of Amino Acids to Specialized Products

Victor W. Rodwell, PhD

BIOMEDICAL IMPORTANCE

Certain proteins contain amino acids that have been post-translationally modified to permit them to perform specific functions. Examples include the carboxylation of glutamate to form γ-carboxyglutamate, which functions in Ca^{2+} binding, the hydroxylation of proline for incorporation into the collagen triple helix, and the hydroxylation of lysine to hydroxylysine whose subsequent modification and cross-linking stabilizes maturing collagen fibers. In addition to serving as the building blocks for protein synthesis, certain amino acids fulfill additional roles as precursors of biologic materials such as heme, purines, pyrimidines, hormones, neurotransmitters, and biologically active peptides. Histamine plays a central role in many allergic reactions. Neurotransmitters derived from amino acids include γ-aminobutyrate, 5-hydroxytryptamine (serotonin), dopamine, norepinephrine and epinephrine. Many drugs used to treat neurologic and psychiatric conditions act by altering the metabolism of these neurotransmitters.

L-α-AMINO ACIDS

Alanine

Alanine serves as a carrier of ammonia and of the carbons of pyruvate from skeletal muscle to liver via the Cori cycle (see Figure 20–4), and together with glycine constitutes a major fraction of the free amino acids in plasma.

Arginine

Figure 30–1 summarizes metabolic fates of arginine. The reaction catalyzed by nitric oxide synthase, a five-electron oxidoreductase with multiple cofactors, converts one nitrogen of the guanidine group of arginine to nitric oxide (NO), an intercellular signaling molecule that serves as a neurotransmitter, smooth muscle relaxant, and vasodilator (see Chapter 49).

$$\text{L-Arginine} + \text{NADPH} + \text{H}^+ + \text{O}_2 \rightarrow \text{NO} + \text{Citrulline} + \text{NADP}^+$$

The guanidino group of arginine is incorporated into creatine, and following conversion to ornithine, its carbon skeleton becomes that of the polyamines putrescine and spermine.

Cysteine

Cysteine participates in the biosynthesis of coenzyme A (see Figure 44–18) by reacting with pantothenate to form 4-phosphopantothenoyl-cysteine (Figure 30–2). Three enzyme-catalyzed reactions convert cysteine to taurine, which can displace the coenzyme A moiety of cholyl-CoA to form the bile acid taurocholic acid (see Figure 26–7). The conversion of cysteine to taurine is initiated by its oxidation to cysteine sulfinate, catalyzed by the nonheme Fe^{2+} enzyme cysteine dioxygenase. Decarboxylation of cysteine sulfinate by cysteine sulfinate decarboxylase forms hypotaurine, whose oxidation by hypotaurine dehydrogenase forms taurine (Figure 30–3).

Glycine

Metabolites and pharmaceuticals excreted as water-soluble glycine conjugates include glycocholic acid (see Chapter 26) and hippuric acid formed from the food additive benzoate (Figure 30–4). Many drugs, drug metabolites, and other compounds with carboxyl groups are excreted in the urine as glycine conjugates. Glycine is incorporated into creatine, and the nitrogen and α-carbon of glycine are incorporated into the pyrrole rings and the methylene bridge carbons of heme (see Chapter 31), and the entire glycine molecule becomes atoms 4, 5, and 7 of purines (Figure 33–1).

Histidine

Decarboxylation of histidine by the pyridoxal 5′-phosphate-dependent enzyme histidine decarboxylase forms histamine (Figure 30–5). A biogenic amine that functions in allergic reactions and gastric secretion, histamine is present in all tissues. Its concentration in the brain hypothalamus varies in accordance with a circadian rhythm. Histidine compounds present in the human body include ergothioneine, carnosine, and dietary anserine (Figure 30–6). While their physiological functions are unknown, carnosine (β-alanyl-histidine) and homocarnosine (γ-aminobutyryl-histidine) are major constituents of excitable tissues, brain, and skeletal muscle. Urinary levels of 3-methylhistidine are unusually low in patients with **Wilson's disease**.

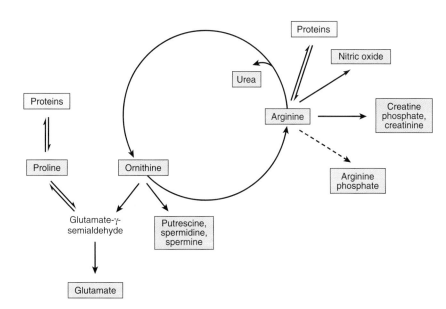

FIGURE 30–1 Arginine, ornithine, and proline metabolism. Reactions with solid arrows all occur in mammalian tissues. Putrescine and spermine synthesis occurs in both mammals and bacteria. Arginine phosphate of invertebrate muscle functions as a phosphagen analogous to creatine phosphate of mammalian muscle.

Methionine

The major nonprotein fate of methionine is conversion to *S*-adenosylmethionine, the principal source of methyl groups in the body. *S*-Adenosylmethionine is synthesized from methionine and ATP, a reaction catalyzed by methionine adenosyltransferase (MAT) (Figure 30–7). Human tissues contain three MAT isozymes (MAT-1 and MAT-3 of liver and MAT-2 of nonhepatic tissues). Although **hypermethioninemia** can result from severely decreased hepatic MAT-1 and MAT-3 activity, if there is residual MAT-1/MAT-3 activity and MAT-2 activity is normal, a high tissue concentration of methionine will assure synthesis of adequate amounts of *S*-adenosylmethionine.

Following decarboxylation of *S*-adenosylmethionine by methionine decarboxylase, three carbons and the α-amino group of methionine contribute to the biosynthesis of the

FIGURE 30–3 Conversion of cysteine to taurine. The reactions are catalyzed by cysteine dioxygenase, cysteine sulfinate decarboxylase, and hypotaurine decarboxylase, respectively.

4–Phospho–pantothenate

Cysteine

CTP

CMP + PP$_i$

4–Phosphopantothenoyl - cysteine

FIGURE 30–2 The reaction catalyzed by phosphopantothenate-cysteine ligase. R-COO⁻ represents 4-phosphopantothenate.

FIGURE 30–4 Biosynthesis of hippurate. Analogous reactions occur with many acidic drugs and catabolites.

FIGURE 30–5 The reaction catalyzed by histidine decarboxylase.

FIGURE 30–6 Derivatives of histidine. Colored boxes surround the components not derived from histidine. The SH group of ergothioneine derives from cysteine.

polyamines spermine and spermidine (Figure 30–8). These polyamines function in cell proliferation and growth, are growth factors for cultured mammalian cells, and stabilize intact cells, subcellular organelles, and membranes. Pharmacologic doses of polyamines are hypothermic and hypotensive. Since they bear multiple positive charges, polyamines associate readily with DNA and RNA. Figure 30–8 summarizes the biosynthesis of polyamines from methionine and ornithine, and Figure 30–9 the catabolism of polyamines.

Serine

Serine participates in the biosynthesis of sphingosine (see Chapter 24), and of purines and pyrimidines, where it provides carbons 2 and 8 of purines and the methyl group of thymine (see Chapter 33). Conversion of serine to homocysteine is catalyzed by cystathionine β-synthase:

$$\text{Serine} + \text{Homocysteine} \rightarrow \text{Cystathionine} + H_2O$$

FIGURE 30–7 Biosynthesis of S-adenosylmethionine, catalyzed by methionine adenosyltransferase (MAT).

Tryptophan

Following hydroxylation of tryptophan to 5-hydroxytryptophan by liver tyrosine hydroxylase, subsequent decarboxylation forms serotonin (5-hydroxytryptamine), a potent vaso-

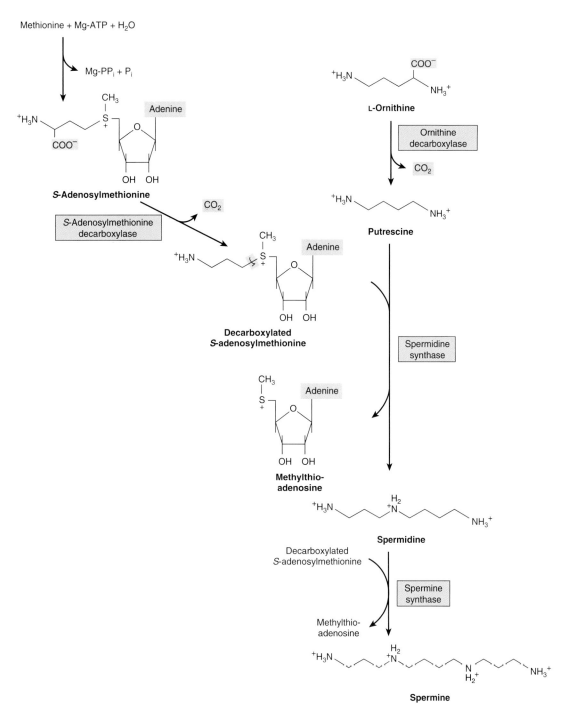

FIGURE 30–8 Intermediates and enzymes that participate in the biosynthesis of spermidine and spermine.

constrictor and stimulator of smooth muscle contraction. Catabolism of serotonin is initiated by monoamine oxidase-catalyzed oxidative deamination to 5-hydroxyindole-3-acetate (Figure 30–10). The psychic stimulation that follows administration of iproniazid results from its ability to prolong the action of serotonin by inhibiting monoamine oxidase. In carcinoid (argentaffinoma), tumor cells overproduce serotonin. Urinary metabolites of serotonin in patients with carcinoid include *N*-acetylserotonin glucuronide and the glycine conjugate of 5-hydroxyindoleacetate. Serotonin and 5-meth-

oxytryptamine are metabolized to the corresponding acids by monoamine oxidase. *N*-Acetylation of serotonin followed by its *O*-methylation in the pineal body forms melatonin. Circulating melatonin is taken up by all tissues, including brain, but is rapidly metabolized by hydroxylation followed by conjugation with sulfate or with glucuronic acid. Kidney tissue, liver tissue, and fecal bacteria all convert tryptophan to tryptamine, then to indole 3-acetate. The principal normal urinary catabolites of tryptophan are 5-hydroxyindoleacetate and indole 3-acetate.

FIGURE 30–9 Catabolism of polyamines. Structures are abbreviated to facilitate presentation.

Tyrosine

Neural cells convert tyrosine to epinephrine and norepinephrine (Figure 30–11). While dopa is also an intermediate in the formation of melanin, different enzymes hydroxylate tyrosine in melanocytes. Dopa decarboxylase, a pyridoxal phosphate-dependent enzyme, forms dopamine. Subsequent hydroxylation by dopamine β-oxidase then forms norepinephrine. In the adrenal medulla, phenylethanolamine-*N*-methyltransferase utilizes *S*-adenosylmethionine to methylate the primary amine of norepinephrine, forming epinephrine (Figure 30–11). Tyrosine is also a precursor of triiodothyronine and thyroxine (see Chapter 41).

Phosphoserine, Phosphothreonine, & Phosphotyrosine

The phosphorylation and dephosphorylation of specific seryl, threonyl or tyrosyl residues of proteins regulate the activity of certain enzymes of lipid and carbohydrate metabolism (see Chapters 9 & 19–26) and of proteins that participate in signal transduction cascades (see Chapter 42).

Sarcosine (*N*-Methylglycine)

The biosynthesis and catabolism of sarcosine (*N*-methylglycine) occur in mitochondria. Formation of sarcosine from dimethylglycine is catalyzed by the flavoprotein dimethylglycine dehydrogenase, which requires reduced pteroylpentaglutamate (TPG)

$$\text{Dimethylglycine} + FADH_2 + H_4TPG + H_2O$$
$$\rightarrow \text{Sarcosine} + N\text{-formyl-TPG}$$

Traces of sarcosine can also arise by methylation of glycine, a reaction catalyzed by *S*-adenosylmethionine:glycine methyltransferase.

$$\text{Glycine} + S\text{-Adenosylmethionine} \rightarrow \text{Sarcosine} +$$
$$S\text{-Adenosylhomocysteine}$$

Catabolism of sarcosine to glycine, catalyzed by the flavoprotein sarcosine dehydrogenase, also requires reduced TPG:

$$\text{Sarcosine} + FAD + H_4TPG + H_2O$$
$$\rightarrow \text{Glycine} + FADH_2 + N\text{-formyl-TPG}$$

The demethylation reactions that form and degrade sarcosine represent important sources of one-carbon units. $FADH_2$ is reoxidized via the electron transport chain (see Chapter 13).

Creatine & Creatinine

Creatinine is formed in muscle from creatine phosphate by irreversible, nonenzymatic dehydration and loss of phosphate (Figure 30–12). Since the 24-h urinary excretion of creatinine is proportionate to muscle mass, it provides a measure of whether a complete 24-h urine specimen has been collected. Glycine, arginine, and methionine all participate in creatine biosynthesis. Synthesis of creatine is completed by methylation of guanidoacetate by *S*-adenosylmethionine (Figure 30–12).

NON-α-AMINO ACIDS

Non-α-amino acids present in tissues in a free form include β-alanine, β-aminoisobutyrate, and γ-aminobutyrate (GABA). β-Alanine is also present in combined form in coenzyme A (see Figure 44–18) and in the β-alanyl dipeptides carnosine, anserine and homocarnosine (see below).

β-Alanine & β-Aminoisobutyrate

β-Alanine and β-aminoisobutyrate are formed during catabolism of the pyrimidines uracil and thymine, respectively (Figure 33–9). Traces of β-alanine also result from the hydrolysis of β-alanyl dipeptides by the enzyme carnosinase. β-Aminoisobutyrate also arises by transamination of methylmalonate semialdehyde, a catabolite of L-valine (see Figure 29–23).

The initial reaction of β-alanine catabolism is transamination to malonate semialdehyde. Subsequent transfer of coenzyme A from succinyl-CoA forms malonyl-CoA semialdehyde, which is then oxidized to malonyl-CoA and decarbox-

FIGURE 30–10 Biosynthesis and metabolism of serotonin and melatonin. ([NH$_4^+$], by transamination; MAO, monoamine oxidase; ~CH$_3$, from S-adenosylmethionine.)

ylated to the amphibolic intermediate acetyl-CoA. Analogous reactions characterize the catabolism of β-aminoisobutyrate. Transamination forms methylmalonate semialdehyde, which is converted to the amphibolic intermediate succinyl-CoA by reactions 8V and 9V of Figure 29–23. Disorders of β-alanine and β-aminoisobutyrate metabolism arise from defects in enzymes of the pyrimidine catabolic pathway. Principal among

these are disorders that result from a total or partial deficiency of dihydropyrimidine dehydrogenase (see Figure 33–9).

β-Alanyl Dipeptides

The β-alanyl dipeptides carnosine and anserine (N-methylcarnosine) (Figure 30–6) activate myosin ATPase, chelate copper,

FIGURE 30–11 Conversion of tyrosine to epinephrine and norepinephrine in neuronal and adrenal cells. (PLP, pyridoxal phosphate.)

FIGURE 30–12 Biosynthesis of creatine and creatinine. Conversion of glycine and the guanidine group of arginine to creatine and creatine phosphate. Also shown is the nonenzymic hydrolysis of creatine phosphate to creatine.

Hydrolysis of carnosine to β-alanine and L-histidine is catalyzed by carnosinase. The heritable disorder carnosinase deficiency is characterized by **carnosinuria**.

Homocarnosine (Figure 30–6), present in human brain at higher levels than carnosine, is synthesized in brain tissue by carnosine synthetase. Serum carnosinase does not hydrolyze homocarnosine. **Homocarnosinosis**, a rare genetic disorder, is associated with progressive spastic paraplegia and mental retardation.

γ-Aminobutyrate

γ-Aminobutyrate (GABA) functions in brain tissue as an inhibitory neurotransmitter by altering transmembrane potential

and enhance copper uptake. β-Alanyl-imidazole buffers the pH of anaerobically contracting skeletal muscle. Biosynthesis of carnosine is catalyzed by carnosine synthetase in a two-stage reaction that involves initial formation of an enzyme-bound acyl-adenylate of β-alanine and subsequent transfer of the β-alanyl moiety to L-histidine.

$$ATP + β\text{-Alanine} \rightarrow β\text{-Alanyl-AMP} + PP_i$$

$$β\text{-Alanyl-AMP} + L\text{-Histidine} \rightarrow Carnosine + AMP$$

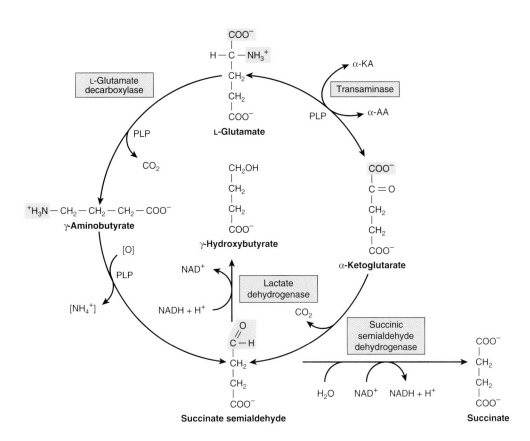

FIGURE 30–13 Metabolism of γ-aminobutyrate. (α-KA, α-keto acids; α-AA, α-amino acids; PLP, pyridoxal phosphate.)

differences. GABA is formed by decarboxylation of glutamate by L-glutamate decarboxylase (Figure 30–13). Transamination of γ-aminobutyrate forms succinate semialdehyde (Figure 30–13), which can be reduced to γ-hydroxybutyrate by L-lactate dehydrogenase, or be oxidized to succinate and thence via the citric acid cycle to CO_2 and H_2O. A rare genetic disorder of GABA metabolism involves a defective GABA aminotransferase, an enzyme that participates in the catabolism of GABA subsequent to its postsynaptic release in brain tissue. Defects in succinic semialdehyde dehydrogenase (Figure 30–13) are responsible for another rare metabolic disorder of γ-aminobutyrate catabolism characterized by **4-hydroxybutyric aciduria**.

SUMMARY

■ In addition to serving structural and functional roles in proteins, α-amino acids participate in a wide variety of other biosynthetic processes.

■ Arginine provides the formamidine group of creatine and the nitrogen of nitric oxide (NO). Via ornithine, arginine provides the skeleton of the polyamines putrescine, spermine and spermidine.

■ Cysteine provides the thioethanolamine portion of coenzyme A, and following its conversion to taurine, part of the bile acid taurocholic acid.

■ Glycine participates in the biosynthesis of heme, purines, creatine, and N-methylglycine (sarcosine). Many drugs and drug metabolites are excreted as glycine conjugates, which increases water solubility for urinary excretion.

■ Decarboxylation of histidine forms the neurotransmitter histamine. Histidine compounds present in the human body include ergothioneine, carnosine, and dietary anserine.

■ S-Adenosylmethionine, the principal source of methyl groups in metabolism, contributes its carbon skeleton to the biosynthesis of the polyamines spermine and spermidine.

■ In addition to its roles in phospholipid and sphingosine biosynthesis, serine provides carbons 2 and 8 of purines and the methyl group of thymine.

■ Key tryptophan metabolites include serotonin and melatonin. Kidney and liver tissue, and also fecal bacteria, convert tryptophan to tryptamine and thence to indole 3-acetate, The principal tryptophan catabolites in urine are indole 3-acetate and 5-hydroxyindoleacetate.

■ Tyrosine forms norepinephrine and epinephrine, and following iodination the thyroid hormones triiodothyronine and thyroxine.

■ The enzyme-catalyzed interconversion of the phospho- and dephospho-forms of peptide bound serine, threonine and tyrosine plays key roles in metabolic regulation, including signal transduction.

■ Glycine, arginine, and S-adenosylmethionine all participate in the biosynthesis of creatine, which as creatine phosphate serves as a major energy reserve in muscle and brain tissue. Excretion in the urine of its catabolite creatine is proportionate to muscle mass.

■ β-Alanine and β-aminoisobutyrate both are present in tissues as free amino acids. β-Alanine also occurs in bound form in coenzyme A, carnosine, anserine, and homocarnosine. Catabolism of β-alanine involves stepwise conversion to acetyl-CoA. Analogous reactions catabolize β-aminoisobutyrate to succinyl-CoA. Disorders of β-alanine and β-aminoisobutyrate metabolism arise from defects in enzymes of pyrimidine catabolism.

■ Decarboxylation of glutamate forms the inhibitory neurotransmitter γ-aminobutyrate (GABA). Two rare metabolic disorders are associated with defects in GABA catabolism.

REFERENCES

Conti M, Beavo J: Biochemistry and physiology of cyclic nucleotide phosphodiesterases: essential components in cyclic nucleotide signaling. Annu Rev Biochem 2007;76:481.

Dominy JE Jr et al: Synthesis of cysteine dioxygenase's amino acid cofactor is regulated by substrate and represents a novel post-translational regulation of activity. J Biol Chem 2008;283:12188.

Joseph CA, Maroney MJ: Cysteine dioxygenase: structure and mechanism. Chem Commun (Camb) 2007;28:3338.

Lindemose S, Nielsen PE, Mollegaard NE: Polyamines preferentially interact with bent adenine tracts in double-stranded DNA. Nucleic Acids Res 2005;33:1790.

Moinard C, Cynober L, de Bandt JP: Polyamines: metabolism and implications in human diseases. Clin Nutr 2005;24:184.

Pearl PL et al: The pediatric neurotransmitter disorders. J Child Neurol 2007;22:606.

Scriver CR et al (editors): *The Metabolic and Molecular Bases of Inherited Disease,* 8th ed. McGraw-Hill, 2001.

Wu F et al: Inhibitory and structural studies of novel coenzyme-substrate analogs of human histidine decarboxylase. FASEB J 2008;22:890.

Porphyrins & Bile Pigments

Robert K. Murray, MD, PhD

BIOMEDICAL IMPORTANCE

The biochemistry of the porphyrins and of the bile pigments is presented in this chapter. These topics are closely related, because heme is synthesized from porphyrins and iron, and the products of degradation of heme are the bile pigments and iron.

Knowledge of the biochemistry of the porphyrins and of heme is basic to understanding the varied functions of **hemoproteins** (see below) in the body. The **porphyrias** are a group of diseases caused by abnormalities in the pathway of biosynthesis of the various porphyrins. Although porphyrias are not very prevalent, physicians must be aware of them. A much more prevalent clinical condition is **jaundice,** due to elevation of bilirubin in the plasma. This elevation is due to overproduction of bilirubin or to failure of its excretion and is seen in numerous diseases ranging from hemolytic anemias to viral hepatitis and to cancer of the pancreas.

METALLOPORPHYRINS & HEMOPROTEINS ARE IMPORTANT IN NATURE

Porphyrins are cyclic compounds formed by the linkage of four pyrrole rings through methyne ($=HC-$) bridges (Figure 31–1). A characteristic property of the porphyrins is the formation of complexes with metal ions bound to the nitrogen atom of the pyrrole rings. Examples are the **iron porphyrins** such as **heme** of hemoglobin and the **magnesium**-containing porphyrin **chlorophyll,** the photosynthetic pigment of plants.

Proteins that contain heme (hemoproteins) are widely distributed in nature. Examples of their importance in humans and animals are listed in Table 31–1.

Natural Porphyrins Have Substituent Side Chains on the Porphin Nucleus

The porphyrins found in nature are compounds in which various **side chains** are substituted for the eight hydrogen atoms numbered in the porphyrin nucleus shown in Figure 31–1. As a simple means of showing these substitutions, Fischer proposed a shorthand formula in which the methyne bridges are omitted

and each pyrrole ring is shown as in Figure 31–2, with the eight substituent positions numbered as indicated. Various porphyrins are represented in Figures 31–2, 31–3, and 31–4.

FIGURE 31–1 The porphyrin molecule. Rings are labeled I, II, III, and IV. Substituent positions on the rings are labeled 1, 2, 3, 4, 5, 6, 7, and 8. The methyne bridges ($=HC-$) are labeled α, β, γ, and δ. The numbering system used is that of Hans Fischer.

TABLE 31–1   Examples of Some Important Human and Animal Hemoproteins[1]

Protein	Function
Hemoglobin	Transport of oxygen in blood
Myoglobin	Storage of oxygen in muscle
Cytochrome *c*	Involvement in electron transport chain
Cytochrome P450	Hydroxylation of xenobiotics
Catalase	Degradation of hydrogen peroxide
Tryptophan pyrrolase	Oxidation of tryptophan

[1]The functions of the above proteins are described in various chapters of this text.

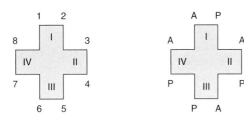

FIGURE 31–2 Uroporphyrin III. (A [acetate] = —CH₂COOH; P [propionate] = —CH₂CH₂COOH.) Note the asymmetry of substituents in ring IV (see text).

The arrangement of the acetate (A) and propionate (P) substituents in the uroporphyrin shown in Figure 31–2 is asymmetric (in ring IV, the expected order of the A and P substituents is reversed). A porphyrin with this type of **asymmetric substitution** is classified as a type III porphyrin. A porphyrin with a completely symmetric arrangement of the substituents is classified as a type I porphyrin. Only types I and III are found in nature, and the type III series is far more abundant (Figure 31–3) and more important because it includes heme.

Heme and its immediate precursor, protoporphyrin IX (Figure 31–4), are both type III porphyrins (ie, the methyl groups are asymmetrically distributed, as in type III copropor-

phyrin). However, they are sometimes identified as belonging to series IX, because they were designated ninth in a series of isomers postulated by Hans Fischer, the pioneer worker in the field of porphyrin chemistry.

HEME IS SYNTHESIZED FROM SUCCINYL-COA & GLYCINE

Heme is synthesized in living cells by a pathway that has been much studied. The two starting materials are **succinyl-CoA,** derived from the citric acid cycle in mitochondria, and the amino acid **glycine.** Pyridoxal phosphate is also necessary in this reaction to "activate" glycine. The product of the condensation reaction between succinyl-CoA and glycine is α-amino-β-ketoadipic acid, which is rapidly decarboxylated to form α-aminolevulinate (ALA) (Figure 31–5). This reaction sequence is catalyzed by ALA synthase, the rate-controlling enzyme in porphyrin biosynthesis in mammalian liver. Synthesis of ALA occurs in **mitochondria.** In the cytosol, two molecules of ALA are condensed by the enzyme **ALA dehydratase** to form two molecules of water and one of **porphobilinogen** (PBG) (Figure 31–5). ALA dehydratase is a zinc-containing enzyme and is sensitive to inhibition by **lead,** as can occur in lead poisoning.

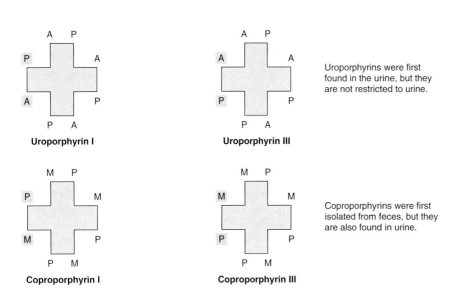

FIGURE 31–3 Uroporphyrins and coproporphyrins. (A, acetate; P, propionate; M, methyl.)

Uroporphyrins were first found in the urine, but they are not restricted to urine.

Coproporphyrins were first isolated from feces, but they are also found in urine.

FIGURE 31–4 Addition of iron to protoporphyrin to form heme. (V [vinyl] = —CH=CH₂.)

FIGURE 31–5 Biosynthesis of porphobilinogen. ALA synthase occurs in the mitochondria, whereas ALA dehydratase is present in the cytosol.

The formation of a cyclic tetrapyrrole—ie, a porphyrin—occurs by condensation of four molecules of PBG (Figure 31–6). These four molecules condense in a head-to-tail manner to form a linear tetrapyrrole, hydroxymethylbilane (HMB). The reaction is catalyzed by uroporphyrinogen I synthase, also named PBG deaminase or HMB synthase. HMB cyclizes spontaneously to form **uroporphyrinogen I** (left-hand side of Figure 31–6) or is converted to **uroporphyrinogen III** by the action of uroporphyrinogen III synthase (right-hand side of Figure 31–6). Under normal conditions, the uroporphyrinogen formed is almost exclusively the III isomer, but in certain of the porphyrias (discussed below), the type I isomers of porphyrinogens are formed in excess.

Note that both of these uroporphyrinogens have the pyrrole rings connected by methylene bridges ($-CH_2-$), which do not form a conjugated ring system. Thus, these compounds are colorless (as are all porphyrinogens). However, the porphyrinogens are readily auto-oxidized to their respective colored porphyrins. These oxidations are catalyzed by light and by the porphyrins that are formed.

Uroporphyrinogen III is converted to coproporphyrinogen III by decarboxylation of all of the acetate (A) groups, which changes them to methyl (M) substituents. The reaction is catalyzed by **uroporphyrinogen decarboxylase,** which is also capable of converting uroporphyrinogen I to coproporphyrinogen I (Figure 31–7). Coproporphyrinogen III then enters the mitochondria, where it is converted to **protoporphyrinogen III** and then to **protoporphyrin III.** Several steps are involved in this conversion. The mitochondrial enzyme **coproporphyrinogen oxidase** catalyzes the decarboxylation and oxidation of two propionic side chains to form proto-porphyrinogen. This enzyme is able to act only on type III coproporphyrinogen, which would explain why type I protoporphyrins do not generally occur in nature. The oxidation of protoporphyrinogen to protoporphyrin is catalyzed by another mitochondrial enzyme, **protoporphyrinogen oxidase.** In mammalian liver, the conversion of coproporphyrinogen to protoporphyrin requires molecular oxygen.

Formation of Heme Involves Incorporation of Iron into Protoporphyrin

The final step in heme synthesis involves the incorporation of ferrous iron into protoporphyrin in a reaction catalyzed by **ferrochelatase (heme synthase),** another mitochondrial enzyme (Figure 31–4).

A summary of the steps in the biosynthesis of the porphyrin derivatives from PBG is given in Figure 31–8. The last three enzymes in the pathway and ALA synthase are located in the mitochondrion, whereas the other enzymes are cytosolic. Both erythroid and nonerythroid ("housekeeping") forms of ALA synthase are found. Heme biosynthesis occurs in most mammalian cells with the exception of mature erythrocytes, which do not contain mitochondria. However, approximately 85% of heme synthesis occurs in erythroid precursor cells in the **bone marrow** and the majority of the remainder in **hepatocytes.**

The porphyrinogens described above are colorless, containing six extra hydrogen atoms as compared with the corresponding colored porphyrins. These **reduced porphyrins** (the porphyrinogens) and not the corresponding porphyrins are the actual intermediates in the biosynthesis of protoporphyrin and of heme.

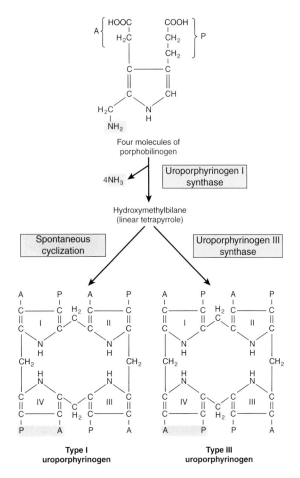

FIGURE 31–6 Conversion of porphobilinogen to uroporphyrinogens. Uroporphyrinogen synthase I is also called porphobilinogen (PBG) deaminase or hydroxymethylbilane (HMB) synthase.

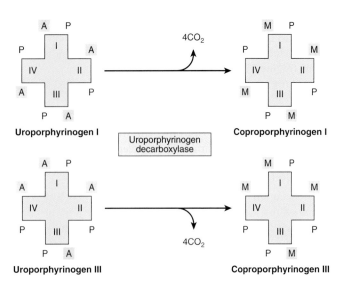

FIGURE 31–7 Decarboxylation of uroporphyrinogens to coproporphyrinogens in cytosol. (A, acetyl; M, methyl; P, propionyl.)

ALA Synthase Is the Key Regulatory Enzyme in Hepatic Biosynthesis of Heme

ALA synthase occurs in both hepatic (ALAS1) and erythroid (ALAS2) forms. The rate-limiting reaction in the synthesis of heme in liver is that catalyzed by ALAS1 (Figure 31–5), a regulatory enzyme. It appears that heme, probably acting through an aporepressor molecule, acts as a negative regulator of the synthesis of ALAS1. This repression-derepression mechanism is depicted diagrammatically in Figure 31–9. Thus, the rate of synthesis of ALAS1 increases greatly in the absence of heme and is diminished in its presence. The turnover rate of ALAS1 in rat liver is normally rapid (half-life about 1 h), a common feature of an enzyme catalyzing a rate-limiting reaction. Heme also affects translation of the enzyme and its transfer from the cytosol to the mitochondrion.

Many **drugs** when administered to humans can result in a marked increase in ALAS1. Most of these drugs are metabolized by a system in the liver that utilizes a specific hemoprotein, **cytochrome P450** (see Chapter 53). During their metabolism, the utilization of heme by cytochrome P450 is greatly increased, which in turn diminishes the intracellular heme concentration. This latter event effects a derepression of ALAS1 with a corresponding increased rate of heme synthesis to meet the needs of the cells.

Several factors affect drug-mediated derepression of ALAS1 in liver—eg, the administration of glucose can prevent it, as can the administration of hematin (an oxidized form of heme).

The importance of some of these regulatory mechanisms is further discussed below when the porphyrias are described.

Regulation of the **erythroid** form of ALAS (ALAS2) differs from that of ALAS1. For instance, it is not induced by the drugs that affect ALAS1, and it does not undergo feedback regulation by heme.

PORPHYRINS ARE COLORED & FLUORESCE

The various porphyrinogens are colorless, whereas the various porphyrins are all colored. In the study of porphyrins or porphyrin derivatives, the **characteristic absorption spectrum** that each exhibits—in both the visible and the ultraviolet regions of the spectrum—is of great value. An example is the absorption curve for a solution of porphyrin in 5% hydrochloric acid (Figure 31–10). Note particularly the sharp absorption band near 400 nm. This is a distinguishing feature of the porphyrin ring and is characteristic of all porphyrins regardless of the side chains present. This band is termed the **Soret band** after its discoverer, the French physicist Charles Soret.

When porphyrins dissolved in strong mineral acids or in organic solvents are illuminated by ultraviolet light, they emit a strong red **fluorescence**. This fluorescence is so characteristic that it is often used to detect small amounts of free porphyrins.

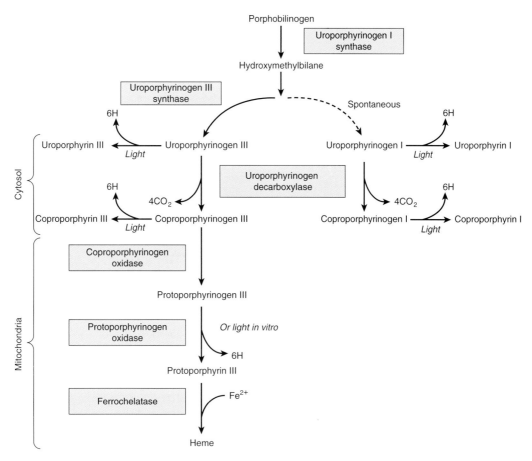

FIGURE 31–8 Steps in the biosynthesis of the porphyrin derivatives from porphobilinogen. Uroporphyrinogen I synthase is also called porphobilinogen deaminase or hydroxymethylbilane synthase.

The double bonds joining the pyrrole rings in the porphyrins are responsible for the characteristic absorption and fluorescence of these compounds; these double bonds are absent in the porphyrinogens.

An interesting application of the photodynamic properties of porphyrins is their possible use in the treatment of certain types of cancer, a procedure called **cancer phototherapy.** Tumors often take up more porphyrins than do normal tissues. Thus, hematoporphyrin or other related compounds are administered to a patient with an appropriate tumor. The tumor is then exposed to an argon laser, which excites the porphyrins, producing cytotoxic effects.

Spectrophotometry Is Used to Test for Porphyrins & Their Precursors

Coproporphyrins and **uroporphyrins** are of clinical interest because they are excreted in increased amounts in the porphyrias. These compounds, when present in urine or feces, can be separated from each other by extraction with appropriate solvent mixtures. They can then be identified and quantified using spectrophotometric methods.

ALA and **PBG** can also be measured in urine by appropriate colorimetric tests.

THE PORPHYRIAS ARE GENETIC DISORDERS OF HEME METABOLISM

The **porphyrias** are a group of disorders due to abnormalities in the pathway of biosynthesis of heme; they can be genetic or acquired. They are not prevalent, but it is important to consider them in certain circumstances (eg, in the differential diagnosis of abdominal pain and of a variety of neuropsychiatric findings); otherwise, patients will be subjected to inappropriate treatments. It has been speculated that King George III had a type of porphyria, which may account for his periodic confinements in Windsor Castle and perhaps for some of his views regarding American colonists. Also, the **photosensitivity** (favoring nocturnal activities) and severe **disfigurement** exhibited by some victims of congenital erythropoietic porphyria have led to the suggestion that these individuals may have been the prototypes of so-called werewolves. No evidence to support this notion has been adduced.

Biochemistry Underlies the Causes, Diagnoses, & Treatments of the Porphyrias

Six major types of **porphyria** have been described, resulting from depressions in the activities of enzymes 3 through 8 shown in Figure 31–9 (see also Table 31–2). Assay of the

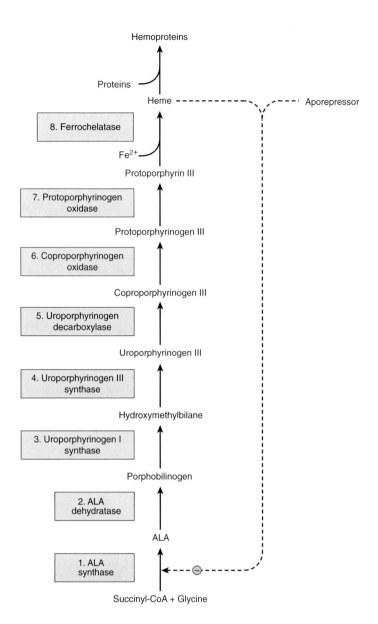

FIGURE 31–9 Intermediates, enzymes, and regulation of heme synthesis. The enzyme numbers are those referred to in column 1 of Table 31–2. Enzymes 1, 6, 7, and 8 are located in mitochondria, the others in the cytosol. Mutations in the gene encoding enzyme 1 causes X-linked sideroblastic anemia. Mutations in the genes encoding enzymes 2–8 cause the porphyrias, though only a few cases due to deficiency of enzyme 2 have been reported. Regulation of hepatic heme synthesis occurs at ALA synthase (ALAS1) by a repression-derepression mechanism mediated by heme and its hypothetical aporepressor. The dotted lines indicate the negative (–) regulation by repression. Enzyme 3 is also called porphobilinogen deaminase or hydroxymethylbilane synthase.

FIGURE 31–10 Absorption spectrum of hematoporphyrin (0.01% solution in 5% HCl).

activity of one or more of these enzymes using an appropriate source (eg, red blood cells) is thus important in making a definitive diagnosis in a suspected case of porphyria. Indi-

viduals with low activities of enzyme 1 (ALAS2) develop anemia, not porphyria (see Table 31–2). Patients with low activities of enzyme 2 (ALA dehydratase) have been reported, but very rarely; the resulting condition is called ALA dehydratase-deficient porphyria.

In general, the porphyrias described are **inherited** in an autosomal dominant manner, with the exception of congenital erythropoietic porphyria, which is inherited in a recessive mode. The precise abnormalities in the genes directing synthesis of the enzymes involved in heme biosynthesis have been determined in some instances. Thus, the use of appropriate gene probes has made possible the prenatal diagnosis of some of the porphyrias.

As is true of most **inborn errors,** the signs and symptoms of porphyria result from either a deficiency of metabolic products beyond the enzymatic block or from an accumulation of metabolites behind the block.

If the enzyme lesion occurs **early** in the pathway prior to the formation of porphyrinogens (eg, enzyme 3 of Figure 31–9,

TABLE 31–2 Summary of Major Findings in the Porphyrias[1]

Enzyme Involved[2]	Type, Class, and OMIM Number	Major Signs and Symptoms	Results of Laboratory Tests
1. ALA synthase (erythroid form)	X-linked sideroblastic anemia[3] (erythropoietic) (OMIM 301300)	Anemia	Red cell counts and hemoglobin decreased
2. ALA dehydratase	ALA dehydratase deficiency (hepatic) (OMIM 125270)	Abdominal pain, neuropsychiatric symptoms	Urinary ALA and coproporphyrin III increased
3. Uroporphyrinogen I synthase[4]	Acute intermittent porphyria (hepatic) (OMIM 176000)	Abdominal pain, neuropsychiatric symptoms	Urinary ALA and PBG increased
4. Uroporphyrinogen III synthase	Congenital erythropoietic (erythropoietic) (OMIM 263700)	Photosensitivity	Urinary, fecal, and red cell uroporphyrin I increased
5. Uroporphyrinogen decarboxylase	Porphyria cutanea tarda (hepatic) (OMIM 176100)	Photosensitivity	Urinary uroporphyrin I increased
6. Coproporphyrinogen oxidase	Hereditary coproporphyria (hepatic) (OMIM 121300)	Photosensitivity, abdominal pain, neuropsychiatric symptoms	Urinary ALA, PBG, and coproporphyrin III and fecal coproporphyrin III increased
7. Protoporphyrinogen oxidase	Variegate porphyria (hepatic) (OMIM 176200)	Photosensitivity, abdominal pain, neuropsychiatric symptoms	Urinary ALA, PBG, and coproporphyrin III and fecal protoporphyrin IX increased
8. Ferrochelatase	Protoporphyria (erythropoietic) (OMIM 177000)	Photosensitivity	Fecal and red cell protoporphyrin IX increased

Abbreviations: ALA, δ-aminolevulinic acid; PBG, porphobilinogen.
[1]Only the biochemical findings in the active stages of these diseases are listed. Certain biochemical abnormalities are detectable in the latent stages of some of the above conditions. Conditions 3, 5, and 8 are generally the most prevalent porphyrias. Condition 2 is rare.
[2]The numbering of the enzymes in this table corresponds to that used in Figure 31–9.
[3]X-linked sideroblastic anemia is not a porphyria but is included here because ALA synthase is involved.
[4]This enzyme is also called PBG deaminase or hydroxymethylbilane synthase.

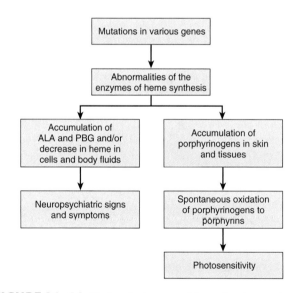

FIGURE 31–11 Biochemical causes of the major signs and symptoms of the porphyrias.

which is affected in acute intermittent porphyria), ALA and PBG will accumulate in body tissues and fluids (Figure 31–11). Clinically, patients complain of abdominal pain and neuropsychiatric symptoms. The precise biochemical cause of these symptoms has not been determined but may relate to elevated levels of ALA or PBG or to a deficiency of heme.

On the other hand, enzyme blocks **later** in the pathway result in the accumulation of the porphyrinogens indicated in

Figures 31–9 and 31–11. Their oxidation products, the corresponding porphyrin derivatives, cause **photosensitivity**, a reaction to visible light of about 400 nm. The porphyrins, when exposed to light of this wavelength, are thought to become "excited" and then react with molecular oxygen to form oxygen radicals. These latter species injure lysosomes and other organelles. Damaged lysosomes release their degradative enzymes, causing variable degrees of skin damage, including scarring.

The porphyrias can be **classified** on the basis of the organs or cells that are most affected. These are generally organs or cells in which synthesis of heme is particularly active. The bone marrow synthesizes considerable hemoglobin, and the liver is active in the synthesis of another hemoprotein, cytochrome P450. Thus, one classification of the porphyrias is to designate them as predominantly either **erythropoietic** or **hepatic;** the types of porphyrias that fall into these two classes are so characterized in Table 31–2. Porphyrias can also be classified as **acute** or **cutaneous** on the basis of their clinical features. Why do specific types of porphyria affect certain organs more markedly than others? A partial answer is that the levels of metabolites that cause damage (eg, ALA, PBG, specific porphyrins, or lack of heme) can vary markedly in different organs or cells depending upon the differing activities of their heme-forming enzymes.

As described above, **ALAS1** is the key regulatory enzyme of the heme biosynthetic pathway in liver. A large number of **drugs** (eg, barbiturates, griseofulvin) induce the enzyme. Most of these drugs do so by inducing cytochrome P450 (see Chapter 53), which uses up heme and thus derepresses (induces) ALAS1. In patients with porphyria, increased activities of ALAS1 result in increased levels of potentially harmful heme precursors prior to the metabolic block. Thus, taking drugs that cause induction of cytochrome P450 (so-called microsomal inducers) can precipitate attacks of porphyria.

The **diagnosis** of a specific type of porphyria can generally be established by consideration of the clinical and family history, the physical examination, and appropriate laboratory tests. The major findings in the six principal types of porphyria are listed in Table 31–2.

High levels of **lead** can affect heme metabolism by combining with SH groups in enzymes such as ferrochelatase and ALA dehydratase. This affects porphyrin metabolism. Elevated levels of protoporphyrin are found in red blood cells, and elevated levels of ALA and of coproporphyrin are found in urine.

It is hoped that **treatment** of the porphyrias at the gene level will become possible. In the meantime, treatment is essentially symptomatic. It is important for patients to avoid drugs that cause induction of cytochrome P450. Ingestion of large amounts of carbohydrates (glucose loading) or administration of hematin (a hydroxide of heme) may repress ALAS1, resulting in diminished production of harmful heme precursors. Patients exhibiting photosensitivity may benefit from administration of β-carotene; this compound appears to lessen production of free radicals, thus diminishing photosensitivity. Sunscreens that filter out visible light can also be helpful to such patients.

CATABOLISM OF HEME PRODUCES BILIRUBIN

Under physiologic conditions in the human adult, 1–2×10^8 erythrocytes are destroyed per hour. Thus, in 1 day, a 70-kg human turns over approximately 6 g of hemoglobin. When hemoglobin is destroyed in the body, **globin** is degraded to its constituent amino acids, which are reused, and the **iron** of heme enters the iron pool, also for reuse. The iron-free **porphyrin** portion of heme is also degraded, mainly in the reticuloendothelial cells of the liver, spleen, and bone marrow.

The catabolism of heme from all of the heme proteins appears to be carried out in the microsomal fractions of cells by a complex enzyme system called **heme oxygenase.** By the time the heme derived from heme proteins reaches the oxygenase system, the iron has usually been oxidized to the ferric form, constituting **hemin.** The heme oxygenase system is substrate-inducible. As depicted in Figure 31–12, the hemin is reduced to heme with NADPH, and, with the aid of more NADPH, oxygen is added to the α-methyne bridge between pyrroles I and II of the porphyrin. The ferrous iron is again oxidized to the ferric form. With the further addition of oxygen, **ferric ion** is released, **carbon monoxide** is produced, and an equimolar quantity of **biliverdin** results from the splitting of the tetrapyrrole ring.

In birds and amphibia, the green biliverdin IX is excreted; in mammals, a soluble enzyme called **biliverdin reductase** reduces the methyne bridge between pyrrole III and pyrrole IV to a methylene group to produce **bilirubin,** a yellow pigment (Figure 31–12).

It is estimated that 1 g of hemoglobin yields 35 mg of bilirubin. The daily bilirubin formation in human adults is approximately 250–350 mg, deriving mainly from hemoglobin but also from ineffective erythropoiesis and from various other heme proteins such as cytochrome P450.

The chemical conversion of heme to bilirubin by reticuloendothelial cells can be observed in vivo as the purple color of the heme in a hematoma is slowly converted to the yellow pigment of bilirubin.

Bilirubin formed in peripheral tissues is **transported to the liver** by plasma albumin. The **further metabolism** of bilirubin occurs primarily in the liver. It can be divided into three processes: (1) uptake of bilirubin by liver parenchymal cells, (2) conjugation of bilirubin with glucuronate in the endoplasmic reticulum, and (3) secretion of conjugated bilirubin into the bile. Each of these processes will be considered separately.

THE LIVER TAKES UP BILIRUBIN

Bilirubin is only **sparingly soluble** in water, but its solubility in plasma is increased by noncovalent binding to albumin. Each molecule of albumin appears to have one high-affinity site and one low-affinity site for bilirubin. In 100 mL of plasma, approximately 25 mg of bilirubin can be tightly bound to albumin at its high-affinity site. Bilirubin in excess of this quantity can be bound only loosely and thus can easily be detached and diffuse into tissues. A number of compounds such as antibiotics and other drugs compete with bilirubin for the high-affinity binding site on albumin. Thus, these compounds can displace bilirubin from albumin and have significant clinical effects.

In the liver, the bilirubin is removed from albumin and taken up at the sinusoidal surface of the hepatocytes by a carrier-mediated saturable system. This **facilitated transport system** has a very large capacity, so that even under pathologic conditions the system does not appear to be rate-limiting in the metabolism of bilirubin.

Since this facilitated transport system allows the equilibrium of bilirubin across the sinusoidal membrane of the hepatocyte, the net uptake of bilirubin will be dependent upon the removal of bilirubin via subsequent metabolic pathways.

Once bilirubin enters the hepatocytes, it can **bind to certain cytosolic proteins,** which help to keep it solubilized prior to conjugation. Ligandin (a member of the family of glutathione S-transferases) and protein Y are the involved proteins. They may also help to prevent efflux of bilirubin back into the blood stream.

Conjugation of Bilirubin with Glucuronic Acid Occurs in the Liver

Bilirubin is **nonpolar** and would persist in cells (eg, bound to lipids) if not rendered water-soluble. Hepatocytes convert bilirubin to a **polar** form, which is readily excreted in the bile, by adding glucuronic acid molecules to it. This process is called **conjugation** and can employ polar molecules other than glucuronic acid (eg, sulfate). Many steroid hormones and

FIGURE 31–12 Schematic representation of the microsomal heme oxygenase system. (Modified, with permission, from Schmid R, McDonough AF in: *The Porphyrins*. Dolphin D [edi.]. Academic Press, 1978. Copyright ©1978. Reprinted with permission from Elsevier.)

drugs are also converted to water-soluble derivatives by conjugation in preparation for excretion (see Chapter 52).

The conjugation of bilirubin is catalyzed by a specific **glucuronosyltransferase.** The enzyme is mainly located in the endoplasmic reticulum, uses UDP-glucuronic acid as the glucuronosyl donor, and is referred to as bilirubin-UGT. Bilirubin monoglucuronide is an intermediate and is subsequently converted to the diglucuronide (Figures 31–13 & 31–14). Most of the bilirubin excreted in the bile of mammals is in the form of bilirubin diglucuronide. However, when bilirubin conjugates exist abnormally in human plasma (eg, in obstruc-

tive jaundice), they are predominantly monoglucuronides. Bilirubin-UGT activity can be **induced** by a number of clinically useful drugs, including phenobarbital. More information about glucuronosylation is presented below in the discussion of inherited disorders of bilirubin conjugation.

Bilirubin Is Secreted into Bile

Secretion of conjugated bilirubin into the bile occurs by an active transport mechanism, which is probably rate-limiting for the entire process of hepatic bilirubin metabolism. The protein

FIGURE 31–13 Structure of bilirubin diglucuronide (conjugated, "direct-reacting" bilirubin). Glucuronic acid is attached via ester linkage to the two propionic acid groups of bilirubin to form an acylglucuronide.

FIGURE 31–14 Conjugation of bilirubin with glucuronic acid. The glucuronate donor, UDP-glucuronic acid, is formed from UDP-glucose as depicted. The UDP-glucuronosyltransferase is also called bilirubin-UGT.

FIGURE 31–15 Diagrammatic representation of the three major processes (uptake, conjugation, and secretion) involved in the transfer of bilirubin from blood to bile. Certain proteins of hepatocytes, such as ligandin (a member of the glutathione S-transferase family of enzymes) and Y protein, bind intracellular bilirubin and may prevent its efflux into the blood stream. The process affected in a number of conditions causing jaundice is also shown.

involved is MRP-2 (multidrug-resistance-like protein 2), also called multispecific organic anion transporter (MOAT). It is located in the plasma membrane of the bile canalicular membrane and handles a number of organic anions. It is a member of the family of ATP-binding cassette (ABC) transporters. The hepatic transport of conjugated bilirubin into the bile is inducible by those same drugs that are capable of inducing the conjugation of bilirubin. Thus, the conjugation and excretion systems for bilirubin behave as a coordinated functional unit.

Figure 31–15 summarizes the three major processes involved in the *transfer of bilirubin* from blood to bile. Sites that are affected in a number of conditions causing jaundice (see below) are also indicated.

Conjugated Bilirubin Is Reduced to Urobilinogen by Intestinal Bacteria

As the conjugated bilirubin reaches the terminal ileum and the large intestine, the glucuronides are removed by specific bacterial enzymes (**β-glucuronidases),** and the pigment is subsequently **reduced** by the fecal flora to a group of colorless tetrapyrrolic compounds called **urobilinogens.** In the terminal ileum and large intestine, a small fraction of the urobilinogens is reabsorbed and reexcreted through the liver to constitute the **enterohepatic urobilinogen cycle.** Under abnormal conditions, particularly when excessive bile pigment is formed or liver disease interferes with this intrahepatic cycle, urobilinogen may also be excreted in the urine.

Normally, most of the colorless urobilinogens formed in the colon by the fecal flora are oxidized there to urobilins (colored compounds) and are excreted in the feces. Darkening of feces upon standing in air is due to the oxidation of residual urobilinogens to urobilins.

HYPERBILIRUBINEMIA CAUSES JAUNDICE

When bilirubin in the blood exceeds 1 mg/dL (17.1 μmol/L), **hyperbilirubinemia** exists. Hyperbilirubinemia may be due to the production of more bilirubin than the normal liver can excrete, or it may result from the failure of a damaged liver to excrete bilirubin produced in normal amounts. In the absence of hepatic damage, obstruction of the excretory ducts of the liver—by preventing the excretion of bilirubin—will also cause hyperbilirubinemia. In all these situations, bilirubin accumulates in the blood, and when it reaches a certain concentration (approximately 2–2.5 mg/dL), it diffuses into the tissues, which then become yellow. That condition is called **jaundice** or **icterus.**

In clinical studies of jaundice, measurement of bilirubin in the serum is of great value. A method for quantitatively assaying the bilirubin content of the serum was first devised by **van den Bergh** by application of **Ehrlich's test** for bilirubin in urine. The Ehrlich reaction is based on the coupling of diazotized sulfanilic acid (Ehrlich's diazo reagent) and bilirubin to produce a reddish-purple azo compound. In the original procedure as described by Ehrlich, methanol was used to provide a solution in which both bilirubin and the diazo regent were soluble. van den Bergh inadvertently omitted the methanol on an occasion when assay of bile pigment in human bile was being attempted. To his surprise, normal development of the color occurred "directly." This form of bilirubin that would react without the addition of methanol was thus termed **"direct-reacting."** It was then found that this same direct reaction would also occur in serum from cases of jaundice due to biliary obstruction. However, it was still necessary to add methanol to detect bilirubin in normal serum or that which was present in excess in serum from cases of hemolytic jaundice where no evidence of obstruction was to be found. To that form of bilirubin which could be measured only after the addition of methanol, the term **"indirect-reacting"** was applied.

It was subsequently discovered that the **indirect bilirubin is "free"** (unconjugated) bilirubin en route to the liver from the reticuloendothelial tissues, where the bilirubin was originally produced by the breakdown of heme porphyrins. Since this bilirubin is not water-soluble, it requires methanol to initiate coupling with the diazo reagent. In the liver, the free bilirubin becomes conjugated with glucuronic acid, and the conjugate, predominantly bilirubin diglucuronide, can then be excreted into the bile. Furthermore, conjugated bilirubin, being water-soluble, can react directly with the diazo reagent, so that the "direct bilirubin" of van den Bergh is actually a bilirubin conjugate (bilirubin glucuronide).

Depending on the type of bilirubin present in plasma—ie, unconjugated or conjugated—hyperbilirubinemia may be classified as **retention hyperbilirubinemia,** due to overproduction, or **regurgitation hyperbilirubinemia,** due to reflux into the bloodstream because of biliary obstruction.

Separation and quantitation of unconjugated bilirubin and the conjugated species can be performed using high-pressure liquid chromatography.

Because of its hydrophobicity, only unconjugated bilirubin can cross the blood-brain barrier into the central nervous system; thus, encephalopathy due to hyperbilirubinemia **(kernicterus)** can occur only in connection with unconjugated bilirubin, as found in retention hyperbilirubinemia. On the other hand, because of its water-solubility, only conjugated bilirubin can appear in urine. Accordingly, **choluric jaundice** (choluria is the presence of bile pigments in the urine) occurs only in regurgitation hyperbilirubinemia, and **acholuric jaundice** occurs only in the presence of an excess of unconjugated bilirubin.

Elevated Amounts of Unconjugated Bilirubin in Blood Occur in a Number of Conditions

Hemolytic Anemias

Hemolytic anemias are important causes of unconjugated hyperbilirubinemia, though unconjugated hyperbilirubinemia is usually only slight (<4 mg/dL; <68.4 µmol/L) even in the event of extensive hemolysis because of the healthy liver's large capacity for handling bilirubin.

Neonatal "Physiologic Jaundice"

This transient condition is the most common cause of unconjugated hyperbilirubinemia. It results from an accelerated hemolysis around the time of **birth** and an immature hepatic system for the uptake, conjugation, and secretion of bilirubin. Not only is the bilirubin-UGT activity reduced, but there probably is reduced synthesis of the substrate for that enzyme, UDP-glucuronic acid. Since the increased amount of bilirubin is unconjugated, it is capable of penetrating the blood-brain barrier when its concentration in plasma exceeds that which can be tightly bound by albumin (20–25 mg/dL). This can result in a hyperbilirubinemic toxic encephalopathy, or **kernicterus,** which can cause mental retardation. Because of the recognized inducibility of this bilirubin-metabolizing system, phenobarbital has been administered to jaundiced neonates and is effective in this disorder. In addition, exposure to blue light (phototherapy) promotes the hepatic excretion of unconjugated bilirubin by converting some of the bilirubin to other derivatives such as maleimide fragments and geometric isomers that are excreted in the bile.

Crigler–Najjar Syndrome, Type I; Congenital Nonhemolytic Jaundice

Type I Crigler–Najjar syndrome is a rare autosomal recessive disorder. It is characterized by severe congenital jaundice (serum bilirubin usually exceeds 20 mg/dL) due to mutations in the gene encoding bilirubin-UGT activity in hepatic tissues. The disease is often fatal within the first 15 months of life. Children with this condition have been treated with phototherapy, resulting in some reduction in plasma bilirubin levels. Phenobarbital has no effect on the formation of bilirubin glucuronides in patients with type I Crigler–Najjar syndrome. A liver transplant may be curative.

It should be noted that the gene encoding human bilirubin-UGT is part of a large UGT gene complex situated on chromosome 2. Many different substrates are subjected to glucuronosylation, so many glucuronosyltransferases are required. The complex contains some 13 substrate-specific first exons, each with its own promoter. Four are pseudogenes, so nine different isoforms with differing glucuronosyltransferase activities are encoded. Exon A1 is that involved with conjugation of bilirubin. In the case of bilirubin, exon A1 is spliced to DNA containing exons 2–5, producing bilirubin-UGT. Other transferases are produced by splicing other first exons (members of A 2–13) to exons 2–5.

Crigler–Najjar Syndrome, Type II

This rare inherited disorder also results from mutations in the gene encoding bilirubin-UGT, but some activity of the enzyme is retained and the condition has a more benign course than type I. Serum bilirubin concentrations usually do not exceed 20 mg/dL. Patients with this condition can respond to treatment with large doses of phenobarbital.

Gilbert Syndrome

Again, this relatively prevalent condition is caused by mutations in the gene encoding bilirubin-UGT. It is more common among males. Approximately 30% of the enzyme's activity is preserved and the condition is entirely harmless.

Toxic Hyperbilirubinemia

Unconjugated hyperbilirubinemia can result from toxin-induced liver dysfunction such as that caused by chloroform, arsphenamines, carbon tetrachloride, acetaminophen, hepatitis virus, cirrhosis, and *Amanita* mushroom poisoning. These acquired disorders are due to hepatic parenchymal cell damage, which impairs conjugation.

Obstruction in the Biliary Tree Is the Most Common Cause of Conjugated Hyperbilirubinemia

Obstruction of the Biliary Tree

Conjugated hyperbilirubinemia commonly results from blockage of the hepatic or common bile ducts, most often due to a gallstone or to cancer of the head of the pancreas (Figure 31–16). Because of the obstruction, bilirubin diglucuronide cannot be excreted. It thus regurgitates into the hepatic veins and lymphatics, and conjugated bilirubin appears in the blood and urine (choluric jaundice). Also, the stools are usually pale in color, and should be examined routinely in any case of jaundice.

The term **cholestatic jaundice** is used to include all cases of extrahepatic obstructive jaundice. It also covers those cases of jaundice that exhibit conjugated hyperbilirubinemia due to micro-obstruction of intrahepatic biliary ductules by swollen, damaged hepatocytes (eg, as may occur in infectious hepatitis).

Dubin–Johnson Syndrome

This benign autosomal recessive disorder consists of conjugated hyperbilirubinemia in childhood or during adult life. The hyperbilirubinemia is caused by mutations in the gene encoding **MRP-2** (see above), the protein involved in the secretion of conjugated bilirubin into bile. The centrilobular

hepatocytes contain an abnormal black pigment that may be derived from epinephrine.

Rotor Syndrome

This is a rare benign condition characterized by chronic conjugated hyperbilirubinemia and normal liver histology. Its precise cause has not been identified.

Some Conjugated Bilirubin Can Bind Covalently to Albumin

When levels of conjugated bilirubin remain high in plasma, a fraction can **bind covalently to albumin** (δ [delta] bilirubin). Because it is bound covalently to albumin, this fraction has a longer half-life in plasma than does conventional conjugated bilirubin. Thus, it remains elevated during the recovery phase of obstructive jaundice after the remainder of the conjugated bilirubin has declined to normal levels; this explains why some patients continue to appear jaundiced after conjugated bilirubin levels have returned to normal.

Urobilinogen & Bilirubin in Urine Are Clinical Indicators

Normally, there are mere traces of urobilinogen in the urine. In **complete obstruction of the bile duct,** no urobilinogen is

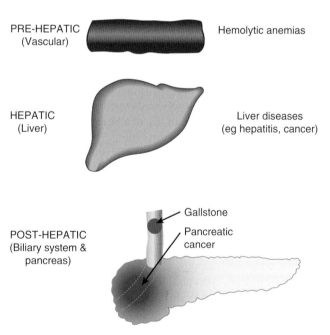

FIGURE 31–16 Diagrammatic representation of some major causes of jaundice. **Prehepatic** indicates events in the blood stream; the major cause would be various forms of hemolytic anemia (see Chapter 52). **Hepatic** signifies events in the liver, such as the various types of hepatitis or other forms of liver disease (eg, cancer). **Posthepatic** refers to events in the biliary tree; the major causes of posthepatic jaundice are obstruction of the common bile duct by a gallstone (biliary calculus) or by cancer of the head of the pancreas.

TABLE 31–3 **Laboratory Results in Normal Patients and Patients with Three Different Causes of Jaundice**

Condition	Serum Bilirubin	Urine Urobilinogen	Urine Bilirubin	Fecal Urobilinogen
Normal	Direct: 0.1–0.4 mg/dL	0–4 mg/24 h	Absent	40–280 mg/24 h
	Indirect: 0.2–0.7 mg/dL			
Hemolytic anemia	↑Indirect	Increased	Absent	Increased
Hepatitis	↑Direct and indirect	Decreased if micro-obstruction is present	Present if micro-obstruction occurs	Decreased
Obstructive jaundice[1]	↑Direct	Absent	Present	Trace to absent

[1]The most common causes of obstructive (posthepatic) jaundice are cancer of the head of the pancreas and a gallstone lodged in the common bile duct. The presence of bilirubin in the urine is sometimes referred to as choluria—therefore, hepatitis and obstruction of the common bile duct cause choluric jaundice, whereas the jaundice of hemolytic anemia is referred to as acholuric. The laboratory results in patients with hepatitis are variable, depending on the extent of damage to parenchymal cells and the extent of micro-obstruction to bile ductules. Serum levels of **alanine aminotransferase** (ALT) and **aspartate aminotransferase** (AST) are usually markedly elevated in hepatitis, whereas serum levels of **alkaline phosphatase** are elevated in obstructive liver disease.

found in the urine, since bilirubin has no access to the intestine, where it can be converted to urobilinogen. In this case, the presence of bilirubin (conjugated) in the urine without urobilinogen suggests obstructive jaundice, either intrahepatic or posthepatic.

In **jaundice secondary to hemolysis,** the increased production of bilirubin leads to increased production of urobilinogen, which appears in the urine in large amounts. Bilirubin is not usually found in the urine in hemolytic jaundice (because unconjugated bilirubin does not pass into the urine), so that the combination of increased urobilinogen and absence of bilirubin is suggestive of hemolytic jaundice. Increased blood destruction from any cause brings about an increase in urine urobilinogen.

Table 31–3 summarizes **laboratory results** obtained on patients with three different causes of jaundice—hemolytic anemia (a prehepatic cause), hepatitis (a hepatic cause), and obstruction of the common bile duct (a posthepatic cause) (see Figure 31–16). Laboratory tests on blood (evaluation of the possibility of a hemolytic anemia and measurement of prothrombin time) and on serum (eg, electrophoresis of proteins; activities of the enzymes ALT, AST, and alkaline phosphatase) are also important in helping to distinguish between prehepatic, hepatic, and posthepatic causes of jaundice.

SUMMARY

- Hemoproteins, such as hemoglobin and the cytochromes, contain heme. Heme is an iron-porphyrin compound (Fe^{2+}-protoporphyrin IX) in which four pyrrole rings are joined by methyne bridges. The eight side groups (methyl, vinyl, and propionyl substituents) on the four pyrrole rings of heme are arranged in a specific sequence.

- Biosynthesis of the heme ring occurs in mitochondria and cytosol via eight enzymatic steps. It commences with formation of δ-aminolevulinate (ALA) from succinyl-CoA and glycine in a reaction catalyzed by ALA synthase, the regulatory enzyme of the pathway.

- Genetically determined abnormalities of seven of the eight enzymes involved in heme biosynthesis result in the inherited porphyrias. Red blood cells and liver are the major sites of metabolic expression of the porphyrias. Photosensitivity and neurologic problems are common complaints. Intake of certain compounds (such as lead) can cause acquired porphyrias. Increased amounts of porphyrins or their precursors can be detected in blood and urine, facilitating diagnosis.

- Catabolism of the heme ring is initiated by the enzyme heme oxygenase, producing a linear tetrapyrrole.

- Biliverdin is an early product of catabolism and on reduction yields bilirubin. The latter is transported by albumin from peripheral tissues to the liver, where it is taken up by hepatocytes. The iron of heme and the amino acids of globin are conserved and reutilized.

- In the liver, bilirubin is made water-soluble by conjugation with two molecules of glucuronic acid and is secreted into the bile. The action of bacterial enzymes in the gut produces urobilinogen and urobilin, which are excreted in the feces and urine.

- Jaundice is due to elevation of the level of bilirubin in the blood. The causes of jaundice can be classified as prehepatic (eg, hemolytic anemias), hepatic (eg, hepatitis), and posthepatic (eg, obstruction of the common bile duct). Measurements of plasma total and nonconjugated bilirubin, of urinary urobilinogen and bilirubin, and of certain serum enzymes as well as inspection and analysis of stool samples help distinguish between these causes.

REFERENCES

Anderson KE et al: Disorders of heme biosynthesis: X-linked sideroblastic anemia and the porphyrias. In: *The Metabolic and Molecular Bases of Inherited Disease,* 8th ed. Chapter 124. Scriver CR et al (eds). McGraw-Hill, 2001.

Chowdhury JR et al: Hereditary jaundice and disorders of bilirubin metabolism. In: *The Metabolic and Molecular Bases of Inherited Disease,* 8th ed. Chapter 125. Scriver CR et al (eds). McGraw-Hill, 2001.

Deacon AC, Whatley SD, Elder GH: Porphyrins and disorders of porphyrin metabolism. In: *Tietz Textbook of Clinical*

Chemistry and Molecular Diagnostics, 4th ed. Chapter 32. Burtis CA, Ashwood ER, Bruns DE (eds). Elsevier Saunders, 2006.

Desnick RJ, Astrin KH: The porphyrias. In: *Harrison's Principles of Internal Medicine,* 17th ed. Fauci AS et al (eds). Chapter 352. McGraw-Hill, 2008.

Dufour DR: Liver disease. In: *Tietz Textbook of Clinical Chemistry and Molecular Diagnostics,* 4th ed. Burtis CA, Ashwood ER, Bruns DE (eds). Chapter 47. Elsevier Saunders, 2006.

Higgins T, Beutler E, Doumas BT: Hemoglobin, iron and bilirubin. In: *Tietz Textbook of Clinical Chemistry and Molecular Diagnostics,* 4th ed. Burtis CA, Ashwood ER, Bruns DE (eds). Chapter 31. Elsevier Saunders, 2006.

Pratt DS, Kaplan MM: Jaundice. In: *Harrison's Principles of Internal Medicine,* 17th ed. Fauci AS et al (eds). Chapter 43. McGraw-Hill, 2008.

Pratt DS, Kaplan MM: Evaluation of liver function. In: *Harrison's Principles of Internal Medicine,* 17th ed. Fauci AS et al (eds). Chapter 296. McGraw-Hill, 2008.

Wolkoff AW: The hyperbilirubinemias. In: *Harrison's Principles of Internal Medicine,* 17th ed. Fauci AS et al (eds). Chapter 297. McGraw-Hill, 2008.

C H A P T E R

32

Nucleotides

Victor W. Rodwell, PhD

BIOMEDICAL IMPORTANCE

In addition to serving as precursors of nucleic acids, purine and pyrimidine nucleotides participate in metabolic functions as diverse as energy metabolism, protein synthesis, regulation of enzyme activity, and signal transduction. When linked to vitamins or vitamin derivatives, nucleotides form a portion of many coenzymes. As the principal donors and acceptors of phosphoryl groups in metabolism, nucleoside tri- and diphosphates such as ATP and ADP are the principal players in the energy transductions that accompany metabolic interconversions and oxidative phosphorylation. Linked to sugars or lipids, nucleosides constitute key biosynthetic intermediates. The sugar derivatives UDP-glucose and UDP-galactose participate in sugar interconversions and in the biosynthesis of starch and glycogen. Similarly, nucleoside-lipid derivatives such as CDP-acylglycerol are intermediates in lipid biosynthesis. Roles that nucleotides perform in metabolic regulation include ATP-dependent phosphorylation of key metabolic enzymes, allosteric regulation of enzymes by ATP, AMP, and CTP, and control by ADP of the rate of oxidative phosphorylation. The cyclic nucleotides cAMP and cGMP serve as the second messengers in

hormonally regulated events, and GTP and GDP play key roles in the cascade of events that characterize signal transduction pathways. Specifically medical applications include the use of synthetic purine and pyrimidine analogs that contain halogens, thiols, or additional nitrogen atoms in the chemotherapy of cancer and AIDS, and as suppressors of the immune response during organ transplantation.

CHEMISTRY OF PURINES, PYRIMIDINES, NUCLEOSIDES, & NUCLEOTIDES

Purines & Pyrimidines Are Heterocyclic Compounds

Purines and pyrimidines are nitrogen-containing **heterocycles,** structures that contain, in addition to carbon, other (hetero) atoms such as nitrogen. Note that the smaller pyrimidine molecule has the *longer* name and the larger purine molecule the *shorter* name, and that their six-atom rings are numbered in opposite directions (Figure 32–1). Purines or pyrimidines with an –NH$_2$ group are weak bases (pK_b values 10–11). The planar character of purines and pyrimidines facilitates their close association, or "stacking," that stabilizes double-stranded DNA (see Chapter 34). The oxo and amino groups of purines and pyrimidines exhibit keto-enol and amine-imine **tautomerism**

FIGURE 32–1 Purine and pyrimidine. The atoms are numbered according to the international system.

FIGURE 32–2 Tautomerism of the oxo and amino functional groups of purines and pyrimidines.

FIGURE 32–3 Ribonucleosides, drawn as the *syn* conformers.

Adenosine   Cytidine   Guanosine   Uridine

(Figure 32–2), although physiologic conditions strongly favor the amino and oxo forms.

Nucleosides Are *N*-Glycosides

Nucleo**sides** are derivatives of purines and pyrimidines that have a sugar linked to a ring nitrogen of a purine or pyrimidine. Numerals with a prime (eg, 2' or 3') distinguish atoms of the sugar from those of the heterocycle. The sugar in **ribonucleosides** is D-ribose, and in **deoxyribonucleosides** is 2-deoxy-D-ribose. Both sugars are linked to the heterocycle by a **β-N-glycosidic bond,** almost always to the N-1 of a pyrimidine or to N-9 of a purine (Figure 32–3).

Nucleotides Are Phosphorylated Nucleosides

Mononucleo**tides** are nucleo**sides** with a phosphoryl group esterified to a hydroxyl group of the sugar. The 3'- and 5'-nucleotides are nucleosides with a phosphoryl group on the 3'- or 5'-hydroxyl group of the sugar, respectively. Since most nucleotides are 5'-, the prefix "5'-" usually is omitted when naming them. UMP and dAMP thus represent nucleotides with a phosphoryl group on C-5 of the pentose. Additional phosphoryl groups, ligated by **acid anhydride bonds** to the phosphoryl group of a mononucleotide, form nucleoside **diphosphates** and **triphosphates** (Figure 32–4).

Heterocylic *N*-Glycosides Exist as *Syn* and *Anti* Conformers

Steric hindrance by the heterocycle dictates that there is no freedom of rotation about the β-N-glycosidic bond of nucleosides or nucleotides. Both therefore exist as noninterconvertible *syn* or *anti* conformers (Figure 32–5). Unlike tautomers, *syn* and *anti* conformers can only be interconverted by cleavage and reformation of the glycosidic bond. Both *syn* and *anti* conformers occur in nature, but the *anti* conformers predominate.

Table 32–1 lists the major purines and pyrimidines and their nucleoside and nucleotide derivatives. Single-letter ab-

Adenosine 5'-monophosphate (AMP)

Adenosine 5'-diphosphate (ADP)

Adenosine 5'-triphosphate (ATP)

FIGURE 32–4 ATP, its diphosphate, and its monophosphate.

Syn   *Anti*

FIGURE 32–5 The *syn* and *anti* conformers of adenosine differ with respect to orientation about the *N*-glycosidic bond.

breviations are used to identify adenine (A), guanine (G), cytosine (C), thymine (T), and uracil (U), whether free or present in nucleosides or nucleotides. The prefix "d" (deoxy) indicates that the sugar is 2'-deoxy-D-ribose (for example, in dATP) (Figure 32–6).

TABLE 32–1 Purine Bases, Ribonucleosides, and Ribonucleotides

Purine or Pyrimidine	X = H	X = Ribose	X = Ribose Phosphate
	Adenine	Adenosine	Adenosine monophosphate (AMP)
	Guanine	Guanosine	Guanosine monophosphate (GMP)
	Cytosine	Cytidine	Cytidine monophosphate (CMP)
	Uracil	Uridine	Uridine monophosphate (UMP)
	Thymine	Thymidine	Thymidine monophosphate (TMP)

Modification of Polynucleotides Can Generate Additional Structures

Small quantities of additional purines and pyrimidines occur in DNA and RNAs. Examples include 5-methylcytosine of bacterial and human DNA, 5-hydroxymethylcytosine of bacterial and viral nucleic acids, and mono- and the di-N-methylated adenine and guanine of mammalian messenger RNAs (Figure 32–7) that function in oligonucleotide recognition and in regulating the half-lives of RNAs. Free nucleotides include

FIGURE 32–6 Structures of AMP, dAMP, UMP, and TMP.

5-Methylcytosine

5-Hydroxymethylcytosine

Dimethylaminoadenine

7-Methylguanine

FIGURE 32–7 Four uncommon but naturally occurring pyrimidines and purines.

**Hypoxanthine
(6-oxopurine)**

**Xanthine
(2,6-dioxopurine)**

**Uric acid
(2,6,8-trioxypurine)**

FIGURE 32–8 Structures of hypoxanthine, xanthine, and uric acid, drawn as the oxo tautomers.

FIGURE 32–9 Caffeine, a trimethylxanthine. The dimethylxanthines theobromine and theophylline are similar but lack the methyl group at N-1 and at N-7, respectively.

hypoxanthine, xanthine, and uric acid (Figure 32–8) that are intermediates in the catabolism of adenine and guanine (see Chapter 33). Methylated heterocycles of plants include the xanthine derivatives caffeine of coffee, theophylline of tea, and theobromine of cocoa (Figure 32–9).

Nucleotides Are Polyfunctional Acids

The phosphoryl groups of nucleosides have pK_a values of about 1.0. Nucleotides therefore bear significant negative charge at physiologic pH. By contrast, the pK_a values of the secondary phosphoryl groups are about 6.2, so these can serve as proton donors or acceptors at pH values approximately two or more units above or below neutrality.

Nucleotides Absorb Ultraviolet Light

The conjugated double bonds of purine and pyrimidine derivatives absorb ultraviolet light. The mutagenic effect of ultraviolet light is due to its absorption by nucleotides in DNA that results in chemical modifications (see Chapter 35). While spectra are pH-dependent, at pH 7.0 all the common nucleotides absorb light at a wavelength close to 260 nm. The concentration of nucleotides and nucleic acids thus often is expressed in terms of "absorbance at 260 nm."

Nucleotides Serve Diverse Physiologic Functions

In addition to their roles as precursors of nucleic acids, ATP, GTP, UTP, CTP and their derivatives each serve unique physiologic functions discussed in other chapters. Selected examples include the role of ATP as the principal biologic transducer of free energy; and the second messenger cAMP (Figure 32–10). The mean intracellular concentration of ATP, the most abundant free nucleotide in mammalian cells, is about 1 mmol/L. Since little cAMP is required, the intracellular cAMP concentration (about 1 nmol/L) is three orders of magnitude below that of ATP. Other examples include adenosine 3′-phosphate-5′-phosphosulfate (Figure 32–11), the sulfate donor for sulfated proteoglycans (see Chapter 48) and for sulfate conjugates of drugs; and the methyl group donor S-adenosylmethionine (Figure 32–12). GTP serves as an allosteric regulator and as

FIGURE 32–10 cAMP, 3′,5′-cyclic AMP, and cGMP, 3′, 5′-cyclic GMP.

Adenine — Ribose — (P) — O — SO_3^{2-}

FIGURE 32–11 Adenosine 3′-phosphate-5′-phosphosulfate.

FIGURE 32–12 *S*-Adenosylmethionine.

an energy source for protein synthesis, and cGMP (Figure 32–10) serves as a second messenger in response to nitric oxide (NO) during relaxation of smooth muscle (see Chapter 49). UDP-sugar derivatives participate in sugar epimerizations and in biosynthesis of glycogen, glucosyl disaccharides, and the oligosaccharides of glycoproteins and proteoglycans (see Chapters 47 & 48). UDP-glucuronic acid forms the urinary glucuronide conjugates of bilirubin (see Chapter 31) and of many drugs, including aspirin. CTP participates in biosynthesis of phosphoglycerides, sphingomyelin, and other substituted sphingosines (see Chapter 24). Finally, many coenzymes incorporate nucleotides as well as structures similar to purine and pyrimidine nucleotides (see Table 32–2).

Nucleoside Triphosphates Have High Group Transfer Potential

Acid anhydrides, unlike phosphate esters, have high group transfer potential. $\Delta G^{0'}$ for the hydrolysis of each of the two terminal (β and γ) phosphoryl groups of nucleoside triphosphates is about -7 kcal/mol (-30 kJ/mol). The high group transfer potential of purine and pyrimidine nucleoside triphosphates permits them to function as group transfer reagents, most frequently of the γ-phosphoryl group. Cleavage of an acid anhydride bond typically is coupled with a highly endergonic process such as covalent bond synthesis—eg, polymerization of nucleoside triphosphates to form a nucleic acid.

SYNTHETIC NUCLEOTIDE ANALOGS ARE USED IN CHEMOTHERAPY

Synthetic analogs of purines, pyrimidines, nucleosides, and nucleotides modified in the heterocyclic ring or in the sugar moiety have numerous applications in clinical medicine. Their toxic effects reflect either inhibition of enzymes essential for nucleic acid synthesis or their incorporation into nucleic acids with resulting disruption of base-pairing. Oncologists employ 5-fluoro- or 5-iodouracil, 3-deoxyuridine, 6-thioguanine and

TABLE 32–2 Many Coenzymes and Related Compounds Are Derivatives of Adenosine Monophosphate

Coenzyme	R	R'	R''	n
Active methionine	Methionine[a]	H	H	0
Amino acid adenylates	Amino acid	H	H	1
Active sulfate	SO_3^{2-}	H	PO_3^{2-}	1
3',5'-Cyclic AMP		H	PO_3^{2-}	1
NAD[b]	Nicotinamide	H	H	2
NADP[b]	Nicotinamide	PO_3^{2-}	H	2
FAD	Riboflavin	H	H	2
Coenzyme A	Pantothenate	H	PO_3^{2-}	2

[a]Replaces phosphoryl group.
[b]R is a B vitamin derivative.

6-mercaptopurine, 5- or 6-azauridine, 5- or 6-azacytidine, and 8-azaguanine (Figure 32–13), which are incorporated into DNA prior to cell division. The purine analog allopurinol, used in treatment of hyperuricemia and gout, inhibits purine biosynthesis and xanthine oxidase activity. Cytarabine is used in chemotherapy of cancer, and azathioprine, which is catabolized to 6-mercaptopurine, is employed during organ transplantation to suppress immunologic rejection (Figure 32–14).

Nonhydrolyzable Nucleoside Triphosphate Analogs Serve as Research Tools

Synthetic, nonhydrolyzable analogs of nucleoside triphosphates (Figure 32–15) allow investigators to distinguish the effects of nucleotides due to phosphoryl transfer from effects mediated by occupancy of allosteric nucleotide-binding sites on regulated enzymes.

DNA & RNA ARE POLYNUCLEOTIDES

The 5'-phosphoryl group of a mononucleotide can esterify a second hydroxyl group, forming a **phosphodiester.** Most commonly, this second hydroxyl group is the 3'-OH of the pentose

FIGURE 32–13 Selected synthetic pyrimidine and purine analogs.

Cytarabine

Azathioprine

FIGURE 32–14 Arabinosylcytosine (cytarabine) and azathioprine.

Parent nucleoside triphosphate

β,γ-Methylene derivative

β,γ-Imino derivative

FIGURE 32–15 Synthetic derivatives of nucleoside triphosphates incapable of undergoing hydrolytic release of the terminal phosphoryl group. (Pu/Py, a purine or pyrimidine base; R, ribose or deoxyribose.) Shown are the parent (hydrolyzable) nucleoside triphosphate (**top**) and the unhydrolyzable β-methylene (**center**) and γ-imino derivatives (**bottom**).

of a second nucleotide. This forms a **dinucleotide** in which the pentose moieties are linked by a 3′,5′-phosphodiester bond to form the "backbone" of RNA and DNA. The formation of a dinucleotide may be represented as the elimination of water between two mononucleotides. Biologic formation of dinucleotides does not, however, occur in this way because the reverse reaction, hydrolysis of the phosphodiester bond, is strongly

favored on thermodynamic grounds. However, despite an extremely favorable ΔG, in the absence of catalysis by **phosphodiesterases**, hydrolysis of the phosphodiester bonds of DNA occurs only over long periods of time. Consequently, DNA persists for considerable periods and has been detected even in fossils. RNAs are far less stable than DNA since the 2′-hydroxyl group of RNA (absent from DNA) functions as a nucleophile during hydrolysis of the 3′,5′-phosphodiester bond.

Posttranslational modification of preformed **polynucleotides** can generate additional structures such as **pseudouri-**

dine, a nucleoside in which D-ribose is linked to C-5 of uracil by a carbon-to-carbon bond rather than by the usual β-N-glycosidic bond. The nucleotide pseudouridylic acid (ψ) arises by rearrangement of a UMP of a preformed tRNA. Similarly, methylation by *S*-adenosylmethionine of a UMP of preformed tRNA forms TMP (thymidine monophosphate), which contains ribose rather than deoxyribose.

Polynucleotides Are Directional Macromolecules

Phosphodiester bonds link the 3'- and 5'-carbons of adjacent monomers. Each end of a nucleotide polymer thus is distinct. We therefore refer to the "5'-end" or the "3'-end" of a polynucleotide, the 5'-end being the one with a free or phosphorylated 5'-hydroxyl.

The base sequence or **primary structure** of a polynucleotide can be represented as shown below. The phosphodiester bond is represented by P or p, bases by a single letter, and pentoses by a vertical line.

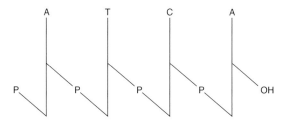

Where all the phosphodiester bonds are 3' → 5', a more compact notation is possible:

pGpGpApTpCpA

This representation indicates that the 5'-hydroxyl—but not the 3'-hydroxyl—is phosphorylated. The most compact representation shows only the base sequence with the 5'-end on the left and the 3'-end on the right. The phosphoryl groups are assumed to be present, but not shown.

SUMMARY

- Under physiologic conditions, the amino and oxo tautomers of purines, pyrimidines, and their derivatives predominate.

- Nucleic acids contain, in addition to A, G, C, T, and U, traces of 5-methylcytosine, 5-hydroxymethylcytosine, pseudouridine (ψ), or N-methylated heterocycles.

- Most nucleosides contain D-ribose or 2-deoxy-D-ribose linked to N-1 of a pyrimidine or to N-9 of a purine by a β-glycosidic bond whose *syn* conformers predominate.

- A primed numeral locates the position of the phosphate on the sugars of mononucleotides (eg, 3'-GMP, 5'-dCMP). Additional phosphoryl groups linked to the first by acid anhydride bonds form nucleoside diphosphates and triphosphates.

- Nucleoside triphosphates have high group transfer potential and participate in covalent bond syntheses. The cyclic phosphodiesters cAMP and cGMP function as intracellular second messengers.

- Mononucleotides linked by 3' → 5'-phosphodiester bonds form polynucleotides, directional macromolecules with distinct 3'- and 5'-ends. For pTpGpTp or TGCATCA, the 5'-end is at the left, and all phosphodiester bonds are 3' → 5'.

- Synthetic analogs of purine and pyrimidine bases and their derivatives serve as anticancer drugs either by inhibiting an enzyme of nucleotide biosynthesis or by being incorporated into DNA or RNA.

REFERENCES

Adams RLP, Knowler JT, Leader DP: *The Biochemistry of the Nucleic Acids,* 11th ed. Chapman & Hall, 1992.

Blackburn GM, Gait MJ: *Nucleic Acids in Chemistry & Biology.* IRL Press, 1990.

Pacher P, Nivorozhkin A, Szabo C: Therapeutic effects of xanthine oxidase inhibitors: renaissance half a century after the discovery of allopurinol. Pharmacol Rev 2006;58:87.

Metabolism of Purine & Pyrimidine Nucleotides

33

Victor W. Rodwell, PhD

BIOMEDICAL IMPORTANCE

Even when humans consume a diet rich in nucleoproteins, dietary purines and pyrimidines are not incorporated directly into tissue nucleic acids. Humans synthesize the nucleic acids, ATP, NAD^+, coenzyme A, etc., from amphibolic intermediates. However, *injected* purine or pyrimidine analogs, including potential anticancer drugs, may be incorporated into DNA. The biosyntheses of purine and pyrimidine oxy- and deoxyribonucleotides (NTPs and dNTPs) are precisely regulated events coordinated by feedback mechanisms that ensure their production in appropriate quantities, and at times that match varying physiologic demand (eg, cell division). Human diseases that involve abnormalities in purine metabolism include gout, Lesch–Nyhan syndrome, adenosine deaminase deficiency, and purine nucleoside phosphorylase deficiency. Diseases of pyrimidine biosynthesis are rarer, but include orotic acidurias. Unlike the urates, the products of pyrimidine catabolism (carbon dioxide, ammonia, β-alanine, and γ-aminoisobutyrate) are highly soluble. One genetic disorder of pyrimidine catabolism is β-hydroxybutyric aciduria, due to total or partial deficiency of the enzyme dihydropyrimidine dehydrogenase. This disorder of pyrimidine catabolism, known also as combined uraciluria-thyminuria, it is also a disorder of β-amino acid biosynthesis, since the formation of β-alanine and of β-aminoisobutyrate is impaired. A nongenetic form can be triggered by administration of the anticancer drug 5-fluorouracil to patients with low levels of dihydropyrimidine dehydrogenase.

PURINES & PYRIMIDINES ARE DIETARILY NONESSENTIAL

Human tissues can synthesize purines and pyrimidines from amphibolic intermediates. Ingested nucleic acids and nucleotides, which therefore are dietarily nonessential, are degraded in the intestinal tract to mononucleotides, which may be absorbed or converted to purine and pyrimidine bases. The purine bases are then oxidized to uric acid, which may be absorbed and excreted in the urine. While little or no dietary purine or pyrimidine is incorporated into tissue nucleic acids, injected compounds are incorporated. The incorporation of

injected [³H]thymidine into newly synthesized DNA thus can be used to measure the rate of DNA synthesis.

BIOSYNTHESIS OF PURINE NUCLEOTIDES

With the exception of parasitic protozoa, all forms of life synthesize purine and pyrimidine nucleotides. Synthesis from amphibolic intermediates proceeds at controlled rates appropriate for all cellular functions. To achieve homeostasis, intracellular mechanisms sense and regulate the pool sizes of nucleotide triphosphates (NTPs), which rise during growth or tissue regeneration when cells are rapidly dividing. Purine and pyrimidine nucleotides are synthesized in vivo at rates consistent with physiologic need. Early investigations of nucleotide biosynthesis first employed birds, and later *Escherichia coli*. Isotopic precursors fed to pigeons established the source of each atom of a purine (Figure 33–1), and initiated study of the intermediates of purine biosynthesis. Avian tissues served as a source of cloned genes that encode enzymes of purine biosynthesis and the regulatory proteins that control the rate of purine biosynthesis.

The three processes that contribute to purine nucleotide biosynthesis are, in order of decreasing importance:

1. Synthesis from amphibolic intermediates (synthesis de novo)
2. Phosphoribosylation of purines
3. Phosphorylation of purine nucleosides.

INOSINE MONOPHOSPHATE (IMP) IS SYNTHESIZED FROM AMPHIBOLIC INTERMEDIATES

Figure 33–2 illustrates the intermediates and the 11 enzyme-catalyzed reactions that convert α-D-ribose 5-phosphate to inosine monophosphate (IMP). In addition to being the first intermediate formed in the de novo pathway for purine biosynthesis, 5-phosphoribosyl 5-pyrophosphate (structure II, Figure 33–2) is an intermediate in the purine salvage pathway, and in the biosynthesis of pyrimidine nucleotides, NAD^+, and

FIGURE 33–1 Sources of the nitrogen and carbon atoms of the purine ring. Atoms 4, 5, and 7 (**blue highlight**) derive from glycine.

$NADP^+$. Separate branches then lead to AMP and GMP (Figure 33–3). Subsequent phosphoryl transfer from ATP converts AMP and GMP to ADP and GDP. Conversion of GDP to GTP involves a second phosphoryl transfer from ATP, whereas conversion of ADP to ATP is achieved primarily by oxidative phosphorylation (see Chapter 13).

Multifunctional Catalysts Participate in Purine Nucleotide Biosynthesis

In prokaryotes, each reaction of Figure 33–2 is catalyzed by a different polypeptide. By contrast, in eukaryotes, the enzymes are polypeptides with multiple catalytic activities whose adjacent catalytic sites facilitate channeling of intermediates between sites. Three distinct multifunctional enzymes catalyze reactions ③, ④, and ⑥; reactions ⑦ and ⑧; and reactions ⑩ and ⑪ of Figure 33–2.

Antifolate Drugs or Glutamine Analogs Block Purine Nucleotide Biosynthesis

The carbons added in reactions ④ and ⑩ of Figure 33–2 are contributed by derivatives of tetrahydrofolate. Purine deficiency states, while rare in humans, generally reflect a deficiency of folic acid. Compounds that inhibit formation of tetrahydrofolates and therefore block purine synthesis have been used in cancer chemotherapy. Inhibitory compounds and the reactions they inhibit include **azaserine** (reaction ⑤, Figure 33–2), **diazanorleucine** (reaction ②, Figure 33–2), **6-mercaptopurine** (reactions ⑬ and ⑭, Figure 33–3), and **mycophenolic acid** (reaction ⑭, Figure 33–3).

"SALVAGE REACTIONS" CONVERT PURINES & THEIR NUCLEOSIDES TO MONONUCLEOTIDES

Conversion of purines, their ribonucleo**sides**, and their deoxyribonucleo**sides** to mononucleo**tides** involves "salvage reactions" that require far less energy than de novo synthesis. The

more important mechanism involves phosphoribosylation by PRPP (structure II, Figure 33–2) of a free purine (Pu) to form a purine 5′-mononucleotide (Pu-RP).

$$Pu + PR\text{-}PP \rightarrow Pu\text{-}RP + PP_i$$

Phosphoryl transfer from ATP, catalyzed by adenosine- and hypoxanthine-phosphoribosyl transferases, converts adenine, hypoxanthine, and guanine to their mononucleotides (Figure 33–4).

A second salvage mechanism involves phosphoryl transfer from ATP to a purine ribonucleo**side** (Pu-R):

$$Pu\text{-}R + ATP \rightarrow PuR\text{-}P + ADP$$

Phosphorylation of the purine nucleotides, catalyzed by adenosine kinase, converts adenosine and deoxyadenosine to AMP and dAMP. Similarly, deoxycytidine kinase phosphorylates deoxycytidine and 2′-deoxyguanosine forming dCMP and dGMP.

Liver, the major site of purine nucleotide biosynthesis, provides purines and purine nucleosides for salvage and for utilization by tissues incapable of their biosynthesis. Human brain tissue has a low level of PRPP glutamyl amidotransferase (reaction ②, Figure 33–2) and hence depends in part on exogenous purines. Erythrocytes and polymorphonuclear leukocytes cannot synthesize 5-phosphoribosylamine (structure III, Figure 33–2) and therefore utilize exogenous purines to form nucleotides.

HEPATIC PURINE BIOSYNTHESIS IS STRINGENTLY REGULATED

AMP & GMP Feedback Regulate PRPP Glutamyl Amidotransferase

Since biosynthesis of IMP consumes glycine, glutamine, tetrahydrofolate derivatives, aspartate, and ATP, it is advantageous to regulate purine biosynthesis. The major determinant of the rate of de novo purine nucleotide biosynthesis is the concentration of PRPP, a function of its rates of synthesis, utilization, and degradation. The rate of PRPP synthesis depends on the availability of ribose 5-phosphate and on the activity of PRPP synthase, an enzyme sensitive to feedback inhibition by AMP, ADP, GMP, and GDP (Figure 33–5).

AMP & GMP Feedback Regulate Their Formation from IMP

Two mechanisms regulate conversion of IMP to ATP and GTP (Figure 33–6). AMP and GMP feedback-inhibit adenylosuccinate synthase and IMP dehydrogenase (reactions ⑫ and ⑭, Figure 33–3), respectively. Furthermore, conversion of IMP to adenylosuccinate en route to AMP requires GTP, and conversion of xanthinylate (XMP) to GMP requires ATP. This cross-regulation between the pathways of IMP metabolism thus serves to decrease synthesis of one purine nucleotide when

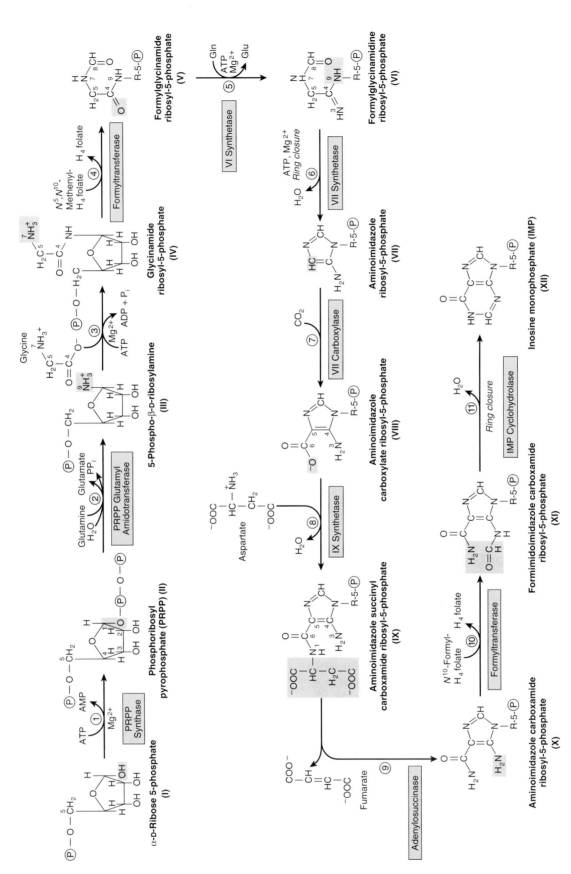

FIGURE 33–2 Purine biosynthesis from ribose 5-phosphate and ATP. See text for explanations. ($\circledP$, PO_3^{2-} or PO_2^-.)

FIGURE 33–3 Conversion of IMP to AMP and GMP.

there is a deficiency of the other nucleotide. AMP and GMP also inhibit hypoxanthine-guanine phosphoribosyltransferase, which converts hypoxanthine and guanine to IMP and GMP (Figure 33–4), and GMP feedback inhibits PRPP glutamyl amidotransferase (reaction 2, Figure 33–2).

REDUCTION OF RIBONUCLEOSIDE DIPHOSPHATES FORMS DEOXYRIBONUCLEOSIDE DIPHOSPHATES

Reduction of the 2′-hydroxyl of purine and pyrimidine ribonucleotides, catalyzed by the **ribonucleotide reductase complex** (Figure 33–7), forms deoxyribonucleoside diphosphates (dNDPs). The enzyme complex is functional only when cells are actively synthesizing DNA. Reduction requires thioredoxin, thioredoxin reductase, and NADPH. The immediate reductant, reduced thioredoxin, is produced by NADPH:thioredoxin reductase (Figure 33–7). Reduction of ribonucleoside diphosphates (NDPs) to deoxyribonucleoside diphosphates (dNDPs) is subject to complex regulatory controls that achieve balanced production of deoxyribonucleotides for synthesis of DNA (Figure 33–8).

BIOSYNTHESIS OF PYRIMIDINE NUCLEOTIDES

Figure 33–9 illustrates the intermediates and enzymes of pyrimidine nucleotide biosynthesis. The catalyst for the initial

reaction is *cytosolic* carbamoyl phosphate synthase II, a different enzyme from the *mitochondrial* carbamoyl phosphate synthase I of urea synthesis (Figure 28–12). Compartmentation thus provides two independent pools of carbamoyl phosphate. PRPP, an early participant in purine nucleotide synthesis (Figure 33–2), is a much later participant in pyrimidine biosynthesis.

Multifunctional Proteins Catalyze the Early Reactions of Pyrimidine Biosynthesis

Five of the first six enzyme activities of pyrimidine biosynthesis reside on multifunctional polypeptides. One such polypeptide catalyzes the first three reactions of Figure 33–9 and ensures efficient channeling of carbamoyl phosphate to pyrimidine biosynthesis. A second bifunctional enzyme catalyzes reactions ⑤ and ⑥ of Figure 33–9.

THE DEOXYRIBONUCLEOSIDES OF URACIL & CYTOSINE ARE SALVAGED

While mammalian cells reutilize few free pyrimidines, "salvage reactions" convert the pyrimidine ribonucleosides uridine and cytidine and the pyrimidine deoxyribonucleosides thymidine and deoxycytidine to their respective nucleotides. ATP-dependent phosphoryltransferases (kinases) catalyze the phosphorylation of the diphosphates of 2′-deoxycytidine, 2′-deoxyguanosine, and 2′-deoxyadenosine to their corresponding nucleoside triphosphates. In addition, orotate phosphoribosyltransferase

FIGURE 33–4 Phosphoribosylation of adenine, hypoxanthine, and guanine to form AMP, IMP, and GMP, respectively.

FIGURE 33–5 Control of the rate of de novo purine nucleotide biosynthesis. Reactions ① and ② are catalyzed by PRPP synthase and by PRPP glutamyl amidotransferase, respectively. Solid lines represent chemical flow. Broken red lines represent feedback inhibition by intermediates of the pathway.

(reaction ⑤, Figure 33–9), an enzyme of pyrimidine nucleotide synthesis, salvages orotic acid by converting it to orotidine monophosphate (OMP).

Methotrexate Blocks Reduction of Dihydrofolate

Reaction ⑫ of Figure 33–9 is the only reaction of pyrimidine nucleotide biosynthesis that requires a tetrahydrofolate derivative. The methylene group of N^5,N^{10}-methylene-tetrahydrofolate is reduced to the methyl group that is transferred, and tetrahydrofolate is oxidized to dihydrofolate. For further pyrimidine synthesis to occur, dihydrofolate must be reduced back to tetrahydrofolate, a reaction catalyzed by dihydrofolate reductase. Dividing cells, which must generate TMP and dihydrofolate, thus are especially sensitive to inhibitors of dihydrofolate reductase such as the anticancer drug **methotrexate.**

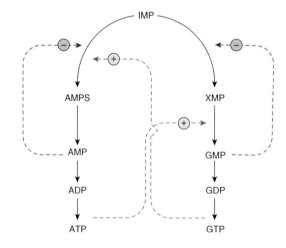

FIGURE 33–6 Regulation of the interconversion of IMP to adenosine nucleotides and guanosine nucleotides. Solid lines represent chemical flow. Broken green lines represent positive feedback loops. Broken red lines represent negative feedback loops.

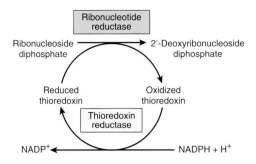

FIGURE 33–7 Reduction of ribonucleoside diphosphates to 2′-deoxyribonucleoside diphosphates.

Certain Pyrimidine Analogs Are Substrates for Enzymes of Pyrimidine Nucleotide Biosynthesis

Orotate phosphoribosyltransferase (reaction ⑤, Figure 33–9) converts the drug **allopurinol** (Figure 32–13) to a nucleotide in which the ribosyl phosphate is attached to N-1 of the pyrimidine ring. The anticancer drug **5-fluorouracil** (Figure 32–13) is also phosphoribosylated by orotate phosphoribosyltransferase.

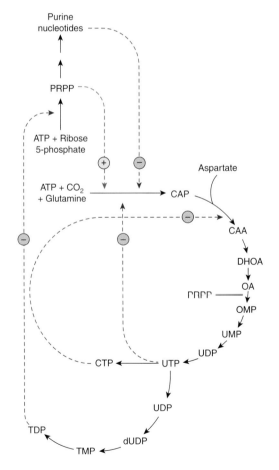

FIGURE 33–8 Regulation of the reduction of purine and pyrimidine ribonucleotides to their respective 2′-deoxyribonucleotides. The broken green line represents a positive feedback loop. Broken red lines represent negative feedback loops.

REGULATION OF PYRIMIDINE NUCLEOTIDE BIOSYNTHESIS

Gene Expression & Enzyme Activity Both Are Regulated

The activities of the first and second enzymes of pyrimidine nucleotide biosynthesis are controlled by allosteric regulation. Carbamoyl phosphate synthase II (reaction ①, Figure 33–9) is inhibited by UTP and purine nucleotides but activated by PRPP. Aspartate transcarbamoylase (reaction ②, Figure 33–9) is inhibited by CTP but activated by ATP (Figure 33–10). In addition, the first three and the last two enzymes of the pathway are regulated by coordinate repression and derepression.

Purine & Pyrimidine Nucleotide Biosynthesis Are Coordinately Regulated

Purine and pyrimidine biosynthesis parallel one another mole for mole, suggesting coordinated control of their biosynthesis. Several sites of cross-regulation characterize purine and pyrimidine nucleotide biosynthesis. PRPP synthase (reaction ①, Figure 33–2), which forms a precursor essential for both processes, is feedback inhibited by both purine and pyrimidine nucleotides.

HUMANS CATABOLIZE PURINES TO URIC ACID

Humans convert adenosine and guanosine to uric acid (Figure 33–11). Adenosine is first converted to inosine by adenosine deaminase. In mammals other than higher primates, uricase converts uric acid to the water-soluble product allantoin. However, since humans lack uricase, the end product of purine catabolism in humans is uric acid.

GOUT IS A METABOLIC DISORDER OF PURINE CATABOLISM

Various genetic defects in PRPP synthetase (reaction ①, Figure 33–2) present clinically as gout. Each defect—eg, an elevated V_{max}, increased affinity for ribose 5-phosphate, or resistance to feedback inhibition—results in overproduction and overexcretion of purine catabolites. When serum urate levels exceed the solubility limit, sodium urate crystalizes in soft tissues and joints and causes an inflammatory reaction, **gouty arthritis.** However, most cases of gout reflect abnormalities in renal handling of uric acid.

OTHER DISORDERS OF PURINE CATABOLISM

While purine deficiency states are rare in human subjects, there are numerous genetic disorders of purine catabolism.

FIGURE 33–9 The biosynthetic pathway for pyrimidine nucleotides.

Hyperuricemias may be differentiated based on whether patients excrete normal or excessive quantities of total urates. Some hyperuricemias reflect specific enzyme defects. Others are secondary to diseases such as cancer or psoriasis that enhance tissue turnover.

rise in intracellular PRPP results in purine overproduction. Mutations that decrease or abolish hypoxanthine-guanine phosphoribosyltransferase activity include deletions, frameshift mutations, base substitutions, and aberrant mRNA splicing.

Lesch–Nyhan Syndrome

Lesch–Nyhan syndrome, an overproduction hyperuricemia characterized by frequent episodes of uric acid lithiasis and a bizarre syndrome of self-mutilation, reflects a defect in **hypoxanthine-guanine phosphoribosyl transferase,** an enzyme of purine salvage (Figure 33–4). The accompanying

von Gierke Disease

Purine overproduction and hyperuricemia in von Gierke disease (**glucose-6-phosphatase deficiency**) occurs secondary to enhanced generation of the PRPP precursor ribose 5-phosphate. An associated lactic acidosis elevates the renal threshold for urate, elevating total body urates.

FIGURE 33–10 Control of pyrimidine nucleotide biosynthesis. Solid lines represent chemical flow. Broken green lines represent positive ⊕, and broken red lines negative ⊖ feedback regulation.

Hypouricemia

Hypouricemia and increased excretion of hypoxanthine and xanthine are associated with **xanthine oxidase deficiency** (Figure 33–11) due to a genetic defect or to severe liver damage. Patients with a severe enzyme deficiency may exhibit xanthinuria and xanthine lithiasis.

Adenosine Deaminase & Purine Nucleoside Phosphorylase Deficiency

Adenosine deaminase deficiency (Figure 33–11) is associated with an immunodeficiency disease in which both thymus-derived lymphocytes (T cells) and bone-marrow-derived lymphocytes (B cells) are sparse and dysfunctional. Patients suffer from severe immunodeficiency. In the absence of enzyme replacement or bone marrow transplantation, infants often succumb to fatal infections. **Purine nucleoside phosphorylase deficiency** is associated with a severe deficiency of T cells but apparently normal B cell function. Immune dysfunctions appear to result from accumulation of dGTP and dATP, which inhibit ribonucleotide reductase and thereby deplete cells of DNA precursors.

CATABOLISM OF PYRIMIDINES PRODUCES WATER-SOLUBLE METABOLITES

Unlike the end products of purine catabolism, the end products of pyrimidine catabolism are the highly water-soluble products CO_2, NH_3, β-alanine, and β-aminoisobutyrate (Figure 33–12). Humans transaminate β-aminoisobutyrate to methylmalonate semialdehyde, which then forms succinyl-CoA (see Figure 20–2).

Excretion of β-aminoisobutyrate increases in leukemia and severe x-ray radiation exposure due to increased

FIGURE 33–11 Formation of uric acid from purine nucleosides by way of the purine bases hypoxanthine, xanthine, and guanine. Purine deoxyribonucleosides are degraded by the same catabolic pathway and enzymes, all of which exist in the mucosa of the mammalian gastrointestinal tract.

destruction of DNA. However, many persons of Chinese or Japanese ancestry routinely excrete β-aminoisobutyrate. Disorders of β-alanine and β-aminoisobutyrate metabolism

FIGURE 33–12 Catabolism of pyrimidines. Hepatic β-ureidopropionase catalyzes the formation of both β-alanine and β-aminoisobutyrate from their pyrimidine precursors.

arise from defects in enzymes of pyrimidine catabolism. These include **β-hydroxybutyric aciduria**, a disorder due to total or partial deficiency of the enzyme **dihydropyrimidine dehydrogenase** (Figure 33–12). The genetic disease reflects an absence of the enzyme. A disorder of pyrimidine catabolism, known also as combined uraciluria-thyminuria, it is also a disorder of β-amino acid metabolism, since the *formation* of β-alanine and of β-aminoisobutyrate is impaired. When due to an inborn error, there are serious neurological complications. A nongenetic form is triggered by the administration of the anti cancer drug 5-fluorouracil (Figure 32–13) to patients with low levels of dihydropyrimidine dehydrogenase.

Pseudouridine Is Excreted Unchanged

Since no human enzyme catalyzes hydrolysis or phosphorolysis of pseudouridine, this unusual nucleoside is excreted unchanged in the urine of normal subjects, and was indeed first isolated from urine.

OVERPRODUCTION OF PYRIMIDINE CATABOLITES IS ONLY RARELY ASSOCIATED WITH CLINICALLY SIGNIFICANT ABNORMALITIES

Since the end products of pyrimidine catabolism are highly water-soluble, pyrimidine overproduction results in few clinical signs or symptoms. In hyperuricemia associated with severe overproduction of PRPP, there is overproduction of pyrimidine nucleotides and increased excretion of β-alanine. Since N^5,N^{10}-methylene-tetrahydrofolate is required for thymidylate synthesis, disorders of folate and vitamin B_{12} metabolism result in deficiencies of TMP.

Orotic Acidurias

The orotic aciduria that accompanies **Reye syndrome** probably is a consequence of the inability of severely damaged mitochondria to utilize carbamoyl phosphate, which then becomes available for cytosolic overproduction of orotic acid. **Type I orotic aciduria** reflects a deficiency of both orotate phosphoribosyltransferase and orotidylate decarboxylase (reactions ⑤ and ⑥, Figure 33–9); the rarer **type II orotic aciduria** is due to a deficiency only of orotidylate decarboxylase (reaction ⑥, Figure 33–9).

Deficiency of a Urea Cycle Enzyme Results in Excretion of Pyrimidine Precursors

Increased excretion of orotic acid, uracil, and uridine accompanies a deficiency in liver mitochondrial ornithine transcarbamoylase (reaction ②, Figure 28–9). Excess carbamoyl phosphate exits to the cytosol, where it stimulates pyrimidine nucleotide biosynthesis. The resulting mild **orotic aciduria** is increased by high-nitrogen foods.

Drugs May Precipitate Orotic Aciduria

Allopurinol (Figure 32–13), an alternative substrate for orotate phosphoribosyltransferase (reaction ⑤, Figure 33–9), competes with orotic acid. The resulting nucleotide product also inhibits orotidylate decarboxylase (reaction ⑥, Figure 33–9), resulting in **orotic aciduria** and **orotidinuria**. 6-Azauridine, following conversion to 6-azauridylate, also competitively inhibits orotidylate decarboxylase (reaction ⑥, Figure 33–9), enhancing excretion of orotic acid and orotidine.

SUMMARY

- Ingested nucleic acids are degraded to purines and pyrimidines. New purines and pyrimidines are formed from amphibolic intermediates and thus are dietarily nonessential.

- Several reactions of IMP biosynthesis require folate derivatives and glutamine. Consequently, antifolate drugs and glutamine analogs inhibit purine biosynthesis.

- Oxidation and amination of IMP forms AMP and GMP, and subsequent phosphoryl transfer from ATP forms ADP and GDP. Further phosphoryl transfer from ATP to GDP forms GTP. ADP is converted to ATP by oxidative phosphorylation. Reduction of NDPs forms dNDPs.

- Hepatic purine nucleotide biosynthesis is stringently regulated by the pool size of PRPP and by feedback inhibition of PRPP-glutamyl amidotransferase by AMP and GMP.

- Coordinated regulation of purine and pyrimidine nucleotide biosynthesis ensures their presence in proportions appropriate for nucleic acid biosynthesis and other metabolic needs.

- Humans catabolize purines to uric acid (pK_a 5.8), present as the relatively insoluble acid at acidic pH or as its more soluble sodium urate salt at a pH near neutrality. Urate crystals are diagnostic of gout. Other disorders of purine catabolism include Lesch–Nyhan syndrome, von Gierke disease, and hypouricemias.

- Since pyrimidine catabolites are water-soluble, their overproduction does not result in clinical abnormalities. Excretion of pyrimidine precursors can, however, result from a deficiency of ornithine transcarbamoylase because excess carbamoyl phosphate is available for pyrimidine biosynthesis.

REFERENCES

Chow EL et al: Reassessing Reye syndrome. Arch Pediatr Adolesc Med 2003;157:1241.

Christopherson RI, Lyons SD, Wilson PK: Inhibitors of de novo nucleotide biosynthesis as drugs. Acc Chem Res 2002;35:961.

Kamal MA, Christopherson RI: Accumulation of 5-phosphoribosyl-1-pyrophosphate in human CCRF-CEM leukemia cells treated with antifolates. Int J Biochem Cell Biol 2004;36:957.

Lipkowitz MS et al: Functional reconstitution, membrane targeting, genomic structure, and chromosomal localization of a human urate transporter. J Clin Invest 2001;107:1103.

Martinez J et al: Human genetic disorders, a phylogenetic perspective. J Mol Biol 2001;308:587.

Moyer RA, John DS: Acute gout precipitated by total parenteral nutrition. J Rheumatol 2003;30:849.

Nofech-Mozes Y et al: Neurologic abnormalities in patients with adenosine deaminase deficiency. Pediatr Neurol 2007;37:218.

Scriver CR et al (editors): *The Metabolic and Molecular Bases of Inherited Disease,* 8th ed. McGraw-Hill, 2001.

Torres RJ, Puig JG: Hypoxanthine-guanine phosophoribosyltransferase (HPRT) deficiency: Lesch–Nyhan syndrome. Orphanet J Rare Dis 2007;2:48.

Wu VC et al: Renal hypouricemia is an ominous sign in patients with severe acute respiratory syndrome. Am J Kidney Dis 2005;45:88.

Nucleic Acid Structure & Function

P. Anthony Weil, PhD

BIOMEDICAL IMPORTANCE

The discovery that genetic information is coded along the length of a polymeric molecule composed of only four types of monomeric units was one of the major scientific achievements of the 20th century. This polymeric molecule, **deoxyribonucleic acid (DNA)**, is the chemical basis of heredity and is organized into genes, the fundamental units of genetic information. The basic information pathway—ie, DNA, which directs the synthesis of RNA, which in turn both directs and regulates protein synthesis—has been elucidated. Genes do not function autonomously; their replication and function are controlled by various gene products, often in collaboration with components of various signal transduction pathways. Knowledge of the structure and function of nucleic acids is essential in understanding genetics and many aspects of pathophysiology as well as the genetic basis of disease.

DNA CONTAINS THE GENETIC INFORMATION

The demonstration that DNA contained the genetic information was first made in 1944 in a series of experiments by Avery, MacLeod, and McCarty. They showed that the genetic determination of the character (type) of the capsule of a specific pneumococcus could be transmitted to another of a different capsular type by introducing purified DNA from the former coccus into the latter. These authors referred to the agent (later shown to be DNA) accomplishing the change as "transforming factor." Subsequently, this type of genetic manipulation has become commonplace. Similar experiments have recently been performed utilizing yeast, cultured plant and mammalian cells, and insect and mammalian embryos as recipients and molecularly cloned DNA as the donor of genetic information.

DNA Contains Four Deoxynucleotides

The chemical nature of the monomeric deoxynucleotide units of DNA—**deoxyadenylate, deoxyguanylate, deoxycytidylate,** and **thymidylate**—is described in Chapter 32. These mono-

meric units of DNA are held in polymeric form by 3′,5′-phosphodiester bonds constituting a single strand, as depicted in Figure 34–1. The informational content of DNA (the genetic code) resides in the sequence in which these monomers—purine and pyrimidine deoxyribonucleotides—are ordered. The polymer as depicted possesses a polarity; one end has a 5′-hydroxyl or phosphate terminal while the other has a 3′-phosphate or hydroxyl terminal. The importance of this polarity will become evident. Since the genetic information resides in the order of the monomeric units within the polymers, there must exist a mechanism of reproducing or replicating this specific information with a high degree of fidelity. That requirement, together with x-ray diffraction data from the DNA molecule and the observation of Chargaff that in DNA molecules the concentration of deoxyadenosine (A) nucleotides equals that of thymidine (T) nucleotides (A = T), while the concentration of deoxyguanosine (G) nucleotides equals that of deoxycytidine (C) nucleotides (G = C), led Watson, Crick, and Wilkins to propose in the early 1950s a model of a double-stranded DNA molecule. The model they proposed is depicted in Figure 34–2. The two strands of this double-stranded helix are held in register by both **hydrogen bonds** between the purine and pyrimidine bases of the respective linear molecules and by **van der Waals** and **hydrophobic interactions** between the stacked adjacent base pairs. The pairings between the purine and pyrimidine nucleotides on the opposite strands are very specific and are dependent upon hydrogen bonding of **A with T** and **G with C** (Figure 34–2).

This common form of DNA is said to be right-handed because as one looks down the double helix, the base residues form a spiral in a clockwise direction. In the double-stranded molecule, restrictions imposed by the rotation about the phosphodiester bond, the favored anticonfiguration of the glycosidic bond (Figure 32–5), and the predominant tautomers (see Figure 32–2) of the four bases (A, G, T, and C) allow A to pair only with T and G only with C, as depicted in Figure 34–3. This base-pairing restriction explains the earlier observation that in a double-stranded DNA molecule the content of A equals that of T and the content of G equals that of C. The two strands of the double-helical molecule, each of which possesses a polarity, are **antiparallel;** ie, one strand runs in the 5′ to 3′ direction and the other in the 3′ to 5′ direction. In the

FIGURE 34–1 A segment of one strand of a DNA molecule in which the purine and pyrimidine bases guanine (G), cytosine (C), thymine (T), and adenine (A) are held together by a phosphodiester backbone between 2′-deoxyribosyl moieties attached to the nucleobases by an *N*-glycosidic bond. Note that the backbone has a polarity (ie, a direction). Convention dictates that a single-stranded DNA sequence is written in the 5′ to 3′ direction (ie, pGpCpTpA, where G, C, T, and A represent the four bases and p represents the interconnecting phosphates).

double-stranded DNA molecules, the genetic information resides in the sequence of nucleotides on one strand, the **template strand.** This is the strand of DNA that is copied during **ribonucleic acid (RNA)** synthesis. It is sometimes referred to as the **noncoding strand.** The opposite strand is considered the **coding strand** because it matches the sequence of the RNA transcript (but containing uracil in place of thymine; see Figure 34–8) that encodes the protein.

The two strands, in which opposing bases are held together by interstrand hydrogen bonds, wind around a central axis in the form of a **double helix.** In the test tube doublestranded DNA can exist in at least six forms (A–E and Z). The B form is usually found under physiologic conditions (low salt, high degree of hydration). A single turn of B-DNA about the long axis of the molecule contains ten base pairs. The distance spanned by one turn of B-DNA is 3.4 nm (34 Å). The width (helical diameter) of the double helix in B-DNA is 2 nm (20 Å).

As depicted in Figure 34–3, three hydrogen bonds, formed by hydrogen bonded to electronegative N or O atoms, hold the deoxyguanosine nucleotide to the deoxycytidine nucleotide, whereas the other pair, the A–T pair, is held together by two hydrogen bonds. Thus, the G–C bonds are more resistant to denaturation, or strand separation, termed "melting," than A–T-rich regions of DNA.

The Denaturation of DNA Is Used to Analyze Its Structure

The double-stranded structure of DNA can be separated into two component strands in solution by increasing the temperature or decreasing the salt concentration. Not only do the two stacks of bases pull apart but the bases themselves unstack while still connected in the polymer by the phosphodiester backbone. Concomitant with this denaturation of the DNA molecule is an increase in the optical absorbance of the purine and pyrimidine bases—a phenomenon referred to as **hyperchromicity** of denaturation. Because of the stacking of the bases and the hydrogen bonding between the stacks, the double-stranded DNA molecule exhibits properties of a rigid rod and in solution is a viscous material that loses its viscosity upon denaturation.

The strands of a given molecule of DNA separate over a temperature range. The midpoint is called the **melting temperature, or T_m.** The T_m is influenced by the base composition of the DNA and by the salt concentration of the solution. DNA rich in G–C pairs, which have three hydrogen bonds, melts at a higher temperature than that rich in A–T pairs, which have two hydrogen bonds. A 10-fold increase of monovalent cation concentration increases the T_m by 16.6°C. The organic solvent

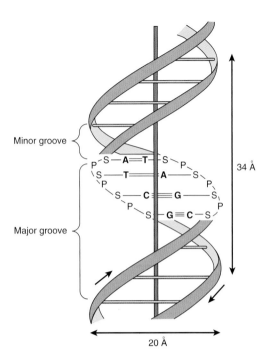

FIGURE 34–2 A diagrammatic representation of the Watson and Crick model of the double-helical structure of the B form of DNA. The horizontal arrow indicates the width of the double helix (20 Å), and the vertical arrow indicates the distance spanned by one complete turn of the double helix (34 Å). One turn of B-DNA includes 10 base pairs (bp), so the rise is 3.4 Å per bp. The central axis of the double helix is indicated by the vertical rod. The short arrows designate the polarity of the antiparallel strands. The major and minor grooves are depicted. (A, adenine; C, cytosine; G, guanine; T, thymine; P, phosphate; S, sugar [deoxyribose].) Hydrogen bonds between A/T and G/C bases indicated by short, red, horizontal lines.

FIGURE 34–3 DNA base pairing between adenosine and thymidine involves the formation of two hydrogen bonds. Three such bonds form between cytidine and guanosine. The broken lines represent hydrogen bonds.

formamide, which is commonly used in recombinant DNA experiments, destabilizes hydrogen bonding between bases, thereby lowering the T_m. This allows the strands of DNA or DNA-RNA hybrids to be separated at much lower temperatures and minimizes the phosphodiester bond breakage that can occur at higher temperatures.

Renaturation of DNA Requires Base Pair Matching

Importantly, separated strands of DNA will renature or reassociate when appropriate physiologic temperature and salt conditions are achieved; this reannealing process is often referred to as **hybridization.** The rate of reassociation depends upon the concentration of the complementary strands. Reassociation of the two complementary DNA strands of a chromosome after transcription is a physiologic example of renaturation (see below). At a given temperature and salt concentration, a particular nucleic acid strand will associate tightly only with a complementary strand. Hybrid molecules will also form under appropriate conditions. For example, DNA will form a hybrid with a complementary DNA (cDNA) or with a cognate messenger RNA (mRNA; see below). When combined with gel electrophoresis techniques that separate nucleic acids by size coupled with radioactive or fluorescent labeling to provide a detectable signal, the resulting analytic techniques are called **Southern (DNA/DNA)** and **Northern (RNA-DNA) blotting,** respectively. These procedures allow for very distinct, high-sensitivity identification of specific nucleic acid species from complex mixtures of DNA or RNA (see Chapter 39).

There Are Grooves in the DNA Molecule

Careful examination of the model depicted in Figure 34–2 reveals a **major groove** and a **minor groove** winding along the molecule parallel to the phosphodiester backbones. In these grooves, proteins can interact specifically with exposed atoms of the nucleotides (via specific hydrophobic and ionic interactions) thereby recognizing and binding to specific nucleotide sequences usually without disrupting the base pairing of the double-helical DNA molecule. As discussed in Chapters 36 and 38, regulatory proteins control the expression of specific genes via such interactions.

DNA Exists in Relaxed & Supercoiled Forms

In some organisms such as bacteria, bacteriophages, many DNA-containing animal viruses, as well as organelles such as mitochondria (see Figure 35–8), the ends of the DNA molecules are joined to create a closed circle with no covalently free ends. This of course does not destroy the polarity of the molecules, but it eliminates all free 3′ and 5′ hydroxyl and phosphoryl groups. Closed circles exist in relaxed or supercoiled forms. Supercoils are introduced when a closed circle is twisted around its own axis or when a linear piece of duplex DNA, whose ends are fixed, is twisted. This energy-requiring process puts the molecule under torsional stress, and the greater the number of supercoils, the greater the stress or torsion (test this by twisting a rubber band). **Negative supercoils** are formed when the molecule is twisted in the direction opposite from the clockwise turns of the right-handed double helix found in B-DNA. Such DNA is said to be underwound. The energy required to achieve this state is, in a sense, stored in the supercoils. The transition to another form that requires energy is thereby facilitated by the underwinding. One such transition is strand separation, which is a prerequisite for DNA replication and transcription. Supercoiled DNA is therefore a preferred form in biologic systems. Enzymes that catalyze topologic changes of DNA are called **topoisomerases.** Topoisomerases can relax or insert supercoils, using ATP as an energy source. Homologs of this enzyme exist in all organisms and are important targets for cancer chemotherapy.

DNA PROVIDES A TEMPLATE FOR REPLICATION & TRANSCRIPTION

The genetic information stored in the nucleotide sequence of DNA serves two purposes. It is the source of information for the synthesis of all protein molecules of the cell and organism, and it provides the information inherited by daughter cells or offspring. Both of these functions require that the DNA molecule serve as a template—in the first case for the transcription of the information into RNA and in the second case for the replication of the information into daughter DNA molecules.

When each strand of the double-stranded parental DNA molecule separates from its complement during replication, each independently serves as a template on which a new complementary strand is synthesized (Figure 34–4). The two newly formed double-stranded daughter DNA molecules, each containing one strand (but complementary rather than identical) from the parent double-stranded DNA molecule, are then sorted between the two daughter cells (Figure 34–5). Each daughter cell contains DNA molecules with information identical to that which the parent possessed; yet in each daughter cell the DNA molecule of the parent cell has been only semiconserved.

THE CHEMICAL NATURE OF RNA DIFFERS FROM THAT OF DNA

Ribonucleic acid (RNA) is a polymer of purine and pyrimidine ribonucleotides linked together by 3′,5′-phosphodiester bonds

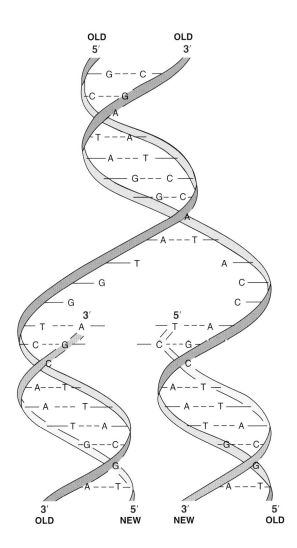

FIGURE 34–4 The double-stranded structure of DNA and the template function of each old strand (orange) on which a new complementary strand (blue) is synthesized.

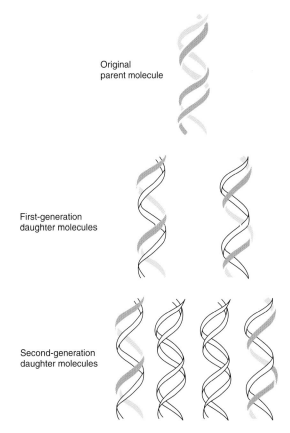

FIGURE 34–5 DNA replication is semiconservative. During a round of replication, each of the two strands of DNA is used as a template for synthesis of a new, complementary strand.

FIGURE 34–6 A segment of a ribonucleic acid (RNA) molecule in which the purine and pyrimidine bases—guanine (G), cytosine (C), uracil (U), and adenine (A)—are held together by phosphodiester bonds between ribosyl moieties attached to the nucleobases by *N*-glycosidic bonds. Note that the polymer has a polarity as indicated by the labeled 3′- and 5′-attached phosphates.

analogous to those in DNA (Figure 34–6). Although sharing many features with DNA, RNA possesses several specific differences:

1. In RNA, the sugar moiety to which the phosphates and purine and pyrimidine bases are attached is ribose rather than the 2′-deoxyribose of DNA.
2. The pyrimidine components of RNA differ from those of DNA. Although RNA contains the ribonucleotides of adenine, guanine, and cytosine, it does not possess thymine except in the rare case mentioned below. Instead of thymine, RNA contains the ribonucleotide of uracil.
3. RNA typically exists as a single strand, whereas DNA exists as a double-stranded helical molecule. However, given the proper complementary base sequence with opposite polarity, the single strand of RNA—as demonstrated in Figure 34–7—is capable of folding back on itself like a hairpin and thus acquiring double-stranded characteristics: G pairing with C, and A with U.
4. Since the RNA molecule is a single strand complementary to only one of the two strands of a gene, its guanine content does not necessarily equal its cytosine content, nor does its adenine content necessarily equal its uracil content.

5. RNA can be hydrolyzed by alkali to 2′,3′ cyclic diesters of the mononucleotides, compounds that cannot be formed from alkali-treated DNA because of the absence of a 2′-hydroxyl group. The alkali lability of RNA is useful both diagnostically and analytically.

Information within the single strand of RNA is contained in its sequence ("primary structure") of purine and pyrimidine nucleotides within the polymer. The sequence is complementary to the template strand of the gene from which it was transcribed. Because of this complementarity, an RNA molecule can bind specifically via the base-pairing rules to its template DNA strand; it will not bind ("hybridize") with the other (coding) strand of its gene. The sequence of the RNA molecule (except for U replacing T) is the same as that of the coding strand of the gene (Figure 34–8).

Nearly All of the Several Species of RNA Are Involved in Some Aspect of Protein Synthesis

Those cytoplasmic RNA molecules that serve as templates for protein synthesis (ie, that transfer genetic information from DNA to the protein-synthesizing machinery) are designated **messenger RNAs,** or **mRNAs.** Many other very abundant cy-

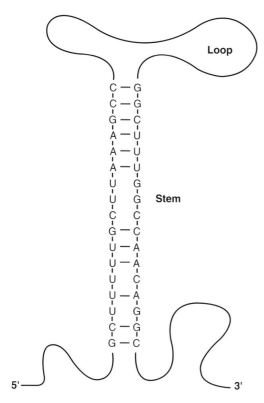

FIGURE 34–7 Diagrammatic representation of the secondary structure of a single-stranded RNA molecule in which a stem loop, or "hairpin," has been formed. Formation of this structure is dependent upon the indicated intramolecular base pairing (colored horizontal lines between bases). Note that A forms hydrogen bonds with U in RNA.

TABLE 34–1 Some of the Species of Small Stable RNAs Found in Mammalian Cells

Name	Length (nucleotides)	Molecules per Cell	Localization
U1	165	1×10^6	Nucleoplasm
U2	188	5×10^5	Nucleoplasm
U3	216	3×10^5	Nucleolus
U4	139	1×10^5	Nucleoplasm
U5	118	2×10^5	Nucleoplasm
U6	106	3×10^5	Perichromatin granules
4.5S	95	3×10^5	Nucleus and cytoplasm
7SK	280	5×10^5	Nucleus and cytoplasm

thesis but play pivotal roles in RNA processing. These relatively small molecules vary in size from 90 to about 300 nucleotides (Table 34–1).

The genetic material for some animal and plant viruses is RNA rather than DNA. Although some RNA viruses never have their information transcribed into a DNA molecule, many animal RNA viruses—specifically, the retroviruses (the HIV virus, for example)—are transcribed by **viral RNA-dependent DNA polymerase, the so-called reverse transcriptase,** to produce a double-stranded DNA copy of their RNA genome. In many cases, the resulting double-stranded DNA transcript is integrated into the host genome and subsequently serves as a template for gene expression and from which new viral RNA genomes and viral mRNAs can be transcribed.

There Exist Several Distinct Classes of RNA

In all prokaryotic and eukaryotic organisms, four main classes of RNA molecules exist: messenger RNA (mRNA), transfer RNA (tRNA), ribosomal RNA (rRNA), and small RNAs. Each differs from the others by abundance, size, function, and general stability.

Messenger RNA (mRNA)

This class is the most heterogeneous in abundance, size and stability; for example, in brewer's yeast specific mRNAs are present in 100's/cell to, on average, ≤0.1/mRNA/cell. As de-

toplasmic RNA molecules (**ribosomal RNAs; rRNAs**) have structural roles wherein they contribute to the formation and function of ribosomes (the organellar machinery for protein synthesis) or serve as adapter molecules (**transfer RNAs; tRNAs**) for the translation of RNA information into specific sequences of polymerized amino acids.

Interestingly, some RNA molecules have intrinsic catalytic activity. The activity of these **ribozymes** often involves the cleavage of a nucleic acid. Two well studied RNA enzymes, or ribozymes, are the peptidyl transferase that catalyzes peptide bond formation on the ribosome and ribozymes involved in the RNA splicing.

In all eukaryotic cells there are **small nuclear RNA (snRNA)** species that are not directly involved in protein syn-

DNA strands:

Coding → **5′**—T G G A A T T G T G A G C G G A T A A C A A T T T C A C A C A G G A A A C A G C T A T G A C C A T G—**3′**
Template → **3′**—A C C T T A A C A C T C G C C T A T T G T T A A A G T G T G T C C T T T G T C G A T A C T G G T A C—**5′**

RNA transcript: **5′**——ppp A U U G U G A G C G G A U A A C A A U U U C A C A C A G G A A A C A G C U A U G A C C A U G **3′**

FIGURE 34–8 The relationship between the sequences of an RNA transcript and its gene, in which the coding and template strands are shown with their polarities. The RNA transcript with a 5′ to 3′ polarity is complementary to the template strand with its 3′ to 5′ polarity. Note that the sequence in the RNA transcript and its polarity is the same as that in the coding strand, except that the U of the transcript replaces the T of the gene.

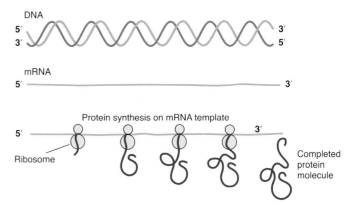

FIGURE 34–9 The expression of genetic information in DNA into the form of an mRNA transcript. This is subsequently translated by ribosomes into a specific protein molecule.

tailed in Chapters 36 & 38, both specific transcriptional and post-transcription mechanisms contribute to this large dynamic range in mRNA content. In mammalian cells mRNA abundance likely varies over a 10^4-fold range. All members of the class function as messengers conveying the information in a gene to the protein synthesizing machinery, where each mRNA serves as a template on which a specific sequence of amino acids is polymerized to form a specific protein molecule, the ultimate gene product (Figure 34–9).

Eukaryotic mRNAs have unique chemical characteristics. The 5′ terminal of mRNA is "capped" by a 7-methylguanosine triphosphate that is linked to an adjacent 2′-*O*-methyl ribonucleoside at its 5′-hydroxyl through the three phosphates (Figure 34–10). The mRNA molecules frequently contain internal 6-methyladenylates and other 2′-*O*-ribose methylated nucleotides. The cap is involved in the recognition of mRNA by the translation machinery, and also helps stabilize the mRNA by preventing the attack of 5′-exonucleases. The protein-synthesizing machinery begins translating the mRNA into proteins beginning downstream of the 5′ or capped terminal. The other end of mRNA molecules, the 3′-hydroxyl terminal, has an attached polymer of adenylate residues 20–250 nucleotides in length. The poly(A) "tail" at the 3′-hydroxyl terminal of mRNAs maintains the intracellular stability of the specific mRNA by preventing the attack of 3′-exonucleases and also facilitates translation (Figure 37–7). A few mRNAs, including those for some histones, do not contain a poly(A) tail. Both the mRNA "cap" and "poly(A) tail" are added post-transcriptionally by non-template-directed enzymes to mRNA precursor molecules (pre-mRNA). mRNA represents 2–5% of total eukaryotic cellular RNA.

In mammalian cells, including cells of humans, the mRNA molecules present in the cytoplasm are not the RNA products immediately synthesized from the DNA template but must be formed by processing from the pre-mRNA before entering the cytoplasm. Thus, in mammalian nuclei, the immediate products of gene transcription (primary transcripts) are very heterogeneous and can be greater than 10- to 50-fold longer than mature mRNA molecules. As discussed in Chapter 36, pre-mRNA molecules are processed to generate the mRNA molecules which then enter the cytoplasm to serve as templates for protein synthesis.

Transfer RNA (tRNA)

tRNA molecules vary in length from 74 to 95 nucleotides. They also are generated by nuclear processing of a precursor molecule (Chapter 36). The tRNA molecules serve as adapters for the translation of the information in the sequence of nucleotides of the mRNA into specific amino acids. There are at least 20 species of tRNA molecules in every cell, at least one (and often several) corresponding to each of the 20 amino acids required for protein synthesis. Although each specific tRNA differs from the others in its sequence of nucleotides, the tRNA molecules as a class have many features in common. The primary structure—ie, the nucleotide sequence—of all tRNA molecules allows extensive folding and intrastrand complementarity to generate a secondary structure that appears in two dimensions like a cloverleaf (Figure 34–11).

All tRNA molecules contain four main arms. The **acceptor arm** terminates in the nucleotides CpCpAOH. These three nucleotides are added posttranscriptionally by a specific nucleotidyl transferase enzyme. The tRNA-appropriate amino acid is attached, or "charged" onto, the 3′-OH group of the A moiety of the acceptor arm (see Figure 37–1). The **D, TψC,** and **extra arms** help define a specific tRNA. tRNAs compose roughly 20% of total cellular RNA.

Ribosomal RNA (rRNA)

A ribosome is a cytoplasmic nucleoprotein structure that acts as the machinery for the synthesis of proteins from the mRNA templates. On the ribosomes, the mRNA and tRNA molecules interact to translate into a specific protein molecule information transcribed from the gene. During periods of active protein synthesis, many ribosomes can be associated with any mRNA molecule to form an assembly called the **polysome** (Figure 37–7).

The components of the mammalian ribosome, which has a molecular weight of about 4.2×10^6 and a sedimentation velocity coefficient of 80S (S = Svedberg units, a parameter sensitive to molecular size and shape) are shown in Table 34–2. The mammalian ribosome contains two major nucleoprotein subunits—a larger one with a molecular weight of 2.8×10^6 (60S) and a smaller subunit with a molecular weight of 1.4×10^6 (40S). The 60S subunit contains a 5S ribosomal RNA (rRNA),

FIGURE 34–10 The cap structure attached to the 5′ terminal of most eukaryotic messenger RNA molecules. A 7-methylguanosine triphosphate (black) is attached at the 5′ terminal of the mRNA (shown in color), which usually also contains a 2′-O-methylpurine nucleotide. These modifications (the cap and methyl group) are added after the mRNA is transcribed from DNA.

a 5.8S rRNA, and a 28S rRNA; there are also more than 50 specific polypeptides. The 40S subunit is smaller and contains a single 18S rRNA and approximately 30 distinct polypeptide chains. All of the ribosomal RNA molecules except the 5S rRNA, which is independently transcribed, are processed from a single 45S precursor RNA molecule in the nucleolus (Chapter 36). The highly methylated ribosomal RNA molecules are packaged in the nucleolus with the specific ribosomal proteins. In the cytoplasm, the ribosomes remain quite stable and capable of many translation cycles. The exact functions of the ribosomal RNA molecules in the ribosomal particle are not fully understood, but they are necessary for ribosomal assembly and also play key roles in the binding of mRNA to ribosomes and its translation. Recent studies indicate that the large rRNA component performs the peptidyl transferase activity and thus is a ribozyme. The ribosomal RNAs (28S + 18S) represent roughly 70% of total cellular RNA.

Small RNA

A large number of discrete, highly conserved, and small RNA species are found in eukaryotic cells; some are quite stable. Most of these molecules are complexed with proteins to form ribonucleoproteins and are distributed in the nucleus, the cytoplasm, or both. They range in size from 20 to 300 nucleotides and are present in 100,000–1,000,000 copies per cell, collectively representing ≤5% of cellular RNA.

FIGURE 34–11 Typical aminoacyl tRNA in which the amino acid (aa) is attached to the 3′ CCA terminal. The anticodon, TψC, and dihydrouracil (D) arms are indicated, as are the positions of the intramolecular hydrogen bonding between these base pairs. Ψ is pseudouridine, an isomer of uridine formed posttranscriptionally. (Reprinted, with permission, from Watson JD: *Molecular Biology of the Gene*, 3rd ed. Copyright © 1976, 1970, 1965, by W.A. Benjamin, Inc., Menlo Park, California.)

Small Nuclear RNAs (snRNAs)

snRNAs, a subset of the small RNAs, are significantly involved in rRNA and mRNA processing and gene regulation. Of the several snRNAs, U1, U2, U4, U5, and U6 are involved in intron removal and the processing of mRNA precursors into mRNA (Chapter 36). The U7 snRNA is involved in production of the correct 3′ ends of histone mRNA—which lacks a poly(A) tail. 7SK RNA associates with several proteins to form a ribonucleoprotein complex, termed P-TEFb, that modulates mRNA gene transcription elongation by RNA polymerase II (see Chapter 36).

Micro-RNAs, miRNAs, and Small Interfering RNAs, siRNAs

One of the most exciting and unanticipated discoveries in the last decade of eukaryotic regulatory biology was the identification and characterization of miRNAs, a class of small RNAs found in most eukaryotes (Chapter 38). Nearly all known **miRNAs and siRNAs cause inhibition of gene expression** by decreasing specific protein production, albeit via distinct mechanisms. miRNAs are typically 21–25 nucleotides in length and are generated by nucleolytic processing of the products of distinct genes/transcription units (see Figure 36–17). miRNA precursors are single stranded but have extensive intramolecular secondary structure. These precursors range in size from about 500 to 1000 nucleotides; the small processed mature **miRNAs typically hybridize, via the formation of imperfect RNA-RNA duplexes within the 3′-untranslated regions** (3′UTRs; see Figure 38–19) of specific target mRNAs, **leading via poorly understood mechanisms to translation arrest.** To date, hundreds of distinct miRNAs have been described in humans. siRNAs are derived by the specific nucleolytic cleavage of larger, double-stranded RNAs to again form small 21–25 nucleotide-long products. These short **siRNAs usually form perfect RNA-RNA hybrids** with their distinct targets potentially anywhere within the length of the mRNA where the complementary sequence exists. Formation of such RNA-RNA duplexes between siRNA and mRNA results in reduced specific protein production because the **siRNA-mRNA complexes are degraded** by dedicated nucleolytic machinery; some or all of this mRNA degradation occurs in specific cytoplasmic organelles termed **P bodies** (Figure 37–11). Both miRNAs and siRNAs represent exciting new **potential targets for therapeutic drug development.** In addition, siRNAs are frequently used to decrease or "knock-down" specific protein levels (via siRNA homology–directed mRNA degradation) in experimental contexts in the laboratory, an extremely useful and powerful alternative to gene-knockout technology (Chapter 39). Interestingly, bacteria also contain small, heterogeneous regulatory RNAs termed sRNAs. Bacterial sRNAs range in size from 50 to 500 nucleotides, and like eukaryotic mi/siRNAs also control a large array of genes. Similarly sRNAs often repress, but sometimes activate protein synthesis by binding to specific mRNA.

TABLE 34–2   Components of Mammalian Ribosomes

		Protein		RNA		
Component	Mass (MW)	Number	Mass	Size	Mass	Bases
40S subunit	1.4×10^6	33	7×10^5	18S	7×10^5	1900
60S subunit	2.8×10^6	50	1×10^6	5S	35,000	120
				5.8S	45,000	160
				28S	1.6×10^6	4700

Note: The ribosomal subunits are defined according to their sedimentation velocity in Svedberg units (40S or 60S). The number of unique proteins and their total mass (MW) and the RNA components of each subunit in size (Svedberg units), mass, and number of bases are listed.

SPECIFIC NUCLEASES DIGEST NUCLEIC ACIDS

Enzymes capable of degrading nucleic acids have been recognized for many years. These nucleases can be classified in several ways. Those which exhibit specificity for DNA are referred to as **deoxyribonucleases.** Those which specifically hydrolyze RNA are **ribonucleases.** Some nucleases degrade both DNA and RNA. Within both of these classes are enzymes capable of cleaving internal phosphodiester bonds to produce either 3′-hydroxyl and 5′-phosphoryl terminals or 5′-hydroxyl and 3′-phosphoryl terminals. These are referred to as **endonucleases.** Some are capable of hydrolyzing both strands of a **double-stranded** molecule, whereas others can only cleave **single strands** of nucleic acids. Some nucleases can hydrolyze only unpaired single strands, while others are capable of hydrolyzing single strands participating in the formation of a double-stranded molecule. There exist classes of endonucleases that recognize specific sequences in DNA; the majority of these are the **restriction endonucleases,** which have in recent years become important tools in molecular genetics and medical sciences. A list of some currently recognized restriction endonucleases is presented in Table 39–2.

Some nucleases are capable of hydrolyzing a nucleotide only when it is present at a terminal of a molecule; these are referred to as **exonucleases.** Exonucleases act in one direction ($3′ \rightarrow 5′$ or $5′ \rightarrow 3′$) only. In bacteria, a $3′ \rightarrow 5′$ exonuclease is an integral part of the DNA replication machinery and there serves to edit—or proofread—the most recently added deoxynucleotide for base-pairing errors.

SUMMARY

- DNA consists of four bases—A, G, C, and T—that are held in linear array by phosphodiester bonds through the 3′ and 5′ positions of adjacent deoxyribose moieties.
- DNA is organized into two strands by the pairing of bases A to T and G to C on complementary strands. These strands form a double helix around a central axis.

- The 3×10^9 base pairs of DNA in humans are organized into the haploid complement of 23 chromosomes. The exact sequence of these 3 billion nucleotides defines the uniqueness of each individual.
- DNA provides a template for its own replication and thus maintenance of the genotype and for the transcription of the roughly 30,000 human genes into a variety of RNA molecules.
- RNA exists in several different single-stranded structures, most of which are directly or indirectly involved in protein synthesis or its regulation. The linear array of nucleotides in RNA consists of A, G, C, and U, and the sugar moiety is ribose.
- The major forms of RNA include messenger RNA (mRNA), ribosomal RNA (rRNA), transfer RNA (tRNA), and small nuclear RNAs (snRNAs; miRNAs). Certain RNA molecules act as catalysts (ribozymes).

REFERENCES

Chapman EJ, Carrington JC: Specialization and evolution of endogenous small RNA pathways. Nature Rev Genetics 2007;8:884.

Green R, Noller HF: Ribosomes and translation. Annu Rev Biochem 1997;66:689.

Guthrie C, Patterson B: Spliceosomal snRNAs. Ann Rev Genet 1988;22:387.

Hopper AK, Shaheen HH: A decade of surprises for tRNA nuclear cytoplasmic dynamics. Trends Cell Biol 2008;18:98.

Hunt T: *DNA Makes RNA Makes Protein.* Elsevier, 1983.

Moore M: From birth to death: the complex lives of eukaryotic mRNAs. Science 2005;309:1514.

Storz G, Altuvia S, Wasserman KM: An abundance of RNA regulators. Ann Rev Biochem 2005;74:199.

Wang G-S, Cooper TA: Splicing in disease: disruption of the splicing code and the decoding machinery. Nature Rev Genetics 2007;8:749.

Watson JD, Crick FHC: Molecular structure of nucleic acids. Nature 1953;171:737.

Watson JD: *The Double Helix.* Atheneum, 1968.

Watson JD et al: *Molecular Biology of the Gene,* 5th ed. Benjamin-Cummings, 2000.

Zamore PD, Haley B: Ribo-gnome: the big world of small RNAs. Science 2005;309:1519.

35

DNA Organization, Replication, & Repair

P. Anthony Weil, PhD

BIOMEDICAL IMPORTANCE*

The genetic information in the DNA of a chromosome can be transmitted by exact replication or it can be exchanged by a number of processes, including crossing over, recombination, transposition, and conversion. These provide a means of ensuring adaptability and diversity for the organism but, when these processes go awry, can also result in disease. A number of enzyme systems are involved in DNA replication, alteration, and repair. Mutations are due to a change in the base sequence of DNA and may result from the faulty replication, movement, or repair of DNA and occur with a frequency of about one in every 10^6 cell divisions. Abnormalities in gene products (either in RNA, protein function, or amount) can be the result of mutations that occur in coding or regulatory-region DNA. A mutation in a germ cell is transmitted to offspring (so-called vertical transmission of hereditary disease). A number of factors, including viruses, chemicals, ultraviolet light, and ionizing radiation, increase the rate of mutation. Mutations often affect somatic cells and so are passed on to successive generations of cells, but only within an organism (ie, horizontally). It is becoming apparent that a number of diseases—and perhaps most cancers—are due to the combined effects of vertical transmission of mutations as well as horizontal transmission of induced mutations.

CHROMATIN IS THE CHROMOSOMAL MATERIAL IN THE NUCLEI OF CELLS OF EUKARYOTIC ORGANISMS

Chromatin consists of very long double-stranded **DNA molecules** and a nearly equal mass of rather small basic proteins termed **histones** as well as a smaller amount of **nonhistone proteins** (most of which are acidic and larger than histones) and a small quantity of **RNA.** The nonhistone proteins include

---

*So far as is possible, the discussion in this chapter and in Chapters 36, 37, & 38 will pertain to mammalian organisms, which are, of course, among the higher eukaryotes. At times it will be necessary to refer to observations in prokaryotic organisms such as bacteria and viruses, but in such cases the information will be of a kind that can be extrapolated to mammalian organisms.

enzymes involved in DNA replication and repair, and the proteins involved in RNA synthesis, processing, and transport to the cytoplasm. The double-stranded DNA helix in each chromosome has a length that is thousands of times the diameter of the cell nucleus. One purpose of the molecules that comprise chromatin, particularly the histones, is to condense the DNA; however, it is important to note that the histones also integrally participate in gene regulation (Chapters 36, 38, & 42). Electron microscopic studies of chromatin have demonstrated dense spherical particles called **nucleosomes,** which are approximately 10 nm in diameter and connected by DNA filaments (Figure 35–1). Nucleosomes are composed of DNA wound around a collection of histone molecules.

Histones Are the Most Abundant Chromatin Proteins

Histones are a small family of closely related basic proteins. **H1 histones** are the ones least tightly bound to chromatin (Figure 35–1) and are, therefore, easily removed with a salt solution, after which chromatin becomes more soluble. The organizational unit of this soluble chromatin is the nucleosome. Nucleosomes contain four types of histones: **H2A, H2B, H3, and H4.** The structures of all four histones—H2A, H2B, H3, and H4, the so-called core histones that form the nucleosome—have been highly conserved between species. This extreme conservation implies that the function of histones is identical in all eukaryotes and that the entire molecule is involved quite specifically in carrying out this function. The carboxyl terminal two-thirds of the histone molecules are hydrophobic, while their amino terminal thirds are particularly rich in basic amino acids. **These four core histones are subject to at least six types of covalent modification:** acetylation, methylation, phosphorylation, ADP-ribosylation, monoubiquitylation, and sumoylation. These histone modifications play an important role in chromatin structure and function, as illustrated in Table 35–1.

The histones interact with each other in very specific ways. **H3 and H4 form a tetramer** containing two molecules of each $(H3\text{-}H4)_2$, while **H2A and H2B form dimers** (H2A-H2B). Under physiologic conditions, these histone oligomers associate to form the **histone octamer** of the composition $(H3\text{-}H4)_2\text{-}(H2A\text{-}H2B)_2$.

FIGURE 35–1 Electron micrograph of nucleosomes (white, ball-shaped) attached to strands of DNA (thin, gray line); see Figure 35–2. (Reproduced, with permission, from Shao Z. Probing nanometer structures with atomic force microscopy. News Physiol Sci, 1999, Aug.;14:142–149. Courtesy of Professor Zhifeng Shao, University of Virginia.)

The Nucleosome Contains Histone & DNA

When the histone octamer is mixed with purified, doubles-tranded DNA under appropriate ionic conditions, the same x-ray diffraction pattern is formed as that observed in freshly isolated chromatin. Electron microscopic studies confirm the existence of reconstituted nucleosomes. Furthermore, the re-constitution of nucleosomes from DNA and histones H2A, H2B, H3, and H4 is independent of the organismal or cellular origin of the various components. Neither the histone H1 nor the nonhistone proteins are necessary for the reconstitution of the nucleosome core.

In the nucleosome, the DNA is supercoiled in a left-handed helix over the surface of the disk-shaped histone octamer (Figure 35–2). The majority of core histone proteins interact with

TABLE 35–1 Possible Roles of Modified Histones

1. Acetylation of histones H3 and H4 is associated with the activation or inactivation of gene transcription.
2. Acetylation of core histones is associated with chromosomal assembly during DNA replication.
3. Phosphorylation of histone H1 is associated with the condensation of chromosomes during the replication cycle.
4. ADP-ribosylation of histones is associated with DNA repair.
5. Methylation of histones is correlated with activation and repression of gene transcription.
6. Monoubiquitylation is associated with gene activation, repression, and heterochromatic gene silencing.
7. Sumoylation of histones (SUMO; small ubiquitin-related modifier) is associated with transcription repression.

the DNA on the inside of the supercoil without protruding, although the amino terminal tails of all the histones probably extend outside of this structure and are available for regulatory covalent modifications (see Table 35–1).

The $(H3-H4)_2$ tetramer itself can confer nucleosome-like properties on DNA and thus has a central role in the formation of the nucleosome. The addition of two H2A-H2B dimers stabilizes the primary particle and firmly binds two additional half-turns of DNA previously bound only loosely to the $(H3-H4)_2$. Thus, 1.75 superhelical turns of DNA are wrapped around the surface of the histone octamer, protecting 146 base pairs of DNA and forming the nucleosome core particle (Figure 35–2). In chroma-tin, core particles are separated by an about 30-bp region of DNA termed "**linker**." Most of the DNA is in a repeating series of these structures, giving the so-called "beads-on-a-string" appearance when examined by electron microscopy (see Figure 35–1).

The assembly of nucleosomes is mediated by one of several nuclear chromatin assembly factors facilitated by histone chap-erones, a group of proteins that exhibit high-affinity histone binding. As the nucleosome is assembled, histones are released from the histone chaperones. Nucleosomes appear to exhibit preference for certain regions on specific DNA molecules, but the basis for this nonrandom distribution, termed **phasing**, is not yet completely understood. Phasing is probably related to the relative physical flexibility of certain nucleotide sequences that are able to accommodate the regions of kinking within the supercoil as well as the presence of other DNA-bound factors that limit the sites of nucleosome deposition.

HIGHER-ORDER STRUCTURES PROVIDE FOR THE COMPACTION OF CHROMATIN

Electron microscopy of chromatin reveals two higher or-ders of structure—the 10-nm fibril and the 30-nm chroma-

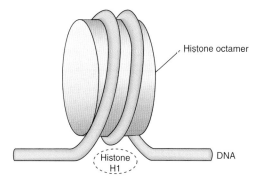

FIGURE 35–2 Model for the structure of the nucleosome, in which DNA is wrapped around the surface of a flat protein cylinder consisting of two each of histones H2A, H2B, H3, and H4 that form the histone octamer. The 146 base pairs of DNA, consisting of 1.75 superhelical turns, are in contact with the histone octamer. This protects the DNA from digestion by a nuclease. The position of histone H1, when it is present, is indicated by the dashed outline at the bottom of the figure.

tin fiber—beyond that of the nucleosome itself. The disk-like nucleosome structure has a 10-nm diameter and a height of 5 nm. The **10-nm fibril** consists of nucleosomes arranged with their edges separated by a small distance (30 bp of DNA) with their flat faces parallel to the fibril axis (Figure 35–3). The 10-nm fibril is probably further supercoiled with six or seven nucleosomes per turn to form the **30-nm chromatin fiber** (Figure 35–3). Each turn of the supercoil is relatively flat, and the faces of the nucleosomes of successive turns would be nearly parallel to each other. H1 histones appear to stabilize the 30-nm fiber, but their position and that of the variable length spacer DNA are not clear. It is probable that nucleosomes can form a variety of packed structures. In order to form a mitotic chromosome, the 30-nm fiber must be compacted in length another 100-fold (see below).

In **interphase chromosomes,** chromatin fibers appear to be organized into 30,000–100,000 bp **loops or domains** anchored in a scaffolding (or supporting matrix) within the nucleus, the

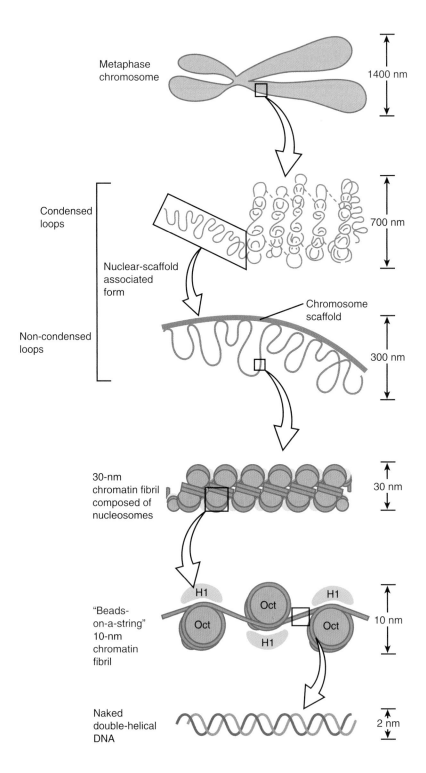

FIGURE 35–3 Shown is the extent of DNA packaging in metaphase chromosomes (top) to noted duplex DNA (bottom). Chromosomal DNA is packaged and organized at several levels as shown (see Table 35–2). Each phase of condensation or compaction and organization (bottom to top) decreases overall DNA accessibility to an extent that the DNA sequences in metaphase chromosomes are almost totally transcriptionally inert. In toto, these five levels of DNA compaction result in nearly a 10^4-fold linear decrease in end-to-end DNA length. Complete condensation and decondensation of the linear DNA in chromosomes occur in the space of hours during the normal replicative cell cycle (see Figure 35–20).

so-called **nuclear matrix**. Within these domains, some DNA sequences may be located nonrandomly. It has been suggested that each looped domain of chromatin corresponds to one or more separate genetic functions, containing both coding and noncoding regions of the cognate gene or genes. This nuclear architecture is likely dynamic, having important regulatory effects upon gene regulation. Recent data suggest that certain genes or gene regions are mobile within the nucleus, moving to discrete loci within the nucleus upon activation (Chapters 36 & 38). Further work will determine if this is a general phenomenon.

SOME REGIONS OF CHROMATIN ARE "ACTIVE" & OTHERS ARE "INACTIVE"

Generally, every cell of an individual metazoan organism contains the same genetic information. Thus, the differences between cell types within an organism must be explained by differential expression of the common genetic information. Chromatin containing active genes (ie, transcriptionally or potentially transcriptionally active chromatin) has been shown to differ in several ways from that of inactive regions. The nucleosome structure of active chromatin appears to be altered, sometimes quite extensively, in highly active regions. DNA in active chromatin contains large regions (about 100,000 bases long) that are relatively more **sensitive to digestion by a nuclease** such as DNase I. DNase I makes single-strand cuts in nearly any segment of DNA (ie, low sequence specificity). It will digest DNA that is not protected, or bound by protein, into its component deoxynucleotides. The sensitivity to DNase I of active chromatin regions reflects only a potential for transcription rather than transcription itself and in several systems can be correlated with a relative lack of 5-methyldeoxycytidine (meC) in the DNA and particular histone covalent modifications (phosphorylation, acetylation, etc; see Table 35–1).

Within the large regions of active chromatin there exist shorter stretches of 100–300 nucleotides that exhibit an even greater (another 10-fold) sensitivity to DNase I. These **hypersensitive sites** probably result from a structural conformation that favors access of the nuclease to the DNA. These regions are often located immediately upstream from the active gene and are the location of interrupted nucleosomal structure caused by the binding of nonhistone regulatory transcription factor proteins (see Chapters 36 & 38). In many cases, it seems that if a gene is capable of being transcribed, it very often has a DNase-hypersensitive site(s) in the chromatin immediately upstream. As noted above, nonhistone regulatory proteins involved in transcription control and those involved in maintaining access to the template strand lead to the formation of hypersensitive sites. Such sites often provide the first clue about the presence and location of a transcription control element.

By contrast, transcriptionally inactive chromatin is densely packed during interphase as observed by electron microscopic studies and is referred to as **heterochromatin;** transcriptionally active chromatin stains less densely and is referred to as **euchromatin.** Generally, euchromatin is replicated earlier than heterochromatin in the mammalian cell cycle (see below). The chromatin in these regions of inactivity is often high in meC content, and histones therein contain relatively lower levels of covalent modifications.

There are two types of heterochromatin: constitutive and facultative. **Constitutive heterochromatin** is always condensed and thus essentially inactive. It is found in the regions near the chromosomal centromere and at chromosomal ends (telomeres). **Facultative heterochromatin** is at times condensed, but at other times it is actively transcribed and, thus, uncondensed and appears as euchromatin. Of the two members of the X chromosome pair in mammalian females, one X chromosome is almost completely inactive transcriptionally and is heterochromatic. However, the heterochromatic X chromosome decondenses during gametogenesis and becomes transcriptionally active during early embryogenesis—thus, it is facultative heterochromatin.

Certain cells of insects, eg, *Chironomus* and *Drosophila*, contain giant chromosomes that have been replicated for multiple cycles without separation of daughter chromatids. These copies of DNA line up side by side in precise register and produce a banded chromosome containing regions of condensed chromatin and lighter bands of more extended chromatin. Transcriptionally active regions of these **polytene chromosomes** are especially decondensed into "**puffs**" that can be shown to contain the enzymes responsible for transcription and to be the sites of RNA synthesis (Figure 35–4). Using highly sensitive fluorescently labeled hybridization probes, specific gene sequences can be mapped, or "painted," within the nuclei of human cells, even without polytene chromosome formation, using FISH (fluorescence in situ hybridization; Chapter 39) techniques.

DNA IS ORGANIZED INTO CHROMOSOMES

At metaphase, mammalian **chromosomes** possess a 2-fold symmetry, with the identical duplicated **sister chromatids** connected at a **centromere,** the relative position of which is characteristic for a given chromosome (Figure 35–5). The centromere is an adenine-thymine (A–T)-rich region containing repeated DNA sequences that range in size from 10^2 (brewers' yeast) to 10^6 (mammals) base pairs. Metazoan centromeres are bound by nucelosomes containing the histone H3 variant protein CENP-A and other specific centromere-binding proteins. This complex, called the **kinetochore,** provides the anchor for the mitotic spindle. It thus is an essential structure for chromosomal segregation during mitosis.

The ends of each chromosome contain structures called **telomeres.** Telomeres consist of short TG-rich repeats. Human telomeres have a variable number of repeats of the sequence 5′-TTAGGG-3′, which can extend for several kilobases. **Telomerase,** a multisubunit RNA-containing complex related to viral RNA-dependent DNA polymerases (reverse transcriptases), is the enzyme responsible for telomere synthesis and thus for maintaining the length of the telomere. Since telomere shortening has been associated with both malignant

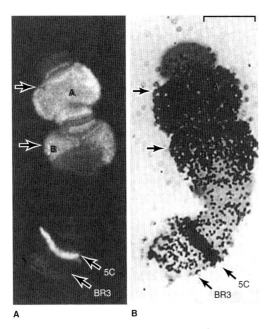

FIGURE 35-4 Illustration of the tight correlation between the presence of RNA polymerase II (Table 36–2) and messenger RNA synthesis. A number of genes, labeled A, B (top), and 5C, but not genes at locus (band) BR3 (5C, BR3, bottom) are activated when *Chironomus tentans* larvae are subjected to heat shock (39°C for 30 min). **(A)** Distribution of RNA polymerase II in isolated chromosome IV from the salivary gland (**at arrows**). The enzyme was detected by immunofluorescence using an antibody directed against the polymerase. The 5C and BR3 are specific bands of chromosome IV, and the arrows indicate puffs. **(B)** Autoradiogram of a chromosome IV that was incubated in ^{3}H-uridine to label the RNA. Note the correspondence of the immunofluorescence and presence of the radioactive RNA (black dots). Bar = 7 μm. (Reproduced, with permission, from Sass H: RNA polymerase B in polytene chromosomes. Cell 1982;28:274. Copyright ©1982. Reprinted with permission from Elsevier.)

FIGURE 35-5 The two sister chromatids of human chromosome 12. The location of the A+T-rich centromeric region connecting sister chromatids is indicated, as are two of the four telomeres residing at the very ends of the chromatids that are attached one to the other at the centromere. (Courtesy of Biophoto Associates/Photo Researchers, Inc.)

TABLE 35-2 **The Packing Ratios of Each of the Orders of DNA Structure**

Chromatin Form	Packing Ratio
Naked double-helical DNA	~1.0
10-nm fibril of nucleosomes	7–10
30-nm chromatin fiber of superhelical nucleosomes	40–60
Condensed metaphase chromosome of loops	8000

transformation and aging, telomerase has become an attractive target for cancer chemotherapy and drug development. Each sister chromatid contains one double-stranded DNA molecule. During interphase, the packing of the DNA molecule is less dense than it is in the condensed chromosome during metaphase. Metaphase chromosomes are nearly completely transcriptionally inactive.

The human haploid genome consists of about 3×10^9 bp and about 1.7×10^7 nucleosomes. Thus, each of the 23 chromatids in the human haploid genome would contain on the average 1.3×10^8 nucleotides in one double-stranded DNA molecule. Therefore the length of each DNA molecule must be compressed about 8000-fold to generate the structure of a condensed metaphase chromosome. In metaphase chromosomes, the 30-nm chromatin fibers are also folded into a series of **looped domains,** the proximal portions of which are anchored to a nonhistone proteinaceous nuclear matrix scaffolding within the nucleus (Figure 35–3). The packing ra-

tios of each of the orders of DNA structure are summarized in Table 35–2.

The packaging of nucleoproteins within chromatids is not random, as evidenced by the characteristic patterns observed when chromosomes are stained with specific dyes such as quinacrine or Giemsa stain (Figure 35–6).

From individual to individual within a single species, the pattern of staining (banding) of the entire chromosome complement is highly reproducible; nonetheless, it differs significantly between species, even those closely related. Thus, the packaging of the nucleoproteins in chromosomes of higher eukaryotes must in some way be dependent upon species-specific characteristics of the DNA molecules.

FIGURE 35–6 A human karyotype (of a man with a normal 46,XY constitution), in which the metaphase chromosomes have been stained by the Giemsa method and aligned according to the Paris Convention. (Courtesy of H Lawce and F Conte.)

A combination of specialized staining techniques and high-resolution microscopy has allowed cytogeneticists to quite precisely map thousands of genes to specific regions of mouse and human chromosomes. With the recent elucidation of the human and mouse genome sequences (among others), it has become clear that many of these visual mapping methods were remarkably accurate.

Coding Regions Are Often Interrupted by Intervening Sequences

The **protein coding regions of DNA,** the transcripts of which ultimately appear in the cytoplasm as single mRNA molecules, are usually **interrupted in the eukaryotic genome by large intervening sequences of nonprotein coding DNA.** Accordingly, the **primary transcripts of DNA, mRNA precursors,** (originally termed **hnRNA** because this species of RNA was quite heterogeneous in size [length] and mostly restricted to the nucleus), contain noncoding intervening sequences of RNA that must be removed in a process which also joins together the appropriate coding segments to form the mature mRNA. Most coding sequences for a single mRNA are interrupted in the genome (and thus in the primary transcript) by at least one—and in some cases as many as 50—noncoding intervening sequences **(introns).** In most cases, the introns are much longer than the coding regions **(exons).** The processing of the primary transcript, which involves precise removal of introns and splicing of adjacent exons, is described in detail in Chapter 36.

The function of the intervening sequences, or introns, is not totally clear. Introns may serve to separate functional domains (exons) of coding information in a form that permits genetic rearrangement by recombination to occur more rapidly than if all coding regions for a given genetic function were contiguous. Such an enhanced rate of genetic rearrangement of functional domains might allow more rapid evolution of biologic function. The relationships among chromosomal DNA, gene clusters on the chromosome, the exon-intron structure of genes, and the final mRNA product are illustrated in Figure 35–7.

MUCH OF THE MAMMALIAN GENOME APPEARS REDUNDANT & MUCH IS NOT HIGHLY TRANSCRIBED

The haploid genome of each human cell consists of 3×10^9 base pairs of DNA subdivided into 23 chromosomes. The entire haploid genome contains sufficient DNA to code for nearly 1.5 million average-sized genes. However, studies of mutation rates and of the complexities of the genomes of higher organisms strongly suggest that humans have significantly fewer than 100,000 proteins encoded by the ~1% of the human genome that is composed of exonic DNA. Indeed current estimates suggest there are ≤25,000 protein-coding genes in humans. This implies that most of the DNA is non-protein-coding—ie, its information is never translated into an amino acid sequence of a protein molecule. Certainly, some of the excess DNA sequences serve to regulate the expression of genes during development, differentiation, and adaptation to the environment, either by serving as binding sites for regulatory proteins or by encoding regulatory RNAs (ie, miRNAs). Some excess clearly makes up the intervening sequences or introns (24% of the total human genome) that split the coding regions of genes, and another portion of the excess appears to be composed of many families of repeated sequences for which clear functions have not yet been defined. A summary of the salient features of the human genome is presented in Chapter 39. Interestingly the ENCODE Project Consortium (Chapter 39) has shown that for the 1% of the genome studied most of the genomic sequence was indeed

FIGURE 35–7 The relationship between chromosomal DNA and mRNA. The human haploid DNA complement of 3×10^9 base pairs (bp) is distributed between 23 chromosomes. Genes are clustered on these chromosomes. An average gene is 2×10^4 bp in length, including the regulatory region (hatched area), which is usually located at the 5′ end of the gene. The regulatory region is shown here as being adjacent to the transcription initiation site (arrow). Most eukaryotic genes have alternating exons and introns. In this example, there are nine exons (dark colored areas) and eight introns (light colored areas). The introns are removed from the primary transcript by the processing reactions, and the exons are ligated together in sequence to form the mature mRNA. (nt, nucleotides.)

transcribed at a low rate. Further research will elucidate the role(s) played by such transcripts.

The DNA in a eukaryotic genome can be divided into different "sequence classes." These are unique-sequence DNA, or nonrepetitive DNA and repetitive-sequence DNA. In the haploid genome, unique-sequence DNA generally includes the single copy genes that code for proteins. The repetitive DNA in the haploid genome includes sequences that vary in copy number from 2 to as many as 10^7 copies per cell.

More Than Half the DNA in Eukaryotic Organisms Is in Unique or Nonrepetitive Sequences

This estimation (and the distribution of repetitive-sequence DNA) is based on a variety of DNA-RNA hybridization techniques and, more recently, on direct DNA sequencing. Similar techniques are used to estimate the number of active genes in a population of unique-sequence DNA. In brewers' yeast (*Saccharomyces cerevisiae*, a lower eukaryote), about two thirds of its 6200 genes are expressed, but only ~1/5 are required for viability under laboratory growth conditions. In typical tissues in a higher eukaryote (eg, mammalian liver and kidney), between 10,000 and 15,000 genes are actively expressed. Different combinations of genes are expressed in each tissue, of course, and how this is accomplished is one of the major unanswered questions in biology.

In Human DNA, at Least 30% of the Genome Consists of Repetitive Sequences

Repetitive-sequence DNA can be broadly classified as moderately repetitive or as highly repetitive. The highly repetitive sequences consist of 5–500 base pair lengths repeated many times in tandem. These sequences are often clustered in centromeres and telomeres of the chromosome and some are

present in about 1–10 million copies per haploid genome. The majority of these sequences are transcriptionally inactive and some of these sequences play a structural role in the chromosome (Figure 35–5; see Chapter 39).

The moderately repetitive sequences, which are defined as being present in numbers of less than 10^6 copies per haploid genome, are not clustered but are interspersed with unique sequences. In many cases, these long interspersed repeats are transcribed by RNA polymerase II and contain caps indistinguishable from those on mRNA.

Depending on their length, moderately repetitive sequences are classified as **long interspersed repeat sequences (LINEs)** or **short interspersed repeat sequences (SINEs)**. Both types appear to be **retroposons;** ie, they arose from movement from one location to another **(transposition)** through an RNA intermediate by the action of reverse transcriptase that transcribes an RNA template into DNA. Mammalian genomes contain 20,000–50,000 copies of the 6–7 kbp LINEs. These represent species-specific families of repeat elements. SINEs are shorter (70–300 bp), and there may be more than 100,000 copies per genome. Of the SINEs in the human genome, one family, the **Alu family,** is present in about 500,000 copies per haploid genome and accounts for 5–6% of the human genome. Members of the human Alu family and their closely related analogs in other animals are transcribed as integral components of mRNA precursors or as discrete RNA molecules, including the well-studied 4.5S RNA and 7S RNA. These particular family members are highly conserved within a species as well as between mammalian species. Components of the short interspersed repeats, including the members of the Alu family, may be mobile elements, capable of jumping into and out of various sites within the genome (see below). These transposition events can have disastrous results, as exemplified by the insertion of Alu sequences into a gene, which, when so mutated, causes neurofibromatosis.

Microsatellite Repeat Sequences

One category of repeat sequences exists as both dispersed and grouped tandem arrays. The sequences consist of 2–6 bp repeated up to 50 times. These **microsatellite sequences** most commonly are found as dinucleotide repeats of AC on one strand and TG on the opposite strand, but several other forms occur, including CG, AT, and CA. The AC repeat sequences occur at 50,000–100,000 locations in the genome. At any locus, the number of these repeats may vary on the two chromosomes, thus providing heterozygosity of the number of copies of a particular microsatellite number in an individual. This is a heritable trait, and because of their number and the ease of detecting them using the **polymerase chain reaction (PCR)** (Chapter 39), such repeats are useful in constructing genetic linkage maps. Most genes are associated with one or more microsatellite markers, so the relative position of genes on chromosomes can be assessed, as can the association of a gene with a disease. Using PCR, a large number of family members can be rapidly screened for a certain **microsatellite polymorphism.** The association of a specific polymorphism with a gene in affected family members—and the lack of this association in unaffected members—may be the first clue about the genetic basis of a disease.

Trinucleotide sequences that increase in number (microsatellite instability) can cause disease. The unstable $p(CGG)_n$ repeat sequence is associated with the fragile X syndrome. Other trinucleotide repeats that undergo dynamic mutation (usually an increase) are associated with Huntington's chorea (CAG), myotonic dystrophy (CTG), spinobulbar muscular atrophy (CAG), and Kennedy disease (CAG).

ONE PERCENT OF CELLULAR DNA IS IN MITOCHONDRIA

The majority of the peptides in mitochondria (about 54 out of 67) are coded by nuclear genes. The rest are coded by genes found in mitochondrial (mt) DNA. Human mitochondria contain two to ten copies of a small circular double-stranded DNA molecule that makes up approximately 1% of total cellular DNA. This mtDNA codes for mt-specific ribosomal and transfer RNAs and for 13 proteins that play key roles in the respiratory chain (Chapter 13). The linearized structural map of the human mitochondrial genes is shown in Figure 35–8. Some of the features of mtDNA are shown in Table 35–3.

An important feature of human mitochondrial mtDNA is that—because all mitochondria are contributed by the ovum during zygote formation—it is transmitted by maternal nonmendelian inheritance. Thus, in diseases resulting from mutations of mtDNA, an affected mother would in theory pass the disease to all of her children but only her daughters would transmit the trait. However, in some cases, deletions in mtDNA occur during oogenesis and thus are not inherited from the mother. A number of diseases have now been shown to be due to mutations of mtDNA. These include a variety of myopathies, neurologic disorders, and some cases of diabetes mellitus.

GENETIC MATERIAL CAN BE ALTERED & REARRANGED

An alteration in the sequence of purine and pyrimidine bases in a gene due to a change—a removal or an insertion—of one or more bases may result in an altered gene product. Such alteration in the genetic material results in a **mutation** whose consequences are discussed in detail in Chapter 37.

Chromosomal Recombination Is One Way of Rearranging Genetic Material

Genetic information can be exchanged between similar or homologous chromosomes. The exchange, or **recombination**

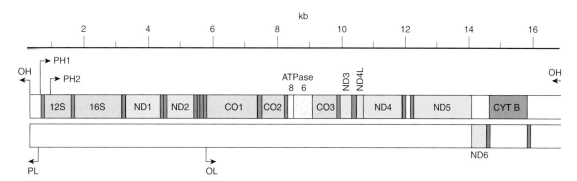

FIGURE 35–8 Maps of human mitochondrial genes. The maps represent the heavy (upper strand) and light (lower map) strands of linearized mitochondrial (mt) DNA, showing the genes for the subunits of NADH-coenzyme Q oxidoreductase (ND1 through ND6), cytochrome *c* oxidase (CO1 through CO3), cytochrome *b* (CYT B), and ATP synthase (ATPase 8 and 6) and for the 12S and 16S ribosomal mt rRNAs. The transfer RNAs are denoted by small blue open boxes. The origin of heavy-strand (OH) and light-strand (OL) replication and the promoters for the initiation of heavy-strand (PH1 and PH2) and light-strand (PL) transcription are indicated by arrows. (Reproduced, with permission, from Moraes CT et al: Mitochondrial DNA deletions in progressive external ophthalmoplegia and Kearns-Sayre syndrome. N Engl J Med 1989;320:1293. Copyright ©1989. Massachusetts Medical Society. All rights reserved.)

TABLE 35-3   Some Major Features of the Structure and Function of Human Mitochondrial DNA

- Is circular, double-stranded, and composed of heavy (H) and light (L) chains or strands

- Contains 16,569 bp

- Encodes 13 protein subunits of the respiratory chain (of a total of about 67)

  Seven subunits of NADH dehydrogenase (complex I)

  Cytochrome *b* of complex III

  Three subunits of cytochrome oxidase (complex IV)

  Two subunits of ATP synthase

- Encodes large (16S) and small (12S) mt ribosomal RNAs

- Encodes 22 mt tRNA molecules

- Genetic code differs slightly from the standard code

  UGA (standard stop codon) is read as Trp

  AGA and AGG (standard codons for Arg) are read as stop codons

- Contains very few untranslated sequences

- High mutation rate (5 to 10 times that of nuclear DNA)

- Comparisons of mtDNA sequences provide evidence about evolutionary origins of primates and other species

Source: Adapted from Harding AE: Neurological disease and mitochondrial genes. Trends Neurol Sci 1991;14:132. Copyright ©1991. Reprinted with permission from Elsevier.

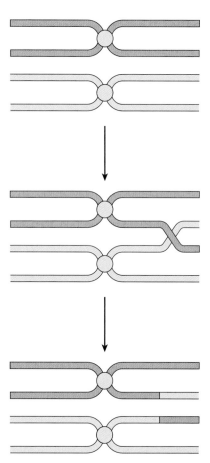

FIGURE 35–9 The process of crossing-over between homologous metaphase chromosomes to generate recombinant chromosomes. See also Figure 35–12.

event, occurs primarily during meiosis in mammalian cells and requires alignment of homologous metaphase chromosomes, an alignment that almost always occurs with great exactness. A process of crossing over occurs as shown in Figure 35–9. This usually results in an equal and reciprocal exchange of genetic information between homologous chromosomes. If the homologous chromosomes possess different alleles of the same genes, the crossover may produce noticeable and heritable genetic linkage differences. In the rare case where the alignment of homologous chromosomes is not exact, the crossing over or recombination event may result in an unequal exchange of information. One chromosome may receive less genetic material and thus a deletion, while the other partner of the chromosome pair receives more genetic material and thus an insertion or duplication (Figure 35–9). Unequal crossing over does occur in humans, as evidenced by the existence of hemoglobins designated Lepore and anti-Lepore (Figure 35–10). The farther apart two sequences are on an individual chromosome, the greater the likelihood of a crossover recombination event. This is the basis for genetic mapping methods. **Unequal crossover** affects tandem arrays of repeated DNAs whether they are related globin genes, as in Figure 35–10, or more abundant repetitive DNA. Unequal crossover through slippage in the pairing can result in expansion or contraction in the copy number of the repeat family and may contribute to the expansion and fixation of variant members throughout the repeat array.

Chromosomal Integration Occurs with Some Viruses

Some bacterial viruses (bacteriophages) are capable of recombining with the DNA of a bacterial host in such a way that the genetic information of the bacteriophage is incorporated in a linear fashion into the genetic information of the host. This integration, which is a form of recombination, occurs by the mechanism illustrated in Figure 35–11. The backbone of the circular bacteriophage genome is broken, as is that of the DNA molecule of the host; the appropriate ends are resealed with the proper polarity. The bacteriophage DNA is figuratively straightened out ("linearized") as it is integrated into the bacterial DNA molecule—frequently a closed circle as well. The site at which the bacteriophage genome integrates or recombines with the bacterial genome is chosen by one of two mechanisms. If the bacteriophage contains a DNA sequence **homologous** to a sequence in the host DNA molecule, then a recombination event analogous to that occurring between homologous chromosomes can occur. However, some bacteriophages synthesize proteins that bind specific sites on bacterial chromosomes to a **nonhomologous** site characteristic of the bacteriophage DNA molecule. Integration occurs at the site and is said to be **"site specific."**

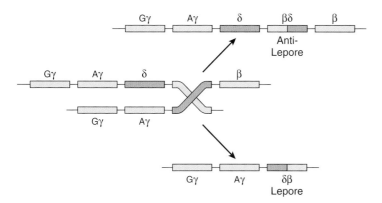

FIGURE 35–10 The process of unequal crossover in the region of the mammalian genome that harbors the structural genes encoding hemoglobins and the generation of the unequal recombinant products hemoglobin delta-beta Lepore and beta-delta anti-Lepore. The examples given show the locations of the crossover regions between amino acid residues. (Redrawn and reproduced, with permission, from Clegg JB, Weatherall DJ: β⁰ Thalassemia: Time for a reappraisal? Lancet 1974;2:133. Copyright ©1974. Reprinted with permission from Elsevier.)

Many animal viruses, particularly the oncogenic viruses—either directly or, in the case of RNA viruses such as HIV that causes AIDS, their DNA transcripts generated by the action of the viral **RNA-dependent DNA polymerase**, or **reverse transcriptase**—can be integrated into chromosomes of the mammalian cell. The integration of the animal virus DNA into the animal genome generally is not "site specific" but does display site preferences.

Transposition Can Produce Processed Genes

In eukaryotic cells, small DNA elements that clearly are not viruses are capable of transposing themselves in and out of the host genome in ways that affect the function of neighboring DNA sequences. These mobile elements, sometimes called **"jumping DNA,"** or jumping genes, can carry flanking regions of DNA and, therefore, profoundly affect evolution. As mentioned above, the Alu family of moderately repeated DNA sequences has structural characteristics similar to the termini

of retroviruses, which would account for the ability of the latter to move into and out of the mammalian genome.

Direct evidence for the transposition of other small DNA elements into the human genome has been provided by the discovery of **"processed genes"** for immunoglobulin molecules, α-globin molecules, and several others. These processed genes consist of DNA sequences identical or nearly identical to those of the messenger RNA for the appropriate gene product. That is, the 5′-nontranslated region, the coding region without intron representation, and the 3′ poly(A) tail are all present contiguously. This particular DNA sequence arrangement must have resulted from the reverse transcription of an appropriately processed messenger RNA molecule from which the intron regions had been removed and the poly(A) tail added. The only recognized mechanism this reverse transcript could have used to integrate into the genome would have been a transposition event. In fact, these "processed genes" have short terminal repeats at each end, as do known transposed sequences in lower organisms. In the absence of their transcription and thus genetic selection for function, many of the processed genes have been randomly altered through evolution so that they now contain nonsense codons that preclude their ability to encode a functional, intact protein (see Chapter 37). Thus, they are referred to as **"pseudogenes."**

Gene Conversion Produces Rearrangements

Besides unequal crossover and transposition, a third mechanism can effect rapid changes in the genetic material. Similar sequences on homologous or nonhomologous chromosomes may occasionally pair up and eliminate any mismatched sequences between them. This may lead to the accidental fixation of one variant or another throughout a family of repeated sequences and thereby homogenize the sequences of the members of repetitive DNA families. This latter process is referred to as **gene conversion.**

Sister Chromatids Exchange

In diploid eukaryotic organisms such as humans, after cells progress through the S phase they contain a tetraploid content of DNA. This is in the form of sister chromatids of chromosome pairs (Figure 35–6). Each of these sister chromatids contains identical genetic information since each is a product of the semi-

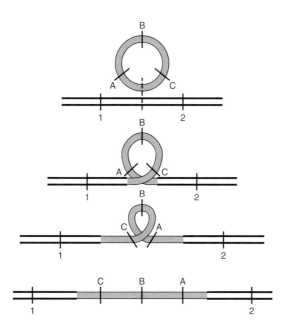

FIGURE 35–11 The integration of a circular genome from a virus (with genes A, B, and C) into the DNA molecule of a host (with genes 1 and 2) and the consequent ordering of the genes.

FIGURE 35–12 Sister chromatid exchanges between human chromosomes. These are detectable by Giemsa staining of the chromosomes of cells replicated for two cycles in the presence of bromodeoxyuridine. The arrows indicate some regions of exchange. (Courtesy of S Wolff and J Bodycote.)

conservative replication of the original parent DNA molecule of that chromosome. Crossing over can occur between these genetically identical sister chromatids. Of course, these **sister chromatid exchanges** (Figure 35–12) have no genetic consequence as long as the exchange is the result of an equal crossover.

Immunoglobulin Genes Rearrange

In mammalian cells, some interesting gene rearrangements occur normally during development and differentiation. For example, in mice the V_L and C_L genes for a single immunoglobulin molecule (see Chapter 38) are widely separated in the germ line DNA. In the DNA of a differentiated immunoglobulin-producing (plasma) cell, the same V_L and C_L genes have been moved physically closer together in the genome and into the same transcription unit. However, even then, this rearrangement of DNA during differentiation does not bring the V_L and C_L genes into contiguity in the DNA. Instead, the DNA contains an interspersed or interruption sequence of about 1200 base pairs at or near the junction of the V and C regions. The interspersed sequence is transcribed into RNA along with the V_L and C_L genes, and the interspersed information is removed from the RNA during its nuclear processing (Chapters 36 & 38).

DNA SYNTHESIS & REPLICATION ARE RIGIDLY CONTROLLED

The primary function of DNA replication is understood to be the provision of progeny with the genetic information pos-

sessed by the parent. Thus, the replication of DNA must be complete and carried out in such a way as to maintain genetic stability within the organism and the species. The process of DNA replication is complex and involves many cellular functions and several verification procedures to ensure fidelity in replication. About 30 proteins are involved in the replication of the *E coli* chromosome, and this process is more complex in eukaryotic organisms. The first enzymologic observations on DNA replication were made by Arthur Kornberg, who described in *E coli* the existence of an enzyme now called DNA polymerase I. This enzyme has multiple catalytic activities, a complex structure, and a requirement for the triphosphates of the four deoxyribonucleosides of adenine, guanine, cytosine, and thymine. The polymerization reaction catalyzed by DNA polymerase I of *E coli* has served as a prototype for all DNA polymerases of both prokaryotes and eukaryotes, even though it is now recognized that the major role of this polymerase is proofreading and repair.

In all cells, replication can occur only from a single-stranded DNA (ssDNA) template. Therefore, mechanisms must exist to target the site of initiation of replication and to unwind the double-stranded DNA (dsDNA) in that region. The replication complex must then form. After replication is complete in an area, the parent and daughter strands must re-form dsDNA. In eukaryotic cells, an additional step must occur. The dsDNA must re-form the chromatin structure, including nucleosomes, that existed prior to the onset of replication. Although this entire process is not completely understood in eukaryotic cells, replication has been quite precisely described in prokaryotic cells, and the general principles are the same in both. The major steps are listed in Table 35–4, illustrated in Figure 35–13, and discussed, in sequence, below. A number of proteins, most with specific enzymatic action, are involved in this process (Table 35–5).

The Origin of Replication

At the **origin of replication (ori),** there is an association of sequence-specific dsDNA-binding proteins with a series of direct repeat DNA sequences. In bacteriophage λ, the oriλ is bound by the λ-encoded O protein to four adjacent sites. In *E coli*, the oriC is bound by the protein dnaA. In both cases, a complex is formed consisting of 150–250 bp of DNA and

TABLE 35–4   Steps Involved in DNA Replication in Eukaryotes

1. Identification of the origins of replication
2. Unwinding (denaturation) of dsDNA to provide an ssDNA template
3. Formation of the replication fork; synthesis of RNA primer
4. Initiation of DNA synthesis and elongation
5. Formation of replication bubbles with ligation of the newly synthesized DNA segments
6. Reconstitution of chromatin structure

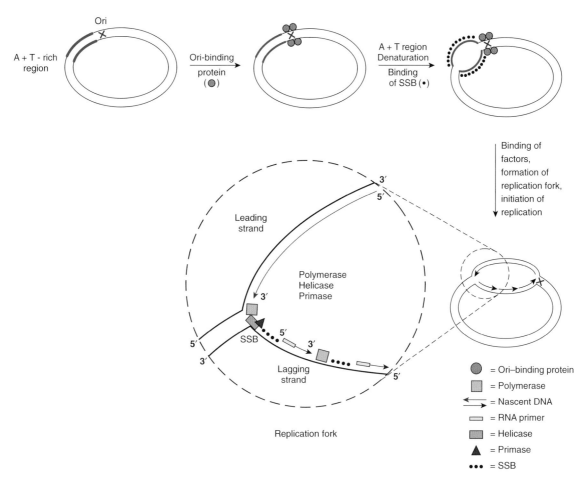

FIGURE 35–13 Steps involved in DNA replication. This figure describes DNA replication in an *E coli* cell, but the general steps are similar in eukaryotes. A specific interaction of a protein (the dnaA protein) to the origin of replication (oriC) results in local unwinding of DNA at an adjacent A+T-rich region. The DNA in this area is maintained in the single-strand conformation (ssDNA) by single-strand-binding proteins (SSBs). This allows a variety of proteins, including helicase, primase, and DNA polymerase, to bind and to initiate DNA synthesis. The replication fork proceeds as DNA synthesis occurs continuously (long red arrow) on the leading strand and discontinuously (short black arrows) on the lagging strand. The nascent DNA is always synthesized in the 5′ to 3′ direction, as DNA polymerases can add a nucleotide only to the 3′ end of a DNA strand.

multimers of the DNA-binding protein. This leads to the local denaturation and unwinding of an adjacent A+T-rich region of DNA. Functionally similar **autonomously replicating sequences (ARS) or replicators** have been identified in yeast cells. The ARS contains a somewhat degenerate 11-bp sequence called the **origin replication element (ORE).** The ORE binds a set of proteins, analogous to the dnaA protein of *E coli*, the group of proteins is collectively called the **origin recognition complex (ORC).** ORC homologs have been found in all eukaryotes examined. The ORE is located adjacent to an approximately 80-bp A+T-rich sequence that is easy to unwind. This is called the **DNA unwinding element (DUE).** The DUE is the origin of replication in yeast and is bound by the MCM protein complex.

Consensus sequences similar to ori or ARS in structure have not been precisely defined in mammalian cells, though several of the proteins that participate in ori recognition and

TABLE 35–5 Classes of Proteins Involved in Replication

Protein	Function
DNA polymerases	Deoxynucleotide polymerization
Helicases	Processive unwinding of DNA
Topoisomerases	Relieve torsional strain that results from helicase-induced unwinding
DNA primase	Initiates synthesis of RNA primers
Single-strand binding proteins	Prevent premature reannealing of dsDNA
DNA ligase	Seals the single strand nick between the nascent chain and Okazaki fragments on lagging strand

function have been identified and appear quite similar to their yeast counterparts in both amino acid sequence and function.

Unwinding of DNA

The interaction of proteins with ori defines the start site of replication and provides a short region of ssDNA essential for initiation of synthesis of the nascent DNA strand. This process requires the formation of a number of protein–protein and protein–DNA interactions. A critical step is provided by a DNA helicase that allows for processive unwinding of DNA. In uninfected *E coli*, this function is provided by a complex of dnaB helicase and the dnaC protein. Single-stranded DNA-binding proteins (SSBs) stabilize this complex. In λ phage-infected *E coli*, the phage protein P binds to dnaB and the P/dnaB complex binds to oriλ by interacting with the O protein. dnaB is not an active helicase when in the P/dnaB/O complex. Three *E coli* heat shock proteins (dnaK, dnaJ, and GrpE) cooperate to remove the P protein and activate the dnaB helicase. In cooperation with SSB, this leads to DNA unwinding and active replication. In this way, the replication of the λ phage is accomplished at the expense of replication of the host *E coli* cell.

Formation of the Replication Fork

A replication fork consists of four components that form in the following sequence: (1) the DNA helicase unwinds a short segment of the parental duplex DNA; (2) a primase initiates synthesis of an RNA molecule that is essential for priming DNA synthesis; (3) the DNA polymerase initiates nascent, daughter strand synthesis; and (4) SSBs bind to ssDNA and prevent premature reannealing of ssDNA to dsDNA. These reactions are illustrated in Figure 35–13.

The polymerase III enzyme (the *dnaE* gene product in *E coli*) binds to template DNA as part of a multiprotein complex that consists of several polymerase accessory factors (β, γ, δ, δ', and τ). DNA polymerases only synthesize DNA in the 5' to 3' direction, and only one of the several different types of polymerases is involved at the replication fork. Because the DNA strands are antiparallel (Chapter 34), the polymerase functions asymmetrically. On the **leading (forward) strand,** the DNA is synthesized continuously. On the **lagging (retrograde) strand,** the DNA is synthesized in short (1–5 kb; see Figure 35–16) fragments, the so-called **Okazaki fragments.** Several Okazaki fragments (up to a thousand) must be sequentially synthesized for each replication fork. To ensure that this happens, the helicase acts on the lagging strand to unwind dsDNA in a 5' to 3' direction. The helicase associates with the primase to afford the latter proper access to the template. This allows the RNA primer to be made and, in turn, the polymerase to begin replicating the DNA. This is an important reaction sequence since DNA polymerases cannot initiate DNA synthesis de novo. The mobile complex between helicase and primase has been called a **primosome.** As the synthesis of an Okazaki fragment is completed and the polymerase is released, a new primer has been synthesized. The same polymerase molecule

remains associated with the replication fork and proceeds to synthesize the next Okazaki fragment.

The DNA Polymerase Complex

A number of different DNA polymerase molecules engage in DNA replication. These share three important properties: (1) **chain elongation,** (2) **processivity,** and (3) **proofreading.** Chain elongation accounts for the rate (in nucleotides per second; ntd/s) at which polymerization occurs. Processivity is an expression of the number of nucleotides added to the nascent chain before the polymerase disengages from the template. The proofreading function identifies copying errors and corrects them. In *E coli*, polymerase III (pol III) functions at the replication fork. Of all polymerases, it catalyzes the highest rate of chain elongation and is the most processive. It is capable of polymerizing 0.5 Mb of DNA during one cycle on the leading strand. Pol III is a large (>1 MDa), multisubunit protein complex in *E coli*. DNA pol III associates with the two identical β subunits of the DNA sliding "clamp"; this association dramatically increases pol III-DNA complex stability, processivity (10 ntd to >50,000 ntd) and rate of chain elongation (20 to 50 ntd/sec) generating the high degree of processivity the enzyme exhibits.

Polymerase I (pol I) and II (pol II) are mostly involved in proofreading and DNA repair. Eukaryotic cells have counterparts for each of these enzymes plus a large number of additional DNA polymerases primarily involved in DNA repair. A comparison is shown in Table 35–6.

In mammalian cells, the polymerase is capable of polymerizing at a rate that is somewhat slower than the rate of polymerization of deoxynucleotides by the bacterial DNA polymerase complex. This reduced rate may result from interference by nucleosomes.

Initiation & Elongation of DNA Synthesis

The initiation of DNA synthesis (Figure 35–14) requires **priming by a short length of RNA,** about 10–200 nucleotides long. In *E coli* this is catalyzed by dnaG (primase), in eukaryotes DNA Pol α synthesizes these RNA primers. The priming process

TABLE 35–6   A Comparison of Prokaryotic and Eukaryotic DNA Polymerases

E coli	Eukaryotic	Function
I		Gap filling following DNA replication, repair, and recombination
II		DNA proofreading and repair
	β	DNA repair
	γ	Mitochondrial DNA synthesis
III	ε	Processive, leading strand synthesis
DnaG	α	Primase
	δ	Processive, lagging strand synthesis

involves nucleophilic attack by the 3′-hydroxyl group of the RNA primer on the phosphate of the first entering deoxynucleoside triphosphate (N in Figure 35–14) with the splitting off of pyrophosphate; this transition to DNA synthesis is catalyzed by the appropriate DNA polymerases (DNA pol III in *E coli*; DNA pol δ and ε in eukaryotes). The 3′-hydroxyl group of the recently attached deoxyribonucleoside monophosphate is then free to carry out a **nucleophilic attack** on the next entering de-

FIGURE 35–14 The initiation of DNA synthesis upon a primer of RNA and the subsequent attachment of the second deoxyribonucleoside triphosphate.

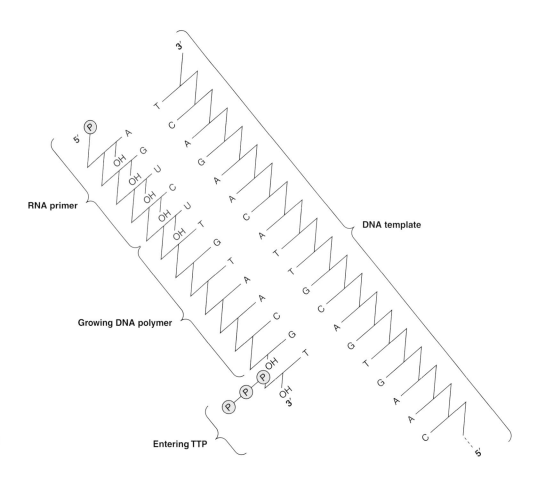

FIGURE 35–15 The RNA-primed synthesis of DNA demonstrating the template function of the complementary strand of parental DNA.

oxyribonucleoside triphosphate (N + 1 in Figure 35–14), again at its α phosphate moiety, with the splitting off of pyrophosphate. Of course, selection of the proper deoxyribonucleotide whose terminal 3′-hydroxyl group is to be attacked is dependent upon **proper base pairing with the other strand** of the DNA molecule according to the rules proposed originally by Watson and Crick (Figure 35–15). When an adenine deoxyribonucleoside monophosphoryl moiety is in the template position, a thymidine triphosphate will enter and its α phosphate will be attacked by the 3′-hydroxyl group of the deoxyribonucleoside monophosphoryl most recently added to the polymer. By this stepwise process, the template dictates which deoxyribonucleoside triphosphate is complementary and by hydrogen bonding holds it in place while the 3′-hydroxyl group of the growing strand attacks and incorporates the new nucleotide into the polymer. These segments of DNA attached to an RNA initiator component are the **Okazaki fragments** (Figure 35–16). In mammals, after many Okazaki fragments are generated, the replication complex begins to remove the RNA primers, to fill in the gaps left by their removal with the proper base-paired deoxynucleotide, and then to seal the fragments of newly synthesized DNA by enzymes referred to as **DNA ligases.**

Replication Exhibits Polarity

As has already been noted, DNA molecules are double stranded and the two strands are antiparallel. The replication of DNA in prokaryotes and eukaryotes occurs on both strands simultaneously. However, an enzyme capable of polymerizing DNA in the 3′ to 5′ direction does not exist in any organism, so

FIGURE 35–16 The discontinuous polymerization of deoxyribonucleotides on the lagging strand; formation of Okazaki fragments during lagging strand DNA synthesis is illustrated. Okazaki fragments are 100–250 nt long in eukaryotes, 1000–2000 bp in prokaryotes.

that both of the newly replicated DNA strands cannot grow in the same direction simultaneously. Nevertheless, the same enzyme does replicate both strands at the same time. The single enzyme replicates one strand ("leading strand") in a continuous manner in the 5′ to 3′ direction, with the same overall forward direction. It replicates the other strand ("lagging strand") discontinuously while polymerizing the nucleotides in short spurts of 150–250 nucleotides, again in the 5′ to 3′ direction, but at the same time it faces toward the back end of the preceding RNA primer rather than toward the unreplicated portion. This process of **semidiscontinuous DNA synthesis** is shown diagrammatically in Figures 35–13 and 35–16.

Formation of Replication Bubbles

Replication proceeds from a single ori in the circular bacterial chromosome, composed of roughly 5×10^6 bp of DNA. This process is completed in about 30 min, a replication rate of 3×10^5 bp/min. The entire mammalian genome replicates in approximately 9 h, the average period required for formation of a tetraploid genome from a diploid genome in a replicating cell. If a mammalian genome (3×10^9 bp) replicated at the same rate as bacteria (ie, 3×10^5 bp/min) from but a single ori, replication would take over 150 h! Metazoan organisms get around this problem using two strategies. First, replication is bidirectional. Second, replication proceeds from multiple origins in each chromosome (a total of as many as 100 in humans). Thus, replication occurs in both directions along all of the chromosomes, and both strands are replicated simultaneously. This replication process generates **"replication bubbles"** (Figure 35–17).

The multiple sites that serve as origins for DNA replication in eukaryotes are poorly defined except in a few animal viruses and in yeast. However, it is clear that initiation is regulated both spatially and temporally, since clusters of adjacent sites initiate replication synchronously. Replication firing, or DNA replication initiation at a replicator/ori, is influenced by a number of distinct properties of chromatin structure that are just beginning to be understood. It is clear, however, that there are more replicators and excess ORC than needed to replicate the mammalian genome within the time of a typical S-phase. Therefore, mechanisms for controlling the excess ORC-bound

replicators must exist. Understanding the control of this process is a major challenge.

During the replication of DNA, there must be a separation of the two strands to allow each to serve as a template by hydrogen bonding its nucleotide bases to the incoming deoxynucleoside triphosphate. The separation of the DNA double helix is promoted by SSBs in *E coli*, a protein termed Replication Protein A (RPA) in eukaryotes. These molecules stabilize the single-stranded structure as the replication fork progresses. The stabilizing proteins bind cooperatively and stoichiometrically to the single strands without interfering with the abilities of the nucleotides to serve as templates (Figure 35–13). In addition to separating the two strands of the double helix, there must be an unwinding of the molecule (once every 10 nucleotide pairs) to allow strand separation. The hexameric DNA β protein complex unwinds DNA in *E coli* while the hexameric MCM complex unwinds eukaryotic DNA. This unwinding happens in segments adjacent to the replication bubble. To counteract this unwinding there are multiple "swivels" interspersed in the DNA molecules of all organisms. The swivel function is provided by specific enzymes that introduce **"nicks" in one strand of the unwinding double helix,** thereby allowing the unwinding process to proceed. The nicks are quickly resealed without requiring energy input, because of the formation of a high-energy covalent bond between the nicked phosphodiester backbone and the nicking-sealing enzyme. The nicking-resealing enzymes are called **DNA topoisomerases.** This process is depicted diagrammatically in Figure 35–18 and there compared with the ATP-dependent resealing carried out by the DNA ligases. Topoisomerases are also capable of unwinding supercoiled DNA. Supercoiled DNA is a higher-ordered structure occurring in circular DNA molecules wrapped around a core, as depicted in Figure 35–19.

There exists in one species of animal viruses (retroviruses) a class of enzymes capable of synthesizing a single-stranded and then a double-stranded DNA molecule from a single-stranded RNA template. This polymerase, RNA-dependent DNA polymerase, or **"reverse transcriptase,"** first synthesizes a DNA-RNA hybrid molecule utilizing the RNA genome as a template. A specific virus-encoded nuclease, **RNase H**, degrades the hybridized template RNA strand, and the remain

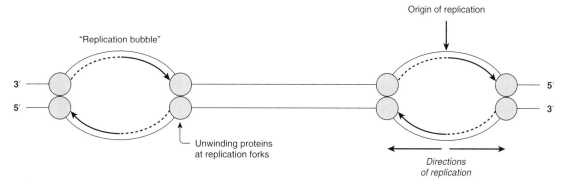

FIGURE 35–17 The generation of "replication bubbles" during the process of DNA synthesis. The bidirectional replication and the proposed positions of unwinding proteins at the replication forks are depicted.

FIGURE 35–18 Comparison of two types of nick-sealing reactions on DNA. The series of reactions at left is catalyzed by DNA topoisomerase I, that at right by DNA ligase; P, phosphate; R, ribose; A, adenine. (Slightly modified and reproduced, with permission, from Lehninger AL: *Biochemistry*, 2nd ed. Worth, 1975. Copyright ©1975 by Worth Publishers. Used with permission from W. H. Freeman and Company.)

ing DNA strand in turn serves as a template to form a double-stranded DNA molecule containing the information originally present in the RNA genome of the animal virus.

Reconstitution of Chromatin Structure

There is evidence that nuclear organization and chromatin structure are involved in determining the regulation and initiation of DNA synthesis. As noted above, the rate of polymerization in eukaryotic cells, which have chromatin and nucleosomes, is slower than that in prokaryotic cells, which have naked DNA. It is also clear that chromatin structure must be re-formed after replication. Newly replicated DNA is rapidly assembled into nucleosomes, and the preexisting and newly assembled histone octamers are randomly distributed to each arm of the replication fork. These reactions are facilitated through the actions of histone chaperone proteins working in concert with chromatin remodeling complexes.

DNA Synthesis Occurs During the S Phase of the Cell Cycle

In animal cells, including human cells, the replication of the DNA genome occurs only at a specified time during the life

FIGURE 35–19 Supercoiling of DNA. A left-handed toroidal (solenoidal) supercoil, at left, will convert to a right-handed interwound supercoil, at right, when the cylindric core is removed. Such a transition is analogous to that which occurs when nucleosomes are disrupted by the high salt extraction of histones from chromatin.

span of the cell. This period is referred to as the synthetic or S phase. This is usually temporally separated from the mitotic, or **M phase**, by nonsynthetic periods referred to as **gap 1 (G1)** and **gap 2 (G2)**, occurring before and after the S phase, respectively (Figure 35–20). Among other things, the cell prepares for DNA synthesis in G1 and for mitosis in G2. The cell regulates the DNA synthesis process by allowing it to occur only at specific times and mostly in cells preparing to divide by a mitotic process.

All eukaryotic cells have gene products that govern the transition from one phase of the cell cycle to another. The **cyclins** are a family of proteins whose concentration increases and decreases throughout the cell cycle—thus their name. The cyclins turn on, at the appropriate time, different **cyclin-dependent protein kinases (CDKs)** that phosphorylate substrates essential for progression through the cell cycle (Figure 35–21). For example, cyclin D levels rise in late G1 phase and allow progression beyond the **start (yeast)** or **restriction point (mammals),** the point beyond which cells irrevocably proceed into the S or DNA synthesis phase.

The D cyclins activate CDK4 and CDK6. These two kinases are also synthesized during G1 in cells undergoing active division. The D cyclins and CDK4 and CDK6 are nuclear proteins that assemble as a complex in late G1 phase. The complex is an active serine-threonine protein kinase. One substrate for this kinase is the retinoblastoma (Rb) protein. Rb is a cell cycle regulator because it binds to and inactivates a transcription factor (E2F) necessary for the transcription of certain genes (histone genes, DNA replication proteins, etc) needed for progression from G1 to S phase. The phosphorylation of Rb by CDK4 or CDK6 results in the release of E2F from Rb-medi-

FIGURE 35–21 Schematic illustration of the points during the mammalian cell cycle during which the indicated cyclins and cyclin-dependent kinases are activated. The thickness of the various colored lines is indicative of the extent of activity.

ated transcription repression—thus, gene activation ensues and cell cycle progression takes place.

Other cyclins and CDKs are involved in different aspects of cell cycle progression (Table 35–7). Cyclin E and CDK2 form a complex in late G1. Cyclin E is rapidly degraded, and the released CDK2 then forms a complex with cyclin A. This sequence is necessary for the initiation of DNA synthesis in S phase. A complex between cyclin B and CDK1 is rate-limiting for the G2/M transition in eukaryotic cells.

Many of the cancer-causing viruses (oncoviruses) and cancer-inducing genes (oncogenes) are capable of alleviating or disrupting the apparent restriction that normally controls the entry of mammalian cells from G1 into the S phase. From the foregoing, one might have surmised that excessive production of a cyclin, loss of a specific CDK inhibitor, or production or activation of a cyclin/CDK at an inappropriate time might result in abnormal or unrestrained cell division. In this context it is noteworthy that the *bcl* oncogene associated with B cell lymphoma appears to be the cyclin D1 gene. Similarly, the

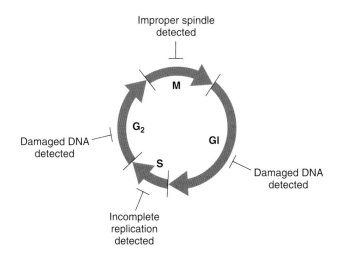

FIGURE 35–20 Mammalian cell cycle and cell cycle checkpoints. DNA, chromosome, and chromosome segregation integrity is continuously monitored throughout the cell cycle. If DNA damage is detected in either the G1 or the G2 phase of the cell cycle, if the genome is incompletely replicated, or if normal chromosome segregation machinery is incomplete (ie, a defective spindle), cells will not progress through the phase of the cycle in which defects are detected. In some cases, if the damage cannot be repaired, such cells undergo programmed cell death (apoptosis).

TABLE 35–7 Cyclins and Cyclin-Dependent Kinases Involved in Cell Cycle Progression

Cyclin	Kinase	Function
D	CDK4, CDK6	Progression past restriction point at G1/S boundary
E, A	CDK2	Initiation of DNA synthesis in early S phase
B	CDK1	Transition from G2 to M

oncoproteins (or transforming proteins) produced by several DNA viruses target the Rb transcription repressor for inactivation, inducing cell division inappropriately.

During the S phase, mammalian cells contain greater quantities of DNA polymerase than during the nonsynthetic phases of the cell cycle. Furthermore, those enzymes responsible for formation of the substrates for DNA synthesis—ie, deoxyribonucleoside triphosphates—are also increased in activity, and their activity will diminish following the synthetic phase until the reappearance of the signal for renewed DNA synthesis. During the S phase, the **nuclear DNA is completely replicated once and only once.** It seems that once chromatin has been replicated, it is marked so as to prevent its further replication until it again passes through mitosis. The molecular mechanisms for this phenomenon are being elucidated.

In general, a given pair of chromosomes will replicate simultaneously and within a fixed portion of the S phase upon every replication. On a chromosome, clusters of replication units replicate coordinately. The nature of the signals that regulate DNA synthesis at these levels is unknown, but the regulation does appear to be an intrinsic property of each individual chromosome.

Enzymes Repair Damaged DNA

The maintenance of the integrity of the information in DNA molecules is of utmost importance to the survival of a particular organism as well as to survival of the species. Thus, it can be concluded that surviving species have evolved mechanisms for repairing DNA damage occurring as a result of either replication errors or environmental insults.

As described in Chapter 34, the major responsibility for the fidelity of replication resides in the specific pairing of nucleotide bases. Proper pairing is dependent upon the presence of the favored tautomers of the purine and pyrimidine nucleotides, but the equilibrium whereby one tautomer is more stable than another is only about 10^4 or 10^5 in favor of that with the greater stability. Although this is not favorable enough to ensure the high fidelity that is necessary, favoring of the preferred tautomers—and thus of the proper base pairing—could be ensured by monitoring the base pairing twice. Such double monitoring does appear to occur in both bacterial and mammalian systems: once at the time of insertion of the deoxyribonucleoside triphosphates, and later by a follow-up energy-requiring mechanism that removes all improper bases which may occur in the newly formed strand. This "proofreading" prevents tautomer-induced misincorporation from occurring more frequently than once every 10^8–10^{10} base pairs of DNA synthesized. The mechanisms responsible for this monitoring mechanism in *E coli* include the 3′ to 5′ exonuclease activities of one of the subunits of the pol III complex and of the pol I molecule. The analogous mammalian enzymes (δ and α) do not seem to possess such a nuclease proofreading function. Other enzymes provide this repair function.

Replication errors, even with a very efficient repair system, lead to the accumulation of mutations. A human has 10^{14}

nucleated cells each with 3×10^9 base pairs of DNA. If about 10^{16} cell divisions occur in a lifetime and 10^{-10} mutations per base pair per cell generation escape repair, there may eventually be as many as one mutation per 10^6 bp in the genome. Fortunately, most of these will probably occur in DNA that does not encode proteins or will not affect the function of encoded proteins and so are of no consequence. In addition, spontaneous and chemically induced damage to DNA must be repaired.

Damage to DNA by environmental, physical, and chemical agents may be classified into four types (Table 35–8). Abnormal regions of DNA, either from copying errors or DNA damage, are replaced by four mechanisms: (1) mismatch repair, (2) base excision-repair, (3) nucleotide excision-repair, and (4) double-strand break repair (Table 35–9). These mechanisms exploit the redundancy of information inherent in the double helical DNA structure. The defective region in one strand can be returned to its original form by relying on the complementary information stored in the unaffected strand.

Mismatch Repair

Mismatch repair corrects errors made when DNA is copied. For example, a C could be inserted opposite an A, or the polymerase could slip or stutter and insert two or more extra unpaired bases. Specific proteins scan the newly synthesized DNA, using adenine methylation within a GATC sequence as the point of reference (Figure 35–22). The template strand is

TABLE 35–8   Types of Damage to DNA

I. Single-base alteration
A. Depurination
B. Deamination of cytosine to uracil
C. Deamination of adenine to hypoxanthine
D. Alkylation of base
E. Insertion or deletion of nucleotide
F. Base-analog incorporation
II. Two-base alteration
A. UV light–induced thymine-thymine (pyrimidine) dimer
B. Bifunctional alkylating agent cross-linkage
III. Chain breaks
A. Ionizing radiation
B. Radioactive disintegration of backbone element
C. Oxidative free radical formation
IV. Cross-linkage
A. Between bases in same or opposite strands
B. Between DNA and protein molecules (eg, histones)

TABLE 35–9 Mechanism of DNA Repair

Mechanism	Problem	Solution
Mismatch repair	Copying errors (single base or two- to five-base unpaired loops)	Methyl-directed strand cutting, exonuclease digestion, and replacement
Base excision–repair	Spontaneous, chemical, or radiation damage to a single base	Base removal by N-glycosylase, abasic sugar removal, replacement
Nucleotide excision–repair	Spontaneous, chemical, or radiation damage to a DNA segment	Removal of an approximately 30-nucleotide oligomer and replacement
Double-strand break repair	Ionizing radiation, chemotherapy, oxidative free radicals	Synapsis, unwinding, alignment, ligation

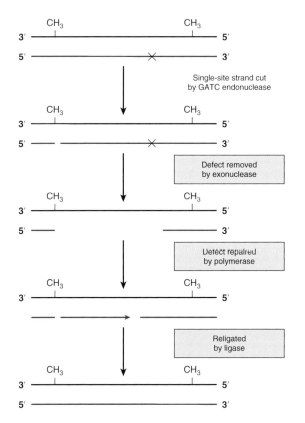

FIGURE 35–22 Mismatch repair of DNA. This mechanism corrects a single mismatch base pair (eg, C to A rather than T to A) or a short region of unpaired DNA. The defective region is recognized by an endonuclease that makes a single-strand cut at an adjacent methylated GATC sequence. The DNA strand is removed through the mutation, replaced, and religated.

methylated, and the newly synthesized strand is not. This difference allows the repair enzymes to identify the strand that contains the errant nucleotide which requires replacement. If a mismatch or small loop is found, a GATC endonuclease cuts the strand bearing the mutation at a site corresponding to the GATC. An exonuclease then digests this strand from the GATC through the mutation, thus removing the faulty DNA. This can occur from either end if the defect is bracketed by two GATC sites. This defect is then filled in by normal cellular enzymes according to base pairing rules. In *E coli*, three proteins (Mut S, Mut C, and Mut H) are required for recognition of the mutation and nicking of the strand. Other cellular enzymes, including ligase, polymerase, and SSBs, remove and replace the strand. The process is somewhat more complicated in mammalian cells, as about six proteins are involved in the first steps.

Faulty mismatch repair has been linked to hereditary non-polyposis colon cancer (HNPCC), one of the most common inherited cancers. Genetic studies linked HNPCC in some families to a region of chromosome 2. The gene located, designated *hMSH2*, was subsequently shown to encode the human analog of the *E coli* MutS protein that is involved in mismatch repair (see above). Mutations of *hMSH2* account for 50–60% of HNPCC cases. Another gene, *hMLH1*, is associated with most of the other cases. *hMLH1* is the human analog of the bacterial mismatch repair gene *MutL*. How does faulty mismatch repair result in colon cancer? The human genes were localized because microsatellite instability was detected. That is, the cancer cells had a microsatellite of a length different from that found in the normal cells of the individual. It appears that the affected cells, which harbor a mutated *hMSH2* or *hMLH1* mismatch repair enzyme, are unable to remove small loops of unpaired DNA, and the microsatellite thus increases in size. Ultimately, microsatellite DNA expansion must affect either the expression or the function of a protein critical in surveillance of the cell cycle in these colon cells.

Base Excision-Repair

The **depurination of DNA,** which happens spontaneously owing to the thermal lability of the purine N-glycosidic bond, occurs at a rate of 5,000–10,000/cell/d at 37°C. Specific enzymes recognize a depurinated site and replace the appropriate purine directly, without interruption of the phosphodiester backbone.

Cytosine, adenine, and guanine bases in DNA spontaneously form uracil, hypoxanthine, or xanthine, respectively. Since none of these normally exist in DNA, it is not surprising that specific **N-glycosylases** can recognize these abnormal bases and remove the base itself from the DNA. This removal marks the site of the defect and allows an **apurinic or apyrimidinic endonuclease** to excise the abasic sugar. The proper base is then replaced by a repair DNA polymerase, and a **ligase** returns the DNA to its original state (Figure 35–23). This series of events is called **base excision-repair.** By a similar series of steps involving initially the recognition of the defect, alkylated bases and base analogs can be removed from DNA

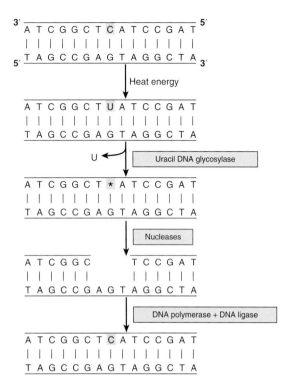

FIGURE 35–23 Base excision-repair of DNA. The enzyme uracil DNA glycosylase removes the uracil created by spontaneous deamination of cytosine in the DNA. An endonuclease cuts the backbone near the defect; then, after an endonuclease removes a few bases, the defect is filled in by the action of a repair polymerase and the strand is rejoined by a ligase. (Courtesy of B Alberts.)

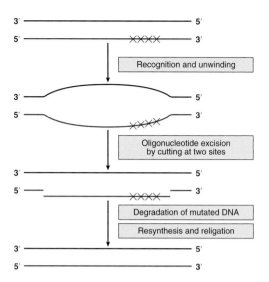

FIGURE 35–24 Nucleotide excision-repair. This mechanism is employed to correct larger defects in DNA and generally involves more proteins than either mismatch or base excision-repair. After defect recognition (indicated by XXXX) and unwinding of the DNA encompassing the defect, an excision nuclease (exinuclease) cuts the DNA upstream and downstream of the defective region. This gap is then filled in by a polymerase (δ/ε in humans) and religated.

and the DNA returned to its original informational content. This mechanism is suitable for replacement of a single base but is not effective at replacing regions of damaged DNA.

Nucleotide Excision-Repair

This mechanism is used to replace regions of damaged DNA up to 30 bases in length. Common causes of such DNA damage include ultraviolet (UV) light, which induces the formation of cyclobutane pyrimidinepyrimidine dimers, and smoking, which causes formation of benzo[a]pyrene-guanine adducts. Ionizing radiation, cancer chemotherapeutic agents, and a variety of chemicals found in the environment cause base modification, strand breaks, cross-linkage between bases on opposite strands or between DNA and protein, and numerous other defects. These are repaired by a process called nucleotide excision-repair (Figure 35–24). This process, which involves more gene products than the two other types of repair, essentially involves the hydrolysis of two phosphodiester bonds on the strand containing the defect. A special excision nuclease (exinuclease), consisting of at least three subunits in *E coli* and 16 polypeptides in humans, accomplishes this task. In eukaryotic cells the enzymes cut between the third to fifth phosphodiester bond 3′ from the lesion, and on the 5′ side the cut is somewhere between the twenty-first and twenty-fifth bonds.

Thus, a fragment of DNA 27–29 nucleotides long is excised. After the strand is removed it is replaced, again by exact base pairing, through the cooperative action of multiple replicative and repair DNA polymerase ($\delta,\varepsilon,\eta,\iota$ in humans), and the ends are joined to the existing strands by DNA ligase.

Xeroderma pigmentosum (XP) is an autosomal recessive genetic disease. The clinical syndrome includes marked sensitivity to sunlight (ultraviolet) with subsequent formation of multiple skin cancers and premature death. In individuals with XP the risk of developing skin cancer is increased 1000- to 2000-fold. The inherited defect seems to involve the repair of damaged DNA, particularly thymine dimers. Cells cultured from patients with xeroderma pigmentosum exhibit low activity for the nucleotide excision-repair process. Seven complementation groups have been identified using hybrid cell analyses, so at least seven gene products (XPA–XPG) are involved. Two of these (XPA and XPC) are involved in recognition and excision. XPB and XPD are helicases and, interestingly, are subunits of the transcription factor TFIIH (see Chapter 36).

Double-Strand Break Repair

The repair of double-strand (ds) breaks is part of the physiologic process of immunoglobulin gene rearrangement. It is also an important mechanism for repairing damaged DNA, such as occurs as a result of ionizing radiation or oxidative free radical generation. Some chemotherapeutic agents destroy cells by causing ds breaks or preventing their repair.

Two proteins are initially involved in the nonhomologous rejoining of a ds break. **Ku**, a heterodimer of 70-kDa and

86-kDa subunits, binds to free DNA ends and has latent ATP-dependent helicase activity. The DNA-bound Ku heterodimer recruits a unique protein kinase, **DNA-dependent protein kinase (DNA-PK).** DNA-PK has a binding site for DNA free ends and another for dsDNA just inside these ends. It therefore allows for the approximation of the two separated ends. The free end DNA-Ku-DNA-PK complex activates the kinase activity in the latter. DNA-PK reciprocally phosphorylates Ku and the other DNA-PK molecule, on the opposing strand, in trans. DNA-PK then dissociates from the DNA and Ku, resulting in activation of the Ku helicase. This results in unwinding of the two ends. The unwound, approximated DNA forms base pairs; the extra nucleotide tails are removed by an exonuclease; and the gaps are filled and closed by DNA ligase. This repair mechanism is illustrated in Figure 35–25.

Some Repair Enzymes Are Multifunctional

Somewhat surprising is the recent observation that DNA repair proteins can serve other purposes. For example, some repair enzymes are also found as components of the large TFIIH complex that plays a central role in gene transcription (Chapter 36). Another component of TFIIH is involved in cell cycle regulation. Thus, three critical cellular processes may be linked through use of common proteins. There is also good evidence that some repair enzymes are involved in gene rearrangements that occur normally.

In patients with **ataxia-telangiectasia,** an autosomal recessive disease in humans resulting in the development of cer-

ebellar ataxia and lymphoreticular neoplasms, there appears to exist an increased sensitivity to damage by x-ray. Patients with **Fanconi anemia,** an autosomal recessive anemia characterized also by an increased frequency of cancer and by chromosomal instability, probably have defective repair of cross-linking damage.

All three of these clinical syndromes are associated with an increased frequency of cancer. It is likely that other human diseases resulting from disordered DNA repair capabilities will be found in the future.

DNA & Chromosome Integrity Is Monitored Throughout the Cell Cycle

Given the importance of normal DNA and chromosome function to survival, it is not surprising that eukaryotic cells have developed elaborate mechanisms to monitor the integrity of the genetic material. As detailed above, a number of complex multisubunit enzyme systems have evolved to repair damaged DNA at the nucleotide sequence level. Similarly, DNA mishaps at the chromosome level are also monitored and repaired. As shown in Figure 35–20, both DNA and chromosomal integrity are continuously monitored throughout the cell cycle. The four specific steps at which this monitoring occurs have been termed **checkpoint controls.** If problems are detected at any of these checkpoints, progression through the cycle is interrupted and transit through the cell cycle is halted until the damage is repaired. The molecular mechanisms underlying detection of DNA damage during the G1 and G2 phases of the cycle are understood better than those operative during S and M phases.

The **tumor suppressor p53,** a protein of MW 53 kDa, plays a key role in both G1 and G2 checkpoint control. Normally a very unstable protein, p53 is a DNA-binding transcription factor, one of a family of related proteins, that is somehow stabilized in response to DNA damage, perhaps by direct p53-DNA interactions. Increased levels of p53 activate transcription of an ensemble of genes that collectively serve to delay transit through the cycle. One of these induced proteins, $p21^{CIP}$, is a potent CDK-cyclin inhibitor (CKI) that is capable of efficiently inhibiting the action of all CDKs. Clearly, inhibition of CDKs will halt progression through the cell cycle (see Figures 35–19 & 35 20). If DNA damage is too extensive to repair, the affected cells undergo **apoptosis** (programmed cell death) in a p53-dependent fashion. In this case, p53 induces the activation of a collection of genes that induce apoptosis. Cells lacking functional p53 fail to undergo apoptosis in response to high levels of radiation or DNA-active chemotherapeutic agents. It may come as no surprise, then, that *p53* is one of the most frequently mutated genes in human cancers. Additional research into the mechanisms of checkpoint control will prove invaluable for the development of effective anticancer therapeutic options.

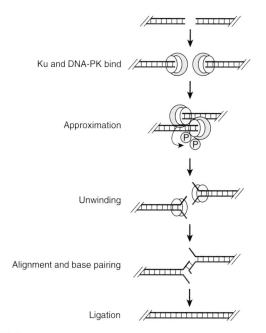

FIGURE 35–25 Double-strand break repair of DNA. The proteins Ku and DNA-dependent protein kinase combine to approximate the two strands and unwind them. The aligned fragments form base pairs; the extra ends are removed, probably by a DNA-PK-associated endo- or exonuclease, and the gaps are filled in; and continuity is restored by ligation.

Labels in figure: Ku and DNA-PK bind / Approximation / Unwinding / Alignment and base pairing / Ligation

SUMMARY

■ DNA in eukaryotic cells is associated with a variety of proteins, resulting in a structure called chromatin.

- Much of the DNA is associated with histone proteins to form a structure called the nucleosome. Nucleosomes are composed of an octamer of histones around which about 150 bp of DNA is wrapped.

- Histones are subject to an extensive array of dynamic covalent modifications that have important regulatory consequences.

- Nucleosomes and higher-order structures formed from them serve to compact the DNA.

- DNA in transcriptionally active regions is sensitive to nuclease attack; some regions are exceptionally sensitive and are often found to contain transcription control sites.

- Highly transcriptionally active DNA (genes) is often clustered in regions of each chromosome. Within these regions, genes may be separated by inactive DNA in nucleosomal structures. In eukaryotes the transcription unit—that portion of a gene that is copied by RNA polymerase—often consists of coding regions of DNA (exons) interrupted by intervening sequences of noncoding DNA (introns).

- After transcription, during RNA processing, introns are removed and the exons are ligated together to form the mature mRNA that appears in the cytoplasm; this process is termed RNA splicing.

- DNA in each chromosome is exactly replicated according to the rules of base pairing during the S phase of the cell cycle.

- Each strand of the double helix is replicated simultaneously but by somewhat different mechanisms. A complex of proteins, including DNA polymerase, replicates the leading strand continuously in the 5′ to 3′ direction. The lagging strand is replicated discontinuously, in short pieces of 150–250 nucleotides, in the 3′ to 5′ direction.

- DNA replication occurs at several sites—called replication bubbles—in each chromosome. The entire process takes about 9 h in a typical cell and only occurs during the S phase of the cell cycle.

- A variety of mechanisms employing different enzymes repair damaged DNA, as after exposure to chemical mutagens or ultraviolet radiation.

REFERENCES

Boutros R et al: CDC25 phosphatases in cancer cells: key players? Good targets? Nat Rev Cancer 2007;7:495.

Dalal Y et al: Structure, dynamics, and evolution of centromeric nucleosomes. Proc Nat Academy of Sciences 2007;104:41.

Encode Consortium: Identification and analysis of functional elements in 1% of the human genome by the ENCODE pilot project. Nature 2007;447:799.

Gilbert DM: In search of the holy replicator. Nature Rev Mol Cell Biol 2004;5:848.

Hillenmeyer ME et al: The chemical genomic portrait of yeast: uncovering a phenotype for all genes. Science 2008;320:362.

Johnson A, O'Donnell M: Cellular DNA replicases: components and dynamics at the replication fork. Ann Rev Biochemistry 2005;74:283.

Krishman KJ et al: What causes mitochondrial DNA deletions in human cells? Nat Genet 2008;40:275.

Lander ES et al: Initial sequencing and analysis of the human genome. Nature 2001;409:860.

Luger K et al: Crystal structure of the nucleosome core particle at 2.8 Å resolution. Nature 1997;389:251.

McCulloch SD, Kunkel TA: The fidelity of DNA synthesis by eukaryotic replicative and translesion synthesis polymerases. Cell Res 2008;18:148.

Michelson RJ, Weinart T: Closing the gaps among a web of DNA repair disorders. Bioessays J 2002;22:966.

Misteli T: Beyond the sequence: cellular organization of genome function. Cell 2007;128:787.

Orr HT, Zoghbi, HY: Trinucleotide repeat disorders. Annu Rev Neurosci 2007;30:375.

Sancar A, Lindsey-Boltz LA, Ünsal-Kacmaz K, Linn S: Molecular mechanisms of mammalian DNA repair and the DNA damage checkpoint. Ann Rev Biochem 2004;73:39.

Sullivan et al: Determining centromere identity: cyclical stories and forking paths. Nat Rev Genet 2001;2:584.

Venter JC et al: The sequence of the human genome. Science 2002;291:1304.

Zilberman D, Henikoff S: Genome wide analysis of DNA methylation patterns. Development 2007;134:3959.

RNA Synthesis, Processing, & Modification

P. Anthony Weil, PhD

BIOMEDICAL IMPORTANCE

The synthesis of an RNA molecule from DNA is a complex process involving one of the group of RNA polymerase enzymes and a number of associated proteins. The general steps required to synthesize the primary transcript are initiation, elongation, and termination. Most is known about initiation. A number of DNA regions (generally located upstream from the initiation site) and protein factors that bind to these sequences to regulate the initiation of transcription have been identified. Certain RNAs—mRNAs in particular—have very different life spans in a cell. It is important to understand the basic principles of messenger RNA synthesis and metabolism, for modulation of this process results in altered rates of protein synthesis and thus a variety of both metabolic and phenotypic changes. This is how all organisms adapt to changes of environment. It is also how differentiated cell structures and functions are established and maintained. The RNA molecules synthesized in mammalian cells are made as precursor molecules that have to be processed into mature, active RNA. Errors or changes in synthesis, processing, splicing, stability, or function of mRNA transcripts are a cause of disease.

RNA EXISTS IN FOUR MAJOR CLASSES

All eukaryotic cells have four major classes of RNA: ribosomal RNA (rRNA), messenger RNA (mRNA), transfer RNA (tRNA), and small RNAs, the small nuclear RNAs and microRNAs (snRNA and miRNA). The first three are involved in protein synthesis, while the small RNAs are involved in mRNA splicing and modulation of gene expression by altering mRNA function. As shown in Table 36–1, these various classes of RNA are different in their diversity, stability, and abundance in cells.

RNA IS SYNTHESIZED FROM A DNA TEMPLATE BY AN RNA POLYMERASE

The processes of DNA and RNA synthesis are similar in that they involve (1) the general steps of initiation, elongation, and termination with 5′ to 3′ polarity; (2) large, multicomponent initiation complexes; and (3) adherence to Watson–Crick base-pairing rules. However, DNA and RNA synthesis do differ in several important ways, including the following: (1) ribonucleotides are used in RNA synthesis rather than deoxyribonucleotides; (2) U replaces T as the complementary base for A in RNA; (3) a primer is not involved in RNA synthesis as RNA polymerases have the ability to initiate synthesis de novo; (4) only portions of the genome are vigorously transcribed or copied into RNA, whereas the entire genome must be copied, once and only once during DNA replication; and (5) there is no highly active, efficient proofreading function during RNA transcription.

The process of synthesizing RNA from a DNA template has been characterized best in prokaryotes. Although in mammalian cells the regulation of RNA synthesis and the processing of the RNA transcripts are different from those in prokaryotes, the process of RNA synthesis per se is quite similar in these two classes of organisms. Therefore, the description of RNA synthesis in prokaryotes, where it is best understood, is applicable to eukaryotes even though the enzymes involved and the regulatory signals, though related, are different.

The Template Strand of DNA Is Transcribed

The sequence of ribonucleotides in an RNA molecule is complementary to the sequence of deoxyribonucleotides in one strand of the double-stranded DNA molecule (Figure 34–8). The strand that is transcribed or copied into an RNA molecule is referred to as the **template strand** of the DNA. The other DNA strand, the **non-template strand**, is frequently referred to as the **coding strand** of that gene. It is called this because, with the exception of T for U changes, it corresponds exactly to the sequence of the RNA primary transcript, which encodes the (protein) product of the gene. In the case of a double-stranded DNA molecule containing many genes, the template strand for each gene will not necessarily be the same strand of the DNA double helix (Figure 36–1). Thus, a given strand of a double-stranded DNA molecule will serve as the template strand for some genes and the coding strand of other genes. Note that the nucleotide sequence of an RNA transcript will be the same (except for U replacing T) as that of the coding strand. The information in the template strand is read out in the 3′ to 5′ direction. Though not shown in Figure 36–1 there are instances of genes embedded within other genes.

TABLE 36–1   Classes of Eukaryotic RNA

RNA	Types	Abundance	Stability
Ribosomal (rRNA)	28S, 18S, 5.8S, 5S	80% of total	Very stable
Messenger (mRNA)	$\sim10^5$ different species	2–5% of total	Unstable to very stable
Transfer (tRNA)	~60 different species	~15% of total	Very stable
Small RNAs			
Small nuclear (snRNA)	~30 different species	≤1% of total	Very stable
Micro (miRNA)	100s–1000	<1% of total	Stable

FIGURE 36–1   This figure illustrates that genes can be transcribed off both strands of DNA. The arrowheads indicate the direction of transcription (polarity). Note that the template strand is always read in the 3′ to 5′ direction. The opposite strand is called the coding strand because it is identical (except for T for U changes) to the mRNA transcript (the primary transcript in eukaryotic cells) that encodes the protein product of the gene.

DNA-Dependent RNA Polymerase Initiates Transcription at a Distinct Site, the Promoter

DNA-dependent RNA polymerase is the enzyme responsible for the polymerization of ribonucleotides into a sequence complementary to the template strand of the gene (see Figures 36–2 & 36–3). The enzyme attaches at a specific site—the promoter—on the template strand. This is followed by initiation of RNA synthesis at the starting point, and the process continues until a termination sequence is reached (Figure 36–3). A **transcription unit** is defined as that region of DNA that includes the signals for transcription initiation, elongation, and termination. The RNA product, which is synthesized in the 5′ to 3′ direction, is the **primary transcript.** Transcription rates vary from gene to gene but can be quite high. An electron micrograph of transcription in action is presented in Figure 36–4. In prokaryotes, this can represent the product of several contiguous genes; in mammalian cells, it usually represents the product of a single gene. If a transcription unit contains only a single gene, then the 5′ termini of the primary RNA transcript and the mature cytoplasmic RNA are identical. **Thus, the starting point of transcription corresponds to the 5′ nucleotide of the mRNA.** This is designated position +1, as is the corresponding nucleotide in the DNA. The numbers increase as the sequence proceeds *downstream* from the start site. This convention makes it easy to locate particular

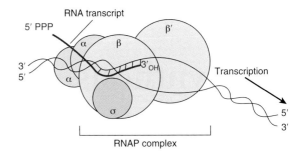

FIGURE 36–2   RNA polymerase (RNAP) catalyzes the polymerization of ribonucleotides into an RNA sequence that is complementary to the template strand of the gene. The RNA transcript has the same polarity (5′ to 3′) as the coding strand but contains U rather than T. *E coli* RNAP consists of a core complex of two α subunits and two β subunits (β and β′). The holoenzyme contains the σ subunit bound to the $\alpha_2\beta\beta'$ core assembly. The ω subunit is not shown. The transcription "bubble" is an approximately 20-bp area of melted DNA, and the entire complex covers 30–75 bp, depending on the conformation of RNAP.

FIGURE 36–3   The transcription cycle in bacteria. Bacterial RNA transcription is described in four steps: **(1) Template binding:** RNA polymerase (RNAP) binds to DNA and locates a promoter (P), which melts the two DNA strands to form a preinitiation complex (PIC). **(2) Chain initiation:** RNAP holoenzyme (core + one of multiple sigma factors) catalyzes the coupling of the first base (usually ATP or GTP) to a second ribonucleoside triphosphate to form a dinucleotide. **(3) Promoter clearance:** RNAP undergoes a conformational change after RNA chain length reaches 10–20 nt and then is able to move away from the promoter, transcribing down the transcription unit. **(4) Chain elongation:** Successive residues are added to the 3′-OH terminus of the nascent RNA molecule. **(5) Chain termination and RNAP release:** The completed RNA chain and RNAP are released from the template. The RNAP holoenzyme re-forms, finds a promoter, and the cycle is repeated.

regions, such as intron and exon boundaries. The nucleotide in the promoter adjacent to the transcription initiation site in the upstream direction is designated –1, and these negative numbers increase as the sequence proceeds *upstream*, away from the initiation site. This provides a conventional way of defining the location of regulatory elements in the promoter.

The primary transcripts generated by RNA polymerase II—one of three distinct nuclear DNA-dependent RNA polymerases in eukaryotes—are promptly capped by 7-methylguanosine triphosphate caps (Figure 34–10) that persist and eventually appear on the 5′ end of mature cytoplasmic mRNA. These caps are necessary for the subsequent processing of the primary transcript to mRNA, for the translation of the mRNA, and for protection of the mRNA against exonucleolytic attack.

Bacterial DNA-Dependent RNA Polymerase Is a Multisubunit Enzyme

The DNA-dependent RNA polymerase (RNAP) of the bacterium *Escherichia coli* exists as an approximately 400 kDa core complex consisting of two identical α subunits, similar but not identical β and β′ subunits, and an ω subunit. The β subunit binds Mg^{2+} ions and composes the catalytic subunit (Figure 36–2). The core RNA polymerase, $\beta\beta'\alpha_2\omega$, often termed E, associates with a specific protein factor (the sigma [σ] factor) to form holoenzyme, $\beta\beta'\alpha_2\omega\sigma$, or Eσ. The σ subunit helps the core enzyme recognize and bind to the specific deoxynucleotide sequence of the promoter region (Figure 36–5) to form the **pre-initiation complex (PIC)**. There are multiple, distinct σ-factor encoding genes in all bacterial species. Sigma factors have a dual role in the process of promoter recognition; σ association with core RNA polymerase decreases its affinity for certain nonpromoter DNA while simultaneously increasing holoenzyme affinity for promoter DNA. The multiple σ-factors, each of which acts as a regulatory protein that modifies the **promoter recognition specificity** of the resulting unique RNA polymerase holoenzyme (ie, $E\sigma_1$, $E\sigma_2$,...). The appearance of different σ factors and their association with Core RNAP forming novel holoenzyme forms can be correlated temporally with various programs of gene expression in prokaryotic systems such as sporulation, growth in various poor nutrient sources, and the response to heat shock.

Mammalian Cells Possess Three Distinct Nuclear DNA-Dependent RNA Polymerases

The properties of mammalian polymerases are described in Table 36–2. Each of these DNA-dependent RNA polymerases is responsible for transcription of different sets of genes. The sizes of the RNA polymerases range from MW 500,000 to MW 600,000. These enzymes exhibit more complex subunit profiles than prokaryotic RNA polymerases. They all have two large

TABLE 36–2 Nomenclature and Properties of Mammalian Nuclear DNA-Dependent RNA Polymerases

Form of RNA Polymerase	Sensitivity to α-Amanitin	Major Products
I	Insensitive	rRNA
II	High sensitivity	mRNA, miRNA, SnRNA
III	Intermediate sensitivity	tRNA, 5s rRNA

subunits and a number of smaller subunits—as many as 14 in the case of RNA pol III. However, the eukaryotic RNA polymerase subunits do exhibit extensive amino acid homologies with prokaryotic RNA polymerases. This homology has been shown recently to extend to the level of three-dimensional structures. The functions of each of the subunits are not yet fully understood.

A peptide toxin from the mushroom *Amanita phalloides,* α-amanitin, is a specific differential inhibitor of the eukaryotic nuclear DNA-dependent RNA polymerases and as such has proved to be a powerful research tool (Table 36–2). α-Amanitin blocks the translocation of RNA polymerase during phosphodiester bond formation.

RNA SYNTHESIS IS A CYCLICAL PROCESS & INVOLVES RNA CHAIN INITIATION, ELONGATION, & TERMINATION

The process of RNA synthesis in bacteria—depicted in Figure 36–3—involves first the binding of the RNA polymerase holoenzyme molecule to the template at the promoter site to form a **preinitiation complex, or PIC.** Binding is followed by a conformational change of the RNAP and unwinding of the DNA around the transcription start site, and the first nucleotide (almost always a purine) then associates with the nucleotide binding site on the β subunit of the enzyme. In the presence of the next appropriate nucleotide bound to the polymerase, RNAP catalyzes the formation of a phosphodiester bond, and the nascent chain is now attached to the polymerization site on the β subunit of RNAP. (The analogy to the A and P sites on the ribosome should be noted; see Figure 37–9.)

Initiation of formation of the RNA molecule at its 5′ end then follows, while elongation of the RNA molecule from the 5′ to its 3′ end continues cyclically, antiparallel to its template. The enzyme polymerizes the ribonucleotides in the specific sequence dictated by the template strand and interpreted by Watson–Crick base-pairing rules. Pyrophosphate is released following each cycle of polymerization. As for DNA synthesis this pyrophosphate (PP_i) is rapidly degraded to 2 mol of inorganic phosphate (P_i) by ubiquitous pyrophosphatases, thereby providing irreversibility on the overall synthetic reaction. In both prokaryotes and eukaryotes, a purine ribonucleotide is usually the first to be polymerized into the RNA molecule. As with eukaryotes, the 5′ triphosphate of the first, initiating nucleotide is maintained in prokaryotic mRNA. After 10–20 nucleotides have been polymerized, RNAP undergoes a second conformational change leading to **promoter clearance.** Once this transition occurs, RNAP physically moves away from the promoter, transcribing down the transcription unit, leading to the next phase of the process, elongation. This decision, to stay at the promoter, "stalled," or transition to elongation appears to be an important regulatory step in eukaryotic pol II transcription.

As the **elongation** complex containing the core RNA polymerase progresses along the DNA molecule, **DNA unwinding** must occur in order to provide access for the appropriate base pairing to the nucleotides of the coding strand. The extent of this transcription bubble (ie, DNA unwinding) is constant throughout transcription and has been estimated to be about 20 base pairs per polymerase molecule. Thus, it appears that the size of the unwound DNA region is dictated by the polymerase and is independent of the DNA sequence in the complex. RNA polymerase has an intrinsic "unwindase" activity that opens the DNA helix. The fact that the DNA double helix must unwind, and the strands part at least transiently for transcription implies some disruption of the nucleosome structure of eukaryotic cells. Topoisomerase both precedes and follows the progressing RNAP to prevent the formation of superhelical tensions that would serve to increase the energy required to unwind the template DNA ahead of RNAP.

Termination of the synthesis of the RNA molecule in bacteria is signaled by a sequence in the template strand of the DNA molecule—a signal that is recognized by a **termination protein, the rho (ρ) factor**. Rho is an ATP-dependent RNA-stimulated helicase that disrupts the nascent RNA-DNA complex. In some cases bacterial RNAP can directly recognize DNA-encoded termination signals without assistance by the rho factor. After termination of synthesis of the RNA, the enzyme separates from the DNA template and probably dissociates to free core enzyme and free factor. With the assistance of another σ factor, the core enzyme then recognizes a promoter at which the synthesis of a new RNA molecule commences. In eukaryotic cells, termination is less well understood but RNA processing, termination, and polyadenylation proteins appear to load onto RNAP II soon after initiation (see below). More than one RNA polymerase molecule may transcribe the same template strand of a gene simultaneously, but the process is phased and spaced in such a way that at any one moment each is transcribing a different portion of the DNA sequence (Figures 36–1 & 36–4).

FIGURE 36–4 Electron photomicrograph of multiple copies of amphibian ribosomal RNA genes in the process of being transcribed. The magnification is about 6000×. Note that the length of the transcripts increases as the RNA polymerase molecules progress along the individual ribosomal RNA genes from transcription start sites (filled circles) to transcription termination sites (open circles). RNA polymerase I (not visualized here) is at the base of the nascent rRNA transcripts. Thus, the proximal end of the transcribed gene has short transcripts attached to it, while much longer transcripts are attached to the distal end of the gene. The arrows indicate the direction (5′ to 3′) of transcription. (Reproduced with permission, from Miller OL Jr, Beatty BR: Portrait of a gene. J Cell Physiol 1969;74[Suppl 1]:225. Copyright © 1969. Reprinted with permission from Wiley-Liss, Inc., a subsidiary of John Wiley & Sons, Inc.)

THE FIDELITY & FREQUENCY OF TRANSCRIPTION IS CONTROLLED BY PROTEINS BOUND TO CERTAIN DNA SEQUENCES

Analysis of the DNA sequence of specific genes has allowed the recognition of a number of sequences important in gene transcription. From the large number of bacterial genes studied, it is possible to construct consensus models of transcription initiation and termination signals.

The question, "How does RNAP find the correct site to initiate transcription?" is not trivial when the complexity of the genome is considered. *E coli* has 4×10^3 transcription initiation sites (ie, gene promoters) in 4.2×10^6 base pairs (bp) of DNA. The situation is even more complex in humans, where as many as 10^5 transcription initiation sites are distributed throughout 3×10^9 bp of DNA. RNAP can bind, with low affinity, to many regions of DNA, but it scans the DNA sequence—at a rate of $\geq 10^3$ bp/s—until it recognizes certain specific regions of DNA to which it binds with higher affinity. These regions are termed promoters, and it is the association of RNAP with promoters that ensures accurate initiation of transcription. The promoter recognition-utilization process is the target for regulation in both bacteria and humans.

Bacterial Promoters Are Relatively Simple

Bacterial promoters are approximately 40 nucleotides (40 bp or four turns of the DNA double helix) in length, a region small enough to be covered by an *E coli* RNA holopolymerase molecule. In a consensus promoter there are two short, conserved sequence elements. Approximately 35-bp upstream of the transcription start site there is a consensus sequence of eight nucleotide pairs (consensus: 5′-TGTTGACA-3′) to which the RNAP binds to form the so-called **closed complex.** More proximal to the transcription start site—about ten nucleotides upstream—is a six-nucleotide-pair A+T-rich sequence (consensus: 5′-TATAAT-3′). These conserved sequence elements together comprise the promoter, and are shown schematically in Figure 36–5. The latter sequence has a low melting temperature because of its lack of GC nucleotide pairs. Thus, the so-

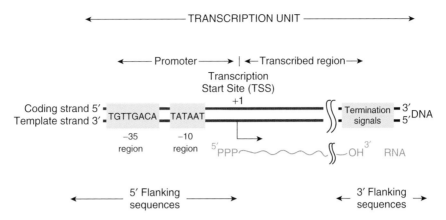

FIGURE 36–5 Bacterial promoters, such as that from *E coli* shown here, share two regions of highly conserved nucleotide sequence. These regions are located 35 and 10 bp upstream (in the 5′ direction of the coding strand) from the start site of transcription, which is indicated as +1. By convention, all nucleotides upstream of the transcription initiation site (at +1) are numbered in a negative sense and are referred to as 5′-flanking sequences. Also by convention, the DNA regulatory sequence elements (TATA box, etc) are described in the 5′ to 3′ direction and as being on the coding strand. These elements function only in double-stranded DNA, however. Note that the transcript produced from this transcription unit has the same polarity or "sense" (ie, 5′ to 3′ orientation) as the coding strand. Termination *cis*-elements reside at the end of the transcription unit (see Figure 36–6 for more detail). By convention the sequences downstream of the site at which transcription termination occurs are termed 3′-flanking sequences.

called **TATA "box"** is thought to ease the dissociation of the two DNA strands so that RNA polymerase bound to the promoter region can have access to the nucleotide sequence of its immediately downstream template strand. Once this process occurs, the combination of RNA polymerase plus promoter is called the **open complex.** Other bacteria have slightly different consensus sequences in their promoters, but all generally have two components to the promoter; these tend to be in the same position relative to the transcription start site, and in all cases the sequences between the two promoter elements have no similarity but still **provide critical spacing functions** that facilitate recognition of −35 and −10 sequences by RNA polymerase holoenzyme. Within a bacterial cell, different sets of genes are often coordinately regulated. One important way that this is accomplished is through the fact that these co-regulated genes share particular −35 and −10 promoter sequences. These unique promoters are recognized by different σ factors bound to core RNA polymerase (ie, Eσ_1, Eσ_2,...).

Rho-dependent transcription **termination signals** in *E coli* also appear to have a distinct consensus sequence, as shown in Figure 36–6. The conserved consensus sequence, which is about 40 nucleotide pairs in length, can be seen to contain a hyphenated or interrupted inverted repeat followed by a series of AT base pairs. As transcription proceeds through the hyphenated, inverted repeat, the generated transcript can form the intramolecular hairpin structure, also depicted in Figure 36–6.

Transcription continues into the AT region, and with the aid of the ρ termination protein the RNA polymerase stops, dissociates from the DNA template, and releases the nascent transcript.

Eukaryotic Promoters Are More Complex

It is clear that the signals in DNA that control transcription in eukaryotic cells are of several types. Two types of sequence elements are promoter-proximal. One of these defines **where transcription is to commence** along the DNA, and the other contributes to the mechanisms that control **how frequently** this event is to occur. For example, in the thymidine kinase gene of the herpes simplex virus, which utilizes transcription factors of its mammalian host for its early gene expression program, there is a single unique transcription start site, and accurate transcription from this start site depends upon a nucleotide sequence located 32 nucleotides upstream from the start site (ie, at −32) (Figure 36–7). This region has the sequence of **TATAAAAG** and bears remarkable similarity to the functionally related **TATA box** that is located about 10 bp upstream from the prokaryotic mRNA start site (Figure 36–5). Mutation or inactivation of the TATA box markedly reduces transcription of this and many other genes that contain this consensus *cis*-active element (see Figures 36–7 & 36–8). The TATA box is usually located 25–30 bp upstream from the transcription start site in mammalian genes that contain it. The consensus sequence for a TATA box is TATAAA, though numerous variations have been characterized. The human TATA box is bound by the 34 kDa **TATA-binding protein (TBP),** which is a subunit in at least two multisubunit complexes, TFIID and SAGA/P-CAF. The non-TBP subunits of TFIID are proteins called **TBP-associated factors (TAFs).** This complex of TBP and TAFs is referred to as TFIID. Binding of the TBP-TAF TFIID complex to the TATA box sequence is thought to represent a first step in the formation of the transcription complex on the promoter.

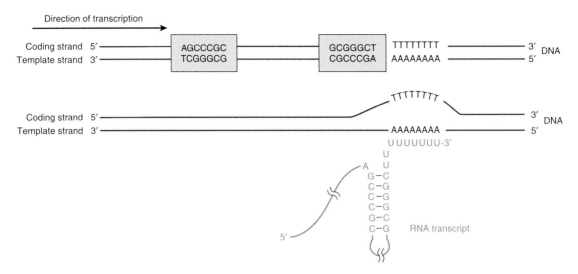

FIGURE 36–6 The predominant bacterial transcription termination signal contains an inverted, hyphenated repeat (the two boxed areas) followed by a stretch of AT base pairs (**top**). The inverted repeat, when transcribed into RNA, can generate the secondary structure in the RNA transcript (**bottom**). Formation of this RNA hairpin causes RNA polymerase to pause and subsequently the ρ termination factor interacts with the paused polymerase and somehow induces chain termination.

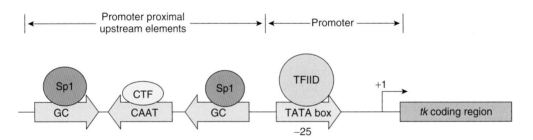

FIGURE 36–7 Transcription elements and binding factors in the herpes simplex virus thymidine kinase *(tk)* gene. DNA-dependent RNA polymerase II (not shown) binds to the region of the TATA box (which is bound by transcription factor TFIID) to form a multicomponent preinitiation complex capable of initiating transcription at a single nucleotide (+1). The frequency of this event is increased by the presence of upstream *cis*-acting elements (the GC and CAAT boxes) located either near to the promoter (promoter proximal) or distant from the promoter (distal elements; Figure 36–8). Proximal and distal *cis* elements are bound by *trans*-acting transcription factors, in this example Sp1 and CTF (also called C/EBP, NF1, NFY). These *cis* elements can function independently of orientation (**arrows**).

Some number of mRNA-encoding genes lack a consensus TATA box. In such instances, additional *cis* elements, an **initiator sequence (Inr)** and/or the so-called **downstream promoter element (DPE),** direct the RNA polymerase II transcription machinery to the promoter and in so doing provide basal transcription starting from the correct site. The Inr element spans the start site (from −3 to +5) and consists of the general consensus sequence TCA_{+1} G/T T T/C (A_{+1} indicates the first nucleotide transcribed). The proteins that bind to Inr in order to direct pol II binding include TFIID. Promoters that have both a TATA box and an Inr may be stronger or more vigorously transcribed than those that have just one of these elements. The DPE has the consensus sequence A/GGA/T CGTG and is localized about 25 bp downstream of the +1 start site. Like the Inr, DPE sequences are also bound by the

TAF subunits of TFIID. In a survey of over 200 eukaryotic genes, roughly 30% contained a TATA box and Inr, 25% contained Inr and DPE, 15% contained all three elements, while ~ 30% contained just the Inr.

Sequences generally, though not always, just upstream from the start site determine how frequently a transcription event occurs. Mutations in these regions reduce the frequency of transcriptional starts 10-fold to 20-fold. Typical of these DNA elements are the GC and CAAT boxes, so named because of the DNA sequences involved. As illustrated in Figure 36–7, each of these boxes binds a specific protein, Sp1 in the case of the GC box and CTF (or C/EPB,NF1,NFY) by the CAAT box; both bind through their distinct **DNA binding domains (DBDs)**. The frequency of transcription initiation is a consequence of these protein-DNA interactions and

FIGURE 36–8 Schematic diagram showing the transcription control regions in a hypothetical mRNA-producing, eukaryotic gene transcribed by RNA polymerase II. Such a gene can be divided into its coding and regulatory regions, as defined by the transcription start site (**arrow**; +1). The coding region contains the DNA sequence that is transcribed into mRNA, which is ultimately translated into protein. The regulatory region consists of two classes of elements. One class is responsible for ensuring basal expression. The "promoter," composed of the TATA box or Inr or DPE elements, directs RNA polymerase II to the correct site (fidelity). In TATA-less promoters, an initiator (Inr) element that spans the initiation site (+1) may direct the polymerase to this site. Another component, the upstream elements, specifies the frequency of initiation; such elements can either be proximal (50–200 bp) or distal (1000–10^5 bp) to the promoter as shown. Among the best studied of the proximal elements is the CAAT box, but several other elements (bound by the transactivator proteins Sp1, NF1, AP1, etc) may be used in various genes. The distal elements enhance or repress expression, several of which mediate the response to various signals, including hormones, heat shock, heavy metals, and chemicals. Tissue-specific expression also involves specific sequences of this sort. The orientation dependence of all the elements is indicated by the arrows within the boxes. For example, the proximal element (the TATA box) must be in the 5′ to 3′ orientation. The upstream elements work best in the 5′ to 3′ orientation, but some of them can be reversed. The locations of some elements are not fixed with respect to the transcription start site. Indeed, some elements responsible for regulated expression can be located either interspersed with the upstream elements, or they can be located downstream from the start site.

complex interactions between particular domains of the transcription factors (distinct from the DBD domains—so-called **activation domains; ADs)** of these proteins and the rest of the transcription machinery (RNA polymerase II, the basal factors **TFIIA, B, D, E, F** and other coregulatory factors such as Mediator, chromatin remodellers and chromatin modifying factors). (See below and Figures 36–9 & 36–10.) The protein-DNA interaction at the TATA box involving RNA polymerase

II and other components of the basal transcription machinery ensures the fidelity of initiation.

Together, the promoter and promoter-proximal *cis*-active upstream elements confer fidelity and frequency of initiation upon a gene. The TATA box has a particularly rigid requirement for both position and orientation. Single-base changes in any of these *cis* elements can have dramatic effects on function by reducing the binding affinity of the cognate *trans* factors (either

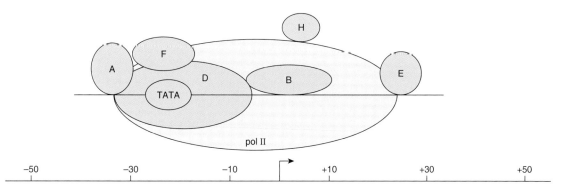

FIGURE 36–9 The eukaryotic basal transcription complex. Formation of the basal transcription complex begins when TFIID binds to the TATA box. It directs the assembly of several other components by protein-DNA and protein-protein interactions; TFIIA, B, E,F, H, and polymerase II (pol II). The entire complex spans DNA from position −30 to +30 relative to the initiation site (+1, marked by bent arrow). The atomic level, x-ray-derived structures of RNA polymerase II alone and of TBP bound to TATA promoter DNA in the presence of either TFIIB or TFIIA have all been solved at 3 Å resolution. The structures of TFIID and TFIIH complexes have been determined by electron microscopy at 30 Å resolution. Thus, the molecular structures of the transcription machinery are beginning to be elucidated. Much of this structural information is consistent with the models presented here.

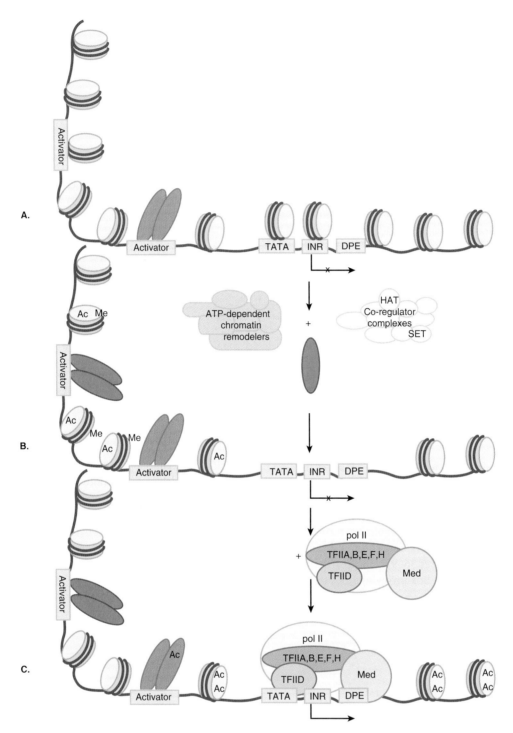

FIGURE 36–10 Nucleosome eviction by chromatin-active coregulators facilitates PIC formation and transcription. Shown in **A**, is an inactive mRNA encoding gene with a single transcription factor bound to its cognate enhancer site (Activator). The enhancer element was nucleosome free and hence available for interaction with this particular activator binding protein. This gene is inactive due to the fact that a portion of the enhancer and the entirety of the promoter are covered by nucleosomes. The enhancer-bound activator transcription factor directly interacts with any of a number of distinct ATP-dependent chromatin remodelers and chromatin-modifying Co-regulators. These coregulators together have the ability to both move and/or remove nucleosomes (ATP-dependent remodelers) as well as to covalently modify nucleosomal histones using intrinsic acetylases (HAT; resulting in acetylation [Ac]) and methylases (SET; resulting in methylation [Me]), carried by subunits of these complexes. The resulting changes in nucleosome position and nucleosome occupancy allow for the binding of additional DNA-binding transactivators (illustrated in **B**) and ultimately the formation of an active PIC and transcription (illustrated in **C**).

TFIID/TBP or Sp1, CTF, and similar factors). The spacing of the TATA box, Inr, and DPE promoter elements is also be critical.

A third class of sequence elements can either increase or decrease the rate of transcription initiation of eukaryotic genes. These elements are called either **enhancers** or **repressors (or silencers),** depending on how they effect RNA synthesis. They have been found in a variety of locations both upstream and downstream of the transcription start site and even within the transcribed protein coding portions of some genes. Enhancers and silencers can exert their effects when located thousands or even tens of thousands of bases away from transcription units located on the same chromosome. Surprisingly, enhancers and silencers can function in an orientation-independent fashion. Literally hundreds of these elements have been described. In some cases, the sequence requirements for binding are rigidly constrained; in others, considerable sequence variation is allowed. Some sequences bind only a single protein, but the majority bind several different proteins. Together these many transfactors binding to promoter distal and proximal *cis*-elements regulate transcription in response to a vast array of biological signals. Such transcriptional regulatory events contribute importantly to control of gene expression.

Specific Signals Regulate Transcription Termination

The **signals for the termination of transcription** by eukaryotic RNA polymerase II are only poorly understood. It appears that the termination signals exist far downstream of the coding sequence of eukaryotic genes. For example, the transcription termination signal for mouse β-globin occurs at several positions 1000–2000 bases beyond the site at which the poly(A) tail will eventually be added. Less is known about the termination process or whether specific termination factors similar to the bacterial ρ factor are involved. However, it is known that formation of the mRNA 3′ terminal, which is generated post-transcriptionally, is somehow coupled to events or structures formed at the time and site of initiation. Moreover, mRNA formation, and in this case 3′end formation depends on a special structure in one of the subunits of RNA polymerase II (the CTD; see below), and this process appears to involve at least two steps as follows. After RNA polymerase II has traversed the region of the transcription unit encoding the 3′ end of the transcript, RNA endonucleases cleave the primary transcript at a position about 15 bases 3′ of the consensus sequence **AAUAAA** that serves in eukaryotic transcripts as a cleavage and polyadenylation signal. Finally, this newly formed 3′ terminal is polyadenylated in the nucleoplasm, as described below.

THE EUKARYOTIC TRANSCRIPTION COMPLEX

A complex apparatus consisting of as many as 50 unique proteins provides accurate and regulatable transcription of eukaryotic genes. The RNA polymerase enzymes (pol I, pol II, and pol III) transcribe information contained in the template strand of DNA into RNA. These polymerases must recognize a specific site in the promoter in order to initiate transcription at the proper nucleotide. In contrast to the situation in prokaryotes, eukaryotic RNA polymerases alone are not able to discriminate between promoter sequences and other regions of DNA; thus, other proteins known as **general transcription factors or GTFs** facilitate promoter-specific binding of these enzymes and formation of the preinitiation complex (PIC). This combination of components can catalyze basal or (non)-unregulated transcription in vitro. Another set of proteins—coactivators, or coregulators—work in conjunction with DNA binding transfactors to help regulate the rate of transcription; all three classes of proteins interact to effect transcription regulation (see below).

Formation of the Pol II Transcription Complex

In bacteria, a σ factor–polymerase holoenzyme complex, Eσ, selectively binds to promoter DNA to form the PIC. The situation is more complex in eukaryotic genes. mRNA-encoding genes, which are transcribed by pol II—are described as an example. In the case of pol II-transcribed genes, the function of σ factors is assumed by a number of proteins. PIC formation requires, in addition to pol II, a number of so-called **general transcription factors (GTFs) termed TFIIA, TFIIB, TFIID, TFIIE, TFIIF, and TFIIH.** These GTFs serve to promote RNA polymerase II transcription on essentially all genes. Some of these GTFs are composed of multiple subunits. **TFIID,** which **binds to the TATA box promoter element,** is the only one of these factors that is independently capable of specific, high affinity binding to promoter DNA. TFIID consists of TATA binding protein (TBP) and 14 TBP-associated factors (TAFs).

TBP binds to the TATA box in the minor groove of DNA (most transcription factors bind in the major groove) and causes an approximately 100-degree bend or kink of the DNA helix. This bending is thought to facilitate the interaction of TBP-associated factors with other components of the transcription initiation complex, the multicomponent eukaryotic promoter and possibly with factors bound to upstream elements. Although initially defined as a component solely required for transcription of pol II gene promoters, TBP, by virtue of its association with distinct, polymerase-specific sets of TAFs, is also an important component of pol I and pol III initiation complexes even if they do not contain TATA boxes.

The binding of TFIID marks a specific promoter for transcription. Of several subsequent in vitro steps, the first is the binding of TFIIA, then TFIIB to the TFIID-promoter complex. This results in a stable ternary complex which is then more precisely located and more tightly bound at the transcription initiation site. This complex then attracts and tethers the pol II–TFIIF complex to the promoter. Addition of TFIIE and TFIIH are the final steps in the assembly of the PIC. TFIIE appears to join the complex with pol II–TFIIF, and TFIIH is then recruited. Each of these binding events extends the size of the complex so that finally about 60 bp (from −30 to +30 relative to +1, the nucleotide from which transcription commences) are covered (Figure 36–9). The PIC is now complete and capable of basal transcription initiated from the correct nucleotide. In genes that lack a TATA box, the same factors are required. In such cases, the Inr or DPE serve to (see Figure 36–8) position the complex for accurate initiation of transcription.

Promoter Accessibility and Hence PIC Formation Is Often Modulated by Nucleosomes

On certain eukaryotic genes the transcription machinery (pol II, etc) cannot access the promoter sequences (ie, TATA-INR-DPE) because these essential promoter elements are wrapped up in nucleosomes (Figure 36–10). Only after transcription factors bind to enhancer DNA upstream of the promoter and recruit chromatin remodeling and modifying coregulatory factors such as the Swi/Snf, SRC-1, p300/CBP (see Chapter 42) or P/CAF factors, are the repressing nucleosomes removed (Figure 36–10). Once the promoter is "open" following nucleosome eviction, pol II and other essential proteins can bind and initiate mRNA gene transcription. Note that the binding of transactivators and coregulators can be sensitive to, and/or directly control the covalent modification status of the histones within the nucleosomes in and around the promoter and enhancer, and thereby increase or decrease the ability of all the other components required for PIC formation to interact with a particular gene. This so called **epigenetic code of histone and protein modifications** can contribute importantly to gene transcription control. Indeed mutations in proteins that catalyze (code writers) or differentially bind (code readers) modified histones can lead to human disease.

Phosphorylation Activates Pol II

Eukaryotic pol II consists of 12 subunits. The two largest subunits, 150 and 190 kDa, are homologous to the bacterial β and β′ subunits. In addition to the increased number of subunits, eukaryotic pol II differs from its prokaryotic counterpart in that it has a series of heptad repeats with consensus sequence Tyr-Ser-Pro-Thr-Ser-Pro-Ser at the carboxyl terminus of the largest pol II subunit. This **carboxyl terminal repeat domain (CTD)** has 26 repeated units in brewers' yeast and 52 units in mammalian cells. The CTD is both a substrate for several kinases, including the kinase component of TFIIH, and a binding site for a wide array of proteins. The CTD has been shown to interact with many RNA processing enzymes as well as nuclear transport proteins. The association of these factors with the CTD of RNA polymerase II (and other components of the basal machinery) thus serves to couple transcription initiation with mRNA splicing, 3′ end formation and transport to the cytoplasm. Pol II is activated when phosphorylated on the Ser and Thr residues and displays reduced activity when the CTD is dephosphorylated. CTD phosphorylation/dephosphorylation is critical for promoter clearance, elongation, termina-

tion, and even appropriate mRNA processing. Pol II lacking the CTD tail is incapable of activating transcription, and cells expressing pol II lacking the CTD are inviable. These results underscore the importance of this domain.

Pol II can associate with other proteins termed **Mediator** or **Med** proteins to form a complex sometimes referred to as the pol II holoenzyme; this complex can form on the promoter or in solution prior to PIC formation (see below). The Med proteins are essential for appropriate regulation of pol II transcription by serving myriad roles, both activating and repressing transcription. Thus Mediator, like TFIID is a transcriptional coregulator (see below). Complex forms of RNA polymerase II holoenzyme (pol II plus Med) have been described in human cells that contain over 30 Med proteins (Med1–Med31).

The Role of Transcription Activators & Coregulators

TFIID was originally considered to be a single protein, TBP. However, several pieces of evidence led to the important discovery that TFIID is actually a complex consisting of TBP and the 14 TAFs. The first evidence that TFIID was more complex than just the TBP molecules came from the observation that TBP binds to a 10-bp segment of DNA, immediately over the TATA box of the gene, whereas native holo-TFIID covers a 35 bp or larger region (Figure 36–9). Second, TBP has a molecular mass of 20–40 kDa (depending on the species), whereas the TFIID complex has a mass of about 1000 kDa. Finally, and perhaps most importantly, TBP supports basal transcription but not the augmented transcription provided by certain activators, eg, Sp1 bound to the GC box. TFIID, on the other hand, supports both basal and enhanced transcription by Sp1, Oct1, AP1, CTF, ATF, etc (Table 36–3). The TAFs are essential

TABLE 36−3   Some of the Transcription Control Elements, Their Consensus Sequences, and the Factors That Bind to Them Which Are Found in Mammalian Genes Transcribed by RNA Polymerase II

Element	Consensus Sequence	Factor
TATA box	TATAAA	TBP/TFIID
CAAT box	CCAATC	C/EBP*, NF-Y*
GC box	GGGCGG	Sp1*
	CAACTGAC	Myo D
	T/CGGA/CN$_5$GCCAA	NF1*
Ig octamer	ATGCAAAT	Oct1, 2, 4, 6*
AP1	TGAG/CTC/AA	Jun, Fos, ATF*
Serum response	GATGCCCATA	SRF
Heat shock	(NGAAN)$_3$	HSF

Note: A complete list would include hundreds of examples. The asterisks mean that there are several members of this family.

TABLE 36–4 **Three Classes of Transcription Factors Involved in mRNA Gene Transcription**

General Mechanisms	Specific Components
Basal components	TBP, TFIIA, B, E, F, and H
Coregulators	TAFs (TBP + TAFs) = TFIID; Mediator, Meds Chromatin modifiers Chromatin remodelers
Activators	SP1, ATF, CTF, AP1, etc

for this activator-enhanced transcription. There are likely several forms of TFIID that differ slightly in their complement of TAFs. Thus different combinations of TAFs with TBP—or one of several recently discovered TBP-like factors (TLFs)—bind to different promoters, and recent reports suggest that this may account for the tissue or cell-selective gene activation noted in various promoters and for the different strengths of certain promoters. TAFs, since they are required for the action of activators, are often called coactivators or coregulators. There are thus three classes of transcription factors involved in the regulation of pol II genes: pol II and GTFs, coregulators, and DNA-binding activator-repressors (Table 36–4). How these classes of

proteins interact to govern both the site and frequency of transcription is a question of central importance. It is thought that coregulators both act as a bridge between the DNA-binding transactivators and pol II/GTFs and modify chromatin.

Two Models Can Explain the Assembly of the Preinitiation Complex

The formation of the PIC described above is based on the sequential addition of purified components as observed through in vitro experiments. An essential feature of this model is that PIC assembly takes place on a DNA template where the transcription proteins all have ready access to DNA. Accordingly, transcription activators, which have autonomous DNA binding and activation domains (see Chapter 38), are thought to function by stimulating PIC formation. Here the TAF or Mediator Complexes are viewed as bridging factors that communicate between the upstream-bound activators, and the GTFs and pol II. This view assumes that there is **stepwise assembly** of the PIC—promoted by various interactions between activators, coactivators, and PIC components, and is illustrated in panel A of Figure 36–11. This model was supported by observations that many of these proteins can indeed bind to one another in vitro.

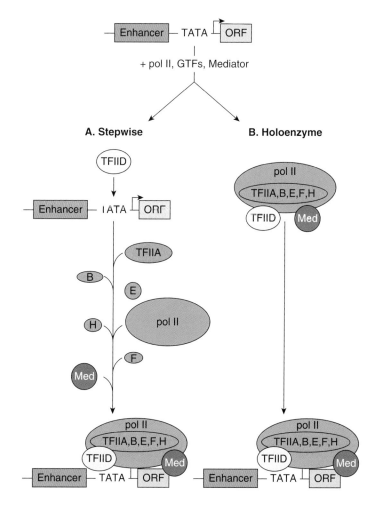

FIGURE 36–11 Models for the formation of a pol II PIC. Shown at top is a typical mRNA encoding transcription unit: enhancer-promoter (TATA)-initiation site (bent arrow) and transcribed region (ORF; open reading frame). PICs have been shown to form by two distinct mechanisms: **(A)** the stepwise binding of GTFs, pol II, and Mediator, or **(B)** by the binding of a single multiprotein complex composed of pol II, Med, and the six GTFs. DNA binding transactivator proteins specifically bind enhancers and in part facilitate PIC formation (or PIC function) by binding directly to the TFIID-TAF subunits or Med subunits of Mediator (not shown, see Figure 36–10); the mechanism(s) by which such protein–protein interactions stimulate transcription remain a subject of intense investigation.

Recent evidence suggests that there is another possible mechanism of PIC formation and thus transcription regulation. First, large preassembled complexes of GTFs and pol II are found in cell extracts, and these complexes can associate with the promoter in a single step. Second, the rate of transcription achieved when activators are added to limiting concentrations of pol II holoenzyme can be matched by increasing the concentration of the pol II holoenzyme in the absence of activators. Thus, at least in vitro, one can establish conditions where activators are not in themselves absolutely essential for PIC formation. These observations led to the **"recruitment" hypothesis,** which has now been tested experimentally. Simply stated, the role of activators and some coactivators may be solely to recruit a preformed holoenzyme-GTF complex to the promoter. The requirement for an activation domain is circumvented when either a component of TFIID or the pol II holoenzyme is artificially tethered, using recombinant DNA techniques, to the DNA binding domain (DBD) of an activator. This anchoring, through the DBD component of the activator molecule, leads to a transcriptionally competent structure, and there is no further requirement for the activation domain of the activator. In this view, the role of activation domains is to direct preformed holoenzyme-GTF complexes to the promoter; they do not assist in PIC assembly (see panel B, Figure 36–11). In this model, the efficiency of the recruitment process directly determines the rate of transcription at a given promoter.

RNA MOLECULES ARE USUALLY PROCESSED BEFORE THEY BECOME FUNCTIONAL

In prokaryotic organisms, the primary transcripts of mRNA-encoding genes begin to serve as translation templates even before their transcription has been completed. This can occur because the site of transcription is not compartmentalized into a nucleus as it is in eukaryotic organisms. Thus, transcription and translation are coupled in prokaryotic cells. Consequently, prokaryotic mRNAs are subjected to little processing prior to carrying out their intended function in protein synthesis. Indeed, appropriate regulation of some genes (eg, the *Trp* operon) relies upon this coupling of transcription and translation. Prokaryotic rRNA and tRNA molecules are transcribed in units considerably longer than the ultimate molecule. In fact, many of the tRNA transcription units encode more than one tRNA molecule. Thus, in prokaryotes the processing of these rRNA and tRNA precursor molecules is required for the generation of the mature functional molecules.

Nearly all eukaryotic RNA primary transcripts undergo extensive processing between the time they are synthesized and the time at which they serve their ultimate function, whether it be as mRNA, miRNAs, or as a component of the translation machinery such as rRNA, 5S RNA, or tRNA. Processing occurs primarily within the nucleus. The processes of **transcription, RNA processing, and even RNA transport from the nucleus are highly coordinated.** Indeed, a transcriptional

coactivator termed SAGA in yeasts and P/CAF in human cells is thought to link transcription activation to RNA processing by recruiting a second complex termed TREX to transcription elongation, splicing, and nuclear export. TREX (transcription-export) represents a likely molecular link between transcription elongation complexes, the RNA splicing machinery, and nuclear export (see Figure 36–12). This coupling presumably dramatically increases both the fidelity and rate of processing and movement of mRNA to the cytoplasm for translation.

The Coding Portions (Exons) of Most Eukaryotic Genes Are Interrupted by Introns

Often interspersed within the amino acid-coding portions **(exons)** of many genes are long sequences of DNA that do not contribute to the genetic information ultimately translated into the amino acid sequence of a protein molecule (see Chapter 35). In fact, these sequences actually interrupt the coding region of structural genes. These **intervening sequences (introns)** exist within most but not all mRNA encoding genes of higher eukaryotes. The intron RNA sequences are cleaved out of the transcript, and the exons of the transcript are appropriately spliced together in the nucleus before the resulting mRNA molecule appears in the cytoplasm for translation (Figures 36–13 & 36–14). One speculation for this exon-intron gene organization is that exons, which often encode an activity domain, or functional module of a protein, represent a convenient means of shuffling genetic information, permitting organisms to quickly test the results of combining novel protein functional domains.

Introns Are Removed & Exons Are Spliced Together

Four different splicing reaction mechanisms for intron removal have been described. The one most frequently used in eukaryotic cells is described below. Although the sequences of nucleotides in the introns of the various eukaryotic transcripts—and even those within a single transcript—are quite heterogeneous, there are reasonably conserved sequences at each of the two exon-intron (splice) junctions and at the branch site, which is located 20–40 nucleotides upstream from the 3' splice site (see consensus sequences in Figure 36–14). A special multicomponent complex, the **spliceosome,** is involved in converting the primary transcript into mRNA. Spliceosomes consist of the primary transcript, five small nuclear RNAs (U1, U2, U4, U5, and U6) and more than 60 proteins, many of which contain conserved "RNP" and "SR" protein motifs. Collectively, the five SnRNAs and RNP-/SR-containing proteins form a **small nuclear ribonucleoprotein termed an SnRNA complex.** It is likely that this penta-snRNP spliceosome forms prior to interaction with mRNA precursors. SnRNPs are thought to position the exon and intron RNA segments for the necessary splicing reactions. The splicing reaction starts with a cut at the junction of the 5'-exon (donor or left) and intron (Figure 36–13).

FIGURE 36–12 RNA Polymerase II–mediated mRNA gene transcription is cotranscriptionally coupled to RNA processing and transport. Shown is RNA pol II actively transcribing an mRNA encoding gene (elongation top to bottom of figure). RNA processing factors (ie, SR/RNP-motif-containing splicing factors as well as polyadenylation and termination factors) interact with the CTD domain of pol II, while mRNA packaging factors such as THO/TREX complex are recruited to the nascent mRNA primary transcript either through direct pol II interactions as shown or through interactions with SR/splicing factors resident on the nascent mRNA. Note that the CTD is not drawn to scale. This conserved domain of the Rpb1 subunit of pol II is in reality 5–10 times the length of the polymerase and thus a significant docking site for RNA processing and transport proteins. In both cases, nascent mRNA chains are thought to be more rapidly and accurately processed due to the rapid recruitment of these many factors to the growing mRNA (precursor) chain. Following appropriate mRNA processing, the mature mRNA is delivered to the nuclear pores dotting the nuclear membrane, where, upon transport through the pores, the mRNAs can be engaged by ribosomes and translated into protein. (Adapted from Jensen et al. [2005]. *Molecular Cell* 11:1129–1138.)

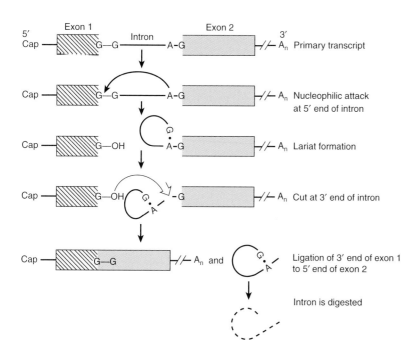

FIGURE 36–13 The processing of the primary transcript to mRNA. In this hypothetical transcript, the 5′ (**left**) end of the intron is cut (↓) and a lariat forms between the G at the 5′ end of the intron and an A near the 3′ end, in the consensus sequence UACUA**A**C. This sequence is called the branch site, and it is the 3′ most **A** that forms the 5′–2′ bond with the G. The 3′ (**right**) end of the intron is then cut (⇓). This releases the lariat, which is digested, and exon 1 is joined to exon 2 at G residues.

FIGURE 36–14 Consensus sequences at splice junctions. The 5′ (donor; **left**) and 3′ (acceptor; **right**) sequences are shown. Also shown is the yeast consensus sequence (UACUA**A**C) for the branch site. In mammalian cells, this consensus sequence is PyNPyPyPuA**P**y, where Py is a pyrimidine, Pu is a purine, and N is any nucleotide. The branch site is located 20–30 nucleotides upstream from the 3′ site. (Copyright © 2005. Reprinted with permission from Elsevier.)

This is accomplished by a nucleophilic attack by an adenylyl residue in the branch point sequence located just upstream from the 3′ end of this intron. The free 5′ terminal then forms a loop or lariat structure that is linked by an unusual 5′–2′ phosphodiester bond to the reactive A in the PyNPyPyPuAPy branch site sequence (Figure 36–14). This adenylyl residue is typically located 20–30 nucleotides upstream from the 3′ end of the intron being removed. The branch site identifies the 3′ splice site. A second cut is made at the junction of the intron with the 3′ exon (donor on right). In this second transesterification reaction, the 3′ hydroxyl of the upstream exon attacks the 5′ phosphate at the downstream exon-intron boundary, and the lariat structure containing the intron is released and hydrolyzed. The 5′ and 3′ exons are ligated to form a continuous sequence.

The snRNAs and associated proteins are required for formation of the various structures and intermediates. U1 within the snRNP complex binds first by base pairing to the 5′ exon-intron boundary. U2 within the snRNP complex then binds by base pairing to the branch site, and this exposes the nucleophilic A residue. U4/U5/U6 within the snRNP complex mediates an ATP-dependent protein-mediated unwinding that results in disruption of the base-paired U4-U6 complex with the release of U4. U6 is then able to interact first with U2, then with U1. These interactions serve to approximate the 5′ splice site, the branch point with its reactive A, and the 3′ splice site. This alignment is enhanced by U5. This process also results in the formation of the loop or lariat structure. The two ends are cleaved, probably by the U2-U6 within the snRNP complex. U6 is certainly essential, since yeasts deficient in this snRNA are not viable. It is important to note that RNA serves as the catalytic agent. This sequence of events is then repeated in genes containing multiple introns. In such cases, a definite pattern is followed for each gene, though the introns are not necessarily removed in sequence—1, then 2, then 3, etc.

Alternative Splicing Provides for Different mRNAs

The processing of mRNA molecules is a site for regulation of gene expression. Alternative patterns of mRNA splicing result from tissue-specific adaptive and developmental control mechanisms. As mentioned above, the sequence of exon-intron splicing events generally follows a hierarchical order for a given gene. The fact that very complex RNA structures are formed during splicing—and that a number of snRNAs and proteins are involved—affords numerous possibilities for a change of this order and for the generation of different mRNAs. Similarly, the use of alternative termination-cleavage polyadenylation sites also results in mRNA variability. Some schematic examples of these processes, all of which occur in nature, are shown in Figure 36–15.

Faulty splicing can cause disease. At least one form of β-thalassemia, a disease in which the β-globin gene of hemoglobin is severely underexpressed, appears to result from a nucleotide change at an exon-intron junction, precluding removal of the intron and therefore leading to diminished or absent synthesis of the β-chain protein. This is a consequence of the fact that the normal translation reading frame of the mRNA is disrupted by a defect in the fundamental process of RNA splicing, underscoring the accuracy that the process of RNA-RNA splicing must maintain.

Alternative Promoter Utilization Provides a Form of Regulation

Tissue-specific regulation of gene expression can be provided by alternative splicing, as noted above, by control elements in

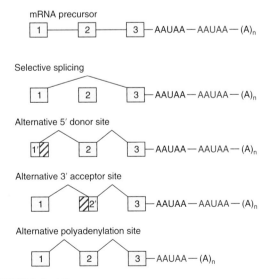

FIGURE 36–15 Mechanisms of alternative processing of mRNA precursors. This form of RNA processing involves the selective inclusion or exclusion of exons, the use of alternative 5′ donor or 3′ acceptor sites, and the use of different polyadenylation sites.

FIGURE 36–16 Alternative promoter use in the liver and pancreatic β cell glucokinase genes. Differential regulation of the glucokinase (*GK*) gene is accomplished by the use of tissue-specific promoters. The β cell *GK* gene promoter and exon 1B are located about 30 kbp upstream from the liver promoter and exon 1L. Each promoter has a unique structure and is regulated differently. Exons 2–10 are identical in the two genes, and the GK proteins encoded by the liver and β cell mRNAs have identical kinetic properties.

the promoter or by the use of alternative promoters. The glucokinase (*GK*) gene consists of ten exons interrupted by nine introns. The sequence of exons 2–10 is identical in liver and pancreatic β cells, the primary tissues in which GK protein is expressed. Expression of the *GK* gene is regulated very differently—by two different promoters—in these two tissues. The liver promoter and exon 1L are located near exons 2–10; exon 1L is ligated directly to exon 2. In contrast, the pancreatic β cell promoter is located about 30 kbp upstream. In this case, the 3′ boundary of exon 1B is ligated to the 5′ boundary of exon 2. The liver promoter and exon 1L are excluded and removed during the splicing reaction (see Figure 36–16). The existence of multiple distinct promoters allows for cell- and tissue-specific expression patterns of a particular gene (mRNA).

Both Ribosomal RNAs & Most Transfer RNAs Are Processed from Larger Precursors

In mammalian cells, the three rRNA molecules (28S, 18S, 5.8S) are transcribed as part of a single large 45S precursor molecule. **The precursor is subsequently processed in the nucleolus** to provide these three RNA components for the ribosome subunits found in the cytoplasm. The rRNA genes are located in the nucleoli of mammalian cells. Hundreds of copies of these genes are present in every cell. This large number of genes is required to synthesize sufficient copies of each type of rRNA to form the 10^7 ribosomes required for each cell replication. Whereas a single mRNA molecule may be copied into 10^5 protein molecules, providing a large amplification, the rRNAs are end products. This lack of amplification requires both a large number of genes and a high transcription rate, typically synchronized with cell growth rate. Similarly, transfer RNAs are often synthesized as precursors, with extra sequences both 5′ and 3′ of the sequences comprising the mature tRNA. A small fraction of tRNAs contain introns.

RNAs CAN BE EXTENSIVELY MODIFIED

Essentially all RNAs are covalently modified after transcription. It is clear that at least some of these modifications are regulatory.

Messenger RNA (mRNA) Is Modified at the 5′ & 3′ Ends

As mentioned above, mammalian mRNA molecules contain a 7-methylguanosine cap structure at their 5′ terminal, and most have a poly(A) tail at the 3′ terminal. The cap structure is added to the 5′ end of the newly transcribed mRNA precursor in the nucleus prior to transport of the mRNA molecule to the cytoplasm. The **5′ cap** of the RNA transcript is required both for efficient translation initiation and protection of the 5′ end of mRNA from attack by 5′ → 3′ exonucleases. The secondary methylations of mRNA molecules, those on the 2′-hydroxy and the N^7 of adenylyl residues, occur after the mRNA molecule has appeared in the cytoplasm.

Poly(A) tails are added to the 3′ end of mRNA molecules in a posttranscriptional processing step. The mRNA is first cleaved about 20 nucleotides downstream from an AAUAA recognition sequence. Another enzyme, **poly(A) polymerase**, adds a poly(A) tail which is subsequently extended to as many as 200 A residues. The **poly(A) tail** appears to protect the 3′ end of mRNA from 3′ → 5′ exonuclease attack. The presence or absence of the poly(A) tail does not determine whether a precursor molecule in the nucleus appears in the cytoplasm, because all poly(A)-tailed nuclear mRNA molecules do not contribute to cytoplasmic mRNA, nor do all cytoplasmic mRNA molecules contain poly(A) tails (histone mRNAs are most notable in this regard). Following nuclear transport cytoplasmic enzymes in mammalian cells can both add and remove adenylyl residues from the poly(A) tails; this process has been associated with an alteration of mRNA stability and translatability.

The size of some cytoplasmic mRNA molecules, even after the poly(A) tail is removed, is still considerably greater than the size required to code for the specific protein for which it is a template, often by a factor of 2 or 3. **The extra nucleotides occur in untranslated (nonprotein coding) regions** both 5′ and 3′ of the coding region; the longest untranslated sequences are usually at the 3′ end. The exact function of **5′ UTR and 3′ UTR** sequences is unknown, but they have been implicated in RNA processing, transport, degradation, and translation; each of these reactions potentially contributes additional levels of control of gene expression. **The micro-RNAs**

typically target sequences within the 3′ UTR. Many of these post-transcriptional events involving mRNAs occur in P bodies (Chapter 37).

Micro-RNAs Are Derived from Large Primary Transcripts Through Specific Nucleolytic Processing

The majority of miRNAs are transcribed by RNA pol II into **primary transcripts** termed **pri-miRNAs.** pri-miRNAs are 5′-capped and 3′-polyadenylated (Figure 36–17). Pri-miRNAs are synthesized from transcription units encoding one or several distinct miRNAs; these transcription units are either located independently in the genome or within the intronic DNA of other genes. miRNA-encoding genes must therefore minimally posess a distinct promoter, coding region and poly-adenylation/termination signals. pri-miRNAs have extensive 2° structure, and this intramolecular structure is maintained following processing by the **Drosha-DGCR8 nuclease**; the portion containing the RNA hairpin is preserved, transported through the nuclear pore and once in the cytoplasm, further processed to a **21 or 22-mer** by the **dicer nuclease.** Ultimately one of the two strands is selected for loading into the **RISC,**

or **RNA induced silencing complex** to form a mature, functional miRNA. SiRNAs are produced similarly. Once in the RISC complex, miRNAs can modulate mRNA function (see Chapter 39).

RNA Editing Changes mRNA After Transcription

The central dogma states that for a given gene and gene product there is a linear relationship between the coding sequence in DNA, the mRNA sequence, and the protein sequence (Figure 35–7). Changes in the DNA sequence should be reflected in a change in the mRNA sequence and, depending on codon usage, in protein sequence. However, exceptions to this dogma have been recently documented. Coding information can be changed at the mRNA level by **RNA editing.** In such cases, the coding sequence of the mRNA differs from that in the cognate DNA. An example is the apolipoprotein B *(apoB)* gene and mRNA. In liver, the single *apoB* gene is transcribed into an mRNA that directs the synthesis of a 100-kDa protein, apoB100. In the intestine, the same gene directs the synthesis of the primary transcript; however, a cytidine deaminase converts a CAA codon in the mRNA to UAA at a single specific

FIGURE 36–17 Biogenesis of miRNAs. miRNA encoding genes are transcribed by RNA pol II into a primary miRNA transcript (pri-miRNA), which is 5′-capped and polyadenylated as is typical of mRNA coding primary transcripts. This pri-miRNA is subjected to processing within the nucleus by the action of the Drosha-DGCR8 nuclease, which trims sequences from both 5′ and 3′ ends generating the pre-miRNA. This partially processed double-stranded RNA is transported through the nuclear pore by exportin-5. The cytoplasmic pre-miRNA is then trimmed further by the action of the multisubunit nuclease termed Dicer, to form the miRNA duplex. One of the two resulting 21–22 nucleotide-long RNA strands is selected, the duplex unwound, and the selected strand loaded into the RISC complex, thereby generating the mature, functional miRNA.

site. Rather than encoding glutamine, this codon becomes a termination signal, and a 48-kDa protein (apoB48) is the result. ApoB100 and apoB48 have different functions in the two organs. A growing number of other examples include a glutamine to arginine change in the glutamate receptor and several changes in trypanosome mitochondrial mRNAs, generally involving the addition or deletion of uridine. The exact extent of RNA editing is unknown, but current estimates suggest that <0.01% of mRNAs are edited in this fashion. Recently, editing of miRNAs has been described suggesting that these two forms of post-transcriptional control mechanisms could cooperatively contribute to gene regulation.

Transfer RNA (tRNA) Is Extensively Processed & Modified

As described in Chapters 34 & 37, the tRNA molecules serve as adapter molecules for the translation of mRNA into protein sequences. The tRNAs contain many modifications of the standard bases A, U, G, and C, including methylation, reduction, deamination, and rearranged glycosidic bonds. Further modification of the tRNA molecules includes nucleotide alkylations and the attachment of the characteristic CpCpA$_{OH}$ terminal at the 3′ end of the molecule by the enzyme nucleotidyl transferase. The 3′ OH of the A ribose is the point of attachment for the specific amino acid that is to enter into the polymerization reaction of protein synthesis. The methylation of mammalian tRNA precursors probably occurs in the nucleus, whereas the cleavage and attachment of CpCpA$_{OH}$ are cytoplasmic functions, since the terminals turn over more rapidly than do the tRNA molecules themselves. Enzymes within the cytoplasm of mammalian cells are required for the attachment of amino acids to the CpCpA$_{OH}$ residues (see Chapter 37).

RNA CAN ACT AS A CATALYST

In addition to the catalytic action served by the snRNAs in the formation of mRNA, several other enzymatic functions have been attributed to RNA. **Ribozymes** are RNA molecules with catalytic activity. These generally involve transesterification reactions, and most are concerned with RNA metabolism (splicing and endoribonuclease). Recently, a ribosomal RNA component was noted to hydrolyze an aminoacyl ester and thus to play a central role in peptide bond function (peptidyl transferases; see Chapter 37). These observations, made using RNA molecules derived from the organelles from plants, yeast, viruses, and higher eukaryotic cells, show that RNA can act as an enzyme, and have revolutionized thinking about enzyme action and the origin of life itself.

SUMMARY

- RNA is synthesized from a DNA template by the enzyme RNA polymerase.
- There are three distinct nuclear DNA-dependent RNA polymerases in mammals: RNA polymerases I, II, and III.

These enzymes catalyze the transcription of rRNA(I), mRNA/miRNAs (II), and tRNA and 5S rRNA-(III) encoding genes.

- RNA polymerases interact with unique *cis*-active regions of genes, termed promoters, in order to form preinitiation complexes (PICs) capable of initiation. In eukaryotes the process of pol II PIC formation requires, in addition to polymerase, multiple general transcription factors (GTFs), TFIIA, B, D, E, F, and H.
- Eukaryotic PIC formation can occur on accessible promoters either step-wise—by the sequential, ordered interactions of GTFs and RNA polymerase with DNA promoters—or in one step by the recognition of the promoter by a pre-formed GTF-RNA polymerase holoenzyme complex.
- Transcription exhibits three phases: initiation, elongation, and termination. All are dependent upon distinct DNA *cis*-elements and modulated by distinct *trans*-acting protein factors.
- The presence of nucleosomes can occlude the binding of transfactors and the transcription machinery thereby inhibiting transcription.
- Most eukaryotic RNAs are synthesized as precursors that contain excess sequences which are removed prior to the generation of mature, functional RNA. These processing steps provide additional potential steps for regulation of RNA synthesis.
- Eukaryotic mRNA synthesis results in a pre-mRNA precursor that contains extensive amounts of excess RNA (introns) that must be precisely removed by RNA splicing to generate functional, translatable mRNA composed of exonic coding and 5′ and 3′ noncoding sequences.
- All steps—from changes in DNA template, sequence, and accessibility in chromatin to RNA stability and translatability—are subject to modulation and hence are potential control sites for eukaryotic gene regulation.

REFERENCES

Bourbon H-M et al: A unified nomenclature for protein subunits of mediator complexes linking transcriptional regulators to RNA polymerase II. Mol Cell 2004;14:553.

Busby S, Ebright RH: Promoter structure, promoter recognition, and transcription activation in prokaryotes. Cell 1994;79:743.

Cramer P, Bushnell DA, Kornberg R: Structural basis of transcription: RNA polymerase II at 2.8 angstrom resolution. Science 2001;292:1863.

Fedor MJ: Ribozymes. Curr Biol 1998;8:R441.

Gott JM, Emeson RB: Functions and mechanisms of RNA editing. Ann Rev Genet 2000;34:499.

Hirose Y, Manley JL: RNA polymerase II and the integration of nuclear events. Genes Dev 2000;14:1415.

Kawauchi J et al: Budding yeast RNA polymerases I and II employ parallel mechanisms of transcriptional termination. Genes Dev 2008;22:1082.

Keaveney M, Struhl K: Activator-mediated recruitment of the RNA polymerase machinery is the predominant mechanism for transcriptional activation in yeast. Mol Cell 1998;1:917.

Kornblihtt AR et al: Multiple links between transcription and splicing. RNA 2004;10:1489.

Lemon B, Tjian R: Orchestrated response: a symphony of transcription factors for gene control. Genes Dev 2000;14:2551.

Lykke-Andersen S, Jensen TH: Overlapping pathways dictate termination of RNA polymerase II. Biochimie 2007;89:1177).

Maniatis T, Reed R: An extensive network of coupling among gene expression machines. Nature 2002;416:499.

Nishikura K: Editor meets silencer: crosstalk between RNA editing and RNA interference. Nat Rev Mol Cell Biol 2006;26:792.

Orphanides G, Reinberg D: A unified theory of gene expression. Cell 2002;108:439.

Price DH: Poised polymerases: on your mark ... get set ... go!. Mol Cell 2008; 30:7.

Reed R, Cheng H: TREX, SR proteins and export of mRNA. Curr Opin Cell Biol 2005;17:269.

Shatkin AJ, Manley JL: The ends of the affair: capping and polyadenylation. Nat Struct Biol 2000;7:838.

Sims RJ, Belotserkovskaya R, Reinberg D: Elongation by RNA polymerase II: the short and long of it. Gene Dev 2004;18:2437.

Stevens SW et al: Composition and functional characterization of the yeast spliceosomal penta-snRNP. Mol Cell 2002;9:31.

Tucker M, Parker R: Mechanisms and control of mRNA decapping in *Saccharomyces cerevisiae*. Ann Rev Biochem 2000;69:571.

West S, Proudfoot N, Dye MJ: Molecular dissection of mammalian RNA polymerase II transcriptional termination. Mol Cell 2008;29:600.

Woychik NA, Hampsey M: The RNA polymerase II machinery: structure illuminates function. Cell 2002;108:453.

Protein Synthesis & the Genetic Code

P. Anthony Weil, PhD

BIOMEDICAL IMPORTANCE

The letters A, G, T, and C correspond to the nucleotides found in DNA. Within the protein coding genes these nucleotides are organized into three-letter code words called **codons,** and the collection of these codons makes up the **genetic code.** It was impossible to understand protein synthesis—or to explain mutations—before the genetic code was elucidated. The code provides a foundation for explaining the way in which protein defects may cause genetic disease and for the diagnosis and perhaps the treatment of these disorders. In addition, the pathophysiology of many viral infections is related to the ability of these infectious agents to disrupt host cell protein synthesis. Many antibacterial drugs are effective because they selectively disrupt protein synthesis in the invading bacterial cell but do not affect protein synthesis in eukaryotic cells.

GENETIC INFORMATION FLOWS FROM DNA TO RNA TO PROTEIN

The genetic information within the nucleotide sequence of DNA is transcribed in the nucleus into the specific nucleotide sequence of an RNA molecule. The sequence of nucleotides in the RNA transcript is complementary to the nucleotide sequence of the template strand of its gene in accordance with the base-pairing rules. Several different classes of RNA combine to direct the synthesis of proteins.

In prokaryotes there is a linear correspondence between the gene, the **messenger RNA (mRNA)** transcribed from the gene, and the polypeptide product. The situation is more complicated in higher eukaryotic cells, in which the primary transcript is much larger than the mature mRNA. The large mRNA precursors contain coding regions (**exons**) that will form the mature mRNA and long intervening sequences (**introns**) that separate the exons. The mRNA is processed within the nucleus, and the introns, which often make up much more of this RNA than the exons, are removed. Exons are spliced together to form mature mRNA, which is transported to the cytoplasm, where it is translated into protein.

The cell must possess the machinery necessary to translate information accurately and efficiently from the nucleotide sequence of an mRNA into the sequence of amino acids of the corresponding specific protein. Clarification of our understanding of this process, which is termed **translation,** awaited deciphering of the genetic code. It was realized early that mRNA molecules themselves have no affinity for amino acids and, therefore, that the translation of the information in the mRNA nucleotide sequence into the amino acid sequence of a protein requires an intermediate adapter molecule. This adapter molecule must recognize a specific nucleotide sequence on the one hand as well as a specific amino acid on the other. With such an adapter molecule, the cell can direct a specific amino acid into the proper sequential position of a protein during its synthesis as dictated by the nucleotide sequence of the specific mRNA. In fact, the functional groups of the amino acids do not themselves actually come into contact with the mRNA template.

THE NUCLEOTIDE SEQUENCE OF AN mRNA MOLECULE CONTAINS A SERIES OF CODONS THAT SPECIFY THE AMINO ACID SEQUENCE OF THE ENCODED PROTEIN

Twenty different amino acids are required for the synthesis of the cellular complement of proteins; thus, there must be at least 20 distinct codons that make up the genetic code. Since there are only four different nucleotides in mRNA, each codon must consist of more than a single purine or pyrimidine nucleotide. Codons consisting of two nucleotides each could provide for only 16 (4^2) specific codons, whereas codons of three nucleotides could provide 64 (4^3) specific codons.

It is now known that each codon consists of a sequence of three nucleotides; ie, **it is a triplet code** (see Table 37–1). The deciphering of the **genetic code** depended heavily on the chemical synthesis of nucleotide polymers, particularly triplets in repeated sequence. These synthetic triplet ribonucleotides were used as mRNAs to program protein synthesis in vitro, allowing investigators to deduce the genetic code.

TABLE 37–1   The Genetic Code[1] (Codon Assignments in Mammalian Messenger RNAs)

First Nucleotide	Second Nucleotide				Third Nucleotide
	U	**C**	**A**	**G**	
U	Phe	Ser	Tyr	Cys	U
	Phe	Ser	Tyr	Cys	C
	Leu	Ser	Term	Term[2]	A
	Leu	Ser	Term	Trp	G
C	Leu	Pro	His	Arg	U
	Leu	Pro	His	Arg	C
	Leu	Pro	Gln	Arg	A
	Leu	Pro	Gln	Arg	G
A	Ile	Thr	Asn	Ser	U
	Ile	Thr	Asn	Ser	C
	Ile[2]	Thr	Lys	Arg[2]	A
	Met	Thr	Lys	Arg[2]	G
G	Val	Ala	Asp	Gly	U
	Val	Ala	Asp	Gly	C
	Val	Ala	Glu	Gly	A
	Val	Ala	Glu	Gly	G

[1]The terms first, second, and third nucleotide refer to the individual nucleotides of a triplet codon. U, uridine nucleotide; C, cytosine nucleotide; A, adenine nucleotide; G, guanine nucleotide; Term, chain terminator codon. AUG, which codes for Met, serves as the initiator codon in mammalian cells and also encodes for internal methionines in a protein. (Abbreviations of amino acids are explained in Chapter 3.)

[2]In mammalian mitochondria, AUA codes for Met and UGA for Trp, and AGA and AGG serve as chain terminators.

THE GENETIC CODE IS DEGENERATE, UNAMBIGUOUS, NONOVERLAPPING, WITHOUT PUNCTUATION, & UNIVERSAL

Three of the 64 possible codons do not code for specific amino acids; these have been termed **nonsense codons.** These nonsense codons are utilized in the cell as **termination signals;** they specify where the polymerization of amino acids into a protein molecule is to stop. **The remaining 61 codons code for 20 amino acids** (Table 37–1). Thus, there must be "degeneracy" in the genetic code—ie, multiple codons must decode the same amino acid. Some amino acids are encoded by several codons; eg, six different codons specify serine. Other amino acids, such as methionine and tryptophan, have a single codon. In general, the third nucleotide in a codon is less important than the first two in determining the specific amino acid to be incorporated, and this accounts for most of the degeneracy of the code. However, for any specific codon, only a single amino acid is indicated; with rare exceptions, the genetic code is **unambiguous**—ie, given a specific codon, only

a single amino acid is indicated. **The distinction between ambiguity and degeneracy is an important concept.**

The unambiguous but degenerate code can be explained in molecular terms. The recognition of specific codons in the mRNA by the tRNA adapter molecules is dependent upon their **anticodon region** and specific base-pairing rules. Each tRNA molecule contains a specific sequence, complementary to a codon, which is termed its anticodon. For a given codon in the mRNA, only a single species of tRNA molecule possesses the proper anticodon. Since each tRNA molecule can be charged with only one specific amino acid, each codon therefore specifies only one amino acid. However, some tRNA molecules can utilize the anticodon to recognize more than one codon. **With few exceptions, given a specific codon, only a specific amino acid will be incorporated—although, given a specific amino acid, more than one codon may be used.**

As discussed below, the reading of the genetic code during the process of protein synthesis does not involve any overlap of codons. **Thus, the genetic code is nonoverlapping.** Furthermore, once the reading is commenced at a specific codon, there is **no punctuation** between codons, and the message is read in a continuing sequence of nucleotide triplets until a translation stop codon is reached.

Until recently, the genetic code was thought to be universal. It has now been shown that the set of tRNA molecules in mitochondria (which contain their own separate and distinct set of translation machinery) from lower and higher eukaryotes, including humans, reads four codons differently from the tRNA molecules in the cytoplasm of even the same cells. As noted in Table 37–1, the codon AUA is read as Met, and UGA codes for Trp in mammalian mitochondria. In addition, in mitochondria, the codons AGA and AGG are read as stop or chain terminator codons rather than as Arg. As a result of these organelle-specific changes in genetic code, mitochondria require only 22 tRNA molecules to read their genetic code, whereas the cytoplasmic translation system possesses a full complement of 31 tRNA species. These exceptions noted, **the genetic code is universal.** The frequency of use of each amino acid codon varies considerably between species and among different tissues within a species. The specific tRNA levels generally mirror these codon usage biases. Thus, a particular abundantly used codon is decoded by a similarly abundant specific tRNA which recognizes that particular codon. Tables of **codon usage** are becoming more accurate as more genes and genomes are sequenced; such information can prove

TABLE 37–2   Features of the Genetic Code

• Degenerate
• Unambiguous
• Nonoverlapping
• Not punctuated
• Universal

FIGURE 37–1 Formation of aminoacyl-tRNA. A two-step reaction, involving the enzyme amino-acyl-tRNA synthetase, results in the formation of aminoacyl-tRNA. The first reaction involves the formation of an AMP-amino acid-enzyme complex. This activated amino acid is next transferred to the corresponding tRNA molecule. The AMP and enzyme are released, and the latter can be reutilized. The charging reactions have an error rate (ie, esterifying the incorrect amino acid on tRNA$_x$) of less than 10^{-4}.

vital for large scale production of proteins for therapeutic purposes (ie, insulin, erythropoietin). Such proteins are often produced in nonhuman cells using recombinant DNA technology (Chapter 39). The main features of the genetic code are listed in Table 37–2.

AT LEAST ONE SPECIES OF TRANSFER RNA (tRNA) EXISTS FOR EACH OF THE 20 AMINO ACIDS

tRNA molecules have extraordinarily similar functions and three-dimensional structures. The adapter function of the tRNA molecules requires the charging of each specific tRNA with its specific amino acid. Since there is no affinity of nucleic acids for specific functional groups of amino acids, this recognition must be carried out by a protein molecule capable of recognizing both a specific tRNA molecule and a specific amino acid. At least 20 specific enzymes are required for these specific recognition functions and for the proper attachment of the 20 amino acids to specific tRNA molecules. The energy requiring process of **recognition and attachment (charging)** proceeds in two steps and is catalyzed by one enzyme for each of the 20 amino acids. These enzymes are termed **aminoacyl-tRNA synthetases.** They form an activated intermediate of aminoacyl-AMP-enzyme complex (Figure 37–1). The specific aminoacyl-AMP-enzyme complex then recognizes a specific tRNA to which it attaches the aminoacyl moiety at the 3′-hydroxyl adenosyl terminal. The charging reactions have an error rate of less than 10^{-4} and so are quite accurate. The amino acid remains attached to its specific tRNA in an ester linkage until it is polymerized at a specific position in the fabrication of a polypeptide precursor of a protein molecule.

The regions of the tRNA molecule referred to in Chapter 34 (and illustrated in Figure 34–11) now become important. The ribothymidine pseudouridine cytidine (TψC) arm is involved in binding of the aminoacyl-tRNA to the ribosomal

surface at the site of protein synthesis. The D arm is one of the sites important for the proper recognition of a given tRNA species by its proper aminoacyl-tRNA synthetase. The acceptor arm, located at the 3′-hydroxyl adenosyl terminal, is the site of attachment of the specific amino acid.

The anticodon region consists of seven nucleotides, and it recognizes the three-letter codon in mRNA (Figure 37–2). The sequence read from the 3′ to 5′ direction in that anticodon loop consists of a variable base–modified purine–XYZ–pyrimidine–pyrimidine-5′. Note that this direction of reading the anticodon is 3′ to 5′, whereas the genetic code in Table 37–1 is read 5′ to 3′, since the codon and the anticodon

FIGURE 37–2 Recognition of the codon by the anticodon. One of the codons for phenylalanine is UUU. tRNA charged with phenylalanine (Phe) has the complementary sequence AAA; hence, it forms a base-pair complex with the codon. The anticodon region typically consists of a sequence of seven nucleotides: variable (N), modified purine (Pu*), X, Y, Z (here, A A A), and two pyrimidines (Py) in the 3′ to 5′ direction.

loop of the mRNA and tRNA molecules, respectively, are **antiparallel** in their complementarity just like all other intermolecular interactions between nucleic acid strands.

The degeneracy of the genetic code resides mostly in the last nucleotide of the codon triplet, suggesting that the base pairing between this last nucleotide and the corresponding nucleotide of the anticodon is not strictly by the Watson–Crick rule. This is called **wobble;** the pairing of the codon and anticodon can "wobble" at this specific nucleotide-to-nucleotide pairing site. For example, the two codons for arginine, AGA and AGG, can bind to the same anticodon having a uracil at its 5′ end (UCU). Similarly, three codons for glycine—GGU, GGC, and GGA—can form a base pair from one anticodon, 3′ CCI 5′ (ie, I can base pair with U, C and A). I is a purine inosine nucleotide generated by deamination of adenine (see Figure 33–2 for structure), another of the peculiar bases often appearing in tRNA molecules.

MUTATIONS RESULT WHEN CHANGES OCCUR IN THE NUCLEOTIDE SEQUENCE

Although the initial change may not occur in the template strand of the double-stranded DNA molecule for that gene, after replication, daughter DNA molecules with mutations in the template strand will segregate and appear in the population of organisms.

Some Mutations Occur by Base Substitution

Single-base changes (**point mutations**) may be **transitions** or **transversions.** In the former, a given pyrimidine is changed to the other pyrimidine or a given purine is changed to the other purine. Transversions are changes from a purine to either of the two pyrimidines or the change of a pyrimidine into either of the two purines, as shown in Figure 37–3.

If the nucleotide sequence of the gene containing the mutation is transcribed into an RNA molecule, then the RNA molecule will of course possess the base change at the corresponding location.

Single-base changes in the mRNA molecules may have one of several effects when translated into protein:

1. There may be no detectable effect because of the degeneracy of the code; such mutations are often referred to as **si-**

lent mutations.** This would be more likely if the changed base in the mRNA molecule were to be at the third nucleotide of a codon. Because of wobble, the translation of a codon is least sensitive to a change at the third position.

2. A **missense effect** will occur when a different amino acid is incorporated at the corresponding site in the protein molecule. This mistaken amino acid—or missense, depending upon its location in the specific protein—might be acceptable, partially acceptable, or unacceptable to the function of that protein molecule. From a careful examination of the genetic code, one can conclude that most single-base changes would result in the replacement of one amino acid by another with rather similar functional groups. This is an effective mechanism to avoid drastic change in the physical properties of a protein molecule. If an acceptable missense effect occurs, the resulting protein molecule may not be distinguishable from the normal one. A partially acceptable missense will result in a protein molecule with partial but abnormal function. If an unacceptable missense effect occurs, then the protein molecule will not be capable of functioning normally.

3. A **nonsense** codon may appear that would then result in the **premature termination** of amino acid incorporation into a peptide chain and the production of only a fragment of the intended protein molecule. The probability is high that a prematurely terminated protein molecule or peptide fragment will not function in its assigned role.

Hemoglobin Illustrates the Effects of Single-Base Changes in Protein Encoding Genes

Some mutations have no apparent effect. The gene system that encodes hemoglobin is one of the best-studied in humans. The lack of effect of a single-base change is demonstrable only by sequencing the nucleotides in the mRNA molecules or cognate genes. The sequencing of a large number of hemoglobin mRNAs and genes from many individuals has shown that the codon for valine at position 67 of the β chain of hemoglobin is not identical in all persons who possess a normally functional β chain of hemoglobin. Hemoglobin Milwaukee has at position 67 a glutamic acid; hemoglobin Bristol contains aspartic acid at position 67. In order to account for the amino acid change by the change of a single nucleotide residue in the codon for amino acid 67, one must infer that the mRNA encoding hemoglobin Bristol possessed a GUU or GUC codon prior to a later change to GAU or GAC, both codons for aspartic acid. However, the mRNA encoding hemoglobin Milwaukee would have to possess at position 67 a codon GUA or GUG in order that a single nucleotide change could provide for the appearance of the glutamic acid codons GAA or GAG. Hemoglobin Sydney, which contains an alanine at position 67, could have arisen by the change of a single nucleotide in any of the four codons for valine (GUU, GUC, GUA, or GUG) to the alanine codons (GCU, GCC, GCA, or GCG, respectively).

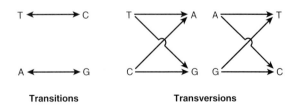

Transitions   **Transversions**

FIGURE 37–3 Diagrammatic representation of transition mutations and transversion mutations.

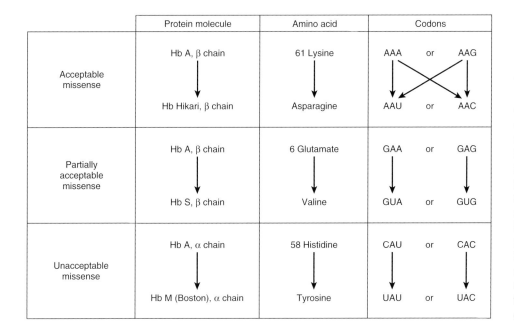

	Protein molecule	Amino acid	Codons			
Acceptable missense	Hb A, β chain → Hb Hikari, β chain	61 Lysine → Asparagine	AAA	or	AAG	
			AAU	or	AAC	
Partially acceptable missense	Hb A, β chain → Hb S, β chain	6 Glutamate → Valine	GAA	or	GAG	
			GUA	or	GUG	
Unacceptable missense	Hb A, α chain → Hb M (Boston), α chain	58 Histidine → Tyrosine	CAU	or	CAC	
			UAU	or	UAC	

FIGURE 37–4 Examples of three types of missense mutations resulting in abnormal hemoglobin chains. The amino acid alterations and possible alterations in the respective codons are indicated. The hemoglobin Hikari β-chain mutation has apparently normal physiologic properties but is electrophoretically altered. Hemoglobin S has a β-chain mutation and partial function; hemoglobin S binds oxygen but precipitates when deoxygenated. Hemoglobin M Boston, an α-chain mutation, permits the oxidation of the heme ferrous iron to the ferric state and so will not bind oxygen at all.

Substitution of Amino Acids Causes Missense Mutations

Acceptable Missense Mutations

An example of an acceptable missense mutation (Figure 37–4, top) in the structural gene for the β chain of hemoglobin could be detected by the presence of an electrophoretically altered hemoglobin in the red cells of an apparently healthy individual. Hemoglobin Hikari has been found in at least two families of Japanese people. This hemoglobin has asparagine substituted for lysine at the 61 position in the β chain. The corresponding transversion might be either AAA or AAG changed to either AAU or AAC. The replacement of the specific lysine with asparagine apparently does not alter the normal function of the β chain in these individuals.

Partially Acceptable Missense Mutations

A partially acceptable missense mutation (Figure 37–4, center) is best exemplified by **hemoglobin S,** which is found in sickle cell anemia. Here glutamic acid, the normal amino acid in position 6 of the β chain, has been replaced by valine. The corresponding single nucleotide change within the codon would be GAA or GAG of glutamic acid to GUA or GUG of valine. Clearly, this missense mutation hinders normal function since it results in sickle cell anemia when the mutant gene is present in the homozygous state. The glutamate-to-valine change may be considered to be partially acceptable because hemoglobin S does bind and release oxygen, albeit abnormally.

Unacceptable Missense Mutations

An unacceptable missense mutation (Figure 37–4, bottom) in a hemoglobin gene generates a nonfunctioning hemoglobin molecule. For example, the hemoglobin M mutations generate molecules that allow the Fe^{2+} of the heme moiety to be oxidized to Fe^{3+}, producing met-hemoglobin. Methemoglobin cannot transport oxygen (see Chapter 6).

Frameshift Mutations Result from Deletion or Insertion of Nucleotides in DNA That Generates Altered mRNAs

The deletion of a single nucleotide from the coding strand of a gene results in an altered reading frame in the mRNA. The machinery translating the mRNA does not recognize that a base was missing, since there is no punctuation in the reading of codons. Thus, a major alteration in the sequence of polymerized amino acids, as depicted in example 1, Figure 37–5, results. Altering the reading frame results in a garbled translation of the mRNA distal to the single nucleotide deletion. Not only is the sequence of amino acids distal to this deletion garbled, but reading of the message can also result in the appearance of a nonsense codon and thus the production of a polypeptide both garbled and prematurely terminated (example 3, Figure 37–5).

If three nucleotides or a multiple of three are deleted from a coding region, the corresponding mRNA when translated will provide a protein from which is missing the corresponding number of amino acids (example 2, Figure 37–5). Because the reading frame is a triplet, the reading phase will not be disturbed for those codons distal to the deletion. If, however, deletion of one or two nucleotides occurs just prior to or within the normal termination codon (nonsense codon), the reading of the normal termination signal is disturbed. Such a deletion might result in reading through the now "mutated" termination signal until another nonsense codon is encountered (example 1, Figure 37–5). Examples of this phenomenon are described in discussions of hemoglobinopathies.

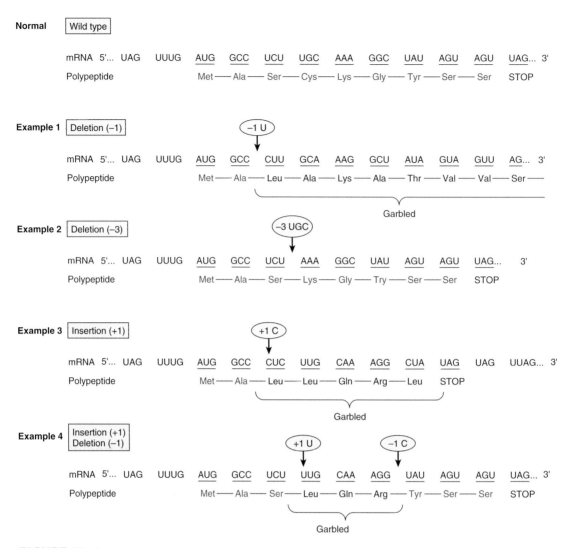

FIGURE 37-5 Examples of the effects of deletions and insertions in a gene on the sequence of the mRNA transcript and of the polypeptide chain translated therefrom. The arrows indicate the sites of deletions or insertions, and the numbers in the ovals indicate the number of nucleotide residues deleted or inserted. Colored type indicates amino acids in correct order.

Insertions of one or two or nonmultiples of three nucleotides into a gene result in an mRNA in which the reading frame is distorted upon translation, and the same effects that occur with deletions are reflected in the mRNA translation. This may result in garbled amino acid sequences distal to the insertion and the generation of a **nonsense codon** at or distal to the insertion, or perhaps reading through the normal termination codon. Following a deletion in a gene, an insertion (or vice versa) can reestablish the proper reading frame (example 4, Figure 37–5). The corresponding mRNA, when translated, would contain a garbled amino acid sequence between the insertion and deletion. Beyond the reestablishment of the reading frame, the amino acid sequence would be correct. One can imagine that different combinations of deletions, of insertions, or of both would result in formation of a protein wherein a portion is abnormal, but this portion is surrounded by the normal amino acid sequences. Such phenomena have been demonstrated convincingly in a number of diseases.

Suppressor Mutations Can Counteract Some of the Effects of Missense, Nonsense, & Frameshift Mutations

The above discussion of the altered protein products of gene mutations is based on the presence of normally functioning tRNA molecules. However, in prokaryotic and lower eukaryotic organisms, abnormally functioning tRNA molecules have been discovered that are themselves the results of mutations. Some of these abnormal tRNA molecules are capable of binding to and decoding altered codons, thereby suppressing the effects of mutations in distinct mutated mRNA encoding structural genes. These **suppressor tRNA molecules,** usually formed as the result of alterations in their anticodon regions, are capable of suppressing certain missense mutations, nonsense mutations, and frameshift mutations. However, since the suppressor tRNA molecules are not capable of distinguishing between a normal codon and one resulting from a gene mutation, their presence in the microbial cell usually results in decreased viability. For instance, the nonsense suppressor tRNA molecules can suppress the normal termination signals to allow a read-through when it is not desirable. Frameshift suppressor tRNA molecules may read a normal codon plus a component of a juxtaposed codon to provide a frameshift, also when it is not desirable. Suppressor tRNA molecules may exist in mammalian cells, since read-through of translation has on occasion been observed.

LIKE TRANSCRIPTION, PROTEIN SYNTHESIS CAN BE DESCRIBED IN THREE PHASES: INITIATION, ELONGATION, & TERMINATION

The general structural characteristics of ribosomes and their self-assembly process are discussed in Chapter 36. These particulate entities serve as the machinery on which the mRNA nucleotide sequence is translated into the sequence of amino acids of the specified protein. The translation of the mRNA commences near its 5′ terminal with the formation of the corresponding amino terminal of the protein molecule. The message is read from 5′ to 3′, concluding with the formation of the carboxyl terminal of the protein. Again, the concept of **polarity** is apparent. As described in Chapter 36, the transcription of a gene into the corresponding mRNA or its precursor first forms the 5′ terminal of the RNA molecule. In prokaryotes, this allows for the beginning of mRNA translation before the transcription of the gene is completed. In eukaryotic organisms, the process of transcription is a nuclear one; mRNA translation occurs in the cytoplasm. This precludes simultaneous transcription and translation in eukaryotic organisms and makes possible the processing necessary to generate mature mRNA from the primary transcript.

Initiation Involves Several Protein–RNA Complexes

Initiation of protein synthesis requires that an mRNA molecule be selected for translation by a ribosome (Figure 37–6). Once the mRNA binds to the ribosome, the latter finds the correct reading frame on the mRNA, and translation begins. This process involves tRNA, rRNA, mRNA, and at least ten eukaryotic initiation factors (eIFs), some of which have multiple (three to eight) subunits. Also involved are GTP, ATP, and amino acids. Initiation can be divided into four steps: (1) dissociation of the ribosome into its 40S and 60S subunits; (2) binding of a ternary complex consisting of met-tRNAⁱ, GTP,

and eIF-2 to the 40S ribosome to form a preinitiation complex; (3) binding of mRNA to the 40S preinitiation complex to form a 43S initiation complex; and (4) combination of the 43S initiation complex with the 60S ribosomal subunit to form the 80S initiation complex.

Ribosomal Dissociation

Two initiation factors, eIF-3 and eIF-1A, bind to the newly dissociated 40S ribosomal subunit. This delays its reassociation with the 60S subunit and allows other translation initiation factors to associate with the 40S subunit.

Formation of the 43S Preinitiation Complex

The first step in this process involves the binding of GTP by eIF-2. This binary complex then binds to met tRNAⁱ, a tRNA specifically involved in binding to the initiation codon AUG. (There are two tRNAs for methionine. One specifies methionine for the initiator codon, the other for internal methionines. Each has a unique nucleotide sequence; both are aminoacylated by the same methionyl tRNA synthetase.) This ternary complex binds to the 40S ribosomal subunit to form the 43S preinitiation complex, which is stabilized by association with eIF-3 and eIF-1A.

eIF-2 is one of two control points for protein synthesis initiation in eukaryotic cells. eIF-2 consists of α, β, and γ subunits. eIF-2α is phosphorylated (on serine 51) by at least four different protein kinases (HCR, PKR, PERK, and GCN2) that are activated when a cell is under stress and when the energy expenditure required for protein synthesis would be deleterious. Such conditions include amino acid and glucose starvation, virus infection, intracellular presence of large quantities of misfolded proteins, serum deprivation, hyperosmolality, and heat shock. PKR is particularly interesting in this regard. This kinase is activated by viruses and provides a host defense mechanism that decreases protein synthesis, including viral protein synthesis, thereby inhibiting viral replication. Phosphorylated eIF-2α binds tightly to and inactivates the GTP-GDP recycling protein eIF-2B. Thus preventing formation of the 43S preinitiation complex and blocking protein synthesis.

Formation of the 43S Initiation Complex

The 5′ terminals of most mRNA molecules in eukaryotic cells are "capped," as described in Chapter 36. This methyl-guanosyl triphosphate cap facilitates the binding of mRNA to the 43S preinitiation complex. A cap binding protein complex, eIF-4F (4F), which consists of eIF-4E (4E) and the eIF-4G (4G)-eIF4A (4A) complex, binds to the cap through the 4E protein. Then eIF-4B (4B) binds and reduces the complex secondary structure of the 5′ end of the mRNA through ATPase and ATP-dependent helicase activities. The association of mRNA with the 43S preinitiation complex to form the 48S initiation complex requires ATP hydrolysis. eIF-3 is a key protein because it binds with high affinity to the 4G component of 4F, and it links this

FIGURE 37–6 Diagrammatic representation of the initiation phase of protein synthesis on an eukaryotic mRNA template containing a 5′ cap (Cap) and 3′ poly(A) terminal [(A)ₙ]. This process proceeds in several steps: (1) activation of mRNA (**right**); (2) formation of the ternary complex consisting of tRNAmetⁱ, initiation factor eIF-2, and GTP (**left**); (3) scanning in the 43S complex to locate the AUG initiator coding, forming the 48S initiation complex (**center**); and (4) formation of the active 80S initiation complex (**bottom, center**). (See text for details.) (GTP, ●; GDP, ○.) The various initiation factors appear in abbreviated form as circles or squares, eg, eIF-3, (③), eIF-4F, (4F), (4F). 4·F is a complex consisting of 4E and 4A bound to 4G (see Figure 37–7). The poly A binding protein, which interacts with the mRNA 3′-poly A tail, is abbreviated PAB. The constellation of protein factors and the 40S ribosomal subunit comprise the 43S preinitiation complex. When bound to mRNA, this forms the 48S preinitiation complex.

complex to the 40S ribosomal subunit. Following association of the 43S preinitiation complex with the mRNA cap, and reduction ("melting") of the secondary structure near the 5′ end of the mRNA through the action of the 4B helicase and ATP, the complex translocates 5′ → 3′ and scans the mRNA for a suitable initiation codon. Generally this is the 5′-most AUG, but the precise initiation codon is determined by so-called **Kozak consensus sequences** that surround the AUG:

$$\overset{\underset{\mid}{-3}\quad\underset{\mid}{-1}\quad\overset{+4}{\mid}}{\text{GCCA/GCC\textbf{AUG}G}}$$

Most preferred is the presence of a purine at positions −3 and +4 relative to the AUG.

Role of the Poly(A) Tail in Initiation

Biochemical and genetic experiments in yeast have revealed that the 3′ poly(A) tail and its binding protein, PAB1, are required for efficient initiation of protein synthesis. Further studies showed that the poly(A) tail stimulates recruitment of the 40S ribosomal subunit to the mRNA through a complex set of interactions. PAB1 (Figure 37–7), bound to the poly(A) tail, interacts with eIF-4G, and 4E subunit of eIF-4F that is bound to the cap. A circular structure is formed that helps direct the 40S ribosomal subunit to the 5′ end of the mRNA and also likely stabilizes mRNAs from exonucleotytic degradation. This helps explain how the cap and poly(A) tail structures have a synergistic effect on protein synthesis. Indeed, differential protein–protein interactions between general and specific mRNA translational repressors and eIF-4E result in m⁷GCap-dependent translation control (Figure 37–8).

Formation of the 80S Initiation Complex

The binding of the 60S ribosomal subunit to the 48S initiation complex involves hydrolysis of the GTP bound to eIF-2 by eIF-5. This reaction results in release of the initiation factors bound to the 48S initiation complex (these factors then are recycled) and the rapid association of the 40S and 60S subunits to form the 80S ribosome. At this point, the met-tRNAⁱ is on the P site of the ribosome, ready for the elongation cycle to commence.

FIGURE 37–7 Schematic illustrating the circularization of mRNA through protein–protein interactions between m⁷G-bound eIF4F and poly A tail-bound PolyA binding protein. eIF4F, composed of eIF4A, 4E, and 4G subunits binds the mRNA 5′-m7G "Cap" (-XpppG⁷ᵐᵉ) upstream of the translation initiation codon (AUG) with high affinity. The eIF4G subunit of the complex also binds Poly A Binding Protein (PAB) with high affinity. Since PAB is bound tightly to the mRNA 3′-poly A tail (OH-AAAAAAA(A)ₙA), circularization results. Shown are multiple 80S ribosomes that are in the process of translating the circularized mRNA into protein (black curlicues), forming a polysome. Upon encountering a termination codon (UAA), translation termination occurs leading to release and dissociation of the 80S ribosome into 60S, 40S subunits and newly translated protein. Dissociated ribosomal subunits can recycle through another round of translation (see Figure 37–6).

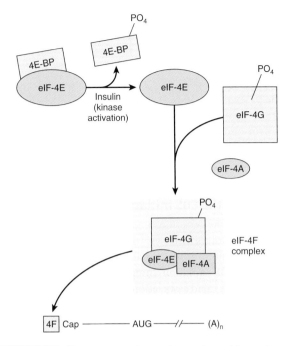

FIGURE 37–8 Activation of eIF-4E by insulin and formation of the cap binding eIF-4F complex. The 4F-cap mRNA complex is depicted as in Figures 37–6 & 37–7. The 4F complex consists of eIF-4E (4E), eIF-4A, and eIF-4G. 4E is inactive when bound by one of a family of binding proteins (4EBPs). Insulin and mitogenic factors (eg, IGF-1, PDGF, inter-leukin-2, and angiotensin II) activate the PI3 kinase/AKT kinase pathways, which activates the mTOR kinase, and results in the phosphorylation of 4E-BP. Phosphorylated 4E-BP dissociates from 4E, and the latter is then able to form the 4F complex and bind to the mRNA cap. These growth peptides also induce phosphorylation of 4G itself by the mTOR and MAP kinase pathways. Phosphorylated 4F binds much more avidly to the cap than does nonphosphorylated 4F.

The Regulation of eIF-4E Controls the Rate of Initiation

The 4F complex is particularly important in controlling the rate of protein translation. As described above, 4F is a complex consisting of 4E, which binds to the m^7G cap structure at the 5' end of the mRNA, and 4G, which serves as a scaffolding protein. In addition to binding 4E, 4G binds to eIF-3, which links the complex to the 40S ribosomal subunit. It also binds 4A and 4B, the ATPase–helicase complex that helps unwind the RNA (Figure 37–8).

4E is responsible for recognition of the mRNA cap structure, a rate-limiting step in translation. This process is further regulated by phosphorylation. Insulin and mitogenic growth factors result in the phosphorylation of 4E on ser 209 (or thr 210). Phosphorylated 4E binds to the cap much more avidly than does the nonphosphorylated form, thus enhancing the rate of initiation. A component of the MAP kinase pathway

(see Figure 42–8) appears to be involved in this phosphorylation reaction.

The activity of 4E is regulated in a second way, and this also involves phosphorylation. A recently discovered set of proteins bind to and inactivate 4E. These proteins include 4E-BP1 (BP1, also known as PHAS-1) and the closely related proteins 4E-BP2 and 4E-BP3. BP1 binds with high affinity to 4E. The [4E]•[BP1] association prevents 4E from binding to 4G (to form 4F). Since this interaction is essential for the binding of 4F to the ribosomal 40S subunit and for correctly positioning this on the capped mRNA, BP-1 effectively inhibits translation initiation.

Insulin and other growth factors result in the phosphorylation of BP-1 at seven unique sites. Phosphorylation of BP-1 results in its dissociation from 4E, and it cannot rebind until critical sites are dephosphorylated. These effects on the activation of 4E explain in part how insulin causes a marked post-transcriptional increase of protein synthesis in liver, adipose tissue, and muscle.

Elongation Is Also a Multistep, Accessory Factor-Facilitated Process

Elongation is a cyclic process on the ribosome in which one amino acid at a time is added to the nascent peptide chain (Figure 37–9). The peptide sequence is determined by the order of the codons in the mRNA. Elongation involves several steps catalyzed by proteins called elongation factors (EFs). These steps are (1) binding of aminoacyl-tRNA to the A site, (2) peptide bond formation, and (3) translocation of the ribosome on the mRNA.

Binding of Aminoacyl-tRNA to the A Site

In the complete 80S ribosome formed during the process of initiation, both the A site (aminoacyl or acceptor site) and E site (deacylated tRNA exit site) are free. The binding of the appropriate aminoacyl-tRNA in the A site requires proper codon recognition. **Elongation factor EF1A** forms a ternary complex with GTP and the entering aminoacyl-tRNA (Figure 37–9). This complex then allows the correct aminoacyl-tRNA to enter the A site with the release of EF1A•GDP and phosphate. GTP hydrolysis is catalyzed by an active site on the ribosome; hydrolysis induces a conformational change in the ribosome concomitantly increasing affinity for the tRNA. As shown in Figure 37–9, EF1A-GDP then recycles to EF1A-GTP with the aid of other soluble protein factors and GTP.

Peptide Bond Formation

The α-amino group of the new aminoacyl-tRNA in the A site carries out a nucleophilic attack on the esterified carboxyl group of the peptidyl-tRNA occupying the P site (peptidyl or polypeptide site). At initiation, this site is occupied by aminoacyl-tRNA met^i. This reaction is catalyzed by a **peptidyltrans-**

TABLE 37–3   Evidence That rRNA Is Peptidyltransferase

- Ribosomes can make peptide bonds even when proteins are removed or inactivated.

- Certain parts of the rRNA sequence are highly conserved in all species.

- These conserved regions are on the surface of the RNA molecule.

- RNA can be catalytic.

- Mutations that result in antibiotic resistance at the level of protein synthesis are more often found in rRNA than in the protein components of the ribosome.

- X-ray crystal structure of large subunit bound to tRNAs suggest detailed mechanism.

ferase, a component of the 28S RNA of the 60S ribosomal subunit. This is another example of ribozyme activity and indicates an important—and previously unsuspected—direct role for RNA in protein synthesis (Table 37–3). Because the amino acid on the aminoacyl-tRNA is already "activated," no further energy source is required for this reaction. The reaction results in attachment of the growing peptide chain to the tRNA in the A site.

Translocation

The now deacylated tRNA is attached by its anticodon to the P site at one end and by the open CCA tail to an **exit (E) site** on the large ribosomal subunit (middle portion of Figure 37–8). At this point, **elongation factor 2 (EF2)** binds to and displaces the peptidyl tRNA from the A site to the P site. In turn, the deacylated tRNA is on the E site, from which it leaves the ribosome. The EF2-GTP complex is hydrolyzed to EF2-GDP, effectively moving the mRNA forward by one codon and leaving the A site open for occupancy by another ternary complex of amino acid tRNA-EF1AGTP and another cycle of elongation.

The charging of the tRNA molecule with the aminoacyl moiety requires the hydrolysis of an ATP to an AMP, equivalent to the hydrolysis of two ATPs to two ADPs and phosphates. The entry of the aminoacyl-tRNA into the A site results in the hydrolysis of one GTP to GDP. Translocation of the newly formed peptidyl-tRNA in the A site into the P site by EF2 similarly results in hydrolysis of GTP to GDP and phosphate. Thus, the energy requirements for the formation of one peptide bond include the equivalent of the hydrolysis of two ATP molecules to ADP and of two GTP molecules to

FIGURE 37–9 Diagrammatic representation of the peptide elongation process of protein synthesis. The small circles labeled n − 1, n, n + 1, etc, represent the amino acid residues of the newly formed protein molecule and corresponding codons in the mRNA. EF1A and EF2 represent elongation factors 1 and 2, respectively. The peptidyl-tRNA, aminoacyl-tRNA, and Exit sites on the ribosome are represented by P site, A site, and E site, respectively.

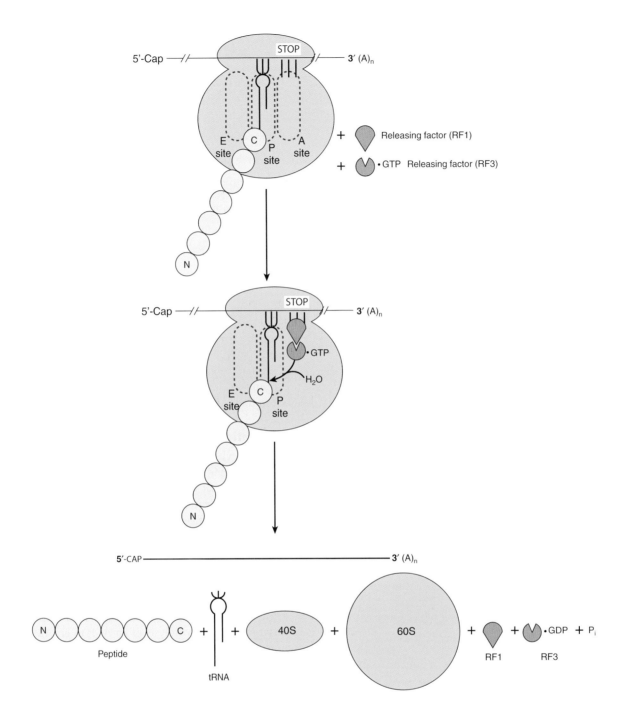

FIGURE 37–10 Diagrammatic representation of the termination process of protein synthesis. The peptidyl-tRNA, aminoacyl-tRNA and exit sites are indicated as P site, A site, and E site, respectively. The termination (stop) codon is indicated by the three vertical bars and stop. Releasing factor RF1 binds to the stop codon. Releasing factor RF3, with bound GTP, binds to RF1. Hydrolysis of the peptidyl-tRNA complex is shown by the entry of H_2O. N and C indicate the amino and carboxyl terminal amino acids of the nascent polypeptide chain, respectively, and illustrate the polarity of protein synthesis.

GDP, or the hydrolysis of four high-energy phosphate bonds. A eukaryotic ribosome can incorporate as many as six amino acids per second; prokaryotic ribosomes incorporate as many as 18 per second. Thus, the energy requiring process of peptide synthesis occurs with great speed and accuracy until a termination codon is reached.

Termination Occurs When a Stop Codon Is Recognized

In comparison to initiation and elongation, termination is a relatively simple process (Figure 37–10). After multiple cycles of elongation culminating in polymerization of the specific amino acids into a protein molecule, the stop or terminating codon of mRNA (UAA, UAG, UGA) appears in the A site. Normally, there is no tRNA with an anticodon capable of recognizing such a termination signal. **Releasing factor RF1** recognizes that a stop codon resides in the A site (Figure 37–10). RF1 is bound by a complex consisting of **releasing factor RF3** with bound GTP. This complex, with the peptidyl transferase, promotes hydrolysis of the bond between the peptide and the tRNA occupying the P site. Thus, a water molecule rather than an amino acid is added. This hydrolysis releases the protein and the tRNA from the P site. Upon hydrolysis and release, the **80S ribosome dissociates** into its 40S and 60S subunits, which are then recycled (Figure 37–7). Therefore, the releasing factors are proteins that hydrolyze the peptidyl-tRNA bond when a stop codon occupies the A site. The mRNA is then released from the ribosome, which dissociates into its component 40S and 60S subunits, and another cycle can be repeated.

Polysomes Are Assemblies of Ribosomes

Many ribosomes can translate the same mRNA molecule simultaneously. Because of their relatively large size, the ribosome particles cannot attach to an mRNA any closer than 35 nucleotides apart. Multiple ribosomes on the same mRNA molecule form a **polyribosome,** or "polysome" (Figure 37–7) In an unrestricted system, the number of ribosomes attached to an mRNA (and thus the size of polyribosomes) correlates positively with the length of the mRNA molecule.

Polyribosomes actively synthesizing proteins can exist as free particles in the cellular cytoplasm or may be attached to sheets of membranous cytoplasmic material referred to as **endoplasmic reticulum.** Attachment of the particulate polyribosomes to the endoplasmic reticulum is responsible for its "rough" appearance as seen by electron microscopy. The proteins synthesized by the attached polyribosomes are extruded into the cisternal space between the sheets of rough endoplasmic reticulum and are exported from there. Some of the protein products of the rough endoplasmic reticulum are packaged by the Golgi apparatus into zymogen particles for eventual export (see Chapter 46) The polyribosomal particles free in the cytosol are responsible for the synthesis of proteins required for intracellular functions.

Nontranslating mRNAs Can Form Ribonucleoprotein Particles That Accumulate in Cytoplasmic Organelles Termed P Bodies

mRNAs, bound by specific packaging proteins and exported from the nucleus as ribonucleoproteins particles (RNPs) sometimes do not immediately associate with ribosomes to be translated. Instead, specific mRNAs can associate with the protein constituents that form P bodies, small dense compartments that incorporate mRNAs as mRNPs (Figure 37–11).

These cytoplasmic organelles are related to similar small mRNA-containing granules found in neurons and certain maternal cells. P bodies are sites of translation repression and mRNA decay. Over 35 distinct proteins have been suggested to reside exclusively or extensively within P bodies. These proteins range from mRNA decapping enzymes, RNA helicases and RNA exonucleases (5′-3′ and 3′-5′), to components involved in miRNA function and mRNA quality control. However, incorporation of an mRNP is not an unequivocal mRNA "death sentence." Indeed, though the mechanisms are not yet fully understood, certain mRNAs appear to be temporarily stored in P bodies and then retrieved and utilized for protein translation. This suggests that an equilibrium exists where the cytoplasmic functions of mRNA (translation and degradation) are controlled by the dynamic interaction of mRNA with polysomes and P bodies.

The Machinery of Protein Synthesis Can Respond to Environmental Threats

Ferritin, an iron-binding protein, prevents ionized iron (Fe^{2+}) from reaching toxic levels within cells. Elemental iron stimulates ferritin synthesis by causing the release of a cytoplasmic protein that binds to a specific region in the 5′ nontranslated region of ferritin mRNA. Disruption of this protein–mRNA interaction activates ferritin mRNA and results in its translation. This mechanism provides for rapid control of the synthesis of a protein that sequesters Fe^{2+}, a potentially toxic molecule.

Many Viruses Co-Opt the Host Cell Protein Synthesis Machinery

The protein synthesis machinery can also be modified in deleterious ways. **Viruses replicate by using host cell processes,** including those involved in protein synthesis. Some viral mRNAs are translated much more efficiently than those of the host cell (eg, encephalomyocarditis virus). Others, such as reovirus and vesicular stomatitis virus, replicate abundantly, and their mRNAs have a competitive advantage over host cell mRNAs for limited translation factors. Other viruses inhibit host cell protein synthesis by preventing the association of mRNA with the 40S ribosome.

Poliovirus and other picornaviruses gain a selective advantage by disrupting the function of the 4F complex. The mRNAs of these viruses do not have a cap structure to direct the binding of the 40S ribosomal subunit (see above). Instead, the 40S ribosomal subunit contacts an **internal ribosomal entry site (IRES)** in a reaction that requires 4G but not 4E. The virus gains a selective advantage by having a protease that attacks 4G and removes the amino terminal 4E binding site. Now the 4E-4G complex (4F) cannot form, so the 40S ribosomal subunit cannot be directed to capped mRNAs. Host cell translation is thus abolished. The 4G fragment can direct binding of the 40S ribosomal subunit to IRES-containing

FIGURE 37–11 The P body is a cytoplasmic organelle that modulates mRNA metabolism. Shown is a photomicrograph of two mammalian cells in which a single distinct protein constituent of the P body has been visualized using the cognate specific fluorescently labeled antibody. P bodies appear as light circles of varying size throughout the cytoplasm. Cell membranes indicated by a solid white line, nuclei by a dashed line. Nuclei were counterstained using a fluorescent dye with different fluorescence excitation/emission spectra from the labeled antibody used to identify P bodies; the nuclear stain intercalates between the DNA base pairs. Modified from http://www.mcb.arizona.edu/parker/WHAT/what.htm. (Used with permission of Dr Roy Parker.)

mRNAs, so viral mRNA translation is very efficient (Figure 37–12). These viruses also promote the dephosphorylation of BP1 (PHAS-1), thereby decreasing cap (4E)-dependent translation.

POSTTRANSLATIONAL PROCESSING AFFECTS THE ACTIVITY OF MANY PROTEINS

Some animal viruses, notably HIV, poliovirus, and hepatitis A virus, synthesize long polycistronic proteins from one long mRNA molecule. The protein molecules translated from these long mRNAs are subsequently cleaved at specific sites to provide the several specific proteins required for viral function. In animal cells, many cellular proteins are synthesized from the mRNA template as a precursor molecule, which then must be modified to achieve the active protein. The prototype is insulin, which is a small protein having two polypeptide chains with interchain and intrachain disulfide bridges. The

molecule is synthesized as a single chain precursor, or **prohormone,** which folds to allow the disulfide bridges to form. A specific protease then clips out the segment that connects the two chains which form the functional insulin molecule (see Figure 41–12).

Many other peptides are synthesized as proproteins that require modifications before attaining biologic activity. Many of the posttranslational modifications involve the removal of amino terminal amino acid residues by specific aminopeptidases. Collagen, an abundant protein in the extracellular spaces of higher eukaryotes, is synthesized as procollagen. Three procollagen polypeptide molecules, frequently not identical in sequence, align themselves in a particular way that is dependent upon the existence of specific amino terminal peptides. Specific enzymes then carry out hydroxylations and oxidations of specific amino acid residues within the procollagen molecules to provide cross-links for greater stability. Amino terminal peptides are cleaved off the molecule to form the final product—a strong, insoluble collagen molecule. Many other posttranslational modifications of proteins oc-

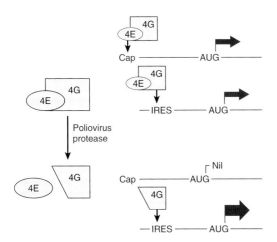

FIGURE 37–12 Picornaviruses disrupt the 4F complex. The 4E-4G complex (4F) directs the 40S ribosomal subunit to the typical capped mRNA (see text). 4G alone is sufficient for targeting the 40S subunit to the internal ribosomal entry site (IRES) of viral mRNAs. To gain selective advantage, certain viruses (eg, poliovirus) express a protease that cleaves the 4E binding site from the amino terminal end of 4G. This truncated 4G can direct the 40S ribosomal subunit to mRNAs that have an IRES but not to those that have a cap. The widths of the arrows indicate the rate of translation initiation from the AUG codon in each example.

FIGURE 37–13 The comparative structures of the antibiotic puromycin (top) and the 3′ terminal portion of tyrosinyl-tRNA (bottom).

cur. Covalent modification by acetylation, phosphorylation, methylation, ubiquitylation, and glycosylation is common, for example.

MANY ANTIBIOTICS WORK BY SELECTIVELY INHIBITING PROTEIN SYNTHESIS IN BACTERIA

Ribosomes in bacteria and in the mitochondria of higher eukaryotic cells differ from the mammalian ribosome described in Chapter 34. The bacterial ribosome is smaller (70S rather than 80S) and has a different, somewhat simpler complement of RNA and protein molecules. This difference is exploited for clinical purposes because many effective antibiotics interact specifically with the proteins and RNAs of prokaryotic ribosomes and thus inhibit protein synthesis. This results in growth arrest or death of the bacterium. The most useful members of this class of antibiotics (eg, tetracyclines, lincomycin, erythromycin, and chloramphenicol) do not interact with components of eukaryotic ribosomes and thus are not toxic to eukaryotes. Tetracycline prevents the binding of aminoacyl-tRNAs to the bacterial ribosome A site. Chloramphenicol and the macrolide class of antibiotics work by binding to 23S rRNA, which is interesting in view of the newly appreciated role of rRNA in peptide bond formation through its peptidyltransferase activity. It should be mentioned that the close similarity between prokaryotic and mitochondrial ribosomes can lead to complications in the use of some antibiotics.

Other antibiotics inhibit protein synthesis on all ribosomes (**puromycin**) or only on those of eukaryotic cells (**cycloheximide**). Puromycin (Figure 37–13) is a structural analog of tyrosinyl-tRNA. Puromycin is incorporated via the A site on the ribosome into the carboxyl terminal position of a peptide but causes the premature release of the polypeptide. Puromycin, as a tyrosinyl-tRNA analog, effectively inhibits protein synthesis in both prokaryotes and eukaryotes. Cycloheximide inhibits peptidyltransferase in the 60S ribosomal subunit in eukaryotes, presumably by binding to an rRNA component.

Diphtheria toxin, an exotoxin of *Corynebacterium diphtheriae* infected with a specific lysogenic phage, catalyzes the ADP-ribosylation of EF-2 on the unique amino acid diphthamide in mammalian cells. This modification inactivates EF-2 and thereby specifically inhibits mammalian protein synthesis. Many animals (eg, mice) are resistant to diphtheria toxin. This resistance is due to inability of diphtheria toxin to cross the cell membrane rather than to insensitivity of mouse EF-2 to diphtheria toxin-catalyzed ADP-ribosylation by NAD.

Ricin, an extremely toxic molecule isolated from the castor bean, inactivates eukaryotic 28S ribosomal RNA by providing the N-glycolytic cleavage or removal of a single adenine.

Many of these compounds—puromycin and cycloheximide in particular—are not clinically useful but have been important in elucidating the role of protein synthesis in the regulation of metabolic processes, particularly enzyme induction by hormones.

SUMMARY

- The flow of genetic information follows the sequence DNA → RNA → protein.
- The genetic information in the structural region of a gene is transcribed into an RNA molecule such that the sequence of the latter is complementary to that in the DNA.
- Ribosomal RNA (rRNA), transfer RNA (tRNA), and messenger RNA (mRNA), are directly involved in protein synthesis; miRNAs regulate mRNA function at the level of translation and/or stability.
- The information in mRNA is in a tandem array of codons, each of which is three nucleotides long.
- The mRNA is read continuously from a start codon (AUG) to a termination codon (UAA, UAG, UGA).
- The open reading frame, or ORF, of the mRNA is the series of codons, each specifying a certain amino acid, that determines the precise amino acid sequence of the protein.
- Protein synthesis, like DNA and RNA synthesis, follows the 5′ to 3′ polarity of mRNA and can be divided into three processes: initiation, elongation, and termination. Mutant proteins arise when single-base substitutions result in codons that specify a different amino acid at a given position, when a stop codon results in a truncated protein, or when base additions or deletions alter the reading frame, so different codons are read.
- A variety of compounds, including several antibiotics, inhibit protein synthesis by affecting one or more of the steps involved in protein synthesis.

REFERENCES

Beckham CJ, Parker R: P bodies, stress granules, and viral life cycles. Cell Host Microbe 2008;3:206.

Crick F et al: The genetic code. Nature 1961;192:1227.

Kapp LD, Lorsch JR: The molecular mechanics of eukaryotic translation. Ann Rev Biochem 2004;73:657.

Kozak M: Structural features in eukaryotic mRNAs that modulate the initiation of translation. J Biol Chem 1991;266:1986.

Parker R, Sheth U: P bodies and the control of mRNA translation and degradation. Mol Cell 2007;25:635.

Sonenberg N, Hinnebusch AG: New modes of translational control in development, behavior, and disease. Mol Cell 2007;28:721.

Steitz TA, Moore PB: RNA, the first macromolecular catalyst: the ribosome is a ribozyme. Trends Biochem Sci 2003;28:411.

Weatherall DJ: Phenotype-genotype relationships in mogenic disease: Lessons from the thalassaemias. Nature Reviews Genetics 2001;2:245.

Regulation of Gene Expression

P. Anthony Weil, PhD

BIOMEDICAL IMPORTANCE

Organisms adapt to environmental changes by altering gene expression. The process of alteration of gene expression has been studied in detail and often involves modulation of gene transcription. Control of transcription ultimately results from changes in the interaction of specific binding regulatory proteins with various regions of DNA in the controlled gene. This can have a positive or negative effect on transcription. Transcription control can result in tissue-specific gene expression, and gene regulation is influenced by hormones, heavy metals, and chemicals. In addition to transcription level controls, gene expression can also be modulated by gene amplification, gene rearrangement, posttranscriptional modifications, RNA stabilization, translational control, protein modification, and protein stabilization. Many of the mechanisms that control gene expression are used to respond to hormones and therapeutic drugs. Thus, a molecular understanding of these processes will lead to development of agents that alter pathophysiologic mechanisms or inhibit the function or arrest the growth of pathogenic organisms.

REGULATED EXPRESSION OF GENES IS REQUIRED FOR DEVELOPMENT, DIFFERENTIATION, & ADAPTATION

The genetic information present in each normal somatic cell of a metazoan organism is practically identical. The exceptions are found in those few cells that have amplified or rearranged genes in order to perform specialized cellular functions or cells that have undergone oncogenic transformation. Expression of the genetic information must be regulated during ontogeny and differentiation of the organism and its cellular components. Furthermore, in order for the organism to adapt to its environment and to conserve energy and nutrients, the expression of genetic information must be cued to extrinsic signals and respond only when necessary. As organisms have evolved, more sophisticated regulatory mechanisms have appeared which provide the organism and its cells with the responsiveness necessary for survival in a complex environment. Mammalian cells possess about 1000 times more genetic information than does the bacterium *Escherichia coli*. Much of this additional genetic information is probably involved in regulation of gene expression during the differentiation of tissues and biologic processes in the multicellular organism and in ensuring that the organism can respond to complex environmental challenges.

In simple terms, there are only two types of gene regulation: **positive regulation** and **negative regulation** (Table 38–1). When the expression of genetic information is quantitatively increased by the presence of a specific regulatory element, regulation is said to be positive; when the expression of genetic information is diminished by the presence of a specific regulatory element, regulation is said to be negative. The element or molecule mediating negative regulation is said to be a negative regulator or **repressor;** that mediating positive regulation is a positive regulator or **activator.** However, a **double negative** has the effect of acting as a positive. Thus, an effector that inhibits the function of a negative regulator will appear to bring about a positive regulation. Many regulated systems that appear to be induced are in fact **derepressed** at the molecular level. (See Chapter 9 for explanation of these terms.)

BIOLOGIC SYSTEMS EXHIBIT THREE TYPES OF TEMPORAL RESPONSES TO A REGULATORY SIGNAL

Figure 38–1 depicts the extent or amount of gene expression in three types of temporal response to an inducing signal. A **type A response** is characterized by an increased extent of gene expression that is dependent upon the continued presence of the inducing signal. When the inducing signal is removed, the amount of gene expression diminishes to its basal level, but the amount repeatedly increases in response to the reappearance of the specific signal. This type of response is commonly observed in prokaryotes in response to sudden changes of the intracellular concentration of a nutrient. It is also observed in many higher organisms after exposure to inducers such as hormones, nutrients, or growth factors (Chapter 42).

A **type B response** exhibits an increased amount of gene expression that is transient even in the continued presence of the regulatory signal. After the regulatory signal has terminated

TABLE 38–1   Effects of Positive and Negative Regulation on Gene Expression

	Rate of Gene Expression	
	Negative Regulation	**Positive Regulation**
Regulator present	Decreased	Increased
Regulator absent	Increased	Decreased

and the cell has been allowed to recover, a second transient response to a subsequent regulatory signal may be observed. This phenomenon of response-desensitization-recovery characterizes the action of many pharmacologic agents, but it is also a feature of many naturally occurring processes. This type of response commonly occurs during development of an organism, when only the transient appearance of a specific gene product is required although the signal persists.

The **type C response** pattern exhibits, in response to the regulatory signal, an increased extent of gene expression that

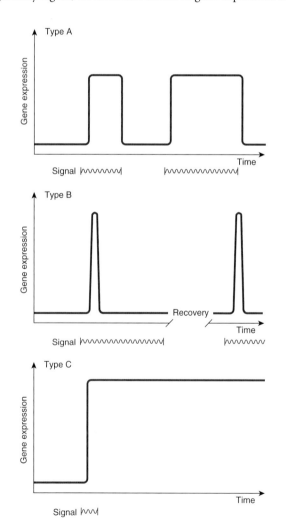

FIGURE 38–1 Diagrammatic representations of the responses of the extent of expression of a gene to specific regulatory signals (such as a hormone as a function of time).

persists indefinitely even after termination of the signal. The signal acts as a trigger in this pattern. Once expression of the gene is initiated in the cell, it cannot be terminated even in the daughter cells; it is therefore an irreversible and inherited alteration. This type of response typically occurs during the development of differentiated function in a tissue or organ.

Prokaryotes Provide Models for the Study of Gene Expression in Mammalian Cells

Analysis of the regulation of gene expression in prokaryotic cells helped establish the principle that information flows from the gene to a messenger RNA to a specific protein molecule. These studies were aided by the advanced genetic analyses that could be performed in prokaryotic and lower eukaryotic organisms. In recent years, the principles established in these early studies, coupled with a variety of molecular biology techniques, have led to remarkable progress in the analysis of gene regulation in higher eukaryotic organisms, including mammals. In this chapter, the initial discussion will center on prokaryotic systems. The impressive genetic studies will not be described, but the physiology of gene expression will be discussed. However, nearly all of the conclusions about this physiology have been derived from genetic studies and confirmed by molecular genetic and biochemical experiments.

Some Features of Prokaryotic Gene Expression Are Unique

Before the physiology of gene expression can be explained, a few specialized genetic and regulatory terms must be defined for prokaryotic systems. In prokaryotes, the genes involved in a metabolic pathway are often present in a linear array called an **operon,** eg, the *lac* operon. An operon can be regulated by a single promoter or regulatory region. The **cistron** is the smallest unit of genetic expression. As described in Chapter 9, some enzymes and other protein molecules are composed of two or more nonidentical subunits. Thus, the "one gene, one enzyme" concept is not necessarily valid. The cistron is the genetic unit coding for the structure of the subunit of a protein molecule, acting as it does as the smallest unit of genetic expression. Thus, the one gene, one enzyme idea might more accurately be regarded as a **one cistron, one subunit** concept. A single mRNA that encodes more than one separately translated protein is referred to as a **polycistronic mRNA.** For example, the polycistronic *lac* operon mRNA is translated into three separate proteins (see below). Operons and polycistronic mRNAs are common in bacteria but not in eukaryotes.

An **inducible gene** is one whose expression increases in response to an **inducer** or **activator,** a specific positive regulatory signal. In general, inducible genes have relatively low basal rates of transcription. By contrast, genes with high basal rates of transcription are often subject to down-regulation by repressors.

The expression of some genes is **constitutive,** meaning that they are expressed at a reasonably constant rate and not known to be subject to regulation. These are often referred to as **housekeeping genes.** As a result of mutation, some inducible gene products become constitutively expressed. A mutation resulting in constitutive expression of what was formerly a regulated gene is called a **constitutive mutation.**

Analysis of Lactose Metabolism in *E coli* Led to the Operon Hypothesis

Jacob and Monod in 1961 described their **operon model** in a classic paper. Their hypothesis was to a large extent based on observations on the regulation of lactose metabolism by the intestinal bacterium *E coli*. The molecular mechanisms responsible for the regulation of the genes involved in the metabolism of lactose are now among the best-understood in any organism. β-Galactosidase hydrolyzes the β-galactoside lactose to galactose and glucose. The structural gene for β-galactosidase *(lacZ)* is clustered with the genes responsible for the permeation of lactose into the cell *(lacY)* and for thiogalactoside transacetylase *(lacA)*. The structural genes for these three enzymes, along with the *lac* promoter and *lac* operator (a regulatory region), are physically associated to constitute the ***lac* operon** as depicted in Figure 38–2. This genetic arrangement of the structural genes and their regulatory genes allows for **coordinate expression** of the three enzymes concerned with lactose metabolism. Each of these linked genes is transcribed into one large polycistronic mRNA molecule that contains multiple independent translation start (AUG) and stop (UAA) codons for each of the three cistrons. Thus, each protein is translated separately, and they are not processed from a single large precursor protein.

It is now conventional to consider that a gene includes regulatory sequences as well as the region that encodes the primary transcript. Although there are many historical exceptions, a gene is generally italicized in lower case and the encoded protein, when abbreviated, is expressed in roman type with the first letter capitalized. For example, the gene *lacI* encodes the repressor protein LacI. When *E coli* is presented with lactose or some specific lactose analogs under appropriate nonrepressing conditions (eg, high concentrations of lactose, no or very low glucose in media; see below), the expression of the activities of β-galactosidase, galactoside permease, and thiogalactoside transacetylase is increased 100-fold to 1000-fold. This is a type A response, as depicted in Figure 38–1. The kinetics of induction can be quite rapid; *lac*-specific mRNAs are fully induced within 5–6 min after addition of lactose to a culture; β-galactosidase protein is maximal within 10 min. Under fully induced conditions, there can be up to 5000 β-galactosidase molecules per cell, an amount about 1000 times greater than the basal, uninduced level. Upon removal of the signal, ie, the inducer, the synthesis of these three enzymes declines.

When *E coli* is exposed to both lactose and glucose as sources of carbon, the organisms first metabolize the glucose and then temporarily stop growing until the genes of the *lac* operon become induced to provide the ability to metabolize lactose as a usable energy source. Although lactose is present

FIGURE 38–2 The positional relationships of the structural and regulatory genes of the *lac* operon. *lacZ* encodes β-galactosidase, *lacY* encodes a permease, and *lacA* encodes a thiogalactoside transacetylase. *lacI* encodes the *lac* operon repressor protein.

from the beginning of the bacterial growth phase, the cell does not induce those enzymes necessary for catabolism of lactose until the glucose has been exhausted. This phenomenon was first thought to be attributable to repression of the *lac* operon by some catabolite of glucose; hence, it was termed catabolite repression. It is now known that catabolite repression is in fact mediated by a **catabolite gene activator protein (CAP)** in conjunction with **cAMP** (Figure 17–5). This protein is also referred to as the cAMP regulatory protein (CRP). The expression of many inducible enzyme systems or operons in *E coli* and other prokaryotes is sensitive to catabolite repression, as discussed below.

The physiology of induction of the *lac* operon is well understood at the molecular level (Figure 38–3). Expression of the normal *lacI* gene of the *lac* operon is constitutive; it is expressed at a constant rate, resulting in formation of the subunits of the **lac repressor.** Four identical subunits with molecular weights of 38,000 assemble into a tetrameric Lac repressor molecule. The LacI repressor protein molecule, the product of *lacI*, has a high affinity (K_d about 10^{-13} mol/L) for the operator locus. The **operator locus** is a region of double-stranded DNA 27 base pairs long with a 2-fold rotational symmetry and an inverted palindrome (indicated by arrows about the dotted axis) in a region that is 21 base pairs long, as shown below:

At any one time, only two of the four subunits of the repressor appear to bind to the operator, and within the 21-base-pair region nearly every base of each base pair is involved in LacI recognition and binding. The binding occurs mostly in the **major groove** without interrupting the base-paired, double-helical nature of the operator DNA. The **operator locus** is between the **promoter site,** at which the DNA-dependent RNA polymerase attaches to commence transcription, and the transcription initiation site of the ***lacZ* gene,** the structural gene for β-galactosidase (Figure 38–2). When attached to the operator locus, the LacI repressor molecule prevents transcription of the distal structural genes, *lacZ, lacY,* and *lacA* by interfering with the binding of RNA polymerase to the promoter; RNA polymerase and LacI repressor cannot be effectively bound to the *lac* operon at the same time. Thus, the LacI repressor molecule is a **negative regulator;** in its pres-

FIGURE 38–3 The mechanism of repression and derepression of the *lac* operon. When either no inducer is present or an inducer is present with glucose **(A),** the constitutively synthesized *lacI* gene products form a repressor tetramer molecule that binds at the operator locus to prevent the initiation of transcription by blocking the efficient binding of RNA polymerase at the promoter locus thus preventing the subsequent transcription of the *lacZ, lacY,* and *lacA* structural genes into a polycistonic mRNA. When inducer is present **(B),** the tetrameric repressor molecules are conformationally altered by inducer and cannot efficiently bind to the operator locus (affinity of binding reduced >1000-fold). In the presence of cAMP and its binding protein (CAP), the RNA polymerase will efficiently transcribe the structural genes *lacZ, lacY,* and *lacA,* and the polycistronic mRNA molecule formed can be translated into the corresponding protein molecules β-galactosidase, permease, and transacetylase, allowing for the catabolism of lactose.

ence (and in the absence of inducer; see below), expression from the *lacZ, lacY,* and *lacA* genes is very, very low. There are normally 20–40 repressor tetramer molecules in the cell, a concentration of tetramer sufficient to effect, at any given time, >95% occupancy of the one *lac* operator element in a bacterium, thus ensuring low (but not zero) basal *lac* operon gene transcription in the absence of inducing signals.

A lactose analog that is capable of inducing the *lac* operon while not itself serving as a substrate for β-galactosidase is an example of a **gratuitous inducer.** An example is isopropylthiogalactoside (IPTG). The addition of lactose or of a gratuitous inducer such as IPTG to bacteria growing on a poorly utilized carbon source (such as succinate) results in prompt induction of the *lac* operon enzymes. Small amounts of the gratuitous inducer or of lactose are able to enter the cell even in the absence of permease. The LacI repressor molecules—both those attached to the operator loci and those free in the cytosol—have a high affinity for the inducer. Binding of the inducer to

repressor molecule induces a conformational change in the structure of the repressor and causes it to dissociate from operator DNA because its affinity for the operator is now 10^3 times lower (K_d about 10^{-9} mol/L) than that of LacI in the absence of IPTG. DNA-dependent RNA polymerase can now bind to the coding strand at the promoter site, and transcription will begin, although this process is relatively inefficient (see below). In such a manner, **an inducer derepresses the *lac* operon** and allows transcription of the structural genes for β-galactosidase, galactoside permease, and thiogalactoside transacetylase. Translation of the polycistronic mRNA can occur even before transcription is completed. Derepression of the *lac* operon allows the cell to synthesize the enzymes necessary to catabolize lactose as an energy source. Based on the physiology just described, IPTG-induced expression of transfected plasmids bearing the *lac* operator-promoter ligated to appropriate bioengineered constructs is commonly used to express mammalian recombinant proteins in *E coli*.

In order for the RNA polymerase to form a PIC at the promoter site most efficiently, there must also be present the **catabolite gene activator protein (CAP)** to which cAMP is bound. By an independent mechanism, the bacterium accumulates cAMP only when it is starved for a source of carbon. In the presence of glucose—or of glycerol in concentrations sufficient for growth—the bacteria will lack sufficient cAMP to bind to CAP because the glucose inhibits adenylyl cyclase, the enzyme that converts ATP to cAMP (see Chapter 41). Thus, in the presence of glucose or glycerol, cAMP-saturated CAP is lacking, so that the DNA-dependent RNA polymerase cannot initiate transcription of the *lac* operon at the maximal rate. However, in the presence of the CAP-cAMP complex, which binds to DNA just upstream of the promoter site, transcription occurs at maximal levels (Figure 38–3). Studies indicate that a region of CAP contacts the RNA polymerase α subunit and facilitates binding of this enzyme to the promoter. Thus, the CAP-cAMP regulator is acting as a **positive regulator** because its presence is required for optimal gene expression. The *lac* operon is therefore controlled by two distinct, ligand-modulated DNA binding *trans* factors; one that acts positively (cAMP-CRP complex) to facilitate productive binding of RNA polymerase to the promoter and one that acts negatively (LacI repressor) that antagonizes RNA polymerase promoter binding. Maximal activity of the *lac* operon occurs when glucose levels are low (high cAMP with CAP activation) and lactose is present (LacI is prevented from binding to the operator).

When the *lacI* gene has been mutated so that its product, LacI, is not capable of binding to operator DNA, the organism will exhibit **constitutive expression** of the *lac* operon. In a contrary manner, an organism with a *lacI* gene mutation that produces a LacI protein which prevents the binding of an inducer to the repressor will remain repressed even in the presence of the inducer molecule, because the inducer cannot bind to the repressor on the operator locus in order to derepress the operon. Similarly, bacteria harboring mutations in their *lac* operator locus such that the operator sequence will not bind a normal repressor molecule constitutively express the *lac* operon genes. Mechanisms of positive and negative regulation comparable to those described here for the *lac* system have been observed in eukaryotic cells (see below).

The Genetic Switch of Bacteriophage Lambda (λ) Provides a Paradigm for Protein-DNA Interactions in Eukaryotic Cells

Like some eukaryotic viruses (eg, herpes simplex, HIV), some bacterial viruses can either reside in a dormant state within the host chromosomes or can replicate within the bacterium and eventually lead to lysis and killing of the bacterial host. Some *E coli* harbor such a "temperate" virus, bacteriophage lambda (λ). When lambda infects an organism of that species it injects its 45,000-bp, double-stranded, linear DNA genome into the cell (Figure 38–4). Depending upon the nutritional state of the cell,

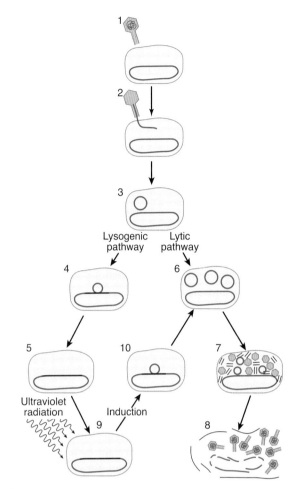

FIGURE 38–4 Infection of the bacterium *E coli* by phage lambda begins when a virus particle attaches itself to specific receptors on the bacterial cell (**1**) and injects its DNA (**dark green line**) into the cell (**2, 3**). Infection can take either of two courses depending on which of two sets of viral genes is turned on. In the lysogenic pathway, the viral DNA becomes integrated into the bacterial chromosome (**red**) (**4, 5**), where it replicates passively as the bacterial DNA and cell divides. This dormant genomically integrated virus is called a prophage, and the cell that harbors it is called a lysogen. In the alternative lytic mode of infection, the viral DNA replicates itself (**6**) and directs the synthesis of viral proteins (**7**). About 100 new virus particles are formed. The proliferating viruses induce lysis of the cell (**8**). A prophage can be "induced" by a DNA damaging agent such as ultraviolet radiation (**9**). The inducing agent throws a switch, so that a different set of genes is turned on. Viral DNA loops out of the chromosome (**10**) and replicates; the virus proceeds along the lytic pathway. (Reproduced, with permission, from Ptashne M, Johnson AD, Pabo CO: A genetic switch in a bacterial virus. Sci Am [Nov] 1982;247:128.)

the lambda DNA will either **integrate** into the host genome (**lysogenic pathway**) and remain dormant until activated (see below), or it will commence **replicating** until it has made about 100 copies of complete, protein-packaged virus, at which point it causes lysis of its host (**lytic pathway**). The newly generated virus particles can then infect other susceptible hosts. Poor growth conditions favor lysogeny while good growth conditions promote the lytic pathway of lambda growth.

When integrated into the host genome in its dormant state, lambda will remain in that state until activated by exposure of its bacterial host to DNA-damaging agents. In response to such a noxious stimulus, the dormant bacteriophage becomes "induced" and begins to transcribe and subsequently translate those genes of its own genome that are necessary for its excision from the host chromosome, its DNA replication, and the synthesis of its protein coat and lysis enzymes. This event acts like a trigger or type C (Figure 38–1) response; ie, once dormant lambda has committed itself to induction, there is no turning back until the cell is lysed and the replicated bacteriophage released. This switch from a dormant or **prophage state** to a **lytic infection** is well understood at the genetic and molecular levels and will be described in detail here; though less well understood at the molecular level, HIV and herpes viruses can behave similarly.

The lytic/lysogenic genetic switching event in lambda is centered around an 80-bp region in its double-stranded DNA genome referred to as the "right operator" (O_R) (Figure 38–5A). The **right operator** is flanked on its left side by the structural gene for the lambda repressor protein, *cI*, and on its right side by the structural gene encoding another regulatory protein called *cro*. When lambda is in its prophage state—ie, integrated into the host genome—the *cI* repressor gene is the *only* lambda gene that is expressed. When the bacteriophage is undergoing lytic growth, the *cI* repressor gene is not expressed, but the *cro* gene—as well as many other genes in lambda—is expressed. That is, **when the repressor gene is on, the *cro* gene is off, and when the *cro* gene is on, the *cI* repressor gene is off.** As we shall see, these two genes regulate each other's expression and thus, ultimately, the decision between lytic and lysogenic growth of lambda. **This decision between repressor gene transcription and *cro* gene transcription is a paradigmatic example of a molecular transcriptional switch.**

The 80-bp lambda right operator, O_R, can be subdivided into three discrete, evenly spaced, 17-bp *cis*-active DNA elements that represent the binding sites for either of two bacteriophage lambda regulatory proteins. Importantly, the nucleotide sequences of these three tandemly arranged sites are similar but not identical (Figure 38–5B). The three related *cis* elements, termed operators O_R1, O_R2, and O_R3, can be bound by either cI or Cro proteins. However, the relative affinities of cI and Cro for each of the sites varies, and this differential binding affinity is central to the appropriate operation of the lambda phage lytic or lysogenic "molecular switch." The DNA region between the *cro* and repressor genes also contains two promoter sequences that direct the binding of RNA polymerase in a specified orientation, where it commences transcribing adjacent genes. One promoter directs RNA polymerase to transcribe in the **rightward direction** and, thus, to transcribe *cro* and other distal genes, while the other promoter directs the transcription of the *cI* **repressor** gene in the **leftward direction** (Figure 38–5B).

The product of the repressor gene, the 236-amino-acid, 27 kDa *cI* **repressor protein,** exists as a **two-domain** molecule in which the **amino terminal domain binds to operator DNA** and the **carboxyl terminal domain promotes the association** of one repressor protein with another to form a dimer. A **dimer** of repressor molecules binds to **operator DNA** much more tightly than does the monomeric form (Figure 38–6A to 38–6C).

The product of the *cro* gene, the 66-amino-acid, 9-kDa **Cro protein,** has a single domain but also binds the operator DNA more tightly as a **dimer** (Figure 38–6D). The Cro protein's single domain mediates both operator binding and dimerization.

In a lysogenic bacterium—ie, a bacterium containing an integrated dormant lambda prophage—the lambda repressor dimer binds **preferentially to O_R1** but in so doing, by a coop-

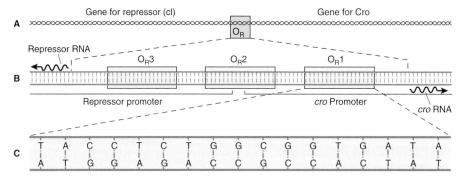

FIGURE 38–5 Right operator (O_R) is shown in increasing detail in this series of drawings. The operator is a region of the viral DNA some 80 base pairs long **(A)**. To its left lies the gene encoding lambda repressor (cI), to its right the gene *(cro)* encoding the regulator protein Cro. When the operator region is enlarged **(B)**, it is seen to include three subregions, O_R1, O_R2, and O_R3, each 17 base pairs long. They are recognition sites to which both repressor and Cro can bind. The recognition sites overlap two promoters—sequences of bases to which RNA polymerase binds in order to transcribe these genes into mRNA (wavy lines), that are translated into protein. Site O_R1 is enlarged **(C)** to show its base sequence. Note that in the O_R region of the lambda chromosome, both strands of DNA act as a template for transcription. (Reproduced, with permission, from Ptashne M, Johnson AD, Pabo CO: A genetic switch in a bacterial virus. Sci Am [Nov] 1982;247:128.)

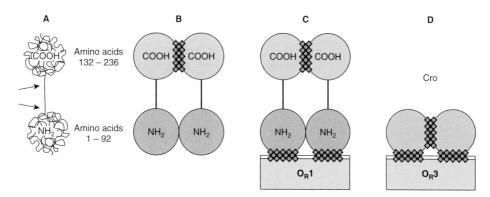

FIGURE 38–6 Schematic molecular structures of cI (lambda repressor, shown in **A, B,** and **C**) and Cro **(D)**. The lambda repressor protein is a polypeptide chain 236 amino acids long. The chain folds itself into a dumbbell shape with two substructures: an amino terminal (NH₂) domain and a carboxyl terminal (COOH) domain. The two domains are linked by a region of the chain that is less structured and susceptible to cleavage by proteases (indicated by the two arrows in **A**). Single repressor molecules (monomers) tend to reversibly associate to form dimers **(B)** A dimer is held together mainly by contact between the carboxyl terminal domains (hatching). Repressor dimers bind to (and can dissociate from) the recognition sites in the operator region; they display the greatest affinity for site O_R1 **(C)**. It is the amino terminal domain of the repressor molecule that makes contact with the DNA (hatching). Cro **(D)** has a single domain with sites that promote dimerization and other sites that promote binding of dimers to operator, preferentially to O_R3. (Reproduced, with permission, from Ptashne M, Johnson AD, Pabo CO: A genetic switch in a bacterial virus. Sci Am [Nov] 1982;247:128.)

erative interaction, enhances the binding (by a factor of 10) of another repressor dimer to O_R2 (Figure 38–7). The affinity of repressor for O_R3 is the least of the three operator subregions. The binding of repressor to O_R1 has two major effects. The occupation of O_R1 by repressor **blocks the binding of RNA polymerase to the rightward promoter** and in that way prevents expression of *cro*. Second, as mentioned above, repressor dimer bound to O_R1 enhances the binding of repressor dimer to O_R2. The binding of repressor to O_R2 has the important added effect of **enhancing the binding of RNA polymerase to the leftward promoter** that overlaps O_R3 and thereby enhances transcription and subsequent expression of the repressor gene. This enhancement of transcription is mediated through direct protein–protein interactions between promoter-bound RNA polymerase and O_R2-bound repressor, much as described above for CAP protein and RNA polymerase on the *lac* operon. Thus, the lambda repressor is both a **negative regulator,** by preventing transcription of *cro,* and a **positive regulator,** by enhancing transcription of its own gene, *cI.* This dual effect of repressor is responsible for the stable state of the dormant lambda bacteriophage; not only does the repressor prevent expression of the genes necessary for lysis, but it also promotes expression of itself to stabilize this state of differentiation. In the event that intracellular repressor protein concentration becomes very high, this excess repressor will bind to O_R3 and by so doing diminish transcription of the repressor gene from the leftward promoter until the repressor concentration drops and repressor dissociates itself from O_R3. Interestingly, similar examples of repressor proteins also having the ability to activate transcription have been observed in eukaryotes.

With such a stable, repressive, cI-mediated, lysogenic state, one might wonder how the lytic cycle could ever be entered. However, this process does occur quite efficiently. When a DNA-damaging signal, such as ultraviolet light, strikes the lysogenic host bacterium, fragments of single-stranded DNA are generated that activate a specific **co-protease** coded by a bacterial gene and referred to as *recA* (Figure 38–7). The activated recA protease hydrolyzes the portion of the repressor protein that connects the amino terminal and carboxyl terminal domains of that molecule (see Figure 38–6A). Such cleavage of the repressor domains causes the **repressor dimers to dissociate,** which in turn causes **dissociation of the repressor molecules from** O_R2 and eventually from O_R1. The effects of removal of repressor from O_R1 and O_R2 are predictable. RNA polymerase immediately has access to the rightward promoter and commences transcribing the *cro gene,* and the enhancement effect of the repressor at O_R2 on leftward transcription is lost (Figure 38–7).

The resulting newly synthesized Cro protein also binds to the operator region as a dimer, but its order of preference is opposite to that of repressor (Figure 38–7). That is, **Cro binds most tightly to** O_R3, but there is no cooperative effect of Cro at O_R3 on the binding of Cro to O_R2. At increasingly higher concentrations of Cro, the protein will bind to O_R2 and eventually to O_R1.

Occupancy of O_R3 by Cro immediately turns off transcription from the leftward *cI* promoter and in that way **prevents any further expression of the repressor gene.** The molecular switch is thus completely "thrown" in the lytic direction. The *cro* gene is now expressed, and the repressor gene is fully

FIGURE 38–7 Configuration of the switch is shown at four stages of the lambda life cycle. The lysogenic pathway (in which the virus remains dormant as a prophage) is selected when a repressor dimer binds to O_R1, thereby making it likely that O_R2 will be filled immediately by another dimer. In the prophage (**top**), the repressor dimers bound at O_R1 and O_R2 prevent RNA polymerase from binding to the rightward promoter and so block the synthesis of Cro (negative control). The repressors also enhance the binding of polymerase to the leftward promoter (positive control), with the result that the repressor gene is transcribed into RNA (wavy line) and more repressor is synthesized, maintaining the lysogenic state. The prophage is induced (**middle**) when ultraviolet radiation activates the protease recA, which cleaves repressor monomers. The equilibrium of free monomers, free dimers, and bound dimers is thereby shifted, and dimers leave the operator sites. RNA polymerase is no longer encouraged to bind to the leftward promoter, so that repressor is no longer synthesized. As induction proceeds, all the operator sites become vacant, and so polymerase can bind to the rightward promoter and Cro is synthesized. During early lytic growth, a single Cro dimer binds to O_R3 (light blue shaded circles), the site for which it has the highest affinity. Consequently, RNA polymerase cannot bind to the leftward promoter, but the rightward promoter remains accessible. Polymerase continues to bind there, transcribing *cro* and other early lytic genes. Lytic growth ensues (**bottom**). (Reproduced, with permission, from Ptashne M, Johnson AD, Pabo CO: A genetic switch in a bacterial virus. Sci Am [Nov] 1982;247:128.)

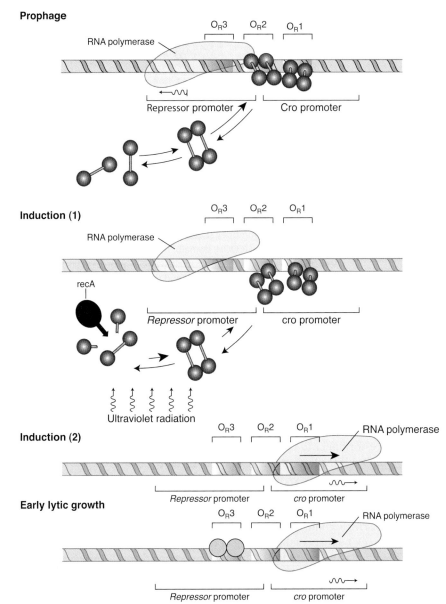

turned off. This event is irreversible, and the expression of other lambda genes begins as part of the lytic cycle. When Cro repressor concentration becomes quite high, it will eventually occupy O_R1 and in so doing reduce the expression of its own gene, a process that is necessary in order to effect the final stages of the lytic cycle.

The three-dimensional structures of Cro and of the lambda repressor protein have been determined by x-ray crystallography, and models for their binding and effecting the above-described molecular and genetic events have been proposed and tested. Both bind to DNA using helix-turn-helix DNA binding domain motifs (see below). To date, this system provides arguably the best understanding of the molecular events involved in gene regulation.

Detailed analysis of the lambda repressor led to the important concept that transcription regulatory proteins have several functional domains. For example, lambda repressor binds to DNA with high affinity. Repressor monomers form dimers, cooperatively interact with each other, and repressor interacts with RNA polymerase. The protein-DNA interface and the three protein-protein interfaces all involve separate and distinct domains of the repressor molecule. As will be noted below (see Figure 38–17), this is a characteristic shared by most (perhaps all) molecules that regulate transcription.

SPECIAL FEATURES ARE INVOLVED IN REGULATION OF EUKARYOTIC GENE TRANSCRIPTION

Most of the DNA in prokaryotic cells is organized into genes, and the templates always have the potential to be transcribed if appropriate positive and negative *trans* factors are activated. A very different situation exists in mammalian cells, in which relatively little of the total DNA is organized into mRNA encoding genes and their associated regulatory regions. The function of the extra DNA is being actively investigated (ie, Chapter 39; the ENCODE Project). More importantly, as described in Chapter 35, the DNA in eukaryotic cells is extensively folded and packed into the protein–DNA complex called chromatin. Histones are an important part of this complex since they both form the structures known as nucleosomes (see Chapter 35) and also factor significantly into gene regulatory mechanisms as outlined below.

Chromatin Remodeling Is an Important Aspect of Eukaryotic Gene Expression

Chromatin structure provides an additional level of control of gene transcription. As discussed in Chapter 35, large regions of chromatin are transcriptionally inactive while others are either active or potentially active. With few exceptions, each cell contains the same complement of genes. The development of specialized organs, tissues, and cells and their function in the intact organism depend upon the differential expression of genes.

Some of this differential expression is achieved by having different regions of chromatin available for transcription in cells from various tissues. For example, the DNA containing the β-globin gene cluster is in **"active" chromatin** in the reticulocyte but in **"inactive" chromatin** in muscle cells. All the factors involved in the determination of active chromatin have not been elucidated. The presence of nucleosomes and of complexes of histones and DNA (see Chapter 35) certainly provides a barrier against the ready association of transcription factors with specific DNA regions. The dynamics of the formation and disruption of nucleosome structure are therefore an important part of eukaryotic gene regulation.

Histone covalent modification is an important determinant of gene activity. Histones are subjected to a wide range of specific post-translational modifications (Table 35–1). These modifications are dynamic and reversible. Histone acetylation and deacetylation are best understood. The surprising discovery that histone acetylase and other enzymatic activities are associated with the coregulators involved in regulation of gene transcription (see Chapter 42) has provided a new concept of gene regulation. Acetylation is known to occur on lysine residues in the amino terminal tails of histone molecules. This modification reduces the positive charge of these tails and decreases the binding affinity of histone for the negatively charged DNA. Such covalent modification of the histones creates new binding sites for additional proteins such as ATP-dependent chromatin remodeling complexes. These complexes can increase accessibility of adjacent DNA sequences by removing nucleosomal histones. Together then coregulators (chromatin modifiers and chromatin remodellers), working in conjunction, can open up gene promoters and regulatory regions, facilitating binding of other *trans* factors and RNA polymerase II

and GTFs (see Figures 36–10, 11). Histone deacetylation catalyzed by transcriptional corepressors would have the opposite effect. Different proteins with specific acetylase and deacetylase activities are associated with various components of the transcription apparatus. The specificity of these processes is under investigation, as are a variety of mechanisms of action. Some specific examples are illustrated in Chapter 42.

There is evidence that the **methylation of deoxycytidine residues** (in the sequence 5′-^mCpG-3′) in DNA may effect changes in chromatin so as to preclude its active transcription, as described in Chapter 35. For example, in mouse liver, only the unmethylated ribosomal genes can be expressed, and there is evidence that many animal viruses are not transcribed when their DNA is methylated. Acute demethylation of deoxycytidine residues in specific regions of steroid hormone inducible genes has been associated with an increased rate of transcription of the gene. However, it is not yet possible to generalize that methylated DNA is transcriptionally inactive, that all inactive chromatin is methylated, or that active DNA is not methylated.

Finally, the binding of specific transcription factors to cognate DNA elements may result in disruption of nucleosomal structure. Many eukaryotic genes have multiple protein-binding DNA elements. The serial binding of transcription factors to these elements—in a combinatorial fashion—may either directly disrupt the structure of the nucleosome, prevent its re-formation, or recruit, via protein-protein interactions, multiprotein coregulator complexes that have the ability to covalently modify or remodel nucleosomes. These reactions result in chromatin-level structural changes that in the end increase DNA accessibility to other factors and the transcription machinery (cf. above).

Eukaryotic DNA that is in an "active" region of chromatin can be transcribed. As in prokaryotic cells, a **promoter** dictates where the RNA polymerase will initiate transcription, but the promoter in mammalian cells (Chapter 36) is more complex. In addition, the *trans*-acting factors generally come from other chromosomes (and so act in *trans*), whereas this consideration is moot in the case of the single chromosome-containing prokaryotic cells. Additional complexity is added by elements or factors that enhance or repress transcription, define tissue-specific expression, and modulate the actions of many effector molecules. Finally, recent results suggest that gene activation and repression might occur when particular genes move into or out of different subnuclear compartments or locations.

Certain DNA Elements Enhance or Repress Transcription of Eukaryotic Genes

In addition to gross changes in chromatin affecting transcriptional activity, certain DNA elements facilitate or enhance initiation at the promoter and hence are termed **enhancers**. **Enhancer elements, which typically contain multiple binding sites for transactivator proteins,** differ from the promoter in notable ways. They can exert their positive influence on transcription even when separated by tens of thousands of base pairs from a promoter; they work when oriented in either direction; and they can work upstream (5′) or downstream (3′) from the promoter. Enhancers are promiscuous; they can stimulate any promoter in the vicinity and may act on more than one promoter. The viral SV40 enhancer can exert an influence on, for example, the transcription of β-globin by increasing its transcription 200-fold in cells containing both the SV40 enhancer and the β-globin gene on the same plasmid (see below and Figure 38–8); in this case the SV40 enhancer β-globin gene was constructed using recombinant DNA technology—see Chapter 39. The enhancer element

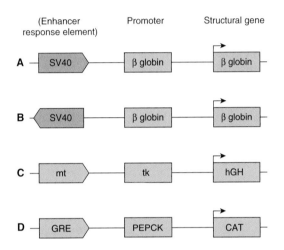

FIGURE 38–8 A schematic illustrating the action of enhancers and other *cis*-acting regulatory elements. These model chimeric genes, all constructed by recombinant DNA techniques (Chapter 39) in vitro, consist of a reporter (structural) gene that encodes a protein that can be readily assayed, and that is not normally produced in the cells to be studied, a promoter that ensures accurate initiation of transcription, and the indicated regulatory elements. In all cases, high-level transcription from the indicated chimeras depends upon the presence of enhancers, which stimulate transcription ≥ 100-fold over basal transcriptional levels (ie, transcription of the same chimeric genes containing just promoters fused to the structural genes). Examples **(A)** and **(B)** illustrate the fact that enhancers (eg, SV40) work in either orientation and upon a heterologous promoter. Example **(C)** illustrates that the metallothionein (mt) regulatory element (which under the influence of cadmium or zinc induces transcription of the endogenous mt gene and hence the metal-binding mt protein) will work through the thymidine kinase (tk) promoter to enhance transcription of the human growth hormone (hGH) gene. The engineered genetic constructions were introduced into the male pronuclei of single-cell mouse embryos and the embryos placed into the uterus of a surrogate mother to develop as transgenic animals. Offspring have been generated under these conditions, and in some the addition of zinc ions to their drinking water effects an increase in liver growth hormone. In this case, these transgenic animals have responded to the high levels of growth hormone by becoming twice as large as their normal litter mates. Example **(D)** illustrates that a glucocorticoid response element (GRE) will work through homologous (PEPCK gene) or heterologous promoters (not shown; ie, tk) promoter, SV40 promoter, β-globin promoter, etc).

does not produce a product that in turn acts on the promoter, since it is active only when it exists within the same DNA molecule as (ie, *cis* to) the promoter. Enhancer binding proteins are responsible for this effect. The exact mechanisms by which these transcription activators work are subject to intensive investigation. Certainly, enhancer binding *trans* factors have been shown to interact with a plethora of other transcription proteins. These interactions include chromatin-modifying coactivators, Mediator, as well as the individual components of the basal RNA polymerase II transcription machinery. Ultimately, *trans* factor-enhancer DNA binding events result in an increase in the binding of the basal transcription machinery to the promoter. Enhancer elements and associated binding proteins often convey nuclease hypersensitivity to those regions where they reside (Chapter 35). A summary of the properties of enhancers is presented in Table 38–2.

One of the best-understood mammalian enhancer systems is that of the β-interferon gene. This gene is induced upon viral infection of mammalian cells. One goal of the cell, once virally infected, is to attempt to mount an antiviral response—if not to save the infected cell, then to help to save the entire organism from viral infection. Interferon production is one mechanism by which this is accomplished. This family of proteins is secreted by virally infected cells. Secreted interferon interacts with neighboring cells to cause an inhibition of viral replication by a variety of mechanisms, thereby limiting the extent of viral infection. The enhancer element controlling induction of the β-interferon gene, which is located between nucleotides −110 and −45, is well characterized. This enhancer is composed of four distinct clustered *cis* elements, each of which is bound by unique *trans* factors. One *cis* element is bound by the transacting factor NF-κB, one by a member of the IRF (interferon regulatory factor) family of *trans* factors, and a third by the heterodimeric leucine zipper factor ATF-2/c-Jun (see below). The fourth factor is the ubiquitous, abundant architectural transcription factor known as HMG I(Y). Upon binding to its degenerate, A+T-rich binding sites, HMG I(Y) induces a significant bend in the DNA. There are four such HMG I(Y) binding sites interspersed throughout the enhancer. These sites play a critical role in forming a particular three-dimensional (3-D) structure, along with the aforementioned three *trans* factors, by inducing a series of critically spaced DNA bends. Consequently, HMG I(Y) induces the cooperative formation

TABLE 38–2   Summary of the Properties of Enhancers

• Work when located long distances from the promoter
• Work when upstream or downstream from the promoter
• Work when oriented in either direction
• Can work with homologous or heterologous promoters
• Work by binding one or more proteins
• Work by facilitating binding of the basal transcription complex to the promoter

of a unique, stereospecific, 3-D structure within which all four factors are active when viral infection signals are sensed by the cell. The structure formed by the cooperative assembly of these four factors is termed the β-interferon enhanceosome (see Figure 38–9), so named because of its obvious structural

FIGURE 38–9 Formation and putative structure of the enhanceosome formed on the human β-interferon gene enhancer. Diagrammatically represented at the top is the distribution of the multiple *cis*-elements (HMG, PRDIV, PRDI-III, PRDII, NRDI) composing the β-interferon gene enhancer. The intact enhancer mediates transcriptional induction of the β-interferon gene (over 100-fold) upon virus infection of human cells. The *cis*-elements of this modular enhancer represent the binding sites for the *trans*-factors HMG I(Y), cJun-ATF-2, IRF3-IRF7, and NF-κB, respectively. The factors interact with these DNA elements in an obligatory, ordered, and highly cooperative fashion as indicated by the arrow. Initial binding of four HMG I(Y) proteins induces sharp DNA bends in the enhancer, causing the entire 70–80 bp region to assume a high level of curvature. This curvature is integral to the subsequent highly cooperative binding of the other *trans*-factors since this enables the DNA-bound factors to make important, direct protein-protein interactions that both contribute to the formation and stability of the enhanceosome and generate a unique three-dimensional surface that serves to recruit chromatin-modifying coregulators that carry enzymatic activities (eg, Swi/Snf: ATPase, chromatin remodeler and P/CAF: histone acetyltransferase) as well as the general transcription machinery (RNA polymerase II and GTFs). Although four of the five *cis*-elements (PRDIV, PRDI-III, PRDII, NRDI) independently can modestly stimulate (~10-fold) transcription of a reporter gene in transfected cells (see Figures 38–10 and 38–12), all five *cis*-elements, in appropriate order, are required to form an enhancer that can appropriately stimulate mRNA gene transcription (ie, ≥100-fold) in response to viral infection of a human cell. This distinction indicates the strict requirement for appropriate enhanceosome architecture for efficient *trans*-activation. Similar enhanceosomes, involving distinct *cis*- and *trans*-factors and coregulators, are proposed to form on many other mammalian genes.

similarity to the nucleosome, also a unique three-dimensional protein-DNA structure that wraps DNA about an assembly of proteins (see Figures 35–1 & 35–2). The enhanceosome, once formed, induces a large increase in β-interferon gene transcription upon virus infection. It is not simply the protein occupancy of the linearly apposed *cis* element sites that induces β-interferon gene transcription—rather, it is the formation of the enhanceosome proper that provides appropriate surfaces for the recruitment of coactivators that results in the enhanced formation of the PIC on the *cis*-linked promoter and thus transcription activation.

The *cis*-acting elements that decrease or **repress** the expression of specific genes have also been identified. Because fewer of these elements have been studied, it is not possible to formulate generalizations about their mechanism of action—though again, as for gene activation, chromatin level covalent modifications of histones and other proteins by (repressor)-recruited multisubunit corepressors have been implicated.

Tissue-Specific Expression May Result From the Action of Enhancers or Repressors

Many genes are now recognized to harbor enhancer or activator elements in various locations relative to their coding regions. In addition to being able to enhance gene transcription, some of these enhancer elements clearly possess the ability to do so in a tissue-specific manner. Thus, the enhancer element associated with the immunoglobulin genes between the J and C regions enhances the expression of those genes preferentially in lymphoid cells. Similarly by fusing known or suspected tissue-specific enhancers to reporter genes (see below) and introducing these chimeric enhancer-reporter constructs microsurgically into single-cell embryo, one can create a transgenic animal (see Chapter 39), and rigorously test whether a given test enhancer truly drives expression in a cell- or tissue-specific fashion. This **transgenic animal** approach has proved useful in studying tissue-specific gene expression.

Reporter Genes Are Used to Define Enhancers & Other Regulatory Elements

By ligating regions of DNA suspected of harboring regulatory sequences to various reporter genes (the **reporter** or **chimeric gene approach**) (Figures 38–8, 38–10, & 38–11), one can determine which regions in the vicinity of structural genes have an influence on their expression. Pieces of DNA thought to harbor regulatory elements are ligated to a suitable reporter gene and introduced into a host cell (Figure 38–10). Basal expression of the reporter gene will be increased if the DNA contains an enhancer. Addition of a hormone or heavy metal to the culture medium will increase expression of the reporter gene if the DNA contains a hormone or metal response element (Figure 38–11). The location of the element can be pinpointed by using progressively shorter pieces of DNA, deletions, or point mutations (Figure 38–11).

FIGURE 38–11 Mapping hormone response elements (HREs) **(A)**, **(B)**, and **(C)** using the reporter gene–transfection approach. A family of reporter genes, constructed as described in Figure 38–10, can be transfected individually into a recipient cell. By analyzing when certain hormone responses are lost in comparison to the 5′ deletion end point, specific hormone-responsive elements can be located.

FIGURE 38–10 The use of reporter genes to define DNA regulatory elements. A DNA fragment from the gene in question—in this example, approximately 2 kb of 5′-flanking DNA and cognate promoter—is ligated into a plasmid vector that contains a suitable reporter gene—in this case, the bacterial enzyme chloramphenicol transferase (CAT). The enzyme luciferase (abbreviated LUC) is another popular reporter gene. Neither LUC nor CAT is present in mammalian cells; hence, detection of these activities in a cell extract means that the cell was successfully transfected by the plasmid. An increase of CAT activity over the basal level, eg, after addition of one or more hormones, means that the region of DNA inserted into the reporter gene plasmid contains functional hormone response elements (HRE). Progressively shorter pieces of DNA, regions with internal deletions, or regions with point mutations can be constructed and inserted to pinpoint the response element (see Figure 38–11 for deletion mapping of the relevant HREs).

This strategy, typically **using transfected cells in culture** (ie, cells induced to take up exogenous DNAs), has led to the identification of hundreds of enhancers, repressors, tissue-specific elements, and hormone, heavy metal, and drug-response elements. The activity of a gene at any moment reflects the interaction of these numerous *cis*-acting DNA elements with their respective *trans*-acting factors. Overall transcriptional output is determined by the balance of positive and negative signaling to the transcription machinery. The challenge now is to figure out how this occurs at the molecular level.

Combinations of DNA Elements & Associated Proteins Provide Diversity in Responses

Prokaryotic genes are often regulated in an on–off manner in response to simple environmental cues. Some eukaryotic genes are regulated in the simple on–off manner, but the process in most genes, especially in mammals, is much more complicated. Signals representing a number of complex environmental stimuli may converge on a single gene. The response of the gene to these signals can have several physiologic characteristics. First, the response may extend over a considerable range. This is accomplished by having additive and synergistic positive responses counterbalanced by negative or repressing effects. In some cases, either the positive or the negative response can be dominant. Also required is a mechanism whereby an effector such as a hormone can activate some genes in a cell while repressing others and leaving still others unaffected. When all of these processes are coupled with tissue-specific element factors, considerable flexibility is afforded. These physiologic variables obviously require an arrangement much more complicated than an on–off switch. The array of DNA elements in a promoter specifies—with associated factors—how a given gene will respond and how long a particular response is maintained. Some simple examples are illustrated in Figure 38–12.

Transcription Domains Can Be Defined by Locus Control Regions & Insulators

The large number of genes in eukaryotic cells and the complex arrays of transcription regulatory factors presents an organizational problem. Why are some genes available for transcription in a given cell whereas others are not? If enhancers can regulate several genes from tens of kilobase distances and are not position- and orientation-dependent, how are they prevented from triggering transcription of all *cis*-linked genes in the vicinity? Part of the solution to these problems is arrived at by having the chromatin arranged in functional units that restrict patterns of gene expression. This may be achieved by having the chromatin form a structure with the nuclear matrix or other physical entity, or compartment within the nucleus.

FIGURE 38–12 Combinations of DNA elements and proteins provide diversity in the response of a gene. Gene A is activated (the width of the arrow indicates the extent) by the combination of activators 1, 2, and 3 (probably with coactivators, as shown in Figure 36–10). Gene B is activated, in this case more effectively, by the combination of 1, 3, and 4; note that 4 does not contact DNA directly in this example. The activators could form a linear bridge that links the basal machinery to the promoter, or this could be accomplished by looping out of the DNA. In either case, the purpose is to direct the basal transcription machinery to the promoter. Gene C is inactivated by the combination of 1, 5, and 3; in this case, factor 5 is shown to preclude the essential binding of factor 2 to DNA, as occurs in example A. If activator 1 helps repressor 5 bind and if activator 1 binding requires a ligand (solid dot), it can be seen how the ligand could activate one gene in a cell (gene A) and repress another (gene C).

Alternatively, some regions are controlled by complex DNA elements called **locus control regions (LCRs).** An LCR—with associated bound proteins—controls the expression of a cluster of genes. The best-defined LCR regulates expression of the globin gene family over a large region of DNA. Another mechanism is provided by **insulators.** These DNA elements, also in association with one or more proteins, prevent an enhancer from acting on a promoter on the other side of an insulator in another transcription domain. Insulators thus serve as transcriptional **boundary elements**.

SEVERAL MOTIFS MEDIATE THE BINDING OF REGULATORY PROTEINS TO DNA

The specificity involved in the control of transcription requires that regulatory proteins bind with high affinity and specificity to the correct region of DNA. Three unique motifs—the **helix-turn-helix,** the **zinc finger,** and the **leucine zipper**—account for many of these specific protein-DNA interactions. Examples of proteins containing these motifs are given in Table 38–3.

Comparison of the binding activities of the proteins that contain these motifs leads to several important generalizations:

1. Binding must be of high affinity to the specific site and of low affinity to other DNA.
2. Small regions of the protein make direct contact with DNA; the rest of the protein, in addition to providing

TABLE 38–3 Examples of Transcription Regulatory Proteins That Contain the Various Binding Motifs

Binding Motif	Organism	Regulatory Protein
Helix-turn-helix	E coli	lac repressor
		CAP
	Phage	λcI, cro, and tryptophan and 434 repressors
	Mammals	Homeobox proteins Pit-1, Oct1, Oct2
Zinc finger	E coli	Gene 32 protein
	Yeast	Gal4
	Drosophila	Serendipity, Hunchback
	Xenopus	TFIIIA
	Mammals	Steroid receptor family, Sp1
Leucine zipper	Yeast	GCN4
	Mammals	C/EBP, fos, Jun, Fra-1, CRE binding protein, c-*myc*, n-*myc*, l-*myc*

the *trans*-activation domains, may be involved in the dimerization of monomers of the binding protein, may provide a contact surface for the formation of heterodimers, may provide one or more ligand-binding sites, or may provide surfaces for interaction with coactivators or corepressors.
3. The protein-DNA interactions are maintained by hydrogen bonds, ionic interactions and van der Waals forces.
4. The motifs found in these proteins are unique; their presence in a protein of unknown function suggests that the protein may bind to DNA.
5. Proteins with the helix-turn-helix or leucine zipper motifs form dimers, and their respective DNA binding sites are symmetric palindromes. In proteins with the zinc finger motif, the binding site is repeated two to nine times. These features allow for cooperative interactions between binding sites and enhance the degree and affinity of binding.

The Helix-Turn-Helix Motif

The first motif described is the **helix-turn-helix**. Analysis of the three-dimensional structure of the lambda Cro transcription regulator has revealed that each monomer consists of three antiparallel β sheets and three α helices (Figure 38–13). The dimer forms by association of the antiparallel β_3 sheets. The α_3 helices form the DNA recognition surface, and the rest of the molecule appears to be involved in stabilizing these structures. The average diameter of an α helix is 1.2 nm, which is the approximate width of the major groove in the B form of DNA.

The DNA recognition domain of each Cro monomer interacts with 5 bp and the dimer binding sites span 3.4 nm, allowing fit into successive half turns of the major groove on the same surface (Figure 38–13). X-ray analyses of the λ cI repressor, CAP (the cAMP receptor protein of *E coli*), tryptophan repressor, and phage 434 repressor, all also display this dimeric helix-turn-helix structure that is present in eukaryotic DNA binding proteins as well (see Table 38–3).

The Zinc Finger Motif

The **zinc finger** was the second DNA binding motif whose atomic structure was elucidated. It was known that the protein TFIIIA, a positive regulator of 5S RNA gene transcription, required zinc for activity. Structural and biophysical analyses revealed that each TFIIIA molecule contains nine zinc ions in a repeating coordination complex formed by closely spaced cysteine-cysteine residues followed 12–13 amino acids later by a histidine–histidine pair (Figure 38–14). In some instances—notably the steroid-thyroid nuclear hormone receptor family—the His-His doublet is replaced by a second Cys-Cys pair. The protein containing zinc fingers appears to lie on one face of the DNA helix, with successive fingers alternatively positioned in one turn in the major groove. As is the case with the recognition domain in the helix-turn-helix protein, each TFIIIA zinc finger contacts about 5 bp of DNA. The importance of this motif in the action of steroid hormones is underscored by an "experiment of nature." A single amino acid mutation in either of the two zinc fingers of the $1,25(OH)_2$-D_3 receptor protein results in resistance to the action of this hormone and the clinical syndrome of rickets.

The Leucine Zipper Motif

Careful analysis of a 30-amino-acid sequence in the carboxyl terminal region of the enhancer binding protein C/EBP revealed a novel structure, **the leucine zipper motif**. As illustrated in Figure 38–15, this region of the protein forms an α helix in which there is a periodic repeat of leucine residues at every seventh position. This occurs for eight helical turns and four leucine repeats. Similar structures have been found in a number of other proteins associated with the regulation of transcription in mammalian and yeast cells. This structure allows two identical or nonidentical monomers (eg, Jun-Jun or Fos-Jun) to "zip together" in a coiled coil and form a tight dimeric complex (Figure 38–15). This protein-protein interaction may serve to enhance the association of the separate DNA binding domains with their target (Figure 38–15).

THE DNA BINDING & TRANSACTIVATION DOMAINS OF MOST REGULATORY PROTEINS ARE SEPARATE & NONINTERACTIVE

DNA binding could result in a general conformational change that allows the bound protein to activate transcription, or these two functions could be served by separate and independent

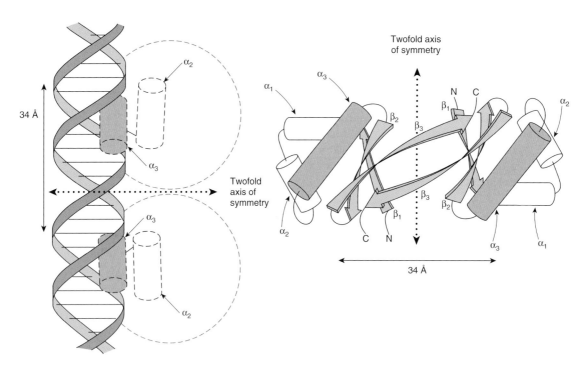

FIGURE 38–13 A schematic representation of the three-dimensional structure of Cro protein and its binding to DNA by its helix-turn-helix motif (**left**). The Cro monomer consists of three antiparallel β sheets (β_1–β_3) and three α-helices (α_1–α_3). The helix-turn-helix motif is formed because the α_3 and α_2 helices are held at about 90 degrees to each other by a turn of four amino acids. The α_3 helix of Cro is the DNA recognition surface (**shaded**). Two monomers associate through the antiparallel β_3 sheets to form a dimer that has a 2-fold axis of symmetry (**right**). A Cro dimer binds to DNA through its α_3 helices, each of which contacts about 5 bp on the same surface of the major groove. The distance between comparable points on the two DNA α-helices is 34 Å, which is the distance required for one complete turn of the double helix. (Courtesy of B Mathews.)

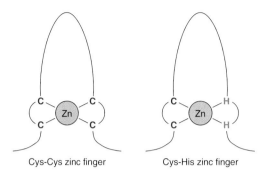

FIGURE 38–14 Zinc fingers are a series of repeated domains (two to nine) in which each is centered on a tetrahedral coordination with zinc. In the case of TFIIIA, the coordination is provided by a pair of cysteine residues (C) separated by 12–13 amino acids from a pair of histidine (H) residues. In other zinc finger proteins, the second pair also consists of C residues. Zinc fingers bind in the major groove, with adjacent fingers making contact with 5 bp along the same face of the helix.

domains. Domain swap experiments suggest that the latter is typically the case.

The *GAL1* gene product is involved in galactose metabolism in yeast. Transcription of this gene is positively regulated by the GAL4 protein, which binds to an upstream activator sequence (UAS), or enhancer, through an amino terminal

domain. The amino terminal 73-amino-acid DNA-binding domain (DBD) of GAL4 was removed and replaced with the DBD of LexA, an *E coli* DNA-binding protein. This domain swap resulted in a molecule that did not bind to the *GAL1* UAS and, of course, did not activate the *GAL1* gene (Figure 38–16). If, however, the *lexA* operator—the DNA sequence normally bound by the *lexA*DBD—was inserted into the promoter region of the *GAL* gene thereby replacing the normal *GAL1*enhancer, the hybrid protein bound to this promoter (at the *lexA* operator) and it activated transcription of *GAL1*. This experiment, which has been repeated a number of times, affords solid evidence that the carboxyl terminal region of GAL4 causes transcriptional activation. These data also demonstrate that the DNA-binding DBD and transactivation domains (ADs) are independent and noninteractive. The hierarchy involved in assembling gene transcription activating complexes includes proteins that bind DNA and transactivate; others that form protein-protein complexes which bridge DNA-binding proteins to transactivating proteins; and others that form protein-protein complexes with components of coregulators or the basal transcription apparatus. A given protein may thus have several modular surfaces or domains that serve different functions (see Figure 38–17). As described in Chapter 36, the primary purpose of these complex assemblies is to facilitate the assembly and/or activity of the basal transcription apparatus on the *cis*-linked promoter.

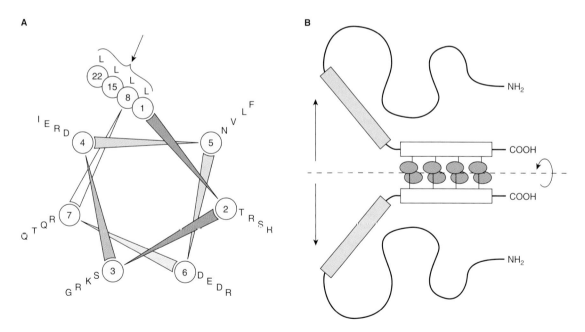

FIGURE 38–15 The leucine zipper motif. **(A)** shows a helical wheel analysis of a carboxyl terminal portion of the DNA binding protein C/EBP. The amino acid sequence is displayed end-to-end down the axis of a schematic α-helix. The helical wheel consists of seven spokes that correspond to the seven amino acids that comprise every two turns of the α-helix. Note that leucine residues (L) occur at every seventh position (in this schematic C/EBP amino acid residues 1, 8, 15, 22; see arrow). Other proteins with "leucine zippers" have a similar helical wheel pattern. **(B)** is a schematic model of the DNA binding domain of C/EBP. Two identical C/EBP polypeptide chains are held in dimer formation by the leucine zipper domain of each polypeptide (denoted by the rectangles and attached ovals). This association is required to hold the DNA binding domains of each polypeptide (the shaded rectangles) in the proper conformation for DNA binding. (Courtesy of S McKnight.)

FIGURE 38–16 Domain-swap experiments demonstrate the independent nature of DNA binding and transcription activation domains. The *GAL1* gene promoter contains an upstream activating sequence (UAS) or enhancer that binds the regulatory protein GAL4 **(A).** This interaction results in a stimulation of *GAL1* gene transcription. A chimeric protein, in which the amino terminal DNA binding domain of GAL4 is removed and replaced with the DNA binding region of the *E coli* protein LexA, fails to stimulate *GAL1* transcription because the LexA domain cannot bind to the UAS **(B).** By contrast, the LexA–GAL4 fusion protein does increase *GAL1* transcription when the *lexA* operator (its natural target) is inserted into the *GAL1* promoter region **(C),** replacing the normal GAL1 UAS.

GENE REGULATION IN PROKARYOTES & EUKARYOTES DIFFERS IN IMPORTANT RESPECTS

In addition to transcription, eukaryotic cells employ a variety of mechanisms to regulate gene expression (Table 38–4). The nuclear membrane of eukaryotic cells physically segregates gene transcription from translation, since ribosomes exist only in the cytoplasm. Many more steps, especially in RNA processing, are involved in the expression of eukaryotic genes than of prokaryotic genes, and these steps provide additional sites for regulatory influences that cannot exist in prokaryotes. These RNA processing steps in eukaryotes, described in detail in Chapter 36, include capping of the 5′ ends of the primary transcripts, addition of a polyadenylate tail to the 3′ ends of transcripts, and excision of intron regions to generate spliced exons in the mature mRNA molecule. To date, analyses of eukaryotic gene expression provide evidence that regulation occurs at the level of **transcription, nuclear RNA processing, mRNA stability, and translation.** In addition, gene amplification and rearrangement influence gene expression.

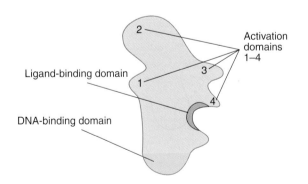

FIGURE 38–17 Proteins that regulate transcription have several domains. This hypothetical transcription factor has a DNA-binding domain (DBD) that is distinct from a ligand-binding domain (LBD) and several activation domains (ADs) (1–4). Other proteins may lack the DBD or LBD and all may have variable numbers of domains that contact other proteins, including co-regulators and those of the basal transcription complex (see also Chapters 41 & 42).

Owing to the advent of recombinant DNA technology, much progress has been made in recent years in the understanding of eukaryotic gene expression. However, because most eukaryotic organisms contain so much more genetic information than do prokaryotes and because manipulation of their genes is so much more difficult, molecular aspects of eukaryotic gene regulation are less well understood than the examples discussed earlier in this chapter. This section briefly describes a few different types of eukaryotic gene regulation.

miRNAs Modulate Gene Expression by Altering mRNA Function

As noted in Chapter 35 the recently discovered class of eukaryotic small RNAs, termed miRNAs, contribute importantly to the control of gene expression. These ~22 nucleotide RNAs regulate the translatability of specific mRNAs by either inhibiting translation or inducing mRNA degradation, though in a few cases miRNAs have been shown to stimulate mRNA function (translation). At least a portion of the miRNA-driven modulation of mRNA activity is thought to occur in the **P body** (Figure 37–11). miRNA action can result in dramatic changes in protein production and hence gene expression. miRNAs have been implicated in numerous human diseases such as heart disease, cancer, muscle wasting, viral infection and diabetes.

miRNAs, like the DNA binding transcription factors described in detail above, are transactive, and once synthesized

TABLE 38–4   Gene Expression Is Regulated by Transcription and in Numerous Other Ways at the RNA Level in Eukaryotic Cells

- Gene amplification
- Gene rearrangement
- RNA processing
- Alternate mRNA splicing
- Transport of mRNA from nucleus to cytoplasm
- Regulation of mRNA stability

and appropriately processed, interact with specific proteins and bind target mRNAs, typically in 3′ untranslated mRNA regions (Figure 36–17). Binding of miRNAs to mRNA targets is directed by normal base pairing rules. In general, if **miRNA–mRNA** base pairing has one or more mismatches, translation of the cognate "target" mRNA is inhibited while if **miRNA-mRNA base pairing** is perfect over all 22 nucleotides, the corresponding mRNA is degraded.

Given the tremendous and ever growing import of miRNAs, many scientists and biotechnology companies are actively studying miRNA biogenesis, transport, and function in hopes of curing human disease. Time will tell the magnitude and universality of miRNA-mediated gene regulation. It is likely that in the near future scientists will unveil the medical significance of these intriguing small RNAs.

Eukaryotic Genes Can Be Amplified or Rearranged During Development or in Response to Drugs

During early development of metazoans, there is an abrupt increase in the need for specific molecules such as ribosomal RNA and messenger RNA molecules for proteins that make up such organs as the eggshell. One way to increase the rate at which such molecules can be formed is to increase the number of genes available for transcription of these specific molecules. Among the repetitive DNA sequences within the genome are hundreds of copies of ribosomal RNA genes. These genes pre-exist repetitively in the DNA of the gametes and thus are transmitted in high copy numbers from generation to generation. In some specific organisms such as the fruit fly (drosophila), there occurs during oogenesis an amplification of a few preexisting genes such as those for the chorion (eggshell) proteins. Subsequently, these amplified genes, presumably generated by a process of repeated initiations during DNA synthesis, provide multiple sites for gene transcription (Figures 36–4 & 38–18).

As noted in Chapter 36, the coding sequences responsible for the generation of specific protein molecules are frequently not contiguous in the mammalian genome. In the case of antibody encoding genes, this is particularly true. As described in detail in Chapter 50, immunoglobulins are composed of two polypeptides, the so-called heavy (about 50 kDa) and light (about 25 kDa) chains. The mRNAs encoding these two protein subunits are encoded by gene sequences that are subjected to extensive DNA sequence-coding changes. These DNA coding changes are integral to generating the requisite recognition diversity central to appropriate immune function.

IgG heavy and light chain mRNAs are encoded by several different segments that are tandemly repeated in the germline. Thus, for example, the IgG light chain is composed of variable (V_L), joining (J_L), and constant (C_L) domains or segments. For particular subsets of IgG light chains, there are roughly 300 tandemly repeated V_L gene coding segments, five tandemly arranged J_L coding sequences, and roughly ten C_L gene coding segments. All of these multiple, distinct coding regions are located in the same region of the same chromosome, and each type of coding segment (V_L, J_L, and C_L) is tandemly repeated in head-to-tail fashion within the segment repeat region. By having multiple V_L, J_L, and C_L segments to choose from, an immune cell has a greater repertoire of sequences to work with to develop both immunologic flexibility and specificity. However, a given functional IgG light chain transcription unit—like all other "normal" mammalian transcription units—contains only the coding sequences for a single protein. Thus, before a particular IgG light chain can be expressed, *single* V_L, J_L, and C_L coding sequences must be recombined to generate a *single,* contiguous transcription unit excluding the multiple nonutilized segments (ie, the other approximately 300 unused V_L segments, the other four unused J_L segments, and the other nine unused C_L segments). This deletion of unused genetic information is accomplished by selective DNA recombination that removes the unwanted coding DNA while retaining the required coding sequences: one V_L, one J_L, and one C_L sequence. (V_L sequences are subjected to additional point mutagenesis to generate even more variability—hence the name.) The newly recombined sequences thus form a single transcription unit that is competent for RNA polymerase II-mediated transcription into a single monocistronic mRNA. Although the IgG genes represent one of the best-studied instances of directed DNA rearrangement modulating gene expression, other cases of gene regulatory DNA rearrangement have been described in the literature. Indeed, as detailed below, drug-induced gene amplification is an important complication of cancer chemotherapy.

In recent years, it has been possible to promote the amplification of specific genetic regions in cultured mammalian cells. In some cases, a several 1000-fold increase in the copy number of specific genes can be achieved over a period of time involving increasing doses of selective drugs. In fact, it has been demonstrated in patients receiving methotrexate for cancer that malignant cells can develop **drug resistance** by increasing the number of genes for dihydrofolate reductase, the target of methotrexate. Gene amplification and deletion events involving 10 to 1,000,000 of bp of DNA such as these occur spontaneously in vivo—ie, in the absence of exogenously supplied selective agents—and these unscheduled extra rounds of replication can become stabilized in the genome under appropriate selective pressures.

FIGURE 38–18 Schematic representation of the amplification of chorion protein genes *s36* and *s38*. (Reproduced, with permission, from Chisholm R: Gene amplification during development. Trends Biochem Sci 1982;7:161. Copyright ©1982. Reprinted with permission from Elsevier.)

Alternative RNA Processing Is Another Control Mechanism

In addition to affecting the efficiency of promoter utilization, eukaryotic cells employ alternative RNA processing to control gene expression. This can result when alternative promoters, intron-exon splice sites, or polyadenylation sites are used. Occasionally, heterogeneity within a cell results, but more commonly the same primary transcript is processed differently in different tissues. A few examples of each of these types of regulation are presented below.

The use of **alternative transcription start sites** results in a different 5′ exon on mRNAs encoding mouse amylase and myosin light chain, rat glucokinase, and drosophila alcohol dehydrogenase and actin. **Alternative polyadenylation sites** in the μ immunoglobulin heavy chain primary transcript result in mRNAs that are either 2700 bases long (μ_m) or 2400 bases long (μ_s). This results in a different carboxyl terminal region of the encoded proteins such that the μ_m protein remains attached to the membrane of the B lymphocyte and the μ_s immunoglobulin is secreted. **Alternative splicing and processing** results in the formation of seven unique α-tropomyosin mRNAs in seven different tissues. It is not clear how these processing-splicing decisions are made or whether these steps can be regulated.

Regulation of Messenger RNA Stability Provides Another Control Mechanism

Although most mRNAs in mammalian cells are very stable (half-lives measured in hours), some turn over very rapidly (half-lives of 10–30 min). In certain instances, mRNA stability is subject to regulation. This has important implications since there is usually a direct relationship between mRNA amount and the translation of that mRNA into its cognate protein. Changes in the stability of a specific mRNA can therefore have major effects on biologic processes.

Messenger RNAs exist in the cytoplasm as ribonucleoprotein particles (RNPs). Some of these proteins protect the mRNA from digestion by nucleases, while others may under certain conditions promote nuclease attack. It is thought that mRNAs are stabilized or destabilized by the interaction of proteins with these various structures or sequences. Certain effectors, such as hormones, may regulate mRNA stability by increasing or decreasing the amount of these proteins.

It appears that **the ends of mRNA molecules are involved in mRNA stability** (Figure 38–19). The 5′ cap structure in eukaryotic mRNA prevents attack by 5′ exonucleases, and the poly(A) tail prohibits the action of 3′ exonucleases. In mRNA molecules with those structures, it is presumed that a single endonucleolytic cut allows exonucleases to attack and digest the entire molecule. Other structures (sequences) in the 5′ untranslated region (5′ UTR), the coding region, and the 3′ UTR are thought to promote or prevent this initial endonucleolytic action (Figure 38–19). A few illustrative examples will be cited.

Deletion of the 5′ UTR results in a 3-fold to 5-fold prolongation of the half-life of c-*myc* mRNA. Shortening the coding region of histone mRNA results in a prolonged half-life. A form of autoregulation of mRNA stability indirectly involves the coding region. Free tubulin binds to the first four amino acids of a nascent chain of tubulin as it emerges from the ribosome. This appears to activate an RNase associated with the ribosome which then digests the tubulin mRNA.

Structures at the 3′ end, including the poly(A) tail, enhance or diminish the stability of specific mRNAs. The absence of a poly(A) tail is associated with rapid degradation of mRNA, and the removal of poly(A) from some RNAs results in their destabilization. Histone mRNAs lack a poly(A) tail but have a sequence near the 3′ terminal that can form a stem-loop structure, and this appears to provide resistance to exonucleolytic attack. Histone H4 mRNA, for example, is degraded in the 3′ to 5′ direction but only after a single endonucleolytic cut occurs about nine nucleotides from the 3′ end in the region of the putative stem-loop structure. Stem-loop structures in the 3′ noncoding sequence are also critical for the regulation, by iron, of the mRNA encoding the transferrin receptor. Stem-loop structures are also associated with mRNA stability in bacteria, suggesting that this mechanism may be commonly employed.

Other sequences in the 3′ ends of certain eukaryotic mRNAs appear to be involved in the destabilization of these mol-

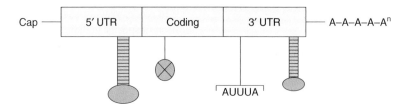

FIGURE 38–19 Structure of a typical eukaryotic mRNA showing elements that are involved in regulating mRNA stability. The typical eukaryotic mRNA has a 5′ noncoding sequence (5′NCS), a coding region, and a 3′NCS. All are capped at the 5′ end, and most have a polyadenylate sequence at the 3′ end. The 5′ cap and 3′ poly(A) tail protect the mRNA against exonuclease attack. Stem-loop structures in the 5′ and 3′ NCS, features in the coding sequence, and the AU-rich region in the 3′ NCS are thought to play roles in mRNA stability.

ecules. Some of this is mediated through the action of specific miRNAs as discussed above. In addition, of particular interest are AU-rich regions, many of which contain the sequence AUUUA. This sequence appears in mRNAs that have a very short half-life, including some encoding oncogene proteins and cytokines. The importance of this region is underscored by an experiment in which a sequence corresponding to the 3′ UTR of the short-half-life colony-stimulating factor (CSF) mRNA, which contains the AUUUA motif, was added to the 3′ end of the β-globin mRNA. Instead of becoming very stable, this hybrid β-globin mRNA now had the short-half-life characteristic of CSF mRNA. Much of this mRNA metabolism occurs in cytoplasmic **P bodies.**

From the few examples cited, it is clear that a number of mechanisms are used to regulate mRNA stability and hence function—just as several mechanisms are used to regulate the synthesis of mRNA. Coordinate regulation of these two processes confers on the cell remarkable adaptability.

SUMMARY

- The genetic constitutions of metazoan somatic cells are nearly all identical.
- Phenotype (tissue or cell specificity) is dictated by differences in gene expression of this complement of genes.
- Alterations in gene expression allow a cell to adapt to environmental changes, developmental cues, and physiological signals.
- Gene expression can be controlled at multiple levels by changes in transcription, RNA processing, localization, and stability or utilization. Gene amplification and rearrangements also influence gene expression.
- Transcription controls operate at the level of protein–DNA and protein–protein interactions. These interactions display protein domain modularity and high specificity.
- Several different classes of DNA-binding domains have been identified in transcription factors.
- Chromatin and DNA modifications contribute importantly in eukaryotic transcription control by modulating DNA accessibility and specifying recruitment of specific coactivators and corepressors to target genes.
- miRNA and siRNAs modulate mRNA translation and stability; these mechanisms complement transcription controls to regulate gene expression.

REFERENCES

Bhaumik SR, Smith E, Shilatifard A: Covalent modifications of histones during development and disease pathogenesis. Nat Struct Mol Biol 2007;14;1008.

Bing L, Carey M, Workman, J: The role of chromatin during transcription. Cell 2007;128:707.

Bird AP, Wolffe AP: Methylation-induced repression—belts, braces and chromatin. Cell 1999;99:451.

Busby S, Ebright RH: Promoter structure, promoter recognition, and transcription activation in prokaryotes. Cell 1994;79:743.

Busby S, Ebright RH: Transcription activation by catabolite activator protein (CAP). J Mol Biol 1999;293:199.

Jacob F, Monod J: Genetic regulatory mechanisms in protein synthesis. J Mol Biol 1961;3:318.

Lemon B, Tjian R: Orchestrated response: a symphony of transcription factors for gene control. Genes Dev 2000; 14:2551.

Letchman DS: Transcription factor mutations and disease. N Engl J Med 1996;334:28.

Merika M, Thanos D: Enhanceosomes. Curr Opin Genet Dev 2001;11:205.

Naar AM, Lemon BD, Tjian R: Transcriptional coactivator complexes. Annu Rev Biochem 2001;70:475.

Narlikar GJ, Fan HY, Kingston RE: Cooperation between complexes that regulate chromatin structure and transcription. Cell 2002;108:475.

Oltz EM: Regulation of antigen receptor gene assembly in lymphocytes. Immunol Res 2001;23:121.

Ptashne M: *A Genetic Switch,* 2nd ed. Cell Press and Blackwell Scientific Publications, 1992.

Roeder RG: Transcriptional regulation and the role of diverse coactivators in animal cells. FEBS Lett 2005;579:909.

Ruthenburg AJ, et al: Multivalent engagement of chromatin modifications by linked binding molecules. Nature Rev Mol Cell Bio 2007;8:983.

Segal E, et al: A genomic code for nucleosome positioning. Nature 2006;442:772.

Sterner DE, Berger SL: Acetylation of histones and transcription-related factors. Microbiol Mol Biol Rev 2000;64:435.

Valencia-Sanchez ME, Liu J, Hannon GJ, Parker R: Control of translation and mRNA degradation by miRNAs and SiRNAs. Genes Dev 2006;20:515.

Weake VM, Workman JL: Histone ubiquitination: triggering gene activity. Mol Cell 2008;29:653.

Yang XJ, Seto E: HATs and HDACs: from structure, function and regulation to nnovel strategies for therapy and prevention Oncogene 2007;26:5310.

Wu R, Bahl CP, Narang SA: Lactose operator-repressor interaction. Curr Top Cell Regul 1978;13:137.

Molecular Genetics, Recombinant DNA, & Genomic Technology

P. Anthony Weil, PhD

BIOMEDICAL IMPORTANCE*

The development of recombinant DNA, high-density DNA microarrays, high-throughput screening, low-cost genome-scale analyses and DNA sequencing and other molecular genetic methodologies has revolutionized biology and is having an increasing impact on clinical medicine. Though much has been learned about human genetic disease from pedigree analysis and study of affected proteins, in many cases where the specific genetic defect is unknown, these approaches cannot be used. The new technologies circumvent these limitations by going directly to the DNA molecule for information. Manipulation of a DNA sequence and the construction of chimeric molecules—so-called genetic engineering—provides a means of studying how a specific segment of DNA works. Novel molecular genetic tools and direct DNA sequencing allow investigators to query and manipulate genomic sequences as well as to examine both cellular mRNA and protein profiles at the molecular level.

Understanding this technology is important for several reasons: (1) It offers a rational approach to understanding the molecular basis of a number of diseases. For example, familial hypercholesterolemia, sickle cell disease, the thalassemias, cystic fibrosis, muscular dystrophy as well as more complex multifactorial diseases like vascular disease, cancer, and diabetes. (2) Human proteins can be produced in abundance for therapy (eg, insulin, growth hormone, tissue plasminogen activator). (3) Proteins for vaccines (eg, hepatitis B) and for diagnostic testing (eg, Ebola and AIDS tests) can be obtained. (4) This technology is used to diagnose existing diseases and predict the risk of developing a given disease and individual response to pharmacological therapeutics. (5) Special techniques have led to remarkable advances in forensic medicine. (6) Gene therapy for potentially curing diseases caused by a single gene deficiency such as sickle cell disease, the thalassemias, adenosine deaminase deficiency, and others may be devised.

---

*See glossary of terms at the end of this chapter.

RECOMBINANT DNA TECHNOLOGY INVOLVES ISOLATION & MANIPULATION OF DNA TO MAKE CHIMERIC MOLECULES

Isolation and manipulation of DNA, including end-to-end joining of sequences from very different sources to make chimeric molecules (eg, molecules containing both human and bacterial DNA sequences in a sequence-independent fashion), is the essence of recombinant DNA research. This involves several unique techniques and reagents.

Restriction Enzymes Cut DNA Chains at Specific Locations

Certain endonucleases—enzymes that cut DNA at specific DNA sequences within the molecule (as opposed to exonucleases, which digest from the ends of DNA molecules)—are a key tool in recombinant DNA research. These enzymes were called **restriction enzymes** because their presence in a given bacterium restricted the growth of certain bacterial viruses called bacteriophages. Restriction enzymes cut DNA of any source into unique, short pieces in a sequence-specific manner—in contrast to most other enzymatic, chemical, or physical methods, which break DNA randomly. These defensive enzymes (hundreds have been discovered) protect the host bacterial DNA from the DNA genome of foreign organisms (primarily infective phages) by specifically inactivating the invading phage DNA by digestion. However, they are present only in cells that also have a companion enzyme which methylates the host DNA, rendering it an unsuitable substrate for digestion by the restriction enzyme. Thus, **site-specific DNA methylases** and restriction enzymes always exist in pairs in a bacterium.

Restriction enzymes are named after the bacterium from which they are isolated. For example, *EcoRI* is from *Escherichia coli,* and *BamHI* is from *Bacillus amyloliquefaciens* (Table 39–1). The first three letters in the restriction enzyme name consist of the first letter of the genus (*E*) and the first two letters of the species (*co*). These may be followed by a

TABLE 39–1 Selected Restriction Endonucleases and Their Sequence Specificities

Endonuclease	Sequence Recognized Cleavage Sites Shown	Bacterial Source
BamHI	↓ GGATCC CCTACC ↑	Bacillus amyloliquefaciens H
BgIII	↓ AGATCT TCTAGA ↑	Bacillus glolbigii
EcoRI	↓ GAATTC CTTAAC ↑	Escherichia coli RY13
EcoRII	↓ CCTGG GGACC ↑	Escherichia coli R245
HindIII	↓ AAGCTT TTCGAA ↑	Haemophilus influenzae R_d
HhaI	↓ GCGC CGCG ↑	Haemophilus haemolyticus
HpaI	↓ GTTAAC CAATTC ↑	Haemophilus parainfluenzae
MstII	↓ CCTnAGG GGAnTCC ↑	Microcoleus strain
PstI	↓ CTGCAG GACGTC ↑	Providencia stuartii 164
TaqI	↓ TCGA AGCT ↑	Thermus aquaticus YTI

Abbreviations: A, adenine; C, cytosine; G, guanine, T, thymine.
Arrows show the site of cleavage; depending on the site, the ends of the resulting cleaved double-stranded DNA are termed sticky ends (*BamHI*) or blunt ends (*HpaI*). The length of the recognition sequence can be 4 bp (*TaqI*), 5 bp (*EcoRII*), 6 bp (*EcoRI*), or 7 bp (*MstII*) or longer. By convention, these are written in the 5′ to 3′ direction for the upper strand of each recognition sequence, and the lower strand is shown with the opposite (ie, 3′ to 5′) polarity. Note that most recognition sequences are palindromes (ie, the sequence reads the same in opposite directions on the two strands). A residue designated n means that any nucleotide is permitted.

strain designation (*R*) and a roman numeral (*I*) to indicate the order of discovery (eg, *EcoRI, EcoRII*). Each enzyme recognizes and cleaves a specific double-stranded DNA sequence that is typically 4–7 bp long. These DNA cuts result in **blunt ends** (eg, *HpaI*) or overlapping (**sticky or cohesive) ends** (eg,

BamHI) (Figure 39–1), depending on the mechanism used by the enzyme. Sticky ends are particularly useful in constructing hybrid or chimeric DNA molecules (see below). If the four nucleotides are distributed randomly in a given DNA molecule, one can calculate how frequently a given enzyme will cut a length of DNA. For each position in the DNA molecule, there are four possibilities (A, C, G, and T); therefore, a restriction enzyme that recognizes a 4-bp sequence cuts, on average, once every 256 bp (4^4), whereas another enzyme that recognizes a 6-bp sequence cuts once every 4096 bp (4^6). A given piece of DNA has a characteristic linear array of sites for the various enzymes dictated by the linear sequence of its bases; hence, a **restriction map** can be constructed. When DNA is digested with a particular enzyme, the ends of all the fragments have the same DNA sequence. The fragments produced can be isolated by electrophoresis on agarose or polyacrylamide gels (see the discussion of blot transfer, below); this is an essential step in DNA cloning as well as various DNA analyses, and a major use of these enzymes.

A number of other enzymes that act on DNA and RNA are an important part of recombinant DNA technology. Many of these are referred to in this and subsequent chapters (Table 39–2).

Restriction Enzymes & DNA Ligase Are Used to Prepare Chimeric DNA Molecules

Sticky-end ligation is technically easy, but some special techniques are often required to overcome problems inherent in this approach. Sticky ends of a vector may reconnect with themselves, with no net gain of DNA. Sticky ends of fragments also anneal so that heterogeneous tandem inserts form. Also, sticky-end sites may not be available or in a convenient position. To circumvent these problems, an enzyme that generates blunt ends can be used. Blunt ends can be ligated directly, however ligation is not directional. Two alternatives thus exist: new ends are added using the enzyme terminal transferase or synthetic sticky ends are added. If poly d(G) is added to the 3′ ends of the vector and poly d(C) is added to the 3′ ends of the foreign DNA using terminal transferase, the two molecules can only anneal to each other, thus circumventing the problems listed above. This procedure is called homopolymer tailing. Alternatively, synthetic blunt-ended duplex oligonucleotide linkers containing the recognition sequence for a convenient restriction enzyme sequence are ligated to the

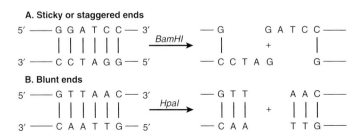

A. Sticky or staggered ends

```
5′ —— G G A T C C —— 3′              — G        G A T C C ——
       | | | | | |        BamHI        |     +         |
3′ —— C C T A G G —— 5′   ———→        — C C T A G        G ——
```

B. Blunt ends

```
5′ —— G T T A A C —— 3′              — G T T      A A C ——
       | | | | | |        HpaI         | | |   +   | | |
3′ —— C A A T T G —— 5′   ———→        — C A A      T T G ——
```

FIGURE 39–1 Results of restriction endonuclease digestion. Digestion with a restriction endonuclease can result in the formation of DNA fragments with sticky, or cohesive, ends **(A)** or blunt ends **(B)**. This is an important consideration in devising cloning strategies.

TABLE 39–2   Some of the Enzymes Used in Recombinant DNA Research

Enzyme	Reaction	Primary Use
Alkaline phosphatase	Dephosphorylates 5' ends of RNA and DNA	Removal of 5'-PO$_4$ groups prior to kinase labeling; also used to prevent self-ligation
BAL 31 nuclease	Degrades both the 3' and 5' ends of DNA	Progressive shortening of DNA molecules
DNA ligase	Catalyzes bonds between DNA molecules	Joining of DNA molecules
DNA polymerase I	Synthesizes double-stranded DNA from single-stranded DNA	Synthesis of double-stranded cDNA; nick translation; generation of blunt ends from sticky ends
DNase I	Under appropriate conditions, produces single-stranded nicks in DNA	Nick translation; mapping of hypersensitive sites; mapping protein-DNA interactions
Exonuclease III	Removes nucleotides from 3' ends of DNA	DNA sequencing; mapping of DNA-protein interactions
λ Exonuclease	Removes nucleotides from 5' ends of DNA	DNA sequencing
Polynucleotide kinase	Transfers terminal phosphate (γ position) from ATP to 5'-OH groups of DNA or RNA	^{32}P end-labeling of DNA or RNA
Reverse transcriptase	Synthesizes DNA from RNA template	Synthesis of cDNA from mRNA; RNA (5' end) mapping studies
S1 nuclease	Degrades single-stranded DNA	Removal of "hairpin" in synthesis of cDNA; RNA mapping studies (both 5' and 3' ends)
Terminal transferase	Adds nucleotides to the 3' ends of DNA	Homopolymer tailing

(Adapted and reproduced, with permission, from Emery AEH: Page 41 in: *An Introduction to Recombinant DNA*. Wiley, 1984. Copyright © 1984 John Wiley & Sons Limited. Reproduced with permission.)

blunt-ended DNA. Direct blunt-end ligation is accomplished using the enzyme bacteriophage T4 DNA ligase. This technique, though less efficient than sticky-end ligation, has the advantage of joining together any pairs of ends. If blunt ends or homopolymer tailing methods are used there is no easy way to retrieve the insert. As an adjunct to the use of restriction endonucleases scientists have utilized specific prokaryotic or eukaryotic recombinases (such as bacterial lox P sites, which are recognized by the CRE recombinase, or yeast FRT sites recognized by the Flp recombinase) to catalyze specific incorporation of two DNA fragments that carry the appropriate recognition sequences. These enzymes catalyze homologous recombination between the relevant recognition sites.

Cloning Amplifies DNA

A **clone** is a large population of identical molecules, bacteria, or cells that arise from a common ancestor. Molecular cloning allows for the production of a large number of identical DNA molecules, which can then be characterized or used for other purposes. This technique is based on the fact that chimeric or hybrid DNA molecules can be constructed in **cloning vectors**—typically bacterial plasmids, phages, or cosmids—which then continue to replicate in a host cell under their own control systems. In this way, the chimeric DNA is amplified. The general procedure is illustrated in Figure 39–2.

Bacterial **plasmids** are small, circular, duplex DNA molecules whose natural function is to confer antibiotic resistance to the host cell. Plasmids have several properties that make them extremely useful as cloning vectors. They exist as single or multiple copies within the bacterium and replicate indepen-

dently from the bacterial DNA. The complete DNA sequence of many plasmids is known; hence, the precise location of restriction enzyme cleavage sites for inserting the foreign DNA is available. Plasmids are smaller than the host chromosome and are therefore easily separated from the latter, and the desired plasmid-inserted DNA is readily removed by cutting the plasmid with the enzyme specific for the restriction site into which the original piece of DNA was inserted.

Phages (bacterial viruses) usually have linear DNA molecules into which foreign DNA can be inserted at several restriction enzyme sites. The chimeric DNA is collected after the phage proceeds through its lytic cycle and produces mature, infective phage particles. A major advantage of phage vectors is that while plasmids accept DNA pieces about 6–10 kb long, phages can accept DNA fragments 10–20 kb long, a limitation imposed by the amount of DNA that can be packed into the phage head.

Larger fragments of DNA can be cloned in **cosmids,** which combine the best features of plasmids and phages. Cosmids are plasmids that contain the DNA sequences, so-called **cos sites,** required for packaging lambda DNA into the phage particle. These vectors grow in the plasmid form in bacteria, but since much of the unnecessary lambda DNA has been removed, more chimeric DNA can be packaged into the particle head. It is not unusual for cosmids to carry inserts of chimeric DNA that are 35–50 kb long. Even larger pieces of DNA can be incorporated into bacterial artificial chromosome (**BAC**), yeast artificial chromosome (**YAC**), or *E coli* bacteriophage P1-based (**PAC**) vectors. These vectors will accept and propagate DNA inserts of several hundred kilobases or more and have largely replaced the plasmid, phage, and cosmid vectors

FIGURE 39–2 Use of restriction nucleases to make new recombinant or chimeric DNA molecules. When inserted back into a bacterial cell (by the process called transformation), typically only a single plasmid is taken up by a single cell, and the plasmid DNA replicates not only itself but also the physically linked new DNA insert. Since recombining the sticky ends, as indicated, regenerates the same DNA sequence recognized by the original restriction enzyme, the cloned DNA insert can be cleanly cut back out of the recombinant plasmid circle with this endonuclease. If a mixture of all of the DNA pieces created by treatment of total human DNA with a single restriction nuclease is used as the source of human DNA, a million or so different types of recombinant DNA molecules can be obtained, each pure in its own bacterial clone. (Modified and reproduced, with permission, from Cohen SN: The manipulation of genes. Sci Am [July] 1975;233:25. Copyright © The Estate of Bunji Tagawa.)

for some cloning and gene mapping applications. A comparison of these vectors is shown in Table 39–3.

Because insertion of DNA into a functional region of the vector will interfere with the action of this region, care must be taken not to interrupt an essential function of the vector. This concept can be exploited, however, to provide a selection technique. For example, a common early plasmid vector **pBR322** has both **tetracycline (tet)** and **ampicillin (amp)** resistance genes. A single *PstI* restriction enzyme site within the amp re-

TABLE 39–3 Cloning Capacities of Common Cloning Vectors

Vector	DNA Insert Size (kb)
Plasmid pBR322	0.01–10
Lambda charon 4A	10–20
Cosmids	35–50
BAC, P1	50–250
YAC	500–3000

sistance gene is commonly used as the insertion site for a piece of foreign DNA. In addition to having sticky ends (Table 39–1 & Figure 39–1), the DNA inserted at this site disrupts the amp resistance gene and makes the bacterium carrying this plasmid amp-sensitive (Figure 39–3). Thus, the parental plasmid, which provides resistance to both antibiotics, can be readily separated from the chimeric plasmid, which is resistant only to tetracycline. YACs contain selection, replication, and segregation functions that work in both bacteria and yeast cells and therefore can be propagated in either organism.

In addition to the vectors described in Table 39–3 that are designed primarily for propagation in bacterial cells, vectors for mammalian cell propagation and insert gene (cDNA)/protein expression have also been developed. These vectors are all based upon various eukaryotic viruses that are composed of RNA or DNA genomes. Notable examples of such **viral vectors** are those utilizing **adenoviral (Ad)**, or **adenovirus-associated viral (AAV)** (DNA-based) and **retroviral** (RNA-based) genomes. Though somewhat limited in the size of DNA sequences that can be inserted, such **mammalian viral cloning vectors** make up for this shortcoming because they will efficiently infect a wide range of different cell types.

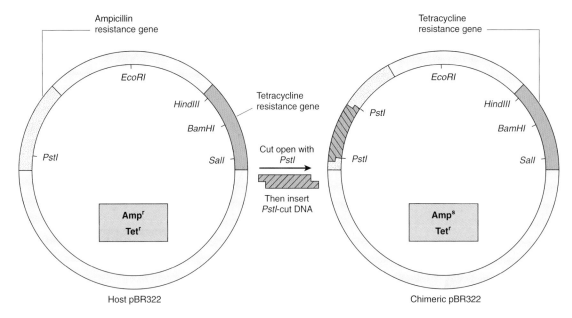

FIGURE 39–3 A method of screening recombinants for inserted DNA fragments. Using the plasmid pBR322, a piece of DNA is inserted into the unique *PstI* site. This insertion disrupts the gene coding for a protein that provides ampicillin resistance to the host bacterium. Hence, cells carrying the chimeric plasmid will no longer survive when plated on a substrate medium that contains this antibiotic. The differential sensitivity to tetracycline and ampicillin can therefore be used to distinguish clones of plasmid that contain an insert. A similar scheme relying upon production of an in-frame fusion of a newly inserted DNA producing a peptide fragment capable of complementing an inactive, deleted form of the enzyme β-galactosidase allows for blue-white colony formation on agar plates containing a dye hydrolyzable by β-galactoside. β-Galactosidase-positive colonies are blue; such colonies contain plasmids in which a DNA was successfully inserted.

For this reason, various mammalian viral vectors are being investigated for use in **gene therapy** and are commonly used for laboratory experiments.

A Library Is a Collection of Recombinant Clones

The combination of restriction enzymes and various cloning vectors allows the entire genome of an organism to be individually packed into a vector. A collection of these different recombinant clones is called a library. A **genomic library** is prepared from the total DNA of a cell line or tissue. A **cDNA library** comprises complementary DNA copies of the population of mRNAs in a tissue. Genomic DNA libraries are often prepared by performing **partial digestion of total DNA** with a restriction enzyme that cuts DNA frequently (eg, a four base cutter such as *TaqI*). The idea is to generate rather large fragments so that most genes will be left intact. The BAC, YAC, and P1 vectors are preferred since they can accept very large fragments of DNA and thus offer a better chance of isolating an intact eukaryotic mRNA-encoding gene on a single DNA fragment.

A vector in which the protein coded by the gene introduced by recombinant DNA technology is actually synthesized is known as an **expression vector.** Such vectors are now commonly used to detect specific cDNA molecules in libraries and to produce proteins by genetic engineering techniques. These vectors are specially constructed to contain very active induc-

ible promoters, proper in-phase translation initiation codons, both transcription and translation termination signals, and appropriate protein processing signals, if needed. Some expression vectors even contain genes that code for protease inhibitors, so that the final yield of product is enhanced. Interestingly as the cost of synthetic DNA synthesis has dropped, many investigators often synthesize an entire cDNA (gene) of interest (in 100–150 nt segments) incorporating the codon preferences of the host used for expression in order to maximize protein production.

Probes Search Libraries or Complex Samples for Specific Genes or cDNA Molecules

A variety of molecules can be used to "probe" libraries in search of a specific gene or cDNA molecule or to define and quantitate DNA or RNA separated by electrophoresis through various gels. Probes are generally pieces of DNA or RNA labeled with a ^{32}P-containing nucleotide—or fluorescently labeled nucleotides (more commonly now). Importantly, neither modification (^{32}P or fluorescent-label) affects the hybridization properties of the resulting labeled nucleic acid probes. The probe must recognize a complementary sequence to be effective. A cDNA synthesized from a specific mRNA can be used to screen either a cDNA library for a longer cDNA or a genomic library for a complementary sequence in the coding region of a gene.

A popular technique for finding specific genes entails taking a short amino acid sequence and, employing the codon usage for that species (see Chapter 37), making an oligonucleotide probe that will detect the corresponding DNA fragment in a genomic library. If the sequences match exactly, probes 15–20 nucleotides long will hybridize. cDNA probes are used to detect DNA fragments on Southern blot transfers and to detect and quantitate RNA on Northern blot transfers. Specific antibodies can also be used as probes provided that the vector used synthesizes protein molecules that are recognized by them.

Blotting & Hybridization Techniques Allow Visualization of Specific Fragments

Visualization of a specific DNA or RNA fragment among the many thousands of "contaminating" molecules in a complex sample requires the convergence of a number of techniques, collectively termed **blot transfer.** Figure 39–4 illustrates the **Southern** (DNA), **Northern** (RNA), and **Western** (protein) blot transfer procedures. (The first is named for the person who devised the technique [Edward Southern], and the other names began as laboratory jargon but are now accepted terms.) These procedures are useful in determining how many copies of a gene are in a given tissue or whether there are any gross alterations in a gene (deletions, insertions, or rearrangements) because the requisite electrophoresis step separates the molecules on the basis of size. Occasionally, if a specific base is changed and a restriction site is altered, these procedures can detect a point mutation. The Northern and Western blot transfer techniques are used to size and quantitate specific RNA and protein molecules, respectively. A fourth hybridization technique, the **Southwestern** blot, examines protein–DNA interactions (not shown). In this method, proteins are separated by electrophoresis, blotted to a membrane, renatured, and analyzed for an interaction with a particular sequence by incubation with a specific labeled DNA probe.

Colony or plaque hybridization is the method by which specific clones are identified and purified. Bacteria are grown as colonies on an agar plate and overlaid with an oriented nitrocellulose filter paper. Cells from each colony stick to the filter and are permanently fixed thereto by heat or UV, which with NaOH treatment also lyses the cells and denatures the DNA so that it is available to hybridize with the probe. A radioactive probe is added to the filter, and (after washing) the hybrid complex is localized by exposing the filter to x-ray film or imaging screen. By matching the spot on the autoradiograph (exposed and developed x-ray film) to a colony, the latter can be picked from the plate. A similar strategy is used to identify fragments in phage libraries. Successive rounds of this procedure result in a clonal isolate (bacterial colony) or individual phage plaque.

All of the hybridization procedures discussed in this section depend on the specific base-pairing properties of complementary nucleic acid strands described above. Perfect matches hybridize readily and withstand high temperatures in the hybridization and washing reactions. Specific complexes

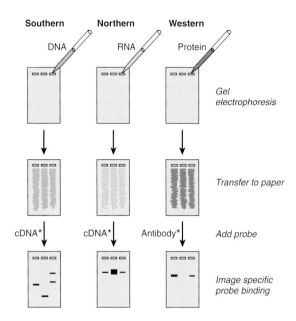

FIGURE 39–4 The blot transfer procedure. In a Southern, or DNA, blot transfer, DNA isolated from a cell line or tissue is digested with one or more restriction enzymes. This mixture is pipetted into a well in an agarose or polyacrylamide gel and exposed to a direct electrical current. DNA, being negatively charged, migrates toward the anode; the smaller fragments move the most rapidly. After a suitable time, the DNA within the gel is denatured by exposure to mild alkali and transferred to nitrocellulose or nylon paper, resulting in an exact replica of the pattern on the gel, by the blotting technique devised by Southern. The DNA is bound to the paper by exposure to heat or UV, and the paper is then exposed to the labeled cDNA probe, which hybridizes to complementary fragments on the filter. After thorough washing, the paper is exposed to x-ray film, which is developed to reveal several specific bands corresponding to the DNA fragment that recognized the sequences in the cDNA probe. The RNA, or Northern, blot is conceptually similar. RNA is subjected to electrophoresis before blot transfer. This requires some different steps from those of DNA transfer, primarily to ensure that the RNA remains intact, and is generally somewhat more difficult. In the protein, or Western, blot, proteins are electrophoresed and transferred to special paper that avidly binds macromolecules and then probed with a specific antibody or other probe molecule. (Asterisks signify labeling, either radioactive or fluorescent.) In the case of Southwestern blotting (see text; not shown), a protein blot similar to that shown above under "Western" is exposed to labeled nucleic acid, and protein-nucleic acid complexes formed are detected by autoradiography.

also form in the presence of low salt concentrations. Less than perfect matches do not tolerate these **stringent conditions** (ie, elevated temperatures and low salt concentrations); thus, hybridization either never occurs or is disrupted during the washing step. Gene families, in which there is some degree of homology, can be detected by varying the stringency of the hybridization and washing steps. Cross-species comparisons of a given gene can also be made using this approach. Hybridization conditions capable of detecting just a single base pair mismatch between probe and target have been devised.

Manual & Automated Techniques Are Available to Determine the Sequence of DNA

The segments of specific DNA molecules obtained by recombinant DNA technology can be analyzed to determine their nucleotide sequence. This method depends upon having a large number of identical DNA molecules. This requirement can be satisfied by cloning the fragment of interest, using the techniques described above. The **manual enzymatic method (Sanger)** employs specific dideoxynucleotides that terminate DNA strand synthesis at specific nucleotides as the strand is synthesized on purified template nucleic acid. The reactions are adjusted so that a population of DNA fragments representing termination at every nucleotide is obtained. By having a radioactive label incorporated at the termination site, one can separate the fragments according to size using polyacrylamide gel electrophoresis. An autoradiograph is made, and each of the fragments produces an image (band) on an x-ray film or imaging plate. These are read in order to give the DNA sequence (Figure 39–5). Another manual method is that of **Maxam and Gilbert,** employs **chemical methods** to cleave the DNA molecules where they contain the specific

nucleotides. Techniques that do not require the use of radioisotopes are employed in automated DNA sequencing. Most commonly employed is an automated procedure in which four different fluorescent labels—one representing each nucleotide—are used. Each emits a specific signal upon excitation by a laser beam of a particular wavelength, and this can be recorded by a computer. The newest DNA sequencing machines use fluorescently labeled nucleotides but detect incorporation using microscopic optics. These machines have reduced the cost of DNA sequencing dramatically, over 100X. These reductions in cost have ushered in the era of personalized genome sequencing. Indeed, using this new technology the sequence of the co-discoverer of the double helix, James Watson, was completely determined.

Oligonucleotide Synthesis Is Now Routine

The automated chemical synthesis of moderately long oligonucleotides (about 100 nucleotides) of precise sequence is now a routine laboratory procedure. Each synthetic cycle takes but a few minutes, so an entire molecule can be made by synthesizing relatively short segments that can then be ligated to one another. Oligonucleotides are now indispensable for DNA se-

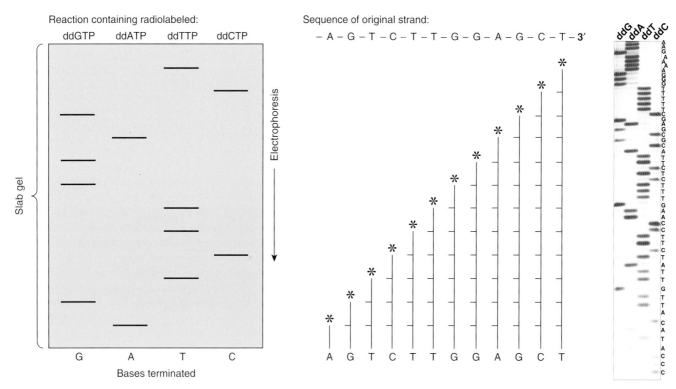

FIGURE 39–5 Sequencing of DNA by the chain termination method devised by Sanger. The ladder-like arrays represent from bottom to top all of the successively longer fragments of the original DNA strand. Knowing which specific dideoxynucleotide reaction was conducted to produce each mixture of fragments, one can determine the sequence of nucleotides from the unlabeled end toward the labeled end (*) by reading up the gel. The base-pairing rules of Watson and Crick (A–T, G–C) dictate the sequence of the other (complementary) strand. (Asterisks signify site of radiolabeling.) Shown (**left, middle**) are the terminated synthesis products of a hypothetical fragment of DNA, sequence shown. An autoradiogram (**right**) of an actual set of DNA sequencing reactions that utilized the four ³²P-labeled dideoxynucleotides indicated at the top of the scanned autoradiogram (ie, dideoxy(dd)G, ddA, ddT, ddC). Electrophoresis was from top to bottom. The deduced DNA sequence is listed on the right side of the gel. Note the log-linear relationship between distance of migration (ie, top to bottom of gel) and DNA fragment length. Current state-of-the-art DNA sequencers no longer utilize gel electrophoresis for fractionation of labeled synthesis products. Moreover in the NGS sequencing platforms, synthesis is followed by monitoring incorporation of the four fluorescently labeled dXTPs.

quencing, library screening, protein–DNA binding assays, the polymerase chain reaction (see below), site-directed mutagenesis, synthetic gene synthesis, and numerous other applications.

The Polymerase Chain Reaction (PCR) Amplifies DNA Sequences

The polymerase chain reaction (PCR) is a method of amplifying a target sequence of DNA. PCR provides a sensitive, selective, and extremely rapid means of amplifying any desired sequence of DNA. Specificity is based on the use of two oligonucleotide primers that hybridize to complementary sequences on opposite strands of DNA and flank the target sequence (Figure 39–6). The DNA sample is first heated to separate the two strands of the template DNA containing the target sequence; the primers, added in vast excess, are allowed to anneal to the DNA; and each strand is copied by a DNA polymerase, starting at the primer sites in the presence of all 4 dXTPs. The two DNA strands each serve as a template for the synthesis of new DNA from the two primers. Repeated cycles of heat denaturation, annealing of the primers to their complementary sequences, and extension of the annealed primers with DNA polymerase result in the exponential amplification of DNA segments of defined length (a doubling at each cycle). Early PCR reactions used an *E coli* DNA polymerase that was destroyed by each heat denaturation cycle. Substitution of a heat-stable DNA polymerase from *Thermus aquaticus* (or the corresponding DNA polymerase from other thermophilic bacteria), an organism that lives and replicates at 70–80°C, obviates this problem and has made possible automation of the reaction, since the polymerase reactions can be run at 70°C. This has also improved the specificity and the yield of DNA.

DNA sequences as short as 50–100 bp and as long as 10 kb can be amplified. Twenty cycles provide an amplification of 10^6 (ie, 2^{20}) and 30 cycles, 10^9 (2^{30}). The PCR allows the DNA in a single cell, hair follicle, or spermatozoon to be amplified and analyzed. Thus, the applications of PCR to forensic medicine are obvious. The PCR is also used (1) to detect infectious agents, especially latent viruses; (2) to make prenatal genetic diagnoses; (3) to detect allelic polymorphisms; (4) to establish precise tissue types for transplants; and (5) to study evolution, using DNA from archeological samples (6) for quantitative RNA analyses after RNA copying and mRNA quantitation by the so-called RT-PCR method (cDNA copies of mRNA generated by a retroviral reverse transcriptase) or (7) to score in vivo protein–DNA occupancy using chromatin immunoprecipitation assays (see below). There are an equal number of applications of PCR to problems in basic science, and new uses are developed every year.

PRACTICAL APPLICATIONS OF RECOMBINANT DNA TECHNOLOGY ARE NUMEROUS

The isolation of a specific (ca. 1000 bp) mRNA-encoding gene from an entire genome requires a technique that will discrimi-

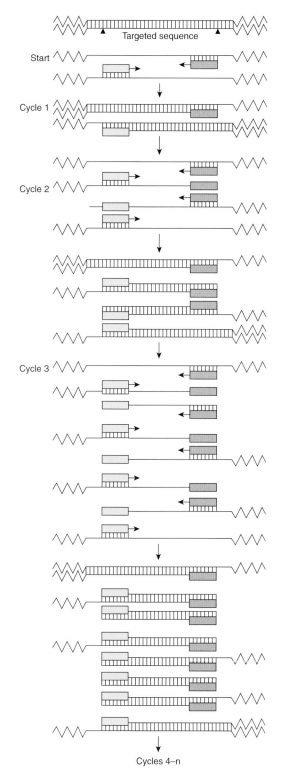

FIGURE 39–6 The polymerase chain reaction is used to amplify specific gene sequences. Double-stranded DNA is heated to separate it into individual strands. These bind two distinct primers that are directed at specific sequences on opposite strands and that define the segment to be amplified. DNA polymerase extends the primers in each direction and synthesizes two strands complementary to the original two. This cycle is repeated several times, giving an amplified product of defined length and sequence. Note that the two primers are present in vast excess.

nate one part in a million. The identification of a regulatory region that may be only 10 bp in length requires a sensitivity of one part in 3×10^8; a disease such as sickle cell anemia is caused by a single base change, or one part in 3×10^9. DNA technology is powerful enough to accomplish all these things.

Gene Mapping Localizes Specific Genes to Distinct Chromosomes

Gene localizing thus can define a map of the human genome. This is already yielding useful information in the definition of human disease. Somatic cell hybridization and in situ hybridization are two techniques used to accomplish this. In **in situ hybridization,** the simpler and more direct procedure, a radioactive probe is added to a metaphase spread of chromosomes on a glass slide. The exact area of hybridization is localized by layering photographic emulsion over the slide and, after exposure, lining up the grains with some histologic identification of the chromosome. **Fluorescence in situ hybridization (FISH), which utilizes fluorescent rather than radioactively labeled probes,** is a very sensitive technique that is also used for this purpose. This often places the gene at a location on a given band or region on the chromosome. Some of the human genes localized using these techniques are listed in Table 39–4. This table represents only a sampling of mapped genes, since thousands of genes have been mapped as a result of the recent sequencing of the human genome. Once the defect is localized to a region of DNA that has the characteristic structure of a gene, a synthetic cDNA copy of the gene can be constructed, which contains only mRNA encoding exons, and expressed in an appropriate vector and its function can be assessed—or the putative peptide, deduced from the open reading frame in the

coding region, can be synthesized. Antibodies directed against this peptide can be used to assess whether this peptide is expressed in normal persons and whether it is absent, or altered in those with the genetic syndrome.

Proteins Can Be Produced for Research & Diagnosis

A practical goal of recombinant DNA research is the production of materials for biomedical applications. This technology has two distinct merits: (1) It can supply large amounts of material that could not be obtained by conventional purification methods (eg, interferon, tissue plasminogen activating factor). (2) It can provide human material (eg, insulin, growth hormone). The advantages in both cases are obvious. Although the primary aim is to supply products—generally proteins—for treatment (insulin) and diagnosis (AIDS testing) of human and other animal diseases and for disease prevention (hepatitis B vaccine), there are other potential commercial applications, especially in agriculture. An example of the latter is the attempt to engineer plants that are more resistant to drought or temperature extremes, more efficient at fixing nitrogen, or that produce seeds containing the complete complement of essential amino acids (rice, wheat, corn, etc).

Recombinant DNA Technology Is Used in the Molecular Analysis of Disease

Normal Gene Variations

There is a normal variation of DNA sequence just as is true of more obvious aspects of human structure. Variations of DNA sequence, **polymorphisms,** occur approximately once in every 500–1000 nucleotides. A recent comparison of the nucleotide sequence of the genome of James Watson, the co-discoverer of DNA structure, identified about 3,300,000 single-nucleotide polymorphisms (SNPs) relative to the "standard" initially sequenced human reference genome. Interestingly, over 80% of the SNPs found in Watson's DNA had already been identified in other individuals. There are also genomic deletions and insertions of DNA (ie, copy number variations; **CNV**) as well as single-base substitutions. In healthy people, these alterations obviously occur in noncoding regions of DNA or at sites that cause no change in function of the encoded protein. This heritable polymorphism of DNA structure can be associated with certain diseases within a large kindred and can be used to search for the specific gene involved, as is illustrated below. It can also be used in a variety of applications in forensic medicine.

Gene Variations Causing Disease

Classic genetics taught that most genetic diseases were due to point mutations which resulted in an impaired protein. This may still be true, but if on reading previous chapters one predicted that genetic disease could result from derangement of any of the steps leading from replication to transcription to RNA processing/transport and protein synthesis, one would have made a proper assessment. This point is again nicely

TABLE 39–4 Localization of Human Genes[1]

Gene	Chromosome	Disease
Insulin	11p15	
Prolactin	6p23-q12	
Growth hormone	17q21-qter	Growth hormone deficiency
α-Globin	16p12-pter	α-Thalassemia
β-Globin	11p12	β-Thalassemia, sickle cell
Adenosine deaminase	20q13-qter	Adenosine deaminase deficiency
Phenylalanine hydroxylase	12q24	Phenylketonuria
Hypoxanthine-guanine phosphoribosyltransferase	Xq26-q27	Lesch-Nyhan syndrome
DNA segment G8	4p	Huntington chorea

[1]This table indicates the chromosomal location of several genes and the diseases associated with deficient or abnormal production of the gene products. The chromosome involved is indicated by the first number or letter. The other numbers and letters refer to precise localizations, as defined in McKusick, Victor A., MD, *Mendelian Inheritance in Man: Catalogs of Autosomal Dominant, Autosomal Recessive, and X-Linked Phenotypes.* Copyright © 1983 Johns Hopkins University Press. Reprinted with permission from the Johns Hopkins University Press.

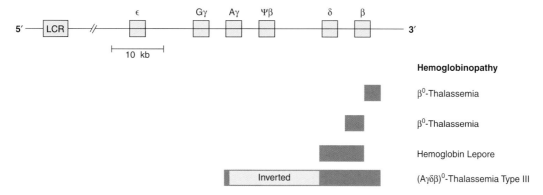

FIGURE 39–7 Schematic representation of the β-globin gene cluster and of the lesions in some genetic disorders. The β-globin gene is located on chromosome 11 in close association with the two γ-globin genes and the δ-globin gene. The β-gene family is arranged in the order 5'-ε-Gγ-Aγ-Ψβ-δ-β-3'. The ε locus is expressed in early embryonic life (as $\alpha_2\varepsilon_2$). The γ genes are expressed in fetal life, making fetal hemoglobin (HbF, $\alpha_2\gamma_2$). Adult hemoglobin consists of HbA ($\alpha_2\beta_2$) or HbA$_2(\alpha_2\delta_2)$. The Ψβ is a pseudogene that has sequence homology with β but contains mutations that prevent its expression. A locus control region (LCR), a powerful enhancer located upstream (5') from the gene, controls the rate of transcription of the entire β-globin gene cluster. Deletions (solid bar) of the β locus cause β-thalassemia (deficiency or absence [β^0] of β-globin). A deletion of δ and β causes hemoglobin Lepore (only hemoglobin is present). An inversion (Aγδβ)^0 in this region (largest bar) disrupts gene function and also results in thalassemia (type III). Each type of thalassemia tends to be found in a certain group of people, eg, the (Aγδβ)^0 deletion inversion occurs in persons from India. Many more deletions in this region have been mapped, and each causes some type of thalassemia.

illustrated by examination of the β-globin gene. This gene is located in a cluster on chromosome 11 (Figure 39–7), and an expanded version of the gene is illustrated in Figure 39–8. Defective production of β-globin results in a variety of diseases and is due to many different lesions in and around the β-globin gene (Table 39–5).

Point Mutations

The classic example is **sickle cell disease,** which is caused by mutation of a single base out of the 3×10^9 in the genome, a T-to-A DNA substitution, which in turn results in an A-to-U change in the mRNA corresponding to the sixth codon of the β-globin gene. The altered codon specifies a different amino acid (valine rather than glutamic acid), and this causes a

structural abnormality of the β-globin molecule. Other point mutations in and around the β-globin gene result in decreased production or, in some instances, no production of β-globin; β-thalassemia is the result of these mutations. (The thalassemias are characterized by defects in the synthesis of hemoglobin subunits, and so β-thalassemia results when there is insufficient production of β-globin.) Figure 39–8 illustrates that point mutations affecting each of the many processes involved in generating a normal mRNA (and therefore a normal protein) have been implicated as a cause of β-thalassemia.

Deletions, Insertions, & Rearrangements of DNA

Studies of bacteria, viruses, yeasts, fruit flies, and now humans show that pieces of DNA can move from one place to

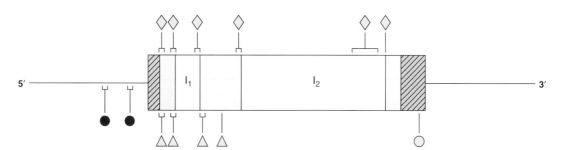

FIGURE 39–8 Mutations in the β-globin gene causing β-thalassemia. The β-globin gene is shown in the 5' to 3' orientation. The cross-hatched areas indicate the 5' and 3' nontranslated regions. Reading from the 5' to 3' direction, the shaded areas are exons 1–3 and the clear spaces are introns 1 (I$_1$) and 2 (I$_2$). Mutations that affect transcription control (●) are located in the 5' flanking-region DNA. Examples of nonsense mutations (△), mutations in RNA processing (◇), and RNA cleavage mutations (○) have been identified and are indicated. In some regions, many distinct mutations have been found. These are indicated by the brackets.

TABLE 39–5   Structural Alterations of the β-Globin Gene

Alteration	Function Affected	Disease
Point mutations	Protein folding	Sickle cell disease
	Transcriptional control	β-Thalassemia
	Frameshift and nonsense mutations	β-Thalassemia
	RNA processing	β-Thalassemia
Deletion	mRNA production	β⁰-Thalassemia
		Hemoglobin Lepore
Rearrangement	mRNA production	β-Thalassemia type III

another within a genome. The deletion of a critical piece of DNA, the rearrangement of DNA within a gene, or the insertion or amplification of a piece of DNA within a coding or regulatory region can all cause changes in gene expression resulting in disease. Again, a molecular analysis of thalassemias produces numerous examples of these processes—particularly deletions—as causes of disease (Figure 39–7). The globin gene clusters seem particularly prone to this lesion. Deletions in the α-globin cluster, located on chromosome 16, cause α-thalassemia. There is a strong ethnic association for many of these deletions, so that northern Europeans, Filipinos, blacks, and Mediterranean peoples have different lesions all resulting in the absence of hemoglobin A and α-thalassemia.

A similar analysis could be made for a number of other diseases. Point mutations are usually defined by sequencing the gene in question, though occasionally, if the mutation destroys or creates a restriction enzyme site, the technique of restriction fragment analysis can be used to pinpoint the lesion. Deletions or insertions of DNA larger than 50 bp can often be detected by the Southern blotting procedure and/or PCR.

Pedigree Analysis

Sickle cell disease again provides an excellent example of how recombinant DNA technology can be applied to the study of human disease. The substitution of T for A in the template strand of DNA in the β-globin gene changes the sequence in the region that corresponds to the sixth codon from

$$\begin{array}{ll} \text{C C T G A G G} & \text{Coding strand} \\ \text{G G A C } \textcircled{T}\text{ C C} & \text{Template strand} \end{array}$$

to

$$\begin{array}{ll} \text{C C T G T G G} & \text{Coding strand} \\ \text{G G A C } \textcircled{A}\text{ C C} & \text{Template strand} \end{array}$$

and destroys a recognition site for the restriction enzyme *MstII* (CCTNAGG; denoted by the small vertical arrows; Table 39–1). Other *MstII* sites 5′ and 3′ from this site (Figure 39–9) are not affected and so will be cut. Therefore, incubation of DNA from normal (AA), heterozygous (AS), and homozygous

(SS) individuals results in three different patterns on Southern blot transfer (Figure 39–9). This illustrates how a DNA pedigree can be established using the principles discussed in this chapter. Pedigree analysis has been applied to a number of genetic diseases and is most useful in those caused by deletions and insertions or the rarer instances in which a restriction endonuclease cleavage site is affected, as in the example cited here. Such analyses are now facilitated by the PCR reaction, which can amplify and hence provide sufficient DNA for analysis from just a few nucleated cells.

Prenatal Diagnosis

If the genetic lesion is understood and a specific probe is available, prenatal diagnosis is possible. DNA from cells collected from as little as 10 mL of amniotic fluid (or by chorionic villus biopsy) can be analyzed by Southern blot transfer. A fetus with the restriction pattern AA in Figure 39–9 does not have sickle cell disease, nor is it a carrier. A fetus with the SS pattern will develop the disease. Probes are now available for this type of analysis of many genetic diseases.

Restriction Fragment Length Polymorphism (RFLP) and Single Nucleotide Polymorphisms (SNPs)

The differences in DNA sequence cited above can result in variations of restriction sites and thus in the length of restriction fragments. Similarly, single nucleotide polymorphisms, or **SNPs**, can be detected by the sensitive PCR method. An inherited difference in the pattern of restriction enzyme digestion enzyme digestion (eg, a DNA variation occurring in more than 1% of the general population) is known as a restriction fragment length polymorphism, or RFLP. Extensive RFLP and SNP maps of the human genome have been constructed. This is proving useful in the Human Genome Analysis Project and is an important component of the effort to understand various single-gene and multigenic diseases. RFLPs result from single-base changes (eg, sickle cell disease) or from deletions or insertions (CNVs) of DNA into a restriction fragment (eg, the thalassemias) and have proved to be useful diagnostic tools. They have been found at known gene loci and in sequences that have no known function; thus, RFLPs may disrupt the function of the gene or may have no apparent biologic consequences. As mentioned above, 80% of the SNPs in the genome of a single known individual had already been mapped independently through the efforts of the SNP-mapping component of the International HapMap Project.

RFLPs and SNPs are inherited, and they segregate in a mendelian fashion. A major use of SNPs/RFLPs is in the definition of inherited diseases in which the functional deficit is unknown. SNPs/RFLPs can be used to establish linkage groups, which in turn, by the process of **chromosome walking,** will eventually define the disease locus. In chromosome walking (Figure 39–10), a fragment representing one end of a long piece of DNA is used to isolate another that overlaps but extends the first. The direction of extension is determined by restriction mapping, and the procedure is repeated sequentially

A. *MstII* restriction sites around and in the β-globin gene

B. Pedigree analysis

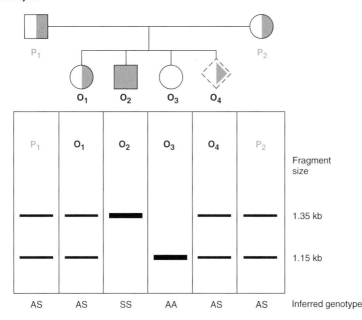

FIGURE 39–9 Pedigree analysis of sickle cell disease. The top part of the figure **(A)** shows the first part of the β-globin gene and the *MstII* restriction enzyme sites in the normal (A) and sickle cell (S) β-globin genes. Digestion with the restriction enzyme *MstII* results in DNA fragments 1.15 kb and 0.2 kb long in normal individuals. The T-to-A change in individuals with sickle cell disease abolishes one of the three *MstII* sites around the β-globin gene; hence, a single restriction fragment 1.35 kb in length is generated in response to *MstII*. This size difference is easily detected on a Southern blot. (The 0.2-kb fragment would run off the gel in this illustration.) **(B)** Pedigree analysis shows three possibilities: AA = normal (open circle); AS = heterozygous (half-solid circles, half-solid square); SS = homozygous (solid square). This approach can allow for prenatal diagnosis of sickle cell disease (dash-sided square). See text.

until the desired sequence is obtained. Collections of mapped, overlapping BAC- or PAC-cloned human genomic DNAs are commercially available. The X chromosome–linked disorders are particularly amenable to the approach of chromosome walking, since only a single allele is expressed. Hence, 20% of the defined RFLPs are on the X chromosome, and a reasonably

complete linkage map of this chromosome exists. The gene for the X-linked disorder, Duchenne-type muscular dystrophy, was found using RFLPs. Similarly, the defect in Huntington disease was localized to the terminal region of the short arm of chromosome 4, and the defect that causes polycystic kidney disease is linked to the α-globin locus on chromosome 16.

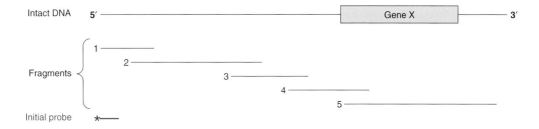

FIGURE 39–10 The technique of chromosome walking. Gene X is to be isolated from a large piece of DNA. The exact location of this gene is not known, but a probe (*——) directed against a fragment of DNA (shown at the 5′ end in this representation) is available, as is a library of clones containing a series of overlapping DNA insert fragments. For the sake of simplicity, only five of these are shown. The initial probe will hybridize only with clones containing fragment 1, which can then be isolated and used as a probe to detect fragment 2. This procedure is repeated until fragment 4 hybridizes with fragment 5, which contains the entire sequence of gene X.

Microsatellite DNA Polymorphisms

Short (2–6 bp), inherited, tandem repeat units of DNA occur about 50,000–100,000 times in the human genome (Chapter 35). Because they occur more frequently—and in view of the routine application of sensitive PCR methods—they are replacing RFLPs as the marker loci for various genome searches.

RFLPs & VNTRs in Forensic Medicine

Variable numbers of tandemly repeated (VNTR) units are one common type of "insertion" that results in an RFLP. The VNTRs can be inherited, in which case they are useful in establishing genetic association with a disease in a family or kindred; or they can be unique to an individual and thus serve as a molecular fingerprint of that person.

Direct Sequencing of Genomic DNA

As noted above, recent advances in DNA sequencing technology, the so-called next generation sequencing (NGS) platforms, have dramatically reduced the per base cost of DNA sequencing. The initial sequence of the human genome cost roughly $350,000,000. The cost of sequencing the same 3×10^9 bp diploid human genome using the new NGS platforms is estimated to be <1% of the original. This dramatic reduction in cost has stimulated various international initiatives to sequence the entire genomes of thousands of individuals of various racial and ethnic backgrounds in order to determine the true extent of DNA/genome polymorphisms present within the population. The resulting cornucopia of genetic information, and the ever decreasing cost of genomic DNA sequencing will dramatically increase our ability to diagnose and, ultimately treat human disease. Obviously, when personal genome sequencing does become commonplace, dramatic changes in the practice of medicine will result.

Gene Therapy and Stem Cell Biology

Diseases caused by deficiency of a single gene product (Table 39–4) are all theoretically amenable to replacement therapy. The strategy is to clone a gene (eg, the gene that codes for adenosine deaminase) into a vector that will readily be taken up and incorporated into the genome of a host cell. Bone marrow precursor cells are being investigated for this purpose because they presumably will resettle in the marrow and replicate there. The introduced gene would begin to direct the expression of its protein product, and this would correct the deficiency in the host cell.

As an alternative to "replacing" defective genes to cure human disease, many scientists are investigating the feasibility of identifying and characterizing pluripotent stem cells that have the ability to differentiate into any cell type in the body. Recent results in this field have shown that adult human somatic cells can readily be converted into apparent pluripotent stem cells by transfection with cDNAs encoding a handful of DNA binding transcription factors. These and other new developments in the fields of gene therapy and stem cell biology promise exciting new potential therapies for curing human disease.

Transgenic Animals

The somatic cell gene replacement therapy described above would obviously not be passed on to offspring. Other strategies to alter germ cell lines have been devised but have been tested only in experimental animals. A certain percentage of genes injected into a fertilized mouse ovum will be incorporated into the genome and found in both somatic and germ cells. Hundreds of transgenic animals have been established, and these are useful for analysis of tissue-specific effects on gene expression and effects of overproduction of gene products (eg, those from the growth hormone gene or oncogenes) and in discovering genes involved in development—a process that heretofore has been difficult to study. The transgenic approach has been used to correct a genetic deficiency in mice. Fertilized ova obtained from mice with genetic hypogonadism were injected with DNA containing the coding sequence for the gonadotropin-releasing hormone (GnRH) precursor protein. This gene was expressed and regulated normally in the hypothalamus of a certain number of the resultant mice, and these animals were in all respects normal. Their offspring also showed no evidence of GnRH deficiency. This is, therefore, evidence of somatic cell expression of the transgene and of its maintenance in germ cells.

Targeted Gene Disruption or Knockout

In transgenic animals, one is adding one or more copies of a gene to the genome, and there is no way to control where that gene eventually resides. A complementary—and much more difficult—approach involves the selective removal of a gene from the genome. Gene knockout animals (usually mice) are made by creating a mutation that totally disrupts the function of a gene. This is then used to replace one of the two genes in an embryonic stem cell that can be used to create a heterozygous transgenic animal. The mating of two such animals will, by mendelian genetics, result in a homozygous mutation in 25% of offspring. Several thousand strains of mice with knockouts of specific genes have been developed. Techniques for disrupting genes in specific cells, tissues, or organs have been developed, so-called conditional, or directed, knockouts. This can be accomplished by taking advantage of particular promoter-enhancer combinations driving expression of DNA recombinases or miRNAs, both of which inactivate gene expression. These methods are particularly useful in cases where gene ablation during early development causes embryonic lethality.

RNA and Protein Profiling, and Protein-DNA Interaction Mapping

The "-omic" revolution of the last decade has culminated in the determination of the nucleotide sequences of entire genomes, including those of budding and fission yeasts, numerous bacteria, the fruit fly, the worm *Caenorhabditis elegans*, plants, the mouse, rat, chicken, monkey and, most notably, humans. Additional genomes are being sequenced at an accelerating pace. The availability of all of this DNA sequence information, coupled with engineering advances, has lead to the

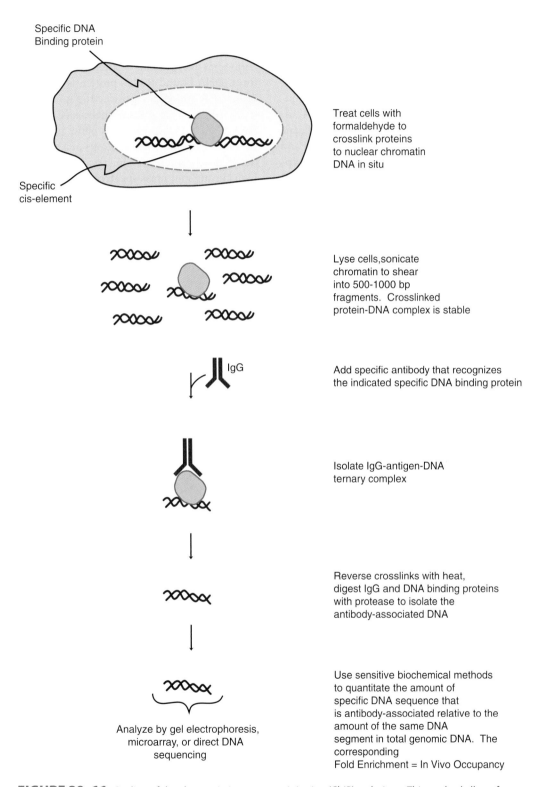

Specific DNA
Binding protein

Specific
cis-element

Treat cells with
formaldehyde to
crosslink proteins
to nuclear chromatin
DNA in situ

Lyse cells,sonicate
chromatin to shear
into 500-1000 bp
fragments. Crosslinked
protein-DNA complex is stable

IgG

Add specific antibody that recognizes
the indicated specific DNA binding protein

Isolate IgG-antigen-DNA
ternary complex

Reverse crosslinks with heat,
digest IgG and DNA binding proteins
with protease to isolate the
antibody-associated DNA

Use sensitive biochemical methods
to quantitate the amount of
specific DNA sequence that
is antibody-associated relative to the
amount of the same DNA
segment in total genomic DNA. The
corresponding
Fold Enrichment = In Vivo Occupancy

Analyze by gel electrophoresis,
microarray, or direct DNA
sequencing

FIGURE 39–11 Outline of the chromatin immunoprecipitation (ChIP) technique. This method allows for
the precise localization of a particular protein (or modified protein if an appropriate antibody is available; eg,
phosphorylated or acetylated histones, transcription factors, etc) on a particular sequence element in living
cells. Depending upon the method used to analyze the immunopurified DNA, quantitative or semi-quantitative
information, at near nucleotide level resolution, can be obtained.

development of several revolutionary methodologies, most of which are based upon **high-density microarray technology** or **NGS next generation sequencing platforms.** In the case of microarrays, it is now possible to deposit thousands of specific, known, definable DNA sequences (more typically now synthetic oligonucleotides) on a glass microscope-style slide in the space of a few square centimeters. By coupling such DNA microarrays with highly sensitive detection of hybridized fluorescently labeled nucleic acid probes derived from mRNA, investigators can rapidly and accurately generate profiles of gene expression (eg, specific cellular mRNA content) from cell and tissue samples as small as 1 g or less. Thus entire **transcriptome information** (the entire collection of cellular mRNAs) for such cell or tissue sources can readily be obtained in only a few days. In the case of NGS sequencing, mRNAs are converted to cDNAs using reverse transcription, and these cDNAs are directly sequenced. Both methods allow for the description of the entire transcriptome. Transcriptome information allows one to predict the collection of proteins that might be expressed in a particular cell, tissue, or organ in normal and disease states based upon the mRNAs present in those cells. Complementing this high-throughput, transcript-profiling method is the recent development of methods to map the loca-

tion, or occupancy of specific proteins bound to discrete sites within living cells. This method, illustrated in Figure 39–11, is termed chromatin immunoprecipitation (ChIP). Proteins are crosslinked in situ in cells or tissues, chromatin isolated, sheared, and specific protein DNA complexes purified using antibodies recognizing a particular protein, or protein isoform. DNA bound to this protein is recovered and analyzed using PCR and either gel electrophoresis, direct sequencing (ChIP-SEQ) or microarray analysis (ChIP-chip). Both ChIP-SEQ and ChIP-chip methods allow investigators to identify the entire genome-wide locations of a single protein throughout all the chromosomes. Finally, methods for high-sensitivity, high-throughput **mass spectrometry of metabolites (metabolomics) and complex protein samples (proteomics)** have been developed. Newer mass spectrometry methods allow one to identify hundreds to thousands of proteins in samples extracted from very small numbers of cells (<1 g). Such analyses can now be used to quantitate the amounts of proteins in two samples. This critical information tells investigators which of the many mRNAs detected in transcript microarray mapping studies are actually translated into protein, generally the ultimate dictator of phenotype. New genetic means for identifying protein–protein interactions and protein function have

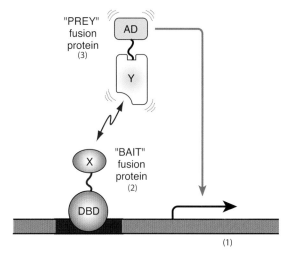

FIGURE 39–12 Overview of two hybrid system for identifying and characterizing protein-protein interactions. Shown are the basic components and operation of the two hybrid system, originally devised by Fields and Song (*Nature* 340:245–246 [1989]) to function in the bakers yeast system. **(1)** A reporter gene, either a selectable marker (ie, a gene conferring prototrophic growth on selective media, or producing an enzyme for which a colony colorimetric assay exists, such as β-galactosidase) that is expressed only when a transcription factor binds upstream to a *cis*-linked enhancer (dark red bar). **(2)** A "bait" fusion protein **(DBD-X)** produced from a chimeric gene expressing a modular DNA binding domain **(DBD;** often derived from the yeast Gal 4 protein or the bacterial Lex A protein, both high-affinity, high-specificity DNA binding proteins) fused in-frame to a protein of interest, here X. In two hybrid experiments, one is testing whether any protein can interact with protein X. Prey protein X may be fused in its entirety or often alternatively just a portion of protein X is expressed in-frame with the DBD. **(3)** A "prey" protein **(Y-AD),** which represents a fusion of a specific protein fused in-frame to a transcriptional activation domain **(AD;** often derived from either the *Herpes simplex* virus VP16 protein or the yeast Gal 4 protein). This system serves as a useful test of protein-protein interactions between proteins X and Y because in the absence of a functional transactivator binding to the indicated enhancer, no transcription of the reporter gene occurs (ie, see Figure 38–16). Thus, one observes transcription only if protein X–protein Y interaction occurs, thereby bringing a functional AD to the *cis*-linked transcription unit, in this case activating transcription of the reporter gene. In this scenario, protein DBD-X alone fails to activate reporter transcription because the X-domain fused to the DBD does not contain an AD. Similarly, protein Y-AD alone fails to activate reporter gene transcription because it lacks a DBD to target the Y-AD protein to the enhancer. Only when both proteins are expressed in a single cell and bind the enhancer and, via DBD-X–Y-AD protein–protein interactions, regenerate a functional transactivator binary "protein," does reporter gene transcription result in activation and mRNA synthesis (line from AD to reporter gene).

also been devised. Systematic genome-wide gene expression knockdown, using SiRNAs (miRNAs), or synthetic lethal genetic interaction screens, have been applied to assess the contribution of individual genes to a variety of processes in model systems (yeast, worms, flies) and mammalian cells (human and mouse). Specific network mappings of protein–protein interactions on a genome-wide basis have been identified using high-throughput variants of the two hybrid interaction test (Figure 39–12). This simple yet powerful method can be performed in bacteria, yeast, or metazoan cells, and allows for detecting specific protein–protein interactions in living cells. Reconstruction experiments indicate that protein–protein interactions with affinities of K_d ~1 μM or tighter can readily be detected with this method. Together, these technologies provide powerful new tools with which to dissect the intricacies of human biology.

Microarray techniques, high-throughput DNA sequencing, two-hybrid genetic knockdown, and mass spectrometric protein and metabolite identification experiments have led to the generation of huge amounts of data. Appropriate data management and interpretation of the deluge of information forthcoming from such studies have relied upon statistical methods; and this new technology, coupled with the flood of DNA sequence information, has led to the development of the fields of **bioinformatics and systems biology,** new disciplines whose goals are to help manage, analyze, and integrate this flood of biologically important information. Future work at the intersection of bioinformatics, transcript-protein profiling, and systems biology will revolutionize our understanding of physiology and medicine.

SUMMARY

- A variety of very sensitive techniques can now be applied to the isolation and characterization of genes and to the quantitation of gene products.

- In DNA cloning, a particular segment of DNA is removed from its normal environment using PCR or one of many restriction endonucleases. This is then ligated into a vector in which the DNA segment can be amplified and produced in abundance.

- Cloned DNA can be used as a probe in one of several types of hybridization reactions to detect other related or adjacent pieces of DNA, or it can be used to quantitate gene products such as mRNA.

- Manipulation of the DNA to change its structure, so-called genetic engineering, is a key element in cloning (eg, the construction of chimeric molecules) and can also be used to study the function of a certain fragment of DNA and to analyze how genes are regulated.

- Chimeric DNA molecules are introduced into cells to make transfected cells or into the fertilized oocyte to make transgenic animals.

- Techniques involving cloned DNA are used to locate genes to specific regions of chromosomes, identify the genes responsible for diseases, study how faulty gene regulation causes disease, diagnose genetic diseases, and increasingly to treat genetic diseases.

REFERENCES

Brass Al et al: Identification of host proteins required for HIV infection through a functional genomic screen. Science 2008;319:921.

Friedman A, Perrimon N: Genome-wide high-throughput screens in functional genomics. Curr Opin Gen Dev 2004;14:470.

Gandhi TKB et al: Analysis of the human protein interactome and comparison with yeast, worm and fly interaction datasets. Nat Genet 2006;38:285.

Martin JB, Gusella JF: Huntington's disease: pathogenesis and management. N Engl J Med 1986:315:1267.

Sambrook J, Fritsch EF, Maniatis T: *Molecular Cloning: A Laboratory Manual.* Cold Spring Harbor Laboratory Press, 1989.

Schlabach MR et al: Cancer proliferation gene discovery through functional genomics. Science 2008;319:620.

Suter B, Kittanakom S, Stagljar I: Interactive proteomics: what lies ahead? Biotechniques 2008;44:681.

Spector DL, Goldman RD, Leinwand LA: *Cells: A Laboratory Manual.* Cold Spring Harbor Laboratory Press, 1998.

Takahashi K et al: Induction of pluripotent stem cells from adult human fibroblasts by defined factors. Cell 2007;131:861.

The ENCODE Project Consortium: Identification and analysis of functional in 1% of the human genome by the ENCODE Pilot Project. Nature 2007;447:799.

Watson JD et al: *Recombinant DNA,* 2nd ed. Scientific American Books. Freeman, 1992.

Weatherall DJ: *The New Genetics and Clinical Practice,* 3rd ed. Oxford Univ Press, 1991.

Werning M et al: In vitro reprogramming of fibroblasts into a pluripotent ES-cell-like state. Nature 2007;448:318.

Wheeler DA et al: The complete genome of an individual by massively parallel DNA sequencing. Nature 2008;451:872.

Wold B, Myers RM: Sequence census methods for functional genomics. Nat Meth 2008;5:19.

GLOSSARY

ARS: Autonomously replicating sequence; the origin of replication in yeast.

Autoradiography: The detection of radioactive molecules (eg, DNA, RNA, protein) by visualization of their effects on photographic or x-ray film.

Bacteriophage: A virus that infects a bacterium.

Blunt-ended DNA: Two strands of a DNA duplex having ends that are flush with each other.

cDNA: A single-stranded DNA molecule that is complementary to an mRNA molecule and is synthesized from it by the action of reverse transcriptase.

Chimeric molecule: A molecule (eg, DNA, RNA, protein) containing sequences derived from two different species.

Clone: A large number of organisms, cells or molecules that are identical with a single parental organism cell or molecule.

Copy number variation (CNV): Change in the copy number of specific genomic regions of DNA between two or more individuals. CNVs can be as large as 10^6 bp of DNA and include deletions or insertions.

Cosmid: A plasmid into which the DNA sequences from bacteriophage lambda that are necessary for the packaging of DNA (cos sites) have been inserted; this permits the plasmid DNA to be packaged in vitro.

ENCODE Project: Encyclopedia of DNA Elements Project; an effort of multiple laboratories throughout the world to provide a detailed, biochemically informative representation of the human genome using high throughput sequencing methods to identify and catalog the functional elements within a single restricted portion (~1%; 30,000,000 bp) of one human chromosome.

Endonuclease: An enzyme that cleaves internal bonds in DNA or RNA.

Epigenetic code: The patterns of modification of chromosomal DNA (ie, cytosine methylation) and nucleosomal histone post-translational modifications. These changes in modification status can lead to dramatic alterations in gene expression. Notably though, the actual underlying DNA sequence involved does not change.

Excinuclease: The excision nuclease involved in nucleotide exchange repair of DNA.

Exon: The sequence of a gene that is represented (expressed) as mRNA.

Exonuclease: An enzyme that cleaves nucleotides from either the 3′ or 5′ ends of DNA or RNA.

Fingerprinting: The use of RFLPs or repeat sequence DNA to establish a unique pattern of DNA fragments for an individual.

FISH: Fluorescence in situ hybridization, a method used to map the location of specific DNA sequences within fixed nuclei.

Footprinting: DNA with protein bound is resistant to digestion by DNase enzymes. When a sequencing reaction is performed using such DNA, a protected area, representing the "footprint" of the bound protein, will be detected because nucleases are unable to cleave the DNA directly bound by the protein.

Hairpin: A double-helical stretch formed by base pairing between neighboring complementary sequences of a single strand of DNA or RNA.

Hybridization: The specific reassociation of complementary strands of nucleic acids (DNA with DNA, DNA with RNA, or RNA with RNA).

Insert: An additional length of base pairs in DNA, generally introduced by the techniques of recombinant DNA technology.

Intron: The sequence of an mRNA-encoding gene that is transcribed but excised before translation. tRNA genes can also contain introns.

Library: A collection of cloned fragments that represents, in aggregate, the entire genome. Libraries may be either genomic DNA (in which both introns and exons are represented) or cDNA (in which only exons are represented).

Ligation: The enzyme-catalyzed joining in phosphodiester linkage of two stretches of DNA or RNA into one; the respective enzymes are DNA and RNA ligases.

Lines: Long interspersed repeat sequences.

Microsatellite polymorphism: Heterozygosity of a certain microsatellite repeat in an individual.

Microsatellite repeat sequences: Dispersed or group repeat sequences of 2–5 bp repeated up to 50 times. May occur at 50–100 thousand locations in the genome.

miRNAs: MicroRNAs, 21–22 nucleotide long RNA species derived from RNA polymerase II transcription units, 500–1500 bp in length via RNA processing. These RNAs, recently discovered, are thought to play crucial roles in gene regulation.

Nick translation: A technique for labeling DNA based on the ability of the DNA polymerase from E coli to degrade a strand of DNA that has been nicked and then to resynthesize the strand; if a radioactive nucleoside triphosphate is employed, the rebuilt strand becomes labeled and can be used as a radioactive probe.

Northern blot: A method for transferring RNA from an agarose or polyacrylamide gel to a nitrocellulose filter, on which the RNA can be detected by a suitable probe.

Oligonucleotide: A short, defined sequence of nucleotides joined together in the typical phosphodiester linkage.

Ori: The origin of DNA replication.

PAC: A high-capacity (70–95 kb) cloning vector based upon the lytic E coli bacteriophage P1 that replicates in bacteria as an extrachromosomal element.

Palindrome: A sequence of duplex DNA that is the same when the two strands are read in opposite directions.

Plasmid: A small, extrachromosomal, circular molecule of DNA that replicates independently of the host DNA.

Polymerase chain reaction (PCR): An enzymatic method for the repeated copying (and thus amplification) of the two strands of DNA that make up a particular gene sequence.

Primosome: The mobile complex of helicase and primase that is involved in DNA replication.

Probe: A molecule used to detect the presence of a specific fragment of DNA or RNA in, for instance, a bacterial colony that is formed from a genetic library or during analysis by blot transfer techniques; common probes are cDNA molecules, synthetic oligodeoxynucleotides of defined sequence, or antibodies to specific proteins.

Proteome: The entire collection of expressed proteins in an organism.

Pseudogene: An inactive segment of DNA arising by mutation of a parental active gene; typically generated by transposition of a cDNA copy of an mRNA.

Recombinant DNA: The altered DNA that results from the insertion of a sequence of deoxynucleotides not previously present into an existing molecule of DNA by enzymatic or chemical means.

Restriction enzyme: An endodeoxynuclease that causes cleavage of both strands of DNA at highly specific sites dictated by the base sequence.

Reverse transcription: RNA-directed synthesis of DNA, catalyzed by reverse transcriptase.

RT-PCR: A method used to quantitate mRNA levels that relies upon a first step of cDNA copying of mRNAs catalyzed by reverse transcriptase prior to PCR amplification and quantitation.

Signal: The end product observed when a specific sequence of DNA or RNA is detected by autoradiography or some other method. Hybridization with a complementary radioactive polynucleotide (eg, by Southern or Northern blotting) is commonly used to generate the signal.

Sines: Short interspersed repeat sequences.

SiRNAs: Silencing RNAs, 21–25 nt in length generated by selective nucleolytic degradation of double-stranded RNAs of cellular or viral origin. SiRNAs anneal to various specific sites within target in RNAs leading to mRNA degradation, hence gene "knockdown."

SNP: Single nucleotide polymorphism. Refers to the fact that single nucleotide genetic variation in genome sequence exists at discrete loci throughout the chromosomes. Measurement of allelic SNP differences is useful for gene mapping studies.

snRNA: Small nuclear RNA. This family of RNAs is best known for its role in mRNA processing.

Southern blot: A method for transferring DNA from an agarose gel to nitrocellulose filter, on which the DNA can be detected by a suitable probe (eg, complementary DNA or RNA).

Southwestern blot: A method for detecting protein-DNA interactions by applying a labeled DNA probe to a transfer membrane that contains a renatured protein.

Spliceosome: The macromolecular complex responsible for precursor mRNA splicing. The spliceosome consists of at least five small nuclear RNAs (snRNA; U1, U2, U4, U5, and U6) and many proteins.

Splicing: The removal of introns from RNA accompanied by the joining of its exons.

Sticky-ended DNA: Complementary single strands of DNA that protrude from opposite ends of a DNA duplex or from the ends of different duplex molecules (see also Blunt-ended DNA, above).

Tandem: Used to describe multiple copies of the same sequence (eg, DNA) that lie adjacent to one another.

Terminal transferase: An enzyme that adds nucleotides of one type (eg, deoxyadenonucleotidyl residues) to the 3′ end of DNA strands.

Transcription: Template DNA-directed synthesis of nucleic acids; typically DNA-directed synthesis of RNA.

Transcriptome: The entire collection of expressed mRNAs in an organism.

Transgenic: Describing the introduction of new DNA into germ cells by its injection into the nucleus of the ovum.

Translation: Synthesis of protein using mRNA as template.

Vector: A plasmid or bacteriophage into which foreign DNA can be introduced for the purposes of cloning.

Western blot: A method for transferring protein to a nitrocellulose filter, on which the protein can be detected by a suitable probe (eg, an antibody).

C H A P T E R

40

Membranes: Structure & Function

Robert K. Murray, MD, PhD & Daryl K. Granner, MD

BIOMEDICAL IMPORTANCE

Membranes are highly fluid, dynamic structures. **Plasma membranes** form closed compartments around cellular protoplasm to define cell boundaries. The plasma membrane has **selective permeabilities** and acts as a barrier, thereby maintaining differences in composition between the inside and outside of the cell. The selective permeabilities for substrates and ions are provided mainly by specific proteins named **transporters** and **ion channels**. The plasma membrane also exchanges material with the extracellular environment by **exocytosis** and **endocytosis**, and there are special areas of membrane structure—**gap junctions**—through which adjacent cells exchange material. In addition, the plasma membrane plays key roles in **cell–cell interactions** and in **transmembrane signaling**.

Membranes also form **specialized compartments** within the cell. Such intracellular membranes help shape many of the morphologically distinguishable structures (organelles), eg, mitochondria, ER, sarcoplasmic reticulum, Golgi complexes, secretory granules, lysosomes, and the nuclear membrane. Membranes localize **enzymes**, function as integral elements in **excitation-response coupling**, and provide sites of **energy transduction**, such as in photosynthesis and oxidative phosphorylation.

Changes in membrane structure (eg, caused by ischemia) can affect water balance and ion flux and therefore every process within the cell. Specific deficiencies or alterations of certain membrane components (eg, caused by mutations genes encoding membrane proteins) lead to a variety of **diseases** (see Table 40–7). In short, normal cellular function depends on normal membranes.

MAINTENANCE OF A NORMAL INTRA- & EXTRACELLULAR ENVIRONMENT IS FUNDAMENTAL TO LIFE

Life originated in an aqueous environment; enzyme reactions, cellular and subcellular processes, and so forth have therefore evolved to work in this milieu. Since mammals live in a gaseous environment, how is the aqueous state maintained? Membranes accomplish this by internalizing and compartmentalizing body water.

The Body's Internal Water Is Compartmentalized

Water makes up about 60% of the lean body mass of the human body and is distributed in two large compartments.

Intracellular Fluid (ICF)

This compartment constitutes two-thirds of total body water and provides a specialized environment for the cell (1) to make, store, and utilize energy; (2) to repair itself; (3) to replicate; and (4) to perform special functions.

Extracellular Fluid (ECF)

This compartment contains about one-third of total body water and is distributed between the plasma and interstitial compartments. The extracellular fluid is a **delivery system**. It brings to the cells nutrients (eg, glucose, fatty acids, amino acids), oxygen, various ions and trace minerals, and a variety of regulatory molecules (hormones) that coordinate the functions of widely separated cells. Extracellular fluid **removes** CO_2, waste products, and toxic or detoxified materials from the immediate cellular environment.

The Ionic Compositions of Intracellular & Extracellular Fluids Differ Greatly

As illustrated in Table 40–1, the **internal environment** is rich in K^+ and Mg^{2+}, and phosphate is its major anion. **Extracellular fluid** is characterized by high Na^+ and Ca^{2+} content, and Cl^- is the major anion. Note also that the concentration of glucose is higher in extracellular fluid than in the cell, whereas the opposite is true for proteins. Why are there such differences? It is thought that the primordial sea in which life originated was rich in K^+ and Mg^{2+}. It therefore follows that enzyme reactions and other biologic processes evolved to function best in that environment—hence the high concentration of these ions within cells. Cells were faced with strong selection pressure as the sea gradually changed to a composition rich in Na^+ and Ca^{2+}. Vast changes would have been required for evolution of a completely new set of biochemical and physiologic machinery; instead, as it happened, **cells developed barriers**—membranes with associated "pumps" such as the Na^+-K^+-ATPase (see below)—to maintain the internal microenvironment.

MEMBRANES ARE COMPLEX STRUCTURES COMPOSED OF LIPIDS, PROTEINS, & CARBOHYDRATE-CONTAINING MOLECULES

We shall mainly discuss the membranes present in eukaryotic cells, although many of the principles described also apply to the membranes of prokaryotes. The various cellular membranes have **different compositions**, as reflected in the **ratio**

TABLE 40–1 Comparison of the Mean Concentrations of Various Molecules Outside and Inside a Mammalian Cell

Substance	Extracellular Fluid	Intracellular Fluid
Na^+	140 mmol/L	10 mmol/L
K^+	4 mmol/L	140 mmol/L
Ca^{2+} (free)	2.5 mmol/L	0.1 µmol/L
Mg^{2+}	1.5 mmol/L	30 mmol/L
Cl^-	100 mmol/L	4 mmol/L
HCO_3^-	27 mmol/L	10 mmol/L
PO_4^{3-}	2 mmol/L	60 mmol/L
Glucose	5.5 mmol/L	0–1 mmol/L
Protein	2 g/dL	16 g/dL

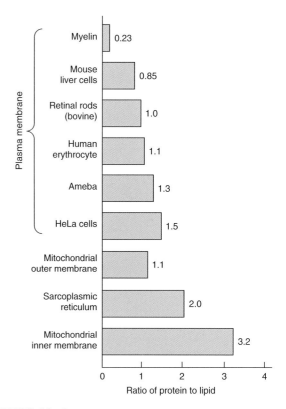

FIGURE 40–1 Ratio of protein to lipid in different membranes. Proteins equal or exceed the quantity of lipid in nearly all membranes. The outstanding exception is myelin, an electrical insulator found on many nerve fibers.

of protein to lipid (Figure 40–1). This is not surprising, given their divergent functions. Membranes are sheet-like enclosed structures consisting of an asymmetric lipid bilayer with distinct inner and outer surfaces. These sheet-like structures are **noncovalent assemblies** that form spontaneously in water due to the amphipathic nature of lipids. Numerous proteins are located in membranes, where they carry out specific functions.

The Major Lipids in Mammalian Membranes Are Phospholipids, Glycosphingolipids, & Cholesterol
Phospholipids

Of the two major phospholipid classes present in membranes, **phosphoglycerides** are the more common and consist of a glycerol backbone to which are attached two fatty acids in ester linkage and a phosphorylated alcohol (Figure 40–2). The fatty acid constituents are usually even-numbered carbon molecules, most commonly containing 16 or 18 carbons. They are unbranched and can be saturated or unsaturated with one or more *cis* double bonds. The simplest phosphoglyceride is phosphatidic acid, which is 1,2-diacylglycerol 3-phosphate, a key intermediate in the formation of other phosphoglycerides (Chapter 24). In most phosphoglycerides present in mem-

Fatty acids

$$R_1 - \overset{\overset{\displaystyle O}{\|}}{C} - O - {}^1CH_2$$
$$R_2 - \overset{\underset{\displaystyle O}{\|}}{C} - O - {}^2CH \qquad \overset{\displaystyle O^-}{\underset{\displaystyle |}{}}$$
$$\phantom{R_2 - C - O - {}^2CH} {}^3CH_2 - O - \overset{\overset{\displaystyle |}{}}{\underset{\underset{\displaystyle O}{\|}}{P}} - O - R_3$$

Glycerol   **Alcohol**

FIGURE 40–2 A phosphoglyceride showing the fatty acids (R₁ and R₂), glycerol, and a phosphorylated alcohol component. In phosphatidic acid, R₃ is hydrogen.

branes, the 3-phosphate is esterified to an alcohol such as ethanolamine, choline, serine, glycerol, or inositol (Chapter 15).

The second major class of phospholipids is composed of sphingomyelin (Figure 15–11), which contains a sphingosine backbone rather than glycerol. A fatty acid is attached by an amide linkage to the amino group of sphingosine, forming **ceramide**. The primary hydroxyl group of sphingosine is esterified to phosphorylcholine, forming sphingomyelin. As the name implies, sphingomyelin is prominent in myelin sheaths.

The amounts and fatty acid compositions of the various phospholipids vary among the different cellular membranes.

Glycosphingolipids

The glycosphingolipids (GSLs) are sugar-containing lipids built on a backbone of ceramide; they include galactosyl- and glucosyl-ceramide (cerebrosides) and the gangliosides. Their structures are described in Chapter 15. They are mainly located in the plasma membranes of cells.

Sterols

The most common sterol in membranes is **cholesterol** (Chapter 15), which resides mainly in the plasma membranes of mammalian cells but can also be found in lesser quantities in mitochondria, Golgi complexes, and nuclear membranes. Cholesterol intercalates among the phospholipids of the membrane, with its hydroxyl group at the aqueous interface and the remainder of the molecule within the leaflet. Its effect on the fluidity of membranes is discussed subsequently.

All of the above lipids can be separated from one another by techniques such as column, thin-layer, and gas-liquid chromatography and their structures established by mass spectrometry and other techniques.

Membrane Lipids Are Amphipathic

All major lipids in membranes contain both hydrophobic and hydrophilic regions and are therefore termed **amphipathic.** If the hydrophobic region were separated from the rest of the molecule, it would be insoluble in water but soluble in oil. Conversely, if the hydrophilic region were separated from the

rest of the molecule, it would be insoluble in oil but soluble in water. The amphipathic nature of a phospholipid is represented in Figure 40–3. Thus, the polar head groups of the phospholipids and the hydroxyl group of cholesterol interface with the aqueous environment; a similar situation applies to the sugar moieties of the GSLs (see below).

Saturated fatty acids have straight tails, whereas **unsaturated fatty acids,** which generally exist in the *cis* form in membranes, make kinked tails (Figure 40–3). As more kinks are inserted in the tails, the membrane becomes less tightly packed and therefore more fluid.

Detergents are amphipathic molecules that are important in biochemistry as well as in the household. The molecular structure of a detergent is not unlike that of a phospholipid. Certain detergents are widely used to solubilize membrane proteins as a first step in their purification. The hydrophobic end of the detergent binds to hydrophobic regions of the proteins, displacing most of their bound lipids. The polar end of the detergent is free, bringing the proteins into solution as detergent–protein complexes, usually also containing some residual lipids.

Membrane Lipids Form Bilayers

The amphipathic character of phospholipids suggests that the two regions of the molecule have incompatible solubilities; however, in a solvent such as water, phospholipids organize themselves into a form that thermodynamically serves the solubility requirements of both regions. A **micelle** (Figure 40–4) is such a structure; the hydrophobic regions are shielded from water, while the hydrophilic polar groups are immersed in the aqueous environment. However, micelles are usually relatively small in size (eg, approximately 200 nm) and thus are limited in their potential to form membranes.

As was recognized in 1925 by Gorter and Grendel, a **bimolecular layer,** or **lipid bilayer,** can also satisfy the thermo-

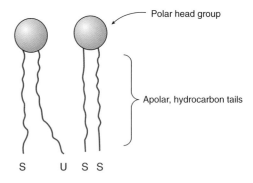

FIGURE 40–3 Diagrammatic representation of a phospholipid or other membrane lipid. The polar head group is hydrophilic, and the hydrocarbon tails are hydrophobic or lipophilic. The fatty acids in the tails are saturated (S) or unsaturated (U); the former are usually attached to carbon 1 of glycerol and the latter to carbon 2. Note the kink in the tail of the unsaturated fatty acid (U), which is important in conferring increased membrane fluidity.

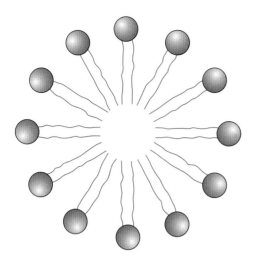

FIGURE 40–4 Diagrammatic cross-section of a micelle. The polar head groups are bathed in water, whereas the hydrophobic hydrocarbon tails are surrounded by other hydrocarbons and thereby protected from water. Micelles are relatively small (compared with lipid bilayers) spherical structures.

dynamic requirements of amphipathic molecules in an aqueous environment. **Bilayers,** not micelles, are indeed the **key structures** in biologic membranes. A bilayer exists as a sheet in which the hydrophobic regions of the phospholipids are protected from the aqueous environment, while the hydrophilic regions are immersed in water (Figure 40–5). Only the ends or edges of the bilayer sheet are exposed to an unfavorable environment, but even these exposed edges can be eliminated by folding the sheet back upon itself to form an enclosed vesicle with no edges. A bilayer can extend over relatively large distances (eg, 1 mm). The closed bilayer provides one of the most essential properties of membranes. It is **impermeable to most water-soluble molecules,** since they would be insoluble in the hydrophobic core of the bilayer.

Lipid bilayers are formed by **self-assembly,** driven by the **hydrophobic effect** (Chapter 2). When lipid molecules come

together in a bilayer, the entropy of the surrounding solvent molecules increases due to the release of immobilized water.

Two questions arise from consideration of the above. First, how many biologic materials are **lipid-soluble** and can therefore readily enter the cell? Gases such as oxygen, CO_2, and nitrogen—small molecules with little interaction with solvents—readily diffuse through the hydrophobic regions of the membrane. The **permeability coefficients** of several ions and of a number of other molecules in a lipid bilayer are shown in Figure 40–6. The three electrolytes shown (Na^+, K^+, and Cl^-) cross the bilayer much more slowly than water. In general, the permeability coefficients of small molecules in a lipid bilayer **correlate with their solubilities in nonpolar solvents.** For instance, **steroids** more readily traverse the lipid bilayer compared with electrolytes. The high permeability coefficient of **water** itself is surprising but is partly explained by its small size and relative lack of charge. Many **drugs** are hydrophobic and can readily cross membranes and enter cells.

The second question concerns **molecules that are not lipid-soluble:** How are the transmembrane concentration gradients for non-lipid-soluble molecules maintained? The answer is that **membranes contain proteins,** many of which are inserted into the lipid bilayer in unique orientations. Such proteins form **channels** for the movement of ions and small molecules and serve as **transporters** for larger molecules that otherwise could not pass the bilayer. These structures are described below.

Membrane Proteins Are Associated with the Lipid Bilayer

Membrane **phospholipids** act as a solvent for membrane proteins, creating an environment in which the latter can function. As described in Chapter 5, the **α-helical structure of proteins** minimizes the hydrophilic character of the peptide bonds themselves. Thus, proteins can be amphipathic and form an integral part of the membrane by having hydrophilic regions protruding at the inside and outside faces of the membrane but connected

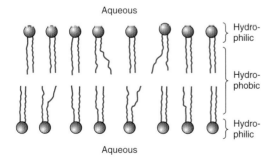

FIGURE 40–5 Diagram of a section of a bilayer membrane formed from phospholipid molecules. The unsaturated fatty acid tails are kinked and lead to more spacing between the polar head groups, hence to more room for movement. This in turn results in increased membrane fluidity. (Slightly modified and reproduced, with permission, from Stryer L: *Biochemistry,* 2nd ed. Freeman, 1981. Copyright ©1981 by W H. Freeman and Company.)

FIGURE 40–6 Permeability coefficients of water, some ions, and other small molecules in lipid bilayer membranes. The permeability coefficient is a measure of the ability of a molecule to diffuse across a permeability barrier. Molecules that move rapidly through a given membrane are said to have a high permeability coefficient. (Slightly modified and reproduced, with permission, from Stryer L: *Biochemistry,* 2nd ed. Freeman, 1981. Copyright © 1981 by W H. Freeman and Company.)

by a hydrophobic region traversing the hydrophobic core of the bilayer. In fact, those portions of membrane proteins that traverse membranes do contain substantial numbers of hydrophobic amino acids and almost invariably have a high α-helical content. For many membranes, a stretch of **approximately 20 amino acids** in an α-helix **will span the bilayer.**

It is possible to calculate whether a particular sequence of amino acids present in a protein is consistent with a **transmembrane location.** This can be done by consulting a table that lists the hydrophobicities of each of the 20 common amino acids and the free energy values for their transfer from the interior of a membrane to water. Hydrophobic amino acids have positive values; polar amino acids have negative values. The total free energy values for transferring successive sequences of 20 amino acids in the protein are plotted, yielding a so-called **hydropathy plot.** Values of over 20 kcal·mol^{-1} are consistent with—but do not prove—the interpretation that the hydrophobic sequence is a transmembrane segment.

Another aspect of the interaction of lipids and proteins is that some proteins are anchored to one leaflet of the bilayer by **covalent linkages to certain lipids. Palmitate** and **myristate** are fatty acids involved in such linkages to specific proteins. A number of other proteins (see Chapter 47) are linked to **glycophosphatidylinositol (GPI) structures.**

Different Membranes Have Different Protein Compositions

The **number of different proteins** in a membrane varies from less than a dozen in the sarcoplasmic reticulum to over 100 in the plasma membrane. Most membrane proteins can be separated from one another using **sodium dodecyl sulfate polyacrylamide gel electrophoresis** (SDS-PAGE), a technique that separates proteins based on their molecular mass. SDS is a powerful detergent that disrupts protein-lipid interactions and thereby solubilizes membrane proteins. SDS also disrupts protein-protein interactions and unfolds proteins. In the absence of SDS, few membrane proteins would remain soluble.

Proteins are the **major functional molecules** of membranes and consist of **enzymes, pumps** and **channels, structural components, antigens** (eg, for histocompatibility), and **receptors** for various molecules. Because every membrane possesses a different complement of proteins, there is no such thing as a typical membrane structure. The enzymatic properties of several different membranes are shown in Table 40–2.

Membranes Are Dynamic Structures

Membranes and their components are **dynamic structures.** The **lipids and proteins** in membranes undergo **turnover,** just as they do in other compartments of the cell. Different lipids have different turnover rates, and the turnover rates of individual species of membrane proteins may vary widely. The membrane itself can turn over even more rapidly than any of its constituents. This is discussed in more detail in the section on endocytosis.

TABLE 40–2 Enzymatic Markers of Different Membranes[1]

Membrane	Enzyme
Plasma	5′-Nucleotidase
	Adenylyl cyclase
	Na$^+$-K$^+$-ATPase
Endoplasmic reticulum	Glucose-6-phosphatase
Golgi apparatus	
Cis	GlcNAc transferase I
Medial	Golgi mannosidase II
Trans	Galactosyl transferase
TGN	Sialyl transferase
Inner mitochondrial membrane	ATP synthase

Abbreviation: TGN, *trans* Golgi network.

[1]Membranes contain many proteins, some of which have enzymatic activity. Some of these enzymes are located only in certain membranes and can therefore be used as markers to follow the purification of these membranes.

Another indicator of the dynamic nature of membranes is that a variety of studies have shown that lipids and certain proteins exhibit **lateral diffusion** in the plane of their membranes. Some proteins do not exhibit lateral diffusion because they are anchored to the underlying actin cytoskeleton. In contrast, the **transverse** movement of lipids across the membrane **(flip-flop)** is extremely slow (see below), and does not occur at all in the case of membrane proteins.

Membranes Are Asymmetric Structures

Proteins have unique orientations in membranes, making the **outside surfaces different from the inside surfaces.** An **inside-outside asymmetry** is also provided by the external location of the carbohydrates attached to membrane proteins. In addition, specific proteins are located exclusively on the outsides or insides of membranes, as in the mitochondrial and plasma membranes.

There are **regional heterogeneities** in membranes. Some, such as occur at the villous borders of mucosal cells, are almost macroscopically visible. Others, such as those at gap junctions, tight junctions, and synapses, occupy much smaller regions of the membrane and generate correspondingly smaller local asymmetries.

There is also inside-outside (transverse) **asymmetry of the phospholipids.** The **choline-containing phospholipids** (phosphatidylcholine and sphingomyelin) are located mainly in the **outer molecular layer;** the **aminophospholipids** (phosphatidylserine and phosphatidylethanolamine) are preferentially located in the **inner leaflet.** Obviously, if this asymmetry is to exist at all, there must be limited transverse mobility (flip-flop) of the membrane phospholipids. In fact, phospholipids in synthetic bilayers exhibit an **extraordinarily slow rate of flip-flop;** the half-life of the asymmetry can be measured in several weeks.

The mechanisms involved in the **establishment of lipid asymmetry** are not well understood. The enzymes involved in the synthesis of phospholipids are located on the cytoplasmic side of microsomal membrane vesicles. Translocases **(flippases)** exist that transfer certain phospholipids (eg, phosphatidylcholine) from the inner to the outer leaflet. Specific **proteins that preferentially bind** individual phospholipids also appear to be present in the two leaflets, contributing to the asymmetric distribution of these lipid molecules. In addition, **phospholipid exchange proteins** recognize specific phospholipids and transfer them from one membrane (eg, the endoplasmic reticulum [ER]) to others (eg, mitochondrial and peroxisomal). There is **further asymmetry** with regard to **GSLs** and also **glycoproteins; the sugar moieties** of these molecules all **protrude outward** from the plasma membrane and are absent from its inner face.

Membranes Contain Integral & Peripheral Proteins

It is useful to classify membrane proteins into two types: **integral** and **peripheral** (Figure 40–7). Most membrane proteins fall into the **integral class**, meaning that they interact extensively with the phospholipids and **require the use of detergents** for their solubilization. Also, they generally **span the bilayer**. Integral proteins are usually **globular** and are themselves **amphipathic**. They consist of two hydrophilic ends separated by an intervening hydrophobic region that traverses the hydrophobic core of the bilayer. As the structures of integral membrane proteins were being elucidated, it became apparent that certain ones (eg, transporter molecules, ion channels, various receptors, and G proteins) **span the bilayer many times**

(see Figure 46–7), whereas others **span the membrane only once**. Integral proteins are also **asymmetrically distributed** across the membrane bilayer. This asymmetric orientation is conferred at the time of their insertion in the lipid bilayer during biosynthesis. The molecular mechanisms involved in insertion of proteins into membranes and the topic of membrane assembly are discussed in Chapter 46.

Peripheral proteins do not interact directly with the hydrophobic cores of the phospholipids in the bilayer and thus **do not require use of detergents** for their release. They are bound to the hydrophilic regions of specific integral proteins and head groups of phospholipids and can be released from them by treatment with **salt solutions of high ionic strength**. For example, **ankyrin**, a peripheral protein, is bound to the inner aspect of the integral protein "band 3" of erythrocyte membrane. **Spectrin**, a cytoskeletal structure within the erythrocyte, is in turn bound to ankyrin and thereby plays an important role in maintenance of the biconcave shape of the erythrocyte.

ARTIFICIAL MEMBRANES MODEL MEMBRANE FUNCTION

Artificial membrane systems can be prepared by appropriate techniques. These systems generally consist of mixtures of one or more phospholipids of natural or synthetic origin that can be treated (eg, by using mild sonication) to form spherical vesicles in which the lipids form a bilayer. Such vesicles, surrounded by a lipid bilayer with an aqueous interior, are termed **liposomes.**

FIGURE 40–7 The fluid mosaic model of membrane structure. The membrane consists of a bimolecular lipid layer with proteins inserted in it or bound to either surface. Integral membrane proteins are firmly embedded in the lipid layers. Some of these proteins completely span the bilayer and are called transmembrane proteins, while others are embedded in either the outer or inner leaflet of the lipid bilayer. Loosely bound to the outer or inner surface of the membrane are the peripheral proteins. Many of the proteins and all the glycolipids have externally exposed oligosaccharide chains. (Reproduced, with permission, from Junqueira LC, Carneiro J: *Basic Histology: Text & Atlas,* 10th ed., McGraw-Hill, 2003.)

Some of the advantages and uses of artificial membrane systems in the study of membrane function are as follows:

1. The **lipid content** of the membranes can be varied, allowing systematic examination of the effects of varying lipid composition on certain functions.

2. **Purified membrane proteins or enzymes** can be incorporated into these vesicles in order to assess what factors (eg, specific lipids or ancillary proteins) the proteins require to reconstitute their function.

3. The environment of these systems can be rigidly controlled and systematically varied (eg, ion concentrations, ligands).

4. When liposomes are formed, they can be made to entrap certain compounds inside themselves, eg, drugs and isolated genes. There is interest in using liposomes to distribute drugs to certain tissues, and if components (eg, antibodies to certain cell surface molecules) could be incorporated into liposomes so that they would be targeted to specific tissues or tumors, the therapeutic impact would be considerable. DNA entrapped inside liposomes appears to be less sensitive to attack by nucleases; this approach may prove useful in attempts at **gene therapy**.

THE FLUID MOSAIC MODEL OF MEMBRANE STRUCTURE IS WIDELY ACCEPTED

The **fluid mosaic model** of membrane structure proposed in 1972 by Singer and Nicolson (Figure 40–7) is now widely accepted. The model is often likened to **icebergs** (membrane proteins) **floating in a sea** of predominantly phospholipid molecules. Early **evidence** for the model was the finding that certain species-specific **integral proteins** (detected by fluorescent labeling techniques) rapidly and randomly redistributed in the plasma membrane of an interspecies hybrid cell formed by the artificially induced fusion of two different parent cells. Biophysical studies of integral proteins showed that they spanned the membrane and had a globular nature. It has subsequently been demonstrated that **phospholipids** also undergo rapid redistribution in the plane of the membrane. This diffusion within the plane of the membrane, termed **lateral diffusion**, can be quite rapid for a phospholipid; in fact, within the plane of the membrane, one molecule of phospholipid can move several micrometers per second.

The **phase changes**—and thus the **fluidity** of membranes—are largely **dependent upon the lipid composition** of the membrane. In a lipid bilayer, the hydrophobic chains of the fatty acids can be highly aligned or ordered to provide a rather stiff structure. As the temperature increases, the hydrophobic side chains undergo a transition from the ordered state (more gel-like or crystalline phase) to a disordered one, taking on a more liquid-like or fluid arrangement. The temperature at which the structure undergoes the transition from ordered to disordered (ie, melts) is called the **"transition temperature"** (**T_m**). The longer and more saturated fatty acid chains interact more strongly with each other via their longer hydrocarbon chains and thus cause higher values of T_m—ie, higher temperatures are required to increase the fluidity of the bilayer. On the other hand, unsaturated bonds that exist in the *cis* configuration tend to increase the fluidity of a bilayer by decreasing the compactness of the side chain packing without diminishing hydrophobicity (Figure 40–3). The phospholipids of cellular membranes generally contain at least one unsaturated fatty acid with at least one cis double bond.

Cholesterol modifies the fluidity of membranes. At temperatures below the T_m, it interferes with the interaction of the hydrocarbon tails of fatty acids and thus increases fluidity. At temperatures above the T_m, it limits disorder because it is more rigid than the hydrocarbon tails of the fatty acids and cannot move in the membrane to the same extent, thus limiting fluidity. At high cholesterol–phospholipid ratios, transition temperatures are altogether indistinguishable.

The **fluidity** of a membrane significantly affects its **functions.** As membrane fluidity increases, so does its **permeability** to water and other small hydrophilic molecules. The lateral mobility of integral proteins increases as the fluidity of the membrane increases. If the active site of an integral protein involved in a given function is exclusively in its hydrophilic regions, changing lipid fluidity will probably have little effect on the activity of the protein; however, if the protein is involved in a transport function in which transport components span the membrane, lipid phase effects may significantly alter the **transport rate**. The **insulin receptor** is an excellent example of altered function with changes in fluidity. As the concentration of unsaturated fatty acids in the membrane is increased (by growing cultured cells in a medium rich in such molecules), fluidity increases. This alters the receptor so that it binds more insulin. At normal body temperature (37°C) the lipid bilayer is in a fluid state.

Lipid Rafts, Caveolae, & Tight Junctions Are Specialized Features of Plasma Membranes

Plasma membranes contain **certain specialized structures** whose biochemical natures have been investigated in some detail.

Lipid rafts are specialized areas of the exoplasmic leaflet of the lipid bilayer enriched in cholesterol, sphingolipids, and certain proteins (see Figure 40–8). They are involved in **signal transduction** and **other processes**. It is thought that clustering certain components of signaling systems closely together may increase the efficiency of their function.

Caveolae may derive from lipid rafts. Many, if not all, contain the protein **caveolin-1,** which may be involved in their formation from rafts. Caveolae are observable by electron microscopy as flask-shaped indentations of the cell membrane facing the cytosol (Figure 40–9). Proteins detected in caveolae include various components of the signal transduction system (eg, the insulin receptor and some G proteins), the folate receptor, and endothelial nitric oxide synthase (eNOS). Caveolae

FIGURE 40–8 Schematic diagram of a lipid raft. Lipid rafts are somewhat thicker than the remainder of the bilayer. They are enriched in sphingolipids (eg, sphingomyelin), glycosphingolipids (eg, the ganglioside GM_1), saturated phospholipids, and cholesterol. They also contain certain GPI-linked proteins (outer leaflet) and acylated and prenylated proteins (inner leaflet). GPI-linked proteins are discussed in Chapter 47. Acylation and prenylation are post-translational modifications of certain membrane proteins.

and lipid rafts are active areas of research, and ideas concerning them and their possible roles in various disorders are rapidly evolving.

Tight junctions are other structures found in surface membranes. They are often located below the apical surfaces of epithelial cells and **prevent the diffusion of macromolecules between cells**. They are composed of **various proteins**, including occludin, various claudins, and junctional adhesion molecules.

Yet **other specialized structures** found in surface membranes include **desmosomes, adherens junctions**, and **microvilli**; their chemical natures and functions are not discussed here. The nature of **gap junctions** is described below.

MEMBRANE SELECTIVITY ALLOWS ADJUSTMENTS OF CELL COMPOSITION & FUNCTION

If the plasma membrane is relatively impermeable, **how do most molecules enter a cell?** How is **selectivity** of this movement

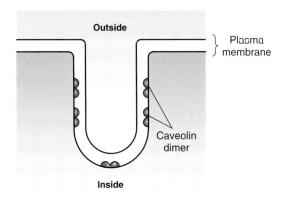

FIGURE 40–9 Schematic diagram of a caveola. A caveola is an invagination in the plasma membrane. The protein caveolin appears to play an important role in the formation of caveolae, and occurs as a dimer. Each caveolin monomer is anchored to the inner leaflet of the plasma membrane by three palmitoyl molecules (not shown).

established? Answers to such questions are important in understanding how cells adjust to a constantly changing extracellular environment. Metazoan organisms also must have **means of communicating** between adjacent and distant cells, so that complex biologic processes can be coordinated. These **signals** must arrive at and be transmitted by the membrane, or they must be **generated** as a consequence of some interaction with the membrane. Some of the major mechanisms used to accomplish these different objectives are listed in Table 40–3.

Passive Diffusion Involving Transporters & Ion Channels Moves Many Small Molecules Across Membranes

Molecules can **passively** traverse the bilayer down electrochemical gradients by **simple diffusion** or by **facilitated diffusion.** This spontaneous movement toward equilibrium contrasts with **active transport,** which **requires energy** because it constitutes movement against an electrochemical gradient. Figure 40–10 provides a schematic representation of these mechanisms. We shall first describe various aspects of passive transport, and then discuss aspects of active transport.

Firstly, let us define the various terms. **Simple diffusion** is the passive flow of a solute from a higher to a lower concentration due to random thermal movement. **Facilitated diffusion** is passive transport of a solute from a higher to a lower concentration mediated by a specific protein transporter. **Active transport** is transport of a solute across a membrane against a concentration gradient, and thus requires energy (frequently derived from the hydrolysis of ATP); a specific transporter (**pump**) is involved.

As mentioned earlier in this chapter, some solutes such as gases can enter the cell by diffusing down an electrochemical gradient across the membrane, and do not require metabolic energy. The **simple diffusion** of a solute across the membrane is limited by the **thermal agitation** of that specific molecule, by the **concentration gradient** across the membrane, and by the **solubility** of that solute (the permeability coefficient, Figure 40–6) in the hydrophobic core of the membrane

TABLE 40–3   Transfer of Material and Information Across Membranes

Cross-membrane movement of small molecules

Diffusion (passive and facilitated)
Active transport

Cross-membrane movement of large molecules

Endocytosis
Exocytosis

Signal transmission across membranes

Cell surface receptors
   1. Signal transduction (eg, glucagon → cAMP)
   2. Signal internalization (coupled with endocytosis, eg, the LDL receptor)
Movement to intracellular receptors (steroid hormones; a form of diffusion)

Intercellular contact and communication

Passive (simple) diffusion is the flow of solute from a higher to a lower concentration due to random thermal movement
Facilitated diffusion is passive transport of a solute from a higher concentration to a lower concentration, mediated by a specific protein transporter
Active transport is transport of a solute across a membrane in the direction of increasing concentration, and thus requires energy (frequently derived from the hydrolysis of ATP); a specific transporter (pump) is involved

The other terms used in this Table are explained later in this Chapter or elsewhere in this text

bilayer. **Solubility** is inversely proportionate to the number of hydrogen bonds that must be broken in order for a solute in the external aqueous phase to become incorporated in the hydrophobic bilayer. Electrolytes, poorly soluble in lipid, do not form hydrogen bonds with water, but they do **acquire a shell of water** from hydration by electrostatic interaction. The size of the shell is directly proportionate to the **charge density** of the electrolyte. Electrolytes with a large charge density have a larger shell of hydration and thus a slower diffusion rate. Na^+, for example, has a higher charge density than K^+. Hydrated Na^+ is therefore larger than hydrated K^+; hence, the latter tends to move more easily through the membrane.

The following factors affect **net diffusion** of a substance: (1) Its **concentration gradient** across the membrane. Solutes move from high to low concentration. (2) The **electrical potential** across the membrane. Solutes move toward the solution that has the opposite charge. The inside of the cell usually has a negative charge. (3) The **permeability coefficient** of the substance for the membrane. (4) The hydrostatic **pressure gradient** across the membrane. Increased pressure will increase the rate and force of the collision between the molecules and the membrane. (5) **Temperature**. Increased temperature will increase particle motion and thus increase the frequency of collisions between external particles and the membrane.

Facilitated diffusion involves either certain **transporters** or **ion channels** (see Figure 40–11). **Other transporters** (mostly ATP-driven) are involved in **active transport**. A multitude of **transporters and channels** exist in biological membranes that route the entry of ions into and out of cells. They are described in the following sections. Table 40–4 summarizes some important points of difference between transporters and ion channels.

Transporters Are Specific Proteins Involved in Facilitated Diffusion & Also Active Transport

Transport systems can be described in a functional sense according to the number of molecules moved and the direction of movement (Figure 40–12) or according to whether movement is toward or away from equilibrium. The following **classification** depends primarily on the former. A **uniport**

FIGURE 40–10 Many small, uncharged molecules pass freely through the lipid bilayer by simple diffusion. Larger uncharged molecules, and some small uncharged molecules, are transferred by specific carrier proteins (transporters) or through channels or pores. Passive transport is always down an electrochemical gradient, toward equilibrium. Active transport is against an electrochemical gradient and requires an input of energy, whereas passive transport does not. (Redrawn and reproduced, with permission, from Alberts B et al: *Molecular Biology of the Cell.* Garland, 1983.)

FIGURE 40–11 A schematic diagram of the two types of membrane transport of small molecules.

system moves one type of molecule bidirectionally. In **cotransport** systems, the transfer of one solute depends upon the stoichiometric simultaneous or sequential transfer of another solute. A **symport** moves two solutes in the same direction. Examples are the proton-sugar transporter in bacteria and the Na^+-sugar transporters (for glucose and certain other sugars) and Na^+-amino acid transporters in mammalian cells. **Antiport** systems move two molecules in opposite directions (eg, Na^+ in and Ca^{2+} out).

Hydrophilic molecules that cannot pass freely through the lipid bilayer membrane do so passively by **facilitated diffusion** or by **active transport**. Passive transport is driven by the transmembrane gradient of substrate. Active transport always occurs against an electrical or chemical gradient, and so it requires energy, usually ATP. Both types of transport involve **specific carrier proteins** (transporters) and both show **specificity** for ions, sugars, and amino acids. Passive and active transport **resemble a substrate–enzyme interaction**. Points of resemblance of both to enzyme action are: (1) There is a specific binding site for the solute. (2) The carrier is saturable, so it has a maximum rate of transport (V_{max}; Figure 40–13). (3) There is a binding constant (K_m) for the solute, and so the whole system has a K_m (Figure 40–13). (4) Structurally similar competitive inhibitors block transport. Transporters are thus like enzymes, but generally do not modify their substrates.

Cotransporters use the gradient of one substrate created by active transport to drive the movement of the other substrate. The Na^+ gradient produced by the Na^+-K^+-ATPase is used to drive the transport of a number of important metabolites. The ATPase is a very important example of **primary transport**, while the Na^+-dependent systems are examples of

secondary transport that rely on the gradient produced by another system. Thus, inhibition of the Na^+-K^+-ATPase in cells also blocks the Na^+-dependent uptake of substances like glucose.

Facilitated Diffusion Is Mediated by a Variety of Specific Transporters

Some specific solutes diffuse down electrochemical gradients across membranes **more rapidly** than might be expected from their size, charge, or partition coefficient. This is because specific transporters are involved. This **facilitated diffusion** exhibits properties distinct from those of simple diffusion. The rate of facilitated diffusion, a uniport system, can be **saturated**; ie, the number of sites involved in diffusion of the specific solutes appears finite. Many facilitated diffusion systems are **stereospecific** but, like simple diffusion, are **driven by the transmembrane electrochemical gradient**.

A **"ping-pong" mechanism** (Figure 40–14) helps explain facilitated diffusion. In this model, the carrier protein exists in two principal conformations. In the **"ping"** state, it is exposed to high concentrations of solute, and molecules of the solute bind to specific sites on the carrier protein. Binding induces a **conformational change** that exposes the carrier to a lower concentration of solute (**"pong"** state). This process is completely **reversible**, and net flux across the membrane depends upon the concentration

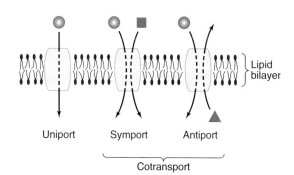

FIGURE 40–12 Schematic representation of types of transport systems. Transporters can be classified with regard to the direction of movement and whether one or more unique molecules are moved. A uniport can also allow movement in the opposite direction, depending on the concentrations inside and outside a cell of the molecule transported. (Redrawn and reproduced, with permission, from Alberts B et al: *Molecular Biology of the Cell*. Garland, 1983.)

TABLE 40–4 Comparison of Transporters and Ion Channels

Transporters	Ion Channels
Bind solute and undergo conformational changes, transferring the solute across the membrane	Form pores in membranes
Involved in passive (facilitated diffusion) and active transport	Involved only in passive transport
Transport is significantly slower than via ion channels	Transport is significantly faster than via transporters

Note: Transporters are also known as carriers or permeases. Active transporters are often called pumps.

FIGURE 40–13 A comparison of the kinetics of carrier-mediated (facilitated) diffusion with passive diffusion. The rate of movement in the latter is directly proportionate to solute concentration, whereas the process is saturable when carriers are involved. The concentration at half-maximal velocity is equal to the binding constant (K_m) of the carrier for the solute. (V_{max}, maximal rate.)

gradient. The **rate** at which solutes enter a cell by facilitated diffusion is determined by the following factors: (1) The concentration gradient across the membrane. (2) The amount of carrier available (this is a key control step). (3) The affinity of the solute-carrier interaction. (4) The rapidity of the conformational change for both the loaded and the unloaded carrier.

Hormones can regulate facilitated diffusion by changing the number of transporters available. **Insulin** via a complex signalling pathway increases glucose transport in fat and muscle by recruiting transporters from an intracellular reservoir. Insulin also enhances amino acid transport in liver and other tissues. One of the coordinated actions of **glucocorticoid hormones** is to enhance transport of amino acids into liver, where the amino acids then serve as a substrate for gluconeogenesis. **Growth hormone** increases amino acid transport in all cells, and **estrogens** do this in the uterus. There are at least five different carrier systems for amino acids in animal cells. Each is specific for a group of closely related amino acids, and most operate as Na^+-symport systems (Figure 40–12).

Ion Channels Are Transmembrane Proteins That Allow the Selective Entry of Various Ions

Natural membranes contain transmembrane channels, pore-like structures composed of proteins that constitute selective **ion channels.** Cation-conductive channels have an average diameter of about 5–8 nm. The **permeability** of a channel depends upon the size, the extent of hydration, and the extent of charge density on the ion. **Specific channels** for Na^+, K^+, Ca^{2+}, and Cl^- have been identified; one such **Na^+** channel is illustrated in Figure 40–15. It is seen to consist of four subunits. Each subunit consists of six α-helical transmembrane domains. The amino and carboxyl terminals are located in the cytoplasm, with both extracellular and intracellular loops being present. The actual pore in the channel through which the ions pass is not shown. A pore constitutes the center (diameter about 5–8 nm) of a structure formed by apposition of the subunits. Ion channels are very **selective,** in most cases permitting the passage of only one type of ion (Na^+, Ca^{2+}, etc). The **selectivity filter** of K^+ channels is made up of a ring of carbonyl groups donated by the subunits. The carbonyls displace bound water from the ion, and thus restrict its size to appropriate precise dimensions for passage through the channel. Many variations on the above structural theme are found, but all ion channels are basically made up of transmembrane subunits that come together to form a central pore through which ions pass selectively.

The membranes of **nerve cells** contain well-studied ion channels that are responsible for the generation of action potentials. The activity of some of these channels is controlled by neurotransmitters; hence, channel activity can be **regulated.**

Ion channels are open transiently and thus are **"gated."** Gates can be controlled by opening or closing. In **ligand-gated channels,** a specific molecule binds to a receptor and opens the channel. **Voltage-gated channels** open (or close) in response to changes in membrane potential. **Mechanically-gated** channels respond to mechanical stimuli (pressure, touch).

Some properties of ion channels are listed in Tables 40–4 & 40–5; other aspects of ion channels are discussed briefly in Chapter 48.

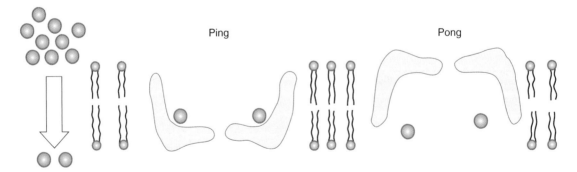

FIGURE 40–14 The "ping-pong" model of facilitated diffusion. A protein carrier (blue structure) in the lipid bilayer associates with a solute in high concentration on one side of the membrane. A conformational change ensues ("ping" to "pong"), and the solute is discharged on the side favoring the new equilibrium. The empty carrier then reverts to the original conformation ("pong" to "ping") to complete the cycle.

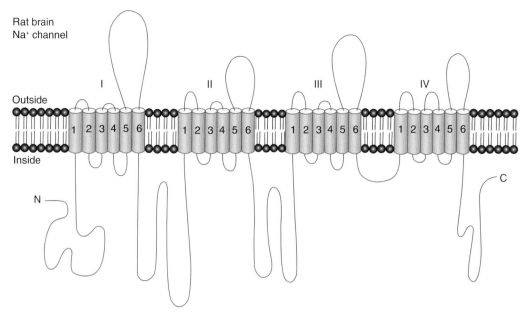

FIGURE 40–15 Diagrammatic representation of the structures of an ion channel (a Na⁺ channel of rat brain). The Roman numerals indicate the four subunits of the channel and the Arabic numerals the α-helical transmembrane domains of each subunit. The actual pore through which the ions (Na⁺) pass is not shown, but is formed by apposition of the various subunits. The specific areas of the subunits involved in the opening and closing of the channel are also not indicated. (After WK Catterall. Modified and reproduced, with permission, from Hall ZW: *An Introduction to Molecular Neurobiology.* Sinauer, 1992.)

Detailed Studies of a K⁺ Channel & of a Voltage-Gated Channel Have Yielded Major Insights into Their Actions

There are at least four features of ion channels that must be elucidated: (1) their overall structures; (2) how they conduct ions so rapidly; (3) their selectivity; and (4) their gating properties. As described below, considerable progress in tackling these difficult problems has been made.

Especial progress has been made by Roderick MacKinnon, who received the Nobel Prize for elucidating the structure and

TABLE 40–5   Some Properties of Ion Channels

- They are composed of transmembrane protein subunits.
- Most are highly selective for one ion; a few are nonselective.
- They allow impermeable ions to cross membranes at rates approaching diffusion limits.
- They can permit ion fluxes of 106–107/s.
- Their activities are regulated.
- The main types are voltage-gated, ligand-gated, and mechanically gated.
- They are usually highly conserved across species.
- Most cells have a variety of Na⁺, K⁺, Ca²⁺, and Cl⁻ channels.
- Mutations in genes encoding them can cause specific diseases.[1]
- Their activities are affected by certain drugs.

[1]Some diseases caused by mutations of ion channels are briefly discussed in Chapter 49.

function of a K⁺ channel (KvAP) present in *Streptomyces lividans*. A variety of techniques were used, including site-directed mutagenesis and x-ray crystallography. The channel is an integral membrane protein composed of four identical subunits, each with two transmembrane segments, creating an inverted teepee-like structure (Figure 40–16). The part of the channels that confers ion selectivity (the **selectivity filter**) measures 12 Å long (a relatively short length of the membrane, so K⁺ does not have far to travel in the membrane) and is situated at the wide end of the inverted teepee. The large, water-filled cavity and helical dipoles shown in Figure 40–16 help overcome the relatively large electrostatic energy barrier for a cation to cross the membrane. The **selectivity filter** is lined with carbonyl oxygen atoms (contributed by a TVGYG sequence), providing a number of sites with which K⁺ can interact. K⁺ ions, which dehydrate as they enter the narrow selectivity filter, fit with proper coordination into the filter, but Na⁺ is too small to interact with the carbonyl oxygen atoms in correct alignment and is rejected. Two K⁺ ions, when close to each other in the filter, repel one another. This repulsion overcomes interactions between K⁺ and the surrounding protein molecule and allows very rapid conduction of K⁺ with high selectivity.

Other studies on a **voltage-gated ion channel** (HvAP) in *Aeropyrum pernix* have revealed many features of its voltage-sensing and voltage-gating mechanisms. This channel is made up of four subunits, each with six transmembrane segments. One of the six segments (S4 and part of S3) is the voltage sensor. It behaves like a **charged paddle** (Figure 40–17), in that it can move through the interior of the membrane transferring four positive charges (due to 4 Arg residues in each subunit) from

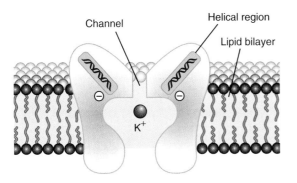

FIGURE 40–16 Schematic diagram of the structure of a K⁺ channel (KvAP) from *Streptomyces lividans*. A single K⁺ is shown in a large aqueous cavity inside the membrane interior. Two helical regions of the channel protein are oriented with their carboxylate ends pointing to where the K⁺ is located. The channel is lined by carboxyl oxygen. (Modified, with permission, from Doyle DA et al: Science 1998;280:69. Copyright © 1998 AAAS.)

one membrane surface to the other in response to changes in voltage. There are four voltage sensors in each channel, linked to the gate. The gate part of the channel is constructed from S6 helices (one from each of the subunits). Movements of this part of the channel in response to changing voltage effectively close the channel or reopen it, in the latter case allowing a current of ions to cross.

Ionophores Are Molecules That Act as Membrane Shuttles for Various Ions

Certain microbes synthesize small cyclic organic molecules, **ionophores, such as valinomycin** that function as shuttles for the movement of ions (K⁺ in the case of valinomycin) across membranes. These ionophores contain hydrophilic centers

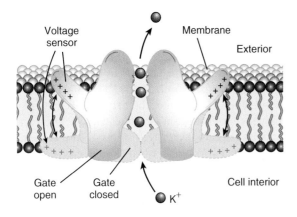

FIGURE 40–17 Schematic diagram of the voltage-gated K⁺ channel of *Aeorpyrum pernix*. The voltage sensors behave like charged paddles that move through the interior of the membrane. Four voltage sensors (only two are shown here) are linked mechanically to the gate of the channel. Each sensor has four positive charges contributed by arginine residues. (Modified, with permission, from Sigworth FJ: Nature 2003;423:21. Copyright © 2003. Macmillan Publishers Ltd.)

that bind specific ions and are surrounded by peripheral hydrophobic regions; this arrangement allows the molecules to dissolve effectively in the membrane and diffuse transversely therein. Others, like the well-studied polypeptide **gramicidin** (an antibiotic), form channels.

Microbial toxins such as **diphtheria toxin** and activated **serum complement components** can produce large pores in cellular membranes and thereby provide macromolecules with direct access to the internal milieu. The toxin α-**hemolysin** (produced by certain species of Streptococcus) consists of seven subunits which come together to form a β-barrel that allows metabolites to leak out of cells, resulting in cell lysis.

Aquaporins Are Proteins That Form Water Channels in Certain Membranes

In certain cells (eg, red cells, cells of the collecting ductules of the kidney), the movement of water by simple diffusion is augmented by movement through **water channels**. These channels are composed of tetrameric transmembrane proteins named **aquaporins**. At least 10 distinct aquaporins (AP-1 to AP-10) have been identified. Crystallographic and other studies have revealed how these channels permit passage of water but exclude passage of ions and protons. In essence, the pores are too narrow to permit passage of ions. Protons are excluded by the fact that the oxygen atom of water binds to two asparagine residues lining the channel, making the water unavailable to participate in a H⁺ relay, and thus preventing entry of protons. Mutations in the gene encoding AP-2 have been shown to be the cause of one type of **nephrogenic diabetes insipidus**. Peter Agre won a Nobel Prize for his work on the structure and function of aquaporins.

ACTIVE TRANSPORT SYSTEMS REQUIRE A SOURCE OF ENERGY

The process of active transport differs from diffusion in that molecules are transported against concentration gradients; hence, **energy is required.** This energy can come from the **hydrolysis of ATP,** from **electron movement,** or from **light.** The maintenance of electrochemical gradients in biologic systems is so important that it consumes approximately 30% of the total energy expenditure in a cell.

As shown in Table 40–6, four major classes of ATP-driven active transporters (**P, F, V, and ABC transporters**) have been recognized. The nomenclature is explained in the legend to the table. The first example of the P class, the **Na⁺K⁺ ATPase**, is discussed below. The **Ca²⁺** ATPase of muscle is discussed in Chapter 48. The second class is referred to as F-type. The most important example of this class is the **mt ATP synthase**, described in Chapter 13. V-type active transporters pump protons into lysosomes and other structures. ABC transporters include the **CFTR protein**, a chloride channel involved in the causation of cystic fibrosis (described later in this chapter and in Chapter 54). Another important member of this class is the **multidrug resistance-1 protein** (MDR-1 protein). This

TABLE 40–6 **Major Types of ATP-Driven Active Transporters**

Type	Example with Subcellular Location
P-type	Ca^{2+} ATPase (SR); Na^+-K^+-ATPase (PM)
F-type	mt ATP synthase of oxidative phosphorylation
V-type	The ATPase that pumps protons into lysosomes and synaptic vesicles
ABC transporter	CFTR protein (PM); MDR-1 protein (PM)

P (in P-type) signifies phosphorylation (these proteins autophosphorylate).
F (in F-type) signifies energy coupling factors.
V (in V-type) signifies vacuolar.
ABC signifies ATP-binding cassette transporter (all have two nucleotide-binding domains and two transmembrane segments).
SR, sarcoplasmic reticulum of muscle; PM, plasma membrane; mt, mitochondrial; CFTR, cystic fibrosis transmembrane regulator protein, a Cl^- transporter, and the protein implicated in the causation of cystic fibrosis (see later in this chapter and also Chapter 54); MDR-1 protein (multidrug resistance-1 protein), a protein that pumps many chemotherapeutic agents out of cancer cells and is thus an important contributor to the resistance of certain cancer cells to treatment.

transporter will pump a variety of drugs, including many anti-cancer agents, out of cells. It is a very important cause of cancer cells exhibiting resistance to chemotherapy, although many other mechanisms are also implicated.

The Na^+-K^+-ATPase of the Plasma Membrane Is a Key Enzyme in Regulating Intracellular Concentrations of Na^+ and K^+

In general, cells maintain a low intracellular Na^+ concentration and a high intracellular K^+ concentration (Table 40–1), along with a net negative electrical potential inside. The pump that maintains these ionic gradients is an **ATPase** that is activated by Na^+ and K^+ (Na^+-K^+-ATPase; see Figure 40–18). It pumps Na^+ out and K^+ into cells. The ATPase is an integral membrane protein that contains a transmembrane domain allowing the passage of ions, and cytosolic domains that couple ATP hydrolysis to transport. It has catalytic centers for both ATP and Na^+ on the cytoplasmic (inner) side of the plasma membrane (PM), with K^+ binding sites located on the extracellular side of the membrane. Phosphorylation by ATP of three Na^+-binding sites on the cytoplasmic surface of the cell induces a conformational change in the protein leading to transfer of three Na^+ ions from the inner to the outer side of the PM. Two molecules of K^+ bind to sites on the protein on the external surface of the PM, resulting in dephosphorylation of the protein and transfer of the K^+ ions across the membrane to the interior. Thus, three Na^+ ions are transported out for every two K^+ ions entering. This creates a charge imbalance between the inside and the outside of the cell, making the inside more negative (an **electrogenic** effect). Ouabain or digitalis (two important cardiac drugs) inhibit this ATPase by binding to the extracellular domain. This enzyme can consume ~30% of cellular energy. The Na^+-K^+-ATPase can be **coupled** to various other transporters, such as those involved in transport of glucose (see below).

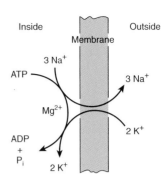

FIGURE 40–18 Stoichiometry of the Na^+-K^+-ATPase pump. This pump moves three Na^+ ions from inside the cell to the outside and brings two K^+ ions from the outside to the inside for every molecule of ATP hydrolyzed to ADP by the membrane-associated ATPase. Ouabain and other cardiac glycosides inhibit this pump by acting on the extracellular surface of the membrane. (Courtesy of R Post.)

TRANSMISSION OF NERVE IMPULSES INVOLVES ION CHANNELS AND PUMPS

The membrane enclosing **neuronal cells** maintains an asymmetry of inside-outside voltage (electrical potential) and is also electrically excitable due to the presence of voltage-gated channels. When appropriately stimulated by a chemical signal mediated by a specific synaptic membrane receptor (see discussion of the transmission of biochemical signals, below), channels in the membrane are opened to allow the rapid influx of Na^+ or Ca^{2+} (with or without the efflux of K^+), so that the voltage difference rapidly collapses and that segment of the membrane is depolarized. However, as a result of the action of the ion pumps in the membrane, the gradient is quickly restored.

When large areas of the membrane are **depolarized** in this manner, the electrochemical disturbance propagates in wave-like form down the membrane, generating a **nerve impulse. Myelin sheets,** formed by Schwann cells, wrap around nerve fibers and provide an electrical insulator that surrounds most of the nerve and greatly speeds up the propagation of the wave (signal) by allowing ions to flow in and out of the membrane only where the membrane is free of the insulation (at the nodes of Ranvier). The myelin membrane has a high content of lipid, accounting for its excellent insulating property. Relatively few proteins are found in the myelin membrane; those present appear to hold together multiple membrane bilayers to form the hydrophobic insulating structure that is impermeable to ions and water. **Certain diseases,** eg, multiple sclerosis and the Guillain-Barré syndrome, are characterized by demyelination and impaired nerve conduction.

TRANSPORT OF GLUCOSE INVOLVES SEVERAL MECHANISMS

A discussion of the **transport of glucose** summarizes many of the points made in this chapter. Glucose must enter cells as the first step in energy utilization. A number of different glucose transporters are involved, varying in different tissues (see Table 20–2). In **adipocytes and skeletal muscle**, glucose enters by a specific transport system that is enhanced by insulin. Changes in transport are primarily due to alterations of V_{max} (presumably from more or fewer transporters), but changes in K_m may also be involved. Glucose transport **in the small intestine** involves some different aspects of the principles of transport discussed above. Glucose and Na^+ bind to different sites on a Na^+-glucose symporter located at the **apical surface**. Na^+ moves into the cell down its electrochemical gradient and "drags" glucose with it (Figure 40–19). Therefore, the greater the Na^+ gradient, the more glucose enters; and if Na^+ in extracellular fluid is low, glucose transport stops. To maintain a steep Na^+ gradient, this Na^+-glucose symporter is dependent on gradients generated by the Na^+-K^+-ATPase, which maintains a low intracellular Na^+ concentration. Similar mechanisms are used to transport **other sugars** as well as **amino acids** across the apical lumen in polarized cells such as are found in the intestine and kidney. The transcellular movement of glucose in this case involves one additional component: a uniport (Figure 40–19) that allows the glucose accumulated within the cell to move across the **basolateral membrane**, and involves **a glucose uniporter** (GLUT2).

The treatment of severe cases of **diarrhea** (such as is found in cholera) makes use of the above information. In **cholera** (see Chapter 54)**,** massive amounts of fluid can be passed as watery stools in a very short time, resulting in severe dehydration and possibly death. Oral rehydration therapy, consisting primarily of NaCl and glucose, has been developed by the World Health Organization (WHO). The transport of glucose and Na^+ across the intestinal epithelium forces (via osmosis) movement of water from the lumen of the gut into intestinal cells, resulting in rehydration. Glucose alone or NaCl alone would not be effective.

CELLS TRANSPORT CERTAIN MACROMOLECULES ACROSS THE PLASMA MEMBRANE BY ENDOCYTOSIS AND EXOCYTOSIS

The process by which cells take up large molecules is called **endocytosis**. Some of these molecules (eg, polysaccharides, proteins, and polynucleotides), when hydrolyzed inside the cell, **yield nutrients**. Endocytosis also provides a mechanism for **regulating** the content of certain membrane components, hormone receptors being a case in point. Endocytosis can be used to learn more about how cells function. DNA from one cell type can be used to transfect a different cell and alter the latter's function or phenotype. A specific gene is often employed in these experiments, and this provides a unique way to study and analyze the regulation of that gene. **DNA transfection** depends upon endocytosis; endocytosis is responsible for the entry of DNA into the cell. Such experiments commonly use calcium phosphate, since Ca^{2+} stimulates endocytosis and precipitates DNA, which makes the DNA a better object for endocytosis. Cells also **release macromolecules** by **exocytosis**. Endocytosis and exocytosis both involve vesicle formation with or from the plasma membrane.

Endocytosis Involves Ingestion of Parts of the Plasma Membrane

Almost all eukaryotic cells are continuously ingesting parts of their plasma membranes. Endocytotic vesicles are generated when segments of the plasma membrane invaginate, enclosing a small volume of extracellular fluid and its contents. The vesicle then pinches off as the fusion of plasma membranes seals the neck of the vesicle at the original site of invagination (Figure 40–20). This vesicle fuses with other membrane structures and thus achieves the transport of its contents to other cellular compartments or even back to the cell exterior. Most endocytotic vesicles fuse with **primary lysosomes** to form **secondary lysosomes,** which contain hydrolytic enzymes and are therefore specialized organelles for intracellular disposal. The macromolecular contents are digested to yield amino acids, simple sugars, or nucleotides, and they are transported out of the vesicles to be reused by the cell. Endocytosis requires (1) energy, usually from the hydrolysis of ATP; (2) Ca^{2+}; and (3) contractile elements in the cell (probably the microfilament system) (Chapter 48).

There are **two** general types of endocytosis. **Phagocytosis** occurs only in specialized cells such as macrophages and granulocytes. Phagocytosis involves the ingestion of large

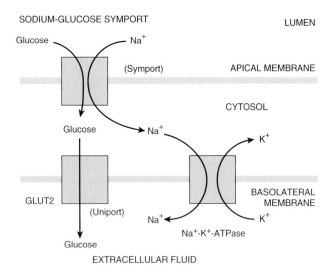

FIGURE 40–19 The transcellular movement of glucose in an intestinal cell. Glucose follows Na^+ across the luminal epithelial membrane. The Na^+ gradient that drives this symport is established by Na^+-K^+ exchange, which occurs at the basal membrane facing the extracellular fluid compartment via the action of the Na^+-K^+-ATPase. Glucose at high concentration within the cell moves "downhill" into the extracellular fluid by facilitated diffusion (a uniport mechanism), via GLUT2 (a glucose transporter, see Table 20–2). The sodium-glucose symport actually carries 2 Na^+ for each glucose.

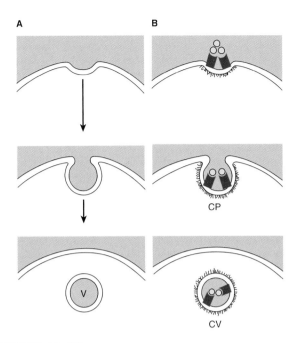

FIGURE 40–20 Two types of endocytosis. An endocytotic vesicle (V) forms as a result of invagination of a portion of the plasma membrane. Fluid-phase endocytosis **(A)** is random and nondirected. Receptor-mediated endocytosis **(B)** is selective and occurs in coated pits (CP) lined with the protein clathrin (the fuzzy material). Targeting is provided by receptors (brown symbols) specific for a variety of molecules. This results in the formation of a coated vesicle (CV).

particles such as viruses, bacteria, cells, or debris. Macrophages are extremely active in this regard and may ingest 25% of their volume per hour. In so doing, a macrophage may internalize 3% of its plasma membrane each minute or the entire membrane every 30 min.

Pinocytosis is a property of all cells and leads to the cellular uptake of fluid and fluid contents. There are **two types**. **Fluid-phase pinocytosis** is a nonselective process in which the uptake of a solute by formation of small vesicles is simply proportionate to its concentration in the surrounding extracellular fluid. The formation of these vesicles is an extremely active process. Fibroblasts, for example, internalize their plasma membrane at about one-third the rate of macrophages. This process occurs more rapidly than membranes are made. The surface area and volume of a cell do not change much, so membranes must be replaced by exocytosis or by being recycled as fast as they are removed by endocytosis.

The other type of pinocytosis, **absorptive pinocytosis,** is a receptor-mediated selective process primarily responsible for the uptake of macromolecules for which there are a finite number of binding sites on the plasma membrane. These high-affinity receptors permit the selective concentration of ligands from the medium, minimize the uptake of fluid or soluble unbound macromolecules, and markedly increase the rate at which specific molecules enter the cell. The vesicles formed during absorptive pinocytosis are derived from invaginations (pits) that are coated on the cytoplasmic side with a filamentous material and are appropriately named **coated pits.** In

many systems, the protein **clathrin** is the filamentous material. It has a three-limbed structure (called a triskelion), with each limb being made up of one light and one heavy chain of clathrin. The polymerization of clathrin into a vesicle is directed by **assembly particles,** composed of four **adapter proteins.** These interact with certain amino acid sequences in the receptors that become cargo, ensuring selectivity of uptake. The lipid **phophatidylinositol 4.5-bisphosphate (PIP$_2$)** (see Chapter 15) also plays an important role in vesicle assembly. In addition, the protein **dynamin,** which both binds and hydrolyzes GTP, is necessary for the pinching off of clathrin-coated vesicles from the cell surface. Coated pits may constitute as much as 2% of the surface of some cells.

As an example, **the low-density lipoprotein** (LDL) molecule and its **receptor** (Chapter 25) are internalized by means of coated pits containing the LDL receptor. These endocytotic vesicles containing LDL and its receptor fuse to lysosomes in the cell. The receptor is released and recycled back to the cell surface membrane, but the apoprotein of LDL is degraded and the cholesteryl esters metabolized. Synthesis of the LDL receptor is regulated by secondary or tertiary consequences of pinocytosis, eg, by metabolic products—such as cholesterol—released during the degradation of LDL. Disorders of the LDL receptor and its internalization are medically important and are discussed in Chapters 25 & 26.

Absorptive pinocytosis of **extracellular glycoproteins** requires that the glycoproteins carry specific carbohydrate recognition signals. These recognition signals are bound by membrane receptor molecules, which play a role analogous to that of the LDL receptor. A galactosyl receptor on the surface of hepatocytes is instrumental in the absorptive pinocytosis of asialoglycoproteins from the circulation (Chapter 47). Acid hydrolases taken up by absorptive pinocytosis in fibroblasts are recognized by their mannose 6-phosphate moieties. Interestingly, the mannose 6-phosphate moiety also seems to play an important role in the intracellular targeting of the acid hydrolases to the lysosomes of the cells in which they are synthesized (Chapter 47).

There is a **dark side** to receptor-mediated endocytosis in that viruses which cause such diseases as hepatitis (affecting liver cells), poliomyelitis (affecting motor neurons), and AIDS (affecting T cells) initiate their damage by this mechanism. Iron toxicity also begins with excessive uptake due to endocytosis.

Exocytosis Releases Certain Macromolecules from Cells

Most cells **release** macromolecules to the exterior by **exocytosis.** This process is also involved in membrane remodeling, when the components synthesized in the Golgi apparatus are carried in vesicles to the plasma membrane. The **signal** for exocytosis is often a hormone which, when it binds to a cell-surface receptor, induces a local and transient change in Ca^{2+} concentration. **Ca^{2+}** triggers exocytosis. Figure 40–21 provides a comparison of the mechanisms of exocytosis and endocytosis.

FIGURE 40–21 A comparison of the mechanisms of endocytosis and exocytosis. Exocytosis involves the contact of two inside-surface (cytoplasmic side) monolayers, whereas endocytosis results from the contact of two outer-surface monolayers.

Molecules released by exocytosis have at least **three fates**: (1) They can attach to the cell surface and become peripheral proteins, eg, antigens. (2) They can become part of the extracellular matrix, eg, collagen and glycosaminoglycans. (3) They can enter extracellular fluid and signal other cells. Insulin, parathyroid hormone, and the catecholamines are all packaged in granules and processed within cells, to be released upon appropriate stimulation.

VARIOUS SIGNALS ARE TRANSMITTED ACROSS MEMBRANES

Specific biochemical signals such as neurotransmitters, hormones, and immunoglobulins bind to specific **receptors** (integral proteins) exposed to the outside of cellular membranes and transmit information through these membranes to the cytoplasm. This process, called **transmembrane signaling** (see Chapter 42), involves the generation of a number of signalling molecules, including cyclic nucleotides, calcium, phosphoinositides, and diacylglycerol. Many of the steps involve **phosphorylation** of various molecules.

GAP JUNCTIONS ALLOW DIRECT FLOW OF MOLECULES FROM ONE CELL TO ANOTHER

Gap junctions are structures that permit **direct transfer of small molecules** (up to ~1200 Da) from one cell to its neighbor. They are composed of a family of proteins called **connexins** that form a hexagonal structure consisting of 12 such proteins. Six connexins form a connexin hemichannel and join to a similar structure in a neighboring cell to make a complete connexon channel (Figure 40–22). One gap junction contains several connexons. Different connexins are found in different tissues. **Mutations** in genes encoding connexins have been found to be associated with a number of conditions, including cardiovascular abnormalities, one type of deafness, and the X-linked form of Charcot-Marie-Tooth disease (a demyelinating neurologic disorder).

MUTATIONS AFFECTING MEMBRANE PROTEINS CAUSE DISEASES

In view of the fact that membranes are located in so many organelles and are involved in so many processes, it is not surprising that **mutations** affecting their protein constitu-

ents should result in many diseases or disorders. Proteins in membranes can be classified as **receptors, transporters, ion channels, enzymes,** and **structural components.** Members of all of these classes are often **glycosylated,** so that mutations affecting this process may alter their function. Examples of diseases or disorders due to abnormalities in membrane proteins are listed in Table 40–7; these mainly reflect mutations in proteins of the **plasma membrane,** with one affecting lysosomal function (I-cell disease). Many genetic diseases or disorders have been ascribed to mutations affecting various proteins involved in the transport of amino acids, sugars, lipids, urate, anions, cations, water, and vitamins across the **plasma membrane.** Mutations in genes encoding proteins in **other membranes** can also have harmful consequences. For example, mutations in genes encoding **mitochondrial membrane proteins** involved in oxidative phosphorylation can cause neurologic and other problems (eg, Leber hereditary optic neuropathy; LHON, **a condition in which some success with gene therapy was reported in 2008**). Membrane proteins can also be affected by conditions other than mutations. Formation of **autoantibodies** to the acetylcholine receptor in skeletal muscle causes myasthenia gravis. **Ischemia** can quickly affect the integrity of various ion channels in membranes. Abnormalities of membrane constituents other than

FIGURE 40–22 Schematic diagram of a gap junction. One connexon is made from two hemiconnexons. Each hemiconnexon is made from six connexin molecules. Small solutes are able to diffuse through the central channel, providing a direct mechanism of cell–cell communication.

TABLE 40–7 Some Diseases or Pathologic States Resulting From or Attributed to Abnormalities of Membranes[1]

Disease	Abnormality
Achondroplasia (OMIM 100800)	Mutations in the gene encoding the fibroblast growth factor receptor 3
Familial hypercholestero- lemia (OMIM 143890)	Mutations in the gene encoding the LDL receptor
Cystic fibrosis (OMIM 219700)	Mutations in the gene encoding the CFTR protein, a Cl⁻ transporter
Congenital long QT syndrome (OMIM 192500)	Mutations in genes encoding ion channels in the heart
Wilson disease (OMIM 277900)	Mutations in the gene encoding a copper-dependent ATPase
I-cell disease (OMIM 252500)	Mutations in the gene encoding GlcNAc phosphotransferase, resulting in absence of the Man 6-P signal for lysosomal localization of certain hydrolases
Hereditary spherocytosis (OMIM 182900)	Mutations in the genes encoding spectrin or other structural proteins in the red cell membrane
Metastasis of cancer cells	Abnormalities in the oligosaccharide chains of membrane glycoproteins and glycolipids are thought to be of importance
Paroxysmal nocturnal hemoglobinuria (OMIM 311770)	Mutation resulting in deficient attachment of the GPI anchor (see Chapter 47) to certain proteins of the red cell membrane

[1]The disorders listed are discussed further in other chapters. The table lists examples of mutations affecting two receptors, one transporter, several ion channels (i.e., congenital long QT syndrome), two enzymes, and one structural protein. Examples of altered or defective glycosylation of glycoproteins are also presented. Most of the conditions listed involve the plasma membrane.

proteins can also be harmful. With regard to **lipids,** excess of cholesterol (eg, in familial hypercholesterolemia), of lyso-phospholipid (eg, after bites by certain snakes, whose venom contains phospholipases), or of glycosphingolipids (eg, in a sphingolipidosis), can all affect membrane function.

Cystic Fibrosis Is Due to Mutations in the Gene Encoding CFTR, a Chloride Transporter

Cystic fibrosis (CF) is a **recessive** genetic disorder prevalent among whites in North America and certain parts of northern Europe. It is characterized by chronic bacterial infections of the airways and sinuses, fat maldigestion due to pancreatic exocrine insufficiency, infertility in males due to abnormal development of the vas deferens, and **elevated levels of chloride in sweat** (> 60 mmol/L). In 1989, it was shown that mutations in a gene encoding a protein named **cystic fibrosis transmembrane regulator protein (CFTR)** were responsible for CF. CFTR is a **cyclic AMP-regulated Cl⁻ transporter**. The major clinical fea-

tures of CF and further information about the gene responsible for CF and about CFTR are presented in Chapter 54.

SUMMARY

- Membranes are complex structures composed of lipids, proteins and carbohydrate-containing molecules,

- The basic structure of all membranes is the lipid bilayer. This bilayer is formed by two sheets of phospholipids in which the hydrophilic polar head groups are directed away from each other and are exposed to the aqueous environment on the outer and inner surfaces of the membrane. The hydrophobic nonpolar tails of these molecules are oriented toward each other, in the direction of the center of the membrane.

- Membranes are dynamic structures. Lipids and certain proteins show rapid lateral diffusion. Flip-flop is very slow for lipids and nonexistent for proteins.

- The fluid mosaic model forms a useful basis for thinking about membrane structure.

- Membrane proteins are classified as integral if they are firmly embedded in the bilayer and as peripheral if they are attached to the outer or inner surface.

- The 20 or so membranes in a mammalian cell have different functions and they define compartments, or specialized environments, within the cell that have specific functions (eg, lysosomes).

- Certain hydrophobic molecules freely diffuse across membranes, but the movement of others is restricted because of their size or charge.

- Various passive and active (usually ATP-dependent) mechanisms are employed to maintain gradients of such molecules across different membranes.

- Certain solutes, eg, glucose, enter cells by facilitated diffusion along a downhill gradient from high to low concentration using specific carrier proteins (transporters).

- The major ATP-driven pumps are classified as P (phosphorylated), F (energy factors), V (vacuolar), and ABC transporters. Member of these classes include the Na⁺-K⁺-ATPase, the Ca²⁺ ATPase of the sarcoplasmic reticulum, the mt ATP synthase, the ATPase acidifying lysosomes, the CFTR protein and the MDR-1 protein.

- Ligand- or voltage-gated ion channels are often employed to move charged molecules (Na⁺, K⁺, Ca²⁺, etc) across membranes.

- Large molecules can enter or leave cells through mechanisms such as endocytosis or exocytosis. These processes often require binding of the molecule to a receptor, which affords specificity to the process.

- Receptors may be integral components of membranes (particularly the plasma membrane). The interaction of a ligand with its receptor may not involve the movement of either into the cell, but the interaction results in the generation of a signal that influences intracellular processes (transmembrane signaling).

- Mutations that affect the structure of membrane proteins (receptors, transporters, ion channels, enzymes, and structural proteins) may cause diseases; examples include cystic fibrosis and familial hypercholesterolemia.

REFERENCES

Alberts B et al: *Molecular Biology of the Cell.* 5th ed. Garland Science, 2008.

Clapham DE: Symmetry, selectivity and the 2003 Nobel Prize. Cell 2003;115:641.

Holland IB et al: *ABC Proteins: From Bacteria to Man.* Academic Press/Elsevier Science, 2003.

Le Roy C, Wrana JL: Clathrin- and non-clathrin-mediated endocytic regulation of cell signaling. Nat Rev Mol Cell Biol 2005;6:112.

Lodish H et al: *Molecular Cell Biology.* 6th ed. WH Freeman & Co, 2008.

Longo N: Inherited defects of membrane transport. In: *Harrison's Principles of Internal Medicine.* 17th ed. Fauci AS et al (editors). Chapter 359. McGraw-Hill, 2008.

Lucero HA, Robbins PW: Lipid rafts: protein association and the regulation of protein activity. Arch Biochem Biophys 2004;426:208.

Pollard TD, Earnshaw WC: *Cell Biology.* 2nd ed. Saunders Elsevier, 2008.

Singer SJ: Some early history of membrane molecular biology. Annu Rev Physiol 2004;66:1.

Vance DE, Vance J (editors): *Biochemistry of Lipids, Lipoproteins and Membranes.* 5th ed. Elsevier, 2008.

Yeagle PL: *The Structure of Biological Membranes.* 2nd ed. CRC Press, 2004.

The Diversity of the Endocrine System

P. Anthony Weil, PhD

ACTH	Adrenocorticotropic hormone		**GH**	Growth hormone
ANF	Atrial natriuretic factor		**IGF-I**	Insulin-like growth factor-I
cAMP	Cyclic adenosine monophosphate		**LH**	Luteotropic hormone
CBG	Corticosteroid-binding globulin		**LPH**	Lipotropin
CG	Chorionic gonadotropin		**MIT**	Monoiodotyrosine
cGMP	Cyclic guanosine monophosphate		**MSH**	Melanocyte-stimulating hormone
CLIP	Corticotropin-like intermediate lobe peptide		**OHSD**	Hydroxysteroid dehydrogenase
			PNMT	Phenylethanolamine-N-methyltransferase
DBH	Dopamine β-hydroxylase		**POMC**	Pro-opiomelanocortin
DHEA	Dehydroepiandrosterone		**SHBG**	Sex hormone-binding globulin
DHT	Dihydrotestosterone		**StAR**	Steroidogenic acute regulatory (protein)
DIT	Diiodotyrosine		**TBG**	Thyroxine-binding globulin
DOC	Deoxycorticosterone		**TEBG**	Testosterone-estrogen-binding globulin
EGF	Epidermal growth factor		**TRH**	Thyrotropin-releasing hormone
FSH	Follicle-stimulating hormone		**TSH**	Thyrotropin-stimulating hormone

BIOMEDICAL IMPORTANCE

The survival of multicellular organisms depends on their ability to adapt to a constantly changing environment. Intercellular communication mechanisms are necessary requirements for this adaptation. The nervous system and the endocrine system provide this intercellular, organism-wide communication. The nervous system was originally viewed as providing a fixed communication system, whereas the endocrine system supplied hormones, which are mobile messages. In fact, there is a remarkable convergence of these regulatory systems. For example, neural regulation of the endocrine system is important in the production and secretion of some hormones; many neurotransmitters resemble hormones in their synthesis, transport, and mechanism of action; and many hormones are synthesized in the nervous system. The word "hormone" is derived from a Greek term that means to arouse to activity. As classically defined, a hormone is a substance that is synthesized in one organ and transported by the circulatory system to act on another tissue. However, this original description is too restrictive because hormones can act on adjacent cells (paracrine action) and on the cell in which they were synthesized (autocrine action) without entering the systemic circulation. A diverse array of hormones—each with distinctive mechanisms of action and properties of biosynthesis, storage, secretion, transport, and metabolism—has evolved to provide homeostatic responses. This biochemical diversity is the topic of this chapter.

THE TARGET CELL CONCEPT

There are about 200 types of differentiated cells in humans. Only a few produce hormones, but virtually all of the 75 trillion cells in a human are targets of one or more of the over 50 known hormones. The concept of the target cell is a useful way of looking at hormone action. It was thought that hormones affected a single cell type—or only a few kinds of cells—and that a hormone elicited a unique biochemical or physiologic action. We now know that a given hormone can affect several different cell types; that more than one hormone can affect a given cell type; and that hormones can exert many different effects in one cell or in different cells. With the discovery of specific cell-surface and intracellular hormone receptors, the definition of a target has been expanded to include any cell in which the hormone (ligand) binds to its receptor, whether or not a biochemical or physiologic response has yet been determined.

Several factors determine the response of a target cell to a hormone. These can be thought of in two general ways: (1) as

TABLE 41–1 Determinants of the Concentration of a Hormone at the Target Cell

The rate of synthesis and secretion of the hormones.
The proximity of the target cell to the hormone source (dilution effect).
The dissociation constants of the hormone with specific plasma transport proteins (if any).
The conversion of inactive or suboptimally active forms of the hormone into the fully active form.
The rate of clearance from plasma by other tissues or by digestion, metabolism, or excretion.

TABLE 41–2 Determinants of the Target Cell Response

The number, relative activity, and state of occupancy of the specific receptors on the plasma membrane or in the cytoplasm or nucleus.
The metabolism (activation or inactivation) of the hormone in the target cell.
The presence of other factors within the cell that are necessary for the hormone response.
Up- or down-regulation of the receptor consequent to the interaction with the ligand.
Postreceptor desensitzation of the cell, including down-regulation of the receptor.

factors that affect the concentration of the hormone at the target cell (Table 41–1) and (2) as factors that affect the actual response of the target cell to the hormone (Table 41–2).

HORMONE RECEPTORS ARE OF CENTRAL IMPORTANCE

Receptors Discriminate Precisely

One of the major challenges faced in making the hormone-based communication system work is illustrated in Figure 41–1. Hormones are present at very low concentrations in the extracellular fluid, generally in the atto- to nanomolar range (10^{-15} to 10^{-9} mol/L). This concentration is much lower than that of the many structurally similar molecules (sterols, amino acids, peptides, proteins) and other molecules that circulate at concentrations in the micro- to millimolar (10^{-6} to 10^{-3}) mol/L range. Target cells, therefore, must distinguish not only between different hormones present in small amounts but also between a given hormone and the 10^6- to 10^9-fold excess of other similar molecules. This high degree of discrimination is provided by cell-associated recognition molecules called receptors. Hormones initiate their biologic effects by binding to specific receptors, and since any effective control system also must provide a means of stopping a response, hormone-induced actions generally but not always terminate when the effector dissociates from the receptor (Figure 38–1).

A target cell is defined by its ability to selectively bind a given hormone to its cognate receptor. Several biochemical features of this interaction are important in order for hormone-receptor interactions to be physiologically relevant: (1) binding should be specific, ie, displaceable by agonist or antagonist; (2) binding should be saturable; and (3) binding should occur within the concentration range of the expected biologic response.

Both Recognition & Coupling Domains Occur on Receptors

All receptors have at least two functional domains. A recognition domain binds the hormone ligand and a second region generates a signal that couples hormone recognition to some intracellular function. Coupling (signal transduction) occurs in two general ways. Polypeptide and protein hormones and the catecholamines bind to receptors located in the plasma membrane and thereby generate a signal that regulates various intracellular functions, often by changing the activity of an enzyme. In contrast, steroid, retinoid, and thyroid hormones interact with intracellular receptors, and it is this ligand–receptor complex that directly provides the signal, generally to specific genes whose rate of transcription is thereby affected.

The domains responsible for hormone recognition and signal generation have been identified in the protein polypeptide and catecholamine hormone receptors. Steroid, thyroid, and retinoid hormone receptors have several functional domains: one site binds the hormone; another binds to specific DNA regions; a third is involved in the interaction with other coregulator proteins that result in the activation (or repression) of gene transcription; and a fourth may specify binding to one or more other proteins that influence the intracellular trafficking of the receptor.

The dual functions of binding and coupling ultimately define a receptor, and it is the coupling of hormone binding to signal transduction—so-called **receptor-effector coupling**—that provides the first step in amplification of the hormonal response. This dual purpose also distinguishes the target cell receptor from the plasma carrier proteins that bind hormone but do not generate a signal (see Table 41–6).

Receptors Are Proteins

Several classes of peptide hormone receptors have been defined. For example, the insulin receptor is a heterotetramer composed of two copies of two different protein subunits ($\alpha_2\beta_2$) linked by multiple disulfide bonds in which the extracellular α subunit binds insulin and the membrane-spanning β subunit transduces the signal through the tyrosine protein kinase domain located in the cytoplasmic portion of this polypeptide. The receptors for insulin-like growth factor I (IGF-I) and epidermal growth factor (EGF) are generally similar in structure to the insulin receptor. The growth hormone and prolactin receptors also span the plasma membrane of target cells but do not con-

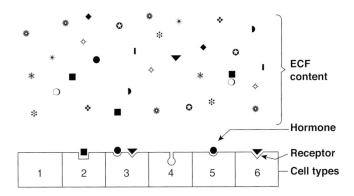

FIGURE 41–1 Specificity and selectivity of hormone receptors. Many different molecules circulate in the extracellular fluid (ECF), but only a few are recognized by hormone receptors. Receptors must select these molecules from among high concentrations of the other molecules. This simplified drawing shows that a cell may have no hormone receptors (1), have one receptor (2+5+6), have receptors for several hormones (3), or have a receptor but no hormone in the vicinity (4).

tain intrinsic protein kinase activity. Ligand binding to these receptors, however, results in the association and activation of a completely different protein kinase signaling pathway, the Jak-Stat pathway. Polypeptide hormone and catecholamine receptors, which transduce signals by altering the rate of production of cAMP through G-proteins, are characterized by the presence of seven domains that span the plasma membrane. Protein kinase activation and the generation of cyclic AMP (cAMP, 3′5′-adenylic acid; see Figure 19–5) is a downstream action of this class of receptor (see Chapter 42 for further details).

A comparison of several different steroid receptors with thyroid hormone receptors revealed a remarkable conservation of the amino acid sequence in certain regions, particularly in the DNA-binding domains. This led to the realization that receptors of the steroid or thyroid type are members of a large superfamily of nuclear receptors. Many related members of this family currently have no known ligand and thus are called orphan receptors. The nuclear receptor superfamily plays a critical role in the regulation of gene transcription by hormones, as described in Chapter 42.

HORMONES CAN BE CLASSIFIED IN SEVERAL WAYS

Hormones can be classified according to chemical composition, solubility properties, location of receptors, and the nature of the signal used to mediate hormonal action within the cell. A classification based on the last two properties is illustrated in Table 41–3, and general features of each group are illustrated in Table 41–4.

The hormones in group I are lipophilic. After secretion, these hormones associate with plasma transport or carrier proteins, a process that circumvents the problem of solubility while prolonging the plasma half-life of the hormone. The relative percentages of bound and free hormone are determined by the amount, binding affinity and binding capacity of the transport protein. The free hormone, which is the biologically active form, readily traverses the lipophilic plasma membrane of all cells and encounters receptors in either the cytosol or nucleus of target cells. The ligand–receptor complex is assumed to be the intracellular messenger in this group.

The second major group consists of water-soluble hormones that bind to the plasma membrane of the target cell. Hormones that bind to the surfaces of cells communicate with intracellular metabolic processes through intermediary molecules called **second messengers** (the hormone itself is the first messenger), which are generated as a consequence of the ligand–receptor interaction. The second messenger concept arose from an observation that epinephrine binds to the plasma membrane of certain cells and increases intracellular cAMP. This was followed by a series of experiments in which cAMP was found to mediate the effects of many hormones. Hormones that clearly employ this mechanism are shown in group II.A of Table 41–3. Atrial natriuretic factor (ANF), uses cGMP as its second messenger (group II.B). Several hormones, many of which were previously thought to affect cAMP, appear to use ionic calcium (Ca^{2+}) or metabolites of complex phosphoinositides (or both) as the intracellular signal. These are shown in group II.C of the table. The intracellular messenger for group II.D is a protein kinase-phosphatase cascade. Several of these have been identified, and a given hormone may use more than one kinase cascade. A few hormones fit into more than one category, and assignments change as new information is brought forward.

DIVERSITY OF THE ENDOCRINE SYSTEM

Hormones Are Synthesized in a Variety of Cellular Arrangements

Hormones are synthesized in discrete organs designed solely for this specific purpose, such as the thyroid (triiodothyronine), adrenal (glucocorticoids and mineralocorticoids), and the pituitary (TSH, FSH, LH, growth hormone, prolactin, ACTH). Some organs are designed to perform two distinct but closely related functions. For example, the ovaries produce mature oocytes and the reproductive hormones estradiol and progesterone. The testes produce mature spermatozoa and testosterone. Hormones are also produced in specialized cells within other organs such as the small intestine (glucagon-like peptide), thyroid (calcitonin), and kidney (angiotensin

TABLE 41–3 Classification of Hormones by Mechanism of Action

I. Hormones that bind to intracellular receptors

Androgens
Calcitriol (1,25[OH]$_2$-D$_3$)
Estrogens
Glucocorticoids
Mineralocorticoids
Progestins
Retinoic acid
Thyroid hormones (T$_3$ and T$_4$)

II. Hormones that bind to cell surface receptors
A. The second messenger is cAMP

α$_2$-Adrenergic catecholamines
β-Adrenergic catecholamines
Adrenocorticotropic hormone (ACTH)
Antidiuretic hormone
Calcitonin
Chorionic gonadotropin, human (CG)
Corticotropin-releasing hormone
Follicle-stimulating hormone (FSH)
Glucagon
Lipotropin (LPH)
Luteinizing hormone (LH)
Melanocyte-stimulating hormone (MSH)
Parathyroid hormone (PTH)
Somatostatin
Thyroid-stimulating hormone (TSH)

B. The second messenger is cGMP

Atrial natriuretic factor
Nitric oxide

C. The second messenger is calcium or phosphatidylinositols (or both)

Acetylcholine (muscarinic)
α$_1$-Adrenergic catecholamines
Angiotensin II
Antidiuretic hormone (vasopressin)
Cholecystokinin
Gastrin
Gonadotropin-releasing hormone
Oxytocin
Platelet-derived growth factor (PDGF)
Substance P
Thyrotropin-releasing hormone (TRH)

D. The second messenger is a kinase or phosphatase cascade

Adiponectin
Chorionic somatomammotropin
Epidermal growth factor
Erythropoietin
Fibroblast growth factor (FGF)
Growth hormone (GH)
Insulin
Insulin-like growth factors I and II
Leptin
Nerve growth factor (NGF)
Platelet-derived growth factor
Prolactin

TABLE 41–4 General Features of Hormone Classes

	Group I	Group II
Types	Steroids, iodo-thyronines, calcitriol, retinoids	Polypeptides, proteins, glycoproteins, catecholamines
Solubility	Lipophilic	Hydrophilic
Transport proteins	Yes	No
Plasma half-life	Long (hours to days)	Short (minutes)
Receptor	Intracellular	Plasma membrane
Mediator	Receptor-hormone complex	cAMP, cGMP, Ca^{2+}, metabolites of complex phosphinositols, kinase cascades

(calcitriol). Examples of this diversity in the approach to hormone synthesis, each of which has evolved to fulfill a specific purpose, are discussed below.

Hormones Are Chemically Diverse

Hormones are synthesized from a wide variety of chemical building blocks. A large series is derived from cholesterol. These include the glucocorticoids, mineralocorticoids, estrogens, progestins, and 1,25(OH)$_2$-D$_3$ (Figure 41–2). In some cases, a steroid hormone is the precursor molecule for another hormone. For example, progesterone is a hormone in its own right but is also a precursor in the formation of glucocorticoids, mineralocorticoids, testosterone, and estrogens. Testosterone is an obligatory intermediate in the biosynthesis of estradiol and in the formation of dihydrotestosterone (DHT). In these examples, described in detail below, the final product is determined by the cell type and the associated set of enzymes in which the precursor exists.

The amino acid tyrosine is the starting point in the synthesis of the catecholamines and of the thyroid hormones tetraiodothyronine (thyroxine; T$_4$) and triiodothyronine (T$_3$) (Figure 41–2). T$_3$ and T$_4$ are unique in that they require the addition of iodine (as I$^-$) for bioactivity. Because dietary iodine is very scarce in many parts of the world, an intricate mechanism for accumulating and retaining I$^-$ has evolved.

Many hormones are polypeptides or glycoproteins. These range in size from thyrotropin-releasing hormone (TRH), a tripeptide, to single-chain polypeptides like adrenocorticotropic hormone (ACTH; 39 amino acids), parathyroid hormone (PTH; 84 amino acids), and growth hormone (GH; 191 amino acids) (Figure 41–2). Insulin is an AB chain heterodimer of 21 and 30 amino acids, respectively. Follicle-stimulating hormone (FSH), luteinizing hormone (LH), thyroid-stimulating hormone (TSH), and chorionic gonadotropin (CG) are glycoprotein hormones of αβ heterodimeric structure. The α chain is identical in all of these hormones,

II). Finally, the synthesis of some hormones requires the parenchymal cells of more than one organ—eg, the skin, liver, and kidney are required for the production of 1,25(OH)$_2$-D$_3$

FIGURE 41–2 Chemical diversity of hormones. **(A)** Cholesterol derivatives. **(B)** Tyrosine derivatives. **(C)** Peptides of various sizes. **(D)** Glycoproteins (TSH, FSH, LH) with common α subunits and unique β subunits.

and distinct β chains impart hormone uniqueness. These hormones have a molecular mass in the range of 25–30 kDa depending on the degree of glycosylation and the length of the β chain.

Hormones Are Synthesized & Modified for Full Activity in a Variety of Ways

Some hormones are synthesized in final form and secreted immediately. Included in this class are the hormones derived from cholesterol. Others such as the catecholamines are synthesized in final form and stored in the producing cells. Others, like insulin, are synthesized from precursor molecules in the producing cell, then are processed and secreted upon a physiologic cue (plasma glucose concentrations). Finally, still others are converted to ac-

tive forms from precursor molecules in the periphery (T_3 and DHT). All of these examples are discussed in more detail below.

MANY HORMONES ARE MADE FROM CHOLESTEROL

Adrenal Steroidogenesis

The adrenal steroid hormones are synthesized from cholesterol. Cholesterol is mostly derived from the plasma, but a small portion is synthesized in situ from acetyl-CoA via mevalonate and squalene. Much of the cholesterol in the adrenal is esterified and stored in cytoplasmic lipid droplets. Upon stimulation of the adrenal by ACTH, an esterase is activated, and the free cholesterol formed is transported into the mitochondrion, where a **cytochrome P450 side chain cleavage enzyme (P450scc)** converts cholesterol to pregnenolone. Cleavage of the side chain involves sequential hydroxylations, first at C_{22} and then at C_{20}, followed by side chain cleavage (removal of the six-carbon fragment isocaproaldehyde) to give the 21-carbon steroid (Figure 41–3, top). An ACTH-dependent **steroidogenic acute regulatory (StAR) protein** is essential for the transport of cholesterol to P450scc in the inner mitochondrial membrane.

All mammalian steroid hormones are formed from cholesterol via pregnenolone through a series of reactions that occur in either the mitochondria or endoplasmic reticulum of the producing cell. Hydroxylases that require molecular oxygen and NADPH are essential, and dehydrogenases, an isomerase, and a lyase reaction are also necessary for certain steps. There is cellular specificity in adrenal steroidogenesis. For instance, 18-hydroxylase and 19-hydroxysteroid dehydrogenase, which are required for aldosterone synthesis, are found only in the zona glomerulosa cells (the outer region of the adrenal cortex), so that the biosynthesis of this mineralocorticoid is confined to this region. A schematic representation of the pathways involved in the synthesis of the three major classes of adrenal steroids is presented in Figure 41–4. The enzymes are shown in the rectangular boxes, and the modifications at each step are shaded.

Mineralocorticoid Synthesis

Synthesis of aldosterone follows the mineralocorticoid pathway and occurs in the zona glomerulosa. Pregnenolone is converted to progesterone by the action of two smooth endoplasmic reticulum enzymes, **3β-hydroxysteroid dehydrogenase (3β-OHSD)** and **Δ^{5,4}-isomerase.** Progesterone is hydroxylated at the C_{21} position to form 11-deoxycorticosterone (DOC), which is an active (Na$^+$-retaining) mineralocorticoid. The next hydroxylation, at C_{11}, produces corticosterone, which has glucocorticoid activity and is a weak mineralocorticoid (it has less than 5% of the potency of aldosterone). In some species (eg, rodents), it is the most potent glucocorticoid. C_{21} hydroxylation is necessary for both mineralocorticoid and glucocorticoid activity, but most steroids with a C_{17} hydroxyl group have more glucocorticoid and less mineralocorticoid action. In the zona glomerulosa, which does not have the smooth endoplasmic reticulum enzyme 17α-hydroxylase, a mitochondrial 18-hydroxylase is present. The **18-hydroxylase (aldosterone synthase)** acts on corticosterone to form 18-hydroxycorticosterone, which is changed to aldosterone by conversion of the 18-alcohol to an aldehyde. This unique distribution of enzymes and the special regulation of the zona glomerulosa by K$^+$ and angiotensin II have led some investigators to suggest that, in addition to the adrenal being two glands, the adrenal cortex is actually two separate organs.

Glucocorticoid Synthesis

Cortisol synthesis requires three hydroxylases located in the fasciculata and reticularis zones of the adrenal cortex that act sequentially on the C_{17}, C_{21}, and C_{11} positions. The first two reactions are rapid, while C_{11} hydroxylation is relatively slow. If the C_{11} position is hydroxylated first, the action of **17α-hydroxylase** is impeded and the mineralocorticoid pathway is followed

FIGURE 41–3 Cholesterol side-chain cleavage and basic steroid hormone structures. The basic sterol rings are identified by the letters A–D. The carbon atoms are numbered 1–21, starting with the A ring.

{AQ}

FIGURE 41–4 Pathways involved in the synthesis of the three major classes of adrenal steroids (mineralocorticoids, glucocorticoids, and androgens). Enzymes are shown in the rectangular boxes, and the modifications at each step are shaded. Note that the 17α-hydroxylase and 17,20-lyase activities are both part of one enzyme, designated P450c17. (Slightly modified and reproduced, with permission, from Harding BW: In: *Endocrinology*, vol 2. DeGroot LJ [editor]. Grune & Stratton, 1979. Copyright © 1979 Elsevier Inc. Reprinted with permission from Elsevier.)

(forming corticosterone or aldosterone, depending on the cell type). 17α-Hydroxylase is a smooth endoplasmic reticulum enzyme that acts upon either progesterone or, more commonly, pregnenolone. 17α-Hydroxyprogesterone is hydroxylated at C_{21} to form 11-deoxycortisol, which is then hydroxylated at C_{11} to form cortisol, the most potent natural glucocorticoid hormone in humans. 21-Hydroxylase is a smooth endoplasmic reticulum enzyme, whereas 11β-hydroxylase is a mitochondrial enzyme. Steroidogenesis thus involves the repeated shuttling of substrates into and out of the mitochondria.

Androgen Synthesis

The major androgen or androgen precursor produced by the adrenal cortex is dehydroepiandrosterone (DHEA). Most 17-hydroxypregnenolone follows the glucocorticoid pathway, but a small fraction is subjected to oxidative fission and removal of the two-carbon side chain through the action of 17,20-lyase. The lyase activity is actually part of the same enzyme (P450c17) that catalyzes 17α-hydroxylation. This is therefore a **dual-function protein.** The lyase activity is important in both the adrenals and the gonads and acts exclusively on 17α-hydroxy-containing molecules. Adrenal androgen production increases markedly if glucocorticoid biosynthesis is impeded by the lack of one of the hydroxylases **(adrenogenital syndrome).** DHEA is really a prohormone, since the actions of 3β-OHSD and $\Delta^{5,4}$-isomerase convert the weak androgen DHEA into the more potent **androstenedione.** Small amounts of androstenedione are also formed in the adrenal by the action of the lyase on 17α-hydroxyprogesterone.

Reduction of androstenedione at the C_{17} position results in the formation of **testosterone,** the most potent adrenal androgen. Small amounts of testosterone are produced in the adrenal by this mechanism, but most of this conversion occurs in the testes.

Testicular Steroidogenesis

Testicular androgens are synthesized in the interstitial tissue by the Leydig cells. The immediate precursor of the gonadal steroids, as for the adrenal steroids, is cholesterol. The rate-limiting step, as in the adrenal, is delivery of cholesterol to the inner membrane of the mitochondria by the transport protein StAR. Once in the proper location, cholesterol is acted upon by the side chain cleavage enzyme P450scc. The conversion of cholesterol to pregnenolone is identical in adrenal, ovary, and testis. In the latter two tissues, however, the reaction is promoted by LH rather than ACTH.

The conversion of pregnenolone to testosterone requires the action of five enzyme activities contained in three proteins: (1) 3β-hydroxysteroid dehydrogenase (3β-OHSD) and $\Delta^{5,4}$-isomerase; (2) 17α-hydroxylase and 17,20-lyase; and (3) 17β-hydroxysteroid dehydrogenase (17β-OHSD). This sequence, referred to as the **progesterone (or Δ^4) pathway,** is shown on the right side of Figure 41–5. Pregnenolone can also be converted to testosterone by the **dehydroepiandrosterone (or Δ^5) pathway,** which is illustrated on the left side of Figure 41–5. The Δ^5 route appears to be most used in human testes.

The five enzyme activities are localized in the microsomal fraction in rat testes, and there is a close functional association between the activities of 3β-OHSD and $\Delta^{5,4}$-isomerase and between those of a 17α-hydroxylase and 17,20-lyase. These enzyme pairs, both contained in a single protein, are shown in the general reaction sequence in Figure 41–5.

Dihydrotestosterone Is Formed from Testosterone in Peripheral Tissues

Testosterone is metabolized by two pathways. One involves oxidation at the 17 position, and the other involves reduction of the A ring double bond and the 3-ketone. Metabolism by the first pathway occurs in many tissues, including liver, and produces 17-ketosteroids that are generally inactive or less active than the parent compound. Metabolism by the second pathway, which is less efficient, occurs primarily in target tissues and produces the potent metabolite dihydrotestosterone (DHT).

The most significant metabolic product of testosterone is DHT, since in many tissues, including prostate, external genitalia, and some areas of the skin, this is the active form of the hormone. The plasma content of DHT in the adult male is about one-tenth that of testosterone, and approximately 400 μg of DHT is produced daily as compared with about 5 mg of testosterone. About 50–100 μg of DHT are secreted by the testes. The rest is produced peripherally from testosterone in a reaction catalyzed by the NADPH-dependent **5α-reductase** (Figure 41–6). Testosterone can thus be considered a prohormone, since it is converted into a much more potent compound (dihydrotestosterone) and since most of this conversion occurs outside the testes. Some estradiol is formed from the peripheral aromatization of testosterone, particularly in males.

Ovarian Steroidogenesis

The estrogens are a family of hormones synthesized in a variety of tissues. 17β-Estradiol is the primary estrogen of ovarian origin. In some species, estrone, synthesized in numerous tissues, is more abundant. In pregnancy, relatively more estriol is produced, and this comes from the placenta. The general pathway and the subcellular localization of the enzymes involved in the early steps of estradiol synthesis are the same as those involved in androgen biosynthesis. Features unique to the ovary are illustrated in Figure 41–7.

Estrogens are formed by the aromatization of androgens in a complex process that involves three hydroxylation steps, each of which requires O_2 and NADPH. The **aromatase enzyme complex** is thought to include a P450 monooxygenase. Estradiol is formed if the substrate of this enzyme complex is testosterone, whereas estrone results from the aromatization of androstenedione.

The cellular source of the various ovarian steroids has been difficult to unravel, but a transfer of substrates between two cell types is involved. Theca cells are the source of androstenedione and testosterone. These are converted by the aromatase enzyme in granulosa cells to estrone and estradiol, respectively. Progesterone, a precursor for all steroid hormones, is produced and secreted by the corpus luteum as an end-product hormone because these cells do not contain the enzymes necessary to convert progesterone to other steroid hormones (Figure 41–8).

Significant amounts of estrogens are produced by the peripheral aromatization of androgens. In human males, the peripheral aromatization of testosterone to estradiol (E_2) accounts for 80% of the production of the latter. In females, adrenal androgens are important substrates, since as much as 50% of the E_2 produced during pregnancy comes from the aromatization of androgens. Finally, conversion of androstenedione to estrone is the major source of estrogens in postmenopausal women. Aromatase activity is present in adipose cells and also in liver, skin, and other tissues. Increased activity of this enzyme may contribute to the "estrogenization" that characterizes such diseases as cirrhosis of the liver, hyperthyroidism, aging, and obesity. Aromatase inhibitors show promise as therapeutic agents in breast cancer and possibly in other female reproductive tract malignancies.

1,25(OH)$_2$-D$_3$ (Calcitriol) Is Synthesized from a Cholesterol Derivative

1,25(OH)$_2$-D$_3$ is produced by a complex series of enzymatic reactions that involve the plasma transport of precursor molecules to a number of different tissues (Figure 41–9). One of these precursors is vitamin D—really not a vitamin, but this common name persists. The active molecule, 1,25(OH)$_2$-D$_3$, is transported to other organs where it activates biologic processes in a manner similar to that employed by the steroid hormones.

FIGURE 41–5 Pathways of testosterone biosynthesis. The pathway on the left side of the figure is called the Δ⁵ or dehydroepiandrosterone pathway; the pathway on the right side is called the Δ⁴ or progesterone pathway. The asterisk indicates that the 17α-hydroxylase and 17,20-lyase activities reside in a single protein, P450c17.

FIGURE 41–6 Dihydrotestosterone is formed from testosterone through action of the enzyme 5α-reductase.

433

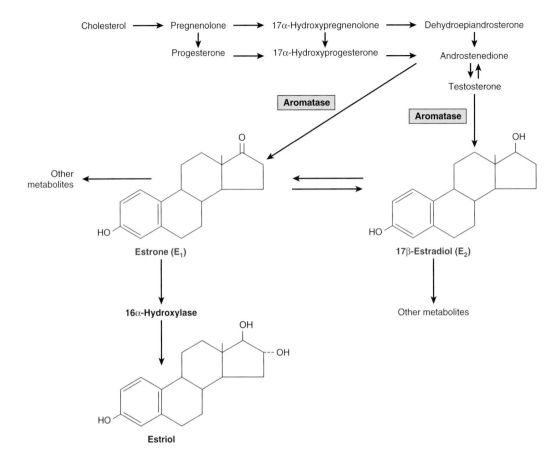

FIGURE 41–7
Biosynthesis of estrogens. (Slightly modified and reproduced, with permission, from Ganong WF: *Review of Medical Physiology*, 21st ed. McGraw-Hill, 2005.)

FIGURE 41–8 Biosynthesis of progesterone in the corpus luteum.

Skin

Small amounts of the precursor for $1,25(OH)_2$-D_3 synthesis are present in food (fish liver oil, egg yolk), but most of the precursor for $1,25(OH)_2$-D_3 synthesis is produced in the malpighian layer of the epidermis from 7 dehydrocholesterol in an ultraviolet light-mediated, nonenzymatic **photolysis** reaction. The extent of this conversion is related directly to the intensity of the exposure and inversely to the extent of pigmentation in the skin. There is an age-related loss of 7-dehydrocholesterol in the epidermis that may be related to the negative calcium balance associated with old age.

Liver

A specific transport protein called the **vitamin D–binding protein** binds vitamin D_3 and its metabolites and moves vitamin D_3 from the skin or intestine to the liver, where it undergoes 25-hydroxylation, the first obligatory reaction in the production of $1,25(OH)_2$-D_3. 25-Hydroxylation occurs in the endoplasmic reticulum in a reaction that requires magnesium, NADPH, molecular oxygen, and an uncharacterized cytoplasmic factor. Two enzymes are involved: an NADPH-dependent cytochrome P450 reductase and a cytochrome P450. This reaction is not regulated, and it also occurs with low efficiency in kidney and intestine. The $25(OH)_2$-D_3 enters the circulation, where it is the major form of vitamin D found in plasma, and is transported to the kidney by the vitamin D–binding protein.

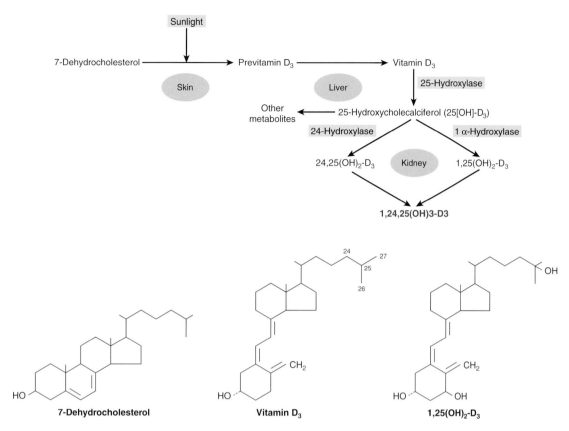

FIGURE 41–9 Formation and hydroxylation of vitamin D_3. 25-Hydroxylation takes place in the liver, and the other hydroxylations occur in the kidneys. 25,26$(OH)_2$-D_3 and 1,25,26$(OH)_3$-D_3 are probably formed as well. The formulas of 7-dehydrocholesterol, vitamin D_3, and 1,25$(OH)_2$-D_3 are also shown. (Modified and reproduced, with permission, from Ganong WF: *Review of Medical Physiology*, 21st ed. McGraw-Hill, 2005.)

Kidney

25$(OH)_2$-D_3 is a weak agonist and must be modified by hydroxylation at position C_1 for full biologic activity. This is accomplished in mitochondria of the renal proximal convoluted tubule by a three-component monooxygenase reaction that requires NADPH, Mg^{2+}, molecular oxygen, and at least three enzymes: (1) a flavoprotein, renal ferredoxin reductase; (2) an iron sulfur protein, renal ferredoxin; and (3) cytochrome P450. This system produces 1,25$(OH)_2$-D_3, which is the most potent naturally occurring metabolite of vitamin D.

CATECHOLAMINES & THYROID HORMONES ARE MADE FROM TYROSINE

Catecholamines Are Synthesized in Final Form & Stored in Secretion Granules

Three amines—dopamine, norepinephrine, and epinephrine—are synthesized from tyrosine in the chromaffin cells of the adrenal medulla. The major product of the adrenal medulla is epinephrine. This compound constitutes about 80% of the catecholamines in the medulla, and it is not made in extramedullary tissue. In contrast, most of the norepinephrine present in organs innervated by sympathetic nerves is made in situ (about 80% of the total), and most of the rest is made in other nerve endings and reaches the target sites via the circulation. Epinephrine and norepinephrine may be produced and stored in different cells in the adrenal medulla and other chromaffin tissues.

The conversion of tyrosine to epinephrine requires four sequential steps: (1) ring hydroxylation; (2) decarboxylation; (3) side-chain hydroxylation to form norepinephrine; and (4) N-methylation to form epinephrine. The biosynthetic pathway and the enzymes involved are illustrated in Figure 41–10.

Tyrosine Hydroxylase Is Rate-Limiting for Catecholamine Biosynthesis

Tyrosine is the immediate precursor of catecholamines, and **tyrosine hydroxylase** is the rate-limiting enzyme in catecholamine biosynthesis. Tyrosine hydroxylase is found in both soluble and particle-bound forms only in tissues that synthesize catecholamines; it functions as an oxidoreductase, with tetrahydropteridine as a cofactor, to convert L-tyrosine to L-dihydroxyphenylalanine (L-**dopa**). As the rate-limiting enzyme, tyrosine hydroxylase is regulated in a variety of ways. The most important mechanism involves feedback inhibi-

FIGURE 41–10 Biosynthesis of catecholamines. (PNMT, phenylethanolamine-*N*-methyltransferase.)

tion by the catecholamines, which compete with the enzyme for the pteridine cofactor. Catecholamines cannot cross the blood–brain barrier; hence, in the brain they must be synthesized locally. In certain central nervous system diseases (eg, Parkinson's disease), there is a local deficiency of dopamine synthesis. L-Dopa, the precursor of dopamine, readily crosses the blood–brain barrier and so is an important agent in the treatment of Parkinson disease.

Dopa Decarboxylase Is Present in All Tissues

This soluble enzyme requires pyridoxal phosphate for the conversion of L-dopa to 3,4-dihydroxyphenylethylamine (**dopamine**). Compounds that resemble L-dopa, such as α-methyldopa, are competitive inhibitors of this reaction. α-Methyldopa is effective in treating some kinds of hypertension.

Dopamine β-Hydroxylase (DBH) Catalyzes the Conversion of Dopamine to Norepinephrine

DBH is a monooxygenase and uses ascorbate as an electron donor, copper at the active site, and fumarate as modulator. DBH is in the particulate fraction of the medullary cells, probably in the secretion granule; thus, the conversion of dopamine to **norepinephrine** occurs in this organelle.

Phenylethanolamine-*N*-Methyltransferase (PNMT) Catalyzes the Production of Epinephrine

PNMT catalyzes the *N*-methylation of norepinephrine to form **epinephrine** in the epinephrine-forming cells of the adrenal medulla. Since PNMT is soluble, it is assumed that norepinephrine-to-epinephrine conversion occurs in the cytoplasm. The synthesis of PNMT is induced by glucocorticoid hormones that reach the medulla via the intra-adrenal portal system. This special system provides for a 100-fold steroid concentration gradient over systemic arterial blood, and this high intra-adrenal concentration appears to be necessary for the induction of PNMT.

T_3 & T_4 Illustrate the Diversity in Hormone Synthesis

The formation of **triiodothyronine (T_3)** and **tetraiodothyronine (thyroxine; T_4)** (see Figure 41–2) illustrates many of the principles of diversity discussed in this chapter. These hormones require a rare element (iodine) for bioactivity; they are synthesized as part of a very large precursor molecule (thyroglobulin); they are stored in an intracellular reservoir (colloid); and there is peripheral conversion of T_4 to T_3, which is a much more active hormone.

The thyroid hormones T_3 and T_4 are unique in that iodine (as iodide) is an essential component of both. In most parts of the world, iodine is a scarce component of soil, and for that reason there is little in food. A complex mechanism has evolved to acquire and retain this crucial element and to convert it into a form suitable for incorporation into organic compounds. At the same time, the thyroid must synthesize thyronine from tyrosine, and this synthesis takes place in thyroglobulin (Figure 41–11).

Thyroglobulin is the precursor of T_4 and T_3. It is a large iodinated, glycosylated protein with a molecular mass of 660 kDa. Carbohydrate accounts for 8–10% of the weight of thyroglobulin and iodide for about 0.2–1%, depending upon the iodine content in the diet. Thyroglobulin is composed of two large subunits. It contains 115 tyrosine residues, each of which is a potential site of iodination. About 70% of the iodide in thyroglobulin exists in the inactive precursors, **monoiodotyrosine (MIT)** and **diiodotyrosine (DIT),** while 30% is in the **iodothyronyl residues,** T_4 and T_3. When iodine supplies are sufficient, the T_4:T_3 ratio is about 7:1. In **iodine deficiency,** this ratio decreases, as does the DIT:MIT ratio. Thyroglobulin, a large molecule of about 5000 amino acids, provides the conformation required for tyrosyl coupling and iodide organification necessary in the

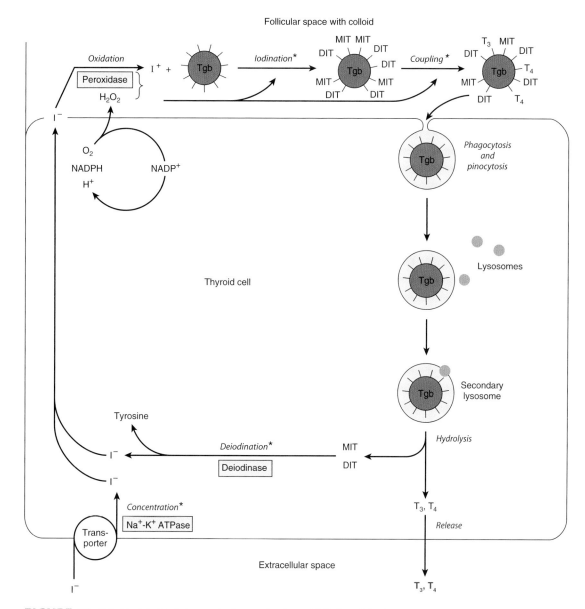

FIGURE 41–11 Model of iodide metabolism in the thyroid follicle. A follicular cell is shown facing the follicular lumen (**top**) and the extracellular space (**bottom**). Iodide enters the thyroid primarily through a transporter (**bottom left**). Thyroid hormone synthesis occurs in the follicular space through a series of reactions, many of which are peroxidase-mediated. Thyroid hormones, stored in the colloid in the follicular space, are released from thyroglobulin by hydrolysis inside the thyroid cell. (Tgb, thyroglobulin; MIT, monoiodotyrosine; DIT, diiodotyrosine; T_3, triiodothyronine; T_4, tetraiodothyronine.) Asterisks indicate steps or processes where inherited enzyme deficiencies cause congenital goiter and often result in hypothyroidism.

formation of the diaminoacid thyroid hormones. It is synthesized in the basal portion of the cell and moves to the lumen, where it is a storage form of T_3 and T_4 in the colloid; several weeks' supply of these hormones exist in the normal thyroid. Within minutes after stimulation of the thyroid by TSH, colloid reenters the cell and there is a marked increase of phagolysosome activity. Various acid proteases and peptidases hydrolyze the thyroglobulin into its constituent amino acids, including T_4 and T_3, which are discharged from the basal portion of the cell (see Figure 41–11). Thyroglobulin is thus a very large prohormone.

Iodide Metabolism Involves Several Discrete Steps

The thyroid is able to concentrate I^- against a strong electrochemical gradient. This is an energy-dependent process and is linked to the Na^+-K^+ ATPase-dependent thyroidal I^- transporter. The ratio of iodide in thyroid to iodide in serum (T:S ratio) is a reflection of the activity of this transporter. This activity is primarily controlled by TSH and ranges from 500:1 in animals chronically stimulated with TSH to 5:1 or less in hypophysectomized animals (no TSH). The T:S ratio in humans on a normal iodine diet is about 25:1.

The thyroid is the only tissue that can oxidize I⁻ to a higher valence state, an obligatory step in I⁻ organification and thyroid hormone biosynthesis. This step involves a heme-containing peroxidase and occurs at the luminal surface of the follicular cell. Thyroperoxidase, a tetrameric protein with a molecular mass of 60 kDa, requires hydrogen peroxide as an oxidizing agent. The H_2O_2 is produced by an NADPH-dependent enzyme resembling cytochrome *c* reductase. A number of compounds inhibit I⁻ oxidation and therefore its subsequent incorporation into MIT and DIT. The most important of these are the thiourea drugs. They are used as antithyroid drugs because of their ability to inhibit thyroid hormone biosynthesis at this step. Once iodination occurs, the iodine does not readily leave the thyroid. Free tyrosine can be iodinated, but it is not incorporated into proteins since no tRNA recognizes iodinated tyrosine.

The coupling of two DIT molecules to form T_4—or of an MIT and DIT to form T_3—occurs within the thyroglobulin molecule. A separate coupling enzyme has not been found, and since this is an oxidative process it is assumed that the same thyroperoxidase catalyzes this reaction by stimulating free radical formation of iodotyrosine. This hypothesis is supported by the observation that the same drugs which inhibit I⁻

oxidation also inhibit coupling. The formed thyroid hormones remain as integral parts of thyroglobulin until the latter is degraded, as described above.

A deiodinase removes I⁻ from the inactive mono and diiodothyronine molecules in the thyroid. This mechanism provides a substantial amount of the I⁻ used in T_3 and T_4 biosynthesis. A peripheral deiodinase in target tissues such as pituitary, kidney, and liver selectively removes I⁻ from the 5′ position of T_4 to make T_3 (see Figure 41–2), which is a much more active molecule. In this sense, T_4 can be thought of as a prohormone, though it does have some intrinsic activity.

Several Hormones Are Made from Larger Peptide Precursors

Formation of the critical disulfide bridges in insulin requires that this hormone be first synthesized as part of a larger precursor molecule, proinsulin. This is conceptually similar to the example of the thyroid hormones, which can only be formed in the context of a much larger molecule. Several other hormones are synthesized as parts of large precursor molecules, not because of some special structural requirement but rather as a mechanism for controlling the available amount of the active hormone. PTH and angiotensin II are examples of this type of regulation. Another interesting example is the POMC

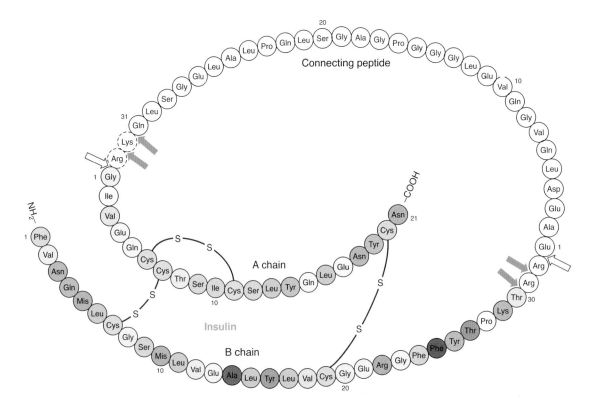

FIGURE 41–12 Structure of human proinsulin. Insulin and C-peptide molecules are connected at two sites by dipeptide links. An initial cleavage by a trypsin-like enzyme (open arrows) followed by several cleavages by a carboxypeptidase-like enzyme (**solid arrows**) results in the production of the heterodimeric (AB) insulin molecule (**colored**) and the C-peptide (**white**).

protein, which can be processed into many different hormones in a tissue-specific manner. These examples are discussed in detail below.

Insulin Is Synthesized as a Preprohormone & Modified Within the β Cell

Insulin has an AB heterodimeric structure with one intrachain (A6–A11) and two interchain disulfide bridges (A7–B7 and A20–B19) (Figure 41–12). The A and B chains could be synthesized in the laboratory, but attempts at a biochemical synthesis of the mature insulin molecule yielded very poor results. The reason for this became apparent when it was discovered that insulin is synthesized as a **preprohormone** (molecular weight approximately 11,500), which is the prototype for peptides that are processed from larger precursor molecules. The hydrophobic 23-amino-acid pre-, or leader, sequence directs the molecule into the cisternae of the endoplasmic reticulum and then is removed. This results in the 9000-MW proinsulin molecule, which provides the conformation necessary for the proper and efficient formation of the disulfide bridges. As shown in Figure 41–12, the sequence of proinsulin, starting from the amino terminal, is B chain—connecting (C) peptide—A chain. The proinsulin molecule undergoes a series of site-specific peptide cleavages that result in the formation of equimolar amounts of mature insulin and C-peptide. These enzymatic cleavages are summarized in Figure 41–12.

Parathyroid Hormone (PTH) Is Secreted as an 84-Amino-Acid Peptide

The immediate precursor of PTH is **proPTH,** which differs from the native 84-amino-acid hormone by having a highly basic hexapeptide amino terminal extension. The primary gene product and the immediate precursor for proPTH is the 115-amino-acid **preproPTH.** This differs from proPTH by having an additional 25-amino-acid amino terminal extension that, in common with the other leader or signal sequences characteristic of secreted proteins, is hydrophobic. The complete structure of preproPTH and the sequences of proPTH and PTH are illustrated in Figure 41–13. PTH_{1-34} has full biologic activity, and the region 25–34 is primarily responsible for receptor binding.

The biosynthesis of PTH and its subsequent secretion are regulated by the plasma ionized calcium (Ca^{2+}) concentration through a complex process. An acute decrease of Ca^{2+} results in a marked increase of PTH mRNA, and this is followed by an increased rate of PTH synthesis and secretion. However, about 80–90% of the proPTH synthesized cannot be accounted for as intact PTH in cells or in the incubation medium of experimental systems. This finding led to the conclusion that most of the proPTH synthesized is quickly degraded. It was later discovered that this rate of degradation decreases when Ca^{2+} concentrations are low, and it increases when Ca^{2+} concentrations are high. A Ca^{2+} receptor on the surface of the parathyroid cell mediates these effects. Very specific fragments of PTH are

FIGURE 41–13
Structure of bovine preproparathyroid hormone. Arrows indicate sites cleaved by processing enzymes in the parathyroid gland (1–5) and in the liver after secretion of the hormone (4–5). The biologically active region of the molecule (**colored**) is flanked by sequence not required for activity on target receptors. (Slightly modified and reproduced, with permission, from Habener JF: Recent advances in parathyroid hormone research. Clin Biochem 1981;14:223. Copyright © 1981. Reprinted with permission from Elsevier.)

generated during its proteolytic digestion (Figure 41–13). A number of proteolytic enzymes, including cathepsins B and D, have been identified in parathyroid tissue. Cathepsin B cleaves PTH into two fragments: PTH$_{1-36}$ and PTH$_{37-84}$. PTH$_{37-84}$ is not further degraded; however, PTH$_{1-36}$ is rapidly and progressively cleaved into di- and tripeptides. Most of the proteolysis of PTH occurs within the gland, but a number of studies confirm that PTH, once secreted, is proteolytically degraded in other tissues, especially the liver, by similar mechanisms.

Angiotensin II Is Also Synthesized from a Large Precursor

The renin-angiotensin system is involved in the regulation of blood pressure and electrolyte metabolism (through production of aldosterone). The primary hormone involved in these processes is angiotensin II, an octapeptide made from angiotensinogen (Figure 41–14). Angiotensinogen, a large α_2-globulin made in liver, is the substrate for renin, an enzyme produced in the juxtaglomerular cells of the renal afferent arteriole. The position of these cells makes them particularly sensitive to blood pressure changes, and many of the physiologic regulators of renin release act through renal baroreceptors. The juxtaglomerular cells are also sensitive to changes of Na^+ and Cl^- concentration in the renal tubular fluid; therefore, any combination of factors that decreases fluid volume (dehydration, decreased blood pressure, fluid or blood loss) or decreases NaCl concentration stimulates renin release. Renal sympathetic nerves that terminate in the juxtaglomerular cells mediate the central nervous system and postural effects on renin release independently of the baroreceptor and salt effects, a mechanism that involves the β-adrenergic receptor. Renin acts upon the substrate angiotensinogen to produce the decapeptide angiotensin I.

Angiotensin-converting enzyme, a glycoprotein found in lung, endothelial cells, and plasma, removes two carboxyl terminal amino acids from the decapeptide angiotensin I to form angiotensin II in a step that is not thought to be rate-limiting. Various nonapeptide analogs of angiotensin I and other compounds act as competitive inhibitors of converting enzyme and are used to treat renin-dependent hypertension. These are referred to as **angiotensin-converting enzyme (ACE) inhibitors.** Angiotensin II increases blood pressure by causing vasoconstriction of the arteriole and is a very potent vasoactive substance. It inhibits renin release from the juxtaglomerular cells and is a potent stimulator of aldosterone production. This results in Na^+ retention, volume expansion, and increased blood pressure.

In some species, angiotensin II is converted to the heptapeptide angiotensin III (Figure 41–14), an equally potent stimulator of aldosterone production. In humans, the plasma level of angiotensin II is four times greater than that of angiotensin III, so most effects are exerted by the octapeptide. Angiotensins II and III are rapidly inactivated by angiotensinases.

Angiotensin II binds to specific adrenal cortex glomerulosa cell receptors. The hormone-receptor interaction does not activate adenylyl cyclase, and cAMP does not appear to mediate the action of this hormone. The actions of angiotensin II, which are to stimulate the conversion of cholesterol to pregnenolone and of corticosterone to 18-hydroxycorticosterone and aldosterone, may involve changes in the concentration of intracellular calcium and of phospholipid metabolites by mechanisms similar to those described in Chapter 42.

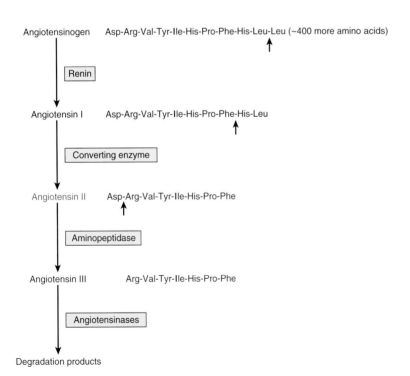

FIGURE 41–14 Formation and metabolism of angiotensins. Small arrows indicate cleavage sites.

Complex Processing Generates the Pro-Opiomelanocortin (POMC) Peptide Family

The POMC family consists of peptides that act as hormones (ACTH, LPH, MSH) and others that may serve as neurotransmitters or neuromodulators (endorphins) (Figure 41–15). POMC is synthesized as a precursor molecule of 285 amino acids and is processed differently in various regions of the pituitary.

The POMC gene is expressed in the anterior and intermediate lobes of the pituitary. The most conserved sequences between species are within the amino terminal fragment, the ACTH region, and the β-endorphin region. POMC or related products are found in several other vertebrate tissues, including the brain, placenta, gastrointestinal tract, reproductive tract, lung, and lymphocytes.

The POMC protein is processed differently in the anterior lobe than in the intermediate lobe. The intermediate lobe of the pituitary is rudimentary in adult humans, but it is active in human fetuses and in pregnant women during late gestation and is also active in many animal species. Processing of the POMC protein in the peripheral tissues (gut, placenta, male reproductive tract) resembles that in the intermediate lobe. There are three basic peptide groups: (1) ACTH, which can give rise to α-MSH and corticotropin-like intermediate lobe peptide (CLIP); (2) β-lipotropin (β-LPH), which can yield γ-LPH, β-MSH, and β-endorphin (and thus α- and γ-endorphins); and (3) a large amino terminal peptide, which generates γ-MSH (not shown). The diversity of these products is due to the many dibasic amino acid clusters that are potential cleavage sites for trypsin-like enzymes. Each of the peptides mentioned is preceded by Lys-Arg, Arg-Lys, Arg-Arg, or Lys-Lys residues. After the prehormone segment is cleaved, the next cleavage, in both anterior and intermediate lobes, is between ACTH and β-LPH, resulting in an amino terminal peptide with ACTH and a β-LPH segment (Figure 41–15). ACTH$_{1–39}$ is subsequently cleaved from the amino terminal peptide, and in the anterior lobe essentially no further cleavages occur. In the intermediate lobe, ACTH$_{1–39}$ is cleaved into α-MSH (residues 1–13) and CLIP (18–39); β-LPH (42–134) is converted to γ-LPH (42–101) and β-endorphin (104–134). β-MSH (84–101) is derived from γ-LPH, while γ-MSH (50–74) is derived from a POMC N-terminal fragment (1–74).

There are extensive additional tissue-specific modifications of these peptides that affect activity. These modifications include phosphorylation, acetylation, glycosylation, and amidation.

Mutations of the α-MSH receptor are linked to a common, early-onset form of obesity. This observation has redirected attention to the POMC peptide hormones.

THERE IS VARIATION IN THE STORAGE & SECRETION OF HORMONES

As mentioned above, the steroid hormones and 1,25(OH)$_2$-D$_3$ are synthesized in their final active form. They are also secreted as they are made, and thus there is no intracellular reservoir of these hormones. The catecholamines, also synthesized in active form, are stored in granules in the chromaffin cells in the adrenal medulla. In response to appropriate neural stimulation, these granules are released from the cell through exocytosis, and the catecholamines are released into the circulation. A several-hour reserve supply of catecholamines exists in the chromaffin cells.

Parathyroid hormone also exists in storage vesicles. As much as 80–90% of the pro PTH synthesized is degraded before it enters this final storage compartment, especially when Ca^{2+} levels are high in the parathyroid cell (see above). PTH is secreted when Ca^{2+} is low in the parathyroid cells, which contain a several-hour supply of the hormone.

The human pancreas secretes about 40–50 units of insulin daily; this represents about 15–20% of the hormone stored in the β cells. Insulin and the C-peptide (see Figure 41–12) are normally secreted in equimolar amounts. Stimuli such as glucose, which provokes insulin secretion, therefore trigger the processing of proinsulin to insulin as an essential part of the secretory response.

A several-week supply of T$_3$ and T$_4$ exists in the thyroglobulin that is stored in colloid in the lumen of the thyroid follicles. These hormones can be released upon stimulation by

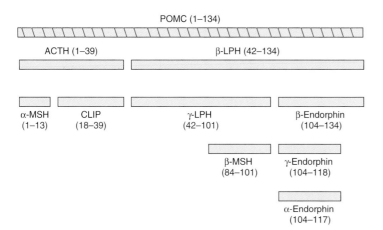

FIGURE 41–15 Products of pro-opiomelanocortin (POMC) cleavage. (MSH, melanocyte-stimulating hormone; CLIP, corticotropin-like intermediate lobe peptide; LPH, lipotropin.)

TABLE 41–5 Diversity in the Storage of Hormones

Hormone	Supply Stored in Cell
Steroids and 1,25(OH)$_2$-D$_3$	None
Catecholamines and PTH	Hours
Insulin	Days
T$_3$ and T$_4$	Weeks

TABLE 41–7 Comparison of T$_4$ and T$_3$ in Plasma

| Total Hormone (µg/dL) | Free Hormone | | | T$_{1/2}$ in Blood (days) |
	Percentage of Total	ng/dL	Molarity	
T$_4$ 8	0.03	~2.24	3.0×10^{-11}	6.5
T$_3$ 0.15	0.3	~0.4	-0.6×10^{-11}	1.5

TSH. This is the most exaggerated example of a prohormone, as a molecule containing approximately 5000 amino acids must be first synthesized, then degraded, to supply a few molecules of the active hormones T$_4$ and T$_3$.

The diversity in storage and secretion of hormones is illustrated in Table 41–5.

SOME HORMONES HAVE PLASMA TRANSPORT PROTEINS

The class I hormones are hydrophobic in chemical nature and thus are not very soluble in plasma. These hormones, principally the steroids and thyroid hormones, have specialized plasma transport proteins that serve several purposes. First, these proteins circumvent the solubility problem and thereby deliver the hormone to the target cell. They also provide a circulating reservoir of the hormone that can be substantial, as in the case of the thyroid hormones. Hormones, when bound to the transport proteins, cannot be metabolized, thereby prolonging their plasma half-life ($t_{1/2}$). The binding affinity of a given hormone to its transporter determines the bound versus free ratio of the hormone. This is important because only the free form of a hormone is biologically active. In general, the concentration of free hormone in plasma is very low, in the range of 10^{-15} to 10^{-9} mol/L. It is important to distinguish between plasma transport proteins and hormone receptors. Both bind hormones but with very different characteristics (Table 41–6).

The hydrophilic hormones—generally class II and of peptide structure—are freely soluble in plasma and do not re-

quire transport proteins. Hormones such as insulin, growth hormone, ACTH, and TSH circulate in the free, active form and have very short plasma half-lives. A notable exception is IGF-I, which is transported bound to members of a family of binding proteins.

Thyroid Hormones Are Transported by Thyroid-Binding Globulin

Many of the principles discussed above are illustrated in a discussion of thyroid-binding proteins. One-half to two-thirds of T$_4$ and T$_3$ in the body is in an extrathyroidal reservoir. Most of this circulates in bound form, ie, bound to a specific binding protein, **thyroxine-binding globulin (TBG).** TBG, a glycoprotein with a molecular mass of 50 kDa, binds T$_4$ and T$_3$ and has the capacity to bind 20 µg/dL of plasma. Under normal circumstances, TBG binds—noncovalently—nearly all of the T$_4$ and T$_3$ in plasma, and it binds T$_4$ with greater affinity than T$_3$ (Table 41–7). The plasma half-life of T$_4$ is correspondingly four to five times that of T$_3$. The small, unbound (free) fraction is responsible for the biologic activity. Thus, in spite of the great difference in total amount, the free fraction of T$_3$ approximates that of T$_4$, and given that T$_3$ is intrinsically more active than T$_4$, most biologic activity is attributed to T$_3$. TBG does not bind any other hormones.

Glucocorticoids Are Transported by Corticosteroid-Binding Globulin

Hydrocortisone (cortisol) also circulates in plasma in protein-bound and free forms. The main plasma binding protein is

TABLE 41–6 Comparison of Receptors with Transport Proteins

Feature	Receptors	Transport Proteins
Concentration	Very low (thousands/cell)	Very high (billions/µL)
Binding affinity	High (pmol/L to nmol/L range)	Low (µmol/L range)
Binding specificity	Very high	Low
Saturability	Yes	No
Reversibility	Yes	Yes
Signal transduction	Yes	No

TABLE 41–8 Approximate Affinities of Steroids for Serum-Binding Proteins

	SHBG[1]	CBG[1]
Dihydrotestosterone	1	>100
Testosterone	2	>100
Estradiol	5	>10
Estrone	>10	>100
Progesterone	>100	~2
Cortisol	>100	~3
Corticosterone	>100	~5

[1]Affinity expressed as K_d (nmol/L).

an α-globulin called **transcortin, or corticosteroid-binding globulin (CBG).** CBG is produced in the liver, and its synthesis, like that of TBG, is increased by estrogens. CBG binds most of the hormone when plasma cortisol levels are within the normal range; much smaller amounts of cortisol are bound to albumin. The avidity of binding helps determine the biologic half-lives of various glucocorticoids. Cortisol binds tightly to CBG and has a $t_{1/2}$ of 1.5–2 h, while corticosterone, which binds less tightly, has a $t_{1/2}$ of less than 1 h (Table 41–8). The unbound (free) cortisol constitutes about 8% of the total and represents the biologically active fraction. Binding to CBG is not restricted to glucocorticoids. Deoxycorticosterone and progesterone interact with CBG with sufficient affinity to compete for cortisol binding. Aldosterone, the most potent natural mineralocorticoid, does not have a specific plasma transport protein. Gonadal steroids bind very weakly to CBG (Table 41–8).

Gonadal Steroids Are Transported by Sex-Hormone-Binding Globulin

Most mammals, humans included, have a plasma β-globulin that binds testosterone with specificity, relatively high affinity, and limited capacity (Table 41–8). This protein, usually called **sex-hormone-binding globulin (SHBG)** or testosterone-estrogen-binding globulin (TEBG), is produced in the liver. Its production is increased by estrogens (women have twice the serum concentration of SHBG as men), certain types of liver disease, and hyperthyroidism; it is decreased by androgens, advancing age, and hypothyroidism. Many of these conditions also affect the production of CBG and TBG. Since SHBG and albumin bind 97–99% of circulating testosterone, only a small fraction of the hormone in circulation is in the free (biologically active) form. The primary function of SHBG may be to restrict the free concentration of testosterone in the serum. Testosterone binds to SHBG with higher affinity than does estradiol (Table 41–8). Therefore, a change in the level of SHBG causes a greater change in the free testosterone level than in the free estradiol level.

Estrogens are bound to SHBG and progestins to CBG. SHBG binds estradiol about five times less avidly than it binds testosterone or DHT, while progesterone and cortisol have little affinity for this protein (Table 41–8). In contrast, progesterone and cortisol bind with nearly equal affinity to CBG, which in turn has little avidity for estradiol and even less for testosterone, DHT, or estrone.

These binding proteins also provide a circulating reservoir of hormone, and because of the relatively large binding capacity they probably buffer against sudden changes in the plasma level. Because the metabolic clearance rates of these steroids are inversely related to the affinity of their binding to SHBG, estrone is cleared more rapidly than estradiol, which in turn is cleared more rapidly than testosterone or DHT.

SUMMARY

- The presence of a specific receptor defines the target cells for a given hormone.

- Receptors are proteins that bind specific hormones and generate an intracellular signal (receptor-effector coupling).

- Some hormones have intracellular receptors; others bind to receptors on the plasma membrane.

- Hormones are synthesized from a number of precursor molecules, including cholesterol, tyrosine per se, and all the constituent amino acids of peptides and proteins.

- A number of modification processes alter the activity of hormones. For example, many hormones are synthesized from larger precursor molecules.

- The complement of enzymes in a particular cell type allows for the production of a specific class of steroid hormone.

- Most of the lipid-soluble hormones are bound to rather specific plasma transport proteins.

REFERENCES

Bain DL et al: Nuclear receptor structure: implications for function. Ann Rev Physiol 2007;69:201.

Bartalina L: Thyroid hormone-binding proteins: update 1994. Endocr Rev 1994;13:140.

Beato M et al: Steroid hormone receptors: many actors in search of a plot. Cell 1995;83:851.

Dai G, Carrasco L, Carrasco N: Cloning and characterization of the thyroid iodide transporter. Nature 1996;379:458.

DeLuca HR: The vitamin D story: a collaborative effort of basic science and clinical medicine. FASEB J 1988;2:224.

Douglass J, Civelli O, Herbert E: Polyprotein gene expression: Generation of diversity of neuroendocrine peptides. Annu Rev Biochem 1984;53:665.

Farooqi IS, O'Rahilly S: Monogenic obesity in humans. Ann Rev Med 2005;56:443.

Miller WL: Molecular biology of steroid hormone biosynthesis. Endocr Rev 1988;9:295.

Nagatsu T: Genes for human catecholamine-synthesizing enzymes. Neurosci Res 1991;12:315.

Russell DW, Wilson JD: Steroid 5 alpha-reductase: two genes/two enzymes. Annu Rev Biochem 1994;63:25.

Russell J et al: Interaction between calcium and 1,25-dihydroxy-vitamin D_3 in the regulation of preproparathyroid hormone and vitamin D receptor mRNA in avian parathyroids. Endocrinology 1993;132:2639.

Steiner DF et al: The new enzymology of precursor processing endoproteases. J Biol Chem 1992;267:23435.

Taguchi A, White M: Insulin-like signaling, nutrient homeostasis, and life span. Ann Rev Physiol 2008;70:191.

Hormone Action & Signal Transduction

P. Anthony Weil, PhD

42

BIOMEDICAL IMPORTANCE

The homeostatic adaptations an organism makes to a constantly changing environment are in large part accomplished through alterations of the activity and amount of proteins. Hormones provide a major means of facilitating these changes. A hormone–receptor interaction results in generation of an intracellular signal that can either regulate the activity of a select set of genes, thereby altering the amount of certain proteins in the target cell, or affect the activity of specific proteins, including enzymes and transporter or channel proteins. The signal can influence the location of proteins in the cell and can affect general processes such as protein synthesis, cell growth, and replication, perhaps through effects on gene expression. Other signaling molecules—including cytokines, interleukins, growth factors, and metabolites—use some of the same general mechanisms and signal transduction pathways. Excessive, deficient, or inappropriate production and release of hormones and of these other regulatory molecules are major causes of disease. Many pharmacotherapeutic agents are aimed at correcting or otherwise influencing the pathways discussed in this chapter.

HORMONES TRANSDUCE SIGNALS TO AFFECT HOMEOSTATIC MECHANISMS

The general steps involved in producing a coordinated response to a particular stimulus are illustrated in Figure 42–1. The stimulus can be a challenge or a threat to the organism, to an organ, or to the integrity of a single cell within that organism. Recognition of the stimulus is the first step in the adaptive response. At the organismic level, this generally involves the nervous system and the special senses (sight, hearing, pain, smell, touch). At the organismic or cellular level, recognition involves physicochemical factors such as pH, O_2 tension, temperature, nutrient supply, noxious metabolites, and osmolarity. Appropriate recognition results in the release of one or more hormones that will govern generation of the necessary adaptive response. For purposes of this discussion, the hormones are categorized as described in Chapter 41, ie, based on the location of their specific cellular receptors and the type of signals generated. Group I hormones interact with an intracellular receptor and group II hormones with

receptor recognition sites located on the extracellular surface of the plasma membrane of target cells. The cytokines, interleukins, and growth factors should also be considered in this latter category. These molecules, of critical importance in homeostatic adaptation, are hormones in the sense that they are produced in specific cells, have the equivalent of autocrine, paracrine, and endocrine actions, bind to cell surface receptors, and activate many of the same signal transduction pathways employed by the more traditional group II hormones.

SIGNAL GENERATION

The Ligand–Receptor Complex Is the Signal for Group I Hormones

The lipophilic group I hormones diffuse through the plasma membrane of all cells but only encounter their specific, high-affinity intracellular receptors in target cells. These receptors can be located in the cytoplasm or in the nucleus of target cells. The hormone–receptor complex first undergoes an **activation reaction.** As shown in Figure 42–2, receptor activation occurs by at least two mechanisms. For example, glucocorticoids diffuse across the plasma membrane and encounter their cognate receptor in the cytoplasm of target cells. Ligand–receptor binding results in a conformational change in the receptor leading to the dissociation of heat shock protein 90 (hsp90). This step appears to be necessary for subsequent nuclear localization of the glucocorticoid receptor. This receptor also contains a nuclear localization sequence that is now free to assist in the translocation from cytoplasm to nucleus. The activated receptor moves into the nucleus (Figure 42–2) and binds with high affinity to a specific DNA sequence called the **hormone response element (HRE).** In the case illustrated, this is a glucocorticoid response element, or GRE. Consensus sequences for HREs are shown in Table 42–1. The DNA-bound, liganded receptor serves as a high-affinity binding site for one or more coactivator proteins, and accelerated gene transcription typically ensues when this occurs. By contrast, certain hormones such as the thyroid hormones and retinoids diffuse from the extracellular fluid across the plasma membrane and go directly into the nucleus. In this case, the cognate receptor is already bound to the HRE (the thyroid hormone response element [TRE], in this

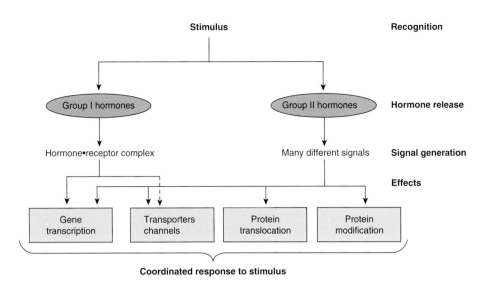

FIGURE 42–1 Hormonal involvement in responses to a stimulus. A challenge to the integrity of the organism elicits a response that includes the release of one or more hormones. These hormones generate signals at or within target cells, and these signals regulate a variety of biologic processes that provide for a coordinated response to the stimulus or challenge. See Figure 42–8 for a specific example.

example). However, this DNA-bound receptor fails to activate transcription because it exists in complex with a corepressor. Indeed, this receptor-corepressor complex serves as an active repressor of gene transcription. The association of ligand with these receptors results in dissociation of the corepressor(s). The

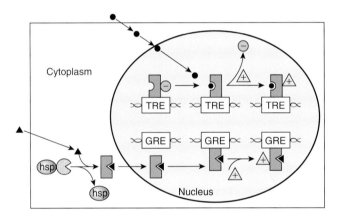

FIGURE 42–2 Regulation of gene expression by two different class I hormones, thyroid hormone and glucocorticoids. Steroid hormones readily gain access to the cytoplasmic compartment of target cells. Glucocorticoid hormones (solid triangles) encounter their cognate receptor (GR) in the cytoplasm, where GR exists in a complex with heat shock protein 90 (hsp). Ligand binding causes dissociation of hsp and a conformational change of the receptor. The receptor-ligand complex then traverses the nuclear membrane and binds to DNA with specificity and high affinity at a glucocorticoid response element (GRE). This event affects the architecture of a number of transcription coregulators (green triangles), and enhanced transcription ensues. By contrast, thyroid hormones and retinoic acid (•) directly enter the nucleus, where their cognate heterodimeric (TR-RXR; see Figure 42–12) receptors are already bound to the appropriate response elements with an associated transcription repressor complex (red circles). Hormone–receptor binds, which again induces conformational changes leading to a reorganization of receptor (TR)-coregulator interactions (ie, molecules such as N-CoR or SMRT [see Table 42–6]). Ligand binding results in dissociation of the repressor complex from the receptor, allowing an activator complex to assemble. The gene is then actively transcribed.

liganded receptor is now capable of binding one or more co-activators with high affinity, resulting in the recruitment of pol II + GTFs and activation of gene transcription. The relationship of hormone receptors to other nuclear receptors and to coregulators is discussed in more detail below.

By selectively affecting gene transcription and the consequent production of appropriate target mRNAs, the amounts of specific proteins are changed and metabolic processes are influenced. The influence of each of these hormones is quite specific; generally, a given hormone affects less than 1% of the genes, mRNA, or proteins in a target cell; sometimes only a few are affected. The nuclear actions of steroid, thyroid, and retinoid hormones are quite well defined. Most evidence suggests

TABLE 42–1 The DNA Sequences of Several Hormone Response Elements (HREs)[1]

Hormone or Effector	HRE	DNA Sequence
Glucocorticoids	GRE	
Progestins	PRE	GGTACA NNN TGTTCT
Mineralocorticoids	MRE	← —→
Androgens	ARE	
Estrogens	ERE	AGGTCA --- TGA/TCCT ← —→
Thyroid hormone	TRE	AGGTCA N3, 4, 5 AGGTCA
Retinoic acid	RARE	—→ —→
Vitamin D	VDRE	
cAMP	CRE	TGACGTCA

[1]Letters indicate nucleotide; N means any one of the four can be used in that position. The arrows pointing in opposite directions illustrate the slightly imperfect inverted palindromes present in many HREs; in some cases these are called "half binding sites" because each binds one monomer of the receptor. The GRE, PRE, MRE, and ARE consist of the same DNA sequence. Specificity may be conferred by the intracellular concentration of the ligand or hormone receptor, by flanking DNA sequences not included in the consensus, or by other accessory elements. A second group of HREs includes those for thyroid hormones, estrogens, retinoic acid, and vitamin D. These HREs are similar except for the orientation and spacing between the half palindromes. Spacing determines the hormone specificity. VDRE (N = 3), TRE (N = 4), and RARE (N = 5) bind to direct repeats rather than to inverted repeats. Another member of the steroid receptor superfamily, the retinoid X receptor (RXR), forms heterodimers with VDR, TR, and RARE, and these constitute the functional forms of these *trans*-acting factors. cAMP affects gene transcription through the CRE.

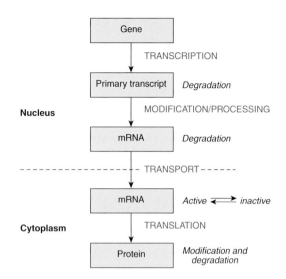

FIGURE 42–3 The "information pathway." Information flows from the gene to the primary transcript to mRNA to protein. Hormones can affect any of the steps involved and can affect the rates of processing, degradation, or modification of the various products.

that these hormones exert their dominant effect on modulating gene transcription, but they—and many of the hormones in the other classes discussed below—can act at any step of the "information pathway," as illustrated in Figure 42–3, to control specific gene expression and, ultimately, a biological response. Direct actions of steroids in the cytoplasm and on various organelles and membranes have also been described. Recently microRNAs have been implicated in mediating some of the diverse actions of the peptide hormone insulin.

GROUP II (PEPTIDE & CATECHOLAMINE) HORMONES HAVE MEMBRANE RECEPTORS & USE INTRACELLULAR MESSENGERS

Many hormones are water-soluble, have no transport proteins (and therefore have a short plasma half-life), and initiate a response by binding to a receptor located in the plasma membrane (Tables 41–3 & 41–4). The mechanism of action of this group of hormones can best be discussed in terms of the **intracellular signals** they generate. These signals include cAMP (cyclic AMP; 3′,5′-adenylic acid; see Figure 19–5), a nucleotide derived from ATP through the action of adenylyl cyclase; cGMP, a nucleotide formed by guanylyl cyclase; Ca^{2+}; and phosphatidylinositides; such molecules are termed second messengers as their synthesis is triggered by the presence of the primary hormone (molecule) binding its receptor. Many of these second messengers affect gene transcription, as described in the previous paragraph; but they also influence a variety of other biologic processes, as shown in Figure 42–3.

G Protein–Coupled Receptors (GPCR)

Many of the group II hormones bind to receptors that couple to effectors through a GTP-binding protein intermediary. These receptors typically have seven hydrophobic plasma membrane-spanning domains. This is illustrated by the seven interconnected cylinders extending through the lipid bilayer in Figure 42–4. Receptors of this class, which signal through guanine nucleotide-bound protein intermediates, are known as **G**

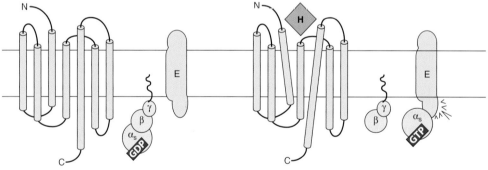

No hormone: inactive effector   Bound hormone (H): active effector

FIGURE 42–4 Components of the hormone receptor–G protein effector system. Receptors that couple to effectors through G proteins (GPCR) typically have seven membrane-spanning domains. In the absence of hormone (**left**), the heterotrimeric G-protein complex (α,β,γ) is in an inactive guanosine diphosphate (GDP)-bound form and is probably not associated with the receptor. This complex is anchored to the plasma membrane through prenylated groups on the βγ subunits (**wavy lines**) and perhaps by myristoylated groups on α subunits (not shown). On binding of hormone (H) to the receptor, there is a presumed conformational change of the receptor—as indicated by the tilted membrane spanning domains—and activation of the G-protein complex. This results from the exchange of GDP with guanosine triphosphate (GTP) on the α subunit, after which α and βγ dissociate. The α subunit binds to and activates the effector (E). E can be adenylyl cyclase, Ca^{2+}, Na$^+$, or Cl$^-$ channels (α$_s$), or it could be a K$^+$ channel (α$_i$), phospholipase Cβ (α$_q$), or cGMP phosphodiesterase (α$_t$). The βγ subunit can also have direct actions on E. (Modified and reproduced, with permission, from Granner DK in: *Principles and Practice of Endocrinology and Metabolism,* 2nd ed. Becker KL [editor]. Lippincott, 1995.)

protein–coupled receptors, or **GPCRs.** To date, hundreds of G protein–linked receptor genes have been identified; this represents the largest family of cell surface receptors in humans. A wide variety of responses are mediated by the GPCRs.

cAMP Is the Intracellular Signal for Many Responses

Cyclic AMP was the first intracellular signal identified in mammalian cells. Several components comprise a system for the generation, degradation, and action of cAMP.

Adenylyl Cyclase

Different peptide hormones can either stimulate (s) or inhibit (i) the production of cAMP from adenylyl cyclase, which is encoded by at least nine different genes (Table 42–2). Two parallel systems, a stimulatory (s) one and an inhibitory (i) one, converge upon a catalytic molecule (C). Each consists of a receptor, R_s or R_i, and a regulatory complex, G_s and G_i. G_s and G_i are each trimers composed of α, β, and γ subunits. Because the α subunit in G_s differs from that in G_i, the proteins, which are distinct gene products, are designated $α_s$ and $α_i$. The α subunits bind guanine nucleotides. The β and γ subunits are always associated (βγ) and appear to function as a heterodimer. The binding of a hormone to R_s or R_i results in a receptor-mediated activation of G, which entails the exchange of GDP by GTP on α and the concomitant dissociation of βγ from α.

The $α_s$ protein has intrinsic GTPase activity. The active form, $α_s$·GTP, is inactivated upon hydrolysis of the GTP to GDP; the trimeric G_s complex (αβγ) is then re-formed and is ready for another cycle of activation. Cholera and pertussis tox-ins catalyze the ADPribosylation of $α_s$ and $α_{i-2}$ (see Table 42–3), respectively. In the case of $α_s$, this modification disrupts the intrinsic GTPase activity; thus, $α_s$ cannot reassociate with βγ and is therefore irreversibly activated. ADP ribosylation of $α_{i-2}$ prevents the dissociation of $α_{i-2}$ from βγ, and free $α_{i-2}$ thus cannot be formed. $α_s$ activity in such cells is therefore unopposed.

There is a large family of G proteins, and these are part of the superfamily of GTPases. The G protein family is classified according to sequence homology into four subfamilies, as illustrated in Table 42–3. There are 21 α, 5 β, and 8 γ subunit genes. Various combinations of these subunits provide a large number of possible αβγ and cyclase complexes.

The α subunits and the βγ complex have actions independent of those on adenylyl cyclase (see Figure 42–4 & Table 42–3). Some forms of $α_i$ stimulate K⁺ channels and inhibit Ca²⁺ channels, and some $α_s$ molecules have the opposite effects. Members of the G_q family activate the phospholipase C group of enzymes. The βγ complexes have been associated with K⁺ channel stimulation and phospholipase C activation. G proteins are involved in many important biologic processes in addition to hormone action. Notable examples include olfaction ($α_{OLF}$) and vision ($α_t$). Some examples are listed in Table 42–3. GPCRs are implicated in a number of diseases and are major targets for pharmaceutical agents.

Protein Kinase

In prokaryotic cells, cAMP binds to a specific protein called catabolite regulatory protein (CRP) that binds directly to DNA and influences gene expression. In eukaryotic cells, cAMP binds to a protein kinase called **protein kinase A (PKA)**, a heterotetrameric molecule consisting of two regulatory subunits (R) and two catalytic subunits (C). cAMP binding results in the following reaction:

$$4\ \text{cAMP} + R_2C_2 \rightleftarrows R_2 \cdot (4\ \text{cAMP}) + 2C$$

The R_2C_2 complex has no enzymatic activity, but the binding of cAMP by R induces dissociation of the R-C complex, thereby activating the latter (Figure 42–5). The active C subunit catalyzes the transfer of the γ phosphate of ATP to a serine or threonine residue in a variety of proteins. The consensus phosphorylation sites are -ArgArg/Lys-X-Ser/Thr- and -Arg-Lys-X-X-Ser-, where X can be any amino acid.

Protein kinase activities were originally described as being "cAMP-dependent" or "cAMP-independent." This classification has changed, as protein phosphorylation is now recognized as being a major regulatory mechanism. Several hundred protein kinases have now been described. The kinases are related in sequence and structure within the catalytic domain, but each is a unique molecule with considerable variability with respect to subunit composition, molecular weight, autophosphorylation, K_m for ATP, and substrate specificity. Both kinase and protein phosphatase activities can be targeted by interaction with specific kinase binding proteins. In the case of PKA, such targeting proteins are termed **AKAPs (A kinase anchoring proteins)**, they serve as scaffolds, which localize PKA near to substrates thereby focusing PKA activity toward

TABLE 42–2 Subclassification of Group II.A Hormones

Hormones That Stimulate Adenylyl Cyclase (H_s)	Hormones That Inhibit Adenylyl Cyclase (H_i)
ACTH	Acetylcholine
ADH	$α_2$-Adrenergics
β-Adrenergics	Angiotensin II
Calcitonin	Somatostatin
CRH	
FSH	
Glucagon	
hCG	
LH	
LPH	
MSH	
PTH	
TSH	

TABLE 42–3   Classes and Functions of Selected G Proteins[1]

Class or Type	Stimulus	Effector	Effect
G_s			
α_s	Glucagon, β-adrenergics	↑Adenylyl cyclase	Glyconeogenesis, lipolysis, glycogenolysis
		↑Cardiac Ca^{2+}, Cl^-, and Na^+ channels	Olfaction
α_{olf}	Odorant	↑Adenylyl cyclase	
G_i			
$\alpha_{i-1,2,3}$	Acetylcholine, α_2-adrenergics	↓Adenylyl cyclase	Slowed heart rate
		↑Potassium channels	
	M_2 cholinergics	↓Calcium channels	
α_0	Opioids, endorphins	↑Potassium channels	Neuronal electrical activity
α_t	Light	↑cGMP phosphodiesterase	Vision
G_q			
α_q	M_1 cholinergics		
	α_1-Adrenergics	↑Phospholipase C-β1	↓Muscle contraction and
α_{11}	α_1-Adrenergics	↑Phospholipase C-β2	↓Blood pressure
G_{12}			
α_{12}	?	Cl^- channel	?

Source: Modified and reproduced, with permission, from Granner DK in: *Principles and Practice of Endocrinology and Metabolism,* 2nd ed. Becker KL (editor). Lippincott, 1995.
[1]The four major classes or families of mammalian G proteins (G_s, G_i, G_q, and G_{12}) are based on protein sequence homology. Representative members of each are shown, along with known stimuli, effectors, and well-defined biologic effects. Nine isoforms of adenylyl cyclase have been identified (isoforms I–IX). All isoforms are stimulated by α_s; α_i isoforms inhibit types V and VI, and α_0 inhibits types I and V. At least 16 different α subunits have been identified.

physiological substrates and facilitating spatiotemporal biological regulation while also allowing for common, shared proteins to elicit specific physiological responses. Multiple AKAPs have been described; they can bind PKA and other kinases as well as phosphatases, phosphodiestases (which hydrolyze cAMP) and protein kinase substrates.

Phosphoproteins

The effects of cAMP in eukaryotic cells are all thought to be mediated by protein phosphorylation-dephosphorylation, principally on serine and threonine residues. The control of any of the effects of cAMP, including such diverse processes as steroidogenesis, secretion, ion transport, carbohydrate and

fat metabolism, enzyme induction, gene regulation, synaptic transmission, and cell growth and replication, could be conferred by a specific protein kinase, by a specific phosphatase, or by specific substrates for phosphorylation. These substrates help define a target tissue and are involved in defining the extent of a particular response within a given cell. For example, the effects of cAMP on gene transcription are mediated by the protein **cyclic AMP response element binding protein (CREB).** CREB binds to a cAMP responsive element (CRE) (see Table 42–1) in its nonphosphorylated state and is a weak activator of transcription. When phosphorylated by PKA, CREB binds the coactivator **CREB-binding protein CBP/p300** (see below) and as a result is a much more potent transcription activator. CBP and the related p300 contain histone acetyltransferase

FIGURE 42–5 Hormonal regulation of cellular processes through cAMP-dependent protein kinase (PKA). PKA exists in an inactive form as an R_2C_2 heterotetramer consisting of two regulatory and two catalytic subunits. The cAMP generated by the action of adenylyl cyclase (activated as shown in Figure 42–4) binds to the regulatory (R) subunit of PKA. This results in dissociation of the regulatory and catalytic subunits and activation of the latter. The active catalytic subunits phosphorylate a number of target proteins on serine and threonine residues. Phosphatases remove phosphate from these residues and thus terminate the physiologic response. A phosphodiesterase can also terminate the response by converting cAMP to 5′-AMP.

activities, and hence serve as chromatin-active transcriptional coregulators (Chapters 36, 38). Interestingly, CBP/p300 can also acetylate certain transcription factors thereby stimulating their ability to bind DNA and modulate transcription.

Phosphodiesterases

Actions caused by hormones that increase cAMP concentration can be terminated in a number of ways, including the hydrolysis of cAMP to 5′-AMP by phosphodiesterases (see Figure 42–5). The presence of these hydrolytic enzymes ensures a rapid turnover of the signal (cAMP) and hence a rapid termination of the biologic process once the hormonal stimulus is removed. There are at least 11 known members of the phosphodiesterase family of enzymes. These are subject to regulation by their substrates, cAMP and cGMP; by hormones; and by intracellular messengers such as calcium, probably acting through calmodulin. Inhibitors of phosphodiesterase, most notably methylated xanthine derivatives such as caffeine, increase intracellular cAMP and mimic or prolong the actions of hormones through this signal.

Phosphoprotein Phosphatases

Given the importance of protein phosphorylation, it is not surprising that regulation of the protein dephosphorylation reaction is another important control mechanism (see Figure 42–5). The phosphoprotein phosphatases are themselves subject to regulation by phosphorylation-dephosphorylation reactions and by a variety of other mechanisms, such as protein-protein interactions. In fact, the substrate specificity of the phosphoserine-phosphothreonine phosphatases may be dictated by distinct regulatory subunits whose binding is regulated hormonally. One of the best-studied roles of regulation by the dephosphorylation of proteins is that of glycogen metabolism in muscle. Two major types of phosphoserine-phosphothreonine phosphatases have been described. Type I preferentially dephosphorylates the β subunit of phosphorylase kinase, whereas type II dephosphorylates the α subunit. Type I phosphatase is implicated in the regulation of glycogen synthase, phosphorylase, and phosphorylase kinase. This phosphatase is itself regulated by phosphorylation of certain of its subunits, and these reactions are reversed by the action of one of the type II phosphatases. In addition, two heat-stable protein inhibitors regulate type I phosphatase activity. Inhibitor-1 is phosphorylated and activated by cAMP-dependent protein kinases; and inhibitor-2, which may be a subunit of the inactive phosphatase, is also phosphorylated, possibly by glycogen synthase kinase-3. Phosphatases that attack phosphotyrosine are also important in signal transduction (see Figure 42–8).

cGMP Is Also an Intracellular Signal

Cyclic GMP is made from GTP by the enzyme guanylyl cyclase, which exists in soluble and membrane-bound forms. Each of these isozymes has unique physiologic properties. The atriopeptins, a family of peptides produced in cardiac atrial

tissues, cause natriuresis, diuresis, vasodilation, and inhibition of aldosterone secretion. These peptides (eg, atrial natriuretic factor) bind to and activate the membrane-bound form of guanylyl cyclase. This results in an increase of cGMP by as much as 50-fold in some cases, and this is thought to mediate the effects mentioned above. Other evidence links cGMP to vasodilation. A series of compounds, including nitroprusside, nitroglycerin, nitric oxide, sodium nitrite, and sodium azide, all cause smooth muscle relaxation and are potent vasodilators. These agents increase cGMP by activating the soluble form of guanylyl cyclase, and inhibitors of cGMP phosphodiesterase (the drug sildenafil [Viagra], for example) enhance and prolong these responses. The increased cGMP activates cGMP-dependent protein kinase (PKG), which in turn phosphorylates a number of smooth muscle proteins. Presumably, this is involved in relaxation of smooth muscle and vasodilation.

Several Hormones Act Through Calcium or Phosphatidylinositols

Ionized calcium is an important regulator of a variety of cellular processes, including muscle contraction, stimulus-secretion coupling, blood clotting cascade, enzyme activity, and membrane excitability. It is also an intracellular messenger of hormone action.

Calcium Metabolism

The extracellular calcium (Ca^{2+}) concentration is about 5 mmol/L and is very rigidly controlled. Although substantial amounts of calcium are associated with intracellular organelles such as mitochondria and the endoplasmic reticulum, the intracellular concentration of free or ionized calcium (Ca^{2+}) is very low: 0.05–10 µmol/L. In spite of this large concentration gradient and a favorable trans-membrane electrical gradient, Ca^{2+} is restrained from entering the cell. A considerable amount of energy is expended to ensure that the intracellular Ca^{2+} is controlled, as a prolonged elevation of Ca^{2+} in the cell is very toxic. A Na^+/Ca^{2+} exchange mechanism that has a high-capacity but low-affinity pumps Ca^{2+} out of cells. There also is a Ca^{2+}/proton ATPase-dependent pump that extrudes Ca^{2+} in exchange for H^+. This has a high affinity for Ca^{2+} but a low capacity and is probably responsible for fine-tuning cytosolic Ca^{2+}. Furthermore, Ca^{2+}-ATPases pump Ca^{2+} from the cytosol to the lumen of the endoplasmic reticulum. There are three ways of changing cytosolic Ca^{2+}: (1) Certain hormones (class II.C, Table 41–3) by binding to receptors that are themselves Ca^{2+} channels, enhance membrane permeability to Ca^{2+} and thereby increase Ca^{2+} influx. (2) Hormones also indirectly promote Ca^{2+} influx by modulating the membrane potential at the plasma membrane. Membrane depolarization opens voltage-gated Ca^{2+} channels and allows for Ca^{2+} influx. (3) Ca^{2+} can be mobilized from the endoplasmic reticulum, and possibly from mitochondrial pools.

An important observation linking Ca^{2+} to hormone action involved the definition of the intracellular targets of Ca^{2+} action. The discovery of a Ca^{2+}-dependent regulator of phosphodiesterase activity provided the basis for a broad understanding of how Ca^{2+} and cAMP interact within cells.

Calmodulin

The calcium-dependent regulatory protein is calmodulin, a 17-kDa protein that is homologous to the muscle protein troponin C in structure and function. Calmodulin has four Ca^{2+} binding sites, and full occupancy of these sites leads to a marked conformational change, which allows calmodulin to activate enzymes and ion channels. The interaction of Ca^{2+} with calmodulin (with the resultant change of activity of the latter) is conceptually similar to the binding of cAMP to PKA and the subsequent activation of this molecule. Calmodulin can be one of numerous subunits of complex proteins and is particularly involved in regulating various kinases and enzymes of cyclic nucleotide generation and degradation. A partial list of the enzymes regulated directly or indirectly by Ca^{2+}, probably through calmodulin, is presented in Table 42–4.

In addition to its effects on enzymes and ion transport, Ca^{2+}/calmodulin regulates the activity of many structural elements in cells. These include the actin-myosin complex of smooth muscle, which is under β-adrenergic control, and various microfilament-mediated processes in noncontractile cells, including cell motility, cell conformation changes, mitosis, granule release, and endocytosis.

Calcium Is a Mediator of Hormone Action

A role for Ca^{2+} in hormone action is suggested by the observations that the effect of many hormones is (1) blunted by Ca^{2+}-free media or when intracellular calcium is depleted; (2) can be mimicked by agents that increase cytosolic Ca^{2+}, such as the Ca^{2+} ionophore A23187; and (3) influences cellular calcium flux. The regulation of glycogen metabolism in liver by vasopressin and β-adrenergic catecholamines provides a good example. This is shown schematically in Figures 19–6 and 19–7.

TABLE 42–4   Enzymes and Proteins Regulated by Calcium or Calmodulin

Adenylyl cyclase
Ca^{2+}-dependent protein kinases
Ca^{2+}-Mg^{2+}-ATPase
Ca^{2+}-phospholipid-dependent protein kinase
Cyclic nucleotide phosphodiesterase
Some cytoskeletal proteins
Some ion channels (eg, L-type calcium channels)
Nitric oxide synthase
Phosphorylase kinase
Phosphoprotein phosphatase 2B
Some receptors (eg, NMDA-type glutamate receptor)

A number of critical metabolic enzymes are regulated by Ca^{2+}, phosphorylation, or both, including glycogen synthase, pyruvate kinase, pyruvate carboxylase, glycerol-3-phosphate dehydrogenase, and pyruvate dehydrogenase.

Phosphatidylinositide Metabolism Affects Ca^{2+}-Dependent Hormone Action

Some signal must provide communication between the hormone receptor on the plasma membrane and the intracellular Ca^{2+} reservoirs. This is accomplished by products of phosphatidylinositol metabolism. Cell surface receptors such as those for acetylcholine, antidiuretic hormone, and $α_1$-type catecholamines are, when occupied by their respective ligands, potent activators of phospholipase C. Receptor binding and activation of phospholipase C are coupled by the G_q isoforms (Table 42–3 & Figure 42–6). Phospholipase C catalyzes the hydrolysis of phosphatidylinositol 4,5-bisphosphate to inositol trisphosphate (IP_3) and 1,2-diacylglycerol (Figure 42–7). Diacylglycerol is itself capable of activating **protein kinase C (PKC),** the activity of which also depends upon Ca^{2+}. IP_3, by interacting with a specific intracellular receptor, is an effective releaser of Ca^{2+} from intracellular storage sites in the endoplasmic reticulum. Thus, the hydrolysis of phosphatidylinositol 4,5-bisphosphate leads to activation of PKC and promotes an increase of cytoplasmic Ca^{2+}. As shown in Figure 42–4, the activation of G proteins can also have a direct action on Ca^{2+} channels. The resulting elevations of cytosolic Ca^{2+} activate Ca^{2+}–calmodulin-dependent kinases and many other Ca^{2+}–calmodulin-dependent enzymes.

Steroidogenic agents—including ACTH and cAMP in the adrenal cortex; angiotensin II, K^+, serotonin, ACTH, and cAMP in the zona glomerulosa of the adrenal; LH in the ovary; and LH and cAMP in the Leydig cells of the testes—have been associated with increased amounts of phosphatidic acid, phosphatidylinositol, and polyphosphoinositides (see Chapter 15) in the respective target tissues. Several other examples could be cited.

The roles that Ca^{2+} and polyphosphoinositide breakdown products might play in hormone action are presented in Figure 42–6. In this scheme the activated protein kinase C can phosphorylate specific substrates, which then alter physiologic processes. Likewise, the Ca^{2+}-calmodulin complex can activate specific kinases. These then modify substrates and thereby alter physiologic responses.

Some Hormones Act Through a Protein Kinase Cascade

Single protein kinases such as PKA, PKC, and Ca^{2+}-calmodulin (CaM)-kinases, which result in the phosphorylation of serine and threonine residues in target proteins, play a very important role in hormone action. The discovery that the EGF receptor contains an intrinsic tyrosine kinase activity that is activated by the binding of the ligand EGF was an important breakthrough. The insulin and IGF-I receptors also contain intrinsic ligand-activated tyrosine kinase activity. Several

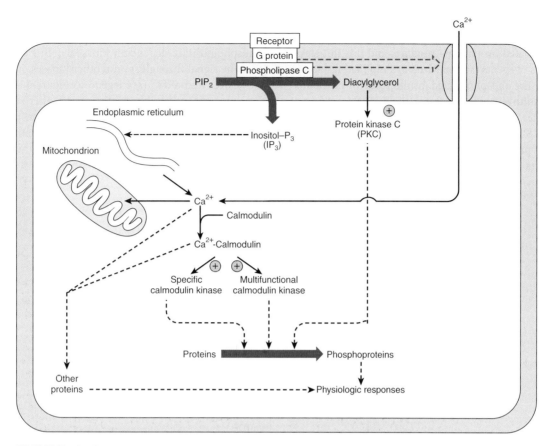

FIGURE 42–6 Certain hormone–receptor interactions result in the activation of phospholipase C. This appears to involve a specific G protein, which also may activate a calcium channel. Phospholipase C results in generation of inositol trisphosphate (IP$_3$), which liberates stored intracellular Ca^{2+}, and diacylglycerol (DAG), a potent activator of protein kinase C (PKC). In this scheme, the activated PKC phosphorylates specific substrates, which then alter physiologic processes. Likewise, the Ca^{2+}-calmodulin complex can activate specific kinases, two of which are shown here. These actions result in phosphorylation of substrates, and this leads to altered physiologic responses. This figure also shows that Ca^{2+} can enter cells through voltage- or ligand-gated Ca^{2+} channels. The intracellular Ca^{2+} is also regulated through storage and release by the mitochondria and endoplasmic reticulum. (Courtesy of JH Exton.)

FIGURE 42–7 Phospholipase C cleaves PIP$_2$ into diacylglycerol and inositol trisphosphate. R$_1$ generally is stearate, and R$_2$ is usually arachidonate. IP$_3$ can be dephosphorylated (to the inactive I-1,4-P$_2$) or phosphorylated (to the potentially active I-1,3,4,5-P$_4$).

receptors—generally those involved in binding ligands involved in growth control, differentiation, and the inflammatory response—either have intrinsic tyrosine kinase activity or are associated with proteins that are tyrosine kinases. Another distinguishing feature of this class of hormone action is that these kinases preferentially phosphorylate tyrosine residues, and tyrosine phosphorylation is infrequent (<0.03% of total amino acid phosphorylation) in mammalian cells. A third distinguishing feature is that the ligand–receptor interaction that results in a tyrosine phosphorylation event initiates a cascade that may involve several protein kinases, phosphatases, and other regulatory proteins.

Insulin Transmits Signals by Several Kinase Cascades

The insulin, epidermal growth factor (EGF), and IGF-I receptors have intrinsic protein tyrosine kinase activities located in their cytoplasmic domains. These activities are stimulated when the receptor binds ligand. The receptors are then autophosphorylated on tyrosine residues, and this initiates a complex series of events (summarized in simplified fashion in Figure 42–8). The phosphorylated insulin receptor next phosphorylates insulin receptor substrates (there are at least four of these molecules, called IRS 1–4) on tyrosine residues. Phosphorylated IRS binds to the Src homology 2 (SH2) domains of a variety of proteins that are directly involved in mediating different effects of insulin. One of these proteins, PI-3 kinase,

links insulin receptor activation to insulin action through activation of a number of molecules, including the kinase PDK1 (phosphoinositide-dependent kinase-1). This enzyme propagates the signal through several other kinases, including PKB (akt), SKG, and aPKC (see legend to Figure 42–8 for definitions and expanded abbreviations). An alternative pathway downstream from PKD1 involves p70S6K and perhaps other as yet unidentified kinases. A second major pathway involves mTOR. This enzyme is directly regulated by amino acid levels and insulin, and is essential for p70S6K activity. This pathway

FIGURE 42–8 Insulin signaling pathways. The insulin signaling pathways provide an excellent example of the "recognition → hormone release → signal generation → effects" paradigm outlined in Figure 42–1. Insulin is released in response to hyperglycemia. Binding of insulin to a target cell-specific plasma membrane receptor results in a cascade of intracellular events. Stimulation of the intrinsic tyrosine kinase activity of the insulin receptor marks the initial event, resulting in increased tyrosine (Y) phosphorylation (Y → Y-P) of the receptor and then one or more of the insulin receptor substrate molecules (IRS 1–4). This increase in phosphotyrosine stimulates the activity of many intracellular molecules such as GTPases, protein kinases, and lipid kinases, all of which play a role in certain metabolic actions of insulin. The two best-described pathways are shown. First, phosphorylation of an IRS molecule (probably IRS-2) results in docking and activation of the lipid kinase, PI-3 kinase, which generates novel inositol lipids that may act as "second messenger" molecules. These, in turn, activate PDK1 and then a variety of downstream signaling molecules, including protein kinase B (PKB or akt), SGK, and aPKC. An alternative pathway involves the activation of p70S6K and perhaps other as yet unidentified kinases. Second, phosphorylation of IRS (probably IRS-1) results in docking of GRB2/mSOS and activation of the small GTPase, p21RAS, which initiates a protein kinase cascade that activates Raf-1, MEK, and the p42/p44 MAP kinase isoforms. These protein kinases are important in the regulation of proliferation and differentiation of several cell types. The mTOR pathway provides an alternative way of activating p70S6K and appears to be involved in nutrient signaling as well as insulin action. Each of these cascades may influence different physiologic processes, as shown. All of the phosphorylation events are reversible through the action of specific phosphatases. For example, the lipid phosphatase PTEN dephosphorylates the product of the PI-3 kinase reaction, thereby antagonizing the pathway and terminating the signal. Representative effects of major actions of insulin are shown in each of the boxes. The asterisk after phosphodiesterase indicates that insulin indirectly affects the activity of many enzymes by activating phosphodiesterases and reducing intracellular cAMP levels. (IGFBP, insulin-like growth factor binding protein; IRS 1–4, insulin receptor substrate isoforms 1–4; PI-3 kinase, phosphatidylinositol 3-kinase; PTEN, phosphatase and tensin homolog deleted on chromosome 10; PKD1, phosphoinositide-dependent kinase; PKB, protein kinase B; SGK, serum and glucocorticoid-regulated kinase; aPKC, atypical protein kinase C; p70S6K, p70 ribosomal protein S6 kinase; mTOR, mammalian target of rapamycin; GRB2, growth factor receptor binding protein 2; mSOS, mammalian son of sevenless; MEK, MAP kinase kinase and ERK kinase; MAP kinase, mitogen-activated protein kinase.)

provides a distinction between the PKB and p70S6K branches downstream from PKD1. These pathways are involved in protein translocation, enzyme activity, and the regulation, by insulin, of genes involved in metabolism (Figure 42–8). Another SH2 domain-containing protein is GRB2, which binds to IRS-1 and links tyrosine phosphorylation to several proteins, the result of which is activation of a cascade of threonine and serine kinases. A pathway showing how this insulin–receptor interaction activates the mitogen-activated protein (MAP) kinase pathway and the anabolic effects of insulin is illustrated in Figure 42–8. The exact roles of many of these docking proteins, kinases, and phosphatases remain to be established.

The Jak/STAT Pathway is Used by Hormones and Cytokines

Tyrosine kinase activation can also initiate a phosphorylation and dephosphorylation cascade that involves the action of several other protein kinases and the counterbalancing actions of phosphatases. Two mechanisms are employed to initiate this cascade. Some hormones, such as growth hormone, prolactin, erythropoietin, and the cytokines, initiate their action by activating a tyrosine kinase, but this activity is not an integral part of the hormone receptor. The hormone–receptor interaction promotes binding and activation of **cytoplasmic protein tyrosine kinases,** such as **Tyk-2, Jak1, or Jak2.**

These kinases phosphorylate one or more cytoplasmic proteins, which then associate with other docking proteins through binding to SH2 domains. One such interaction results in the activation of a family of cytosolic proteins called **signal transducers and activators of transcription (STATs).**

The phosphorylated STAT protein dimerizes and translocates into the nucleus, binds to a specific DNA element such as the interferon response element, and activates transcription. This is illustrated in Figure 42–9. Other SH2 docking events may result in the activation of PI-3 kinase, the MAP kinase pathway (through SHC or GRB2), or G protein–mediated activation of phospholipase C (PLCγ) with the attendant production of diacylglycerol and activation of protein kinase C. It is apparent that there is a potential for "cross-talk" when different hormones activate these various signal transduction pathways.

The NF-κB Pathway Is Regulated by Glucocorticoids

The transcription factor **NF-κB is a heterodimeric complex** typically composed of two **subunits termed p50 and p65** (Figure 42–10). Normally, NF-κB is kept sequestered in the cytoplasm in a transcriptionally inactive form by members of the **Inhibitor of NF-κB (IκB) family**. Extracellular stimuli such as proinflammatory cytokines, reactive oxygen species, and mitogens lead to activation of the IκB kinase complex, IKK, which is a heterohexameric structure consisting of α, β, and γ subunits. IKK phosphorylates IκB on two serine residues, and this targets IκB for ubiquitination and subsequent degradation by the proteasome. Following IκB degradation, free NF-κB translocates to the nucleus, where it binds to a number of gene promoters and activates transcription, particularly of genes involved in the **inflammatory response.** Transcriptional regulation by NF-κB is mediated by a variety of coactivators such as CREB binding protein (CBP), as described below (see Figure 42–13).

FIGURE 42–9 Initiation of signal transduction by receptors linked to Jak kinases. The receptors (R) that bind prolactin, growth hormone, interferons, and cytokines lack endogenous tyrosine kinase. Upon ligand binding, these receptors dimerize and an associated protein (Jak1, Jak2, or TYK) is phosphorylated. Jak-P, an active kinase, phosphorylates the receptor on tyrosine residues. The STAT proteins associate with the phosphorylated receptor and then are themselves phosphorylated by Jak-P. STAT Ⓟ dimerizes, translocates to the nucleus, binds to specific DNA elements, and regulates transcription. The phosphotyrosine residues of the receptor also bind to several SH2 domain-containing proteins (X-SH2). This results in activation of the MAP kinase pathway (through SHC or GRB2), PLCγ, or PI-3 kinase.

FIGURE 42–10 Regulation of the NF-κB pathway. NF-κB consists of two subunits, p50 and p65, which when present in the nucleus regulate transcription of the multitude of genes important for the inflammatory response. NF-κB is restricted from entering the nucleus by IκB, an inhibitor of NF-κB. IκB binds to—and masks—the nuclear localization signal of NF-κB. This cytoplasmic protein is phosphorylated by an IKK complex which is activated by cytokines, reactive oxygen species, and mitogens. Phosphorylated IκB can be ubiquitinylated and degraded, thus releasing its hold on NF-κB. Glucocorticoids, potent anti-inflammatory agents, are thought to affect at least three steps in this process (1, 2, 3), as described in the text.

Glucocorticoid hormones are therapeutically useful agents for the treatment of a variety of inflammatory and immune diseases. Their anti-inflammatory and immunomodulatory actions are explained in part by the inhibition of NF-κB and its subsequent actions. Evidence for three mechanisms for the inhibition of NF-κB by glucocorticoids has been presented: (1) Glucocorticoids increase IκB mRNA, which leads to an increase of IκB protein and more efficient sequestration of NF-κB in the cytoplasm. (2) The glucocorticoid receptor competes with NF-κB for binding to coactivators. (3) The glucocorticoid receptor directly binds to the p65 subunit of NF-κB and inhibits its activation (Figure 42–10).

HORMONES CAN INFLUENCE SPECIFIC BIOLOGIC EFFECTS BY MODULATING TRANSCRIPTION

The signals generated as described above have to be translated into an action that allows the cell to effectively adapt to a challenge (Figure 42–1). Much of this adaptation is accomplished through alterations in the rates of transcription of specific genes. Many different observations have led to the current view of how hormones affect transcription. Some of these are as follows: (1) Actively transcribed genes are in regions of "open" chromatin (defined by a susceptibility to the enzyme DNase I), which allows for the access of transcription factors to DNA. (2) Genes have regulatory regions, and transcription factors bind to these to modulate the frequency of transcription initiation. (3) The hormone–receptor complex can be one of these transcription factors. The DNA sequence to which this binds is called a hormone response element (HRE; see Table 42–1 for examples). (4) Alternatively, other hormone-generated signals can modify the location, amount, or activity of transcription factors and thereby influence binding to the regulatory or response element. (5) Members of a large superfamily of nuclear receptors act with—or in a manner analogous to—hormone receptors. (6) These nuclear receptors interact with another large group of coregulatory molecules to effect changes in the transcription of specific genes.

Several Hormone Response Elements (HREs) Have Been Defined

Hormone response elements resemble enhancer elements in that they are not strictly dependent on position and location or orientation. They generally are found within a few hundred nucleotides upstream (5′) of the transcription initiation site, but they may be located within the coding region of the gene, in introns. HREs were defined by the strategy illustrated in Figure 38–11. The consensus sequences illustrated in Table 42–1 were arrived at through analysis of many genes regulated by a given hormone using simple, heterologous reporter systems (see Figure 38–10). Although these simple HREs bind the hormone–receptor complex more avidly than surrounding DNA—or DNA from an unrelated source—and confer hormone responsiveness to a reporter gene, it soon became apparent that the regulatory circuitry of natural genes must be much more complicated. Glucocorticoids, progestins, mineralocorticoids, and androgens have vastly different physiologic actions. How could the specificity required for these effects be achieved through regulation of gene expression by the same HRE (Table 42–1)? Questions like this have led to experiments which have allowed for elaboration of a very complex model of transcription regulation. For example, the HRE must associate with other DNA elements (and associated binding proteins) to function optimally. The extensive sequence similarity noted between steroid hormone receptors, particularly in their DNA-binding domains, led to discovery of the **nuclear receptor superfamily** of proteins. These—and a large number of **coregulator proteins**—allow for a wide variety of DNA–protein and protein–protein interactions and the specificity

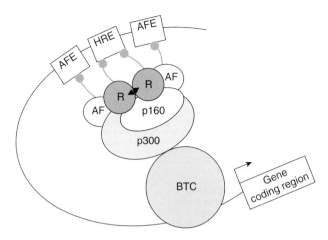

FIGURE 42–11 The hormone response transcription unit. The hormone response transcription unit is an assembly of DNA elements and bound proteins that interact, through protein-protein interactions, with a number of coactivator or corepressor molecules. An essential component is the hormone response element which binds the ligand (▲)-bound receptor (R). Also important are the accessory factor elements (AFEs) with bound transcription factors. More than two dozen of these accessory factors (AFs), which are often members of the nuclear receptor superfamily, have been linked to hormone effects on transcription. The AFs can interact with each other, with the liganded nuclear receptors, or with coregulators. These components communicate with the basal transcription complex (BTC) through a coregulator complex that can consist of one or more members of the p160, corepressor, mediator-related, or CBP/p300 families (see Table 42–6). Recall (Chapters 36, 38) that many of the transcription coregulators carry intrinsic enzymatic activities, which covalently modify the DNA, transcription proteins, and the histones present in the nucleosomes (not shown here) in and around the enhancer (HRE, AFE) and promoter. Collectively the hormone, hormone receptor, chromatin, DNA and transcription machinery integrate and process hormone signals to regulate transcription in a physiological fashion.

necessary for highly regulated physiologic control. A schematic of such an assembly is illustrated in Figure 42–11.

There Is a Large Family of Nuclear Receptor Proteins

The nuclear receptor superfamily consists of a diverse set of transcription factors that were discovered because of a sequence similarity in their DNA-binding domains. This family, now with more than 50 members, includes the nuclear hormone receptors discussed above, a number of other receptors whose ligands were discovered after the receptors were identified, and many putative or orphan receptors for which a ligand has yet to be discovered.

These nuclear receptors have several common structural features (Figure 42–12). All have a centrally located **DNA-binding domain (DBD)** that allows the receptor to bind with high affinity to a response element. The DBD contains two zinc finger binding motifs (see Figure 38–14) that direct binding either as homodimers, as heterodimers (usually with a retinoid X receptor [RXR] partner), or as monomers. The

target response element consists of one or two DNA half-site consensus sequences arranged as an inverted or direct repeat. The spacing between the latter helps determine binding specificity. Thus, in general, a direct repeat with three, four, or five nucleotide spacer regions specifies the binding of the vitamin D, thyroid, and retinoic acid receptors, respectively, to the same consensus response element (Table 42–1). A multifunctional **ligand-binding domain (LBD)** is located in the carboxyl terminal half of the receptor. The LBD binds hormones or metabolites with selectivity and thus specifies a particular biologic response. The LBD also contains domains that mediate the binding of heat shock proteins, dimerization, nuclear localization, and transactivation. The latter function is facilitated by the carboxyl terminal transcription activation function (**AF-2 domain),** which forms a surface required for the interaction with coactivators. A highly variable **hinge region** separates the DBD from the LBD. This region provides flexibility to the receptor, so it can assume different DNA-binding conformations. Finally, there is a highly variable amino terminal region that contains another transactivation domain referred to as **AF-1.** The AF-1 domain likely provides for distinct physiologic functions through the binding of different coregulator proteins. This region of the receptor, through the use of different promoters, alternative splice sites, and multiple translation initiation sites, provides for receptor isoforms that share DBD and LBD identity but exert different physiologic responses because of the association of various coregulators with this variable amino terminal AF-1 domain.

It is possible to sort this large number of receptors into groups in a variety of ways. Here they are discussed according to the way they bind to their respective DNA elements (Figure 42–12). Classic hormone receptors for glucocorticoids (GR), mineralocorticoids (MR), estrogens (ER), androgens (AR), and progestins (PR) bind as homodimers to inverted repeat sequences. Other hormone receptors such as thyroid (TR), retinoic acid (RAR), and vitamin D (VDR) and receptors that bind various metabolite ligands such as PPAR α, β, and γ, FXR, LXR, PXR/SXR, and CAR bind as heterodimers, with retinoid X receptor (RXR) as a partner, to direct repeat sequences (see Figure 42–12 & Table 42–5). Another group of orphan receptors that as yet have no known ligand bind as homodimers or monomers to direct repeat sequences.

As illustrated in Table 42–5, the discovery of the nuclear receptor superfamily has led to an important understanding of how a variety of metabolites and xenobiotics regulate gene expression and thus the metabolism, detoxification, and elimination of normal body products and exogenous agents such as pharmaceuticals. Not surprisingly, this area is a fertile field for investigation of new therapeutic interventions.

A Large Number of Nuclear Receptor Coregulators Also Participate in Regulating Transcription

Chromatin remodeling (histone modifications, DNA methylation), transcription factor modification by various enzyme

FIGURE 42–12 The nuclear receptor superfamily. Members of this family are divided into six structural domains (A–F). Domain A/B is also called AF-1, or the modulator region, because it is involved in activating transcription. The C domain consists of the DNA-binding domain (DBD). The D region contains the hinge, which provides flexibility between the DBD and the ligand-binding domain (LBD, region E). The C terminal part of region E contains AF-2, another domain important for transactivation. The F region is poorly defined. The functions of these domains are discussed in more detail in the text. Receptors with known ligands, such as the steroid hormones, bind as homodimers on inverted repeat half-sites. Other receptors form heterodimers with the partner RXR on direct repeat elements. There can be nucleotide spacers of one to five bases between these direct repeats (DR1–5). Another class of receptors for which ligands have not been determined (orphan receptors) bind as homodimers to direct repeats and occasionally as monomers to a single half-site.

activities, and the communication between the nuclear receptors and the basal transcription apparatus are accomplished by protein-protein interactions with one or more of a class of coregulator molecules. The number of these coregulator molecules now exceeds 100, not counting species variations and splice variants. The first of these to be described was the **CREB-binding protein, CBP.** CBP, through an amino terminal domain, binds to phosphorylated serine 137 of CREB

and mediates transactivation in response to cAMP. It thus is described as a coactivator. CBP and its close relative, p300, interact directly or indirectly with a number of signaling molecules, including activator protein-1 (AP-1), signal transducers and activators of transcription (STATs), nuclear receptors, and CREB (Figure 42–13). **CBP/p300** also binds to the p160 family of coactivators described below and to a number of other proteins, including viral transcription factor Ela, the p90[rsk]

TABLE 42–5 Nuclear Receptors with Special Ligands[1]

Receptor		Partner	Ligand	Process Affected
Peroxisome	PPAR$_\alpha$	RXR (DR1)	Fatty acids	Peroxisome proliferation
Proliferator-activated	PPAR$_\beta$ PPAR$_\gamma$		Fatty acids Fatty acids Eicosanoids, thiazolidinediones	Lipid and carbohydrate metabolism
Farnesoid X	FXR	RXR (DR4)	Farnesol, bile acids	Bile acid metabolism
Liver X	LXR	RXR (DR4)	Oxysterols	Cholesterol metabolism
Xenobiotic X	CAR	RXR (DR5)	Androstanes Phenobarbital Xenobiotics	Protection against certain drugs, toxic metabolites, and xenobiotics
	PXR	RXR (DR3)	Pregnanes Xenobiotics	

[1]Many members of the nuclear receptor superfamily were discovered by cloning, and the corresponding ligands were subsequently identified. These ligands are not hormones in the classic sense, but they do have a similar function in that they activate specific members of the nuclear receptor superfamily. The receptors described here form heterodimers with RXR and have variable nucleotide sequences separating the direct repeat binding elements (DR1–5). These receptors regulate a variety of genes encoding cytochrome p450s (CYP), cytosolic binding proteins, and ATP-binding cassette (ABC) transporters to influence metabolism and protect cells against drugs and noxious agents.

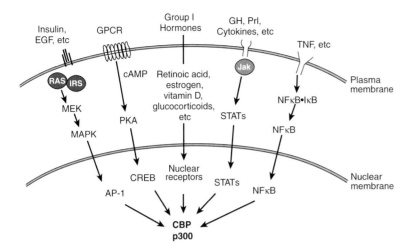

FIGURE 42–13 Several signal transduction pathways converge on CBP/p300. Many ligands that associate with membrane or nuclear receptors eventually converge on CBP/p300. Several different signal transduction pathways are employed. (EGF, epidermal growth factor; GH, growth hormone; Prl, prolactin; TNF, tumor necrosis factor; other abbreviations are expanded in the text.)

protein kinase, and RNA helicase A. It is important to note, as mentioned above, that **CBP/p300 also has intrinsic histone acetyltransferase (HAT) activity.** Some of the many actions of CBP/p300, which appear to depend on intrinsic enzyme activities and its ability to serve as a scaffold for the binding of other proteins, are illustrated in Figure 42–11. Other coregulators serve similar functions.

Several other families of coactivator molecules have been described. Members of the **p160 family of coactivators**, all of about 160 kDa, include (1) SRC-1 and NCoA-1; (2) GRIP 1, TIF2, and NCoA-2; and (3) p/CIP, ACTR, AIB1, RAC3, and TRAM-1 (Table 42–6). The different names for members within a subfamily often represent species variations or minor splice variants. There is about 35% amino acid identity between members of the different subfamilies. The p160 coactivators share several properties. They (1) bind nuclear receptors in an agonist- and AF-2 transactivation domain-dependent manner; (2) have a conserved amino terminal basic helix-loop-helix (bHLH) motif (see Chapter 38); (3) have a weak carboxyl terminal transactivation domain and a stronger amino terminal transactivation domain in a region that is required for the CBP/p160 interaction; (4) contain at least three of the **LXXLL motifs** required for protein-protein interaction with other coactivators; and (5) often have HAT activity. The role of HAT is particularly interesting, as mutations of the HAT domain disable many of these transcription factors. Current thinking holds that these HAT activities acetylate histones and result in remodeling of chromatin into a transcription-efficient environment; however, other protein substrates for HAT-mediated acetylation have been reported. Histone acetylation/deacetylation thus plays a critical role in gene expression.

A small number of proteins, including NCoR and SMRT, comprise the **corepressor family.** They function, at least in part, as described in Figure 42–2. Another family includes the TRAPs, DRIPs, and ARC (Table 42–6). These proteins represent subunits of the Mediator (Chapter 36) and range in size from 80 kDa to 240 kDa and are thought to link the nuclear receptor-coactivator complex to RNA polymerase II and the other components of the basal transcription apparatus.

TABLE 42–6 Some Mammalian Coregulator Proteins

I. 300-kDa family of coactivators		
A. CBP		CREB-binding protein
B. p300		Protein of 300 kDa
II. 160-kDa family of coactivators		
A. SRC-1		Steroid receptor coactivator 1
	NCoA-1	Nuclear receptor coactivator 1
B. TIF2		Transcriptional intermediary factor 2
	GRIP1	Glucocorticoid receptor-interacting protein
	NCoA-2	Nuclear receptor coactivator 2
C. p/CIP		p300/CBP cointegrator-associated protein 1
	ACTR	Activator of the thyroid and retinoic acid receptors
	AIB	Amplified in breast cancer
	RAC3	Receptor-associated coactivator 3
	TRAM-1	TR activator molecule 1
III. Corepressors		
A. NCoR		Nuclear receptor corepressor
B. SMRT		Silencing mediator for RXR and TR
IV. Mediator-related proteins		
A. TRAPs		Thyroid hormone receptor-associated proteins
B. DRIPs		Vitamin D receptor-interacting proteins
C. ARC		Activator-recruited cofactor

The exact role of these coactivators is presently under intensive investigation. Many of these proteins have intrinsic enzymatic activities. This is particularly interesting in view of the fact that acetylation, phosphorylation, methylation, sumoylation, and ubiquitination—as well as proteolysis and cellular translocation—have been proposed to alter the activity of some of these coregulators and their targets.

It appears that certain combinations of coregulators—and thus different combinations of activators and inhibitors—are responsible for specific ligand-induced actions through various receptors. Furthermore, these interactions on a given promoter are dynamic. In some cases, complexes consisting of as many as 47 transcription factors have been observed on a single gene.

SUMMARY

- Hormones, cytokines, interleukins, and growth factors use a variety of signaling mechanisms to facilitate cellular adaptive responses.

- The ligand–receptor complex serves as the initial signal for members of the nuclear receptor family.

- Class II hormones, which bind to cell surface receptors, generate a variety of intracellular signals. These include cAMP, cGMP, Ca^{2+}, phosphatidylinositides, and protein kinase cascades.

- Many hormone responses are accomplished through alterations in the rate of transcription of specific genes.

- The nuclear receptor superfamily of proteins plays a central role in the regulation of gene transcription.

- Nuclear receptors, which may have hormones, metabolites, or drugs as ligands, bind to specific DNA elements as homodimers or as heterodimers with RXR. Some—orphan receptors—have no known ligand but bind DNA and influence transcription.

- Another large family of coregulator proteins remodel chromatin, modify other transcription factors, and bridge the nuclear receptors to the basal transcription apparatus.

REFERENCES

Arvanitakis L et al: Constitutively signaling G-protein-coupled receptors and human disease. Trends Endocrinol Metab 1998;9:27

Aylon Y, Oren M: Living with p53, Dying of p53. Cell 2007;130:597.

Beene DL, Scott JD: A-kinase anchoring proteins take shape. Current Opinion in Cell Biol 2007;19:192.

Darnell JE Jr, Kerr IM, Stark GR: Jak-STAT pathways and transcriptional activation in response to IFNs and other extracellular signaling proteins. Science 1994;264:1415.

Fantl WJ, Johnson DE, Williams LT: Signalling by receptor tyrosine kinases. Annu Rev Biochem 1993;62:453.

Hanoune J, Defer N: Regulation and role of adenylyl cyclase isoforms. Annu Rev Pharmacol Toxicol 2001;41:145.

Jaken S: Protein kinase C isozymes and substrates. Curr Opin Cell Biol 1996;8:168.

Lee C-H, Olson P, Evans RM: Mini-review: Lipid metabolism, metabolic diseases and peroxisome proliferators-activated receptor. Endocrinology 2003;144:2201.

Métivier R et al: Cyclical DNA methylation of a transcriptionally active promter. Nature 2008;452:45.

Métivier R, Reid G, Gannon F: Transcription in four dimensions: nuclear receptor-directed initiation of gene expression. EMBO Journal 2006;7:161.

Montminy M: Transcriptional regulation by cyclic AMP. Annu Rev Biochem 1997;66:807

Morris AJ, Malbon CC: Physiological regulation of G protein-linked signaling. Physiol Rev 1999;79:1373.

O'Malley B: Coregulators: from whence came these "master genes." Mol Endocrinology 2007;21:1009.

Rosenfeld MG, Lunyak VV, Glass CK: Sensors and signals: a coactivator/corepressor/epigenetic code for integrating signal-dependent programs of transcriptional response. Genes and Dev 2006;20:1405.

Sonoda J, Pei L, Evans RM: Nuclear receptors: decoding metabolic disease. Fed of European Biochem Soc 2007;582:2.

Tang X, Tang G, Ozcan S: Role of microRNAs in diabetes. Biochim Biophys Acta 2008;1779:697.

Walton KM, Dixon JE: Protein tyrosine phosphatases. Annu Rev Biochem 1993;62:101.

Nutrition, Digestion, & Absorption

David A. Bender, PhD

BIOMEDICAL IMPORTANCE

In addition to water, the diet must provide metabolic fuels (mainly carbohydrates and lipids), protein (for growth and turnover of tissue proteins), fiber (for bulk in the intestinal lumen), minerals (containing elements with specific metabolic functions), and vitamins and essential fatty acids (organic compounds needed in smaller amounts for other metabolic and physiologic functions). The polysaccharides, triacylglycerols, and proteins that make up the bulk of the diet must be hydrolyzed to their constituent monosaccharides, fatty acids, and amino acids, respectively, before absorption and utilization. Minerals and vitamins must be released from the complex matrix of food before they can be absorbed and utilized.

Globally, **undernutrition** is widespread, leading to impaired growth, defective immune systems, and reduced work capacity. By contrast, in developed countries, there is excessive food consumption (especially of fat), leading to obesity, and the development of cardiovascular disease and some forms of cancer. Deficiencies of vitamin A, iron, and iodine pose major health concerns in many countries, and deficiencies of other vitamins and minerals are a major cause of ill health. In developed countries nutrient deficiency is rare, although there are vulnerable sections of the population at risk. Intakes of minerals and vitamins that are adequate to prevent deficiency may be inadequate to promote optimum health and longevity.

Excessive secretion of gastric acid, associated with *Helicobacter pylori* infection, can result in the development of gastric and duodenal **ulcers;** small changes in the composition of bile can result in crystallization of cholesterol as **gallstones;** failure of exocrine pancreatic secretion (as in **cystic fibrosis**) leads to undernutrition and steatorrhea. **Lactose intolerance** is the result of lactase deficiency, leading to diarrhea and intestinal discomfort. Absorption of intact peptides that stimulate antibody responses cause **allergic reactions** and **celiac disease** is an allergic reaction to wheat gluten.

DIGESTION & ABSORPTION OF CARBOHYDRATES

The digestion of carbohydrates is by hydrolysis to liberate oligosaccharides, then free mono- and disaccharides. The increase in blood glucose after a test dose of a carbohydrate compared with that after an equivalent amount of glucose (as glucose or from a reference starchy food) is known as the **glycemic index.** Glucose and galactose have an index of 1, as do lactose, maltose, isomaltose, and trehalose, which give rise to these monosaccharides on hydrolysis. Fructose and the sugar alcohols are absorbed less rapidly and have a lower glycemic index, as does sucrose. The glycemic index of starch varies between near 1 to near 0 as a result of variable rates of hydrolysis, and that of nonstarch polysaccharides is 0. Foods that have a low glycemic index are considered to be more beneficial since they cause less fluctuation in insulin secretion. Resistant starch and nonstarch polysaccharides provide substrates for bacterial fermentation in the large intestine, and the resultant butyrate and other short chain fatty acids provide a significant source of fuel for intestinal enterocytes. There is some evidence that butyrate also has antiproliferative activity, and so provides protection against colorectal cancer.

Amylases Catalyze the Hydrolysis of Starch

The hydrolysis of starch is catalyzed by salivary and pancreatic amylases, which catalyze random hydrolysis of $\alpha(1\rightarrow4)$ glycoside bonds, yielding dextrins, then a mixture of glucose, maltose, and maltotriose and small branched dextrins (from the branchpoints in amylopectin).

Disaccharidases Are Brush Border Enzymes

The disaccharidases, maltase, sucrase-isomaltase (a bifunctional enzyme catalyzing hydrolysis of sucrose and isomaltose), lactase, and trehalase are located on the brush border of the intestinal mucosal cells, where the resultant monosaccharides and those arising from the diet are absorbed. Congenital deficiency of lactase occurs rarely in infants, leading to lactose intolerance and failure to thrive when fed on breast milk or normal infant formula. Congenital deficiency of sucrase-isomaltase occurs among the Inuit, leading to sucrose intolerance, with persistent diarrhea and failure to thrive when the diet contains sucrose.

In most mammals, and most human beings, lactase activity begins to fall after weaning, and is almost completely lost by late adolescence, leading to **lactose intolerance.** Lactose remains in the intestinal lumen, where it is a substrate for bacterial fermentation to lactate, resulting in abdominal discomfort and diarrhea after consumption of relatively large amounts. In two population groups, people of north European origin and nomadic tribes of sub-Saharan Africa and Arabia, lactase persists after weaning and into adult life. Marine mammals secrete a high-fat milk that contains no carbohydrate, and their pups lack lactase.

There Are Two Separate Mechanisms for the Absorption of Monosaccharides in the Small Intestine

Glucose and galactose are absorbed by a sodium-dependent process. They are carried by the same transport protein (SGLT 1), and compete with each other for intestinal absorption (Figure 43–1). Other monosaccharides are absorbed by carrier-mediated diffusion. Because they are not actively transported, fructose and sugar alcohols are only absorbed down their concentration gradient, and after a moderately high intake, some may remain in the intestinal lumen, acting as a substrate for bacterial fermentation.

DIGESTION & ABSORPTION OF LIPIDS

The major lipids in the diet are triacylglycerols and, to a lesser extent, phospholipids. These are hydrophobic molecules, and have to be hydrolyzed and emulsified to very small droplets (micelles) before they can be absorbed. The fat-soluble vitamins, A, D, E, and K, and a variety of other lipids (including cholesterol) are absorbed dissolved in the lipid micelles. Absorption of the fat-soluble vitamins is impaired on a very low fat diet.

Hydrolysis of triacylglycerols is initiated by lingual and gastric lipases, which attack the *sn*-3 ester bond forming 1,2-diacylglycerols and free fatty acids, aiding emulsification. Pancreatic lipase is secreted into the small intestine, and re-

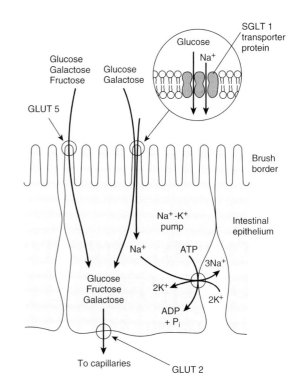

FIGURE 43–1 Transport of glucose, fructose, and galactose across the intestinal epithelium. The SGLT 1 transporter is coupled to the Na^+-K^+ pump, allowing glucose and galactose to be transported against their concentration gradients. The GLUT 5 Na^+-independent facilitative transporter allows fructose, as well as glucose and galactose, to be transported down their concentration gradients. Exit from the cell for all sugars is via the GLUT 2 facilitative transporter.

quires a further pancreatic protein, colipase, for activity. It is specific for the primary ester links—ie, positions 1 and 3 in triacylglycerols—resulting in 2-monoacylglycerols and free fatty acids as the major end products of luminal triacylglycerol digestion. Monoacylglycerols are poor substrates for hydrolysis, so that less than 25% of ingested triacylglycerol is completely hydrolyzed to glycerol and fatty acids (Figure 43–2). Bile salts, formed in the liver and secreted in the bile, permit emulsification of the products of lipid digestion into micelles together with phospholipids and cholesterol from the bile. Because the micelles are soluble, they allow the products of digestion, including the fat-soluble vitamins, to be transported through the aqueous environment of the intestinal lumen and permit close contact with the brush border of the mucosal cells, allowing uptake into the epithelium. The bile salts pass on to the ileum, where most are absorbed into the **enterohepatic circulation** (Chapter 26). Within the intestinal epithelium, 1-monoacylglycerols are hydrolyzed to fatty acids and glycerol and 2-monoacylglycerols are reacylated to triacylglycerols via the **monoacylglycerol pathway.** Glycerol released in the intestinal lumen is not reutilized but passes into the portal vein; glycerol released within the epithelium is reutilized for triacylglycerol synthesis via the normal phosphatidic acid pathway (Chapter 24). Long-chain fatty acids are esterified to yield to triacylglycerol in the mucosal cells and together with the other products

FIGURE 43–2 Digestion and absorption of triacylglycerols. The values given for percentage uptake may vary widely but indicate the relative importance of the three routes shown.

of lipid digestion, secreted as chylomicrons into the lymphatics, entering the bloodstream via the thoracic duct (Chapter 25). Short- and medium-chain fatty acids are mainly absorbed into the hepatic portal vein as free fatty acids.

Cholesterol is absorbed dissolved in lipid micelles, and is mainly esterified in the intestinal mucosa before being incorporated into chylomicrons. Unesterified cholesterol and other sterols are actively transported out of the mucosal cells into the intestinal lumen. Plant sterols and stanols (in which the B ring is saturated) compete with cholesterol for esterification, but are poor substrates. They therefore reduce the absorption of cholesterol, and act to lower serum cholesterol.

DIGESTION & ABSORPTION OF PROTEINS

Few bonds are accessible to the proteolytic enzymes that catalyze hydrolysis of peptide bonds, without prior denaturation of dietary proteins (by heat in cooking and by the action of gastric acid).

Several Groups of Enzymes Catalyze the Digestion of Proteins

There are two main classes of proteolytic digestive enzymes (**proteases**), with different specificities for the amino acids forming the peptide bond to be hydrolyzed. **Endopeptidases** hydrolyze peptide bonds between specific amino acids throughout the molecule. They are the first enzymes to act, yielding a larger number of smaller fragments. Pepsin in the gastric juice catalyzes hydrolysis of peptide bonds adjacent to aromatic and branched-chain amino acids and methionine. Trypsin, chymotrypsin, and elastase are secreted into the small intestine by the pancreas. Trypsin catalyzes hydrolysis of lysine and arginine esters, chymotrypsin esters of aromatic amino acids, and elastase esters of small neutral aliphatic amino acids. **Exopeptidases** catalyze the hydrolysis of peptide bonds, one at a time, from the ends of peptides. **Carboxypeptidase**s, secreted in the pancreatic juice, release amino acids from the free carboxyl terminal; **aminopeptidases,** secreted by the intestinal mucosal cells, release amino acids from the amino terminal. **Dipeptidases** and **tripeptidases** in the brush border of intestinal mucosal cells catalyze the hydrolysis of di- and tripeptides, which are not substrates for amino- and carboxypeptidases.

The proteases are secreted as inactive **zymogens;** the active site of the enzyme is masked by a small region of the peptide chain that is removed by hydrolysis of a specific peptide bond. Pepsinogen is activated to pepsin by gastric acid and by activated pepsin (autocatalysis). In the small intestine, trypsinogen, the precursor of trypsin, is activated by enteropeptidase, which is secreted by the duodenal epithelial cells; trypsin can then activate chymotrypsinogen to chymotrypsin, proelastase to elastase, procarboxypeptidase to carboxypeptidase, and proaminopeptidase to aminopeptidase.

Free Amino Acids & Small Peptides Are Absorbed by Different Mechanisms

The end product of the action of endopeptidases and exopeptidases is a mixture of free amino acids, di- and tripeptides, and oligopeptides, all of which are absorbed. Free amino acids are absorbed across the intestinal mucosa by sodium-dependent active transport. There are several different amino acid transporters, with specificity for the nature of the amino acid side-chain (large or small, neutral, acidic or basic). The various amino acids carried by any one transporter compete with each other for absorption and tissue uptake. Dipeptides and tripeptides enter the brush border of the intestinal mucosal cells, where they are hydrolyzed to free amino acids, which are then transported into the hepatic portal vein. Relatively large peptides may be absorbed intact, either by uptake into mucosal epithelial cells (transcellular) or by passing between epithelial cells (paracellular). Many such peptides are large enough to stimulate antibody formation—this is the basis of **allergic reactions** to foods.

DIGESTION & ABSORPTION OF VITAMINS & MINERALS

Vitamins and minerals are released from food during digestion, although this is not complete, and the availability of vitamins and minerals depends on the type of food and, especially for minerals, the presence of chelating compounds. The fat-soluble vitamins are absorbed in the lipid micelles that are the result of fat digestion; water-soluble vitamins and most mineral salts are absorbed from the small intestine either by active transport or by carrier-mediated diffusion followed by binding to intracellular proteins to achieve concentrative uptake. Vitamin B_{12} absorption requires a specific transport protein, intrinsic factor (Chapter 44); calcium absorption is dependent on vitamin D; zinc absorption probably requires a zinc-binding ligand secreted by the exocrine pancreas, and the absorption of iron is limited (see below).

Calcium Absorption Is Dependent on Vitamin D

In addition to its role in regulating calcium homeostasis, vitamin D is required for the intestinal absorption of calcium. Synthesis of the intracellular calcium-binding protein, **calbindin,** required for calcium absorption, is induced by vitamin D, which also affects the permeability of the mucosal cells to calcium, an effect that is rapid and independent of protein synthesis.

Phytic acid (inositol hexaphosphate) in cereals binds calcium in the intestinal lumen, preventing its absorption. Other minerals, including zinc, are also chelated by phytate. This is mainly a problem among people who consume large amounts of unleavened whole-wheat products; yeast contains an enzyme, **phytase**, that dephosphorylates phytate, so rendering it inactive. High concentrations of fatty acids in the intestinal lumen, as a result of impaired fat absorption, can also reduce calcium absorption, by forming insoluble calcium salts; a high intake of oxalate can sometimes cause deficiency, since calcium oxalate is insoluble.

Iron Absorption Is Limited and Strictly Controlled, but Enhanced by Vitamin C and Alcohol

Although iron deficiency is a common problem, about 10% of the population are genetically at risk of iron overload **(hemochromatosis),** and in order to reduce the risk of adverse effects of nonenzymic generation of free radicals by iron salts, absorption is strictly regulated. Inorganic iron is transported into the mucosal cell by a proton-linked divalent metal ion transporter, and accumulated intracellularly by binding to **ferritin.** Iron leaves the mucosal cell via a transport protein ferroportin, but only if there is free **transferrin** in plasma to bind to. Once transferrin is saturated with iron, any that has accumulated in the mucosal cells is lost when the cells are shed. Expression of the ferroportin gene (and possibly also that for the divalent metal ion transporter) is downregulated by hepcidin, a peptide secreted by the liver when body iron reserves are adequate. In response to hypoxia, anemia or hemorrhage, the synthesis of hepcidin is reduced, leading to increased synthesis of ferroportin and increased iron absorption (Figure 43–3). As a result of this mucosal barrier, only about 10% of dietary iron is absorbed, and only 1–5% from many plant foods (Chapter 50).

Inorganic iron is absorbed in the Fe^{2+} (reduced) state, hence the presence of reducing agents enhances absorption. The most effective compound is **vitamin C,** and while intakes of 40–80 mg of vitamin C/day are more than adequate to meet requirements, an intake of 25–50 mg per meal enhances iron absorption, especially when iron salts are used to treat iron deficiency anemia. Alcohol and fructose also enhance iron absorption. Heme iron from meat is absorbed separately, and is considerably more available than inorganic iron. However, the absorption of both inorganic and heme iron is impaired by calcium—a glass of milk with a meal significantly reduces iron availability.

ENERGY BALANCE: OVER- & UNDERNUTRITION

After the provision of water, the body's first requirement is for metabolic fuels—fats, carbohydrates, amino acids from proteins (Table 16–1). Food intake in excess of energy expenditure leads to **obesity,** while intake less than expenditure leads to emaciation and wasting, **marasmus,** and **kwashiorkor.** Both obesity and severe undernutrition are associated with increased mortality. The body mass index = weight (in kg)/height^2 (in m) is commonly used as a way of expressing relative obesity; a desirable range is between 20 and 25.

Energy Requirements Are Estimated by Measurement of Energy Expenditure

Energy expenditure can be determined directly by measuring heat output from the body, but is normally estimated indirectly from the consumption of oxygen. There is an energy expenditure of 20 kJ/liter of oxygen consumed, regardless of whether the fuel being metabolized is carbohydrate, fat, or protein (Table 16–1).

Measurement of the ratio of the volume of carbon dioxide produced: volume of oxygen consumed **(respiratory quotient, RQ)** is an indication of the mixture of metabolic fuels being oxidized (Table 16–1).

A more recent technique permits estimation of total energy expenditure over a period of 1–2 weeks, using dual isotopically labeled water, $^2H_2^{18}O$. ^2H is lost from the body only in water, while ^{18}O is lost in both water and carbon dioxide; the difference in the rate of loss of the two labels permits estimation of total carbon dioxide production, and hence oxygen consumption and energy expenditure.

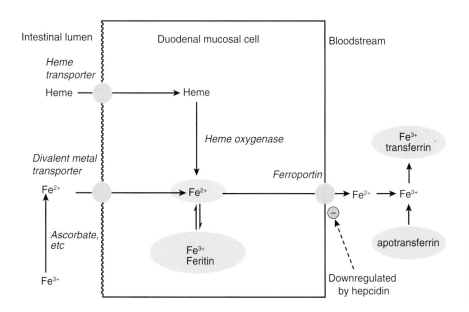

FIGURE 43–3 Absorption of iron. Hepcidin secreted by the liver downregulates synthesis of ferroportin and limits iron absorption.

Basal metabolic rate (BMR) is the energy expenditure by the body when at rest, but not asleep, under controlled conditions of thermal neutrality, measured about 12 h after the last meal, and depends on weight, age, and gender. Total energy expenditure depends on the basal metabolic rate, the energy required for physical activity and the energy cost of synthesizing reserves in the fed state. It is therefore possible to calculate an individual's energy requirement from body weight, age, gender, and level of physical activity. Body weight affects BMR because there is a greater amount of active tissue in a larger body. The decrease in BMR with increasing age, even when body weight remains constant, is the result of muscle tissue replacement by adipose tissue, which is metabolically less active. Similarly, women have a significantly lower BMR than do men of the same body weight because women's bodies contain proportionally more adipose tissue.

Energy Requirements Increase with Activity

The most useful way of expressing the energy cost of physical activities is as a multiple of BMR. Sedentary activities use only about 1.1–1.2 × BMR. By contrast, vigorous exertion, such as climbing stairs, cross-country walking uphill, etc., may use 6–8 × BMR.

Ten Percent of the Energy Yield of a Meal May Be Expended in Forming Reserves

There is a considerable increase in metabolic rate after a meal (**diet-induced thermogenesis**). A small part of this is the energy cost of secreting digestive enzymes and of active transport of the products of digestion; the major part is the result of synthesizing reserves of glycogen, triacylglycerol, and protein.

There Are Two Extreme Forms of Undernutrition

Marasmus can occur in both adults and children, and occurs in vulnerable groups of all populations. **Kwashiorkor** affects only children, and has been reported only in developing countries. The distinguishing feature of kwashiorkor is that there is fluid retention, leading to edema, and fatty infiltration of the liver. Marasmus is a state of extreme emaciation; it is the outcome of prolonged negative energy balance. Not only have the body's fat reserves been exhausted, but there is wastage of muscle as well, and as the condition progresses there is loss of protein from the heart, liver, and kidneys. The amino acids released by the catabolism of tissue proteins are used as a source of metabolic fuel and as substrates for gluconeogenesis to maintain a supply of glucose for the brain and red blood cells (Chapter 20). As a result of the reduced synthesis of proteins, there is impaired immune response and more risk from infections. Impairment of cell proliferation in the intestinal mucosa

occurs, resulting in reduction in surface area of the intestinal mucosa, and reduction in absorption of such nutrients as are available.

Patients with Advanced Cancer and AIDS Are Malnourished

Patients with advanced cancer, HIV infection and AIDS, and a number of other chronic diseases are frequently undernourished, a condition called **cachexia.** Physically, they show all the signs of marasmus, but there is considerably more loss of body protein than occurs in starvation. The secretion of cytokines in response to infection and cancer increases the catabolism of tissue protein by the ATP-dependent ubiquitin-proteasome pathway (Figure 45–8), so increasing energy expenditure. This differs from marasmus, in which protein synthesis is reduced, but catabolism in unaffected. Patients are **hypermetabolic,** ie, a considerable increase in basal metabolic rate. In addition to activation of the ubiquitin-proteasome pathway of protein catabolism, three other factors are involved. Many tumors metabolize glucose anaerobically to release lactate. This is then used for gluconeogenesis in the liver, which is energy consuming with a net cost of 6 ATP for each mol of glucose cycled (see Figure 20–4). There is increased stimulation of **uncoupling proteins** by **cytokines** leading to thermogenesis and increased oxidation of metabolic fuels. **Futile cycling of lipids** occurs because hormone sensitive lipase is activated by a proteoglycan secreted by tumors resulting in liberation of fatty acids from adipose tissue and ATP-expensive re-esterification to triacylglycerols in the liver, which are exported in VLDL.

Kwashiorkor Affects Undernourished Children

In addition to the wasting of muscle tissue, loss of intestinal mucosa and impaired immune responses seen in marasmus, children with **kwashiorkor** show a number of characteristic features. The defining characteristic is **edema,** associated with a decreased concentration of plasma proteins. In addition, there is enlargement of the liver as a result of accumulation of fat. It was formerly believed that the cause of kwashiorkor was a lack of protein, with a more or less adequate energy intake, however, analysis of the diets of affected children shows that this is not so. Children with kwashiorkor are less stunted than those with marasmus and the edema begins to improve early in treatment, when the child is still receiving a low protein diet.

Very commonly, an infection precipitates kwashiorkor. Superimposed on general food deficiency, there is probably a deficiency of the antioxidant nutrients such as zinc, copper, carotene, and vitamins C and E. The **respiratory burst** in response to infection leads to the production of oxygen and halogen **free radicals** as part of the cytotoxic action of stimulated macrophages. This added oxidant stress may well trigger the development of kwashiorkor (see Chapter 54).

PROTEIN & AMINO ACID REQUIREMENTS

Protein Requirements Can Be Determined by Measuring Nitrogen Balance

The state of protein nutrition can be determined by measuring the dietary intake and output of nitrogenous compounds from the body. Although nucleic acids also contain nitrogen, protein is the major dietary source of nitrogen and measurement of total nitrogen intake gives a good estimate of protein intake (mg N × 6.25 = mg protein, as N is 16% of most proteins). The output of N from the body is mainly in urea and smaller quantities of other compounds in urine, undigested protein in feces; significant amounts may also be lost in sweat and shed skin. The difference between intake and output of nitrogenous compounds is known as **nitrogen balance.** Three states can be defined. In a healthy adult, nitrogen balance is in **equilibrium,** when intake equals output, and there is no change in the total body content of protein. In a growing child, a pregnant woman, or a person in recovery from protein loss, the excretion of nitrogenous compounds is less than the dietary intake and there is net retention of nitrogen in the body as protein—**positive nitrogen balance.** In response to trauma or infection, or if the intake of protein is inadequate to meet requirements, there is net loss of protein nitrogen from the body—**negative nitrogen balance.** Except when replacing protein losses, nitrogen equilibrium can be maintained at any level of protein intake above requirements. A high intake of protein does not lead to positive nitrogen balance; although it increases the rate of protein synthesis, it also increases the rate of protein catabolism, so that nitrogen equilibrium is maintained, albeit with a higher rate of protein turnover.

The continual catabolism of tissue proteins creates the requirement for dietary protein, even in an adult who is not growing; although some of the amino acids released can be re-utilized, much is used for gluconeogenesis in the fasting state. Nitrogen balance studies show that the average daily requirement is 0.66 g of protein/kg body weight (0.825 allowing for individual variation), approximately 55 g/day, or 0.825% of energy intake. Average intakes of protein in developed countries are of the order of 80–100 g/day, ie, 14–15% of energy intake. Because growing children are increasing the protein in the body, they have a proportionally greater requirement than adults and should be in positive nitrogen balance. Even so, the need is relatively small compared with the requirement for protein turnover. In some countries, protein intake is inadequate to meet these requirements, resulting in stunting of growth.

There Is a Loss of Body Protein in Response to Trauma & Infection

One of the metabolic reactions to a major trauma, such as a burn, a broken limb, or surgery, is an increase in the net catabolism of tissue proteins, both in response to cytokines and glucocorticoid hormones, and as a result of excessive utilization of threonine and cysteine in the synthesis of **acute-phase proteins**. As much as 6–7% of the total body protein may be lost over 10 days. Prolonged bed rest results in considerable loss of protein because of atrophy of muscles. Protein catabolism may be increased in response to cytokines, and without the stimulus of exercise it is not completely replaced. Lost protein is replaced during **convalescence,** when there is positive nitrogen balance. A normal diet is adequate to permit this replacement.

The Requirement Is Not Just for Protein, but for Specific Amino Acids

Not all proteins are nutritionally equivalent. More of some is needed to maintain nitrogen balance than others because different proteins contain different amounts of the various amino acids. The body's requirement is for amino acids in the correct proportions to replace tissue proteins. The amino acids can be divided into two groups: **essential** and **nonessential**. There are nine essential or indispensable amino acids, which cannot be synthesized in the body: histidine, isoleucine, leucine, lysine, methionine, phenylalanine, threonine, tryptophan, and valine. If one of these is lacking or inadequate, then regardless of the total intake of protein, it will not be possible to maintain nitrogen balance, since there will not be enough of that amino acid for protein synthesis.

Two amino acids, cysteine and tyrosine, can be synthesized in the body, but only from essential amino acid precursors—cysteine from methionine and tyrosine from phenylalanine. The dietary intakes of cysteine and tyrosine thus affect the requirements for methionine and phenylalanine. The remaining 11 amino acids in proteins are considered to be nonessential or dispensable, since they can be synthesized as long as there is enough total protein in the diet. If one of these amino acids is omitted from the diet, nitrogen balance can still be maintained. However, only three amino acids, alanine, aspartate, and glutamate, can be considered to be truly dispensable; they are synthesized from common metabolic intermediates (pyruvate, oxaloacetate, and ketoglutarate, respectively). The remaining amino acids are considered as nonessential, but under some circumstances the requirement may outstrip the capacity for synthesis.

SUMMARY

- Digestion involves hydrolyzing food molecules into smaller molecules for absorption through the gastrointestinal epithelium. Polysaccharides are absorbed as monosaccharides, triacylglycerols as 2-monoacylglycerols, fatty acids and glycerol, and proteins as amino acids.

- Digestive disorders arise as a result of (1) enzyme deficiency, eg, lactase and sucrase; (2) malabsorption, eg, of glucose and galactose as a result of defects in the Na^+-glucose cotransporter (SGLT 1); (3) absorption of unhydrolyzed polypeptides leading to immune responses, eg, as in celiac disease; and (4) precipitation of cholesterol from bile as gallstones.

▪ In addition to water, the diet must provide metabolic fuels (carbohydrate and fat) for body growth and activity, protein for synthesis of tissue proteins, fiber for roughage, minerals for specific metabolic functions, certain polyunsaturated fatty acids of the n-3 and n-6 families, and vitamins, organic compounds needed in small amounts for other essential functions.

▪ Twenty different amino acids are required for protein synthesis, of which nine are essential in the human diet. The quantity of protein required is affected by protein quality, energy intake, and physical activity.

▪ Undernutrition occurs in two extreme forms: marasmus, in adults and children, and kwashiorkor in children. Overnutrition leads to excess energy intake, and is associated with diseases such as obesity, noninsulin-dependent diabetes mellitus, atherosclerosis, cancer, and hypertension.

REFERENCES

Bender DA, Bender AE: *Nutrition: A Reference Handbook.* Oxford University Press, 1997.

Fuller MF, Garllick PJ: Human amino acid requirements: can the controversy be resolved? Ann Rev Nutr 1994;14:217.

Geissler C, Powers HJ: *Human Nutrition.* 11th ed. Elsevier, 2005.

Institute of Medicine: *Dietary Reference Intakes for Energy, Carbohydrate, Fiber, Fat, Fatty Acids, Cholesterol, Protein, and Amino Acids (Macronutrients).* National Academies Press, 2002.

Pencharz PB, Ball RO: Different approaches to define individual amino acid requirements. Ann Rev Nutr 2003;23:101.

Royal College of Physicians: *Nutrition and Patients—A Doctor's Responsibility.* Royal College of Physicians, 2002.

Swallow DM: Genetic influences on carbohydrate digestion. Nutr Res Rev 2003;16:37

World Health Organization Technical Report Series 894: *Obesity—Preventing and Managing the Global Epidemic.* WHO, 2000.

World Health Organization Technical Report Series 916: *Diet and the Prevention of Chronic Diseases.* WHO, 2003.

Micronutrients: Vitamins & Minerals

44

David A. Bender, PhD

BIOMEDICAL IMPORTANCE

Vitamins are a group of organic nutrients, required in small quantities for a variety of biochemical functions that, generally, cannot be synthesized by the body and must therefore be supplied in the diet.

The lipid-soluble vitamins are hydrophobic compounds that can be absorbed efficiently only when there is normal fat absorption. Like other lipids, they are transported in the blood in lipoproteins or attached to specific binding proteins. They have diverse functions—eg, vitamin A, vision and cell differentiation; vitamin D, calcium and phosphate metabolism, and cell differentiation; vitamin E, anti-oxidant; and vitamin K, blood clotting. As well as dietary inadequacy, conditions affecting the digestion and absorption of the lipid-soluble vitamins, such as steatorrhea and disorders of the biliary system, can all lead to deficiency syndromes, including night blindness and xerophthalmia (vitamin A); rickets in young children and osteomalacia in adults (vitamin D); neurologic disorders and hemolytic anemia of the newborn (vitamin E); and hemorrhagic disease of the newborn (vitamin K). Toxicity can result from excessive intake of vitamins A and D. Vitamin A and the carotenes (many of which are precursors of vitamin A), and vitamin E are antioxidants (Chapter 45) and have possible roles in prevention of atherosclerosis and cancer.

The water-soluble vitamins are composed of the B vitamins and vitamin C; they function mainly as enzyme cofactors. Folic acid acts as a carrier of one-carbon units. Deficiency of a single vitamin of the B complex is rare, since poor diets are most often associated with **multiple deficiency states.** Nevertheless, specific syndromes are characteristic of deficiencies of individual vitamins, eg, beriberi (thiamin); cheilosis, glossitis, seborrhea (riboflavin); pellagra (niacin); megaloblastic anemia, methylmalonic aciduria, and pernicious anemia (vitamin B_{12}); megaloblastic anemia (folic acid); and scurvy (vitamin C).

Inorganic mineral elements that have a function in the body must be provided in the diet. When the intake is insufficient, deficiency signs may arise, eg, anemia (iron), and cretinism and goiter (iodine). Excessive intakes may be toxic.

The Determination of Micronutrient Requirements Depends on the Criteria of Adequacy Chosen

For any nutrient, there is a range of intakes between that which is clearly inadequate, leading to **clinical deficiency disease,** and that which is so much in excess of the body's metabolic capacity that there may be signs of **toxicity.** Between these two extremes is a level of intake that is adequate for normal health and the maintenance of metabolic integrity. Individuals do not all have the same requirement for nutrients, even when calculated on the basis of body size or energy expenditure. There is a range of individual requirements of up to 25% around the mean. Therefore, in order to assess the adequacy of diets, it is necessary to set a reference level of intake high enough to ensure that no one either suffers from deficiency or is at risk of toxicity. If it is assumed that individual requirements are distributed in a statistically normal fashion around the observed mean requirement, then a range of $\pm 2 \times$ the standard deviation (SD) around the mean includes the requirements of 95% of the population. Reference or recommended intakes are therefore set at the average requirement plus $2 \times$ SD, and so meet or exceed the requirements of 97.5% of the population.

THE VITAMINS ARE A DISPARATE GROUP OF COMPOUNDS WITH A VARIETY OF METABOLIC FUNCTIONS

A vitamin is defined as an organic compound that is required in the diet in small amounts for the maintenance of normal metabolic integrity. Deficiency causes a specific disease, which is cured or prevented only by restoring the vitamin to the diet (Table 44–1). However, **vitamin D,** which is formed in the skin from 7-dehydrocholesterol on exposure to sunlight, and **niacin,** which can be formed from the essential amino acid tryptophan, do not strictly comply with this definition.

TABLE 44–1 The Vitamins

Vitamin		Functions	Deficiency disease
Lipid-Soluable			
A	Retinol, β-carotene	Visual pigments in the retina; regulation of gene expression and cell differentiation (β-carotene is an antioxidant)	Night blindness, xerophthalmia; keratinization of skin
D	Calciferol	Maintenance of calcium balance; enhances intestinal absorption of Ca^{2+} and mobilizes bone mineral; regulation of gene expression and cell differentiation	Rickets = poor mineralization of bone; osteomalacia = bone demineralization
E	Tocopherols, tocotrienols	Antioxidant, especially in cell membranes; roles in cell signaling	Extremely rare—serious neurologic dysfunction
K	Phylloquinone: menaquinones	Coenzyme in formation of γ-carboxy-glutamate in enzymes of blood clotting and bone matrix	Impaired blood clotting, hemorrhagic disease
Water-Soluable			
B_1	Thiamin	Coenzyme in pyruvate and α-ketoglutarate dehydrogenases, and transketolase; regulates Cl^- channel in nerve conduction	Peripheral nerve damage (beriberi) or central nervous system lesions (Wernicke-Korsakoff syndrome)
B_2	Riboflavin	Coenzyme in oxidation and reduction reactions (FAD and FMN); prosthetic group of flavoproteins	Lesions of corner of mouth, lips, and tongue, seborrheic dermatitis
Niacin	Nicotinic acid, nicotinamide	Coenzyme in oxidation and reduction reactions, functional part of NAD and NADP; role in intracellular calcium regulation and cell signaling	Pellagra—photosensitive dermatitis, depressive psychosis
B_6	Pyridoxine, pyridoxal, pyridoxamine	Coenzyme in transamination and decarboxylation of amino acids and glycogen phosphorylase; modulation of steroid hormone action	Disorders of amino acid metabolism, convulsions
	Folic acid	Coenzyme in transfer of one-carbon fragments	Megaloblastic anemia
B_{12}	Cobalamin	Coenzyme in transfer of one-carbon fragments and metabolism of folic acid	Pernicious anemia = megaloblastic anemia with degeneration of the spinal cord
	Pantothenic acid	Functional part of CoA and acyl carrier protein: fatty acid synthesis and metabolism	Peripheral nerve damage (nutritional melalgia or "burning foot syndrome")
H	Biotin	Coenzyme in carboxylation reactions in gluconeogenesis and fatty acid synthesis; role in regulation of cell cycle	Impaired fat and carbohydrate metabolism, dermatitis
C	Ascorbic acid	Coenzyme in hydroxylation of proline and lysine in collagen synthesis; antioxidant; enhances absorption of iron	Scurvy—impaired wound healing, loss of dental cement, subcutaneous hemorrhage

LIPID-SOLUBLE VITAMINS

TWO GROUPS OF COMPOUNDS HAVE VITAMIN A ACTIVITY

Retinoids comprise **retinol, retinaldehyde,** and **retinoic acid** (prefovrmed vitamin A, found only in foods of animal origin); carotenoids, found in plants, are composed of carotenes and related compounds; many are precursors of vitamin A, as they can be cleaved to yield retinaldehyde, then retinol and retinoic acid (Figure 44–1). The α-, β-, and γ-carotenes and cryptoxanthin are quantitatively the most important provitamin A carotenoids. Although it would appear that one molecule of β-carotene should yield two of retinol, this is not so in practice; 6 μg of β-carotene is equivalent to 1 μg of preformed retinol. The total amount of vitamin A in foods is therefore expressed as micrograms of retinol equivalents. β-Carotene and other provitamin A carotenoids are cleaved in the intestinal mucosa by carotene dioxygenase, yielding retinaldehyde, which is reduced to retinol, esterified and secreted in chylomicrons together with esters formed from dietary retinol. The intestinal activity of carotene dioxygenase is low, so that a relatively large proportion of ingested β-carotene may appear in the circulation unchanged. While the principal site of carotene dioxy-

FIGURE 44–1 β-Carotene and the major vitamin A vitamers. Asterisk shows the site of cleavage of β-carotene by carotene dioxygenase, to yield retinaldehyde.

genase attack is the central bond of β-carotene, asymmetric cleavage may also occur, leading to the formation of 8′-, 10′-, and 12′-apo-carotenals, which are oxidized to retinoic acid, but cannot be used as sources of retinol or retinaldehyde.

Vitamin A Has a Function in Vision

In the retina, retinaldehyde functions as the prosthetic group of the light-sensitive opsin proteins, forming **rhodopsin** (in rods) and **iodopsin** (in cones). Any one cone cell contains only one type of opsin, and is sensitive to only one color. In the pigment epithelium of the retina, all-*trans*-retinol is isomerized to 11-*cis*-retinol and oxidized to 11-*cis*-retinaldehyde. This reacts with a lysine residue in opsin, forming the holo-protein rhodopsin. As shown in Figure 44–2, the absorption of light by rhodopsin causes isomerization of the retinaldehyde from 11-*cis* to all-*trans*, and a conformational change in opsin. This results in the release of retinaldehyde from the protein, and the initiation of a nerve impulse. The formation of the initial excited form of rhodopsin, bathorhodopsin, occurs within picoseconds of illumination. There is then a series of conformational changes leading to the formation of metarhodopsin II, which initiates a guanine nucleotide amplification cascade and then a nerve impulse. The final step is hydrolysis to release all-*trans*-retinaldehyde and opsin. The key to initiation of the visual cycle is the availability of 11-*cis*-retinaldehyde, and hence vitamin A. In deficiency, both the time taken to adapt to darkness and the ability to see in poor light are impaired.

Retinoic Acid Has a Role in the Regulation of Gene Expression and Tissue Differentiation

A major role of vitamin A is in the control of cell differentiation and turnover. All-*trans*-retinoic acid and 9-*cis*-retinoic acid (Figure 44–1) regulate growth, development, and tissue differ-

FIGURE 44–2 The role of retinaldehyde in the visual cycle.

entiation; they have different actions in different tissues. Like the thyroid and steroid hormones and vitamin D, retinoic acid binds to nuclear receptors that bind to response elements of DNA and regulate the transcription of specific genes. There are two families of nuclear retinoid receptors: the retinoic acid receptors (RAR) bind all-*trans*-retinoic acid or 9-*cis*-retinoic acid, and the retinoid X receptors (RXR) bind 9-*cis*-retinoic acid. Retinoid X receptors also form dimers with vitamin D, thyroid, and other a nuclear acting hormone receptors. Deficiency of vitamin A impairs vitamin D function because of lack of 9-*cis*-retinoic acid to form receptor dimers, while excessive vitamin A also impairs vitamin D function, because of formation of RXR-homodimers, meaning that there are not enough RXR available to form heterodimers with the vitamin D receptor.

Vitamin A Deficiency Is a Major Public Health Problem Worldwide

Vitamin A deficiency is the most important preventable cause of blindness. The earliest sign of deficiency is a loss of sensitivity to green light, followed by impairment to adapt to dim light, followed by night blindness. More prolonged deficiency leads to **xerophthalmia:** keratinization of the cornea and blindness. Vitamin A also has an important role in differentiation of immune system cells, and even mild deficiency leads to increased susceptibility to infectious diseases. Also, the synthesis of retinol binding protein is reduced in response to infection (it is a negative **acute phase protein**), decreasing the circulating concentration of the vitamin, and further impairing immune responses.

Vitamin A Is Toxic in Excess

There is only a limited capacity to metabolize vitamin A, and excessive intakes lead to accumulation beyond the capacity of binding proteins, so that unbound vitamin A causes tissue damage. Symptoms of toxicity affect the central nervous system (headache, nausea, ataxia, and anorexia, all associated with increased cerebrospinal fluid pressure); the liver (hepatomegaly with histologic changes and hyperlipidemia); calcium homeostasis (thickening of the long bones, hypercalcemia, and calcification of soft tissues); and the skin (excessive dryness, desquamation, and alopecia).

VITAMIN D IS REALLY A HORMONE

Vitamin D is not strictly a vitamin, since it can be synthesized in the skin, and under most conditions that is the major source of the vitamin. Only when sunlight exposure is inadequate is a dietary source required. Its main function is in the regulation of calcium absorption and homeostasis; most of its actions are mediated by way of nuclear receptors that regulate gene expression. It also has a role in regulating cell proliferation and differentiation. There is evidence that intakes considerably higher than are required to maintain calcium homeostasis reduce the risk of insulin resistance, obesity and the metabolic syndrome, as well as various cancers. Deficiency, leading to rickets in children and osteomalacia in adults, continues to be a problem in northern latitudes, where sunlight exposure is inadequate.

Vitamin D Is Synthesized in the Skin

7-Dehydrocholesterol (an intermediate in the synthesis of cholesterol that accumulates in the skin) undergoes a nonenzymic

reaction on exposure to ultraviolet light, yielding previtamin D (Figure 44–3). This undergoes a further reaction over a period of hours to form cholecalciferol, which is absorbed into the bloodstream. In temperate climates, the plasma concentration of vitamin D is highest at the end of summer and lowest at the end of winter. Beyond latitudes about 40° north or south there is very little ultraviolet radiation of the appropriate wavelength in winter.

Vitamin D Is Metabolized to the Active Metabolite, Calcitriol, in Liver & Kidney

Cholecalciferol, either synthesized in the skin or from food, undergoes two hydroxylations to yield the active metabolite, 1,25-dihydroxyvitamin D or calcitriol (Figure 44–4). Ergocalciferol from fortified foods undergoes similar hydroxylation to yield ercalcitriol. In the liver, cholecalciferol is hydroxylated to form the 25-hydroxy-derivative, calcidiol. This is released into the circulation bound to a vitamin D binding globulin, which is the main storage form of the vitamin. In the kidney, calcidiol undergoes either 1-hydroxylation to yield the active metabolite 1,25-dihydroxy-vitamin D (calcitriol), or 24-hydroxylation to yield a probably inactive metabolite, 24,25-dihydroxyvitamin D (24-hydroxycalcidiol).

Vitamin D Metabolism Is Both Regulated by and Regulates Calcium Homeostasis

The main function of vitamin D is in the control of calcium homeostasis, and in turn, vitamin D metabolism is regulated by factors that respond to plasma concentrations of calcium and phosphate. Calcitriol acts to reduce its own synthesis by inducing the 24-hydroxylase and repressing the 1-hydroxylase in the kidney. The principal function of vitamin D is to maintain the plasma calcium concentration. Calcitriol achieves this in three ways: it increases intestinal absorption of calcium; it reduces excretion of calcium (by stimulating resorption in the distal renal tubules); and it mobilizes bone mineral. In addition, calcitriol is involved in insulin secretion, synthesis and secretion of parathyroid and thyroid hormones, inhibition of production of interleukin by activated T-lymphocytes and of immunoglobulin by activated B-lymphocytes, differentiation of monocyte precursor cells, and modulation of cell proliferation. In most of these actions, it acts like a steroid hormone, binding to nuclear receptors and enhancing gene expression, although it also has rapid effects on calcium transporters in the intestinal mucosa. For further details of the role of calcitriol in calcium homeostasis, see Chapter 47.

FIGURE 44–3 The synthesis of vitamin D in the skin.

FIGURE 44–4 Metabolism of vitamin D.

Vitamin D Deficiency Affects Children & Adults

In the vitamin D deficiency disease **rickets,** the bones of children are undermineralized as a result of poor absorption of calcium. Similar problems occur as a result of deficiency during the adolescent growth spurt. **Osteomalacia** in adults results from the demineralization of bone, especially in women who have little exposure to sunlight, especially after several pregnancies. Although vitamin D is essential for prevention and treatment of osteomalacia in the elderly, there is little evidence that it is beneficial in treating **osteoporosis.**

Vitamin D Is Toxic in Excess

Some infants are sensitive to intakes of vitamin D as low as 50 μg/day, resulting in an elevated plasma concentration of calcium. This can lead to contraction of blood vessels, high blood pressure, and **calcinosis**—the calcification of soft tissues. Although excess dietary vitamin D is toxic, excessive exposure to sunlight does not lead to vitamin D poisoning, because there is a limited capacity to form the precursor, 7-dehydrocholesterol, and prolonged exposure of previtamin D to sunlight leads to formation of inactive compounds.

VITAMIN E DOES NOT HAVE A PRECISELY DEFINED METABOLIC FUNCTION

No unequivocal unique function for vitamin E has been defined. It acts as a lipid-soluble antioxidant in cell membranes, where many of its functions can be provided by synthetic antioxidants, and is important in maintaining the fluidity of cell membranes. It also has a (relatively poorly defined) role in cell signaling. Vitamin E is the generic descriptor for two families

FIGURE 44–5 Vitamin E vitamers. In α-tocopherol and tocotrienol R_1, R_2, and R_3 are all —CH_3 groups. In the β-vitamers R_2 is H, in the γ-vitamers R_1 is H, and in the Δ-vitamers R_1 and R_2 are both H.

of compounds, the **tocopherols** and the **tocotrienols** (Figure 44–5). The different vitamers have different biologic potency; the most active is D-α-tocopherol, and it is usual to express vitamin E intake in terms of milligrams D-α-tocopherol equivalents. Synthetic DL-α-tocopherol does not have the same biologic potency as the naturally occurring compound.

Vitamin E Is the Major Lipid-Soluble Antioxidant in Cell Membranes and Plasma Lipoproteins

The main function of vitamin E is as a chain-breaking, free-radical-trapping antioxidant in cell membranes and plasma lipoproteins by reacting with the lipid peroxide radicals formed by peroxidation of polyunsaturated fatty acids (Chapter 45). The tocopheroxyl radical product is relatively unreactive, and ultimately forms nonradical compounds. Commonly, the tocopheroxyl radical is reduced back to tocopherol by reaction with vitamin C from plasma (Figure 44–6). The resultant monodehydroascorbate radical then undergoes enzymic or nonenzymic reaction to yield ascorbate and dehydroascorbate, neither of which is a radical.

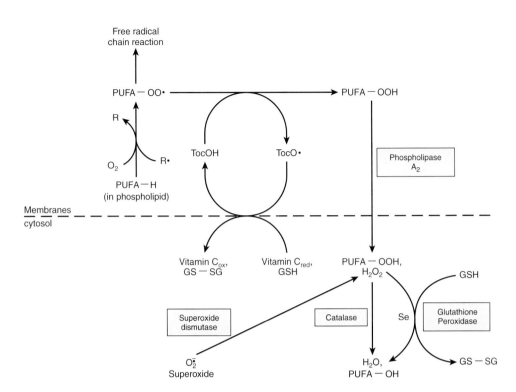

FIGURE 44–6 Interaction between antioxidants in the lipid phase (cell membranes) and the aqueous phase (cytosol). (R•, free radical; PUFA-OO•, peroxyl radical of polyunsaturated fatty acid in membrane phospholipid; PUFA-OOH, hydroxyperoxy polyunsaturated fatty acid in membrane phospholipid, released into the cytosol as hydroxyperoxy polyunsaturated fatty acid by the action of phospholipase A₂; PUFA-OH, hydroxy polyunsaturated fatty acid; Toc-OH vitamin E [α-tocopherol]; TocO•, tocopheroxyl radical; Se, selenium; SSH, reduced glutathione; GS-SG, oxidized glutathione, which is reduced to GSH after reaction with NADPH, catalyzed by glutathione reductase; PUFA-H, polyunsaturated fatty acid.)

Vitamin E Deficiency

In experimental animals, vitamin E deficiency results in resorption of fetuses and testicular atrophy. Dietary deficiency of vitamin E in humans is unknown, although patients with severe fat malabsorption, cystic fibrosis, and some forms of chronic liver disease suffer deficiency because they are unable to absorb the vitamin or transport it, exhibiting nerve and muscle membrane damage. Premature infants are born with inadequate reserves of the vitamin. The erythrocyte membranes are abnormally fragile as a result of peroxidation, leading to hemolytic anemia.

VITAMIN K IS REQUIRED FOR SYNTHESIS OF BLOOD CLOTTING PROTEINS

Vitamin K was discovered as a result of investigations into the cause of a bleeding disorder, hemorrhagic (sweet clover) disease of cattle and of chickens fed on a fat-free diet. The missing factor in the diet of the chickens was vitamin K, while the cattle feed contained **dicumarol,** an antagonist of the vitamin. Antagonists of vitamin K are used to reduce blood coagulation in patients at risk of thrombosis; the most widely used is **warfarin.**

Three compounds have the biological activity of vitamin K (Figure 44–7): **phylloquinone,** the normal dietary source, found in green vegetables; **menaquinones,** synthesized by intestinal bacteria, with differing lengths of side-chain; and **menadione** and menadiol diacetate, synthetic compounds that can be metabolized to phylloquinone. Menaquinones are absorbed to some extent, but it is not clear to what extent they are biologically active as it is possible to induce signs of vitamin K deficiency simply by feeding a phylloquinone-deficient diet, without inhibiting intestinal bacterial action.

FIGURE 44–7 The vitamin K vitamers. Menadiol (or menadione) and menadiol diacetate are synthetic compounds that are converted to menaquinone in the liver.

Vitamin K Is the Coenzyme for Carboxylation of Glutamate in Postsynthetic Modification of Calcium-Binding Proteins

Vitamin K is the cofactor for the carboxylation of glutamate residues in the post-synthetic modification of proteins to form the unusual amino acid γ-carboxyglutamate (Gla) (Figure 44–8). Initially, vitamin K hydroquinone is oxidized to the epoxide, which activates a glutamate residue in the protein substrate to a carbanion, which reacts nonenzymically with carbon dioxide to form γ-carboxyglutamate. Vitamin K epoxide is reduced to the quinone by a warfarin-sensitive reductase, and the quinone is reduced to the active hydroquinone by either the same warfarin-sensitive reductase or a warfarin-insensitive quinone reductase. In the presence of warfarin, vitamin K epoxide cannot be reduced, but accumulates and is excreted. If enough vitamin K (as the quinone) is provided in the diet, it can be reduced to the active hydroquinone by the warfarin-insensitive enzyme, and carboxylation can continue, with stoichiometric utilization of vitamin K and excretion of the epoxide. A high dose of vitamin K is the antidote to an overdose of warfarin.

Prothrombin and several other proteins of the blood clotting system (Factors VII, IX, and X, and proteins C and S, Chapter 50) each contain 4–6 γ-carboxyglutamate residues. γ-Carboxyglutamate chelates calcium ions, and so permits the binding of the blood clotting proteins to membranes. In vita-min K deficiency, or in the presence of warfarin, an abnormal precursor of prothrombin (preprothrombin) containing little or no γ-carboxyglutamate, and incapable of chelating calcium, is released into the circulation.

Vitamin K Is Also Important in Synthesis of Bone Calcium-Binding Proteins

Treatment of pregnant women with warfarin can lead to fetal bone abnormalities (fetal warfarin syndrome). Two proteins that contain γ-carboxyglutamate are present in bone, osteocalcin, and bone matrix Gla protein. Osteocalcin also contains hydroxyproline, so its synthesis is dependent on both vitamins K and C; in addition, its synthesis is induced by vitamin D. The release into the circulation of osteocalcin provides an index of vitamin D status.

WATER-SOLUBLE VITAMINS

VITAMIN B₁ (THIAMIN) HAS A KEY ROLE IN CARBOHYDRATE METABOLISM

Thiamin has a central role in energy-yielding metabolism, and especially the metabolism of carbohydrates (Figure 44–9). **Thiamin diphosphate** is the coenzyme for three multi-enzyme complexes that catalyze oxidative decarboxylation reactions: pyruvate dehydrogenase in carbohydrate metabolism (Chapter 17); α-ketoglutarate dehydrogenase in the citric acid cycle (Chapter 17); and the branched-chain keto-acid dehydrogenase involved in the metabolism of leucine, isoleucine, and valine (Chapter 29). In each case, the thiamin diphosphate provides a reactive carbon on the thiazole moiety that forms a carbanion, which then adds to the carbonyl group, eg, pyruvate. The addition compound is then decarboxylated, eliminating CO_2. Thiamin diphosphate is also the coenzyme for transketolase, in the pentose phosphate pathway (Chapter 21).

Thiamin triphosphate has a role in nerve conduction; it phosphorylates, and so activates, a chloride channel in the nerve membrane.

Thiamin Deficiency Affects the Nervous System & the Heart

Thiamin deficiency can result in three distinct syndromes: a chronic peripheral neuritis, **beriberi,** which may or may not be associated with **heart failure** and **edema;** acute pernicious (fulminating) beriberi (shoshin beriberi), in which heart failure and metabolic abnormalities predominate, without peripheral neuritis; and **Wernicke encephalopathy** with **Korsakoff psychosis,** which is associated especially with alcohol and narcotic abuse. The role of thiamin diphosphate in pyruvate dehydrogenase means that in deficiency there is impaired conversion of pyruvate to acetyl CoA. In subjects on a relatively high carbohydrate diet, this results in increased plasma

FIGURE 44–8 The role of vitamin K in the synthesis of γ-carboxyglutamate.

FIGURE 44–9 Thiamin, thiamin diphosphate, and the carbanion form.

concentrations of lactate and pyruvate, which may cause life-threatening **lactic acidosis.**

Thiamin Nutritional Status Can Be Assessed by Erythrocyte Transketolase Activation

The activation of apo-transketolase (the enzyme protein) in erythrocyte lysate by thiamin diphosphate added in vitro has become the accepted index of thiamin nutritional status.

VITAMIN B₂ (RIBOFLAVIN) HAS A CENTRAL ROLE IN ENERGY-YIELDING METABOLISM

Riboflavin provides the reactive moieties of the coenzymes **flavin mononucleotide (FMN)** and **flavin adenine dinucleotide (FAD)** (Figure 44–10). FMN is formed by ATP-dependent phosphorylation of riboflavin, whereas FAD is synthesized by further reaction with ATP in which its AMP moiety is transferred to FMN. The main dietary sources of riboflavin are milk and dairy products. In addition, because of its intense yellow color, riboflavin is widely used as a food additive.

Flavin Coenzymes Are Electron Carriers in Oxidoreduction Reactions

These include the mitochondrial respiratory chain, key enzymes in fatty acid and amino acid oxidation, and the citric acid cycle. Reoxidation of the reduced flavin in oxygenases and mixed-function oxidases proceeds by way of formation of the flavin radical and flavin hydroperoxide, with the intermediate generation of superoxide and perhydroxyl radicals and

hydrogen peroxide. Because of this, flavin oxidases make a significant contribution to the total oxidant stress in the body (Chapter 45).

Riboflavin Deficiency Is Widespread but Not Fatal

Although riboflavin is centrally involved in lipid and carbohydrate metabolism, and deficiency occurs in many countries, it is not fatal, because there is very efficient conservation of tissue riboflavin. Riboflavin released by the catabolism of enzymes is rapidly incorporated into newly synthesized enzymes. Deficiency is characterized by cheilosis, desquamation and inflammation of the tongue, and a seborrheic dermatitis. Riboflavin nutritional status is assessed by measurement of the activation of erythrocyte glutathione reductase by FAD added in vitro.

NIACIN IS NOT STRICTLY A VITAMIN

Niacin was discovered as a nutrient during studies of **pellagra.** It is not strictly a vitamin since it can be synthesized in the body from the essential amino acid tryptophan. Two compounds, **nicotinic acid** and **nicotinamide,** have the biologic activity of niacin; its metabolic function is as the nicotinamide ring of the coenzymes **NAD** and **NADP** in oxidation/reduction reactions (Figure 44–11). Some 60 mg of tryptophan is equivalent to 1 mg of dietary niacin. The niacin content of foods is expressed as

$$\text{mg Niacin equivalents} = \text{mg Preformed niacin} + 1/60 \times \text{mg Tryptophan}$$

Because most of the niacin in cereals is biologically unavailable, this is discounted.

FIGURE 44–10 Riboflavin and the coenzymes flavin mononucleotide (FMN) and flavin adenine dinucleotide (FAD).

FIGURE 44–11 Niacin (nicotinic acid and nicotinamide).

NAD Is the Source of ADP-Ribose

In addition to its coenzyme role, NAD is the source of ADP-ribose for the **ADP-ribosylation** of proteins and polyADP-ribosylation of nucleoproteins involved in the **DNA repair mechanism.** Cyclic ADP-ribose and nicotinic acid adenine dinucleotide, formed from NAD, act to increase intracellular calcium in response to neurotransmitters and hormones.

Pellagra Is Caused by Deficiency of Tryptophan & Niacin

Pellagra is characterized by a photosensitive dermatitis. As the condition progresses, there is dementia and possibly diarrhea. Untreated pellagra is fatal. Although the nutritional etiology of pellagra is well established, and tryptophan or niacin prevents or cures the disease, additional factors, including deficiency of riboflavin or vitamin B₆, both of which are required for synthesis of nicotinamide from tryptophan, may be important. In most outbreaks of pellagra, twice as many women as men are affected, probably the result of inhibition of tryptophan metabolism by estrogen metabolites.

Pellagra Can Occur as a Result of Disease Despite an Adequate Intake of Tryptophan & Niacin

A number of genetic diseases that result in defects of tryptophan metabolism are associated with the development of pellagra, despite an apparently adequate intake of both tryptophan and niacin. **Hartnup disease** is a rare genetic condition in which there is a defect of the membrane transport mechanism for tryptophan, resulting in large losses as a result of intestinal malabsorption and failure of the renal reabsorption mechanism. In **carcinoid syndrome,** there is metastasis of a primary liver tumor of enterochromaffin cells, which synthesize 5-hydroxytryptamine. Overproduction of 5-hydroxytryptamine may account for as much as 60% of the body's tryptophan metabolism, causing pellagra because of the diversion away from NAD synthesis.

Niacin Is Toxic in Excess

Nicotinic acid has been used to treat hyperlipidemia when of the order of 1–6 g/day are required, causing dilatation of blood vessels and flushing, along with skin irritation. Intakes of both nicotinic acid and nicotinamide in excess of 500 mg/day also cause liver damage.

VITAMIN B₆ IS IMPORTANT IN AMINO ACID & GLYCOGEN METABOLISM & IN STEROID HORMONE ACTION

Six compounds have vitamin B₆ activity (Figure 44–12): **pyridoxine, pyridoxal, pyridoxamine,** and their 5′-phosphates. The active coenzyme is pyridoxal 5′-phosphate. Some 80% of the body's total vitamin B₆ is pyridoxal phosphate in muscle, mostly associated with glycogen phosphorylase. This is not available in deficiency, but is released in starvation, when glycogen reserves become depleted, and is then available, especially in liver and kidney, to meet increased requirement for gluconeogenesis from amino acids.

Vitamin B₆ Has Several Roles in Metabolism

Pyridoxal phosphate is a coenzyme for many enzymes involved in amino acid metabolism, especially transamination and decarboxylation. It is also the cofactor of glycogen phosphorylase, where the phosphate group is catalytically important. In addition, B₆ is important in steroid hormone action. Pyridoxal phosphate removes the hormone-receptor complex from DNA binding, terminating the action of the hormones. In vitamin B₆ deficiency, there is increased sensitivity to the actions of low concentrations of estrogens, androgens, cortisol, and vitamin D.

Vitamin B₆ Deficiency Is Rare

Although clinical deficiency disease is rare, there is evidence that a significant proportion of the population have marginal vitamin B₆ status. Moderate deficiency results in abnormalities of tryptophan and methionine metabolism. Increased sensitiv-

FIGURE 44–12 Interconversion of the vitamin B₆ vitamers.

ity to steroid hormone action may be important in the development of **hormone-dependent cancer** of the breast, uterus, and prostate, and vitamin B_6 status may affect the prognosis.

Vitamin B_6 Status Is Assessed by Assaying Erythrocyte Transaminases

The most widely used method of assessing vitamin B_6 status is by the activation of erythrocyte transaminases by pyridoxal phosphate added in vitro, expressed as the activation coefficient.

In Excess, Vitamin B_6 Causes Sensory Neuropathy

The development of sensory neuropathy has been reported in patients taking 2–7 g of pyridoxine per day for a variety of reasons (there is some slight evidence that it is effective in treating **premenstrual syndrome**). There was some residual damage after withdrawal of these high doses; other reports suggest that intakes in excess of 200 mg/d are associated with neurologic damage.

VITAMIN B_{12} IS FOUND ONLY IN FOODS OF ANIMAL ORIGIN

The term "vitamin B_{12}" is used as a generic descriptor for the **cobalamins**—those **corrinoids** (cobalt-containing compounds possessing the corrin ring) having the biologic activity of the vitamin (Figure 44–13). Some corrinoids that are

growth factors for microorganisms not only have no vitamin B_{12} activity, but may also be anti-metabolites of the vitamin. Although it is synthesized exclusively by microorganisms, for practical purposes vitamin B_{12} is found only in foods of animal origin, there being no plant sources of this vitamin. This means that strict vegetarians (vegans) are at risk of developing B_{12} deficiency. The small amounts of the vitamin formed by bacteria on the surface of fruits may be adequate to meet requirements, but preparations of vitamin B_{12} made by bacterial fermentation are available.

Vitamin B_{12} Absorption Requires Two Binding Proteins

Vitamin B_{12} is absorbed bound to **intrinsic factor,** a small glycoprotein secreted by the parietal cells of the gastric mucosa. Gastric acid and pepsin release the vitamin from protein binding in food and make it available to bind to **cobalophilin,** a binding protein secreted in the saliva. In the duodenum, cobalophilin is hydrolyzed, releasing the vitamin for binding to intrinsic factor. **Pancreatic insufficiency** can therefore be a factor in the development of vitamin B_{12} deficiency, resulting in the excretion of cobalophilin-bound vitamin B_{12}. Intrinsic factor binds only the active vitamin B_{12} vitamers and not other corrinoids. Vitamin B_{12} is absorbed from the distal third of the ileum via receptors that bind the intrinsic factor-vitamin B_{12} complex, but not free intrinsic factor or free vitamin.

There Are Three Vitamin B_{12}– Dependent Enzymes

Methylmalonyl CoA mutase, leucine aminomutase, and **methionine synthase** (Figure 44–14) are vitamin B_{12}–dependent enzymes. Methylmalonyl CoA is formed as an intermediate in the catabolism of valine and by the carboxylation of propionyl CoA arising in the catabolism of isoleucine, cholesterol, and, rarely, fatty acids with an odd number of carbon atoms or directly from propionate, a major product of microbial fermentation in the rumen. It undergoes a vitamin B_{12}–dependent rearrangement to succinyl CoA, catalyzed by methylmalonyl CoA

FIGURE 44–13 Vitamin B_{12}. Four coordination sites on the central cobalt atom are chelated by the nitrogen atoms of the corrin ring, and one by the nitrogen of the dimethylbenzimidazole nucleotide. The sixth coordination site may be occupied by: CN^- (cyanocobalamin), OH^- (hydroxocobalamin), H_2O (aquocobalamin, $—CH_3$ (methyl cobalamin) or 5′-deoxyadenosine (adenosylcobalamin).

FIGURE 44–14 Homocysteine and the "folate trap." Vitamin B_{12} deficiency leads to impairment of methionine synthase, resulting in accumulation of homocysteine and trapping folate as methyltetrahydrofolate.

mutase (Figure 20–2). The activity of this enzyme is greatly reduced in vitamin B_{12} deficiency, leading to an accumulation of methylmalonyl CoA and urinary excretion of methylmalonic acid, which provides a means of assessing vitamin B_{12} nutritional status.

Vitamin B_{12} Deficiency Causes Pernicious Anemia

Pernicious anemia arises when vitamin B_{12} deficiency impairs the metabolism of folic acid, leading to functional folate deficiency that disturbs erythropoiesis, causing immature precursors of erythrocytes to be released into the circulation (megaloblastic anemia). The most common cause of pernicious anemia is failure of the absorption of vitamin B_{12} rather than dietary deficiency. This can be the result of failure of intrinsic factor secretion caused by autoimmune disease affecting parietal cells or from production of anti-intrinsic factor antibodies. There is irreversible degeneration of the spinal cord in pernicious anemia, as a result of failure of methylation of one arginine residue on myelin basic protein. This is the result of methionine deficiency in the central nervous system, rather than secondary folate deficiency.

THERE ARE MULTIPLE FORMS OF FOLATE IN THE DIET

The active form of folic acid (pteroyl glutamate) is tetrahydrofolate (Figure 44–15). The folates in foods may have up to seven additional glutamate residues linked by γ-peptide bonds. In addition, all of the one-carbon substituted folates in Figure 44–15 may also be present in foods. The extent to which the different forms of folate can be absorbed varies, and folate intakes are calculated as dietary folate equivalents – the sum of μg food folates + 1.7 × μg of folic acid (used in food enrichment).

Tetrahydrofolate Is a Carrier of One-Carbon Units

Tetrahydrofolate can carry one-carbon fragments attached to N-5 (formyl, formimino, or methyl groups), N-10 (formyl) or bridging N-5–N-10 (methylene or methenyl groups). 5-Formyl-tetrahydrofolate is more stable than folate, and is therefore used pharmaceutically (known as **folinic acid**), and the synthetic (racemic) compound **(leucovorin).** The major point of entry for one-carbon fragments into substituted folates is methylene-tetrahydrofolate (Figure 44–16), which is formed by the reaction of glycine, serine, and choline with tetrahydrofolate. Serine is the most important source of substituted folates for biosynthetic reactions, and the activity of serine hydroxymethyltransferase is regulated by the state of folate substitution and the availability of folate. The reaction is reversible, and in liver it can form serine from glycine as a substrate for gluconeogenesis. Methylene-, methenyl-, and

FIGURE 44–15 Tetrahydrofolic acid and the one-carbon substituted folates.

10-formyl-tetrahydrofolates are interconvertible. When one-carbon folates are not required, the oxidation of formyl-tetrahydrofolate to yield carbon dioxide provides a means of maintaining a pool of free folate.

Inhibitors of Folate Metabolism Provide Cancer Chemotherapy, Antibacterial, & Antimalarial Drugs

The methylation of deoxyuridine monophosphate (dUMP) to thymidine monophosphate (TMP), catalyzed by thymidylate synthase, is essential for the synthesis of DNA. The one-carbon fragment of methylene-tetrahydrofolate is reduced to a methyl group with release of dihydrofolate, which is then reduced back to tetrahydrofolate by **dihydrofolate reductase.** Thymidylate synthase and dihydrofolate reductase are especially active in tissues with a high rate of cell division. **Methotrexate,** an analog of 10-methyl-tetrahydrofolate, inhibits dihydrofolate reductase and has been exploited as an anti-cancer drug. The dihydrofolate reductases of some bacteria and parasites differ from the human enzyme; inhibitors of these enzymes can be used as antibacterial drugs (eg, **trimethoprim**) and antimalarial drugs (eg, **pyrimethamine**).

Vitamin B_{12} Deficiency Causes Functional Folate Deficiency—the "Folate Trap"

When acting as a methyl donor, S-adenosyl methionine forms homocysteine, which may be remethylated by methyl-tetrahy-

FIGURE 44–16 Sources and utilization of one-carbon substituted folates.

drofolate catalyzed by methionine synthase, a vitamin B_{12}–dependent enzyme (Figure 44–14). As the reduction of methylene-tetrahydrofolate to methyl-tetrahydrofolate is irreversible and the major source of tetrahydrofolate for tissues is methyltetrahydrofolate, the role of methionine synthase is vital, and provides a link between the functions of folate and vitamin B_{12}. Impairment of methionine synthase in vitamin B_{12} deficiency results in the accumulation of methyltetrahydrofolate—the "folate trap." There is therefore functional deficiency of folate, secondary to the deficiency of vitamin B_{12}.

Folate Deficiency Causes Megaloblastic Anemia

Deficiency of folic acid itself or deficiency of vitamin B_{12}, which leads to functional folic acid deficiency, affects cells that are dividing rapidly because they have a large requirement for thymidine for DNA synthesis. Clinically, this affects the bone marrow, leading to megaloblastic anemia.

Folic Acid Supplements Reduce the Risk of Neural Tube Defects & Hyperhomocysteinemia, & May Reduce the Incidence of Cardiovascular Disease & Some Cancers

Supplements of 400 μg/day of folate begun before conception result in a significant reduction in the incidence of **spina bifida** and other **neural tube defects.** Because of this, there is mandatory enrichment of flour with folic acid in many countries. Elevated blood homocysteine is a significant risk factor for **atherosclerosis, thrombosis,** and **hypertension.** The condition is the result of an impaired ability to form methyltetrahydrofolate by methylene-tetrahydrofolate reductase, causing functional folate deficiency, resulting in failure to remethylate homocysteine to methionine. People with an abnormal variant of methylene-tetrahydrofolate reductase that occurs in 5–10% of the population do not develop hyperhomocysteinemia if they have a relatively high intake of folate, but it is not yet known whether this affects the incidence of cardiovascular disease.

There is also evidence that low folate status results in impaired methylation of CpG islands in DNA, which is a factor in the development of colorectal and other cancers. A number of studies suggest that folate supplementation or food enrichment may reduce the risk of developing some cancers.

Folate Enrichment of Foods May Put Some People at Risk

Folate supplements will rectify the megaloblastic anemia of vitamin B_{12} deficiency but may hasten the development of the (irreversible) nerve damage found in vitamin B_{12} deficiency. There is also antagonism between folic acid and the anticonvulsants used in the treatment of epilepsy.

DIETARY BIOTIN DEFICIENCY IS UNKNOWN

The structures of biotin, biocytin, and carboxybiotin (the active metabolic intermediate) are shown in Figure 44–17. Biotin is widely distributed in many foods as biocytin (ε-amino-biotinyllysine), which is released on proteolysis. It is synthe-

FIGURE 44–17 Biotin, biocytin, and carboxy-biocytin.

FIGURE 44–18 Pantothenic acid and coenzyme A. Asterisk shows site of acylation by fatty acids.

sized by intestinal flora in excess of requirements. Deficiency is unknown, except among people maintained for many months on total parenteral nutrition, and a very small number who eat abnormally large amounts of uncooked egg white, which contains avidin, a protein that binds biotin and renders it unavailable for absorption.

Biotin Is a Coenzyme of Carboxylase Enzymes

Biotin functions to transfer carbon dioxide in a small number of reactions: acetyl-CoA carboxylase (Figure 23–1), pyruvate carboxylase (Figure 20–1), propionyl-CoA carboxylase (Figure 20–2), and methylcrotonyl-CoA carboxylase. A holocarboxylase synthetase catalyzes the transfer of biotin onto a lysine residue of the apo-enzyme to form the biocytin residue of the holoenzyme. The reactive intermediate is 1-N-carboxy-biocytin, formed from bicarbonate in an ATP-dependent reaction. The carboxy group is then transferred to the substrate for carboxylation.

Biotin also has a role in regulation of the cell cycle, acting to biotinylate key nuclear proteins.

AS PART OF CoA & ACP, PANTOTHENIC ACID ACTS AS A CARRIER OF ACYL GROUPS

Pantothenic acid has a central role in acyl group metabolism when acting as the pantetheine functional moiety of coenzyme A or acyl carrier protein (ACP) (Figure 44–18). The pantetheine moiety is formed after combination of pantothenate with cysteine, which provides the–SH prosthetic group of CoA and ACP. CoA takes part in reactions of the citric acid cycle (Chapter 17), fatty acid oxidation (Chapter 22), acetylations and cholesterol synthesis (Chapter 26). ACP participates in fatty acid synthesis (Chapter 23). The vitamin is widely distributed in all food-stuffs, and deficiency has not been unequivocally reported in humans except in specific depletion studies.

ASCORBIC ACID IS A VITAMIN FOR ONLY SOME SPECIES

Vitamin C (Figure 44–19) is a vitamin for humans and other primates, the guinea pig, bats, passeriform birds, and most fishes and invertebrates; other animals synthesize it as an intermediate in the uronic acid pathway of glucose metabolism (Figure 21-4). In those species for which it is a vitamin, there is a block in the pathway as a result of absence of gulonolactone oxidase. Both ascorbic acid and dehydroascorbic acid have vitamin activity.

Vitamin C Is the Coenzyme for Two Groups of Hydroxylases

Ascorbic acid has specific roles in the copper-containing hydroxylases and the α-ketoglutarate-linked iron-containing hydroxylases. It also increases the activity of a number of other enzymes in vitro, although this is a nonspecific reducing action. In addition, it has a number of nonenzymic effects as a result of its action as a reducing agent and oxygen radical quencher (Chapter 45).

Dopamine β-hydroxylase is a copper-containing enzyme involved in the synthesis of the catecholamines (norepinephrine and epinephrine), from tyrosine in the adrenal medulla and central nervous system. During hydroxylation the Cu^+ is oxidized to Cu^{2+}; reduction back to Cu^+ specifically requires ascorbate, which is oxidized to monodehydroascorbate.

FIGURE 44–19 Vitamin C.

A number of peptide hormones have a carboxy terminal amide that is derived from a terminal glycine residue. This glycine is hydroxylated on the α-carbon by a copper-containing enzyme, **peptidylglycine hydroxylase,** which, again, requires ascorbate for reduction of Cu^{2+}.

A number of iron-containing, ascorbate-requiring hydroxylases share a common reaction mechanism, in which hydroxylation of the substrate is linked to decarboxylation of α-ketoglutarate. Many of these enzymes are involved in the modification of precursor proteins. **Proline** and **lysine hydroxylases** are required for the postsynthetic modification of **procollagen** to **collagen,** and proline hydroxylase is also required in formation of **osteocalcin** and the C1q component of **complement.** Aspartate β-hydroxylase is required for the postsynthetic modification of the precursor of protein C, the vitamin K–dependent protease that hydrolyzes activated factor V in the blood-clotting cascade (Chapter 50). Trimethyllysine and γ-butyrobetaine hydroxylases are required for the synthesis of carnitine.

Vitamin C Deficiency Causes Scurvy

Signs of vitamin C deficiency include skin changes, fragility of blood capillaries, gum decay, tooth loss, and bone fracture, many of which can be attributed to deficient collagen synthesis.

There May Be Benefits from Higher Intakes of Vitamin C

At intakes above about 100 mg/day, the body's capacity to metabolize vitamin C is saturated, and any further intake is excreted in the urine. However, in addition to its other roles, vitamin C enhances the absorption of iron, and this depends on the presence of the vitamin in the gut. Therefore, increased intakes may be beneficial. There is very little good evidence that high doses of vitamin C prevent the common cold, although they may reduce the duration and severity of symptoms.

MINERALS ARE REQUIRED FOR BOTH PHYSIOLOGIC & BIOCHEMICAL FUNCTIONS

Many of the essential minerals (Table 44–2) are widely distributed in foods, and most people eating a mixed diet are likely to receive adequate intakes. The amounts required vary from grams per day for sodium and calcium, through milligrams per day (eg, iron, zinc), to micrograms per day for the trace elements. In general, mineral deficiencies occur when foods come from one region where the soil may be deficient in some minerals (eg, iodine and selenium, deficiencies of both of which occur in many areas of the world); when foods come from a variety of regions, mineral deficiency is less likely to occur. However, iron deficiency is a general problem, because if iron losses from the body are relatively high (eg, from heavy

TABLE 44–2   Classification of Minerals According to Their Function

Function	Mineral
Structural function	Calcium, magnesium, phosphate
Involved in membrane function	Sodium, potassium
Function as prosthetic groups in enzymes	Cobalt, copper, iron, molybdenum, selenium, zinc
Regulatory role or role in hormone action	Calcium, chromium, iodine, magnesium, manganese, sodium, potassium
Known to be essential, but function unknown	Silicon, vanadium, nickel, tin
Have effects in the body, but essentiality is not established	Fluoride, lithium
May occur in foods and known to be toxic in excess	Aluminum, arsenic, antimony, boron, bromine, cadmium, cesium, germanium, lead, mercury, silver, strontium

menstrual blood loss), it is difficult to achieve an adequate intake to replace losses. Foods grown on soil containing high levels of selenium cause toxicity, and excessive intakes of sodium cause hypertension in susceptible people.

SUMMARY

- Vitamins are organic nutrients with essential metabolic functions, generally required in small amounts in the diet because they cannot be synthesized by the body. The lipid-soluble vitamins (A, D, E, and K) are hydrophobic molecules requiring normal fat absorption for their absorption and the avoidance of deficiency.

- Vitamin A (retinol), present in meat, and the provitamin (β-carotene), found in plants, form retinaldehyde, utilized in vision, and retinoic acid, which acts in the control of gene expression.

- Vitamin D is a steroid prohormone yielding the active hormone calcitriol, which regulates calcium and phosphate metabolism; deficiency leads to rickets and osteomalacia.

- Vitamin E (tocopherol) is the most important antioxidant in the body, acting in the lipid phase of membranes protecting against the effects of free radicals.

- Vitamin K functions as cofactor of a carboxylase that acts on glutamate residues of precursor proteins of clotting factors and bone proteins to enable them to chelate calcium.

- The water-soluble vitamins of the B complex act as enzyme cofactors. Thiamin is a cofactor in oxidative decarboxylation of α-keto acids and of transketolase in the pentose phosphate pathway. Riboflavin and niacin are important cofactors in oxidoreduction reactions, present in flavoprotein enzymes and in NAD and NADP, respectively.

- Pantothenic acid is present in coenzyme A and acyl carrier protein, which act as carriers for acyl groups in metabolic reactions.

- Vitamin B$_6$ as pyridoxal phosphate, is the coenzyme for several enzymes of amino acid metabolism, including the transaminases, and of glycogen phosphorylase. Biotin is the coenzyme for several carboxylase enzymes.
- Vitamin B$_{12}$ and folate provide one-carbon residues for DNA synthesis and other reactions; deficiency results in megaloblastic anemia.
- Vitamin C is a water-soluble antioxidant that maintains vitamin E and many metal cofactors in the reduced state.
- Inorganic mineral elements that have a function in the body must be provided in the diet. When intake is insufficient, deficiency may develop, and excessive intakes may be toxic.

REFERENCES

Bender DA, Bender AE: *Nutrition: A Reference Handbook.* Oxford University Press, 1997.

Bender DA: *Nutritional Biochemistry of the Vitamins,* 2nd edition. Cambridge University Press, 2003.

Department of Health: *Dietary Reference Values for Food Energy and Nutrients for the United Kingdom.* Her Majesty's Stationery Office, 1991.

FAO/WHO: *Human Vitamin and Mineral Requirements: Report of a Joint FAO/WHO Expert Consultation: Bankok, Thailand.* Food and Nutrition Division of the United Nations Food and Agriculture Organization, 2000.

Geissler C, Powers HJ: *Human Nutrition,* 11th edition. Elsevier, 2005.

Institute of Medicine: *Dietary Reference Intakes for Calcium, Phosphorus, Magnesium, Vitamin D and Fluoride.* National Academy Press, 1997.

Institute of Medicine: *Dietary Reference Values for Thiamin, Riboflavin, Niacin, Vitamin B6, Folate, Vitamin B12, Pantothenic Acid, Biotin and Choline.* National Academy Press, 2000.

Institute of Medicine: *Dietary Reference Values for Vitamin C, Vitamin E, Selenium and Carotenoids.* National Academy Press, 2000.

Institute of Medicine: *Dietary Reference Intakes for Vitamin A, Vitamin K, Arsenic, Boron, Chromium, Copper, Iodine, Iron, Manganese, Molybdenum, Nickel, Silicon, Vanadium and Zinc.* National Academy Press, 2001.

Scientific Advisory Committee on Nutrition of the Food Standards Agency: *Folate and Disease Prevention.* The Stationery Office, 2006.

Free Radicals and Antioxidant Nutrients

David A. Bender, PhD

BIOMEDICAL IMPORTANCE

Free radicals are formed in the body under normal conditions. They cause damage to nucleic acids, proteins, and lipids in cell membranes and plasma lipoproteins. This can cause **cancer**, **atherosclerosis** and **coronary artery disease**, and **autoimmune diseases**. Epidemiological and laboratory studies have identified a number of protective **antioxidant nutrients**: selenium, vitamins C and E, β-carotene, and a variety of polyphenolic compounds derived from plant foods. Many people take supplements of one or more antioxidant nutrients. However, intervention trials show little benefit of antioxidant supplements except among people who were initially deficient, and many trials of β-carotene and vitamin E have shown increased mortality among those taking the supplements.

Free Radical Reactions Are Self-Perpetuating Chain Reactions

Free radicals are highly reactive molecular species with an unpaired electron; they persist for only a very short time (of the order of 10^{-9}–10^{-12} sec) before they collide with another molecule and either abstract or donate an electron in order to achieve stability. In so doing, they generate a new radical from the molecule with which they collided. The only way in which a free radical can be quenched, so terminating this chain reaction, is if two radicals react together, when the unpaired electrons can become paired in one or other of the parent molecules. This is a rare occurrence, because of the very short half-life of an individual radical and the very low concentrations of radicals in tissues.

The most damaging radicals in biological systems are oxygen radicals (sometimes called reactive oxygen species)—especially superoxide, $\cdot O_2^-$, hydroxyl, $\cdot OH$, and perhydroxyl, $\cdot O_2H$. Tissue damage caused by oxygen radicals is often called oxidative damage, and factors that protect against oxygen radical damage are known as antioxidants.

Radicals Can Damage DNA, Lipids, and Proteins

Interaction of radicals with bases in DNA can lead to chemical changes that, if not repaired (Chapter 35) may be inherited in daughter cells. Radical damage to unsaturated fatty acids in cell membranes and plasma lipoproteins leads to the formation of lipid peroxides, then highly reactive dialdehydes that can chemically modify proteins and nucleic acid bases. Proteins are also subject to direct chemical modification by interaction with radicals. Oxidative damage to tyrosine residues in proteins can lead to the formation of dihydroxyphenylalanine that can undergo non-enzymic reactions leading to further formation of oxygen radicals.

The total body radical burden can be estimated by measuring the products of lipid peroxidation. Lipid peroxides can be measured by the ferrous oxidation in xylenol orange (FOX) assay. Under acidic conditions, they oxidize Fe^{2+} to Fe^{3+}, which forms a chromophore with xylenol orange. The dialdehydes formed from lipid peroxides can be measured by reaction with thiobarbituric acid, when they form a red fluorescent adduct—the results of this assay are generally reported as total thiobarbituric acid reactive substances, TBARS. Peroxidation of n-6 polyunsaturated fatty acids leads to the formation of pentane, and of n-3 polyunsaturated fatty acids to ethane, both of which can be measured in exhaled air.

Radical Damage May Cause Mutations, Cancer, Autoimmune Disease, and Atherosclerosis

Radical damage to DNA in germ-line cells in ovaries and testes can lead to heritable mutations; in somatic cells the result may be initiation of cancer. The dialdehydes formed as a result of radical-induced lipid peroxidation in cell membranes can also modify bases in DNA.

Chemical modification of amino acids in proteins, either by direct radical action or as a result of reaction with the products of radical-induced lipid peroxidation, leads to proteins that are recognized as non-self by the immune system. The resultant antibodies will also cross-react with normal tissue proteins, so initiating autoimmune disease.

Chemical modification of the proteins or lipids in plasma low density lipoprotein (LDL) leads to abnormal LDL that is not recognized by the liver LDL receptors, and so is not cleared by the liver. The modified LDL is taken up by macrophage scavenger receptors. Lipid engorged macrophages in-

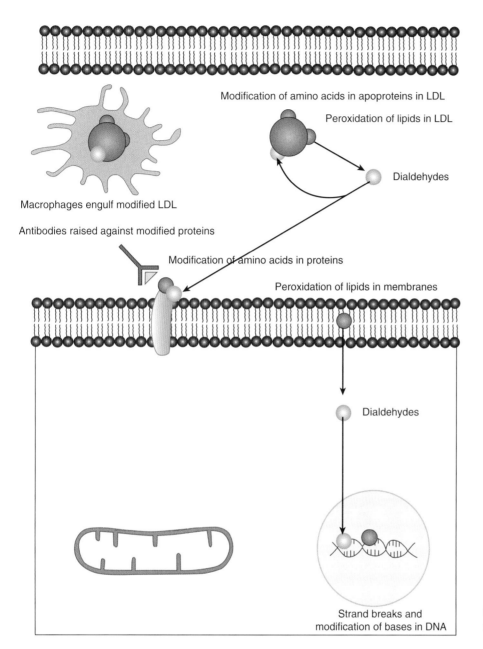

FIGURE 45–1 Tissue damage by radicals.

filtrate under blood vessel endothelium (especially when there is already some damage to the endothelium), and are killed by the high content of unesterified cholesterol they have accumulated. This occurs in the development of atherosclerotic plaques which, in extreme cases, can more or less completely occlude a blood vessel.

There Are Multiple Sources of Oxygen Radicals in the Body

Ionizing radiation (x-rays and UV) can lyse water, leading to the formation of hydroxyl radicals. Transition metal ions, including Cu^+, Co^{2+}, Ni^{2+}, and Fe^{2+} can react non-enzymically with oxygen or hydrogen peroxide, again leading to the formation of hydroxyl radicals. Nitric oxide (the endothelium-

derived relaxation factor) is itself a radical, and, more importantly, can react with superoxide to yield peroxynitrite, which decays to form hydroxyl radicals.

The respiratory burst of activated macrophages (Chapter 52) is increased utilisation of glucose via the pentose phosphate pathway (Chapter 21) to reduce $NADP^+$ to NADPH, and increased utilisation of oxygen to oxidise NADPH to produce oxygen (and halogen) radicals as cytotoxic agents to kill phagocytosed microorganisms. The respiratory burst oxidase (NADPH oxidase) is a flavoprotein that reduces oxygen to superoxide: $NADPH + 2O_2 \rightarrow NADP^+ + 2O_2^{\cdot-} + 2H^+$. Plasma markers of radical damage to lipids increase considerably in response to even a mild infection.

The oxidation of reduced flavin coenzymes in the mitochondrial (Chapter 13) and microsomal electron transport

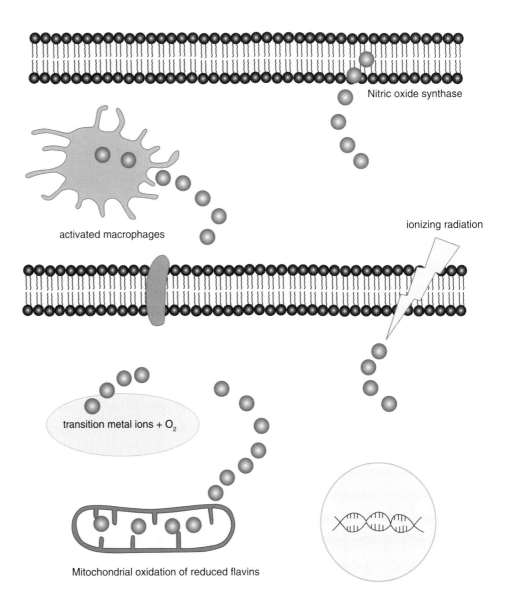

FIGURE 45–2 Sources of radicals.

chains proceeds through a series of steps in which the flavin semiquinone radical is stabilized by the protein to which it is bound, and forms oxygen radicals as transient intermediates. Although the final products are not radicals, because of the unpredictable nature of radicals there is considerable "leakage" of radicals, and some 3–5% of the daily consumption of 30 mol of oxygen by an adult human being is converted to singlet oxygen, hydrogen peroxide, and superoxide, perhydroxyl and hydroxyl radicals, rather then undergoing complete reduction to water. This results in daily production of about 1.5 mol of reactive oxygen species.

There Are Various Mechanisms of Protection Against Radical Damage

The metal ions that undergo non-enzymic reaction to form oxygen radicals are not normally free in solution, but are bound to either the proteins for which they provide the pros-

thetic group, or to specific transport and storage proteins, so that they are unreactive. Iron is bound to transferrin, ferritin and hemosiderin, copper to ceruloplasmin, and other metal ions are bound to metallothionein. This binding to transport proteins that are too large to be filtered in the kidneys also prevents loss of metal ions in the urine.

Superoxide is produced both accidentally and also as the reactive oxygen species required for a number of enzyme-catalyzed reactions. A family of superoxide dismutases catalyze the reaction between superoxide and water to yield oxygen and hydrogen peroxide: $O_2^{\cdot -} + H_2O \rightarrow O_2 + H_2O_2$. The hydrogen peroxide is then removed by catalase and various peroxidases: $2H_2O_2 \rightarrow 2H_2O + O_2$. Most enzymes that produce and require superoxide are in the peroxisomes, together with superoxide dismutase, catalase, and peroxidases.

The peroxides that are formed by radical damage to lipids in membranes and plasma lipoproteins are reduced to fatty acids by glutathione peroxidase, a selenium-dependent enzyme (hence the importance of adequate selenium intake to maximize antioxidant activity), and the oxidized glutathione is reduced by NADPH-dependent glutathione reductase (Figure 21–3). Lipid peroxides are also reduced to fatty acids by reaction with vitamin E, forming the relatively stable tocopheroxyl radical, which persist long enough to undergo reduction back to tocopherol by reaction with vitamin C at the surface of the cell or lipoprotein (Figure 44–6). The resultant monodehydroascorbate radical then undergoes enzymic reduction back to ascorbate or a non-enzymic reaction of 2 mol of monodehydroascorbate to yield 1 mol each of ascorbate and dehydroascorbate.

Ascorbate, uric acid and a variety of polyphenols derived from plant foods act as water-soluble radical trapping antioxidants, forming relatively stable radicals that persist long enough to undergo reaction to non-radical products. Ubiquinone and carotenes similarly act as lipid-soluble radical-trapping antioxidants in membranes and plasma lipoproteins.

Antioxidants Can Also Be Pro-Oxidants

Although ascorbate is an anti-oxidant, reacting with superoxide and hydroxyl to yield monodehydroascorbate and hydrogen peroxide or water, it can also be a source of superoxide radicals by reaction with oxygen, and hydroxyl radicals by reaction with Cu^{2+} ions (Table 45-1). However, these pro-oxidant actions require relatively high concentrations of ascorbate that are unlikely to be reached in tissues, since once the plasma concentration of ascorbate reaches about 30 mmol/L, the renal threshold is reached, and at intakes above about 100–120 mg/day the vitamin is excreted in the urine quantitatively with intake.

A considerable body of epidemiological evidence suggested that carotene is protective against lung and other cancers. However, two major intervention trials in the 1990s showed an increase in death from lung (and other) cancer among people given supplements of β-carotene. The problem is that although β-carotene is indeed a radical trapping antioxidant under conditions of low partial pressure of oxygen, as in most tissues, at high partial pressures of oxygen (as in the lungs) and especially in high concentrations, β-carotene is an autocatalytic pro-oxidant, and hence can initiate radical damage to lipids and proteins.

Epidemiological evidence also suggests that vitamin E is protective against atherosclerosis and cardiovascular disease. However, meta-analysis of intervention trials with vitamin E shows increased mortality among those taking (high dose) supplements. These trials have all used α-tocopherol, and it is possible that the other vitamers of vitamin E that are present in foods, but not the supplements, may be important. In vitro, plasma lipoproteins form less cholesterol ester hydroperoxide when incubated with sources of low concentrations of perhydroxyl radicals when vitamin E has been removed than when it is present. The problem seems to be that vitamin E acts as an antioxidant by forming a stable radical that persists long enough to undergo metabolism to non-radical products. This

TABLE 45–1 Antioxidant and Pro-Oxidant Roles of Vitamin C

Antioxidant roles:
Ascorbate + $\cdot O_2^-$ → H_2O_2 + monodehydroascorbate; catalase and peroxidases catalyze the reaction: $2H_2O_2$ → $2H_2O + O_2$
Ascorbate + $\cdot OH$ → H_2O + monodehydroascorbate
Pro-oxidant roles:
Ascorbate + O_2 → $\cdot O_2^-$ + monodehydroascorbate
Ascorbate + Cu^{2+} → Cu^+ + monodehydroasacorbate
$Cu^+ + H_2O_2$ → $Cu^{2+} + OH^- + \cdot OH$

means that the radical also persists long enough to penetrate deeper in to the lipoprotein, causing further radical damage, rather than interacting with a water-soluble antioxidant at the surface of the lipoprotein.

SUMMARY

- Free radicals are highly reactive molecular species with an unpaired electron. They can react with, and modify, proteins, nucleic acids and fatty acids in cell membranes and plasma lipoproteins.

- Radical damage to lipids and proteins in plasma lipoproteins is a factor in the development of atherosclerosis and coronary artery disease; radical damage to nucleic acids may induce heritable mutations and cancer; radical damage to proteins may lead to the development of auto-immune diseases.

- Oxygen radicals as a result of exposure to ionising radiation, non-enzymic reactions of transition metal ions, the respiratory burst of activated macrophages, and the normal oxidation of reduced flavin coenzymes.

- Protection against radical damage is afforded by enzymes that remove superoxide ions and hydrogen peroxide, enzymic reduction of lipid peroxides linked to oxidation of glutathione, non-enzymic reaction of lipid peroxides with vitamin E, and reaction of radicals with compounds such as vitamins C and E, carotene, ubiquinone, uric acid, and dietary polyphenols that form relatively stable radicals that persist long enough to undergo reaction to non-radical products.

- Except in people who were initially deficient, intervention trials of vitamin E and β-carotene have generally shown increased mortality among those taking the supplements. β-Carotene is only an antioxidant at low concentrations of oxygen; at higher concentrations of oxygen it is an autocatalytic pro-oxidant. Vitamin E forms a stable radical that is capable of either undergoing reaction with water-soluble antioxidants or penetrating further into lipoproteins and tissues, so increasing radical damage.

REFERENCES

Asplund K: Antioxidant vitamins in the prevention of cardiovascular disease: a systematic review. J Intern Med 2002;251:372.

Bjelakovic G, Nikolova D, Gluud LL, Simonetti RG, Gluud C: Mortality in randomised trials of antioxidant supplements for primary and secondary prevention. JAMA 2007;297:842.

Burton G, Ingold K: β-Carotene, an unusual type of lipid antioxidant. Science 1984;224:569.

Carr A, Frei B: Does vitamin C act as a pro-oxidant under physiological conditions? FASEB J 1999;13:1007.

Halliwell B, Gutteridge JMC, Cross CE: Free radicals, antioxidants and human disease: where are we now? J Lab Clin Med 1992;119:598.

Imlay JA: Pathways of oxidative damage. Ann Rev Microbiol 2003;57:395.

Imlay JA: Cellular Defenses against superoxide and hydrogen peroxide. Ann Rev Biochem 2008;77: (doi:10.1146/annurev. biochem.77.061606.161055).

Klaunig JE, Kamendulis LM: The role of oxidative stress in carcinogenesis. Ann Rev Pharm Tox 2004;44:239.

Miller ER, Pastor-Barriuso R, et al: Meta-analysis: high-dosage vitamin E supplementation may increase all-cause mortality. Ann Intern Med 2005;142:37.

Omenn GS, Goodman GE, et al: Effects of a combination of beta carotene and vitamin A on lung cancer and cardiovascular disease. N Engl J Med 1996;334:1150.

Various authors: Symposium: Antioxidant vitamins and β-carotene in disease prevention. Amer J Clin Nutr 1995;62, suppl 6:12995–15405.

Various authors: Symposium Proceedings: Molecular Mechanisms of protective effects of vitamin E in atherosclerosis. J Nutr 2001;131:366–397.

46

Intracellular Traffic & Sorting of Proteins

Robert K. Murray, MD, PhD

BIOMEDICAL IMPORTANCE

Proteins must travel from polyribosomes, where they are synthesized, to many different sites in the cell to perform their particular functions. Some are destined to be components of specific organelles, others for the cytosol or for export, and yet others will be located in the various cellular membranes. Thus, there is considerable **intracellular traffic of proteins.** A major insight was the recognition by Blobel and others that for proteins to attain their proper locations, they generally contain **information** (a signal or coding sequence) that **targets** them appropriately. Once a number of the signals were defined (see Table 46–1), it became apparent that **certain diseases** result from mutations that affect these signals. In this chapter we discuss the intracellular traffic of proteins and their sorting and briefly consider some of the disorders that result when abnormalities occur.

MANY PROTEINS ARE TARGETED BY SIGNAL SEQUENCES TO THEIR CORRECT DESTINATIONS

The protein biosynthetic pathways in cells can be considered to be **one large sorting system.** Many proteins carry **signals** (usually but not always specific sequences of amino acids) that direct them to their destination, thus ensuring that they will end up in the appropriate membrane or cell compartment; these signals are a fundamental component of the sorting system. Usually the signal sequences are recognized and interact with complementary areas of other proteins that serve as receptors for those containing the signals.

A **major sorting decision** is made early in protein biosynthesis, when specific proteins are synthesized either on **free** or on **membrane-bound polyribosomes.** This results in two sorting branches, called the **cytosolic branch** and the **rough endoplasmic reticulum (RER) branch** (Figure 46–1). This sorting occurs because proteins synthesized on membrane-bound polyribosomes contain a **signal peptide** that mediates their attachment to the membrane of the ER. Further details on the signal peptide are given below. Proteins synthesized on

free polyribosomes lack this particular signal peptide and are delivered into the cytosol. There they are directed to mitochondria, nuclei, and peroxisomes by specific signals—or remain in the cytosol if they lack a signal. Any protein that contains a targeting sequence that is subsequently removed is designated as a **preprotein.** In some cases a second peptide is also removed, and in that event the original protein is known as a **preproprotein** (eg, preproalbumin; Chapter 50).

Proteins synthesized and sorted in the **rough ER branch** (Figure 46–1) include many destined for various membranes (eg, of the ER, Golgi apparatus [GA], plasma membrane [PM]) and for secretion. Lysosomal enzymes are also included. These various proteins may thus reside in the membranes or lumen of the ER, or follow the major transport route of intracellular proteins to the GA. The entire pathway of ER → GA→ plasma membrane is often called the **secretory** or **exocytotic pathway.** Events along this route will be given special attention. Proteins destined for the GA, the PM, certain other sites, or for secretion are carried in **transport vesicles** (Figure 46–2); a brief description of the formation of these important particles will be given subsequently. Certain other proteins destined for secretion are carried in **secretory vesicles** (Figure 46–2). These are prominent in the pancreas and certain other glands. Their mobilization and discharge are regulated and often referred to as **"regulated secretion,"** whereas the secretory pathway involving transport vesicles is called **"constitutive."** Passage of enzymes to the lysosomes using the mannose 6-phosphate signal is described in Chapter 47.

The Golgi Apparatus Is Involved in Glycosylation & Sorting of Proteins

The **GA** plays two major roles in membrane synthesis. First, it is involved in the **processing of the oligosaccharide chains** of membrane and other N-linked glycoproteins and also contains enzymes involved in O-glycosylation (see Chapter 47). Second, it is involved in the **sorting** of various proteins prior to their delivery to their appropriate intracellular destinations. All parts of the GA participate in the first role, whereas the **trans** Golgi network (TGN) is particularly involved in the second and is very rich in vesicles.

TABLE 46–1   Some Sequences or Molecules That Direct Proteins to Specific Organelles

Targeting Sequence or Compound	Organelle Targeted
Signal peptide sequence	Membrane of ER
Amino terminal KDEL sequence (Lys-Asp-Glu-Leu) in ER-resident proteins in COPI vesicles	Luminal surface of ER
Di-acidic sequences (eg, Asp-X-Glu) in membrane proteins in COPII vesicles	Golgi membranes
Amino terminal sequence (20–80 residues)	Mitochondrial matrix
NLS (eg, Pro_2-Lys_3-Arg-Lys-Val)	Nucleus
PTS (eg, Ser-Lys-Leu)	Peroxisome
Mannose 6-phosphate	Lysosome

Abbreviations: NLS, nuclear localization signal; PTS, peroxisomal-matrix targeting sequence.

FIGURE 46–1 Diagrammatic representation of the two branches of protein sorting occurring by synthesis on (1) cytosolic and (2) membrane-bound polyribosomes. The mitochondrial proteins listed are encoded by nuclear genes; one of the signals used in further sorting of mitochondrial matrix proteins is listed in Table 46–1. (ER, endoplasmic reticulum; GA, Golgi apparatus.)

A Wide Variety of Experimental Techniques Have Been Used to Investigate Trafficking and Sorting

Approaches that have afforded major insights to the processes described in this chapter include (1) electron microscopy; (2) use of yeast mutants; (3) subcellular fractionation; (4) application of recombinant DNA techniques (eg, mutating or eliminating particular sequences in proteins, or fusing new sequences onto them); and (5) development of in vitro systems (eg, to study translocation in the ER and mechanisms of vesicle formation); (6) use of fluorescent tags to follow the movement of proteins; and (7) structural studies on certain proteins, particularly by x-ray crystallography.

The sorting of proteins belonging to **the cytosolic branch** referred to above is described next, starting with mitochondrial proteins.

THE MITOCHONDRION BOTH IMPORTS & SYNTHESIZES PROTEINS

Mitochondria contain many proteins. Thirteen polypeptides (mostly membrane components of the electron transport chain) are encoded by the **mitochondrial (mt) genome** and synthesized in that organelle using its own protein synthesizing system. However, the majority (at least several hundred) are encoded by **nuclear genes,** are synthesized outside the mitochondria on cytosolic **polyribosomes,** and must be imported. **Yeast cells** have proved to be a particularly useful system for analyzing the mechanisms of import of mitochondrial proteins, partly because it has proved possible to generate a variety of **mutants** that have illuminated the fundamental processes involved. Most progress has been made in the study of proteins present in **the mitochondrial matrix**, such as the F_1 ATPase subunits. Only the pathway of import of matrix proteins will be discussed in any detail here.

Matrix proteins must pass from cytosolic polyribosomes through the **outer** and **inner mitochondrial membranes** to reach their destination. Passage through the two membranes is called **translocation.** They have an amino terminal leader sequence **(presequence),** about 20–50 amino acids in length (see Table 46–1), which is not highly conserved but is amphipathic and contains many hydrophobic and positively charged amino acids (eg, Lys or Arg). The presequence is equivalent to a signal peptide mediating attachment of polyribosomes to membranes of the ER (see below), but in this instance **targeting proteins to the matrix**; if the leader sequence is cleaved off, potential matrix proteins will not reach their destination. Some general features of the passage of a protein from the cytosol to the mt matrix are shown in Figure 46–3.

Translocation occurs **posttranslationally,** after the matrix proteins are released from the cytosolic polyribosomes. Interactions with a number of cytosolic proteins that act as **chaperones** (see below) and as **targeting factors** occur prior to translocation.

Two distinct **translocation complexes** are situated in the outer and inner mitochondrial membranes, referred to (respectively) as TOM (translocase-of-the-outer membrane) and TIM (translocase-of-the-inner membrane). Each complex has been analyzed and found to be composed of a number of proteins, some of which act as **receptors (eg, Tom20/22)** for the incoming proteins and others as **components (eg, Tom40) of the transmembrane pores** through which these proteins must pass. Proteins must be in the **unfolded state** to pass through the complexes, and this is made possible by **ATP-dependent binding to several chaperone proteins.** The roles of chaperone proteins in protein folding are discussed later in this chapter. In mitochondria, they are involved in translocation, sorting, folding, assembly, and degradation of imported proteins. A **proton-motive force** across the inner membrane is required for import; it is made up of the **electric potential** across the membrane (inside negative) and the **pH gradient**

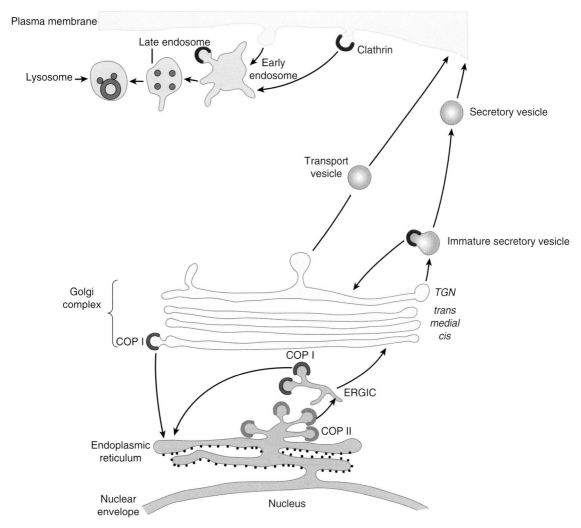

FIGURE 46–2 Diagrammatic representation of the rough ER branch of protein sorting. Newly synthesized proteins are inserted into the ER membrane or lumen from membrane-bound polyribosomes (small black circles studding the cytosolic face of the ER). Proteins that are transported out of the ER are carried in COPII vesicles to the *cis*-Golgi (anterograde transport). Movement of proteins through the Golgi appears to be mainly by cisternal maturation. In the TGN, the exit side of the Golgi, proteins are segregated and sorted. Secretory proteins accumulate in secretory vesicles (regulated secretion), from which they are expelled at the plasma membrane. Proteins destined for the plasma membrane or those that are secreted in a constitutive manner are carried out to the cell surface in as yet to be characterized transport vesicles (constitutive secretion). Clathrin-coated vesicles are involved in endocytosis, carrying cargo to late endosomes and to lysosomes. Mannose 6-phosphate (not shown; see Chapter 47) acts as a signal for transporting enzymes to lysosomes. COPI vesicles are involved in retrieving proteins from the Golgi to the ER (retrograde transport) and may be involved in some intra-Golgi. transport. The ERGIC/VTR compartment appears to be a site mainly for concentrating cargo destined for retrograde transport into COPI vesicles. (TGN, *trans*-Golgi network; ERGIC/VTR, ER-Golgi intermediate complex or vesicular tubule clusters.) (Courtesy of E Degen.)

(see Chapter 13). The positively charged leader sequence may be helped through the membrane by the negative charge in the matrix. The presequence is split off in the matrix by a **matrix-processing protease (MPP).** Contact with **other chaperones** present in the matrix is essential to complete the overall process of import. Interaction with mt-Hsp70 (mt = mitochondrial; Hsp = heat shock protein; 70 = ~70 kDa) ensures proper import into the matrix and prevents misfolding or aggrega-

tion, while interaction with the mt-Hsp60-Hsp10 system ensures proper folding. The interactions of imported proteins with the above chaperones require **hydrolysis of ATP** to drive them.

The details of how preproteins are translocated have not been fully elucidated. It is possible that the electric potential associated with the inner mitochondrial membrane causes a conformational change in the unfolded preprotein being trans-

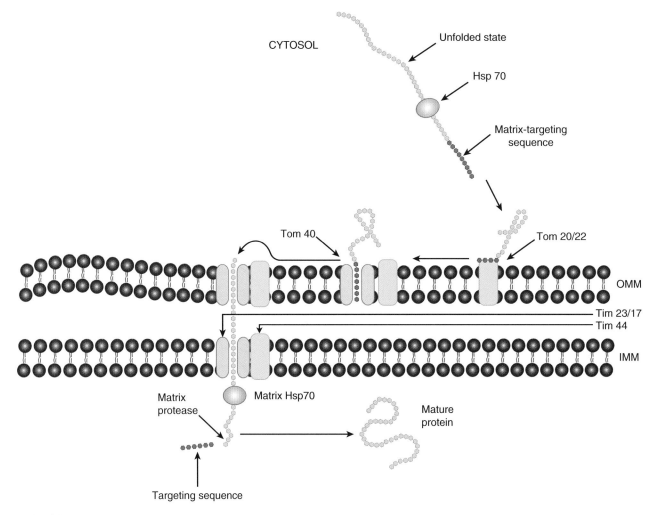

FIGURE 46–3 Schematic representation of the entry of a protein into the mitochondrial matrix. The unfolded protein synthesized on cytosolic poyribosomes and containing a matrix-targeting sequence interacts with the cytosolic chaperone Hsp 70. The protein next interacts with the mt outer membrane receptor Tom 20/22, and is transferred to the neighboring import channel Tom 40 (Tom, translocon of the outer membrane). The protein is then translocated across the channel; the channel on the inner mt membrane is largely composed of Tim 23 and Tim 17 proteins (Tim, translocon of the inner membrane). On the inside of the inner mt membrane, it interacts with the matrix chaperone Hsp 70, which in turn interacts with membrane protein Tim 44. The hydrolysis of ATP by mt Hsp70 probably helps drive the translocation, as does the electronegative interior of the matrix. The targeting sequence is subsequently cleaved by the matrix processing enzyme, and the imported protein assumes its final shape, or may interact with an mt chaperonin prior to this. At the site of translocation, the outer and inner mt membranes are in close contact. (Modified, with permission, from Lodish H, et al: *Molecular Cell Biology*, 6th ed. W.H. Freeman & Co., 2008.)

located and that this helps to pull it across. Furthermore, the fact that the matrix is more negative than the intermembrane space may "attract" the positively charged amino terminal of the preprotein to enter the matrix. Close apposition at **contact sites** between the outer and inner membranes is necessary for translocation to occur.

The above describes the major pathway of proteins destined for the mitochondrial matrix. However, certain proteins insert into the **outer mitochondrial membrane** facilitated by the TOM complex. Others stop in the **intermembrane space,** and some insert into the **inner membrane.** Yet others proceed into the matrix and then return to the inner membrane

or intermembrane space. A number of proteins contain two signaling sequences—one to enter the mitochondrial matrix and the other to mediate subsequent relocation (eg, into the inner membrane). Certain mitochondrial proteins do not contain presequences (eg, cytochrome *c*, which locates in the inter membrane space), and others contain **internal presequences.** Overall, proteins employ a variety of mechanisms and routes to attain their final destinations in mitochondria.

General features that apply to the import of proteins into organelles, including mitochondria and some of the other organelles to be discussed below, are summarized in Table 46–2.

TABLE 46–2   Some General Features of Protein Import to Organelles

- Import of a protein into an organelle usually occurs in three stages: recognition, translocation, and maturation.

- Targeting sequences on the protein are recognized in the cytoplasm or on the surface of the organelle.

- The protein is generally unfolded for translocation, a state maintained in the cytoplasm by chaperones.

- Threading of the protein through a membrane requires energy and organellar chaperones on the *trans* side of the membrane.

- Cycles of binding and release of the protein to the chaperone result in pulling of its polypeptide chain through the membrane.

- Other proteins within the organelle catalyze folding of the protein, often attaching cofactors or oligosaccharides and assembling them into active monomers or oligomers.

Source: Data from McNew JA, Goodman JM: The targeting and assembly of peroxisomal proteins: some old rules do not apply. Trends Biochem Sci 1998;21:54. Reprinted with permission from Elsevier.

LOCALIZATION SIGNALS, IMPORTINS, & EXPORTINS ARE INVOLVED IN TRANSPORT OF MACROMOLECULES IN & OUT OF THE NUCLEUS

It has been estimated that more than a million macromolecules per minute are transported between the nucleus and the cytoplasm in an active eukaryotic cell. These macromolecules include histones, ribosomal proteins and ribosomal subunits, transcription factors, and mRNA molecules. The transport is bidirectional and occurs through the **nuclear pore complexes** (NPCs). These are complex structures with a mass approximately 15 times that of a ribosome and are composed of aggregates of about 30 different proteins. The minimal diameter of an NPC is approximately 9 nm. Molecules smaller than about 40 kDa can pass through the channel of the NPC by **diffusion**, but **special translocation mechanisms** exist for larger molecules. These mechanisms are under intensive investigation, but some important features have already emerged.

Here we shall mainly describe **nuclear import** of certain macromolecules. The general picture that has emerged is that proteins to be imported (cargo molecules) carry a **nuclear localization signal (NLS).** One example of an NLS is the amino acid sequence $(Pro)_2$-$(Lys)_3$-Arg-Lys-Val (see Table 46–1), which is markedly rich in basic lysine residues. Depending on which NLS it contains, a cargo molecule interacts with one of a family of soluble proteins called **importins,** and the complex **docks** transiently at the NPC. Another family of proteins called **Ran** plays a critical regulatory role in the interaction of the complex with the NPC and in its translocation through the NPC. Ran proteins are small monomeric nuclear **GTPases** and, like other GTPases, exist in either GTP-bound or GDP-bound states. They are themselves regulated by **guanine nucleotide exchange factors** (GEFs), which are located

in the nucleus, and Ran **guanine-activating proteins** (GAPs), which are predominantly cytoplasmic. The GTP-bound state of Ran is favored in the nucleus and the GDP-bound state in the cytoplasm. The conformations and activities of Ran molecules vary depending on whether GTP or GDP is bound to them (the GTP-bound state is active; see discussion of G proteins in Chapter 42). The **asymmetry** between nucleus and cytoplasm—with respect to which of these two nucleotides is bound to Ran molecules—is thought to be crucial in understanding the roles of Ran in transferring complexes unidirectionally across the NPC. When **cargo molecules** are **released inside the nucleus**, the **importins recirculate to the cytoplasm** to be used again. Figure 46–4 summarizes some of the principal features in the above process.

Proteins similar to importins, referred to as **exportins,** are involved in the export of many macromolecules (various protein, tRNA molecules, ribosomal subunits and certain mRNA molecules) from the nucleus. Cargo molecules for export carry **nuclear export signals (NESs).** Ran proteins are involved in this process also, and it is now established that the processes of import and export share a number of common features. The family of importins and exportins are referred to as **karyopherins.**

Another system is involved in the translocation of the majority of **mRNA molecules**. These are exported from the nucleus to the cytoplasm as ribonucleoprotein (RNP) complexes attached to a protein named **mRNP exporter**. This is a heterodimeric molecule (ie, composed of 2 different subunits, TAP and Nxt-1) which carries RNP molecules through the NPC. Ran is not involved. This system appears to use the hydrolysis of **ATP** by an RNA helicase (Dbp5) to drive translocation.

Other **small monomeric GTPases** (eg, ARF, Rab, Ras, and Rho) are important in various cellular processes such as vesicle formation and transport (ARF and Rab; see below), certain growth and differentiation processes (Ras), and formation of the actin cytoskeleton. A process involving GTP and GDP is also crucial in the transport of proteins across the membrane of the ER (see below).

PROTEINS IMPORTED INTO PEROXISOMES CARRY UNIQUE TARGETING SEQUENCES

The **peroxisome** is an important organelle involved in aspects of the metabolism of many molecules, including fatty acids and other lipids (eg, plasmalogens, cholesterol, bile acids), purines, amino acids, and hydrogen peroxide. The peroxisome is bounded by a single membrane and contains more than 50 enzymes; catalase and urate oxidase are marker enzymes for this organelle. Its proteins are **synthesized on cytosolic polyribosomes** and fold prior to import. The pathways of import of a number of its proteins and enzymes have been studied, some being **matrix components** (see Figure 46–5) and others **mem-**

FIGURE 46–4 Simplified representation of the entry of a protein into the nucleoplasm. As shown in the top left-hand side of the figure, a cargo molecule in the cytoplasm via its NLS interacts to form a complex with an importin. (This may be either importin α or both importin α and importin β.) This complex next interacts with Ran·GDP and traverses the NPC into the nucleoplasm. In the nucleoplasm, Ran·GDP is converted to Ran·GTP by GEF, causing a conformational change in Ran resulting in the cargo molecule being released. The importin-Ran·GTP complex then leaves the nucleoplasm via the NPC to return to the cytoplasm. In the cytoplasm, due to the action of GTP-activating protein (GAP), which converts GTP to GDP, the importin is released to participate in another import cycle. The Ran·GTP is the active form of the complex, with the Ran·GDP form being considered inactive. Directionality is believed to be conferred on the overall process by the dissociation of Ran·GTP in the nucleoplasm. (C, cargo molecule; I, importin; NLS, nuclear localizing signal; NPC, nuclear pore complex; GEF, guanine nucleotide exchange factor; GAP, GTPase activating factor.) (Modified, with permission, from Lodish H, et al: *Molecular Cell Biology*, 6th ed. W.H. Freeman & Co., 2008.)

brane components. At least two **peroxisomal-matrix targeting sequences (PTSs)** have been discovered. One, **PTS1**, is a tripeptide (ie, Ser-Lys-Leu [SKL], but variations of this sequence have been detected) located at the carboxyl terminal of a number of matrix proteins, including catalase. Another, **PTS2**, is at the N-terminus and has been found in at least four matrix proteins (eg, thiolase). Neither of these two sequences is cleaved after entry into the matrix. Proteins containing PTS1 sequences **form complexes** with a cytosolic receptor protein **(Pex5)** and proteins containing PTS2 sequences complex with another receptor protein. The resulting complexes then interact with a membrane receptor complex, **Pex2/10/12**, which translocates them into the matrix. Proteins involved in further transport of proteins into the matrix are also present. Pex5 is re-cycled to the cytosol. Most peroxisomal **membrane proteins** have been found to contain neither of the above two targeting sequences, but apparently contain others. The import

system can handle **intact oligomers** (eg, tetrameric catalase). Import of **matrix proteins** requires **ATP**, whereas import of **membrane proteins** does **not**.

Most Cases of Zellweger Syndrome Are Due to Mutations in Genes Involved in the Biogenesis of Peroxisomes

Interest in import of proteins into peroxisomes has been stimulated by studies on **Zellweger syndrome.** This condition is apparent at birth and is characterized by **profound neurologic impairment**, victims often dying within a year. The number of peroxisomes can vary from being almost normal to being virtually absent in some patients. Biochemical findings include an accumulation of very-long-chain fatty acids, abnormalities of the synthesis of bile acids, and a marked reduction of plasmalogens. The condition is believed to be due to **mutations**

FIGURE 46–5 Schematic representation of the entry of a protein into the peroxisomal matrix. The protein to be imported into the matrix is synthesized on cytosolic polyribosomes, assumes its folded shape prior to import, and contains a C-terminal peroxisomal targeting sequence (PTS). It interacts with cytosolic receptor protein Pex5, and the complex then interacts with a receptor on the peroxisomal membrane, Pex14. In turn, the protein-Pex 14 complex passes to the Pex 2/10/12 complex on the peroxisomal membrane and is translocated. Pex 5 is returned to the cytosol. The protein retains its PTS in the matrix. (Modified, with permission, from Lodish H, et al: *Molecular Cell Biology*, 6th ed. W.H. Freeman & Co., 2008.)

in genes encoding certain proteins—so called **peroxins**—involved in various steps of **peroxisome biogenesis** (such as the import of proteins described above), or in genes encoding **certain peroxisomal enzymes** themselves. Two closely related conditions are **neonatal adrenoleukodystrophy** and **infantile Refsum disease.** Zellweger syndrome and these two conditions represent a **spectrum** of overlapping features, with Zellweger syndrome being the **most severe** (many proteins affected) and infantile Refsum disease the least severe (only one or a few proteins affected). Table 46–3 lists these and related conditions.

THE SIGNAL HYPOTHESIS EXPLAINS HOW POLYRIBOSOMES BIND TO THE ENDOPLASMIC RETICULUM

As indicated above, the **rough ER branch** is the second of the two branches involved in the synthesis and sorting of proteins. In this branch, proteins are synthesized on **membrane-bound polyribosomes** and **translocated into the lumen** of the rough ER prior to further sorting (Figure 46–2).

The **signal hypothesis** was proposed by Blobel and Sabatini partly to explain the distinction between free and membrane-bound polyribosomes. They found that proteins synthesized on membrane-bound polyribosomes contained a peptide exten-

sion **(signal peptide)** at their amino terminals which mediated their attachment to the membranes of the ER. As noted above, proteins whose entire synthesis occurs on free polyribosomes

TABLE 46–3   Disorders Due to Peroxisomal Abnormalities

	OMIM Number[1]
Zellweger syndrome	214100
Neonatal adrenoleukodystrophy	202370
Infantile Refsum disease	266510
Hyperpipecolic academia	239400
Rhizomelic chondrodysplasia punctata	215100
Adrenoleukodystrophy	300100
Pseudoneonatal adrenoleukodystrophy	264470
Pseudo-Zellweger syndrome	261515
Hyperoxaluria type 1	259900
Acatalasemia	115500
Glutaryl-CoA oxidase deficiency	231690

Source: Reproduced, with permission, from Seashore MR, Wappner RS: *Genetics in Primary Care & Clinical Medicine.* Appleton & Lange, 1996.
[1]OMIM = *Online Mendelian Inheritance in Man.* Each number specifies a reference in which information regarding each of the above conditions can be found.

TABLE 46–4   Some Properties of Signal Peptides

- Usually, but not always, located at the amino terminal
- Contain approximately 12–35 amino acids
- Methionine is usually the amino terminal amino acid
- Contain a central cluster of hydrophobic amino acids
- Contain at least one positively charged amino acid near their amino terminal
- Usually cleaved off at the carboxyl terminal end of an Ala residue by signal peptidase

lack this signal peptide. An important aspect of the signal hypothesis was that it suggested—as turns out to be the case—that **all ribosomes have the same structure** and that the distinction between membrane-bound and free ribosomes depends solely on the former carrying proteins that have signal peptides. Much evidence has confirmed the original hypothesis. Because many membrane proteins are synthesized on membrane-bound polyribosomes, the signal hypothesis plays an important role in **concepts of membrane assembly**. Some **characteristics of signal peptides** are summarized in Table 46–4.

Figure 46–6 illustrates the principal features in relation to the passage of **a secreted protein** through the membrane

of the ER. It incorporates features from the original signal hypothesis and from subsequent work. The mRNA for such a protein encodes an amino terminal **signal peptide** (also variously called a leader sequence, a transient insertion signal, a signal sequence, or a presequence). The signal hypothesis proposed that the protein is inserted into the ER membrane at the same time as its mRNA is being translated on polyribosomes, so-called **cotranslational insertion.** As the signal peptide emerges from the large subunit of the ribosome, it is recognized by a **signal recognition particle (SRP)** that blocks further translation after about 70 amino acids have been polymerized (40 buried in the large ribosomal subunit and 30 exposed). The block is referred to as **elongation arrest.** The SRP contains **six proteins** and has a **7S RNA** associated with it that is closely related to the Alu family of highly repeated DNA sequences (Chapter 35). The SRP-imposed block is not released until the SRP-signal peptide-polyribosome complex has bound to the so-called **docking protein** (SRP-R, a receptor for the SRP) on the ER membrane; the SRP thus guides the signal peptide to the SRP-R and prevents premature folding and expulsion of the protein being synthesized into the cytosol.

The **SRP-R** is an integral membrane protein composed of **α and β subunits**. The α subunit **binds GDP** and the β subunit spans the membrane. When the SRP-signal peptide complex interacts with the receptor, the **exchange of GDP for GTP is**

FIGURE 46–6 Diagram of the signal hypothesis for the transport of secreted proteins across the ER membrane. The ribosomes synthesizing a protein move along the messenger RNA specifying the amino acid sequence of the protein. (The messenger is represented by the line between 5′ and 3′.) The codon AUG marks the start of the message for the protein; the hatched lines that follow AUG represent the codons for the signal sequence. As the protein grows out from the larger ribosomal subunit, the signal sequence is exposed and bound by the signal recognition particle (SRP). Translation is blocked until the complex binds to the "docking protein," also designated SRP-R (represented by the black bar) on the ER membrane. There is also a receptor (red bar) for the ribosome itself. The interaction of the ribosome and growing peptide chain with the ER membrane results in the opening of a channel through which the protein is transported to the interior space of the ER. During translocation, the signal sequence of most proteins is removed by an enzyme called the "signal peptidase," located at the luminal surface of the ER membrane. The completed protein is eventually released by the ribosome, which then separates into its two components, the large and small ribosomal subunits. The protein ends up inside the ER. See text for further details. (Slightly modified and reproduced, with permission, from Marx JL: Newly made proteins zip through the cell. Science 1980;207:164. Copyright ©1980 by the American Association for the Advancement of Science.)

stimulated. This form of the receptor (with GTP bound) has a **high affinity for the SRP** and thus **releases the signal peptide**, which binds to the translocation machinery (translocon) also present in the ER membrane. The α subunit then **hydrolyzes its bound GTP**, restoring GDP and completing a GTP-GDP cycle. The **unidirectionality** of this cycle helps drive the interaction of the polyribosome and its signal peptide with the ER membrane in the forward direction.

The **translocon** consists of three membrane proteins (the Sec61 complex) that form a **protein-conducting channel** in the ER membrane through which the newly synthesized protein may pass. The channel appears to be **open only when a signal peptide is present**, preserving conductance across the ER membrane when it closes. The conductance of the channel has been measured experimentally.

The insertion of the signal peptide into the conducting channel, while the other end of the parent protein is still attached to ribosomes, is termed "**cotranslational insertion.**" The process of elongation of the remaining portion of the protein probably facilitates passage of the nascent protein across the lipid bilayer as the ribosomes remain attached to the membrane of the ER. Thus, the rough (or ribosome-studded) ER is formed. It is important that the protein be kept in an **unfolded state** prior to entering the conducting channel—otherwise, it may not be able to gain access to the channel.

Ribosomes remain attached to the ER during synthesis of signal peptide-containing proteins but are **released** and **dissociated** into their two types of subunits when the process is completed. The signal peptide is hydrolyzed by **signal peptidase,** located on the luminal side of the ER membrane (Figure 46–6), and then is apparently rapidly degraded by proteases.

Cytochrome P450 (Chapter 53), an integral ER membrane protein, does not completely cross the membrane. Instead, it resides in the membrane with its signal peptide intact. Its passage through the membrane is prevented by a sequence of amino acids called a **halt-** or **stop-transfer signal.**

Secretory proteins and **soluble proteins destined for organelles distal to the ER** completely traverse the membrane bilayer and are discharged into the lumen of the ER. *N*-Glycan chains, if present, are added (Chapter 47) as these proteins traverse the inner part of the ER membrane—a process called "**cotranslational glycosylation.**" Subsequently, the proteins are found in the **lumen of the Golgi apparatus,** where further changes in glycan chains occur (Figure 47–9) prior to intracellular distribution or secretion. There is strong evidence that the **signal peptide** is involved in the process of protein **insertion into ER membranes.** Mutant proteins, containing altered signal peptides in which a hydrophobic amino acid is replaced by a hydrophilic one, are not inserted into ER membranes. Nonmembrane proteins (eg, α-globin) to which signal peptides have been attached by genetic engineering can be inserted into the lumen of the ER or even secreted.

There is evidence that the ER membrane is involved in **retrograde transport** of various molecules from the ER lumen **to the cytosol.** These molecules include unfolded or misfolded glycoproteins, glycopeptides, and oligosaccharides. At least some of these molecules are **degraded in proteasomes** (see below). Whether the translocon is involved in retrotranslocation is not clear; one or more other channels may be involved. Whatever the case, there is **two-way traffic** across the ER membrane.

PROTEINS FOLLOW SEVERAL ROUTES TO BE INSERTED INTO OR ATTACHED TO THE MEMBRANES OF THE ENDOPLASMIC RETICULUM

The routes that proteins follow to be inserted into the membranes of the ER include the following.

Cotranslational Insertion

Figure 46–7 shows a variety of ways in which proteins are distributed in the plasma membrane. In particular, the **amino terminals** of certain proteins (eg, the LDL receptor) can be seen to be on the extracytoplasmic face, whereas for other proteins (eg, the asialoglycoprotein receptor) the **carboxyl terminals** are on this face. To explain these dispositions, one must consider the initial biosynthetic events at the ER membrane. The **LDL receptor** enters the ER membrane in a manner analogous to a secretory protein (Figure 46–6); it partly traverses the ER membrane, its signal peptide is cleaved, and its amino terminal protrudes into the lumen. However, it is **retained in the membrane** because it contains a highly hydrophobic segment, the **halt-** or **stop-transfer signal.** This sequence forms the single transmembrane segment of the protein and is its membrane-anchoring domain. The small patch of ER membrane in which the newly synthesized LDL receptor is located subsequently **buds off** as a component of a **transport vesicle.** As described below in the discussion of asymmetry of proteins and lipids in membrane assembly, the disposition of the receptor in the ER membrane is preserved in the vesicle, which eventually fuses with the plasma membrane. In contrast, the **asialoglycoprotein receptor** possesses an **internal insertion sequence,** which inserts into the membrane but is **not cleaved.** This acts as an **anchor,** and its carboxyl terminal is extruded through the membrane. The **more complex disposition** of the **transporters** (eg, for glucose) can be explained by the fact that **alternating** transmembrane α-helices act as **uncleaved insertion sequences** and as **halt-transfer signals**, respectively. Each pair of helical segments is inserted as a **hairpin.** Sequences that determine the structure of a protein in a membrane are called **topogenic sequences.** As explained in the legend to Figure 46–7, the above three proteins are examples of **type I, type II,** and **type IV** transmembrane proteins.

Synthesis on Free Polyribosomes & Subsequent Attachment to the Endoplasmic Reticulum Membrane

An example is **cytochrome b_5,** which enters the ER membrane spontaneously.

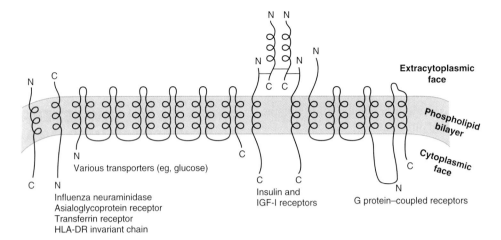

FIGURE 46–7 Variations in the way in which proteins are inserted into membranes. This schematic representation, which illustrates a number of possible orientations, shows the segments of the proteins within the membrane as α helices and the other segments as lines. The LDL receptor, which crosses the membrane once and has its amino terminal on the exterior, is called a type I transmembrane protein. The asialoglycoprotein receptor, which also crosses the membrane once but has its carboxyl terminal on the exterior, is called a type II transmembrane protein. Cytochrome P450 (not shown) is an example of a type III transmembrane protein; its disposition is similar to type I proteins, but does not contain a cleavable signal sequence. The various transporters indicated (eg, glucose) cross the membrane a number of times and are called type IV transmembrane proteins; they are also referred to as polytopic membrane proteins. (N, amino terminal; C, carboxyl terminal.) (Adapted, with permission, from Wickner WT, Lodish HF: Multiple mechanisms of protein insertion into and across membranes. Science 1985;230:400. Copyright ©1985 by the American Association for the Advancement of Science.)

Retention at the Luminal Aspect of the Endoplasmic Reticulum by Specific Amino Acid Sequences

A number of proteins possess the amino acid sequence **KDEL** (Lys-Asp-Glu-Leu) at their carboxyl terminal (see Table 46–1). KDEL-containing proteins first travel to the **GA** in COPII transport vesicles (see below), interact there with a specific KDEL receptor protein, and then **return in COPI transport vesicles to the ER**, where they dissociate from the receptor.

Retrograde Transport from the Golgi Apparatus

Certain other **non-KDEL-containing proteins** destined for the membranes of the ER also pass to the Golgi and then return, by **retrograde vesicular transport,** to the ER to be inserted therein (see below).

The foregoing paragraphs demonstrate that a **variety of routes** are involved in assembly of the proteins of the ER membranes; a similar situation probably holds for other membranes (eg, the mitochondrial membranes and the plasma membrane). Precise targeting sequences have been identified in some instances (eg, KDEL sequences).

The topic of membrane biogenesis is discussed further later in this chapter.

CHAPERONES ARE PROTEINS THAT PREVENT FAULTY FOLDING & UNPRODUCTIVE INTERACTIONS OF OTHER PROTEINS

Molecular chaperones have been referred to previously in this Chapter. A number of important properties of these proteins are listed in Table 46–5, and the names of some of particular importance in the ER are listed in Table 46–6. Basically, they **stabilize unfolded or partially folded intermediates,** allowing them time to fold properly, and prevent inappropriate interactions, thus combating the formation of nonfunctional structures. Most chaperones exhibit **ATPase activity** and bind ADP and ATP. This activity is important for their effect on protein folding. The ADP-chaperone complex often has a high affinity for the unfolded protein, which, when bound, stimulates release of ADP with replacement by ATP. The ATP-chaperone complex, in turn, releases segments of the protein that have folded properly, and the **cycle** involving ADP and ATP binding is repeated until the protein is released.

Chaperonins are the second major class of chaperones. They form complex **barrel-like structures** in which an unfolded protein is retained, giving it time and suitable conditions in which to fold properly. The mtGroEL chaperonin has been much studied. It is polymeric, has two ring-like struc-

TABLE 46–5 Some Properties of Chaperone Proteins

- Present in a wide range of species from bacteria to humans

- Many are so-called heat shock proteins (Hsp)

- Some are inducible by conditions that cause unfolding of newly synthesized proteins (eg, elevated temperature and various chemicals)

- They bind to predominantly hydrophobic regions of unfolded proteins and prevent their aggregation

- They act in part as a quality control or editing mechanism for detecting misfolded or otherwise defective proteins

- Most chaperones show associated ATPase activity, with ATP or ADP being involved in the protein–chaperone interaction

- Found in various cellular compartments such as cytosol, mitochondria, and the lumen of the endoplasmic reticulum

TABLE 46–6 Some Chaperones and Enzymes Involved in Folding That Are Located in the Rough Endoplasmic Reticulum

- BiP (immunoglobulin heavy chain binding protein)

- GRP94 (glucose-regulated protein)

- Calnexin

- Calreticulin

- PDI (protein disulfide isomerase)

- PPI (peptidyl prolyl *cis-trans* isomerase)

tures, each composed of seven identical subunits, and again ATP is involved in its action.

Several examples of chaperones were introduced above when the sorting of mitochondrial proteins was discussed. The **immunoglobulin heavy chain binding protein (BiP)** is located in the lumen of the ER. This protein **promotes proper folding by preventing aggregation** and will bind abnormally folded immunoglobulin heavy chains and certain other proteins and prevent them from leaving the ER. Another important chaperone is **calnexin,** a calcium-binding protein located in the ER membrane. This protein binds a wide variety of proteins, including major histocompatibility complex (MHC) antigens and a variety of plasma proteins. As described in Chapter 47, calnexin binds the monoglycosylated species of glycoproteins that occur during processing of glycoproteins, retaining them in the ER until the glycoprotein has folded properly. **Calreticulin,** which is also a calcium-binding protein, has properties similar to those of calnexin; it is not membrane-bound. Chaperones are not the only proteins in the ER lumen that are concerned with proper folding of proteins. Two **enzymes** are present that play an active role in folding. **Protein disulfide isomerase (PDI)** promotes rapid **formation and** reshuffling of disulfide bonds until the correct set is achieved. **Peptidyl prolyl isomerase (PPI)** accelerates folding of proline-containing proteins by catalyzing the cis-trans isomerization of X-Pro bonds, where X is any amino acid residue.

ACCUMULATION OF MISFOLDED PROTEINS IN THE ENDOPLASMIC RETICULUM CAN INDUCE THE UNFOLDED PROTEIN RESPONSE (UPR)

Maintenance of **homeostasis in the ER** is important for normal cell function. When the unique environment within the lumen of the ER is perturbed (eg, changes in ER Ca^{2+}, alterations of redox status, exposure to various toxins or some viruses), this can lead to reduced protein folding capacity and the accumulation of misfolded proteins. The accumulation of misfolded proteins in the ER is referred to as ER stress. The cell has evolved a mechanism termed the unfolded protein response (UPR) to sense the levels of misfolded proteins and initiate intracellular signaling mechanisms to compensate for the stress conditions and restore ER homeostasis. The UPR is initiated by ER stress sensors which are transmembrane proteins embedded in the ER membrane. Activation of these stress sensors causes three principal effects: transient inhibition of translation to reduce the amount of newly synthesized proteins and induction of a transcriptional response that leads to increased expression of ER chaperones and of proteins involved in degradation of misfolded ER proteins (discussed below). Therefore, the UPR increases the ER folding capacity and prevents a buildup of unproductive and potentially toxic protein products, in addition to other responses to restore cellular homeostasis. However, if impairment of folding persists, cell death pathways (apoptosis) are activated. A more complete understanding of the UPR is likely to provide new approaches to treating diseases in which ER stress and defective protein folding occur (see Table 46–7).

MISFOLDED PROTEINS UNDERGO ENDOPLASMIC RETICULUM–ASSOCIATED DEGRADATION (ERAD)

Misfolded proteins occur in many genetic diseases (eg, see Table 46–7). Proteins that misfold in the ER are selectively **transported back across the ER (retrotranslocation or dislocation)** to enter **proteasomes** present in the cytosol. The precise route by which the misfolded proteins pass back across the ER membrane is still under investigation. If a channel is involved, it does not appear to be the translocon (Sec61 complex) described earlier, although it may contain some of its components. The energy for translocation appears to be at least partly supplied by **p97**, an AAA-ATPase (one of a family of **A**TPases **A**ssociated with various cellular **A**ctivities). **Chaperones** present in the lumen of the ER (eg, BiP) and in the cytosol help target misfolded proteins to proteasomes. Prior to entering proteasomes, most proteins are **ubiquitinated** (see the next paragraph) and are escorted to proteasomes by polyubiquitin-binding proteins. Ubiquitin ligases are present in the ER membrane. The above process is referred to as ERAD and is outlined in Figure 46–8.

TABLE 46–7   Some Conformational Diseases That Are Caused by Abnormalities in Intracellular Transport of Specific Proteins and Enzymes Due to Mutations[1]

Disease	Affected Protein
α_1-Antitrypsin deficiency with liver disease (OMIM 107400)	α_1-Antitrypsin
Chediak-Higashi syndrome (OMIM 214500)	Lysosomal trafficking regulator
Combined deficiency of factors V and VIII (OMIM 227300)	ERGIC53, a mannose-binding lectin
Cystic fibrosis (OMIM 219700)	CFTR
Diabetes mellitus [some cases] (OMIM 147670)	Insulin receptor (α-subunit)
Familial hypercholesterolemia, autosomal dominant (OMIM 143890)	LDL receptor
Gaucher disease (OMIM 230800)	β-Glucosidase
Hemophilia A (OMIM 306700) and B (OMIM 306900)	Factors VIII and IX
Hereditary hemochromatosis (OMIM 235200)	HFE
Hermansky-Pudlak syndrome (OMIM 203300)	AP-3 adaptor complex β3A subunit
I-cell disease (OMIM 252500)	N-acetylglucosamine 1-phospho-transferase
Lowe oculocerebrorenal syndrome (OMIM 309000)	PIP$_2$ 5-phosphatase
Tay-Sachs disease (OMIM 272800)	β-Hexosaminidase
von Willebrand disease (OMIM 193400)	von Willebrand factor

Abbreviation: PIP$_2$, phosphatidylinositol 4,5-bisphosphate.
Note: Readers should consult textbooks of medicine or pediatrics for information on the clinical manifestations of the conditions listed.
[1]See Schroder M, Kaufman RJ: The Mammalian Unfolded Protein Response. Annu Rev Biochem 2005;74, 739 and OlkonnenV, Ikonen E: Genetic defects of intracellular membrane transport. N Eng J Med 2000;343: 10095.

UBIQUITIN IS A KEY MOLECULE IN PROTEIN DEGRADATION

There are two major pathways of protein degradation in eukaryotes. One involves **lysosomal proteases** and does not require ATP. The other pathway involves **ubiquitin** and is ATP-dependent. It plays the major role in the degradation of proteins, and is particularly associated with **disposal of misfolded proteins and regulatory enzymes that have short half-lives.** Research on ubiquitin has expanded rapidly, and it is known to be involved in **cell cycle regulation** (degradation of cyclins), **DNA repair**, **activation of NFκB** (see Chapter 50), **muscle wasting**, **viral infections**, and **many other** important physiologic and pathologic processes. Ubiquitin is a **small** (76 amino acids), **highly conserved protein** that plays a key role in **marking** various proteins for subsequent **degradation in proteasomes**. The mechanism of attachment of ubiquitin to a target protein (eg, a misfolded form of CFTR, the protein in-

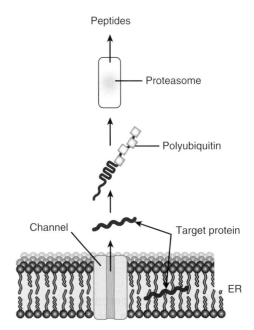

FIGURE 46–8 Schematic diagram of the events in ERAD. A target protein (which may be misfolded or normally folded) undergoes retrograde transport through the ER membrane into the cytosol, where it is subjected to polyubiquitination. Following polyubiquitination, it enters a proteasome, inside which it is degraded to small peptides that exit and may have several fates. Liberated ubiquitin molecules are recycled. The precise route by which misfolded proteins pass back through the ER membrane is not as yet known; a channel may exist (as shown in the figure), but that has not apparently been established.

volved in the causation of cystic fibrosis; see Chapters 40 & 54) is shown in Figure 46–9 and involves **three enzymes**: an **activating enzyme**, a **conjugating enzyme**, and a **ligase**. There are a number of types of conjugating enzymes, and, surprisingly, some hundreds of different ligases. It is the latter enzyme that confers substrate specificity. Once the molecule of ubiquitin is attached to the protein, a number of others are also attached, resulting in a **polyubiquitinated target protein**. It has been estimated that a **minimum of four ubiquitin molecules** must be attached to commit a target molecule to degradation in a proteasome. Ubiquitin can be **cleaved** from a target protein by deubiquitinating **enzymes** and the liberated ubiquitin can be reused.

Ubiquitinated Proteins Are Degraded in Proteasomes

Polyubiquitinated target proteins enter the proteasomes, located in the cytosol. The proteasome is a relatively **large cylindrical structure** and is composed of some **50 subunits.** The proteasome has a hollow **core**, and one or two **caps** that play a regulatory role. Target proteins are **unfolded** by ATPases present in the proteasome caps. Proteasomes can hydrolyze a very wide variety of peptide bonds. Target proteins pass into the core to be degraded to small peptides, which then exit the proteasome (Figure 46–8) to be further degraded by cytosolic peptidases. Both normally and abnormally folded proteins

FIGURE 46–9 Sequence of reactions in addition of ubiquitin to a target protein. In the reaction catalyzed by E1, the C-terminal COO⁻ group of ubiquitin is linked in a thioester bond to an SH group of E1. In the reaction catalyzed by E2, the activated ubiquitin is transferred to an SH group of E2. In the reaction catalyzed by E3, ubiquitin is transferred from E2 to an ε-amino group on a lysine of the target protein. Additional rounds of ubiquitination then build up the polyubiquitin chain. (Ub, ubiquitin; E1, activating enzyme; E2, conjugating enzyme; E3, ligase; LYS ^^^^ Pr, target protein.)

are substrates for the proteasome. Liberated ubiquitin molecules are recycled. The proteasome plays an important role in **presenting small peptides** produced by **degradation of various viruses** and other molecules to **major histocompatibility class I molecules**, a key step in antigen presentation to T lymphocytes.

TRANSPORT VESICLES ARE KEY PLAYERS IN INTRACELLULAR PROTEIN TRAFFIC

Proteins that are synthesized on membrane-bound polyribosomes and are destined for the GA or PM reach these sites inside **transport vesicles.** Those vesicles involved in **anterograde transport** (COPII) from the ER to the GA and in **retrograde transport** (COPI) from the GA to the ER are clathrin-free. Transport and secretory vesicles carrying cargo from the GA to the PM are also clathrin-free. The vesicles involved in endocytosis (see discussions of the LDL receptor in Chapters 25 & 26) are coated with clathrin, as are certain vesicles carrying cargo to lysosomes. For the sake of clarity, the non-clathrin-coated vesicles are referred to in this text as **transport vesicles.** Table 46–8 summarizes **the types** and **functions** of the major vesicles identified to date.

TABLE 46–8 Some Types of Vesicles and Their Functions

Vesicle	Function
COPI	Involved in intra-GA transport and retrograde transport from the GA to the ER
COPII	Involved in export from the ER to either ERGIC or the GA
Clathrin	Involved in transport in post-GA locations including the PM, TGN and endosomes
Secretory vesicles	Involved in regulated secretion from organs such as the pancreas (eg, secretion of insulin)
Vesicles from the TGN to the PM	They carry proteins to the PM and are also involved in constitutive secretion

Abbreviations: GA, Golgi apparatus; ER, endoplasmic reticulum; ERGIC, ER-GA intermediate compartment; PM, plasma membrane; TGN, *trans*-Golgi network.
Note: Each vesicle has its own set of coat proteins. Clathrin is associated with various adapter proteins (APs), eg, AP-1, AP-2 and AP-3, forming different types of clathrin vesicles. These various clathrin vesicles have different intracellular targets. The proteins of secretory vesicles and vesicles involved in transport from the GA to the PM are not well characterized, nor are the mechanisms involved in their formations and fates.

Model of Transport Vesicles Involves SNAREs & Other Factors

Vesicles lie at the heart of intracellular transport of many proteins. Significant progress has been made in understanding the events involved in vesicle formation and transport. This has transpired because of the use of a number of approaches. In particular, the use by Schekman and colleagues of **genetic approaches for studying vesicles in yeast** and the development by Rothman and colleagues of **cell-free systems** to study vesicle formation have been crucial. For instance, it is possible to observe, by electron microscopy, budding of vesicles from Golgi preparations incubated with cytosol and ATP. The overall mechanism is complex, with its own **nomenclature** (Table 46–9), and involves a variety of cytosolic and membrane proteins, GTP, ATP, and accessory factors. **Budding, tethering, docking,** and **membrane fusion** are key steps in the life cycles of vesicles with Sar, ARF, and the Rab GTPases (see below) acting as **molecular switches.**

There are common general steps in transport vesicle formation, vesicle targeting and fusion with a target membrane, irrespective of the membrane the vesicle forms from or its intracellular destination. The nature of the coat proteins, GTPases and targeting factors differ depending on where the vesicle forms from and its eventual destination. Transport from the ER to the Golgi is the best studied example and will be used to illustrate these steps. **Anterograde vesicular transport** from the ER to the Golgi involves **COPII vesicles** and the process can be considered to occur in eight steps (Figure 46–10). The basic concept is that each transport vesicle is loaded with specific cargo and also one or more **v-SNARE**

TABLE 46–9   Some Factors Involved in the Formation of Non-Clathrin-Coated Vesicles and Their Transport

- ARF: ADP-ribosylation factor, a GTPase involved in formation of COPI and also clathrin-coated vesicles.

- Coat proteins: A family of proteins found in coated vesicles. Different transport vesicles have different complements of coat proteins.

- GTP-γ-S: A nonhydrolyzable analog of GTP, used to test the involvement of GTP in biochemical processes.

- NEM: N-Ethylmaleimide, a chemical that alkylates sulfhydryl groups and inactivates NSF.

- NSF: NEM-sensitive factor, an ATPase.

- Sar1: A GTPase that plays a key role in assembly of COPII vesicles.

- Sec 12: A guanine nucleotide exchange factor (GERF) that interconverts Sar1.GDP and Sar1.GTP.

- α-SNAP: Soluble NSF attachment protein. Along with NSF, this protein is involved in dissociation of SNARE complexes.

- SNARE: SNAP receptor. SNAREs are key molecules in the fusion of vesicles with acceptor membranes.

- t-SNARE: Target SNARE.

- v-SNARE: Vesicle SNARE.

- Rab proteins: A family of Ras-related proteins (monomeric GTPases) first observed in rat brain. They are active when GTP is bound. Different Rab molecules dock different vesicles to acceptor membranes.

- Rab effector proteins: A family of proteins that interact with Rab molecules; some act to tether vesicles to acceptor membranes.

proteins that direct targeting. Each target membrane bears one or more **complementary t-SNARE proteins** with which the former interact, mediating SNARE protein-dependent vesicle-membrane fusion. In addition, **Rab proteins** also help direct the vesicles to specific membranes and are involved in tethering, prior to vesicle docking at a target membrane.

Step 1: **Budding** is initiated when **Sar1** is activated by binding **GTP**, which is exchanged for GDP via the action of **Sec12**. This causes a conformational change in Sar1·GTP, embedding it in the ER membrane to form a focal point for vesicle assembly.

Step 2: Various **coat proteins** bind to **Sar1·GTP**. In turn, membrane cargo proteins bind to the coat proteins and soluble cargo proteins inside vesicles bind to receptor regions of the former. Additional coat proteins are assembled to **complete bud formation**. Coat proteins promote budding, contribute to the curvature of buds and also help sort proteins.

Step 3: The **bud pinches off**, completing formation of the coated vesicle. The curvature of the ER membrane and protein–protein and protein–lipid interactions in the bud facilitate pinching off from ER exit sites.

Step 4: **Coat disassembly** (involving **dissociation of Sar1** and the **shell** of coat proteins) follows **hydrolysis of bound GTP to GDP** by Sar1, promoted by a specific coat protein.

Sar1 thus plays key roles in both assembly and dissociation of the coat proteins. Uncoating is necessary for fusion to occur.

Step 5: **Vesicle targeting** is achieved by attachment of **Rab** molecules to vesicles. Rab·GDP molecules in the cytosol are converted to Rab·GTP molecules by a specific guanine nucleotide exchange factor and these attach to the vesicles. The Rab. GTP molecules subsequently interact with **Rab effector proteins** on membranes to **tether** the vesicle to the membranes.

Step 6: **v-SNAREs pair with cognate t-SNAREs** in the target membrane to **dock** the vesicles and initiate fusion. Generally one v-SNARE in the vesicle pairs with three t-SNAREs on the acceptor membrane to form a tight **four-helix bundle**.

Step 7: **Fusion** of the vesicle with the acceptor membrane occurs once the v- and t-SNARES are closely aligned. After vesicle fusion and release of contents occurs, GTP is hydrolyzed to GDP, and the Rab·GDP molecules are released into the cytosol. When a SNARE on one membrane interacts with a SNARE on another membrane, linking the two membranes, this is referred to as a trans-SNARE complex or a SNARE pin. Interactions of SNARES on the same membrane form a cis-SNARE complex. In order to **dissociate the four-helix bundle** between the v- and t-SNARES so that they can be re-used, two additional proteins are required. These are an **ATPase** (NSF) and α-**SNAP**. NSF hydrolyzes ATP and the energy released dissociates the four-helix bundle making the SNARE proteins available for another round of membrane fusion.

Step 8: Certain components are **recycled** (eg, Rab, possibly v-SNAREs).

During the above cycle, SNARES, tethering proteins, Rab and other proteins all **collaborate** to deliver a vesicle and its contents to the appropriate site.

COPI, COPII, and Clathrin-Coated Vesicles Have Been Most Studied

The following points clarify and expand on the previous section.

1. As indicated in Table 46–8, there are **a number of different types of vesicles**. Other types of vesicles may remain to be discovered. Here we focus mainly on COPII, COPI and clathrin-coated vesicles. Each of these types has a different complement of proteins in its coat. The details of assembly for COPI and clathrin-coated vesicles are somewhat different from those described above. For example, **Sar1** is the protein involved in step 1 of formation of COPII vesicles, whereas **ARF** is involved in the formation of COPI and clathrin-coated vesicles. However, the principles concerning assembly of these different types are generally similar.

2. Regarding **selection** of cargo molecules by vesicles, this appears to be primarily **a function of the coat proteins** of vesicles. **Cargo molecules** via their sorting signals may interact with coat proteins either **directly** or via **intermediary proteins** that attach to coat proteins, and they then become enclosed in their appropriate vesicles. A number of **signal se-**

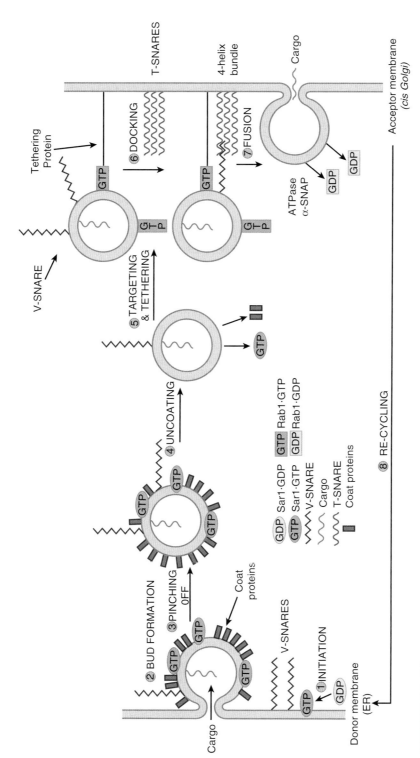

FIGURE 46–10 Model of the steps in a round of anterograde transport involving COPII vesicles. The cycle starts in the bottom left-hand side of the figure, where two molecules of Sar1 are represented as small ovals containing GDP. The steps in the cycle are described in the text. The various components are briefly described in Table 46–7. The roles of Rab and Rab effector proteins (see text) in the overall process are not dealt with in this figure. (Adapted, with permission, from Rothman JE: Mechanisms of intracellular protein transport. Nature 1994;372:55. Courtesy of E Degen.)

quences on cargo molecules have been identified (see Table 46–1). For example KDEL sequences direct certain ER resident proteins in retrograde flow to the ER in COPI vesicles. Di-acidic sequences (eg, Asp-X-Glu) and short hydrophobic sequences on membrane proteins are involved in interactions with coat proteins of COPII vesicles.

Proteins in the **apical** or **basolateral** areas of the plasma membranes of polarized epithelial cells can be transported to these sites in **transport vesicles** budding from the TGN. Different Rab proteins likely direct some vesicles to apical regions and others to basolateral regions. In certain cells, proteins are first directed to the basolateral membrane, then endocytosed and transported across the cell by **transcytosis** to the apical region. Yet another mechanism for sorting proteins to the apical region (or in some cases to the basolateral region) involves the **glycosylphosphatidylinositol (GPI) anchor** described in Chapter 47. This structure is also often present in **lipid rafts** (see Chapter 40).

Not all cargo molecules may have a sorting signal. Some highly abundant secretory proteins travel to various cellular destinations in transport vesicles by **bulk flow**; ie, they enter into transport vesicles at the same concentration that they occur in the organelle. The precise extent of bulk flow is not clearly known, although it appears that most proteins are actively sorted (concentrated) into transport vesicles and bulk flow is used by only a select group of cargo proteins.

3. Once proteins in the secretory pathway reach the *cis*-Golgi from the ER in vesicles, they can travel through the GA to the *trans*-Golgi **in vesicles,** or by a process called **cisternal maturation**, or perhaps in some cases by **simple diffusion**. A former view was that the GA is essentially a static organelle, allowing vesicular flow from one static cisterna to the next. There is now, however, evidence to support the view that the cisternae move and transform into one another (ie, cisternal maturation). In this model, vesicular elements from the ER fuse with one another to help form the *cis*-Golgi, which in turn can move forward to become the medial Golgi, etc. COPI vesicles return Golgi enzymes (eg, glycosyltransferases) back from distal cisternae of the GA to more proximal (eg, *cis*) cisternae.

4. Vesicles move through cells along **microtubules** or along **actin filaments**.

5. The fungal metabolite **brefeldin A prevents GTP from binding to ARF**, and thus inhibits formation of COPI vesicles. In its presence, the Golgi apparatus appears to **collapse into the ER**. It may do this by inhibiting the guanine nucleotide exchanger involved in formation of COPI vesicles. Brefeldin A has thus proven to be a useful tool for examining some aspects of Golgi structure and function.

6. **GTP-γ-S** (a nonhydrolyzable analog of GTP often used in investigations of the role of GTP in biochemical processes) **blocks disassembly of the coat** from coated vesicles, leading to a build-up of coated vesicles, facilitating their study.

7. As mentioned above, a family of Ras-like proteins, called the **Rab protein family,** are required in several steps of intracellular protein transport and also in regulated secretion and endocytosis. (Ras proteins are involved in cell signaling via receptor tyrosine kinases). Like Ras, Rab proteins are **small monomeric GTPases** that attach to the cytosolic faces of membranes (via **geranylgeranyl** lipid anchors). They **attach in the GTP-bound state** to the budding vesicle and are also present on acceptor membranes. Rab proteins interact with **Rab effector proteins**, that have various roles, such as involvement in tethering and in membrane fusion.

8. The fusion of **synaptic vesicles** with the plasma membrane of **neurons** involves a series of events similar to that described above. For example, one v-SNARE is designated **synaptobrevin** and two t-SNAREs are designated **syntaxin** and **SNAP 25** (synaptosome-associated protein of 25 kDa). **Botulinum B toxin** is one of the most lethal toxins known and the most serious cause of food poisoning. One component of this toxin is a **protease** that appears to **cleave only synaptobrevin**, thus **inhibiting release of acetylcholine** at the neuromuscular junction and possibly proving fatal, depending on the dose taken.

9. Although the above model refers to **non-clathrin-coated vesicles,** it appears likely that many of the events described above apply, at least in principle, to clathrin-coated vesicles.

10. Some proteins are further subjected to **further processing by proteolysis** while inside either transport or secretory vesicles. For example, **albumin** is synthesized by hepatocytes as **preproalbumin** (see Chapter 50). Its signal peptide is removed, converting it to **proalbumin**. In turn, proalbumin, while inside transport vesicles, is converted to **albumin** by action of **furin** (Figure 46–11). This enzyme cleaves a hexapeptide from proalbumin immediately C-terminal to a dibasic amino acid site (ArgArg). The resulting mature albumin is secreted into the plasma. Hormones such as **insulin** (see Chapter 41) are subjected to similar proteolytic cleavages while inside secretory vesicles.

THE ASSEMBLY OF MEMBRANES IS COMPLEX

There are many cellular membranes, each with its own specific features. No satisfactory scheme describing the assembly of

Signal peptidase   Furin

Preproalbumin ⟶ Signal peptide + Proalbumin ⟶ Hexapeptide + Albumin

FIGURE 46–11 Cleavage of preproalbumin to proalbumin and of the latter to albumin. Furin cleaves proalbumin at the C-terminal end of a basic dipeptide (ArgArg).

any one of these membranes is available. How various proteins are initially inserted into the membrane of the ER has been discussed above. The transport of proteins, including membrane proteins, to various parts of the cell inside vesicles has also been described. Some general points about membrane assembly remain to be addressed.

Asymmetry of Both Proteins & Lipids Is Maintained During Membrane Assembly

Vesicles formed from membranes of the ER and Golgi apparatus, either naturally or pinched off by homogenization, exhibit **transverse asymmetries** of both lipid and protein. These **asymmetries are maintained** during fusion of transport vesicles with the plasma membrane. The **inside** of the vesicles after fusion becomes the **outside of the plasma membrane**, and the cytoplasmic side of the vesicles remains the cytoplasmic side of the membrane (Figure 46–12). Since the transverse asymmetry of the membranes already exists in the vesicles of the ER well before they are fused to the plasma membrane, a major problem of membrane assembly becomes understanding how the integral proteins are inserted into the lipid bilayer of the ER. This problem was addressed earlier in this chapter.

Phospholipids are the major class of lipid in membranes. The enzymes responsible for the synthesis of phospholipids reside in the cytoplasmic surface of the cisternae of the ER. As phospholipids are synthesized at that site, they probably self-assemble into thermodynamically stable bimolecular layers, thereby expanding the membrane and perhaps promoting the detachment of so-called lipid vesicles from it. It has been proposed that these vesicles travel to other sites, donating their lipids to other membranes; however, little is known about this matter. As indicated above, cytosolic proteins that take up phospholipids from one membrane and release them to another (ie, **phospholipid exchange proteins**) have been demonstrated; they probably play a role in contributing to the specific lipid composition of various membranes.

It should be noted that the **lipid compositions** of the ER, Golgi and plasma membrane differ, the latter two membranes containing **higher amounts of cholesterol, sphingomyelin and glycosphingolipids**, and **less phosphoglycerides** than does the ER. Sphingolipids pack more densely in membranes than do phosphoglycerides. These differences affect the structures and functions of membranes. For example, the **thickness of the bilayer** of the GA and PM is greater than that of the ER, which affects what particular transmembrane proteins are found in these organelles. Also, **lipid rafts** (see earlier discussion) are believed to be formed in the GA

Lipids & Proteins Undergo Turnover at Different Rates in Different Membranes

It has been shown that the half-lives of the lipids of the ER membranes of rat liver are generally shorter than those of its

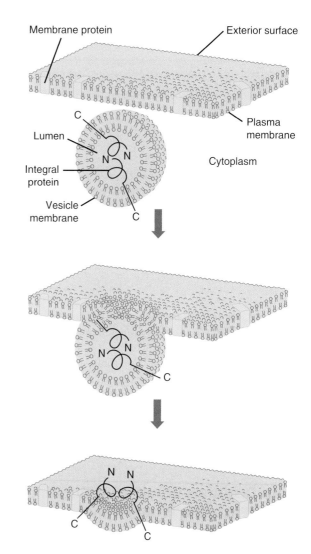

FIGURE 46–12 Fusion of a vesicle with the plasma membrane preserves the orientation of any integral proteins embedded in the vesicle bilayer. Initially, the amino terminal of the protein faces the lumen, or inner cavity, of such a vesicle. After fusion, the amino terminal is on the exterior surface of the plasma membrane. That the orientation of the protein has not been reversed can be perceived by noting that the other end of the molecule, the carboxyl terminal, is always immersed in the cytoplasm. The lumen of a vesicle and the outside of the cell are topologically equivalent. (Redrawn and modified, with permission, from Lodish HF, Rothman JE: The assembly of cell membranes. Sci Am [Jan] 1979;240:43.)

proteins, so that the **turnover rates of lipids and proteins are independent.** Indeed, different lipids have been found to have different half-lives. Furthermore, the half-lives of the proteins of these membranes vary quite widely, some exhibiting short (hours) and others long (days) half-lives. Thus, individual lipids and proteins of the ER membranes appear to be inserted into it relatively independently; this is the case for many other membranes.

The biogenesis of membranes is thus a complex process about which much remains to be learned. One indication of

TABLE 46–10   Some Major Features of Membrane Assembly

- Lipids and proteins are inserted independently into membranes.

- Individual membrane lipids and proteins turn over independently and at different rates.

- Topogenic sequences (eg, signal [amino terminal or internal] and stop-transfer) are important in determining the insertion and disposition of proteins in membranes.

- Membrane proteins inside transport vesicles bud off the endoplasmic reticulum on their way to the Golgi; final sorting of many membrane proteins occurs in the trans-Golgi network.

- Specific sorting sequences guide proteins to particular organelles such as lysosomes, peroxisomes, and mitochondria.

the complexity involved is to consider the number of **post-translational modifications** that membrane proteins may be subjected to prior to attaining their mature state. These include disulfide formation, proteolysis, assembly into multimers, glycosylation, addition of a glycophosphatidylinositol (GPI) anchor, sulfation on tyrosine or carbohydrate moieties, phosphorylation, acylation, and prenylation—a list that is not complete. Nevertheless, significant progress has been made; Table 46–10 summarizes some of the major features of membrane assembly that have emerged to date.

Various Disorders Result from Mutations in Genes Encoding Proteins Involved in Intracellular Transport

Some disorders reflecting abnormal **peroxisomal** function and abnormalities of protein synthesis in the **ER** and of the synthesis of **lysosomal proteins** have been listed earlier in this Chapter (see Table 46–3 and Table 46–7, respectively). Many other mutations affecting intracellular protein transport to various organelles have been reported, but are not included here. The elucidation of the causes of these various **conformational disorders** has contributed significantly to our understanding of **molecular pathology**. Apart from the possibility of **gene therapy**, it is hoped that attempts to restore at least a degree of normal folding to misfolded proteins by **administration to affected individuals of small molecules** that interact specifically with such proteins will be of therapeutic benefit. This is an active area of research.

SUMMARY

- Many proteins are targeted to their destinations by signal sequences. A major sorting decision is made when proteins are partitioned between cytosolic and membrane-bound polyribosomes by virtue of the absence or presence of a signal peptide.

- Pathways of protein import into mitochondria, nuclei, peroxisomes, and the endoplasmic reticulum are described.

- Numerous proteins synthesized on membrane-bound polyribosomes proceed to the Golgi apparatus and the plasma membrane in transport vesicles.

- Many glycosylation reactions occur in compartments of the Golgi, and proteins are further sorted in the *trans*-Golgi network.

- The role of chaperone proteins in the folding of proteins is presented and the unfolded protein response is described.

- Endoplasmic reticulum–associated degradation (ERAD) is briefly described and the key role of ubiquitin in protein degradation is shown.

- A model describing budding and attachment of transport vesicles to a target membrane is summarized.

- Certain proteins (eg, precursors of albumin and insulin) are subjected to proteolysis while inside transport vesicles, producing the mature proteins.

- Small GTPases (eg, Ran, Rab) and guanine nucleotide-exchange factors play key roles in many aspects of intracellular trafficking.

- The complex process of membrane assembly is discussed briefly. Asymmetry of both lipids and proteins is maintained during membrane assembly.

- Many disorders have been shown to be due to mutations in genes that affect the folding of various proteins. These conditions are often referred to as conformational diseases. Apart from gene therapy, the development of small molecules that interact with misfolded proteins and help restore at least some of their function is an important area of research.

REFERENCES

Alberts B et al: *Molecular Biology of the Cell.* 5th ed. Garland Science, 2008. (An excellent textbook of cell biology, with comprehensive coverage of trafficking and sorting).

Alder NN, Johnson AE: Cotranslational membrane protein biogenesis at the endoplasmic reticulum. J Biol Chem 2004;279:22787.

Bonifacino JS, Glick BS: The mechanisms of vesicle budding and fusion. Cell 2004;116:153.

Dalbey RE, von Heijne G (editors): *Protein Targeting, Transport and Translocation.* Academic Press, 2002.

Ellgaard L, Helenius A: Quality control in the endoplasmic reticulum. Nat Rev Mol Cell Biol 2003;4:181.

Koehler CM: New developments in mitochondrial assembly. Annu Rev Cell Dev Biol 2004;20:309.

Lai E, Teodoro T, Volchuk A: Endoplasmic reticulum stress: Signaling the unfolded protein response. Physiology 2007;22:193.

Lee MCS et al: Bi-directional protein transport between the ER and Golgi. Annu Rev Cell Dev Biol 2004;20:87.

Lodish H et al: *Molecular Cell Biology.* 6th ed. WH Freeman & Co., 2008. (An excellent textbook of cell biology, with comprehensive coverage of trafficking and sorting).

Owen DJ, Collins BM, Evans PR: Adaptors for clathrin coats: structure and function. Annu Rev Cell Dev Biol 2004;20:153.

Pelham HRB: Maturation of Golgi cisternae directly observed. Trends Biochem Sci 2006;31:601.

Pollard TD, Earnshaw WC, Lippincott-Schwartz J: *Cell Biology*. 2nd ed. WB Saunders, 2007. (An excellent textbook of cell biology, with comprehensive coverage of trafficking and sorting).

Romisch K: Endoplasmic-reticulum–associated degradation. Annu Rev Cell Dev Biol 2005;21:435.

Schroder M, Kaufman RJ: The mammalian unfolded protein response. Annu Rev Biochem 2005;74:739.

Trombetta ES, Parodi AJ: Quality control and protein folding in the secretory pathway. Annu Rev Cell Dev Biol 2003;19:649.

Van Meer G, Sprong H: Membrane lipids and vesicular traffic. Curr Opin Cell Biol 2004;16:373.

Wiedemann N, Frazier AE, Pfanner N: The protein import machinery of mitochondria. J Biol Chem 2004;279:14473.

Zaidiu SK et al: Intranuclear trafficking: organization and assembly of regulatory machinery for combinatorial biological control. J Biol Chem 2004;279:43363.

Glycoproteins

Robert K. Murray, MD, PhD

BIOMEDICAL IMPORTANCE

Glycobiology is the study of the roles of sugars in health and disease. The **glycome** is the entire complement of sugars, whether free or present in more complex molecules, of an organism. **Glycomics,** an analogous term to genomics and proteomics, is the comprehensive study of glycomes, including genetic, physiologic, pathologic, and other aspects.

One major class of molecules included in the glycome is **glycoproteins.** These are proteins that contain oligosaccharide chains (glycans) covalently attached to their polypeptide backbones. It has been estimated that approximately 50% of eukaryotic proteins have sugars attached, so that **glycosylation** (enzymic attachment of sugars) is the most frequent post-translational modification of proteins. Nonenzymic attachment of sugars to proteins can also occur, and is referred to as **glycation.** This process can have serious pathologic consequences (eg, in poorly controlled diabetes mellitus). Glycoproteins are one class of **glycoconjugate** or **complex carbohydrate**—equivalent terms used to denote molecules containing one or more carbohydrate chains covalently linked to protein (to form glycoproteins or proteoglycans) or lipid (to form glycolipids). (**Proteoglycans** are discussed in Chapter 48 and **glycolipids** in Chapter 15.) Almost all the **plasma proteins** of humans—with the notable exception of albumin—are glycoproteins. Many **proteins of cellular membranes** (Chapter 40) contain substantial amounts of carbohydrate. A number of the **blood group substances** are glycoproteins, whereas others are glycosphingolipids. Certain **hormones** (eg, chorionic gonadotropin) are glycoproteins. A major problem in cancer is **metastasis,** the phenomenon whereby cancer cells leave their tissue of origin (eg, the breast), migrate through the bloodstream to some distant site in the body (eg, the brain), and grow there in an unregulated manner, with catastrophic results for the affected individual. Many cancer researchers think that alterations in the structures of glycoproteins and other glycoconjugates on the surfaces of cancer cells are important in the phenomenon of metastasis.

GLYCOPROTEINS OCCUR WIDELY & PERFORM NUMEROUS FUNCTIONS

Glycoproteins occur in most organisms, from bacteria to humans. Many viruses also contain glycoproteins, some of which have been much investigated, in part because they often play key roles in viral attachment to cells (eg, HIV-1 and influenza A virus). Numerous proteins with diverse functions are glycoproteins (Table 47–1); their carbohydrate content ranges from 1% to over 85% by weight.

Many studies have been conducted in an attempt to define the precise roles oligosaccharide chains play in the functions of glycoproteins. Table 47–2 summarizes results from such studies. Some of the functions listed are firmly established; others are still under investigation.

OLIGOSACCHARIDE CHAINS ENCODE BIOLOGIC INFORMATION

An enormous number of glycosidic linkages can be generated between sugars. For example, three different hexoses may be linked to each other to form over 1000 different trisaccharides. The conformations of the sugars in oligosaccharide chains vary depending on their linkages and proximity to other molecules with which the oligosaccharides may interact. It is now established that certain oligosaccharide chains encode **biologic information** and that this depends upon their constituent sugars, their sequences, and their linkages. For instance, mannose 6-phosphate residues target newly synthesized lysosomal enzymes to that organelle (see later). The biologic information that sugars contain is expressed via interactions between specific sugars, either free or in glycoconjugates, and proteins (such as lectins; see below) or other molecules. These interactions lead to changes of cellular activity. Thus, deciphering the so-called "**sugar code of life**" (one of the principal aims of glycomics) entails elucidating all of the interactions that sugars and sugar-containing molecules participate in, and also the results of these interactions on cellular behavior. This will not be an easy task, considering the diversity of glycans found in cells.

TECHNIQUES ARE AVAILABLE FOR DETECTION, PURIFICATION, STRUCTURAL ANALYSIS, & SYNTHESIS OF GLYCOPROTEINS

A variety of methods used in the detection, purification, and structural analysis of glycoproteins are listed in Table 47–3.

TABLE 47-1 Some Functions Served by Glycoproteins

Function	Glycoproteins
Structural molecule	Collagens
Lubricant and protective agent	Mucins
Transport molecule	Transferrin, ceruloplasmin
Immunologic molecule	Immunoglobulins, histocompatibility antigens
Hormone	Chorionic gonadotropin, thyroid-stimulating hormone (TSH)
Enzyme	Various, eg, alkaline phosphatase
Cell attachment-recognition site	Various proteins involved in cell-cell (eg, sperm-oocyte), virus-cell, bacterium-cell, and hormone-cell interactions
Antifreeze	Certain plasma proteins of cold-water fish
Interact with specific carbohydrates	Lectins, selectins (cell adhesion lectins), antibodies
Receptor	Various proteins involved in hormone and drug action
Affect folding of certain proteins	Calnexin, calreticulin
Regulation of development	Notch and its analogs, key proteins in development
Hemostasis (and thrombosis)	Specific glycoproteins on the surface membranes of platelets

TABLE 47-2 Some Functions of the Oligosaccharide Chains of Glycoproteins

- Modulate physicochemical properties, eg, solubility, viscosity, charge, conformation, denaturation, and binding sites for various molecules, bacteria viruses and some parasites
- Protect against proteolysis, from inside and outside of cell
- Affect proteolytic processing of precursor proteins to smaller products
- Are involved in biologic activity, eg, of human chorionic gonadotropin (hCG)
- Affect insertion into membranes, intracellular migration, sorting and secretion
- Affect embryonic development and differentiation
- May affect sites of metastases selected by cancer cells

Source: Adapted from Schachter H: Biosynthetic controls that determine the branching and heterogeneity of protein-bound oligosaccharides. Biochem Cell Biol 1986;64:163.

The conventional methods used to purify proteins and enzymes are also applicable to the purification of glycoproteins. Once a glycoprotein has been purified, the use of **mass spectrometry** and **high-resolution NMR spectroscopy** can often

TABLE 47-3 Some Important Methods Used to Study Glycoproteins

Method	Use
Periodic acid–Schiff reagent	Detects glycoproteins as pink bands after electrophoretic separation.
Incubation of cultured cells with a radioactive sugar	Leads to detection of glycoproteins as radioactive bands after electrophoretic separation.
Treatment with appropriate endo- or exoglycosidase or phospholipases	Resultant shifts in electrophoretic migration help distinguish among proteins with N-glycan, O-glycan, or GPI linkages and also between high mannose and complex N-glycans.
Sepharose-lectin column chromatography	To purify glycoproteins or glycopeptides that bind the particular lectin used.
Compositional analysis following acid hydrolysis	Identifies sugars that the glycoprotein contains and their stoichiometry.
Mass spectrometry	Provides information on molecular mass, composition, sequence, and sometimes branching of a glycan chain.
NMR spectroscopy	To identify specific sugars, their sequence, linkages, and the anomeric nature of glycosidic linkages.
Methylation (linkage) analysis	To determine linkages between sugars.
Amino acid or cDNA sequencing	Determination of amino acid sequence.

identify the structures of its glycan chains. Analysis of glycoproteins can be complicated by the fact that they often exist as **glycoforms;** these are proteins with identical amino acid sequences but somewhat different oligosaccharide compositions. Although linkage details are not stressed in this chapter, it is critical to appreciate that the precise natures of the linkages between the sugars of glycoproteins are of fundamental importance in determining the structures and functions of these molecules.

Impressive advances are also being made in **synthetic chemistry,** allowing synthesis of complex glycans that can be tested for biologic and pharmacologic activity. In addition, methods have been developed that use simple organisms, such as yeasts, to secrete human glycoproteins of therapeutic value (eg, erythropoietin) into their surrounding medium.

EIGHT SUGARS PREDOMINATE IN HUMAN GLYCOPROTEINS

About 200 monosaccharides are found in nature; however, only eight are commonly found in the oligosaccharide chains of glycoproteins (Table 47–4). Most of these sugars were described in Chapter 14. N-Acetylneuraminic acid (NeuAc) is usually found at the termini of oligosaccharide chains, attached

TABLE 47–4   The Principal Sugars Found in Human Glycoproteins[1]

Sugar	Type	Abbreviation	Nucleotide Sugar	Comments
Galactose	Hexose	Gal	UDP-Gal	Often found subterminal to NeuAc in N-linked glycoproteins. Also found in the core trisaccharide of proteoglycans.
Glucose	Hexose	Glc	UDP-Glc	Present during the biosynthesis of N-linked glycoproteins but not usually present in mature glycoproteins. Present in some clotting factors.
Mannose	Hexose	Man	GDP-Man	Common sugar in N-linked glycoproteins.
N-Acetylneuraminic acid	Sialic acid (nine C atoms)	NeuAc	CMP-NeuAc	Often the terminal sugar in both N- and O-linked glycoproteins. Other types of sialic acid are also found, but NeuAc is the major species found in humans. Acetyl groups may also occur as O-acetyl species as well as N-acetyl.
Fucose	Deoxyhexose	Fuc	GDP-Fuc	May be external in both N- and O-linked glycoproteins or internal, linked to the GlcNAc residue attached to Asn in N-linked species. Can also occur internally attached to the OH of Ser (eg, in t-PA and certain clotting factors).
N-Acetylgalactosamine	Aminohexose	GalNAc	UDP-GalNAc	Present in both N- and O-linked glycoproteins.
N-Acetylglucosamine	Aminohexose	GlcNAc	UDP-GlcNAc	The sugar attached to the polypeptide chain via Asn in N-linked glycoproteins; also found at other sites in the oligosaccharides of these proteins. Many nuclear proteins have GlcNAc attached to the OH of Ser or Thr as a single sugar.
Xylose	Pentose	Xyl	UDP-Xyl	Xyl is attached to the OH of Ser in many proteoglycans. Xyl in turn is attached to two Gal residues, forming a link trisaccharide. Xyl is also found in t-PA and certain clotting factors.

[1]Structures of glycoproteins are illustrated in Chapter 14.

to subterminal galactose (Gal) or N-acetylgalactosamine (Gal-NAc) residues. The other sugars listed are generally found in more internal positions. **Sulfate** is often found in glycoproteins, usually attached to Gal, GalNAc, or GlcNAc.

NUCLEOTIDE SUGARS ACT AS SUGAR DONORS IN MANY BIOSYNTHETIC REACTIONS

It is important to understand that in most biosynthetic reactions, it is not the free sugar or phosphorylated sugar that is involved in such reactions, but rather the corresponding **nucleotide sugar**. The first nucleotide sugar to be reported was uridine diphosphate glucose (UDP-Glc); its structure is shown in Figure 19–2. The common nucleotide sugars involved in the biosynthesis of glycoproteins are listed in Table 47–4; the reasons some contain UDP and others guanosine diphosphate (GDP) or cytidine monophosphate (CMP) are not clear. Many of the glycosylation reactions involved in the biosynthesis of glycoproteins utilize these compounds (see below). The **anhydro nature** of the linkage between the phosphate group and the sugars is of the high-energy, high-group-transfer-potential type (Chapter 11). The sugars of these compounds are thus **"activated"** and can

be transferred to suitable acceptors provided appropriate transferases are available.

Most nucleotide sugars are formed in the cytosol, generally from reactions involving the corresponding nucleoside triphosphate. CMP-sialic acids are formed in the nucleus. Formation of uridine diphosphate galactose (UDP-Gal) requires the following two reactions in mammalian tissues:

Because many glycosylation reactions occur within the lumen of the Golgi apparatus, **carrier systems** (permeases, transporters) are necessary to transport nucleotide sugars across the Golgi membrane. Systems transporting UDP-Gal, GDP-Man, and CMP-NeuAc into the cisternae of the Golgi apparatus have been described. They are **antiport** systems; ie, the influx of one molecule of nucleotide sugar is balanced by the efflux of one molecule of the corresponding nucleotide (eg,

UMP, GMP, or CMP) formed from the nucleotide sugars. This mechanism ensures an adequate concentration of each nucleotide sugar inside the Golgi apparatus. UMP is formed from UDP-Gal in the above process as follows:

EXO- & ENDOGLYCOSIDASES FACILITATE STUDY OF GLYCOPROTEINS

A number of **glycosidases** of defined specificity have proved useful in examining structural and functional aspects of glycoproteins (Table 47–5). These enzymes act at either external (exoglycosidases) or internal (endoglycosidases) positions of oligosaccharide chains. Examples of exoglycosidases are **neuraminidases** and **galactosidases;** their sequential use removes terminal NeuAc and subterminal Gal residues from most glycoproteins. **Endoglycosidases F** and **H** are examples of the latter class; these enzymes cleave the oligosaccharide chains at specific GlcNAc residues close to the polypeptide backbone (ie, at internal sites; Figure 47–5) and are thus useful in releasing large oligosaccharide chains for structural analyses. A glycoprotein can be treated with one or more of the above glycosidases to analyze the effects on its biologic behavior of removal of specific sugars.

THE MAMMALIAN ASIALOGLYCOPROTEIN RECEPTOR IS INVOLVED IN CLEARANCE OF CERTAIN GLYCOPROTEINS FROM PLASMA BY HEPATOCYTES

Experiments performed by Ashwell and his colleagues in the early 1970s played an important role in focusing attention on the functional significance of the oligosaccharide chains of glycoproteins. They treated rabbit ceruloplasmin (a plasma protein; see Chapter 50) with neuraminidase in vitro. This procedure exposed subterminal Gal residues that were normally masked by terminal NeuAc residues. Neuraminidase-treated radioactive ceruloplasmin was found to disappear rapidly from the circulation, in contrast to the slow clearance of the untreated protein. Very significantly, when the Gal residues exposed to treatment with neuraminidase were removed by treatment with a galactosidase, the clearance rate of the protein returned to normal. Further studies demonstrated that liver cells contain a **mammalian asialoglycoprotein receptor** that recognizes the Gal moiety of many desialylated plasma proteins and leads to their endocytosis. This work indicated that an individual sugar, such as Gal, could play an important role in governing at least one of the biologic properties (ie, time of residence in the circulation) of certain glycoproteins. This greatly strengthened the concept that oligosaccharide chains could contain biologic information.

LECTINS CAN BE USED TO PURIFY GLYCOPROTEINS & TO PROBE THEIR FUNCTIONS

Lectins are **carbohydrate-binding proteins** that agglutinate cells or precipitate glycoconjugates; a number of lectins are themselves glycoproteins. Immunoglobulins that react with sugars are not considered lectins. Lectins contain at least two sugar-binding sites; proteins with a single sugar-binding site will not agglutinate cells or precipitate glycoconjugates. The specificity of a lectin is usually defined by the sugars that are best at inhibiting its ability to cause agglutination or precipitation. Enzymes, toxins, and transport proteins can be classified as lectins if they bind carbohydrate. Lectins were first discovered in plants and microbes, but many lectins of animal origin are now known. The mammalian asialoglycoprotein receptor described above is an important example of an animal lectin. Some important lectins are listed in Table 47–6. Much current research is centered on the roles of various animal lectins in the mechanisms of action of glycoproteins, some of which are discussed below (eg, with regard to the selectins).

Numerous lectins have been purified and are commercially available; three plant lectins that have been widely used experimentally are listed in Table 47–7. Among many uses, lectins have been employed to purify specific glycoproteins, as tools for probing the glycoprotein profiles of cell surfaces, and as reagents for generating mutant cells deficient in certain enzymes involved in the biosynthesis of oligosaccharide chains.

THERE ARE THREE MAJOR CLASSES OF GLYCOPROTEINS

Based on the nature of the linkage between their polypeptide chains and their oligosaccharide chains, glycoproteins can be divided into three major classes (Figure 47–1): (1) those containing an **O-glycosidic linkage** (ie, O-linked), involv-

TABLE 47–5 Some Glycosidases Used to Study the Structure and Function of Glycoproteins[1]

Enzymes	Type
Neuraminidases	Exoglycosidase
Galactosidases	Exo-or endoglycosidase
Endoglycosidase F	Endoglycosidase
Endoglycosidase H	Endoglycosidase

[1]The enzymes are available from a variety of sources and are often specific for certain types of glycosidic linkages and also for their anomeric natures. The sites of action of endoglycosidases F and H are shown in Figure 47–5. F acts on both high-mannose and complex oligosaccharides, whereas H acts on the former.

TABLE 47–6 Some Important Lectins

Lectins	Examples or Comments
Legume lectins	Concanavalin A, pea lectin
Wheat germ agglutinin	Widely used in studies of surfaces of normal cells and cancer cells
Ricin	Cytotoxic glycoprotein derived from seeds of the castor plant
Bacterial toxins	Heat-labile enterotoxin of *E coli* and cholera toxin
Influenza virus hemagglutinin	Responsible for host-cell attachment and membrane fusion
C-type lectins	Characterized by a Ca^{2+}-dependent carbohydrate recognition domain (CRD); includes the mammalian asialoglycoprotein receptor, the selectins, and the mannose-binding protein
S-type lectins	β-Galactoside-binding animal lectins with roles in cell-cell and cell-matrix interactions
P-type lectins	Mannose 6-P receptor
I-type lectins	Members of the immunoglobulin super-family, eg, sialoadhesin mediating adhesion of macrophages to various cells

TABLE 47–7 Three Plant Lectins and the Sugars with Which They Interact[1]

Lectin	Abbreviation	Sugars
Concanavalin A	ConA	Man and Glc
Soybean lectin		Gal and GalNAc
Wheat germ agglutinin	WGA	Glc and NeuAc

[1]In most cases, lectins show specificity for the anomeric nature of the glycosidic linkage (α or β); this is not indicated in the table.

ing the hydroxyl side chain of serine or threonine and a sugar such as *N*-acetylgalactosamine (GalNAc-Ser[Thr]); (2) those containing an **N-glycosidic linkage** (ie, N-linked), involving the amide nitrogen of asparagine and *N*-acetylglucosamine (GlcNAc-Asn); and (3) those linked to the carboxyl terminal amino acid of a protein via a phosphoryl-ethanolamine moiety joined to an oligosaccharide (glycan), which in turn is linked via glucosamine to phosphatidylinositol (PI). This latter class is referred to as **glycosylphosphatidylinositol-anchored** (**GPI-anchored,** or **GPI-linked**) glycoproteins. It is involved in directing certain glycoproteins to the apical or basolateral areas of the plasma membrane of certain polarized epithelial cells (see Chapter 46 and below). Other minor classes of glycoproteins also exist.

The number of oligosaccharide chains attached to one protein can vary from one to 30 or more, with the sugar chains ranging from one or two residues in length to much larger structures. Many proteins contain more than one type of sugar chain; for instance, **glycophorin,** an important red cell mem-

brane glycoprotein (Chapter 52), contains both O- and N-linked oligosaccharides.

GLYCOPROTEINS CONTAIN SEVERAL TYPES OF O-GLYCOSIDIC LINKAGES

At least four subclasses of O-glycosidic linkages are found in human glycoproteins: (1) The **GalNAcSer(Thr)** linkage shown in Figure 47–1 is the predominant linkage. Two typical oligosaccharide chains found in members of this subclass are shown in Figure 47–2. Usually a Gal or a NeuAc residue is attached to the GalNAc, but many variations in the sugar compositions and lengths of such oligosaccharide chains are found. This type of linkage is found in **mucins** (see below). (2) **Proteoglycans** contain a **Gal-Gal-Xyl-Ser** trisaccharide (the so-called link trisaccharide). (3) **Collagens** contain a **Gal-hydroxylysine (Hyl)** linkage. (Subclasses [2] and [3] are discussed further in Chapter 48.) (4) Many **nuclear proteins** (eg, certain transcription factors) and **cytosolic proteins** contain side chains consisting of a single GlcNAc attached to a serine or threonine residue (**GlcNAc-Ser[Thr]**).

Mucins Have a High Content of O-Linked Oligosaccharides & Exhibit Repeating Amino Acid Sequences

Mucins are glycoproteins with two major characteristics: (1) a high content of **O-linked oligosaccharides** (the carbohydrate content of mucins is generally more than 50%); and (2) the presence of **repeating amino acid sequences** (tandem repeats) in the center of their polypeptide backbones, to which the O-glycan chains are attached in clusters (Figure 47–3). These sequences are rich in serine, threonine, and proline. Although O-glycans predominate, mucins often contain a number of N-glycan chains. Both **secretory** and **membrane-bound** mucins occur. The former are found in the mucus present in the secretions of the gastrointestinal, respiratory, and reproductive tracts. **Mucus** consists of about 94% water and 5% mucins, with the remainder being a mixture of various cell molecules, electrolytes, and remnants of cells. Secretory mucins generally have an oligomeric structure and thus often have a very high molecular mass. The oligomers are composed of monomers linked by disulfide bonds. Mucus exhibits a high **viscosity** and often forms a **gel.** These qualities are functions of its content of mucins. The high content of O-glycans confers an extended structure on mucins. This is in part explained by steric interactions between their GalNAc moieties and adjacent amino acids, resulting in a chain-stiffening effect so that the conformations of mucins often become those of rigid rods. Intermolecular noncovalent interactions between various sugars on neighboring glycan chains contribute to gel formation. The high content of **NeuAc** and **sulfate** residues found in many mucins confers a negative charge on them. With regard to function, mucins help **lubricate** and form a **protective physical barrier** on epithelial surfaces. Membrane-bound mucins participate in various **cell-**

A

CH$_2$OH

OH — C — O

C — OH

C — H

H — N

C=O

CH$_3$

α — O — CH$_2$ — C

Ser

B

CH$_2$OH

C — O

HO — C

OH

H — N

C=O

CH$_3$

β

C — N — C — CH$_2$ — C

Asn

C

NH$_2$

Protein

▽ Glycine

○ Ethanolamine

P

■ Mannose

Ethanolamine ○ — P — ■ Mannose

■ Mannose

△ Glucosamine

⬡ Inositol

PI-PLC — P ← Additional fatty acid

Plasma membrane

FIGURE 47–1 Depictions of **(A)** an O-linkage (*N*-acetylgalactosamine to serine), **(B)** an N-linkage (*N*-acetylglucosamine to asparagine), and **(C)** a glycosylphosphatidylinositol (GPI) linkage. The GPI structure shown is that linking acetylcholinesterase to the plasma membrane of the human red blood cell. The carboxyl terminal amino acid is glycine joined in amide linkage via its COOH group to the NH$_2$ group of phosphorylethanolamine, which in turn is joined to a mannose residue. The core glycan contains three mannose and one glucosamine residues. The glucosamine is linked to inositol, which is attached to phosphatidic acid. The site of action of PI-phospholipase C (PI-PLC) is indicated. The structure of the core glycan is shown in the text. This particular GPI contains an extra fatty acid attached to inositol and also an extra phosphorylethanolamine moiety attached to the middle of the three mannose residues. Variations found among different GPI structures include the identity of the carboxyl terminal amino acid, the molecules attached to the mannose residues, and the precise nature of the lipid moiety.

cell interactions (eg, involving selectins; see below). The density of oligosaccharide chains makes it difficult for **proteases** to approach their polypeptide backbones, so that mucins are often resistant to their action. Mucins also tend to "mask" certain surface antigens. Many cancer cells form excessive amounts of mucins; perhaps the mucins may mask certain surface antigens on such cells and thus protect the cells from immune surveillance. Mucins also carry cancer-specific peptide and carbohy-

drate epitopes (an epitope is a site on an antigen recognized by an antibody, also called an antigenic determinant). Some of these epitopes have been used to stimulate an immune response against cancer cells.

A

NeuAc $\xrightarrow{\alpha 2,6}$ GalNAc $\longrightarrow$ Ser(Thr)

B

Gal $\xrightarrow{\beta 1,3}$ GalNAc $\longrightarrow$ Ser(Thr)

↑α 2,3   ↑α 2,6

NeuAc   NeuAc

FIGURE 47–2 Structures of two O-linked oligosaccharides found in **(A)** submaxillary mucins and **(B)** fetuin and in the sialoglycoprotein of the membrane of human red blood cells. (Modified and reproduced, with permission, from Lennarz WJ: *The Biochemistry of Glycoproteins and Proteoglycans.* Plenum Press, 1980. Reproduced with kind permission from Springer Science and Business Media.)

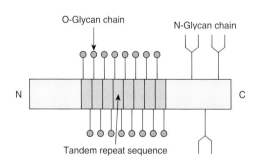

FIGURE 47–3 Schematic diagram of a mucin. O-glycan chains are shown attached to the central region of the extended polypeptide chain and N-glycan chains to the carboxyl terminal region. The narrow rectangles represent a series of tandem repeat amino acid sequences. Many mucins contain cysteine residues whose SH groups form interchain linkages; these are not shown in the figure. (Adapted, with permission, from Strous GJ, Dekker J: Mucin-type glycoproteins. Crit Rev Biochem Mol Biol 1992;27:57. Copyright ©1992. Reproduced by permission of Taylor & Francis Group, LLC.)

The **genes** encoding the polypeptide backbones of a number of mucins derived from various tissues (eg, pancreas, small intestine, trachea and bronchi, stomach, and salivary glands) have been cloned and sequenced. These studies have revealed new information about the polypeptide backbones of mucins (size of tandem repeats, potential sites of N-glycosylation, etc.) and ultimately should reveal aspects of their genetic control. Some important properties of mucins are summarized in Table 47–8.

The Biosynthesis of O-Linked Glycoproteins Uses Nucleotide Sugars

The polypeptide chains of O-linked and other glycoproteins are encoded by mRNA species; because most glycoproteins are membrane-bound or secreted, they are generally translated on membrane-bound polyribosomes (Chapter 37). Hundreds of different oligosaccharide chains of the O-glycosidic type exist. These glycoproteins are built up by the **stepwise donation of sugars from nucleotide sugars,** such as UDP-GalNAc, UDP-Gal, and CMP-NeuAc. The enzymes catalyzing this type of reaction are membrane-bound **glycoprotein glycosyltransferases.** Generally, synthesis of one specific type of linkage requires the activity of a correspondingly specific transferase. The factors that determine which specific serine and threonine residues are glycosylated have not been identified but are probably found in the peptide structure surrounding the glycosylation site. The enzymes assembling O-linked chains are located in the Golgi apparatus, sequentially arranged in an

TABLE 47–8   Some Properties of Mucins

- Found in secretions of the gastrointestinal, respiratory, and reproductive tracts and also in membranes of various cells.

- Exhibit high content of O-glycan chains, usually containing NeuAc.

- Contain repeating amino acid sequences rich in serine, threonine, and proline.

- Extended structure contributes to their high viscoelasticity.

- Form protective physical barrier on epithelial surfaces, are involved in cell–cell interactions, and may contain or mask certain surface antigens.

TABLE 47–9   Summary of Main Features of O-Glycosylation

- Involves a battery of membrane-bound glycoprotein glycosyltransferases acting in a stepwise manner; each transferase is generally specific for a particular type of linkage.

- The enzymes involved are located in various subcompartments of the Golgi apparatus.

- Each glycosylation reaction involves the appropriate nucleotide sugar.

- Dolichol-P-P-oligosaccharide is not involved, nor are glycosidases; and the reactions are not inhibited by tunicamycin.

- O-Glycosylation occurs posttranslationally at certain Ser and Thr residues.

assembly line with terminal reactions occurring in the *trans*-Golgi compartments.

The major features of the biosynthesis of O-linked glycoproteins are summarized in Table 47–9.

N-LINKED GLYCOPROTEINS CONTAIN AN Asn-GLcNAc LINKAGE

N-Linked glycoproteins are distinguished by the presence of the Asn-GlcNAc linkage (Figure 47–1). It is the major class of glycoproteins and has been much studied, since the most readily accessible glycoproteins (eg, plasma proteins) mainly belong to this group. It includes both **membrane-bound** and **circulating** glycoproteins. The principal difference between this and the previous class, apart from the nature of the amino acid to which the oligosaccharide chain is attached (Asn vs Ser or Thr), concerns their biosynthesis.

Complex, Hybrid, & High-Mannose Are the Three Major Classes of N-Linked Oligosaccharides

There are three major classes of N-linked oligosaccharides: **complex, hybrid,** and **high-mannose** (Figure 47–4). Each type shares a common pentasaccharide, $Man_3GlcNAc_2$—shown within the boxed area in Figure 47–4 and depicted also in Figure 47–5—but they differ in their outer branches. The presence of the **common pentasaccharide** is explained by the fact that all three classes share an initial common mechanism of biosynthesis. Glycoproteins of the complex type generally contain terminal NeuAc residues and underlying Gal and GlcNAc residues, the latter often constituting the disaccharide *N*-acetyllactosamine. Repeating *N*-acetyllactosamine units—$[Gal\beta1–3/4GlcNAc\beta 1–3]_n$ (poly-*N*-acetyllactosaminoglycans)—are often found on N-linked glycan chains. I/i blood group substances belong to this class. The majority of complex-type oligosaccharides contain two, three, or four outer branches (Figure 47–4), but structures containing five branches have also been described. The oligosaccharide branches are often referred to as **antennae,** so that bi-, tri-, tetra-, and penta-antennary structures may all be found. A bewildering number of chains of the complex type exist, and that indicated in Figure 47–4 is only one of many. Other complex chains may terminate in Gal or Fuc. High-mannose oligosaccharides typically have two to six additional Man residues linked to the pentasaccharide core. Hybrid molecules contain features of both of the two other classes.

The Biosynthesis of N-Linked Glycoproteins Involves Dolichol-P-P-Oligosaccharide

Leloir and his colleagues described the occurrence of a **dolichol-pyrophosphate-oligosaccharide (Dol-P-P-oligosaccharide),** which subsequent research showed to play a key role in the biosynthesis of N-linked glycoproteins. The oligosaccharide chain of this compound generally has the structure

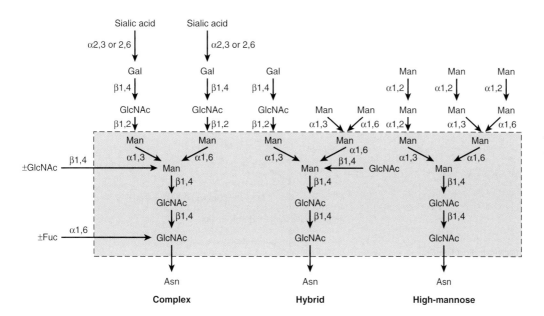

FIGURE 47–4 Structures of the major types of asparagine-linked oligosaccharides. The boxed area encloses the pentasaccharide core common to all N-linked glycoproteins. (Reproduced, with permission, from Kornfeld R, Kornfeld S: Assembly of asparagine-linked oligosaccharides. Annu Rev Biochem 1985;54:631. Copyright © 1985 by Annual Reviews. Reprinted with permission.)

FIGURE 47–5 Schematic diagram of the pentasaccharide core common to all N-linked glycoproteins and to which various outer chains of oligosaccharides may be attached. The sites of action of endoglycosidases F and H are also indicated.

R-GlcNAc$_2$Man$_9$Glc$_3$ (R = DolP-P). The sugars of this compound are first assembled on the Dol-P-P backbone, and the oligosaccharide chain is then transferred en bloc to suitable Asn residues of acceptor apoglycoproteins during their synthesis on membrane-bound polyribosomes. All N-glycans have a common pentasaccharide core structure (Figure 47–5).

To form **high-mannose** chains, only the Glc residues plus certain of the peripheral Man residues are removed. To form an oligosaccharide chain of the **complex type,** the Glc residues and four of the Man residues are removed by glycosidases in the endoplasmic reticulum and Golgi. The sugars characteristic of complex chains (GlcNAc, Gal, NeuAc) are added by the action of individual glycosyltransferases located in the Golgi apparatus. The phenomenon whereby the glycan chains of N-linked glycoproteins are first partially degraded and then in some cases rebuilt is referred to as **oligosaccharide processing. Hybrid chains** are formed by partial processing, forming complex chains on one arm and Man structures on the other arm.

Thus, the initial steps involved in the biosynthesis of the N-linked glycoproteins differ markedly from those involved

in the biosynthesis of the O-linked glycoproteins. The former involves Dol-P-P-oligosaccharide; the latter, as described earlier, does not.

The process of N-glycosylation can be broken down into two stages: (1) assembly of Dol-P-P-oligosaccharide and transfer of the oligosaccharide; and (2) processing of the oligosaccharide chain.

Assembly & Transfer of Dolichol-P-P-Oligosaccharide

Polyisoprenol compounds exist in both bacteria and eukaryotic cells. They participate in the synthesis of bacterial polysaccharides and in the biosynthesis of N-linked glycoproteins and GPI anchors. The polyisoprenol used in eukaryotic tissues is **dolichol,** which is, next to rubber, the longest naturally occurring hydrocarbon made up of a single repeating unit. Dolichol is composed of 17–20 repeating isoprenoid units (Figure 47–6).

Before it participates in the biosynthesis of Dol-PP-oligosaccharide, dolichol must first be phosphorylated to form dolichol phosphate (Dol-P) in a reaction catalyzed by **dolichol kinase** and using ATP as the phosphate donor.

Dolichol-P-P-GlcNAc (Dol-P-P-GlcNAc) is the key lipid that acts as an acceptor for other sugars in the assembly of Dol-P-P-oligosaccharide. It is synthesized in the membranes of the endoplasmic reticulum from Dol-P and UDP-GlcNAc in the following reaction, catalyzed by GlcNAc-P transferase:

$$Dol\text{-}P + UDP\text{-}GlcNAc \rightarrow Dol\text{-}P\text{-}P\text{-}GlcNAc + UMP$$

The above reaction—which is the first step in the assembly of Dol-P-P-oligosaccharide—and the other later reactions are

FIGURE 47–6 The structure of dolichol. The phosphate in dolichol phosphate is attached to the primary alcohol group at the left-hand end of the molecule. The group within the brackets is an isoprene unit (n = 17–20 isoprenoid units).

$$HO-CH_2-CH_2-\underset{\underset{CH_3}{|}}{\overset{\overset{H}{|}}{C}}-CH_2\left[CH_2-CH=\underset{}{\overset{\overset{CH_3}{|}}{C}}-CH_2\right]_n CH_2-CH=\overset{\overset{CH_3}{|}}{C}-CH_3$$

FIGURE 47–7 Pathway of biosynthesis of dolichol-P-P-oligosaccharide. The specific linkages formed are indicated in Figure 47–8. Note that the first five internal mannose residues are donated by GDP-mannose, whereas the more external mannose residues and the glucose residues are donated by dolichol-P-mannose and dolichol-P-glucose. (UDP, uridine diphosphate; Dol, dolichol; P, phosphate; UMP, uridine monophosphate; GDP, guanosine diphosphate; M, mannose; G, glucose.)

summarized in Figure 47–7. The essential features of the subsequent steps in the assembly of Dol-P-P-oligosaccharide are as follows:

1. A second GlcNAc residue is added to the first, again using UDP-GlcNAc as the donor.
2. Five Man residues are added, using GDP-mannose as the donor.
3. Four additional Man residues are next added, using Dol-P-Man as the donor. Dol-P-Man is formed by the following reaction:

$$\text{Dol-P} + \text{GDP-Man} \rightarrow \text{Dol-P-Man} + \text{GDP}$$

4. Finally, the three peripheral glucose residues are donated by Dol-P-Glc, which is formed in a reaction analogous to that just presented except that Dol-P and UDP-Glc are the substrates.

It should be noted that the first seven sugars (two GlcNAc and five Man residues) are donated by nucleotide sugars, whereas the last seven sugars (four Man and three Glc residues) added are donated by dolichol-sugars. The net result is assembly of the compound illustrated in Figure 47–8 and referred to in shorthand as Dol-P-P-GlcNAc$_2$Man$_9$Glc$_3$.

The oligosaccharide linked to dolichol-P-P is transferred en bloc to form an N-glycosidic bond with one or more specific Asn residues of an acceptor protein emerging from the luminal surface of the membrane of the endoplasmic reticulum. The reaction is catalyzed by **oligosaccharide: protein transferase,** a membrane-associated enzyme complex. The transferase will recognize and transfer any substrate with the general structure Dol-P-P-(GlcNAc)$_2$-R, but it has a strong preference for the Dol-P-P-GlcNAc$_2$Man$_9$Glc$_3$ structure. Glycosylation occurs at the Asn residue of an Asn-XSer/Thr tripeptide

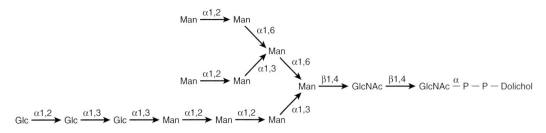

FIGURE 47–8 Structure of dolichol-P-P-oligosaccharide. (From Li E, et al: Structure of the lipid-linked oligosaccharide precursor of the complex-type oligosaccharides of the vesicular stomatitis virus G protein. J Biol Chem 1978;253:7762.)

sequence, where X is any amino acid except proline, aspartic acid, or glutamic acid. A tripeptide site contained within a β turn is favored. Only about one-third of the Asn residues that are potential acceptor sites are actually glycosylated, suggesting that factors other than the tripeptide are also important.

The acceptor proteins are of both the secretory and integral membrane class. Cytosolic proteins are rarely glycosylated. The transfer reaction and subsequent processes in the glycosylation of N-linked glycoproteins, along with their subcellular locations, are depicted in Figure 47–9. The other product of the oligosaccharide: protein transferase reaction is dolichol-PP, which is subsequently converted to dolichol-P by a phosphatase. The dolichol-P can serve again as an acceptor for the synthesis of another molecule of Dol-P-P-oligosaccharide.

Processing of the Oligosaccharide Chain

1. Early Phase. The various reactions involved are indicated in Figure 47–9. The oligosaccharide: protein transferase catalyzes reaction 1 (see above). Reactions 2 and 3 involve the removal of the terminal Glc residue by glucosidase I and of the next two Glc residues by glucosidase II, respectively. In the case of **high-mannose** glycoproteins, the process may stop here, or up to four

Man residues may also be removed. However, to form **complex** chains, additional steps are necessary, as follows. Four external Man residues are removed in reactions 4 and 5 by at least two different mannosidases. In reaction 6, a GlcNAc residue is added to the Man residue of the Man 1–3 arm by GlcNAc transferase I. The action of this latter enzyme permits the occurrence of reaction 7, a reaction catalyzed by yet another mannosidase (Golgi α-mannosidase II) and which results in a reduction of the Man residues to the core number of three (Figure 47–5).

An important additional pathway is indicated in reactions I and II of Figure 47–9. This involves enzymes destined for **lysosomes.** Such enzymes are targeted to the lysosomes by a specific chemical marker. In reaction I, a residue of GlcNAc-1-P is added to carbon 6 of one or more specific Man residues of these enzymes. The reaction is catalyzed by a GlcNAc phosphotransferase, which uses UDPGlcNAc as the donor and generates UMP as the other product:

$$\text{UDP-GlcNAc + Man — Protein} \xrightarrow{\boxed{\begin{array}{c}\text{GlcNAc}\\\text{PHOSPHO-}\\\text{TRANSFERASE}\end{array}}} \text{GlcNAc-1-P-6-Man — Protein + UMP}$$

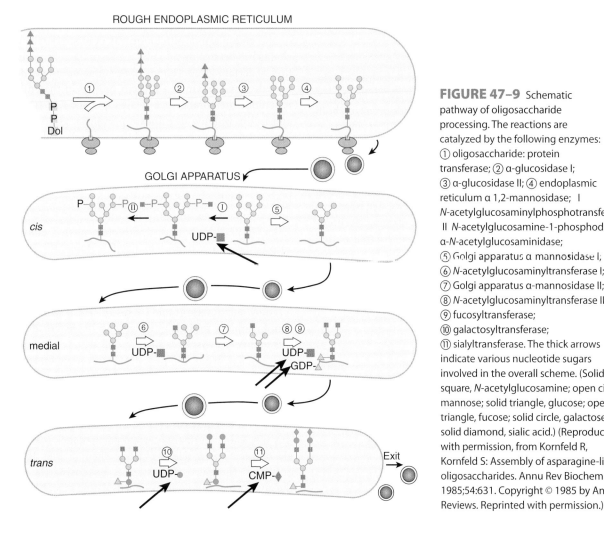

FIGURE 47–9 Schematic pathway of oligosaccharide processing. The reactions are catalyzed by the following enzymes: ① oligosaccharide: protein transferase; ② α-glucosidase I; ③ α-glucosidase II; ④ endoplasmic reticulum α 1,2-mannosidase; I N-acetylglucosaminylphosphotransferase; II N-acetylglucosamine-1-phosphodiester α-N-acetylglucosaminidase; ⑤ Golgi apparatus α mannosidase I; ⑥ N-acetylglucosaminyltransferase I; ⑦ Golgi apparatus α-mannosidase II; ⑧ N-acetylglucosaminyltransferase II; ⑨ fucosyltransferase; ⑩ galactosyltransferase; ⑪ sialyltransferase. The thick arrows indicate various nucleotide sugars involved in the overall scheme. (Solid square, N-acetylglucosamine; open circle, mannose; solid triangle, glucose; open triangle, fucose; solid circle, galactose; solid diamond, sialic acid.) (Reproduced, with permission, from Kornfeld R, Kornfeld S: Assembly of asparagine-linked oligosaccharides. Annu Rev Biochem 1985;54:631. Copyright © 1985 by Annual Reviews. Reprinted with permission.)

In reaction II, the GlcNAc is removed by the action of a phosphodiesterase, leaving the Man residues phosphorylated in the 6 position:

$$\text{GlcNAc-1-P-6-Man — Protein} \xrightarrow{\boxed{\text{PHOSPHO-DIESTERASE}}} \text{P-6-Man — Protein + GlcNAc}$$

Man 6-P receptors, located in the Golgi apparatus, bind the Man 6-P residues of these enzymes and direct them to the lysosomes. Fibroblasts from patients with **I-cell disease** (see below) are severely deficient in the activity of the GlcNAc phosphotransferase.

2. **Late Phase.**   To assemble a typical complex oligosaccharide chain, additional sugars must be added to the structure formed in reaction 7. Hence, in reaction 8, a second GlcNAc is added to the peripheral Man residue of the other arm of the bi-antennary structure shown in Figure 47–9; the enzyme catalyzing this step is GlcNAc transferase II. Reactions 9, 10, and 11 involve the addition of Fuc, Gal, and NeuAc residues at the sites indicated, in reactions catalyzed by fucosyl, galactosyl, and sialyl transferases, respectively. The assembly of poly-*N*-acetyl-lactosamine chains requires additional GlcNAc transferases.

The Endoplasmic Reticulum & Golgi Apparatus Are the Major Sites of Glycosylation

As indicated in Figure 47–9, the endoplasmic reticulum and the Golgi apparatus are the major sites involved in glycosylation processes. The assembly of Dol-P-P-oligosaccharide occurs on both the cytoplasmic and luminal surfaces of the ER membranes. Addition of the oligosaccharide to protein occurs in the rough endoplasmic reticulum during or after translation. Removal of the Glc and some of the peripheral Man residues also occurs in the endoplasmic reticulum. The Golgi apparatus is composed of *cis*, medial, and *trans* cisternae; these can be separated by appropriate centrifugation procedures. Vesicles containing glycoproteins bud off in the endoplasmic reticulum and are transported to the *cis*-Golgi. Various studies have shown that the enzymes involved in glycoprotein processing show differential locations in the cisternae of the Golgi. As indicated in Figure 47–9, Golgi α-mannosidase I (catalyzing reaction 5) is located mainly in the *cis*-Golgi, whereas GlcNAc transferase I (catalyzing reaction 6) appears to be located in the medial Golgi, and the fucosyl, galactosyl, and sialyl transferases (catalyzing reactions 9, 10, and 11) are located mainly in the *trans*-Golgi. The major features of the biosynthesis of N-linked glycoproteins are summarized in Table 47–10 and should be contrasted with those previously listed (Table 47–9) for O-linked glycoproteins.

Some Glycan Intermediates Formed During N-Glycosylation Have Specific Functions

The following are a number of specific functions of N-glycan chains that have been established or are under investigation:

TABLE 47–10   Summary of Main Features of N-Glycosylation

- The oligosaccharide $Glc_3Man_9(GlcNAc)_2$ is transferred from dolichol-P-P-oligosaccharide in a reaction catalyzed by oligosaccharide:protein transferase, which is inhibited by tunicamycin.
- Transfer occurs to specific Asn residues in the sequence AsnX-Ser/Thr, where X is any residue except Pro, Asp, or Glu.
- Transfer can occur cotranslationally in the endoplasmic reticulum.
- The protein-bound oligosaccharide is then partially processed by glucosidases and mannosidases; if no additional sugars are added, this results in a high-mannose chain.
- If processing occurs down to the core heptasaccharide $(Man_5[GlcNAc]_2)$, complex chains are synthesized by the addition of GlcNAc, removal of two Man, and the stepwise addition of individual sugars in reactions catalyzed by specific transferases (eg, GlcNAc, Gal, NeuAc transferases) that employ appropriate nucleotide sugars.

(**1**) The involvement of the **mannose 6-P signal** in targeting of certain lysosomal enzymes is clear (see above and discussion of I-cell disease, below). (**2**) It is likely that the large N-glycan chains present on newly synthesized glycoproteins may assist in keeping these proteins in a **soluble state** inside the lumen of the endoplasmic reticulum. (**3**) One species of N-glycan chains has been shown to play a role in the folding and retention of certain glycoproteins in the lumen of the endoplasmic reticulum. **Calnexin** is a protein present in the endoplasmic reticulum membrane that acts as a chaperone (Chapter 46) and lectin. Binding to calnexin prevents a glycoprotein from aggregating. It has been found that calnexin will bind specifically to a number of glycoproteins (eg, the influenza virus hemagglutinin [HA]) that possess the **monoglycosylated core structure**. This species is the product of reaction 2 shown in Figure 47–9, but from which the terminal glucose residue has been removed, leaving only the innermost glucose attached. Calnexin and the bound glycoprotein form a complex with **ERp57**, a homolog of protein disulfide isomerase (PDI), which catalyzes disulfide bond interchange, facilitating proper folding. The bound glycoprotein is released from its complex with calnexin-ERp57 when the sole remaining glucose is hydrolyzed by glucosidase II and **leaves the ER if properly folded**. If **not properly folded**, an ER **glucosyltransferase** recognizes this and **re-glucosylates** the glycoprotein, which **re-binds** to the calnexin-Erp57 complex. If now properly folded, the glycoprotein is again de-glucosylated and leaves the ER. If not capable of proper folding, it is **translocated out of the ER** into the cytoplasm, where it is **degraded** (compare Figure 46–8). This so-called **calnexin cycle** is illustrated in Figure 47–10. In this way, calnexin retains certain partly folded (or misfolded) glycoproteins and releases them when further folding has occurred. The glucosyltransferase, by **sensing** the folding of the glycoprotein and only re-glucosylating misfolded proteins, is a key component of the cycle. The calnexin cycle is an important component of the **quality control systems** operating

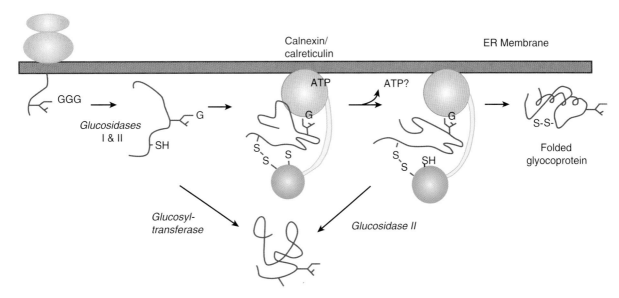

FIGURE 47–10 Model of the calnexin cycle. As a nascent (growing) polypeptide chain enters the ER, certain Asn residues are glycosylated by addition of $Glc_3Man_9GlcNAc_2$ (see text). The outermost two molecules of glucose are removed via the actions of glucosidases I and II. This exposes the innermost molecule of glucose, which is recognized by the lectin sites of calnexin and calreticulin. In their ATP-bound state, calnexin and calreticulin bind to the monoglucosylated oligosaccharide (via their lectin sites) as well as to hydrophobic segments of the unfolded glycoprotein (via their polypeptide binding or chaperone sites). Glycoprotein dissociation involves the action of glucosidase II to remove the terminal glucose and also a change in affinity of the polypeptide binding site. After dissociation, if folding does not occur rapidly, the glycoprotein is re-glucosylated by an ER glucosyltransferase, which acts only on non-native protein conformers (conformer = a protein in one of several possible conformations). The re-glucosylated glycoprotein can then re-bind to the ATP form of calnexin/calreticulin. Thus, both the glucosyltransferase and calnexin/calreticulin act as folding sensors. This cycle of binding and release has three functions: it prevents glycoprotein aggregation; it retains non-native conformers in the ER until a native structure is acquired (quality control); and binding to calnexin/calreticulin brings ERp57 into proximity with the non-native glycoprotein. ERp57 catalyzes disulfide bond formation and isomerization within the glycoprotein substrate, assisting it to assume its native conformation. If the glycoprotein is not capable of proper folding, it is translocated out of the ER into the cytoplasm for proteosomal degradation (compare Figure 46–8). Calreticulin, a soluble ER protein, plays a similar role to calnexin. (G, glucose.) (Figure and legend generously supplied by Dr D B Williams, and modified slightly with his permission.)

in the lumen of the ER. The soluble ER protein **calreticulin** performs a similar function.

Several Factors Regulate the Glycosylation of Glycoproteins

It is evident that glycosylation of glycoproteins is a complex process involving a large number of enzymes. It has been estimated that some 1% of the human genome may be involved with glycosylation events. Another index of its complexity is that more than ten distinct GlcNAc transferases involved in glycoprotein biosynthesis have been reported, and others are theoretically possible. Multiple species of the other glycosyltransferases (eg, sialyltransferases) also exist. Controlling factors of the first stage of N-linked glycoprotein biosynthesis (ie, **oligosaccharide assembly and transfer**) include (1) the presence of suitable acceptor sites in proteins, (2) the tissue level of Dol-P, and (3) the activity of the oligosaccharide: protein transferase.

Some factors known to be involved in the regulation of **oligosaccharide processing** are listed in Table 47–11. Two of the points listed merit further comment. First, **species varia-** tions among processing enzymes have assumed importance in relation to production of glycoproteins of therapeutic use by means of recombinant DNA technology. For instance, **recombinant erythropoietin** (epoetin alfa; EPO) is sometimes administered to patients with certain types of chronic anemia in order to stimulate erythropoiesis. The half-life of EPO in plasma is influenced by the nature of its glycosylation pattern, with certain patterns being associated with a short half-life, appreciably limiting its period of therapeutic effectiveness. It is thus important to harvest EPO from host cells that confer a pattern of glycosylation consistent with a normal half-life in plasma. Second, there is great interest in analysis of the activities of glycoprotein-processing enzymes in various types of **cancer cells.** These cells have often been found to synthesize different oligosaccharide chains (eg, they often exhibit greater branching) from those made in control cells. This could be due to cancer cells containing different patterns of glycosyltransferases from those exhibited by corresponding normal cells, due to specific gene activation or repression. The differences in oligosaccharide chains could affect adhesive interactions between cancer cells and their normal parent tissue cells, contributing to metastasis. If a correlation could be found be-

TABLE 47–11   Some Factors Affecting the Activities of Glycoprotein Processing Enzymes

Factor	Comment
Cell type	Different cell types contain different profiles of processing enzymes.
Previous enzyme	Certain glycosyltransferases act only on an oligosaccharide chain if it has already been acted upon by another processing enzyme.[1]
Development	The cellular profile of processing enzymes may change during development if their genes are turned on or off.
Intracellular location	For instance, if an enzyme is destined for insertion into the membrane of the ER (eg, HMG-CoA reductase), it may never encounter Golgi-located processing enzymes.
Protein conformation	Differences in conformation of different proteins may facilitate or hinder access of processing enzymes to identical oligosaccharide chains.
Species	Same cells (eg, fibroblasts) from different species may exhibit different patterns of processing enzymes.
Cancer	Cancer cells may exhibit processing enzymes different from those of corresponding normal cells.

[1]For example, prior action of GlcNAc transferase I is necessary for the action of Golgi α-mannosidase II.

tween the activity of particular processing enzymes and the **metastatic properties** of cancer cells, this could be important as it might permit synthesis of drugs to inhibit these enzymes and, secondarily, metastasis.

The genes encoding many glycosyltransferases have already been **cloned**, and others are under study. Cloning has revealed new information on both protein and gene structures. The latter should also cast light on the mechanisms involved in their **transcriptional control**, and **gene knockout studies** are being used to evaluate the biologic importance of various glycosyltransferases.

Tunicamycin Inhibits N- But Not O-Glycosylation

A number of compounds are known to inhibit various reactions involved in glycoprotein processing. **Tunicamycin, deoxynojirimycin,** and **swainsonine** are three such agents. The reactions they inhibit are indicated in Table 47–12. These agents can be used experimentally to inhibit various stages of glycoprotein biosynthesis and to study the effects of specific alterations upon the process. For instance, if cells are grown in the presence of tunicamycin, no glycosylation of their normally N-linked glycoproteins will occur. In certain cases, lack of glycosylation has been shown to increase the susceptibility of these proteins to proteolysis. Inhibition of glycosylation does not appear to have a consistent effect upon the secretion

TABLE 47–12   Three Inhibitors of Enzymes Involved in the *N*-Glycosylation of Glycoproteins and Their Sites of Action

Inhibitor	Site of Action
Tunicamycin	Inhibits GlcNAc-P transferase, the enzyme catalyzing addition of GlcNAc to dolichol-P, the first step in the biosynthesis of oligosaccharide-P-P-dolichol
Deoxynojirimycin	Inhibitor of glucosidases I and II
Swainsonine	Inhibitor of mannosidase II

of glycoproteins that are normally secreted. The inhibitors of glycoprotein processing listed in Table 47–12 do not affect the biosynthesis of O-linked glycoproteins. The extension of O-linked chains can be prevented by GalNAc-benzyl. This compound competes with natural glycoprotein substrates and thus prevents chain growth beyond GalNAc.

SOME PROTEINS ARE ANCHORED TO THE PLASMA MEMBRANE BY GLYCOSYLPHOSPHATIDYLINOSITOL STRUCTURES

Glycosylphosphatidylinositol (GPI)-linked glycoproteins comprise the third major class of glycoprotein. The GPI structure (sometimes called a "sticky foot") involved in linkage of the enzyme acetylcholinesterase (ACh esterase) to the plasma membrane of the red blood cell is shown in Figure 47–1. GPI-linked proteins are anchored to the outer leaflet of the plasma membrane by the fatty acids of phosphatidylinositol (PI). The PI is linked via a GlcN moiety to a glycan chain that contains various sugars (eg, Man, GlcN). In turn, the oligosaccharide chain is linked via phosphorylethanolamine in an amide linkage to the carboxyl terminal amino acid of the attached protein. The core of most GPI structures contains one molecule of phosphorylethanolamine, three Man residues, one molecule of GlcN, and one molecule of phosphatidylinositol, as follows:

$$\text{Ethanolamine-phospho} \rightarrow 6\text{Man}\alpha1 \rightarrow$$
$$2\text{Man}\alpha1 \rightarrow 6\text{Man}\alpha1 \rightarrow \text{GlN}\alpha1 \rightarrow$$
$$6 - myo - \text{inositol} - 1 - \text{phospholipid}$$

Additional constituents are found in many GPI structures; for example, that shown in Figure 47–1 contains an extra phosphorylethanolamine attached to the middle of the three Man moieties of the glycan and an extra fatty acid attached to GlcN. The functional significance of these variations among structures is not understood. This type of linkage was first detected by the use of bacterial PI-specific phospholipase C (PI-PLC), which was found to release certain proteins from the plasma membrane of cells by splitting the bond indicated in Figure 47–1. Examples of some proteins that are anchored by this type of linkage are given in Table 47–13. At least three possible functions of this type of linkage have been suggested: (1) The GPI anchor may allow greatly enhanced **mobility** of a protein

TABLE 47–13 *Some* GPI-Linked Proteins

- Acetylcholinesterase (red cell membrane)
- Alkaline phosphatase (intestinal, placental)
- Decay-accelerating factor (red cell membrane)
- 5'-Nucleotidase (T lymphocytes, other cells)
- Thy-1 antigen (brain, T lymphocytes)
- Variable surface glycoprotein (*Trypanosoma brucei*)

in the plasma membrane compared with that observed for a protein that contains transmembrane sequences. This is perhaps not surprising, as the GPI anchor is attached only to the outer leaflet of the lipid bilayer, so that it is freer to diffuse than a protein anchored via both leaflets of the bilayer. Increased mobility may be important in facilitating rapid responses to appropriate stimuli. (2) Some GPI anchors may connect with **signal transduction** pathways. (3) It has been shown that GPI structures can **target** certain proteins to apical domains and also basolateral domains of the plasma membrane of certain polarized epithelial cells. The biosynthesis of GPI anchors is complex and begins in the endoplasmic reticulum. The GPI anchor is assembled independently by a series of enzyme-catalyzed reactions and then transferred to the carboxyl terminal end of its acceptor protein, accompanied by cleavage of the preexisting carboxyl terminal hydrophobic peptide from that protein. This process is sometimes called **glypiation**. An acquired defect in an early stage of the biosynthesis of the GPI structure has been implicated in the causation of **paroxysmal nocturnal hemoglobinuria** (see later).

ADVANCED GLYCATION END-PRODUCTS (AGEs) ARE THOUGHT TO BE IMPORTANT IN THE CAUSATION OF TISSUE DAMAGE IN DIABETES MELLITUS

Glycation refers to non-enzymatic attachment of sugars (mainly glucose) to amino groups of proteins and also to other molecules (eg, DNA, lipids). Glycation is distinguished from **glycosylation** because the latter involves enzyme-catalyzed attachment of sugars. When glucose attaches to a protein, in-

termediate products formed include **Schiff bases**. These can further re-arrange by the **Amadori rearrangement** to ketoamines (see Figure 47–11). The overall series of reactions is known as the **Maillard reaction**. These reactions are involved in the **browning** of certain foodstuffs that occurs on storage or processing (eg, heating). The end-products of glycation reactions are termed **advanced glycation end-products (AGEs)**.

The major medical interest in AGEs has been in relation to them causing **tissue damage in diabetes mellitus**, in which the level of blood glucose is often consistently elevated, promoting increased glycation. At constant time intervals, the extent of glycation is more or less proportional to the blood glucose level. It has also been suggested that AGEs are involved in other processes, such as **aging**.

Glycation of collagen and other proteins in the ECM alters their properties (eg, increasing the **cross-linking of collagen**). Cross-linking can lead to accumulation of various plasma proteins in the walls of blood vessels; in particular, accumulation of **LDL** can contribute to **atherogenesis**. AGEs appear to be involved in both **microvascular** and **macrovascular** damage in diabetes mellitus. Also endothelial cells and macrophages have AGE receptors on their surfaces. Uptake of glycated proteins by these receptors can activate the transcription factor **NF-kB** (see Chapter 50), generating a variety of **cytokines** and **pro-inflammatory molecules**. It is thus believed that AGEs are one significant contributor to some of the pathologic finding found in diabetes mellitus (see Figure 47–12). **Aminoguanidine**, an inhibitor of the formation of AGEs, may be of benefit in reducing the organ and tissue complications of diabetes.

Non-enzymatic attachment of glucose to **hemoglobin A** present in red blood cells (ie, formation of **HbA$_{1c}$**) occurs in normal individuals and is increased in patients with diabetes mellitus whose blood sugar levels are elevated. As discussed in Chapter 6, measurement of HbA$_{1c}$ has become a very important part of the **management of patients with diabetes mellitus**.

GLYCOPROTEINS ARE INVOLVED IN MANY BIOLOGIC PROCESSES & IN MANY DISEASES

As listed in Table 47–1, glycoproteins have many different functions; some have already been addressed in this chapter and

FIGURE 47–11 Formation of AGEs from glucose. Glucose is shown interacting with the amino group of hemoglobin (Hb) forming a Schiff base. This is subject to the Amadori rearrangement, forming a ketoamine. Further rearrangements can occur, leading to other AGEs.

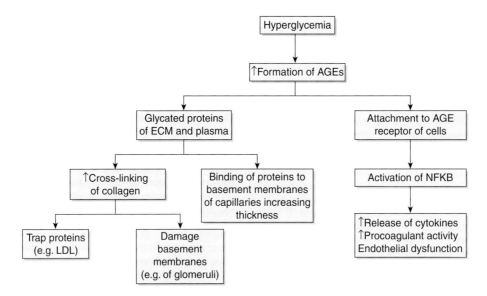

FIGURE 47–12 Some consequences of the formation of AGEs. Hyperglycemia (eg, occurring in poorly controlled diabetes) leads to the formation of AGEs. These can occur in proteins of the ECM or plasma. In the ECM, they can cause increased cross-linking of collagen, which can trap proteins such as LDL (contributing to atherogenesis) and damage basement membranes in the kidneys and other sites. Thickening of basement membranes can also occur by binding of glycated proteins to them. AGEs can attach to AGE receptors on cells, activating NFkB (see Chapter 50), which has several consequences (as shown). Damage to renal basement membranes, thickening of these membranes in capillaries and endothelial dysfunction are found in ongoing uncontrolled diabetes mellitus.

others are described elsewhere in this text (eg, transport molecules, immunologic molecules, and hormones). Here, their involvement in two specific processes—**fertilization** and **inflammation**—will be briefly described. In addition, the **bases of a number of diseases** that are due to abnormalities in the synthesis and degradation of glycoproteins will be summarized.

Glycoproteins Are Important in Fertilization

To reach the plasma membrane of an oocyte, a sperm has to traverse the **zona pellucida (ZP),** a thick, transparent, noncellular envelope that surrounds the oocyte. The zona pellucida contains three glycoproteins of interest, ZP1–3. Of particular note is ZP3, an O-linked glycoprotein that functions as a receptor for the sperm. A protein on the sperm surface, possibly galactosyl transferase, interacts specifically with oligosaccharide chains of ZP3; in at least certain species (eg, the mouse), this interaction, by transmembrane signaling, induces the **acrosomal reaction,** in which enzymes such as proteases and hyaluronidase and other contents of the acrosome of the sperm are released. Liberation of these enzymes helps the sperm to pass through the zona pellucida and reach the plasma membrane (PM) of the oocyte. In hamsters, it has been shown that another glycoprotein, PH-30, is important in both the binding of the PM of the sperm to the PM of the oocyte and also in the subsequent fusion of the two membranes. These interactions enable the sperm to enter and thus fertilize the oocyte. It may

be possible to **inhibit fertilization** by developing drugs or antibodies that interfere with the normal functions of ZP3 and PH-30 and which would thus act as contraceptive agents.

Selectins Play Key Roles in Inflammation & in Lymphocyte Homing

Leukocytes play important roles in many inflammatory and immunologic phenomena. The first steps in many of these phenomena are interactions between circulating leukocytes and **endothelial cells** prior to passage of the former out of the circulation. Work done to identify specific molecules on the surfaces of the cells involved in such interactions has revealed that leukocytes and endothelial cells contain on their surfaces specific lectins, called **selectins,** that participate in their intercellular adhesion. Features of the three major classes of selectins are summarized in Table 47–14. Selectins are single-chain Ca^{2+}-binding transmembrane proteins that contain a number of domains (Figure 47–13). Their amino terminal ends contain the lectin domain, which is involved in binding to specific carbohydrate ligands.

The adhesion of neutrophils to endothelial cells of postcapillary venules can be considered to occur in four stages, as shown in Figure 47–14. The initial baseline stage is succeeded by **slowing or rolling** of the neutrophils, mediated by selectins. Interactions between L-selectin on the neutrophil surface and CD34 and GlyCAM-1 or other glycoproteins on the endothelial surface are involved. These particular interactions are initially

TABLE 47–14 Some Molecules Involved in Leukocyte-Endothelial Cell Interactions

Molecule	Cell	Ligands
Selectins		
L-selectin	PMN, lymphs	CD34, Gly-CAM-[1], sialyl-Lewis[x], and others
P-selectin	EC, platelets	P-selectin glycoprotein ligand-1 (PSGL-1), sialyl-Lewis[x], and others
E-selectin	EC	Sialyl-Lewis[x] and others
Integrins		
LFA-1	PMN, lymphs	ICAM-1, ICAM-2
(CD11a/CD18)		
Mac-1	PMN	ICAM-1 and others
(CD11b/CD18)		
Immunoglobulin superfamily		
ICAM-1	Lymphs, EC	LFA-1, Mac-1
ICAM-2	Lymphs, EC	LFA-1
PECAM-1	EC, PMN, lymphs	Various platelets

Source: Modified, with permission, from Albelda SM, Smith CW, Ward PA: Adhesion molecules and inflammatory injury. FASEB J 1994;8:504.

Abbreviations: PMN, polymorphonuclear leukocytes; EC, endothelial cell; lymphs, lymphocytes; CD, cluster of differentiation; ICAM, intercellular adhesion molecule; LFA-1, lymphocyte function-associated antigen-1; PECAM-1, platelet endothelial cell adhesion cell molecule-1.

[1]These are ligands for lymphocyte L-selectin; the ligands for neutrophil L-selectin have not apparently been identified.

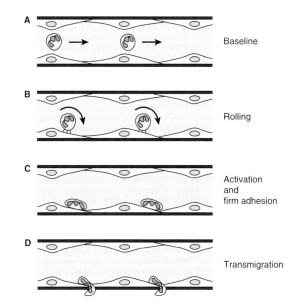

FIGURE 47–14 Schematic diagram of neutrophil-endothelial cell interactions. **(A)** Baseline conditions: Neutrophils do not adhere to the vessel wall. **(B)** The first event is the slowing or rolling of the neutrophils within the vessel (venule) mediated by selectins. **(C)** Activation occurs, resulting in neutrophils firmly adhering to the surfaces of endothelial cells and also assuming a flattened shape. This requires interaction of activated CD18 integrins on neutrophils with ICAM-1 on the endothelium. **(D)** The neutrophils then migrate through the junctions of endothelial cells into the interstitial tissue; this requires involvement of PECAM-1. Chemotaxis is also involved in this latter stage. (Reproduced, with permission, from Albelda SM, Smith CW, Ward PA: Adhesion molecules and inflammatory injury. FASEB J 1994;8;504.)

FIGURE 47–13 Schematic diagram of the structure of human L-selectin. The extracellular portion contains an amino terminal domain homologous to C-type lectins and an adjacent epidermal growth factor-like domain. These are followed by a variable number of complement regulatory-like modules (numbered circles) and a trans-membrane sequence (black diamond). A short cytoplasmic sequence (red rectangle) is at the carboxyl terminal. The structures of P- and E-selectin are similar to that shown except that they contain more complement-regulatory modules. The numbers of amino acids in L-, P-, and E- selectins, as deduced from the cDNA sequences, are 385, 789, and 589, respectively. (Reproduced, with permission, from Bevilacqua MP, Nelson RM: Selectins. J Clin Invest 1993;91:370.)

short-lived, and the overall binding is of relatively low affinity, permitting rolling. However, during this stage, **activation** of the neutrophils by various chemical mediators (discussed below) occurs, resulting in a change of shape of the neutrophils and firm adhesion of these cells to the endothelium. An additional set of **adhesion molecules** is involved in firm adhesion, namely, LFA-1 and Mac-1 on the neutrophils and ICAM-1

and ICAM-2 on endothelial cells. LFA-1 and Mac-1 are CD11/CD18 integrins (see Chapter 52 for a discussion of integrins), whereas ICAM-1 and ICAM-2 are members of the immunoglobulin superfamily. The fourth stage is **transmigration** of the neutrophils across the endothelial wall. For this to occur, the neutrophils insert pseudopods into the junctions between endothelial cells, squeeze through these junctions, cross the basement membrane, and then are free to migrate in the extravascular space. Platelet-endothelial cell adhesion molecule-1 (PECAM-1) has been found to be localized at the junctions of endothelial cells and thus may have a role in transmigration. A variety of biomolecules have been found to be involved in **activation** of neutrophil and endothelial cells, including tumor necrosis factor, various interleukins, platelet activating factor (PAF), leukotriene B$_4$, and certain complement fragments. These compounds stimulate various signaling pathways, resulting in changes in cell shape and function, and some are also chemotactic. One important functional change is **recruitment of selectins** to the cell surface, as in some cases selectins are stored in granules (eg, in endothelial cells and platelets).

The precise chemical nature of some of the ligands involved in selectin-ligand interactions has been determined. All three selectins bind **sialylated and fucosylated oligosaccharides,** and in particular all three bind **sialyl-Lewis^x**

FIGURE 47–15 Schematic representation of the structure of sialyl-Lewis^x.

(Figure 47–15), a structure present on both glycoproteins and glycolipids. Whether this compound is the actual ligand involved in vivo is not established. Sulfated molecules, such as the sulfatides (Chapter 15), may be ligands in certain instances. This basic knowledge is being used in attempts to synthesize compounds that block selectin-ligand interactions and thus may inhibit the inflammatory response. Approaches include administration of specific monoclonal antibodies or of chemically synthesized analogs of sialyl-Lewis^x, both of which bind selectins. **Cancer cells** often exhibit sialyl-Lewis^x and other selectin ligands on their surfaces. It is thought that these ligands play a role in the invasion and metastasis of cancer cells.

Abnormalities in the Synthesis of Glycoproteins Underlie Certain Diseases

Table 47–15 lists a number of conditions in which abnormalities in the synthesis of glycoproteins are of importance. As mentioned above, many **cancer cells** exhibit different profiles

TABLE 47–15 Some Diseases Due to or Involving Abnormalities in the Biosynthesis of Glycoproteins

Disease	Abnormality
Cancer	Increased branching of cell surface glycans or presentation of selectin ligands may be important in metastasis.
Congenital disorders of glycosylation[1]	See Table 47–16.
HEMPAS[2] (OMIM 224100)	Abnormalities in certain enzymes (eg, mannosidase II and others) involved in the biosynthesis of N-glycans, particularly affecting the red blood cell membrane.
Leukocyte adhesion deficiency, type II (OMIM 266265)	Probably mutations affecting a Golgi-located GDP-fucose transporter, resulting in defective fucosylation.
Paroxysmal nocturnal hemoglobinuria (OMIM 311770)	Acquired defect in biosynthesis of the GPI[3] structures of decay accelerating factor (DAF) and CD59.
I-cell disease (OMIM 252500)	Deficiency of GlcNAc phosphotransferase, resulting in abnormal targeting of certain lysosomal enzymes.

[1]The OMIM number for congenital disorder of glycosylation type Ia is 212065.
[2]Hereditary erythroblastic multinuclearity with a positive acidified serum lysis test (congenital dyserythropoietic anemia type II). This is a relatively mild form of anemia. It reflects at least in part the presence in the red cell membranes of various glycoproteins with abnormal N-glycan chains, which contribute to the susceptibility to lysis.
[3]Glycosylphosphatidylinositol.

of oligosaccharide chains on their surfaces, some of which may contribute to metastasis.

The **congenital disorders of glycosylation (CDG)** are a group of disorders of considerable current interest. The major features of these conditions are summarized in Table 47–16.

Leukocyte adhesion deficiency (LAD) II is a rare condition probably due to mutations affecting the activity of a Golgi-located GDP-fucose transporter. It can be considered a congenital disorder of glycosylation. The absence of fucosylated ligands for selectins leads to a marked decrease in neutrophil rolling. Subjects suffer life-threatening, recurrent bacterial infections and also psychomotor and mental retardation. The condition appears to respond to oral fucose.

Hereditary erythroblastic multinuclearity with a positive acidified lysis test (HEMPAS)—congenital dyserythropoietic anemia type II—is another disorder in which abnormalities in the processing of N-glycans are thought to be involved. Some cases have been claimed to be due to defects in alpha–mannosidase II.

Paroxysmal nocturnal hemoglobinuria (PNH) is an acquired mild anemia characterized by the presence of hemoglobin in urine due to hemolysis of red cells, particularly during sleep. This latter phenomenon may reflect a slight drop in plasma pH during sleep, which increases susceptibility to lysis by the complement system (Chapter 50). The basic defect in paroxysmal nocturnal hemoglobinuria is the acquisition of somatic mutations in the *PIG-A* (for phosphatidylinositol glycan class A) gene of certain hematopoietic cells. The product of this gene appears to be the enzyme that links glucosamine to phosphatidylinositol in the GPI structure (Figure 47–1). Thus, proteins that are anchored by a GPI linkage are deficient in the red cell membrane. Two proteins are of particular interest:

TABLE 47–16 Major Features of the Congenital Disorders of Glycosylation

- Autosomal recessive disorders

- Multisystem disorders that have probably not been recognized in the past

- Generally affect the central nervous system, resulting in psychomotor retardation and other features

- Type I disorders are due to mutations in genes encoding enzymes (eg, phosphomannomutase-2 [PMM-2], causing CDG Ia) involved in the synthesis of dolichol-P-P-oligosaccharide

- Type II disorders are due to mutations in genes encoding enzymes (eg, GlcNAc transferase-2, causing CDG IIa) involved in the processing of N-glycan chains

- At least 15 distinct disorders have been recognized

- Isoelectric focusing of transferrin is a useful biochemical test for assisting in the diagnosis of these conditions; truncation of the oligosaccharide chains of this protein alters its iso-electric focusing pattern

- Oral mannose has proved of benefit in the treatment of CDG Ia

Abbreviation: CDG, congenital disorder of glycosylation.

decay accelerating factor (DAF) and another protein designated CD59. They normally interact with certain components of the complement system (Chapter 50) to prevent the hemolytic actions of the latter. However, when they are deficient, the complement system can act on the red cell membrane to cause hemolysis. A monoclonal antibody to C5, a terminal component of the complement system, has proven useful in the management of PNH. by inhibiting the complement cascade. PNH can be diagnosed relatively simply, as the red cells are much more sensitive to hemolysis in normal serum acidified to pH 6.2 (Ham's test); the complement system is activated under these conditions, but normal cells are not affected. Figure 47–16 summarizes the etiology of PNH.

Study of the congenital muscular dystrophies (CMDs) has revealed that certain of them (eg, the Walker-Warburg syndrome, muscle-eye-brain disease, Fukuyama CMD) are the result of defects in the synthesis of glycans in the protein α-dystroglycan (α-DG). This protein protrudes from the surface membrane of muscle cells and interacts with laminin-2 (merosin) in the basal lamina (see Figure 49–11). If the glycans of α-DG are not correctly formed (as a result of mutations in genes encoding certain glycosyltransferases), this results in defective interaction of α-DG with laminin, which in turn leads to the development of a CMD.

Rheumatoid arthritis is associated with an alteration in the glycosylation of circulating immunoglobulin G (IgG) molecules (Chapter 50), such that they lack galactose in their Fc regions and terminate in GlcNAc. Mannose-binding protein (MBP, not to be confused with the mannose 6-P receptor), a C-lectin synthesized by liver cells and secreted into the circulation, binds mannose, GlcNAc, and certain other sugars. It can thus bind agalactosyl IgG molecules, which subsequently activate the complement system (see Chapter 50), contributing to chronic inflammation in the synovial membranes of joints.

MBP can also bind the above sugars when they are present on the surfaces of certain bacteria, fungi, and viruses, preparing these pathogens for opsonization or for destruction by the complement system. This is an example of innate immunity, not involving immunoglobulins or T lymphocytes. Deficiency of this protein in young infants as a result of mutation renders them very susceptible to recurrent infections.

I-Cell Disease Results from Faulty Targeting of Lysosomal Enzymes

As indicated above, Man 6-P serves as a chemical marker to target certain lysosomal enzymes to that organelle. Analysis of cultured fibroblasts derived from patients with I-cell (inclusion cell) disease played a large part in revealing the above role of Man 6-P. I-cell disease is an uncommon condition characterized by severe progressive psychomotor retardation and a variety of physical signs, with death often occurring in the first decade. Cultured cells from patients with I-cell disease were found to lack almost all of the normal lysosomal enzymes; the lysosomes thus accumulate many different types of undegraded molecules, forming inclusion bodies. Samples of plasma from patients with the disease were observed to contain very high activities of lysosomal enzymes; this suggested that the enzymes were being synthesized but were failing to reach their proper intracellular destination and were instead being secreted. Cultured cells from patients with the disease were noted to take up exogenously added lysosomal enzymes obtained from normal subjects, indicating that the cells contained a normal receptor on their surfaces for endocytic uptake of lysosomal enzymes. In addition, this finding suggested that lysosomal enzymes from patients with I-cell disease might lack a recognition marker. Further studies revealed that lysosomal enzymes from normal individuals carried the Man 6-P recognition marker described above, which interacted with a specific intracellular protein, the Man 6-P receptor. Cultured cells from patients with I-cell disease were then found to be deficient in the activity of the cis-Golgi-located GlcNAc phosphotransferase, explaining how their lysosomal enzymes failed to acquire the Man 6-P marker. It is now known that there are two Man 6-P receptor proteins, one of high (275 kDa) and one of low (46 kDa) molecular mass. These proteins are lectins, recognizing Man 6-P. The former is cation-independent and also binds IGF-II (hence it is named the Man 6-P–IGF-II receptor), whereas the latter is cation-dependent in some species and does not bind IGF-II. It appears that both receptors function in the intracellular sorting of lysosomal enzymes into clathrin-coated vesicles, which occurs in the trans-Golgi subsequent to synthesis of Man 6-P in the cis-Golgi. These vesicles then leave the Golgi and fuse with a prelysosomal compartment. The low pH in this compartment causes the lysosomal enzymes to dissociate from their receptors and subsequently enter into lysosomes. The receptors are recycled and reused. Only the smaller receptor functions in the endocytosis of extracellular lysosomal enzymes, which is a minor pathway for lysosomal location. Not all cells employ the Man 6-P receptor to target their lysosomal enzymes (eg, hepatocytes use a different but undefined pathway); furthermore, not all lysosomal enzymes are targeted by this mechanism. Thus, biochemical

FIGURE 47–16 Scheme of causation of paroxysmal nocturnal hemoglobinuria (OMIM 311770).

FIGURE 47–17 Summary of the causation of I-cell disease (OMIM 252500).

TABLE 47–17   Major Features of Some Diseases[1] Due to Deficiencies of Glycoprotein Hydrolases[2]

• Usually exhibit mental retardation or other neurologic abnormalities, and in some disorders coarse features or visceromegaly (or both)
• Variations in severity from mild to rapidly progressive
• Autosomal recessive inheritance
• May show ethnic distribution (eg, aspartylglycosaminuria is common in Finland)
• Vacuolization of cells observed by microscopy in some disorders
• Presence of abnormal degradation products (eg, oligosaccharides that accumulate because of the enzyme deficiency) in urine, detectable by TLC and characterizable by GLC-MS
• Definitive diagnosis made by assay of appropriate enzyme, often using leukocytes
• Possibility of prenatal diagnosis by appropriate enzyme assays
• No definitive treatment at present

[1]α-Mannosidosis, β-mannosidosis, fucosidosis, sialidosis, aspartylglycosaminuria, and Schindler disease.
[2]OMIM numbers: α-mannosidosis, 248500; β-mannosidosis, 248510; fucosidosis, 230000; sialidosis, 256550; aspartylglycosaminuria, 208400; Schindler disease, 609241.

investigations of I-cell disease not only led to elucidation of its basis but also contributed significantly to knowledge of how newly synthesized proteins are targeted to specific organelles, in this case the lysosome. Figure 47–17 summarizes the causation of I-cell disease.

Pseudo-Hurler polydystrophy is another genetic disease closely related to I-cell disease. It is a **milder condition**, and patients may survive to adulthood. Studies have revealed that the **GlcNAc phosphotransferase** involved in I-cell disease has several domains, including a catalytic domain and a domain that specifically recognizes and interacts with lysosomal enzymes. It has been proposed that the defect in pseudo-Hurler polydystrophy lies in the latter domain, and the retention of some catalytic activity results in a milder condition.

Genetic Deficiencies of Glycoprotein Lysosomal Hydrolases Cause Diseases Such as α-Mannosidosis

Glycoproteins, like most other biomolecules, undergo both synthesis and degradation (ie, turnover). Degradation of the oligosaccharide chains of glycoproteins involves a battery of lysosomal hydrolases, including α-neuraminidase, β-galactosidase, β-hexosaminidase, α- and β-mannosidases, α-N-acetylgalactosaminidase, α-fucosidase, endo- β-N-acetyl-glucosaminidase, and aspartylglucosaminidase. The sites of action of the last two enzymes are indicated in the legend to Figure 47–5. Genetically determined defects of the activities of these enzymes can occur, resulting in abnormal degradation of glycoproteins. The accumulation in tissues of such degraded glycoproteins can lead to various diseases. Among the best-recognized of these diseases are mannosidosis, fucosidosis, sialidosis, aspartylglycosaminuria, and Schindler disease, due respectively to deficiencies of α-mannosidase, α-fucosidase, α-neuraminidase, aspartylglucosaminidase, and α-N-acetyl-galactosaminidase. These diseases, which are relatively un-

common, have a variety of manifestations; some of their major features are listed in Table 47–17. The fact that patients affected by these disorders all show signs referable to the **central nervous system** reflects the importance of glycoproteins in the development and normal function of that system.

THE GLYCANS OF GLYCOCONJUGATES ARE INVOLVED IN THE BINDING OF VIRUSES, BACTERIA, & CERTAIN PARASITES TO HUMAN CELLS

A principal feature of glycans, and one that explains many of their biologic actions, is that they **bind** specifically to a variety of molecules such as proteins or other glycans. One reflection of this is their ability to bind certain viruses, many bacteria and some parasites.

Influenza virus A binds to cell surface glycoprotein receptor molecules containing NeuAc via a protein named **hemagglutinin (H)**. It also possesses a **neuraminidase (N)** that plays a key role in allowing elution of newly synthesized progeny from infected cells. If this process is inhibited, spread of the viruses is markedly diminished. Inhibitors of this enzyme (eg, zanamivir, oseltamivir) are now available for use in treating patients with influenza. Influenza viruses are classified according to the type of hemagglutinin and neuraminidase that they possess. There are at least 16 types of hemagglutinin and 9 types of neuraminidase. Thus, **avian influenza virus** is classified as **H5N1**. There is great interest in how this virus attaches to human cells, in view of the possibility of a pandemic

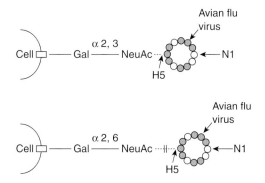

FIGURE 47–18 Schematic representation of binding of the avian influenza virus (H5N1) to a respiratory epithelial cell. The viral hemagglutin (HA) mediates its entry to cells by binding to a glycan on the cell surface that is terminated by the disaccharide galactose → α 2,3-NeuAc. It will not bind to a glycan terminated by galactose → α 2,6-NeuAc, which is the type predominantly found in the human respiratory tract. If the viral HA alters via mutation to be able to bind to the latter disaccharide, this could greatly increase its pathogenicity for humans. (H5, hemagglutinin type 5; N1, neuraminidase type 1.)

FIGURE 47–19 Attachment of *Helicobacter pylori* to epithelial cells of the stomach. Adhesin, a protein present in the tail of *H pylori*, interacts with two different glycans (structures shown below) present in glycoproteins on the surface of gastric epithelial cells. This provides an attachment site for the bacterium. Subsequently it liberates molecules, such as ammonia, that contribute to initiating peptic ulceration. (A) NeuAcα2,3Galβ1,4—Protein (Neuraminyl-galactose); (B) Fucα1,2Galβ1,3GlcNAc—Protein (Lewis^B substance).

occurring. It has been found that the virus preferentially attaches to glycans terminated by the disaccharide **galactose → α 2,3-NeuAc** (Figure 47–18). However, the predominant disaccharide terminating glycans in cells of the human respiratory tract is **galactose → α 2,6-NeuAc**. If a change in the structure of the viral hemagglutinin (due to mutation) occurs that allows it to bind to the latter disaccharide, this could greatly increase the potential infectivity of the virus, possibly resulting in very serious consequences.

Human immunodeficiency virus type 1 (HIV-1), thought by most to be the cause of AIDS, attaches to cells via one of its surface glycoproteins (gp120) and uses another surface glycoprotein (gp 41) to fuse with the host cell membrane. **Antibodies** to gp 120 develop during infection by HIV-1, and there has been interest in using the protein as a vaccine. One major problem with this approach is that the structure of gp 120 can change relatively rapidly, allowing the virus to escape from the neutralizing activity of antibodies directed against it.

Helicobacter pylori is believed to be the major cause of **peptic ulcers**. Studies have shown that this bacterium binds to at least two different glycans present on the surfaces of epithelial cells in the stomach (see Figure 47–19). This allows it to establish a stable attachment site to the stomach lining, and subsequent secretion of ammonia and other molecules by the bacterium are believed to initiate ulceration.

Similarly, many **bacteria that cause diarrhea** are also known to attach to surface cells of the intestine via glycans present in glycoproteins or glycolipids.

The basic cause of **cystic fibrosis** (CF) is mutations in the gene encoding CFTR (see Chapters 40 & 54). A major problem in this disease is recurring lung infections by bacteria such as *Pseudomonas aeruginosa*. In CF, a relative dehydration of re-

spiratory secretions occurs secondary to changes in electrolyte composition in the airway as a result of mutations in CFTR. Bacteria such as *P aeruginosa* attach to the sugar chains of mucins and find the dehydrated environment in the bronchioles a favorable location in which to multiply.

The attachment of *Plasmodium falciparum*—one of the types of plasmodia causing **malaria**—to human cells is mediated by a GPI present on the surface of the parasite.

Various researchers are **analyzing the surfaces of viruses, bacteria, parasites** and **human cells** to determine which molecules are involved in attachment. It is important to define the precise nature of the interactions between invading organisms and host cells, as this will hopefully lead to the development of drugs or other agents that will specifically inhibit attachment.

THE PACE OF RESEARCH IN GLYCOMICS IS ACCELERATING

Research on glycoconjugates in the past has been hampered by the lack of availability of suitable technics to determine the structures of glycans. However, appropriate analytical technics are now available (some of which are listed in Table 47–3), as are powerful new genetic technics (eg, knock-outs and knock-downs using RNAi molecules). It is certain that research in glycomics will not only provide a wealth of structural information on glyconconjugates, helping to disclose "the sugar code of life," but will also uncover many new important biologic interactions that are sugar-dependent and will provide targets for drug and other therapies.

SUMMARY

- Glycoproteins are widely distributed proteins—with diverse functions—that contain one or more covalently linked carbohydrate chains.

- The carbohydrate components of a glycoprotein range from 1% to more than 85% of its weight and may be simple or very complex in structure. Eight sugars are mainly found in the sugar chains of human glycoproteins: xylose, fucose, galactose, glucose, mannose, N-acetylgalactosamine, N-acetylglucosamine and N-acetylneuraminic acid.

- At least certain of the oligosaccharide chains of glycoproteins encode biologic information; they are also important to glycoproteins in modulating their solubility and viscosity, in protecting them against proteolysis, and in their biologic actions.

- The structures of oligosaccharide chains can be elucidated by gas-liquid chromatography, mass spectrometry, and high-resolution NMR spectrometry.

- Glycosidases hydrolyze specific linkages in oligosaccharides and are used to explore both the structures and functions of glycoproteins.

- Lectins are carbohydrate-binding proteins involved in cell adhesion and many other biologic processes.

- The major classes of glycoproteins are O-linked (involving an OH of serine or threonine), N-linked (involving the N of the amide group of asparagine), and glycosylphosphatidylinositol (GPI)-linked.

- Mucins are a class of O-linked glycoproteins that are distributed on the surfaces of epithelial cells of the respiratory, gastrointestinal, and reproductive tracts.

- The endoplasmic reticulum and Golgi apparatus play a major role in glycosylation reactions involved in the biosynthesis of glycoproteins.

- The oligosaccharide chains of O-linked glycoproteins are synthesized by the stepwise addition of sugars donated by nucleotide sugars in reactions catalyzed by individual specific glycoprotein glycosyltransferases.

- In contrast, the synthesis of N-linked glycoproteins involves a specific dolichol-P-P-oligosaccharide and various glycotransferases and glycosidases. Depending on the enzymes and precursor proteins in a tissue, it can synthesize complex, hybrid, or high-mannose types of N-linked oligosacccharides.

- Glycoproteins are implicated in many biologic processes. For instance, they have been found to play key roles in fertilization and inflammation.

- A number of diseases involving abnormalities in the synthesis and degradation of glycoproteins have been recognized. Glycoproteins are also involved in many other diseases, including influenza, AIDS, rheumatoid arthritis, cystic fibrosis and peptic ulcer.

- Developments in the new field of glycomics are likely to provide much new information on the roles of sugars in health and disease and also indicate targets for drug and other types of therapies.

REFERENCES

Burtis CA, Ashwood ER, Bruns DE (editors): *Tietz Textbook of Clinical Chemistry and Molecular Diagnostics.* 4th ed. Elsevier Saunders, 2006. (Chapter 25 contains a good discussion of glycated hemoglobin).

Chandrasekeran A, Srinivasan A, Raman R et al: Glycan topology determines human adaptation of avian H5N1 virus hemagglutinin. Nat Biotechnology 2008;26:107.

Freeze HH: Congenital disorders of glycosylation: CDG-I, CDG-II, and beyond. Curr Mol Med 2007;7:389.

Helenius A, Aebi M: Roles of N-linked glycans in the endoplasmic reticulum. Annu Rev Biochem 2004;73:1019.

Kornfeld R, Kornfeld S: Assembly of asparagine-linked oligosaccharides. Annu Rev Biochem 1985;54:631.

Michele DE, Campbell KP: Dystrophin-glycoproteins complex: post-translational processing and dystroglycan function. J Biol Chem 2003;278:15457.

Ohtsubo K, Marth JD: Glycosylation in cellular mechanisms of health and disease. Cell 2006;126:855.

Pilobelli KT, Mahal LK: Deciphering the glycocode: the complexity and analytical challenge of glycomics. Curr Opin Chem Biol 2007;11:300.

Ramasamy R et al: Receptor for advanced glycation end products: fundamental roles in the inflammatory response: winding the way to the pathogenesis of endothelial dysfunction and atherosclerosis. Ann NY Acad Sci 2008;1126:7.

Roseman S: Reflections on glycobiology. J Biol Chem 2001;276:41527.

Schachter H: The clinical relevance of glycobiology. J Clin Invest 2001;108:1579.

Scriver CR et al (editors): *The Metabolic and Molecular Bases of Inherited Disease.* 8th ed. McGraw-Hill, 2001. (Various chapters in this text and its up-dated online version [see References for Chapter 1] give in-depth coverage of topics such as I-cell disease and disorders of glycoprotein degradation.)

Sharon N, Lis H: History of lectins: from hemagglutinins to biological recognition molecules. Glycobiology 2004;14:53R.

Taylor ME, Drickamer K: *Introduction to Glycobiology.* Oxford University Press, 2003.

Varki A et al (editors): *Essentials of Glycobiology.* 2nd ed. Cold Spring Harbor Laboratory Press, 2008.

Von Itzstein M, Plebanski M, Cooke RM, Coppel RL: Hot, sweaty and sticky: the glycobiology of *Plasmodium falciparum.* Trends Parasitol 2008; 24:210.

Werz DB, Seeberger PH: Carbohydrates are the next frontier in pharmaceutical research. Chemistry 2005; 11:3194.

48

The Extracellular Matrix

Robert K. Murray, MD, PhD & Frederick W. Keeley, PhD

BIOMEDICAL IMPORTANCE

Most mammalian cells are located in tissues where they are surrounded by a complex **extracellular matrix (ECM)** often referred to as **"connective tissue."** The ECM contains three major classes of biomolecules: (1) the **structural proteins,** collagen, elastin, and fibrillin-1; (2) certain **specialized proteins** such as fibronectin and laminin; and (3) **proteoglycans,** whose chemical natures are described below. The ECM has been found to be involved in many normal and pathologic processes—eg, it plays important roles in development, in inflammatory states, and in the spread of cancer cells. Involvement of certain components of the ECM has been documented in both **rheumatoid arthritis** and **osteoarthritis.** Several diseases (eg, osteogenesis imperfecta and a number of types of the Ehlers-Danlos syndrome) are due to genetic disturbances of the synthesis of collagen. Specific components of proteoglycans (the glycosaminoglycans; GAGs) are affected in the group of genetic disorders known as the **mucopolysaccharidoses.** Changes occur in the ECM during the **aging process.** This chapter describes the basic biochemistry of the three major classes of biomolecules found in the ECM and illustrates their biomedical significance. Major biochemical features of two specialized forms of ECM—bone and cartilage—and of a number of diseases involving them are also briefly considered.

COLLAGEN IS THE MOST ABUNDANT PROTEIN IN THE ANIMAL WORLD

Collagen, the major component of most connective tissues, constitutes approximately 25% of the protein of mammals. It provides an extracellular framework for all metazoan animals and exists in virtually every animal tissue. At least 28 distinct types of collagen made up of over 30 distinct polypeptide chains (each encoded by a separate gene) have been identified in human tissues. Although several of these are present only in small proportions, they may play important roles in determining the physical properties of specific tissues. In addition, a number of proteins (eg, the C1q component of the complement system, pulmonary surfactant proteins SPA and SP-D) that are not classified as collagens have collagen-like domains

in their structures; these proteins are sometimes referred to as "noncollagen collagens."

Table 48–1 summarizes information on many of the types of collagens found in human tissues; the nomenclature used to designate types of collagen and their genes is described in the footnote.

In Table 48–2, the types of collagen listed in Table 48–1 are subdivided into a number of classes based primarily on the structures they form. In this chapter, we shall be primarily concerned with the fibril-forming **collagens I and II,** the major collagens of skin and bone and of cartilage, respectively. However, mention will be made of some of the other collagens.

COLLAGEN TYPE I IS COMPOSED OF A TRIPLE HELIX STRUCTURE & FORMS FIBRILS

All collagen types have a **triple helical structure.** In some collagens, the entire molecule is triple helical, whereas in others the triple helix may involve only a fraction of the structure. Mature collagen type I, containing approximately 1,000 amino acids, belongs to the former type; in it, each polypeptide subunit or alpha chain is twisted into a left-handed polyproline helix of three residues per turn (Figure 48–1). Three of these alpha chains are then wound into a **right-handed superhelix,** forming a rodlike molecule 1.4 nm in diameter and about 300 nm long. A striking characteristic of collagen is the occurrence of **glycine** residues at every third position of the triple helical portion of the alpha chain. This is necessary because glycine is the only amino acid small enough to be accommodated in the limited space available down the central core of the triple helix. This **repeating structure,** represented as $(Gly-X-Y)_n$, is an absolute requirement for the formation of the triple helix. While X and Y can be any other amino acids, about 100 of the X positions are proline and about 100 of the Y positions are hydroxyproline. Proline and hydroxyproline confer **rigidity** on the collagen molecule. **Hydroxyproline** is formed by the posttranslational hydroxylation of peptide-bound proline residues catalyzed by the enzyme **prolyl hydroxylase,** whose cofactors are **ascorbic acid** (vitamin C) and α-ketoglutarate.

TABLE 48–1   Types of Collagen and Their Genes[1]

Type	Genes	Tissue
I	COL1A1, COL1A2	Most connective tissues, including bone
II	COL2A1	Cartilage, vitreous humor
III	COL3A1	Extensible connective tissues such as skin, lung, and the vascular system
IV	COL4A1–COL4A6	Basement membranes
V	COL5A1–COL5A3	Minor component in tissues containing collagen I
VI	COL6A1–COL6A3	Most connective tissues
VII	COL7A1	Anchoring fibrils
VIII	COL8A1–COL8A2	Endothelium, other tissues
IX	COL9A1–COL9A3	Tissues containing collagen II
X	COL10A1	Hypertrophic cartilage
XI	COL11A1, COL11A2, COL2A1	Tissues containing collagen II
XII	COL12A1	Tissues containing collagen I
XIII	COL13A1	Many tissues
XIV	COL14A1	Tissues containing collagen I
XV	COL15A1	Many tissues
XVI	COL16A1	Many tissues
XVII	COL17A1	Skin hemidesmosomes
XVIII	COL18A1	Many tissues (eg, liver, kidney)
XIX	COL19A1	Rhabdomyosarcoma cells

Source: Adapted slightly from Prockop DJ, Kivirrikko KI: Collagens: molecular biology, diseases, and potentials for therapy. Annu Rev Biochem 1995;64:403. Copyright © 1995 by Annual Reviews, www.annualreviews.org. Reprinted with permission.
[1]The types of collagen are designated by Roman numerals. Constituent procollagen chains, called proα chains, are numbered using Arabic numerals, followed by the collagen type in parentheses. For instance, type I procollagen is assembled from two proα1(I) and one proα2(I) chains. It is thus a heterotrimer, whereas type 2 procollagen is assembled from three proα1(II) chains and is thus a homotrimer. The collagen genes are named according to the collagen type, written in Arabic numerals for the gene symbol, followed by an A and the number of the proα chain that they encode. Thus, the COL1A1 and COL1A2 genes encode the α1 and α2 chains of type I collagen, respectively. At least 28 types of collagen have now been recognized.

TABLE 48–2   Classification of Collagens, Based Primarily on the Structures That They Form

Class	Type
Fibril-forming	I, II, III, V, and XI
Network-like	IV, VIII, X
FACITs[1]	IX, XII, XIV, XVI, XIX
Beaded filaments	VI
Anchoring fibrils	VII
Transmembrane domain	XIII, XVII
Others	XV, XVIII

Source: Based on Prockop DJ, Kivirrikko KI: Collagens: molecular biology, diseases, and potentials for therapy. Annu Rev Biochem 1995;64:403. Copyright © 1995 by Annual Reviews. Reprinted with permission.
[1]FACITs = fibril-associated collagens with interrupted triple helices. Additional collagens to those listed above have been recognized.

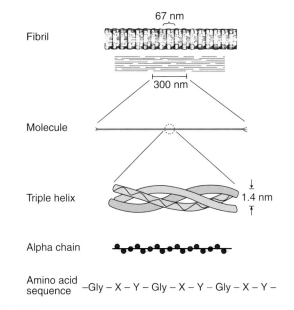

FIGURE 48–1 Molecular features of collagen structure from primary sequence up to the fibril. Each individual polypeptide chain is twisted into a left-handed helix of three residues (Gly-X-Y) per turn, and all of these chains are then wound into a right-handed superhelix. (Slightly modified and reproduced, with permission, from Eyre DR: Collagen: Molecular diversity in the body's protein scaffold. Science 1980;207:1315. Copyright © 1980 by the American Association for the Advancement of Science.)

Lysines in the Y position may also be posttranslationally modified to hydroxylysine through the action of **lysyl hydroxylase,** an enzyme with similar cofactors. Some of these hydroxylysines may be further modified by the addition of galactose or galactosyl-glucose through an **O-glycosidic linkage,** a glycosylation site that is unique to collagen.

Collagen types that form long rod-like fibers in tissues are assembled by lateral association of these triple helical units into a **"quarter staggered" alignment** such that each is displaced longitudinally from its neighbor by slightly less than one-quarter of its length (Figure 48–1, upper part). This arrangement is responsible for the banded appearance of these fibers in connective tissues. Collagen fibers are further stabilized by the formation of **covalent cross-links,** both within and between the triple helical units. These cross-links form through the action of **lysyl oxidase,** a copper-dependent enzyme that oxidatively deaminates the ε-amino groups of certain lysine and hydroxylysine residues, yielding reactive aldehydes. Such aldehydes can form aldol condensation products with other lysine- or hydroxylysine-derived aldehydes or form

Schiff bases with the ε-amino groups of unoxidized lysines or hydroxylysines. These reactions, after further chemical rearrangements, result in the stable covalent cross-links that are important for the tensile strength of the fibers. Histidine may also be involved in certain cross-links.

Several collagen types do not form fibrils in tissues (Table 48–2). They are characterized by interruptions of the triple helix with stretches of protein lacking Gly-X-Y repeat sequences. These non-Gly-X-Y sequences result in areas of globular structure interspersed in the triple helical structure.

Type IV collagen, the best-characterized example of a collagen with discontinuous triple helices, is an important component of **basement membranes,** where it forms a mesh-like network.

Collagen Undergoes Extensive Posttranslational Modifications

Newly synthesized collagen undergoes extensive **posttranslational modification** before becoming part of a mature extracellular collagen fiber (Table 48–3). Like most secreted proteins, collagen is synthesized on ribosomes in a precursor form, **preprocollagen,** which contains a leader or signal sequence that directs the polypeptide chain into the lumen of the endoplasmic reticulum. As it enters the endoplasmic reticulum, this leader sequence is enzymatically removed. **Hydroxylation** of proline and lysine residues and **glycosylation** of hydroxylysines in the **procollagen** molecule also take place at this site. The procollagen molecule contains polypeptide extensions **(extension peptides)** of 20–35 kDa at both its amino and carboxyl terminal ends, neither of which is present in mature collagen. Both extension peptides contain cysteine residues. While the amino terminal propeptide forms only intrachain disulfide bonds, the carboxyl terminal propeptides form both intrachain and interchain disulfide bonds. Formation of

these disulfide bonds assists in the **registration** of the three collagen molecules to form the triple helix, winding from the carboxyl terminal end. After formation of the triple helix, no further hydroxylation of proline or lysine or glycosylation of hydroxylysines can take place. **Self-assembly** is a cardinal principle in the biosynthesis of collagen.

Following **secretion** from the cell by way of the Golgi apparatus, extracellular enzymes called **procollagen aminoproteinase** and **procollagen carboxyproteinase** remove the extension peptides at the amino and carboxyl terminal ends, respectively. Cleavage of these propeptides may occur within crypts or folds in the cell membrane. Once the propeptides are removed, the triple helical collagen molecules, containing approximately 1000 amino acids per chain, **spontaneously assemble** into collagen fibers. These are further stabilized by the formation of **inter- and intrachain cross-links** through the action of lysyl oxidase, as described previously.

The same cells that secrete collagen also secrete **fibronectin,** a large glycoprotein present on cell surfaces, in the extracellular matrix, and in blood (see below). Fibronectin binds to aggregating precollagen fibers and alters the kinetics of fiber formation in the pericellular matrix. Associated with fibronectin and procollagen in this matrix are the **proteoglycans** heparan sulfate and chondroitin sulfate (see below). In fact, **type IX collagen,** a minor collagen type from cartilage, contains attached proteoglycan chains. Such interactions may serve to regulate the formation of collagen fibers and to determine their orientation in tissues.

Once formed, collagen is relatively **metabolically stable.** However, its breakdown is increased during starvation and various inflammatory states. Excessive production of collagen occurs in a number of conditions, eg, hepatic cirrhosis.

TABLE 48–3 Order and Location of Processing of the Fibrillar Collagen Precursor

Intracellular
1. Cleavage of signal peptide
2. Hydroxylation of prolyl residues and some lysyl residues; glycosylation of some hydroxylysyl residues
3. Formation of intrachain and interchain S–S bonds in extension peptides
4. Formation of triple helix
Extracellular
1. Cleavage of amino and carboxyl terminal propeptides
2. Assembly of collagen fibers in quarter-staggered alignment
3. Oxidative deamination of ε-amino groups of lysyl and hydroxylysyl residues to aldehydes
4. Formation of intra- and interchain cross-links via Schiff bases and aldol condensation products

A Number of Genetic Diseases Result from Abnormalities in the Synthesis of Collagen

About 30 genes encode the collagens, and their pathway of biosynthesis is complex, involving at least eight enzyme-catalyzed posttranslational steps. Thus, it is not surprising that a number of diseases (Table 48–4) are due to **mutations in collagen genes** or in **genes encoding some of the enzymes** involved in these posttranslational modifications. The diseases affecting bone (eg, osteogenesis imperfecta) and cartilage (eg, the chondrodysplasias) will be discussed later in this chapter.

Ehlers-Danlos syndrome comprises a group of inherited disorders whose principal clinical features are hyperextensibility of the skin, abnormal tissue fragility, and increased joint mobility. The clinical picture is variable, reflecting underlying extensive genetic heterogeneity. At least 10 types have been recognized, most but not all of which reflect a variety of lesions in the synthesis of collagen. **Type IV** is the most serious because of its tendency for spontaneous rupture of arteries or the bowel, reflecting abnormalities in type III collagen. Patients with **type VI,** due to a deficiency of lysyl hydroxylase, exhibit marked joint hypermobility and a tendency to ocular rupture. A deficiency of procollagen N-proteinase, causing

TABLE 48–4   Diseases Caused by Mutations in Collagen Genes or by Deficiencies in the Activities of Posttranslational Enzymes Involved in the Biosynthesis of Collagen

Gene or Enzyme	Disease[1]
COL1A1, COL1A2	Osteogenesis imperfecta, type 1[2] (OMIM 166200) Osteoporosis[3] (OMIM 166710) Ehlers-Danlos syndrome type VII autosomal dominant (OMIM 130060)
COL2A1	Severe chondrodysplasias Osteoarthritis[3] (OMIM 165720)
COL3A1	Ehlers-Danlos syndrome type IV (OMIM 130050)
COL4A3–COL4A6	Alport syndrome (including both autosomal and X-linked forms) (OMIM 104200)
COL7A1	Epidermolysis bullosa, dystrophic (OMIM 131750)
COL10A1	Schmid metaphysial chondrodysplasia (OMIM 156500)
Lysyl hydroxylase	Ehlers-Danlos syndrome type VI (OMIM 225400)
Procollagen N-proteinase	Ehlers-Danlos syndrome type VII autosomal recessive (OMIM 225410)
Lysyl hydroxylase	Menkes disease[4] (OMIM 309400)

Source: Adapted from Prockop DJ, Kivirrikko KI: Collagens: molecular biology, diseases, and potentials for therapy. Annu Rev Biochem 1995;64:403. Copyright © 1995 by Annual Reviews. Reprinted with permission.

[1]Genetic linkage to collagen genes has been shown for a few other conditions not listed here.

[2]At least four types of osteogenesis imperfecta are recognized; the great majority of mutations in all types are in the COL1A1 and COL1A2 genes.

[3]At present applies to only a relatively small number of such patients.

[4]Secondary to a deficiency of copper (Chapter 50).

formation of abnormal thin, irregular collagen fibrils, results in **type VIIC,** manifested by marked joint hypermobility and soft skin.

Alport syndrome is the designation applied to a number of genetic disorders (both X-linked and autosomal) affecting the structure of **type IV** collagen fibers, the major collagen found in the basement membranes of the renal glomeruli (see discussion of laminin, below). Mutations in several genes encoding type IV collagen fibers have been demonstrated. The presenting sign is hematuria, and patients may eventually develop end-stage renal disease. Electron microscopy reveals characteristic abnormalities of the structure of the basement membrane and lamina densa.

In **epidermolysis bullosa,** the skin breaks and blisters as a result of minor trauma. The dystrophic form is due to mutations in *COL7A1,* affecting the structure of **type VII** collagen. This collagen forms delicate fibrils that anchor the basal lamina to collagen fibrils in the dermis. These anchoring fibrils have been shown to be markedly reduced in this form of the disease, probably resulting in the blistering. Epidermolysis

bullosa simplex, another variant, is due to mutations in keratin 5 (Chapter 49).

Scurvy affects the structure of collagen. However, it is due to **a deficiency of ascorbic acid** (Chapter 44), and is not a genetic disease. Its major signs are bleeding gums, subcutaneous hemorrhages, and poor wound healing. These signs reflect impaired synthesis of collagen due to **deficiencies of prolyl and lysyl hydroxylases**, both of which require ascorbic acid as a cofactor.

In **Menkes disease** deficiency of copper results in defective cross-linking of collagen and elastin by the copper-dependent enzyme lysyl oxidase. (Menkes disease is discussed in Chapter 50.)

ELASTIN CONFERS EXTENSIBILITY & RECOIL ON LUNG, BLOOD VESSELS, & LIGAMENTS

Elastin is a connective tissue protein that is responsible for properties of extensibility and elastic recoil in tissues. Although not as widespread as collagen, elastin is present in large amounts, particularly in tissues that require these physical properties, eg, lung, large arterial blood vessels, and some elastic ligaments. Smaller quantities of elastin are also found in skin, ear cartilage, and several other tissues. In contrast to collagen, there appears to be only one genetic type of elastin, although variants arise by alternative splicing (Chapter 36) of the hnRNA for elastin. Elastin is synthesized as a soluble monomer of ~70 kDa called **tropoelastin.** Some of the prolines of tropoelastin are hydroxylated to **hydroxyproline** by prolyl hydroxylase, though hydroxylysine and glycosylated hydroxylysine are not present. Unlike collagen, tropoelastin is not synthesized in a pro- form with extension peptides. Furthermore, elastin does not contain repeat Gly-X-Y sequences, triple helical structure, or carbohydrate moieties.

After secretion from the cell, certain lysyl residues of tropoelastin are oxidatively deaminated to aldehydes by **lysyl oxidase,** the same enzyme involved in this process in collagen. However, the major cross-links formed in elastin are the **desmosines,** which result from the condensation of three of these lysine-derived aldehydes with an unmodified lysine to form a tetrafunctional cross-link unique to elastin. Once cross-linked in its mature, extracellular form, elastin is highly insoluble and **extremely stable** and has a very low turnover rate. Elastin exhibits a variety of random coil conformations that permit the protein to stretch and subsequently recoil during the performance of its physiologic functions.

Table 48–5 summarizes the main differences between collagen and elastin.

Deletions in the elastin gene (located at 7q11.23) have been found in approximately 90% of subjects with **Williams-Beuren syndrome** (OMIM 194050), a developmental disorder affecting connective tissue and the central nervous system. The mutations, by affecting synthesis of elastin, probably play a causative role in the **supravalvular aortic stenosis** often found

TABLE 48-5 Major Differences Between Collagen and Elastin

Collagen	Elastin
1. Many different genetic types	One genetic type
2. Triple helix	No triple helix; random coil conformations permitting stretching
3. $(Gly-X-Y)_n$ repeating structure	No $(Gly-X-Y)_n$ repeating structure
4. Presence of hydroxylysine	No hydroxylysine
5. Carbohydrate-containing	No carbohydrate
6. Intramolecular aldol cross-links	Intramolecular desmosine cross-links
7. Presence of extension peptides during biosynthesis	No extension peptides present during biosynthesis

FIGURE 48-2 Probable sequence of events in the causation of the major signs exhibited by patients with Marfan syndrome (OMIM 154700).

in this condition. Fragmentation or, alternatively, a decrease of elastin is found in conditions such as pulmonary emphysema, cutis laxa, and aging of the skin.

MARFAN SYNDROME IS DUE TO MUTATIONS IN THE GENE FOR FIBRILLIN-1, A PROTEIN PRESENT IN MICROFIBRILS

Marfan syndrome is a relatively prevalent inherited disease affecting connective tissue; it is inherited as an autosomal dominant trait. It affects the **eyes** (eg, causing dislocation of the lens, known as ectopia lentis), the **skeletal system** (most patients are tall and exhibit long digits [arachnodactyly] and hyperextensibility of the joints), and the **cardiovascular system** (eg, causing weakness of the aortic media, leading to dilation of the ascending aorta). Abraham Lincoln may have had this condition. Most cases are caused by mutations in the gene (on chromosome 15) for **fibrillin-1**; missense mutations have been detected in several patients with Marfan syndrome. This results in abnormal fibrillin and/or lower amounts being deposited in the ECM. There is evidence that the cytokine **TGF-β** normally binds to fibrillin-1, and if this binding is decreased (due to lower amounts of fibrillin-1), this can lead to an excess of the cytokine. The excess of TGF-β may contribute to the pathology (eg, in the aorta and aortic valve) found in the syndrome. This finding may lead to the development of therapies for the condition using drugs that antagonize TGF-β (eg, Losartan).

Fibrillin-1 is a large glycoprotein (about 350 kDa) that is a structural component of microfibrils, 10- to 12-nm fibers found in many tissues. It is secreted (subsequent to a proteolytic cleavage) into the extracellular matrix by fibroblasts and becomes incorporated into the insoluble **microfibrils,** which appear to provide a **scaffold** for deposition of elastin. Of special relevance to Marfan syndrome, fibrillin-1 is found in the zonular fibers of the **lens,** in the **periosteum,** and associated with elastin fi-

bers in the **aorta** (and elsewhere); these locations respectively explain the ectopia lentis, arachnodactyly, and cardiovascular problems found in the syndrome. Other proteins (eg, emelin and two microfibril-associated proteins) are also present in microfibrils. It appears likely that abnormalities of them may cause other connective tissue disorders. A gene for another fibrillin—**fibrillin-2**—exists on chromosome 5; mutations in this gene are linked to causation of **congenital contractural arachnodactyly** (OMIM 121050), but not to Marfan syndrome. Fibrillin-2 may be important in deposition of microfibrils early in development. The probable sequence of events leading to Marfan syndrome is summarized in Figure 48–2.

FIBRONECTIN IS AN IMPORTANT GLYCOPROTEIN INVOLVED IN CELL ADHESION & MIGRATION

Fibronectin is a major glycoprotein of the extracellular matrix, also found in a soluble form in plasma. It consists of two identical subunits, each of about 230 kDa, joined by two disulfide bridges near their carboxyl terminals. The gene encoding fibronectin is very large, containing some 50 exons; the RNA produced by its transcription is subject to considerable alternative splicing, and as many as 20 different mRNAs have been detected in various tissues. Fibronectin contains three types of repeating motifs (I, II, and III), which are organized into functional **domains** (at least seven); functions of these domains include binding **heparin** (see below) and fibrin, collagen, DNA, and cell surfaces (Figure 48–3). The amino acid sequence of the fibronectin receptor of fibroblasts has been derived, and the protein is a member of the transmembrane integrin class of proteins (Chapter 51). The **integrins** are heterodimers, containing various types of α and β polypeptide chains. Fibronectin contains an Arg-Gly-Asp (RGD) sequence that binds to the receptor. The **RGD sequence** is shared by a number of other proteins present in the ECM that bind to integrins present in cell surfaces. Synthetic peptides containing the RGD sequence

FIGURE 48–3 Schematic representation of fibronectin. Seven functional domains of fibronectin are represented; two different types of domain for heparin, cell-binding, and fibrin are shown. The domains are composed of various combinations of three structural motifs (I, II, and III), not depicted in the figure. Also not shown is the fact that fibronectin is a dimer joined by disulfide bridges near the carboxyl terminals of the monomers. The approximate location of the RGD sequence of fibronectin, which interacts with a variety of fibronectin integrin receptors on cell surfaces, is indicated by the arrow. (Redrawn after Yamada KM: Adhesive recognition sequences. J Biol Chem 1991;266:12809.)

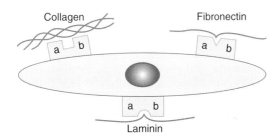

FIGURE 48–4 Schematic representation of a cell interacting through various integrin receptors with collagen, fibronectin, and laminin present in the ECM. (Specific subunits are not indicated.) (Redrawn after Yamada KM: Adhesive recognition sequences. J Biol Chem 1991;266:12809.)

inhibit the binding of fibronectin to cell surfaces. Figure 48–4 illustrates the interaction of collagen, fibronectin, and laminin, all major proteins of the ECM, with a typical cell (eg, fibroblast) present in the matrix.

The fibronectin receptor interacts indirectly with **actin** microfilaments (Chapter 49) present in the cytosol (Figure 48–5). A number of proteins, collectively known as **attachment proteins,** are involved; these include talin, vinculin, an actin-filament capping protein, and α-actinin. Talin interacts with the receptor and vinculin, whereas the latter two interact with actin. The interaction of fibronectin with its receptor provides one route whereby the **exterior of the cell can communicate with the interior** and thus affect cell behavior. Via the interaction with its cell receptor, fibronectin plays an important role in the **adhesion** of cells to the ECM. It is also involved in **cell migration** by providing a binding site for cells and thus helping them to steer their way through the ECM. The amount of fibronectin around many **transformed cells** is sharply reduced, partly explaining their faulty interaction with the ECM.

LAMININ IS A MAJOR PROTEIN COMPONENT OF RENAL GLOMERULAR & OTHER BASAL LAMINAS

Basal laminas are specialized areas of the ECM that surround epithelial and some other cells (eg, muscle cells); here

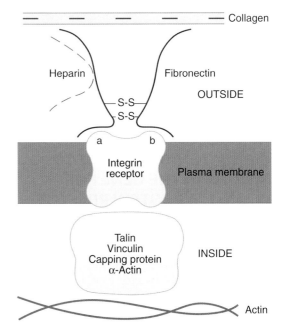

FIGURE 48–5 Schematic representation of fibronectin interacting with an integrin fibronectin receptor situated in the exterior of the plasma membrane of a cell of the ECM and of various attachment proteins interacting indirectly or directly with an actin microfilament in the cytosol. For simplicity, the attachment proteins are represented as a complex.

we discuss only the laminas found in the **renal glomerulus.** In that structure, the basal lamina is contributed by two separate sheets of cells (one endothelial and one epithelial), each disposed on opposite sides of the lamina; these three layers make up the **glomerular membrane.** The primary components of the basal lamina are three proteins—laminin, entactin, and type IV collagen—and the GAG **heparin** or **heparan sulfate.** These components are synthesized by the underlying cells.

Laminin (about 850 kDa, 70 nm long) consists of three distinct elongated polypeptide chains (A, B_1, and B_2) linked together to form an elongated cruciform shape (see Figure 49–11). It has potential binding sites for type IV collagen, heparin, and integrins on cell surfaces. The collagen inter-

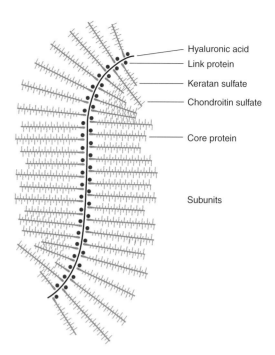

FIGURE 48–7 Schematic representation of the proteoglycan aggrecan. (Reproduced, with permission, from Lennarz WJ: *The Biochemistry of Glycoproteins and Proteoglycans*. Plenum Press, 1980. Reproduced with kind permission from Springer Science and Business Media.)

FIGURE 48–6 Darkfield electron micrograph of a proteoglycan aggregate in which the proteoglycan subunits and filamentous backbone are particularly well extended. (Reproduced, with permission, from Rosenberg L, Hellman W, Kleinschmidt AK: Electron microscopic studies of proteoglycan aggregates from bovine articular cartilage. J Biol Chem 1975;250:1877.)

acts with laminin (rather than directly with the cell surface), which in turn interacts with integrins or other laminin receptor proteins, thus anchoring the lamina to the cells. **Entactin,** also known as "nidogen," is a glycoprotein containing an RGD sequence; it binds to laminin and is a major cell attachment factor. The relatively thick basal lamina of the renal glomerulus has an important role in **glomerular filtration,** regulating the passage of large molecules (most plasma proteins) across the glomerulus into the renal tubule. The glomerular membrane allows small molecules, such as **inulin** (5.2 kDa), to pass through as easily as water. On the other hand, only a small amount of the protein **albumin** (69 kDa), the major plasma protein, passes through the normal glomerulus. This is explained by two sets of facts: (1) The **pores** in the glomerular membrane are large enough to allow molecules up to about 8 nm to pass through. (2) Albumin is smaller than this pore size, but it is prevented from passing through easily by the **negative charges** of heparan sulfate and of certain sialic acid-containing glycoproteins present in the lamina. These negative charges repel albumin and most plasma proteins, which are negatively charged at the pH of blood. The normal structure of the glomerulus may be severely damaged in certain types of **glomerulonephritis** (eg, caused by antibodies directed against various components of the glomerular membrane). This alters the pores and the amounts and dispositions of the negatively charged macromolecules referred to above, and relatively massive amounts of albumin (and of certain other plasma proteins) can pass through into the urine, resulting in severe **albuminuria.**

PROTEOGLYCANS & GLYCOSAMINOGLYCANS

The Glycosaminoglycans Found in Proteoglycans Are Built Up of Repeating Disaccharides

Proteoglycans are proteins that contain covalently linked glycosaminoglycans. At least 30 have been characterized and given names such as syndecan, betaglycan, serglycin, perlecan, aggrecan, versican, decorin, biglycan, and fibromodulin. They vary in tissue distribution, nature of the core protein, attached glycosaminoglycans, and function. The proteins bound covalently to glycosaminoglycans are called **"core proteins";** they have proved difficult to isolate and characterize, but the use of recombinant DNA technology is beginning to yield important information about their structures. The amount of **carbohydrate** in a proteoglycan is usually much greater than is found in a glycoprotein and may comprise up to 95% of its weight. Figures 48–6 & 48–7 show the general structure of one particular proteoglycan, **aggrecan,** the major type found in cartilage. It is very large (about 2×10^3 kDa), with its overall

FIGURE 48–8 Summary of structures of glycosaminoglycans and their attachments to core proteins. (GlcUA, D-glucuronic acid; IdUA, L-iduronic acid; GlcN, D-glucosamine; GalN, D-galactosamine; Ac, acetyl; Gal, D-galactose; Xyl, D-xylose; Ser, L-serine; Thr, L-threonine; Asn, L-asparagine; Man, D-mannose; NeuAc, N-acetylneuraminic acid.) The summary structures are qualitative representations only and do not reflect, for example, the uronic acid composition of hybrid glycosaminoglycans such as heparin and dermatan sulfate, which contain both L-iduronic and D-glucuronic acid. Neither should it be assumed that the indicated substituents are always present, eg, whereas most iduronic acid residues in heparin carry a 2'-sulfate group, a much smaller proportion of these residues are sulfated in dermatan sulfate. The presence of link trisaccharides (Gal-Gal-Xyl) in the chondroitin sulfates, heparin, and heparan and dermatan sulfates is shown. (Slightly modified and reproduced, with permission, from Lennarz WJ: *The Biochemistry of Glycoproteins and Proteoglycans.* Plenum Press, 1980. Reproduced with kind permission from Springer Science and Business Media.)

structure resembling that of a bottle brush. It contains a long strand of hyaluronic acid (one type of GAG) to which link proteins are attached **noncovalently.** In turn, these latter interact noncovalently with core protein molecules from which chains of other GAGs (keratan sulfate and chondroitin sulfate in this case) project. More details on this macromolecule are given when cartilage is discussed below.

There are at least seven **glycosaminoglycans (GAGs):** hyaluronic acid, chondroitin sulfate, keratan sulfates I and II, heparin, heparan sulfate, and dermatan sulfate. A GAG is an unbranched polysaccharide made up of repeating disaccharides, one component of which is always an **amino sugar** (hence the name GAG), either D-glucosamine or D-galactosamine. The other component of the repeating disaccharide (except in the case of keratan sulfate) is a **uronic acid,** either L-glucuronic acid (GlcUA) or its 5'-epimer, L-iduronic acid (IdUA). With the exception of hyaluronic acid, all the GAGs contain **sulfate groups,** either as O-esters or as N-sulfate (in heparin and heparan sulfate). Hyaluronic acid affords another exception because there is no clear evidence that it is attached covalently to protein, as the definition of a proteoglycan given above specifies. Both GAGs and proteoglycans have proved difficult to work with, partly because of their complexity. However, they are major components of the ECM; they have a number of important biologic roles; and they are involved in a number of disease processes—so that interest in them is increasing rapidly.

Biosynthesis of Glycosaminoglycans Involves Attachment to Core Proteins, Chain Elongation, & Chain Termination

Attachment to Core Proteins

The linkage between GAGs and their core proteins is generally one of three types.

1. An **O-glycosidic bond** between **xylose** (Xyl) and **Ser,** a bond that is unique to proteoglycans. This linkage is formed by transfer of a Xyl residue to Ser from UDP-xylose. Two residues of Gal are then added to the Xyl residue, forming a **link trisaccharide,** Gal-Gal-XylSer. Further chain growth of the GAG occurs on the terminal Gal.

2. An **O-glycosidic bond** forms between **GalNAc** (N-acetylgalactosamine) and **Ser (Thr)** (Figure 47–1A), present in keratan sulfate II. This bond is formed by donation to Ser (or Thr) of a GalNAc residue, employing UDP-GalNAc as its donor.

3. An **N-glycosylamine bond** between **GlcNAc** (N-acetylglucosamine) and the amide nitrogen of **Asn,** as found in N-linked glycoproteins (Figure 47–1B). Its synthesis is believed to involve dolichol-P-P-oligosaccharide.

The synthesis of the core proteins occurs in the **endoplasmic reticulum,** and formation of at least some of the above linkages also occurs there. Most of the later steps in the bio-

TABLE 48–6 Major Properties of the Glycosaminoglycans

GAG	Sugars	Sulfate[1]	Linkage of Protein	Location
HA	GlcNAc, GlcUA	Nil	No firm evidence	Synovial fluid, vitreous humor, loose connective tissue
CS	GalNAc, GlcUA	GalNAc	Xyl-Ser; associated with HA via link proteins	Cartilage, bone, cornea
KS I	GlcNAc, Gal	GlcNAc	GlcNAc-Asn	Cornea
KS II	GlcNAc, Gal	Same as KS I	GalNAc-Thr	Loose connective tissue
Heparin	GlcN, IdUA	GlcN	Ser	Mast cells
		GlcN		
		IdUA		
Heparan sulfate	GlcN, GlcUA	GlcN	Xyl-Ser	Skin fibroblasts, aortic wall
Dermatan sulfate	GalNAc, IdUA, (GlcUA)	GalNAc	Xyl-Ser	Wide distribution
		IdUa		

[1]The sulfate is attached to various positions of the sugars indicated (see Figure 48–8).
Note that all of the GAGs (except the keratan sulfers) contain a uronic add (glucuronic or iduronic acid).

synthesis of GAG chains and their subsequent modifications occur in the **Golgi apparatus.**

Chain Elongation

Appropriate **nucleotide sugars** and highly specific Golgi-located **glycosyltransferases** are employed to synthesize the oligosaccharide chains of GAGs. The **"one enzyme, one linkage"** relationship appears to hold here, as in the case of certain types of linkages found in glycoproteins. The enzyme systems involved in chain elongation are capable of high-fidelity reproduction of complex GAGs.

Chain Termination

This appears to result from (1) **sulfation,** particularly at certain positions of the sugars, and (2) the **progression** of the growing GAG chain away from the membrane site where catalysis occurs.

Further Modifications

After formation of the GAG chain, **numerous chemical modifications** occur, such as the introduction of sulfate groups onto GalNAc and other moieties and the epimerization of GlcUA to IdUA residues. The enzymes catalyzing sulfation are designated **sulfotransferases** and use **3′-phosphoadenosine-5′-phosphosulfate** [PAPS; active sulfate] (see Figure 32–11) as the sulfate donor. These Golgi-located enzymes are highly specific, and distinct enzymes catalyze sulfation at different positions (eg, carbons 2, 3, 4, and 6) on the acceptor sugars. An **epimerase** catalyzes conversions of glucuronyl to iduronyl residues.

The Various Glycosaminoglycans Exhibit Differences in Structure & Have Characteristic Distributions

The seven GAGs named above **differ** from each other in a number of the following properties: amino sugar composition, uronic acid composition, linkages between these components, chain length of the disaccharides, the presence or absence of sulfate groups and their positions of attachment to the constituent sugars, the nature of the core proteins to which they are attached, the nature of the linkage to core protein, their tissue and subcellular distribution, and their biologic functions.

The structures (Figure 48–8) and the distributions of each of the GAGs will now be briefly discussed. The major features of the seven GAGs are summarized in Table 48–6.

Hyaluronic Acid

Hyaluronic acid consists of an unbranched chain of repeating disaccharide units containing GlcUA and GlcNAc. Hyaluronic acid is present in bacteria and is widely distributed among various animals and tissues, including synovial fluid, the vitreous body of the eye, cartilage, and loose connective tissues.

Chondroitin Sulfates (Chondroitin 4-Sulfate & Chondroitin 6-Sulfate)

Proteoglycans linked to **chondroitin sulfate** by the Xyl-Ser O-glycosidic bond are prominent components of **cartilage** (see below). The repeating disaccharide is similar to that found in hyaluronic acid, containing GlcUA but with **GalNAc** replacing GlcNAc. The GalNAc is substituted with **sulfate** at either its 4′ or its 6′ position, with approximately one sulfate being present per disaccharide unit.

Keratan Sulfates I & II

As shown in Figure 48–8, the keratan sulfates consist of repeating **Gal-GlcNAc** disaccharide units containing **sulfate** attached to the 6′ position of GlcNAc or occasionally of Gal. Type I is abundant in **cornea,** and type II is found along with chondroitin sulfate attached to hyaluronic acid in **loose connective tissue.** Types I and II have different attachments to protein (Figure 48–8).

FIGURE 48–9 Structure of heparin. The polymer section illustrates structural features typical of heparin; however, the sequence of variously substituted repeating disaccharide units has been arbitrarily selected. In addition, non-O-sulfated or 3-O-sulfated glucosamine residues may also occur. (Modified, redrawn, and reproduced, with permission, from Lindahl U et al: Structure and biosynthesis of heparin-like polysaccharides. Fed Proc 1977;36:19.)

Heparin

The repeating disaccharide contains **glucosamine** (GlcN) and either of the two uronic acids (Figure 48–9). Most of the amino groups of the GlcN residues are **N-sulfated,** but a few are acetylated. The GlcN also carries a sulfate attached to carbon 6.

Approximately 90% of the uronic acid residues are **IdUA.** Initially, all of the uronic acids are GlcUA, but a 5′-epimerase converts approximately 90% of the GlcUA residues to IdUA after the polysaccharide chain is formed. The protein molecule of the heparin proteoglycan is unique, consisting exclusively of serine and glycine residues. Approximately two-thirds of the serine residues contain GAG chains, usually of 5–15 kDa but occasionally much larger. Heparin is found in the granules of **mast cells** and also in liver, lung, and skin.

Heparan Sulfate

This molecule is present on many **cell surfaces** as a proteoglycan and is extracellular. It contains **GlcN** with fewer N-sulfates than heparin, and, unlike heparin, its predominant uronic acid is **GlcUA.**

Dermatan Sulfate

This substance is widely distributed in animal tissues. Its structure is similar to that of chondroitin sulfate, except that in place of a GlcUA in β-1,3 linkage to GalNAc it contains an **IdUA** in an α-1,3 linkage to **GalNAc.** Formation of the IdUA occurs, as in heparin and heparan sulfate, by 5′-epimerization of GlcUA. Because this is regulated by the degree of sulfation and because sulfation is incomplete, dermatan sulfate contains **both** IdUA-GalNAc and GlcUA-GalNAc disaccharides.

Deficiencies of Enzymes That Degrade Glycosaminoglycans Result in Mucopolysaccharidoses

Both **exo-** and **endoglycosidases** degrade GAGs. Like most other biomolecules, GAGs are subject to **turnover,** being both synthesized and degraded. In adult tissues, GAGs generally exhibit relatively **slow** turnover, their half-lives being days to weeks.

Understanding of the degradative pathways for GAGs, as in the case of glycoproteins (Chapter 47) and glycosphingo-

lipids (Chapter 24), has been greatly aided by elucidation of the specific enzyme deficiencies that occur in certain **inborn errors of metabolism.** When GAGs are involved, these inborn errors are called **mucopolysaccharidoses** (Table 48–7).

Degradation of GAGs is carried out by a battery of **lyso-somal hydrolases.** These include certain **endoglycosidases,** various **exoglycosidases,** and **sulfatases,** generally acting in sequence to degrade the various GAGs. A number of them are indicated in Table 48–7.

The **mucopolysaccharidoses** share a common mechanism of causation, as illustrated in Figure 48–10. They are usually inherited in an **autosomal recessive** manner, with **Hurler** and **Hunter syndromes** being perhaps the most widely studied. None is common. **General features** of these conditions are summarized in Table 48–8 and **laboratory tests** of use in their diagnosis are summarized in Table 48–9. In some cases, **a family history** of a mucopolysaccharidosis is obtained.

The term **"mucolipidosis"** was introduced to denote diseases that combined features common to both mucopolysaccharidoses and sphingolipidoses (Chapter 24). Three mucolipidoses are listed in Table 48–7. In **sialidosis** (mucolipidosis I, ML-I), various oligosaccharides derived from glycoproteins and certain gangliosides can accumulate in tissues. **I-cell disease** (ML-II) and **pseudo-Hurler polydystrophy** (ML-III) are described in Chapter 47. The term "mucolipidosis" is retained because it is still in relatively widespread clinical usage, but it is not appropriate for these two latter diseases since the mechanism of their causation involves **mislocation** of certain lysosomal enzymes. Genetic defects of the catabolism of the oligosaccharide chains of glycoproteins (eg, mannosidosis, fucosidosis) are also described in Chapter 47. Most of these defects are characterized by increased excretion of various fragments of glycoproteins in the urine, which accumulate because of the metabolic block, as in the case of the mucolipidoses.

Hyaluronidase is one important enzyme involved in the catabolism of both hyaluronic acid and chondroitin sulfate. It is a widely distributed endoglycosidase that cleaves hexosaminidic linkages. From hyaluronic acid, the enzyme will generate a tetrasaccharide with the structure (GlcUA-β-1,3-GlcNAc-β-1,4)$_2$, which can be degraded further by a β-glucuronidase and β-N-acetylhexosaminidase. Surprisingly, only one case of an apparent genetic deficiency of this enzyme appears to have been reported (OMIM 601492).

TABLE 48–7   Biochemical Defects and Diagnostic Tests in Mucopolysaccharidoses (MPS) and Mucolipidoses (ML)

Name	Alternative Designation[1,2]	Enzymatic Defect	Urinary Metabolites
Mucopolysaccharidoses			
Hurler (OMIM 607014), Scheie (OMIM 607016), Hurler-Scheie (OMIM 607015)	MPS I	α-L-Iduronidase	Dermatan sulfate, heparan sulfate
Hunter (OMIM 309900)	MPS II	Iduronate sulfatase	Dermatan sulfate, heparan sulfate
Sanfilippo A (OMIM 252900)	MPS IIIA	Heparan sulfate N-sulfatase (sulfamidase)	Heparan sulfate
Sanfilippo B (OMIM 252920)	MPS IIIB	α-N-Acetylglucosaminidase	Heparan sulfate
Sanfilippo C (OMIM 252930)	MPS IIIC	α-Glucosaminide N-acetyltransferase	Heparan sulfate
Sanfilippo D (OMIM 252940)	MPS IIID	N-Acetylglucosamine 6-sulfatase	Heparan sulfate
Morquio A (OMIM 253000)	MPS IVA	Galactosamine 6-sulfatase	Keratan sulfate, chondroitin 6-sulfate
Morquio B (OMIM 253010)	MPS IVB	β-Galactosidase	Keratan sulfate
Maroteaux-Lamy (OMIM 253200)	MPS VI	N-Acetylgalactosamine 4-sulfatase (arylsulfatase B)	Dermatan sulfate
Sly (OMIM 253220)	MPS VII	β-Glucuronidase	Dermatan sulfate, heparan sulfate, chondroitin 4-sulfate, chondroitin 6-sulfate
Mucolipidoses			
Sialidosis (OMIM 256550)	ML I	Sialidase (neuraminidase)	Glycoprotein fragments
I-cell disease (OMIM 252500)	ML II	N-acetylglucosamine-1-phosphotransferase (acid hydrolases thus lack phosphomannosyl residues)	Glycoprotein fragments
Pseudo-Hurler polydystrophy (OMIM 252600)	ML III	As for ML II but deficiency is incomplete	Glycoprotein fragments

Source: Modified and reproduced, with permission, from DiNatale P, Neufeld EF: The biochemical diagnosis of mucopolysaccharidoses, mucolipidoses and related disorders. In: *Perspectives in Inherited Metabolic Diseases,* vol 2. Barr B et al (editors). Editiones Ermes (Milan), 1979.
[1]Fibroblasts, leukocytes, tissues, amniotic fluid cells, or serum can be used for the assay of many of the above enzymes. Patients with these disorders exhibit a variety of clinical findings that may include cloudy corneas, mental retardation, stiff joints, cardiac abnormalities, hepatosplenomegaly, and short stature, depending on the specific disease and its severity.
[2]The term MPS V is no longer used. The existence of MPS VIII (suspected glucosamine 6-sulfatase deficiency: OMIM 253230) has not been confirmed. At least one case of hyaluronidase deficiency (MPS IX; OMIM 601492) has been reported.

FIGURE 48–10 Simplified scheme of causation of a mucopolysaccharidosis, such as Hurler syndrome (OMIM 607014), in which the affected enzyme is α-L-iduronidase. Marked accumulation of the GAGs in the tissues mentioned in the figure could cause hepatomegaly, splenomegaly, disturbances of growth, coarse facies, and mental retardation, respectively.

Proteoglycans Have Numerous Functions

As indicated above, **proteoglycans** are remarkably complex molecules and are found in **every tissue** of the body, mainly in the ECM or "ground substance." There they are associated with each other and also with the other major structural components of the matrix, collagen and elastin, in quite specific manners. Some proteoglycans bind to collagen and others to elastin. These interactions are important in determining the structural organization of the matrix. Some proteoglycans (eg, decorin) can also **bind growth factors** such as TGF-β, modulating their effects on cells. In addition, some of them interact with certain **adhesive proteins** such as fibronectin and laminin (see above), also found in the matrix. The GAGs

TABLE 48–8   Summary of the Major Features of the Mucopolysaccharidoses

They exhibit a chronic progressive course.

They affect a number of organ systems (ie, they are multisystem disorders).

Many patients exhibit organomegaly (eg, hepato- and splenomegaly may be present).

Patients often exhibit dysostosis multiplex (characterized by severe abnormalities in the development of cartilage and bone, and also mental retardation).

Patients often exhibit abnormal facies (facial appearance).

Other signs sometimes found are abnormalities of hearing, of vision, of the cardiovascular system, and of mental development.

TABLE 48–9   Some Laboratory Tests Used in the Diagnosis of a Mucopolysaccharidosis

Urinalysis for presence of increased amounts of GAGs.

Assays of suspected enzymes in white blood cells, fibroblasts or possibly serum.

Tissue biopsy with subsequent analysis of GAGs by electrophoresis.

Use of specific gene tests.

Prenatal diagnosis can now be performed in at least certain cases using amniotic fluid cells or chorionic villus biopsy.

TABLE 48–10   Some Functions of Glycosaminoglycans and Proteoglycans

- Act as structural components of the ECM

- Have specific interactions with collagen, elastin, fibronectin, laminin, and other proteins such as growth factors

- As polyanions, bind polycations and cations

- Contribute to the characteristic turgor of various tissues

- Act as sieves in the ECM

- Facilitate cell migration (HA)

- Have role in compressibility of cartilage in weight-bearing (HA, CS)

- Play role in corneal transparency (KS I and DS)

- Have structural role in sclera (DS)

- Act as anticoagulant (heparin)

- Are components of plasma membranes, where they may act as receptors and participate in cell adhesion and cell-cell interactions (eg, HS)

- Determine charge-selectiveness of renal glomerulus (HS)

- Are components of synaptic and other vesicles (eg, HS)

Abbreviations: ECM, extracellular matrix; HA, hyaluronic acid; CS, chondroitin sulfate; KS I, keratan sulfate I; DS, dermatan sulfate; HS, heparan sulfate.

TABLE 48–11   The Principal Proteins Found in Bone[1]

Proteins	Comments
Collagens	
Collagen type I	Approximately 90% of total bone protein. Composed of two α1(I) and one α2(I) chains.
Collagen type V	Minor component.
Noncollagen proteins	
Plasma proteins	Mixture of various plasma proteins.
Proteoglycans[2] CS-PG I (biglycan)	Contains two GAG chains; found in other tissues.
CS-PG II (decorin)	Contains one GAG chain; found in other tissues.
CS-PG III	Bone-specific.
Bone SPARC[3] protein (osteonectin)	Not bone-specific.
Osteocalcin (bone Gla protein)	Contains γ-carboxyglutamate residues that bind to hydroxyapatite. Bone-specific.
Osteopontin	Not bone-specific. Glycosylated and phosphorylated.
Bone sialoprotein	Bone-specific. Heavily glycosylated, and sulfated on tyrosine.
Bone morphogenetic proteins (BMPs)	A family (eight or more) of secreted proteins with a variety of actions on bone; many induce ectopic bone growth.
Osteoprotegerin	Inhibits osteoclastogenesis

[1]Various functions have been ascribed to the noncollagen proteins, including roles in mineralization; however, most of them are still speculative. It is considered unlikely that the noncollagen proteins that are not bone-specific play a key role in mineralization. A number of other proteins are also present in bone, including a tyrosine-rich acidic matrix protein (TRAMP), some growth factors (eg, TGFβ), and enzymes involved in collagen synthesis (eg, lysyl oxidase).
[2]CS-PG, chondroitin sulfate–proteoglycan; these are similar to the dermatan sulfate PGs (DS-PGs) of cartilage (Table 48–13).
[3]SPARC, secreted protein acidic and rich in cysteine.

present in the proteoglycans are **polyanions** and hence bind polycations and cations such as Na+ and K+. This latter ability attracts water by osmotic pressure into the extracellular matrix and contributes to its turgor. GAGs also **gel** at relatively low concentrations. Because of the long extended nature of the polysaccharide chains of GAGs and their ability to gel, the proteoglycans can act as **sieves,** restricting the passage of large macromolecules into the ECM but allowing relatively free diffusion of small molecules. Again, because of their extended structures and the huge macromolecular aggregates they often form, they occupy a **large volume** of the matrix relative to proteins.

Some Functions of Specific GAGs & Proteoglycans

Hyaluronic acid is especially high in concentration in embryonic tissues and is thought to play an important role in permitting **cell migration** during morphogenesis and wound repair. Its ability to attract water into the extracellular matrix and thereby "loosen it up" may be important in this regard. The high concentrations of hyaluronic acid and chondroitin sulfates present in **cartilage** contribute to its compressibility (see below).

Chondroitin sulfates are located at sites of calcification in endochondral **bone** and are also found in **cartilage.** They are also located inside certain **neurons** and may provide an endoskeletal structure, helping to maintain their shape.

Both **keratan sulfate I** and **dermatan sulfate** are present in the **cornea.** They lie between collagen fibrils and play a critical role in corneal transparency. Changes in proteoglycan composition found in corneal scars disappear when the cornea heals. The presence of dermatan sulfate in the **sclera** may also play a role in maintaining the overall shape of the eye. Keratan sulfate I is also present in **cartilage.**

Heparin is an important **anticoagulant.** It binds with factors IX and XI, but its most important interaction is with **plasma antithrombin** (discussed in Chapter 51). Heparin can also bind specifically to **lipoprotein lipase** present in capillary walls, causing a release of this enzyme into the circulation.

Certain proteoglycans (eg, **heparan sulfate**) are associated with the plasma membrane of cells, with their core proteins actually spanning that membrane. In it they may act as **receptors** and may also participate in the mediation of **cell growth** and **cell–cell communication.** The attachment of cells to their substratum in culture is mediated at least in part by heparan sulfate. This proteoglycan is also found in the **basement membrane of the kidney** along with type IV collagen and laminin (see above), where it plays a major role in determining the charge selectiveness of glomerular filtration.

Proteoglycans are also found in **intracellular locations** such as the nucleus; their function in this organelle has not been elucidated. They are present in some storage or secretory granules, such as the chromaffin granules of the adrenal medulla. It has been postulated that they play a role in release of the contents of such granules. The various functions of GAGs are summarized in Table 48–10.

Associations with Major Diseases & with Aging

Hyaluronic acid may be important in permitting **tumor cells to migrate** through the ECM. Tumor cells can induce fibroblasts to synthesize greatly increased amounts of this GAG, thereby perhaps facilitating their own spread. Some tumor cells have less heparan sulfate at their surfaces, and this may play a role in the **lack of adhesiveness** that these cells display.

The intima of the **arterial wall** contains hyaluronic acid and chondroitin sulfate, dermatan sulfate, and heparan sulfate proteoglycans. Of these proteoglycans, dermatan sulfate binds plasma low-density lipoproteins. In addition, dermatan sulfate appears to be the major GAG synthesized by arterial smooth muscle cells. Because it is these cells that proliferate in **atherosclerotic lesions** in arteries, dermatan sulfate may play an important role in development of the atherosclerotic plaque.

In various types of **arthritis,** proteoglycans may act as **autoantigens,** thus contributing to the pathologic features of these conditions. The amount of chondroitin sulfate in cartilage diminishes with age, whereas the amounts of keratan sulfate and hyaluronic acid increase. These changes may contribute to the development of **osteoarthritis,** as may increased activity of the enzyme aggrecanase, which acts to degrade aggrecan. Changes in the amounts of certain GAGs in the skin are also observed with **aging** and help to account for the characteristic changes noted in this organ in the elderly.

An exciting new phase in proteoglycan research is opening up with the findings that mutations that affect individual proteoglycans or the enzymes needed for their synthesis alter the regulation of **specific signaling pathways** in *Drosophila* and *Caenorhabditis elegans,* thus affecting **development;** it already seems likely that similar effects exist in mice and humans.

BONE IS A MINERALIZED CONNECTIVE TISSUE

Bone contains both **organic** and **inorganic** material. The **organic** matter is mainly **protein.** The principal proteins of bone are listed in Table 48–11; **type I collagen** is the major protein, comprising 90–95% of the organic material. Type V collagen is also present in small amounts, as are a number of noncollagen proteins, some of which are relatively specific to bone. The **inorganic** or mineral component is mainly crystalline **hydroxyapatite**—$Ca_{10}(PO_4)_6(OH)_2$—along with sodium, magnesium, carbonate, and fluoride; approximately 99% of the body's calcium is contained in bone (Chapter 44). Hydroxyapatite confers on bone the strength and resilience required by its physiologic roles.

Bone is a **dynamic structure** that undergoes continuing cycles of remodeling, consisting of resorption followed by deposition of new bone tissue. This remodeling permits bone to adapt to both physical (eg, increases in weight-bearing) and hormonal signals.

The major cell types involved in bone resorption and deposition are **osteoclasts** and **osteoblasts** (Figure 48–11). The former are associated with resorption and the latter with deposition of bone. Osteocytes are descended from osteoblasts; they also appear to be involved in maintenance of bone matrix but will not be discussed further here.

Osteoclasts are multinucleated cells derived from pluripotent hematopoietic stem cells. Osteoclasts possess an apical membrane domain, exhibiting a ruffled border that plays a key role in bone resorption (Figure 48–12). A proton-translocating **ATPase** expels protons across the ruffled border into

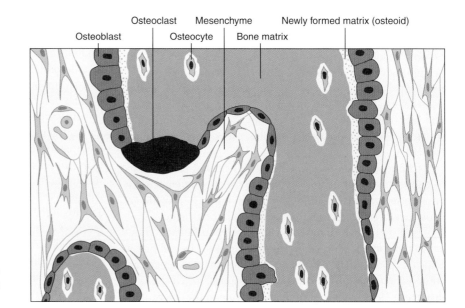

FIGURE 48–11 Schematic illustration of the major cells present in membranous bone. Osteoblasts (lighter color) are synthesizing type I collagen, which forms a matrix that traps cells. As this occurs, osteoblasts gradually differentiate to become osteocytes. (Reproduced, with permission, from Junqueira LC, Carneiro J: *Basic Histology: Text & Atlas,* 10th ed. McGraw-Hill, 2003.)

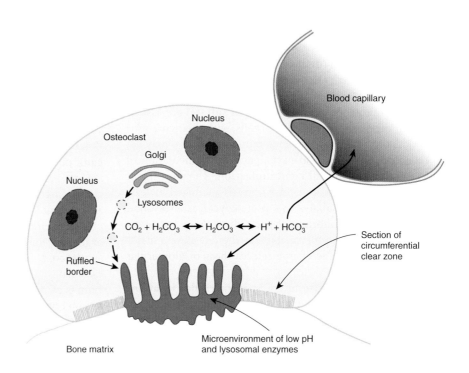

FIGURE 48–12 Schematic illustration of some aspects of the role of the osteoclast in bone resorption. Lysosomal enzymes and hydrogen ions are released into the confined microenvironment created by the attachment between bone matrix and the peripheral clear zone of the osteoclast. The acidification of this confined space facilitates the dissolution of calcium phosphate from bone and is the optimal pH for the activity of lysosomal hydrolases. Bone matrix is thus removed, and the products of bone resorption are taken up into the cytoplasm of the osteoclast, probably digested further, and transferred into capillaries. The chemical equation shown in the figure refers to the action of carbonic anhydrase II, described in the text. (Reproduced, with permission, from Junqueira LC, Carneiro J: *Basic Histology: Text & Atlas,* 10th ed. McGraw-Hill, 2003.)

the resorption area, which is the microenvironment of low pH shown in the figure. This lowers the local pH to 4.0 or less, thus increasing the solubility of hydroxyapatite and allowing demineralization to occur. Lysosomal acid proteases are released that digest the now accessible matrix proteins. **Osteoblasts**—mononuclear cells derived from pluripotent mesenchymal precursors—synthesize most of the proteins found in bone (Table 48–11) as well as various growth factors and cytokines. They are responsible for the deposition of new bone matrix (osteoid) and its subsequent mineralization. Osteoblasts **control mineralization** by regulating the passage of calcium and phosphate ions across their surface membranes. The lat-

ter contain **alkaline phosphatase,** which is used to generate phosphate ions from organic phosphates. The mechanisms involved in mineralization are not fully understood, but several factors have been implicated. Alkaline phosphatase contributes to mineralization, but in itself is not sufficient. Small vesicles (matrix vesicles) containing calcium and phosphate have been described at sites of mineralization, but their role is not clear. **Type I collagen** appears to be necessary, with mineralization being first evident in the gaps between successive molecules. Recent interest has focused on **acidic phosphoproteins,** such as bone sialoprotein, acting as sites of nucleation. These proteins contain motifs (eg, poly-Asp and poly-Glu

TABLE 48–12 Some Metabolic and Genetic Diseases Affecting Bone and Cartilage

Disease	Comments
Dwarfism	Often due to a deficiency of growth hormone, but has many other causes.
Rickets	Due to a deficiency of vitamin D during childhood.
Osteomalacia	Due to a deficiency of vitamin D during adulthood.
Hyperparathyroidism	Excess parathormone causes bone resorption.
Osteogenesis imperfecta (eg, OMIM 166200)	Due to a variety of mutations in the *COL1A1* and *COL1A2* genes affecting the synthesis and structure of type I collagen.
Osteoporosis (OMIM 166710)	Commonly postmenopausal or in other cases is more gradual and related to age; a small number of cases are due to mutations in the *COL1A1* and *COL1A2* genes and possibly in the vitamin D receptor gene
Osteoarthritis	A small number of cases are due to mutations in the *COL1A* genes
Several chondrodysplasias	Due to mutations in *COL2A1* genes
Pfeiffer syndrome[1] (OMIM 101600)	Mutations in the gene encoding fibroblast growth receptor 1 (FGFR1)
Jackson-Weiss (OMIM 123150) and Crouzon (OMIM 123500) syndromes[1]	Mutations in the gene encoding FGFR2
Achondroplasia (OMIM 100800) and thanatophoric dysplasia[2] (OMIM 187600)	Mutations in the gene encoding FGFR3

[1]The Pfeiffer, Jackson-Weiss, and Crouzon syndromes are craniosynostosis syndromes; craniosynostosis is a term signifying premature fusion of sutures in the skull.
[2]Thanatophoric (Gk *thanatos* "death" + *phoros* "bearing") dysplasia is the most common neonatal lethal skeletal dysplasia, displaying features similar to those of homozygous achondroplasia.

stretches) that bind calcium and may provide an initial scaffold for mineralization. Some macromolecules, such as certain proteoglycans and glycoproteins, can also act as **inhibitors** of nucleation.

It is estimated that approximately 4% of compact bone is **renewed annually** in the typical healthy adult, whereas approximately 20% of trabecular bone is replaced.

Many factors are involved in the **regulation of bone metabolism,** only a few of which will be mentioned here (see case no. 15 on osteoporosis, Chapter 54). Some **stimulate osteoblasts** (eg, parathyroid hormone and 1,25-dihydroxycholecalciferol) and others **inhibit** them (eg, corticosteroids).

Parathyroid hormone and 1,25-dihydroxycholecalciferol also stimulate osteoclasts, whereas calcitonin and estrogens inhibit them.

BONE IS AFFECTED BY MANY METABOLIC & GENETIC DISORDERS

A number of the more important examples of metabolic and genetic disorders that affect bone are listed in Table 48–12.

Osteogenesis imperfecta (brittle bones) is characterized by abnormal fragility of bones. The scleras are often abnormally thin and translucent and may appear blue owing to a deficiency of connective tissue. **Four types** of this condition (mild, extensive, severe, and variable) have been recognized, of which the extensive type occurring in the newborn is the most ominous. Affected infants may be born with multiple fractures and not survive. Over 90% of patients with osteogenesis imperfecta have mutations in the *COL1A1* and *COL1A2* genes, encoding proα1(I) and proα2(I) chains, respectively. Over 100 mutations in these two genes have been documented and include partial gene deletions and duplications. Other mutations affect RNA splicing, and the most frequent type results in the **replacement of glycine** by another bulkier amino acid, affecting formation of the triple helix. In general, these mutations result in decreased expression of collagen or in structurally abnormal pro chains that assemble into **abnormal fibrils**, weakening the overall structure of bone. When one abnormal chain is present, it may interact with two normal chains, but folding may be prevented, resulting in enzymatic degradation of all of the chains. This is called **"procollagen suicide"** and is an example of a dominant negative mutation, a result often seen when a protein consists of multiple different subunits.

Osteopetrosis (marble bone disease), characterized by **increased bone density**, is due to inability to resorb bone. One form occurs along with renal tubular acidosis and cerebral calcification. It is due to mutations in the gene (located on chromosome 8q22) encoding **carbonic anhydrase II** (CA II), one of four isozymes of carbonic anhydrase present in human tissues. The reaction catalyzed by carbonic anhydrase is shown below:

$$CO_2 + H_2O \leftrightarrow H_2CO_3 \leftrightarrow H^+ + HCO_3^-$$

In osteoclasts involved in bone resorption, CA II apparently provides protons to neutralize the OH^- ions left inside the cell when H^+ ions are pumped across their ruffled borders (see above). Thus, **if CA II is deficient in activity** in osteoclasts, normal bone resorption does not occur, and osteopetrosis results. The mechanism of the cerebral calcification is not clear, whereas the renal tubular acidosis reflects deficient activity of CA II in the renal tubules.

Osteoporosis (see Case History no. 15 in Chapter 54) is a generalized progressive reduction in bone tissue mass per unit volume causing skeletal weakness. The ratio of **mineral** to **organic elements** is unchanged in the remaining normal bone. Fractures of various bones, such as the head of the femur, occur very easily and represent a huge burden to both the affected patients and to the health care budget of society. Among other factors, **estrogens** and the cytokines **interleukins-1 and -6** appear to be intimately involved in the causation of osteoporosis.

THE MAJOR COMPONENTS OF CARTILAGE ARE TYPE II COLLAGEN & CERTAIN PROTEOGLYCANS

The **principal proteins** of hyaline cartilage (the major type of cartilage) are listed in Table 48–13. **Type II collagen** is the principal protein (Figure 48–13), and a number of other minor types of collagen are also present. In addition to these components, elastic cartilage contains elastin and fibroelastic cartilage contains type I collagen. Cartilage contains a number of **proteoglycans,** which play an important role in its compressibility. **Aggrecan** (about 2×10^3 kDa) is the major proteoglycan. As shown in Figure 48–14, it has a very complex structure, containing several GAGs (hyaluronic acid, chondroitin sulfate, and keratan sulfate) and both link and core proteins. The core protein contains three domains: A, B, and C. The hyaluronic acid binds noncovalently to domain A of the core protein as well as to the link protein, which stabilizes the hyaluronate–core protein interactions. The keratan sulfate chains are located in domain B, whereas the chondroitin sulfate chains are located in domain C; both of these types of GAGs are bound covalently to the core protein. The core protein also contains both O- and N-linked oligosaccharide chains.

The other proteoglycans found in cartilage have simpler structures than aggrecan.

Chondronectin is involved in the attachment of type II collagen to chondrocytes.

TABLE 48–13   The Principal Proteins Found in Cartilage

Proteins	Comments
Collagen proteins	
Collagen type II	90–98% of total articular cartilage collagen. Composed of three α 1(II) chains.
Collagens V, VI, IX, X, XI	Type IX cross-links to type II collagen. Type XI may help control diameter of type II fibrils.
Noncollagen proteins	
Proteoglycans	
Aggrecan	The major proteoglycan of cartilage.
Large non-aggregating proteoglycan	Found in some types of cartilage.
DS-PG I (biglycan)[1]	Similar to CS-PG I of bone.
DS-PG II (decorin)	Similar to CS-PG II of bone.
Chondronectin	May play role in binding type II collagen to surface of cartilage.
Anchorin C II	May bind type II collagen to surface of chondrocyte.

[1]The core proteins of DS-PG I and DS-PG II are homologous to those of CS-PG I and CS-PG II found in bone (Table 48–11). A possible explanation is that osteoblasts lack the epimerase required to convert glucuronic acid to iduronic acid, the latter of which is found in dermatan sulfate.

Cartilage is an avascular tissue and obtains most of its nutrients from synovial fluid. It exhibits slow but continuous **turnover.** Various **proteases** (eg, collagenases and stromalysin) synthesized by chondrocytes can **degrade collagen** and the other proteins found in cartilage. Interleukin-1 (IL-1) and tumor necrosis factor α (TNFα) appear to stimulate the production of such proteases, whereas transforming growth factor β (TGFβ) and insulin-like growth factor 1 (IGF-I) generally exert an anabolic influence on cartilage.

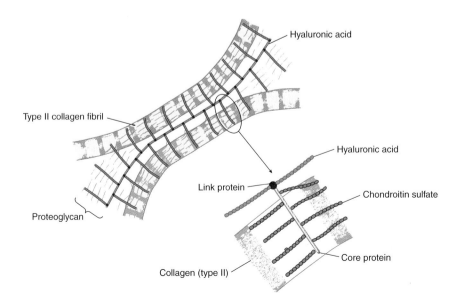

FIGURE 48–13 Schematic representation of the molecular organization in cartilage matrix. Link proteins noncovalently bind the core protein (lighter color) of proteoglycans to the linear hyaluronic acid molecules (darker color). The chondroitin sulfate side chains of the proteoglycan electrostatically bind to the collagen fibrils, forming a cross-linked matrix. The oval outlines the area enlarged in the lower part of the figure. (Reproduced, with permission, from Junqueira LC, Carneiro J: *Basic Histology: Text & Atlas*, 10th ed. McGraw-Hill, 2003.)

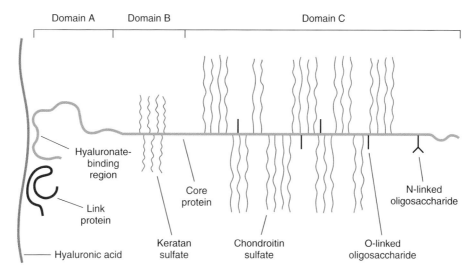

FIGURE 48–14 Schematic diagram of the aggrecan from bovine nasal cartilage. A strand of hyaluronic acid is shown on the left. The core protein (about 210 kDa) has three major domains. Domain A, at its amino terminal end, interacts with approximately five repeating disaccharides in hyaluronate. The link protein interacts with both hyaluronate and domain A, stabilizing their interactions. Approximately 30 keratan sulfate chains are attached, via GalNAc-Ser linkages, to domain B. Domain C contains about 100 chondroitin sulfate chains attached via Gal-Gal-Xyl-Ser linkages and about 40 O-linked oligosaccharide chains. One or more N-linked glycan chains are also found near the carboxyl terminal of the core protein. (Reproduced, with permission, from Moran LA et al: *Biochemistry,* 2nd ed. Neil Patterson Publishers, 1994. Copyright © 1994. Reprinted with permission of Pearson Education, Inc.)

THE MOLECULAR BASES OF THE CHONDRODYSPLASIAS INCLUDE MUTATIONS IN GENES ENCODING TYPE II COLLAGEN & FIBROBLAST GROWTH FACTOR RECEPTORS

Chondrodysplasias are a mixed group of hereditary disorders affecting cartilage. They are manifested by shortlimbed dwarfism and numerous skeletal deformities. A number of them are due to a variety of mutations in the *COL2A1* gene, leading to abnormal forms of type II collagen. One example is **Stickler syndrome,** manifested by degeneration of joint cartilage and of the vitreous body of the eye.

The best-known of the chondrodysplasias is **achondroplasia,** the most common cause of **short-limbed dwarfism**. Affected individuals have short limbs, normal trunk size, macrocephaly, and a variety of other skeletal abnormalities. The condition is often inherited as an autosomal dominant trait, but many cases are due to new mutations. The molecular basis of achondroplasia is outlined in Figure 48–15. Achondroplasia is not a collagen disorder but is due to mutations in the gene encoding **fibroblast growth factor receptor 3 (FGFR3). Fibroblast growth factors** are a family of at least nine proteins that affect the growth and differentiation of cells of mesenchymal and neuroectodermal origin. Their **receptors** are transmembrane proteins and form a subgroup of the family of receptor tyrosine kinases. FGFR3 is one member of this subgroup of

FIGURE 48–15 Simplified scheme of the causation of achondroplasia (OMIM 100800). In most cases studied so far, the mutation has been a G to A transition at nucleotide 1138. In a few cases, the mutation was a G to C transversion at the same nucleotide. This particular nucleotide is a real "hot spot" for mutation. Both mutations result in replacement of a Gly residue by an Arg residue in the transmembrane segment of the receptor. A few cases involving replacement of Gly by Cys at codon 375 have also been reported.

four and mediates the actions of FGF3 on cartilage. In almost all cases of achondroplasia that have been investigated, the mutations were found to involve nucleotide 1138 and resulted in substitution of arginine for glycine (residue number 380) in the transmembrane domain of the protein, rendering it inactive. No such mutation was found in unaffected individuals.

Rather amazingly, other mutations in the same gene can result in **hypochondroplasia, thanatophoric dysplasia** (types I and II) and the **SADDAN phenotype** (severe achondroplasia with developmental delay and acanthosis nigricans [the latter is a brown to black hyperpigmentation of the skin]).

As indicated in Table 48–12, **other skeletal dysplasias** (including certain craniosynostosis syndromes) are also due to mutations in genes encoding FGF receptors. Another type of skeletal dysplasia (diastrophic dysplasia) has been found to be due to mutation in a sulfate transporter. Thus, thanks to recombinant DNA technology, a new era in understanding of skeletal dysplasias has begun.

SUMMARY

- The major components of the ECM are the structural proteins collagen, elastin, and fibrillin-1; a number of specialized proteins (eg, fibronectin and laminin); and various proteoglycans.

- Collagen is the most abundant protein in the animal kingdom; approximately 28 types have been isolated. All collagens contain greater or lesser stretches of triple helix and the repeating structure $(Gly-X-Y)_n$.

- The biosynthesis of collagen is complex, featuring many posttranslational events, including hydroxylation of proline and lysine.

- Diseases associated with impaired synthesis of collagen include scurvy, osteogenesis imperfecta, Ehlers-Danlos syndrome (many types), and Menkes disease.

- Elastin confers extensibility and elastic recoil on tissues. Elastin lacks hydroxylysine, Gly-X-Y sequences, triple helical structure, and sugars but contains desmosine and isodesmosine cross-links not found in collagen.

- Fibrillin-1 is located in microfibrils. Mutations in the gene encoding fibrillin-1 cause Marfan syndrome. The cytokine TGF-β appears to contribute to the cardiovascular pathology.

- The glycosaminoglycans (GAGs) are made up of repeating disaccharides containing a uronic acid (glucuronic or iduronic) or hexose (galactose) and a hexosamine (galactosamine or glucosamine). Sulfate is also frequently present.

- The major GAGs are hyaluronic acid, chondroitin 4- and 6-sulfates, keratan sulfates I and II, heparin, heparan sulfate, and dermatan sulfate.

- The GAGs are synthesized by the sequential actions of a battery of specific enzymes (glycosyltransferases, epimerases, sulfotransferases, etc) and are degraded by the sequential action of lysosomal hydrolases. Genetic deficiencies of the latter result in mucopolysaccharidoses (eg, Hurler syndrome).

- GAGs occur in tissues bound to various proteins (linker proteins and core proteins), constituting proteoglycans. These structures are often of very high molecular weight and serve many functions in tissues.

- Many components of the ECM bind to proteins of the cell surface named integrins; this constitutes one pathway by which the exteriors of cells can communicate with their interiors.

- Bone and cartilage are specialized forms of the ECM. Collagen I and hydroxyapatite are the major constituents of bone. Collagen II and certain proteoglycans are major constituents of cartilage.

- The molecular causes of a number of heritable diseases of bone (eg, osteogenesis imperfecta) and of cartilage (eg, the chondrodystrophies) are being revealed by the application of recombinant DNA technology.

REFERENCES

Brodsky B, Persikov AV: Molecular structure of the collagen triple helix. Adv Prot Chem 2005;70:301.

Fauci AS et al (editors): *Harrison's Principles of Internal Medicine,* 17th ed. McGrawHill, 2008. (Chapter 357, *Heritable Disorders of Connective Tissue;* Chapter 355, *Lysosomal Storage Diseases;* Chapter 326, *Osteoarthritis;* Chapter 346, *Bone and Mineral Metabolism in Health and Disease;* Chapter 349, *Paget Disease and Other Dysplasias of Bone*).

Gelb BD: Marfan's syndrome and related disorders—more tightly connected than we thought. N Engl J Med 2006;354:2250.

Hacker U, Nybakken K, Perrimon N: Heparan sulfate proteoglycans: the sweet side of development. Nat Rev Mol Cell Biol 2005;6:530.

Handel TM et al: Regulation of protein function by glycosaminoglycans—as exemplified by chemokines. Annu Rev Biochem 2005;74:385.

Iozzo RV: Basement membrane proteoglycans: from cellar to ceiling. Nat Rev Mol Cell Biol 2005;6:645.

Karsenty G: An aggrecanase and osteoarthritis. N Engl J Med 2005;353:522.

Khosla S, Westendorf JJ, Oursler MJ: Building bone to reverse osteoporosis and repair fractures. J Clin Invest 2008; 118:421.

Marx J: All in the stroma: cancer's Cosa Nostra. Science 2008;320:38. (An article that indicates the importance of the ECM in the behavior of cancer cells).

Scriver CR et al (editors): *The Metabolic and Molecular Bases of Inherited Disease,* 8th ed. McGraw-Hill, 2001. (This comprehensive four-volume text and the updated online version [see Chapter 1] contain chapters on disorders of collagen biosynthesis and structure, Marfan syndrome, the mucopolysaccharidoses, achondroplasia, Alport syndrome, and craniosynostosis syndromes.)

Seeman E, Delams PD: Bone quality——the material and structural basis of bone strength and fragility. N Engl J Med 2006; 354:2250.

Zuscik MJ, Hilton MJ, Zhang X et al: Regulation of chondrogenesis and chondrocyte differentiation by stress. J Clin Invest 2008;118:429.

Muscle & the Cytoskeleton

Robert K. Murray, MD, PhD

BIOMEDICAL IMPORTANCE

Proteins play an important role in **movement** at both the organ (eg, skeletal muscle, heart, and gut) and cellular levels. In this chapter, the roles of specific proteins and certain other key molecules (eg, Ca^{2+}) in **muscular contraction** are described. A brief coverage of **cytoskeletal proteins** is also presented.

Knowledge of the molecular bases of a number of conditions that affect muscle has advanced greatly in recent years. Understanding of the molecular basis of **Duchenne-type muscular dystrophy** was greatly enhanced when it was found that it was due to mutations in the gene encoding dystrophin (see case history no. 7 in Chapter 54). Significant progress has also been made in understanding the molecular basis of **malignant hyperthermia**, a serious complication for some patients undergoing certain types of anesthesia. **Heart failure** is a very common medical condition, with a variety of causes; its rational therapy requires understanding of the biochemistry of heart muscle. One group of conditions that cause heart failure are the **cardiomyopathies**, some of which are genetically determined. **Nitric oxide** (NO) has been found to be a major regulator of smooth muscle tone. Many widely used **vasodilators**—such as nitroglycerin, used in the treatment of angina pectoris—act by increasing the formation of NO. Muscle, partly because of its mass, plays major roles in the **overall metabolism** of the body.

MUSCLE TRANSDUCES CHEMICAL ENERGY INTO MECHANICAL ENERGY

Muscle is the major biochemical **transducer** (machine) that converts potential (chemical) energy into kinetic (mechanical) energy. Muscle, the largest single tissue in the human body, makes up somewhat less than 25% of body mass at birth, more than 40% in the young adult, and somewhat less than 30% in the aged adult. We shall discuss aspects of the three types of muscle found in vertebrates: **skeletal, cardiac,** and **smooth.** Both skeletal and cardiac muscle appear **striated** upon microscopic observation; smooth muscle is **nonstriated.** Although skeletal muscle is under voluntary nervous control, the control of both cardiac and smooth muscle is involuntary.

The Sarcoplasm of Muscle Cells Contains ATP, Phosphocreatine, & Glycolytic Enzymes

Striated muscle is composed of multinucleated muscle fiber cells surrounded by an electrically excitable plasma membrane, the **sarcolemma.** An individual muscle fiber cell, which may extend the entire length of the muscle, contains a bundle of many **myofibrils** arranged in parallel, embedded in intracellular fluid termed **sarcoplasm.** Within this fluid is contained glycogen, the high-energy compounds ATP and phosphocreatine, and the enzymes of glycolysis.

The Sarcomere Is the Functional Unit of Muscle

An overall view of voluntary muscle at several levels of organization is presented in Figure 49–1.

When the **myofibril** is examined by electron microscopy, alternating dark and light bands (anisotropic bands, meaning birefringent in polarized light; and isotropic bands, meaning not altered by polarized light) can be observed. These bands are thus referred to as **A and I bands,** respectively. The central region of the A band (the H band) appears less dense than the rest of the band. The I band is bisected by a very dense and narrow **Z line** (Figure 49–2).

The **sarcomere** is defined as the region between two Z lines (Figures 49–1 & 49–2) and is repeated along the axis of a fibril at distances of 1500–2300 nm depending upon the state of contraction.

The **striated** appearance of voluntary and cardiac muscle in light microscopic studies results from their high degree of organization, in which most muscle fiber cells are aligned so that their sarcomeres are in parallel register (Figure 49–1).

Thick Filaments Contain Myosin; Thin Filaments Contain Actin, Tropomyosin, & Troponin

When *myofibrils* are examined by electron microscopy, it appears that each one is constructed of two types of longitudinal filaments. One type, the **thick filament,** confined to the

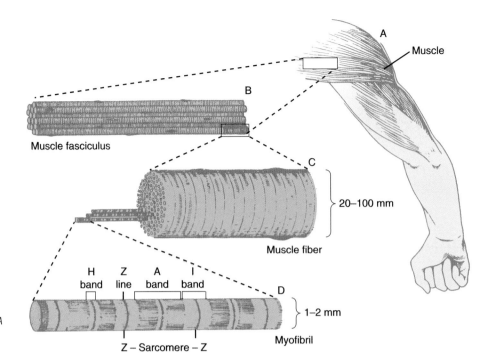

FIGURE 49–1 The structure of voluntary muscle. The sarcomere is the region between the Z lines. (Drawing by Sylvia Colard Keene. Reproduced, with permission, from Bloom W, Fawcett DW: *A Textbook of Histology*, 10th ed. Saunders, 1975.)

A band, contains chiefly the protein myosin. These filaments are about 16 nm in diameter and arranged in cross-section as a hexagonal array (Figure 49–2, center; right-hand cross-section).

The **thin filament** (about 7 nm in diameter) lies in the I band and extends into the A band but not into its H zone (Figure 49–2). Thin filaments contain the proteins actin, tropomyosin, and troponin (Figure 49–3). In the A band, the thin filaments are arranged around the thick (myosin) filament as a secondary hexagonal array. Each thin filament lies symmetrically between three thick filaments (Figure 49–2, center; mid cross-section), and each thick filament is surrounded symmetrically by six thin filaments.

The thick and thin filaments interact via **cross-bridges** that emerge at intervals of 14 nm along the thick filaments. As depicted in Figure 49–2, the cross-bridges (drawn as arrowheads at each end of the myosin filaments, but not shown extending fully across to the thin filaments) have opposite polarities at the two ends of the thick filaments. The two poles of the thick filaments are separated by a 150-nm segment (the M band, not labeled in the figure) that is free of projections.

The Sliding Filament Cross-Bridge Model Is the Foundation on Which Current Thinking About Muscle Contraction Is Built

This model was proposed independently in the 1950s by Henry Huxley and Andrew Huxley and their colleagues. It was largely based on careful morphologic observations on resting, extended, and contracting muscle. Basically, when muscle contracts, there is no change in the lengths of the thick and thin filaments, but the H zones and the I bands shorten (see legend to Figure 49–2). Thus, the arrays of interdigitat-

ing filaments must **slide past one another** during contraction. **Cross-bridges** that link thick and thin filaments at certain stages in the contraction cycle generate and sustain the tension. The tension developed during muscle contraction is proportionate to the filament overlap and to the number of cross-bridges. Each cross-bridge head is connected to the thick filament via a flexible fibrous segment that can bend outward from the thick filament. This flexible segment facilitates contact of the head with the thin filament when necessary but is also sufficiently pliant to be accommodated in the interfilament spacing.

ACTIN & MYOSIN ARE THE MAJOR PROTEINS OF MUSCLE

The mass of a muscle is made up of 75% water and more than 20% protein. The two major proteins are actin and myosin.

Monomeric **G-actin** (43 kDa; G, globular) makes up 25% of muscle protein by weight. At physiologic ionic strength and in the presence of Mg^{2+}, G-actin polymerizes noncovalently to form an insoluble double helical filament called F-actin (Figure 49–3). The **F-actin** fiber is 6–7 nm thick and has a pitch or repeating structure every 35.5 nm.

Myosins constitute a family of proteins, with at least 12 classes having been identified in the human genome. The myosin discussed in this chapter is **myosin-II,** and when myosin is referred to in this text, it is this species that is meant unless otherwise indicated. Myosin-I is a monomeric species that binds to cell membranes. It may serve as a linkage between microfilaments and the cell membrane in certain locations.

Myosin contributes 55% of muscle protein by weight and forms the **thick filaments.** It is an asymmetric hexamer with

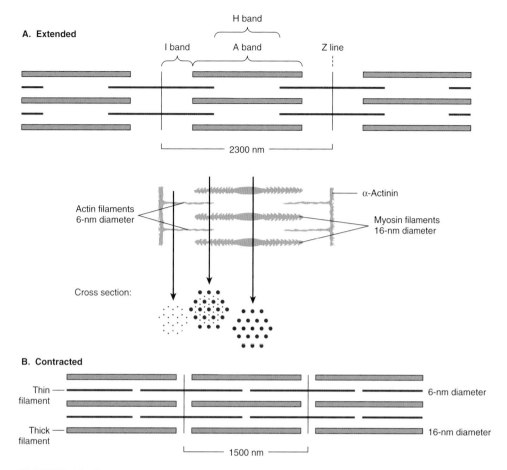

FIGURE 49–2 Arrangement of filaments in striated muscle. **(A)** Extended. The positions of the I, A, and H bands in the extended state are shown. The thin filaments partly overlap the ends of the thick filaments, and the thin filaments are shown anchored in the Z lines (often called Z disks). In the lower part of Figure 49–2A, "arrowheads," pointing in opposite directions, are shown emanating from the myosin (thick) filaments. Four actin (thin) filaments are shown attached to two Z lines via α-actinin. The central region of the three myosin filaments, free of arrowheads, is called the M band (not labeled). Cross-sections through the M bands, through an area where myosin and actin filaments overlap and through an area in which solely actin filaments are present, are shown. **(B)** Contracted. The actin filaments are seen to have slipped along the sides of the myosin fibers toward each other. The lengths of the thick filaments (indicated by the A bands) and the thin filaments (distance between Z lines and the adjacent edges of the H bands) have not changed. However, the lengths of the sarcomeres have been reduced (from 2300 nm to 1500 nm), and the lengths of the H and I bands are also reduced because of the overlap between the thick and thin filaments. These morphologic observations provided part of the basis for the sliding filament model of muscle contraction.

a molecular mass of approximately 460 kDa. Myosin has a fibrous tail consisting of two intertwined helices. Each helix has a globular head portion attached at one end (Figure 49–4). The hexamer consists of one pair of **heavy (H) chains** each of approximately 200 kDA molecular mass, and two pairs of **light (L) chains** each with a molecular mass of approximately 20 kDa. The L chains differ, one being called the **essential** light chain and the other the **regulatory** light chain. Skeletal muscle myosin binds actin to form **actomyosin** (actin-myosin), and its intrinsic ATPase activity is markedly enhanced in this complex. Isoforms of myosin exist whose amounts can vary in different anatomic, physiologic, and pathologic situations.

The structures of actin and of the head of myosin have been determined by x-ray crystallography; these studies have confirmed a number of earlier findings concerning their structures and have also given rise to much new information.

Limited Digestion of Myosin with Proteases Has Helped to Elucidate Its Structure & Function

When myosin is digested with **trypsin,** two myosin fragments (meromyosins) are generated. **Light meromyosin** (LMM)

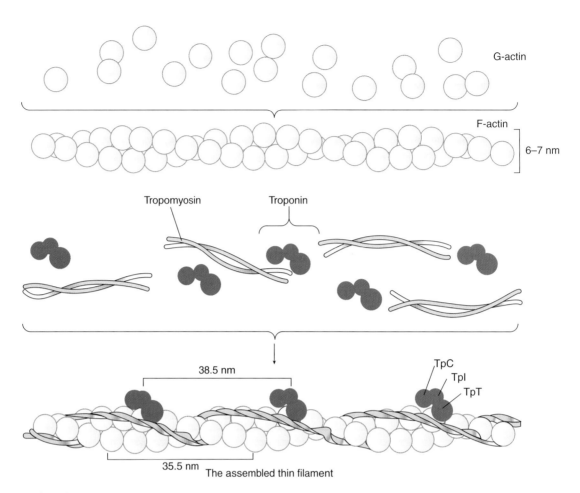

FIGURE 49–3 Schematic representation of the thin filament, showing the spatial configuration of its three major protein components: actin, myosin, and tropomyosin. The upper panel shows individual molecules of G-actin. The middle panel shows actin monomers assembled into F-actin. Individual molecules of tropomyosin (two strands wound around one another) and of troponin (made up of its three subunits) are also shown. The lower panel shows the assembled thin filament, consisting of F-actin, tropomyosin, and the three subunits of troponin (TpC, TpI, and TpT).

consists of aggregated, insoluble α-helical fibers from the tail of myosin (Figure 49–4). LMM exhibits no ATPase activity and does not bind to F-actin.

Heavy meromyosin (HMM; molecular mass about 340 kDa) is a soluble protein that has both a fibrous portion and a globular portion (Figure 49–4). It exhibits **ATPase** activity and binds to F-actin. Digestion of HMM with **papain** generates two subfragments, S-1 and S-2. The S-2 fragment is fibrous in character, has no ATPase activity, and does not bind to F-actin.

S-1 (molecular mass approximately 115 kDa) does exhibit ATPase activity, binds L chains, and in the absence of ATP will bind to and decorate actin with "arrowheads" (Figure 49–5). Both S-1 and HMM exhibit **ATPase** activity, which is accelerated 100- to 200-fold by complexing with F-actin. As discussed below, F-actin greatly enhances the rate at which myosin ATPase releases its products, ADP and P_i. Thus, although **F-actin** does not affect the hydrolysis step per se, its ability

to **promote release** of the products produced by the ATPase activity greatly accelerates the overall rate of catalysis.

CHANGES IN THE CONFORMATION OF THE HEAD OF MYOSIN DRIVE MUSCLE CONTRACTION

How can hydrolysis of ATP produce macroscopic movement? Muscle contraction essentially consists of the cyclic **attachment and detachment** of the S-1 head of myosin to the F-actin filaments. This process can also be referred to as the making and breaking of crossbridges. The attachment of actin to myosin is followed by **conformational changes** which are of particular importance in the S-1 head and are dependent upon which nucleotide is present (ADP or ATP). These changes result in the **power stroke,** which drives movement of actin filaments past myosin filaments. The energy for the

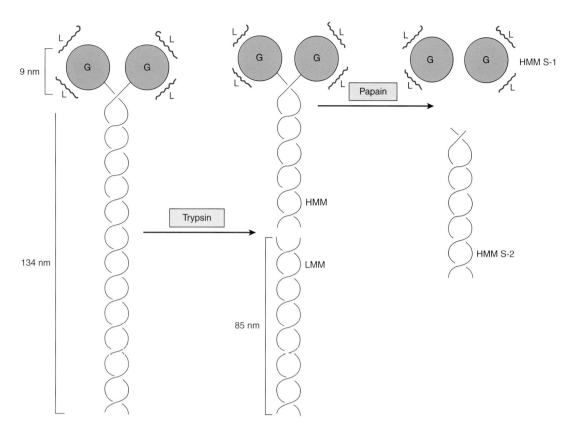

FIGURE 49–4 Diagram of a myosin molecule showing the two intertwined α-helices (fibrous portion), the globular region or head (G), the light chains (L), and the effects of proteolytic cleavage by trypsin and papain. The globular region (myosin head) contains an actin-binding site and an L chain-binding site and also attaches to the remainder of the myosin molecule.

FIGURE 49–5 The decoration of actin filaments with the S-1 fragments of myosin to form "arrowheads." (Courtesy of JA Spudich.)

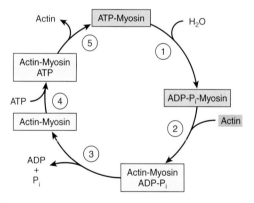

FIGURE 49–6 The hydrolysis of ATP drives the cyclic association and dissociation of actin and myosin in five reactions described in the text. (Modified from Stryer L: *Biochemistry*, 2nd ed. Freeman, 1981. Copyright © 1981 by W. H. Freeman and Company.)

power stroke is ultimately supplied by **ATP**, which is hydrolyzed to ADP and P$_i$. However, the power stroke itself occurs as a result of **conformational changes** in the myosin head when **ADP** leaves it.

The major biochemical events occurring during one cycle of muscle contraction and relaxation can be represented in the five steps shown in Figure 49–6 are as follows:

1. In the **relaxation phase** of muscle contraction, the S-1 head of myosin hydrolyzes ATP to ADP and P$_i$, but these products remain bound. The resultant ADPP$_i$-myosin complex has been energized and is in a so-called high-energy conformation.

2. When **contraction** of muscle is stimulated (via events involving Ca^{2+}, troponin, tropomyosin, and actin, which

are described below), actin becomes accessible and the S-1 head of myosin finds it, binds it, and forms the actin-myosin-ADP-P$_i$ complex indicated.

3. Formation of this complex **promotes the release of P$_i$,** which initiates the power stroke. This is followed by release of ADP and is accompanied by a large conformational change in the head of myosin in relation to its tail (Figure 49–7), pulling actin about 10 nm toward the center of the sarcomere. This is the **power stroke.** The myosin is now in a so-called low-energy state, indicated as actin-myosin.

4. Another molecule of ATP binds to the S-1 head, forming an actin-myosin-ATP complex.

5. Myosin-ATP has a low affinity for actin, and **actin is thus released.** This last step is a key component of relaxation and is dependent upon the binding of ATP to the actin-myosin complex.

Another cycle then commences with the hydrolysis of ATP (step 1 of Figure 49–6), re-forming the high-energy conformation.

FIGURE 49–7 Representation of the active cross-bridges between thick and thin filaments. This diagram was adapted by AF Huxley from HE Huxley: The mechanism of muscular contraction. Science 1969;164:1356. The latter proposed that the force involved in muscular contraction originates in a tendency for the myosin head (S-1) to rotate relative to the thin filament and is transmitted to the thick filament by the S-2 portion of the myosin molecule acting as an inextensible link. Flexible points at each end of S-2 permit S-1 to rotate and allow for variations in the separation between filaments. The present figure is based on HE Huxley's proposal but also incorporates elastic (the coils in the S-2 portion) and stepwise-shortening elements (depicted here as four sites of interaction between the S-1 portion and the thin filament). (See Huxley AF, Simmons RM: Proposed mechanism of force generation in striated muscle. Nature [Lond] 1971;233:533.) The strengths of binding of the attached sites are higher in position 2 than in position 1 and higher in position 3 than position 2. The myosin head can be detached from position 3 with the utilization of a molecule of ATP; this is the predominant process during shortening. The myosin head is seen to vary in its position from about 90° to about 45°, as indicated in the text. (S-1, myosin head; S-2, portion of the myosin molecule; LMM, light meromyosin) (see legend to Figure 49–4). (Reproduced from Huxley AF: Muscular contraction. J Physiol 1974;243:1. By kind permission of the author and the *Journal of Physiology*.)

Thus, hydrolysis of ATP is used to drive the cycle, with the actual power stroke being the conformational change in the S-1 head that occurs upon the release of ADP. The **hinge regions** of myosin (referred to as flexible points at each end of S-2 in the legend to Figure 49–7) permit the large range of movement of S-1 and also allow S-1 to find actin filaments.

If intracellular **levels of ATP drop** (eg, after death), ATP is not available to bind the S-1 head (step 4 above), **actin does not dissociate,** and relaxation (step 5) does not occur. This is the explanation for **rigor mortis,** the stiffening of the body that occurs after death.

Calculations have indicated that the **efficiency** of contraction is about 50%; that of the internal combustion engine is less than 20%.

Tropomyosin & the Troponin Complex Present in Thin Filaments Perform Key Functions in Striated Muscle

In striated muscle, there are two other proteins that are minor in terms of their mass but important in terms of their function. **Tropomyosin** is a fibrous molecule that consists of two chains, alpha and beta, that attach to F-actin in the groove between its filaments (Figure 49–3). Tropomyosin is present in all muscular and muscle-like structures. The **troponin complex** is unique to striated muscle and consists of three polypeptides. **Troponin T** (TpT) binds to tropomyosin as well as to the other two troponin components. **Troponin I** (TpI) inhibits the F-actin-myosin interaction and also binds to the other components of troponin. **Troponin C** (TpC) is a calcium-binding polypeptide that is structurally and functionally analogous to **calmodulin,** an important calcium-binding protein widely distributed in nature. Four molecules of calcium ion are bound per molecule of troponin C or calmodulin, and both molecules have a molecular mass of 17 kDa.

Ca²⁺ Plays a Central Role in Regulation of Muscle Contraction

The contraction of muscles from all sources occurs by the general mechanism described above. Muscles from different organisms and from different cells and tissues within the same organism may have different molecular mechanisms responsible for the regulation of their contraction and relaxation. In all systems, **Ca²⁺** plays a key regulatory role. There are two general mechanisms of **regulation** of muscle contraction: **actin-based** and **myosin-based.** The former operates in skeletal and cardiac muscle, the latter in smooth muscle.

Actin-Based Regulation Occurs in Striated Muscle

Actin-based regulation of muscle occurs in vertebrate skeletal and cardiac muscles, both striated. In the general mechanism described above (Figure 49–6), the only potentially limiting factor in the cycle of muscle contraction might be ATP. The skeletal muscle system is **inhibited** at rest; this inhibition is relieved to activate contraction. The inhibitor of striated muscle is the **troponin system,** which is bound to tropomyosin and F-actin in the thin filament (Figure 49–3). In striated muscle, there is no control of contraction unless the tropomyosin-troponin systems are present along with the actin and myosin filaments. As described above, **tropomyosin** lies along the groove of F-actin, and the three components of **troponin**—TpT, TpI, and TpC—are bound to the F-actin–tropomyosin complex. TpI prevents binding of the myosin head to its F-actin attachment site either by altering the conformation of F-actin via the tropomyosin molecules or by simply rolling tropomyosin into a position that directly blocks the sites on F-actin to which the myosin heads attach. Either way prevents activation of the myosin ATPase that is mediated by binding of the myosin head to F-actin. Hence, the TpI system blocks the contraction cycle at step 2 of Figure 49–6. This accounts for the **inhibited state** of relaxed striated muscle.

The Sarcoplasmic Reticulum Regulates Intracellular Levels of Ca²⁺ in Skeletal Muscle

In the sarcoplasm of resting muscle, the concentration of Ca^{2+} is 10^{-8} to 10^{-7} mol/L. The resting state is achieved because Ca^{2+}

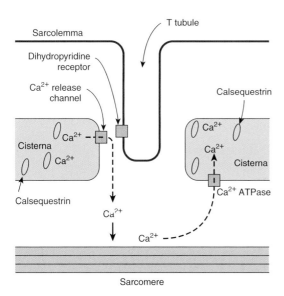

FIGURE 49–8 Diagram of the relationships among the sarcolemma (plasma membrane), a T tubule, and two cisternae of the sarcoplasmic reticulum of skeletal muscle (not to scale). The T tubule extends inward from the sarcolemma. A wave of depolarization, initiated by a nerve impulse, is transmitted from the sarcolemma down the T tubule. It is then conveyed to the Ca^{2+} release channel (ryanodine receptor), perhaps by interaction between it and the dihydropyridine receptor (slow Ca^{2+} voltage channel), which are shown in close proximity. Release of Ca^{2+} from the Ca^{2+} release channel into the cytosol initiates contraction. Subsequently, Ca^{2+} is pumped back into the cisternae of the sarcoplasmic reticulum by the Ca^{2+} ATPase (Ca^{2+} pump) and stored there, in part bound to calsequestrin.

is pumped into the sarcoplasmic reticulum through the action of an active transport system, called the Ca^{2+} ATPase (Figure 49–8), initiating relaxation. The sarcoplasmic reticulum is a network of fine membranous sacs. Inside the sarcoplasmic reticulum, Ca^{2+} is bound to a specific Ca^{2+}-binding protein designated **calsequestrin.** The sarcomere is surrounded by an excitable membrane (the T tubule system) composed of transverse (T) channels closely associated with the sarcoplasmic reticulum.

When the sarcolemma is excited by a **nerve impulse,** the signal is transmitted into the T tubule system and a **Ca²⁺ release channel** in the nearby sarcoplasmic reticulum opens, releasing Ca^{2+} from the sarcoplasmic reticulum into the sarcoplasm. The concentration of Ca^{2+} in the sarcoplasm rises rapidly to 10^{-5} mol/L. The Ca^{2+}-binding sites on *TpC* in the thin filament are quickly occupied by Ca^{2+}. The TpC-4Ca²⁺ interacts with TpI and TpT to alter their interaction with tropomyosin. Accordingly, tropomyosin moves out of the way or alters the conformation of F-actin so that the myosin head–ADP-P$_i$ (Figure 49–6) can interact with F-actin to start the contraction cycle.

The Ca^{2+} release channel is also known as the **ryanodine receptor** (RYR). There are two isoforms of this receptor, RYR1 and RYR2, the former being present in skeletal muscle and the latter in heart muscle and brain. **Ryanodine** is a plant alkaloid that binds to RYR1 and RYR2 specifically and modulates their activities. The Ca^{2+} release channel is a homotetramer made up of four subunits of kDa 565. It has transmembrane sequences at its carboxyl terminal, and these probably form the Ca^{2+} channel. The remainder of the protein protrudes into the cytosol, bridging the gap between the sarcoplasmic reticulum and the transverse tubular membrane. The channel is ligand-gated, Ca^{2+} and ATP working synergistically in vitro, although how it operates in vivo is not clear. A possible sequence of events leading to opening of the channel is shown in Figure 49–9. The channel lies very close to the **dihydropyridine receptor** (DHPR; a voltage-gated slow K type Ca^{2+} channel) of the transverse tubule system (Figure 49–8). Experiments in vitro employing an affinity column chromatography approach have

FIGURE 49–9 Possible chain of events leading to opening of the Ca^{2+} release channel. As indicated in the text, the Ca^{2+} voltage channel and the Ca^{2+} release channel have been shown to interact with each other in vitro via specific regions in their polypeptide chains. (DHPR, dihydropyridine receptor; RYR1, ryanodine receptor 1.)

indicated that a 37-amino-acid stretch in RYR1 interacts with one specific loop of DHPR.

Relaxation occurs when sarcoplasmic Ca^{2+} falls below 10^{-7} mol/L owing to its resequestration into the sarcoplasmic reticulum by Ca^{2+} ATPase. TpC-$4Ca^{2+}$ thus loses its Ca^{2+}. Consequently, **troponin,** via interaction with tropomyosin, **inhibits** further myosin head and F-actin interaction, and in the presence of ATP the myosin head detaches from the F-actin.

Thus, Ca^{2+} controls skeletal muscle contraction and relaxation by an allosteric mechanism mediated by TpC, TpI, TpT, tropomyosin, and F-actin.

A **decrease** in the concentration of **ATP** in the sarcoplasm (eg, by excessive usage during the cycle of contraction-relaxation or by diminished formation, such as might occur in ischemia) has two major effects: (1) The Ca^{2+} **ATPase** (Ca^{2+} pump) in the sarcoplasmic reticulum ceases to maintain the low concentration of Ca^{2+} in the sarcoplasm. Thus, the interaction of the myosin heads with F-actin is promoted. (2) The ATP-dependent **detachment of myosin heads** from F-actin cannot occur, and rigidity (contracture) sets in. The condition of **rigor mortis,** following death, is an extension of these events.

Muscle contraction is a delicate dynamic balance of the attachment and detachment of myosin heads to F-actin, subject to fine regulation via the nervous system.

Table 49–1 summarizes the overall events in contraction and relaxation of skeletal muscle.

TABLE 49–1   Sequence of Events in Contraction and Relaxation of Skeletal Muscle

Steps in contraction
1. Discharge of motor neuron
2. Release of transmitter (acetylcholine) at motor endplate
3. Binding of acetylcholine to nicotinic acetylcholine receptors
4. Increased Na^+ and K^+ conductance in endplate membrane
5. Generation of endplate potential
6. Generation of action potential in muscle fibers
7. Inward spread of depolarization along T tubules
8. Release of Ca^{2+} from terminal cisterns of sarcoplasmic reticulum and diffusion to thick and thin filaments
9. Binding of Ca^{2+} to troponin C, uncovering myosin binding sites of actin
10. Formation of cross-linkages between actin and myosin and sliding of thin on thick filaments, producing shortening
Steps in relaxation
1. Ca^{2+} pumped back into sarcoplasmic reticulum
2. Release of Ca^{2+} from troponin
3. Cessation of interaction between actin and myosin

Source: Reproduced, with permission, from Ganong WF: *Review of Medical Physiology,* 21st ed. McGraw-Hill, 2003.

Mutations in the Gene Encoding the Ca^{2+} Release Channel Are One Cause of Human Malignant Hyperthermia

Some genetically predisposed patients experience a severe reaction, designated **malignant hyperthermia,** on exposure to certain anesthetics (eg, halothane) and depolarizing skeletal muscle relaxants (eg, succinylcholine). The reaction consists primarily of rigidity of skeletal muscles, hypermetabolism, and high fever. A **high cytosolic concentration of Ca^{2+}** in skeletal muscle is a major factor in its causation. Unless malignant hyperthermia is recognized and treated immediately, patients may die acutely of ventricular fibrillation or survive to succumb subsequently from other serious complications. Appropriate treatment is to stop the anesthetic and administer the drug **dantrolene** intravenously. Dantrolene is a skeletal muscle relaxant that acts to inhibit release of Ca^{2+} from the sarcoplasmic reticulum into the cytosol, thus preventing the increase of cytosolic Ca^{2+} found in malignant hyperthermia.

Malignant hyperthermia also occurs in **swine.** Susceptible animals homozygous for malignant hyperthermia respond to stress with a fatal reaction (**porcine stress syndrome**) similar to that exhibited by humans. If the reaction occurs prior to slaughter, it affects the quality of the pork adversely, resulting in an inferior product. Both events can result in considerable economic losses for the swine industry.

The finding of a high level of cytosolic Ca^{2+} in muscle in malignant hyperthermia suggested that the condition might be caused by abnormalities of the Ca^{2+} ATPase or of the **Ca^{2+} release channel.** No abnormalities were detected in the former, but sequencing of cDNAs for the latter protein proved insightful, particularly in swine. All cDNAs from **swine** with malignant hyperthermia so far examined have shown a substitution of T for C1843, resulting in the substitution of Cys for Arg^{615} in the Ca^{2+} release channel. The mutation affects the function of the channel in that it opens more easily and remains open longer; the net result is massive release of Ca^{2+} into the cytosol, ultimately causing sustained muscle contraction.

The picture is more complex in **humans,** since malignant hyperthermia exhibits **genetic heterogeneity.** Members of a number of families who suffer from malignant hyperthermia have not shown genetic linkage to the *RYR1* gene. Some humans susceptible to malignant hyperthermia have been found to exhibit the same mutation found in swine, and others have a variety of point mutations at different loci in the *RYR1* gene. Certain families with malignant hypertension have been found to have mutations affecting the **DHPR.** Figure 49–10 summarizes the probable chain of events in malignant hyperthermia. The major promise of these findings is that, once additional mutations are detected, it will be possible to **screen,** using suitable DNA probes, for individuals at risk of developing malignant hyperthermia during anesthesia. Current screening tests (eg, the in vitro caffeine-halothane test) are relatively unreliable. Affected individuals could then be given **alternative anesthetics,** which would not endanger their lives. It should also be possible, if desired, to eliminate malignant

FIGURE 49–10 Simplified scheme of the causation of malignant hyperthermia (OMIM 145600). Many different point mutations have been detected in the *RYR1* gene, some of which are associated with central core disease (OMIM 117000). It is estimated that at least 50% of families with members who have malignant hyperthermia are linked to the *RYR1* gene. Some individuals with mutations in the gene encoding DHPR have also been detected; it is possible that mutations in other genes for proteins involved in certain aspects of muscle metabolism will also be found.

hyperthermia from swine populations using suitable breeding practices.

Another condition due to mutations in the *RYR1* gene is **central core disease.** This is a rare myopathy presenting in infancy with hypotonia and proximal muscle weakness. Electron microscopy reveals an absence of mitochondria in the center of many type I (see below) muscle fibers. Damage to mitochondria induced by high intracellular levels of Ca^{2+} secondary to abnormal functioning of *RYR1* appears to be responsible for the morphologic findings.

MUTATIONS IN THE GENE ENCODING DYSTROPHIN CAUSE DUCHENNE MUSCULAR DYSTROPHY

A number of **additional proteins** play various roles in the structure and function of muscle. They include titin (the largest protein known), nebulin, α-actinin, desmin, dystrophin, and calcineurin. Some properties of these proteins are summarized in Table 49–2.

Dystrophin is of special interest. As discussed in case no. 9 of Chapter 54, mutations in the gene encoding this protein have been shown to be the cause of **Duchenne muscular dystrophy** and the milder **Becker muscular dystrophy**. They are also implicated in some cases of **dilated cardiomyopathy** (see below). As shown in Figure 49–11, dystrophin forms part of a large complex of proteins that attach to or interact with the plasmalemma. Dystrophin links the actin cytoskeleton to the ECM and appears to be needed for assembly of the synaptic junction. Impairment of these processes by formation of defective dystrophin is thought to be critical in the causation of Duchenne muscular dystrophy. Mutations in the genes en-

TABLE 49–2   Some Other Important Proteins of Muscle

Protein	Location	Comment or Function
Titin	Reaches from the Z line to the M line	Largest protein in body. Role in relaxation of muscle.
Nebulin	From Z line along length of actin filaments	May regulate assembly and length of actin filaments.
α-Actinin	Anchors actin to Z lines	Stabilizes actin filaments.
Desmin	Lies alongside actin filaments	Attaches to plasma membrane (plasmalemma).
Dystrophin	Attached to plasmalemma	Deficient in Duchenne muscular dystrophy. Mutations of its gene can also cause dilated cardiomyopathy.
Calcineurin	Cytosol	A calmodulin-regulated protein phosphatase. May play important roles in cardiac hypertrophy and in regulating amounts of slow and fast twitch muscles.
Myosin-binding protein C	Arranged transversely in sarcomere A-bands	Binds myosin and titin. Plays a role in maintaining the structural integrity of the sarcomere.

coding some of the components of the **sarcoglycan complex** shown in Figure 49–11 are responsible for **limb-girdle** and certain **other congenital forms** of muscular dystrophy.

Mutations in genes encoding **several glycosyltransferases** involved in the synthesis of the sugar chains of **α-dystroglycan** have been found to be the cause of certain types of **congenital muscular dystrophy** (see Chapter 47).

CARDIAC MUSCLE RESEMBLES SKELETAL MUSCLE IN MANY RESPECTS

The general picture of muscle contraction in the heart resembles that of skeletal muscle. Cardiac muscle, like skeletal muscle, is **striated** and uses the actin-myosin-tropomyosin-troponin system described above. Unlike skeletal muscle, cardiac muscle exhibits **intrinsic rhythmicity,** and individual myocytes communicate with each other because of its syncytial nature. The **T tubular system** is more developed in cardiac muscle, whereas the **sarcoplasmic reticulum** is less extensive and consequently the intracellular supply of Ca^{2+} for contraction is less. Cardiac muscle thus relies on **extracellular Ca^{2+}** for contraction; if isolated cardiac muscle is deprived of Ca^{2+}, it ceases to beat within approximately 1 min, whereas skeletal muscle can continue to contract without an extracellular

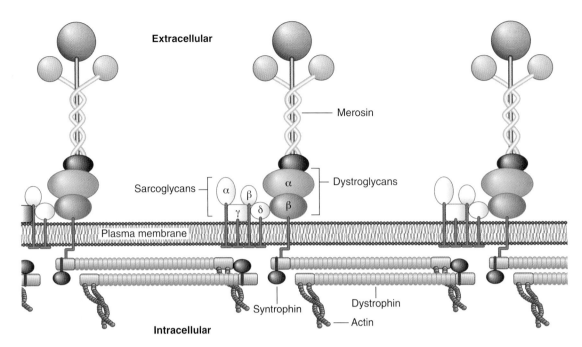

FIGURE 49–11 Organization of dystrophin and other proteins in relation to the plasma membrane of muscle cells. Dystrophin is part of a large oligomeric complex associated with several other protein complexes. The dystroglycan complex consists of α-dystroglycan, which associates with the basal lamina protein merosin (also named laminin-2), and α-dystroglycan, which binds α-dystroglycan and dystrophin. Syntrophin binds to the carboxyl terminal of dystrophin. The sarcoglycan complex consists of four transmembrane proteins: α-, β-, β-, and δ-sarcoglycan. The function of the sarcoglycan complex and the nature of the interactions within the complex and between it and the other complexes are not clear. The sarcoglycan complex is formed only in striated muscle, and its subunits preferentially associate with each other, suggesting that the complex may function as a single unit. Mutations in the gene encoding dystrophin cause Duchenne and Becker muscular dystrophy. Mutations in the genes encoding the various sarcoglycans have been shown to be responsible for limb-girdle dystrophies (eg, OMIM 604286) and mutations in genes encoding other muscle proteins cause other types of muscular dystrophy. Mutations in genes encoding certain glycosyltransferases involved in the synthesis of the glycan chains of α-dystroglycan are responsible for certain congenital muscular dystrophies (see Chapter 47). (Reproduced, with permission, from Duggan DJ et al: Mutations in the sarcoglycan genes in patients with myopathy. N Engl J Med 1997;336:618. Copyright © 1997 Massachusetts Medical Society. All rights reserved.)

source of Ca^{2+} for a longer period. **Cyclic AMP** plays a more prominent role in cardiac than in skeletal muscle. It modulates intracellular levels of Ca^{2+} through the activation of protein kinases; these enzymes phosphorylate various transport proteins in the sarcolemma and sarcoplasmic reticulum and also in the troponin-tropomyosin regulatory complex, affecting intracellular levels of Ca^{2+} or responses to it. There is a rough correlation between the phosphorylation of TpI and the increased contraction of cardiac muscle induced by catecholamines. This may account for the **inotropic effects** (increased contractility) of β-adrenergic compounds on the heart. Some differences among skeletal, cardiac, and smooth muscle are summarized in Table 49–3.

Ca²⁺ Enters Myocytes Via Ca²⁺ Channels & Leaves Via the Na⁺-Ca²⁺ Exchanger & the Ca²⁺ ATPase

As stated above, **extracellular Ca^{2+}** plays an important role in contraction of cardiac muscle but not in skeletal muscle. This

means that Ca^{2+} both enters and leaves myocytes in a regulated manner. We shall briefly consider three transmembrane proteins that play roles in this process.

Ca²⁺ Channels

Ca^{2+} enters myocytes via these channels, which allow entry only of Ca^{2+} ions. The major portal of entry is the L-type (long-duration current, large conductance) or **slow Ca^{2+}** channel, which is voltage-gated, opening during depolarization induced by spread of the cardiac action potential and closing when the action potential declines. These channels are equivalent to the dihydropyridine receptors of skeletal muscle (Figure 49–8). Slow Ca^{2+} channels are **regulated** by cAMP-dependent protein kinases (stimulatory) and cGMP-protein kinases (inhibitory) and are blocked by so-called calcium channel blockers (eg, verapamil). **Fast** (or T, transient) Ca^{2+} channels are also present in the plasmalemma, though in much lower numbers; they probably contribute to the early phase of increase of myoplasmic Ca^{2+}.

The resultant increase of Ca^{2+} in the myoplasm acts on the Ca^{2+} release channel of the sarcoplasmic reticulum to open

TABLE 49–3   Some Differences Among Skeletal, Cardiac, and Smooth Muscle

Skeletal Muscle	Cardiac Muscle	Smooth Muscle
1. Striated.	1. Striated.	1. Nonstriated.
2. No syncytium.	2. Syncytial.	2. Syncytial.
3. Small T tubules.	3. Large T tubules.	3. Generally rudimentary T tubules.
4. Sarcoplasmic reticulum well developed and Ca^{2+} pump acts rapidly.	4. Sarcoplasmic reticulum present and Ca^{2+} pump acts relatively rapidly.	4. Sarcoplasmic reticulum often rudimentary and Ca^{2+} pump acts slowly.
5. Plasmalemma lacks many hormone receptors.	5. Plasmalemma contains a variety of receptors (eg, α- and β-adrenergic).	5. Plasmalemma contains a variety of receptors (eg, α- and β-adrenergic).
6. Nerve impulse initiates contraction.	6. Has intrinsic rhythmicity.	6. Contraction initiated by nerve impulses, hormones, etc.
7. Extracellular fluid Ca^{2+} not important for contraction.	7. Extracellular fluid Ca^{2+} important for contraction.	7. Extracellular fluid Ca^{2+} important for contraction.
8. Troponin system present.	8. Troponin system present.	8. Lacks troponin system; uses regulatory head of myosin.
9. Caldesmon not involved.	9. Caldesmon not involved.	9. Caldesmon is important regulatory protein.
10. Very rapid cycling of the cross-bridges.	10. Relatively rapid cycling of the cross-bridges.	10. Slow cycling of the cross-bridges permits slow, prolonged contraction and less utilization of ATP.

it. This is called **Ca^{2+}-induced Ca^{2+} release** (CICR). It is estimated that approximately 10% of the Ca^{2+} involved in contraction enters the cytosol from the extracellular fluid and 90% from the sarcoplasmic reticulum. However, the former 10% is important, as the rate of increase of Ca^{2+} in the myoplasm is important, and entry via the Ca^{2+} channels contributes appreciably to this.

Ca^{2+}-Na^+ Exchanger

This is the principal route of **exit** of Ca^{2+} from myocytes. In resting myocytes, it helps to maintain a low level of free intracellular Ca^{2+} by exchanging one Ca^{2+} for three Na^+. The energy for the uphill movement of Ca^{2+} out of the cell comes from the downhill movement of Na^+ into the cell from the plasma. This exchange contributes to relaxation, but may run in the reverse direction during excitation. Because of the Ca^{2+}-Na^+ exchanger, anything that causes intracellular Na^+ (Na^+_i) to rise will secondarily cause Ca^{2+}_i to rise, causing more forceful contraction. This is referred to as a **positive inotropic effect.** One example is when the drug **digitalis** is used to treat heart failure. Digitalis inhibits the sarcolemmal Na^+-K^+ ATPase, diminishing exit of Na^+ and thus increasing Na^+_i. This in turn causes Ca^{2+} to increase, via the Ca^{2+}-Na^+ exchanger. The increased Ca^{2+}_i results in increased force of cardiac contraction (see Figure 49–12), of benefit in heart failure.

Ca^{2+} ATPase

This Ca^{2+} pump, situated in the sarcolemma, also contributes to Ca^{2+} exit but is believed to play a relatively minor role as compared with the Ca^{2+}-Na^+ exchanger.

It should be noted that there are a variety of **ion channels** (Chapter 40) in most cells, for Na^+, K^+, Ca^{2+}, etc. Many

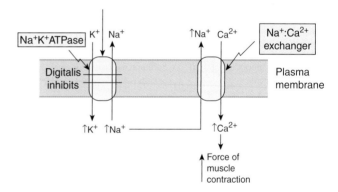

FIGURE 49–12 Scheme of how the drug digitalis (used in the treatment of certain cases of heart failure) increases cardiac contraction. Digitalis inhibits the Na^+K^+ ATPase (see Chapter 40). This results in less Na^+ being pumped out of the cardiac myocyte and leads to an increase of the intracellular concentration of Na^+. In turn, this stimulates the Na^+-Ca^{2+} exchanger so that more Na^+ is exchanged outward, and more Ca^{2+} enters the myocyte. The resulting increased intracellular concentration of Ca^{2+} increases the force of muscular contraction.

of them have been cloned in recent years and their dispositions in their respective membranes worked out (number of times each one crosses its membrane, location of the actual ion transport site in the protein, etc). They can be classified as indicated in Table 49–4. Cardiac muscle is rich in ion channels, and they are also important in skeletal muscle. Mutations in genes encoding ion channels have been shown to be responsible for a number of relatively rare conditions affecting muscle. These and other diseases due to mutations of ion channels have been termed **channelopathies;** some are listed in Table 49–5.

TABLE 49–4   Major Types of Ion Channels Found in Cells

Type	Comment
External ligand-gated	Open in response to a specific extracellular molecule, eg, acetylcholine.
Internal ligand-gated	Open or close in response to a specific intracellular molecule, eg, a cyclic nucleotide.
Voltage-gated	Open in response to a change in membrane potential, eg, Na$^+$, K$^+$, and Ca^{2+} channels in heart.
Mechanically gated	Open in response to change in mechanical pressure.

TABLE 49–5   Some Disorders (Channelopathies) Due to Mutations in Genes Encoding Polypeptide Constituents of Ion Channels

Disorder[1]	Ion Channel and Major Organs Involved
Central core disease (OMIM 117000)	Ca^{2+} release channel (RYR1) Skeletal muscle
Cystic fibrosis (OMIM 219700)	CFTR (Cl$^-$ channel) Lungs, pancreas
Hyperkalemic periodic paralysis (OMIM 170500)	Sodium channel Skeletal muscle
Hypokalemic periodic paralysis (OMIM 170400)	Slow Ca^{2+} voltage channel (DHPR) Skeletal muscle
Malignant hyperthermia (OMIM 145600)	Ca^{2+} release channel (RYR1) Skeletal muscle
Myotonia congenita (OMIM 160800)	Chloride channel Skeletal muscle

Source: Data in part from Ackerman NJ, Clapham DE: Ion channels—basic science and clinical disease. N Engl J Med 1997;336:1575.
[1]Other channelopathies include the long QT syndrome (OMIM 192500); pseudoaldosteronism (Liddle syndrome, OMIM 177200); persistent hyperinsulinemic hypoglycemia of infancy (OMIM 601820); hereditary X-linked recessive type II nephrolithiasis of infancy (Dent syndrome, OMIM 300009); and generalized myotonia, recessive (Becker disease, OMIM 255700). The term "myotonia" signifies any condition in which muscles do not relax after contraction.

Inherited Cardiomyopathies Are Due to Disorders of Cardiac Energy Metabolism or to Abnormal Myocardial Proteins

An **inherited cardiomyopathy** is any structural or functional abnormality of the ventricular myocardium due to an inherited cause. There are nonheritable types of cardiomyopathy, but these will not be described here. As shown in Table 49–6, the causes of inherited cardiomyopathies fall into two broad classes: (1) disorders of **cardiac energy metabolism,** mainly reflecting mutations in genes encoding enzymes or proteins involved in fatty acid oxidation (a major source of energy for the myocardium) and oxidative phosphorylation; and (2) mu-

TABLE 49–6   Biochemical Causes of Inherited Cardiomyopathies[1]

Cause	Proteins or Process Affected
Inborn errors of fatty acid oxidation	Carnitine entry into cells and mitochondria Certain enzymes of fatty acid oxidation
Disorders of mitochondrial oxidative phosphorylation	Proteins encoded by mitochondrial genes Proteins encoded by nuclear genes
Abnormalities of myocardial contractile and structural proteins	β-Myosin heavy chains, troponin, tropomyosin, dystrophin

Source: Based on Kelly DP, Strauss AW: Inherited cardiomyopathies. N Engl J Med 1994;330:913.
[1]Mutations (eg, point mutations, or in some cases deletions) in the genes (nuclear or mitochondrial) encoding various proteins, enzymes, or tRNA molecules are the fundamental causes of the inherited cardiomyopathies. Some conditions are mild, whereas others are severe and may be part of a syndrome affecting other tissues.

tations in genes encoding proteins involved in or **affecting myocardial contraction,** such as myosin, tropomyosin, the troponins, and cardiac myosin-binding protein C. Mutations in the genes encoding these latter proteins cause familial hypertrophic cardiomyopathy, which will now be discussed.

Mutations in the Cardiac β-Myosin Heavy Chain Gene Are One Cause of Familial Hypertrophic Cardiomyopathy

Familial hypertrophic cardiomyopathy is one of the most frequent hereditary cardiac diseases. Patients exhibit hypertrophy—often massive—of one or both ventricles, starting early in life, and not related to any extrinsic cause such as hypertension. Most cases are transmitted in an autosomal dominant manner; the rest are sporadic. Until recently, its cause was obscure. However, this situation changed when studies of one affected family showed that a **missense mutation** (ie, substitution of one amino acid by another) in the β-**myosin heavy chain gene** was responsible for the condition. Subsequent studies have shown a number of missense mutations in this gene, all coding for highly conserved residues. Some individuals have shown other mutations, such as formation of an α/β-myosin heavy chain hybrid gene. Patients with familial hypertrophic cardiomyopathy can show great variation in clinical picture. This in part reflects **genetic heterogeneity;** ie, mutation in a number of **other genes** (eg, those encoding cardiac actin, tropomyosin, cardiac troponins I and T, essential and regulatory myosin light chains, cardiac myosin-binding protein C, titin, and mitochondrial tRNA-glycine and tRNA-isoleucine) may also cause familial hypertrophic cardiomyopathy. In addition, mutations at different sites in the gene for β-myosin heavy chain may affect the function of the protein to a greater or lesser extent. The missense mutations are clustered in the head and head-rod regions of myosin heavy chain. One hypothesis is that the mutant polypeptides ("poison polypeptides") cause formation of abnormal myofibrils, eventually resulting in compensatory hypertrophy. Some mutations alter the **charge** of the amino acid (eg, substitution of arginine for glutamine), presumably affecting the **conformation** of the protein more markedly and thus affecting its function. Patients with these mutations have a significantly shorter life expectancy than patients in whom the mutation produced no alteration in charge. Thus, definition of the precise mutations involved in the genesis of FHC may prove to be of important prognostic value; it can be accomplished by appropriate use of the polymerase chain reaction on genomic DNA obtained from one sample of blood lymphocytes. Figure 49–13 is a simplified scheme of the events causing familial hypertrophic cardiomyopathy.

Another type of cardiomyopathy is termed **dilated cardiomyopathy.** Mutations in the genes encoding dystrophin, muscle LIM protein (so called because it was found to contain a cysteine-rich domain originally detected in three proteins: Lin-II, Isl-1, and Mec-3), the cyclic response-element binding protein (CREB), desmin and lamin have been implicated in the causation of this condition. The first two proteins help organize the contractile apparatus of cardiac muscle cells, and CREB is involved in the regulation of a number of genes in these cells. Current research is not only elucidating the molecular causes of the cardiomyopathies but is also disclosing mutations that cause **cardiac developmental disorders** (eg, septal defects) and **arrhythmias** (eg, due to mutations affecting ion channels).

Ca²⁺ Also Regulates Contraction of Smooth Muscle

While all muscles contain actin, myosin, and tropomyosin, only vertebrate **striated** muscles contain the **troponin system.** Thus, the mechanisms that regulate contraction must differ in various contractile systems.

Smooth muscles have molecular structures similar to those in striated muscle, but the sarcomeres are not aligned so as to generate the striated appearance. Smooth muscles contain α-actinin and tropomyosin molecules, as do skeletal muscles. They **do not have the troponin system,** and the light chains of smooth muscle myosin molecules differ from those of striated muscle myosin. Regulation of smooth muscle contraction is **myosin-based,** unlike striated muscle, which is actin-based. However, like striated muscle, smooth muscle contraction is **regulated by Ca²⁺.**

Phosphorylation of Myosin Light Chains Initiates Contraction of Smooth Muscle

When smooth muscle myosin is bound to F-actin in the absence of other muscle proteins such as tropomyosin, there is **no detectable ATPase activity.** This absence of activity is quite unlike the situation described for striated muscle myosin and F-actin, which has abundant ATPase activity. Smooth muscle myosin contains **light chains** that prevent the binding of the myosin head to F-actin; they **must be phosphorylated** before they allow F-actin to activate myosin ATPase. The ATPase activity then attained hydrolyzes ATP about 10-fold more slowly than the corresponding activity in skeletal muscle. The phosphate on the myosin light chains may form a chelate with the Ca²⁺ bound to the tropomyosin-TpC-actin complex, leading to an increased rate of formation of cross-bridges between the myosin heads and actin. The phosphorylation of light chains **initiates** the attachment-detachment contraction cycle of smooth muscle.

Myosin Light Chain Kinase Is Activated by Calmodulin-4Ca²⁺ & Then Phosphorylates the Light Chains

Smooth muscle sarcoplasm contains a **myosin light chain kinase** that is calcium-dependent. The Ca²⁺ activation of myosin light chain kinase requires binding of **calmodulin-4Ca²⁺** to its kinase subunit (Figure 49–14). The calmodulin-4Ca²⁺-activated light chain kinase phosphorylates the light chains, which then ceases to inhibit the myosin–F-actin interaction. The contraction cycle then begins.

Another non-Ca²⁺-dependent pathway exists in smooth muscle for initiating contraction. This involves **Rho kinase,** which is activated by a variety of stimuli (not shown in Figure 49–14). This enzyme phosphorylates myosin light chain phosphatase, inhibiting it, and thus increasing the phosphorylation of the light chain. Rho kinase also directly phosphorylates the light chain of myosin. Both of these actions increase the contraction of smooth muscle.

Smooth Muscle Relaxes When the Concentration of Ca²⁺ Falls Below 10⁻⁷ Molar

Relaxation of smooth muscle occurs when sarcoplasmic Ca²⁺ falls below 10⁻⁷ mol/L. The **Ca²⁺** dissociates from calmodulin,

FIGURE 49–13 Simplified scheme of the causation of familial hypertrophic cardiomyopathy (OMIM 192600) due to mutations in the gene encoding β-myosin heavy chain. Mutations in genes encoding other proteins (see text) can also cause this condition.

The figure shows boxes connected by downward arrows:

- Predominantly missense mutations in the β-myosin heavy chain gene on chromosome 14
- Mutant polypeptide chains ("poison polypeptides") that lead to formation of defective myofibrils
- Compensatory hypertrophy of one or both cardiac ventricles
- Cardiomegaly and various cardiac signs and symptoms, including sudden death

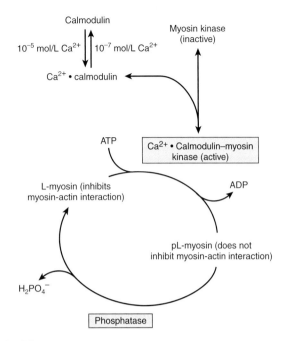

FIGURE 49–14 Regulation of smooth muscle contraction by Ca^{2+}. pL-myosin is the phosphorylated light chain of myosin; L-myosin is the dephosphorylated light chain. (Adapted, with permission, from Adelstein RS, Eisenberg R: Regulation and kinetics of actin-myosin ATP interaction. Annu Rev Biochem 1980;49:921. Copyright © 1980 by Annual Reviews, www.annualreviews.org.)

which in turn dissociates from the myosin light chain kinase, inactivating the kinase. No new phosphates are attached to the p-light chain, and **light chain protein phosphatase,** which is continually active and calcium-independent, removes the existing phosphates from the light chains. Dephosphorylated myosin p-light chain then inhibits the binding of myosin heads

to F-actin and the ATPase activity. The myosin head detaches from the F-actin in the presence of ATP, but it cannot reattach because of the presence of dephosphorylated plight chain; hence, **relaxation** occurs.

Table 49–7 summarizes and compares the regulation of actin-myosin interactions (activation of myosin ATPase) in striated and smooth muscles.

The myosin light chain kinase is not directly affected or activated by **cAMP.** However, cAMP-activated protein kinase can phosphorylate the myosin light chain kinase (not the light chains themselves). The phosphorylated myosin light chain kinase exhibits a significantly lower affinity for calmodulin-Ca^{2+} and thus is less sensitive to activation. Accordingly, an **increase in cAMP dampens the contraction response** of smooth muscle to a given elevation of sarcoplasmic Ca^{2+}. This molecular mechanism can explain the relaxing effect of β-adrenergic stimulation on smooth muscle.

Another protein that appears to play a Ca^{2+}-dependent role in the regulation of smooth muscle contraction is **caldesmon** (87 kDa). This protein is ubiquitous in smooth muscle and is also found in nonmuscle tissue. At low concentrations of Ca^{2+}, it binds to tropomyosin and actin. This **prevents interaction of actin with myosin,** keeping muscle in a relaxed state. At higher concentrations of Ca^{2+}, Ca^{2+}-calmodulin binds caldesmon, **releasing it from actin.** The latter is then free to bind to myosin, and contraction can occur. Caldesmon is also subject to phosphorylation-dephosphorylation; when phosphorylated, it cannot bind actin, again freeing the latter to interact with myosin. Caldesmon may also participate in organizing the structure of the contractile apparatus in smooth muscle. Many of its effects have been demonstrated in vitro, and its physiologic significance is still under investigation.

As noted in Table 49–3, slow cycling of the cross-bridges permits slow prolonged contraction of smooth muscle (eg, in

TABLE 49–7   Actin-Myosin Interactions in Striated and Smooth Muscle

	Striated Muscle	Smooth Muscle (and Nonmuscle Cells)
Proteins of muscle filaments	Actin Myosin Tropomyosin Troponin (TpI, TpT, TpC)	Actin Myosin[1] Tropomyosin
Spontaneous interaction of F-actin and myosin alone (spontaneous activation of myosin ATPase by F-actin)	Yes	No
Inhibitor of F-actin–myosin interaction (inhibitor of F-actin–dependent activation of ATPase)	Troponin system (TpI)	Unphosphorylated myosin light chain
Contraction activated by	Ca^{2+}	Ca^{2+}
Direct effect of Ca^{2+}	$4Ca^{2+}$ bind to TpC	$4Ca^{2+}$ bind to calmodulin
Effect of protein-bound Ca^{2+}	TpC $4Ca^{2+}$ antagonizes TpI inhibition of F-actin–myosin interaction (allows F-actin activation of ATPase)	Calmodulin $4Ca^{2+}$ activates myosin light chain kinase that phosphorylates myosin p-light chain. The phosphorylated p-light chain no longer inhibits F-actin–myosin interaction (allows F-actin activation of ATPase)

[1]Light chains of myosin are different in striated and smooth muscles.

viscera and blood vessels) with less utilization of ATP compared with striated muscle. The ability of smooth muscle to maintain force at reduced velocities of contraction is referred to as the **latch state;** this is an important feature of smooth muscle, and its precise molecular bases are under study.

Nitric Oxide Relaxes the Smooth Muscle of Blood Vessels & Also Has Many Other Important Biologic Functions

Acetylcholine is a vasodilator that acts by causing relaxation of the smooth muscle of blood vessels. However, it does not act directly on smooth muscle. A key observation was that if **endothelial cells** were stripped away from underlying smooth muscle cells, acetylcholine no longer exerted its vasodilator effect. This finding indicated that vasodilators such as acetylcholine initially interact with the endothelial cells of small blood vessels via receptors. The receptors are coupled to the phosphoinositide cycle, leading to the intracellular release of Ca^{2+} through the action of inositol trisphosphate. In turn, the elevation of Ca^{2+} leads to the liberation of **endothelium-derived relaxing factor (EDRF),** which diffuses into the adjacent smooth muscle. There, it reacts with the heme moiety of a soluble guanylyl cyclase, resulting in activation of the latter, with a consequent elevation of intracellular levels of **cGMP** (Figure 49–15). This in turn stimulates the activities of certain cGMP-dependent protein kinases, which probably phosphorylate specific muscle proteins, causing relaxation; however, the details are still being clarified. The important coronary artery vasodilator **nitroglycerin,** widely used to relieve angina pectoris, acts to increase intracellular release of EDRF and thus of cGMP.

Quite unexpectedly, EDRF was found to be the gas **nitric oxide (NO).** NO is formed by the action of the enzyme NO synthase, which is cytosolic. The endothelial and neuronal forms of NO synthase are activated by Ca^{2+} (Table 49–8). The substrate is **arginine,** and the products are citrulline and NO:

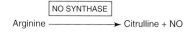

NO SYNTHASE
Arginine ——————→ Citrulline + NO

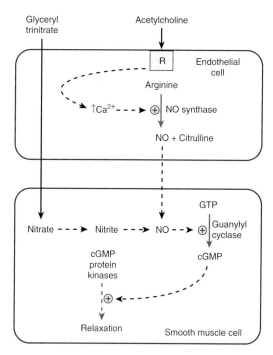

FIGURE 49–15 Diagram showing formation in an endothelial cell of nitric oxide (NO) from arginine in a reaction catalyzed by NO synthase. Interaction of an agonist (eg, acetylcholine) with a receptor (R) probably leads to intracellular release of Ca^{2+} via inositol trisphosphate generated by the phosphoinositide pathway, resulting in activation of NO synthase. The NO subsequently diffuses into adjacent smooth muscle, where it leads to activation of guanylyl cyclase, formation of cGMP, stimulation of cGMP protein kinases, and subsequent relaxation. The vasodilator nitroglycerin is shown entering the smooth muscle cell, where its metabolism also leads to formation of NO.

NO synthase catalyzes a five-electron oxidation of an amidine nitrogen of arginine. L-Hydroxyarginine is an intermediate that remains tightly bound to the enzyme. NO synthase is a very complex enzyme, employing five redox cofactors: NADPH, FAD, FMN, heme, and tetrahydrobiopterin. NO can also be formed from **nitrite,** derived from vasodilators such as glyceryl trinitrate during their metabolism. NO has a very short half-life (approximately 3–4 seconds) in tissues because

TABLE 49–8   Summary of the Nomenclature of the NO Synthases and of the Effects of Knockout of Their Genes in Mice

Subtype	Name[1]	Comments	Result of Gene Knockout in Mice[2]
1	nNOS	Activity depends on elevated Ca^{2+}; first identified in neurons; calmodulin-activated	Pyloric stenosis, resistant to vascular stroke, aggressive sexual behavior (males)
2	iNOS[3]	Independent of elevated Ca^{2+}; prominent in macrophages	More susceptible to certain types of infection
3	eNOS	Activity depends on elevated Ca^{2+}; first identified in endothelial cells	Elevated mean blood pressure

Source: Adapted from Snyder SH: NO. Nature 1995;377:196.
[1]n, neuronal; i, inducible; e, endothelial.
[2]Gene knockouts were performed by homologous recombination in mice. The enzymes are characterized as neuronal, inducible (macrophage), and endothelial because these were the sites in which they were first identified. However, all three enzymes have been found in other sites, and the neuronal enzyme is also inducible. Each gene has been cloned, and its chromosomal location in humans has been determined.
[3]iNOS is Ca^{2+}-independent but binds calmodulin very tightly.

it reacts with oxygen and superoxide. The product of the reaction with superoxide is **peroxynitrite** (ONOO⁻), which decomposes to form the highly reactive OH⁻ radical. NO is inhibited by hemoglobin and other heme proteins, which bind it tightly. **Chemical inhibitors of NO synthase** are now available that can markedly decrease formation of NO. Administration of such inhibitors to animals and humans leads to vasoconstriction and a marked elevation of blood pressure, indicating that NO is of major importance in the maintenance of blood pressure in vivo. Another important cardiovascular effect is that by increasing synthesis of cGMP, it acts as an **inhibitor of platelet aggregation** (Chapter 51).

Since the discovery of the role of NO as a vasodilator, there has been intense experimental interest in this molecule. It has turned out to have a variety of physiologic roles, involving virtually every tissue of the body (Table 49–9). Three major isoforms of NO synthase have been identified, each of which has

been cloned, and the chromosomal locations of their genes in humans have been determined. Gene knockout experiments have been performed on each of the three isoforms and have helped establish some of the postulated functions of NO.

To summarize, research in the past decade has shown that NO plays an important role in many physiologic and pathologic processes.

SEVERAL MECHANISMS REPLENISH STORES OF ATP IN MUSCLE

The **ATP** required as the constant energy source for the contraction-relaxation cycle of muscle can be generated (1) by glycolysis, using blood glucose or muscle glycogen, (2) by oxidative phosphorylation, (3) from creatine phosphate, and (4) from two molecules of ADP in a reaction catalyzed by adenylyl kinase (Figure 49–16). The amount of ATP in skeletal muscle is only sufficient to provide energy for contraction for a few seconds, so that ATP must be constantly renewed from one or more of the above sources, depending upon metabolic conditions. As discussed below, there are at least **two distinct types of fibers** in skeletal muscle, one predominantly active in **aerobic** conditions and the other in **anaerobic** conditions; not unexpectedly, they use each of the above sources of energy to different extents.

Skeletal Muscle Contains Large Supplies of Glycogen

The sarcoplasm of skeletal muscle contains large stores of **glycogen,** located in granules close to the I bands. The release of glucose from glycogen is dependent on a specific muscle **glycogen phosphorylase** (Chapter 19), which can be activated by Ca²⁺, epinephrine, and AMP. To generate glucose 6-phosphate for glycolysis in skeletal muscle, glycogen phosphorylase b

TABLE 49–9 Some Physiologic Functions and Pathologic Involvements of Nitric Oxide (NO)

- Vasodilator, important in regulation of blood pressure

- Involved in penile erection; sildenafil citrate (Viagra) affects this process by inhibiting a cGMP phosphodiesterase

- Neurotransmitter in the brain and peripheral autonomic nervous system

- Role in long-term potentiation

- Role in neurotoxicity

- Low level of NO involved in causation of pylorospasm in infantile hypertrophic pyloric stenosis

- May have role in relaxation of skeletal muscle

- May constitute part of a primitive immune system

- Inhibits adhesion, activation, and aggregation of platelets

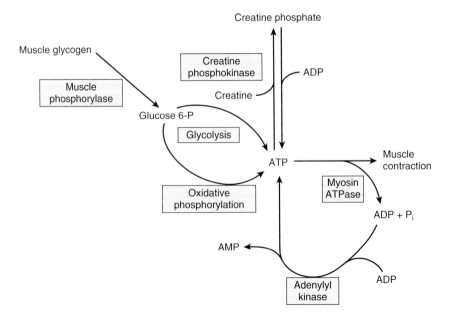

FIGURE 49–16 The multiple sources of ATP in muscle.

must be activated to phosphorylase a via phosphorylation by phosphorylase b kinase (Chapter 19). Ca^{2+} promotes the activation of phosphorylase b kinase, also by phosphorylation. Thus, Ca^{2+} both initiates muscle contraction and activates a pathway to provide necessary energy. The hormone **epinephrine** also activates glycogenolysis in muscle. **AMP**, produced by breakdown of ADP during muscular exercise, can also activate phosphorylase b without causing phosphorylation. Muscle glycogen phosphorylase b is inactive in **McArdle disease,** one of the glycogen storage diseases (Chapter 19).

Under Aerobic Conditions, Muscle Generates ATP Mainly by Oxidative Phosphorylation

Synthesis of ATP via **oxidative phosphorylation** requires a supply of oxygen. Muscles that have a high demand for oxygen as a result of sustained contraction (eg, to maintain posture) store it attached to the heme moiety of **myoglobin.** Because of the heme moiety, muscles containing myoglobin are red, whereas muscles with little or no myoglobin are white. **Glucose,** derived from the blood glucose or from endogenous glycogen, and **fatty acids** derived from the triacylglycerols of adipose tissue are the principal substrates used for aerobic metabolism in muscle.

Creatine Phosphate Constitutes a Major Energy Reserve in Muscle

Creatine phosphate prevents the rapid depletion of ATP by providing a readily available high-energy phosphate that can be used to regenerate ATP from ADP. Creatine phosphate is formed from ATP and creatine (Figure 49–16) at times when the muscle is relaxed and demands for ATP are not so great. The enzyme catalyzing the phosphorylation of creatine is **creatine kinase** (CK), a muscle-specific enzyme with clinical utility in the detection of acute or chronic diseases of muscle.

SKELETAL MUSCLE CONTAINS SLOW (RED) & FAST (WHITE) TWITCH FIBERS

Different types of fibers have been detected in skeletal muscle. One classification subdivides them into type I (slow twitch), type IIA (fast twitch-oxidative), and type IIB (fast twitch-glycolytic). For the sake of simplicity, we shall consider only two types: type I (slow twitch, oxidative) and type II (fast twitch, glycolytic) (Table 49–10). The **type I** fibers are red because they contain myoglobin and mitochondria; their metabolism is aerobic, and they maintain relatively sustained contractions. The **type II** fibers, lacking myoglobin and containing few mitochondria, are white: they derive their energy from anaerobic glycolysis and exhibit relatively short durations of contraction. The **proportion** of these two types of fibers varies among the muscles of the body, depending on function (eg, whether or not a muscle is involved in sustained contraction, such as maintaining posture). The proportion also varies with train-

TABLE 49–10 Characteristics of Type I and Type II Fibers of Skeletal Muscle

	Type I Slow Twitch	Type II Fast Twitch
Myosin ATPase	Low	High
Energy utilization	Low	High
Mitochondria	Many	Few
Color	Red	White
Myoglobin	Yes	No
Contraction rate	Slow	Fast
Duration	Prolonged	Short

ing; for example, the number of type I fibers in certain leg muscles increases in athletes training for marathons, whereas the number of type II fibers increases in sprinters.

A Sprinter Uses Creatine Phosphate & Anaerobic Glycolysis to Make ATP, Whereas a Marathon Runner Uses Oxidative Phosphorylation

In view of the two types of fibers in skeletal muscle and of the various energy sources described above, it is of interest to compare their involvement in a sprint (eg, 100 meters) and in the marathon (42.2 km; just over 26 miles) (Table 49–11).

The major sources of energy in the **100-m sprint** are **creatine phosphate** (first 4–5 sec) and then **anaerobic glycolysis,** using muscle glycogen as the source of glucose. The two main sites of metabolic control are at **glycogen phosphorylase** and at **PFK-1.** The former is activated by Ca^{2+} (released from the sarcoplasmic reticulum during contraction), epinephrine, and AMP. PFK-1 is activated by AMP, P_i, and NH_3. Attesting to the efficiency of these processes, the flux through glycolysis can increase as much as 1000-fold during a sprint.

In contrast, in the **marathon, aerobic metabolism** is the principal source of ATP. The major fuel sources are **blood**

TABLE 49–11 Types of Muscle Fibers and Major Fuel Sources Used by a Sprinter and by a Marathon Runner

Sprinter (100 m)	Marathon Runner
Type II (glycolytic) fibers are used predominantly	Type I (oxidative) fibers are used predominantly
Creatine phosphate is the major energy source during the first 4–5 sec	ATP is the major energy source throughout
Glucose derived from muscle glycogen and metabolized by anaerobic glycolysis is the major fuel source	Blood glucose and free fatty acids are the major fuel sources
Muscle glycogen is rapidly depleted	Muscle glycogen is slowly depleted

glucose and **free fatty acids,** largely derived from the breakdown of triacylglycerols in adipose tissue, stimulated by epinephrine. Hepatic glycogen is degraded to maintain the level of blood glucose. Muscle glycogen is also a fuel source, but it is degraded much more gradually than in a sprint. It has been calculated that the amounts of glucose in the blood, of glycogen in the liver, of glycogen in muscle, and of triacylglycerol in adipose tissue are sufficient to supply muscle with energy during a marathon for 4 min, 18 min, 70 min, and approximately 4000 min, respectively. However, the rate of oxidation of fatty acids by muscle is slower than that of glucose, so that oxidation of glucose and of fatty acids are both major sources of energy in the marathon.

A number of procedures have been used by athletes to counteract muscle fatigue and inadequate strength. These include **carbohydrate loading, soda (sodium bicarbonate) loading, blood doping** (administration of red blood cells), and ingestion of **creatine** and **androstenedione.** Their rationales and efficacies will not be discussed here.

SKELETAL MUSCLE CONSTITUTES THE MAJOR RESERVE OF PROTEIN IN THE BODY

In humans, **skeletal muscle protein** is the major non-fat source of stored energy. This explains the very large losses of muscle mass, particularly in adults, resulting from prolonged caloric undernutrition.

The study of **tissue protein breakdown** in vivo is difficult, because amino acids released during intracellular breakdown of proteins can be extensively reutilized for protein synthesis within the cell, or the amino acids may be transported to other organs where they enter anabolic pathways. However, actin and myosin are methylated by a posttranslational reaction, forming **3-methylhistidine.** During intracellular breakdown of actin and myosin, 3-methylhistidine is released and excreted into the urine. The urinary output of the methylated amino acid provides a reliable index of the rate of myofibrillar protein breakdown in the musculature of human subjects.

Various features of muscle metabolism, most of which are dealt with in other chapters of this text, are summarized in Table 49–12.

THE CYTOSKELETON PERFORMS MULTIPLE CELLULAR FUNCTIONS

Non-muscle cells perform mechanical work, including self-propulsion, morphogenesis, cleavage, endocytosis, exocytosis, intracellular transport, and changing cell shape. These cellular functions are carried out by an extensive intracellular network of filamentous structures constituting the **cytoskeleton.** The cell cytoplasm is not a sac of fluid, as once thought. Essentially all eukaryotic cells contain three types of filamentous structures: **actin filaments** (also known as microfilaments), **microtubules,** and **intermediate filaments.** Each type of filament can be distinguished biochemically and by the electron microscope.

Some properties of these three structures are summarized in Tables 49–13 & 49–14.

Non-Muscle Cells Contain Actin That Forms Microfilaments

G-actin is present in most if not all cells of the body. With appropriate concentrations of magnesium and potassium chloride, it spontaneously polymerizes to form double helical **F-actin** filaments like those seen in muscle. There are at least

TABLE 49–12 Summary of Major Features of the Biochemistry of Skeletal Muscle Related to Its Metabolism[1]

- Skeletal muscle functions under both aerobic (resting) and anaerobic (eg, sprinting) conditions, so both aerobic and anaerobic glycolysis operate, depending on conditions.

- Skeletal muscle contains myoglobin as a reservoir of oxygen.

- Skeletal muscle contains different types of fibers primarily suited to anaerobic (fast twitch fibers) or aerobic (slow twitch fibers) conditions.

- Actin, myosin, tropomyosin, troponin complex (TpT, TpI, and TpC), ATP, and Ca^{2+} are key constituents in relation to contraction.

- The Ca^{2+} ATPase, the Ca^{2+} release channel, and calsequestrin are proteins involved in various aspects of Ca^{2+} metabolism in muscle.

- Insulin acts on skeletal muscle to increase uptake of glucose.

- In the fed state, most glucose is used to synthesize glycogen, which acts as a store of glucose for use in exercise; "preloading" with glucose is used by some long-distance athletes to build up stores of glycogen.

- Epinephrine stimulates glycogenolysis in skeletal muscle, whereas glucagon does not because of absence of its receptors.

- Skeletal muscle cannot contribute directly to blood glucose because it does not contain glucose-6-phosphatase.

- Lactate produced by anaerobic metabolism in skeletal muscle passes to liver, which uses it to synthesize glucose, which can then return to muscle (the Cori cycle).

- Skeletal muscle contains phosphocreatine, which acts as an energy store for short-term (seconds) demands.

- Free fatty acids in plasma are a major source of energy, particularly under marathon conditions and in prolonged starvation.

- Skeletal muscle can utilize ketone bodies during starvation.

- Skeletal muscle is the principal site of metabolism of branched chain amino acids, which are used as an energy source.

- Proteolysis of muscle during starvation supplies amino acids for gluconeogenesis.

- Major amino acids emanating from muscle are alanine (destined mainly for gluconeogenesis in liver and forming part of the glucose-alanine cycle) and glutamine (destined mainly for the gut and kidneys).

[1]This table brings together material from various chapters in this book.

TABLE 49–13 Some Properties of Microfilaments and Microtubules

	Microfilaments	Microtubules
Protein(s)	Actin	α-and β-tubulins
Diameter	8–9 nm	25 nm
Functions	Structural, motility	Structural, motility, polarity

Note: Some properties of intermediate filaments are described in Table 49–14.

TABLE 49–14 Classes of Intermediate Filaments of Eukaryotic Cells and Their Distributions

Proteins	Molecular Mass	Distributions
Lamins		
A, B, and C	65–75 kDa	Nuclear lamina
Keratins		
Type I (acidic)	40–60 kDa	Epithelial cells, hair, nails
Type II (basic)	50–70 kDa	
Vimentin-like		
Vimentin	54 kDa	Various mesenchymal cells
Desmin	53 kDa	Muscle
Glial fibrillary acid protein	50 kDa	Glial cells
Peripherin	66 kDa	Neurons
Neurofilaments		
Low (L), medium (M), and high (H)[1]	60–130 kDa	Neurons

Note: Intermediate filaments have an approximate diameter of 10 nm and have various functions. For example, keratins are distributed widely in epithelial cells and adhere via adapter proteins to desmosomes and hemidesmosomes. Lamins provide support for the nuclear membrane.
[1]Refers to their molecular masses.

two types of actin in nonmuscle cells: β-actin and γ-actin. Both types can coexist in the same cell and probably even copolymerize in the same filament. In the cytoplasm, **F-actin** forms **microfilaments** of 7–9.5 nm that frequently exist as bundles of a tangled-appearing meshwork. These bundles are prominent just underlying the plasma membrane of many cells and are there referred to as **stress fibers.** The stress fibers disappear as cell motility increases or upon malignant transformation of cells by chemicals or oncogenic viruses.

Although not organized as in muscle, actin filaments in nonmuscle cells interact with **myosin** to cause cellular movements.

Microtubules Contain α- & β-Tubulins

Microtubules, an integral component of the cellular cytoskeleton, consist of cytoplasmic tubes 25 nm in diameter and often of extreme length (see Figure 49–17). Microtubules are neces-

sary for the formation and function of the **mitotic spindle** and thus are present in all eukaryotic cells. They are also involved in the intracellular movement of endocytic and exocytic **vesicles** and form the major structural components of **cilia** and **flagella.** Microtubules are a major component of **axons** and **dendrites,** in which they maintain structure and participate in the axoplasmic flow of material along these neuronal processes.

Microtubules are cylinders of 13 longitudinally arranged protofilaments, each consisting of dimers of **α-tubulin** and **β-tubulin,** closely related proteins of approximately 50 kDa molecular mass. The tubulin dimers assemble into protofilaments and subsequently into sheets and then cylinders. A microtubule-organizing center, located around a pair of centrioles, nucleates the growth of new microtubules. A third species of tubulin, **γ-tubulin,** appears to play an important role in this assembly. **GTP** is required for assembly. A variety of proteins are associated with microtubules (**microtubule-associated proteins [MAPs],** one of which is **tau**) and play important roles in microtubule assembly and stabilization. Microtubules are in a state of dynamic instability, constantly assembling and disassembling. They exhibit **polarity** (plus and minus ends); this is important in their growth from centrioles and in their ability to direct intracellular movement. For instance, in axonal transport, the protein **kinesin,** with a myosin-like ATPase activity, uses hydrolysis of ATP to move vesicles down the axon toward the positive end of the microtubular formation. Flow of materials in the opposite direction, toward the negative end, is powered by **cytosolic dynein,** another protein with ATPase activity. Similarly, **axonemal dyneins** power ciliary and flagellar movement. Another protein, **dynamin,** uses GTP and is involved in endocytosis. Kinesins, dyneins, dynamin, and myosins are referred to as **molecular motors.**

An absence of dynein in cilia and flagella results in immotile cilia and flagella, leading to male sterility, situs inversus and chronic respiratory infection, a condition known as **Kartagener syndrome** (OMIM 244400). Mutations in genes affecting the synthesis of dynein have been detected in individuals with this syndrome.

Certain **drugs** bind to microtubules and thus interfere with their assembly or disassembly. These include **colchicine** (used for treatment of acute gouty arthritis), **vinblastine** (a vinca alkaloid used for treating certain types of cancer), **paclitaxel** (Taxol) (effective against ovarian cancer), and **griseofulvin** (an antifungal agent).

Intermediate Filaments Differ from Microfilaments & Microtubules

An intracellular fibrous system exists of filaments with an axial periodicity of 21 nm and a diameter of 8–10 nm that is intermediate between that of microfilaments (6 nm) and microtubules (23 nm). At least four classes of **intermediate filaments** are found, as indicated in Table 49–14.

They are all elongated, fibrous molecules, with a central rod domain, an amino terminal head, and a carboxyl terminal tail. They form a structure like a rope, and the mature fila-

FIGURE 49–17 Schematic representation of microtubules. The upper left-hand corner shows a drawing of microtubules as seen in the electron microscope following fixation with tannic acid in glutaraldehyde. The unstained tubulin subunits are delineated by the dense tannic acid. Cross sections of tubules reveal a ring of 13 subunits of dimers arranged in a spiral. Changes in microtubule length are due to the addition or loss of individual tubulin subunits. Characteristic arrangements of microtubules (not shown here) are found in centrioles, basal bodies, cilia and flagellae. (Reproduced, with permission, from Junqueirai LC, Carneiro J, Kelley RO: *Basic Histology,* 7th ed. Appleton & Lange, 1992.)

ments are composed of tetramers packed together in a helical manner. They are important structural components of cells, and most are **relatively stable** components of the cytoskeleton, not undergoing rapid assembly and disassembly and not disappearing during mitosis, as do actin and many microtubular filaments.

An important exception to this is provided by the **lamins,** which, subsequent to phosphorylation, disassemble at mitosis and reappear when it terminates. **Lamins** form a meshwork in apposition to the inner nuclear membrane.

Mutations in the gene encoding **lamin A** and lamin C cause Hutchinson-Gilford progeria syndrome **(progeria)** [OMIM 176670], characterized by the appearance of **accelerated aging** and other features. A farnesylated form (see Figure 26–2 for the structure of farnesyl) of prelamin A accumulates in the condition, because the site of normal proteolytic action to cleave off the farnesylated portion of lamin A is altered by mutation. Lamin A is an important component of the structural scaffolding that maintains the integrity of the nucleus of a cell. It appears that the accumulation of the farnesylated prelamin A makes nuclei unstable, altering their shape, and somehow this predisposes to the development of signs of premature aging. Experiments in mice have indicated that administration of a farnesyltransferase inhibitor may ameliorate the development of misshapen nuclei. Children affected by this condition often die in their teens of atherosclerosis. A brief scheme of the causation of progeria is shown in Figure 49–18.

Keratins form a large family, with about 30 members being distinguished. As indicated in Table 49–14, two major types of keratins are found; all individual keratins are **heterodimers** made up of one member of each class.

Vimentins are widely distributed in mesodermal cells, and desmin, glial fibrillary acidic protein, and peripherin are related to them. All members of the vimentin-like family can copolymerize with each other.

FIGURE 49–18 Scheme of the causation of progeria (Hutchinson-Gilford syndrome, OMIM 176670).

Intermediate filaments are very prominent in nerve cells; neurofilaments are classified as low, medium, and high on the basis of their molecular masses. The **distribution of intermediate filaments** in normal and abnormal (eg, cancer) cells can be studied by the use of immunofluorescent techniques, using antibodies of appropriate specificities. These antibodies to specific intermediate filaments can also be of use to pathologists in helping to decide the origin of certain dedifferentiated malignant tumors. These tumors may still retain the type of intermediate filaments found in their cell of origin.

A number of **skin diseases,** mainly characterized by blistering, have been found to be due to mutations in genes encoding **various keratins.** Two of these disorders are epidermolysis bullosa simplex (OMIM 131800) and epidermolytic palmoplantar keratoderma (OMIM 144200). The **blistering** found in these disorders probably reflects a diminished capacity of various layers of the skin to resist mechanical stresses due to abnormalities in keratin structure.

SUMMARY

- The myofibrils of skeletal muscle contain thick and thin filaments. The thick filaments contain myosin. The thin filaments contain actin, tropomyosin, and the troponin complex (troponins T, I, and C).

- The sliding filament cross-bridge model is the foundation of current thinking about muscle contraction. The basis of this model is that the interdigitating filaments slide past one another during contraction and cross-bridges between myosin and actin generate and sustain the tension.

- The hydrolysis of ATP is used to drive movement of the filaments. ATP binds to myosin heads and is hydrolyzed to ADP and P_i by the ATPase activity of the actomyosin complex.

- Ca^{2+} plays a key role in the initiation of muscle contraction by binding to troponin C. In skeletal muscle, the sarcoplasmic reticulum regulates distribution of Ca^{2+} to the sarcomeres, whereas inflow of Ca^{2+} via Ca^{2+} channels in the sarcolemma is of major importance in cardiac and smooth muscle.

- Many cases of malignant hyperthermia in humans are due to mutations in the gene encoding the Ca^{2+} release channel.

- A number of differences exist between skeletal and cardiac muscle; in particular, the latter contains a variety of receptors on its surface.

- Some cases of familial hypertrophic cardiomyopathy are due to missense mutations in the gene coding for β-myosin heavy chain. Mutations in genes encoding a number of other proteins have also been detected.

- Smooth muscle, unlike skeletal and cardiac muscle, does not contain the troponin system; instead, phosphorylation of myosin light chains initiates contraction.

- Nitric oxide is a regulator of vascular smooth muscle; blockage of its formation from arginine causes an acute elevation of blood pressure, indicating that regulation of blood pressure is one of its many functions.

- Duchenne-type muscular dystrophy is due to mutations in the gene, located on the X chromosome, encoding the protein dystrophin.

- Two major types of muscle fibers are found in humans: white (anaerobic) and red (aerobic). The former are particularly used in sprints and the latter in prolonged aerobic exercise. During a sprint, muscle uses creatine phosphate and glycolysis as energy sources; in the marathon, oxidation of fatty acids is of major importance during the later phases.

- Non-muscle cells perform various types of mechanical work carried out by the structures constituting the cytoskeleton.

These structures include actin filaments (microfilaments), microtubules (composed primarily of α-tubulin and β-tubulin), and intermediate filaments. The latter include lamins, keratins, vimentin-like proteins, and neurofilaments. Mutations in the gene encoding lamin A cause progeria, a condition characterized by the appearance of premature aging. Mutations in genes for certain keratins cause a number of skin diseases.

REFERENCES

Alberts B et al: *Molecular Biology of the Cell*. 5th ed. Garland Science, 2008. (Contains excellent coverage of muscle and the cytoskeleton).

Brosnan JT, Brosnan ME: Creatine: endogenous metabolite, dietary and therapeutic supplement. Annu Rev Nutr 2007;27:241.

De La Cruz EM, Ostap EM: Relating biochemistry and function in the myosin superfamily. Curr Opin Cell Biol 2004;16:61.

Herrman H, Aebi U: Intermediate filaments: molecular structure, assembly mechanism, and integration into functionally distinct intracellular scaffolds. Annu Rev Biochem 2004;73:749.

Lodish H et al: *Molecular Cell Biology,* 6th ed. WH Freeman & Co, 2008. (Contains excellent coverage of muscle and the cytoskeleton).

Matthews KD, Moore SA: Multicore myopathy, central core disease, malignant hyperthermia susceptibility, and RYR1 mutations: one disease with many faces? Arch Neurol 2004;61:106.

Murad F: Nitric oxide and cyclic GMP in cell signalling and drug development. N Engl J Med 2006; 355:2003.

Murphy RT, Starling RC: Genetics and cardiomyopathy: where are we now? Cleve Clin J Med 2005;72:465.

Neubauer S: The failing heart—an engine out of fuel. N Engl J Med 2007; 356:1140.

Palmer KJ, Watson P, Stephens DJ: The role of microtubules in transport between the endoplasmic reticulum and Golgi apparatus in mammalian cells. Biochem Soc Symp 2005;72:1.

Pollard TD, Earnshaw WC: *Cell Biology,* 2nd ed. Saunders, 2008. (Contains excellent coverage of muscle and the cytoskeleton).

Scriver CR et al (editors): *The Metabolic and Molecular Bases of Inherited Disease,* 8th ed. McGraw-Hill, 2001. (This comprehensive four-volume text and its updated online edition [see Chapter 1]contain chapters on malignant hyperthermia, channelopathies, hypertrophic cardiomyopathy, the muscular dystrophies, and disorders of intermediate filaments.)

Webb RC: Smooth muscle contraction and relaxation. Adv Physiol Educ 2003;27:201.

Wehrens XHT, Lehnart SE, Marks AR: Intracellular calcium release and cardiac disease. Annu Rev Physiol 2005;67:69.

Plasma Proteins & Immunoglobulins

Robert K. Murray, MD, PhD

BIOMEDICAL IMPORTANCE

The fundamental role of blood in the maintenance of **homeostasis** and the ease with which blood can be obtained have meant that the study of its constituents has been of central importance in the development of biochemistry and clinical biochemistry. The basic properties of a number of **plasma proteins,** including the **immunoglobulins** (antibodies), are described in this chapter. Changes in the amounts of various plasma proteins and immunoglobulins occur in many diseases and can be monitored by electrophoresis or other suitable procedures. As indicated in an earlier chapter, alterations of the activities of certain **enzymes** found in plasma are of diagnostic use in a number of pathologic conditions. Plasma proteins involved in blood coagulation are discussed in Chapter 51.

THE BLOOD HAS MANY FUNCTIONS

The functions of blood—except for specific cellular ones such as oxygen transport and cell-mediated immunologic defense—are carried out by plasma and its constituents (Table 50–1).

Plasma consists of water, electrolytes, metabolites, nutrients, proteins, and hormones. The water and electrolyte composition of plasma is practically the same as that of all extracellular fluids. Laboratory determinations of levels of Na^+, K^+, Ca^{2+}, Mg^{2+}, Cl^-, HCO_3^-, $Paco_2$, and of blood pH are important in the management of many patients.

PLASMA CONTAINS A COMPLEX MIXTURE OF PROTEINS

The concentration of total protein in human plasma is approximately 7.0–7.5 g/dL and comprises the major part of the solids of the plasma. The proteins of the plasma are actually a complex mixture that includes not only simple proteins but also conjugated proteins such as **glycoproteins** and various types of **lipoproteins.** Use of proteomic techniques is allowing the isolation and characterization of previously unknown plasma proteins, some present in very small amounts (eg, detected in hemodialysis fluid and in the plasma of patients with cancer), thus expanding **the plasma proteome.** Thousands of **antibod-**ies are present in human plasma, although the amount of any one antibody is usually quite low under normal circumstances. The relative dimensions and molecular masses of some of the most important plasma proteins are shown in Figure 50–1.

The **separation** of individual proteins from a complex mixture is frequently accomplished by the use of solvents or electrolytes (or both) to remove different protein fractions in accordance with their solubility characteristics. This is the basis of the so-called salting-out methods, which find some usage in the determination of protein fractions in the clinical laboratory. Thus, one can separate the proteins of the plasma into three major groups—**fibrinogen, albumin,** and **globulins**—by the use of varying concentrations of sodium or ammonium sulfate.

The most common method of analyzing plasma proteins is by **electrophoresis.** There are many types of electrophoresis, each using a different supporting medium. In clinical laboratories, **cellulose acetate** is widely used as a supporting medium. Its use permits resolution, after staining, of plasma proteins into five bands, designated albumin, α_1, α_2, β, and γ fractions, respectively (Figure 50–2). The stained strip of cellulose acetate (or other supporting medium) is called an electrophoretogram. The amounts of these five bands can be conveniently quantified by use of densitometric scanning machines. Characteristic changes in the amounts of one or more of these five bands are found in many diseases.

The Concentration of Protein in Plasma Is Important in Determining the Distribution of Fluid Between Blood & Tissues

In arterioles, the **hydrostatic pressure** is about 37 mm Hg, with an interstitial (tissue) pressure of 1 mm Hg opposing it. The **osmotic pressure** (oncotic pressure) exerted by the plasma proteins is approximately 25 mm Hg. Thus, a net outward force of about 11 mm Hg drives fluid out into the interstitial spaces. In venules, the hydrostatic pressure is about 17 mm Hg, with the oncotic and interstitial pressures as described above; thus, a net force of about 9 mm Hg attracts water back into the circulation. The above pressures are often referred to as the **Starling forces.** If the concentration of plasma proteins is markedly diminished (eg, due to severe protein malnutrition),

TABLE 50–1 Major Functions of Blood

1. **Respiration**—transport of oxygen from the lungs to the tissues and of CO_2 from the tissues to the lungs

2. **Nutrition**—transport of absorbed food materials

3. **Excretion**—transport of metabolic waste to the kidneys, lungs, skin, and intestines for removal

4. Maintenance of the normal **acid-base balance** in the body

5. Regulation of **water balance** through the effects of blood on the exchange of water between the circulating fluid and the tissue fluid

6. Regulation of **body temperature** by the distribution of body heat

7. **Defense** against infection by the white blood cells and circulating antibodies

8. Transport of **hormones** and regulation of metabolism

9. Transport of **metabolites**

10. **Coagulation**

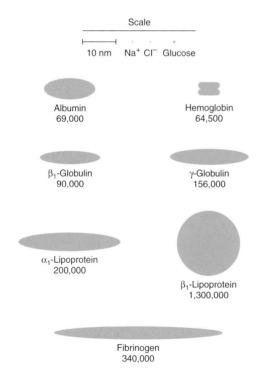

FIGURE 50–1 Relative dimensions and approximate molecular masses of protein molecules in the blood (Oncley).

fluid is not attracted back into the intravascular compartment and accumulates in the extravascular tissue spaces, a condition known as **edema.** Edema has many causes; protein deficiency is one of them.

Plasma Proteins Have Been Studied Extensively

Because of the relative ease with which they can be obtained, plasma proteins have been studied extensively in both humans and animals. Considerable information is available about the biosynthesis, turnover, structure, and functions of the major plasma proteins. Alterations of their amounts and of their metabolism in many disease states have also been investigated. In recent years, many of the genes for plasma proteins have been cloned and their structures determined.

The preparation of **antibodies** specific for the individual plasma proteins has greatly facilitated their study, allowing the precipitation and isolation of pure proteins from the complex mixture present in tissues or plasma. In addition, the use of **isotopes** has made possible the determination of their pathways of biosynthesis and of their turnover rates in plasma.

The following generalizations have emerged from studies of plasma proteins.

Most Plasma Proteins Are Synthesized in the Liver

This has been established by experiments at the whole animal level (eg, hepatectomy) and by use of the isolated perfused liver preparation, of liver slices, of liver homogenates, and of in vitro translation systems using preparations of mRNA extracted from liver. However, the γ-globulins are synthesized in plasma cells and certain plasma proteins are synthesized in other sites, such as endothelial cells.

Plasma Proteins Are Generally Synthesized on Membrane-Bound Polyribosomes

They then traverse the major secretory route in the cell (rough endoplasmic membrane → smooth endoplasmic membrane → Golgi apparatus → secretory vesicles) prior to entering the plasma. Thus, most plasma proteins are synthesized as **preproteins** and initially contain amino terminal signal peptides (Chapter 46). They are usually subjected to various posttranslational modifications (proteolysis, glycosylation, phosphorylation, etc) as they travel through the cell. Transit times through the hepatocyte from the site of synthesis to the plasma vary from 30 min to several hours or more for individual proteins.

Most Plasma Proteins Are Glycoproteins

Accordingly, they generally contain either N- or O-linked oligosaccharide chains, or both (Chapter 47). Albumin is the major exception; it does not contain sugar residues. The oligosaccharide chains have various functions (Table 47–2). Removal of terminal sialic acid residues from certain plasma proteins (eg, ceruloplasmin) by exposure to neuraminidase can markedly shorten their half-lives in plasma (Chapter 47).

Many Plasma Proteins Exhibit Polymorphism

A **polymorphism** is a mendelian or monogenic trait that exists in the population in at least two phenotypes, neither of which is rare (ie, neither of which occurs with frequency of less than 1–2%). The ABO blood group substances (Chapter 52) are the best-known examples of human polymorphisms. Human plasma proteins that exhibit polymorphism include α₁-antitrypsin, haptoglobin, transferrin, ceruloplasmin, and

FIGURE 50–2 Technique of cellulose acetate zone electrophoresis. **(A)** A small amount of serum or other fluid is applied to a cellulose acetate strip. **(B)** Electrophoresis of sample in electrolyte buffer is performed. **(C)** Separated protein bands are visualized in characteristic positions after being stained. **(D)** Densitometer scanning from cellulose acetate strip converts bands to characteristic peaks of albumin, α_1-globulin, β_2-globulin, β-globulin, and γ-globulin. (Reproduced, with permission, from Parslow TG et al [editors]: *Medical Immunology,* 10th ed. McGraw-Hill, 2001.)

immunoglobulins. The polymorphic forms of these proteins can be distinguished by different procedures (eg, various types of electrophoresis or isoelectric focusing), in which each form may show a characteristic migration. Analyses of these human polymorphisms have proved to be of genetic, anthropologic, and clinical interest.

Each Plasma Protein Has a Characteristic Half-Life in the Circulation

The **half-life** of a plasma protein can be determined by labeling the isolated pure protein with ^{131}I or Cr^{51} under mild, nondenaturing conditions. The labeled protein is freed of unbound free isotope and its specific activity (disintegrations per minute per milligram of protein) determined. A known amount of the radioactive protein is then injected into a normal adult subject, and samples of blood are taken at various time intervals for determinations of radioactivity. The values for radioactivity are plotted against time, and the half-life of the protein (the time for the radioactivity to decline from its peak value to one-half of its peak value) can be calculated from the resulting graph, discounting the times for the injected protein to equilibrate (mix) in the blood and in the extravascular spaces. The half-lives obtained for albumin and haptoglobin in normal healthy adults are approximately 20 and 5 days, respectively. In certain diseases, the half-life of a protein may be markedly altered. For instance, in some gastrointestinal diseases such as regional ileitis (Crohn disease), considerable amounts of plasma proteins, including albumin, may be lost into the bowel through the inflamed intestinal mucosa. Patients with this condition have a **protein-losing gastroenteropathy,** and the half-life of injected iodinated albumin in these subjects may be reduced to as little as 1 day.

The Levels of Certain Proteins in Plasma Increase During Acute Inflammatory States or Secondary to Certain Types of Tissue Damage

These proteins are called **"acute-phase proteins"** (or reactants) and include **C-reactive protein** (CRP, so named because it reacts with the C polysaccharide of pneumococci), α_1-antitrypsin, haptoglobin, α_1-acid glycoprotein, and **fibrinogen.** The elevations of the levels of these proteins vary from as little as 50% to as much as 1000-fold in the case of CRP. Their levels are also usually elevated during chronic inflammatory states and in patients with cancer. These proteins are believed to play a role in the body's response to inflammation. For example, C-reactive protein can stimulate the classic complement pathway, and α_1-antitrypsin can neutralize certain proteases released during the acute inflammatory state. CRP is used as a marker of tissue injury, infection, and inflammation, and there is considerable interest in its use as a predictor of certain types of cardiovascular conditions secondary to atherosclerosis. The cytokine (= a protein made by cells that affects the behavior of other cells) interleukin-1 (IL-1), a polypeptide released from mononuclear phagocytic cells, is the principal—but not the sole—stimulator of the synthesis of the majority of acute phase reactants by hepatocytes. Additional molecules such as IL-6 are involved, and they as well as IL-1 appear to work at the level of gene transcription.

Nuclear factor kappa-B (NFkB) is a transcription factor that has been involved in the stimulation of synthesis of certain of the acute phase proteins. This important factor is also involved in the expression of many cytokines, chemokines, growth factors, and cell adhesion molecules implicated in immunologic phenomena. Normally it exists in an inactive form in the cytosol but is activated and translocated to the nucleus via the action of a number of molecules (eg, interleukin-1) produced in processes such as inflammation, infection, and radiation injury.

Table 50–2 summarizes the functions of many of the plasma proteins. The remainder of the material in this chapter presents basic information regarding selected plasma proteins: albumin, haptoglobin, transferrin, ceruloplasmin, α_1-antitrypsin, α_2-macroglobulin, the immunoglobulins, and the complement system. The lipoproteins are discussed in Chapter 25. New information is constantly forthcoming on plasma proteins and their variants (including those discussed here), as the techniques of proteomics, particularly sensitive new methods of determining proteins sequences by mass spectrometry (see Chapter 4), are applied to their study. A number of laboratories are participating in efforts to determine the complete human plasma protein proteome. It is believed that this will shed further light on **genetic variations** in humans and also provide many new **biomarkers** to aid in the diagnosis of many diseases. (A biomarker has been defined as a characteristic that is objectively measured and evaluated as an indicator of normal biologic processes, pathogenic processes, or pharmacologic responses to a therapeutic intervention.)

Albumin Is the Major Protein in Human Plasma

Albumin (69 kDa) is the major protein of human plasma (3.4–4.7 g/dL) and makes up approximately 60% of the total plasma protein. About 40% of albumin is present in the plasma, and the other 60% is present in the extracellular space. The liver produces about 12 g of albumin per day, representing about 25% of total hepatic protein synthesis and half its secreted protein. Albumin is initially synthesized as a **preproprotein.** Its **signal peptide** is removed as it passes into the cisternae of the rough endoplasmic reticulum, and a **hexapeptide** at the resulting amino terminal is subsequently cleaved off farther along the secretory pathway (see Figure 46–11). The synthesis of albumin is depressed in a variety of diseases, particularly those of the liver. The plasma of patients with **liver disease** often shows a decrease in the ratio of albumin to globulins (decreased albumin-globulin ratio). The synthesis of albumin decreases relatively early in conditions of protein malnutrition, such as kwashiorkor.

Mature human albumin consists of one polypeptide chain of 585 amino acids and contains 17 disulfide bonds. By the use of proteases, albumin can be subdivided into three **domains,** which have different functions. Albumin has an ellipsoidal shape, which means that it does not increase the viscosity of the plasma as much as an elongated molecule such as fibrinogen does. Because of its relatively low molecular mass (about 69 kDa) and high concentration, albumin is thought to be responsible for 75–80% of the **osmotic pressure** of human plasma. Electrophoretic studies have shown that the plasma of certain humans lacks albumin. These subjects are said to exhibit **analbuminemia.** One cause of this condition is a mutation that affects splicing. Subjects with analbuminemia show only moderate edema, despite the fact that albumin is the major determinant of plasma osmotic pressure. It is thought that the amounts of the other plasma proteins increase and compensate for the lack of albumin.

Another important function of albumin is its ability to **bind various ligands.** These include free fatty acids (FFA), calcium, certain steroid hormones, bilirubin, and some of the plasma tryptophan. In addition, albumin appears to play an important role in transport of copper in the human body (see below). A variety of drugs, including sulfonamides, penicillin G, dicumarol, and aspirin, are bound to albumin; this finding has important pharmacologic implications.

Preparations of human albumin have been widely used in the treatment of hemorrhagic shock and of burns. However, some recent studies question the value of this therapy.

TABLE 50–2 Some Functions of Plasma Proteins

Function	Plasma Proteins
Antiproteases	Antichymotrypsin α_1-Antitrypsin (α_1-antiproteinase) α_2-Macroglobulin Antithrombin
Blood clotting	Various coagulation factors, fibrinogen
Enzymes	Function in blood, eg, coagulation factors, cholinesterase Leakage from cells or tissues, eg, aminotransferases
Hormones	Erythropoietin[1]
Immune defense	Immunoglobulins, complement proteins, β_2-microglobulin
Involvement in inflammatory responses	Acute phase response proteins (eg, C-reactive protein, α_1-acid glycoprotein [orosomucoid])
Oncofetal	α_1-Fetoprotein (AFP)
Transport or binding proteins	Albumin (various ligands, including bilirubin, free fatty acids, ions [Ca^{2+}], metals [eg, Cu^{2+}, Zn^{2+}], metheme, steroids, other hormones, and a variety of drugs) Ceruloplasmin (contains Cu^{2+}; albumin probably more important in physiologic transport of Cu^{2+}) Corticosteroid-binding globulin (transcortin) (binds cortisol) Haptoglobin (binds extracorpuscular hemoglobin) Lipoproteins (chylomicrons, VLDL, LDL, HDL) Hemopexin (binds heme) Retinol-binding protein (binds retinol) Sex-hormone-binding globulin (binds testosterone, estradiol) Thyroid-binding globulin (binds T_4, T_3) Transferrin (transport iron) Transthyretin (formerly prealbumin; binds T_4 and forms a complex with retinol-binding protein)

[1]Various other protein hormones circulate in the blood but are not usually designated as plasma proteins. Similarly, ferritin is also found in plasma in small amounts, but it too is not usually characterized as a plasma protein.

Haptoglobin Binds Extracorpuscular Hemoglobin, Preventing Free Hemoglobin from Entering the Kidney

Haptoglobin (Hp) is a plasma glycoprotein that binds extra-corpuscular hemoglobin (Hb) in a tight noncovalent complex (Hb-Hp). The amount of haptoglobin in human plasma ranges from 40 mg to 180 mg of hemoglobin-binding capacity per deciliter. Approximately 10% of the hemoglobin that is degraded each day is released into the circulation and is thus extracorpuscular. The other 90% is present in old, damaged red blood cells, which are degraded by cells of the histiocytic system. The molecular mass of hemoglobin is approximately 65 kDa, whereas the molecular mass of the simplest polymorphic form of haptoglobin (Hp 1-1) found in humans is approximately 90 kDa. Thus, the Hb-Hp complex has a molecular mass of approximately 155 kDa. Free hemoglobin passes through the glomerulus of the kidney, enters the tubules, and tends to precipitate therein (as can happen after a massive incompatible blood transfusion, when the capacity of haptoglobin to bind hemoglobin is grossly exceeded) (Figure 50–3). However, the Hb-Hp complex is too large to pass through the glomerulus. The function of Hp thus appears to be to prevent loss of free hemoglobin into the kidney. This conserves the valuable iron present in hemoglobin, which would otherwise be lost to the body.

Human haptoglobin exists in **three polymorphic forms,** known as Hp 1-1, Hp 2-1, and Hp 2-2. Hp 1-1 migrates in starch gel electrophoresis as a single band, whereas Hp 2-1 and Hp 2-2 exhibit much more complex band patterns. Two genes, designated Hp^1 and Hp^2, direct these three phenotypes, with

Hp 2-1 being the heterozygous phenotype. It has been suggested that the haptoglobin polymorphism may be associated with the prevalence of many inflammatory diseases.

The levels of haptoglobin in human plasma vary and are of some diagnostic use. Low levels of haptoglobin are found in patients with **hemolytic anemias.** This is explained by the fact that whereas the half-life of haptoglobin is approximately 5 days, the half-life of the Hb-Hp complex is about 90 min, the complex being rapidly removed from plasma by hepatocytes. Thus, when haptoglobin is bound to hemoglobin, it is cleared from the plasma about 80 times faster than normally. Accordingly, the level of haptoglobin falls rapidly in situations where hemoglobin is constantly being released from red blood cells, such as occurs in hemolytic anemias. Haptoglobin is an acute phase protein, and its plasma level is elevated in a variety of inflammatory states.

Haptoglobin-related protein is another protein found in human plasma. It bears a high degree of homology to haptoglobin and it appears to bind hemoglobin. Its level is elevated in some patients with cancers, although the significance of this is not understood.

Certain other plasma proteins **bind heme** but not hemoglobin. **Hemopexin** is a β_1-globulin that binds free heme. **Albumin** will bind some metheme (ferric heme) to form methemalbumin, which then transfers the metheme to hemopexin.

Absorption of Iron from the Small Intestine Is Tightly Regulated

Transferrin (Tf) is a plasma protein that plays a central role in transporting iron around the body to sites where it is needed. Before we discuss it further, certain aspects of iron metabolism will be reviewed.

Iron is important in the human body because of its occurrence in many hemoproteins such as hemoglobin, myoglobin, and the cytochromes. It is ingested in the diet either as heme or nonheme iron (Figure 50–4); as shown, these different forms involve separate pathways. Absorption of iron in the proximal duodenum is tightly regulated, as there is no physiologic pathway for its excretion from the body. Under normal circumstances, the body guards its content of iron zealously,

A.   Hb → Kidney → Excreted in urine or precipitates in
(MW 65,000)   tubules; iron is lost to body

B.   Hb   +   Hp → Hb : Hp complex ↮ Kidney
(MW 65,000)   (MW 90,000) │ (MW 155,000)
                          Catabolized by liver cells;
                          iron is conserved and reused

FIGURE 50–3 Different fates of free hemoglobin and of the hemoglobin-haptoglobin complex.

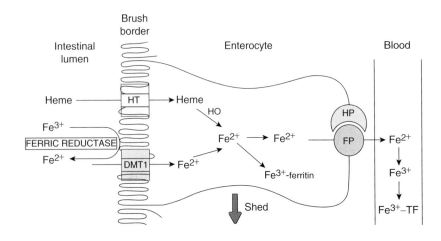

FIGURE 50–4 Absorption of iron. Fe^{3+} is converted to Fe^{2+} by ferric reductase, and Fe^{2+} is transported into the enterocyte by the apical membrane iron transporter DMT1. Heme is transported into the enterocyte by a separate heme transporter (HT), and heme oxidase (HO) releases Fe^{2+} from the heme. Some of the intracellular Fe^{2+} is converted to Fe^{3+} and bound by ferritin. The remainder binds to the basolateral Fe^{2+} transporter (FP) and is transported into the bloodstream, aided by hephaestin (HP). In plasma, Fe^{3+} is bound to the iron transport protein transferrin (TF). (Reproduced, with permission, from Ganong WF: *Review of Medical Physiology,* 21st ed. McGraw-Hill, 2003.)

TABLE 50-3 Distribution of Iron in a 70-kg Adult Male[1]

Transferrin	3–4 mg
Hemoglobin in red blood cells	2500 mg
In myoglobin and various enzymes	300 mg
In stores (ferritin and hemosiderin)	1000 mg
Absorption	1 mg/d
Losses	1 mg/d

[1]In an adult female of similar weight, the amount in stores would generally be less (100–400 mg) and the losses would be greater (1.5–2 mg/d).

so that a healthy adult male loses only about 1 mg/d, which is replaced by absorption. Adult females are more prone to states of iron deficiency because some may lose excessive blood during menstruation. The amounts of iron in various body compartments are shown in Table 50–3.

Enterocytes in the proximal duodenum are responsible for absorption of iron. Incoming iron in the Fe^{3+} state is reduced to Fe^{2+} by a **ferrireductase** present on the surface of enterocytes (Figure 50–4). Vitamin C in food also favors reduction of ferric iron to ferrous iron. The transfer of iron from the apical surfaces of enterocytes into their interiors is performed by a proton-coupled **divalent metal transporter (DMT1).** This protein is not specific for iron, as it can transport a wide variety of divalent cations.

A recently discovered peptide (25 amino acids, synthesized by liver cells) named **hepcidin** appears to play an important role in iron metabolism. It down-regulates the intestinal absorption and placental transfer of iron and also the release of iron from macrophages, possibly by interaction with ferroportin. When plasma levels of iron are high, synthesis of hepcidin increases; the opposite occurs when plasma levels of iron are low. It may play an important role in hereditary hemochromatosis (see case no. 10, Chapter 54) and also in the iron deficiency anemia seen in chronic inflammatory conditions. Another recently discovered protein named **hemojuvelin** may act by modulating the expression of hepcidin.

Once inside an enterocyte, iron can either be stored as **ferritin** or transferred across the basolateral membrane into the plasma, where it is carried by transferrin (see below). Passage across the basolateral membrane appears to be carried out by another protein, **ferroportin.** This protein may interact with the copper-containing protein **hephaestin,** a protein similar to ceruloplasmin (see below). Hephaestin is thought to have a ferroxidase activity, which is important in the release of iron from cells. Thus, Fe^{2+} is converted back to Fe^{3+}, the form in which it is transported in the plasma by transferrin.

The **overall regulation** of iron absorption is complex and not well understood; it appears that **hepcidin** plays a key role. Regulation occurs at the level of the enterocyte, where further absorption of iron is blocked (likely by hepcidin) if a sufficient amount has been taken up (so-called dietary regulation exerted by "mucosal block"). It also appears to be responsive to

the overall requirement of erythropoiesis for iron (erythropoietic regulation). Absorption is excessive in hereditary hemochromatosis (see case no 10, Chapter 54).

Transferrin Shuttles Iron to Sites Where It Is Needed

Transferrin (Tf) is a β_1-globulin with a molecular mass of approximately 76 kDa. It is a glycoprotein and is synthesized in the liver. About 20 polymorphic forms of transferrin have been found. It plays a central role in the body's metabolism of iron because it transports iron (2 mol of Fe^{3+} per mole of Tf) in the circulation to sites where iron is required, eg, from the gut to the bone marrow and other organs. Approximately 200 billion red blood cells (about 20 mL) are catabolized per day, releasing about 25 mg of iron into the body—most of which will be transported by transferrin.

There are **receptors** (**TfR1** and **TfR2**) on the surfaces of many cells for transferrin. It binds to these receptors and is internalized by receptor-mediated endocytosis (compare the fate of LDL; Chapter 25). The acid pH inside the lysosome causes the iron to dissociate from the protein. The dissociated iron leaves the endosome via DMT1 to enter the cytoplasm. Unlike the protein component of LDL, apoTf is not degraded within the lysosome. Instead, it remains associated with its receptor, returns to the plasma membrane, dissociates from its receptor, reenters the plasma, picks up more iron, and again delivers the iron to needy cells. Normally, the iron bound to Tf turns over 10–20 times a day.

Abnormalities of the glycosylation of transferrin occur in the **congenital disorders of glycosylation** (Chapter 47) and in **chronic alcohol abuse.** Their detection by, for example, isoelectric focusing is used to help diagnose these conditions.

Iron Deficiency Anemia Is Extremely Prevalent

Attention to iron metabolism is **particularly important in women** for the reason mentioned above. Additionally, in **pregnancy,** allowances must be made for the growing fetus. **Older people** with poor dietary habits ("tea and toasters") may develop iron deficiency. Iron deficiency anemia due to inadequate intake, inadequate utilization, or excessive loss of iron is one of the most prevalent conditions seen in medical practice.

The concentration of transferrin in plasma is approximately 300 mg/dL. This amount of transferrin can bind 300 µg of iron per deciliter, so that this represents the **total iron-binding capacity** of plasma. However, the protein is normally only one-third saturated with iron. In **iron deficiency anemia,** the protein is even less saturated with iron, whereas in conditions of storage of excess iron in the body (eg, hemochromatosis) the saturation with iron is much greater than one-third.

Ferritin Stores Iron in Cells

Ferritin is another protein that is important in the metabolism of iron. Under normal conditions, it stores iron that can be called upon for use as conditions require. In conditions of excess iron (eg, hemochromatosis), body stores of iron are greatly increased and much more ferritin is present in the tissues, such as the liver and spleen. Ferritin contains approximately 23% iron, and **apoferritin** (the protein moiety free of iron) has a molecular mass of approximately 440 kDa. Ferritin is composed of 24 subunits of 18.5 kDa, which surround in a micellar form some 3000 to 4500 ferric atoms. Normally, there is a little ferritin in human plasma. However, in patients with excess iron, the amount of ferritin in plasma is markedly elevated. The amount of ferritin in plasma can be conveniently measured by a sensitive and specific radioimmunoassay and serves as an index of body iron stores.

Synthesis of the **transferrin receptor (TfR)** and that of **ferritin** are reciprocally linked to cellular iron content. When iron levels are high, ferritin is synthesized to store iron and no further uptake of iron is required, so the TfR is not synthesized. Conversely, when iron levels are low, ferritin is not synthesized and the TfR is synthesized in order to promote uptake of iron from transferrin. The mechanisms involved have been studied in detail. Different untranslated sequences of the mRNAs for both proteins (named **iron-response elements, IREs**) interact with a cytosolic protein sensitive to variations in levels of cellular iron (**IRE-binding protein, IRE-BP**) (see Figure 50–5).

The IREs form hairpin loops on different parts of the two mRNAs. When **iron levels are high** (Ai and Bi in Figure 50–5), iron binds to the IRE-BP as 4Fe-4S cluster, which which prevents the IRE-BP binding to the IREs on the respective mRNAs. The cells thus translate stored ferritin mRNA and synthesize ferritin, while the TfR mRNA is degraded. In contrast, when **levels of iron are low**, iron is not bound significantly to the IRE-BP, and the latter can now bind to the IREs. Under these conditions, the ferritin mRNA is apparently stored as an inactive form, whereas the TfR mRNA is stabilized from degradation, translated and increased synthesis of receptors occurs. Thus, the needs of the cell for iron are satisfied. This is an important example of control of expression of proteins at the **translational** level. One other aspect of the IRE-BP is of interest. When iron levels are high, the protein acts as a cytoplasmic **aconitase** (see Chapter 17); when iron levels drop, the protein undergoes extensive conformational change and can act as an IRE-BP.

Hemosiderin is a somewhat ill-defined molecule; it appears to be a partly degraded form of ferritin but still containing iron. It can be detected by histologic stains (eg, Prussian blue) for iron, and its presence is determined histologically when excessive storage of iron occurs.

Various Laboratory Tests Are Used to Assess and Many Proteins Are Involved in the Metabolism of Iron

Table 50–4 summarizes laboratory tests useful in the assessment of patients with abnormalities of iron metabolism and Table 50–5 lists many of the proteins involved in iron metabolism.

Ceruloplasmin Binds Copper, & Low Levels of This Plasma Protein Are Associated With Wilson Disease

Ceruloplasmin (about 160 kDa) is an α_2-globulin. It has a blue color because of its high copper content and carries 90% of the copper present in plasma. Each molecule of ceruloplasmin binds six atoms of copper very tightly, so that the copper is not readily exchangeable. **Albumin** carries the other ~10% of the plasma copper, but binds the metal less tightly than does ceruloplasmin. Albumin thus donates its copper to tissues more readily than ceruloplasmin and appears to be more important than ceruloplasmin in copper transport in the human

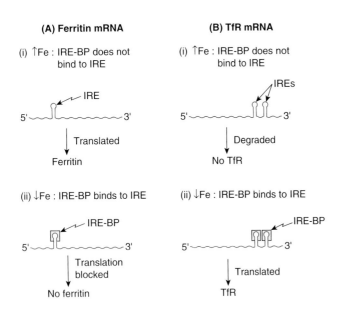

FIGURE 50–5 Schematic representation of the reciprocal relationship between synthesis of ferritin and the transferrin receptor (TfR). The mRNA for ferritin is represented on the left, and that for TfR on the right of the diagram. At high concentrations of iron, the iron bound to the IRE-BP prevents that protein from binding the IREs on either type of mRNA. The mRNA for ferritin is able to be translated under these circumstances, and ferritin is synthesized. On the other hand, when the IRE-BP is not able to bind to the IRE on the mRNA for TfR, that mRNA is degraded. In contrast, at low concentrations of iron, the IRE-BP is able to bind to the IREs on both types of mRNA. In the case of the ferritin mRNA, this prevents it from being translated, and it is apparently stored. In the case of the mRNA for TfR, binding of the IRE-BP prevents that mRNA from being degraded, it is translated, and TfR is synthesized.

TABLE 50–4 Laboratory Tests for Assessing Patients with Disorders of Iron Metabolism

- Red blood cell count and estimation of hemoglobin
- Determinations of plasma iron, total iron-binding capacity (TIBC), and % transferrin saturation
- Determination of ferritin in plasma by radioimmunoassay
- Prussian blue stain of tissue sections
- Determination of amount of iron (μg/g) in a tissue biopsy

TABLE 50–5 Some Proteins Involved in Iron Metabolism

- Ceruloplasmin (ferroxidase activity)
- DMT1
- Ferrireductase (cytochrome *b* reductase I)
- Ferritin
- Ferroportin
- Heme transporter
- Hemojuvelin
- Hepcidin
- Hephaestin
- HFE
- Iron-responsive element-binding protein
- Transferrin
- Transferrin receptors 1 and 2

Note: Further information on most of these proteins is provided in the text or in Chapter 54 (case no. 10) or can be accessed online at OMIM.

TABLE 50–6 Some Important Enzymes That Contain Copper

- Amine oxidase
- Copper-dependent superoxide dismutase
- Cytochrome oxidase
- Tyrosinase

body. Ceruloplasmin exhibits a copper-dependent **oxidase** activity, but its physiologic significance has not been clarified apart from possible involvement in the oxidation of Fe^{2+} in transferrin to Fe^{3+}. The amount of ceruloplasmin in plasma is decreased in liver disease. In particular, low levels of ceruloplasmin are found in **Wilson disease** (hepatolenticular degeneration), a disease due to abnormal metabolism of copper. In order to clarify the description of Wilson disease, we shall first consider **the metabolism of copper** in the human body and then **Menkes disease,** another condition involving abnormal copper metabolism.

Copper Is a Cofactor for Certain Enzymes

Copper is an essential trace element. It is required in the diet because it is the metal cofactor for a variety of enzymes (see Table 50–6). Copper plays important roles in cellular respiration (cytochrome *c* oxidase), iron homeostasis (ceruloplasmin), melanin formation (tyrosinase), neurotransmitter production (various enzymes), synthesis of connective tissue (lysyl oxidase) and protection against oxidants (eg, superoxide dismutase). It accepts and donates electrons and is involved in reactions involving dismutation, hydroxylation, and oxygenation. However, **excess copper** can cause problems because it can oxidize proteins and lipids, bind to nucleic acids, and enhance the production of free radicals. It is thus important to have mechanisms that will maintain the amount of copper in the body within normal limits. The body of the normal adult contains about 100 mg of copper, located mostly in bone, liver, kidney, and muscle. The daily intake of copper is about 2–4 mg, with about 50% being absorbed in the stomach and upper small intestine and the remainder excreted in the feces. Copper is carried to the liver **bound to albumin,** taken up by liver cells, and part of it is excreted in the bile. Copper also leaves the liver attached to **ceruloplasmin,** which is synthesized in that organ.

The Tissue Levels of Copper & of Certain Other Metals Are Regulated in Part by Metallothioneins

Metallothioneins are a group of small proteins (about 6.5 kDa), found in the cytosol of cells, particularly of liver, kidney, and intestine. They have a high content of cysteine and can **bind copper, zinc, cadmium,** and **mercury.** The SH groups of cysteine are involved in binding the metals. Acute intake (eg, by injection) of copper and of certain other metals increases the amount (induction) of these proteins in tissues, as does administration of certain hormones or cytokines. These proteins may function to store the above metals in a nontoxic form and are involved in their overall metabolism in the body. Sequestration of copper also diminishes the amount of this metal available to generate free radicals.

Menkes Disease Is Due to Mutations in the Gene Encoding a Copper-Binding P-Type ATPase

Menkes disease ("kinky" or "steely" hair disease) is a disorder of copper metabolism. It is X-linked, affects only male infants, involves the nervous system, connective tissue, and vasculature, and is usually fatal in infancy. Early diagnosis is important, because injections of copper may be effective if the condition is treated promptly. In 1993, it was reported that the basis of Menkes disease was mutations in the gene (the *ATP7A* gene) for a **copper-binding P-type ATPase** (the ATP7A protein). Interestingly, the enzyme showed structural similarity to certain metal-binding proteins in microorganisms. This ATPase is thought to be responsible for directing the efflux of copper from cells. When altered by mutation, copper is not mobilized normally from the intestine, in which it accumulates, as it does in a variety of other cells and tissues, from which it cannot exit. Despite the accumulation of copper, the activities of many copper-dependent enzymes are decreased, perhaps because of a defect of its incorporation into the apoenzymes. Normal liver expresses very little of the ATPase, which explains the absence of hepatic involvement in Menkes disease. This work led to the suggestion that liver might contain a different copper-binding ATPase, which could be involved in the causation of Wilson disease. As described below, this turned out to be the case.

Wilson Disease Is Also Due to Mutations in a Gene Encoding a Copper-Binding P-Type ATPase

Wilson disease is a genetic disease in which copper fails to be excreted in the bile and accumulates in liver, brain, kidney, and red blood cells. It can be regarded as an inability to maintain a near-zero copper balance, resulting in **copper toxicosis.** The increase of copper in liver cells appears to inhibit the coupling of copper to apoceruloplasmin and leads to low levels of ceruloplasmin in plasma. As the amount of copper accumulates, patients may develop a hemolytic anemia, chronic liver disease (cirrhosis, hepatitis), and a neurologic syndrome owing to accumulation of copper in the basal ganglia and other centers. A frequent clinical finding is the **Kayser-Fleischer ring.** This is a green or golden pigment ring around the cornea due to deposition of copper in Descemet's membrane. The major laboratory tests of copper metabolism are listed in Table 50–7. If Wilson disease is suspected, a **liver biopsy** should be performed; a value for liver copper of over 250 μg/g dry weight along with a plasma level of ceruloplasmin of under 20 mg/dL is diagnostic.

The cause of Wilson disease was also revealed in 1993, when it was reported that a variety of mutations in a gene encoding a **copper-binding P-type ATPase** (ATP7B protein) were responsible. The gene (*ATP7B*) is estimated to encode a protein of 1411 amino acids, which is highly homologous to the product of the gene affected in Menkes disease. In a manner not yet fully explained, a nonfunctional ATPase causes defective excretion of copper into the bile, a reduction of incorporation of copper into apoceruloplasmin, and the accumulation of copper in liver and subsequently in other organs such as brain.

Treatment for Wilson disease consists of a diet low in copper along with lifelong administration of **penicillamine,** which chelates copper, is excreted in the urine, and thus depletes the body of the excess of this mineral.

Another condition involving ceruloplasmin is **aceruloplasminemia.** In this genetic disorder, levels of ceruloplasmin are very low and consequently its ferroxidase activity is markedly deficient. This leads to failure of release of iron from cells, and iron accumulates in certain brain cells, hepatocytes, and pancreatic islet cells. Affected individuals show **severe neurologic signs** and have diabetes mellitus. Use of a chelating agent or administration of plasma or ceruloplasmin concentrate may be beneficial.

TABLE 50–7   Major Laboratory Tests Used in the Investigation of Diseases of Copper Metabolism

Test	Normal Adult Range
Serum copper	10–22 μmol/L
Ceruloplasmin	200–600 mg/L
Urinary copper	<1 μmol/24 h
Liver copper	20–50 μg/g dry weight

Source: Based on Gaw A et al: *Clinical Biochemistry.* Churchill Livingstone, 1995. Copyright © 1995 Elsevier Ltd. Reprinted with permission from Elsevier.

Deficiency of α₁-Antiproteinase (α₁-Antitrypsin) Is Associated with Emphysema & One Type of Liver Disease

α_1-**Antiproteinase** (about 52 kDa) was formerly called α_1-antitrypsin, and this name is retained here. It is a single-chain protein of 394 amino acids, contains three oligosaccharide chains, and is the major component (> 90%) of the α_1 fraction of human plasma. It is synthesized by hepatocytes and macrophages and is the principal **serine protease inhibitor** (**serpin,** or **Pi**) of human plasma. It inhibits trypsin, elastase, and certain other proteases by forming complexes with them. At least 75 **polymorphic forms** occur, many of which can be separated by electrophoresis. The major genotype is MM, and its phenotypic product is PiM. There are two areas of clinical interest concerning α_1-antitrypsin. A deficiency of this protein has a role in certain cases (approximately 5%) of **emphysema.** This occurs mainly in subjects with the **ZZ genotype,** who synthesize PiZ, and also in PiSZ heterozygotes, both of whom secrete considerably less protein than PiMM individuals. Considerably less of this protein is secreted as compared with PiM. When the amount of α_1-antitrypsin is deficient and polymorphonuclear white blood cells increase in the lung (eg, during pneumonia), the affected individual lacks a countercheck to proteolytic damage of the lung by proteases such as elastase (Figure 50–6). It is of considerable interest that a particular **methionine** (residue 358) of α-antitrypsin is involved in its binding to proteases. **Smoking** oxidizes this methionine to methionine sulfoxide and thus inactivates it. As a result, affected molecules of α_1-antitrypsin no longer neutralize proteases. This is particularly devastating in patients (eg, PiZZ phenotype) who already have low levels of α_1-antitrypsin. The further diminution in α_1-antitrypsin brought about by smoking results in increased proteolytic destruction of lung tissue, accelerating the development of emphysema. **Intravenous administration of α_1-antitrypsin** (augmentation therapy) has been used as an adjunct in the treatment of patients with emphysema due to α_1-antitrypsin deficiency. Attempts are being made, using the techniques of protein engineering, to replace methionine 358 by another residue that would not be subject to oxidation. The resulting "mutant" α_1-antitrypsin would thus afford protection against proteases for a much longer period of time than would native α_1-antitrypsin. Attempts are also being made to develop **gene therapy** for this condition. One approach is to use a modified adenovirus (a pathogen of the respiratory tract) into which the gene for α_1-antitrypsin has been inserted. The virus would then be introduced into the respiratory tract (eg, by an aerosol). The hope is that pulmonary epithelial cells would express the gene and secrete α_1-antitrypsin locally. Experiments in animals have indicated the feasibility of this approach.

Deficiency of α_1-antitrypsin is also implicated in one type of **liver disease** (α_1-antitrypsin deficiency liver disease). In this condition, molecules of the **ZZ phenotype** accumulate and aggregate in the cisternae of the endoplasmic reticulum of hepatocytes. Aggregation is due to formation of **polymers** of mutant

A. Active elastase + α_1–AT → Inactive elastase: α_1–AT complex → No proteolysis of lung → No tissue damage

B. Active elastase + ↓ or no α_1–AT → Active elastase → Proteolysis of lung → Tissue damage

FIGURE 50–6 Scheme illustrating **(A)** normal inactivation of elastase by α_1-antitrypsin and **(B)** situation in which the amount of α_1-antitrypsin is substantially reduced, resulting in proteolysis by elastase and leading to tissue damage.

α_1-antitrypsin, the polymers forming via a strong interaction between a specific loop in one molecule and a prominent β-pleated sheet in another (loop-sheet polymerization). By mechanisms that are not understood, **hepatitis** results with consequent **cirrhosis** (accumulation of massive amounts of collagen, resulting in fibrosis). It is possible that administration of a synthetic peptide resembling the loop sequence could inhibit loop-sheet polymerization. Diseases such as α_1-antitrypsin deficiency, in which cellular pathology is primarily caused by the presence of aggregates of aberrant forms of individual proteins, have been named **conformational diseases** (see also Chapter 46). Most appear to be due to the formation of β sheets by conformationally unstable proteins, which in turn leads to formation of aggregates. Other members of this group of conditions include Alzheimer disease, Parkinson disease, and Huntington disease.

At present, severe α_1-antitrypsin deficiency liver disease can be successfully treated by **liver transplantation.** In the future, introduction of the gene for normal α_1-antitrypsin into hepatocytes may become possible, but this would not stop production of the PiZ protein. Figure 50–7 is a scheme of the causation of this disease.

α_2-Macroglobulin Neutralizes Many Proteases & Targets Certain Cytokines to Tissues

α_2-**Macroglobulin** is a large plasma glycoprotein (720 kDa) made up of four identical subunits of 180 kDa. It comprises 8–10% of the total plasma protein in humans. Approximately 10% of the **zinc** in plasma is transported by α_2-macroglobulin,

FIGURE 50–8 An internal cyclic thiol ester bond, as found in α_2-macroglobulin. AA_x and AA_y are neighboring amino acids to cysteine and glutamine.

the remainder being transported by albumin. The protein is synthesized by a variety of cell types, including monocytes, hepatocytes, and astrocytes. It is the major member of a group of plasma proteins that include complement proteins C3 and C4. These proteins contain a unique **internal cyclic thiol ester bond** (formed between a cysteine and a glutamine residue, see Figure 50–8) and for this reason have been designated as the **thiol ester plasma protein family.** This bond is highly reactive and is involved in some of the biologic actions of α_2-macroglobulin.

α_2-Macroglobulin binds many proteinases and is thus an important **panproteinase inhibitor.** The α_2-macroglobulin-proteinase complexes are rapidly cleared from the plasma by a receptor located on many cell types. In addition, α_2-macroglobulin binds many **cytokines** (platelet-derived growth factor, transforming growth factor-β, etc) and appears to be involved in targeting them toward particular tissues or cells. Once taken up by cells, the cytokines can dissociate from α_2-macroglobulin and subsequently exert a variety of effects on cell growth and function. The binding of proteinases and cytokines by α_2-macroglobulin involves different mechanisms that will not be considered here.

Amyloidosis Occurs by the Deposition of Fragments of Various Plasma Proteins in Tissues

Amyloidosis is the accumulation of various insoluble fibrillar proteins between the cells of tissues to an extent that affects function. The accumulation is generally due to either **increased production** of certain proteins or **accumulation of mutated forms** of other proteins (see below). One or more organs or tissues may be affected, and the clinical picture depends on the sites and extent of deposition of amyloid fibrils. The fibrils generally represent proteolytic fragments of various plasma proteins and possess a **β-pleated sheet structure.** The

---

GAG to AAG mutation in exon 5 of gene for α_1-AT on chromosome 14

↓

Results in GLU^{342} to Lys^{342} substitution in α_1-AT, causing formation of PiZZ

↓

PiZZ accumulates in cisternae of endoplasmic reticulum and aggregates via loop-sheet polymerization

↓

Leads to hepatitis (mechanism unknown) and cirrhosis in ~10% ot ZZ homozygotes

FIGURE 50–7 Scheme of causation of α_1-antitrypsin-deficiency liver disease. The mutation shown causes formation of PiZZ (OMIM 107400). (α_1-AT, α_1-antitrypsin.)

term "amyloidosis" is a misnomer, as it was originally thought that the fibrils were starch-like in nature.

Amyloidosis is now generally classified as AX, where A represents amyloidosis and X the protein in the fibrils. However, this system will not be used here. A simple **classification** of amyloidosis is shown in Table 50–8. **Primary** amyloidosis is usually due to a monoclonal plasma cell disorder in which the protein that accumulates is a fragment of a **light chain** (see below) of an immunoglobulin. **Secondary** amyloidosis usually occurs secondary to chronic infections or cancer and is due to accumulation of degradation products of **serum amyloid A** (SAA). Increased synthesis of SAA occurs in chronic inflammatory states due to elevated levels of certain inflammatory cytokines that stimulate the liver to produce more of this protein. **Familial** amyloidosis results from accumulation of mutated forms of certain plasma proteins, particularly **transthyretin** (see Table 50–2). Over 80 mutated forms of this protein have been documented. Other plasma proteins can also accumulate in other rare types of familial amyloidosis. In patients undergoing long-term chronic dialysis, the plasma protein β_2-**microglobulin** can accumulate, because it is retained in the plasma by dialysis membranes. Accumulation of an amyloid type protein is believed to be a crucial factor in the causation of Alzheimer disease (see Case 2, Chapter 54). In all, at least 20 different proteins have been implicated in the different types of amyloidosis. The precise factors that determine the deposition of proteolytic fragments in tissues await elucidation. Amyloid fibrils generally have a **P component** associated with them, which is derived from **serum amyloid P component,** a plasma protein closely related to C-reactive protein. Tissue sections containing amyloid fibrils interact with **Congo red dye** and display striking green birefringence when viewed by polarizing microscopy. Deposition of amyloid occurs in patients with a variety of disorders; **treatment of the underlying disorder** should be provided if possible.

In general, experimental approaches to the treatment of amyloidosis can be considered under three headings: (1) preventing production of the precursor protein; (2) stabilizing the structures of precursor proteins so that they do not convert to β–pleated sheet structures; and (3) destabilizing amyloid fibrils so that they re-convert to their normal conformations. For instance, regarding the third approach, several **small ligands** bind avidly to amyloid fibrils. For example, **iodinated anthra-**cycline binds specifically and with high affinity to all natural amyloid fibrils and promotes their disaggregation in vitro. Another similar approach has been the development of the drug **eprodisate.** Amyloid fibrils bind to glycosaminoglycans (see Chapter 48) in tissues. Eprodisate binds to the GAGs, and thus disrupts the binding of the fibrils to these molecules. It is hoped that molecules affecting any of the three processes just mentioned may prove useful in the treatment of amyloidosis.

PLASMA IMMUNOGLOBULINS PLAY A MAJOR ROLE IN THE BODY'S DEFENSE MECHANISMS

The immune system of the body consists of three major components: **B lymphocytes**, **T lymphocytes** and **the innate immune system.** The B lymphocytes are mainly derived from bone marrow cells in higher animals and from the bursa of Fabricius in birds. The T lymphocytes are of thymic origin. The **B cells** are responsible for the synthesis of circulating, humoral antibodies, also known as **immunoglobulins.** The **T cells** are involved in a variety of important **cell-mediated immunologic processes** such as graft rejection, hypersensitivity reactions, and defense against malignant cells and many viruses. The **innate immune system** defends against infection in a non-specific manner and unlike B cells and T cells is **not adaptive.** It contains a variety of cells such as phagocytes, neutrophils, natural killer cells and others. Case number 1 in Chapter 54 describes one condition in which there is a genetic deficiency of T cells due to mutation in the gene encoding adenosine deaminase. There are a variety of other conditions in which **various components of the immune system are deficient due to mutations.** Most of these are characterized by **recurrent infections**, which must be treated vigorously by, for example, administration of immunoglobulins (if these are deficient) and appropriate antibiotics.

This section considers only the plasma immunoglobulins, which are synthesized mainly in **plasma cells.** These are specialized cells of B cell lineage that synthesize and secrete immunoglobulins into the plasma in response to exposure to a variety of **antigens.**

All Immunoglobulins Contain a Minimum of Two Light & Two Heavy Chains

Immunoglobulins contain a minimum of two identical light (L) chains (23 kDa) and two identical heavy (H) chains (53–75 kDa), held together as a tetramer (L_2H_2) by disulfide bonds. The structure of IgG is shown in Figure 50–9; it is **Y-shaped,** with binding of antigen occurring at both tips of the Y. Each chain can be divided conceptually into specific domains, or regions, that have structural and functional significance. The half of the light (L) chain toward the carboxyl terminal is referred to as the **constant region** (C_L), while the amino terminal half is the **variable region** of the light chain (V_L). Approximately one-quarter of the heavy (H) chain at the amino terminals is referred to as its **variable region** (V_H), and the other three-quarters of the heavy chain are referred to as the **constant**

TABLE 50–8   A Classification of Amyloidosis

Type	Protein Implicated
Primary	Principally light chains of immunoglobulins
Secondary	Serum amyloid A (SAA)
Familial	Transthyretin; also rarely apolipoprotein A-1, cystatin C, fibrinogen, gelsolin, lysozyme
Alzheimer disease	Amyloid β peptide (see Chapter 54, case no. 2)
Dialysis-related	β_2-microglobulin

Note: Proteins other than these listed have also been implicated in amyloidosis.

regions (C_H1, C_H2, C_H3) of that H chain. The portion of the immunoglobulin molecule that binds the specific antigen is formed by the amino terminal portions (variable regions) of both the H and L chains—ie, the V_H and V_L domains. The domains of the protein chains consist of two sheets of antiparallel distinct stretches of amino acids that bind antigen.

As depicted in Figure 50–9, digestion of an immunoglobulin by the enzyme **papain** produces two antigen-binding fragments **(Fab)** and one crystallizable fragment **(Fc)**, which is responsible for functions of immunoglobulins other than direct binding of antigens. Because there are two Fab regions, IgG molecules bind two molecules of antigen and are termed **divalent.** The site on the antigen to which an antibody binds is termed an **antigenic determinant,** or **epitope.** The area in which papain cleaves the immunoglobulin molecule—ie, the region between the C_H1 and C_H2 domains—is referred to as the **"hinge region."** The hinge region confers **flexibility** and allows both Fab arms to move independently, thus helping them to bind to antigenic sites that may be variable distances apart (eg, on bacterial surfaces). Fc and hinge regions differ in the different classes of antibodies, but the overall model of antibody structure for each class is similar to that shown in Figure 50–9 for IgG.

All Light Chains Are Either Kappa or Lambda in Type

There are two general types of light chains, **kappa (κ)** and **lambda (λ),** which can be distinguished on the basis of structural differences in their C_L regions. A given immunoglobulin molecule always contains two κ or two λ light chains—never a mixture of κ and λ. In humans, κ the chains are more frequent than chains in immunoglobulin molecules.

The Five Types of Heavy Chain Determine Immunoglobulin Class

Five classes of H chain have been found in humans (Table 50–9), distinguished by differences in their C_H **regions.** They are designated λ, α, μ δ, and ε. The μ and ε chains each have four C_H domains rather than the usual three. The type of H chain determines the class of immunoglobulin and thus its effector function. There are thus five immunoglobulin classes: **IgG, IgA, IgM, IgD,** and **IgE.** The biologic functions of these five classes are summarized in Table 50–10.

No Two Variable Regions Are Identical

The **variable regions** of immunoglobulin molecules consist of the V_L and V_H domains and are quite heterogeneous. In fact, no two variable regions from different humans have been found to have identical amino acid sequences. However, amino acid analyses have shown that the variable regions are comprised of **relatively invariable regions** and other **hypervariable regions** (Figure 50–10). L chains have three hypervariable regions (in V_L) and H chains have four (in V_H). These **hypervariable regions** comprise the **antigen-binding site** (located at the tips of the Y shown in Figure 50–9) and dictate the amazing specificity of antibodies. For this reason, hypervariable regions are also termed **comple-**

FIGURE 50–9 Structure of IgG. The molecule consists of two light (L) chains and two heavy (H) chains. Each light chain consists of a variable (V_L) and a constant (C_L) region. Each heavy chain consists of a variable region (V_H) and a constant region that is divided into three domains (C_H1, C_H2, and C_H3). The C_H2 domain contains the complement-binding site and the C_H3 domain contains a site that attaches to receptors on neutrophils and macrophages. The antigen-binding site is formed by the hypervariable regions of both the light and heavy chains, which are located in the variable regions of these chains (see Figure 50–10). The light and heavy chains are linked by disulfide bonds, and the heavy chains are also linked to each other by disulfide bonds. (Reproduced, with permission, from Parslow TG et al [editors]: *Medical Immunology,* 10th ed. McGraw-Hill, 2001.)

TABLE 50-9   Properties of Human Immunoglobulins

Property	IgG	IgA	IgM	IgD	IgE
Percentage of total immunoglobulin in serum (approximate)	75	15	9	0.2	0.004
Serum concentration (mg/dL) (approximate)	1000	200	120	3	0.05
Sedimentation coefficient	7S	7S or 11S[1]	19S	7S	8S
Molecular weight (× 1000)	150	170 or 400[1]	900	180	190
Structure	Monomer	Monomer or dimer	Monomer or dimer	Monomer	Monomer
H-chain symbol	γ	A	μ	δ	ε
Complement fixation	+	—	+	—	—
Transplacental passage	+	—	—	?	—
Mediation of allergic responses	—	—	—	—	+
Found in secretions	—	+	—	—	—
Opsonization	+	—	—[2]	—	—
Antigen receptor on B cell	—	—	+	?	—
Polymeric form contains J chain	—	+	+	—	—

Source: Reproduced, with permission, from Levinson W, Jawetz E: *Medical Microbiology and Immunology,* 7th ed. McGraw-Hill, 2002.
[1]The 11S form is found in secretions (eg, saliva, milk, tears) and fluids of the respiratory, intestinal, and genital tracts.
[2]IgM opsonizes indirectly by activating complement. This produces C3b, which is an opsonin.

TABLE 50-10   Major Functions of Immunoglobulins

Immunoglobulin	Major Functions
IgG	Main antibody in the secondary response. Opsonizes bacteria, making them easier to phagocytose. Fixes complement, which enhances bacterial killing. Neutralizes bacterial toxins and viruses. Crosses the placenta.
IgA	Secretory IgA prevents attachment of bacteria and viruses to mucous membranes. Does not fix complement.
IgM	Produced in the primary response to an antigen. Fixes complement. Does not cross the placenta. Antigen receptor on the surface of B cells.
IgD	Found on the surfaces of B cells where it acts as a receptor for antigen.
IgE	Mediates immediate hypersensitivity by causing release of mediators from mast cells and basophils upon exposure to antigen (allergen). Defends against worm infections by causing release of enzymes from eosinophils. Does not fix complement. Main host defense against helminthic infections.

Source: Reproduced, with permission, from Levinson W, Jawetz E: *Medical Microbiology and Immunology,* 7th ed. McGraw-Hill, 2002.

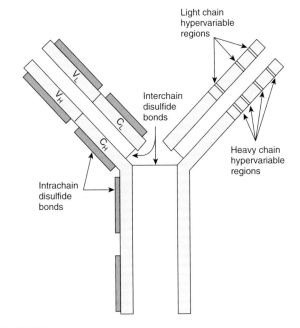

FIGURE 50-10 Schematic model of an IgG molecule showing approximate positions of the hypervariable regions in heavy and light chains. The antigen-binding site is formed by these hypervariable regions. The hypervariable regions are also called complementarity-determining regions (CDRs). (Modified and reproduced, with permission, from Parslow TG et al [editors]: *Medical Immunology,* 10th ed. McGraw-Hill, 2001.)

mentarity-determining regions (CDRs). About five to ten amino acids in each hypervariable region (CDR) contribute to the antigen-binding site. CDRs are located on small loops of the variable domains, the surrounding polypeptide regions between the hypervariable regions being termed **framework regions.** CDRs from both V_H and V_L domains, brought together by folding of the polypeptide chains in which they are contained, form a single hypervariable surface comprising the **antigen-binding site.** Various combinations of H and L chain CDRs can give rise to many antibodies of different specificities, a feature that contributes to the tremendous diversity of antibody molecules and is termed **combinatorial diversity.** Large antigens interact with all of the CDRs of an antibody, whereas small ligands may interact with only one or a few CDRs that form a pocket or groove in the antibody molecule. The essence of antigen-antibody interactions is **mutual complementarity** between the surfaces of CDRs and epitopes. The interactions between antibodies and antigens involve **noncovalent forces and bonds** (electrostatic and van der Waals forces and hydrogen and hydrophobic bonds).

The Constant Regions Determine Class-Specific Effector Functions

The **constant regions** of the immunoglobulin molecules, particularly the C_H2 and C_H3 (and C_H4 of IgM and IgE), which constitute the Fc fragment, are responsible for the **class-specific effector functions** of the different immunoglobulin

molecules (Table 50–9, bottom part), eg, complement fixation or transplacental passage.

Some immunoglobulins such as immune IgG exist only in the basic tetrameric structure, while others such as IgA and IgM can exist as higher order polymers of two, three (IgA), or five (IgM) tetrameric units (Figure 50–11).

The L chains and H chains are synthesized as separate molecules and are subsequently assembled within the B cell or plasma cell into mature immunoglobulin molecules, all of which are **glycoproteins.**

Both Light & Heavy Chains Are Products of Multiple Genes

Each immunoglobulin **light chain** is the product of at least three separate structural genes: a **variable region** (V_L) gene, a **joining region** (**J**) gene (bearing no relationship to the J chain of IgA or IgM), and a **constant region** (C_L) gene. Each **heavy chain** is the product of at least **four** different genes: a **variable region** (V_H) gene, a **diversity region** (**D**) gene, a **joining region** (**J**) gene, and a **constant region** (C_H) gene. Thus, the "one gene, one protein" concept is not valid. The molecular mechanisms responsible for the generation of the single im-

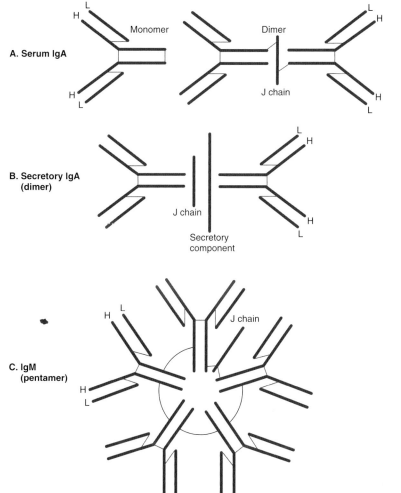

FIGURE 50–11 Schematic representation of serum IgA, secretory IgA, and IgM. Both IgA and IgM have a J chain, but only secretory IgA has a secretory component. Polypeptide chains are represented by thick lines; disulfide bonds linking different polypeptide chains are represented by thin lines. (Reproduced, with permission, from Parslow TG et al [editors]: *Medical Immunology*, 10th ed. McGraw-Hill, 2001.)

munoglobulin chains from multiple structural genes are discussed in Chapters 35 & 38.

Antibody Diversity Depends on Gene Rearrangements

Each person is capable of generating antibodies directed against perhaps 1 million different antigens. The generation of such immense **antibody diversity** depends upon a number of factors including the existence of multiple gene segments (V, C, J, and D segments), their recombinations (see Chapters 35 & 38), the combinations of different L and H chains, a high frequency of somatic mutations in immunoglobulin genes, and **junctional diversity**. The latter reflects the addition or deletion of a random number of nucleotides when certain gene segments are joined together, and introduces an additional degree of diversity. Thus, the above factors ensure that **a vast number of antibodies** can be synthesized from several hundred gene segments.

Class (Isotype) Switching Occurs During Immune Responses

In most humoral immune responses, antibodies with identical specificity but of different classes are generated in a specific chronologic order in response to the immunogen (immunizing antigen). For instance, antibodies of the IgM class normally precede molecules of the IgG class. The switch from one class to another is designated **"class or isotype switching,"** and its molecular basis has been investigated extensively. A single type of immunoglobulin light chain can combine with an antigen-specific μ chain to generate a specific IgM molecule. Subsequently, the same antigen-specific light chain combines with a γ chain with an identical V_H region to generate an IgG molecule with antigen specificity identical to that of the original IgM molecule. The same light chain can also combine with an α heavy chain, again containing the identical V_H region, to form an IgA molecule with identical antigen specificity. These three classes (IgM, IgG, and IgA) of immunoglobulin molecules against the same antigen have **identical variable domains** of both their light (V_L) chains and heavy (V_H) chains and are said to share an **idiotype**. (Idiotypes are the antigenic determinants formed by the specific amino acids in the hypervariable regions.) The **different classes** of these three immunoglobulins (called **isotypes**) are thus determined by their **different C_H** regions, which are combined with the same antigen-specific V_H regions.

Both Over- & Underproduction of Immunoglobulins May Result in Disease States

Disorders of immunoglobulins include **increased production** of specific classes of immunoglobulins or even specific immunoglobulin molecules, the latter by clonal tumors of plasma cells called myelomas. **Multiple myeloma** is a neoplastic condition; electrophoresis of serum or urine will usually reveal a large increase of one particular immunoglobulin or one particular light chain (the latter termed a Bence Jones protein). **Decreased production** may be restricted to a single class of immunoglobulin molecules (eg, IgA or IgG) or may involve underproduction of all classes of immunoglobulins (IgA, IgD, IgE, IgG, and IgM). A severe reduction in synthesis of an immunoglobulin class due to a genetic abnormality can result in a serious immunodeficiency disease—eg, **agammaglobulinemia,** in which production of IgG is markedly affected—because of impairment of the body's defense against microorganisms.

Hybridomas Provide Long-Term Sources of Highly Useful Monoclonal Antibodies

When an antigen is injected into an animal, the resulting antibodies are **polyclonal,** being synthesized by a mixture of B cells. Polyclonal antibodies are directed against a number of different sites (epitopes or determinants) on the antigen and thus are **not monospecific.** However, by means of a method developed by Kohler and Milstein, almost limitless amounts of a single monoclonal antibody specific for one epitope can be obtained.

The method involves **cell fusion,** and the resulting permanent cell line is called a **hybridoma.** Typically, B cells are obtained from the spleen of a mouse (or other suitable animal) previously injected with an antigen or mixture of antigens (eg, foreign cells). The B cells are mixed with mouse **myeloma cells** and exposed to polyethylene glycol, which causes cell fusion. A summary of the principles involved in generating hybridoma cells is given in Figure 50–12. Under the conditions used, only the hybridoma cells multiply in cell culture. This involves plating the hybrid cells into hypoxanthine-aminopterin-thymidine (HAT)-containing medium at a concentration such that each dish contains approximately one cell. Thus, a **clone** of hybridoma cells multiplies in each dish. The culture medium is harvested and screened for antibodies that react with the original antigen or antigens. If the immunogen is a mixture of many antigens (eg, a cell membrane preparation), an individual culture dish will contain a clone of hybridoma cells synthesizing a monoclonal antibody to one specific antigenic determinant of the mixture. By harvesting the media from many culture dishes, a battery of monoclonal antibodies can be obtained, many of which are specific for individual components of the immunogenic mixture. The hybridoma cells can be frozen and stored and subsequently thawed when more of the antibody is required; this ensures its long-term supply. The hybridoma cells can also be grown in the abdomen of mice, providing relatively large supplies of antibodies. Attempts to produce **human** monoclonal antibodies are underway.

Because of their **specificity,** monoclonal antibodies have become extremely **useful reagents** in many areas of biology and medicine. For example, they can be used to measure the amounts of many individual proteins (eg, plasma proteins), to determine the nature of infectious agents (eg, types of bacte-

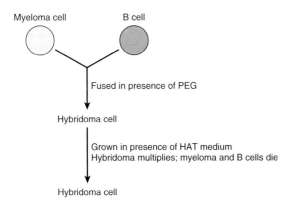

FIGURE 50–12 Scheme of production of a hybridoma cell. The myeloma cells are immortalized, do not produce antibody, and are HGPRT⁻ (rendering the salvage pathway of purine synthesis [see Chapter 33] inactive). The B cells are not immortalized, each produces a specific antibody, and they are HGPRT⁺. Polyethylene glycol (PEG) stimulates cell fusion. The resulting hybridoma cells are immortalized (via the parental myeloma cells), produce antibody, and are HGPRT⁺ (both latter properties gained from the parental B cells). The B cells will die in the medium because they are not immortalized. In the presence of HAT, the myeloma cells will also die, since the aminopterin in HAT suppresses purine synthesis by the de novo pathway by inhibiting reutilization of tetrahydrofolate (see Chapter 33). However, the hybridoma cells will survive, grow (because they are HGPRT⁺), and—if cloned—produce monoclonal antibody. (HAT, hypoxanthine, aminopterin, and thymidine; HGPRT, hypoxanthine-guanine phosphoribosyl transferase.)

ria), and to subclassify both normal (eg, lymphocytes) and tumor cells (eg, leukemic cells). In addition, they are being used to direct therapeutic agents to tumor cells and also to accelerate removal of drugs from the circulation when they reach toxic levels (eg, digoxin).

For **therapeutic use in humans**, monoclonal antibodies made in mice can be **humanized**. This can be achieved by attaching the complementarity-determining regions [CDRs] (the sites that bind antigens) onto appropriate sites in a human immunoglobulin molecule. This produces an antibody that is very similar to a human antibody, thus markedly **lessening immunogenicity** and the chances of an anaphylactic reaction.

The Complement System Comprises Some 20 Plasma Proteins & Is Involved in Cell Lysis, Inflammation, & Other Processes

Plasma contains approximately 20 proteins that are members of **the complement system**. This system was discovered when it was observed that addition of fresh serum containing antibodies directed to a bacterium caused its **lysis**. Unlike antibodies, the factor was **labile** when heated at 56°C. Subsequent work has resolved the proteins of the system and how they function; most have been cloned and sequenced. The complement system is involved in the ability to **lyse** various cells, but also in aspects of **inflammation** (eg, chemotaxis and phagocytosis) and in the **clearance of antigen–antibody complexes**

from the circulation. Deficiencies of various components of the system due to mutations cause **complement deficiency disorders**. The details of this system are relatively complex, and a textbook of immunology should be consulted. The basic concept is that the **normally inactive proteins** of the system, when triggered by a stimulus, become **activated by proteolysis** and interact in a specific sequence with one or more of the other proteins of the system. This results in **cell lysis** and generation of **peptide or polypeptide fragments** that are involved in aspects of inflammation. The complement system resembles blood coagulation (Chapter 51) in that it involves both **conversion of inactive precursors to active products by proteases** and **a cascade with amplification.**

SUMMARY

- Plasma contains many proteins with a variety of functions. Most are synthesized in the liver and are glycosylated.
- Albumin, which is not glycosylated, is the major protein and is the principal determinant of intravascular osmotic pressure; it also binds many ligands, such as drugs and bilirubin.
- Haptoglobin binds extracorpuscular hemoglobin, prevents its loss into the kidney and urine, and hence preserves its iron for reutilization.
- Transferrin binds iron, transporting it to sites where it is required. Ferritin provides an intracellular store of iron. Iron deficiency anemia is a very prevalent disorder. Hereditary hemochromatosis is a genetic disease involving excess absorption of iron; it is discussed in Chapter 54 (Case 10). A number of laboratory tests are available for assessing the status of iron (eg, excess or deficiency) in the human body, and many different proteins are involved in different aspects of its metabolism.
- Ceruloplasmin contains substantial amounts of copper, but albumin appears to be more important with regard to its transport. Both Wilson disease and Menkes disease, which reflect abnormalities of copper metabolism, have been found to be due to mutations in genes encoding copper-binding P-type ATPases.
- α₁-Antitrypsin is the major serine protease inhibitor of plasma, in particular inhibiting the elastase of neutrophils. Genetic deficiency of this protein is a cause of emphysema and can also lead to liver disease.
- α₂-Macroglobulin is a major plasma protein that neutralizes many proteases and targets certain cytokines to specific organs.
- Immunoglobulins play a key role in the defense mechanisms of the body, as do proteins of the complement system. Some of the principal features of these proteins are described.

REFERENCES

Adamson JW: Iron deficiency and other hypoproliferative anemias, Chapter 98. In: Fauci AS et al (editors): *Harrison's Principles of Internal Medicine,* 17th ed. McGraw-Hill, 2008.

Adamson JW, Longo DL: Anemia and polycythemia, Chapter 58. In: Fauci AS et al (editors): *Harrison's Principles of Internal Medicine,* 17th ed. McGraw-Hill, 2008.

Chua AC, Graham RM, Trinder D, Olnyk JK: The regulation of cellular iron metabolism. Crit Rev Clin Lab Sci 2007;44:413.

Craig WY, Ledue TB, Ritchie RF: *Plasma Proteins: Clinical Utility and Interpretation.* Foundation for Blood Research, 2008.

Haynes BF, Soderberg KA, Fauci AS: Introduction to the immune system, Chapter 308. In: Fauci AS et al (editors): *Harrison's Principles of Internal Medicine,* 17th ed. McGraw-Hill, 2008.

Higgins T, Beutler E, Doumas BT: Hemoglobin, iron, and bilirubin, Chapter 31. In: Burtis CA, Ashwood EA, Bruns DE (editors): *Tietz Textbook of Clinical Chemistry and Molecular Diagnostics,* 4th ed. Elsevier Saunders, 2006.

Janeway CA Jr et al: *Immunobiology,* 6th ed. Garland Science Publishing, 2005.

Johnson AM et al: Amino acids, peptides, and proteins, Chapter 20. In: Burtis CA, Ashwood EA, Bruns DE (editors): *Tietz Textbook of Clinical Chemistry and Molecular Diagnostics,* 4th ed. Elsevier Saunders, 2006. (Includes discussions of plasma proteins, complement proteins and immunoglobulins; C-reactive protein is discussed in Chapter 26 of the same text).

Levinson W: *Review of Medical Microbiology and Immunology,* 10th ed. Appleton & Lange, 2008.

Rajkumar SV, Gertz MA: Advances in the treatment of amyloidosis. N Engl J Med 356:2423, 2007.

Schaller H, Gerber S, Kaempfer U et al: *Human Blood Plasma Proteins: Structure and Function.* Wiley, 2008.

Tall AR: C-reactive protein reassessed. N Engl J Med 2004;350:1450.

51

Hemostasis & Thrombosis

Peter L. Gross, MD, Robert K. Murray, MD, PhD
& Margaret L. Rand, PhD

BIOMEDICAL IMPORTANCE

Basic aspects of the proteins of the blood coagulation system and of fibrinolysis are described in this chapter. Some fundamental aspects of platelet biology are also presented. Hemorrhagic and thrombotic states can cause serious medical emergencies, and thromboses in the coronary and cerebral arteries are major causes of death in many parts of the world. Rational management of these conditions requires a clear understanding of the bases of blood coagulation, fibrinolysis, and platelet aggregation.

HEMOSTASIS & THROMBOSIS HAVE THREE COMMON PHASES

Hemostasis is the cessation of bleeding from a cut or severed vessel, whereas thrombosis occurs when the endothelium lining blood vessels is damaged or removed (eg, upon rupture of an atherosclerotic plaque). These processes encompass blood clotting (coagulation) and involve blood vessels, platelet aggregation, and plasma proteins that cause formation or dissolution of platelet aggregates.

In hemostasis, there is initial vasoconstriction of the injured vessel, causing diminished blood flow distal to the injury. Then hemostasis and thrombosis share three phases:

1. Formation of a loose and temporary platelet aggregate at the site of injury. Platelets bind to collagen at the site of vessel wall injury, and form thromboxane A_2 and release ADP, which activate other platelets flowing by the vicinity of the injury. (The mechanism of platelet activation is described below.) Thrombin, formed during coagulation at the same site, causes further platelet activation. Upon activation, platelets change shape and, in the presence of fibrinogen, aggregate to form the hemostatic plug (in hemostasis) or thrombus (in thrombosis).
2. Formation of a fibrin mesh that binds to the platelet aggregate, forming a more stable hemostatic plug or thrombus.
3. Partial or complete dissolution of the hemostatic plug or thrombus by plasmin.

There Are Three Types of Thrombi

Three types of thrombi or clots are distinguished. All three contain **fibrin** in various proportions.

1. The **white** thrombus is composed of platelets and fibrin and is relatively poor in erythrocytes. It forms at the site of an injury or abnormal vessel wall, particularly in areas where blood flow is rapid (arteries).
2. The **red** thrombus consists primarily of red cells and fibrin. It morphologically resembles the clot formed in a test tube and may form in vivo in areas of retarded blood flow or stasis (eg, veins) with or without vascular injury, or it may form at a site of injury or in an abnormal vessel in conjunction with an initiating platelet plug.
3. A third type is a disseminated **fibrin deposit** in very small blood vessels or capillaries.

We shall first describe the coagulation pathway leading to the formation of fibrin. Then we shall briefly describe some aspects of the involvement of platelets and blood vessel walls in the overall process. This separation of clotting factors and platelets is artificial, since both play intimate and often mutually interdependent roles in hemostasis and thrombosis, but it facilitates description of the overall processes involved.

Both Extrinsic & Intrinsic Pathways Result in the Formation of Fibrin

Two pathways lead to fibrin clot formation: the extrinsic and the intrinsic pathways. These pathways are not independent, as previously thought. However, this artificial distinction is retained in the following text to facilitate their description.

Initiation of the fibrin clot in response to tissue injury is carried out by the extrinsic pathway. The intrinsic pathway is activated by negatively charged surfaces in vitro, eg, glass. Both pathways lead to activation of prothrombin to thrombin and the thrombin-catalyzed cleavage of fibrinogen to form the fibrin clot. The pathways are complex and involve many different proteins (Figures 51–1 & 51–2; Table 51–1). In general, as shown in Table 51–2, these proteins can be classified into five types: (1) zymogens of serine-dependent proteases, which become activated during the process of coagulation; (2) co-

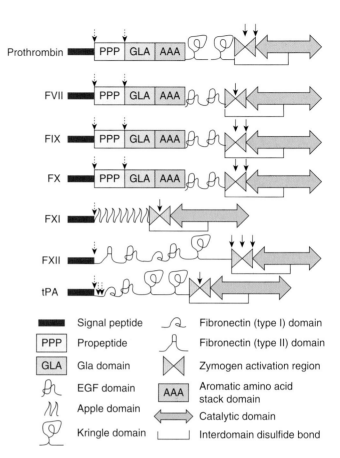

FIGURE 51–1 The pathways of blood coagulation, with the extrinsic pathway indicated at the top left and the intrinsic pathway at the top right. The pathways converge in the activation of factor Xa and culminate in the formation of cross-linked fibrin. Complexes of tissue factor and factor VIIa activate not only factor X (extrinsic Xase (tenase)) but also factor IX in the intrinsic pathway (dotted arrow). In addition, thrombin feedback activates at the sites indicated (dashed arrows); factor Xa feedback activates factor VII to VIIa (not shown). The three predominant complexes, extrinsic Xase, intrinsic Xase and prothrombinase, are indicated in the arrows; the reactions require anionic procoagulant phospholipid membrane and calcium. Activated proteases are in solid-outlined boxes; active cofactors are in dash-outlined boxes and inactive factors are not in boxes. (PK, prekallikrein; HK, HMW kininogen.)

FIGURE 51–2 The structural domains of selected proteins involved in coagulation and fibrinolysis. The domains are as identified at the bottom of the figure, and include signal peptide, propeptide, Gla domain, epidermal growth factor (EGF) domain, apple domain, kringle domain, fibronectin (types I and II) domain, the zymogen activation region, aromatic amino acid stack, and the catalytic domain. Interdomain disulfide bonds are indicated, but numerous intradomain disulfide bonds are not. Sites of proteolytic cleavage in synthesis or activation are indicated by arrows (dashed and solid, respectively). FVII, factor VII; FIX, factor IX; FX, factor X, FXI; factor XI; FXII, factor XII; tPA, tissue plasminogen activator. (Adapted, with permission, from Furie B, Furie BC: The molecular basis of blood coagulation. Cell 1988;53:505.)

factors; (3) fibrinogen; (4) a transglutaminase, which stabilizes the fibrin clot; and (5) regulatory and other proteins.

The Extrinsic Pathway Leads to Activation of Factor X

The extrinsic pathway involves tissue factor, factors VII and X, and Ca^{2+} and results in the production of factor Xa. It is initiated at the site of tissue injury with the exposure of **tissue factor** (Figure 51–1) on activated endothelial cells and monocytes. Tissue factor interacts with and activates factor VII (53 kDa, a zymogen containing vitamin K–dependent γ-carboxyglutamate [Gla] residues; see Chapter 44), synthesized in the liver. It should be noted that in the Gla-containing zymogens (factors II, VII, IX, and X), the Gla residues in the amino terminal regions of the molecules serve as high-affinity binding sites for Ca^{2+}. Tissue factor acts as a cofactor for factor VIIa (by convention, activated clotting factors are referred to by use of the suffix a), enhancing its enzymatic activity to activate factor X. The association of tissue factor and factor VIIa is

called **tissue factor complex.** The reaction by which factor Xa is activated requires the assembly of components, termed the extrinsic tenase complex, on a membrane surface; these components are Ca^{2+}, tissue factor complex, and factor X. Factor VIIa cleaves an Arg-Ile bond in factor X (56 kDa) to produce the two-chain serine protease, factor Xa. Tissue factor and factor VIIa also activate factor IX in the intrinsic pathway. Indeed, **the formation of complexes between tissue factor and factor VIIa is now considered to be the key process involved in initiation of blood coagulation in vivo. Tissue factor pathway inhibitor (TFPI)** is a major physiologic inhibitor of coagulation. It is a protein that circulates in the blood associated with lipoproteins. TFPI directly inhibits factor Xa by binding to the enzyme near its active site. This factor Xa-TFPI complex then inhibits the factor VIIa-tissue factor complex.

TABLE 51–1 Numerical System for Nomenclature of Blood Clotting Factors

Factor	Common Name
I	Fibrinogen ⎫ These factors are usually referred
II	Prothrombin ⎭ to by their common names
III	Tissue factor ⎫ These factors are usually not
IV	Ca²⁺ ⎭ referred to as coagulation factors
V	Proaccelerin, labile factor, accelerator (Ac-) globulin
VII[1]	Proconvertin, serum prothrombin conversion accelerator (SPCA), cothromboplastin
VIII	Antihemophilic factor A, antihemophilic globulin (AHG)
IX	Antihemophilic factor B, Christmas factor, plasma thromboplastin component (PTC)
X	Stuart-Prower factor
XI	Plasma thromboplastin antecedent (PTA)
XII	Hageman factor
XIII	Fibrin stabilizing factor (FSF), fibrinoligase

Note: The numbers indicate the order in which the factors have been discovered and bear no relationship to the order in which they act.
[1]There is no factor VI.

The Intrinsic Pathway Also Leads to Activation of Factor X

The activation of factor Xa is the major site where the intrinsic and extrinsic pathways converge (Figure 51–1). The intrinsic pathway (Figure 51–1) involves factors XII, XI, IX, VIII, and X as well as prekallikrein, high molecular-weight (HMW) kininogen, Ca²⁺, and phospholipid. It results in the production of factor Xa that is cleaved by the tenase complex of the intrinsic pathway. Activation of factor X provides an important link between the intrinsic and extrinsic pathways.

This pathway can be initiated with the "contact phase" in which prekallikrein, HMW kininogen, factor XII, and factor XI are exposed to a negatively charged activating surface. Kaolin can be used for in vitro tests as an initiator of the intrinsic pathway. When the components of the contact phase assemble on the activating surface, factor XII is activated to factor XIIa upon proteolysis by kallikrein. This factor XIIa, generated by kallikrein, attacks prekallikrein to generate more kallikrein, setting up a reciprocal activation. Factor XIIa, once formed, activates factor XI to XIa and also releases bradykinin (a nonapeptide with potent vasodilator action) from HMW kininogen.

Factor XIa in the presence of Ca²⁺ activates factor IX (55 kDa, a Gla-containing zymogen), to the serine protease, factor IXa. This in turn also cleaves an Arg-Ile bond in factor X to produce factor Xa. This latter reaction requires the assembly of components, called **the intrinsic tenase complex,** on a membrane surface: Ca²⁺ and factor VIIIa, as well as factors IXa and X.

Factor VIII (330 kDa), a glycoprotein, is not a protease precursor but a cofactor that serves as a receptor for factors IXa and X on the platelet surface. Factor VIII is activated by minute quantities of thrombin to form factor VIIIa, which is in turn inactivated upon further cleavage by thrombin.

The role of the initial steps of the intrinsic pathway in initiating coagulation has been called into question because patients with a hereditary deficiency of factor XII, prekallikrein or HMW kininogen do not exhibit bleeding problems. Similarly, patients with a deficiency of factor XI may not have bleeding problems. The intrinsic pathway largely serves to amplify factor Xa and ultimately thrombin formation, through feedback mechanisms (see below). The intrinsic pathway may also be important in **fibrinolysis** (see below), since kallikrein, factor XIIa, and factor XIa can cleave plasminogen and kallikrein can activate single chain urokinase.

Factor Xa Leads to Activation of Prothrombin to Thrombin

Factor Xa, produced by either the extrinsic or the intrinsic pathway, activates **prothrombin** (factor II) to **thrombin** (factor IIa), which then converts fibrinogen to fibrin (Figure 51–1).

The activation of prothrombin, like that of factor X, occurs on a membrane surface and requires the assembly of a **prothrombinase complex,** consisting of Ca²⁺, factor Va, factor Xa, and prothrombin. The assembly of the prothrombinase and tenase complexes takes place on the membrane surface of platelets activated to expose the acidic (anionic) phospholipid **phosphatidylserine,** which is normally on the internal side of the plasma membrane of resting, nonactivated platelets.

Factor V (330 kDa), a glycoprotein with homology to factor VIII and ceruloplasmin, is synthesized in the liver, spleen, and kidney and is found in platelets as well as in plasma. It functions as a cofactor in a manner similar to that of factor VIII in the tenase complex. When activated to factor Va by traces of thrombin, it binds specifically to the platelet membrane (Figure 51–3) and forms a complex with factor Xa and prothrombin. It is subsequently inactivated by further action of thrombin, thereby providing a means of limiting the activation of prothrombin to thrombin. **Prothrombin** (72 kDa; Figure 51–3) is a single-chain glycoprotein synthesized in the liver. The amino terminal region of prothrombin (Figure 51–2) contains ten Gla residues, and the serine-dependent active protease site is in the catalytic domain close to the carboxyl terminal region of the molecule. Upon binding to the complex of factors Va and Xa on the platelet membrane (Figure 51–3), prothrombin is cleaved by factor Xa at two sites to generate the active, two-chain thrombin molecule, which is then released from the platelet surface.

Conversion of Fibrinogen to Fibrin Is Catalyzed by Thrombin

Fibrinogen (factor I, 340 kDa; see Figures 51–1 & 51–4; Tables 51–1 & 51–2) is a soluble plasma glycoprotein that consists of three nonidentical pairs of polypeptide chains (Aα,Bβ,γ)₂ covalently linked by disulfide bonds. The Bβ and γ chains contain

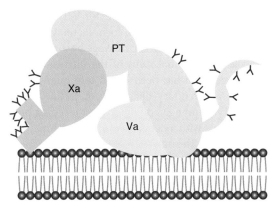

FIGURE 51–3 Diagrammatic representation (not to scale) of the binding of factors Va, Xa and prothrombin (PT) to the plasma membrane of the activated platelet. A central theme in blood coagulation is the assembly of protein complexes on membrane surfaces. Gamma-carboxyglutamate residues (indicated by Y) on vitamin K–dependent proteins bind calcium and contribute to the exposure of membrane binding sites on these proteins. (Adapted, with permission, from Furie B, Furie BC: The molecular basis of blood coagulation. Cell 1988:53:505.)

asparagine-linked complex oligosaccharides. All three chains are synthesized in the liver; the three genes are on the same chromosome, and their expression is coordinately regulated in humans. The amino terminal regions of the six chains are held in close proximity by a number of disulfide bonds, while the carboxyl terminal regions are spread apart, giving rise to a highly asymmetric, elongated molecule (Figure 51–4). The Aα and Bβ portions of the A and B chains, designated **fibrinopeptide A (FPA) and B (FPB),** respectively, at the amino terminal ends of the chains, bear excess negative charges as a result of the presence of aspartate and glutamate residues, as well as an unusual tyrosine O-sulfate in FPB. These negative charges contribute to the solubility of fibrinogen in plasma and also serve to prevent aggregation by causing electrostatic repulsion between fibrinogen molecules.

Thrombin (34 kDa), a serine protease formed by the prothrombinase complex, hydrolyzes the four Arg-Gly bonds between the fibrinopeptides and the α and β portions of the Aα and Bβ chains of fibrinogen (Figure 51–5A). The release of the fibrinopeptides by thrombin generates fibrin monomer, which has the subunit structure $(\alpha, \beta, \gamma)_2$. Since FPA and FPB contain only 16 and 14 residues, respectively, the fibrin molecule retains 98% of the residues present in fibrinogen. The removal

FIGURE 51–4 Diagrammatic representation (not to scale) of fibrinogen showing pairs of Aα, Bβ, and γ chains linked by disulfide bonds. (FPA, fibrinopeptide A; FPB, fibrinopeptide B.)

TABLE 51–2 The Functions of the Proteins Involved in Blood Coagulation

Zymogens of serine proteases	
Factor XII	Binds to negatively charged surface, eg, kaolin, glass; activated by HMW kininogen and kallikrein
Factor XI	Activated by factor XIIa
Factor IX	Activated by factor XIa
Factor VII	Activated by thrombin
Factor X	Activated on surface of activated platelets by tenase complex (Ca^{2+}, factors VIIIa and IXa) and by factor VIIa in presence of tissue factor and Ca^{2+}
Factor II	Activated on surface of activated platelets by prothrombinase complex (Ca^{2+}, factors Va and Xa) [Factors II, VII, IX, and X are Gla-containing zymogens] (Gla = γ-carboxyglutamate)
Cofactors	
Factor VIII	Activated by thrombin; factor VIIIa is a cofactor in the activation of factor X by factor IXa
Factor V	Activated by thrombin; factor Va is a cofactor in the activation of prothrombin by factor Xa
Tissue factor (factor III)	A glycoprotein expressed on the surface of stimulated endothelial cells and monocytes to act as a cofactor for factor VIIa
Fibrinogen	
Factor I	Cleaved by thrombin to form fibrin clot
Thiol-dependent transglutaminase	
Factor XIII	Activated by thrombin; stabilizes fibrin clot by covalent cross-linking
Regulatory and other proteins	
Protein C	Activated to protein Ca by thrombin bound to thrombomodulin; then degrades factors VIIIa and Va
Protein S	Acts as a cofactor of protein C; both proteins contain Gla (γ-carboxyglutamate) residues
Thrombomodulin	Protein on the surface of endothelial cells; binds thrombin, which then activates protein C

of the fibrinopeptides exposes binding sites that allow the molecules of fibrin monomers to aggregate spontaneously in a regularly staggered array, forming an insoluble fibrin clot. It is the formation of this insoluble fibrin polymer that traps platelets, red cells, and other components to form the white or red thrombi. This initial fibrin clot is rather weak, held together only by the noncovalent association of fibrin monomers.

In addition to converting fibrinogen to fibrin, thrombin also converts factor XIII to factor XIIIa. This factor is a highly specific **transglutaminase** that covalently cross-links fibrin molecules by forming peptide bonds between the amide groups of glutamine and the ε-amino groups of lysine residues (Figure 51–5B), yielding a more stable fibrin clot with increased resistance to proteolysis.

A

Thrombin

Fibrinopeptide
(A or B)

Fibrin chain
(α or β)

B

Fibrin — CH₂ — CH₂ — CH₂ — CH₂ — NH₃⁺
(Lysyl)

H₂N — C — CH₂ — CH₂ — Fibrin
(Glutaminyl)

NH₄⁺

Factor XIIIa (Transglutaminase)

Fibrin — CH₂ — CH₂ — CH₂ — CH₂ — NH — C — CH₂ — CH₂ — Fibrin

FIGURE 51–5 Formation of a fibrin clot. **(A)** Thrombin-induced cleavage of Arg-Gly bonds of the Aα and Bβ chains of fibrinogen to produce fibrinopeptides (left-hand side) and the α and β chains of fibrin monomer (right-hand side). **(B)** Cross-linking of fibrin molecules by activated factor XIII (factor XIIIa).

Levels of Circulating Thrombin Must Be Carefully Controlled or Clots May Form

Once active thrombin is formed in the course of hemostasis or thrombosis, its concentration must be carefully controlled to prevent further fibrin formation or platelet activation. This is achieved in two ways. Thrombin circulates as its inactive precursor, prothrombin, which is activated as the result of a cascade of enzymatic reactions, each converting an inactive zymogen to an active enzyme and leading finally to the conversion of prothrombin to thrombin (Figure 51–1). At each point in the cascade, **feedback mechanisms** produce a delicate balance of activation and inhibition. The concentration of factor XII in plasma is approximately 30 μg/mL, while that of fibrinogen is 3 mg/mL, with intermediate clotting factors increasing in concentration as one proceeds down the cascade, showing that the clotting cascade provides amplification. The second means of controlling thrombin activity is the inactivation of any thrombin formed by **circulating inhibitors,** the most important of which is antithrombin (see below).

The Activity of Antithrombin, an Inhibitor of Thrombin, Is Increased by Heparin

Four naturally occurring thrombin inhibitors exist in normal plasma. The most important is **antithrombin,** which contributes approximately 75% of the antithrombin activity. Antithrombin can also inhibit the activities of factors IXa, Xa, XIa, XIIa, and VIIa complexed with tissue factor. α₂-**Macroglobulin** contributes most of the remainder of the antithrombin activity, with **heparin cofactor II** and α₁-**antitrypsin** acting as minor inhibitors under physiologic conditions.

The endogenous activity of antithrombin is greatly potentiated by the presence of sulfated glycosaminoglycans (heparans) (Chapter 48). These bind to a specific cationic site of antithrombin, inducing a conformational change and promoting its binding to thrombin as well as to its other substrates. This is the basis for the use of **heparin,** a derivatized heparan, in clinical medicine to inhibit coagulation. The anticoagulant effects of heparin can be antagonized by strongly cationic polypeptides such as **protamine,** which bind strongly to heparin, thus inhibiting its binding to antithrombin.

Low-molecular-weight heparins (LMWHs), derived from enzymatic or chemical cleavage of unfractionated heparin, are finding increasing clinical use. They can be administered subcutaneously at home, have greater bioavailability than unfractionated heparin, and do not need frequent laboratory monitoring.

Individuals with inherited deficiencies of antithrombin are prone to develop venous thrombosis, providing evidence that antithrombin has a physiologic function and that the coagulation system in humans is normally in a dynamic state.

Thrombin is involved in an additional regulatory mechanism that operates in coagulation. It combines with **thrombomodulin,** a glycoprotein present on the surfaces of endothelial cells. The complex activates **protein C** on the **endothelial protein C receptor.** In combination with **protein S,** activated protein C (APC) degrades factors Va and VIIIa, limiting their actions in coagulation. A genetic deficiency of either protein C or protein S can cause venous thrombosis. Furthermore, patients with **factor V Leiden** (which has a glutamine residue in place of an arginine at position 506) have an increased risk of venous thrombotic disease because factor V Leiden is resistant to inactivation by APC. This condition is termed APC resistance.

Coumarin Anticoagulants Inhibit the Vitamin K–Dependent Carboxylation of Factors II, VII, IX, & X

The coumarin drugs (eg, warfarin), which are used as anticoagulants, inhibit the vitamin K–dependent carboxylation of

Glu to Gla residues (see Chapter 44) in the amino terminal regions of factors II, VII, IX, and X and also proteins C and S. These proteins, all of which are synthesized in the liver, are dependent on the Ca^{2+}-binding properties of the Gla residues for their normal function in the coagulation pathways. The coumarins act by inhibiting the reduction of the quinone derivatives of vitamin K to the active hydroquinone forms (Chapter 44). Thus, the administration of vitamin K will bypass the coumarin-induced inhibition and allow the post-translational modification of carboxylation to occur. Reversal of coumarin inhibition by vitamin K requires 12–24 h, whereas reversal of the anticoagulant effects of heparin by protamine is almost instantaneous.

Heparin and warfarin are widely used in the treatment of thrombotic and thromboembolic conditions, such as deep vein thrombosis and pulmonary embolus. Heparin is administered first, because of its prompt onset of action, whereas warfarin takes several days to reach full effect. Their effects are closely monitored by use of appropriate tests of coagulation (see below) because of the risk of producing hemorrhage.

There Are Several Hereditary Bleeding Disorders, Including Hemophilia A

Inherited deficiencies of the clotting system that result in bleeding are found in humans. The most common is deficiency of factor VIII, causing **hemophilia A,** an X chromosome-linked disease that has played a major role in the history of the royal families of Europe. **Hemophilia B** is due to a deficiency of factor IX; its clinical features are almost identical to those of hemophilia A, but the conditions can be separated on the basis of specific assays that distinguish between the two factors.

The gene for human factor VIII has been cloned and is one of the largest so far studied, measuring 186 kb in length and containing 26 exons. A variety of mutations in the factor VIII and IX genes have been detected leading to diminished activities of the factor VIII and IX proteins; these include partial gene deletions and point and missense mutations. **Prenatal diagnosis** by DNA analysis after chorionic villus sampling is now possible.

In the past, treatment for patients with hemophilia A and B consisted of administration of **cryoprecipitates** (enriched in factor VIII) prepared from individual donors or lyophilized factor VIII or IX **concentrates** prepared from very large plasma pools. It is now possible to prepare factors VIII and IX by **recombinant DNA technology.** Such preparations are free of contaminating viruses (eg, hepatitis A, B, C, or HIV-1) found in human plasma, but are expensive; their use may increase if cost of production decreases.

The most common hereditary bleeding disorder is **von Willebrand disease,** with a prevalence of up to 1% of the population. It results from a deficiency or defect in **von Willebrand factor,** a large multimeric glycoprotein that is secreted by endothelial cells into the plasma, where it stabilizes factor VIII. von Willebrand factor also promotes platelet adhesion at sites of vessel wall injury (see below).

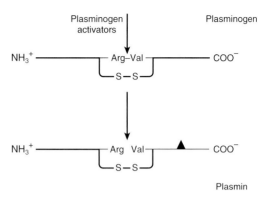

FIGURE 51–6 Activation of plasminogen. The same Arg-Val bond is cleaved by all plasminogen activators to give the two-chain plasmin molecule. The solid triangle indicates the serine residue of the active site. The two chains of plasmin are held together by a disulfide bridge.

Fibrin Clots Are Dissolved by Plasmin

As stated above, the coagulation system is normally in a state of dynamic equilibrium in which fibrin clots are constantly being laid down and dissolved. This latter process is termed **fibrinolysis. Plasmin,** the serine protease mainly responsible for degrading fibrin and fibrinogen, circulates in the form of its inactive zymogen, **plasminogen** (90 kDa), and any small amounts of plasmin that are formed in the fluid phase under physiologic conditions are rapidly inactivated by the fast-acting plasmin inhibitor, α_2-antiplasmin. Plasminogen binds to fibrin and thus becomes incorporated in clots as they are produced; since plasmin that is formed when bound to fibrin is protected from α_2-antiplasmin, it remains active. **Activators of plasminogen** of various types are found in most body tissues, and all cleave the same Arg-Val bond in plasminogen to produce the two-chain serine protease, plasmin (Figure 51–6). The specificity of plasmin for fibrin is another mechanism to regulate fibrinolysis. Via one of its kringle domains, plasmin (ogen) specifically binds lysine residues on fibrin and so is increasingly incorporated into the fibrin mesh as it cleaves it. (Kringle domains [Figure 51–2]) are common protein motifs of about 100 amino acid residues in length, that have a characteristic covalent structure defined by a pattern of 3 disulfide bonds.) Thus, the carboxypeptidase **TAFIa (activated thrombin activatable fibrinolysis inhibitor)** (Figure 51–7), which removes terminal lysines, is able to inhibit fibrinolysis. Thrombin activates TAFI to TAFIa, thereby inhibiting fibrinolysis during clot formation.

Tissue plasminogen activator (t-PA) (Figure 51–2) is a serine protease that is released into the circulation from vascular endothelium under conditions of injury or stress and is catalytically inactive unless bound to fibrin. Upon binding to fibrin, t-PA cleaves plasminogen within the clot to generate plasmin, which in turn digests the fibrin to form soluble degradation products and thus dissolves the clot. Neither plasmin nor the plasminogen activator can remain bound to these degradation products, and so they are released into the fluid phase, where they are inactivated by their natural inhibitors.

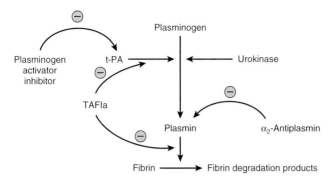

FIGURE 51–7 Initiation of fibrinolysis by the activation of plasmin. Scheme of sites of action of tissue plasminogen activator (t-PA), urokinase, plasminogen activator inhibitor, α_2-antiplasmin and thrombin-activatable fibrinolysis inhibitor (TAFIa) (the last three proteins exert inhibitory actions).

Prourokinase is the precursor of a second activator of plasminogen, **urokinase.** Originally isolated from urine, it is now known to be synthesized by cell types such as monocytes and macrophages, fibroblasts, and epithelial cells. Its main action is probably in the degradation of extracellular matrix. Figure 51–7 indicates the sites of action of five proteins that influence the formation and action of plasmin.

Recombinant t-PA & Streptokinase Are Used as Clot Busters

Alteplase, t-PA produced by recombinant DNA technology, is used therapeutically as a fibrinolytic agent, as is **streptokinase.** However, the latter is less selective than t-PA, activating plasminogen in the fluid phase (where it can degrade circulating fibrinogen) as well as plasminogen that is bound to a fibrin clot. The amount of plasmin produced by therapeutic doses of streptokinase may exceed the capacity of the circulating α_2-antiplasmin, causing fibrinogen as well as fibrin to be degraded and resulting in the bleeding often encountered during fibrinolytic therapy. Because of its relative **selectivity** for degrading fibrin, recombinant t-PA has been widely used to restore the patency of coronary arteries following thrombosis. If administered early enough, before irreversible damage of heart muscle occurs (about 6 h after onset of thrombosis), t-PA can significantly reduce the mortality rate from myocardial damage following coronary thrombosis. Streptokinase has also been widely used in the treatment of coronary thrombosis, but has the disadvantage of being antigenic.

t-PA has also been used in the treatment of ischemic stroke, peripheral arterial occlusion, and pulmonary embolism.

There are a number of disorders, including cancer and sepsis, in which **the concentrations of plasminogen activators increase.** In addition, the antiplasmin activities contributed by α_1-antitrypsin and α_2-antiplasmin may be impaired in diseases such as cirrhosis. Since certain bacterial products, such as streptokinase, are capable of activating plasminogen,

they may be responsible for the diffuse hemorrhage sometimes observed in patients with disseminated bacterial infections.

Activation of Platelets Involves Stimulation of the Polyphosphoinositide Pathway

Platelets normally circulate in an unstimulated disk-shaped form. During hemostasis or thrombosis, they become activated and help form hemostatic plugs or thrombi. Three major steps are involved: (1) adhesion to exposed collagen in blood vessels, (2) release (exocytosis) of the contents of their storage granules, and (3) aggregation.

Platelets adhere to collagen via specific receptors on the platelet surface, including the glycoprotein complexes GPIa–IIa ($\alpha 2\beta 1$ integrin; Chapter 52) and GPIb–IX–V, and GPVI. The binding of GPIb–IX–V to collagen is mediated via von Willebrand factor; this interaction is especially important in platelet adherence to the subendothelium under conditions of high shear stress that occur in small vessels and partially stenosed arteries.

Platelets adherent to collagen change shape and spread out on the subendothelium. They release the contents of their storage granules (the dense granules and the alpha granules); secretion is also stimulated by thrombin.

Thrombin, formed from the coagulation cascade, is the most potent activator of platelets and initiates activation by interacting with its receptors PAR (protease activated receptor)-1, PAR-4, and GPIb–IX–V on the platelet plasma membrane (Figure 51–8A). The further events leading to platelet activation upon binding to PAR-1 and PAR-4 are examples of **transmembrane signaling,** in which a chemical messenger outside the cell generates effector molecules inside the cell. In this instance, thrombin acts as the external chemical messenger (stimulus or agonist). The interaction of thrombin with its G protein–coupled receptors stimulates the activity of an intracellular **phospholipase Cβ.** This enzyme hydrolyzes the membrane phospholipid phosphatidylinositol 4,5-bisphosphate (PIP$_2$, a polyphosphoinositide) to form the two internal effector molecules, 1,2-diacylglycerol and 1,4,5-inositol trisphosphate.

Hydrolysis of PIP$_2$ is also involved in the action of many hormones and drugs. Diacylglycerol stimulates **protein kinase C,** which phosphorylates the protein **pleckstrin** (47 kDa). This results in aggregation and release of the contents of the storage granules. ADP released from dense granules can also activate platelets, resulting in aggregation of additional platelets. IP$_3$ causes release of Ca^{2+} into the cytosol mainly from the dense tubular system (or residual smooth endoplasmic reticulum from the megakaryocyte), which then interacts with calmodulin and myosin light chain kinase, leading to phosphorylation of the light chains of myosin. These chains then interact with actin, causing changes of platelet shape.

Collagen-induced activation of a platelet phospholipase A$_2$ by increased levels of cytosolic Ca^{2+} results in liberation of arachidonic acid from platelet phospholipids, leading to the

A

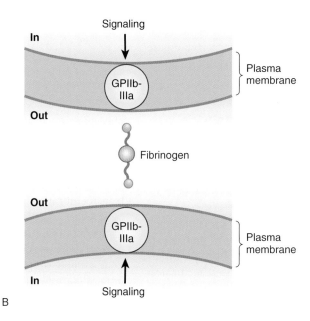

B

FIGURE 51–8 **(A)** Diagrammatic representation of platelet activation by collagen, thrombin, thromboxane A$_2$ and ADP, and inhibition by prostacyclin. The external environment, the plasma membrane, and the inside of a platelet are depicted from top to bottom. *Elevation of Ca^{2+} levels inside the platelet and activation of protein kinase C result in further signaling events, leading to change of platelet shape, release of the contents of the storage granules, and aggregation. (AC, adenylyl cyclase; cAMP, cyclic AMP; COX-1, cyclooxgenase-1; DAG, 1,2-diacylglycerol; GP, glycoprotein; IP, prostacyclin receptor; IP$_3$, inositol 1,4,5-trisphosphate; P2Y$_1$, P2Y$_{12}$, purinoceptors; PAR, protease activated receptor; PIP$_2$, phosphatidylinositol 4,5-bisphosphate; PKC, protein kinase C; PL, phospholipid; PLA$_2$, phospholipase A$_2$; PLCβ, phospholipase Cβ; PLCγ, phospholipase Cγ; TP, thromboxane A$_2$ receptor; TxA$_2$, thromboxane A$_2$; VWF, von Willebrand factor.) The G proteins that are involved are not shown. **(B)** Diagrammatic representation of platelet aggregation mediated by fibrinogen binding to activated GPIIb-IIIa molecules on adjacent platelets. Signaling events initiated by all aggregating agents transform GPIIb-IIIa from its resting state to an activated form that can bind fibrinogen.

formation of **thromboxane A$_2$** (Chapter 23), which in turn, in a G protein–coupled receptor-mediated fashion, can further activate phospholipase C, promoting platelet aggregation.

Activated platelets, besides forming a platelet aggregate, are required, via the newly expressed anionic phospholipid phosphatidylserine on the membrane surface, for acceleration of the activation of coagulation factors X and II (Figure 51–1).

All of the aggregating agents, including thrombin, collagen, ADP, and others such as platelet-activating factor, modify via signaling pathways the platelet surface glycoprotein complex GPIIb-IIIa (αIIbβ3; Chapter 52) so that fibrinogen can bind to it on the activated platelet surface (Figure 51–8B). Molecules of divalent fibrinogen then link adjacent activated platelets to each other, forming a platelet aggregate. Some agents, including epinephrine, serotonin, and vasopressin, exert synergistic effects with other aggregating agents.

Endothelial Cells Synthesize Prostacyclin & Other Compounds That Affect Clotting & Thrombosis

The endothelial cells in the walls of blood vessels make important contributions to the overall regulation of hemostasis and thrombosis. As described in Chapter 23, these cells synthesize **prostacyclin** (PGI_2), a potent inhibitor of platelet aggregation, opposing the action of thromboxane A_2. Prostacyclin acts by stimulating the activity of adenylyl cyclase in the surface membranes of platelets. The resulting increase of intraplatelet cAMP opposes the increase in the level of intracellular Ca^{2+} produced by IP_3 and thus inhibits platelet activation (Figure 51–8). Endothelial cells play other roles in the regulation of thrombosis. For instance, these cells possess an ADPase, which hydrolyzes ADP, and thus opposes its aggregating effect on platelets. In addition, these cells appear to synthesize heparan sulfate, an anticoagulant, and they also synthesize plasminogen activators, which may help dissolve thrombi. Table 51–3 lists some molecules produced by endothelial cells that affect thrombosis and fibrinolysis. Nitric oxide (endothelium-derived relaxing factor) is discussed in Chapter 49.

Analysis of the mechanisms of uptake of atherogenic lipoproteins, such as LDL, by endothelial, smooth muscle, and monocytic cells of arteries, along with detailed studies of how these lipoproteins damage such cells is a key area of study in elucidating the mechanisms of **atherosclerosis** (Chapter 26).

TABLE 51–3 Molecules Synthesized by Endothelial Cells That Play a Role in the Regulation of Thrombosis and Fibrinolysis

Molecule	Action
ADPase (CD39, an ectoenzyme)	Degrades ADP (an aggregating agent of platelets) to AMP + Pi
Nitric oxide (NO)	Inhibits platelet adhesion and aggregation by elevating levels of cGMP
Prostacyclin (PGI_2, a prostaglandin)	Inhibits platelet aggregation by increasing levels of cAMP
Thrombomodulin (a glycoprotein)	Binds protein C, which is then cleaved by thrombin to yield activated protein C; this in combination with protein S degrades factors Va and VIIIa, limiting their actions
Endothelial protein C receptor (EPCR, a glycoprotein)	Facilitates protein C activation by the thrombin-thrombomodulin complex
Tissue plasminogen activator (t-PA, a protease)	Activates plasminogen to plasmin, which digests fibrin; the action of t-PA is opposed by plasminogen activator inhibitor-1 (PAI-1)

Source: Adapted from Wu KK: Endothelial cells in hemostasis, thrombosis and inflammation. Hosp Pract (Off Ed) 1992 Apr; 27:145.

Aspirin Is an Effective Antiplatelet Drug

Certain drugs (antiplatelet drugs) inhibit platelet responses. The most commonly used antiplatelet drug is aspirin (acetylsalicylic acid), which irreversibly acetylates and thus inhibits the platelet cyclooxygenase system (COX-1) involved in formation of thromboxane A_2 (Chapter 15), a potent aggregator of platelets and also a vasoconstrictor. Platelets are very sensitive to aspirin; as little as 30 mg/d (one regular aspirin tablet contains 325 mg) effectively eliminates the synthesis of thromboxane A_2. Aspirin also inhibits production of prostacyclin (PGI_2, which opposes platelet aggregation and is a vasodilator) by endothelial cells, but unlike platelets, these cells regenerate cyclooxygenase within a few hours. Thus, the overall balance between thromboxane A_2 and prostacyclin can be shifted in favor of the latter, opposing platelet aggregation. Indications for treatment with aspirin thus include management of angina, evolving myocardial infarction, transient cerebral ischemic attacks, acute ischemic stroke, severe carotid artery stenosis, and primary prevention of atherothrombotic disease.

Other antiplatelet drugs include clopidogrel, a specific inhibitor of the $P2Y_{12}$ receptor for ADP, and antagonists of ligand binding to GPIIb–IIIa (eg, abciximab) that interfere with fibrinogen binding and thus platelet aggregation.

Laboratory Tests Measure Coagulation, Thrombolysis, & Platelet Aggregation

A number of laboratory tests are available to measure the phases of hemostasis described above. The tests include platelet count, bleeding time, platelet aggregation, activated partial thromboplastin time (aPTT or PTT), prothrombin time (PT), thrombin time (TT), concentration of fibrinogen, fibrin clot stability, and measurement of fibrin degradation products. The platelet count quantitates the number of platelets, the bleeding time is an overall test of platelet and vessel wall function, and platelet aggregation measures responses to specific aggregating agents. aPTT is a measure of the intrinsic pathway and PT of the extrinsic pathway. PT is used to measure the effectiveness of oral anticoagulants such as warfarin, and aPTT is used to monitor heparin therapy. The reader is referred to a textbook of hematology for a discussion of these tests.

SUMMARY

- Hemostasis and thrombosis are complex processes involving coagulation factors, platelets, and blood vessels.
- Many coagulation factors are zymogens of serine proteases, becoming activated during the overall process.
- Both extrinsic and intrinsic pathways of coagulation exist, the former initiated in vivo by tissue factor. The pathways converge at factor Xa, ultimately resulting in thrombin-catalyzed conversion of fibrinogen to fibrin, which is strengthened by covalent cross-linking, catalyzed by factor XIIIa.
- Genetic disorders that lead to bleeding occur; the principal disorders involve factor VIII (hemophilia A), factor IX

(hemophilia B), and von Willebrand factor (von Willebrand disease).

- Antithrombin is an important natural inhibitor of coagulation; genetic deficiency of this protein can result in thrombosis.

- For their activity, factors II, VII, IX, and X and proteins C and S require vitamin K–dependent γ-carboxylation of certain glutamate residues, a process that is inhibited by the anticoagulant warfarin.

- Fibrin is dissolved by plasmin. Plasmin exists as an inactive precursor, plasminogen, which can be activated by tissue plasminogen activator (t-PA). Both t-PA and streptokinase are widely used to treat early thrombosis in the coronary arteries.

- Thrombin and other agents cause platelet aggregation, which involves a variety of biochemical and morphologic events. Stimulation of phospholipase C and the polyphosphoinositide pathway is a key event in platelet activation, but other processes are also involved.

- Aspirin is an important antiplatelet drug that acts by inhibiting production of thromboxane A_2.

REFERENCES

Hoffman R et al (editors): *Hematology: Basic Principles and Practice,* 4th ed. Elsevier Churchill Livingston, 2005.

Israels LG, Israels ED: *Mechanisms in Hematology,* 3rd ed. Core Health Sciences Inc, 2002. (This text has many excellent illustrations of basic mechanisms in hematology.)

Kasper DL et al (editors): *Harrison's Principles of Internal Medicine,* 17th ed. McGraw-Hill, 2008.

Red & White Blood Cells

Robert K. Murray, MD, PhD

BIOMEDICAL IMPORTANCE

Blood cells have been studied intensively because they are obtained easily, because of their functional importance, and because of their involvement in many disease processes. The structure and function of **hemoglobin,** the **porphyrias, jaundice,** and aspects of **iron metabolism** are discussed in previous chapters. Table 52–1 summarizes the causes of a number of important diseases affecting red blood cells; some are discussed in this chapter, and the remainder are discussed elsewhere in this text. **Anemia** is a very prevalent condition with many causes. The **discovery of the causes of certain types of anemias** (eg, of pernicious anemia [a form of B_{12} deficient anemia] and of sickle cell anemia) has been an area where the reciprocal relationship between medicine and biochemistry referred to in Chapter 1 has been extremely beneficial. The World Health Organization (WHO) defines **anemia** as a hemoglobin level of <130 g/L in men and <120g/L in females. There are many causes of anemia; only the most prevalent or biochemically relevant are mentioned here. A simplified classification of the causes of anemia is given in Table 52–2. Certain of the blood group systems, present on the membranes of erythrocytes and other blood cells, are of extreme importance in relation to blood transfusion and tissue transplantation Every organ in the body can be affected by inflammation; neutrophils play a central role in acute inflammation, and other white blood cells, such as lymphocytes, play important roles in chronic inflammation. Leukemias, defined as malignant neoplasms of blood-forming tissues, can affect precursor cells of any of the major classes of white blood cells; common types are acute and chronic myelocytic leukemia, affecting precursors of the neutrophils; and acute and chronic lymphocytic leukemias. Knowledge of the molecular mechanisms involved in the causation of the leukemias is increasing rapidly, but is not discussed in this text. Combination chemotherapy, using combinations of various chemotherapeutic agents, all of which act at one or more biochemical loci, has been remarkably effective in the treatment of certain of these types of leukemias. Understanding the role of red and white cells in health and disease requires a knowledge of certain fundamental aspects of their biochemistry.

ALL BLOOD CELLS DERIVE FROM HEMATOPOIETIC STEM CELLS

Figure 52–1 summarizes the derivation of the various types of blood cells from **hematopoietic stem cells.** The first solid evidence for the existence of stem cells, and in particular hematopoietic stem cells, was reported from studies done in mice by Ernest McCulloch and James Till in 1963. In recent years, interest in stem cells has grown enormously, and they are now of interest to almost every area of medicine and the health sciences. A stem cell is a cell with a unique capacity to produce unaltered daughter cells (ie, **self-renewal**) and to generate specialized cell types (**potency**). Stem cells may be **totipotent** (capable of producing all the cells in an organism), **pluripotent** (able to differentiate into cells of any of the three germ layers), **multipotent** (produce only cells of a closely related family) or **unipotent** (produce only one type of cell). Stem cells are also classified as **embryonic** and **adult;** the latter are more limited in their capabilities to differentiate than the former, although genetic approaches to overcoming this restriction are becoming available.

As shown in Figure 52–1, **red blood cells** and **platelets** share a common pathway of differentiation until the stage of megakaryocyte erythroid progenitors. Cells of **lymphoid origin** branch off at the stage of multipotent progenitors, and **other white blood cells** at the stage of the common myeloid progenitors. Each pathway is **regulated** by various factors (eg, stem cell factor, thrombopoietin, various interleukins, erythropoietin, etc), and key specific **transcription factors** (not indicated in the Figure) are also involved at the stages indicated.

Stem cell factor is a cytokine that plays an important role in the proliferation of hematopoietic stem cells and some of their progeny. **Thrombopoietin** is a glycoprotein that is important in regulating the production of platelets by the bone marrow. **Interleukins** are cytokines produced by leukocytes; they regulate various aspects of hematopoiesis and of the immune system.

THE RED BLOOD CELL IS SIMPLE IN TERMS OF ITS STRUCTURE & FUNCTION

The **major functions of the red blood cell** are relatively simple, consisting of delivering oxygen to the tissues and of helping in

TABLE 52–1   Summary of the Causes of Some Important Disorders Affecting Red Blood Cells

Disorder	Sole or Major Cause
Iron deficiency anemia	Inadequate intake or excessive loss of iron
Methemoglobinemia	Intake of excess oxidants (various chemicals and drugs) Genetic deficiency in the NADH-dependent methemoglobin reductase system (OMIM 250800) Inheritance of HbM (OMIM 141900)
Sickle cell anemia (OMIM 603903)	Sequence of codon 6 of the β chain changed from GAG in the normal gene to GTG in the sickle cell gene, resulting in substitution of valine for glutamic acid
α-Thalassemias (OMIM 141800)	Mutations in the α-globin genes, mainly unequal crossing-over and large deletions and less commonly nonsense and frameshift mutations
β-Thalassemia (OMIM 141900)	A very wide variety of mutations in the β-globin gene, including deletions, nonsense and frameshift mutations, and others affecting every aspect of its structure (eg, splice sites, promoter mutants)
Megaloblastic anemias Deficiency of vitamin B_{12}	Decreased absorption of B_{12}, often due to a deficiency of intrinsic factor, normally secreted by gastric parietal cells
Deficiency of folic acid	Decreased intake, defective absorption, or increased demand (eg, in pregnancy) for folate
Hereditary spherocytosis[1] (OMIM 182900)	Deficiencies in the amount or in the structure of α or β spectrin, ankyrin, band 3 or band 4.1
Glucose-6-phosphate dehydrogenase (G6PD) deficiency[1] (OMIM 305900)	A variety of mutations in the gene (X-linked) for G6PD, mostly single point mutations
Pyruvate kinase (PK) deficiency[1] (OMIM 266200)	A variety of mutations in the gene for the R (red cell) isozyme of PK
Paroxysmal nocturnal hemoglobinuria[1] (OMIM 311770)	Mutations in the PIG-A gene, affecting synthesis of GPI-anchored proteins

[1]The last four disorders cause hemolytic anemias, as do a number of the other disorders listed. Most of the above conditions are discussed in other chapters of this text. OMIM numbers apply only to disorders with a genetic basis.

the disposal of carbon dioxide and protons formed by tissue metabolism. Thus, it has a much simpler structure than most human cells, being essentially composed of a membrane surrounding a solution of hemoglobin (this protein forms about 95% of the intracellular protein of the red cell). There are no intracellular organelles, such as mitochondria, lysosomes, or Golgi apparatus. Human red blood cells, like most red cells of animals, are nonnucleated. However, the red cell is not metabolically inert. **ATP** is synthesized from **glycolysis** and is im-

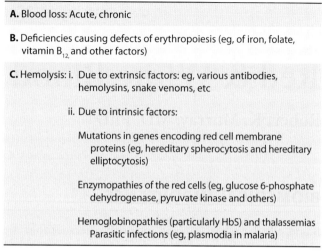

TABLE 52–2   A Brief Classification of the Causes of Anemia

A. Blood loss: Acute, chronic

B. Deficiencies causing defects of erythropoiesis (eg, of iron, folate, vitamin B_{12}, and other factors)

C. Hemolysis: i. Due to extrinsic factors: eg, various antibodies, hemolysins, snake venoms, etc

ii. Due to intrinsic factors:

Mutations in genes encoding red cell membrane proteins (eg, hereditary spherocytosis and hereditary elliptocytosis)

Enzymopathies of the red cells (eg, glucose 6-phosphate dehydrogenase, pyruvate kinase and others)

Hemoglobinopathies (particularly HbS) and thalassemias Parasitic infections (eg, plasmodia in malaria)

Note: Figure 52–3 also indicates the causes of hemolytic anemias. In anemia, the red cells may be **larger** than normal (macrocytes, as in folate and B12 deficiencies, of **normal size** (normocytes, as in acute blood loss or bone marrow failure) or **smaller** than normal (microcytes, as in iron deficiency anemia). They may also **stain** more intensely than usual (hyperchromic), normally (normochromic) or paler than usual (hypochromic). These differences in intensity of staining qualitatively reflect higher, normal or lower contents of hemoglobin.

portant in processes that help the red blood cell maintain its biconcave shape and also in the regulation of the **transport of ions** (eg, by the Na^+-K^+ ATPase and the anion exchange protein [see below]) and of **water** in and out of the cell. The biconcave shape increases the surface-to-volume ratio of the red blood cell, thus facilitating gas exchange. The red cell contains cytoskeletal components (see below) that play an important role in determining its shape.

About Two Million Red Blood Cells Enter the Circulation per Second

The **life span** of the normal red blood cell is 120 days; this means that slightly less than 1% of the population of red cells (200 billion cells, or 2 million per second) is replaced daily. The new red cells that appear in the circulation still contain ribosomes and elements of the endoplasmic reticulum. The RNA of the ribosomes can be detected by suitable stains (such as cresyl blue), and cells containing it are termed reticulocytes; they normally number about 1% of the total red blood cell count. The life span of the red blood cell can be dramatically shortened in a variety of **hemolytic anemias.** The number of reticulocytes is markedly increased in these conditions, as the bone marrow attempts to compensate for rapid breakdown of red blood cells by increasing the amount of new, young red cells in the circulation.

Erythropoietin Regulates Production of Red Blood Cells

Human **erythropoietin (EPO)** is a glycoprotein of 166 amino acids (molecular mass about 34 kDa). Its amount in plasma

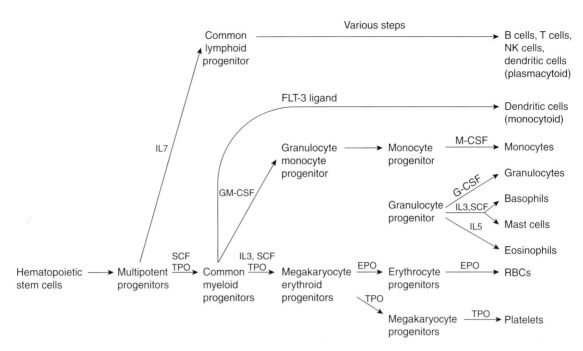

FIGURE 52–1 Simplified scheme of differentiation of red blood cells and other blood cells from the hematopoietic stem cell. Sites of action of interleukins (IL-7, IL-3, and IL-5), stem cell factor (SCF), thrombopoietin (TPO), FLT-3 ligand (a growth factor), granulocyte-macrophage colony-stimulating factor (GM-CSF), erythropoietin (EPO), monocyte colony-stimulating factor (M-CSF), and granulocyte colony-stimulating factor (G-CSF) are shown. Sites of action of important transcription factors are not shown. Various steps in the development of lymphoid cells (top part of the Figure) have been omitted and abbreviated to one step. (Modified, with permission, from Scadden DT, Longo DL in Fauci AS et al [editors], *Harrison's Principles of Internal Medicine,* 17th ed, Chapter 68, McGraw-Hill, 2008.)

can be measured by radioimmunoassay. It is the major regulator of human erythropoiesis (Figure 52–1). As shown in the figure, earlier stages in the development of red blood cells involve stem cell factor, thrombopoietin and interleukin-3. EPO is synthesized mainly by the kidney and is released in response to hypoxia into the bloodstream, in which it travels to the bone marrow. There it interacts with progenitors of red blood cells via a specific receptor. The receptor is a transmembrane protein consisting of two different subunits and a number of domains. It is not a tyrosine kinase, but it stimulates the activities of specific members of this class of enzymes involved in downstream signal transduction.

The availability of a cDNA for EPO has made it possible to produce substantial amounts of this hormone for analysis and for therapeutic purposes; previously the isolation of erythropoietin from human urine yielded very small amounts of the protein. The major use of **recombinant EPO** has been in the treatment of a small number of **anemic states,** such as that due to renal failure. As described in Chapter 47, attempts have been made to prolong the half-life of EPO (thus lengthening its activity) in the circulation by altering the nature of its sugar chains.

MANY GROWTH FACTORS REGULATE PRODUCTION OF WHITE BLOOD CELLS

A large number of **hematopoietic growth factors** have been identified in recent years in addition to erythropoietin. This area

of study adds to knowledge about the differentiation of blood cells, provides factors that may be useful in treatment, and also has implications for understanding of the abnormal growth of blood cells (eg, the leukemias). Like erythropoietin, most of the growth factors isolated have been glycoproteins, are very active in vivo, and in vitro interact with their target cells via specific cell surface receptors, and ultimately (via intracellular signals) affect gene expression, thereby promoting differentiation. Many have been cloned, permitting their production in relatively large amounts. Two of particular interest are **granulocyte-** and **granulocyte-macrophage colony-stimulating factors** (G-CSF and GM-CSF, respectively). As indicated in Figure 52–1, G-CSF is relatively specific, inducing mainly granulocytes, whereas GM-CSF induces a wider variety of white blood cells. When the production of neutrophils is severely depressed, this condition is referred to as **neutropenia.** It is particularly likely to occur in patients treated with certain chemotherapeutic regimens and after bone marrow transplantation. These patients are liable to develop overwhelming infections. G-CSF has been administered to such patients to boost production of neutrophils.

THE RED BLOOD CELL HAS A UNIQUE & RELATIVELY SIMPLE METABOLISM

Various aspects of the **metabolism of the red cell,** many of which are discussed in other chapters of this text, are summarized in Table 52–3.

TABLE 52–3   Summary of Important Aspects of the Metabolism of the Red Blood Cell

- The RBC is highly dependent upon glucose as its energy source; its membrane contains high affinity glucose transporters.

- Glycolysis, producing lactate, is the site of production of ATP.

- Because there are no mitochondria in RBCs, there is no production of ATP by oxidative phosphorylation.

- The RBC has a variety of transporters that maintain ionic and water balance.

- Production of 2,3-bisphosphoglycerate, by reactions closely associated with glycolysis, is important in regulating the ability of Hb to transport oxygen.

- The pentose phosphate pathway is operative in the RBC (it metabolizes about 5–10% of the total flux of glucose) and produces NADPH; hemolytic anemia due to a deficiency of the activity of glucose-6-phosphate dehydrogenase is common.

- Reduced glutathione (GSH) is important in the metabolism of the RBC, in part to counteract the action of potentially toxic peroxides; the RBC can synthesize GSH and requires NADPH to return oxidized glutathione (G-S-S-G) to the reduced state.

- The iron of Hb must be maintained in the ferrous state; ferric iron is reduced to the ferrous state by the action of an NADH-dependent methemoglobin reductase system involving cytochrome b_5 reductase and cytochrome b_5.

- Synthesis of glycogen, fatty acids, protein, and nucleic acids does not occur in the RBC; however, some lipids (eg, cholesterol) in the red cell membrane can exchange with corresponding plasma lipids.

- The RBC contains certain enzymes of nucleotide metabolism (eg, adenosine deaminase, pyrimidine nucleotidase, and adenylyl kinase); deficiencies of these enzymes are involved in some cases of hemolytic anemia.

- When RBCs reach the end of their life span, the globin is degraded to amino acids (which are reutilized in the body), the iron is released from heme and also reutilized, and the tetrapyrrole component of heme is converted to bilirubin, which is mainly excreted into the bowel via the bile.

TABLE 52–4   Some Properties of the Glucose Transporter of the Membrane of the Red Blood Cell (GLUT1)

- It accounts for about 2% of the protein of the membrane of the RBC.

- It exhibits specificity for glucose and related D-hexoses (L-hexoses are not transported).

- The transporter functions at approximately 75% of its V_{max} at the physiologic concentration of blood glucose, is saturable and can be inhibited by certain analogs of glucose.

- Some twelve similar but distinct glucose transporters have been detected to date in mammalian tissues, of which the red cell transporter is one.

- It is not dependent upon insulin, unlike the corresponding carrier in muscle and adipose tissue.

- Its complete amino acid sequence (492 amino acids) has been determined.

- It transports glucose when inserted into artificial liposomes.

- It is estimated to contain 12 transmembrane helical segments.

- It functions by generating a gated pore in the membrane to permit passage of glucose; the pore is conformationally dependent on the presence of glucose and can oscillate rapidly (about 900 times/s).

Reticulocytes Are Active in Protein Synthesis

The mature red blood cell cannot synthesize protein. **Reticulocytes** are active in protein synthesis. Once reticulocytes enter the circulation, they lose their intracellular organelles (ribosomes, mitochondria, etc) within about 24 h, becoming young red blood cells and concomitantly losing their ability to synthesize protein. Extracts of rabbit reticulocytes (obtained by injecting rabbits with a chemical—phenylhydrazine—that causes a severe hemolytic anemia, so that the red cells are almost completely replaced by reticulocytes) are widely used as an in vitro system for synthesizing proteins. Endogenous mRNAs present in these reticulocytes are destroyed by use of a nuclease, whose activity can be inhibited by addition of Ca^{2+}. The system is then programmed by adding purified mRNAs or whole-cell extracts of mRNAs, and radioactive proteins are synthesized in the presence of ^{35}S-labeled L-methionine or other radiolabeled amino acids. The radioactive proteins synthesized are separated by SDS-PAGE and detected by radioautography.

The Red Blood Cell Has a Glucose Transporter in Its Membrane

The entry rate of glucose into red blood cells is far greater than would be calculated for simple diffusion. Rather, it is an example of **facilitated diffusion** (Chapter 40). The specific protein involved in this process is called the **glucose transporter** (GLUT1) or glucose permease. Some of its properties are summarized in Table 52–4. The process of entry of glucose into red blood cells is of major importance because it is the major fuel supply for these cells. About twelve different but related glucose transporters have been isolated from various human tissues; unlike the red cell transporter, some of these are insulin-dependent (eg, in muscle and adipose tissue). There is considerable interest in the latter types of transporter because defects in their recruitment from intracellular sites to the surface of skeletal muscle cells may help explain the **insulin resistance** displayed by patients with type 2 diabetes mellitus.

Superoxide Dismutase, Catalase, & Glutathione Protect Blood Cells from Oxidative Stress & Damage

Several powerful **oxidants** are produced during the course of metabolism, in both blood cells and most other cells of the body. These include superoxide ($O_2^{-\cdot}$), hydrogen peroxide (H_2O_2), peroxyl radicals (ROO$^\cdot$), and hydroxyl radicals (OH$^\cdot$)

TABLE 52–5   Reactions of Importance in Relation to Oxidative Stress in Blood Cells and Various Tissues

1. Production of superoxide (by-product of various reactions)	$O_2 + e^- \rightarrow O_2^{\cdot-}$
2. NADPH oxidase	$2\,O_2 + NADPH \rightarrow 2\,O_2^{\cdot-} + NADP + H^+$
3. Superoxide dismutase	$O_2^{\cdot-} + O_2^{\cdot-} + 2\,H^+ \rightarrow H_2O_2 + O_2$
4. Catalase	$H_2O_2 \rightarrow 2\,H_2O + O_2$
5. Myeloperoxidase	$H_2O_2 + X^- + H^+ \rightarrow HOX + H_2O$ ($X^- = Cl^-, Br^-, SCN^-$)
6. Glutathione peroxidase (Se-dependent)	$2\,GSH + R\text{-}O\text{-}OH \rightarrow GSSG + H_2O + ROH$
7. Fenton reaction	$Fe^{2+} + H_2O_2 \rightarrow Fe^{3+} + OH^{\cdot} + OH^-$
8. Iron-catalyzed Haber-Weiss reaction	$O_2^{\cdot-} + H_2O_2 \rightarrow O_2 + OH^{\cdot} + OH^-$
9. Glucose-6-phosphate dehydrogenase (G6PD)	$G6P + NADP \rightarrow 6\,Phosphogluconate + NADPH + H^+$
10. Glutathione reductase	$G\text{-}S\text{-}S\text{-}G + NADPH + H^+ \rightarrow 2\,GSH + NADP$

and are referred to as **reactive oxygen species (ROS). Free radicals** are atoms or groups of atoms that have an unpaired electron (see Chapters 15 & 45). OH· is a particularly reactive molecule and can react with proteins, nucleic acids, lipids, and other molecules to alter their structure and produce tissue damage. The reactions listed in Table 52–5 play an important role in forming these oxidants and in disposing of them; each of these reactions will now be considered in turn.

Superoxide is formed (reaction 1) in the red blood cell by the auto-oxidation of hemoglobin to methemoglobin (approximately 3% of hemoglobin in human red blood cells has been calculated to auto-oxidize per day); in other tissues, it is formed by the action of enzymes such as cytochrome P450 reductase and xanthine oxidase. When stimulated by contact with bacteria, **neutrophils** exhibit a **respiratory burst** (see below) and produce superoxide in a reaction catalyzed by NADPH oxidase (reaction 2). Superoxide spontaneously dismutates to form H_2O_2 and O_2; however, the rate of this same reaction is speeded up tremendously by the action of the enzyme **superoxide dismutase** (reaction 3). **Hydrogen peroxide** is subject to a number of fates. The enzyme **catalase,** present in many types of cells, converts it to H_2O and O_2 (reaction 4). Neutrophils possess a unique enzyme, **myeloperoxidase,** that uses H_2O_2 and halides to produce hypohalous acids (reaction 5); this subject is discussed further below. The selenium-containing enzyme **glutathione peroxidase** (Chapter 21) will also act on reduced glutathione (GSH) and H_2O_2 to produce oxidized glutathione (GSSG) and H_2O (reaction 6); this enzyme can also use other peroxides as substrates. OH· and OH$^-$ can be formed from H_2O_2 in a nonenzymatic reaction catalyzed by Fe^{2+} (the **Fenton reaction,** reaction 7). $O_2^{\cdot-}$ and and H_2O_2 are the substrates in the iron-catalyzed **Haber-Weiss reaction** (reaction 8), which also produces OH· and OH$^-$. Superoxide can release iron ions from ferritin. Thus, production of OH· may be one of the mechanisms involved in tissue injury due to iron overload in hemochromatosis (see case no. 10 Chapter 54).

Chemical compounds and reactions capable of generating potential toxic oxygen species can be referred to as **pro-oxidants.** On the other hand, compounds and reactions disposing of these species, scavenging them, suppressing their formation, or opposing their actions are **antioxidants** and include compounds such as NADPH, GSH, ascorbic acid, and vitamin E. In a normal cell, there is an appropriate pro-oxidant: antioxidant balance. However, this balance can be shifted toward the pro-oxidants when production of oxygen species is increased greatly (eg, following ingestion of certain chemicals or drugs) or when levels of antioxidants are diminished (eg, by inactivation of enzymes involved in disposal of oxygen species and by conditions that cause low levels of the antioxidants mentioned above). This state is called **"oxidative stress"** (see Chapter 45) and can result in serious cell damage if the stress is massive or prolonged.

ROS are now thought to play an important role in many types of **cellular injury** (eg, resulting from administration of various toxic chemicals or from ischemia), some of which can result in cell death. Indirect evidence supporting a role for these species in generating cell injury is provided if administration of an enzyme such as superoxide dismutase or catalase is found to protect against cell injury in the situation under study.

Deficiency of Glucose-6-Phosphate Dehydrogenase Is Frequent in Certain Areas & Is an Important Cause of Hemolytic Anemia

NADPH, produced in the reaction catalyzed by the X-linked glucose-6-phosphate dehydrogenase (Table 52–5, reaction 9) in the **pentose phosphate pathway** (Chapter 21), plays a key role in supplying reducing equivalents in the red cell and in other cells such as the hepatocyte. Because the pentose phosphate pathway is virtually its sole means of producing NADPH, the red blood cell is very sensitive to oxidative damage if the function of this pathway is impaired (eg, by enzyme deficiency). One function of NADPH is to reduce GSSG to GSH, a reaction catalyzed by glutathione reductase (reaction 10).

Deficiency of the activity of **glucose-6-phosphate dehydrogenase,** owing to mutation, is extremely frequent in some regions of the world (eg, tropical Africa, the Mediterranean, certain parts of Asia, and in North America among blacks). It is the most common of all enzymopathies (diseases caused by abnormalities of enzymes), and some 140 genetic variants of the enzyme have been distinguished; at least 400 million people are estimated to have a variant gene. It is thought that an abnormal form of this enzyme confers resistance to malaria. The disorder resulting from deficiency of glucose-6-phosphate dehydrogenase is **hemolytic anemia.** When an abnormal form of an enzyme causes pathology, that is referred to as an **enzymopathy.** Consumption of broad beans (*Vicia faba*) by individuals deficient in activity of the enzyme can precipitate an acute attack of hemolytic anemia because they contain potential oxidants. In addition, a number of drugs (eg, the antimalarial drug **primaquine** [the condition caused by intake of primaquine is called **primaquine-sensitive hemolytic anemia**] and **sulfonamides**) and chemicals (eg, naphthalene) precipitate an attack, because their intake leads to generation of H_2O_2 or O_2^-. Normally, H_2O_2 is disposed of by **catalase** and **glutathione peroxidase** (Table 52–5, reactions 4 and 6), the latter causing increased production of GSSG. GSH is regenerated from GSSG by the action of the enzyme **glutathione reductase,** which depends on the availability of NADPH (reaction 10). The red blood cells of individuals who are deficient in the activity of glucose-6-phosphate dehydrogenase cannot generate sufficient NADPH to regenerate GSH from GSSG, which in turn impairs their ability to dispose of H_2O_2 and of oxygen radicals. These compounds can cause oxidation of critical SH groups in proteins and possibly peroxidation of lipids in the membrane of the red cell, causing lysis of the red cell membrane. Some of the SH groups of hemoglobin become oxidized, and the protein precipitates inside the red blood cell, forming **Heinz bodies,** which stain purple with cresyl violet. The presence of Heinz bodies indicates that red blood cells have been subjected to oxidative stress. Figure 52–2 summarizes the possible chain of events in hemolytic anemia due to deficiency of glucose-6-phosphate dehydrogenase.

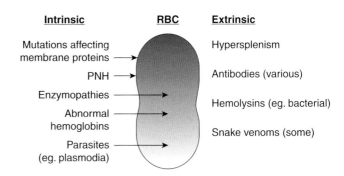

FIGURE 52–3 Schematic diagram of some causes of hemolytic anemias. Extrinsic causes are causes outside the red cell; they include hypersplenism, various antibodies, certain bacterial hemolysins and some snake venoms. Causes intrinsic to the red cells include mutations affecting the structures of membrane proteins (eg, in hereditary spherocytosis and hereditary elliptocytosis), PNH (paroxysmal nocturnal hemoglobinuria, see Chapter 47), enzymopathies, abnormal hemoglobins and certain parasites (eg, plasmodia causing malaria).

Hemolytic Anemias Are Caused by Abnormalities Outside, Within or Inside the Red Cell Membrane

Various causes of hemolytic anemias are summarized in Figure 52–3. Causes **outside the membrane** (ie, extrinsic) include **hypersplenism,** a condition in which the spleen is enlarged from a variety of causes and red blood cells become sequestered in it. **Various antibodies** (eg, transfusion reactions and anti-Rh antibodies, the presence in plasma of warm and cold antibodies that lyse red blood cells) also fall in this class, as do **hemolysins** released by various infectious agents, such as certain bacteria (eg, certain strains of *E coli* and clostridia). Some snakes release **venoms** that act to lyse the red cell membrane (eg, via the action of phospholipases or proteinases).

Causes **within the membrane** (intrinsic) include abnormalities of proteins. The most important conditions are **hereditary spherocytosis** and **hereditary elliptocytosis,** principally caused by abnormalities in the amount or structure of spectrin (see below). **Paroxysmal nocturnal hemoglobinuria** is discussed in Chapter 47.

Causes **inside the red blood cell** (also intrinsic) include **hemoglobinopathies** and **enzymopathies.** Sickle cell anemia and thalassemias are the most prevalent hemoglobinopathies. Abnormalities of enzymes in the pentose phosphate pathway and in glycolysis are the most frequent enzymopathies involved, particularly the former. Deficiency of **glucose-6-phosphate dehydrogenase** is prevalent in certain parts of the world and is a frequent cause of hemolytic anemia (see above). Deficiency of **pyruvate kinase** is not frequent, but it is the second commonest enzyme deficiency resulting in hemolytic anemia; the mechanism appears to be due to impairment of glycolysis, resulting in decreased formation of ATP, affecting various aspects of membrane integrity. **Parasitic** infections (eg, the plasmodia causing malaria) are also important causes of hemolytic anemias in certain geographic areas.

FIGURE 52–2 Summary of probable events causing hemolytic anemia due to deficiency of the activity of glucose-6-phosphate dehydrogenase (G6PD) (OMIM 305900).

TABLE 52–6 Laboratory Investigations That Assist in the Diagnosis of Hemolytic Anemia

General tests and findings
Increased nonconjugated (indirect) bilirubin
Shortened red cell survival time as measured by injection of autologous ^{51}Cr-labeled red cells
Reticulocytosis
Low level of plasma haptoglobin
Specific tests and findings
Hb electrophoresis (eg, HbS)
Red cell enzymes (eg, G6PD or pyruvate kinase deficiency)
Osmotic fragility (eg, hereditary spherocytosis)
Coombs test[1]
Cold agglutinins

[1]The direct Coombs test detects the presence of antibodies on red cells, whereas the indirect test detects the presence of circulating antibodies to antigens present on red cells.

Laboratory investigations that aid in the diagnosis of hemolytic anemia are listed in Table 52–6.

Methemoglobin Is Useless in Transporting Oxygen

The ferrous iron of hemoglobin is susceptible to oxidation by superoxide and other oxidizing agents, forming **methemoglobin,** which cannot transport oxygen. Only a very small amount of methemoglobin is present in normal blood, as the red blood cell possesses an effective system (the NADH-cytochrome b_5 methemoglobin reductase system) for reducing heme Fe^{3+} back to the Fe^{2+} state. This system consists of **NADH** (generated by glycolysis), a flavoprotein named **cytochrome b_5** reductase (also known as methemoglobin reductase), and **cytochrome b_5**. The Fe^{3+} of methemoglobin is reduced back to the Fe^{2+} state by the action of reduced cytochrome b_5:

$$Hb-Fe^{3+} + Cyt\,b_{5red} \rightarrow Hb-Fe^{2+} + Cyt\,b_{5ox}$$

Reduced cytochrome b_5 is then regenerated by the action of cytochrome b_5 reductase:

$$Cyt\,b_{5ox} + NADH \rightarrow Cyt\,b_{5red} + NAD$$

Methemoglobinemia Is Inherited or Acquired

Methemoglobinemia can be classified as either **inherited** or **acquired** by ingestion of certain drugs and chemicals. Neither type occurs frequently, but physicians must be aware of them. The inherited form is usually due to deficient activity of **cytochrome b_5** reductase, but mutations can also affect the activity of cytochrome b_5. Certain **abnormal hemoglobins** (eg, HbM) are also

rare causes of methemoglobinemia. In HbM, mutation changes the amino acid residue to which heme is attached, thus altering its affinity for oxygen and favoring its oxidation. Ingestion of **certain drugs** (eg, sulfonamides) or **chemicals** (eg, aniline) can cause acquired methemoglobinemia. Cyanosis (bluish discoloration of the skin and mucous membranes due to increased amounts of deoxygenated hemoglobin in arterial blood, or in this case due to increased amounts of methemoglobin) is usually the presenting sign in both types and is evident when over 10% of hemoglobin is in the "met" form. Diagnosis is made by spectroscopic analysis of blood, which reveals the characteristic absorption spectrum of methemoglobin. Additionally, a sample of blood containing methemoglobin cannot be fully reoxygenated by flushing oxygen through it, whereas normal deoxygenated blood can. Electrophoresis can be used to confirm the presence of an abnormal hemoglobin. Ingestion of **methylene blue** or **ascorbic acid** (both reducing agents) is used to treat mild methemoglobinemia due to enzyme deficiency. Acute massive methemoglobinemia (due to ingestion of chemicals) should be treated by intravenous injection of methylene blue.

MORE IS KNOWN ABOUT THE MEMBRANE OF THE HUMAN RED BLOOD CELL THAN ABOUT THE SURFACE MEMBRANE OF ANY OTHER HUMAN CELL

A variety of biochemical approaches have been used to study the membrane of the red blood cell. These include analysis of membrane proteins by SDS-PAGE, the use of specific enzymes (proteinases, glycosidases, and others) to determine the location of proteins and glycoproteins in the membrane, and various techniques to study both the lipid composition and disposition of individual lipids. Morphologic (eg, electron microscopy, freeze-fracture electron microscopy) and other techniques (eg, use of antibodies to specific components) have also been widely used. When red blood cells are lysed under specific conditions, their membranes will reseal in their original orientation to form **ghosts** (right-side-out ghosts). By altering the conditions, ghosts can also be made to reseal with their cytosolic aspect exposed on the exterior (inside-out ghosts). Both types of ghosts have been useful in analyzing the disposition of specific proteins and lipids in the membrane. In recent years, cDNAs for many proteins of this membrane have become available, permitting the deduction of their amino sequences and domains. All in all, more is known about the membrane of the red blood cell than about any other membrane of human cells (Table 52–7).

Analysis by SDS-PAGE Resolves the Proteins of the Membrane of the Red Blood Cell

When the membranes of red blood cells are analyzed by **SDS-PAGE,** about ten major proteins are resolved (Figure 52–4),

TABLE 52–7   Summary of Biochemical Information About the Membrane of the Human Red Blood Cell

- The membrane is a bilayer composed of about 50% lipid and 50% protein.

- The major lipid classes are phospholipids and cholesterol; the major phospholipids are phosphatidylcholine (PC), phosphatidylethanolamine (PE), and phosphatidylserine (PS) along with sphingomyelin (Sph).

- The choline-containing phospholipids, PC and Sph, predominate in the outer leaflet and the amino-containing phospholipids (PE and PS) in the inner leaflet.

- Glycosphingolipids (GSLs) (neutral GSLs, gangliosides, and complex species, including the ABO blood group substances) constitute about 5–10% of the total lipid.

- Analysis by SDS-PAGE shows that the membrane contains about 10 major proteins and more than 100 minor species.

- The major proteins (which include spectrin, ankyrin, the anion exchange protein, actin, and band 4.1) have been studied intensively, and the principal features of their disposition (eg, integral or peripheral), structure, and function have been established.

- Many of the proteins are glycoproteins (eg, the glycophorins) containing O- or N-linked (or both) oligosaccharide chains located on the external surface of the membrane.

TABLE 52–8   Principal Proteins of the Red Cell Membrane

Band Number[1]	Protein	Integral (I) or Peripheral (P)	Approximate Molecular Mass (kDa)
1	Spectrin (α)	P	240
2	Spectrin (β)	P	220
2.1	Ankyrin	P	210
2.2	Ankyrin	P	195
2.3	Ankyrin	P	175
2.6	Ankyrin	P	145
3	Anion exchange protein	I	100
4.1	Unnamed	P	80
5	Actin	P	43
6	Glyceraldehyde-3-phosphate dehydrogenase	P	35
7	Tropomyosin	P	29
8	Unnamed	P	23
	Glycophorins A, B, and C	I	31, 23, and 28

Source: Adapted from Lux DE, Becker PS: Disorders of the red cell membrane skeleton: hereditary spherocytosis and hereditary elliptocytosis. Chapter 95 in: *The Metabolic Basis of Inherited Disease,* 6th ed. Scriver CR et al (editors). McGraw-Hill, 1989.
[1]The band number refers to the position of migration on SDS-PAGE (see Figure 52–4). The glycophorins are detected by staining with the periodic acid-Schiff reagent. A number of other components (eg, 4.2 and 4.9) are not listed. Native spectrin is α₂β₂.

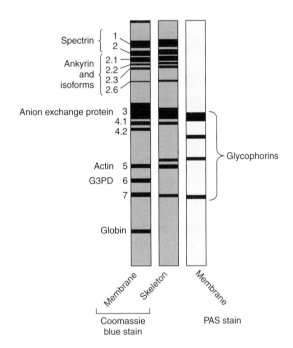

FIGURE 52–4 Diagrammatic representation of the major proteins of the membrane of the human red blood cell separated by SDS-PAGE. The bands detected by staining with Coomassie blue are shown in the two left-hand channels, and the glycoproteins detected by staining with periodic acid-Schiff (PAS) reagent are shown in the right-hand channel. (Reproduced, with permission, from Beck WS, Tepper RI: Hemolytic anemias III: membrane disorders. In: *Hematology,* 5th ed. Beck WS [editor]. The MIT Press, 1991.)

several of which have been shown to be **glycoproteins.** Their migration on SDS-PAGE was used to name these proteins, with the slowest migrating (and hence highest molecular mass) being designated band 1 or **spectrin.** All these major proteins have been isolated, most of them have been identified, and considerable insight has been obtained about their functions (Table 52–8). Many of their amino acid sequences also have been established. In addition, it has been determined which are integral or peripheral membrane proteins, which are situated on the external surface, which are on the cytosolic surface, and which span the membrane (Figure 52–5). Many minor components can also be detected in the red cell membrane by use of sensitive staining methods or two-dimensional gel electrophoresis. One of these is the glucose transporter described above.

The Major Integral Proteins of the Red Blood Cell Membrane Are the Anion Exchange Protein & the Glycophorins

The **anion exchange protein (band 3)** is a transmembrane glycoprotein, with its carboxyl terminal end on the external surface of the membrane and its amino terminal end on the

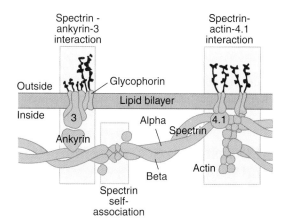

FIGURE 52–5 Diagrammatic representation of the interaction of cytoskeletal proteins with each other and with certain integral proteins of the membrane of the red blood cell. (Reproduced, with permission, from Beck WS, Tepper RI: Hemolytic anemias III: membrane disorders. In: *Hematology*, 5th ed. Beck WS [editor]. The MIT Press, 1991.)

cytoplasmic surface. It is an example of a **multipass** membrane protein, extending across the bilayer approximately fourteen times. It probably exists as a dimer in the membrane, in which it forms a tunnel, permitting the exchange of chloride for bicarbonate. Carbon dioxide, formed in the tissues, enters the red cell as bicarbonate, which is exchanged for chloride in the lungs, where carbon dioxide is exhaled. The amino terminal end binds many proteins, including hemoglobin, proteins 4.1 and 4.2, ankyrin, and several glycolytic enzymes. Purified band 3 has been added to lipid vesicles in vitro and has been shown to perform its transport functions in this reconstituted system.

Glycophorins A, B, and C are also transmembrane glycoproteins but of the **single-pass** type, extending across the membrane only once. A is the major glycophorin, is made up of 131 amino acids, and is heavily glycosylated (about 60% of its mass). Its amino terminal end, which contains 16 oligosaccharide chains (15 of which are O-glycans), extrudes out from the surface of the red blood cell. Approximately 90% of the sialic acid of the red cell membrane is located in this protein. Its transmembrane segment (23 amino acids) is α-helical. The carboxyl terminal end extends into the cytosol and binds to protein 4.1, which in turn binds to spectrin. **Polymorphism** of this protein is the basis of the MN blood group system (see below). Glycophorin A contains binding sites for influenza virus and for *Plasmodium falciparum,* the cause of one form of malaria. Intriguingly, the function of red blood cells of individuals who lack glycophorin A does not appear to be affected.

Spectrin, Ankyrin, & Other Peripheral Membrane Proteins Help Determine the Shape & Flexibility of the Red Blood Cell

The red blood cell must be able to squeeze through some tight spots in the microcirculation during its numerous passages around the body; the sinusoids of the spleen are of special importance in this regard. For the red cell to be easily and reversibly **deformable,** its membrane must be both fluid and flexible; it should also preserve its biconcave shape, since this facilitates gas exchange. Membrane **lipids** help determine membrane fluidity. Attached to the inner aspect of the membrane of the red blood cell are a number of **peripheral cytoskeletal proteins** (Table 52–8) that play important roles in respect to preserving shape and flexibility; these will now be described.

Spectrin is the major protein of the cytoskeleton. It is composed of two polypeptides: spectrin 1 (α chain) and spectrin 2 (β chain). These chains, measuring approximately 100 nm in length, are aligned in an antiparallel manner and are loosely intertwined, forming a dimer. Both chains are made up of segments of 106 amino acids that appear to fold into triple-stranded α-helical coils joined by nonhelical segments. One dimer interacts with another, forming a head-to-head tetramer. The overall shape confers **flexibility** on the protein and in turn on the membrane of the red blood cell. At least four **binding sites** can be defined in spectrin: (1) for self-association, (2) for ankyrin (bands 2.1, etc), (3) for actin (band 5), and (4) for protein 4.1.

Ankyrin is a pyramid-shaped protein that **binds spectrin.** In turn, ankyrin binds tightly to band 3, securing attachment of spectrin to the membrane. Ankyrin is sensitive to proteolysis, accounting for the appearance of bands 2.2, 2.3, and 2.6, all of which are derived from band 2.1.

Actin (band 5) exists in red blood cells as short, double-helical filaments of F-actin. The tail end of spectrin dimers binds to actin. Actin also binds to protein 4.1.

Protein 4.1, a globular protein, binds tightly to the tail end of spectrin, near the actin-binding site of the latter, and thus is part of a protein 4.1-spectrinactin ternary complex. Protein 4.1 also binds to the integral proteins, glycophorins A and C, thereby attaching the ternary complex to the membrane. In addition, protein 4.1 may interact with certain membrane phospholipids, thus connecting the lipid bilayer to the cytoskeleton.

Certain other proteins (4.9, adducin, and tropomyosin) also participate in **cytoskeletal assembly.**

Abnormalities in the Amount or Structure of Spectrin Cause Hereditary Spherocytosis & Elliptocytosis

Hereditary spherocytosis is a genetic disease, transmitted as an autosomal dominant, that affects about 1:5000 North Americans. It is characterized by the presence of spherocytes (spherical red blood cells, with a low surface-to-volume ratio) in the peripheral blood, by a **hemolytic anemia** (see Figure 52–3), and by splenomegaly. The spherocytes are not as deformable as are normal red blood cells, and they are subject to destruction in the spleen, thus greatly shortening their life in the circulation. Hereditary spherocytosis is **curable by splenectomy** because the spherocytes can persist in the circulation if the spleen is absent.

The spherocytes are much more susceptible to osmotic lysis than are normal red blood cells. This is assessed in the

FIGURE 52–6 Summary of the causation of hereditary spherocytosis (OMIM 182900). Approximately 50% of cases are due to abnormalities in ankyrin and 25% to abnormalities in spectrin.

osmotic fragility test, in which red blood cells are exposed in vitro to decreasing concentrations of NaCl. The physiologic concentration of NaCl is 0.85 g/dL. When exposed to a concentration of NaCl of 0.5 g/dL, very few normal red blood cells are hemolyzed, whereas approximately 50% of spherocytes would lyse under these conditions. The explanation is that the spherocyte, being almost circular, has little potential extra volume to accommodate additional water and thus lyses readily when exposed to a slightly lower osmotic pressure than is normal.

One cause of hereditary spherocytosis (Figure 52–6) is a deficiency in the amount of **spectrin** or abnormalities of its structure, so that it no longer tightly binds the other proteins with which it normally interacts. This weakens the membrane and leads to the spherocytic shape. Abnormalities of **ankyrin** and of **bands 3, 4.1,** and **4.2** are involved in other cases.

Hereditary elliptocytosis is a genetic disorder that is similar to hereditary spherocytosis except that affected red blood cells assume an elliptic, disk-like shape, recognizable by microscopy. It is also due to abnormalities in **spectrin;** some cases reflect abnormalities of band **4.1** or of **glycophorin C.**

THE BIOCHEMICAL BASES OF THE ABO BLOOD GROUP SYSTEM HAVE BEEN ESTABLISHED

Approximately 30 human **blood group systems** have been recognized, the best known of which are the **ABO, Rh (Rhesus),** and **MN** systems. The term "**blood group**" applies to a defined system of red blood cell antigens (blood group substances) controlled by a genetic locus having a variable number of alleles (eg, A, B, and O in the ABO system). The term "**blood type**" refers to the antigenic phenotype, usually recognized by the use of appropriate antibodies. For purposes of blood transfusion, it is particularly important to know the basics of the **ABO** and **Rh** systems. However, knowledge of blood group systems is also of biochemical, genetic, immuno-

logic, anthropologic, obstetric, pathologic, and forensic interest. Here, we shall discuss only some key features of the **ABO system.** From a biochemical viewpoint, the major interests in the ABO substances have been in isolating and determining their structures, elucidating their pathways of biosynthesis, and determining the natures of the products of the A, B, and O genes.

The ABO System Is of Crucial Importance in Blood Transfusion

This system was first discovered by Landsteiner in 1900 when investigating the basis of compatible and incompatible transfusions in humans. The membranes of the red blood cells of most individuals contain one blood group substance of type A, type B, type AB, or type O. Individuals of **type A** have anti-B antibodies in their plasma and will thus agglutinate type B or type AB blood. Individuals of **type B** have anti-A antibodies and will agglutinate type A or type AB blood. **Type AB** blood has neither anti-A nor anti-B antibodies and has been designated the **universal recipient. Type O** blood has neither A nor B substances and has been designated the **universal donor.** The explanation of these findings is related to the fact that the body does not usually produce antibodies to its own constituents. Thus, individuals of type A do not produce antibodies to their own blood group substance, A, but do possess antibodies to the foreign blood group substance, B, possibly because similar structures are present in microorganisms to which the body is exposed early in life. Since individuals of type O have neither A nor B substances, they possess antibodies to both these foreign substances. The above description has been simplified considerably; eg, there are two subgroups of type A: A_1 and A_2.

The genes responsible for production of the ABO substances are present on the long arm of chromosome 9. There are **three alleles,** two of which are codominant (A and B) and the third (O) recessive; these ultimately determine the four phenotypic products: the A, B, AB, and O substances.

The ABO Substances Are Glycosphingolipids & Glycoproteins Sharing Common Oligosaccharide Chains

The **ABO substances** are complex oligosaccharides present in most cells of the body and in certain secretions. On membranes of red blood cells, the oligosaccharides that determine the specific natures of the ABO substances appear to be mostly present in **glycosphingolipids,** whereas in secretions the same oligosaccharides are present in **glycoproteins.** Their presence in secretions is determined by a gene designated *Se* (for **secretor**), which codes for a specific **fucosyl (Fuc) transferase** in secretory organs, such as the exocrine glands, but which is not active in red blood cells. Individuals of *SeSe* or *Sese* genotypes secrete A or B antigens (or both), whereas individuals of the *sese* genotype do not secrete A or B substances, but their red blood cells can express the A and B antigens.

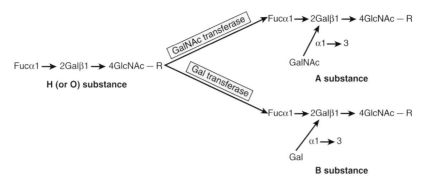

FIGURE 52–7 Diagrammatic representation of the structures of the H, A, and B blood group substances. R represents a long complex oligosaccharide chain, joined either to ceramide where the substances are glycosphingolipids, or to the polypeptide backbone of a protein via a serine or threonine residue where the substances are glycoproteins. Note that the blood group substances are biantennary; ie, they have two arms, formed at a branch point (not indicated) between the GlcNAc—R, and only one arm of the branch is shown. Thus, the H, A, and B substances each contain two of their respective short oligosaccharide chains shown above. The AB substance contains one type A chain and one type B chain.

H Substance Is the Biosynthetic Precursor of Both the A & B Substances

The ABO substances have been isolated and their structures determined; simplified versions, showing only their non-reducing ends, are presented in Figure 52–7. It is important to first appreciate the structure of the **H substance,** since it is the precursor of both the A and B substances and is the blood group substance found in persons of type O. H substance itself is formed by the action of a **fucosyltransferase,** which catalyzes the addition of the terminal fucose in $\alpha 1 \rightarrow 2$ linkage onto the terminal Gal residue of its precursor:

$$GDP\text{-}Fuc + Gal\text{-}\beta\text{-}R \rightarrow Fuc\text{-}\alpha 1,2\text{-}Gal\text{-}\beta\text{-}R + GDP$$
$$\quad\quad\quad\quad Precursor \quad\quad\quad\quad\quad\quad H\,substance$$

The H locus codes for this fucosyltransferase. The *h* allele of the H locus codes for an inactive fucosyltransferase; therefore, individuals of the *hh* genotype cannot generate H substance, the precursor of the A and B antigens. Thus, individuals of the *hh* genotype will have red blood cells of type O, even though they may possess the enzymes necessary to make the A or B substances (see below). They are referred to as being Bombay phenotype (O_h).

The *A* Gene Encodes a GalNAc Transferase, the *B* Gene a Gal Transferase, & the *O* Gene an Inactive Product

In comparison with H substance (Figure 52–7), **A substance** contains an additional GalNAc and **B substance** an additional Gal, linked as indicated. Anti-A antibodies are directed to the additional GalNAc residue found in the A substance, and anti-B antibodies are directed toward the additional Gal residue

found in the B substance. Thus, GalNAc is the **immunodominant sugar** (ie, the one determining the specificity of the antibody formed) of blood group A substance, whereas Gal is the immunodominant sugar of the B substance. In view of the structural findings, it is not surprising that A substance can be synthesized in vitro from O substance in a reaction catalyzed by a GalNAc transferase, employing UDP-GalNAc as the sugar donor. Similarly, blood group B can be synthesized from O substance by the action of a Gal transferase, employing UDP-Gal. It is crucial to appreciate that the product of the *A* gene is the **GalNAc transferase** that adds the terminal GalNAc to the O substance. Similarly, the product of the *B* gene is the **Gal transferase** adding the Gal residue to the O substance. Individuals of **type AB** possess both enzymes and thus have two oligosaccharide chains (Figure 52–6), one terminated by a GalNAc and the other by a Gal. Individuals of type O apparently synthesize an inactive protein, detectable by immunologic means; thus, H substance is their ABO blood group substance.

In 1990, a study using cloning and sequencing technology described the nature of the differences between the glycosyltransferase products of the *A, B,* and *O* genes. A difference of four nucleotides is apparently responsible for the distinct specificities of the A and B glycosyltransferases. On the other hand, the *O* allele has a single base-pair mutation, causing a **frameshift mutation** resulting in a protein lacking transferase activity.

NEUTROPHILS HAVE AN ACTIVE METABOLISM & CONTAIN SEVERAL UNIQUE ENZYMES & PROTEINS

The major biochemical features of **neutrophils** are summarized in Table 52–9. Prominent features are active aerobic

TABLE 52–9 Summary of Major Biochemical Features of Neutrophils

- Active glycolysis
- Active pentose phosphate pathway
- Moderate oxidative phosphorylation
- Rich in lysosomes and their degradative enzymes
- Contain certain unique enzymes (eg, myeloperoxidase and NADPH oxidase) and proteins
- Contain CD 11/CD18 integrins in plasma membrane

glycolysis, active pentose phosphate pathway, moderately active oxidative phosphorylation (because mitochondria are relatively sparse), and a high content of lysosomal enzymes. Many of the enzymes listed in Table 52–5 are also of importance in the oxidative metabolism of neutrophils (see below). Table 52–10 summarizes the functions of some proteins that are relatively unique to neutrophils.

Neutrophils Are Key Players in the Body's Defense Against Bacterial Infection

Neutrophils are motile phagocytic cells of the innate immune system that play a key role in acute inflammation. When bacteria enter tissues, a number of phenomena result that are collectively known as the "acute inflammatory response." They include (1) increase of vascular permeability, (2) entry of activated neutrophils into the tissues, (3) activation of platelets, and (4) spontaneous subsidence (resolution) if the invading microorganisms have been dealt with successfully.

A variety of molecules are released from cells and plasma proteins during acute inflammation whose net overall effect is to increase vascular permeability, resulting in tissue edema (Table 52–11).

In acute inflammation, neutrophils are recruited from the bloodstream into the tissues to help eliminate the foreign invaders. The neutrophils are attracted into the tissues by **chemotactic factors,** including complement fragment C5a, small peptides derived from bacteria (eg, N-formyl-methionyl-leucyl-phenylalanine), and a number of leukotrienes. To

TABLE 52–10 Some Important Enzymes and Proteins of Neutrophils[1]

Enzyme or Protein	Reaction Catalyzed or Function	Comment
Myeloperoxidase (MPO)	$H_2O_2 + X^-$ (halide) $+ H^+ \rightarrow HOX + H_2O$ (where $X^- = Cl^-$, HOX = hypochlorous acid)	Responsible for the green color of pus Genetic deficiency can cause recurrent infections
NADPH oxidase	$2O_2 + NADPH \rightarrow 2O_2^{\cdot-} + NADP + H^+$	Key component of the respiratory burst Deficient in chronic granulomatous disease
Lysozyme	Hydrolyzes link between N-acetylmuramic acid and N-acetyl-D-glucosamine found in certain bacterial cell walls	Abundant in macrophages
Defensins	Basic antibiotic peptides of 20–33 amino acids	Apparently kill bacteria by causing membrane damage
Lactoferrin	Iron-binding protein	May inhibit growth of certain bacteria by binding iron and may be involved in regulation of proliferation of myeloid cells
CD11a/CD18, CD11b/CD18, CD11c/CD18[2]	Adhesion molecules (members of the integrin family)	Deficient in leukocyte adhesion deficiency type I (OMIM 116920)
Receptors for Fc fragments of IgGs	Bind Fc fragments of IgG molecules	Target antigen-antibody complexes to myeloid and lymphoid cells, eliciting phagocytosis and other responses

[1]The expression of many of these molecules has been studied during various stages of differentiation of normal neutrophils and also of corresponding leukemic cells employing molecular biology techniques (eg, measurements of their specific mRNAs). For the majority, cDNAs have been isolated and sequenced, amino acid sequences deduced, genes have been localized to specific chromosomal locations, and exons and intron sequences have been defined. Some important proteinases of neutrophils are listed in Table 52–13.

[2]CD = cluster of differentiation. This refers to a uniform system of nomenclature that has been adopted to name surface markers of leukocytes. A specific surface protein (marker) that identifies a particular lineage or differentiation stage of leukocytes and that is recognized by a group of monoclonal antibodies is called a member of a cluster of differentiation. The system is particularly helpful in categorizing subclasses of lymphocytes. Many CD antigens are involved in cell-cell interactions, adhesion, and transmembrane signaling.

TABLE 52–11 Sources of Biomolecules With Vasoactive Properties Involved in Acute Inflammation

Mast Cells and Basophils	Platelets	Neutrophils	Plasma Proteins
Histamine	Serotonin	Platelet-activating factor (PAF)	C3a, C4a, and C5a from the complement system
		Eicosanoids (various prostaglandins and leukotrienes)	Bradykinin and fibrin degradation products from the coagulation system

reach the tissues, circulating neutrophils must pass through the capillaries. To achieve this, they marginate along the vessel walls and then adhere to endothelial (lining) cells of the capillaries.

Integrins Mediate Adhesion of Neutrophils to Endothelial Cells

Adhesion of neutrophils to endothelial cells employs specific adhesive proteins **(integrins)** located on their surface and also specific receptor proteins in the endothelial cells. (See also the discussion of selectins in Chapter 47.)

The **integrins** are a superfamily of surface proteins present on a wide variety of cells. They are involved in the **adhesion** of cells to other cells or to specific components of the extracellular matrix. They are **heterodimers,** containing an α and a β subunit linked noncovalently. The subunits contain extracellular, transmembrane, and intracellular segments. The **extracellular segments** bind to a variety of ligands such as specific proteins of the extracellular matrix and of the surfaces of other cells. These ligands often contain ArgGly-Asp (R-G-D) sequences. The **intracellular domains** bind to various proteins of the cytoskeleton, such as actin and vinculin. The integrins are proteins that **link the outsides of cells to their insides,** thereby helping to integrate responses of cells (eg, movement, phagocytosis) to changes in the environment.

Three **subfamilies** of integrins were recognized initially. Members of each subfamily were distinguished by containing a common β subunit, but they differed in their subunits. However, more than three β subunits have now been identified, and the classification of integrins has become rather complex. Some integrins of specific interest with regard to neutrophils are listed in Table 52–12.

A deficiency of the $β_2$ subunit (also designated CD18) of LFA-1 and of two related integrins found in neutrophils and macrophages, Mac-1 (CD11b/CD18) and p150,95 (CD11c/CD18), causes **type 1 leukocyte adhesion deficiency,** a disease characterized by recurrent bacterial and fungal infec-

tions. Among various results of this deficiency, the adhesion of affected white blood cells to endothelial cells is diminished, and lower numbers of neutrophils thus enter the tissues to combat infection.

Once having passed through the walls of small blood vessels, the neutrophils migrate toward the highest concentrations of the chemotactic factors, encounter the invading bacteria, and attempt to attack and destroy them. The neutrophils must be **activated** in order to turn on many of the metabolic processes involved in phagocytosis and killing of bacteria.

Activation of Neutrophils Is Similar to Activation of Platelets & Involves Hydrolysis of Phosphatidylinositol Bisphosphate

The mechanisms involved in platelet activation are discussed in Chapter 51 (see Figure 51–8). The process involves interaction of the stimulus (eg, thrombin) with a receptor, activation of G proteins, stimulation of phospholipase C, and liberation from phosphatidylinositol bisphosphate of inositol triphosphate and diacylglycerol. These two second messengers result in an elevation of intracellular Ca^{2+} and activation of protein kinase C. In addition, activation of phospholipase A_2 produces arachidonic acid that can be converted to a variety of biologically active eicosanoids.

The process of **activation of neutrophils** is essentially similar. They are activated, via specific receptors, by interaction with bacteria, binding of chemotactic factors, or antibody-antigen complexes. The resultant **rise in intracellular Ca^{2+}** affects many processes in neutrophils, such as assembly of microtubules and the actin-myosin system. These processes are respectively involved in secretion of contents of granules and in motility, which enables neutrophils to seek out the invaders. The activated neutrophils are now ready to destroy the invaders by mechanisms that include production of active derivatives of oxygen.

TABLE 52–12 **Examples of Integrins That Are Important in the Function of Neutrophils, of Other White Blood Cells, and of Platelets[1]**

Integrin	Cell	Subunit	Ligand	Function
VLA-1 (CD49a)	WBCs, others	α1β1	Collagen, laminin	Cell-ECM adhesion
VLA-5 (CD49e)	WBCs, others	α5β1	Fibronectin	Cell-ECM adhesion
VLA-6 (CD49f)	WBCs, others	α6β1	Laminin	Cell-ECM adhesion
LFA-1 (CD11a)	WBCs	αLβ2	ICAM-1	Adhesion of WBCs
Glycoprotein IIb/IIIa	Platelets	αIIbβ3	ICAM-2	
Fibrinogen, fibronectin, von Willebrand factor	Platelet adhesion and aggregation			

[1]LFA-1, lymphocyte function-associated antigen 1; VLA, very late antigen; CD, cluster of differentiation; ICAM, intercellular adhesion molecule; ECM, extracellular matrix. A deficiency of LFA-1 and related integrins is found in type I leukocyte adhesion deficiency (OMIM 116920). A deficiency of platelet glycoprotein IIb/IIIa complex is found in Glanzmann thrombasthenia (OMIM 273800), a condition characterized by a history of bleeding, a normal platelet count, and abnormal clot retraction. These findings illustrate how fundamental knowledge of cell surface adhesion proteins is shedding light on the causation of a number of diseases.

The Respiratory Burst of Phagocytic Cells Involves NADPH Oxidase & Helps Kill Bacteria

When neutrophils and other phagocytic cells engulf bacteria, they exhibit a rapid increase in oxygen consumption known as **the respiratory burst.** This phenomenon reflects the rapid utilization of oxygen (following a lag of 15–60 seconds) and production from it of large amounts of **reactive derivatives,** such as O_2^-, H_2O_2, OH·, and OCl⁻ (hypochlorite ion). Some of these products are potent microbicidal agents.

The **electron transport chain system** responsible for the respiratory burst (named NADPH oxidase) is composed of several components. One is **cytochrome b_{558}**, located in the plasma membrane; it is a heterodimer, containing two polypeptides of 91 kDa and 22 kDa. When the system is activated (see below), two cytoplasmic polypeptides of 47 kDa and 67 kDa are recruited to the plasma membrane and, together with cytochrome b_{558}, form **the NADPH oxidase** responsible for the respiratory burst. The reaction catalyzed by NADPH oxidase, involving formation of superoxide anion, is shown in Table 52–5 (reaction 2). This system catalyzes the one-electron reduction of oxygen to superoxide anion. The NADPH is generated mainly by the pentose phosphate cycle, whose activity increases markedly during phagocytosis.

The above reaction is followed by the spontaneous production (by spontaneous dismutation) of **hydrogen peroxide** from two molecules of superoxide:

$$O_2^- + O_2^- + 2H^+ \rightarrow H_2O_2 + O_2$$

The **superoxide ion** is discharged to the outside of the cell or into phagolysosomes, where it encounters ingested bacteria. Killing of bacteria within phagolysosomes appears to depend on the combined action of elevated pH, superoxide ion, or further oxygen derivatives (H_2O_2, OH·, and HOCl [hypochlorous acid; see below]) and on the action of certain bactericidal peptides (defensins) and other proteins (eg, cathepsin G and certain cationic proteins) present in phagocytic cells. Any superoxide that enters the cytosol of the phagocytic cell is converted to H_2O_2 by the action of **superoxide dismutase,** which catalyzes the same reaction as the spontaneous dismutation shown above. In turn, H_2O_2 is used by myeloperoxidase (see below) or disposed of by the action of glutathione peroxidase or catalase.

NADPH oxidase is inactive in resting phagocytic cells and is **activated** upon contact with various ligands (complement fragment C5a, chemotactic peptides, etc) with receptors in the plasma membrane. The events resulting in activation of the oxidase system have been much studied and are similar to those described above for the process of activation of neutrophils. They involve **G proteins,** activation of **phospholipase C,** and generation of **inositol 1,4,5-triphosphate** (IP₃). The last mediates a transient increase in the level of cytosolic Ca²⁺, which is essential for induction of the respiratory burst. **Diacylglycerol** is also generated and induces the translocation of protein kinase C into the plasma membrane from the cytosol,

FIGURE 52–8 Simplified scheme of the sequence of events involved in the causation of chronic granulomatous disease (OMIM 306400). Mutations in any of the genes for the four polypeptides involved (two are components of cytochrome b_{558} and two are derived from the cytoplasm) can cause the disease. The polypeptide of 91 kDa is encoded by a gene in the X chromosome; approximately 60% of cases of chronic granulomatous disease are X-linked, with the remainder being inherited in an autosomal recessive fashion.

where it catalyzes the **phosphorylation** of various proteins, some of which are components of the oxidase system. A second pathway of activation not involving Ca²⁺ also operates.

Mutations in the Genes for Components of the NADPH Oxidase System Cause Chronic Granulomatous Disease

The importance of the **NADPH oxidase system** was clearly shown when it was observed that the respiratory burst was defective in **chronic granulomatous disease,** a relatively uncommon condition characterized by recurrent infections and widespread granulomas (chronic inflammatory lesions) in the skin, lungs, and lymph nodes. The granulomas form as attempts to wall off bacteria that have not been killed, owing to genetic deficiencies in the NADPH oxidase system. The disorder is due to mutations in the genes encoding the four polypeptides that constitute the NADPH oxidase system. Some patients have responded to treatment with gamma interferon, which may increase transcription of the 91-kDa component if it is affected. Attempts are being made to develop gene therapy for this condition. The probable sequence of events involved in the causation of chronic granulomatous disease is shown in Figure 52–8.

Neutrophils Contain Myeloperoxidase, Which Catalyzes the Production of Chlorinated Oxidants

The enzyme **myeloperoxidase,** present in large amounts in neutrophil granules and responsible for the green color of pus, can act on H_2O_2 to produce hypohalous acids:

$$H_2O_2 + X^- + H^+ \xrightarrow{\text{MYELOPEROXIDASE}} HOX + H_2O$$

(X⁻ = Cl⁻, Br⁻, I⁻ or SCN⁻; HOCl = hypochlorous acid)

The H_2O_2 used as substrate is generated by the NADPH oxidase system. Cl^- is the halide usually employed, since it is present in relatively high concentration in plasma and body fluids. **HOCl,** the active ingredient of household liquid bleach, is a powerful oxidant and is highly microbicidal. When applied to normal tissues, its potential for causing damage is diminished because it reacts with primary or secondary amines present in neutrophils and tissues to produce various nitrogen-chlorine derivatives; these **chloramines** are also oxidants, though less powerful than HOCl, and act as microbicidal agents (eg, in sterilizing wounds) without causing tissue damage.

The Proteinases of Neutrophils Can Cause Serious Tissue Damage if Their Actions Are Not Checked

Neutrophils contain a number of proteinases (Table 52–13) that can hydrolyze elastin, various types of collagens, and other proteins present in the extracellular matrix. Such enzymatic action, if allowed to proceed unopposed, can result in serious damage to tissues. Most of these proteinases are **lysosomal enzymes** and exist mainly as inactive precursors in normal neutrophils. Small amounts of these enzymes are released into normal tissues, with the amounts increasing markedly during inflammation. The activities of elastase and other proteinases are normally kept in check by a number of **antiproteinases** (also listed in Table 52–13) present in plasma and the extracellular fluid. Each of them can combine—usually forming a noncovalent complex—with one or more specific proteinases and thus cause inhibition. In Chapter 50 it was shown that a genetic deficiency of α_1-**antiproteinase inhibitor** (α_1-antitrypsin) permits elastase to act unopposed and digest pulmonary tissue, thereby participating in the causation of emphysema. α_2-**Macroglobulin** is a plasma protein that plays an important role in the body's defense against excessive action of proteases; it combines with and thus neutralizes the activities of a number of important proteases (Chapter 50).

TABLE 52–13   **Proteinases of Neutrophils and Antiproteinases of Plasma and Tissues[1]**

Proteinases	Antiproteinases
Elastase	α_1-Antiproteinase (α_1-antitrypsin)
Collagenase	α_2-Macroglobulin
Gelatinase	Secretory leukoproteinase inhibitor
Cathepsin G	α_1-Antichymotrypsin
Plasminogen activator	Plasminogen activator inhibitor–1 Tissue inhibitor of metalloproteinase

[1]The table lists some of the important proteinases of neutrophils and some of the proteins that can inhibit their actions. Most of the proteinases listed exist inside neutrophils as precursors. The proteinases listed can digest many proteins of the extracellular matrix, causing tissue damage. The overall balance of proteinase:antiproteinase action can be altered by activating the precursors of the proteinases, or by inactivating the antiproteinases. The latter can be caused by proteolytic degradation or chemical modification, eg, Met-358 of α_1-antiproteinase inhibitor is oxidized by cigarette smoke.

When increased amounts of chlorinated oxidants are formed during inflammation, they affect the proteinase: antiproteinase equilibrium, tilting it in favor of the former. For instance, certain of the proteinases listed in Table 52–13 are **activated** by HOCl, whereas certain of the antiproteinases are **inactivated** by this compound. In addition, the tissue inhibitor of metalloproteinases and α_1-antichymotrypsin can be hydrolyzed by activated elastase, and α_1-antiproteinase inhibitor can be hydrolyzed by activated collagenase and gelatinase. In most circumstances, **an appropriate balance** of proteinases and antiproteinases is achieved. However, in certain instances, such as in the lung when α_1-antiproteinase inhibitor is deficient or when large amounts of neutrophils accumulate in tissues because of inadequate drainage, considerable **tissue damage** can result from the unopposed action of proteinases.

RECOMBINANT DNA TECHNOLOGY HAS HAD A PROFOUND IMPACT ON HEMATOLOGY

Recombinant DNA technology has had a major impact on many aspects of hematology. The bases of the **thalassemias** and of many **disorders of coagulation** (Chapter 51) have been greatly clarified by investigations using cloning and sequencing. The study of oncogenes and chromosomal translocations has advanced understanding of the **leukemias.** As discussed above, cloning techniques have made available therapeutic amounts of **erythropoietin** and **other growth factors.** Deficiency of **adenosine deaminase,** which affects lymphocytes particularly, is the first disease to be treated by gene therapy (see case no. 1, Chapter 54). Like many other areas of biology and medicine, hematology has been and will continue to be revolutionized by this technology.

SUMMARY

- Anemias are very prevalent conditions. Major causes include blood loss, deficiencies of iron, folate and vitamin B_{12} and various factors causing hemolysis.

- The red blood cell is simple in terms of its structure and function, consisting principally of a concentrated solution of hemoglobin surrounded by a membrane.

- The production of red cells is regulated by erythropoietin, whereas other growth factors (eg, granulocyte and granulocyte-macrophage colony-stimulating factors) regulate the production of white blood cells.

- The red cell contains a battery of cytosolic enzymes, such as superoxide dismutase, catalase, and glutathione peroxidase, to dispose of powerful oxidants (ROS) generated during its metabolism.

- Genetically determined deficiency of the activity of glucose-6-phosphate dehydrogenase, which produces NADPH, is an important cause of hemolytic anemia.

- Methemoglobin is unable to transport oxygen; both genetic and acquired causes of methemoglobinemia are recognized.

- Considerable information has accumulated concerning the proteins and lipids of the red cell membrane. A number of cytoskeletal proteins, such as spectrin, ankyrin, and actin, interact with specific integral membrane proteins to help regulate the shape and flexibility of the membrane.

- Deficiency of spectrin results in hereditary spherocytosis and hereditary elliptocytosis, both causes of hemolytic anemia.

- The ABO blood group substances in the red cell membrane are complex glycosphingolipids; the immunodominant sugar of A substance is N-acetyl-galactosamine, whereas that of the B substance is galactose. O substance does not contain either of these two sugar residues in the particular linkages found in the A and B substances.

- Neutrophils play a major role in the body's defense mechanisms. Integrins on their surface membranes determine specific interactions with various cell and tissue components.

- Leukocytes are activated on exposure to bacteria and other stimuli; NADPH oxidase plays a key role in the process of activation (the respiratory burst). Mutations in this enzyme and associated proteins cause chronic granulomatous disease.

- The proteinases of neutrophils can digest many tissue proteins; normally, this is kept in check by a battery of antiproteinases. However, this defense mechanism can be overcome in certain circumstances, resulting in extensive tissue damage.

- The application of recombinant DNA technology is revolutionizing the field of hematology.

REFERENCES

Fauci AS et al (editors): *Harrison's Principles of Internal Medicine,* 17th ed. McGraw-Hill, 2008. (Chapters 58, 61, & 98–108 deal with various blood disorders. Chapters 66–68 deal with various aspects of hematopoietic and other stem cells).

ffrench-Constant C, Colognato H: Integrins: versatile integrators of extracellular signals. Trends Cell Biol 2004;14:678.

Hoffman R et al (editors): *Hematology: Basic Principles and Practice,* 4th ed. Elsevier Churchill Livingston, 2005.

Israels LG, Israels ED: *Mechanisms in Hematology,* 3rd ed. Core Health Sciences Inc, 2002.

Scriver CR et al (editors): *The Molecular Bases of Inherited Disease,* 8th ed. McGraw-Hill, 2001. (This text is now available online and updated as *The Online Metabolic & Molecular Bases of Inherited Disease* at www.ommbid.com Subscription is required, although access may be available via university and hospital libraries and other sources). A number of the chapters concern topics described in this chapter.

Yonekawa K, Harlan JM: Targeting leukocyte integrins in human diseases. J Leukoc Biol 2005;77:129.

53

Metabolism of Xenobiotics

Robert K. Murray, MD, PhD

BIOMEDICAL IMPORTANCE

Increasingly, humans are subjected to exposure to various foreign chemicals (**xenobiotics**)—drugs, food additives, pollutants, etc. The situation is well summarized in the following quotation from Rachel Carson: "As crude a weapon as the cave man's club, the chemical barrage has been hurled against the fabric of life." **Understanding how xenobiotics are handled at the cellular level** is important in learning how to cope with the chemical onslaught, and thus helping to **preserve the environment.** For example, building on such information, attempts are being made to modify microorganisms by introducing genes that encode various enzymes involved in metabolizing specific xenobiotics to harmless products. These modified organisms will then be used to help dispose of various pollutants that contaminate the planet.

Knowledge of the metabolism of xenobiotics is basic to a rational understanding of pharmacology and therapeutics, pharmacy, toxicology, management of cancer, and drug addiction. All these areas involve administration of, or exposure to, xenobiotics.

HUMANS ENCOUNTER THOUSANDS OF XENOBIOTICS THAT MUST BE METABOLIZED BEFORE BEING EXCRETED

A **xenobiotic** (Gk *xenos* "stranger") is a compound that is foreign to the body. The principal classes of xenobiotics of medical relevance are **drugs**, **chemical carcinogens**, and **various compounds** that have found their way into our environment by one route or another, such as polychlorinated biphenyls (PCBs) and certain insecticides. More than 200,000 manufactured environmental chemicals exist. Most of these compounds are subject to metabolism (chemical alteration) in the human body, with the liver being the main organ involved; occasionally, a xenobiotic may be excreted unchanged. **At least 30 different enzymes** catalyze reactions involved in xenobiotic metabolism; however, this chapter will only cover a selected group of them.

It is convenient to consider the metabolism of xenobiotics in two phases. In **phase 1,** the major reaction involved is **hydroxylation,** catalyzed mainly by members of a class of enzymes referred to as **monooxygenases** or **cytochrome P450s.** Hydroxylation may terminate the action of a drug, though this is not always the case. In addition to hydroxylation, these enzymes catalyze **a wide range of reactions**, including those involving deamination, dehalogenation, desulfuration, epoxidation, peroxygenation, and reduction. Reactions involving hydrolysis (eg, catalyzed by esterases) and certain other non-P450-catalyzed reactions also occur in phase 1.

In **phase 2,** the hydroxylated or other compounds produced in phase 1 are converted by specific enzymes to **various polar metabolites** by **conjugation** with glucuronic acid, sulfate, acetate, glutathione, or certain amino acids, or by **methylation.**

The overall purpose of the two phases of metabolism of xenobiotics is to increase their **water solubility (polarity)** and thus **excretion** from the body. Very hydrophobic xenobiotics would persist in adipose tissue almost indefinitely if they were not converted to more polar forms. In certain cases, phase 1 metabolic reactions convert xenobiotics from **inactive** to **biologically active** compounds. In these instances, the original xenobiotics are referred to as "prodrugs" or "procarcinogens." In other cases, additional phase 1 reactions (eg, further hydroxylation reactions) convert the active compounds to less active or inactive forms prior to conjugation. In yet other cases, it is the conjugation reactions themselves that convert the active products of phase 1 reactions to less active or inactive species, which are subsequently excreted in the urine or bile. In a very few cases, conjugation may actually increase the biologic activity of a xenobiotic.

The term "**detoxification**" is sometimes used for many of the reactions involved in the metabolism of xenobiotics. However, the term is not always appropriate because, as mentioned above, in some cases the reactions to which xenobiotics are subject actually **increase** their biologic activity and toxicity.

ISOFORMS OF CYTOCHROME P450 HYDROXYLATE A MYRIAD OF XENOBIOTICS IN PHASE 1 OF THEIR METABOLISM

Hydroxylation is the chief reaction involved in phase 1. The responsible enzymes are called **monooxygenases** or **cytochrome P450s.** It is estimated that there are some 57 cytochrome P450 genes present in humans. The reaction catalyzed by a monooxygenase (cytochrome P450) is as follows:

$$RH + O_2 + NADPH + H^+ \rightarrow R\text{-}OH + H_2O + NADP$$

RH above can represent a very wide variety of xenobiotics, including drugs, carcinogens, pesticides, petroleum products, and pollutants (such as a mixture of PCBs). In addition, **endogenous compounds,** such as certain steroids, eicosanoids, fatty acids, and retinoids, are also substrates. The substrates are generally **lipophilic** and are rendered more **hydrophilic** by hydroxylation.

Cytochrome P450 is considered the **most versatile biocatalyst** known. The actual reaction mechanism is complex and has been briefly described previously (Figure 12–6). It has been shown by the use of $^{18}O_2$ that one atom of oxygen enters R–OH and one atom enters water. This dual fate of the oxygen accounts for the former naming of monooxygenases as **"mixed-function oxidases."** The reaction catalyzed by cytochrome P450 can also be represented as follows:

Reduced cytochrome P450   Oxidized cytochrome P450

$$RH + O_2 \rightarrow R\text{-}OH + H_2O$$

Cytochrome P450 is so named because the enzyme was discovered when it was noted that preparations of **microsomes** that had been chemically reduced and then exposed to carbon monoxide **exhibited a distinct peak at 450 nm.** Microsomes contain fragments of the endoplasmic reticulum, where much of the P450 content of cells is located (see below). Among reasons that this enzyme is important is the fact that **approximately 50% of the common drugs humans ingest are metabolized by isoforms of cytochrome P450**; these enzymes also act on various carcinogens and pollutants. The major cytochrome P450s in drug metabolism are members of the CYP1, CYP2, and CYP3 families (see below).

Isoforms of Cytochrome P450 Make Up a Superfamily of Heme-Containing Enzymes

The following are important points concerning cytochrome P450s.

1. Because of **the large number of isoforms** (about 150) that have been discovered, it became important to have a **systematic nomenclature** for isoforms of P450 and for their genes. This is now available and in wide use and is based on structural homology. The abbreviated root symbol CYP denotes a cytochrome P450. This is followed by an Arabic number designating the **family;** cytochrome P450s are included in the same family if they exhibit 40% or more amino acid sequence identity. The Arabic number is followed by a capital letter indicating the **subfamily,** if two or more members exist; P450s are in the same subfamily if they exhibit greater than 55% sequence identity. The **individual** P450s are then arbitrarily assigned Arabic numerals. Thus, CYP1A1 denotes a cytochrome P450 that is a member of family 1 and subfamily A and is the first individual member of that subfamily. The nomenclature for the **genes** encoding cytochrome P450s is identical to that described above except that italics are used; thus, the gene encoding CYP1A1 is *CYP1A1.*

2. Like **hemoglobin,** they are hemoproteins.

3. They are **widely distributed across species,** including bacteria.

4. They are present in highest amount in **liver cells** and enterocytes but are probably present in all tissues. In liver and most other tissues, they are present mainly in the **membranes of the smooth endoplasmic reticulum,** which constitute part of the **microsomal fraction** when tissue is subjected to subcellular fractionation. In hepatic microsomes, cytochrome P450s can comprise as much as 20% of the total protein. P450s are found in most tissues, though often in low amounts compared with liver. In the **adrenal,** they are found in **mitochondria** as well as in the endoplasmic reticulum; the various hydroxylases present in that organ play an important role in cholesterol and steroid biosynthesis. The mitochondrial cytochrome P450 system differs from the microsomal system in that it uses an NADPH-linked flavoprotein, **adrenodoxin reductase,** and a nonheme iron-sulfur protein, **adrenodoxin.** In addition, the specific P450 isoforms involved in steroid biosynthesis are generally much more restricted in their substrate specificity.

5. At least six different species of cytochrome P450 are present in the endoplasmic reticulum of human liver, each with wide and somewhat overlapping **substrate specificities** and acting on both xenobiotics and endogenous compounds. The genes for many isoforms of P450 (from both humans and animals such as the rat) have been isolated and studied in detail in recent years. The combination of there being **a number of different types** and **each having a relatively wide substrate specificity** explains why the cytochrome P450 family can metabolize thousands of different chemicals.

6. **NADPH,** not NADH, is involved in the reaction mechanism of cytochrome P450. The enzyme that uses NADPH to yield the reduced cytochrome P450, shown at the left-hand side of the above equation, is called **NADPH-cytochrome P450 reductase.** Electrons are transferred from NADPH to NADPH-cytochrome P450 reductase and then to cytochrome P450. This leads to the **reductive activation of molecular oxygen,** and one atom of oxygen is subsequently inserted into the substrate. **Cytochrome b_5,** another hemoprotein found in the membranes of the smooth endoplasmic reticulum (Chapter 12), may be involved as an electron donor in some cases.

7. **Lipids** are also components of the cytochrome P450 system. The preferred lipid is **phosphatidylcholine,** which

is the major lipid found in membranes of the endoplasmic reticulum.

8. Most isoforms of cytochrome P450 are **inducible.** For instance, the administration of phenobarbital or of many other drugs causes hypertrophy of the smooth endoplasmic reticulum and a three- to fourfold increase in the amount of cytochrome P450 within 4–5 days. The mechanism of induction has been studied extensively and in most cases involves **increased transcription of mRNA** for cytochrome P450. However, certain cases of induction involve **stabilization of mRNA**, **enzyme stabilization**, or **other mechanisms** (eg, an effect on translation).

Induction of cytochrome P450 has important clinical implications, since it is a biochemical mechanism of **drug interaction.** A drug interaction has occurred when the effects of one drug are altered by prior, concurrent, or later administration of another. As an illustration, consider the situation when a patient is taking the anticoagulant **warfarin** to prevent blood clotting. This drug is metabolized by **CYP2C9**. Concomitantly, the patient is started on **phenobarbital** (an inducer of this P450) to treat a certain type of epilepsy, but the dose of warfarin is not changed. After 5 days or so, the level of CYP2C9 in the patient's liver will be elevated three- to fourfold. This in turn means that warfarin will be **metabolized much more quickly than before**, and its **dosage will have become inadequate.** Therefore, the **dose must be increased** if warfarin is to be therapeutically effective. To pursue this example further, **a problem** could arise later on **if the phenobarbital is discontinued** but the increased dosage of warfarin stays the same. The patient will be at risk of bleeding, since the high dose of warfarin will be even more active than before, because the level of CYP2C9 will decline once phenobarbital has been stopped.

Another example of enzyme induction involves **CYP2E1,** which is induced by consumption of **ethanol.** This is a matter for concern, because this P450 metabolizes certain widely used solvents and also components found in tobacco smoke, many of which are established **carcinogens.** Thus, if the activity of CYP2E1 is elevated by induction, this may increase the risk of carcinogenicity developing from exposure to such compounds.

9. Certain isoforms of cytochrome P450 **(eg, CYP1A1)** are particularly involved in the metabolism of polycyclic aromatic hydrocarbons (PAHs) and related molecules; for this reason they were formerly called **aromatic hydrocarbon hydroxylases (AHHs).** This enzyme is important in the metabolism of PAHs and in carcinogenesis produced by these agents. For example, in the lung it may be involved in the conversion of inactive PAHs (procarcinogens), inhaled by smoking, to active carcinogens by hydroxylation reactions. Smokers have higher levels of this enzyme in some of their cells and tissues than do nonsmokers. Some reports have indicated that the activity of this enzyme may be elevated (induced) in **the placenta** of a woman who smokes, thus potentially altering the quantities of metabolites of PAHs (some of which could be harmful) to which the fetus is exposed.

10. Certain cytochrome P450s exist in **polymorphic forms** (genetic isoforms), some of which exhibit low catalytic activity. These observations are one important explanation for the variations in drug responses noted among many patients. One P450 exhibiting polymorphism is **CYP2D6,** which is involved in the metabolism of **debrisoquin** (an antihypertensive drug; see Table 53–2) and **sparteine** (an antiarrhythmic and oxytocic drug). Certain polymorphisms of CYP2D6 cause poor metabolism of these and a variety of other drugs so that they can accumulate in the body, resulting in untoward consequences. Another interesting polymorphism is that of **CYP2A6,** which is involved in the metabolism of **nicotine** to conitine. Three *CYP2A6* alleles have been identified: a wild type and two null or inactive alleles. It has been reported that individuals with the null alleles, who have impaired metabolism of nicotine, are apparently protected against becoming tobacco-dependent smokers (Table 53–2). These individuals smoke less, presumably because their blood and brain concentrations of nicotine remain elevated longer than those of individuals with the wild-type allele. It has been speculated that inhibiting CYP2A6 may be a novel way to help prevent and to treat smoking.

Table 53–1 summarizes some principal features of cytochrome P450s.

CONJUGATION REACTIONS PREPARE XENOBIOTICS FOR EXCRETION IN PHASE 2 OF THEIR METABOLISM

In phase 1 reactions, xenobiotics are generally converted to more polar, hydroxylated derivatives. In phase 2 reactions, these derivatives are conjugated with molecules such as glucuronic acid, sulfate, or glutathione. This renders them even more water-soluble, and they are eventually excreted in the urine or bile.

Five Types of Phase 2 Reactions Are Described Here

Glucuronidation

The glucuronidation of bilirubin is discussed in Chapter 31; the reactions whereby xenobiotics are glucuronidated are essentially similar. UDP-glucuronic acid is the glucuronyl donor, and a variety of glucuronosyltransferases, present in both the endoplasmic reticulum and cytosol, are the catalysts. Molecules such as 2-acetylaminofluorene (a carcinogen), aniline, benzoic acid, meprobamate (a tranquilizer), phenol, and many steroids are excreted as glucuronides. The glucuronide may be attached to oxygen, nitrogen, or sulfur groups of the substrates. Glucuronidation is probably the most frequent conjugation reaction.

Sulfation

Some alcohols, arylamines, and phenols are sulfated. The **sulfate donor** in these and other biologic sulfation reactions (eg,

TABLE 53–1   Some Properties of Human Cytochrome P450s

- Involved in phase I of the metabolism of innumerable xenobiotics, including perhaps 50% of the drugs administered to humans; they may increase, decrease or not affect the activities of various drugs.
- Involved in the metabolism of many endogenous compounds (eg, steroids)
- All are hemoproteins
- Often exhibit broad substrate specificity, thus acting on many compounds; consequently, different P450s may catalyze formation of the same product
- Extremely versatile catalysts, perhaps catalyzing about 60 types of reactions
- However, basically they catalyze reactions involving introduction of one atom of oxygen into the substrate and one into water
- Their hydroxylated products are more water-soluble than their generally lipophilic substrates, facilitating excretion
- Liver contains highest amounts, but found in most if not all tissues, including small intestine, brain, and lung
- Located in the smooth endoplasmic reticulum or in mitochondria (steroidogenic hormones)
- In some cases, their products are mutagenic or carcinogenic
- Many have a molecular mass of about 55 kDa
- Many are inducible, resulting in one cause of drug interactions
- Many are inhibited by various drugs or their metabolic products, providing another cause of drug interactions
- Some exhibit genetic polymorphisms, which can result in atypical drug metabolism
- Their activities may be altered in diseased tissues (eg, cirrhosis), affecting drug metabolism
- Genotyping the P450 profile of patients (eg, to detect polymorphisms) may in the future permit individualization of drug therapy

sulfation of steroids, glycosaminoglycans, glycolipids, and glycoproteins) is **adenosine 3′-phosphate-5′-phosphosulfate (PAPS)** (Chapter 24); this compound is called "active sulfate."

Conjugation With Glutathione

Glutathione (γ-glutamyl-cysteinylglycine) is a **tripeptide** consisting of glutamic acid, cysteine, and glycine (Figure 3–3). Glutathione is commonly abbreviated GSH (because of the sulfhydryl group of its cysteine, which is the business part of the molecule). A number of potentially toxic electrophilic xenobiotics (such as certain carcinogens) are conjugated to the nucleophilic GSH in reactions that can be represented as follows:

$$R + GSH \rightarrow R-S-G$$

where R = an electrophilic xenobiotic. The enzymes catalyzing these reactions are called **glutathione S-transferases** and are present in high amounts in liver cytosol and in lower amounts

in other tissues. A variety of glutathione S-transferases are present in human tissue. They exhibit different substrate specificities and can be separated by electrophoretic and other techniques. If the potentially toxic xenobiotics were not conjugated to GSH, they would be free to combine covalently with DNA, RNA, or cell protein and could thus lead to serious cell damage. GSH is therefore an important **defense mechanism** against certain toxic compounds, such as some drugs and carcinogens. If the levels of GSH in a tissue such as liver are **lowered** (as can be achieved by the administration to rats of certain compounds that react with GSH), then that tissue can be shown to be more susceptible to injury by various chemicals that would normally be conjugated to GSH. Glutathione conjugates are subjected to **further metabolism** before excretion. The glutamyl and glycinyl groups belonging to glutathione are removed by specific enzymes, and an acetyl group (donated by acetyl-CoA) is added to the amino group of the remaining cysteinyl moiety. The resulting compound is a **mercapturic acid,** a conjugate of L-acetylcysteine, which is then excreted in the urine.

Glutathione has other important functions in human cells apart from its role in xenobiotic metabolism.

1. It participates in the decomposition of potentially toxic **hydrogen peroxide** in the reaction catalyzed by glutathione peroxidase (Chapter 21).
2. It is an important **intracellular reductant,** helping to maintain essential SH groups of enzymes in their reduced state. This role is discussed in Chapter 21, and its involvement in the hemolytic anemia caused by deficiency of glucose-6-phosphate dehydrogenase is discussed in Chapters 21 and 52.
3. A metabolic cycle involving GSH as a carrier has been implicated in the **transport of certain amino acids** across membranes in the kidney. The first reaction of the cycle is shown below.

$$\text{Amino acid} + \text{GSH} \rightarrow \gamma\text{-Glutamyl amino acid} + \text{Cysteinylglycine}$$

This reaction helps transfer certain amino acids across the plasma membrane, the amino acid being subsequently hydrolyzed from its complex with GSH and the GSH being resynthesized from cysteinylglycine. The enzyme catalyzing the above reaction is **γ-glutamyltransferase (GGT).** It is present in the plasma membrane of renal tubular cells and bile ductule cells, and in the endoplasmic reticulum of hepatocytes. The enzyme has diagnostic value because it is released into the blood from hepatic cells in **various hepatobiliary diseases**.

Other Reactions

The two most important other reactions are acetylation and methylation.

1. **Acetylation**—Acetylation is represented by

$$X + \text{Acetyl-CoA} \rightarrow \text{Acetyl-X} + \text{CoA}$$

where X represents a xenobiotic. As for other acetylation reactions, **acetyl-CoA** (active acetate) is the acetyl donor. These reactions are catalyzed by **acetyltransferases** present in the cytosol of various tissues, particularly liver. The drug **isoniazid,** used in the treatment of tuberculosis, is subject to acetylation. **Polymorphic types** of acetyltransferases exist, resulting in individuals who are classified as **slow or fast acetylators,** and influence the rate of clearance of drugs such as isoniazid from blood. Slow acetylators are more subject to certain toxic effects of isoniazid because the drug persists longer in these individuals.

2. Methylation—A few xenobiotics are subject to methylation by methyltransferases, employing *S*-adenosylmethionine (Figure 29–18) as the methyl donor.

THE ACTIVITIES OF XENOBIOTIC-METABOLIZING ENZYMES ARE AFFECTED BY AGE, SEX, & OTHER FACTORS

Various factors affect the activities of the enzymes metabolizing xenobiotics. The activities of these enzymes may differ substantially among **species.** Thus, for example, the possible **toxicity** or **carcinogenicity** of xenobiotics cannot be extrapolated freely from one species to another. There are significant differences in enzyme activities among individuals, many of which appear to be due to **genetic factors.** The activities of some of these enzymes vary according to **age** and **sex.**

Intake of various xenobiotics such as phenobarbital, PCBs, or certain hydrocarbons can cause **enzyme induction.** It is thus important to know whether or not an individual has been exposed to these inducing agents in evaluating biochemical responses to xenobiotics. (It is always important when taking a clinical history to ask whether the patient has been taking any drugs or other therapeutic preparations). Metabolites of certain xenobiotics can **inhibit** or **stimulate** the activities of xenobiotic-metabolizing enzymes. Again, this can affect the doses of certain drugs that are administered to patients. Various **diseases** (eg, cirrhosis of the liver) can affect the activities of drug-metabolizing enzymes, sometimes necessitating adjustment of dosages of various drugs for patients with these disorders.

RESPONSES TO XENOBIOTICS INCLUDE PHARMACOLOGIC, TOXIC, IMMUNOLOGIC, & CARCINOGENIC EFFECTS

Xenobiotics are metabolized in the body by the reactions described above. When the xenobiotic is a drug, phase 1 reactions may produce its active form or may diminish or terminate its action if it is pharmacologically active in the body without prior metabolism. The diverse effects produced by drugs com-

TABLE 53–2 Some Important Drug Reactions Due to Mutant or Polymorphic Forms of Enzymes or Proteins[1]

Enzyme or Protein Affected	Reaction or Consequence
Glucose-6-phosphate dehydrogenase (G6PD) [mutations] (OMIM 305900)	Hemolytic anemia following ingestion of drugs such as primaquine
Ca^{2+} release channel (ryanodine receptor) in the sarcoplasmic reticulum [mutations] (OMIM 180901)	Malignant hyperthermia (OMIM 145600) following administration of certain anesthetics (eg, halothane)
CYP2D6 [polymorphisms] (OMIM 124030)	Slow metabolism of certain drugs (eg, debrisoquin), resulting in their accumulation
CYP2A6 [polymorphisms] (OMIM 122720)	Impaired metabolism of nicotine, resulting in protection against becoming a tobacco-dependent smoker

[1]G6PD deficiency is discussed in Chapters 21 & 52 and malignant hyperthermia in Chapter 49. At least one gene other than that encoding the ryanodine receptor is involved in certain cases of malignant hypertension. Many other examples of drug reactions based on polymorphism or mutation are available.

prise the area of study of pharmacology; here it is important to appreciate that drugs act primarily through biochemical mechanisms. Table 53–2 summarizes four important reactions to drugs that reflect **genetically determined differences** in enzyme and protein structure among individuals—part of the field of study known as **pharmacogenetics.** This area of science has been defined as **the study of the contribution of genetic factors to variation in drug response and toxicity.**

Polymorphisms that affect drug metabolism can occur in any of the **enzymes** involved in drug metabolism (including cytochrome P450s), in **transporters** and in **receptors.**

Certain xenobiotics are very toxic even at low levels (eg, cyanide). On the other hand, there are few xenobiotics, including drugs, that do not exert some toxic effects if sufficient amounts are administered. The **toxic effects of xenobiotics** cover a wide spectrum, but the major effects can be considered under three general headings (Figure 53–1).

The first is **cell injury** (cytotoxicity), which can be severe enough to result in cell death. There are many mechanisms by which xenobiotics injure cells. The one considered here is **covalent binding to cell macromolecules** of reactive species of xenobiotics produced by metabolism. These macromolecular targets include **DNA, RNA,** and **protein.** If the macromolecule to which the reactive xenobiotic binds is essential for short-term cell survival, eg, a protein or enzyme involved in some critical cellular function such as oxidative phosphorylation or regulation of the permeability of the plasma membrane, then severe effects on cellular function could become evident quite rapidly.

Second, the reactive species of a xenobiotic may bind to a protein, altering its **antigenicity.** The xenobiotic is said to act as a **hapten,** ie, a small molecule that by itself does not stimulate antibody synthesis but will combine with antibody once formed. The resulting **antibodies** can then damage the cell by

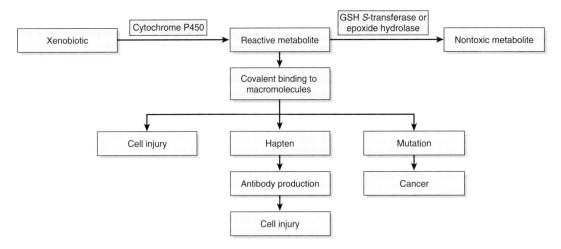

FIGURE 53–1 Simplified scheme showing how metabolism of a xenobiotic can result in cell injury, immunologic damage, or cancer. In this instance, the conversion of the xenobiotic to a reactive metabolite is catalyzed by a cytochrome P450, and the conversion of the reactive metabolite (eg, an epoxide) to a nontoxic metabolite is catalyzed either by a GSH S-transferase or by epoxide hydrolase.

several immunologic mechanisms that grossly perturb normal cellular biochemical processes.

Third, reactions of activated species of chemical carcinogens with **DNA** are thought to be of great importance in **chemical carcinogenesis.** Some chemicals (eg, benzo[a]pyrene) require activation by monooxygenases in the endoplasmic reticulum to become carcinogenic (they are thus called **indirect carcinogens**). The activities of the monooxygenases and of other xenobiotic-metabolizing enzymes present in the endoplasmic reticulum thus help to determine whether such compounds become carcinogenic or are "detoxified." Other chemicals (eg, various alkylating agents) can react directly (direct carcinogens) with DNA without undergoing intracellular chemical activation.

The enzyme **epoxide hydrolase** is of interest because it can exert a protective effect against certain carcinogens. The products of the action of certain monooxygenases on some procarcinogen substrates are **epoxides.** Epoxides are highly reactive and mutagenic or carcinogenic or both. Epoxide hydrolase—like cytochrome P450, also present in the membranes of the endoplasmic reticulum—acts on these compounds, converting them into much less reactive dihydrodiols. The reaction catalyzed by epoxide hydrolase can be represented as follows:

$$\underset{\textbf{Epoxide}}{-\overset{|}{\underset{\diagdown\!\!\!O\!\!\!\diagup}{C}}-\overset{|}{C}-} + H_2O \longrightarrow \underset{\textbf{Dihydrodiol}}{-\overset{|}{\underset{HO}{C}}-\overset{|}{\underset{OH}{C}}-}$$

PHARMACOGENOMICS WILL DRIVE THE DEVELOPMENT OF NEW & SAFER DRUGS

As indicated above, **pharmacogenetics** is **the study of the contribution of genetic factors to variation in drug re-**

sponse and toxicity. As a result of the progress made in sequencing the human genome, a new field of study—**pharmacogenomics**—has developed recently. It has been defined as **the use of genomic information and technologies to optimize the discovery and development of drug targets and drugs.** It builds on pharmacogenetics, but covers a wider sphere of activity. Information from genomics, proteomics, bioinformatics, and other disciplines such as biochemistry and toxicology will be integrated to make possible the synthesis of newer and safer drugs. As the sequences of all our genes and their encoded proteins are determined, this will

FIGURE 53–2 Simplified scheme of some approaches to the development of new drugs.

reveal many new **targets for drug actions.** It will also reveal **polymorphisms** (this term is briefly discussed in Chapter 50) of enzymes and proteins **related to drug metabolism, action, and toxicity.** Microarrays capable of detecting them will be constructed, permitting **screening of individuals** for potentially harmful polymorphisms prior to the start of drug therapy. Already **gene chips** are available for analyzing certain P450 genotypes (eg, for CYP2D6, whose gene product is involved in the metabolism of many antidepressants, antipsychotics, β-blockers, and some chemotherapeutic agents). Figure 53–2 summarizes some approaches to developing new drugs. Major thrusts of new drug development are to **enhance treatment** and to **provide safer, personalized drugs**, taking into account polymorphisms and other **genetic** and **environmental** factors. It has been estimated that some 100,000 deaths from adverse drug reactions occur each year in the United States alone. It is hoped that new information provided by studies in the various areas indicated in Figure 53–2 and in other areas will translate into successful therapies and also **eventually** into a new era of personalized therapeutics. However, much work remains to be done before this is achievable.

- Many cytochrome P450s are inducible. This has important implications in phenomena such as drug interaction.
- Mitochondrial cytochrome P450s also exist and are involved in cholesterol and steroid biosynthesis. They use a non-heme iron-containing sulfur protein, adrenodoxin, not required by microsomal isoforms.
- Cytochrome P450s, because of their catalytic activities, play major roles in the reactions of cells to chemical compounds and in chemical carcinogenesis.
- Phase 2 reactions are catalyzed by enzymes such as glucuronosyltransferases, sulfotransferases, and glutathione S-transferases, using UDP-glucuronic acid, PAPS (active sulfate), and glutathione, respectively, as donors.
- Glutathione not only plays an important role in phase 2 reactions but is also an intracellular reducing agent and is involved in the transport of certain amino acids into cells.
- Xenobiotics can produce a variety of biologic effects, including pharmacologic responses, toxicity, immunologic reactions, and cancer.
- Catalyzed by the progress made in sequencing the human genome, the new field of pharmacogenomics offers the promise of being able to make available a host of new rationally designed, safer drugs.

SUMMARY

- Xenobiotics are chemical compounds foreign to the body, such as drugs, food additives, and environmental pollutants; more than 200,000 have been identified.
- Xenobiotics are metabolized in two phases. The major reaction of phase 1 is hydroxylation catalyzed by a variety of monooxygenases, also known as the cytochrome P450s. In phase 2, the hydroxylated species are conjugated with a variety of hydrophilic compounds such as glucuronic acid, sulfate, or glutathione. The combined operation of these two phases renders lipophilic compounds into water-soluble compounds that can be eliminated from the body.
- Cytochrome P450s catalyze reactions that introduce one atom of oxygen derived from molecular oxygen into the substrate, yielding a hydroxylated product. NADPH and NADPH-cytochrome P450 reductase are involved in the complex reaction mechanism.
- All cytochrome P450s are hemoproteins and generally have a wide substrate specificity, acting on many exogenous and endogenous substrates. They represent the most versatile biocatalyst known.
- Some 57 cytochrome P450 genes are found in human tissue.
- Cytochrome P450s are generally located in the endoplasmic reticulum of cells and are particularly enriched in liver.

REFERENCES

Austin CP: The impact of the completed human genome sequence on the development of novel therapeutics for human disease. Annu Rev Med 2004;55:1.

Evans WE, McLeod HL: Pharmacogenomics—drug disposition, drug targets, and side effects. N Engl J Med 2003;348:538.

Human Cytochrome P450 (CYP) Allele Nomenclature Committee. http://www.imm.ki.se/CYPalleles/

Ingelman-Sundberg M: Pharmacogenomic biomarkers for prediction of severe adverse drug reactions. N Engl J Med 2008;358:637.

Kalant H, Grant DM, Mitchell J (eds): *Principles of Medical Pharmacology*, 7th ed. Saunders Elsevier, 2007. (Chapters 4 [Drug Biotransformation by Riddick DS] and 10 [Pharmacogenetics and Pharmacogenomics by Grant DM and Kalow W] are particularly relevant to this Chapter).

Katzung BG (editor): *Basic & Clinical Pharmacology*, 10th ed. McGraw-Hill, 2006.

Lee C, Morton CC: Structural genomic variation and personalized medicine. N Engl J Med 2008;358:740.

Reilly J-J, Blakey JD: Why pharmacogenetics is something worth talking about. Student BMJ 2006;14:265.

Shurin SB, Nabel EG: Pharmacogenomics—ready for prime time? N Engl J Med 2008:358:1061.

Wilkinson GR: Drug metabolism and variability among patients in drug response. N Engl J Med 2005;350:2211.

54

Biochemical Case Histories

Robert K. Murray, MD, PhD & Peter L. Gross, MD

INTRODUCTION

In this final chapter, 16 case histories are presented and discussed. They illustrate the importance of a knowledge of Biochemistry for **the understanding of disease**. Of course, as has been shown throughout the text, biochemistry is also crucial for the **understanding of health and wellness**.

Most of the diseases discussed here are **prevalent**, or relatively prevalent, in a global sense. (Prevalence is the proportion of persons in a given population that has a particular disease at a point or interval of time.) However, two (xeroderma pigmentosum and severe combined immunodeficiency disease due to deficiency of adenosine deaminase [ADA]) are relatively rare. They are included because they illustrate two crucial biologic facts: the importance of **DNA repair** and of the **immune system** as protective mechanisms. In addition, ADA deficiency is the first disease for which **gene therapy** was performed in humans.

The **reference values** for laboratory tests cited in the cases below may differ from these listed by laboratories with which the reader may be familiar. This is because reference values from different laboratories vary somewhat, in part due to different methodologies. In this chapter, laboratory results are generally given as **SI units** (Systeme International d'Unites). **Appendix I** lists the majority of the normal values for the lab tests referred to in this chapter, and gives both SI and "conventional" values (as widely used in the United States).

The **doses of drugs** administered in the treatment of the cases described here are not generally given; the reader should check these out on her/his own.

CASE 1: ADENOSINE DEAMINASE (ADA) DEFICIENCY CAUSING SEVERE COMBINED IMMUNODEFICIENCY DISEASE (SCID)

Causation

Genetic (due to mutations in the gene encoding ADA). Deficiency of ADA affects the immune system.

History and Physical Examination

A little girl aged 11 months was brought by her parents to a children's hospital. She had had a number of attacks of pneumonia and thrush (oral infection usually due to *Candida albicans*) since birth. The major findings of a thorough workup were very low levels of circulating lymphocytes (ie, severe lymphopenia) and low levels of circulating immunoglobulins. The attending pediatrician suspected SCID.

Laboratory Findings

Analysis of a sample of red blood cells revealed a very low activity of ADA and also a very high level (about 50 times normal) of dATP. This confirmed the diagnosis of SCID due to deficiency of ADA, the enzyme that converts adenosine to inosine (Chapter 33):

$$Adenosine \rightarrow Inosine + NH_4^+$$

Treatment

Appropriate **antibiotic** treatment was commenced and the child was given periodic injections of **immune globulin**. In addition, she was started on weekly intramuscular injections of **bovine ADA conjugated to polyethylene glycol**. Bovine ADA is relatively non-immunogenic, and conjugation to polyethylene glycol prolongs its half-life. It has been shown to be beneficial in the treatment of ADA deficiency. Her parents were informed that **bone marrow transplantation** was the most appropriate therapy, but the treatment was declined. In view of reports of successes with ADA gene therapy, this treatment was offered, and the parents consented. The treatment was approved by the hospital Ethics Committee. Lymphocytes and mononuclear cells were isolated from her blood using a gradient of Ficoll (a neutral highly branched polysaccharide). They were then grown in the presence of interleukin-2 (to stimulate cell division) and infected with a modified retrovirus containing inserts encoding ADA and also a gene (the NeoR gene) encoding an enzyme governing neomycin resistance, which was used to show that gene transfer had been achieved. An alternative at the present time would be to use **bone marrow stem cells** (which have been reported to give good increases of both B and T cells), but these were not available at the time of treatment. The autologous gene-treated cells were then injected intravenously. The child received similar injections once a month over the next year, and in addition continued to receive polyethylene glycol-conjugated ADA. Measurement of the activity of ADA revealed a sustained elevation (about 20% of normal) of the enzyme in circulating lymphocytes after 6 months of treatment; analyses using the PCR technique with NeoR probes revealed that approximately the same percentage of lymphocytes contained inserted genetic material.

Discussion

Deficiency of the activity of ADA is inherited as an **autosomal recessive** condition. It accounts for approximately 15% of cases of SCID; other causes involve mutations in a variety of genes affecting the function of cells of the immune system. Most of the mutations in the gene for ADA so far detected have been base substitutions, though deletions have also been detected. These mutations result in diminished activity or stability of ADA. The block of activity of ADA results in accumulation of adenosine, which in turn results in accumulation of deoxyadenosine and **dATP**. Elevated levels of the latter are **toxic**, particularly to **T lymphocytes**, which normally exhibit a high activity of ADA. Thus, lymphocytes are injured or killed, resulting in **impairment of both cell and humoral immunity**, because impairment of T cell function can secondarily affect B-cell function.

Adenosine deaminase deficiency has become quite prominent because it is the first disease to be treated by somatic cell gene therapy. Several patients have been treated by protocols similar to that described above, which is an example of ex vivo (the lymphocytes and mononuclear cells were removed from the body prior to insertion of the ADA gene) gene therapy. One reason for selecting ADA deficiency as a suitable condition for somatic gene therapy was that cells that express the gene for ADA would have **a selective advantage for growth** over uncorrected cells. Gene therapy is discussed briefly in Chapter 39. Important points concerning gene therapy include that the **level of expression** of the affected protein should be sufficient to sustain normal function, that ideally the inserted gene should show **normal regulation** and that **no significant undesirable side effects** should occur (eg, cancer due to insertional mutagenesis). In relation to the last point, the gene that is being delivered may insert into a gene that is essential for normal cell growth, and if this occurs that may result in the cell becoming cancerous (an example of insertional mutagenesis), as has been noted in some instances of gene therapy.

A simplified scheme of the events involved in the causation of ADA deficiency is given in Figure 54–1. Safe and effective treatment of ADA deficiency by gene therapy has been reported recently (see references).

CASE 2: ALZHEIMER DISEASE (AD)

Before studying this case, the reader is advised to consult the material in Chapter 5 on AD.

FIGURE 54–1 Summary of probable events in the causation of SCID due to ADA deficiency (OMIM 102700).

Causation

Deposition of amyloid β peptide ($A\beta_{42}$) in certain parts of the brain is believed by many neuroscientists to be one major cause of AD. It is thought that this 42-amino-acid peptide, occurring as beta sheets, oligomerizes and is deposited around neurons; the oligomers may be toxic to neurones. Deposition of $A\beta_{42}$ may be due to excessive formation or decreased removal of the peptide. In certain cases of familial AD, specific genes have been identified (eg, these encoding amyloid precursor protein [APP], presenilins 1 and 2, and apolipoprotein E4), which affect the production or removal of $A\beta_{42}$.

History and Physical Examination

A 72-year-old woman who lived on her own was found wandering around her neighborhood at 2 a.m. Her husband had died 3 years previously and her one son lived some distance away. The lady was confused and taken to the hospital. The son was notified and came immediately to see his mother. She was not able at the time of admission to give a clear history. The son volunteered that she had been diagnosed by a neurologist as having early AD, but had refused to go into a nursing home. She had home help during the day, and had not previously wandered out of her home. Sometimes a lady friend visited and spent the night at her house. In fact, she had appeared relatively normal prior to the present situation, as her son spoke to her on the phone every day of the week. However, her short-term memory had become worse in recent months, and he had become concerned about her. She was on a medication (donezepil) for AD. Otherwise she had no significant medical history. She was kept in the hospital for a couple of days, during which time her family doctor and the neurologist were consulted.

Treatment

There is no specific treatment for AD at the present time. Donezepil and several other drugs that are used in the management of AD **inhibit the activity of cholinesterase**, an enzyme that hydrolyzes acetylcholine (ACh) to acetate and choline. They are used because some studies had shown lower than normal levels of ACh in specimens of brains from subjects who had died of AD. They appear to produce a modest improvement of brain function and memory in some patients. Memantine, a drug that **antagonizes N-methyl-D-aspartate receptors**, may cause some slowing of the progression of AD. Symptoms such as depression, agitation, anxiety, and insomnia can be treated by appropriate drugs and antipsychotics may be required if psychosis intervenes. A possible preventive role of **omega-3 fatty acids** is under study. However, overall, there is as yet no effective therapy for AD. This patient was kept on donepezil and admitted to a nursing home that specialized in caring for patients with AD. Apart from high-quality basic care, the nursing home offered a variety of programs, including music therapy and exercise programs.

Discussion

AD is a progressive neurodegenerative condition in which **decline of general cognitive function** occurs, usually accompanied by **affective** and **behavioral disturbances**. At least 2 million people in the United States suffer from AD, and its prevalence is likely to increase as more people live longer. Some cases have a **familial (genetic)** basis, but the great majority (~90%) appear to be **sporadic**. AD is the commonest cause of **dementia**, which can be defined as a progressive decline in intellectual functions, due to an organic cause, that interferes substantially with an individual's activities. AD places a tremendous burden on families and on the health care system, as, sooner or later, most patients cannot look after themselves. The usual age at onset is over 65 years, but the disease can have an early onset (eg, in the 40s), particularly in cases where there is a genetic predisposition (see below). Survival ranges from 2 to 20 years. It is estimated that about 40% of people over 85 years of age have variable degrees of AD. **Loss of short-term memory** is often the first sign. AD usually **progresses** inexorably, and many patients are eventually completely incapacitated.

The **diagnosis** is usually one of exclusion. A complete neurologic exam is necessary and also a recognized mental status exam. Other forms of dementia (Lewy body, vascular, etc) must be excluded, as must other organic and psychiatric problems; various lab tests may thus be indicated to do this. In certain cases an MRI or CT scan may be indicated; these will usually reveal variable degrees of cortical atrophy and enlargement of ventricles if AD is present. Considerable research is underway to develop laboratory tests (eg, on blood or cerebrospinal fluid) that will assist in making the unequivocal diagnosis of AD.

The basic **pathologic picture** is of a degenerative process characterized by the death and consequent loss of cells in certain areas of the brain (eg, the cortex, hippocampus and certain other sites). **Apoptosis** (a programmed type of cell death in which various mechanisms, particularly the activities of proteolytic enzyme known as caspases, are activated within a cell leading to rapid cell death) may be involved in the cell death occurring in AD. At the microscopic level, **neuritic plaques** containing aggregated amyloid β peptide (Aβ, a peptide of 42 amino acids, occurring in beta sheets) surrounded

by nerve cells containing **neurofibrillary tangles** (paired helical filaments formed from a hyperphosphorylated form of the microtubule associated protein, **tau**) are hallmarks. Deposits of $A\beta_{12}$ are frequent in small blood vessels.

Intensive research is under way to determine the cause of AD. Particular interest has focused on the presence of $A\beta_{42}$, the major constituent of the plaques found in AD. The term "amyloid" refers to a group of diverse extracellular protein deposits found in many different diseases (see Chapter 50). Amyloid proteins usually stain blue with iodine, like starch, which accounts for the name (amylo denotes starch). The **amyloid cascade hypothesis** proposes that deposition of $A\beta_{42}$ is the cause of the pathologic changes observed in the brains of the victims of AD and that other changes, such as neurofibrillary tangles and vascular alterations are secondary. $A\beta_{42}$ is derived from a larger precursor protein named **amyloid precursor protein (APP)**, whose gene is located on chromosome 21 close to the area affected in Down syndrome (trisomy 21). Individuals with Down syndrome who survive to age 50 often suffer from AD.

As shown in Figure 54–2, APP is a transmembrane protein which can be cleaved by proteases known as secretases. In step 1, APP is cleaved by either β-secretase or α-secretase to produce small nontoxic products. Then in step 2, cleavage of the β-secretase product by γ-secretase results in either the toxic $A\beta_{42}$ (containing 42 amino acids) or the non-toxic $A\beta_{40}$ peptide. Cleavage of the α-secretase product by γ-secretase produces the non-toxic P3 peptide. When split off from its parent protein,

Step 1: Cleavage by either α- or β-secretase

Step 2: Cleavage by γ-secretase

FIGURE 54–2 Simplified scheme of the formation of $A\beta_{42}$. Amyloid precursor protein (APP) is digested by β-, α-, and γ-secretases. A key initial step (step 1) is the digestion by either β-secretase or α-secretase, producing smaller nontoxic products. Cleavage of the β-secretase product by γ-secretase (step 2) results in either the toxic $A\beta_{42}$ (containing 42 amino acids) or the nontoxic $A\beta_{40}$ peptide. Cleavage of the α-secretase product by γ-secretase produces the nontoxic P3 peptide. Excess production of $A\beta_{42}$ is a key initiator of cellular damage in Alzheimer disease (AD). Among research efforts on AD have been attempts to develop therapies to reduce accumulation of $A\beta_{42}$ by inhibiting β- or γ-secretases, promoting α-secretase activity or clearing $A\beta_{42}$ by use specific antibodies. (Reproduced, with permission, from Fauci AS et al [eds,] *Harrison's Principles of Internal Medicine,* 17th ed, McGraw-Hill, 2008, p. 2542.)

Aβ$_{42}$ forms an insoluble extracellular deposit. **Aggregation** of Aβ$_{42}$, produced by its **oligomerization** and formation of **beta sheets,** is thought by some to be a key event in causing AD.

Mutations in certain genes have been found in some patients with AD (familial AD). These mutations often predispose to early onset AD. One of these genes is that encoding APP. Table 54–1 summarizes some aspects of the principal genes so far discovered. In general, the effects of the products of these genes are to enhance deposition of amyloid or to diminish its removal. Precise dissection of their mechanisms of action is underway.

A second part of the amyloid cascade hypothesis is that Aβ or Aβ-containing fragments are directly or indirectly **neurotoxic.** There is evidence that exposure of neurons to Aβ can increase their intracellular concentrations of **Ca^{2+}**. Some **protein kinases,** including that involved in phosphorylation of **tau,** are regulated by levels of Ca^{2+}. Thus, increase of Ca^{2+} may lead to hyperphosphorylation of tau and formation of the paired helical filaments present in the neurofibrillary tangles. Interference with **synaptic function** is also probable, perhaps secondary to neuronal damage.

Further research may reveal unexpected developments that alter the validity of the amyloid cascade theory as presented above.

Work on AD has shown the probable importance of an **abnormally folded peptide** in the causation of this important brain disease. It is hoped that further research on AD may result in drugs that will prevent, arrest, or even reverse AD. For example, it may be possible to **develop small molecules** that either prevent formation or deposition of Aβ$_{42}$, prevent its aggregation, or accelerate its clearance. In addition, it is possible that specific antibodies to Aβ$_{42}$ or tau could prevent their putative toxic actions.

AD is one of the so-called **conformational diseases** (Chapters 46 & 50), in which abnormally folded proteins play a central role in the causation of a disease. Other examples of these diseases are cystic fibrosis (this chapter), alpha$_1$-antitrypsin disease (Chapter 50) and the prion diseases (Chapter 5).

FIGURE 54–3 A tentative scheme of the possible sequence of events in at least certain cases of AD.

The study of **various neurodegenerative diseases** is providing dramatic evidence of **the importance of protein structure and function** in their causation. For example, abnormal forms of the protein huntingtin play an important role in Huntington disease, abnormalities of α-synuclein play a role in some cases of Parkinson disease, and prions have been found to play key roles in the causation of bovine spongiform encephalopathy (BSE) and Creutzfeldt-Jacob disease. The application of genomic and proteomic techniques is also beginning to throw light on the **causation of major psychiatric disorders,** such as bipolar disease and schizophrenia. The importance of genetic and biochemical approaches in understanding disease processes has never been more clear. A simplified scheme of the causation of AD is shown in Figure 54–3.

CASE 3: CHOLERA

Causation

Infection by *Vibrio cholerae.*

History and Physical Examination

A 21-year-old female medical student working in a developing country suddenly began to pass profuse watery stools almost continuously. She soon started to vomit, her general condition declined abruptly, and she was rushed to the local village hospital. On admission, she was cyanotic, skin turgor was poor, blood pressure was 70/50 mm Hg (normal 120/80 mm Hg), and her pulse was rapid and weak. The doctor on duty diagnosed cholera, took a stool sample, and started treatment immediately.

Treatment

Treatment consisted of **intravenous** administration of a solution made up in the hospital containing 5 g NaCl, 4 g NaHCO$_3$,

TABLE 54–1 Some Genes Involved in Familial Types of Alzheimer Disease (AD)

Gene	Type of AD	Chromosome	Protein Product
APP	AD1, Familial (OMIM 104300)	21	APP
APOE4	AD2, Late onset (OMIM 104310)	19	ApoE4
PS1	AD3, Early onset (OMIM 104311)	14	Presenilin 1
PS2	AD4, Familial (OMIM 606889)	1	Presenilin 2

In general, the products of these genes act by increasing the production of Aβ$_{42}$ (APP, PS1 and PS2) or by decreasing its clearance (APOE4). Presenilins 1 and 2 may be involved in the action of γ-secretase. (APP, amyloid precursor protein; OMIM, Online Mendelian Inheritance in Man catalog number; APOE4, apolipoprotein E4; PS, presenilin.)

and 1 g KCl per liter of pyrogen-free distilled water. This solution was initially given rapidly (100 mL per h) until her blood pressure and pulse rate were normalized. She was also given the antibiotic **doxycycline**. On the second day, she was able to take the **oral rehydration solution** recommended by the World Health Organization (WHO) for the treatment of cholera, consisting of 20 g of glucose, 3.5 g of NaCl, trisodium citrate and dihydrate 2.9 g (or 2.5 g NaHCO$_3$), and 1.5 g KCl per liter of drinking water.

She took amounts moderately exceeding the volume of her daily stools. **Solid food** was reinstituted on the fourth day after admission. She continued to recover rapidly and was discharged 7 days after admission.

Discussion

Cholera is an important infectious disease endemic in certain Asian countries and other parts of the world. Fecal contamination of water and food is the principal method of transmission. It is due to *Vibrio cholerae*, a bacterium that secretes a protein **enterotoxin**. The toxin is actually encoded by a bacteriophage (CTXφ) resident in *V cholerae*. The enterotoxin is made up of **one A subunit** (composed of one A1 and one A2 peptide joined by a disulfide link) and **five B subunits** and has a molecular mass of approximately 84 kDa. In the small intestine, the toxin **attaches** by means of the **B subunits** binding to the **ganglioside GM1** (Figure 15–13) present in the plasma membrane of mucosal cells (Figure 54–4). The A subunit then dissociates, and the **A1 peptide** passes across to the inner aspect of the plasma membrane. It catalyzes the **ADP-ribosylation** (using **NAD$^+$** as donor) of the GTP-binding regulatory component (G$_s$) of adenylate cyclase, upregulating the activity of this enzyme. Thus, **adenylyl cyclase** becomes chronically activated (Chapter 42). This results in an elevation of **cAMP**, which activates protein kinase A (PKA). This in turn via phosphorylation of CFTR and of a Na$^+$-H$^+$ exchanger leads to inhibition of

FIGURE 54–4 Diagrammatic representation of the mechanism of action of cholera toxin (CT) attaches to the plasma membrane via interaction of its B subunits with the ganglioside GM$_1$. The A subunit crosses the membrane and catalyzes the addition of the ADP-ribose component of NAD$^+$ to the G protein involved in stimulating adenylyl cyclase (NAD$^+$ → ADP-ribose + nicotinamide). The addition of ADP-ribose to the G protein locks adenylyl cyclase in its active conformation, increasing the intracellular level of cAMP. This leads to phosphorylation of several membrane transporters, in turn resulting in accumulation of the ions shown and of water in the intestinal lumen, thus producing often massive diarrhea. (Reproduced, with permission, from Nester EW et al, Microbiology: A Human Perspective, 5th ed. McGraw Hill, 2007.)

absorption of Na^+ and enhancement of secretion of Cl^-. Thus, massive amounts of NaCl accumulate inside the lumen of the intestine, attracting water by osmosis and contributing to the liquid stools characteristic of cholera.

Cholera toxin may also affect other molecules involved in intestinal secretion (eg, prostaglandins and nerve histamine receptors). The histologic structure of the small intestine remains remarkably unaffected, despite the loss of large amounts of Na^+, Cl^- water, HCO_3 and K^+. It is the loss of these constituents that results in the marked fluid loss (dehydration), low blood volume, acidosis, and K^+ depletion found in serious cases of cholera, and which can prove fatal unless appropriate replacement therapy (as described above) is begun immediately. A person suffering from cholera can lose as much as 1 L of fluid per hour.

The recognition and easy availability of appropriate replacement fluids, such as oral rehydration solution, has led to tremendous improvement in the treatment of cholera. It should be emphasized that **glucose** is an essential component of oral rehydration solution (Chapter 40); whereas the cholera toxin inhibits absorption of Na^+ by intestinal cells, it does not inhibit glucose-facilitated Na^+ transport into these cell, so glucose will be absorbed and used to supply energy.

Figure 54–5 summarizes the mechanisms involved in the causation of the diarrhea of cholera.

FIGURE 54–5 Summary of mechanisms involved in the causation of the diarrhea of cholera.

CASE 4: COLORECTAL CANCER

Causation

Environmental and genetic (most if not all cancers are thought to originate by the occurrence of mutations in key genes regulating cell growth [oncogenes and tumor suppressor genes]). The mutations may be inherited, but much more often various environmental influences (chemicals, radiation, and some viruses) are involved.

History, Physical Examination, and Results of Tests

A 62-year-old male consulted his family physician. He had noted that he had passed some fresh bright red blood in his stools several times in the previous 3 months, which he attributed to hemorrhoids. Over the previous 12 months his appetite had decreased and he had lost over 10 lbs. He had always been in good health until the past year, and was not on any medications. He had no other complaints.

In view of the history of rectal bleeding, weight loss, and the patient's age, his physician suspected that he might have colorectal cancer and requested the patient to submit three consecutive daily specimens of feces for the fecal occult blood test. Shortly thereafter the physician received a report indicating that the results were positive. He also ordered a complete blood count and estimations of levels of serum iron, iron-binding capacity, and ferritin. The results showed a microcytic anemia (see Chapter 52), often found in patients with colorectal cancer because of bleeding from the tumor. A rectal examination was negative. No abnormalities were noted in chest x-rays.

The physician arranged a consult 4 days later with a gastroenterologist. Colonoscopy was performed 1 week later. This revealed the presence of a moderately large tumor (approximately 5×6 cm) in the middle of the transverse colon. Measurement of carcinoembryonic antigen (CEA), a biomarker for colorectal cancer (see below for further comments and also Chapter 7), was ordered. It was elevated (20 µg/L, normal 0–3 µg/L). Surgery was scheduled 2 weeks later, when the tumor was resected and end-to-end anastomosis performed. The regional lymph nodes were also excised and submitted along with the tumor specimen to the pathology lab. No local invasion by the tumor was noted, and no tumor was visible elsewhere in the abdomen, including the liver. The subsequent pathology report described the tumor as a relatively well-differentiated adenocarcinoma, invading the muscular mucosa. No tumor cells were noted in the lymph glands; no distant metastases were noted at the time of surgery. The TNM stage was T1N0M0 (cancer limited to the mucosa and submucosa, with an approximate 5-year survival rate of >90% [T = tumor, N = nodes, M = metastases]). (The interested reader should check the staging of tumors of the large intestine in a textbook of pathology.) In view of these findings, no chemotherapy or radiation therapy was considered necessary. Determination of CEA several weeks after surgery showed it had declined to normal levels. The patient was advised to return for follow-up at regular intervals, when, among other tests, samples of blood for measurements of CEA were taken; they remained at normal levels. A follow-up colonoscopy was performed 3 years after the operation; no tumor was seen in the colon. The patient was alive and well at 5 years after operation.

Discussion

The biochemistry of cancer is a huge subject; only a few topics will be covered here, particularly (1) oncogenes and tumor suppressor genes and (2) the use of CEA as a tumor marker.

Colorectal cancer is the second most common cancer in the United States, lung cancer being number one. It can occur anywhere in the large intestine, although the rectum is the most common site. Some 95% of malignant tumors in the large intestine are adenocarcinomas (cancers of epithelial origin arising from glandular structures). About 10% of colorectal cancers occur in the transverse colon. In this case, although the tumor was moderately large, no extension from the primary site of the tumor occurred, no local nodes were involved, and no distant metastases had occurred. This was fortunate for the patient, as it meant there was an excellent prognosis and also he did not have to be subjected to chemotherapy or radiotherapy.

Most colorectal adenocarcinomas originate from **adenomatous polyps**. A polyp is a growth, usually benign, protruding from a mucous membrane. There are a variety of types. The one of interest here is the adenomatous polyp. Most tumors of the colon arise from such polyps, although the majority of such polyps do not progress to cancer.

There a number of well-defined **genetic syndromes** that predispose to colorectal cancer. The most common is **hereditary nonpolyposis colorectal cancer** (HNPCC), in which mutations in various genes involved in DNA mismatch repair are involved (see Chapter 35). Another relatively rare condition is **familial adenomatous polyposis** (adenomatous polyposis coli) in which hundreds or thousands of polyps appear in the colon and rectum.

The *APC* gene is located on chromosome 5q21 and many mutations have been described. Overall, it has been estimated that approximately 20% of colorectal cancers have a genetic basis.

Various **environmental factors** have been proposed as being involved in the causation of colorectal cancer. These include diets rich in **saturated fat**, high in **calories**, low in **calcium**, and low in **fiber**. How exactly each of these proposed factor oper-

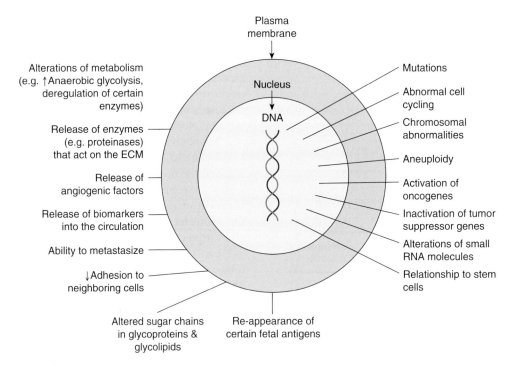

FIGURE 54–6 Some biochemical and genetic changes occurring in human cancer cells. Many biochemical changes are observed in cancer cells, only a few of which are shown here. The roles of mutations in activating oncogenes and inactivating tumor suppressor genes are discussed in the text. Alterations of small RNA molecules have recently been reported, and the relationship of stem cells to cancer cells is under study. Chromosomal abnormalities (eg, aneuploidy, a chromosome complement that is not an exact multiple of 23) are often evident in cancer cells. Many changes in metabolism have been observed; for example, cancer cells often exhibit a high rate of anaerobic glycolysis. Changes in plasma membrane constituents (eg, alteration of the sugar chains of various glycoproteins and glycolipids) have been detected and may be of importance in relation to metastasis. Alterations in the interactions of cancer cells with constituents of the ECM may also be important in relation to the ability of cancer cells to spread to other sites. Various molecules can pass out of cancer cells and can be detected in the blood as tumor markers.

ates is the subject of ongoing research. For example, dietary fat appears to enhance the production of cholesterol and bile acids by the liver. When bile acids are excreted into the bowel, bacterial enzymes may act on them to convert them to secondary bile acids, which are thought to be tumor promoters. A **tumor promoter** is a molecule that along with an **initiator** (ie, a molecule that causes a mutation in DNA) leads a cell to become cancerous. **Inflammatory bowel disease** (eg, ulcerative colitis) is another predisposing factor to colorectal cancer.

The **essence of cancer cells** (ie, malignant tumor cells) is that they exhibit deregulation of many control mechanisms involved in cell growth and cell division, so that they grow more rapidly than their normal cell counterparts. Another crucial feature of cancer cells is that they **spread** to other sites of the body (ie, they metastasize). **Benign** tumor cells exhibit abnormal growth, but do not metastasize. Malignant tumors of epithelial tissue are called carcinomas, and those of soft tissues are called sarcomas. Some important features of cancer cells are summarized in Figure 54–6.

Over the past 30 years or so, major advances have been made in understanding how cancer cells develop and grow. Two key findings were the discovery of oncogenes and tumor suppressor genes. An **oncogene** can be defined as an altered gene (by mutation) whose product acts in a dominant manner to accelerate cell growth or cell division, contributing to cancer development. Oncogenes are generally derived by mutation from normal cellular proto-oncogenes. Factors that cause mutation include chemicals and radiation. A **tumor suppressor gene** produces a protein product that normally suppresses cell growth or cell division When such a gene is altered (eg, by mutation), the inhibitory effect of its product is lost or diminished, also leading to increased cell growth and/or division. The study of the genes of certain viruses that cause cancer (eg, Rous sarcoma virus) was of great importance in opening up knowledge in this area. Many oncogenes and tumor suppressor genes in human and other animals have now been identified.

Studies of **the development of colon cancers** by Vogelstein and others have led to important insights into the roles of such genes in human cancers. These workers analyzed various oncogenes, tumor suppressor genes and genes closely associated with the actions of the products of the two former types of genes in samples of normal colonic epithelium, in dysplastic epithelium, in various stages of adenomatous polyps and in adenocarcinomas. Dysplasia is a pre-neoplastic condition, characterized by abnormal development of epithelium. Some of their major results are summarized in Figure 54–7. It can be seen that certain genes were found to be mutated at relatively specific stages of the total sequence shown. Functions of the various genes identified are listed in Table 54–2. The overall sequence of changes can vary somewhat from that shown, and other genes may also be involved. Similar studies have been performed on a number of other human tumors, revealing somewhat different patterns of activation of oncogenes and mutations of tumor suppressor genes. Further mutations in these and other genes are involved in the phenomenon of **tumor progression**, a phenomenon whereby tumor cells be-

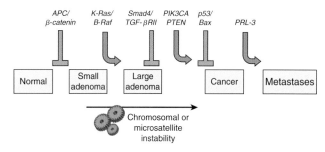

FIGURE 54–7 Multistep genetic changes associated with the development of colorectal cancers. Mutations in the *APC* gene initiate the formation of adenomas. One sequence of mutations in an oncogene and in various tumor suppressor genes that can result in further progression to large adenomas and cancer is indicated. Patients with familial adenomatous polyposis (OMIM 175100) inherit mutations in the *APC* gene and develop numerous dysplastic aberrant crypt foci (ACF), some of which progress as they acquire the other mutations indicated in the figure. The tumors from patients with hereditary nonpolyposis colon cancer (OMIM 120435) go through a similar though not identical series of mutations; mutations in the mismatch repair system speed up this process. *K-RAS* is an oncogene and the other specific genes indicated are tumor suppressor genes. The chromosomal locations of the various genes indicated are known. The sequence of events shown here is not invariable in the development of all colorectal cancers. A variety of other genetic alterations have been described in a small fraction of advanced colorectal cancers. These may be responsible for the heterogeneity of biologic and clinical properties observed among different cases. Instability of chromosomes and microsatellites (see Chapter 35) occurs in many tumors, and likely involves mutations in a considerable number of genes. (Reproduced, with permission, from Bunz F, Kinzler, KW, Vogelstein B: Colorectal Tumors, Fig. 48–2, *The Online Metabolic & Molecular Bases of Inherited Disease,* www.ommbid.com)

come selected for fast growth rate and ability to spread. Studies such as these just described show that **cancer is truly a genetic disease**, but in a somewhat different sense from the general meaning of the phrase, insofar as many of the gene alterations are due to somatic mutations.

There is also evidence that **epigenetic mechanisms** (eg, methylation/demethylation of cytosine in specific genes and acetylation of histones H3 and H4 affecting gene transcription) may play a role in carcinogenesis. One area of interest regarding such changes are that they are potentially reversible.

Another area of great interest at present is the implication of **stem cells** in carcinogenesis. This topic will not be addressed here, but is attracting much research.

CEA is a glycoprotein present in the plasma membranes of many cells. It is released into the plasma in a number of conditions, in which it can measured by a radioimmunoassay. Levels of CEA are increased in serum in colorectal cancer, but also in other alimentary and non-alimentary cancers, and in certain non-cancerous conditions.

Less than 50% of individuals with localized colorectal cancer have elevated levels of CEA, so it is not a useful screening test for this condition. Its main use is to **monitor the effects of treatment** and to detect **recurrence** of cancers such

TABLE 54–2   Some Genes Associated with Colorectal Carcinogenesis

Gene[1]	Action of Encoded Protein
APC (OMIM 611731)	Antagonizes Wnt[2] signaling; if mutated, Wnt signaling is enhanced, stimulating cell growth
β-Catenin (OMIM 116806)	Encodes β-catenin, a protein present in adherens junctions, which are important in the integrity of epithelial layers
K-RAS (OMIM 601599)	Involved in tyrosine kinase signaling
BRAF (OMIM 164757)	A serine/threonine kinase
Smad4 (OMIM 600993)	Affects signaling by transforming growth factor-beta (TGF-β)
TGF-βRII	Acts as a receptor for TGF-β[3]
PIK3CA (OMIM 171834)	Acts as a catalytic subunit of phosphatidylinositol 3-kinase
PTEN (OMIM 601728)	A protein-tyrosine phosphatase with an area of homology to tensin, a protein that interacts with actin filaments at focal adhesions
p53 (OMIM 191170)	The product, p53, is involved in reaction to DNA damage and is also a transcription factor for many genes involved in cell division; p53 is sometimes called the "guardian of the genome"
Bax (OMIM 600040)	Acts to induce cell death (apoptosis)[4]
PRL-3 (OMIM 606449)	A protein-tyrosine phosphatase

Abbreviations: *APC*, adenomatous polyposis coli gene; *K-RAS*, Kirsten-Ras-associated gene; *BRAF*, the human homolog of an avian protooncogene; *Smad4,* the homolog of a gene found in Drosophila*; PIK3CA*, the catalytic subunit of phosphatidylinositol 3-kinase; *PTEN*, a protein-tyrosine phosphatase and tensin homolog; p53, a polypeptide of molecular mass 53,000; *Bax*, BCL2-associated X protein (BCL-2 is a repressor of apoptosis); *PRL3*, a protein-tyrosine phosphatase with homology to PRL1, another protein-tyrosine phosphatase found in regenerating liver.
Note: The various genes listed are either oncogenes, tumor suppressor genes or genes whose products are closely associated with the products of these two types of genes. The cumulative effects of mutations in the genes listed are to drive colonic epithelial cells to proliferate and eventually become cancerous. They achieve this mainly via effects on various signaling pathways affecting cellular proliferation. Other genes and proteins not listed here are also involved. This table and Figure 54–7 vividly show the importance of a detailed knowledge of cell signaling for understanding the genesis of cancer.
[1]K-RAS and BRAF are oncogenes; the other genes listed are either tumor suppressor genes or genes whose products are associated with the actions of the products of tumor suppressor genes.
[2]The Wnt family of secreted glycoproteins is involved in a variety of developmental processes. Tensin is a protein that interacts with actin filaments at focal adhesions.
[3]TGF-β is a polypeptide that regulates proliferation and differentiation in many cell types.
[4]A protooncogene is the normal counterpart of an oncogene. Apoptosis is programmed cell death.

FIGURE 54–8 Simplified scheme of the multistep causation of colorectal cancer.

as colorectal cancer, by following changes in its level. In the present case, the level of CEA remained low for 5 years after surgery, suggesting no recurrence of cancer had occurred. A goal (probably unattainable) of research in this area would be to develop highly specific biomarkers for very early colorectal cancer and for other very early tumors that would be positive in 100% of cases and negative in 100% of normal individuals!

Figure 54–8 summarizes some major factors leading to the development of colorectal cancer.

CASE 5: CYSTIC FIBROSIS (CF)

Causation

Genetic (mutations in the gene encoding the cystic fibrosis transmembrane regulator [CFTR] protein).

History and Physical Examination

A 1-year-old girl, an only child of Caucasian background, was brought to the clinic at the Hospital for Sick Children by her mother. She had been feverish for the past 24 h and was coughing frequently. The mother stated that her daughter had experienced three attacks of "bronchitis" since birth, each of which had been treated with antibiotics by their family physician. The mother had also noted that her daughter had been passing somewhat bulky, foul-smelling stools for the past several months and was not gaining weight as expected. In view of the history of pulmonary and gastrointestinal problems, the attending physician suspected that the patient might have CF, although no family history of this condition was elicited.

Laboratory Findings

Chest x-rays showed signs consistent with bronchopneumonia. Culture of sputum revealed predominantly *Pseudomonas aeruginosa*. Fecal fat was increased. A quantitative pilocarpine iontopheresis sweat test was performed, and the sweat Cl^- was 70 mmol/L (>60 mmol/L is abnormal); the test was repeated a week later with similar results.

Treatment

The child was given an appropriate **antibiotic** and referred to the **cystic fibrosis clinic** for further care. A **comprehensive program** was instituted to look after all aspects of her health, including psychosocial considerations. She was started on a **pancreatic enzyme preparation** (given with each meal) and placed on a **high-calorie diet** supplemented with **multivitamins** and **vitamin E**. **Postural drainage** was begun for the thick pulmonary secretions. Subsequent infections were treated promptly with appropriate antibiotics and with an aerosolized recombinant preparation of **human DNase** that digests the DNA of microorganisms present in the respiratory tract. At age 6 years, she had grown normally, had been relatively free of infection for a year, was attending school and making satisfactory progress. Serious chronic cases of CF in which the lungs are severely compromised are candidates for **lung transplants**, although the efficacy of this treatment has been challenged recently.

Research on **gene therapy** for CF is under examination (eg, using recombinant viruses encoding the CFTR protein). Another line of research is investigating whether **small molecules** can be found for clinical use that help abnormally folded CFTR molecules re-fold into at least partially active molecules.

Discussion

CF is a **prevalent** and usually serious genetic disease among whites in North America. It affects approximately 1:2500 individuals and is inherited as an **autosomal recessive** disease; about one person in 25 is a carrier. It is a disease of the **exocrine glands**, with the **respiratory and gastrointestinal tracts** being most affected. A diagnostic hallmark is the presence of **high amounts of NaCl in sweat**, reflecting an underlying abnormality in exocrine gland function (see below). **Pilocarpine iontophoresis** has generally been used to allow collection of sufficient amounts of sweat for analysis. Iontophoresis is a process by which drugs are introduced into the body (in this case the skin) via use of an electrical current. Its use is diminishing as the availability of specific genetic probes increases.

The **classic presentation** of CF is that of a young child with a history of recurrent pulmonary infection and signs of exocrine insufficiency (eg, fatty, bulky stools due to a lack of pancreatic lipase), as in the present case. However, the disease is **clinically heterogeneous**, which at least partly reflects heterogeneity at the molecular level (see below). Approximately 15% of patients may have sufficient pancreatic function to be classified as "pancreatic sufficient."

For reasons related to abnormalities in Cl^- and Na^+ transport (see below), the **pancreatic ducts** and the ducts of certain other exocrine glands become filled with **viscous mucus**, which leads to their **obstruction**. This mucus is also present in the **bronchioles**, leading to their obstruction; this favors the growth of certain bacteria (eg, *Staphylococcu aureus* and *P aeruginosa*) that cause **recurrent bronchopulmonary infections**, eventually seriously compromising lung function. In turn, the pulmonary disease can lead to right ventricular hypertrophy and possible heart failure. Patients usually die of **pulmonary infection** or **heart failure**. In recent years, more patients have been living into their 30s and later, as the condition is now diagnosed earlier and appropriate comprehensive therapy started. Sometimes, problems due to lack of pancreatic secretions can be present **at birth**, the infants presenting with intestinal obstruction due to very thick meconium **(meconium ileus)**. Other patients, less severely affected, may not be diagnosed until they are in their teens or later. CF also affects the **genital tract** and most males and many females are **infertile**.

In 1989, results of a collaborative program between Canadian and American scientists revealed the nature of the genetic lesion in the majority of sufferers from CF. The gene involved in CF was the first to be cloned solely on its position determined by **linkage analysis (positional cloning)** and constituted an enormous amount of painstaking labor and a tremendous triumph for **"reverse genetics."** By reverse genetics is meant that the gene was isolated based on its map location, and not with the availability of chromosomal rearrangements or deletions (in contrast to, for example, the isolation of the gene involved in Duchenne muscular dystrophy). The success of this Herculean effort showed that, at least in theory, the molecular basis of any genetic disease could be revealed by similar approaches. More recent advances (eg, outcomes of the Human Genome Project) have further facilitated identification of "disease genes."

Table 54–3 summarizes the major strategies used in detecting the gene involved in the causation of CF. The protein product of the gene encodes an integral membrane protein of approximately 170kDa. It was named **CFTR** and was found

TABLE 54–3 Some Strategies Used in Isolating the Gene Involved in CF

- From study of a large number of families with CF, assignment of the gene to chromosome 7 was made by demonstrating linkage to several RFLPs on that chromosome.

- Further narrowing to a smaller region of chromosome 7 was accomplished by use of additional RFLPs.

- Chromosome jumping and chromosome walking were used to isolate clones.

- The affected region was sequenced by looking for mutations in DNA that were not present in DNA from normal individuals, for exons expressed as mRNAs in tissues affected by CF (eg, pancreas and lung), for sequences conserved across species, and for an open reading frame (indicating an expressed protein).

TABLE 54–4   Some Characteristics of the Gene for the CFTR Protein and of the Protein Itself

- About 250,000-bp gene on chromosome 7.
- 25 exons.
- mRNA of 6129 bp.
- Transmembrane protein of 1480 amino acids.
- CFTR contains two NBFs and one regulatory domain.
- Commonest mutation in CF is deletion of ΔF508 present in the first NBF.
- CFTR is a cAMP-responsive chloride transporter.
- Shows homology to other proteins that use ATP to effect transport across membranes (eg, P-glycoprotein).

Abbreviations: CFTR, cystic fibrosis transmembrane regulatory protein; F, phenylalanine; NBF, nucleotide-binding fold.

to be a **cAMP-responsive chloride transporter**, helping to explain the high chloride content of sweat found in patients with CF.

Some of the characteristics of the **gene** and of the **CFTR** protein are listed in Table 54–4. The major mutation initially located in the gene was deletion of three bases encoding phenylalanine residue 508 (ΔF508); in North America, approximately 70% of CF carriers have this mutation. Subsequent work has revealed **over 1000 different mutations** in the gene. A variety of different types of mutations have been found, including small deletions, insertions, and missense and nonsense mutations. Because of the importance of early diagnosis, in certain countries, **all newborn infants** are now undergoing genetic screening for CF. Elsewhere, techniques to detect deletion of ΔF508 and a number of the other most frequent mutations are being used to **confirm the diagnosis** of CF, to **detect carriers** and in **prenatal diagnosis**.

The **CFTR protein** (Figure 54–9) consists of two similar halves, each containing six transmembrane regions and a

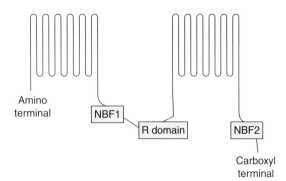

FIGURE 54–9 Diagram of the structure of the CFTR protein (not to scale). The protein contains 12 transmembrane segments, two nucleotide-binding folds or domains (NBF1 and NBF2), and one regulatory (R) domain. NBF1 and NBF2 bind ATP and couple its hydrolysis to transport of Cl⁻; they are called ATP-binding cassettes (ABCs), a feature found in many membrane transporters. Phe 508, the major locus of mutations in CF, is located in NBF1.

TABLE 54–5   Some Mechanisms by Which Mutations May Affect the CFTR Protein

I	Reduce or abolish its synthesis
II	Block its intracellular processing
III	Alters its regulation of chloride flux
IV	Alters conductance of chloride channel

nucleotide (ATP)-binding fold (NBF). The two halves of the molecule are joined by a regulatory domain. F508 is located in NBF1. The protein shows similarities in structure to certain other proteins that use ATP to transport molecules across cells membranes (eg, P-glycoprotein, involved in resistance to certain cancer chemotherapeutic agents).

Normally, CFTR is synthesized on bound polyribosomes and exported to the plasma membrane, where it functions. Mutations can affect CFTR is a number of ways, summarized briefly in Table 54–5. Many mutations affect the **folding** of the protein, markedly reducing its function; this classifies CF as a **conformational disease** (see Chapter 46 and the discussion of Alzheimer disease in this chapter). Mutations affect many other proteins in a similar manner to those summarized in Table 54–5, affecting their synthesis, processing, or function.

Figure 54–10 summarizes some of the mechanisms involved in the causation of CF.

FIGURE 54–10 Summary of possible mechanisms involved in cells in the airways of individuals with cystic fibrosis (OMIM 219700) who have pulmonary pathology. In individuals of Caucasian origin, 70% of the mutations occur at one locus, resulting in deletion of ΔF508 from the CTR protein. However, over 1000 mutations have been identified in the *CFTR* gene. Basically, the CFTR protein acts normally as a cAMP-regulated transporter involved in secretion of Cl⁻, but in addition normally inhibits absorption of Na⁺ by a Na⁺ channel. The viscosity of the mucus in the pancreatic ductules is also increased, leading to their obstruction. The details of how abnormalities of CFTR affect ion transport in the pancreas are somewhat different than in the lung.

CASE 6: DIABETIC KETOACIDOSIS (DKA)

Causation

Endocrine (due to deficiency of insulin).

History and Physical Examination

A 14-year-old girl was admitted to a children's hospital in coma. Her mother stated that the girl had been in good health until approximately 2 weeks previously, when she developed a sore throat and moderate fever. She subsequently lost her appetite and generally did not feel well. Several days before admission she began to complain of undue thirst and also started to get up several times during the night to urinate. Her family doctor was out of town and her mother felt reluctant to contact another physician. However, on the day of admission the girl had started to vomit, had become drowsy and difficult to arouse, and accordingly had been brought to the emergency department. On examination she was dehydrated, her skin was cold, she was breathing in a deep sighing manner (Kussmaul respiration), and her breath had a fruity odor. Her blood pressure was 90/60 and her pulse rate 115/min. She could not be aroused. A diagnosis of type 1 diabetes mellitus (formerly called insulin-dependent) with resulting ketoacidosis and coma (DKA) was made by the intern on duty.

Laboratory Findings

The admitting diagnosis was confirmed by the laboratory findings, kindly supplied by Dr. ML Halperin:

Plasma or serum results (normal levels in parentheses, SI Units):
- Glucose, 50 mmol/L (4.2–6.1 mmol/L)
- Ketoacids ++++ (trace)
- Bicarbonate, 6 mmol/L (22–30 mmol/L)
- Urea nitrogen, 15 mM (2.5–7.1 mmol/L)
- Arterial blood pH, 7.07 (7.35–7.45)
- Na^+, 136 (136–146 mmol/L)
- Cl^-, 100 (102–109 mmol/L)
- P_{CO_2} 2.7 (4.3–6.0 kPa [or 32–45 mm Hg])
- Anion gap, 31 (7–16 mmol/L)
  (The anion gap is calculated from plasma $Na^+ - [Cl^- + HCO_3^-]$.)
- Potassium, 5.5 mmol/L (3.5–5.0 mmol/L)
- Creatinine, 200 μmol/L (44–80 μmol/L)
- Albumin 50 g/L (41–53 g/L)
- Osmolality, 325 (275–295 mOsm/kg serum water)
- Hematocrit, 0.500 (0.354–0.444)

Urine results:
- Glucose, ++++ (normal −)
- Ketoacids, ++++ (normal −)

Treatment

The most important initial measures in treatment of diabetes ketoacidosis are intravenous administration of insulin and saline solution. This patient was given intravenous insulin (10 units/h) added to 0.9% NaCl. Glucose was withheld until the level of plasma glucose fell below 15 mM. Insulin and glucose facilitate entry of K^+ into cells. KCl was also administered cautiously, with plasma K^+ levels monitored every hour initially. Continual monitoring of K^+ levels is extremely important in the management of diabetic ketoacidosis because inadequate management of K^+ balance is the main cause of death. Bicarbonate is not needed routinely, but may be required if acidosis is very severe.

Discussion

The precise **cause** of type 1 (insulin-dependent) diabetes mellitus has not been elucidated, and is under intense investigation. Genetic, environmental and immunologic factors have all been implicated. A very tentative scheme of the chains of events is the following. Patients with this type of diabetes have a **genetic susceptibility** (a large number of genes, including histocompatibility genes located on chromosome 6, have been implicated), which may predispose to a **viral infection** (eg, by coxsackie or rubella viruses). The infection and consequent inflammatory reaction may alter the antigenicity of the surface of the pancreatic B cells and set up an autoimmune reaction involving both cytotoxic antibodies and T lymphocytes. This leads eventually to widespread destruction of beta cells, resulting in type I diabetes mellitus. Perhaps the **sore throat** this patient had several weeks before admission reflected the initiating viral infection.

The marked **hyperglycemia**, **glucosuria**, **ketonemia** and **ketonuria** confirmed the diagnosis of DKA. The **low pH** indicated a severe acidosis due to the greatly increased production of acetoacetic acid and β-hydroxybutyric acid. The low levels of **bicarbonate** and P_{CO_2} confirmed the presence of a metabolic acidosis with partial respiratory compensation (the hyperventilation). Calculation of the **anion gap** is useful in a number of metabolic situations. In this case it is elevated because of the presence of excess ketoacids in the blood. There are a number of other causes of elevation of the anion gap, including lacticacidosis and intoxication by methanol, ethylene glycol, and salicylates.

The elevated values of **urea** and **creatinine** indicated some renal impairment (due to diminished renal perfusion because of low blood volume secondary to dehydration), dehydration, and increased degradation of protein. A high plasma level of **potassium** is often found in DKA owing to a lowered uptake of potassium by cells in the absence of insulin. Thus, the clinical picture in DKA reflects the abnormalities in the metabolism of carbohydrate, lipid, and protein that occur when plasma levels of insulin are sharply reduced. The increased **osmolality** of plasma due to hyperglycemia also contributes to the development of coma in diabetic ketoacidosis. It should be apparent that the rational treatment of a patient with DKA depends on thorough familiarity with the actions of insulin.

A general scheme of the events occurring in DKA is given in Figure 54–11.

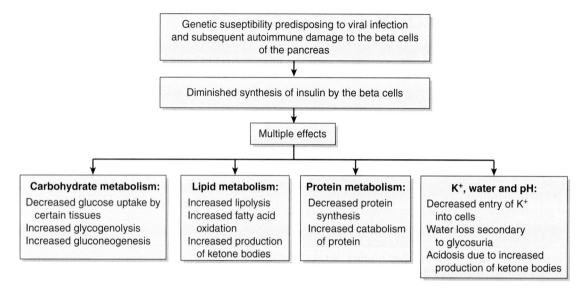

FIGURE 54–11 Summary of some mechanisms involved in the causation of the ketoacidosis of type 1 diabetes mellitus (OMIM 222100).

CASE 7: DUCHENNE MUSCULAR DYSTROPHY (DMD)

Causation

Genetic (mutations in the gene encoding the protein dystrophin).

History and Physical Examination

A 4-year-old boy was brought to a children's hospital clinic. His mother was concerned because she had noticed that her son was walking awkwardly, fell over frequently, and had difficulty climbing stairs. There were no siblings, but the mother had a brother who died at age 19 of muscular dystrophy. The pediatrician on call noted muscular weakness in both the pelvic and shoulder girdle. Modest enlargement of the calf muscles was also noted. Because of the muscle weakness and its distribution, the pediatrician made a provisional diagnosis of DMD.

Laboratory and Other Findings

The activity in serum of creatine kinase (CK) was markedly increased. It was decided to proceed directly to mutation analysis using a sample of the patient's lymphocytes. This showed a large deletion in the gene for dystrophin, confirming the diagnosis of DMD. This saved the patient from being tested by electromyography and also from having a muscle biopsy performed; these tests, along with Western blotting for detection of dystrophin, were routinely performed prior to the availability of mutation analysis, and still may be performed in certain circumstances.

Discussion

The family history, the typical distribution of muscular weakness, the elevation of the activity in serum of CK, and the results of mutation analysis confirmed the provisional diagnosis of DMD. This is a severe **X-linked** degenerative disease of muscle. It has a prevalence of approximately 1:3500 live male births. It affects **young boys**, who first show loss of strength in their proximal muscles, leading to a waddling gait, difficulty in standing up, and eventually very severe weakness. Death generally occurs from respiratory insufficiency.

The cause of DMD was revealed in 1986–1987. Various studies led to localization of the defect to the middle of the short arm of the X chromosome and to subsequent identification of a segment of DNA that was **deleted** in patients with DMD. Using the corresponding non-deleted segment from normal individuals, a cDNA was isolated derived by reverse transcription from a transcript (mRNA) of 14 kb that was expressed in fetal and adult skeletal muscle. This was cloned and the protein product was identified as **dystrophin**, a 400-kDa red-shaped protein of 3685 amino acids. Dystrophin was absent or markedly reduced in electrophoretograms of extracts of muscle from patients with DMD and from mice with an X-linked muscular dystrophy. Antibodies against dystrophin were used to study its localization in muscle; it is associated with the **sarcolemma** (plasma membrane) of normal muscle and was absent or markedly deficient in patients with DMD. A less severe reduction in the amount of dystrophin, or a reduction in its size, is the cause of **Becker muscular dystrophy**, a milder type of muscular dystrophy. Whereas the gene **deletions** and **duplications** found in DMD tend to cause frameshift mutations, the same types of mutations in Becker MD are generally in-frame, and thus synthesis of dystrophin is not as affected in the latter.

Dystrophin appears to have four domains, two of which are similar to domains present in α-actinin (another muscle protein) and one to a domain in spectrin, a protein of the cytoskeleton of the red blood cell. As shown in Figure 49–11, dystrophin interacts with actin, syntrophin, and β-dystroglycan. Its **function** is not clear, but it may play a role in transmembrane signaling and in stabilizing the cytoskeleton and sarcolemma.

FIGURE 54–12 Summary of mechanisms involved in the causation of Duchenne muscular dystrophy (OMIM 310200).

Deficiency of dystrophin may affect the integrity of the sarcolemma, resulting in increased osmotic fragility of dystrophic muscle or permitting excessive influx of Ca^{2+}. The **gene** coding for dystrophin is the largest human gene recognized to date (~2500 kb, 79 exons), which helps explain the observation that approximately one-third of cases of DMD are new mutations. Attempts are being made to **produce dystrophin** by recombinant DNA technology and perhaps eventually administer it to patients. The availability of **probes** for dystrophin facilitates **prenatal diagnosis** of DMD by chorionic villus sampling or amniocentesis. The demonstration of lack of dystrophin as a cause of DMD has been one of the major accomplishments of the application of molecular biology to the study of human diseases.

There are many **different types** of muscular dystrophy. The molecular causes of many of them have been elucidated. Not surprisingly, perhaps, they have been found to be due to a variety of mutations in genes that encode specific muscle proteins, such as these shown in Figure 49–11 and others not shown. The various muscular dystrophies can be classified on the basis of their clinical features (eg, affecting the limb girdle, etc), or increasingly on the basis of the genes or proteins affected by the causative mutations. Dystrophin also occurs in heart muscle and the brain. Its occurrence in the former can result in **cardiomyopathy.** The absence of dystrophin in brain results in an IQ of less than 75 being observed in some 25% of boys with DMD.

Treatment

At present, no specific therapy for DMD exists. Treatment in this case was thus essentially **symptomatic.** He was enrolled in a special **muscular dystrophy clinic,** started on **prednisone** (which can slow down the progress of DMD for a few years) and encouraged to undertake mild **exercise.** A **physiotherapist** was available when needed. **Portable respirators** have proven very useful when breathing is affected. The mother was advised to seek genetic **counseling** regarding future pregnancies. At different times, a variety of therapeutic measures intended to benefit patients with DMD have been used. These include the use of myoblast transfer (to supply dystrophin), antisense oligonucleotides (to skip mutated dystrophin gene exons), CoE Q10 (to possibly increase muscle strength), creatine monohydrate (to perhaps help build up muscle mass), pentoxyfylline (anti-inflammatory), and gentamicin (may ignore premature stop codons in the dystrophin gene). None appears to have been very successful. A small trial using dystrophin gene transfer was scheduled to start in the United States in January 2008.

Figure 54–12 summarizes the mechanisms involved in the causation of DMD.

CASE 8: ETHANOL INTOXICATION, ACUTE

Causation

Chemical (due to excess intake of ethanol).

History and Physical Examination

A 52-year-old man was admitted to the emergency department in a coma. Apparently, he had become increasingly depressed after the death of his wife one month previously. Before her death he had been a moderate drinker, but his consumption of alcohol had increased markedly over the last few weeks. He had also been eating poorly. His married daughter had dropped round to see him on Sunday morning and had found him unconscious on the living room couch. Two empty bottles of rye whiskey were found on the living room table. On examination, he could not be roused, his breathing was deep and noisy, alcohol could be smelled on his breath, and his temperature was 35.5°C (normal; 36.3–37.1°C). The diagnosis on admission was coma due to excessive intake of alcohol.

Laboratory Findings

The pertinent laboratory results were alcohol 500 mg/dL, glucose 2.7 mmol/L (normal: 4.2–6.1), lactate 8.0 mmol/L (normal: 0.5–1.6), and blood pH of 7.21 (normal: 7.35–7.45).

These results were consistent with the admitting diagnosis, accompanied by a metabolic acidosis.

Treatment

He was started on intravenous normal saline, and then, because of the very high level of blood alcohol and the coma, it was decided to start **hemodialysis** immediately. This directly eliminates the toxic ethanol from the body but is only required in very serious cases of ethanol toxicity. In this case, the level of blood alcohol fell rapidly and the patient regained consciousness later the same day. Intravenous **glucose** (5%) was administered after dialysis was stopped to counteract the hypoglycemia that this patient exhibited. The patient made a good recovery and was referred for **psychiatric counseling**.

Discussion

Excessive consumption of alcohol is a major health problem in most societies. The present case deals with the acute, toxic

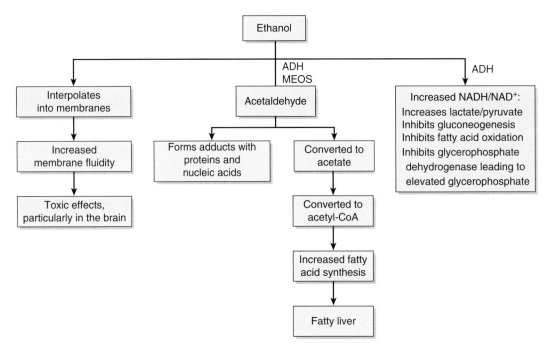

FIGURE 54–13 Summary of some mechanisms involved in acute ethanol toxicity. (ADH, alcohol dehydrogenase; MEOS, microsomal ethanol oxidizing system, involving cytochrome P-450 species CYP2E1.)

effects of a very large intake of ethanol. A related problem, which is not discussed here but which has many biochemical aspects, is the development of **liver cirrhosis** in individuals who maintain a high intake of ethanol (eg, 80 g of absolute ethanol daily for more than 10 years).

From a biochemical viewpoint, the major question concerning the present case is how does ethanol produce its diverse acute effects, including coma, lacticacidosis, and hypoglycemia? The clinical viewpoint is how best to treat this condition.

The metabolism of ethanol was decribed in Chapter 25; it occurs mainly in the liver and involves two routes. The first and major route uses **alcohol dehydrogenase** and **acetaldehyde dehydrogenase**, converting ethanol via acetadehyde to acetate (Chapter 25), which is then converted to acetyl-CoA. NADH + H$^+$ are produced in both of these reactions. The intracellular **ratio of NADH/NAD$^+$** can thus be increased appreciably by ingestion of large amounts of ethanol. In turn, this can affect the Keq of a number of important metabolic reactions that use these two cofactors. High levels of NADH favor **formation of lactate** from pyruvate, accounting for the lacticacidosis. This diminishes the concentration of pyruvate (required for the **pyruvate carboxylase** reaction, Chapter 17) and thus **inhibits gluconeogenesis**. In severe cases, when liver glycogen is depleted and no longer available for glycogenolysis, **hypoglycemia** results. The second route involves a **microsomal cytochrome P450** (microsomal ethanol oxidizing system), also producing acetaldehyde (Chapter 25). **Acetaldehyde** is a highly reactive molecule and can form adducts with proteins, nucleic acids, and other molecules. It appears likely that its ability to react with various molecules is involved in the causation of the toxic effects of ethanol. Ethanol also appears to be able to **interpolate into**

biologic membranes, expanding them and increasing their fluidity. When the membranes affected are excitable, this results in alterations of their action potentials, impairs active transport across them, and also affects neurotransmitter release. All of these depress cerebral function and, if severe enough, can produce coma and death from respiratory paralysis.

Figure 54–13 summarizes some of the major mechanisms involved in the causation of toxicity by ethanol.

CASE 9: GOUT, ACUTE

Before studying this case, the reader is advised to consult the material in Chapter 33 dealing with uric acid.

Causation

Deposition of crystals of monosodium urate (MSU) in one or more joints and various tissues. The great majority of cases (~90%) are associated with decreased renal excretion of MSU, but in certain cases, increased production of MSU is involved, and increased dietary intake of purines may play a role.

History and Physical Examination

A moderately obese 64-year-old male appeared at the emergency department complaining of severe pain of 12-h duration in his left big toe. He stated that he regularly had at least 2–3 drinks of scotch whisky every evening after work. He had no other significant medical history. On examination, his left big toe was found to be red and markedly swollen around the metacarpophalangeal joint, and exquisitely sensitive. There was no evidence of arthritis elsewhere. Because of the history and location of the affected joint, the intern on duty suspected that the patient was having an attack of acute gout. She ordered a number of lab tests, including a white cell count, determination of serum uric acid, and x-ray examination of the affected joint. The serum level of uric acid, was 0.61 mmol/L (normal, 0.18–0.41 mmol/L in males); the white cell count was at the high end of normal. The x-ray findings were non-specific; no indication of chronic arthritis was evident. Under local anesthesia, arthrocentesis was performed on the affected joint, and a small amount of synovial fluid withdrawn and sent to the laboratory for detection of cells and crystals. Typical needle-shaped crystals of MSU showing negative birefringence were detected in the synovial fluid, as were neutrophils.

Treatment

The patient was started on a suitable dose of a **non-steroidal anti-inflammatory drug** (NSAID) to relieve the acute inflammation and pain. He was referred to a rheumatologist for ongoing treatment; she continued administration of the NSAID. Several months later he had another acute attack of joint pain, this time in his right knee. His plasma level of uric acid was 0.57 mmol/L. **Daily excretion of uric acid** was measured and found to be ~9.0 mmol [~1500 mg] (normal 3.6–5.4 mmol/day). The rheumatologist decided to start him on long-term therapy with **allopurinol**, a drug used to decrease formation of uric acid by inhibiting xanthine oxidase, the enzyme responsible for formation of uric acid from xanthine (see Chapter 33). In addition, the patient was referred to a dietitian regarding losing weight, advised to drink plenty of fluids, to markedly limit his intake of alcohol, to restrict intake of purine-rich foods (eg, anchovies and red meat), and started on a regular exercise program. Until the present, the patient has had no further attacks of acute arthritis.

Discussion

Uric acid is formed from purine nucleosides (eg, adenosine and guanosine) produced by the breakdown of nucleic acids and other molecules (eg, ATP), and in humans is the end-product of purine catabolism. The daily synthesis rate is estimated to be about 1.8 mmol (~300 mg), with a total body pool of approximately 7.2 mmol (1200 mg in adult males, and about one-half that in females). In individuals with gout, the total body pool can be as large as 180 mmol (30,000 mg).

The enzyme involved in formation of uric acid is xanthine oxidase (see Chapter 33). Humans do not possess the peroxisomal enzyme uricase (urate oxidase), which is involved in the degradation of uric acid to allantoin. Approximately 70% of uric acid is excreted by the kidneys, and the rest by the gut. Uric acid exhibits **antioxidant properties** (see Chapter 45); the possible significance of this is under investigation.

Gout is a type of arthritis, acute or chronic, due to deposition of crystals of MSU usually in relatively avascular areas, such as cartilage and tissues around joints, and also where body temperature is lower (eg, the ears, distal ends of the limbs). When crystals of MSU are deposited in synovial fluid, they elicit an inflammatory reaction. In acute gout, this consists mainly of neutrophils. The inflammatory reaction causes the characteristic signs and symptoms of heat, pain, swelling and redness experienced in acute gout. It is generally important to ascertain that the characteristic crystals of MSU are actually present in the synovial fluid of an affected joint, as other crystals (eg, calcium pyrophosphate) may cause signs and symptoms similar to gout. One joint is usually affected initially (ie, monoarticular arthritis), often the metacarpophalangeal joint of the big toe, as in this case. One factor helping to account for this is that the temperatures of the joints of the lower extremities are lower than elsewhere in the body.

MSU has a solubility in plasma of ~0.42 mmol/L at 37°C. It is much more soluble than uric acid, which is the major ionic species below pH 5.75 (the pKa value for the dissociation of uric acid to urate). This difference is particularly important in **urine**, where **calculi** of uric acid may form at acidic values of pH. When the above concentration is exceeded, plasma is supersaturated with respect to MSU. The concentration at which precipitation of MSU in tissues and joints occurs appears to vary, for unknown reasons. Before treatments were available to prevent chronic gout occurring (eg, allopurinol, see below), large aggregates of MSU would accumulate in various tissues; these are called **tophi**, and can attain a considerable size. Tophi may still occur if gout is not diagnosed and treated early.

Gout is usually preceded and accompanied by **hyperuricemia** (plasma uric acid level >0.41 mmol/L). The sequence shown in Figure 54–14 is often involved. Chronic gout can be prevented if appropriate treatment is instituted following an attack of acute gout. Hyperuricemia is much commoner in **males**, although its incidence in females increases after menopause. It should be noted that approximately 30% of people experiencing an attack of gout may have normal levels of MSU in the plasma. **Hyperuricemia** is caused by decreased renal excretion, increased production, or increased intake of uric acid. **Decreased renal excretion** is involved in the great majority of cases of gout, and genetic factors are likely involved. Many kidney diseases affect renal excretion, as does acidosis caused by various metabolic conditions. A variety of drugs (eg, certain diuretics and also salicylates) interfere with excretion of uric acid. Handling of MSU by the kidneys is complex, including phases of glomerular filtration, reabsorption, secretion, and further reabsoption in various parts of the renal tubule. The precise contributions of alterations of these phases to the causation of hyperuricemia have not been clearly defined as yet. **Increased production** can occur due to certain enzyme abnormalities (eg, deficiency of hypoxanthine-guanine phosphoribosyl transferase [HGPRT] and over-activity of PRPP synthetase), although these are uncommon (see Chapter 33). In Lesch-Nyhan syndrome, mutations of the gene encoding

FIGURE 54–14 One common sequence of events leading to chronic gout.

HGPRT are involved (see Chapter 33), and gout can be a feature. Death of cancer cells caused by chemotherapy leads to degradation of their nucleic acids, and thus to increased formation of purines. **Increased intake** can occur via ingestion of purine-rich foods, such as sweetbreads and certain meats, although this is not thought to be a major contributor to elevation of serum uric acid.

The role of **intake of alcohol** in precipitating gout has long been recognized. Ingestion of ethanol can lead to formation of lactic acid, which inhibits secretion of uric acid. In addition, ethanol appears to promote breakdown of ATP, leading to increased production of purines from which uric acid is formed. Also, the solubility of MSU is markedly diminished as the pH in tissues drops, a situation favored by increased production of lactic acid.

Renal stones (urate calculi) develop frequently in patients with chronic gout; the risk of these are lessened by therapy with allopurinol.

For **treatment** of the acute inflammation and pain of acute gout, an **NSAID** is generally used. **Colchicine** is also effective in blocking inflammation caused by MSU crystals. It is known to bind to free tubulin causing the depolymerization of microtubules (Chapter 49); this may prevent the movement of neutrophils into an area containing crystals of MSU. However, it may cause nausea and vomiting. A **corticosteroid** or **ACTH** may also be employed for their anti-inflammatory effects. For long-term management, intended to prevent or reverse any complications that may have arisen, **allopurinol** is used to chronically inhibit the production of uric acid from xanthine.

Uricosuric drugs (which increase the rate of excretion of uric acid) may be used instead of allopurinol if under-excretion of urate is involved; these include probenecid, sulfinpyrazone, and benzbromarone. Any **accompanying conditions** should be addressed (eg, obesity, hypertension, hypertriglyceridemia, alcoholism, renal disease). If gout is treated early, it is compatible with a normal life-span.

A simplified scheme of the causation of acute gout is shown in Figure 54–15.

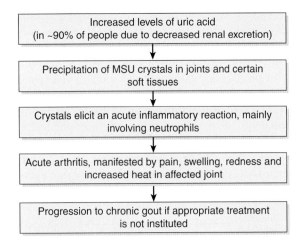

FIGURE 54–15 Simplified scheme of some of the events involved in the causation of gout.

CASE 10: HEREDITARY HEMOCHROMATOSIS

Causation

Genetic (due to mutations in the *HFE* gene or certain other genes whose protein products affect the metabolism of iron).

History and Physical Examination

A 50-year-old man visited his family doctor complaining of fatigue, low libido, and moderate generalized joint pains of approximately 1 year's duration. The joint pains were mostly in the fingers, wrists, hips, knees, and ankles. His parents, both dead, were born in Scotland but emigrated to Canada in early adulthood. The patient had no siblings and did not smoke or drink. He occasionally took acetaminophen for his joint pains, but otherwise was not receiving any medication. An uncle had died of liver cancer about 10 years previously. In addition to stiffness and slight swelling over some joints, the physician noted a grayish skin pigmentation, most prominent in exposed parts, and for that reason referred the patients to an internist, who also noted that the liver edge was firm and palpable just below the costal margin. The internist suspected hereditary hemochromatosis and ordered appropriate laboratory tests as well as x-rays of the hands, hips, knees, and ankles.

Laboratory Findings

Normal reference values in parentheses:

- Hb, 120 g/L (133–162 g/L, males)
- RBC, 4.6×10^{12}/L (4.30–5.60×10^{12}/L, males)
- Glucose (fasting), 5 mmol/L (4.2–6.1 mmol/L)
- Alanine aminotransferase [ALT], 1.8 μkat/L or 105 units/L (0.12–0.70 μkat/L or 7–41 units/L)
- Plasma iron, 50 μmol/L (7–25 μmol/L)
- Total iron-binding capacity, 55 μmol/L (45–73 μmol/L)
- Transferrin saturation with iron 82% (16–35%)
- Serum ferritin, 3200 μg/L, (29–248 μg/L, males)

X-rays of the joints showed loss of articular cartilage, narrowing of joint spaces, and diffuse demineralization.

In view of the above findings, it was decided to perform a **liver biopsy**. Histologic examination revealed moderate periportal fibrosis. Hemosiderin (aggregates of ferritin micelles) was visible as golden brown granules in both parenchymal and bile duct epithelial cells; with Prussian blue staining, iron was markedly visible in these cells. Quantitative measurement of iron in the biopsy material revealed a marked elevation of iron (8100 μg per g dry weight; normal: 300–1400 μg). The laboratory and other findings were consistent with a diagnosis of hereditary hemochromatosis.

Treatment

The treatment of hereditary hemochromatosis is relatively simple, consisting of withdrawing blood from the patient until the excess iron in the body is brought down to near normal values. This is accomplished initially by weekly **phlebotomy** of approximately 500 mL of blood (250 mg of iron). The efficacy of treatment is assessed by monthly monitoring of serum ferritin and transferrin saturation. Once these parameters have reached satisfactory levels, phlebotomy is required only once every 3 months.

Chelation therapy (eg, with deferoxamine) is rarely required and is much more expensive, as well as less effective than phlebotomy. Foods high in iron should be avoided.

Complications such as hepatic cirrhosis, diabetes mellitus, infertility, and cardiac problems must be treated appropriately. **Screening** of family members is recommended in order to give them early treatment if necessary. If the condition is recognized early (eg, before hepatic cirrhosis develops), the prognosis is excellent. **Hepatocellular carcinoma** may develop in patients with cirrhosis. The **arthropathy** of hemochromatosis does not respond well to treatment and may continue relentlessly despite normalization of body iron parameters.

Discussion

The key feature of hemochromatosis is an **increase of total body iron** sufficient to cause tissue damage. Total body iron ranges between 2.5 g and 3.5 g in normal adults; in hemochromatosis, it usually exceeds 15 g. A combination of hepatomegaly, skin pigmentation, diabetes mellitus, heart disease, arthropathy, and hypogonadism should suggest the condition. The finding of elevated levels of **transferrin saturation** and of serum **ferritin** are the most useful tests for **diagnosis**. Because **cirrhosis** is a major determinant of clinical outcome, a **liver biopsy** is important if the diagnosis of hemochromatosis is not made early in the course of the disease. It is critical to **diagnose the condition early** to prevent complications such as cirrhosis and liver cancer. To achieve early diagnosis, a **high index of suspicion** is necessary.

Hemochromatosis may be **hereditary** or **secondary** to a variety of causes (eg, conditions requiring blood transfusions and dietary overload in parts of Africa where beer is brewed in iron pots). In hereditary hemochromatosis, **absorption** of iron from the small intestine is greatly increased (see below). There is no pathway for disposing of excess iron and also the solubility of ferric iron is low. The accumulated iron damages organs such as the liver, pancreatic islets, and heart; it is thought to play an important role in tissue damage (eg, cirrhosis), perhaps in part due to effects on free radical production (see Chapter 45). Melanin and various amounts of iron accumulate in the skin, accounting for the slate-blue color often seen. The precise cause of melanin accumulation is not clear. The frequent coexistence of diabetes mellitus (due to islet damage) and the skin pigmentation led to use of the term **bronze diabetes** for hereditary hemochromatosis.

Hereditary hemochromatosis is a prevalent **autosomal recessive** disorder. It is common in Europe (carrier frequency of approximately 1:10), particularly in Ireland and Scotland,

and emigrants from these countries have contributed to dissemination of the affected gene around the world. Since 1976 it has been known that there is an association between HLA antigens and hereditary hemochromatosis. In 1996, Feder and colleagues **isolated a gene**, now known as **HFE**, located on chromosome 6 (6p21.3) about 3–5 telomeric from the major histocompatibility complex genes. The encoded product was found to be related to MHC class 1 antigens. *HFE* has been found to exhibit three missense mutations in individuals with primary hemochromatosis. The more frequent mutation was one that changed cysteinyl residue 282 to a tyrosyl residue (CY282Y), disturbing the structure of the protein; the other mutation changed histidyl residue 63 to an aspartyl residue (H63D). The incidence of these two mutations varies in different ethnic groups, CY282Y being less frequent in Italians than in northern Europeans. A third less frequent mutation (S65C) has also been found, but has not as yet been studied in much detail. A small group of individuals are compound heterozygotes (C282Y/H63D). Some patients with primary hemochromatosis have none of the above mutations, perhaps because other mutations in *HFE* occur, or because one or more other genes involved in iron metabolism is involved (eg, genes encoding hepcidin, hemojuvelin, ferroportin; transferrin receptor 2 and ferritin, see Chapter 50).

Penetrance of hemochromatosis is low, and is lower in women than in men. Penetrance is the fraction of individuals with a genotype known to cause a disease who have signs or symptoms of the disease.

Genetic testing has been evaluated but is not presently recommended, except in family members. Tests for *HFE* mutations in individuals with elevated serum iron concentrations may be useful.

How the protein product (HFE) of the *HFE* gene functions in iron absorption has not as yet been elucidated. HFE has been shown to be located in the surface of the cells in the crypts of the small intestine, where it associates with β_2-microglobulin and the transferrin receptor-1 (TfR1). This interaction is disrupted when the CY282Y mutation occurs, probably because HFE fails to reach the cell surface. It is postulated that this in some manner results in up-regulation of the **divalent metal transporter (DMT1)** (see Figure 50–4), leading to increased uptake of iron from the small intestine by villus cells.

Hepcidin appears to be a key player in iron transport. It is a 25 amino acid peptide synthesized by liver cells and secreted into the blood. It **inhibits iron absorption by enterocytes** and also **release of iron from macrophages**. It does this by **inhibiting ferroportin**, present in both of these cells. Ferroportin is the protein in these cells responsible for export of iron into the blood. The level of hepcidin is decreased in hereditary hemochromatosis, allowing increased iron absorption. Levels of hepcidin appear to be regulated by HFE, by hemojuvelin and by transferrin receptor 2 (TfR2), although the details of these controls have yet to be discovered.

A tentative scheme of the main events in the causation of hereditary hemochromatosis is given in Figure 54–16.

FIGURE 54–16 Tentative scheme of the main events in causation of hereditary hemochromatosis (OMIM 235200). The two principal mutations are CY282Y and H63D (see text). Mutations in genes other than *HFE* also cause hemochromatosis. Figure 50–4, showing various proteins involved in iron transport, should be consulted.

CASE 11: HYPOTHYROIDISM, PRIMARY

Before studying this case, the reader is advised to consult the material in Chapter 41 dealing with the thyroid gland.

Causation

Primary hypothyroidism is a state due to deficiency of the thyroid hormones, usually due to impaired function, damage to or surgical removal of the thyroid gland. Specific causes are discussed below.

History, Physical Examination, and Laboratory Tests

A 57-year-old woman visited her family physician complaining of chronic fatigue and sluggishness for a number of years; this was her first visit to her doctor for 5 years. On questioning, a history of constipation and feeling cold (cold intolerance) was elicited. She had two adult children; her last menstrual period had occurred some seven years previously. A sister had pernicious anemia and a maternal aunt had had "a thyroid problem."

On examination, the patient was moderately obese. She answered questions slowly, with little change of expression, her voice sounded coarse, and her tongue appeared moderately swollen. Some puffiness around her cheeks was also evident. Palpation of the neck revealed that her thyroid gland was rather firm in consistency, and modestly enlarged. Her blood pressure was mildly elevated and her deep tendon reflexes were delayed. Some clinical findings in a case of hypothyroidism are summarized in Figure 54–17. On the basis of the history and clinical examination, her doctor suspected that the lady had

hypothyroidism. She ordered various tests to investigate this possibility; the following results were these of relevance:

- Thyroid stimulating hormone (TSH): 20 mIU/L (normal range, 0.34–4.25 mIU/L)
- Thyroxine (T_4), free: 4.0 pmol/L (normal 10.3–21.9 pmol/L)
- Anti-thyroperoxidase (anti-TPO) antibodies: ++++ (normal trace)
- Cholesterol, total: 6.20 mmol/L (normal <5.17 mmol/L)
- Chest x-ray: Revealed a small pericardial effusion
- ECG: Revealed bradycardia and low voltage complexes, but no evidence of ischemia or arrhythmias
- Hemoglobin and RBC count: Results consistent with a mild normocytic anemia

Treatment

The clinical history, physical examination, and lab results were all consistent with primary hypothyroidism. Accordingly, the patient was started on a low dose of thyroxine (T_4). It is important to begin therapy with a small dose of T_4, as larger doses can precipitate serious cardiac events, due to the changes in me-

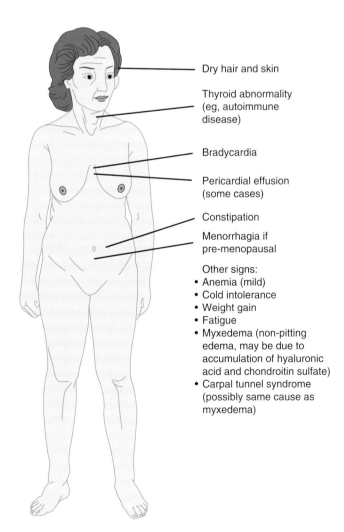

Dry hair and skin

Thyroid abnormality (eg, autoimmune disease)

Bradycardia

Pericardial effusion (some cases)

Constipation

Menorrhagia if pre-menopausal

Other signs:
- Anemia (mild)
- Cold intolerance
- Weight gain
- Fatigue
- Myxedema (non-pitting edema, may be due to accumulation of hyaluronic acid and chondroitin sulfate)
- Carpal tunnel syndrome (possibly same cause as myxedema)

FIGURE 54–17 Some of the major signs of hypothyroidism.

tabolism caused by administration of the hormone. The dosage of T_4 is gradually increased at 6–8 week intervals, up to a maximum of approximately 125 μg. Assessment of progress is made by assays of TSH, which should eventually decline to normal range and be sustained there. Regular assessments are important. Once started, T_4 therapy is generally continued for life.

Discussion

Primary hypothyroidism is a relatively **prevalent** condition, and is probably the most common endocrine problem (excluding diabetes mellitus) seen in clinical practice. The most common cause worldwide is **deficient intake of iodine**. In North America (as in the present case) and other developed countries, a major cause is **Hashimoto disease**, an autoimmune condition affecting the thyroid. **Other causes** include ^{131}I ablation of the thyroid, surgical resection of the thyroid, and the use of drugs for treating hyperthyroidism. It is commoner in **females** than males. In this case, the diagnosis was relatively easy because of the classical history and clinical findings. However, it often has an **insidious onset**, developing gradually over years, and may not be considered. A case can be made for routine assay of TSH in everyone over 35 years of age, but there is as yet no consensus regarding this.

Secondary hypothyroidism is much less common, and is due to decreased secretion of TSH by various pathologic conditions affecting the pituitary gland. Pathology in the **hypothalamus** can cause **tertiary** hypothyroidism, due to decreased secretion of the hypothalamic thyroid-releasing hormone (TRH). **Congenital** hypothyroidism is usually due to various blocks in manufacture of the thyroid hormones and can result in **cretinism** if not diagnosed early and treated appropriately. All newborns in North America and many other countries are routinely screened for levels of TSH at birth.

Detection of increased levels of serum **TSH** is the most useful test for hypothyroidism. As levels of circulating thyroid hormones (T_4, T_3) drop due to thyroid destruction in Hashimoto disease, feedback inhibition on the pituitary declines and levels of TSH rise.

The presence of **elevated TSH** and **decreased T_4** are highly indicative of hypothyroidism. In **Hashimoto disease** the thyroid gland becomes heavily infiltrated by **lymphocytes** and other inflammatory cells, which gradually destroy and replace much of the gland resulting in a progressive decrease of secretion of the thyroid hormones (see Chapter 41), causing the hypothyroid state. The lymphocytes include activated CD4[+] T cells specific for various thyroid antigens. **Various auto-antibodies** can be detected in serum from patients with Hashimoto disease; among these often measured at present are **anti-thyroperoxidase (anti-TPO) antibodies**, which serve as markers of Hashimoto disease. Very often there is a **family history** of the disease or of other autoimmune conditions, indicating a **genetic** contribution. In the present case, a sister of the patient's mother had a family history of "thyroid disease" and a sister of the patient had pernicious anemia, another autoimmune condition.

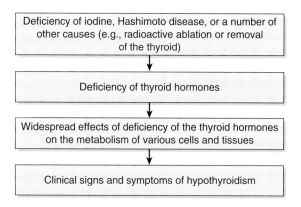

FIGURE 54–18 Simplified scheme of the causation of primary hypothyroidism.

It is important to consider hypothyroidism as a diagnosis, as **early treatment** can make a huge difference to a patient's quality of life.

A simplified scheme of the causation of hypothyroidism is given in Figure 54–18.

CASE 12: KWASHIORKOR, ONE TYPE OF PROTEIN-ENERGY MALNUTRITION (PEM)

Before studying this case, the reader is advised to consult the material in Chapters 43 and 44 on nutrition. Chapter 43 contains information on energy balance, kwashiorkor, marasmus, and essential amino acids.

Causation

Nutritional deficiency primarily of protein in the case of kwashiorkor, but usually accompanied by deficiencies of other nutrients. Deficiency of energy intake causes marasmus. Some patients have features of both kwashiorkor and marasmus.

History and Physical Examination

A 2-year-old African girl was brought by her mother to the outpatient department of the local hospital. The mother had four children, the youngest of whom was 3 months of age and still being breast-fed. The father had broken his leg in an accident during the previous year and had not been able to work since. Family income was thus low, and they were not able to buy milk and meat on a regular basis. Their main subsistence food was a starchy gruel, high in carbohydrate and low in protein, and even that had been in short supply recently. The mother stated that the daughter had been eating poorly for the past several months, had intermittent diarrhea for that time, had recently developed a cough and fever, and had become very irritable, weak, and apathetic.

On examination, she was found to be both underweight for her height and small for her age. Her temperature was

40.5°C. Her midarm circumference was a little below normal. Her skin was flaky and her hair dry, brittle, and easily pluckable. The abdomen was distended and the liver moderately enlarged. Peripheral edema was evident. Rales were heard over the lower lobes of both lungs.

The doctor on duty made the diagnosis of kwashiorkor, diarrhea, pneumonia, and possible bacteremia.

Laboratory Findings

Samples of blood were taken for analysis. Results were subsequently reported as hemoglobin 6.0 g/dL (normal for a 2 year old: 11–14 g/dL), total serum protein 4.4 g/dL (normal 6.0–8.0 g/dL), and albumin 2.2 g/dL (normal 3.5–5.5 g/dL). Stool and blood samples were taken for culture; a gram-negative anaerobe was later reported in both. The neutrophil count was elevated (consistent with a bacterial pneumonia) and her lymphocyte count was markedly depressed. Chest x-rays revealed mottled opacities in the lower lobes of both lungs, consistent with bilateral acute bronchopneumonia.

Treatment

In many cases it is better not to treat children with mild to moderate kwashiorkor in the hospital, because this only increases the chance of infection. However, in view of the fever, weakness, drowsiness, and severe edema, this patient was admitted. She was immediately started on an appropriate **antibiotic** and **intravenous saline-dextrose** infusion. Sadly, her condition worsened and she died approximately 12 h after admission. Autopsy findings were compatible with kwashiorkor and also revealed severe fatty liver and bilateral bronchopneumonia.

Discussion

PEM is the commonest nutritional disorder in many parts of the world. As many as one billion people suffer from various degrees of severity of PEM. It is due to inadequate dietary intakes of protein (causing kwashiorkor) or energy (causing marasmus). It is almost always accompanied by deficiencies of other nutrients (eg, vitamins, minerals, etc). Children and the elderly are particularly susceptible, but it may occur at any age.

PEM may be defined as **primary** (due directly to deficiency of protein and energy intake), or **secondary** (due to increased nutrient needs, decreased absorption of nutrients, or increased loss of nutrients). This was a case of primary kwashiorkor.

Many of the features of primary PEM represent **adaptations** to the dietary deficiencies of energy and protein. For example, physical activity decreases in the face of deficient intake of nutrients. Stores of glycogen in muscle and liver are only capable of supplying energy for a short time (a day or two), so fat stores are mobilized to produce energy.

Eventually, when these are exhausted, protein is catabolized (mainly in muscles) to supply amino acids and energy.

Thus patients with PEM exhibit little activity, have diminished or no body stores of fat, and show muscle wasting, depending on the severity of the condition.

PEM has been classified as **edematous** (kwashiorkor) or **nonedematous** (marasmus). The precise cause of the edema in kwashiorkor in still under study. Hypoalbuminemia (due to deficient supply of amino acids to synthesize the protein) is likely one contributing factor (see Chapter 50), although this is not settled. Increased vascular permeability may also be important. Deficiency of the amino acid methionine, a precursor of cysteine, may also contribute. Cysteine is one of the three amino acids present in glutathione, the body's major antioxidant. If the tissues levels of glutathione decline, this could result in free radicals damaging various molecules and tissues (see Chapter 45) and perhaps damaging cell membranes, increasing their permeability.

Classically, **kwashiorkor** is one end of the spectrum of PEM, in which the essential feature is **protein deficiency**, with relatively adequate energy intake. **Marasmus** is at the other end, and is due to severe and prolonged restriction not only of protein but of all **food**. However, many cases have features of both marasmus and kwashiorkor, a state called **marasmic kwashiorkor**; this patient exhibited primarily signs of kwashiorkor. The **hallmarks of kwashiorkor** are a hypoalbuminemia, fragile skin (eg, poor wound healing, ulcers), easy hair pluckability, and edema (see Figure 54–19). Kwashiorkor is the word used by members of the Ga tribe in Ghana to describe

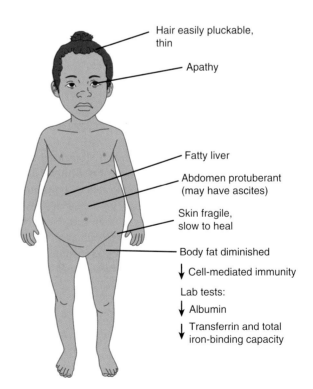

Hair easily pluckable, thin

Apathy

Fatty liver

Abdomen protuberant (may have ascites)

Skin fragile, slow to heal

Body fat diminished

↓ Cell-mediated immunity

Lab tests:

↓ Albumin

↓ Transferrin and total iron-binding capacity

FIGURE 54–19 Some of the major signs of kwashiorkor.

TABLE 54–6 Some Differences Between Kwashiorkor and Marasmus

	Kwashiorkor	Marasmus
Cause	Decreased protein intake	Decreased energy intake
Onset	Rapid (eg, weeks), often associated with stress such as infections	Gradual (months to years)
Hypolbuminemia	Present and may be severe	Mild if present
Muscle wasting	Absent or mild	May be very severe
Body fat	Diminished	Absent

Note: Patients with marasmic kwashiorkor exhibit variable combinations of the above features.

"the sickness the older child gets when the next child is born." It follows weaning from breast milk and exposure to a diet low in protein and high in carbohydrate. **Fatty liver** is often found in kwashiorkor because deficiency of dietary protein depresses the synthesis of apolipoproteins in the liver, resulting in accumulation of triglycerides. The poor skin and hair seen in kwashiorkor are mainly due to protein deficiency. **Hypoalbuminemia** is a common feature. While deficiency of protein can cause hypoalbuminemia, chronic inflammation can also contribute by suppressing synthesis of albumin. Total iron-binding capacity and levels of transferrin are also depressed.

Hormones may be important in the generation of PEM. The exposure to a relatively high intake of carbohydrate is thought by some to keep levels of insulin high and levels of epinephrine and cortisol low in kwashiorkor, as opposed to marasmus. The combination of low insulin and high cortisol greatly favors **catabolism of muscle**; thus, **muscle wasting** is

Low protein intake

↓

Deficiency of essential amino acids

↓

Decreased synthesis of proteins

↓

Defects in skin and hair quality; fatty liver (due to decreased synthesis of apolipoproteins); edema (due to protein deficiency and other factors); low levels of hemoglobin, albumin, transferrin and cells of the immune system

↓

Poor resistance to infection

- -

In comparison, a child with severe marasmus would show marked loss of muscle mass.

FIGURE 54–20 Summary of some of the factors involved in the causation of kwashiorkor.

greater in marasmus than in kwashiorkor. In addition, because of the lower levels of epinephrine, fat is not mobilized to the same extent in kwashiorkor. The **immune system** is impaired in PEM, particularly T-cell function. Individuals are thus very susceptible to **infections** (eg, causing diarrhea), and infections worsen the situation by placing a higher metabolic demand on the body (eg, through fever). Some differences between kwashiorkor and marasmus are summarized in Table 54–6.

PEM is entirely **preventable** by a well-balanced diet containing adequate amounts of the major macronutrients, micronutrients, vitamins, and minerals.

Figure 54–20 summarizes some of the mechanisms involved in kwashiorkor.

CASE 13: MYOCARDIAL INFARCTION (MI)

Etiology

Lack of oxygen and various metabolites (due to blockage of blood flow in a coronary artery to an area of myocardium). Genetic and other factors predispose to this situation.

History and Physical Examination

A 46-year-old businessman was admitted to the emergency department of his local hospital complaining of severe retrosternal pain of 1-h duration. He had previously been admitted to the hospital once for treatment of a small MI, but despite this he continued to smoke heavily. He had been advised to take a mainly vegetarian diet, restrict his salt intake, join an exercise program, and prescribed an HMG-CoA reductase inhibitor (his total cholesterol and LDL cholesterol had been elevated and HDL depressed) and a combination of a thiazide diuretic and an ACE (angiotensin-converting enzyme) inhibitor for moderate hypertension. He was also taking one aspirin (81 mg) per day. His blood pressure was 150/90 Hg (before this incident it had been 140/80 mm Hg; it was likely elevated because of stress), pulse was 60/min, and he was sweating profusely. There was no evidence of cardiac failure. His father had died at age 52 from a "heart attack," and one of his two siblings had had an MI at age 49. Because of the admitting diagnosis of probable MI, he was given morphine to relieve his pain and apprehension and also for its coronary dilator effect, and immediately transferred to a cardiac care unit, where continuous electrocardiographic monitoring was started at once.

Laboratory Findings

The initial ECG showed S-T segment elevation and other changes in certain leads, indicative of an acute anterior transmural left ventricular infarction. Blood was taken initially and at regular intervals thereafter for measurement of troponin T; on admission the level was within normal limits, but had risen 8-fold by 6 h after admission.

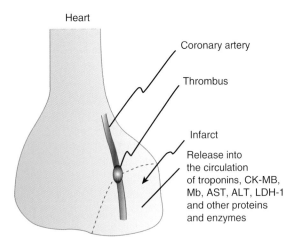

FIGURE 54–21 Diagrammatic representation of a coronary artery thrombus resulting in release of various proteins and enzymes into the circulation from an area of myocardial infarction (MI). Various proteins and enzymes are released from infarcted tissue at different rates, and they exhibit different half-lives in the circulation. At the present time, troponins are widely used to assist in the diagnosis of MI, but the other enzymes shown are still used to variable extents, and were more widely used in the past. (Mb, myoglobin; CK-MB, the MB isozyme of creatine kinase; AST, aspartate aminotransferase; ALT, alanine aminotransferase; LDH-1, the heart muscle isozyme of lactate dehydrogenase.)

Some proteins and enzymes that have been used in the diagnosis of a myocardial infarction are indicated in Figure 54–21. The use of enzymes and proteins in the diagnosis of MI and other diseases is discussed in Chapter 7. Levels of total cholesterol and the LDL/HDL cholesterol ratio were within normal limits (<5.17 mmol/L and 4:1, respectively), and triacylglycerols were 1.50 mmol/L (normal <2.26 mmol/L).

Treatment

The attending cardiologist, after reviewing all aspects of the case, decided to administer tissue plasminogen activator (t-PA) intravenously because of the diagnosis of anterior transmural MI. Some 1.5 h had elapsed since the onset of symptoms. Chest pain began to disappear after 12 h, and the patient felt increasingly comfortable. He was discharged from hospital 7 days later under the care of his family doctor. He was instructed to continue his medication, to attend a cardiac rehabilitation program, and to stop smoking.

Discussion

An MI is generally caused by an occlusive thrombus lying in close proximity to an unstable atherosclerotic plaque which has often ruptured recently. The rupture of the plaque helps generate the thrombus. Usually the diagnosis can be made from the clinical history, the electrocardiographic results, and serial measurements of a cardiac biomarker, such as **troponin T**. Measurement of this protein has replaced that of CK-MB in many hospitals (see Chapter 7).

Major aims of treatment are to prevent death from cardiac arrhythmias by administration of appropriate drugs and to limit the size of the infarction. In this case, the decision was made to limit the size of the infarct by **administration of t-PA**, which can dissolve or limit the growth of the thrombus (see Chapter 51) if given up to 12 h of onset of symptoms, although preferably sooner. An alternative would have been **percutaneous coronary intervention** (PCI), consisting of percutaneous transluminal coronary angioplasty (PTCA), with or without insertion of a stent.

Atherosclerosis, **coronary thrombosis** and **MI** are described here very briefly; a textbook of pathology should be consulted for detailed descriptions. Atherosclerosis consists of patchy plaques in the intima of medium and large arteries. **Endothelial dysfunction** is believed to play an important role in the genesis of atherosclerosis. Platelets and fibrin can deposit on the luminal aspect of an artery and monoclonally derived smooth muscle cells present in the medial layer of artery grow into the intimal lesion, attracted by growth factors released by macrophages and platelets (eg, platelet-derived growth factor). Plasma lipoproteins, glycosaminoglycans, collagen, and calcium subsequently accumulate in a lesion called a **fatty streak**. The presence of **oxidized LDL** in atherosclerotic lesions appears to be particularly important, in that it encourages recruitment of macrophages (inflammatory cells) and stimulates the release of growth factors. **Inflammation** is thus thought to be a key factor in atherosclerosis, as reflected by accumulation of macrophages and lymphocytes. The elevated plasma level of **C-reactive protein** (CRP) (Chapter 50) is also a reflection of chronic inflammation.

As the above processes progress, the fatty streak evolves into an **intimal plaque** Inflammation and hemorrhage into the plaque can occur, leading to rupture of its surface and exposure of its underlying constituents to the blood. Platelets will adhere to the exposed collagen, and a **thrombus** is initiated (Chapter 51).

Risk factors for atherosclerosis include, age, family history, male sex, high levels of LDL and low levels of HDL, hypertension, diabetes mellitus, and smoking. This patient had elevated levels of total cholesterol and LDL and depressed levels of HDL before starting dietary and drug treatment.

If the thrombus in a coronary artery occludes approximately 90% of the vessel wall, blood flow through the affected vessel may cease (**total ischemia**) and the oxygen supply of the affected area of mycocardium will be rapidly compromised. The normal metabolism of the myocardium is **aerobic**, with most of its ATP being derived from oxidative phosphorylation. The anoxia secondary to total ischemia results in a switch to **anaerobic glycolysis**, which generates only about one-tenth of the ATP produced by oxidative phosphorylation. Not only does this switch in metabolism occur, but the flow of substrates into the myocardium via the blood and the removal of metabolic products from it are also greatly reduced. This accumulation of intracellular metabolites increases the intracellular oncotic pressure, resulting in cell swelling, affecting the permeability of the plasma membrane. Thus, the affected myocardium exhibits depletion of ATP, accumulation of lactic acid, development of severe acidosis, and marked reduction of

FIGURE 54–22 Summary of mechanisms involved in the causation of an acute myocardial infarction. The arrows do not in all cases imply a strict causal relationship.

contractile force. The **precise metabolic changes that commit a cell to dying** are being investigated in many laboratories; this is a very important area of research. Changes under study include depletion of ATP, activation of intracellular phospholipases (resulting in damage to cellular membranes), activation of proteases, and accumulation of intracellular Ca^{2+}.

Some of the mechanisms involved in the causation of an acute MI are summarized in Figure 54–22.

CASE 14: OBESITY

Before studying this case history the reader is advised to consult the material on triacylglycerols and adipose tissue in Chapter 25 and the material on nutrition in Chapters 43 and 44.

Causation

Many factors contribute to obesity (genetic, environmental, cultural, etc). However, the central issue is that energy intake exceeds energy expenditure, resulting in the storage of triacylglycerols in adipose tissue.

History, Physical Examination, and Laboratory Findings

A 30-year-old woman visited her family doctor complaining of being markedly overweight. She was married, but had no children. She also complained that her periods were irregular. She gave no significant past history and was not on any drugs. She stated that she had always tended to being overweight, and that her mother and two sisters were also overweight. She had a sedentary job in an office and did not exercise regularly. In addition, she said that she had "a healthy appetite," and both she and her husband (who was also overweight) often consumed a variety of fast foods. Many obese individuals deny that they eat excessively, and it is difficult to measure food consumption precisely in ordinary medical practice.

On examination, she was obviously overweight (200 lb, 91 kg) for her size (5′4″, 163 cm). Her body mass index (BMI = weight kg/height m^2) was calculated from a table and found to be ~34. A BMI of 25–29.9 kg/m^2 indicates overweight and a BMI of >30 kg/m^2 indicates obesity. Other clinical indicators of obesity include waist–hip ratio and skin fold thickness. If fat accumulates around the abdomen this confers an apple shape, whereas fat accumulating around the buttocks confers a pear shape. The former is more serious, as abdominal (visceral) fat upon lipolysis can release fatty acids into the portal vein, leading to fat deposition in the liver and muscles. More precise tools for measuring obesity are also available (eg, bioelectrical impedance analysis and underwater weighing). Her blood pressure was 140/95 and her total cholesterol was 6.1 mmol/L (borderline high). Urinalysis was negative. Her fasting blood glucose was 6.6 mmol/L (borderline high).

Treatment

Her physician advised her that she was obese, but not morbidly obese (BMI >40). The laboratory results indicated that her levels of cholesterol and blood glucose were borderline high, as was her blood pressure. He told her that further elevation of these values would predispose her to **heart disease, diabetes mellitus, hypertension,** and **the metabolic syndrome.** The latter is characterized by excess abdominal fat, high blood glucose (resistance to insulin), abnormal blood lipids (increased LDL and decreased HDL), and high blood pressure. He also pointed out a number of **other complications** to which obesity predisposed her (eg, reproductive problems such as irregular periods, gallbladder disease, deep vein thrombosis, sleep apnea, etc). He indicated to the patient that treatment of obesity was not easy, and that she would have to undergo a **permanent change in lifestyle** if she wished therapy to be successful and weight loss to be maintained. The **major approaches to treatment** of obesity were summarized: (1) diet, (2) exercise, (3) behavioral therapy, (4) drugs (eg, to suppress appetite centrally, or acting as an inhibitor of lipase activity in the gut, thus reducing absorption of fatty acids), and (5) surgery (eg, laparoscopic adjustable gastric banding and other procedures) in very severe cases.

He stated that in her case he believed satisfactory progress could be made, at least initially, through use of the first two lines of therapy. He told the patient that he seldom recommended drugs for treatment of obesity, and that surgery was generally restricted to people with morbid obesity (BMI >40) who had not responded to other approaches. He also encouraged her to get her husband involved because of his weight problem, and because it would be mutually beneficial if he also were on a weight reduction program.

Regarding **diet**, her doctor (who had a special interest in implementing healthful nutrition in his practice) discussed general features of a suitable diet and also the advantages of **regular daily exercise**. Specific dietary recommendations that he made are listed in Table 54–7.

She subsequently joined a weight loss organization and found that it stressed changes in behavior (eg, developing sensible eating habits, rewards for good results, group counseling) and also provided support and encouragement from the other members. The patient was very keen to lose weight, and over

the next year she closely followed the various advices given. She lost 34 lbs over that period of time. She felt much better and also stated that her periods had regularized and she was hoping to get pregnant. In addition, her blood pressure and her levels of total cholesterol and blood glucose were down to normal values. She was absolutely determined to continue on the weight loss program, as was her husband. Her sole regret was that she had not lost even more weight. Many patients under treatment for obesity eventually regain lost weight, for a variety of reasons.

Discussion

Obesity is a very prevalent condition, which is on the increase. Almost one-third of adults in the Unites States are obese, and, alarmingly, more and more children are obese. Some talk of an epidemic of obesity in Western society. Dissecting all the factors contributing to this is complex, but **increased energy intake** and **decreased energy expenditure** play important roles. The increased intake of fast foods and high calorie soft drinks and TV watching by "couch potatoes" are important contributors.

While the dangers of obesity have been widely circulated, it is of interest that some obese individuals feel that they are being unfairly targeted by health professionals and that the dangers of obesity have been over-emphasized. It can be debated that it is better to be somewhat obese and fit, than to be of normal weight and unfit. In addition, certain individuals claim that they enjoy being obese.

While obesity itself is relatively easy to recognize and define (eg, we can somewhat arbitrarily use a BMI >30), pinning down the specific factors that contribute to individual cases is not that easy.

In the present case, **a number of contributory factors** are apparent. For example, she led a sedentary lifestyle, she frequently ate fast foods, she did not exercise regularly, etc. However, what was her precise caloric intake? What was her precise energy expenditure? Were the various mechanisms controlling appetite (see Figure 54–23) working properly? What role if any did genetic factors play in her obesity (she gave a family history of obesity)? These factors are not easy to quantify for a physician in her/his office.

Because of its medical importance, much research is going on presently in the field of obesity. This spans basic research on the adipocyte and signaling mechanisms to epidemiologic studies on various populations and their susceptibility to obesity. Here we shall only touch briefly on three areas: (1) regulation of appetite and food intake; (2) some genetic aspects; and (3) some aspects of energy expenditure.

Figure 54–23 summarizes a body of knowledge concerning the **regulation of appetite**. The **hypothalamus** plays a key role in the central regulation of appetite. Factors that both increase and suppress appetite are shown. Psychologic, neural, and cultural factors all play a role. A variety of peptides, affecting specific areas of the hypothalamus, are indicated. In addition, levels of circulating metabolites (eg, glucose) and of

TABLE 54–7   Summary of Dietary Advice Regarding Weight Loss

Acquire information regarding general nutrition and calories and study food labels.

Start a low calorie diet (about 1200 calories/day), containing appropriate amounts of carbohydrates, proteins and fats.

Eat more frequent, smaller meals.

Reduce intake of simple sugars and refined carbohydrates, and increase intake of complex carbohydrates (grains, etc) and low glycemic index foods.

Reduce intake of red and processed meat.

Reduce intake of saturated fats, and increase intake of monounsaturated and polyunsaturated fats and of omega-3 fatty acids.

Reduce salt intake.

Increase consumption of fresh fruits, vegetables, legumes, nuts, and low-fat dairy foods.

Markedly reduce intake of fast foods and calorie-rich soft drinks.

Avoid any of the fad diets.

Drink good quality water.

Take vitamin and mineral supplements.

Consult a dietitian for further details on diet and nutrition

Join an organization that specializes in advising people on weight loss and runs a daily exercise program.

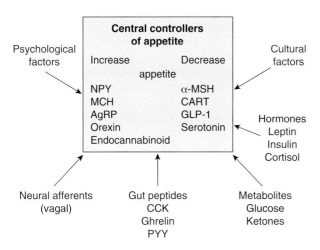

FIGURE 54–23 Factors that regulate appetite through effects on central neural circuits. Factors that increase or decrease appetite are listed. (NPY, neuropeptide Y; MCH, melanin-concentrating hormone; Ag RP, Agouti-related peptide; MSH, melanocyte-stimulating hormone; CART, cocaine- and amphetamine-related transcript; GLP-1, glucagon-related peptide-1, CCK, cholecystokinin; PYY, peptide YY.) The reader is advised to consult a textbook of Physiology for descriptions of the actions of these various molecules. (Reproduced, with permission, from Harrison's *Principles of Internal Medicine*, 17th ed, Fauci AS et al [eds], McGraw-Hill, 2008.)

hormones affect the hypothalamic centers. Regarding hormones, for example, individuals with Cushing syndrome (high levels of cortisol or related hormones) are obese and display a characteristic distribution of body fat. Particular interest has focused on **leptin**, a polypeptide released from adipocytes that acts mainly on the hypothalamus. Elevated levels of leptin decrease food intake and also increase energy expenditure. Leptin was discovered through studies of genetically obese mice (ob/ob). In humans, leptin is the product of the *OB* gene. The levels of leptin are elevated in most obese humans, suggesting that in some manner they may be resistant to its action.

Genetic influences on obesity have been recognized. Identical twins tend to have similar body weights. Occasional cases of mutations in the genes encoding leptin and the leptin receptor have been identified. Cases of obese individuals having mutations in the gene encoding proopiomelanortin (POMC), which is processed to form α-melanocyte hormone (α–MSH), a powerful appetite suppressant, have been described. In addition, mutations in the gene encoding the type 4 receptor for α-MSH and in the genes encoding two proteolytic enzymes involved in the conversion of POMC to α-MSH have been reported.

Children with the Prader-Willi syndrome (due to deletion of a part of chromosome 15; affected individuals overeat, among other signs) and with the Laurence-Moon-Biedl syndrome (an autosomal recessive genetic disorder) are characteristically obese. It should be stressed that the great majority of obese individuals do not apparently have mutations in the genes referred to above. However, it is likely that multiple

FIGURE 54–24 Simplified scheme of the causation of obesity.

subtle genetic factors (eg, single nucleotide polymorphisms, SNPs) influence obesity.

Regarding **energy expenditure**, like food intake, this is difficult to measure precisely except in research facilities. A sophisticated method for measuring energy expenditure using doubly labeled water is described in Chapter 44, as are the concepts of basal metabolic rate (BMR) and other factors involved in daily energy expenditure. It appears that most obese individuals have a higher energy expenditure than people of normal weight, because their lean body mass is increased. Another important issue is whether obese individuals actually eat more than non-obese individuals. It appears likely that they do, although many deny this. In the present case, it seems likely from her history that the patient did overeat.

An issue that has been the subject of debate is the possible role of **variations in diet-induced thermogenesis** (see Chapter 25) in contributing to obesity. Brown adipose tissue contains a mitochondrial protein known as **thermogenin** (uncoupling protein–1) that dissipates energy as heat. While brown adipose tissue is not a prominent component of adults (unlike newborns), it does occur. Also it appears to be diminished in amount in at least certain obese individuals, which could mean that they dissipate less energy as heat than non-obese individuals and have more available for other purposes.

Also two other uncoupling proteins are known to exist in human tissues, although their overall contribution to diet-induced thermogenesis is not clearly established. Thus, many factors can contribute to obesity and it is not easy to assess their contribution in most cases seen in clinical practice. At the present time, it is reasonable to treat obesity primarily by reducing intake of food, eating a healthful diet, increasing physical activity and offering appropriate support and encouragement.

Figure 54–24 summarizes some major factors involved in the causation of obesity.

CASE 15: OSTEOPOROSIS, PRIMARY (POSTMENOPAUSAL)

Before studying this case history the reader is advised to peruse the material in Chapter 48 on bone and the material in Chapter 44 on vitamin D.

Causation

Osteoporosis is loss of bone mass with preservation of the normal ratio of organic matrix (mostly proteins) to mineral. A variety of factors (endocrine, nutritional, lack of physical activity, etc) contribute to its development. In the postmenopausal type of osteoporosis dealt with here, estrogen deficiency is the major factor.

History, Physical Examination, and Investigations

A 64-year-old woman presented at emergency department, having tripped in her garden and apparently fallen rather gently onto her right forearm. Nevertheless, she suspected that she had fractured a bone in her arm, because of the pain and swelling just above her right wrist joint. X-rays showed a fracture of the distal end of the radius, with displacement. The radius also showed moderate reduction of radiodensity, suggestive of osteoporosis. The fracture was reduced, an appropriate cast was applied and she was asked to report to her family physician 2 weeks later. The emergency department physician gave her a note to give to her own physician, which mentioned that, because of the mildness of her fall, the resultant fracture, and the reduced radiodensity of her radius, he suspected that she might have osteoporosis. The patient attended her family doctor 2 weeks later. She had four adult children. Her last menstrual period had been some 5 years previously, and she had only attended her physician very irregularly over the years for the occasional minor ailment. She was not on any medication, did not take any vitamin or mineral supplements, and had never taken hormonal treatment for menopause. She ate very few fruits and vegetables, and overall she consumed a high carbohydrate diet along with liberal amounts of fried foods. She smoked one pack of cigarettes a day and also had several drinks of vodka each evening. In addition, she rarely exercised. She complained of chronic lower backache, but otherwise gave no significant history. In view of the fracture and the suggestion of osteoporosis, her family doctor ordered dual-energy x-ray absorptiometry (DEXA) of the lumbar spine and hip areas to assess bone density. He also ordered x-rays of the lower spine because of the history of lower back pain. In addition, he ordered determinations of serum Ca, P, alkaline phosphatase, 25-hydroxyvitamin D, parathyroid hormone, and complete urinalysis (including 24-h calcium) to check for any other bone disease (eg, due to vitamin D deficiency, hyperparathyroidism, or multiple myeloma). The results of her DEXA scan showed a marked reduction (over 3 standard deviations; over 2.5 is diagnostic for osteoporosis) from the average value in 25-year-old women of her race, compatible with severe osteoporosis. The x-rays of her lower spine showed decreased radiodensity, but no fractures. Results of her blood and urine tests were within normal limits, suggesting that she had no other serious bone disorder.

When bone resorption occurs, there is increased production of **collagen cross-link products**. These include N-telopeptide, deoxypyridinoline and C-telopeptide. Also, in some cases of osteoporosis, **increased bone formation** occurs; bone alkaline phosphatase and osteocalcin are markers of this. These various markers are not in themselves diagnostic of osteoporosis, but can be measured in serum or urine as **indicators of response to therapy**. For example, a 30% decrease of these markers would suggest successful therapy. They were not measured in this case, and in fact are not measured in many clinical laboratories.

Treatment

The history, DEXA scan, and the results of the other tests were consistent with a diagnosis of relatively severe osteoporosis. She was advised to start on an exercise program immediately at a gym, initially under the supervision of a personal trainer. She was also referred to a dietitian to change her dietary habits; recommendations included eating regular, daily portions of fresh fruits and vegetables, and consuming a more balanced diet (eg, reducing the starchy and fried foods). It was recommended that she stop smoking and cut down on her consumption of alcohol, as both of these may contribute to osteoporosis. She was also started on appropriate doses of calcium citrate and vitamin D. In addition to the above, her physician recommended that she start on a bisphosphonate (eg, alendronate or risedronate), giving detailed instructions as to how to take the drug. These two drugs, N-containing bisphosphonates, are taken up by osteoclasts and inhibit formation of farnesyl diphosphate. This in turn inhibits **prenylation of certain proteins** (see Figure 26–2) and their attachment to the plasma membrane, negatively affecting osteoclast activity and thus inhibiting bone resorption. Other drugs that may be used in the treatment of osteoporosis in selected cases include salmon calcitonin, estrogen, selective estrogen modulators (eg, raloxifene), and parathyroid hormone. Estrogen was formerly widely prescribed early in menopause to reduce osteoporosis, and appeared to be relatively effective. However, the Women's Health Initiative Study indicated that the risks of estrogen therapy outweighed the benefits in this situation.

The patient was followed up over the next few years. She lost considerable weight and maintained her exercise program. In general, she felt much more healthy than she had done in the previous 20 years. Her fracture healed satisfactorily, no further loss of bone mass occurred, but normal bone mass was not restored. She was advised on how to take precautions to avoid falls and it was also recommended that she wear hip pads.

Discussion

Osteoporosis can be defined as **reduction of bone mass or density**. Often it is first detected after a fracture, as the loss of bone tissue predisposes to fractures. The World Health Organization (WHO) has suggested that osteoporosis exists when **bone density falls 2.5 standard deviations or more below the mean for young healthy adults of the same race and gender**. In the United States, some 8 million women and 2 million

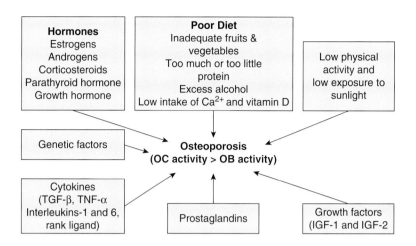

FIGURE 54–25 Simplified scheme of various factors involved in the causation of osteoporosis. (OC, osteoclastic, OB, osteoblastic, IGF-1 and IGF-2, insulin-like growth factors 1 and 2; TGF-β, transforming growth factor β; TNF-α, tumor necrosis factor α.) The hormones listed have a variety of effects on bone. IGF-1 and IGF-2 have anabolic effects on bone, but may also stimulate turnover of bone. RANK-ligand is a cytokine that is involved in communication between osteoblasts and osteoclasts; when it interacts with osteoclasts, it increases their activity. The other cytokines are made by osteoblasts. Their synthesis is increased or decreased by estrogen deficiency, with the overall effect of extending the life span of osteoclasts (by inhibiting apoptosis).

men have osteoporosis, and many others are at risk of developing it.

Two terms related to osteoporosis are osteopenia and osteomalacia. **Osteopenia** is decreased bone mass, embracing both osteoporosis and osteomalacia. In **osteoporosis**, bone mass decreases due to decreased bone formation and increased resorption, but a normal ratio of bone mineral (hydroxyapatite) to bone matrix (mostly collagen type 1) is preserved. **Osteomalacia** is also an example of osteopenia, but in it there is decreased mineralization; its commonest cause is deficiency of vitamin D.

Primary osteoporosis can be divided into 3 types: **idiopathic** (uncommon, and occurs in children and young adults), **postmenopausal,** and **involutional** (elderly) types. This case is an example of **postmenopausal osteoporosis**, and decline in estrogen levels is a major factor in its causation. This type can also occur in males due to a decline of serum testosterone, which acts to increase osteoclastic activity. **Involutional** osteoporosis occurs in older individuals and is due to the decline of the number of osteoblasts with age. Postmenopausal and involutional osteoporosis may coexist.

Secondary osteoporosis is relatively uncommon, and may be due to a variety of conditions (eg, chronic renal disease, various drugs [especially corticosteroids], a number of endocrine disorders, malabsorption syndrome, etc).

FIGURE 54–26 Simplified scheme of some major factors in the causation of osteoporosis.

Many factors are involved in the causation of osteoporosis (see Figure 54–25). An in-depth understanding of them requires a knowledge of bone **modeling**, of a variety of **cytokines**, of the actions of **a number of hormones**, and of various **nutritional** and **genetic factors**, among other considerations. (Cytokines are a heterogeneous group of proteins released by various cells and which have autocrine or paracrine effects.) Here we only indicate the complexity of a complete understanding of osteoporosis by briefly mentioning the main players. The central point is that osteoporosis occurs when **bone resorption** (osteoclastic [OC]) activity) **exceeds bone formation** (osteoblastic [OB] activity). In regards to post-menopausal osteoporosis in particular, one important consideration is that **estrogen loss** appears to increase the secretion of a number of cytokines that lead to recruitment of osteoclasts. Also, estrogen loss diminishes the secretion of certain other cytokines that promote osteoblastic activity. The overall effect is thus an imbalance between osteoclasts and osteoblasts, in favor of the former.

Figure 54–26 is a simplified schematic representation of the causation of osteoporosis.

CASE 16: XERODERMA PIGMENTOSUM (XP) (PIGMENTED DRY SKIN)

The reader should consult the material in Chapter 35 on DNA repair and on XP before reading this case.

Causation

Genetic (a mutation in a gene directing synthesis of a DNA repair enzyme in the nucleotide excision repair pathway) and physical (exposure to ultraviolet [UV] irradiation).

History and Physical Examination

An 8-year-old boy, an only child, presented at a dermatology clinic with a skin tumor on his right cheek. He had always avoided exposure to sunlight because it made his skin blister. His skin had scattered areas of hyperpigmentation and other areas where it looked mildly atrophied. There was no family history of a similar disorder. Because of the presence of a skin tumor at such a young age, the history of avoidance of sunlight, and the other milder skin lesions, the dermatologist made a provisional diagnosis of XP.

Laboratory Findings

Histologic examination of the excised tumor showed that it was a **squamous cell carcinoma** (a common type of tumor skin cancer in older people, but very unusual in a boy of this age). A small piece of skin was removed for **preparing fibroblasts** to be grown in tissue culture. A research laboratory in the hospital specialized in radiation biology and was set up to measure the amount of **thymine dimers** (see below) formed following exposure to UV light. The patient's fibroblasts and control fibroblasts were exposed to UV light, and cell samples were taken at 8-h intervals for a total of 32 h postirradiation. Extracts of DNA were prepared and the numbers of dimers remaining at each time point indicated were determined. Whereas only 24% of the dimers formed persisted in DNA extracted from the normal cells at 32 h, approximately 95% were found in the extract from the patient's cells at 32 h. This showed that the UV-induced lesions had not been repaired, and thus confirmed the diagnosis of XP.

Discussion

XP is a rare **autosomal recessive** condition in which the mechanisms for repair of DNA subsequent to damage by UV irradiation are defective. This arises because of mutations in the genes encoding components of the **nucleotide excision pathway** of DNA repair (nucleotide excision repair, NER; see Chapter 35). The major damage inflicted on DNA by UV irradiation is the formation of thymine (pyrimidine) dimers (also known as cyclobutane pyrimidine dimers), where covalent bonds are formed between carbon atoms 5 and 5 and 6 and 6 of adjacent intrachain thymine residues, resulting in the dimers. Other types of damage can also occur.

NER has two sub-pathways: global genome repair and transcription-coupled repair. The former scans the whole genome and removes damaged areas rapidly. The second is linked to transcription, operates slowly, and removes damage from the transcribed strand of DNA.

Detailed analyses of NER involved in the removal of thymine dimers have been performed. The pathway is highly conserved across species, indicating its importance. In general, the pathway involves four major steps, all involving various enzymes: (1) **recognition** of damaged DNA; (2) **excision** of damaged region; (3) **filling in** of the gap by DNA polymerase; and (4) **ligation** of the filled in area. In *E coli*, an endonucleolytic cleavage, catalyzed by a specific **endonuclease** (also called an excinuclease), occurs on either side of the damage, releasing a 12- to 13-base-pair oligonucleotide. The polymerization step involves **DNA polymerase** and the final step is sealing by **DNA ligase**.

The NER pathway operates in humans, and its details are still being elucidated. It appears to be generally similar to the pathway in *E coli*. The most noticeable difference is that a much larger oligonucleotide (about 30 bases) is excised in humans. A simplified scheme of the initial steps in the human pathway is shown in Figure 35–24. The products of at least seven genes (these encoding XPA through XPG) have been implicated in NER in humans, and all have been cloned. Mutations in any one of these genes cause XP. In the boy whose situation is discussed here, the specific gene involved was not determined.

Because the responsible genes have been cloned, **prenatal diagnosis** of XP is now possible using appropriate probes. The NER pathway is also involved in processes other than DNA repair, such as recombination, replication, and transcription. The involvement of the seven genes mentioned above in DNA repair was originally shown in the following manner. It was observed that when cultured cells from individuals with XP were cocultured with cells from other individuals under conditions in which **cell fusion** occurred, the defect in DNA repair could sometimes be corrected. This indicated that one set of cells was providing a normal gene product to the other, thus correcting the defect. In this manner, at least **seven complementation groups**, corresponding to the seven genes and their products mentioned above, have been recognized.

If UV damage is not repaired, **mutations in DNA** will result, chromosomal abnormalities may occur and **cancer** may ensue. Patients with XP often suffer from a variety of skin cancers from an early age.

As described in Chapter 35, there are **other pathways** of DNA repair operative in humans. They are all important in preserving the integrity of DNA, and abnormalities of some of them have been associated with cancer (eg, mismatch repair).

FIGURE 54–27 Summary of mechanisms involved in the causation of xeroderma pigmentosum (OMIM 278730 & other entries). Also compare with Figure 35–24, showing the mechanism of nucleotide-excision repair.

TABLE 54–8 Some Major Challenges Facing Medicine and Allied Health Sciences[1]

Topic	Challenge
Aging	Understanding its molecular bases and perhaps modifying some of its effects
Various types of arthritis and osteoporosis	Understanding their molecular bases (eg, further study of the roles of the ECM in their causation) and improving present therapies
Cancers	Understanding their molecular bases (eg, additional studies on oncogenes, tumor suppressor genes and mechanism of metastasis), developing better biomarkers for earlier diagnosis, and improving present therapies
Cardiovascular diseases (eg, myocardial infarctions and strokes)	Understanding their molecular bases (eg, increased knowledge of atherosclerosis and thrombosis) and improving present therapies
Chronic neurodegenerative diseases (eg, Alzheimer disease)	Understanding their molecular bases (eg, additional insights into the roles of various proteins in their causation) and improving present therapies
Diabetes mellitus	Obtaining further insights as to its causes and effects (eg, obtaining a complete picture of all aspects of insulin action and of mechanisms of tissue damage such as glycation) and improving therapy
Environmental medicine	Requires a concerted effort among scientists and health workers to prevent further degradation of the environment and forestall potentially serious health effects
Genetic diseases	Establishing their molecular bases and developing gene therapy and other treatments (eg, use of small molecules to help affected proteins fold properly)
Infections, including AIDS and tropical diseases	Understanding their molecular bases, preventing their spread (eg, by increased knowledge of the biochemistry of microorganisms and of mechanisms of their attachment to cells) and improving present therapies
Nutrition	Improving the standard world-wide and combatting problems such as protein-energy-malnutrition and obesity
Poverty	Encouragement of a global attack on poverty, which is a root cause of many physical and mental disorders
Psychiatric diseases	Understanding their molecular bases (eg, determining which genes are involved in diseases such as schizophrenia and bipolar disorders) and improving present therapies
Wellness	Educating populations regarding its maintenance (cellular health) and instituting measures (eg, nutrition, exercise, vaccines, mental health, avoidance of toxins) to help prevent illnesses

[1]Biochemical and various related approaches (genetic, cell biologic, immunologic, pathologic, pharmacologic, etc) will be critical in addressing many of these challenges, as will public health approaches.

The parents of the boy in the present case were told that he would have to be **watched closely** throughout life for the development of new skin cancers. In addition, he was advised to **avoid sunlight** and to use an appropriate **sunscreen ointment**. Although XP is a rare condition, the existence of a variety of mechanisms for repairing DNA following exposure to different types of irradiation and to chemical damage is of **great protective importance**. Without their existence, life on this planet would be fraught with even more danger that it presently is! For instance, it has been estimated that patients with XP have a 1000-fold greater chance of developing skin cancer than do normal individuals.

Figure 54–27 summarizes the mechanisms involved in the causation of XP.

EPILOG

Remarkable progress has been made in biochemistry and in related fields such as genetics and cell biology. Many of the discoveries in these disciplines have had great impact on medicine and related life sciences. Crucial insights into the natures of many diseases have emerged from studies of their molecular bases. New drugs and other treatments are constantly being developed based on such discoveries. However, it is obvious that there are still many major challenges facing medical science. Table 54–8 summarizes some of them. The authors of this text and other biochemists believe that the application of biochemical, genetic, and allied approaches to these problems and to others not listed will pay rich dividends from which people all across the globe will benefit. Hopefully some of the readers of this text will contribute to such endeavors.

REFERENCES

Aiuti A et al: Gene therapy for immunodeficiency due to adenosine deaminase deficiency. N Engl J Med 2009;360:447.

Axford JS, O'Callaghan C (editors). *Medicine.* 2nd ed. Blackwell Science, 2004. (Contains fundamental coverage of many of the conditions described here).

Beers MH, Porter RS, Jones TV et al (editors): *The Merck Manual of Diagnosis and Therapy.* 18th ed. Merck Research Laboratories, 2006. (This book is available free online at http://merck.com/mmpe/index.html and contains coverage of many of the conditions discuused here).

Burtis CA, Ashwood ER, Bruns DE (editors): *Tietz Textbook of Clinical Chemistry and Molecular Diagnostics.* 4th ed. Elsevier Saunders, 2006. (Chapters 23 [Tumor Markers] and 49 [Mineral and Bone Metabolism] are of particular interest to the contents of this Chapter.)

Doherty GM, Way LW (editors): *Current Surgical Diagnosis & Treatment.* 12th ed. Lange Medical Books/McGraw-Hill, 2006. (Chapter 30 contains a discussion of colorectal cancer).

Eckel RH: Nonsurgical management of obesity in adults. N Engl J Med 2008;358:1941.

Fauci AS et al (editors): *Harrison's Principles of Internal Medicine.* 17th ed. McGraw-Hill, 2008. (Contains comprehensive descriptions of many of the conditions described here).

Kohn DB, Candotti F: Gene therapy fulfilling its promise. N Engl J Med 2009;360:518. (Discusses aspects of the successful treatment of ADA deficiency.)

Rakel RE, Bope ET (editors): *Conn's Current Therapy.* Saunders Elsevier, 2007. (Various Chapters contain discussions of many of the conditions described here).

Riordan JR: CFTR function and prospects for therapy. Annu Rev Biochem 2008;77:70.

Scientific American. Volume 18, no. 3, 2008. (A special edition on various aspects of cancer).

Scriver CR et al (editors): *The Metabolic and Molecular Bases odf Inherited Disease.* 8th ed. McGraw-Hill, 2001. (The online version of this text contains comprehensive up-dated descriptions of many of the conditions discussed in this Chapter).

Shils ME, Shile M, Ross AC et al (editors): *Modern Nutrition in Health and Disease.* 10th ed. Lippincott Williams & Wilkins, 2006. (Contains comprehensive coverage of nutritional topics, including PEM [Chapter 57] and obesity [Chapters 63 & 64]).

Tannock IF, Hill RP, Bristow RG, Harrington L (editors): *The Basic Science of Oncology.* 4th ed. McGraw-Hill. 2005. (Various Chapters contain discussions of DNA repair and properties of cancer cells).

Appendix I

SELECTED LABORATORY RESULTS

Results are listed in both SI units (SI = Systeme International d'Unites) and "conventional" units. The results are relevant to understanding the cases described in Chapter 54. Laboratory results can vary from laboratory to laboratory; the reader should check with her/his local laboratory their reference values. (The results listed here are from *Harrison's Principles of Internal Medicine,* Fauci AS et al [editors], Appendix: Laboratory Values of Clinical Importance, by Kratz A et al, 17th edition, McGraw-Hill, 2008, with permission.)

Blood Analyte or Measurement	SI Units	Conventional Units
Alanine aminotransferase (ALT)	0.12–0.70 µkat/L	7–41 U/L
Albumin Female Male	41–53 g/L 40–50 g/L	4.1–5.3 g/dL 4.0–5.0 g/dL
Anion gap	7–16 mmol/L	7–16 mmol/L
Bicarbonate	22–30 mmol/L	22–30 meq/L
Calcium	2.2–2.6 mmol/L	8.7–10.2 mg/dL
Carcinoembryonic antigen (CEA)	0.0–3 µg/L	0.0–3.0 ng/mL
Chloride	102–109 mmol/L	102–109 meq/L
Cholesterol, total (these levels are considered desirable)	<5.17 mmol/L	<200 mg/dL
Cholesterol, LDL (considered optimal)	< 2.59 mmol/L	< 100 mg/dL
Cholesterol, HDL	1.03–1.55 mmol/L	40–60 mg/dL
C-reactive protein	0.20–3.0 mg/L	0.20–3.0 mg/L
Creatinine Female Male	44–80 µmol/L 53–106 µmol/L	0.5–0.9 ng/mL 0.6–1.2 ng/mL
Ferritin Female Male	10–150 µg/L 29–248 µg/L	10–150 ng/mL 29–248 ng/mL
Glucose (fasting)	4.2–6.1 mmol/L	75–110 mg/dL
Hematocrit Female Male	0.354–0.444 0.388–0.464	35.4–44.4 38.8–46.4
Hemoglobin Female Male	120–158 g/L 133–162 g/L	12.0–15.8 g/dL 13.3–16.2 g/dL

Blood Analyte or Measurement	SI Units	Conventional Units
Iron	7–25 µmol/L	41–141 µg/dL
Iron-binding capacity	45–73 µmol/L	251–406 µg/dL
Iron saturation	0.16–0.35	16–35%
Lactate (arterial blood)	0.5–1.6 mmol/L	4.5–14.4 mg/dL
Magnesium	0.62–0.95 mmol/L	1.5–2.3 mg/dL
Osmolality	275–295 mOsmol/kg serum water	275–295 mOsmol/kg serum water
pH (arterial)	7.35–7.45	7.3–7.45
Phosphorus, inorganic	0.81–1.4 mmol/L	2.5–4.3 mg/dL
Potassium	3.5–5.0 mmol/L	3.5–5.0 meq/L
Protein, total	67–86 g/L	6.7–8.6 g/dL
Red blood cells Female Male	$4.00–5.20 \times 10^{12}$/L $4.30–5.60 \times 10^{12}$/L	$4.00–5.20 \times 10^{6}$/mm³ $4.30–5.60 \times 10^{6}$/mm³
Sodium	136–146 mmol/L	136–146 meq/
Thyroxine Free Total	10.3–21.9 pmol/L 70–151 nmol/L	0.8–1.7 ng/dL 5.4–11.7 µg/dL
Transferrin	2.0–4.0 g/L	200–400 mg/dL
Triglycerides (fasting)	0.34–2.26 mmol/L	30–200 mg/dL
Troponin T	0–0.1 µg/L	0–0.1 ng/mL
TSH	0.34–4.25 mIU/L	0.34 –4.25 µIU/mL
Urea nitrogen	2.5–7.1 mmol/L	7–20 mg/dL
Uric acid Female Male	0.15–0.33 mmol/L 0.18–0.41 mmol/L	2.5–5.6 mg/dL 3.1–7.0 mg/dL

Appendix II

SELECTED WORLDWIDE WEB SITES

The following is a list of web sites that readers may find useful. The sites have been visited at various times by one or more of the authors. Most are located in the United States, but many provide extensive links to international sites and to databases (eg, for protein and nucleic acid sequences) and online journals. RKM would be grateful if readers who find other useful sites would notify him of their URLs by e-mail (rmurray6745@ rogers.com) so that they may be considered for inclusion in future editions of this text.

Readers should note that URLs may change or cease to exist.

ACCESS TO THE BIOMEDICAL LITERATURE

High Wire Press: http://highwire.stanford.edu/
(Extensive lists of various classes of journals—biology, medicine, etc—and offers also the most extensive list of journals with free online access.)
National Library of Medicine: http://www.nlm.nih.gov/
(Free access to Medline via PubMed.)

GENERAL RESOURCE SITES

The Biology Project (from the University of Arizona): http:// www. biology.arizona.edu/default.html
(Contains excellent biochemical coverage of enzymes, membranes, etc.)
Harvard University Department of Molecular & Cellular Biology Links: http://mcb.harvard.edu/BioLinks.html
(Contains many useful links.)

SITES ON SPECIFIC TOPICS

American Heart Association: http://www.americanheart.org/
(Valuable information on nutrition, on the role of various biomolecules—eg, cholesterol, lipoproteins—in heart disease, and on the major cardiovascular diseases.)

Cancer Genome Anatomy Project (CGAP): http://www.cgap. nci. nih.gov/
(An interdisciplinary program to generate the information and technical tools to decipher the molecular anatomy of the cancer cell.)
Carbohydrate Chemistry and Glycobiology: A Web Tour: http:// sciencemag.org/feature/data/carbohydrates.dtl
(Contains links to organic chemistry, carbohydrate chemistry and glycobiology.)
European Bioinformatics Institute: http://www.ebi.ac.uk/
(Maintains the EMBL Nucleotide and SWISS-PROT databases as well as other databases.)
GeneCards: http://www.genecards.org/
(A database of human genes, their products, and their involvements in disease; from the Weizmann Institute of Science.)
GeneTests: http://www.geneclinics.org/
(A medical genetics information resource with comprehensive articles on many genetic diseases.)
Genes and Disease: http://www.ncbi.nlm.nih.gov/disease/
(Coverage of the genetic bases of many different diseases.)
Howard Hughes Medical Institute: http://www.hhmi.org/
(An excellent site for following current biomedical research. Contains a comprehensive Research News Archive.)
Human Gene Mutation Database: http://www.hgmd.cf.ac.uk/ ac/index.php
(An extensive tabulation of mutations in human genes from the Institute of Medical Genetics in Cardiff, Wales.)
Human Genome Project Information: http://www.doege-nomes. org/
(From the U.S. Department of Energy; also contains general information on genomics and on microbial genomes.)
J. Craig Venter Institute: http://www.tigr.org/index.shtml
(Contains sequences of bacterial genomes and other information).
Karolinska Institute: Diseases and Disorders: http://www.mic. ki.se/Diseases/C18.html
(Contains extensive links pertaining to nutritional and metabolic diseases.)

Lipids Online: http://lipidsonline.org/
(A resource from Baylor College of Medicine for health care practitioners with an interest in atherosclerosis, dyslipidemias, and lipid management.)

MITOMAP: http://www.mitomap.org/
(A human mitochondrial genome database.)

National Center for Biotechnology Information: http://ncbi.nlm.nih.gov/
(Information on molecular biology and how molecular processes affect human health and disease.)

National Human Genome Research Institute: http://www.genome.gov/
(Extensive information about the Human Genome Project and subsequent work.)

National Institutes of Health: http://www.nih.gov/
(Includes links to the separate Institutes and Centers that constitute NIH, covering a wide range of biomedical research.)

Office of Rare Diseases: http://rarediseases.info.nih.gov
(Access to information on more than 7,000 rare diseases, including current research.)

OMIM (Online Mendelian Inheritance in Man): http://www.ncbi.nlm.nih.gov/sites/entrez?db=omim
(A fantastically comprehensive resource on human genetic diseases, initiated by Dr Victor A. McKusick, considered by many to be the father of modern human genetics.)

Protein Data Bank: http://www.rcsb.org/pdb/home/home.do
(A worldwide repository for the processing and distribution of three-dimensional biologic macromolecular structure data.)

Society for Endocrinology: http://www.endocrinology.org/
(The site aims to advance education and research in endocrinology for the public benefit.)

Society for Neuroscience: http://www.sfn.org
(Contains useful information on a variety of topics in neuroscience.)

The Broad Institute: http://www.broad.mit.edu/
(The Broad Institute is a research collaboration of MIT, Harvard and its affiliated hospitals, and the Whitehead Institute and was created to bring the power of genomics to medicine.)

The Endocrine Society: http://www.endo-society.org/
(The mission of this Society is to advance excellence in endocrinology and promote its essential role as an integrative force in scientific research and medical practice).

The Protein Kinase Resource: http://www.kinasenet.org/
(Information on the protein kinase family of enzymes.)

The UCSD-Nature Signaling Gateway: http://www.signalling-gateway.org/
(A comprehensive resource for anyone interested in signal transduction.)

The Wellcome Trust Sanger Institute: http://www.sanger.ac.uk/
(A genome research center whose purpose is to increase knowledge of genomes, particularly through large-scale sequencing and analysis.)

BIOCHEMICAL JOURNALS AND REVIEWS

The following is a partial list of biochemistry journals and review series and of some biomedical journals that contain biochemical articles. Biochemistry and biology journals now usually have Web sites, often with useful links, and some journals are fully accessible without charge. The reader can obtain the URLs for the following by using a search engine.

- *Annual Reviews of Biochemistry, Cell and Developmental Biology, Genetics, Genomics and Human Genetics*
- *Archives of Biochemistry and Biophysics (Arch Biochem Biophys)*
- *Biochemical and Biophysical Research Communications (Biochem Biophys Res Commun)*
- *Biochemical Journal (Biochem J)*
- *Biochemistry (Biochemistry)*
- *Biochemistry (Moscow) (Biochemistry [Mosc])*
- *Biochimica et Biophysica Acta (Biochim Biophys Acta)*
- *Biochimie (Biochimie)*
- *European Journal of Biochemistry (Eur J Biochem)*
- *Indian Journal of Biochemistry and Biophysics (Indian J Biochem Biophys)*
- *Journal of Biochemistry (Tokyo) (J Biochem [Tokyo])*
- *Journal of Biological Chemistry (J Biol Chem)*
- *Journal of Clinical Investigation (J Clin Invest)*
- *Journal of Lipid Research (J Lipid Res)*
- *Nature (Nature)*
- *Nature Genetics (Nat Genet)*
- *Proceedings of the National Academy of Sciences USA (Proc Natl Acad Sci USA)*
- *Science (Science)*
- *Trends in Biochemical Sciences (Trends Biochem Sci)*

Index

Note: Page numbers followed by f indicate figures; and page numbers followed by t indicate tables.

A band, 545, 546f, 547f
A blood group substance, 603, 603f
A cyclins, 329, 329f, 329t
A gene, GalNAc transferase encoded by, 603
A (aminoacyl/acceptor) site, aminoacyl-tRNA
    binding to, in protein synthesis,
    362, 363f
A1 peptide of *V cholerae* enterotoxin, 620
ABC-1. *See* ATP-binding cassette transporter-1
Abetalipoproteinemia, 214, 232t
Abnormally folded peptide in Alzheimer
    disease, 619
ABO blood group system, biochemical basis of,
    602–603, 603f
Absolute stereospecificity of enzymes, 51
Absorption, 459–466
Absorption chromatography, for protein/peptide
    purification, 22
Absorption spectra, of porphyrins, 275, 276f
Absorptive pinocytosis, 421
ACAT (acyl-CoA:cholesterol acyltransferase),
    225
Accelerated aging, 586
Accelerator (Ac-) globulin (factor V), 585, 586t,
    586f
Acceptor arm, of tRNA, 308, 310f,
    355, 355f
Acceptor (A/aminoacyl) site, aminoacyl-tRNA
    binding to, in protein synthesis,
    362, 363f
Accessory factor elements (AFEs),
    455f
Accumulation of mutated proteins, 575
Aceruloplasminemia, 574
ACEs. *See* Angiotensin-converting enzyme
    inhibitors
Acetal links, 116
Acetaldehyde, 630
Acetaldehyde dehydrogenase, 630
Acetic acid, 122t
    pK/pK_a value of, 13t
Acetoacetate, 187, 187f
    in tyrosine catabolism, 258, 261f
Acetoacetyl-CoA synthetase, in mevalonate syn-
    thesis, 224, 224f
Acetone, 187
Acetyl (acyl)-malonyl enzyme, 193, 195f
Acetyl-CoA, 132, 132f, 137
    carbohydrate metabolism and, 132, 132f
    catabolism of, 143–145, 144f, 145f. *See also*
        Citric acid cycle
    cholesterol synthesis and, 224–225, 225f, 226f,
        227f

fatty acid oxidation to, 132, 133f, 184–187, 185f,
    186f
formation of, 254–257, 253f–257f
lipogenesis and, 194–195, 195f, 196f
    as fatty acid building block, 195–196
in lipogenesis regulation, 196
in platelet-activating factor synthesis, 208f
pyruvate dehydrogenase regulated by, 154–155,
    154f, 197
pyruvate oxidation to, 147, 147f, 153–155, 153f,
    154f, 155t
xenobiotic metabolism and, 613
Acetyl-CoA carboxylase, 198
    in lipogenesis regulation, 193, 194f, 197, 198,
        198f
Acetyl hexosamines, in glycoproteins, 119t
Acetyl transacylase, 194, 194f, 195f
Acetylation
    in covalent modification, mass increases and,
        26t
    of xenobiotics, 612–613
Acetylcholine, inhibiting release of, 502
N-Acetylgalactosamine (GalNAc), in glycopro-
    teins, 507–508, 508t
N-Acetylglucosamine (GlcNAc), in glycoproteins,
    508t
N-Acetylglucosamine phosphotransferase (Gl-
    cNAc phosphotransferase)
    in I-cell disease, 523
    in pseudo-Hurler polydystrophy, 524
N-Acetylglutamate, in urea biosynthesis,
    243, 244f
N-Acetylneuraminic acid, 119, 119f, 180, 181f
    in gangliosides, 209, 210f
    in glycoproteins, 180, 181f, 506–507, 507t
    in mucins, 510, 511f
Acetyltransferases, xenobiotic metabolism and, 613
Acholuric jaundice, 281
Achondroplasia, 423t, 541t, 543, 543f
Acid anhydride bonds, 286
Acid anhydrides, group transfer potential for, 289,
    289f, 289t
Acid-base balance, ammonia metabolism in, 243
Acid-base catalysis, 53
    HIV protease in, 54, 55f
Acid phosphatase, diagnostic significance
    of, 59t
Acidemia, isovaleric, 257, 258
Acidosis
    lactic. *See* Lactic acidosis
    metabolic, ammonia in, 243
Acids
    conjugate, 10

molecular structure affecting strength of, 12, 13t
polyfunctional, nucleotides as, 288
as proton donors, 10
strong, 10
weak. *See* Weak acids
Aciduria
    dicarboxylic, 191
    methylmalonic, 167
    orotic, 300
    urocanic, 250
Aconitase (aconitate hydratase), 143, 572
ACP. *See* Acyl carrier protein
Acrosomal reaction, glycoproteins in, 520
ACTH. *See* Adrenocorticotropic hormone
Actin, 546
    decoration of, 548, 549f
    fibronectin receptor interacting with, 532, 532f
    in muscle contraction, 546, 547f, 549–550, 549f,
        550f
    regulation of striated muscle and, 550–551
    in nonmuscle cells, 562–563
    in red cell membranes, 600–601, 600f, 600t
    in striated *vs.* smooth muscle, 558t
    structure of, 547
    F-Actin, 546, 548, 548f
        in nonmuscle cells, 562
    G-Actin, 546, 548f
        in nonmuscle cells, 562
Actin-filament capping protein, 532, 532f
Actin (thin) filaments, 502, 545–546, 547f, 548f
Activated protein C, in blood coagulation, 587
Activated sugars, 508
Activating enzyme, 498
Activation energy, 64, 65
Activation energy barrier, enzymes affecting, 65
Activation reaction, 444, 445f
Activator-recruited cofactor (ARC), 457, 457t
Activators
    in regulation of gene expression, 369, 371. *See
        also* Enhancers/enhancer elements
    transcription, 344–345, 344t
Active chromatin, 315, 316f, 377
Active site, 53, 53f. *See also* Catalytic site
Active sulfate (adenosine 3′-phosphate-5′-
    phosphosulfate), 288, 288f, 612
Active transport, 413, 414f, 414t, 415, 416f, 419t
    in bilirubin secretion, 279–289, 280f
Actomyosin, 547
ACTR coactivator, 457, 457t
Acute fatty liver of pregnancy, 191
Acute inflammatory response, neutrophils in, 604
Acute phase proteins, 465, 568–569, 569t
    negative, vitamin A as, 470

Acyl carrier protein (ACP), 193, 194f
  synthesis of, from pantothenic acid, 193, 479
Acyl-CoA dehydrogenase, 100, 185, 186f
  medium-chain, deficiency of, 191
Acyl-CoA synthetase (thiokinase)
  in fatty acid activation, 184, 185f
  in triacylglycerol synthesis, 207, 220, 220f
Acyl-CoA:cholesterol acyltransferase (ACAT), 225
Acylcarnitine, 184, 185f
Acylglycerol, 205
Acylglycerol metabolism, 205–208
  catabolism, 205
  clinical aspects of, 209–211
  synthesis, 205–208, 206f
    in endoplasmic reticulum, 135, 136f
ADA. See Adenosine deaminase
Adapter proteins, in absorptive pinocytosis, 421
Additional proteins in muscle, 553
Adenine, 287f, 287t
Adenine nucleotide transporter, 109f, 110
Adenomatous polyps, 622
Adenosine, 286f, 287t
  base pairing of in DNA, 302, 304f
  in uric acid formation, 297, 300f
Adenosine 3′-phosphate-5′-phosphosulfate, 288,
      288f, 612
Adenosine deaminase
  deficiency of. See Adenosine deaminase defi-
      ciency (ADA)
  localization of gene for, 396t
Adenosine deaminase deficiency (ADA), 299, 607
  case study, 616–617
Adenosine diphosphate. See ADP
Adenosine monophosphate. See AMP
Adenosine triphosphate. See ATP
S-Adenosylmethionine, 257, 257f, 263, 264f, 288,
      289f, 289t
  biosynthesis of, 264f
Adenylic acid, as second messenger, 446
Adenylyl cyclase, 620
  in cAMP-dependent signal transduction, 447,
      447t
  cAMP derived from, 159
  in lipolysis, 221f, 222
Adenylyl kinase (myokinase), 96
  in gluconeogenesis regulation, 168
  as source of ATP in muscle, 560, 560f
Adhesion molecules, 520–521, 521t. See also Cell
      adhesion
Adipocytes, 222
Adipose tissue, 121, 131, 220–221, 221f
  brown, 222f, 223
  in fasting state, 140
  glucose uptake into, 138
  metabolism in, 139f, 141t, 220–221, 221f
ADP, 286f
  free energy of catabolism capture by, 95f, 108
  free energy of hydrolysis of, 94t
  myosin, muscle contraction and, 549, 549f
  in platelet activation, 590, 590f
  in respiratory control, 147
  structure of, 95f
ADP-chaperone complex, 496. See also Chaperones
ADP-ribose, NAD as source of, 475
ADP-ribosylation, 475, 620
ADPase, 591, 591t
Adrenal cortical hormones. See also Mineralo-
      corticoids; specific hormone and
      Glucocorticoids
  synthesis of, 429–432, 430f, 431f

Adrenal gland, cytochrome P450 isoforms in, 610
Adrenal medulla, catecholamines produced in, 435
Adrenergic receptors, in glycogenolysis, 160
Adrenocorticotropic hormone (ACTH), 427, 428,
      429f, 441, 441f, 632
Adrenodoxin, 610
Adrenodoxin reductase, 610
Adrenogenital syndrome, 431
Adrenoleukodystrophy, neonatal, 493, 493t
Adult stem cells, 593
Advanced glycation end-products (AGEs), 519,
      519f, 520f
Aerobic glycolysis, 152
  as muscle ATP source, 560, 560f, 561, 561t
Aerobic metabolism, 638
Aerobic respiration, citric acid cycle and, 143
AF-2 domain, 455
Affective and behavioral disturbances in
      Alzheimer disease, 618
Affinity of SRP–R, for SRP, 495
Affinity chromatography
  for protein/peptide purification, 23
  in recombinant fusion protein purification,
      60–61, 60f
AFP. See Alpha-fetoprotein
Agammaglobulinemia, 580
Age, xenobiotic-metabolizing enzymes affected
      by, 613
AGEs. See Advanced glycation end-products
Aggrecan, 533, 542, 542t, 543f
Aggregates, formation of after denaturation, 39
Aggregation prevention, 497
Aging, glycosaminoglycans and, 539
Aging process, extracellular matrix in, 527
Aglycone, 116
AHG. See Antihemophilic factor A/globulin
AIB1 coactivator, 457, 457t
ALA. See Aminolevulinate
ALA synthase, 274
Alanine, 15t, 240, 262
  pI of, 17
  in pyruvate formation, 251
  synthesis of, 235, 236f
α-Alanine, 251
β-Alanine, 266–267
Alanine (alanine-pyruvate) aminotransferase
      (ALT/SGPT), 59
  diagnostic significance of, 59t
  in urea synthesis, 241–242, 242f
Albumin, 502, 566, 567f, 569, 569t
  conjugated bilirubin binding to, 282
  copper binding to, 573
  free fatty acids in combination with, 184, 213,
      213t, 569
  glomerular membrane permeability to, 533
Albuminuria, 533
Alcohol, ethyl. See Ethanol
Alcohol dehydrogenase, 630
  in fatty liver, 219
Alcoholism
  cirrhosis and, 218
  fatty liver and, 218–219
  transferrin glycosylation in, 571
Aldehyde dehydrogenase, 99
  in fatty liver, 219
Aldolase A, deficiency of, 155
Aldolase B, 179, 179f
  deficiency of, 182
Aldolases, in glycolysis, 150, 151f
Aldose-ketose isomerism, 114f, 115f

Aldose reductase, 179, 179f, 182
Aldoses, 113, 114t, 115f
  ring structure of, 114f
Aldosterone
  binding of, 443
  synthesis of, 430, 431f
    angiotensin affecting, 440
Aldosterone synthase (18-hydroxylase), in steroid
      synthesis, 430–431, 431f
Alimentary pentosuria, 181
Alkaline phosphatase
  in bone mineralization, 540
  isozymes of, diagnostic significance of, 59t
  in recombinant DNA technology, 390t
Alkalosis, metabolic, ammonia in, 243
Alkaptonuria, 254
Allergic reactions, peptide absorption causing,
      459
Allopurinol, 289, 290f, 297, 300, 631, 632
Allosteric activators, 167
Allosteric effectors/modifiers, 22f, 137
  in gluconeogenesis regulation, 167–168
  negative, 77. See also Feedback inhibition
  second messengers as, 79
Allosteric enzymes, 77–78, 137
  aspartate transcarbamoylase as model of, 78
Allosteric properties of hemoglobin, 44–48
Allosteric regulation, of enzymatic catalysis,
      77–78, 78f, 137, 138f
  gluconeogenesis regulation and, 167–168
Allosteric site, 78
Alpha-adrenergic receptors, in glycogenolysis,
      160
Alpha-amino acids. See also Amino acids
  genetic code specifying, 14, 15t–16t
  in proteins, 14, 16
Alpha-amino nitrogen. See Amino acid nitrogen
Alpha anomers, 114–115
Alpha-fetoprotein, 569t
Alpha-globin gene, localization of, 396t
Alpha helix, 32–33
Alpha-lipoproteins. See also High-density
      lipoproteins
  familial deficiency of, 232t
Alpha-R groups, amino acid properties affected
      by, 18
Alpha thalassemias, 49
Alpha-tocopherol. See Tocopherol
7α-Hydroxylase, sterol, 228
α-Linolenic acid, for essential fatty acid deficiency, 200
α-Aminoadipate, 255f
α-Aminoadipate, δ-semialdehyde, 255f
α₁-Antiproteinase, in emphysema, 574–575
α₁-Antitrypsin, in emphysema, 574–575
α₁-Antitrypsin liver disease, 574–575
α₂-Macroglobulin, 575
Alport syndrome, 530, 530t
ALT. See Alanine aminotransferase
Alteplase (tissue plasminogen activator/t-PA),
      584f, 588, 589, 591t
Altitude, high, adaptation to, 48
Alu family, 318
Alzheimer disease, amyloid in, 576t, 617–619,
      618f, 619f, 619t
  case study, 617–619, 619f
  causation, 617
  history and physical examination, 617
  treatment, 617–618
Amadori rearrangement, to ketoamines, 519
Ambiguity, genetic code and, 354

Amino acid carbon skeletons, catabolism of, 248–260
  acetyl-CoA formation and, 254–257, 254f–258f
  branched-chain, 257–258, 259f, 260f
    disorders of, 258
  pyruvate formation and, 250–254, 250f–252f
  transamination in initiation of, 248–250, 248f–250f
Amino acid nitrogen
  catabolism of, 239–247
    in amino acid carbon skeleton catabolism, 248–250, 248f–250f
  end products of, 240–241
    urea as, 243–245, 244f
  L-glutamate dehydrogenase in, 242, 243f
  transamination of, 241–242, 241f, 242f
L-Amino acid oxidase, 99
  in nitrogen metabolism, 242, 242f
Amino acid sequences. *See also* Protein sequencing
  determination of, for glycoproteins, 507t
  primary structure determined by, 18
  repeating, in mucins, 510–512, 511f
Amino acids, 2, 14–20, 15t–16t, 242f. *See also* Peptides
  absorption of, 462
  ammonia removal from, 242, 242f
  analysis/identification of, 20
  biosynthesis, 235
  blood glucose and, 169
  branched chain, catabolism of, 257–258, 259f, 260f
    disorders of, 258
  in catalysis, conservation of, 56, 57t
  chemical reactions of, functional groups dictating, 18–20
  in citric acid cycle, 137
  deamination of. *See* Deamination
  deficiency of, 234, 465
  excitatory. *See* Aspartate; Glutamate
  glucogenic, 137
  in gluconeogenesis, 146, 146f
  interconvertability of, 137
  interorgan exchange maintaining the circulating levels of, 240
  keto acid replacement of in diet, 238
  ketogenic, 137
  melting point of, 18
  metabolism of, 132, 132f, 133, 133f. *See also* Amino acid carbon skeletons; Amino acid nitrogen
    pyridoxal phosphate in, 475–476
  net charge of, 16–17, 17f
  nutritionally essential, 133, 234–235
  nutritionally nonessential, 133
    synthesis of, 234–238
  in peptides, 14, 19, 19f
  pK/pKa values of, 15t–16t, 17–18, 17f
    environment affecting, 17–18, 18t
  products derived from, 262–270. *See also* specific product
  properties of, 14–18
  protein degradation and, 239–240, 240f
  in proteins, 14, 15
  requirements for, 465
  sequence of, in primary structure, 18
  solubility point of, 18
  substitutions of, missense mutations caused by, 357, 357f
  synthesis of, 234–238

in carbohydrate metabolism, 132
  citric acid cycle in, 146, 146f
  transamination of. *See* Transamination
  transporter/carrier systems for
    glutathione and, 612
    hormones affecting, 416
Amino sugars (hexosamines), 116, 117f
  glucose as precursor of, 180, 181f
  in glycosaminoglycans, 119, 180, 181f
  in glycosphingolipids, 180, 181f
  interrelationships in metabolism of, 181f
Aminoacyl residues, 18–19
  peptide structure and, 19
Aminoacyl (A/acceptor) site, aminoacyl-tRNA binding to, 362, 363f
Aminoacyl-tRNA, in protein synthesis, 362
Aminoacyl-tRNA synthetases, 355, 355f
Aminoguanidine, 519
Aminolevulinate (ALA), 272, 273f
  in porphyria, 277
Aminolevulinate dehydratase, 272, 273f
  in porphyria, 276, 277t
Aminolevulinate synthase, 272, 273f, 274, 276f
  in porphyria, 276, 276f, 277t
Aminopeptidases, 462
Aminophospholipids, membrane asymmetry and, 410
Aminoproteinase, procollagen, 529
Aminotransferases (transaminases), 146f, 147
  diagnostic significance of, 59t
  in urea biosynthesis, 241–242, 242f
Ammonia
  in acid-base balance, 243
  detoxification of, 239
  excess of, 245
  glutamine synthase fixing, 242, 243f
  nitrogen removed as, 242, 242f
Ammonia intoxication, 242, 245
Ammonium ion, pK/pKa value of, 12t
Amobarbital, on oxidative phosphorylation, 103
AMP, 286f, 287f, 287t, 296f
  coenzyme derivatives of, 289t
  cyclic. *See* Cyclic AMP
  free energy of hydrolysis of, 94t
  IMP conversion to, 293, 295f
  feedback regulation of, 293, 295, 295f
  PRPP glutamyl amidotransferase regulated by, 293
  structure of, 95f, 287f
Amp resistance genes, 391, 392f
Amphibolic pathways/processes, 131
  citric acid cycle and, 146
Amphipathic helices, 33
Amphipathic lipids, 129, 129f
  in lipoproteins, 212, 213f
  in membranes, 129, 129f, 408, 408f
Amphipathic molecules, folding and, 7
Ampicillin resistance genes, 391, 392f
Amplification, gene, in gene expression regulation, 385, 385f
Amylases
  diagnostic significance of, 59t
  in hydrolysis of starch, 459–460
Amyloid β peptide (Aβ42), 617, 618f
  aggregation of, 619
Amyloid cascade hypothesis, 618
Amyloid precursor proteins, 40, 618, 618f
  in Alzheimer disease, 40
Amyloidosis, 575–576, 576t

Amylopectin, 118, 118f, 460
Amylopectinosis, 160t
Amylose, 117, 118f
Anabolic pathways/anabolism, 93, 131. *See also* Endergonic reaction; Metabolism
Anaerobic glycolysis, 149, 150, 150f, 638
  as muscle ATP source, 560f, 561–562, 561t
Analbuminemia, 569
Anaphylaxis, slow-reacting substance of, 202
Anaplerotic reactions, in citric acid cycle, 146
Anchor, 495
Anchorin, in cartilage, 542t
Andersen disease, 160t
Androgen response element, 445t
Androgens
  estrogens produced from, 432, 434f
  receptors for, 455
  synthesis of, 431–432, 431f, 433f
Androstenedione
  estrone produced from, 432, 434f
  synthesis of, 431, 433f
Anemias, 49, 593
  Fanconi, 333
  hemolytic, 149, 155, 594, 599, 599t
    glucose-6-phosphate dehydrogenase deficiency causing, 174, 180–181, 597, 598, 598f
    haptoglobin levels in, 570
    hyperbilirubinemia/jaundice in, 281, 283t
    pentose phosphate pathway/glutathione peroxidase and, 177, 177f, 180–181
    primaquine-sensitive, 598
    red cell membrane abnormalities causing, 598
  iron deficiency, 463, 480, 571, 594t
  megaloblastic
    folate deficiency causing, 468t, 478
    vitamin B12 deficiency causing, 468t, 477, 478, 494t
  pernicious, 468t, 477
  recombinant erythropoietin for, 517, 595
  sickle cell. *See* Sickle cell disease
Angelman syndrome, 240
Angiotensin-converting enzyme, 440, 440f
Angiotensin-converting enzyme inhibitors, 459
Angiotensin II, 427–428, 440, 440f
  synthesis of, 440, 440f
Angiotensin III, 440, 440f
Angiotensinogen, 440, 440f
Anion exchange protein, 600–601, 600f, 600t
Anion gap, 627
Ankyrin, 600f, 600t, 601
Anomeric carbon atom, 114–115
Anserine, 262, 264f, 266
Antennae (oligosaccharide branches), 512
Anterior pituitary gland hormones, blood glucose affected by, 171
Anterograde transport (COPII), 499, 501f
Anti conformers, 286, 286f
Antibiotic treatment, 616, 625
Antibiotics, 636
  amino sugars in, 116
  bacterial protein synthesis affected by, 367–368
  folate inhibitors as, 477
Antibodies, 525, 566, 567, 598. *See also* Immunoglobulins
  monoclonal, hybridomas in production of, 580–581, 581f
  in xenobiotic cell injury, 613, 614t

Antibody diversity, 576
    DNA/gene rearrangement and, 580
Antichymotrypsin, 569t
Anticoagulants, coumarin, 587–588
Anticodon region, of tRNA, 354, 355, 355f
Antifolate drugs, purine nucleotide synthesis affected by, 293
Antigenic determinant (epitope), 35, 577
Antigenicity, xenobiotics altering, cell injury and, 613, 614f
Antigens, 576
Antihemophilic factor A/globulin (factor VIII), 585, 586t
    deficiency of, 588
Antihemophilic factor B (factor IX), 586t
    coumarin drugs affecting, 588
    deficiency of, 588
Antimalarial drugs, folate inhibitors as, 477
Antimycin A, on respiratory chain, 108, 109f
Antioxidant nutrients, 482, 485
Antioxidants, 102, 128, 597, 597t
    retinoids and carotenoids as, 128, 468t
    vitamin C as, 128
    vitamin E as, 102, 128, 471, 472f
Antioxidant properties, uric acid, 631
Antiparallel beta sheet, 34, 34f
Antiparallel loops, mRNA and tRNA, 356
Antiparallel strands, DNA, 302
Antiport systems, 415, 415f
    for nucleotide sugars, 508
α_1-Antiproteinase, in emphysema and liver disease, 574–575
Antiproteinases, 607, 607t
Antithrombin/antithrombin III, 569t, 587
    heparin binding to, 539, 587
Anti-thyroperoxidase (anti-TPO) antibodies, 635
α_1-Antitrypsin, in emphysema, 574–575
α_1-Antitrypsin liver disease, 574
APC. See Activated protein C
APC resistance, 587
Apical proteins, 502
Apo A-I, 213, 213t, 228
    deficiencies of, 232t
Apo A-II, 213, 213t
    lipoprotein lipase affected by, 215
Apo A-IV, 213, 213t
Apo B-48, 213, 213t
Apo B-100, 213, 213t
    in LDL metabolism, 216, 216f
    regulation of, 225
Apo B-100 receptor
    in LDL metabolism, 216
Apo C-I, 213, 213t
Apo C-II, 213, 213t
    in lipoprotein lipase activity, 215
Apo C-III, 213, 213t
    lipoprotein lipase affected by, 215
Apo D, 213, 213t
Apo E, 213, 213t, 215
Apo E receptor
    in chylomicron remnant uptake, 215, 216f
    in LDL metabolism, 215, 216, 217f
Apo-transketolase activation, in thiamin nutritional status assessment, 474
Apoferritin, 572
Apolipoproteins/apoproteins, 212–233
    distribution of, 212–213, 213t
    hemoglobin, oxygenation affecting, 46
Apomyoglobin, hindered environment for heme iron and, 43–44, 44f

Apoproteins. See Apolipoproteins/apoproteins
Apoptosis, 208, 618
    p53 and, 333
APP. See Amyloid precursor proteins
Apurinic endonuclease, in base excisionrepair, 331
Apyrimidinic endonuclease, in base excision-repair, 331
Aquaporins, 418
D-Arabinose, 115f, 116t
Arabinosyl cytosine (cytarabine), 289, 290
Arachidonic acid/arachidonate, 122, 123f, 198, 198f
    eicosanoid formation and, 200, 201f, 202f, 203f
    for essential fatty acid deficiency, 200
ARC, 457, 457t
ARE. See Androgen response element
Argentaffinoma (carcinoid), serotonin in, 265
Arginase
    disorders of, 246
    in urea synthesis, 249f
Arginine, 16t, 262
    catabolism of, 249f, 250
    metabolism of, 263f
    in urea synthesis, 249, 250
Argininosuccinase
    deficiency of, 246
    in urea synthesis, 244f, 246
Argininosuccinate, in urea synthesis, 243, 244, 244f
Argininosuccinate lyase deficiency, 246
Arginosuccinate synthase, 246
    deficiency of, 246
Argininosuccinicaciduria, 246
Aromatase enzyme complex, 432, 434f
ARS (autonomously replicating sequences), 323, 403
Arsenate, oxidation and phosphorylation affected by, 150
Arsenite, oxidation and phosphorylation affected by, 155
Arterial wall intima, proteoglycans in, 539
Arthritis
    gouty, 297
    proteoglycans in, 539
    rheumatoid, glycosylation alterations in, 523
Arthropathy, of hemochromatosis, 633
Artificial membranes, 411–412
Ascorbate, 177, 178f, 485
Ascorbic acid (vitamin C), 174, 468, 479–480, 479f
    as antioxidant, 128
    in collagen synthesis, 41, 480, 527
    deficiency of, 490t, 480, 530
        collagen affected in, 41, 480, 530
Asialoglycoprotein receptors
    in cotranslational insertion, 495, 496f
    in glycoprotein clearance, 509
Asn-GlcNAc linkage
    in glycoproteins, 512
    in glycosaminoglycans, 535
Asparaginase, in amino acid nitrogen catabolism, 242–243, 243f
Asparagine, 15t
    in amino acid nitrogen catabolism, 248, 249
    synthesis of, 235–236, 236f
Asparagine synthetase, 235, 236f
Aspartate
    catabolism of, 248, 249f
    synthesis of, 235, 236f
    in urea synthesis, 244
Aspartate 102, in covalent catalysis, 55, 55f

Aspartate aminotransferase (AST/SGOT), diagnostic significance of, 59, 59t
Aspartate transcarbamoylase, 78
    in pyrimidine synthesis, 297, 298f
Aspartic acid, 15t
    pI of, 17
Aspartic protease family, in acid-base catalysis, 54, 55f
Aspartylglycosaminuria, 524, 524t
Aspirin
    antiplatelet actions of, 591
    cyclooxygenase affected by, 200
    prostaglandin synthesis affected by, 193
Assembly particles, in absorptive pinocytosis, 421
AST. See Aspartate aminotransferase
Asthma, leukotrienes in, 122
Asymmetric substitution, in porphyrins, 272, 272f
Asymmetry
    importin binding and, 491
    inside-outside, 410
    lipid and protein, membrane assembly and, 503, 503f
    in membranes, 410
Ataxia-telangiectasia, 333
ATCase. See Aspartate transcarbamoylase
Atherogenesis, 519
Atherosclerosis, 212, 591, 638
    cholesterol and, 126, 224, 230
    HDL and, 217
    hyperhomocysteinemia and, folic acid supplements in prevention of, 478
    LDL plasma concentration and, 216
    lysophosphatidylcholine (lysolecithin) and, 125
    risk factors for, 638
Atlas of Protein Sequence and Structure, 86
Atorvastatin, 232
ATP, 94–96, 286f, 288, 491, 492
    in active transport, 418–419, 419f
    in coupling, 94, 95–96
    fatty acid oxidation producing, 185, 187
    from free energy of catabolism, 95f, 106
    free energy of hydrolysis of, 94, 94t, 94t
    in free energy transfer from exergonic to endergonic processes, 94, 94f
    hydrolysis of
        in muscle contraction, 549, 549f
        by NSF, 500
    inorganic pyrophosphate production and, 96
    in mitochondrial protein synthesis and import, 488
    in muscle/muscle contraction, 545, 549, 549f
        decrease in availability of, 552
        sources of, 560–561, 560f, 561t
    in purine synthesis, 293, 294f
    from respiratory control, 108, 108t
    respiratory control in maintenance of supply of, 147
    structure of, 95f
    synthesis of
        in citric acid cycle, 144f, 145, 155, 155t
        glucose oxidation yielding, 155, 155t
        respiratory chain electron transport in, 106, 107f, 108f
ATP/ADP cycle, 95, 95f
ATP-binding cassette transporter, 217, 217f
ATP-chaperone complex, 516. See also Chaperones
ATP-citrate lyase, 147, 147f
    acetyl-CoA for lipogenesis and, 196
ATP synthase, 106, 107f, 108f

ATPase, 418, 500
   in active transport, 419, 419f
   chaperones exhibiting activity of, 496
   copper-binding P-type, mutations in gene for
      Menkes disease caused by, 573
      Wilson disease caused by, 574
Atractyloside, on respiratory chain, 109, 109f
Attachment proteins, 532, 532f
Autoimmune diseases, 482
Auto-oxidation. See Peroxidation
Autoantibodies, in myasthenia gravis, 439, 635
Autonomously replicating sequences (ARS), 323,
      403
Autoradiography, definition of, 403
Autosomal recessive condition, 617, 625, 633
Autotrophic organisms, 94
Avian influenza virus, 524, 525f
Avidin, biotin deficiency caused by, 479
Axial ratios, 31
Axonemal dyneins, 563
5- or 6-Azacytidine, 289
8-Azaguanine, 289, 290f
Azathioprine, 289, 290f
5- or 6-Azauridine, 289, 290f

β-Alanine, 266–267
β Alanyl dipeptides, 267–268
β-Alanyl-imidazole, 268
β-Aminoisobutyrate, 266, 267
B blood group substance, 603, 603f
β-Carotene, 485
B cyclins, 329, 329f, 329t
B gene, Gal transferase encoded by, 603
B lymphocytes, 576
   in hybridoma production, 580, 580f
B subunit of SRP-R, 494
B vitamins. See Vitamin B complex
β₂-microglobulin, 576
BAC vector. See Bacterial artificial chromosome
      (BAC) vector
Bacteria
   intestinal, in bilirubin deconjugation, 280
   transcription cycle in, 336
Bacterial artificial chromosome (BAC) vector,
      390, 391t
Bacterial artificial chromosomes (BACs), 85
Bacterial promoters, in transcription, 338–339,
      340f
Bacteriophage, definition of, 403
BAL. See Dimercaprol
BAL 31 nuclease, in recombinant DNA technol-
      ogy, 390t
Balanced chemical equations, 62
BamHI, 388, 389t
Barbiturates, on respiratory chain, 108, 109f
Barrel-like structures, 496
Barth syndrome,
Basal lamina, laminin as component of, 532–533
Basal metabolic rate, 464
Base excision-repair of DNA, 330, 331t, 331–332,
      332f
Base pairing in DNA, 8, 302–303, 304f
   matching of for renaturation, 304
   recombinant DNA technology and, 388
   replication/synthesis and, 324–326, 325f
Base substitution, mutations occurring by, 356,
      356f
Basement membranes, collagen in, 529
Bases
   conjugate, 10

   as proton acceptors, 10
   strong, 10
   weak, 10
Basic Local Alignment Search Tool (BLAST), 88
Basolateral proteins, 502
Becker muscular dystrophy, 553, 628
Bence Jones protein, 580
Bends (protein conformation), 34–35, 35f
Benign tumor, 623
Beriberi, 468t, 473
β-Alanine, 266–267
Beta anomers, 114–115
Beta-endorphins, 441, 441f
Beta-globin gene
   localization of, 396t
   recombinant DNA technology in detection of
      variations in, 396, 396t, 397f
β-Hydroxybutyric acid, 292, 300f
Beta-lipoproteins, 212. See also Low-density
      lipoproteins
β₂-Microglobulin, 576
Beta-oxidation of fatty acids, 184–187, 185f, 186f
   ketogenesis regulation and, 190, 190f
   modified, 186f, 187
Beta sheet, 33–34, 619
Beta subunits of hemoglobin, myoglobin and,
      45–46
Beta thalassemias, 49–50
BFU-E. See Burst-forming unit-erythroid
BgIII, 389t
BHA. See Butylated hydroxyanisole
BHT. See Butylated hydroxytoluene
Bi-Bi reactions, 72–73, 73f
   Michaelis-Menten kinetics and, 73, 73f
Bicarbonate, 627
   in extracellular and intracellular fluid, 407t
Biglycan
   in bone, 538t
   in cartilage, 542t
Bilayers, lipid, 408–409, 409f
   membrane proteins and, 409–410
Bile, bilirubin secretion into, 279–280, 280f
Bile acids (salts), 228–229
   enterohepatic circulation of, 229
   in lipid digestion and absorption, 460, 461f
   secondary, 229, 231f
   synthesis of, 228–229, 231f
      regulation of, 229, 231f
Bile pigments, 278–280. See also Bilirubin
Biliary obstruction, hyperbilirubinemia/jaundice
      caused by, 282, 283t
Bilirubin
   accumulation of (hyperbilirubinemia), 280–282,
      283t
   conjugated
      binding to albumin and, 282
      reduction of to urobilinogen, 280
   conjugation of, 278–279, 280f
   fecal, in jaundice, 283t
   heme catabolism producing, 278, 279f
   liver uptake of, 278–280, 280f
   normal values for, 283t
   secretion of into bile, 279–280, 280f
   unconjugated, disorders occurring in, 281–282
   urine, in jaundice, 282–283, 283t
Biliverdin, 278, 279f
Biliverdin reductase, 278
Bimolecular membrane layer, 408. See also Lipid
      bilayer
Binding by glycans of resuses, 524

Binding constant, Michaelis constant (K_m)
      approximating, 69
Binding proteins, 442t, 443, 569t
Biochemical case histories, 616
   adenosine deaminase deficiency, 616–617
   Alzheimer disease, 617–619
   cholera, 619–621
   colorectal cancer, 621–624
   cystic fibrosis, 624–626
   diabetic ketoacidosis, 627–628
   Duchenne muscular dystrophy, 628–629
   ethanol intoxication, acute, 629–630
   gout, acute, 630–632
   hereditary hemochromatosis, 632–634
   hypothyroidism, 634–635
   Kwashiorkor, protein-energy malnutrition
      (PEM), 635–637
   myocardial infarction, 637–639
   obesity, 639–641
   osteoporosis, primary, (postmenopausal),
      641–643
   xeroderma pigmentosum, 643–645
Biochemistry, 3
   as basis of health/disease, 2–4, 3t
   definition of, 1
   Human Genome Project and, 3–4, 4f
   methods and preparations used in, 1, 2t
   relationship of to medicine, 1–4, 3f
Biocytin, 478, 478f
Bioenergetics, 92–96. See also ATP
Bioengineering, 4
Bioethics, 4
Bioinformatics, 4–5, 84–91, 403
   computational biology, 87
   computer-aided drug design, 89
   definition of, 5, 85–86
   genomes and medicine, 85
   genomic resources for, 86–87
   Human Genome Project in, 84–85
   protein function and, 29
   proteins, identification of, 88
   "unknown proteins", identification of, 88
   virtual cells, 89–90
Biologic oxidation. See Oxidation
Biologic information, 506
Biologic membranes, 630
Biology, 3
Biomarkers, 569
Biomolecules. See also specific type
   stabilization of, 7
   water affecting structure of, 7–8, 7t
Biophysics, 4
Biotechnology, 4
Biotin, 468t, 478, 478f
   deficiency of, 468t, 478
   in malonyl-CoA synthesis, 193, 194f
   as prosthetic group, 52
BiP. See Immunoglobulin heavy chain binding
      protein
1,3-Bisphosphoglycerate (BPG), free energy of
      hydrolysis of, 94t
2,3-Bisphosphoglycerate (BPG), T structure of
      hemoglobin stabilized by, 48, 48f
Bisphosphoglycerate mutase, in glycolysis in
      erythrocytes, 152f, 153
2,3-Bisphosphoglycerate phosphatase, in erythro-
      cytes, 152f, 153
BLAST, 88
blastn, 88
blastp, 88

blastx, 88
Blindness, vitamin A deficiency causing, 470
Blistering, 564
Blood, functions of, 566, 567t
Blood cells, 593–608. *See also* Erythrocytes; Neutrophils; Platelets
Blood clotting. *See* Coagulation
Blood coagulation, 586f. *See also* Coagulation (blood); Coagulation factors
Blood glucose. *See* Glucose, blood
Blood group substances, 603, 603f
    glycoproteins as, 506, 602
Blood group systems, 593, 602, 603f
Blood plasma. *See* Plasma
Blood type, 602–603
Blood vessels, nitric oxide affecting, 559–560, 559f, 559t, 560t
Blot transfer techniques, 393–394, 393f
Blunt end ligation/blunt-ended DNA, 389–390, 389f, 403
BMR. *See* Basal metabolic rate
Body mass index, 463
Body water. *See* Water
Bohr effect, 47, 48f
    in hemoglobin M, 49
Bonds. *See* specific types
Bone, 540f, 539–541
    metabolic and genetic disorders involving, 541–542, 541t
    proteins in, 539, 540t
Bone Gla protein, 538t
Bone marrow, heme synthesis in, 273
Bone marrow stem cells, 616
Bone marrow transplantation, 616
Bone matrix Gla protein, 473
Bone morphogenetic proteins, 538t
Bone sialoprotein, 538t
Bone SPARC protein, 538t
Botany, 1
Botulinum B toxin, 502
Bovine ADA conjugated to polyethylene glycol, 616
Bovine spongiform encephalopathy, 40
BPG. *See* 1,3-Bisphosphoglycerate (BPG); 2,3-Bisphosphoglycerate (BPG)
Bradykinin, in inflammation, 604t
Brain, metabolism in, 141t
    glucose as necessity for, 137
Branch point, 157
Branched chain amino acids, catabolism of, 240, 257–258, 259f, 260f
    disorders of, 258
Branched chain ketonuria (maple syrup urine disease), 258
Branching enzymes absence of, 160t
    in glycogen biosynthesis, 157, 159f
Brefeldin A, 502
Brittle bones (osteogenesis imperfecta), 541t
Broad beta disease, 232t
Bronze diabetes, 633
Brown adipose tissue, 222f, 223
Browning of certain foodstuffs, 519
Brush border enzymes, 460
BSE. *See* Bovine spongiform encephalopathy
Budding of vesicles, 499, 500
Buffers
    Henderson-Hasselbalch equation describing behavior of, 11–12, 12f
    weak acids and their salts as, 12, 12f
"Bulk flow", of membrane proteins, 502

Brst-forming unit-erythroid, 595f
Butylated hydroxyanisole (BHA), as antioxidant/food preservative, 128
Butylated hydroxytoluene (BHT), as antioxidant/food preservative, 128
Butyric acid, 122t

C-peptide, 438f, 439
C-reactive protein, 568, 569t, 638
C regions/segments. *See* Constant regions/segments
C-terminal-binding domain, 35
C_{20} polyunsaturated acids, eicosanoids formed from, 200, 201f, 202f
Ca^{2+}-ATPase, 449
Ca^{2+}-Na+ exchanger, 449, 555
Cachexia, 138
    cancer, 140, 151, 464
CADD. *See* Computer-aided drug design
Caffeine, 288, 288f
    hormonal regulation of lipolysis and, 222
Caged subtratres, 38
Calbindin, 462
Calcidiol (25-hydroxycholecalciferol), in vitamin D metabolism, 470, 471f
Calciferol. *See* Vitamin D
Calcineurin, 553t
Calcinosis, 471
Calcitonin, 427
Calcitriol (1,25(OH)$_2$-D$_3$), 428, 429f, 470
    calcium concentration regulated by, 470
    storage/secretion of, 442, 442t
    synthesis of, 432, 434–435, 435f, 470, 471f
Calcium, 480t, 622
    absorption of, 462–463
        vitamin D metabolism and, 462, 470
    in blood coagulation, 584f, 585, 585t
    in bone, 539
    in extracellular fluid, 407, 407t
    in intracellular fluid, 407, 407t
    iron absorption affected by, 463
    in malignant hyperthermia, 552–553, 335f
    metabolism of, 449–450
        vitamin D metabolism and, 470
    in muscle contraction, 550
        in cardiac muscle, 554–555
        phosphorylase activation and, 160
        sarcoplasmic reticulum and, 549f, 551
        in smooth muscle, 557–559
    in platelet activation, 589, 590f
    as second messenger, 427, 428t, 446, 450, 450t
        phosphatidylinositide metabolism affecting, 450, 451f
    vitamin D metabolism affected by, 470
Calcium ATPase, 449, 555
Calcium-binding proteins, vitamin K and glutamate carboxylation and postsynthetic modification and, 473, 473f
    synthesis and, 473, 587–588
Calcium/calmodulin, 450
Calcium/calmodulin-sensitive phosphorylase kinase, in glycogenolysis, 159–160
Calcium channels, 449. *See also* Calcium release channel
    in cardiac muscle, 554–555
Calcium-induced calcium release, in cardiac muscle, 551
Calcium pump, 449, 555
Calcium release channel, 551, 551f
    dihydropyridine receptor and, 551, 551f

mutations in gene for, malignant hyperthermia caused by, 552–553, 574f, 613t
Calcium-sodium exchanger, 449, 555
Calculi, 631
Caldesmon, 558
Calmodulin, 450, 450t, 550
    muscle phosphorylase and, 160, 161f
Calmodulin-4 Ca^{2+}, in smooth muscle contraction, 557, 558f
Calnexin, 497, 516, 517f
Calories, 622
Calreticulin, 497, 517
Calsequestrin, 551, 551f
cAMP, 620, 626. *See also* Cyclic AMP
Cancer, 482
Cancer cachexia, 140, 151, 464
Cancer/cancer cells, 522. *See also* Carcinogenesis/carcinogens
    colorectal, 621
    cyclins and, 329
    essence of, 623
    glycoproteins and, 506, 517, 522–523, 522t
    hormone-dependent, vitamin B_6 deficiency and, 475–476
    membrane abnormalities and, 423t
    mucins produced by, 511
    spread of, 623
Cancer chemotherapy
    combination, 593
    folate inhibitors in, 477
    neutropenia caused by, 595
    synthetic nucleotide analogs in, 289, 290f
Cancer Genome Atlas, 87
Cancer phototherapy, porphyrins in, 275
CAP. *See* Catabolite gene activator protein
Caps, 498
Caproic acid, 122t
Carbamates, hemoglobin, 47
Carbamoyl phosphate
    excess, 300
    free energy of hydrolysis of, 94t
    in urea synthesis, 243, 244f, 245
Carbamoyl phosphate synthetase I, 243
    deficiency of, 245
    in urea synthesis, 243, 244f
Carbamoyl phosphate synthetase II, in pyrimidine synthesis, 297, 298f
Carbohydrate-binding proteins, 509
Carbohydrate complex. *See* specific types
Carbohydrates, 113–120. *See also* Glucose; Sugars; specific types
    in cell membranes, 120
    cell surface, glycolipids and, 119
    classification of, 113, 114t
    complex (glycoconjugate). *See also* specific types
        glycoproteins as, 506
    digestion and absorption of, 459–460, 460f
    in fatty acid synthesis, 137
    interconvertibility of, 137
    isomerism of, 113–115, 114f
    in lipoproteins, 120
    metabolism of, 132, 132f, 133f
        diseases associated with, 113
        vitamin B_1 (thiamin) in, 473, 474f
    in proteoglycans, 533
    very low, weight loss from diets with, 173
Carbon dioxide
    citric acid cycle in production of, 143–144, 145f
    enters red cell as bicarbonate, 601
    transport of, by hemoglobin, 47, 48f

Carbon monoxide
   heme catabolism producing, 278
   on oxidative phosphorylation, 103
   on respiratory chain, 108, 109f
Carbon skeleton, amino acid. *See* Amino acid
         carbon skeletons
Carbonic acid, p*K*/p*Ka* value of, 13t
Carbonic anhydrase, in osteopetrosis, 541
Carboxybiotin, 478, 478f
Carboxyl terminal repeat domain, 344
Carboxylase enzymes, biotin as coenzyme of, 479
Carboxypeptidases, 462
Carboxyproteinase, procollagen, 529
Carcinogenesis/carcinogens, 614
   chemical, 614
      cytochrome P450 induction and, 611
   indirect, 614
Carcinoid (argentaffinoma), serotonin in, 265
Carcinoid syndrome, 475
Cardiac developmental defects, 557
Cardiac glycosides, 116
Cardiac muscle, 545, 553–560, 555t, 556t
Cardiac troponins, 60
Cardiolipin, 125, 125f
   synthesis of, 205–207, 206f, 207f
Cardiomyopathies, 545, 556, 556t, 629
Cargo proteins/molecules, 491, 500
   in export, 491
   in import, 491, 492f
Carnitine
   deficiency of, 184, 191
   in fatty acid transport, 184, 185f
Carnitine-acylcarnitine translocase, 184, 185f
Carnitine palmitoyltransferase, deficiency of, 184
Carnitine palmitoyltransferase-I, 184, 185f
   deficiency of, 191
   in ketogenesis regulation, 190, 190f
Carnitine palmitoyltransferase-II, 184, 185f
   deficiency of, 191
Carnitine system, 110
Carnosinase deficiency, 268
Carnosine, 268
Carnosinuria, 268
Carotene, 485
Carotene dioxygenase, 468, 469f
Carotenoids, 468–469, 469f. *See also* Vitamin A
Carrier proteins/systems, 415, 416f
   for nucleotide sugars, 508
Cartilage, 535, 542, 542f, 542t, 543f
   chondrodysplasia affecting, 542
Catabolic pathways/catabolism, 93, 131. *See also*
         Exergonic reaction; Metabolism;
         specific substances
   energy captured in, from respiratory chain, 95f,
         106, 107f
Catabolite gene activator protein (cyclic AMP
         regulatory protein), 371, 373
Catabolite regulatory protein, 447
Catalase, 100, 595–597
   as antioxidant, 128, 597t
   in nitrogen metabolism, 242, 242f
Catalysis/catalytic reactions (enzymatic). *See also*
         Metabolism
   acid-base, 53
      HIV protease in, 54, 55f
   at active site, 53–54, 53f
   Bi-Bi, 72–73, 73f
      Michaelis-Menten kinetics and, 73, 73f
   coenzymes/cofactors in, 52, 53f
   conservation of residues and, 56, 57t

covalent, 54, 54f, 65
   chymotrypsin in, 54–56, 55f, 65
   fructose-2,6-bisphosphatase in, 55f, 56
double displacement, 72, 73f
enzyme detection facilitated by, 56–58, 58f
equilibrium constant and, 65
isozymes and, 56
kinetics of, 65–74
   activation energy and, 64, 65
   balanced equations and, 62
   competitive *vs.* noncompetitive inhibition
         and, 70, 70f, 71f
   in drug development, 73–74
   factors affecting rates of, 64f, 66, 66f
   free energy changes and, 62–63
   initial velocity and, 66
   multiple substrates and, 72–73, 73f
   substrate concentration and, 66–67, 67f
   models of, 67–70, 68f, 69f, 70f
   transition states and, 63–64
mechanisms of, 53–54, 54f
   prosthetic groups/cofactors/coenzymes in,
         52, 53f
   site-directed mutagenesis in study of, 61
oxaloacetate and, 143
ping-pong, 72–73, 73f
prosthetic groups in, 52, 53f
by proximity, 53
regulation of, 75–82, 135, 138f
   active and passive processes in, 75, 76f
   allosteric, 77–78, 78f, 135, 138f
   compartmentation in, 75–76
   covalent, 77, 79, 80–81, 80f
   enzyme quantity and, 76–77
   feedback inhibition and, 77–78, 78f, 135
   feedback regulation and, 78–79, 135
   metabolite flow and, 75, 76f
   Michaelis constant (*Km*) in, 75, 76f
   phosphorylation-dephosphorylation in, 80f,
         81t
   proteolysis in, 79, 80f
   second messengers in, 79
RNA and, 351
sequential (single) displacement, 72, 73f
specificity of, 51, 52f
by strain, 53–54
substrate concentration affecting rate of, 66–67,
         67f
   Hill model of, 69–70, 70f
   Michaelis-Menten model of, 67–68, 68f
Catalytic constant (*k*cat), 68
Catalytic efficiency (*k*cat/*K*m), 68–69
Catalytic residues, conserved, 56, 57t
Catalytic site, 77. *See also* Active site
Cataracts, diabetic, 182
Catecholamines. *See also* specific type
   receptors for, 427
   storage/secretion of, 442, 442t
   synthesis of, 435–441, 436f
Cathepsins, in acid-base catalysis, 54
Cation. *See also* specific cations
   membrane penetration by, 110
Caveolae, 412
Caveolin-1, 412
CBG. *See* Corticosteroid-binding globulin
CBP/CBP/p300 (CREB-binding protein), 448, 453,
         455f, 456–457, 457t
CD11a-c/CD18, in neutrophils, 604t, 605
CD18, 605
CD49a/e/f, 605t

CD59, 523
CDK-cyclin inhibitor/CDKI, DNA/chromosome
         integrity and, 333
CDKs. *See* Cyclin-dependent protein kinases
cDNA, 403
cDNA library, 392, 403
cDNA sequencing, in glycoprotein analysis, 507t
CDRs. *See* Complementarity-determining regions
CEA, 623
Celiac disease, 459
Cell, 1
   in macromolecule transport, 420–422, 421f,
         422f
Cell adhesion
   fibronectin in, 531–532, 532f
   glycosphingolipids in, 209
   integrins in, 605, 605t
   selectins in, 520–521, 520f, 520t
Cell biology, 1
Cell-cell communication, via gap junctions, 422,
         422f
Cell-cell interactions, 406
   mucins in, 510–511
Cell cycle, S phase of, DNA synthesis during,
         328–330, 329f, 329t
Cell-cycle regulation, 498
Cell death, 208
Cell-free systems, vesicles studied in, 499
Cell fusion, 580, 644
Cell injury
   oxygen species causing, 597, 597t
   xenobiotics causing, 613, 614f
Cell lysis, complement in, 581
Cell-mediated immunity, 576
Cell membrane. *See* Plasma membrane
Cell migration, fibronectin in, 531–532
Cell recognition, glycosphingolipids in, 209
Cell sap. *See* Cytosol
Cell surface carbohydrates, glycolipids and, 126
Cell surfaces, heparan sulfate on, 536
Cellulose, 118
Cellulose acetate zone electrophoresis, 566, 568f
Central core disease, 553, 556t
Central nervous system, 524
   glucose as metabolic necessity for, 137
Centromere, 315, 316f
Cephalin (phosphatidylethanolamine), 124, 125f
   membrane asymmetry and, 410
   synthesis of, 205, 206f
Ceramide, 125, 126f, 208–209, 210f
   synthesis of, 208–209, 210f
Cerebrohepatorenal (Zellweger) syndrome, 191,
         492–493, 493t
Cerebrosides, 208
Ceruloplasmin, 569, 572–573
   deficiency of, 574
   diagnostic significance of, 59t, 573
Cervonic acid, 123t
CFTR. *See* Cystic fibrosis transmembrane
         regulator
CFU-E. *See* Colony-forming unit-erythroid
Chain elongation. *See also* Elongation
   by DNA polymerase, 324
   in glycosaminoglycan synthesis, 535
   in transcription cycle, 336, 336f
Chain initiation. *See also* Initiation
   in transcription cycle, 336, 336f
Chain termination. *See also* Termination
   in glycosaminoglycan synthesis, 535
   in transcription cycle, 336, 336f

Channeling, in citric acid cycle, 143
Channelopathies, 555, 556t
Channels, membrane, 409
Chaperones, 39, 496–497, 497t
    ATP-dependent protein binding to, 488, 496
    ATPase activity of, 496
    histone, 313
    in protein sorting, 488, 497t
Chaperonins, 39, 496
Charged paddle, 417, 418f
Charging, in protein synthesis, 355, 355f
Checkpoint controls, 333
Chèdiak-Higashi syndrome, 498t
Chelation therapy, 633
Chemical carcinogenesis/carcinogens, 609, 614
Chemiosmotic theory, 106
    on respiratory control, 107f, 109
Chemotactic factors, 604
Chemotherapy, cancer
    combination, 593
    folate inhibitors in, 477
    neutropenia caused by, 595
    synthetic nucleotide analogs in, 289, 290f
Chenodeoxycholic acid, 228, 231f
Chenodeoxycholyl CoA, 228, 231f
Chimeric gene approach, 378f, 379, 380f
Chimeric molecules, 388–395, 403
    restriction enzymes and DNA ligase in preparation of, 389–390, 391f
Chips, gene array, protein expression and, 29
Chitin, 119, 119f
Chloramines, 607
Chloride
    in extracellular and intracellular fluid, 407, 407t
    permeability coefficient of, 409f
Chlorophyll, 271
Cholecalciferol (vitamin D₃)
    skin synthesis of, 432, 434–435, 435f, 470, 471f
    in vitamin D metabolism, 470, 470f, 471f
Cholera
    case study, 619–621, 620f
    glucose transport in treatment of, 420
    toxin, 209
Cholestatic jaundice, 282
Cholesterol, 126, 127, 127f, 212, 462, 503
    in bile acid synthesis, 228–229, 231f
    in calcitriol (1,25(OH)₂-D₃) synthesis, 428, 429f, 470
    dietary, 224
    excess of. See Hypercholesterolemia
    excretion of, 228–229, 231f
    in hormone synthesis, 429–435, 429f, 430f
    in lipoprotein, 212, 213f
    in membranes, 408
        fluid mosaic model and, 412
    metabolism of, 133, 134f
        clinical aspects of, 230–233, 232t
        diurnal variations in, 225
        high-density lipoproteins in, 216–217, 217f
    plasma levels of
        atherosclerosis and coronary heart disease and, 230
        dietary changes affecting, 230–231
        drug therapy affecting, 232
        lifestyle changes affecting, 231–232
        normal, 232
    synthesis of, 224–225, 225f, 226f, 227f
        acetyl-CoA in, 132, 133f, 224–225, 225f, 226f, 227f
        carbohydrate metabolism and, 132

HMG-CoA reductase in regulation of, 225, 228f
    in tissues, 127, 127f
        factors affecting balance of, 225–227, 229f
    transport of, 227–228, 230f
        reverse, 217, 217f, 224, 228
Cholesteryl ester hydrolase, 225
Cholesteryl ester transfer protein, 228, 229f
Cholesteryl esters, 127, 212, 228
    in lipoprotein core, 212, 213f
Cholic acid, 228
Choline, 124, 125f
    deficiency of, fatty liver and, 218
    in glycine synthesis, 236, 236f
    membrane asymmetry and, 410
Cholinesterase, 618
Cholinesterase activity inhibition, 617
Choluric jaundice, 281
Cholyl CoA, in bile acid synthesis, 228, 231f
Chondrodysplasias, 541t, 543–544, 543f
Chondroitin sulfates, 119, 119f, 529, 534f, 535, 535t
    functions of, 539
Chondronectin, 542, 542t
Chorionic gonadotropin, human (hCG), 428
Christmas factor (factor IX), 584, 584f, 585t
    coumarin drugs affecting, 588–589
    deficiency of, 589
Chromatids
    nucleoprotein packing in, 315, 316t, 317f
    sister, 315, 316f
        exchanges between, 321–322, 322f
Chromatin, 312–315, 313f, 313t
    active vs. inactive regions of, 314f, 315
    higher order structure/compaction of, 313–315, 314f
    reconstitution of, in DNA replication, 328
    remodeling of in gene expression, 377
Chromatography. See also specific type
    affinity
        for protein/peptide purification, 23
        for recombinant fusion protein purification, 60–61, 60f
        for protein/peptide purification, 21–24
        Sepharose-lectin column, for glycoprotein analysis, 507t
Chromium, 480t
Chromosomal integration, 320–321, 321f
Chromosomal recombination, 319–320, 320f, 321f
Chromosomal transposition, 321
Chromosome walking, 398, 400f
Chromosomes, 315–317, 316f, 316t, 317f, 318f
    integrity of, monitoring, 333
    interphase, chromatin fibers in, 314
    metaphase, 314f, 316, 316t
    polytene, 315, 316f
Chronic granulomatous disease, 606, 606f
Chyle, 214
Chylomicron remnants, 213t, 215, 215f, 203
    liver uptake of, 215
Chylomicrons, 134, 138, 212, 213t
    apolipoproteins of, 213, 213t
    metabolism of, 134, 135f, 214–216, 216f
    in triacylglycerol transport, 215, 215f, 216f
Chymotrypsin, 462
    conserved residues and, 57t
    in covalent catalysis, 54–56, 55f
    in digestion, 462
Chymotrypsinogen, 462

cI repressor protein/cI repressor gene, 374, 374f, 375f, 376f
CICR. See Calcium-induced calcium release
Cirrhosis of liver, 143, 218, 633
Cisternal maturation, 502
Cistron, 370
Citrate
    in citric acid cycle, 143, 144f
    in lipogenesis regulation, 197
Citrate synthase, 143, 145f
Citric acid, pK/pKa value of, 13t
Citric acid cycle (Krebs/tricarboxylic acid cycle), 95, 104, 137, 143–144, 144f, 145f
    ATP generated by, 144f, 145, 154–155, 155t
    carbon dioxide liberated by, 143–144, 145f
    deamination and, 146–147
    gluconeogenesis and, 146, 146f, 165–167, 166f
    in metabolism, 132, 132f, 133f, 135, 136f, 143, 146–147, 146f
        amino acid, 132f, 133f
        carbohydrate, 132, 132f, 146, 146f
        lipid/fatty acid, 132, 133f, 147, 147f
        at subcellular level, 135, 136f
    in mitochondria, 135, 136f
    reducing equivalents liberated by, 143–144, 145f
    regulation of, 147
    respiratory chain substrates provided by, 143, 144f
    transamination and, 146f147
    vitamins in, 146
Citrulline, in urea synthesis, 243–244
Citrullinemia, 246
CJD. See Creutzfeldt-Jakob disease
CK. See Creatine kinase
Cl. See Chloride
Class B scavenger receptor B1, 217, 217f
Class (isotype) switching, 580
Classic pathway, of complement activation, 581
Clathrin, 421, 421f
Clathrin-coated vesicles, 500, 502
Clathrin-free vesicles, 499
Clearance of antigen–antibody complexes, 581
Cleavage
    of preproalbumin, to proalbumin, 502f
    in protein sequencing, 25
    of ubiquitin, 498
Clinical deficiency disease. See also specific diseases
    vitamin, 467
CLIP, 441, 441f
Clofibrate, 232
Clones
    definition of, 403
    library of, 392, 403
    in monoclonal antibody production, 580
Cloning, 390–392, 391f, 391t, 392f, 518
Cloning vectors, 390–392, 391f, 391t, 392f, 405
Closed complex, 338
Clotting factors, 585t. See also specific type under Factor
    vitamin K in synthesis of, 472–473, 472f
CMP, 287t
CMP-NeuAc, 508, 508t
CO. See Carbon monoxide
CO₂. See Carbon dioxide
CoA. See Coenzyme A
Coactivators, transcription, 345, 345t
Coagulation (blood), 583–592
    endothelial cell products in, 591, 591t

extrinsic pathway of, 583–584, 584f, 586t
fibrin formation in, 583–584, 584f
final common pathway in, 609f
intrinsic pathway of, 583–584, 584f
laboratory tests in evaluation of, 591
pathways of, 587f
prostaglandins in, 193
proteins involved in, 585, 586t. *See also* Coagu-
    lation factors
vitamin K in, 472, 472f
    coumarin anticoagulants affecting,
    587–588
Coagulation factors, 586t. *See also* specific type
    under Factor
vitamin K in synthesis of, 473, 473f
Coat proteins
function of, 500
recruitment of, 500, 501f
Coated pits, in absorptive pinocytosis, 437,
    437f
Coating, vesicle, 499–500, 501f
brefeldin A affecting, 502
Cobalamin (vitamin B$_{12}$), 468t, 476–477, 477f
absorption of, 476
    intrinsic factor in, 462, 476
deficiency of, 468t, 477
    functional folate deficiency and, 477–478
in methylmalonic aciduria, 167
Cobalophilin, 476
Cobalt, 480t
in vitamin B$_{12}$, 476
Cobamide, coenzymes derived from, 52
Coding regions, 317, 318f
in recombinant DNA technology, 392
Coding strand, 303
in RNA synthesis, 335, 336f, 340f
Codon usage tables, 354–355
Codons, 353, 354t
amino acid sequence of encoded protein speci-
    fied by, 353
nonsense, 354
Coenzyme A, synthesis of from pantothenic acid,
    479, 479f
Coenzymes, 52
in catalysis, 52, 53f
nucleotide derivatives, 289, 289t
Cofactors, 52
in blood coagulation, 585, 586t, 587
in catalysis, 52, 53f
in citric acid cycle regulation, 147
Cognitive function, decline of, 618
Colchicine, 563, 632
Colipase, 460
Collagen, 40–41, 366, 527–530, 528t
in bone, 538t, 539
in cartilage, 542–543, 542t, 543f
classification of, 527, 528t
cross-linking of, 519, 642
elastin differentiated from, 531t
fibril formation by, 527–530, 528f, 529t
genes for, 527, 528t
    diseases caused by mutations in, 41,
    529–530, 530t
chondrodysplasias, 541t, 543–544
osteogenesis imperfecta, 541, 541t
maturation/synthesis of, 40–41
    ascorbic acid (vitamin C) in, 41, 479
    disorders of, 41
O-glycosidic linkage in, 510
in platelet activation, 590, 590f, 591

posttranslational modification of, 529, 529t
triple helix structure of, 40–41, 41f, 527–530,
    528f
type I, 527–529, 528t, 538t
type II, 527, 542t
type IV, 528t, 529, 530t
type V, 538t, 539, 542t
type VI, 529, 530t, 542t
type VII, 530, 530t
type IX, 519, 542t, 542t
types of, 527, 528t
Collision-induced dissociation, in mass
    spectrometry, 27
Collision (kinetic) theory, 64
Colon cancer. *See* Colorectal cancer
Colony-forming unit-erythroid, 594, 595f
Colony hybridization, 393. *See also* Hybridization
Colorectal cancer
case study, 621–624, 623f, 624f
development of, 623
mismatch repair genes in, 330–331
Column chromatography, for protein/peptide
    purification, 21, 23f
Combination chemotherapy, 593
Combinatorial chemistry, 57
Combinatorial diversity, 579
Compartmentation, 75–76
Competitive inhibition, noncompetitive inhibition
    differentiated from, 70–72, 70f,
    71f, 72f
Complement, 569t, 581
in inflammation, 581, 604t
Complement deficiency disorders, 581
Complementarity
of DNA, 306, 307f
    recombinant DNA technology and, 392
of RNA, 308, 310f
Complementarity-determining regions, 578f,
    581
Complementary DNA (cDNA), 403
Complementary DNA (cDNA) library, 392,
    403
Complex carbohydrate, 506
Complex (glycoconjugate) carbohydrates, glyco-
    proteins as, 506
Complex oligosaccharide chains, 512, 513f
formation of, 512, 513
Computational biology, 84–91
definition of, 87
genomes and medicine, 85
genomic resources for, 86–87
Human Genome Project in, 84–85
Computer-aided drug design (CADD), 89
Concanavalin A (ConA), 120, 510t
Conformation. *See also* specific substances
polypeptide/protein, 22f
native, 38
Conformational diseases, 575, 619, 626
Conformational disorders, 504
Congenital contractural arachnodactyly, 531
Congenital disorders of glycosylation (CDG), 522,
    522t, 571
Congenital forms of muscular dystrophy, 553
Congenital hypothyroidism, 635
Congenital long QT syndrome, 423t
Congenital muscular dystrophy, 553
Congenital nonhemolytic jaundice (type I Crigler-
    Najjar syndrome), 281–282
Congo red dye, 576
Conjugate acid, 10

Conjugate base, 10
Conjugating enzyme, 498
Conjugation
of bilirubin, 278–279, 280f
of xenobiotics, 609, 611–613
Connecting (C)-peptide, 438f, 439
Connective tissue, 527
bone as, 539–541
keratan sulfate I in, 535
Connexin, 422, 422f
Consensus sequences, 346, 347f
Kozak, 361
Conservation of energy, 95
Conserved residues, 56, 57t
Constant regions/segments, 576
gene for, 579
    DNA rearrangement and, 322, 385, 580
immunoglobulin heavy chain, 576–577,
    577f
immunoglobulin light chain, 322, 386, 576–577,
    577f
Constitutive gene expression, 370, 373
Constitutive heterochromatin, 315
Constitutive mutation, 371
Constitutive secretion, 487
Conversion of inactive precursors, to active
    products
by proteases and a cascade with amplification,
    581
Contact sites, 490
Contractility/contraction. *See* Muscle
    contraction
Cooperative binding
hemoglobin, 46
    Bohr effect and, 47, 48f
Hill equation describing, 69–70, 70f
COPI vesicles, 496, 499, 499t, 500, 502
COPII vesicles, 499, 499t, 500, 502
Coplanar atoms, partial double-bond character
    and, 19, 19f
Copper, 480t
ceruloplasmin in binding of, 572–573
as cofactor, 573, 573t
enzymes containing, 573t
excess, 573
in Menkes disease, 573
metallothioneins in regulation of, 573
in oxidases, 98
tests for disorders of metabolism of, 574t
in Wilson disease, 572, 573, 574
Copper-binding P-type ATPase, mutations in
    gene for
Menkes diseases caused by, 573
Wilson disease caused by, 573
Copper toxicosis, 574. *See also* Wilson disease
Coproporphyrinogen I, 273, 274f, 275f
Coproporphyrinogen III, 273, 274f, 275f
Coproporphyrinogen oxidase, 273, 274f, 275f
in porphyria, 277t
Coproporphyrins, 272f, 275
spectrophotometry in detection of, 275
Coprostanol (coprosterol), 228
Core proteins, 533, 534–535, 542, 542f, 543f
in glycosaminoglycan synthesis, 551
Coregulator proteins, 455–458, 457t
Corepressors, 457, 457t
Cori cycle, 170, 170f
Cori disease, 160t
Cornea, keratan sulfate I in, 535, 537t
Coronary artery disease, 482

Coronary (ischemic) heart disease. *See also* Atherosclerosis
    cholesterol and, 230
Coronary thrombosis, 638
Corrinoids, 476. *See also* Cobalamin
Corticosteroid, 632
Corticosteroid-binding globulin (CBG/transcortin), 442–443, 442t, 569t
Corticosterone
    binding of, 442t, 443
    synthesis of, 430–431, 431f
Corticotropin. *See* Adrenocorticotropic hormone
Cortisol, 429f, 430f
    binding of, 442–443, 442t
    synthesis of, 430–431, 431f
Cos sites, 390
Cosmids, 390, 391t, 403
Cothromboplastin (factor VII), 584, 584f, 585t
    coumarin drugs affecting, 587–588
Cotranslational glycosylation, 495
Cotranslational insertion, 494, 495, 496f
Cotransport systems, 404–405, 416f
Coulomb's law, 6
Coumarin, 587–588
Coupling, 93–94, 93f
    ATP in, 94, 95–96
    hormone receptor-effector, 426
Coupling domains, on hormone receptors, 426
Covalent bonds
    biologic molecules stabilized by, 7, 7t
    membrane lipid-protein interaction and, 410
    xenobiotic cell injury and, 613, 614f
Covalent catalysis, 54, 54f, 65
    chymotrypsin in, 54–56, 55f, 65
    fructose-2,6-bisphosphatase in, 56, 56f
Covalent cross-links, collagen, 528
Covalent modification
    mass spectrometry in detection of, 26–28, 26t, 27f
    in protein maturation, 40
    in regulation of enzymatic catalysis, 77, 79, 80–81, 80f. *See also* Phosphorylation; Proteolysis
        gluconeogenesis regulation and, 167
        irreversible, 79, 80f
        metabolite flow and, 80–81
        reversible, 80, 80f, 81t
Coxibs, 200
CPT-I. *See* Carnitine palmitoyltransferase-I
CRE. *See* Cyclic AMP response element
Creatine, 266, 268f
Creatine kinase, diagnostic significance of, 59, 59t, 628
Creatine phosphate, 263f, 266, 268f
    free energy of hydrolysis of, 94t
    in muscle, 560f, 561, 561t
Creatine phosphate shuttle, 111f, 112
Creatinine, 266, 268f
CREB, 448
CREB-binding protein, 448, 456
Cretinism, 635
Creutzfeldt-Jakob disease, 39–40
Crigler-Najjar syndrome
    type I (congenital nonhemolytic jaundice), 281–282
    type II, 282
Cro protein/cro gene, 374, 374f, 375f, 376f
    binding of to DNA, by helix-turn-helix motif, 381–382, 382f
Cross-bridges, 546, 550f
Cross-links, covalent in collagen, 528

Crossing-over, in chromosomal recombination, 320, 320f, 321f
Crouzon syndrome, 541t
CRP. *See* C-reactive protein; Catabolite regulatory protein; Cyclic AMP regulatory protein
Cryoprecipitates, recombinant DNA technology in production of, 588
Cryptoxanthin, 468
Crystallography, x-ray, protein structure demonstrated by, 37–38
CS-PG I/II/III, in bone, 542t
CT. *See* Calcitonin
CTD. *See* Carboxyl terminal repeat domain
CTP, 289
    in phosphorylation, 96
Cyanide
    on oxidative phosphorylation, 103
    on respiratory chain, 108, 109f
Cyclic AMP, 159, 160f, 288, 288f, 289t, 447–449, 447t, 448f
    adenylyl cyclase affecting, 159, 447, 448t
    in cardiac muscle regulation, 554
    in gluconeogenesis, 168–169, 169f
    in glycogen metabolism regulation, 160–162, 160f, 161f, 162f
    as second messenger, 159, 427, 428t, 446, 447–449, 447t, 448f
    smooth muscle contraction affected by, 557
Cyclic AMP-dependent protein kinase, 36f. *See also* Protein kinases
Cyclic AMP regulatory protein (catabolite gene activator protein), 371, 373
Cyclic AMP response element, 445t, 448–449
Cyclic AMP response element binding protein, 448
Cyclic GMP, 288f, 289
    role in smooth muscle, 559f
    as second messenger, 289, 427, 428t, 446, 449
Cyclic 3',5'-nucleotide phosphodiesterase, in lipolysis, 222
Cyclin-dependent protein kinases, 329, 329f, 329t
    inhibition of, DNA/chromosome integrity and, 333
Cyclins, 328–330, 329f, 329t
Cycloheximide, 367
Cyclooxygenase, 200
    as "suicide enzyme," 200–201
Cyclooxygenase pathway, 200–201, 202f, 203f
CYP nomenclature, for cytochrome P450 isoforms, 610
CYP2A6, polymorphism of, 611, 613t
CYP2C9, in warfarin-phenobarbital interaction, 611
CYP2D6, polymorphism of, 611, 613t, 615
CYP2E1, enzyme induction and, 611
Cysteine, 15t, 262
    conversion to taurine, 263f
    metabolism of, 251f, 252, 254
        abnormalities of, 248–250, 252f
    in pyruvate formation, 252, 251f, 252f
    requirements for, 465
    synthesis of, 236, 237f
Cystic fibrosis, 423, 423t, 459, 525, 556t
    case study, 624–626, 625t
Cystic fibrosis transmembrane regulator (CFTR), 423, 525, 625, 626
    degradation of, 498
Cystine reductase, 251f, 252
Cystinosis (cystine storage disease), 252

Cystinuria (cystine-lysinuria), 252
Cytarabine (arabinosyl cytosine), 289
Cytidine, 286f, 287t
Cytidine monophosphate (CMP), 287t
Cytidine monophosphate N-acetylneuraminic acid (CMP-NeuAc), 508, 508t
Cytidine triphosphate (CTP), 289
    in phosphorylation, 96
Cytochrome b_5, 101, 495, 599
Cytochrome b_H, 104
Cytochrome b_L, 104
Cytochrome c oxidase, 103, 105f, 106f
Cytochrome c_1, 104
Cytochrome oxidase/cytochrome a_3, 98
Cytochrome P450, 495, 610
Cytochrome P450-dependent microsomal ethanol oxidizing system, 218–220
Cytochrome P450 side chain cleavage enzyme (P450scc), 430, 430f, 432
Cytochrome P450 system, 98, 101, 102f, 609
    ALA synthase affected by, 274, 278
    enzyme induction and, 274, 611
    genes encoding, nomenclature for, 610
    isoforms of, 610–611
    membrane insertion, 495
    in metabolism of xenobiotics, 609–611, 612t
    mitochondrial, 101
    nomenclature system for, 610
    in xenobiotic cell injury, 613, 614f
Cytochromes, as dehydrogenases, 100
Cytokines, 519, 643
    α_2-macroglobulin binding of, 575
    in cachexia, 140
Cytosine, 287t
    base pairing of in DNA, 302, 303, 304f
    deoxyribonucleosides of, in pyrimidine synthesis, 295–297, 298f
Cytoskeleton/cytoskeletal proteins, 545, 562–564
    red cell, 600, 600f, 600
Cytosol, 495
    ALA synthesis in, 272, 273f
    glycolysis in, 135, 136f, 149
    lipogenesis in, 193–196, 194f, 195f
    pentose phosphate pathway reactions in, 174
    pyrimidine synthesis in, 297, 298f
Cytosolic branch, for protein sorting, 487, 488, 488f
Cytosolic dynein, 563
Cytosolic proteins, O-glycosidic linkages in, 510
Cytotoxicity, xenobiotic, 613, 614f

D-amino acids, free, 14
D arm, of tRNA, 308, 310, 355, 355f
D cyclins, 329, 329f, 329t
    cancer and, 329
D isomerism, 113–114, 114f
DAF. *See* Decay accelerating factor
dAMP, 287f
Dantrolene, for malignant hyperthermia, 552
Database of Genotype and Phenotype (dbGAP), 87
Databases, 86
DATP, 617
DBD. *See* DNA binding domains
dbGAP. *See* Database of Genotype and Phenotype
Deamination, 133, 134f
    citric acid cycle in, 145
    liver in, 134

Debranching enzymes
    absence of, 160t
    in glycogenolysis, 158–159, 159f
Debrisoquin, CYP2D6 in metabolism of, 611
Decarboxylation of S-adenosylmethionine,
        263
Decay accelerating factor, 523
Decorin
    in bone, 538t
    in cartilage, 542t
Defensins, 604t
Deficiency, in cis-Golgi-located GlcNAc phospho-
        transferase, 523
Degeneracy, of genetic code, 354
Degradation, of virus, 499
Dehydrocholesterol, in vitamin D metabolism,
        470, 470f
Dehydroepiandrosterone (DHEA), synthesis of,
        431, 431f
Dehydrogenases, 98, 99–100, 100f
    in enzyme detection, 58, 58f
    nicotinamide coenzyme-dependent, 99, 100f
    in respiratory chain, 99–100
    riboflavin-dependent, 99–100
Deletions, DNA, recombinant DNA technology
        in detection of, 397–398, 398t
Dementia, 618
Denaturation
    DNA structure analysis and, 303–304
    protein refolding and, 39
    temperature and, 66
Deoxynojirimycin, 518, 518t
Deoxy sugars, 116, 117f
Deoxyadenylate, 302
Deoxycholic acid, synthesis of, 229
Deoxycorticosterone
    binding of, 443
    synthesis of, 430, 431f
11-Deoxycortisol, synthesis of, 431, 431f
Deoxycytidine residues, methylation of, gene
        expression affected by, 377
Deoxycytidylate, 302
Deoxyguanylate, 302
Deoxyhemoglobin, proton binding by, 47, 48f
Deoxyhemoglobin A, "sticky patch" receptor on,
        49f
Deoxyhemoglobin S, "sticky patch" receptor on,
        49f
Deoxynucleotides, 302–303, 303f, 304f
Deoxyribonucleases (DNase)/DNase I, 311
    active chromatin and, 315
Deoxyribonucleic acid. See DNA
Deoxyribonucleoside diphosphates (dNDPs),
        reduction of NDPs to, 295, 297f
Deoxyribonucleosides, 286
    in pyrimidine synthesis, 295–296
Deoxyribose, 113, 116, 117f
3-Deoxyuridine, 289
Dephosphorylation. See also Phosphorylation
    in covalent modification, 80–81, 81f
Depolarization, in nerve impulse transmission,
        419
Depurination, DNA, base excision-repair
        and, 331
Dermatan sulfate, 534f, 535t, 536
    functions of, 539
Desmin, 553t, 563t
Desmosines, 530
Desmosterol, in cholesterol synthesis, 225, 227f
Detergents, 408

Detoxification, 609
    cytochrome P450 system in, 101, 102f, 610–611,
        612t
Dextrinosis, limit, 160t
Dextrins, 118
Dextrose, 114
DHA. See Docosahexaenoic acid
DHEA. See Dehydroepiandrosterone
DHPR. See Dihydropyridine receptor
DHT. See Dihydrotestosterone
Diabetes mellitus, 113, 172, 639
    fatty liver and, 218
    free fatty acid levels in, 214
    hyperglycemia in, 141
    insulin resistance and, 596
    ketosis/ketoacidosis in, 191
    lipid transport and storage disorders and, 213
    lipogenesis in, 193
    as metabolic disease, 131
    management of patients with, 519
Diabetic cataract, 182
Diabetic ketoacidosis, case study, 627–628
Diacylglycerol, 124, 460, 461f
    in calcium-dependent signal transduction, 450,
        451f
    formation of, 206f
    in platelet activation, 589, 590f
    in respiratory burst, 606
Diacylglycerol acyltransferase, 206f, 207
Diagnostic enzymology, 58–60, 59t
Diagnostic window, 59
Diarrhea
    bacteria causing, glycan binding of, 525
    severe, glucose transport in treatment of,
        420
Dicarboxylic aciduria, 191
Dicer nuclease, 350
Dicumarol (4-hydroxydicoumarin), 472
Dielectric constant, of water, 6
Diet. See also Nutrition
    blood glucose regulation and, 170
    cholesterol levels affected by, 230–231
    hepatic VLDL secretion and, 218, 219f
    high-fat, fatty liver and, 218
    very low carbohydrate, weight loss from, 173
Diet-induced thermogenesis, 223, 464
Diethylenetriaminepentaacetate (DTPA), as pre-
        ventive antioxidant, 128
Diffusion
    facilitated, 413, 414t, 414f, 415–416, 41f
        of bilirubin, 278
        of glucose. See also Glucose transporters
        insulin affecting, 416
        in red cell membrane, 596
            hormones in regulation of, 416
            "Ping-Pong" model of, 415, 416f
    net, 414, 414f
    passive, 413, 414t, 414f
    simple, 413, 414t, 414f
Digestion, 459–466
Digitalis, 418
    Ca^{2+}-Na^+ exchanger in action of, 554–556
    Na^+-K^+ ATPase affected by, 419, 554–556
Dihydrobiopterin, defect in synthesis of, 254
Dihydrobiopterin reductase, defect in, 254
Dihydrofolate/dihydrofolate reductase, methotrex-
        ate affecting, 296, 477
Dihydrolipoyl dehydrogenase, 153f, 154
Dihydrolipoyl transacetylase, 153f, 154
Dihydropyridine receptor, 551, 551f

Dihydrotestosterone, 432, 433f
    binding of, 442t
Dihydroxyacetone, 115f
Dihydroxyacetone phosphate, in glycolysis, 205,
        206f
1,25-Dihydroxyvitamin D3. See Calcitriol
24,25-Dihydroxyvitamin D3 (24-hydroxycalci-
        diol), in vitamin D metabolism,
        470, 471f
Diiodotyrosine (DIT), 436, 437f, 438
Dilated cardiomyopathy, 553, 557
Dimercaprol, on respiratory chain, 108, 109f
Dimeric proteins, 35
Dimers
    Cro protein, 374, 375f
    histone, 313
    lambda repressor (cI) protein, 374, 375f
Dimethylallyl diphosphate, in cholesterol synthe-
        sis, 224, 226f
Dimethylaminoadenine, 288f
2,4-Dinitrophenol, 109
Dinucleotide, 290
Dioxygenases, 101
Dipalmitoyl lecithin, 124
Dipeptidases, 462
Diphosphates, nucleoside, 286, 286f
Diphosphatidylglycerol. See Cardiolipin
Diphtheria toxin, 367, 418
Dipoles, water forming, 6, 7f
Disaccharidases, 460
Disaccharides, 113, 117, 117f, 117t. See also
        specific type
Disassociation
    of Sar1, 500
    of four-helix bundle, 500
Discovery of causes of certain anemia types, 593
Disease, 1. See also Biochemical case histories;
        specific diseases
    biochemical basis of, 2, 3t
    conformational, 498t
    genes, tracking, 3
    Human Genome Project and, 3–4
    major causes of, 3t
Dislocation, 497
Displacement reactions
    double, 72, 73f
    sequential (single), 72, 73f
Dissociation, 523
    of water, 9
Dissociation constant, 9
    Michaelis constant (K_m) and, 69
    in pH calculation, 10
    of weak acids, 10–11, 12
Distal histidine (histidine E7), in oxygen binding,
        43, 44f
Disulfide bonds, protein folding and, 39
DIT. See Diiodotyrosine
Diurnal rhythm, in cholesterol synthesis, 225
Divalent metal transporter (DMT1), 571, 633
Diversity
    antibody, 579, 580
    combinatorial, 579
    in gene expression, 380, 381f
    junctional, 580
Diversity segment, DNA rearrangement and, 580
Dixon plot, 71, 72f
DNA, 302–305, 312–333, 336f
    base excision-repair of, 331t, 331–332, 332f
    base pairing in, 8, 302–303, 304f
        matching of for renaturation, 304

DNA (*Continued*)
  recombinant DNA technology and, 388–395
  replication/synthesis and, 326, 326f
  binding to regulatory proteins of, motifs for, 381–382, 381t, 382f, 383f
  blunt-ended, 389–90, 389f, 403
  in chromatin, 312–315, 313f, 313t, 314f, 316f
  chromosomal, 315–317, 316f, 316t, 317f, 318f
    relationship of to mRNA, 318f
  coding regions of, 317, 318f
  complementarity of, 305, 305f
    recombinant DNA technology and, 388–395
  damage to, 330, 330t
    repair of, 330–333, 331t
    ADP-ribosylation for, 475
  deletions in, recombinant DNA technology in detection of, 397–398, 398t
  depurination of, base excision-repair and, 331
  double-strand break repair of, 331t, 332–333, 333f
  double-stranded, 302–303
  flanking sequence, 395
  genetic information contained in, 302–305
  grooves in, 304, 304f
  insertions in, recombinant DNA technology in detection of, 397–398
  integrity of, monitoring, 333
  "jumping," 321
  mismatch repair of, 330–331, 331t, 331f
  mitochondrial, 319, 319f, 320t
  mutations in, 312, 319–322, 320f, 321f, 322f. *See also* Mutations
  in nucleosomes, 313, 314f
  nucleotide excision-repair of, 332, 332f
  rearrangements of
    in antibody diversity, 321, 580
    recombinant DNA technology in detection of, 397–398, 398t
  recombinant. *See* Recombinant DNA/ recombinant DNA technology
  relaxed form of, 304–305
  renaturation of, base pair matching and, 304
  repair of, 330–333, 331t, 498
  repetitive-sequence, 318
  replication/synthesis of, 305, 305f, 322–333, 322f, 323f, 324t, 329t
    DNA polymerase complex in, 323t, 324
    DNA primer in, 323t, 324–325, 325f, 326f
    initiation of, 324–325, 325f, 326f
    origin of, 322
    polarity of, 326–327
    proteins involved in, 323t
    reconstitution of chromatin structure and, 328
    repair during, 330–333, 331t
    replication bubble formation and, 327–328, 327f, 328f
    replication fork formation and, 323f, 324
    ribonucleoside diphosphate reduction and, 295, 297f
    in S phase of cell cycle, 328–330, 329f
    semiconservative nature of, 305, 305f
    semidiscontinuous, 323f, 327, 326f
    unwinding and, 324
  in RNA synthesis, 335–337, 336f, 337t
  stabilization of, 8
  structure of, 302–305, 303f, 304f
    denaturation in analysis of, 303–304
    double-helical, 8, 302–303, 304f
    recombinant DNA technology and, 388
  supercoiled, 305, 327, 328t

  transcription of, 305
  transposition of, 321
  unique-sequence (nonrepetitive), 318
  unwinding of, 324
    RNA synthesis and, 337–338
  xenobiotic cell injury and, 613
DNA binding domains, 382, 383f, 454
DNA binding motifs, 381–382, 381t, 382f, 383f
DNA-dependent protein kinase, in doublestrand break repair, 332–333
DNA-dependent RNA polymerases, 336–337, 336f, 337t
DNA elements, gene expression affected by, 377, 378f, 378t, 379f
  diversity and, 380, 381f
DNA fingerprinting, 404
DNA footprinting, 404
DNA helicase, 323f, 323t, 324
DNA ligase, 323t, 326
  in recombinant DNA technology, 389–390, 390t, 391f
DNA-PK. *See* DNA-dependent protein kinase
DNA polymerases, 322–324, 323f, 323t
  in recombinant DNA technology, 390t
DNA primase, 323f, 323t, 324
DNA probes, 392–393, 404
  library searched with, 392–393
  in porphyria diagnosis, 276
DNA-protein interactions, bacteriophage lambda as paradigm for, 373–376, 373f, 374f, 375f, 376f
DNA sequences amplification of by PCR, 395, 395f
  determination of, 394, 394f
  protein sequencing and, 25–26
DNA topoisomerases, 305, 323t, 327, 328f
DNA transfection, endocytosis in, 420
DNA transfection, identification of enhancers/ regulatory elements and, 379–380
DNA unwinding element, 324
DNase (deoxyribonuclease)/DNase I, 311
  active chromatin and, 315
  in recombinant DNA technology, 390t
dNDPs. *See* Deoxyribonucleoside diphosphates
DOC. *See* Deoxycorticosterone
Docking, 499
  in nuclear import, 491
Docking programs, molecular, 38
Docking protein, 494, 494f
Docosahexaenoic acid, 200
Dolichol, 128, 128f, 513, 513f
  in cholesterol synthesis, 225, 226f
  in N-glycosylation, 514
Dolichol kinase, 513
Dolichol-P-P-GlcNAc, 513–515
Dolichol-P-P-oligosaccharide (dolicholpyrophosphate-oligosaccharide), 512, 513, 514f
  in N-glycosylation, 513, 514f
Dolichol phosphate, 513
Domains. *See also* specific type
  albumin, 569
  carboxyl terminal repeat, 344
  chromatin, 314, 316, 316f
  coupling, on hormone receptors, 426
  DNA binding, 382, 383f, 454
  fibronectin, 531, 532f
  protein, 35
  Src homology 2 (SH2)
    in insulin signal transmission, 452, 42f
    in Jak/STAT pathway, 453

  trans-activation, of regulatory proteins, 382–383, 384f
  transcription, 380–381
L-Dopa, 435, 436, 436f
Dopa decarboxylase, 266, 268f, 436, 436f
Dopamine, 436, 436f. *See also* Catecholamines
  synthesis of, 266, 268f, 436, 436f
Dopamine β-hydroxylase, 476
Double displacement reactions, 72, 73f
Double helix, of DNA structure, 8, 302–303, 304f
  recombinant DNA technology and, 388
Double reciprocal plot
  inhibitor evaluation and, 71, 71f
  K_m and V_{max} estimated from, 68, 68fDoublestrand break repair of DNA, 331t, 332–333, 333f
Double-stranded DNA, 302–303, 312
  unwinding
    for replication, 324
    RNA synthesis and, 337–338
Downstream promoter element, 340, 341f
Doxycycline, antibiotic, 620
DPE. *See* Downstream promoter element
DRIPs, 457, 457t
Drosha-DGCR8 nuclease, 350
Drug detoxification/interactions, cytochromes P450 and, 101, 102f, 611
Drug development
  enzyme kinetics, mechanism, and inhibition in, 73–74
  pharmacogenetics and, 614–615
  RNA targets for, 310
Drug discovery, enzyme assays suitable for "high-throughput" screening in, 57
Drug metabolism, in vivo, 74
Drug resistance, gene amplification in, 385
Drugs, 609
  dose, 616
  as enzyme inhibitors, 73
dsDNA. *See* Double-stranded DNA
DS-PG I/DS-PG II, in cartilage, 542t
DTPA, as preventive antioxidant, 128
Dubin-Johnson syndrome, 282
Duchenne muscular dystrophy, 399, 553
  case study, 628–629
DUE. *See* DNA unwinding element
Dwarfism, 541t, 543
α-dystroglycan, 553
Dynamin, in absorptive pinocytosis, 421, 563
Dyneins, 563
Dysbetalipoproteinemia, familial, 232t
Dyslipoproteinemias, 232–233, 232t
Dysplasia, 623
Dystrophin, 545, 553, 553f, 628, 629
  mutation in gene for, in muscular dystrophy, 553, 553f

E coli, lactose metabolism in, operon hypothesis and, 371–373, 371f, 372f
E coli bacteriophage P1-based (PAC) vector, 390, 391t, 404
E cyclins, 329, 329f, 329t
E-selectin, 521t
E (exit) site, in protein synthesis, 363, 363f
E_0. *See* Redox (oxidation-reduction) potential
Eact. *See* Activation energy
ECF. *See* Extracellular fluid
ECM. *See* Extracellular matrix
EcoRI, 388, 389t, 391f
EcoRII, 389t

Edema
    in kwashiorkor, 463, 464
    plasma protein concentration and, 567
    in thiamin deficiency, 473
Edematous. See Kwashiorkor
Edman reaction, for peptide/protein sequencing,
        25, 26f
Edman reagent (phenylisothiocyanate), in protein
        sequencing, 25, 26f
EDRF. See Endothelium-derived relaxing factor
EDTA, as preventive antioxidant, 128
EFA. See Essential fatty acids
EFs. See Elongation factors
EGF. See Epidermal growth factor
Egg white, uncooked, biotin deficiency caused
        by, 479
Ehlers-Danlos syndrome, 41, 234, 527, 529, 530t
Eicosanoids, 122, 193, 200, 201f, 604
Eicosapentaenoic acid, 198f
eIF-4E complex, in protein synthesis, 362
eIFs, in protein synthesis, 359
80S initiation complex, in protein synthesis, 361,
        361f
Elaidic acid, 122, 123t, 123f
Elastase, in digestion, 462
Elastin, 530–531, 531t
Electrogenic effect, 418
Electron carriers, flavin coenzymes as, 474
Electron flow, through respiratory chain, 104,
        105f, 106f
Electron movement, in active transport, 418
Electron-transferring flavoprotein, 100, 185
Electron transport chain system, 606. See also
        Respiratory chain
Electrophiles, 8
Electrophoresis
    for plasma protein analysis, 566
    polyacrylamide, for protein/peptide purifica-
        tion, 24, 24f, 25f
    two-dimensional, protein expression and, 29
Electrospray ionization, 27, 28f
    in mass spectrometry, 27
Electrostatic bonds/interactions, 8. See also Salt
        (electrostatic) bonds
    oxygen binding rupturing, Bohr effect protons
        and, 47, 48f
ELISAs. See Enzyme-linked immunoassays
Elliptocytosis, hereditary, 602
Elongase, 196, 197f
    in polyunsaturated fatty acid synthesis, 199,
        199f
Elongation
    chain
        in fatty acid synthesis, 196, 197f
        in transcription cycle, 336, 336f
    in DNA synthesis, 324–326
    in glycosaminoglycan synthesis, 535
    in protein synthesis, 362–364, 363f
    in RNA synthesis, 336, 336f, 337–338
Elongation arrest, 494
Elongation factor 2, in protein synthesis, 363, 363f
Elongation factor EF1A, in protein synthesis, 362,
        363f
Elongation factors, in protein synthesis, 362–364,
        363f
Emaciation, 131
Embryonic stem cells, 593
Emelin, 531
Emphysema
    α_1-antitrypsin deficiency in, 574

α_1-antitrypsin (α_1-antiproteinase) for, 574–575
Emulsions, amphipathic lipids forming, 129, 129f
Emtricitabine, 73
Encephalopathies
    hyperbilirubinemia causing (kernicterus), 281
    from inherited mitochondrial defects, 103
    spongiform (prion diseases), 40
    Wernicke, 473
ENCODE Project, 87
Endergonic reaction, 93
    coupling and, 93–94, 93f
        ATP in, 94, 95–96
Endocrine system, 425–443, 627. See also
        Hormones
    diversity of, 427–429
Endocytosis, 420–421, 421f
    receptor-mediated, 421, 421f
Endoglycosidase F, 509
Endoglycosidase H, 509
Endoglycosidases, 536
    in glycoprotein analysis, 509, 509t
Endonucleases, 311, 404
    apurinic and apyrimidinic, in base excision-
        repair, 331
    restriction, 311, 388–390, 389t, 389f, 403
        in recombinant DNA technology, 389–390,
        389t, 389f, 390t, 391f
Endopeptidases, 462
Endoplasmic reticulum, 365
    accumulation of misfolded proteins in, 497
    acylglycerol synthesis and, 135, 136f
    fatty acid chain elongation in, 196, 197f
    rough
        glycosylation in, 515f, 516
        in protein sorting, 487, 488f, 489f
        protein synthesis and, 365
        routes of protein insertion into, 495–496, 496f
        signal hypothesis of polyribosome binding to,
        493–496, 494t, 494f
    smooth, cytochrome P450 isoforms in, 610
Endoplasmic reticulum–associated degradation
        (ERAD)
    of misfolded proteins, 497, 498f
Endorphins, 441, 441f
Endothelial cells, 520
    in clotting and thrombosis, 591, 591t
    neutrophil interaction and
        integrins in, 521t, 605, 605t
        selectins in, 520–522, 521t, 521f
Endothelial dysfunction, 638
Endothelium-derived relaxing factor, 559. See also
        Nitric oxide
Energy
    activation, 63, 65
    conservation of, 95
    free. See Free energy
    in muscle, creatine phosphate as reserve for,
        560f, 561
    nutritional requirement for, 463
    transduction of
        membranes in, 406
        in muscle, 545–546
Energy balance, 463–464
Energy capture, 93f, 94–95
Energy expenditure, 463–464
Energy transfer, 93f, 94–95
Enhanceosome, 379, 379f
Enhancers/enhancer elements, 343
    in gene expression, 377, 378f, 378t
        tissue-specific expression and, 379

recombinant DNA technology and, 388
    reporter genes in definition of, 377–379, 380f,
        381f
Enolase, in glycolysis, 150, 151f
Entactin, in basal lamina, 533
Enterocytes, iron absorption in, 571
Enterohepatic circulation, 229
    lipid absorption and, 460
Enterohepatic urobilinogen cycle, 280
Enteropeptidase, 462
Enterotoxin, 620
Enthalpy, 92
Entrez Gene, 87
Entropy, 92
Enzyme induction, 613
    cytochrome P450 and, 274, 611
    in gluconeogenesis regulation, 167, 168t
Enzyme inhibitors, drugs as, 73
Enzyme-linked immunoassays (ELISAs), 57
Enzyme-substrate (ES) complex, 53
Enzymes, 8
    active sites of, 52–53, 53f
    analysis, aiding diagnosis, 58–59, 59t
    assisting in diagnosis of myocardial infarction,
        59–60
    assay of, 58, 58f
    branching, in glycogen biosynthesis, 157, 159f
    catalytic activity of, 51, 52f. See also Catalysis/
        catalytic reactions (enzymatic)
        detection facilitated by, 56–58, 58f
        kinetics of, 65–74. See also Kinetics (enzyme)
    in drug development, 73–74
        regulation of, 75–82, 135, 138f
        RNA and, 351
        specificity of, 51, 52f
    classification of, 51–52
    debranching
        absence of, 160t
        in glycogenolysis, 159, 159f
    degradation of, control of, 77
    in disease diagnosis/prognosis, 58, 59t, 59f,
        566
    in DNA repair, 330, 330t, 333
    hydrolysis rate affected by, 8
    irreversible inhibition ("poisoning") of, 72
    isosteric, 78
    isozymes and, 56
    kinetics of, 62–74. See also Kinetics (enzyme)
    mechanisms of action of, 51–61
    membranes in localization of, 406
    metal-activated, 52
    plasma, diagnostic significance of, 58, 59t
    quantity of, catalytic capacity affected by,
        76–77
    recombinant DNA technology in study of,
        60–61, 60f
    regulatory, 135, 138f
    restriction. See Restriction endonucleases/
        enzymes
    sophisticated control networks and, 81–82
    specificity of, 51, 52f
    substrates affecting conformation of, 54, 55f
Enzymology, single molecule, 56–57, 57f
Enzymopathies, 598
Epidermal growth factor (EGF), receptor for,
        37f, 426
Epigenetic code of histone and protein modifica-
        tion, 344
Epigenetic mechanisms, 623

Epimerases
    in galactose metabolism, 180, 180f
    in glycosaminoglycan synthesis, 535
    in pentose phosphate pathway, 174, 176f
Epimers, 115, 115f
Epinephrine, 429f, 436, 436f. *See also*
        Catecholamines
    blood glucose affected by, 171
    in gluconeogenesis regulation, 167
    in lipogenesis regulation, 197
    synthesis of, 266, 268f, 435–436, 436f
Epitope (antigenic determinant), 35, 577
Epoxide hydrolase, 614
Epoxides, 614
Eprodisate, 576
Equilibrium constant (K_{eq}), 65
    in enzymatic catalysis, 65
    free energy changes and, 63
ER. *See* Estrogens
Ercalcitriol, 470
ERE. *See* Estrogen response element
eRF. *See* Releasing factors
Ergocalciferol (vitamin D$_2$), 470
Ergosterol, 127, 127f
ERp57, 516
Erythrocyte aminotransferases, in vitamin B$_6$
        status assessment, 476
Erythrocyte transketolase activation, in thiamin
        nutritional status assessment, 474
Erythrocytes, 593, 595–602
    2,3-bisphosphoglycerate pathway in, 152f, 153
    disorders affecting, 593, 594t
    erythropoietin in regulation of, 594–595, 595f
    glucose-6-phosphate dehydrogenase deficiency
        affecting, 597, 598, 598f
    glucose as metabolic necessity for, 137
    glycolysis in, 152f, 153
    hemoglobin S "sticky patch" affecting, 49
    hemolysis of, pentose phosphate pathway/glu-
        tathione peroxidase and, 177, 177f,
        180–181
    life span of, 594
    membranes of, 599–602, 600f, 600t, 601f
        glucose transporter of, 596, 596t
        hemolytic anemias and, 598–599, 599t
    metabolism of, 141t, 595–599, 596t
        oxidants produced during, 596–597, 597t
    recombinant DNA technology in study of, 607
    structure and function of, 593–595
Erythroid ALA synthase (ALAS2), 274
    in porphyria, 277t
Erythropoiesis, 594–595, 595f
Erythropoietin/recombinant erythropoietin (epo-
        etin alfa/EPO), 517, 569, 594–595,
        595f
    cloning of, 607
D-Erythrose, 115f
ES. *See* enzyme-substrate (ES) complex
Escherichia coli, lactose metabolism in, operon
        hypothesis and, 371–373, 371f,
        372f
Escherichia coli bacteriophage P1-based (PAC)
        vector, 390, 391t, 404
Essential amino acids. *See* Nutritionally essential
        amino acids
Essential fatty acids, 193, 198, 198f, 200
    abnormal metabolism of, 202
    deficiency of, 200, 201
    prostaglandin production and, 193
Essential fructosuria, 174, 182

Essential pentosuria, 174, 181
Estriol, synthesis of, 432, 434f
Estrogen response element, 445t
Estrogens, 542
    on amino acid transport, 416
    binding of, 442t, 443
    receptors for, 455
    synthesis of, 432, 434f
Estrone
    binding of, 442t
    synthesis of, 432, 434f
Ethanol, 630
    CYP2E1 induction and, 611
    fatty liver and, 218–220
    iron absorption and, 463
    thiamin deficiency and, 473
    transferrin glycosylation with chronic abuse
        of, 571
Ethanol intoxication, acute, case study, 629–630
Ether lipids, biosynthesis of, 208f
Ethylenediaminetetraacetate (EDTA), as preven-
        tive antioxidant, 128
Euchromatin, 315
Eukaryotic cell cycle, versatility of, 81–82, 82f
Eukaryotic gene expression, 376–381, 384–387,
        384t. *See also* Gene expression
    chromatic remodeling in, 377
    diversity of, 380, 381f
    DNA elements affecting, 377–379, 378f, 378t,
        379f
    DNA-protein interactions in, bacteriophage
        lambda as paradigm for, 373–376,
        373f, 374f, 375f, 376f
    locus control regions and insulators in,
        380–381
    prokaryotic gene expression compared with,
        384–387, 384t
    reporter genes and, 379–380, 380f
    tissue-specific, 379
Eukaryotic promoters, in transcription, 339–343,
        340f, 341f, 342f
Eukaryotic transcription complex, 343–346, 344t,
        345t
Exchange diffusion systems, 110
Exchange transporters, 110–112, 110f, 111f
Excitation-response coupling, membranes in, 406
Exergonic reaction, 92, 93
    coupling and, 93–94, 93f
    ATP in, 94, 95–96
Exinuclease, in DNA repair, 332, 332f
Exit (E) site, in protein synthesis, 363, 363f
Exocrine glands, 625
Exocytosis, 420, 421–422, 422f
Exocytotic (secretory) pathway, 487
Exoglycosidases, 536
    in glycoprotein analysis, 507t, 508t, 509
Exons, 317, 353, 404
    interruptions in. *See* Introns
    splicing, 346–348, 405
        alternative, in regulation of gene expression,
        348, 348f, 386
Exonucleases, 311, 404
    in recombinant DNA technology, 390t
Exopeptidases, 462
Exportins, 491
Expression vector, 392
Extra arm, of tRNA, 308, 310f
Extracellular environment, membranes in mainte-
        nance of, 406–407, 407t
Extracellular fluid (ECF), 406–407, 407t

Extracellular lysosomal enzymes, 523
Extracellular matrix, 527–544. *See also* Matrix;
        specific components
Extramitochondrial system, fatty acid synthesis
        in, 193
Extrinsic pathway of blood coagulation, 583–584,
        584f
Eye, fructose and sorbitol in, diabetic cataract
        and, 182
Ezetimibe, for hypercholesterolemia, 232

F-Actin, 546, 548, 548f
    in non-muscle cells, 562
Fab region, 577, 577f
Fabry disease, 210t
Facilitated diffusion/transport system, 413, 414f,
        414t, 415–416, 416f
    for bilirubin, 278
    for glucose. *See also* Glucose transporters
        insulin affecting, 416
        in red cell membrane, 596
    hormones in regulation of, 416
    "Ping-Pong" model of, 415, 416f
    and transporters, 415–416
Factor I (fibrinogen), 566, 585t, 586t
    conversion of to fibrin, 585–586
Factor II (prothrombin), 585, 585t, 586t
    coumarin drugs affecting, 473, 587–588
    vitamin K in synthesis of, 473
Factor III (tissue factor), 584, 584f, 585t, 586t
Factor IV. *See* Calcium
Factor V (proaccelerin/labile factor/accelerator
        globulin), 585, 585t, 586f, 586t
Factor V Leiden, 587
Factor VII (proconvertin/serum prothrombin
        conversion accelerator/cothrombo-
        plastin), 584, 584f, 585t, 586t
    coumarin drugs affecting, 587–588
Factor VIII (antihemophilic factor A/globulin),
        584f, 585, 585t, 586t
    deficiency of, 588
Factor VIII concentrates, recombinant DNA tech-
        nology in production of, 588
Factor IX (antihemophilic factor B/Christmas
        factor/plasma thromboplastin com-
        ponent), 584f, 585, 585t, 586t
    coumarin drugs affecting, 587–588
    deficiency of, 588
Factor X (Stuart-Prower factor), 584f, 585, 585t,
        586t
    activation of, 584f, 584–585
    coumarin drugs affecting, 587–588
Factor Xa, prothrombin to thrombin activation
        by, 585
Factor XI (plasma thromboplastin antecedent),
        584f, 585, 585t, 586t
    deficiency of, 585
Factor XII (Hageman factor), 584f, 585, 585t,
        586t
Factor XIII (fibrin stabilizing factor/fibrinoligase),
        586t
Facultative heterochromatin, 315
FAD. *See* Flavin adenine dinucleotide
FADH2, fatty acid oxidation yielding, 185
Familial adenomatous polyposis, 622
Familial amyloidosis, 576
Familial hypertrophic cardiomyopathy, 556–557,
        557f
Fanconi anemia, 333
Farber disease, 210t

Farnesoid X receptor, in bile acid synthesis regulation, 229
Farnesyl diphosphate, in cholesterol/polyisoprenoid synthesis, 224, 225, 226f
Fast acetylators, 613
Fast (white) twitch fibers, 561, 561t
Fasting state, metabolic fuels in, 131, 137–140, 140f, 140t
Fat tissue. See Adipose tissue
Fatal infantile mitochondrial myopathy and renal dysfunction, 112
Fatigue (muscle), 149
Fats, 121. See also Lipids
    diets high in, fatty liver and, 225
    metabolism of, 132, 132f, 133f, 134, 135f
Fatty acid-binding protein, 184, 214
Fatty acid chains, elongation of, 196, 197f
Fatty acid elongase system, 196, 197f
    in polyunsaturated fatty acid synthesis, 199, 199f
Fatty acid oxidase, 185, 186f
Fatty acid synthase, 69, 193
Fatty acid synthase complex, 193–195, 194f, 195f, 198
Fatty acid synthesis, carbohydrates in, 137
Fatty acid-transport protein, membrane, 214
Fatty acids, 2, 121–121
    activation of, 184, 185f
    calcium absorption affected by, 462
    eicosanoids formed from, 193, 200, 201f, 202f
    essential, 193, 198, 198f, 200
        abnormal metabolism of, 202
        deficiency of, 200, 201–202
        prostaglandin production and, 193
    free. See Free fatty acids
    interconvertibility of, 137
    in membranes, 408
    metabolism of, 132, 133f
    nomenclature of, 121, 121f
    oxidation of, 184. See also Ketogenesis
        acetyl-CoA release and, 132, 133f, 184–187, 185f, 186f
        clinical aspects of, 191
        hypoglycemia caused by impairment of, 191
        in mitochondria, 184, 185f
    physical/physiologic properties of, 124
    saturated, 121, 122, 122t
    synthesis of, 193–198, 194f, 195f. See also Lipogenesis
        carbohydrate metabolism and, 132
        citric acid cycle in, 147, 147f
        extramitochondrial, 193
    trans, 122, 200
    transport of, carnitine in, 184, 185f
    triacylglycerols (triglycerides) as storage form of, 124, 124f
    unesterified (free). See Free fatty acids
    unsaturated. See Unsaturated fatty acids
Fatty liver, 637
    alcoholism and, 218–220
    nonalcoholic fatty liver desease, 218
    nonalcaholic hepatic steatosis, 218
    of pregnancy, 191
    triacylglycerol metabolism imbalance and, 218
Fatty streak, 638
Favism, 181
Fc fragment, 577, 577f
    receptors for, in neutrophils, 604t
Fe. See Iron
Fed state, metabolic fuels in, 131, 137–139, 140t
Feedback inhibition, in allosteric regulation, 77–78, 78f, 137

Feedback regulation
    in allosteric regulation, 78–78, 137
    of circulating thrombin levels, 587
Fenton reaction, 597
Ferric iron, 278
    in methemoglobinemia, 49
Ferrireductase, 571
Ferritin, 463, 571–572, 633
    protein synthesis affected by, 365
Ferritin receptor, 572
Ferrochelatase (heme synthase), 272t, 273
    in porphyria, 277t
Ferroportin, 463f, 571
Ferrous iron
    incorporation of into protoporphyrin, 272t, 273
    in oxygen transport, 43–44
Fertilization, glycoproteins in, 520
Fetal hemoglobin, P_{50} of, 46
Fetal warfarin syndrome, 473
FFA. See Free fatty acids
FGFs. See Fibroblast growth factors
Fiber, 622
Fibrillin, 527, 531
    Marfan syndrome caused by mutations in gene for, 531, 531f
Fibrils, collagen, 527–530, 528f, 529t
Fibrin
    dissolution of by plasmin, 588–589, 588f
    formation of, 583–584, 584t
        thrombin in, 585–586, 587t
    in thrombi, 583
Fibrin deposit, 583
Fibrin mesh, formation of, 583
Fibrin split products, in inflammation, 604t
Fibrin stabilizing factor (factor XIII), 586t
Fibrinogen, 586f
Fibrinogen (factor I), 566, 585t, 586t
    conversion of to fibrin, 585–586
Fibrinoligase (factor XIII), 585t, 586t
Fibrinolysis, 588
Fibrinopeptides A and B, 586, 587f
Fibroblast growth factor receptor 3, achondroplasia caused by mutation in gene for, 541t, 543, 543f
Fibroblast growth factor receptors, chondrodysplasias caused by mutation in gene for, 541t, 543, 543f
Fibroblast growth factors (FGFs), 543–544
Fibronectin, 527, 529, 531–532, 532f
Fibrous proteins, 31
    collagen as, 40
Figlu. See Formiminoglutamate
Fingerprinting, DNA, 404
FISH. See Fluorescence in situ hybridization
Fish-eye disease, 232t
5′ cap, mRNA modification and, 349
Flavin adenine dinucleotide (FAD), 98, 289t, 474
    in citric acid cycle, 146
Flavin mononucleotide (FMN), 52, 98, 474
Flavoproteins
    electron-transferring, 100
    as oxidases, 98–99, 99f
    in respiratory chain complexes, 100, 103
Flip-flop, 410
Flip-flop, phospholipid, membrane asymmetry and, 410
Flippases, membrane asymmetry and, 411
Fluid mosaic model, 411f, 412–413
Fluid-phase pinocytosis, 421, 421f
Fluidity, membrane, 412–413

Fluorescence, of porphyrins, 274–275, 276f
Fluorescence in situ hybridization (FISH), 396
    in gene mapping, 396t
Fluoride, 480t
    in glycolysis, 150
1-Fluoro-2,4-dinitrobenzene (Sanger reagent), for polypeptide sequencing, 25
Fluoroacetate, 143, 145f
5-Fluorouracil, 290f, 297
Fluvastatin, 232
Flux-generating reaction, 137
FMN. See Flavin mononucleotide
Folate. See Folic acid
Folate trap, 476f, 477–478
Folding
    polar and charged group positioning and, 7
    protein, 22f, 38–39, 39f
        after denaturation, 39
Folic acid (folate/pteroylglutamic acid), 468t, 477–478, 477f, 478f
    coenzymes derived from, 52
    deficiency of, 250, 468t, 478
        functional, 477–478
    forms of in diet, 477–478, 477f, 478f
    inhibitors of metabolism of, 477
    supplemental, 478
Folinic acid, 477
Follicle-stimulating hormone (FSH), 425, 428t, 429f
Footprinting, DNA, 404
Forbes disease, 160t
Forensic medicine
    polymerase chain reaction (PCR) in, 395
    restriction fragment length polymorphisms (RFLPs) in, 400
    variable numbers of tandemly repeated units (VNTRs) in, 400
Forkhead transcription factor, 37f
Formic acid, pK/pK_a value of, 13t
Formiminoglutamate, in histidine catabolism, 250, 250f
Formyl-tetrahydrofolate, 477, 477f, 478f
Four-helix bundle, 500
Fourier synthesis, 37
43S initiation complex, in protein synthesis, 359–361, 360f
43S preinitiation complex, in protein synthesis, 359, 360f
FPA/FPB. See Fibrinopeptides A and B
Fractions, 23f
Frameshift mutations, 357–358, 358f
    ABO blood group and, 603
Framework regions, 577
Free amino acids, absorption of, 462
Free energy
    changes in, 92
        chemical reaction direction and, 62–63
        coupling and, 93–94, 93f
        enzymes affecting, 65
        equilibrium state and, 62–63
        redox potential and, 98, 99t
        transition states and, 63–64
    of hydrolysis of ATP, 94, 94t
Free fatty acids, 121, 184, 212, 213t
    in fatty liver, 218
    glucose metabolism affecting, 220–221
    insulin affecting, 221
    ketogenesis regulation and, 189–191, 190f
    lipogenesis affected by, 197, 198f
    metabolism of, 213–214
        starvation and, 139–140, 140f, 140t

Free polyribosomes, protein synthesis on, 487, 495. *See also* Polyribosomes
Free radicals (reactive oxygen species). *See also* Antioxidants
  causing damage, 482–483, 483f
  hydroperoxidases in protection against, 100
  in kwashiorkor, 464
  lipid peroxidation producing, 128–129, 128f
  multiple sources of oxygen, 483–484
  in oxygen toxicity, 102, 597, 597t
  protection mechanisms against damage, 484–485
  as self-perpetuating chain reactions, 482
  xenobiotic cell injury and, 613–614, 614f
D-Fructofuranose, 115f
Fructokinase, 179, 179f
  deficiency of, 182
D-Fructopyranose, 115f
Fructose
  absorption of, 460, 460f
  in diabetic cataract, 182
  glycemic index of, 459
  hepatic
    hypertriacylglycerolemia/hypercholesterolemia/ hyperuricemia and, 182
    metabolism affected by, 178–179, 179f
  iron absorption affected by, 463
  metabolism of, 178–179, 179f
    defects in, 182
  pyranose and furanose forms of, 115f
D-Fructose, 116t
Fructose-1 deficiency, 182
Fructose-1,6-bisphosphatase, 176
Fructose-1,6-bisphosphate
  in gluconeogenesis, 165, 166f
  in glycolysis, 150, 151f
Fructose-2,6-bisphosphatase, 168–168, 169f
  in covalent catalysis, 56, 56f
Fructose-2,6-bisphosphate, 168–168, 169f
Fructose 6-phosphate
  free energy of hydrolysis of, 94t
  in gluconeogenesis, 165, 166f
  in glycolysis, 150, 151f
Fructose intolerance, hereditary, 182
Fructosuria, essential, 174, 182
FSF. *See* Fibrin stabilizing factor
FSH. *See* Follicle-stimulating hormone
Fucose, in glycoproteins, 508t
Fucosidosis, 524, 524t
Fucosylated oligosaccharides, selectins binding, 521
Fucosyltransferase/fucosyl (Fuc) transferase, 603
Fuels, metabolic. *See* Metabolic fuels
Fumarase (fumarate hydratase), 144, 145f
Fumarate, 144, 145f
  in tyrosine catabolism, 253f, 254f
  in urea synthesis, 244
Fumarylacetoacetate, in tyrosine catabolism, 253f, 254
Fumarylacetoacetate hydrolase, defect at, in tyrosinemia, 254
Functional groups
  amino acid chemical reactions affected by, 18–20
  amino acid properties affected by, 18
  physiologic significance of, 10–11
  pK of, medium affecting, 12–13
Furanose ring structures, 114, 114f
Furin, 502
Fusion, of vesicles, 500

Fusion proteins, recombinant, in enzyme study, 60–61, 60f
FXR. *See* Farnesoid X receptor

G-Actin, 546, 548f, 562
  in nonmuscle cells, 562
G-CSF. *See* Granulocyte colony-stimulating factor
G protein-coupled receptors (GPCRs), 446–447, 446f
G proteins, 447, 448t
  in calcium-dependent signal transduction, 450, 451f
  in cAMP-dependent signal transduction, 447, 448t
  in Jak/STAT pathway, 453
  in respiratory burst, 606
GAGs. *See* Glycosaminoglycans
Gal-Gal-Xyl-Ser trisaccharide, 510
Gal-hydroxylysine (Hyl) linkage, 510
Gal transferase, 603, 603f
Galactokinase, 179, 180f
  inherited defects in, 182
Galactosamine, 180, 181f
D-Galactosamine (chondrosamine), 116
Galactose, 113, 179–180, 180f, 525
  absorption of, 460, 460f
  glycemic index of, 459
  in glycoproteins, 508t
  metabolism of, 179–180, 180f
    enzyme deficiencies and, 182
D-Galactose, 115f, 116t
Galactose → α 2,6-NeuAc, 525
Galactose 1-phosphate uridyl transferase, 180, 180f
Galactosemia, 113, 174, 182
Galactosidases, in glycoprotein analysis, 509
Galactoside, 116
Galactosylceramide, 126, 126f, 208, 210t
GalCer. *See* Galactosylceramide
Gallstones, 459
  cholesterol, 224
GalNAc, in glycoproteins, 508, 508t
GalNAc-Ser(Thr) linkage
  in glycoproteins, 510, 511f
  in glycosaminoglycans, 534
GalNAc transferase, in ABO system, 603, 603f
Gamma-globulin, 567f
γ-Hydroxybutyrate metabolism of, 269, 269f
Ganglioside GM, 620
Ganglioside GM1, 620
Gangliosides, 126
  amino sugars in, 116, 181f
  sialic acids in, 119
  synthesis of, 209, 210f
Gap junctions, 422
  schematic diagram of, 422f
GAPs. *See* Guanine activating proteins
Gastric lipase, 460
Gastroenteropathy, protein-losing, 568
Gated ion channels, 416, 417–418
Gaucher disease, 210t
GDH. *See* Glutamate dehydrogenase/L-glutamate dehydrogenase
GDP, 494
GDP-Fuc, 508t
GDP-Man, 508, 508t
GEFs. *See* Guanine nucleotide exchange factors
Gel electrophoresis, polyacrylamide, for protein/peptide purification, 23–24, 24f, 25f

Gel filtration, for protein/peptide purification, 21–22, 24f
Gemfibrozil, 232
GenBank, 86
GenBank, UniProt, and Protein Database (PDB), 86
Gender, xenobiotic-metabolizing enzymes affected by, 613
Gene. *See* Genes; Genome
Gene amplification, in gene expression regulation, 384–385, 385f
Gene array chips, protein expression and, 29
Gene coding for dystrophin, 629
Gene conversion, 321
Gene deletions and duplications, 628
Gene disruption/knockout, targeted, 400
Gene expression
  constitutive, 370, 371
  miRNA and siRNA inhibition of, 310
  in pyrimidine nucleotide synthesis, regulation of, 297
  regulation of, 369–387
    alternative splicing and, 348, 348f, 386
    eukaryotic transcription and, 376–381
    hormones in, 445f
    negative *vs.* positive, 369, 370t, 371–372, 375
    in prokaryotes *vs.* eukaryotes, 384–387, 384t
    regulatory protein DNA binding and trans-activation domains and, 382–383, 384f
    regulatory protein-DNA binding motifs and, 381–382, 381t, 382f, 383f
    retinoic acid in, 469
    temporal responses and, 369–370, 370f
Gene knockout studies, 518
Gene mapping, 317, 396, 396t
Gene products, diseases associated with deficiency of, 396t
Gene therapy, 4, 400, 504, 625
  and level of expression, 617
  for urea biosynthesis defects, 246
Gene transcription GeneCards database, 87. *See* Transcription
General acid/base catalysis, 53
Genes, 626
  alteration of, 319–322, 321f, 322f
  amplification of, in gene expression regulation, 319–322, 321f, 322f
  disease causing, recombinant DNA technology in detection of, 396, 397f, 398t
  heterogeneous nuclear RNA processing in regulation of, 349
  housekeeping, 371
  immunoglobulin, DNA rearrangement and, 322, 386, 580
    double-strand break repair and, 332–333
  inducible, 370
  knockout, 400
  previously unknown, 3
  processed, 321
  reporter, 379–380, 380f
  targeted disruption of, 400
  variations in
    normal, recombinant DNA techniques for identification of, 396
Genetic code, 302, 353–368, 354t
  features of, 354, 354t
Genetic contribution, 635
Genetic counseling, 629
Genetic diseases. *See also* specific diseases
  diagnosis of
    enzymes in, 60

recombinant DNA technology in, 396–400, 397f, 398t, 399f
gene therapy for, 400
Genetic influences, on obesity, 641
Genetic linkage. *See* Linkage analysis
Genetic susceptibility, 627
Genetic syndromes, 622
Genetic testing, 633
Genetic variations, 569
Genetically determined differences, 613
Genetics, 1
  molecular, 1
  xenobiotic-metabolizing enzymes affected by, 613
Genevan system, for fatty acid nomenclature, 121
Genome
  and medicines, in bioinformatics, 85
  redundancy in, 317–319
  removal of gene from (targeted gene disruption/knockout), 400
Genomic library, 392, 404
Genomic resources, 86–87
Genomic technology, 388. *See also* Recombinant DNA/recombinant DNA technology
Genomics, 84
  protein sequencing and, 28
Genomics revolution, 85
Genital tract, 625
Geometric isomerism, of unsaturated fatty acids, 122, 123f
Geranyl diphosphate, in cholesterol synthesis, 224, 226f
Geranylgeranyl, in vesicle coating, 502
Ghosts, red cell membrane analysis and, 599
Gibbs change in free energy, 92
Gibbs free energy/Gibbs energy. *See* Free energy
Gilbert syndrome, 282
GK (glucokinase) gene, regulation of, 349, 349f
GlcCer. *See* Glucosylceramide
GlcNAc phosphotransferase, 523, 524
GlcNAc. *See* N-Acetylglucosamine (GlcNAc)
Glial fibrillary acid protein, 563t
Glibenclamide. *See* Glyburide
Globin, 278
Globular proteins, 31
Globulins, 566
Glomerular filtration, basal lamina in, 533
Glomerular membrane, laminin in, 532–533
Glomerulonephritis, 533
Glomerulus, renal, laminin in, 532–533
Glucagon, 131, 160, 171
  in fasting state, 139
  in gluconeogenesis regulation, 167
  in lipogenesis regulation, 197, 198f
Glucagon/insulin ratio, in ketogenesis regulation, 190
Glucagon-like peptide, 427
Glucan (glucosan), 117
Glucan transferase, in glycogenolysis, 158, 158f, 159f
Glucocorticoid receptor-interacting protein (GRIP1 coactivator), 457, 457t
Glucocorticoid response element (GRE), 444, 445f, 445t
Glucocorticoids, 428. *See also* specific type
  in amino acid transport, 116
  blood glucose affected by, 171
  in lipolysis, 221f, 222
  receptors for, 455

synthesis of, 430–431, 431f
transport of, 442–443, 442t
D-Glucofuranose, 114f
Glucogenic amino acids, 137
Glucokinase, 168t
  in blood glucose regulation, 170, 171f
  in glycogen biosynthesis, 157, 158f, 168t
  in glycolysis, 150, 151f, 168t
Glucokinase gene, regulation of, 349, 349f
Gluconeogenesis, 131, 132, 133, 165–173, 166f, 630
  blood glucose regulation and, 169–172, 170f, 171f
  citric acid cycle in, 146–147, 146f, 165–167, 166f
  energy cost of, in weight loss from very low carbohydrate diets, 173
  in glycolysis, 149–153, 151f, 152f, 165–167, 166f
    regulation of, 152–153
  regulation of, 167–170, 168t, 169f
    allosteric modification in, 167–168
    covalent modification in, 167
    enzyme induction/repression in, 167, 168t
    fructose 2,6-bisphosphate in, 168–169, 169f
    substrate (futile) cycles in, 169
    thermodynamic barriers to glycolysis and, 165–167, 166f
Gluconeogenic amino acid, 240
Gluconolactone hydrolase, 174, 176f
D-Glucopyranose, 114f
Glucosamine, 117f, 180, 181f
  in heparin, 536
Glucosan (glucan), 117
Glucose, 113–120, 629
  absorption of, 459, 460, 460f
  as amino sugar precursor, 180, 181f
  epimers of, 115, 115f
  in extracellular and intracellular fluid, 407, 407t
  furanose forms of, 114, 114f
  galactose conversion to, 179–180, 180f
  glycemic index of, 459
  in glycogen biosynthesis, 157, 158f
  in glycoproteins, 508t
  insulin secretion and, 170–171, 172
  interconvertibility of, 137
  isomers of, 113–115, 114f
  as metabolic necessity, 137
  permeability coefficient of, 409f
  pyranose forms of, 114, 114f
  renal threshold for, 172
  structure of, 113, 114f
  transport of, 169–170, 171f, 419–420, 420f, 460, 460f
    insulin affecting, 415–416
  uptake of, 138
D-Glucose, 114f, 115f, 116t
L-Glucose, 114f
Glucose-alanine cycle, 170
Glucose, blood
  normal, 157
  regulation of
    clinical aspects of, 172–173, 172f
    diet/gluconeogenesis/glycogenolysis in, 170–172, 170f, 171f
    glucagon in, 171
    glucokinase in, 170, 171f
    glycogen in, 157
    insulin in, 170–171
    limits of, 169
    metabolic and hormonal mechanisms in, 170, 171, 171t

Glucose metabolism, 132–133, 132f, 134f, 135f, 149–153, 151f, 152f, 169, 170f. *See also* Gluconeogenesis; Glycolysis
  ATP generated by, 155, 155t
  in fed state, 137–139
  free fatty acids and, 220–221
  insulin affecting, 170–171, 172
  by pentose phosphate pathway, 132, 174–177, 175f, 176f, 177f
  starvation and, 139–140, 140f, 140t
Glucose 1-phosphate
  free energy of hydrolysis of, 94t
  in gluconeogenesis, 165–167, 166f
Glucose-6-phosphatase
  deficiency of, 160t, 298
  in gluconeogenesis, 168t
  in glycogenolysis, 159
Glucose 6-phosphate, 159
  free energy of hydrolysis of, 94t
  in gluconeogenesis, 164, 166f
  in glycogen biosynthesis, 157, 158f
  in glycolysis, 150, 151f
Glucose-6-phosphate dehydrogenase
  deficiency of, 174, 180–181, 597–598, 598, 598f, 633t
  in pentose phosphate pathway, 174, 175f, 176f
Glucose synthesis, fatty acids and, 137
Glucose tolerance, 172, 172f
Glucose transporters, 170, 171t
  in blood glucose regulation, 170, 171, 221
  insulin affecting, 416
  red cell membrane, 596, 596t
Glucoside, 116
Glucosuria, 172, 627
Glucosylceramide, 126, 208, 210f
Glucosyltransferase, 516
D-Glucuronate, 115, 116f
Glucuronate/glucuronic acid, 177, 178f
  bilirubin conjugation with, 278–279, 280f
Glucuronidation
  of bilirubin, 278–279, 280f
  of xenobiotics, 611
Glucuronides, 174
GLUT 1-4. *See* Glucose transporters
Glutamate
  carboxylation of, vitamin K as cofactor for, 473, 473f
  catabolism of, 248, 248f, 249f
  in proline synthesis, 236, 237f
  synthesis of, 235, 235f
  transamination and, 241–242, 241f, 242f
  in urea biosynthesis, 241–242, 241f, 242f
Glutamate aminotransferase, 241–242
Glutamate/aspartate transporter, 111, 111f
L-Glutamate decarboxylase, 266, 269f
Glutamate dehydrogenase/L-glutamate dehydrogenase, 235, 235f
  in nitrogen metabolism, 242–243, 242f
Glutamate r-semialdehyde, 237f
Glutamic acid, 15t
Glutaminase, in amino acid nitrogen catabolism, 242–243
Glutaminase reaction, 243f
Glutamine, 15t, 240
  in amino acid nitrogen catabolism, 242–243
  catabolism of, 248, 248f, 249f
  synthesis of, 235, 235f, 243f
Glutamine analogs, purine nucleotide synthesis affected by, 293
Glutamine synthetase/synthase, 235, 235f, 242, 243f

Glutamyl amidotransferase, PRPP, regulation of, 293, 294f

γ-Glutamyl phosphate, 235

Glutaric acid, pK/pKₐ value of, 13t

Glutathione
  as antioxidant, 597, 597t
  in conjugation of xenobiotics, 612
  as defense mechanism, 612
  functions of, 612

Glutathione peroxidase, 100, 177, 177f, 181, 597, 597t

Glutathione reductase, erythrocyte
  pentose phosphate pathway and, 177, 177f
  riboflavin status and, 474

Glutathione S-transferases, 612
  in enzyme study, 60f, 61

Glyburide (glibenclamide), 191

N-Glycan chains, 495

Glycan intermediates, formation of during N-glycosylation, 516–517

Glycans, virus and bacteria bound by, 524–525

Glycation, 506, 519

Glycemic index, 118, 459

Glyceraldehyde (glycerose), D and L isomers of, 114f

Glyceraldehyde 3-phosphate
  in glycolysis, 150, 151f
  oxidation of, 150, 152f

Glyceraldehyde 3-phosphate dehydrogenase
  in glycolysis, 150, 151f
  in red cell membranes, 600t

Glycerol, 124
  permeability coefficient of, 409f
  synthesis of, 167

Glycerol-3-phosphate
  acylglycerol biosynthesis and, 205, 206f
  electron transfer via, 104
  free energy of hydrolysis of, 94t
  triacylglycerol esterification and, 220, 220f

Glycerol-3-phosphate acyltransferase, 206f, 207

Glycerol-3-phosphate dehydrogenase, 206f, 207
  mitochondrial, 100

Glycerol ether phospholipids, synthesis of, 207, 208f

Glycerol kinase, 205, 206f, 220

Glycerol moiety, of triacylglycerols, 132

Glycerol phosphate pathway, 206f

Glycerophosphate shuttle, 110f, 111

Glycerophospholipids, 121

Glycerose (glyceraldehyde), D and L isomers of, 114f

Glycine, 15t, 262
  catabolism of, pyruvate formation and, 250–251, 251f
  in collagen, 527
  in heme synthesis, 270, 279–282, 273f, 283f, 276f
  synthesis of, 236, 236f

Glycine cleavage complex, 250–251

Glycinuria, 251

Glycobiology, 506

Glycocalyx, 120

Glycochenodeoxycholic acid, synthesis of, 231f

Glycocholic acid, synthesis of, 231f

Glycoconjugate, 506

Glycoconjugate (complex) carbohydrates, glycoproteins as, 506

Glycoforms, 507

Glycogen, 111, 118, 118f
  in carbohydrate metabolism, 132, 133f, 165–167

carbohydrate storage and, 157, 158t
metabolism of, 155–159. See also Glycogenesis; Glycogenolysis
  branching in, 157, 159f
  clinical aspects of, 160t, 163
  regulation of
cyclic AMP in, 160–162, 160f, 161f, 162f
glycogen synthase and phosphorylase in, 162–163, 163f
  in starvation, 140–141
muscle, 138, 157, 158t, 560–561, 560f
synthesis of, 137

Glycogen phosphorylase, 157–158, 158f, 560
  pyridoxal phosphate as cofactor for, 475
  regulation of, 162, 162f, 163f

Glycogen storage diseases, 113, 157, 160t, 163

Glycogen synthase, in glycogen metabolism, 157, 158f, 165, 168t
  regulation of, 160–162, 162f, 163f

Glycogen synthase a, 162, 162f

Glycogen synthase b, 162, 162f

Glycogenesis, 132, 157, 158f
  regulation of
cyclic AMP in, 160f, 161f, 162–163, 162f
enzymes in, 168t
glycogen synthase and phosphorylase in, 162, 162f, 163f

Glycogenin, 157, 158f

Glycogenolysis, 134, 157–159, 158f
  blood glucose regulation and, 169–171, 170f, 171f
  cyclic AMP in regulation of, 160f, 161f, 162–163, 162f
  cyclic AMP-independent, 160, 162
  debranching enzymes in, 157, 159f
  glycogen synthase and phosphorylase in regulation of, 162–163, 162f, 163f

Glycolipid storage diseases, 205

Glycolipids (glycosphingolipids), 121, 125–126, 126f
  ABO blood group and, 602–603
  amino sugars in, 180, 181f
  galactose in synthesis of, 179–180, 180f

Glycolysis, 95, 132, 132f, 149–156, 150f
  aerobic, 150
  anaerobic, 149, 150, 150f
    as muscle ATP source, 560f, 561, 561t
  ATP generated by, 154–155, 155t
  clinical aspects of, 155
  in erythrocytes, 152f, 153
  glucose utilization/gluconeogenesis and, 149–152, 151f, 152f, 165–167, 166f. See also Gluconeogenesis
  pathway of, 149–153, 151f, 152f
  pyruvate oxidation and, 146, 147f, 153f, 154–155, 154f, 155t
  regulation of, 154–155
    enzymes in, 168t
    fructose 2,6-bisphosphate in, 168–169, 169f
    gluconeogenesis and, 152–153, 167–169, 168t, 169f
  at subcellular level, 135, 136f
  thermodynamic barriers to reversal of, 165–167

Glycolytic enzymes, in muscle, 545

Glycome, 506

Glycomics, 4, 506, 525

Glycophorins, 120, 510, 600–601, 600t, 602f

Glycoprotein glycosyltransferases, 512, 518

Glycoprotein IIb-IIIa, in platelet activation, 590, 605t

Glycoproteins, 31, 119, 119t, 429f, 506–526, 566, 567. See also specific type and Plasma proteins
  amino sugars in, 116, 180, 181f
  asialoglycoprotein receptor in clearance of, 509
  as blood group substances, 506, 602, 603
  carbohydrates in, 119t
  classes of, 509–510, 511f
  complex, 512, 513f
    formation of, 512, 513
  diseases associated with abnormalities of, 522–523, 522t, 523f
  extracellular, absorptive pinocytosis of, 421
  in fertilization, 520
  functions of, 506, 507t, 519–524, 521t
  galactose in synthesis of, 179–180, 180f
  glycosylphosphatidylinositol-anchored, 510, 511f, 518–519, 519t
  high-mannose, 512–513, 513f
    formation of, 512, 513
  hybrid, 512–513, 513f
    formation of, 512
  immunoglobulins as, 580
  membrane asymmetry and, 411
  N-linked, 510, 511f, 512–518
  nucleotide sugars, 508–509, 508t
  O-linked, 510–512, 512t, 513f, 513f
  oligosaccharide chains of, 506
  red cell membrane, 600, 601f
  sugars in, 507–508, 508t
  techniques for study of, 506–507, 507t
    asialoglycoprotein receptor in, 509
    glycosidases in, 509, 509t
    lectins in, 509, 510t
  in zona pellucida, 520

Glycosaminoglycans, 119, 119f, 534–537, 538t. See also specific type
  amino sugars in, 116
  deficiencies of enzymes in degradation of, 536–537, 537f
  disease associations of, 539
  distributions of, 535–536, 534f, 535t, 536f
  functions of, 537–539, 538t
  structural differences among, 535–536, 534f, 535t, 536f
  synthesis of, 533–535

Glycosidases, in glycoprotein analysis, 509, 509t

Glycosides, 116

N-Glycosides, heterocyclic, 286

N-Glycosidic linkage, 510

Glycosphingolipids (glycolipids), 121, 125–126, 126f, 208–209, 210f, 503
  ABO blood group and, 602
  amino sugars in, 180, 181f
  galactose in synthesis of, 179–180, 180f
  in membranes, 411
    membrane asymmetry and, 410

N-Glycosylases, in base excision-repair, 331, 332f

Glycosylated hemoglobin (HbA₁c), 50

Glycosylation, 506, 519
  of collagen, 528
  congenital disorders of, 522, 522t, 571
  cotranslational, 495
  in covalent modification, mass increases and, 26t
  Golgi apparatus in, 487
  inhibitors of, 518, 518t
  nucleotide sugars in, 508–509, 508t

N-Glycosylation, 512–518, 513f, 514f, 515f, 516t
  dolichol-P-P-oligosaccharide in, 512–516, 513f

in endoplasmic reticulum, 515f, 516
glycan intermediates formed during, 516–517
in Golgi apparatus, 515f, 516
inhibition of, 518, 518t
regulation of, 517–518, 518t
tunicamycin affecting, 518, 518t
O-Glycosylation, 512, 512t
Glycosylphosphatidylinositol (GPI), 502
Glycosylphosphatidylinositol-anchored (GPI-anchored/GPI-linked) glycoproteins, 510, 511f, 518–519, 519t
in paroxysmal nocturnal hemoglobinuria, 522, 523f
Glycosyltransferases, 553
Glycosyltransferases, glycoprotein, 512, 518
Glypiation, 519
GM-CSF. See Granulocyte-macrophage colony-stimulating factor
GM$_1$ ganglioside, 126, 126f
GM$_3$ ganglioside, 126
GMP, 287t, 296f
cyclic, 288f, 289f
as second messenger, 289, 427, 428t, 446, 449
IMP conversion to, 292, 294f
feedback-regulation of, 293, 296f
PRPP glutamyl amidotransferase regulated by, 293
Golgi apparatus, 487
appearing to collapse into ER, 502
core protein synthesis in, 534–535
glycosylation and, 487, 515f, 516
lumen of, 495
in membrane synthesis, 487
in protein sorting, 487, 489f, 487
proteins destined for membrane of, 487, 495
retrograde transport from, 495, 496
in VLDL formation, 219f
Gout, acute, case study, 630–632
Gout, chronic, 631, 631f
Gout/gouty arthritis, 297, 631
GPCRs. See G protein-coupled receptors
GPI-anchored/linked glycoproteins, 510, 511f, 518–519, 519t
in paroxysmal nocturnal hemoglobinuria, 519, 522, 523f
GPIIb-IIIa, in platelet activation, 590, 605t
Granulocyte-colony stimulating factor, 595
Granulocyte-macrophage colony-stimulating factor, 595
Granulomatous disease, chronic, 606, 606f
Granulosa cells, hormones produced by, 432
Gratuitous inducers, 372
GRE. See Glucocorticoid response element
GRIP1 coactivator, 457, 457t
Griseofulvin, 563
Group transfer potential, 94, 95f
of nucleoside triphosphates, 288f, 289, 289f, 289t
Group transfer reactions, 8–9
Growth factors, hematopoietic, 595. See also specific growth factors
Growth hormone, 426, 427
amino acid transport affected by, 416
localization of gene for, 396t
receptor for, 426–427
GSH. See Glutathione
GSLs. See Glycosphingolipids
GST (glutathione S-transferase) tag, in enzyme study, 60f, 61

GTP, 288–289, 494, 495, 500
binding proteins, 225
cyclic GMP formed from, 449
in phosphorylation, 96
GTPases, 447, 491
small monomeric, 491, 502
GTP-bound state, 502
Guanine, 287t
Guanine activating proteins, 491, 492f
Guanine nucleotide exchange factors, 491, 492f
Guanosine, 286f, 287t
base pairing of in DNA, 302, 303, 304f
in uric acid formation, 297, 299f
Guanosine diphosphate fucose (GDP-Fuc), 508t
Guanosine diphosphate mannose (GDPMan), 508, 508t
Guanosine monophosphate. See GMP
Guanylyl cyclase, 449
L-Gulonolactone oxidase, 177
Gyrate atrophy of retina, 250

H bands, 545, 546f, 547f
H blood group substance, 603, 603f
H chains. See Heavy chains
H substance, ABO blood group and, 603
H1 histones, 312, 313f
H2A histones, 312, 313f
H2B histones, 312, 313f
H2S. See Hydrogen sulfide
H3 histones, 312, 313f
H4 histones, 312, 313f
H5N1, 524
Haber-Weiss reaction, 597
Hageman factor (factor XII), 584f, 585, 585t
Hairpin, 306, 307f, 404, 495
Half-life
enzyme, 239
protein, 239
plasma protein, 568
Halt-transfer signal, 495
Haplotype, 87
Haplotype map (HapMap), 87
HapMap database, 86–87
Hapten, in xenobiotic cell injury, 613, 614f
Haptoglobin, 569t, 570, 570f
Haptoglobin-related protein, 570
Hartnup disease, 257, 475
Hashimoto disease, 635
HAT activity. See Histone acetyltransferase activity
Haworth projection, 113, 114f
HbA (hemoglobin A), P$_{50}$ of, 46
HbA$_{1c}$ (glycosylated hemoglobin), 50
HbF (fetal hemoglobin), P$_{50}$ of, 46
HbM (hemoglobin M), 49, 357, 599
HbS (hemoglobin S), 49, 49f, 357
hCG. See Human chorionic gonadotropin
HDL. See High-density lipoproteins
Health, 1
normal biochemical processes as basis of, 2–4, 3t
Heart
developmental defects of, 557
metabolism in, 141t
thiamin deficiency affecting, 473–474
Heart disease, coronary (ischemic), 639. See also Atherosclerosis
cholesterol and, 230
Heart failure, 125, 545
in thiamin deficiency, 473–474
Heat, from respiratory chain, 106–108
Heat-shock proteins, as chaperones, 39, 497

Heavy chains
immunoglobulin, 576–577, 577f
genes producing, 602
myosin, 556–557
familial hypertrophic cardiomyopathy caused by mutations in gene for, 556–557, 557f
Heavy meromyosin, 548, 549f
Heinz bodies, 598
Helicases, DNA, 323f, 323t, 324
Helicobacter pylori
glycan binding of, 525, 525f
ulcers associated with, 459
Helix
double, of DNA structure, 8, 303, 304f
triple, of collagen structure, 40–41, 41f, 527–530, 528f
Helix-loop-helix motifs, 35
Helix-turn-helix motif, 381–382, 381t, 382f
Hemagglutinin, influenza virus, 524
calnexin binding to, 516, 517f
Hematology, recombinant DNA technology affecting, 607
Hematopoietic growth factors, 595
Hematopoietic stem cells, 593
Heme, 41, 44f, 271
catabolism of, bilirubin produced by, 278, 279f
in proteins, 271. See also Heme proteins
synthesis of, 272–274, 273f, 274f, 275f, 276f
disorders of (porphyrias), 268, 271, 276f, 277t
Heme binding, 570
Heme iron, 278, 593
absorption of, 463, 570–571, 570f
hindered environment for, 43–44, 44f
Heme oxygenase system, 278, 279f
Heme proteins (hemoproteins), 271t, 272. See also Hemoglobin; Myoglobin
cytochrome P450 isoforms as, 610
Heme synthase (ferrochelatase), 273, 274f
in porphyria, 277t
Hemiacetal, 113
Hemiconnexin, 422, 422f
Hemin, 278, 279f
Hemochromatosis, 463
arthropathy of, 633
case study 632–634
hereditary, 633
penetrance of, 633
secondary, 633
Hemodialysis, 629
Hemoglobin, 43–50, 567f
allosteric properties of, 44–48
bilirubin synthesis and, 278, 279f
in carbon dioxide transport, 47, 48f
extracorpuscular, haptoglobin binding of, 569t, 570
glycosylated (HbA$_{1c}$), 50
mutations, 48–49, 356
oxygen affinities (P$_{50}$) and, 46, 46f
oxygen dissociation curve for, 44, 45f
in oxygen transport, 43–44
oxygenation of
conformational changes and, 46–47, 46f, 47f
apoprotein, 46
2,3-bisphosphoglycerate stabilizing, 48, 48f
high altitude adaptation and, 48
mutant hemoglobins and, 48–49
in proton transport, 47
tetrameric structure of, 45
changes in during development, 45, 45f

Hemoglobin A (HbA), P_{50} of, 46
Hemoglobin A (HbA$_{1c}$), 519
Hemoglobin A$_{1c}$ (glycosylated hemoglobin), 50
Hemoglobin Bristol, 356
Hemoglobin Chesapeake, 49
Hemoglobin F (fetal hemoglobin), P_{50} of, 46
Hemoglobin Hikari, 357, 357f
Hemoglobin M, 49, 357, 599
Hemoglobin Milwaukee, 356
Hemoglobin S, 49, 49f, 357
Hemoglobin Sydney, 356
Hemoglobinopathies, 48–49, 598
Hemoglobinuria, paroxysmal nocturnal, 423t, 519,
        522, 522t, 523f
Hemojuvelin, 593
Hemolysins, 598
Hemolytic anemias, 149, 155, 180, 594, 598–599,
        599t
    causes, 598–599
    glucose-6-phosphate dehydrogenase deficiency
        causing, 174, 180–181, 597–598,
        598f, 607
    haptoglobin levels in, 570
    hyperbilirubinemia/jaundice in, 281, 283t
    inside red blood cells, 598
    outside cell membrane, 598
    pentose phosphate pathway/glutathione
        peroxidase and, 177, 177f, 180–181
    primaquine-sensitive, 598
    red cell membrane abnormalities causing,
        598–599
    within membrane, 598
Hemopexin, 569t
Hemophilia A, 588
Hemophilia B, 588
Hemoproteins. See Heme proteins
Hemosiderin, 572
Hemostasis, 583–592. See also Coagulation
    laboratory tests in evaluation of, 591
    pathways of, 607f
    phases of, 583
HEMPAS. See Hereditary erythroblastic multi-
        nuclearity with a positive acidified
        lysis test
Henderson-Hasselbalch equation, 11–12
Heparan sulfate, 529, 533, 534, 534f, 535f, 536,
        537t, 539
    in basal lamina, 532
    clotting/thrombosis affected by, 591, 591t
Heparin, 119, 119f, 535t, 536, 536f, 587
    antithrombin III activity affected by, 539, 588
    in basal lamina, 532
    binding of, fibronectin in, 532, 532f
    functions of, 539
    lipoprotein and hepatic lipases affected by, 215
Heparin cofactor II, as thrombin inhibitor,
        587
Hepatic ALA synthase (ALAS1), 274
    in porphyria, 277, 277t
Hepatic lipase, 215
    in chylomicron remnant uptake, 216, 216f
    deficiency of, 232t
Hepatic portal system, 169
    in metabolite circulation, 134, 134f
Hepatic purine biosynthesis, 293, 295, 296f
Hepatitis, 143
    jaundice in, 282f, 283t
Hepatobiliary diseases, 612
Hepatocellular carcinoma, 633
Hepatocytes

glycoprotein clearance from, asialoglycoprotein
        receptor in, 509
    heme synthesis in, 274
        ALA synthase in regulation of, 274, 276f
Hepatolenticular degeneration (Wilson disease),
        423t, 572, 573, 574
    ceruloplasmin levels in, 572–573
    gene mutations in, 423t, 574
Hepcidin, 463f, 571, 633
    inhibits iron absorption by enterocytes, 633
    ferroportin, 633
    release of iron from macrophages, 633
Hephaestin, 571
Heptoses, 113, 114t
Hereditary elliptocytosis, 598, 602
Hereditary erythroblastic multinuclearity with
        a positive acidified lysis test
        (HEMPAS), 522, 522t
Hereditary hemochromatosis, 571
    case study, 632–634
Hereditary nonpolyposis colon cancer, 622
    mismatch repair genes in, 331
Hereditary spherocytosis, 423t, 594t, 598,
        601–602, 602f
Hermansky-Pudlak syndrome, 498t
Hers disease, 160t
Heterochromatin, 315
Heterodimer, 35
Heterogeneous nuclear RNA (hnRNA)
    processing of, gene regulation and, 346
Heterotrophic organisms, 94
Hexapeptide, in albumin synthesis, 569
Hexokinase, 168t
    in blood glucose regulation, 170, 171f
    in fructose metabolism, 178–179, 179f
    in glycogen biosynthesis, 157, 168t
    in glycolysis, 149–150, 151f, 168t
        as flux-generating reaction, 137
        regulation and, 152
Hexosamines (amino sugars), 116, 117f
    glucose as precursor of, 180, 181f
    in glycosaminoglycans, 119, 180, 181f
    in glycosphingolipids, 180, 181f
    interrelationships in metabolism of, 181f
Hexose monophosphate shunt. See Pentose phos-
        phate pathway
Hexoses, 113, 114t
    in glycoproteins, 119t
    metabolism of, 174–177, 175f, 176f, 177f. See
        also Pentose phosphate pathway
        clinical aspects of, 180–181
        physiologic importance of, 115, 116t
HFE gene, 633
HFE mutations, in hemochromatosis, 573f
HGP. See Human Genome Project
HGRPT. See Hypoxanthine-guanine phosphoribo-
        syl transferase
HhaI, 389t
High altitude, adaptation to, 48
High-calorie diet, 625
High-density lipoproteins, 212, 213t
    apolipoproteins of, 216–217, 213t
    atherosclerosis and, 216–217, 230
    cycles, 217
    metabolism of, 216–217, 217f
    ratio of to low-density lipoproteins, 230
    receptor for, 216–217, 217f
High-density microarray technology, 402
High-energy phosphates, 94. See also ATP
    in energy capture and transfer, 94, 94f, 94t, 95f

as "energy currency" of cell, 95–96, 95f, 96f, 106
    symbol designating, 94
High-mannose oligosaccharides, 506–526, 513f
    formation of, 513, 516
High-molecular-weight kininogen, 584f, 585
High-performance liquid chromatography,
        reversed phase, for protein/peptide
        purification, 23
"High-throughput" screening, 57
Hill coefficient, 70
Hill equation, 69–70, 70f
Hindered environment, for heme iron, 43–44, 44f
HindIII, 389t
Hinge region, 455, 456f
    immunoglobulin, 576–577, 577f
Hippuric acid/hippurate, synthesis of, 262, 264f
Histamine, 604t
    formation of, 262
Histidase, impaired, 250
Histidine, 16t, 262, 264f
    catabolism of, 250, 250f
    conserved residues and, 57t
    decarboxylation of, 262, 264f
    in oxygen binding, 43, 44f
    requirements for, 465
Histidine E7, in oxygen binding, 43, 44f
Histidine F8
    in oxygen binding, 43, 44f
    replacement of in hemoglobin M, 49
Histidine 57, in covalent catalysis, 55, 55f
Histidinemia, 250
Histone acetyltransferase activity, of coactivators,
        457
Histone chaperones, 313
Histone covalent modification, 377
Histone dimer, 312, 313
Histone octamer, 312, 313, 313f
Histone tetramer, 312, 313
Histones, 312–313, 313f, 313t
HIV-I, glycoproteins in attachment of, 525
HIV protease, in acid-base catalysis, 54, 55f
HMG-CoA. See 3-Hydroxy-3-methylglutaryl-CoA
        (HMG-CoA)
HMM. See Heavy meromyosin
hMSH1/hMSH2, in colon cancer, 331
HNCC. See Hereditary nonpolyposis colon cancer
hnRNA. See Heterogeneous nuclear RNA
Holocarboxylase synthetase, biotin as coenzyme
        of, 479
Homeostasis
    blood in maintenance of, 566
    in ER, 497
    hormone signal transduction in regulation of,
        444, 445f
Homocarnosine, 262, 264f, 268
Homocarnosinosis, 268
Homocysteine
    in cysteine and homoserine synthesis, 236, 237f
    functional folate deficiency and, 476f, 478
Homocystinurias, 252
    vitamin B$_{12}$ deficiency/functional folate defi-
        ciency and, 476f, 478
Homodimers, 35
Homogentisate, in tyrosine catabolism, 253f, 254
Homogentisate dioxygenase/oxidase, 101
    deficiency of, in alkaptonuria, 253f, 254

Homology, 88
    conserved residues and, 56, 57t
    in protein classification, 31

Homology modeling, 38
Homopolymer tailing, 389
Homoserine, synthesis of, 236, 237f
Hormone-dependent cancer, vitamin B$_6$ deficiency and, 476
Hormone response elements, 379, 380f, 444–445, 445t, 454–455, 455f
Hormone response transcription unit, 455f
Hormone-sensitive lipase, 220, 220f
    insulin affecting, 221
Hormones, 637. See also specific hormones
    in blood glucose regulation, 170
    classification of, 427, 428t
    facilitated diffusion regulated by, 415–416
    glycoproteins as, 506
    lipid metabolism regulated by, 221–223, 221f
    in metabolic control, 137, 138f
    receptors for, 426–427, 427f, 455
        proteins as, 426–427
        recognition and coupling domains on, 426
        specificity/selectivity of, 426, 427f
    signal transduction and, 444–458
        intracellular messengers and, 446–454, 448t, 450t
        response to stimulus and, 444, 445f
        signal generation and, 444–445, 445f, 445t, 446f
        transcription modulation and, 454–458, 456f, 457f, 457t
    stimulus recognition by, 444, 445f
    storage/secretion of, 441–442, 442t
    synthesis of
        chemical diversity of, 428–429, 429f
        cholesterol in, 429–435, 429f, 430f
        peptide precursors and, 438–439
        specialization of, 427
        tyrosine in, 421, 428, 429f
    target cells for, 425–426, 426t
    transport of, 442–443, 442t
    vitamin D as, 470–471
Housekeeping genes, 371
Hp. See Haptoglobin
HpaI, 389t
HPETE. See Hydroperoxides
HPLC. See High-performance liquid chromatography
HREs. See Hormone response elements
hsp60/hsp70, as chaperones, 37
5-HT (5-hydroxytryptamine). See Serotonin
Human chorionic gonadotropin (hCG), 428
Human DNase, 625
Human evolution, 3
Human gene mutation database, 87
Human Genome Project, 3–4, 4–5
    areas of current interest, 5f
    implications, 3
Human immunodeficiency virus (HIV-I), glycoproteins in attachment of, 525
Human plasma protein proteome, 569
Hunter syndrome, 536, 537t
Hurler-Scheie syndrome, 537t
Hurler syndrome, 537t
Hyaluronic acid, 119, 119f, 534f, 535, 535t
    disease associations and, 539
    functions of, 539
Hyaluronidase, 536
Hybrid glycoproteins, 512, 513f
    formation of, 512
Hybridization, 304, 393, 396, 404
    in situ, in gene mapping, 396, 396t

Hybridomas, 580–581, 581f
Hydrocortisone. See Cortisol
Hydrogen bonds, 6, 7f
    in DNA, 302, 303, 304f
Hydrogen ion concentration. See also pH
    enzyme-catalyzed reaction rate affected by, 66, 66f
Hydrogen peroxide
    glutathione in decomposition of, 612
    as hydroperoxidase substrate, 100
    production of in respiratory burst, 606
Hydrogen sulfide, on respiratory chain, 108–109, 109f
Hydrolases, 52
    cholesteryl ester, 225
    fumarylacetoacetate, defect at, in tyrosinemia, 254
    gluconolactone, 174, 176f
    lysosomal, deficiencies of, 524, 524t
Hydrolysis (hydrolytic reactions), 8. See also specific reaction
    of bound GTP to GDP, 500
    free energy of, 94, 94t
    in glycogenolysis, 158–159, 158f, 159f
    of triacylglycerols, 205
Hydropathy plot, 410
Hydroperoxidases, 98, 100
Hydroperoxides, formation of, 201, 203f
Hydrophilic compounds, hydroxylation producing, 610
Hydrophilic portion of lipid molecule, 129, 129f
Hydrophobic effect, in lipid bilayer self-assembly, 409
Hydrophobic interaction chromatography, for protein/peptide purification, 22–23
Hydrophobic interactions, 7–8
Hydrophobic membrane domains, 35
Hydrophobic portion of lipid molecule, 129, 129f
Hydrostatic pressure, 566
3-Hydroxy-3-methylglutaryl-CoA (HMGCoA)
    in ketogenesis, 187–189, 188f
    in mevalonate synthesis, 224, 225f
3-Hydroxy-3-methylglutaryl-CoA (HMGCoA) lyase
    deficiency of, 191
    in ketogenesis, 188f, 189
3-Hydroxy-3-methylglutaryl-CoA (HMGCoA) reductase
    cholesterol synthesis controlled by, 225, 228f
    in mevalonate synthesis, 224, 225f
3-Hydroxy-3-methylglutaryl-CoA (HMGCoA) synthase
    in ketogenesis, 188f, 189
    in mevalonate synthesis, 224, 225f
l(+)-3-Hydroxyacyl-CoA dehydrogenase, 185, 186f
3-Hydroxyanthranilate dioxygenase/oxygenase, 101
Hydroxyapatite, 539
d-3-Hydroxybutyrate dehydrogenase, 187, 187f
24-Hydroxycalcidiol (24,25-dihydroxyvitamin D$_3$), in vitamin D metabolism, 470, 471f
25-Hydroxycholecalciferol (calcidiol), in vitamin D metabolism, 470, 471f
4-Hydroxydicoumarin (dicumarol), 472
18-Hydroxylase, in steroid synthesis, 430, 431f
21-Hydroxylase, in steroid synthesis, 431, 431f
27-Hydroxylase, sterol, 228
Hydroxylase cycle, 101, 102f
Hydroxylases, 101

in steroid synthesis, 430–431, 431f
Hydroxylation
    in collagen processing, 528
    in covalent modification, mass increases and, 26t
    of xenobiotics, 610–611, 612t
Hydroxylysine, synthesis of, 237–238
5-Hydroxymethylcytosine, 287, 288f
17-Hydroxypregnenolone, 431, 431f
17-Hydroxyprogesterone, 431f
17α-Hydroxyprogesterone, 431–432
Hydroxyproline
    catabolism of, 252f, 254
    synthesis of, 237–238, 238f, 527–528
    tropoelastin hydroxylation and, 530
4-Hydroxyproline dehydrogenase, defect in, in hyperhydroxyprolinemia, 254
15-Hydroxyprostaglandin dehydrogenase, 200–201
5-Hydroxytryptamine. See Serotonin
Hyperalphalipoproteinemia, familial, 232t
Hyperammonemia type 2, 244f, 245, 246
Hyperargininemia, 246
Hyperbilirubinemia, 280–283, 283t
Hypercholesterolemia, 212
    familial, 1, 232t, 423t
        LDL receptor deficiency in, 216, 423t
        from fructose loading of liver, 182
Hyperchromicity of denaturation, 303
Hyperglycemia, 627. See also Diabetes mellitus
    in diabetes mellitus, 141
    glucagon causing, 171
    insulin release in response to, 452f
Hyperhomocysteinemia, folic acid supplements in prevention of, 478
Hyperhydroxyprolinemia, 254
Hyperkalemic periodic paralysis, 556t
Hyperlacticacidemia, 219
Hyperlipidemia, niacin for, 475
Hyperlipoproteinemias, 212, 232, 232t
    familial, 232t
Hyperlysinemia, periodic, 254
Hypermetabolism, 149, 464
Hypermethioninemia, 263
Hyperornithinemia, hyperammonemia, and homocitrullinuria syndrome (HHH syndrome), 244f, 245
Hyperornithinemia-hyperammonemia syndrome, 250
Hyperoxaluria, primary, 251
Hyperparathyroidism, bone and cartilage affected in, 541t
Hyperphenylalaninemias, 254
Hyperprolinemias, types I and II, 249–250
Hypersensitive sites, chromatin, 315
Hypersplenism, in hemolytic anemia, 598
Hypertension, 639
    hyperhomocysteinemia and, folic acid supplements in prevention of, 478
Hyperthermia, malignant, 545, 552–553, 553f, 556t
Hypertriacylglycerolemia
    in diabetes mellitus, 212
    familial, 232t
    from fructose loading of liver, 182
Hypertrophic cardiomyopathy, familial, 556–557, 557f
Hyperuricemia, 298, 631
    from fructose loading of liver, 182
    males in, 631
Hypervariable regions, 577, 578f

Hypoalbuminemia, 637
Hypochondroplasia, 543
Hypoglycemia, 165, 630
    fatty acid oxidation and, 184, 191
    fructose-induced, 182
    insulin excess causing, 172
    during pregnancy and in neonate, 172
Hypoglycemic effect
    of glucagon, 171
Hypoglycin, 184, 191
Hypokalemic periodic paralysis, 556t
Hypolipidemic drugs, 232
Hypolipoproteinemia, 212, 232, 232t
Hypothalamus, 635, 640
Hypothyroidism, 125
    primary, case study, 634–635
Hypouricemia, 299
Hypoxanthine, 288, 288f
Hypoxanthine-guanine phosphoribosyl trans-
        ferase (HRPT)
    defect of in Lesch-Nyhan syndrome, 298
    localization of gene for, 396t
Hypoxia, lactate production and, 149, 150–152,
        150f

I. See Iodine/iodide
I bands, 545, 546f, 547f
I-cell disease, 422, 423t, 516, 522t, 523–524, 524f,
        536, 537t
    causation of, 524f
Ibuprofen, 193
    cyclooxygenases affected by, 200
IC$_{50}$, 71
ICAM-1, 521, 521t
ICAM-2, 521, 521t
ICF. See Intracellular fluid
Icterus (jaundice), 271, 280–283, 283t
IDDM. See Insulin-dependent diabetes mellitus
Idiotypes, 580
IDL. See Intermediate-density lipoproteins
IEF. See Isoelectric focusing
IgA, 577, 578t, 579f
IgD, 577, 578t
IgE, 577, 578t
IGF-I receptor, 426
IgG, 577, 577f, 578t
    deficiency of, 580
    hypervariable regions of, 577, 578f
IgM, 577, 578t, 579f
Immune globulin, 616
Immune response, class/isotype switching and, 580
Immune system, 637
Immunity, innate, 523
Immunogenicity, lessening, 581
Immunoglobulin genes, 580
    DNA rearrangement and, 322, 385, 580
        double-strand break repair and, 332
Immunoglobulin heavy chain binding protein, 497
Immunoglobulin heavy chains, 576–577, 577f
    genes producing, 579
Immunoglobulin light chains, 576, 577f
    in amyloidosis, 575–576
    genes producing, 579
        DNA rearrangement and, 322, 385, 580
Immunoglobulins, 566, 569t, 576–581, 578t. See
        also specific type under Ig
    class switching and, 580
    classes of, 577, 578t
    diseases caused by over- and underproduction
        of, 580

functions of, 578t, 579
    genes for. See Immunoglobulin genes
    hybridomas as sources of, 580–581, 581f
    structure of, 576–577, 577f, 578f, 579f
Immunology, 1
IMP (inosine monophosphate)
    conversion of to AMP and GMP, 293, 295f
        feedback regulation of, 293, 295f
    synthesis of, 292–293, 294f, 295f, 296f
Impairment of both cell and humoral immunity,
        617
Importins, 491, 492f
In situ hybridization, 396
In situ hybridization/fluorescence in situ hybrid-
        ization, 396
    in gene mapping, 396t
Inactive chromatin, 315, 377
Inborn errors of metabolism, 1, 248, 536
Inclusion cell (I-cell) disease, 422, 423t, 516, 522t,
        523–524, 524f
Increased bone density, 541
Increased production, 575
Indole, permeability coefficient of, 409f
Indomethacin, cyclooxygenases affected by, 198
Induced fit model, 54, 54f
Inducers
    enzyme synthesis affected by, 77
        in gluconeogenesis regulation, 167
    gratuitous, 372
    in regulation of gene expression, 370
Inducible gene, 370
Infantile Refsum disease, 191, 493, 493t
Infection, 637
    neutrophils in, 604–605, 604t
    protein loss and, 465
    respiratory burst in, 606
Inflammation, 193, 520, 581, 593, 632, 638
    acute phase proteins in, 568–569, 569t
    complement in, 581
    neutrophils in, 604–605, 604t
        integrins and, 521t, 604t, 605
        selectins and, 520–522, 521f, 521t
    prostaglandins in, 193
    selectins in, 520–522, 521f, 521t
Inflammatory bowel disease, 623
Influenza virus
    hemagglutinin in, calnexin binding to, 516
    neuraminidase in, 524
Information pathway, 446, 446f
Inhibition
    competitive vs. noncompetitive, 70–72, 70f,
        71f
    feedback, in allosteric regulation, 77–78, 78f
    mechanism-based, 72
    irreversible, 72
    tightly bound, 71–72
Inhibitor-1, 161f, 162, 163, 163f
Initial velocity, 66
    inhibitors affecting, 71, 71f
Initiation
    in DNA synthesis, 324, 325f, 326f, 327f
    in protein synthesis, 359, 360f
    in RNA synthesis, 336, 336f, 337
Initiation complexes, in protein synthesis, 360f, 362
Initiator sequence, 340, 341f, 623
Innate immunity, 523, 576
Inner mitochondrial membrane, 103, 488
    protein insertion in, 490
    relative impermeability of, exchange transport-
        ers and, 110–112, 110f, 111f

Inorganic pyrophosphatase, in fatty acid activa-
        tion, 96
Inosine monophosphate (IMP)
    conversion of to AMP and GMP, 293, 295f
        feedback-regulation of, 293, 295f
    synthesis of, 292–293, 294f, 295f, 296f
Inositol hexaphosphate (phytic acid), calcium
        absorption affected by, 462
Inositol trisphosphate, 124, 450, 451f
    in platelet activation, 589, 590f
    in respiratory burst, 606
Inotropic effects, 554
Inr. See Initiator sequence
Insert/insertions, DNA, 404
    recombinant DNA technology in detection of,
        397–398
Inside-outside asymmetry, membrane, 410
Insulators, 380–381
    nonpolar lipids as, 121
Insulin, 131, 428, 438, 438f, 502
    adipose tissue metabolism affected by, 222–223
    in blood glucose regulation, 170–171
    deficiency of, 172. See also Diabetes mellitus
    free fatty acids affected by, 221
    gene for, localization of, 398t
    glucagon opposing actions of, 171
    in glucose transport, 416
    in glycolysis, 150, 167
    initiation of protein synthesis affected by, 362,
        362f
    in lipogenesis regulation, 197
    in lipolysis regulation, 197, 221–222, 221f
    on metabolic fuel reserves, 137
    phosphorylase b affected by, 162
    receptor for, 426, 450–451, 452f
    signal transmission by, 451–452, 452f
    storage/secretion of, 441, 442t
    synthesis of, 438f, 439
Insulin-dependent diabetes mellitus (IDDM/type 1),
        172. See also Diabetes mellitus
Insulin/glucagon ratio, in ketogenesis regulation,
        190
Insulin-like growth factor I receptor, 426
Insulin resistance, 596
Integral proteins, 31, 411, 411f
    as receptors, 422
    red cell membrane, 600–601, 600t, 601f
Integration, chromosomal, 320–321, 321f
Integrins, neutrophil interactions and, 521t, 604t,
        605
Interleukins, 593
Interleukins-1 and -6, 542
Intermediary proteins and cargo molecules, 500
Intermediate-density lipoproteins, 213t, 215, 216, 230
Intermediate filaments, 563–564, 563t
Intermembrane space, proteins in, 490
Intermittent branched-chain ketonuria, 258
Internal presequences, 490
Internal ribosomal entry site, 365, 367f
Interphase chromosomes, chromatin fibers in, 314
Intervening sequences. See Introns
Intestinal bacteria, in bilirubin conjugation, 280
Intimal plaque, 638
Intracellular environment, membranes in mainte-
        nance of, 406–407, 407t
Intracellular fluid (ICF), 406, 407, 407t
Intracellular membranes, 406
Intracellular messengers, 446–454, 448t, 450t.
        See also specific type and Second
        messengers

Intracellular signals, 446–454
Intracellular traffic, 487–504. *See also* Protein
        sorting
    disorders due to mutations in genes encoding,
        504, 504t
    transport vesicles in, 499–502
Intravenous saline-dextrose, 636
Intravenous treatment, 619
Intrinsic factor, 462, 476
    in pernicious anemia, 477
Intrinsic pathway of blood coagulation, 583–584,
        584f, 585
Intrinsic tenase complex, 585
Introns (intervening sequences), 317, 346–348,
        347f, 353, 404
    removal of from primary transcript, 346–348,
        347f
Inulin, 118
Inulin, glomerular membrane permeability to, 533
"Invert sugar," 117
Iodinated anthracycline, 576
Iodine/iodide, 480t
    deficiency of, 436
    in thyroid hormone synthesis, 437–438, 437f
5-Iodo-2′-deoxyuridine, 290f
Iodopsin, 469
Iodothyronyl residues, 436. *See also* Thyroxine;
        Triiodothyronine
5-Iodouracil, 289
Ion channels, 406, 416, 417f, 417t, 556t
    in cardiac muscle, 555, 556t
    diseases associated with disorders of, 556t, 556t
Ion exchange, 22
Ion exchange chromatography, for protein/peptide
        purification, 22
Ion product, 9
Ionizing radiation, nucleotide excisionrepair of
        DNA damage caused by, 332
Ionophores, 110, 418
IP3. *See* Inositol trisphosphate
IPTG. *See* Isopropylthiogalactoside
IRES. *See* Internal ribosomal entry site
Iron, 480t
    absorption of, 463, 570–571, 570f, 571t, 633
        in hemochromatosis, 463
        vitamin C and ethanol affecting, 463, 480
    deficiency of, 480
    distribution of, 571t
    ferrous, in oxygen transport, 43–44
    heme, 278, 570
        absorption of, 463, 270f, 271
        hindered environment for, 43–44, 44f
        in methemoglobinemia, 49
    incorporation of into protoporphyrin, 272f, 273
    metabolism of, 570f, 571
        disorders of, 575t
    nonheme, 570
    transferrin in transport of, 570, 570f, 571t
Iron-binding capacity, total, 571
Iron deficiency/iron deficiency anemia, 463, 480,
        571
Iron overload, 463
Iron porphyrins, 271
Iron response elements, 572
Iron-responsive element-binding protein, 572
Iron-sulfur proteins, in respiratory chain com-
        plexes, 103–104, 105f
Irreversible covalent modifications, 79, 80f
Irreversible inhibition, enzyme, 72
IRS 1-4, in insulin signal transmission, 452, 452f

Ischemia, 149, 422
Islets of Langerhans, insulin produced by, 170
Isocitrate dehydrogenase, 143, 145f
    in NADPH production, 194, 196f
Isoelectric focusing, for protein/peptide purifica-
        tion, 24, 25f
Isoelectric pH (pI), amino acid net charge and, 17
Isoenzymes. *See* Isozymes
Isoleucine, 15t
    catabolism of, 258, 259f, 260f
    interconversion of, 238
    requirements for, 465
Isomaltose, 117t
Isomerases, 52
    in steroid synthesis, 430, 431f, 432, 433f
Isomerism
    geometric, of unsaturated fatty acids, 122, 123f
    of steroids, 127, 127f
    of sugars, 113–115, 114f, 115f
Isomorphous displacement, 37
Isoniazid, acetylation of, 613
Isopentenyl diphosphate, in cholesterol synthesis,
        224, 226f
Isoprene units, polyprenoids synthesized from,
        128, 128f
Isoprenoids, synthesis of, 224, 227f
    in cholesterol synthesis, 226f
Isopropylthiogalactoside, 372
Isoprostanes (prostanoids), 122, 129
    cyclooxygenase pathway in synthesis of,
        200–201, 201f, 202f
Isosteric enzymes, 78
Isothermic systems, biologic systems as, 92
Isotopes. *See also* specific types
    in plasma protein analysis, 567
Isotype (class) switching, 580
Isotypes, 580
Isovaleric acidemia, 258
Isovaleryl-CoA dehydrogenase, in isovaleric
        acidemia, 258
Isozymes, 56

J chain, 579f
Jackson-Weiss syndrome, 541t
JAK kinases, 427, 453, 453f
Jak-STAT pathway, 427, 453, 453f
Jamaican vomiting sickness, 191
Jaundice (icterus), 271, 280–283, 283t
Joining region, gene for, 579
    DNA rearrangement and, 385, 580
"Jumping DNA," 321
Junctional diversity, 580
Juxtaglomerular cells, in renin-angiotensin system,
        440

K. See Dissociation constant; Potassium
k. See Rate constant
K_{cat}. *See* Catalytic constant
K⁺ channel, 417–418
Kappa light chains, 577
Kartagener syndrome, 563
Karyopherins, 491
Karyotype, 317f
Kayser-Fleischer ring, 594
K_{cat}/K_m. *See* Catalytic efficiency
K_d. *See* Dissociation constant
KDEL-containing proteins, 488t, 496
K_{eq}. *See* Equilibrium constant
Keratan sulfates, 534t, 535
    functions of, 538

Keratins, 563t, 564
Kernicterus, 281
Ketoacidosis, 184, 191
    in diabetes mellitus, 141
3-Ketoacyl synthase, 193, 195f
3-Ketoacyl-CoA thiolase deficiency, 191
Ketoamines, 519
Ketogenesis, 135, 135f, 184–192. *See also* Fatty
        acids, oxidation of
    high rates of fatty acid oxidation and, 187–189,
        187f, 188f
    HMG-CoA in, 187–189, 188f
    regulation of, 189–191, 190f
Ketogenic amino acids, 137
α-Ketoglutarate, in amino acid carbon skeleton
        catabolism, 248f
Ketoglutarate transporter, 110, 110f
Ketone bodies, 132, 135, 135f, 184, 187, 187f, 188f
    in fasting state, 140
    free fatty acids as precursors of, 189
    as fuel for extrahepatic tissues, 189, 189f
    in starvation, 139–140, 140f, 140t
Ketonemia, 189, 191
Ketonuria, 191, 627
    branched chain (maple syrup urine disease),
        258
Ketoses (sugars), 113, 114t
Ketosis, 184, 189, 191
    in cattle
        fatty liver and, 218
        lactation and, 191
    in diabetes mellitus, 141, 191
    ketoacidosis caused by, 191
    in lactation, 141
    nonpathologic, 191
    in starvation, 191
Kidney
    in fasting state, 140
    glycogenolysis in, 159
    metabolism in, 141t
    in renin-angiotensin system, 440
    vitamin D metabolism in, 470
    vitamin D3 synthesis in, 435, 435f
Kinases, protein. *See* Protein kinases
Kinesin, 563
Kinetic (collision) theory, 64
Kinetics (enzyme), 62–74. *See also* Catalysis/cata-
        lytic reactions (enzymatic)
    activation energy affecting, 63–64, 65
    balanced equations and, 62
    competitive *vs.* noncompetitive inhibition and,
        70–72, 70f, 71f
    in drug development, 73–74
    factors affecting reaction rate and, 64–65, 64f,
        66f
    free energy changes affecting, 62–63
    initial velocity and, 66
    multisubstrate enzymes and, 72, 73f
    saturation, 67f, 69
        sigmoid (Hill equation), 69–70, 69f
    substrate concentration and, 66–67, 67f
        models of effects of, 67–70, 68f, 69f
    transition states and, 63–64
Kinetochore, 315
Kininogen, high-molecular-weight, 584f, 585
Kinky hair disease (Menkes disease), 573
K_m. *See* Michaelis constant
Knockout genes, 400
Korsakoff psychosis, 473

Kozak consensus sequences, 361
Krabbe disease, 210t
Krebs cycle. *See* Citric acid cycle
Ku, in double-strand break repair, 332–333, 333f
K_w. *See* Ion product
Kwashiorkor (edematous), 234, 463, 464, 636–637
  hallmarks of, 636
  protein-energy malnutrition (PEM), case study, 635–637
Kynureninase, 256f, 257
Kynurenine-anthranilate pathway, for tryptophan catabolism, 256f, 257
Kynurenine formylase, 256f, 257

L-amino acids
  in proteins, 14–16
L chains. *See* Light chains
L-Dopa, 435, 436f
L-Iduronate, 115, 116f
L isomerism, 113–114, 114f
L-tryptophan dioxygenase (tryptophan pyrrolase), 101
l-type calcium channel, 554
Labile factor (factor V), 585, 585f, 585t
lac operon, 370, 371–373, 371f, 372f
lac repressor, 371, 372f
lacA gene, 371, 371f, 372f
lacI gene, 371, 372, 372f
Lactase, 460
  deficiency of (lactose/milk intolerance), 113, 459, 460
Lactate
  anaerobic glycolysis and, 149, 150–152, 150f
  hypoxia and, 150–152, 150f
Lactate dehydrogenase, 35, 36
  in anaerobic glycolysis, 150
  isozymes, 59
Lactate dehydrogenase isozymes, 59, 150
  diagnostic significance of, 59, 59t, 59f
Lactate formation, 630
Lactation, ketosis in, 141
Lactic acid, pK/pKa value of, 12t
Lactic acid cycle, 170, 170f
Lactic acidosis, 149
  from inherited mitochondrial defects, 103
  pyruvate metabolism and, 155
  thiamin deficiency and, 474
Lactoferrin, 604t
Lactogenic hormone. *See* Prolactin
Lactose, 117, 117f, 117t, 179
  galactose in synthesis of, 179–180, 180f
  metabolism of, operon hypothesis and, 371–373, 371f, 372f
Lactose (milk) intolerance, 113, 459, 460
Lactose synthase, 180, 180f
Lactulose, 117t
lacY gene, 371, 371f, 372, 372f
lacZ gene, 371, 371f, 372, 372f
Lagging (retrograde) strand, in DNA replication, 323f, 324, 327
Lambda light chains, 577
Lambda repressor (cI) protein/gene, 373–376, 374f, 375f, 376f
Lamin A, 586
Laminin, 527, 531f, 532–533
Lamins, 563t, 564, 586
Langerhans, islets of, insulin produced by, 170
Latch state, 559
Laue x-ray crystallography, 38

Lauric acid, 122t
Laws of thermodynamics, 92–93
  hydrophobic interactions and, 7
LBD. *See* Ligand-binding domain
LCAT. *See* Lecithin:cholesterol acyltransferase
LCRs. *See* Locus control regions
LDH. *See* Lactate dehydrogenase isozymes
LDL. *See* Low-density lipoproteins
LDL:HDL cholesterol ratio, 230
Lead poisoning, ALA dehydratase inhibition and, 272, 278
Leader sequence. *See* Signal peptide
Leading (forward) strand, in DNA replication, 323f, 324, 327
Lecithin. *See also* Phosphatidylcholine
  metabolism of, 209f
Lecithin:cholesterol acyltransferase (LCAT), 207, 216, 217f, 228
  familial deficiency of, 232t
Lecithins (phosphatidylcholines), 124, 125f
  membrane asymmetry and, 410
  synthesis of, 205, 206f
Lectins, 120, 509, 510t, 641
  in glycoprotein analysis, 507t, 509, 510t
  recognizing Man 6-P, 523
Leiden factor V, 587
Lens of eye, fructose and sorbitol in, diabetic cataract and, 182
Leptin, 222
Lesch-Nyhan syndrome, 298, 631
Leucine, 15t
  catabolism of, 258, 259f, 260f
  interconversion of, 238
  requirements for, 465
Leucine aminomutase, 476
Leucine zipper motif, 381, 381t, 382, 383f
Leucovorin, 477
Leukemias, 593, 607
Leukocyte adhesion deficiency
  type I, 605
  type II, 522, 522t
Leukocytes, 520, 605
  growth factors regulating production of, 595
  recombinant DNA technology in study of, 608
Leukodystrophy, metachromatic, 210t
Leukotriene A4, 123f
Leukotrienes, 122, 123f, 193, 200
  clinical significance of, 202–203
  lipoxygenase pathway in formation of, 200, 201, 201f, 203f
LFA-1, 521, 521t, 605, 605t
LH. *See* Luteinizing hormone
Library, 392, 404
Lifestyle changes, cholesterol levels affected by, 231–232
Ligand-binding domain, 455
Ligand-gated channels, 416, 556t
Ligand-receptor complex, in signal generation, 444–446
Ligands, 89
Ligases, 52, 498
  DNA, 323t, 326, 327, 328f
Ligation, 404
Light, in active transport, 418
Light chains, 576
  immunoglobulin, 577, 577f
    in amyloidosis, 576
    genes producing, 579
  DNA rearrangement and, 322, 385, 580

myosin, 546
  in smooth muscle contraction, 557
Light meromyosin, 548, 549f
Limit dextrinosis, 160t
LINEs. *See* Long interspersed repeat sequences
Lines, definition of, 404
Lineweaver-Burk plot
  inhibitor evaluation and, 71, 71f
  K_m and V_{max} estimated from, 68, 68f
Lingual lipase, 460
Link trisaccharide, in glycosaminoglycan synthesis, 534
Linkage analysis (positional cloning), 625
  in glycoprotein study, 507t
N-linked glycoproteins synthesis, dolichol-P-P-oligosaccharide in, 512–515, 514f
Linoleic acid/linoleate, 123t, 198, 198f, 200
  in essential fatty acid deficiency, 199
  synthesis of, 199f
α-Linolenic acid, for essential fatty acid deficiency, 200
Lipases
  diagnostic significance of, 59t
  in digestion, 460, 461f
  in triacylglycerol metabolism, 205, 220, 220f, 460, 461f
Lipid bilayer, 409–410, 409f
  membrane proteins and, 409–410
Lipid compositions, of ER, 503
Lipid core, of lipoprotein, 212
Lipid droplets, 220
Lipid rafts, 412, 502, 503
Lipid storage disorders (lipidoses), 210–211, 210t
Lipidomics, 4
Lipids, 121–130. *See also* specific type
  amphipathic, 129, 129f
  asymmetry of, membrane assembly and, 503, 503f
  classification of, 121
  complex, 121
  in cytochrome P450 system, 610–611
  derived, 121
  digestion and absorption of, 460, 461f
  disorders associated with abnormalities of, 422
  fatty acids, 121–124
  glycolipids, 121, 125–126, 126f
  interconvertibility of, 137
  in membranes, 407–408
    ratio of to protein, 407, 407f
  metabolism of, 132, 132f, 133f, 134, 135f. *See also* Lipolysis
    in fed state, 137–139
    in liver, 218, 219f
  neutral, 121
  peroxidation of, 128–129, 128f
  phospholipids, 121, 124–125, 125f
  precursor, 121
  simple, 121
  steroids, 126–128, 126f, 127f, 128f
  transport and storage of, 212–223
    adipose tissue and, 220–221, 220f
    brown adipose tissue and, 222f, 223
    clinical aspects of, 218–220
    fatty acid deficiency and, 201–202
    as lipoproteins, 212–213, 213f, 213t
    liver in, 218, 219f
    triacylglycerols (triglycerides), 124, 124f
    turnover of, membranes and, 503–504

Lipogenesis, 134, 135, 137, 193–198, 194f, 195f, 222–223
  acetyl-CoA for, 195–196
  fatty acid synthase complex in, 193–194, 194f, 195f
  malonyl-CoA production in, 193, 194f
  NADPH for, 194–195, 195f, 196f
  regulation of, 196–198, 198f
    enzymes in, 168t, 193–194, 194f, 197–198
    nutritional state in, 196–197
    short- and long-term mechanisms in, 197–198
Lipolysis, 134, 135f, 221–222, 221f. See also Lipids
  hormone-sensitive lipase in, 220, 220f
  hormones affecting, 221–222, 221f
  insulin affecting, 197
  triacylglycerol, 205
Lipophilic compounds, cytochrome P450 isoforms in hydroxylation of, 610–611
Lipoprotein lipase, 134, 135f, 215, 215f, 216f
  familial deficiency of, 232t
  involvement in remnant uptake, 215–216, 215f
Lipoprotein(a) excess, familial, 232t
Lipoproteins, 31, 121, 134, 212–223, 213t, 213f, 566, 569t. See also specific type
  carbohydrates in, 120
  in cholesterol transport, 227–228, 230f
  classification of, 212, 213t
  deficiency of, fatty liver and, 218
  disorders of, 232, 232t
  remnant, 213t, 215, 215f
    liver uptake of, 215–216
Liposomes, 411
  amphipathic lipids forming, 129, 129f
  artificial membranes and, 411–412
Lipotropic factor, 218
Lipoxins, 122, 123f, 193, 200
  clinical significance of, 202–203
  lipoxygenase pathway in formation of, 200, 201, 201f, 203f
Lipoxygenase, 129, 201, 203f
  reactive species produced by, 129
5-Lipoxygenase, 201, 203f
Lipoxygenase pathway, 200, 201, 201f, 203f
Liquid chromatography, 21, 23f
  high-performance reversed-phase, for peptide separation, 23
Lithium, 480t
Lithocholic acid, synthesis of, 231f
Liver
  angiotensinogen made in, 440
  bilirubin uptake by, 278–280, 280f
  liver biopsy, 633
  cirrhosis of, 143, 218, 630
  cytochrome P450 isoforms in, 610
  in fasting state, 137–140
  fatty
    alcoholism and, 220
    of pregnancy, 191
    triacylglycerol metabolism imbalance and, 218
  fructose overload and, 182
  glucose uptake into, 138
  glycogen in, 157, 158t
  glycogenolysis in, 159
  heme synthesis in, 274
    ALA synthase in regulation of, 274, 276f
  ketone bodies produced by, 187, 187f, 188f, 189
  metabolism in, 132, 133f, 134f, 135f, 141t, 143

  fatty acid oxidation, ketogenesis and, 187–189, 187f, 188f
  fructose, 178–179, 179f
  glucose, 166f, 169–170, 170f
  fructose 2,6-bisphosphate in regulation of, 168–169, 169f
  glycogen, 157–159, 158f, 160–162
  lipid, 218, 219f
  phosphorylase in, control of, 159
  plasma protein synthesis in, 134, 567
  vitamin D metabolism in, 470
  vitamin D_3 synthesis in, 434, 435f, 471f
Liver phosphorylase deficiency, 160t
LMM. See Light meromyosin
Lock and key model, 54
Locus control regions, 380–381
Long interspersed repeat sequences (LINEs), 318
Looped domains, chromatin, 315, 316, 316f
Loops (protein conformation), 34–35
Loose connective tissue, keratan sulfate I in, 535
Low-density lipoprotein receptor-related protein (LRP), 213
  in chylomicron remnant uptake, 215f, 215–216
Low-density lipoproteins, 212, 213t, 230, 519
  apolipoproteins of, 212–213, 213t
  metabolism of, 216, 216f
  ratio of to high-density lipoproteins, atherosclerosis and, 230
  receptors for, 216
    in chylomicron remnant uptake, 215–216, 215f
    in cotranslational insertion, 495, 496f
    regulation of, 225–227
Low-energy phosphates, 94
Low molecular weight heparins (LMWHs), 587
Low pH, 523, 627
LRP. See Low-density lipoprotein receptor-related protein
LTs. See Leukotrienes
Lung surfactant, 124, 205
  deficiency of, 124, 209–210
Luteinizing hormone (LH), 427, 428, 429f
LXs. See Lipoxins
LXXLL motifs, nuclear receptor coregulators, 457
Lyases, 52
  in steroid synthesis, 431, 431f, 433f
Lymphocyte homing, selectins in, 520–522, 521f, 521t
Lymphocytes, 635. See also B lymphocytes; T lymphocytes
  recombinant DNA technology in study of, 607
Lymphoid origin cells, 593
Lysine, 16t
  catabolism of, 255, 255f
  pI of, 17
  requirements for, 465
Lysine hydroxylase, vitamin C as coenzyme for, 480
Lysis, cell, complement in, 581
Lysogenic pathway, 373, 373f
Lysolecithin (lysophosphatidylcholine), 125, 125f
  metabolism of, 207–208, 209f
Lysophosphatidylcholine. See Lysolecithin
Lysophospholipase, 207, 209f
Lysophospholipids, 125, 125f
Lysosomal degradation pathway, defect in lipidoses, 211
Lysosomal enzymes, 607

  in I-cell disease, 422, 423t, 523–524, 524f
Lysosomal hydrolases, deficiencies of, 524, 524t
Lysosomal proteases, in protein degradation, 498
Lysosomal proteins, 504
Lysosomes
  in endocytosis, 420
  in oligosaccharide processing, 515
  protein entry into
    disorders associated with defects in, 504, 504t
Lysozyme, 34f, 35, 604t
Lysyl hydroxylase
  diseases caused by deficiency of, 530t
  in hydroxylysine synthesis, 238, 528
Lysyl oxidase, 528, 530
Lytic pathway, 373, 373f
D-Lyxose, 115f

Mac-1, 521, 521t
α_2-Macroglobulin, 575
Macromolecules, cellular transport of, 420–422, 421f, 422f
Mad cow disease (bovine spongiform encephalopathy), 38
Magnesium, 480t
  in chlorophyll, 271
  in extracellular and intracellular fluid, 407, 407t
Maillard reaction, 519
Major groove, in DNA, 304f, 304
  operon model and, 371
Major histocompatibility class I molecules, 499
Major sorting decision, 487
Malate, 144, 145f
Malate dehydrogenase, 144, 145f
Malate shuttle, 111, 111f
MALDI. See Matrix-assisted laser-desorption and ionization
Maleylacetoacetate, in tyrosine catabolism, 253f, 254
Malic enzyme, in NADPH production, 194–195, 196f
Malignancy/malignant cells. See Cancer/cancer cells
Malignant hyperthermia, 545, 552–553, 553f, 556t
Malonate
  on respiratory chain, 109, 109f
  succinate dehydrogenase inhibition by, 70, 70f
Malonyl-CoA, in fatty acid synthesis, 193, 194f
Malonyl transacylase, 193, 194f, 195f
Maltase, 460
Maltose, 117, 117f, 117t
Mammalian target of rapamycin (mTOR), in insulin signal transmission, 452, 452f
Mammotropin. See Prolactin
Man 6-P receptor proteins, 523
Mannosamine, 180, 181f
D-Mannosamine, 116
D-Mannose, 115f, 116t
Mannose, in glycoproteins, 508t
Mannose 6-phosphate/mannose 6-P signal, 516
  in I-cell disease, 523–524, 524f
  in protein flow, 488t
Mannose-binding protein, deficiency of, 523
Mannosidosis, 524, 524t
Manual enzymatic method
  for DNA sequencing, 394

MAP (mitogen-activated protein) kinase
  in insulin signal transmission, 452f, 453
  in Jak/STAT pathway, 453
Maple syrup urine disease (branched-chain keto-
    nuria), 258
Marasmic kwashiorkor, 636
Marasmus (nonedematous), 92, 234, 463, 464, 636
Marble bone disease (osteopetrosis), 541
Marfan syndrome, fibrillin mutations causing,
    531, 531f
Maroteaux-Lamy syndrome, 537t
Mass spectrometry, 26–28, 27f
  covalent modifications detected by, 26–28, 26t,
    27f, 28f
  for glycoprotein analysis, 507, 507t
  tandem, 28
  tandem, detection of metabolic diseases, 246
  transcript-protein profiling and, 402
Mast cells, heparin in, 536
Matrix
  extracellular, 527–544. See also specific
    component
  mitochondrial, 103, 143
Matrix-assisted laser-desorption (MALDI), in
    mass spectrometry, 27, 28f
Matrix-processing protease (MPP), 489
Matrix proteins, 488, 492
  diseases caused by defects in import of, 492
Maxam and Gilbert's method, for DNA sequenc-
    ing, 394
Maximal velocity (V_{max})
  allosteric effects on, 78
  inhibitors affecting, 71, 71f
Michaelis-Menten equation in determination of,
    67–70, 68f
  substrate concentration and, 66–67, 67f
McArdle disease/syndrome, 160t, 561
Mechanically gated ion channels, 556t
Mechanism-based inhibition, 72
Mediator-related proteins, 457, 457t
Medicine
  preventive, biochemical research affecting, 2
  relationship of to biochemistry, 1–4, 3f
Meconium ileus, 625
Medium-chain acyl-CoA dehydrogenase, defi-
    ciency of, 191
Megaloblastic anemia
  folate deficiency causing, 468t, 477, 594t
  vitamin B12 deficiency causing, 468t, 477, 478,
    594t
Melanocyte-stimulating hormone (MSH), 441,
    441f
MELAS, 112
Melatonin, biosynthesis and metabolism of, 267f
Melting point, of amino acids, 18
Melting temperature/transition temperature, 303,
    412
Membrane assembly, 494
Membrane attack complex, 581
Membrane-bound polyribosomes, 493
Membrane fatty acid–transport protein, 214
Membrane fusion, 499
Membrane proteins, 409–410, 410t, 492, 506. See
    also Glycoproteins
  association of with lipid bilayer, 411
  integral, 31, 411, 411f
  mutations affecting, diseases caused by,
    422–423, 422f, 423t
  peripheral, 411, 411f
  red cell, 599–602, 600f, 600t

structure of, dynamic, 410–411
Membrane transport, 413–418, 414t, 414f, 415f,
    416f, 417f, 418f. See also specific
    mechanism
Membranes, 406–426
  artificial, 411–412
  assembly of, 502–504, 503f, 504t
  asymmetry of, 407, 410–411
  bilayers of, 409–410, 409f
    membrane protein association and, 409–410
  biogenesis of, 502–504, 503f, 504t
  cholesterol in, 408
    fluid mosaic model and, 412
  depolarization of, in nerve impulse transmis-
    sion, 419
  function of, 406–412
  fluidity affecting, 412
  glycosphingolipids in, 408
  Golgi apparatus in synthesis of, 499
  intracellular, 406
  lipids in, 407–408
    amphipathic, 129, 129f, 408, 408f
    mutations affecting, diseases caused by,
      422–423, 422f, 423t
  phospholipids in, 124–125, 125f, 407–408, 408f
  plasma. See Plasma membrane
  protein:lipid ratio in, 407, 407f
  proteins in, 409–410, 410t t. See also Membrane
    proteins
  red cell, 599–602, 600f, 600t
    hemolytic anemias and, 598, 599t
  selectivity of, 406, 413, 414f, 414t, 415f, 415t,
    416f, 417f, 417t, 418f
  sterols in, 408
  structure of, 407–411, 407f
    asymmetry and, 410–411
    fluid mosaic model of, 412–413, 413f
Menadiol, 472, 472f
Menadiol diacetate, 472, 472f
Menadione, 472. See also Vitamin K
Menaquinone, 468t, 472, 472f. See also Vitamin K
Menkes disease, 530, 573
Menkes syndrome, 234
MEOS. See Cytochrome P450-dependent mi-
    crosomal ethanol oxidizing system
6-Mercaptopurine, 289, 290f
Mercapturic acid, 612
Mercuric ions, pyruvate metabolism affected by,
    155
Meromyosin
  heavy, 548, 549f
  light, 547, 549f
Messenger RNA (mRNA), 306, 307–308, 308f,
    309f, 335, 336t, 354, 361f. See also
    RNA
  alternative splicing and, 348, 348f, 386
  codon assignments in, 353, 354t
  editing of, 350–351
  exporter, 491
  modification of, 351
  molecules, 491
  nontranslating, 365
  nucleotide sequence of, 353
    mutations caused by changes in, 356–359,
      356f, 357f, 358f
  polycistronic, 370
  relationship of to chromosomal DNA, 318f
  stability of, regulation of gene expression and,
    386–387, 386f
  transcription starting point and, 336

Metabolic acidosis, ammonia in, 243
Metabolic alkalosis, ammonia in, 243
Metabolic fuels, 137–140. See also Digestion
  clinical aspects of, 140–141
  diet providing, 463, 464
  diurnal requirements for, 131
  in fed and starving states, 137–140, 139f, 140f,
    140t
  interconvertability of, 137–140
  in normal adult, 131
  provision of, 131–142. See also Metabolism
Metabolic syndrome, 639
Metabolic pathway/metabolite flow, 133. See also
    specific type and specific types and
    Metabolism
  flux-generating reactions in, 137
  nonequilibrium reactions in, 137
  regulation of, 75–76, 76f, 137, 138f
    covalent modification in, 80
  unidirectional nature of, 75, 75f
Metabolism, 93, 131–142, 141t, 612. See also
    Catalysis/catalytic reactions (enzy-
    matic); Metabolic pathway/metabo-
    lite flow; specific types
  blood circulation and, 134–135, 134f, 135f
  of drugs, in vivo, 74
  group transfer reactions in, 8
  inborn errors of, 1, 248
  integration of, metabolic fuels and, 137–140
  regulation of, 76, 76f, 137, 138f
    allosteric and hormonal mechanisms in,
      77–78, 78f, 137, 138f
    enzymes in, 137, 138f
  allosteric regulation and, 77–78, 78f, 137, 138f
  compartmentation and, 75–76
  control of quantity in, 76–77
  covalent modification and, 77, 79, 80–81, 80f
  rate-limiting reactions and, 76
  at subcellular level, 135, 136f
  at tissue and organ levels, 133, 134f, 135f, 141t
  of xenobiotics, 609–615
Metabolite flow, 75
Metabolomics, 4
Metachromatic leukodystrophy, 210t
Metal-activated enzymes, 52
Metal ions, in enzymatic reactions, 52
Metalloenzymes, 52
Metalloflavoproteins, 98
Metalloproteins, 31
Metallothioneins, 573
Metaphase chromosomes, 314f, 315, 316t
Metastasis
  glycoproteins and, 506, 517–518, 522, 522t
  membrane abnormalities and, 423t
Methacrylyl-CoA, catabolism of, 259f
Methemoglobin, 49, 357, 599
Methemoglobinemia, 49, 599
Methionine, 15t, 263, 264f
  of α_1-antitrypsin, 574
  active (S-adenosylmethionine), 257, 257f, 263,
    264f, 288, 289f, 291t
  catabolism of, 257, 257f, 258f
  in folate trap, 476f
  requirements for, 465
Methionine adenosyltransferase (MAT), 263, 264f
Methionine synthase, 476, 476f, 478
Methotrexate, 296, 477
  dihydrofolate/dihydrofolate reductase affected
    by, 296, 477
Methyl pentose, in glycoproteins, 119t

Methyl-tetrahydrofolate, in folate trap, 477–478, 477f
Methylation
    in covalent modification, mass increases and, 26t
    of deoxycytidine residues, gene expression affected by, 377
    in glycoprotein analysis, 507t
    of xenobiotics, 609, 613
5-Methylcytosine, 287, 288f
Methylene tetrahydrofolate, 477, 477f
    in folate trap, 476f, 478
7-Methylguanine, 288f
Methylhistidine, 561
    in Wilson disease, 262
Methylmalonicaciduria, 167
Methylmalonyl-CoA, accumulation of in vitamin B_{12} deficiency, 476
Methylmalonyl-CoA isomerase (mutase), in propionate metabolism, 165–167, 166f, 167f, 476
Methylmalonyl-CoA mutase (isomerase), 166, 167f, 476–477
Methylmalonyl-CoA racemase, in propionate metabolism, 166, 167f
Mevalonate, synthesis of, 224, 227f
    in cholesterol synthesis, 225f, 226f
Mg. See Magnesium
MI. See Myocardial infarction
mi and small interfering (si) RNA, 310
Micelles, 408, 409, 409f
    amphipathic lipids forming, 129, 129f, 408, 409f
    in lipid absorption, 460
Michaelis constant (Km), 67
    allosteric effects on, 78
    binding constant approximated by, 69
    enzymatic catalysis rate and, 68, 68f, 75, 76f
    inhibitors affecting, 71, 71f
    Michaelis-Menten equation in determination of, 67–70, 68f
Michaelis-Menten equation, 67–68
    Bi-Bi reactions and, 73, 73f
    regulation of metabolite flow and, 75, 76f
Micro (mi) RNAs, 310
Microbial toxins, 418
Microbiology, 1
Microfilaments, 562–563, 563t
Micronutrients. See also specific micronutrients
    determination of requirements for, 465
    minerals, 480, 480t
        digestion and absorption of, 462
    vitamins. See Vitamins
Micro-RNAs, 349–350, 384–385
Microsatellite instability, 319
Microsatellite polymorphisms (DNA), 319, 400, 404
Microsatellite repeat sequences, 319, 404
Microsomal cytochrome P450, 630
Microsomal elongase system, 196, 197f
Microsomal fraction, cytochrome P450 isoforms in, 610
Microtubules, 502, 563, 563t
Microvascular and macrovascular damage, in diabetes mellitus, 519
Migration, cell, fibronectin in, 531–532
Milk (lactose) intolerance, 113, 460, 463
Mineral to organic elements ratio, 541
Mineralocorticoid response element, 445t
Mineralocorticoids, 428
    receptors for, 455
    synthesis of, 428, 431f

Minerals, 2, 480, 480t
    digestion and absorption of, 462–463
Minor groove, in DNA, 304, 304f
Misfolded proteins, accumulation of in endoplasmic reticulum, 497
Mismatch repair of DNA, 330–331, 331t, 331f
    colon cancer and, 331
Missense mutations, 356, 357, 357f
    familial hypertrophic cardiomyopathy caused by, 556–557, 557f
MIT. See Monoiodotyrosine
Mitchell chemiosmotic theory. See Chemiosmotic theory
Mitochondria
    ALA synthesis in, 272, 273f
    citric acid cycle in, 132, 132f, 135f, 135, 136f, 143, 144, 147, 147f
    fatty acid oxidation in, 184, 185f
    high-energy phosphate transport from, 111f, 112
    ion transport in, 111
    protein synthesis and import by, 488, 488t
    respiratory chain in. See Respiratory chain
Mitochondrial cytochrome P450, 101, 610. See also Cytochrome P450 system
Mitochondrial DNA, 319, 319f, 320t
Mitochondrial encephalopathy, lactic acidosis, and stroke (MELAS), 112
Mitochondrial genome, 488
Mitochondrial glycerol-3-phosphate dehydrogenase, 100
Mitochondrial matrix, 488, 490f
Mitochondrial membrane proteins, mutations of, 439
Mitochondrial membranes
    enzymes as markers of compartment separated by, 103
    protein insertion in, 489
    structure of, 103, 104f
Mitochondrial oxidation, of reduced flavins, 484f
Mitochondrial protein complexes, in respiratory chain, 99t, 103, 105f, 106f
Mitogen-activated protein (MAP) kinase
    in insulin signal transmission, 541, 452f
    in Jak/STAT pathway, 453
Mitotic spindle, microtubules in formation of, 563
Mixed-function oxidases, 101, 610. See also Cytochrome P450 system
ML. See Mucolipidoses
MOAT. See Multispecific organic anion transporter
Modeling, molecular, in protein structure analysis, 38
Molecular biology, 1. See also Recombinant DNA/recombinant DNA technology in primary structure determination, 25–26
Molecular chaperones. See Chaperones
Molecular diagnostic tests, 4
Molecular docking programs, 38, 89
Molecular dynamics, 38
Molecular genetics, 1, 388. See also Recombinant DNA/recombinant DNA technology
Molecular modeling, in protein structure analysis, 38
Molecular motors, 563
Molecular pathology, 504
Molecular replacement, 38

Molecular switches, 499
Molybdenum, 480t
Monoacylglycerol acyltransferase, 206f, 207
Monoacylglycerol pathway, 206f, 207, 460, 461f
2-Monoacylglycerols, 206f, 207
Monoclonal antibodies
    hybridomas in production of, 580–581, 581f
    and therapeutic use in humans, 581
Monoglycosylated core structure, calnexin binding and, 516
Monoiodotyrosine (MIT), 436, 437f, 438
Monomeric proteins, 35
Mononucleotides, 286
    "salvage" reactions and, 293, 294f, 295f
Monooxygenases, 101. See also Cytochrome P450 system
    in metabolism of xenobiotics, 610
Monosaccharides, 113. See also specific type and Glucose
    absorption of, 460, 460f
    physiologic importance of, 115, 116t
Monounsaturated fatty acids, 122, 123t. See also Fatty acids; Unsaturated fatty acids
    dietary, cholesterol levels affected by, 230
    synthesis of, 198–199, 199f
Morquio syndrome, 537t
MPP. See Matrix-processing peptidase
MPS. See Mucopolysaccharides
MRE. See Mineralocorticoid response element
mRNA. See Messenger RNA
MRP-2, in bilirubin secretion, 282
MSH. See Melanocyte-stimulating hormone
MstII, 389t
    in sickle cell disease, 398, 398t, 399f
mtDNA. See Mitochondrial DNA
mTOR, in insulin signal transmission, 452, 452f
Mucins, 510–511, 510t
    genes for, 511
    O-glycosidic linkages in, 510–511, 511f
    repeating amino acid sequences in, 510–511, 511f
Mucolipidoses, 536, 537t
Mucopolysaccharides, 119, 119f
Mucopolysaccharidoses, 536–537, 537t, 538f
    extracellular matrix in, 527
Mucoproteins. See Glycoproteins
Mucus, 510
Multidrug resistance-like protein 2 (MRP-2), in bilirubin secretion, 280
Multifactorial diseases, bioinformatics and, 84
Multipass membrane protein, anion exchange protein as, 600t, 601, 601f
Multiple deficiency states, vitamin, 467
Multiple myeloma, 580
Multiple sclerosis, 210
Multiple sequence alignment, 88
Multiple sulfatase deficiency, 211
Multipotent stem cells, 593
Multisite phosphorylation, in glycogen metabolism, 163
Multispecific organic anion transporter (MOAT), in bilirubin secretion, 280
Muscle, 546f. See also Cardiac muscle; Skeletal muscle
    ATP in, 545, 550, 560f, 561
    catabolism of, 637
    contraction of. See Muscle contraction
    in energy transduction, 545–546, 546f, 547f, 548f

Muscle (*Continued*)
  in fasting state, 139–140
  fibers in, 545
  glucose uptake into, 138
  glycogen in, 157, 158t
    in fasting state, 139–140
  metabolism in, 133–135, 133f, 141t, 562t
    glycogen, 157
    lactate production and, 151
  phosphorylase in, control of, 159
  as protein reserve, 562
  proteins of, 553t. *See also* Actin; Myosin; Titin
  striated, 545
Muscle contraction, 545, 547f, 548–553, 552t
  ATP hydrolysis in, 549f, 550
  in cardiac muscle, 553–560
  regulation of
    actin-based, 550–551
    calcium in, 550
    in cardiac muscle, 553–560
    sarcoplasmic reticulum and, 551–552, 551f
    in smooth muscle, 558, 558f
      myosin-based, 557
      myosin light chain kinase in, 557, 558f
  relaxation phase of, 549, 551, 552t
    in smooth muscle
    calcium in, 557–558
    nitric oxide in, 559–560, 559f
  sliding filament cross-bridge model of, 546, 547f
    in smooth muscle, 557–559
  tropomyosin and troponin in, 550
Muscle fatigue, 149
Muscle phosphorylase
  absence of, 160t
  activation of
    calcium/muscle contraction and, 161
    cAMP and, 159, 161f
Muscle wasting, 498, 637
Muscular dystrophy, Duchenne, 545, 553, 557f,
    629
Mutagenesis, site-directed, in enzyme study, 61
Mutations, 312, 319, 320f, 321f, 322f, 492, 626
  base substitution, 356, 356f
  constitutive, 371
  frameshift, 358, 358f
    ABO blood group and, 603
  gene conversion and, 321, 619
  integration and, 320–321, 321f
  of membrane proteins, diseases caused by,
      422–423, 423t
  missense, 356, 357, 357f
    familial hypertrophic cardiomyopathy
      caused by, 556–557, 557f
  mRNA nucleotide sequence changes causing,
      355f, 356–358, 357f, 358f
  nonsense, 358–359
  point, 359
    recombinant DNA technology in detection
      of, 396, 397f
  recombination and, 320, 320f, 321f
  sister chromatid exchanges and, 321–322, 322f
  suppressor, 358
  transition, 356, 356f
  transposition and, 321
  transversion, 356, 356f
Myasthenia gravis, 439
Myelin sheets, 419
Myeloma, 580
Myeloma cells, hybridomas grown from, 580, 581f
Myeloperoxidase, 597, 604t, 606–607

Myocardial infarction (MI), 638
  case study, 637–639
  causation of, 639f
  enzymes assisting in diagnosis of, 59–60
Myocardial infarction, lactate dehydrogenase
    isoenzymes in diagnosis of, 58f, 59–60, 59t
Myofibrils, 545, 546, 546f, 547f
Myoglobin, 43–50
  oxygen dissociation curve for, 44, 45f
  oxygen stored by, 43, 44, 45f, 561
Myoglobinuria, 49
Myokinase (adenylyl kinase), 96
  in gluconeogenesis regulation, 168
  as source of ATP in muscle, 560f, 561
Myopathy, from inherited mitochondrial defects,
    103
Myophosphorylase deficiency, 160t
Myosin, 546–548, 549f
  in muscle contraction, 546, 547f, 549, 549f,
      550f
    regulation of smooth muscle contraction
      and, 557
  in striated *vs.* smooth muscle, 558t
  structure and function of, 547–548, 549f
Myosin-binding protein C, 553t
Myosin (thick) filaments, 546, 547f
Myosin head, 547, 549f
  conformational changes in, in muscle contrac-
      tion, 548–553
Myosin heavy chains, 547
  familial hypertrophic cardiomyopathy
    caused by mutations in gene for, 556–557, 557f
Myosin light chain kinase, 557–558, 558f
Myosin light chains, 547
  in smooth muscle contraction, 557
Myotonia congenita, 556t
Myristic acid, 122t
Myristylation
  in covalent modification, mass increases and,
      26t

N-acetylneuraminic acid, 119, 119f, 180, 181f
  in gangliosides, 209, 210f
  in glycoproteins, 180, 181f, 508t
  in mucins, 507, 511f
N-glycan chains, 495
N-linked glycoproteins, 510, 511f, 512–518
  classes of, 512–513, 513f
  synthesis of, 512–514, 513f, 514f, 515f, 516f
    in endoplasmic reticulum and Golgi
      apparatus, 516, 516t
    glycan intermediates formed during,
      516–517
    regulation of, 516–517, 518t
    tunicamycin affecting, 518, 518t
Na. *See* Sodium
Na⁺-Ca²⁺ exchanger, 449
Na⁺-K⁺ ATPase, 419, 419f
  in glucose transport, 420, 420f
NaCl, high amounts in sweat, 625
NAD⁺ (nicotinamide adenine dinucleotide), 99,
    475, 475f
  absorption spectrum of, 58, 58f
  in citric acid cycle, 147
  as coenzyme, 99, 100f, 289t
  of plasma membrane, 418
NAD(P)⁺-dependent dehydrogenases, in enzyme
    detection, 57–58
NADH
  absorption spectrum of, 58, 58f

extramitochondrial oxidation of, substrate
    shuttles in, 111, 110f
  fatty acid oxidation yielding, 185
  in pyruvate dehydrogenase regulation, 154–155,
      154f
NADH dehydrogenase, 100
NADH-Q oxidoreductase, 103, 105f, 106f
  as electron acceptor, 104, 104f, 145f
NADP⁺ (nicotinamide adenine dinucleotide phos-
    phate), 99, 474
  as coenzyme, 99, 100f, 289t
  in pentose phosphate pathway, 174, 175f, 176f
NADPH
  in cytochrome P450 reactions, 102f, 610
  intramitochondrial, proton-translocating
    transhydrogenase in, 110–111
  for lipogenesis, 195f, 194, 196f
  pentose phosphate pathway and, 174, 175f,
      175f, 180
NADPH-cytochrome P450 reductase, 610
NADPH oxidase, 604t, 606
  chronic granulomatous disease associated with
    mutations in, 606, 606f
Nanotechnology, 57
National Center for Biotechnology Information
    (NCBI), 87
Native conformation, protein, 38
NCoA-1/NCoA-2 coactivators, 457, 457t
NCoR, 457, 457t
NDPs. *See* Ribonucleoside diphosphates
Nebulin, 553t
NEFA (nonesterified fatty acids). *See* Free
fatty acids
Negative nitrogen balance, 465
Negative regulators, of gene expression, 369, 370t,
    371, 375
Negative supercoils, DNA, 304
NEM-sensitive factor (NSF), 500, 500t
Neonatal adrenoleukodystrophy, 493, 493t
Neonatal (physiologic) jaundice, 281
Neonatal tyrosinemia, 254
Nephrogenic diabetes insipidus, 418
Nerve cells. *See* Neurons
Nerve impulses, 419
Nervous system
  glucose as metabolic necessity for, 137
  thiamin deficiency affecting, 473–474
NESs. *See* Nuclear export signals
Net charge, of amino acid, 16–17, 17f
Net diffusion, 414
NeuAc. *See* N-Acetylneuraminic acid
Neural tube defects, folic acid supplements
    in prevention of, 478
Neuraminic acid, 119, 126
Neuraminidases
  deficiency of, 524, 524t
  in glycoprotein analysis, 509
  influenza virus, 524
Neuritic plaques, 618
Neurodegenerative diseases, 619
Neurofibrillary tangles, 618
Neurofilaments, 563t
Neurologic diseases, protein conformation altera-
    tions and, 39–40
Neurologic impairment, profound, 492
Neurologic signs, severe, 574
Neurons, membranes of
  impulses transmitted along, 419
  ion channels in, 415f, 416, 417f
  synaptic vesicle fusion with, 502

Neuropathy, sensory, in vitamin B6 excess, 476
Neurotoxicity, 619
Neutral lipids, 121
Neutropenia, 595
Neutrophils, 603–607
    activation of, 521, 605
    biochemical features of, 604t
    enzymes and proteins of, 604t
    in infection, 604–605
    in inflammation, 604–605, 604t
        integrins and, 605, 605t
        selectins and, 520–521, 521t, 521f
    proteinases of, 607, 607t
    respiratory burst and, 606
NFκB activation, 498, 519
Niacin, 467, 468t, 474–475, 475f. See also Nicotin-
        amide; Nicotinic acid
    deficiency of, 468t, 475
    excess/toxicity of, 475
Nick translation, 404
Nickel, 480t
Nicks/nick-sealing, in DNA replication, 327
Nicotine, 611
Nicotinamide, 468t, 475, 475f. See also Niacin
    coenzymes derived from, 52
        dehydrogenases and, 99, 100f
    excess/toxicity of, 475
Nicotinamide adenine dinucleotide (NAD⁺), 99,
        475, 475f
    absorption spectrum of, 58, 58f
    in citric acid cycle, 146
    as coenzyme, 99, 100f, 289t
Nicotinamide adenine dinucleotide phosphate
        (NADP⁺), 99, 474
    as coenzyme, 99, 100f, 289t
    in pentose phosphate pathway, 174, 175f, 176f
Nicotinic acid, 468t, 474, 475, 475f. See also Niacin
    as hypolipidemic drug, 232
NIDDM. See Non-insulin dependent diabetes
        mellitus
Nidogen (entactin), in basal lamina, 533
Niemann-Pick C-like 1 protein, 232
Niemann-Pick disease, 210t
Night blindness, vitamin A deficiency causing,
        168t, 470
Nitric oxide, 545, 559–560, 559f, 560t, 591t
    clotting/thrombosis affected by, 591, 591t
Nitric oxide synthases, 559–560, 559f, 559t
Nitrite, nitric oxide formation from, 559–560
Nitrogen balance, 465
Nitroglycerin, 559
NLS. See Nuclear localization signal
NMR. See Nuclear magnetic resonance (NMR)
        spectroscopy
NO. See Nitric oxide
NO synthase, 559–560, 559f, 559t
Nomenclature, 499
Non–clathrin-coated vesicles, 499
Noncoding regions, in recombinant DNA technol-
        ogy, 388, 389f
Noncoding strand, 303
Noncompetitive inhibition, competitive
    inhibition differentiated from, 70–72, 70f, 71f, 72f
Noncovalent assemblies, in membranes, 407
Noncovalent forces
    in biomolecule stabilization, 8
    peptide conformation and, 19–20
Nonedematous. See Marasmus
Nonequilibrium reactions, 137
    citric acid cycle regulation and, 147

glycolysis regulation and, 152–153, 165–167
Nonesterified fatty acids. See Free fatty acids
Nonheme iron, 570
Nonhistone proteins, 312
Non-insulin dependent diabetes mellitus
        (NIDDM/type 2), 172
Nonoxidative phase, of pentose phosphate path-
        way, 174–177
Nonrepetitive (unique-sequence) DNA, 318
Nonsense codons, 354, 356, 357, 358–359
Nonsense mutations, 359
Nonsteroidal anti-inflammatory drugs (NSAID),
        631, 632
    cyclooxygenase affected by, 200
    prostaglandins synthesis, 193
Norepinephrine, 429f, 435, 436f. See also
        Catecholamines
    synthesis of, 266, 268f, 435, 436f
    in thermogenesis, 222f, 223
Normal regulation and gene, 617
Normally inactive proteins, 581
Northern blot transfer procedure, 304, 393, 393f,
        398
NOS, 559–560, 559f, 559t
NPCs. See Nuclear pore complexes
NSAID. See Nonsteroidal anti-inflammatory
        drugs
NSF. See NEM-sensitive factor
Nuclear export signals, 491
Nuclear genes, proteins encoded by, 488
Nuclear localization signal (NLS), 488f, 491, 492t
Nuclear magnetic resonance (NMR) spectroscopy
    for glycoprotein analysis, 507, 507t
    protein structure demonstrated by, 38
Nuclear pore complexes, 491
Nuclear proteins, O-glycosidic linkages in, 510
Nuclear receptor coactivators (NCoA-1/NCoA-2),
        457, 457t
Nuclear receptor corepressor (NCoR), 457, 457t
Nuclear receptor superfamily, 427, 455, 456t, 457f
Nucleases, 8, 311
    active chromatin and, 315
Nucleic acids. See also DNA; RNA
    bases of, 286, 287t
    dietarily nonessential, 292
    digestion of, 311
    structure and function of, 302–311
Nucleolytic processing, of RNA, 350
Nucleophile, water as, 8–9
Nucleoplasm, 492f
Nucleoproteins, packing of, 316, 316t, 317f
Nucleosidases (nucleoside phosphorylases),
        purine, deficiency of, 299
Nucleoside diphosphate kinase, 96
Nucleoside triphosphates
    group transfer potential of, 289, 290f, 289f, 289t
    nonhydrolyzable analogs of, 289, 290f
    in phosphorylation, 96
    in transfer of high-energy phosphate, 96
Nucleosides, 285–291, 287t
Nucleosomes, 312, 313, 313f
    eviction, 342f
    and promoter accessibility, 344
Nucleotide bending fold. See Rossmann fold
Nucleotide excision-repair of DNA, 332, 332f, 644
Nucleotide sugars, in glycoprotein biosynthesis,
        507, 508t, 512, 512t
Nucleotides, 285–291, 287t. See also Purine; Py-
        rimidines/pyrimidine nucleotides

adenylyl kinase (myokinase) in interconversion
        of, 96
    as coenzymes, 289, 289t
    DNA, deletion/insertion of, frameshift
        mutations and, 358, 358f
    metabolism of, 292–301
    in mRNA, 353–354
    mutations caused by changes in, 356–358, 356f,
        357f, 358f
    physiologic functions of, 288
    as polyfunctional acids, 288
    polynucleotides, 289–291
    synthetic analogs of, in chemotherapy, 289, 290f
    ultraviolet light absorbed by, 288
Nucleus (cell), importins and exportins in
        transport and, 491, 489f
Nutrition, 459–466. See also Diet
    biochemical research affecting, 2
    lipogenesis regulated by, 194
Nutritional deficiencies, 459
    in AIDS and cancer, 464
Nutritionally essential amino acids, 133, 242t, 465.
        See also Amino acids
Nutritionally essential fatty acids, 198. See also
        Fatty acids
    abnormal metabolism of, 202
    deficiency of, 200, 202
Nutritionally nonessential amino acids, 133, 234,
        235t, 465
    synthesis of, 234–238
Nutrigenomics, 4
O blood group gene, 603
O blood group substance, 603, 603f
O-glycosidic linkage, 509
    of collagen, 528
    of proteoglycans, 533–534
O-linked glycoproteins, 510, 512, 511, 512t
    synthesis of, 510–511, 512t
O-linked oligosaccharides, in mucins, 510–511,
        511f
1,25(OH)2-D3. See Calcitriol
Obesity, 92, 131, 212, 459, 463
    case study, 639–641
    lipogenesis and, 193
Octamers, histone, 313, 313f
Oculocerebrorenal syndrome, 498t
Okazaki fragments, 324, 326f
Oleic acid, 122, 122f, 123t, 123f, 198t
    synthesis of, 199, 199f
Oligomerization, 619
Oligomers, import of by peroxisomes, 491
Oligomycin, on oxidation and phosphorylation,
        109, 109f
Oligonucleotide
    definition of, 404
    in primary structure determination, 26
Oligosaccharide branches (antennae), 512
Oligosaccharide chains
    glycoprotein, 506, 507t, 567
        in N-glycosylation, 513, 515f
    regulation of, 517–518
        sugars in, 507, 508t
    glycosaminoglycans, 535
Oligosaccharide processing, 487, 495, 513, 514f, 515
    Golgi apparatus in, 495
    regulation of, 517–518, 518t
Oligosaccharide:protein transferase, 515
Oligosaccharides, 113
    O-linked, in mucins, 510–511, 511f
Omega-3 fatty acids, 618

OMP (orotidine monophosphate), 296, 298f
Oncogenes, 1, 623
    cyclins and, 329
Oncoproteins, Rb protein and, 330
Oncotic (osmotic) pressure, 566, 569
Oncoviruses, cyclins and, 329
Open complex, 339
Online mendelian inheritance in man database, 87
Open reading frame, 28
Operator locus, 371, 372f
Operon/operon hypothesis, 370, 371–373, 371f, 372f
Optical activity/isomer, 114
OR. See Right operator
Oral rehydration solution, 620
ORC. See Origin replication complex
ORE. See Origin replication element
ORF. See open reading frame
Ori (origin of replication), 322–323, 323f, 404
Origin of replication (ori), 322–323, 323f, 404
Origin replication complex, 323
Origin replication element, 323
Ornithine, 262
    catabolism of, 249f, 250
    metabolism of, 263f
    in urea synthesis, 244, 245, 246
Ornithine-citrulline antiporter, defective, 249
Ornithine transcarbamoylase/L-ornithine
    transcarbamoylase deficiency of, 246, 297
    in urea synthesis, 246
Ornithine transporter disorders, 246
Orotate phosphoribosyltransferase, 295, 297, 298f
Orotic aciduria, 300
Orotidine monophosphate (OMP), 296, 298f
Orotidinuria, 301
Orphan receptors, 427, 455
Osmolality, 627
Osmotic fragility test, 602
Osmotic (oncotic) pressure, 566, 569
Osteoarthritis, 527, 530t
    proteoglycans in, 539
Osteoblasts, 539–540, 540f
Osteocalcin, 473, 480, 538t
Osteoclasts, 539, 541, 540f
Osteocytes, 539, 540f
Osteogenesis imperfecta (brittle bones), 234, 541,
    541f
Osteoid, 540, 540f
Osteomalacia, 468t, 470, 471, 541t, 643
Osteonectin, 538t
Osteopenia, 643
Osteopetrosis (marble bone disease), 541
Osteopontin, 538t
Osteoporosis, 471, 541–542, 541t, 643, 643f
    primary (postmenopausal), case study, 641–643
Ouabain, 116, 418
    Na+-K+ ATPase affected by, 419
Outer mitochondrial membrane, 103, 488
    protein insertion in, 490
Ovary, hormones produced by, 427, 432, 433f, 434f
Overnutrition, 466
Oxaloacetate
    in amino acid carbon skeleton catabolism, 248,
        248f, 249f
    in aspartate synthesis, 235, 236f
    in citric acid cycle, 135, 136f, 143, 144f, 146,
        146f, 147
Oxidases, 98–99, 99f. See also specific type
    ceruloplasmin as, 572–573
    copper in, 98
    flavoproteins as, 98–99, 99f

mixed-function, 101, 610. See also Cytochrome
        P450 system
Oxidation, 98
    definition of, 98
    dehydrogenases in, 99–100, 100f
    fatty acid, 184. See also Ketogenesis
        acetyl-CoA release and, 132, 132f, 184–187,
            185f, 186f
        clinical aspects of, 191
        hypoglycemia caused by impairment of, 191
        in mitochondria, 184, 185f
    hydroperoxidases in, 100
    oxidases in, 98, 99f, 100f
    oxygen toxicity and, 102, 597, 597t
    oxygenases in, 101, 102f
    redox potential and, 98, 99t
Oxidation-reduction (redox) potential, 98, 99t
Oxidative deamination, 242f
Oxidative phase, of pentose phosphate pathway,
    174, 175f, 176f
Oxidative phosphorylation, 95, 132. See also Phos-
        phorylation; Respiratory chain
    ATP generation by, 106
    enzymes as markers of compartments separated
        by mitochondrial membranes in, 103
    muscle generation of ATP by, 560f, 561,
        561–562, 561t
    respiration and, via ATP, 103
    at respiratory chain level, 106, 155t
Oxidative stress, 597
Oxidized LDL, 638
Oxidoreductase, 35
    c, 103, 105f, 106f, 104
    NADH-Q, 103, 105f, 106f
        as electron acceptor, 104, 104f, 145f
Oxidoreductases, 52, 98. See also specific type
Oxidosqualene:lanosterol cyclase, 224, 227f
Oxygen
    binding, 45, 45f. See also Oxygenation
        Bohr effect and, 47, 48f
        histidines F8 and E7 in, 43, 44f
    debt, 152
    hemoglobin affinities (P_{50}) for, 46, 46f
    myoglobin in storage of, 43, 44, 45f, 561
    reductive activation of, 610
    transport of, ferrous iron in, 43
Oxygen dissociation curve, for myoglobin and
        hemoglobin, 44, 45f
Oxygen radicals. See Free radicals
Oxygen toxicity, superoxide free radical and, 102,
        597, 597t. See also Free radicals
Oxygenases, 98, 101–102
Oxygenation of hemoglobin
    conformational changes and, 46–47, 46f, 47f
        2,3-bisphosphoglycerate stabilizing, 48, 48f
        apoprotein, 46
    high altitude adaptation and, 48
    mutant hemoglobins and, 49
Oxysterols, 129

P bodies, 387, 365, 366f
p/CIP coactivator, 457, 457t
P component, in amyloidosis, 576
p-Hydroxyphenylpyruvate, in tyrosine catabolism,
        254, 253f
P-selectin, 521t
P50, hemoglobin affinity for oxygen and, 46, 46f
p53 protein/p53 gene, 333
p97, 497
p160 coactivators, 456, 457t

p300 coactivator/CPB/p300, 448, 453, 455f, 457,
        457t
P450 cytochrome. See Cytochrome P450 system
P450scc (cytochrome P450 side chain cleavage
        enzyme), 429–430, 430f, 432
PAC (P1-based) vector, 390, 391t, 404
Paclitaxel (Taxol), 563
Paddle, charged, 417–418, 418f
PAF. See Platelet-activating factor
PAGE. See Polyacrylamide gel electrophoresis
Pain, prostaglandins in, 193
Palindrome, 404
Palmitate, 193
Palmitic acid, 122t
Palmitoleic acid, 123t, 198f
    synthesis of, 198
Palmitoylation, in covalent modification, mass
        increases and, 26t
Pancreatic ducts and obstruction, 625
Pancreatic enzyme preparation, 625
Pancreatic insufficiency, in vitamin B12 deficiency,
        476
Pancreatic islets, insulin produced by, 170
Pancreatic lipase, 460, 461f
Pancreatic sufficient, 625
Panproteinase inhibitor, 575
Pantothenic acid, 193, 468t, 479, 479f
    in citric acid cycle, 146
    coenzymes derived from, 52
    deficiency of, 468t
Papain, immunoglobulin digestion by, 577
PAPS. See Adenosine 3'-phosphate-5'-
        phosphosulfate
Parallel beta sheet, 33f, 34
Parasitic infections, 598
Parathyroid hormone (PTH), 428, 439–440, 439f
    storage/secretion of, 441–442, 442t
    synthesis of, 439–440, 439f
Paroxysmal nocturnal hemoglobinuria, 423t, 519,
        522–523, 522t, 523f, 598
Partition chromatography, for protein/peptide
        purification, 21
Passive diffusion/transport, 413, 414t, 414f
Pathologic picture and AD, 618
Pathology, 1
Pathway regulation, 593
pBR322, 391, 391t, 392f
PCR. See Polymerase chain reaction
PDH. See Pyruvate dehydrogenase
PDI. See Protein disulfide isomerase
PECAM-1, 521, 521t
Pedigree analysis, 398, 399f
Pellagra, 468t, 475
PEM (protein-energy malnutrition), 637
    primary, 636
    secondary, 636
Penicillamine, for Wilson disease, 574
Pentasaccharide, in N-linked glycoproteins,
        512–513, 513f
Pentose phosphate pathway, 132, 174–177, 175f,
        176f, 177f
    cytosol as location for reactions of, 174
    enzymes of, 168t
    erythrocyte hemolysis and, 180–181, 597
    impairment of, 180–181
    NADPH produced by, 174, 175f, 176f
        for lipogenesis, 194–195, 195f, 196f
    nonoxidative phase of, 174–176
    oxidative phase of, 174, 175f, 176f
    ribose produced by, 174, 175f

Pentoses, 113, 114t
  in glycoproteins, 119t
  physiologic importance of, 115, 116t
Pentosuria, essential, 174, 181
PEPCK. See Phosphoenolpyruvate carboxykinase
Pepsin, 462
  in acid-base catalysis, 54
Pepsinogen, 462
Peptic ulcers
  Helicobacter pylori in, 459
    glycan binding of, 525, 525f
Peptidases, in protein degradation, 239, 240f
Peptide bonds, 19. See also Peptides
  formation of, 8, 362–363
  partial double-bond character of, 19, 19f
  on secondary conformations, 32
Peptides, 14–20, 429f. See also Amino acids;
Proteins
  absorption of, 462
  amino acids in, 14, 19, 19f
  as hormone precursors, 438–439
  intracellular messengers used by, 446–454, 448t,
    450t
  as polyelectrolytes, 19
  purification of, 21–24
Peptidyl prolyl isomerase, 497
Peptidylglycine hydroxylase, vitamin C as coen-
    zyme for, 480
Peptidyltransferase, 362–363, 363t
Percutaneous coronary intervention (PCI), 638
Percutaneous transluminal coronary angioplasty
    (PTCA), 638
Perilipin, 222
Periodic acid-Schiff reagent, in glycoprotein
    analysis, 507t
Periodic hyperlysinemia, 255
Periodic paralysis
  hyperkalemic, 556t
  hypokalemic, 556t
Peripheral proteins, 411, 411f
Peripherin, 563t
Permeability coefficients, of substances in lipid
    bilayer, 409, 409f
Pernicious anemia, 468t, 477
Peroxidases, 100, 200
Peroxidation, lipid, free radicals produced by,
    128–129, 128f
Peroxides, 484–485
Peroxins, 493
Peroxisomal abnormalities, disorders due to, 493t,
    504
Peroxisomal enzymes, 493
Peroxisomal-matrix targeting sequences (PTS),
    488t, 492, 493f
Peroxisomes, 101, 491
  absence/abnormalities of, 492, 493t
    in Zellweger syndrome, 191, 492
  biogenesis of, 492
  in fatty acid oxidation, 187
Pex2/10/12, 492
Pex5, 492
Pfeiffer syndrome, 541t
PFK-1. See Phosphofructokinase
    (phosphofructokinase-1)
PGHS. See Prostaglandin H synthase
PGIs. See Prostacyclins
PGs. See Prostaglandins
pH, 9–13. See also Acid-base balance
  amino acid net charge and, 16–17, 17f
  buffering and, 11–12, 12f. See also Buffers

calculation of, 9–10
  definition of, 9
  enzyme-catalyzed reaction rate affected by, 66,
    66f
  isoelectric, amino acid net charge and, 17
Phage lambda, 373–376, 373f, 374f, 375f, 376f
Phages, in recombinant DNA technology, 390
Phagocytic cells, respiratory burst of, 606
Phagocytosis, 420–421
Pharmacogenetics, 613, 614
Pharmacogenomics, 4, 614–615
Pharmacology, 1
Pharmacy, 1
Phasing, nucleosome, 313
Phenobarbital, warfarin interaction and, cyto-
    chrome P450 induction affecting,
    611
Phenylalanine, 16t
  catabolism of, 254–255, 254f
  in phenylketonuria, 254, 254f
  requirements for, 465
  in tyrosine synthesis, 237, 237f
Phenylalanine hydroxylase, 37f
  defect in, 254
  localization of gene for, 396t
  in tyrosine synthesis, 237, 237f
Phenylethanolamine-N-methyltransferase
    (PNMT), 436, 436f
Phenylisothiocyanate (Edman reagent), in protein
    sequencing, 25, 26f
Phenylketonuria, 254
Phi angle, 32, 32f
Phlebotomy, 633
Phosphagens, 95, 95f
Phosphatases
  acid, diagnostic significance of, 59t
  alkaline
    in bone mineralization, 540
    isozymes of, diagnostic significance
      of, 59t
    in recombinant DNA technology, 390t
Phosphate transporter, 110, 110f
Phosphates/phosphorus, 480t
  in extracellular and intracellular fluid, 407t
  free energy of hydrolysis of, 94–95, 94t
  high-energy, 94. See also ATP
    in energy capture and transfer, 94–95, 94f,
      94t, 95f
    as "energy currency" of cell, 95–96, 95f, 96f,
      106
    symbol designating, 94
  low-energy, 94
Phosphatidate, 206f, 207
  in triacylglycerol synthesis, 205, 206f, 207
Phosphatidate phosphohydrolase, 206f, 207
Phosphatidic acid, 124, 125f, 407, 408f
Phosphatidic acid pathway, 460, 461f
Phosphatidylcholines (lecithins), 124, 125f
  membrane asymmetry and, 411
  synthesis of, 205, 206f
Phosphatidylethanolamine (cephalin), 124,
    125f
  membrane asymmetry and, 411
  synthesis of, 205, 206f
Phosphatidylglycerol, 125, 125f
Phosphatidylinositol 3-kinase (PI-3 kinase)
  in insulin signal transmission, 452, 452f
  in Jak/STAT pathway, 453
Phosphatidylinositol 4
  in platelet activation, 590f

Phosphatidylinositol 4,5-bisphosphate (PIP2), 124,
    450, 451f
  in absorptive pinocytosis, 421
  in platelet activation, 589
Phosphatidylinositol bisphosphate, in neutrophil
    activation, 605
Phosphatidylinositol/phosphatidylinositide, 124,
    125f
  GPI-linked glycoproteins and, 536. See also
    Glycosylphosphatidylinositol-
    anchored GPIanchored/GPI-linked)
    glycoproteins
  metabolism of, 450, 451f
  as second messenger/second messenger precur-
    sor, 124, 125f, 427, 428t, 446, 450,
    450t, 451f
  synthesis of, 205, 206f
Phosphatidylserine, 124, 125f
  membrane asymmetry and, 410
Phosphocreatine, in muscle, 545
Phosphodiester, 289–290
Phosphodiesterases, 290
  in calcium-dependent signal transduction,
    449–450
  in cAMP-dependent signal transduction, 448f,
    449
  cAMP hydrolyzed by, 159
Phosphoenolpyruvate
  free energy of hydrolysis of, 94t
  in gluconeogenesis, 146, 146f
Phosphoenolpyruvate carboxykinase (PEPCK),
    146, 146f
  in gluconeogenesis regulation, 146, 146f, 165,
    166f
Phosphoenolpyruvate carboxylase, 168t
  in gluconeogenesis, 168t
Phosphofructokinase (phosphofructokinase-1), 168t
  in gluconeogenesis regulation, 168
  in glycolysis, 150, 151f, 168t
    regulation and, 152
  muscle, deficiency of, 155, 160t
Phosphofructokinase-2, 168–169, 169f
6-Phosphofructo-2-kinase/fructose-2, 6-bisphos-
    phatase, 37f
Phosphoglucomutase, in glycogen biosynthesis,
    177, 158f
6-Phosphogluconate dehydrogenase, 174, 175f, 176f
3-Phosphoglycerate
  in glycolysis, 152, 151f
  in serine synthesis, 236, 236f
Phosphoglycerate kinase, in glycolysis, 150, 151f
  in erythrocytes, 152f, 153, 154
Phosphoglycerate mutase, in glycolysis, 150, 151f
Phosphoglycerides, in membranes, 407–408, 408f
Phosphoglycerols
  lysophospholipids in metabolism of, 125, 125f
  synthesis of, 205, 206f
Phosphohexose isomerase, in glycolysis, 150, 151f
Phosphoinositide-dependent kinase-1 (PDK1), in
    insulin signal transmission, 452
Phospholipase A1, 207, 209f
Phospholipase A2, 207, 209f
  in platelet activation, 589, 590f
Phospholipase C, 207, 209f
  in calcium-dependent signal transduction, 450,
    451f
  in Jak/STAT pathway, 453
  in respiratory burst, 606
Phospholipase Cβ, 589
Phospholipase D, 207, 209f

Phospholipases
  in glycoprotein analysis, 507t
  in phosphoglycerol degradation and remodeling, 207–208, 209f
Phospholipids, 121, 212
  digestion and absorption of, 460–462, 461f
  glycerol ether, synthesis of, 207, 208f
  in lipoprotein lipase activity, 215
  in membranes, 124–125, 125f, 407–408, 408f, 409, 503
    membrane asymmetry and, 427, 503
  in multiple sclerosis, 210
  as second messenger precursors, 205
  synthesis of, 206f
Phosphoprotein phosphatases, in cAMP-dependent signal transduction, 448f, 449
Phosphoproteins, in cAMP-dependent signal transduction, 448f,
Phosphoprotein phosphatases, in cAMP-dependent signal transduction, 449
Phosphoric acid, pK/pKa value of, 13t
Phosphorus. See Phosphates/phosphorus
Phosphorylase
  activation of, cAMP and, 160
  in glycogen metabolism, 161–162, 158f
    regulation of, 162, 162f, 163f
    liver, 159
      deficiency of, 160t
    muscle, 159
      absence of, 160t
      activation of
    calcium/muscle contraction and, 160
    cAMP and, 161f
Phosphorylase a, 162, 161f
Phosphorylase b, 162, 161f
Phosphorylase kinase
  calcium/calmodulin-sensitive, in glycogenolysis, 161, 162
  deficiency of, 160t
  protein phosphatase-1 affecting, 162
Phosphorylase kinase a, 160, 161f
Phosphorylase kinase b, 159, 161f
Phosphorylation, protein
  in covalent modification, 79, 80, 80f, 81t
    mass increases and, 26t
  multisite, in glycogen metabolism, 163
  oxidative. See Oxidative phosphorylation
  in respiratory burst, 606
  at substrate level, 107f, 106
  versatility of, 81, 81t, 82f
Phosphatidylcholine metabolism, 209f
Phosphotriose isomerase, 150
Photolysis reaction, in vitamin D synthesis, 434
Photosensitivity, in porphyria, 275
Phototherapy, cancer, porphyrins in, 275
Phylloquinone, 468t, 475, 472f. See also Vitamin K
Physiologic (neonatal) jaundice, 281
Physiology, 1
Phytanic acid, Refsum disease caused by accumulation of, 191
Phytase, 462
Phytic acid (inositol hexaphosphate), calcium absorption affected by, 462
Pi, 574
  in muscle contraction, 549, 549f
pI (isoelectric pH), amino acid net charge and, 17
PI-3 kinase
  in insulin signal transmission, 452, 452f
  in Jak/STAT pathway, 453
PIC. See Preinitiation complex

PIG-A gene, mutations of in paroxysmal nocturnal hemoglobinuria, 522, 523f
Pilocarpine, 625
"Ping-Pong" mechanism, in facilitated diffusion, 415, 416f
Ping-pong reactions, 72–73, 73f
Pinocytosis, 421
PIP₂ (phosphatidylinositol 4,5-bisphosphate), 124, 450, 451f
  in absorptive pinocytosis, 421
  in platelet activation, 590
Pituitary hormones, 427. See also specific type
  blood glucose affected by, 171
pK/pKa
  of amino acids, 15t–16t, 17, 17f
    environment affecting, 17–18, 18t
  medium affecting, 12–13
  of weak acids, 10–13, 13t, 17
PKA. See Protein kinase A
PKB. See Protein kinase B
PKC. See Protein kinase C
PKU. See Phenylketonuria
Placenta, 611
  estriol synthesis by, 432
Plaque hybridization, 393. See also Hybridization
Plasma, 566, 627
  analysis of enzymes in, 58
Plasma cells, immunoglobulins synthesized in, 576
Plasma enzymes. See also Enzymes
  diagnostic significance of, 59t
Plasma lipoproteins. See Lipoproteins
Plasma membrane, 406, 414–422, 415f, 503. See also Membranes
  carbohydrates in, 120
  mutations in, diseases caused by, 422–423, 423t
Plasma proteins, 506, 566–576, 567f, 569t. See also specific type and Glycoproteins
  in bone, 538t
  concentration of, 571
  electrophoresis for analysis of, 566, 568f
  functions of, 569, 569t
  half-life of, 568
  in inflammation, 604t
  polymorphism of, 567
  synthesis of
    in liver, 134, 567
    on polyribosomes, 567
  transport, 442t, 443, 569t
Plasma proteome, 566
Plasma thromboplastin antecedent (PTA/factor XI), 584f, 585, 585t
  deficiency of, 585
Plasma thromboplastin component (PTC/factor IX), 584f, 585, 585t
  coumarin drugs affecting, 587–588
  deficiency of, 588
Plasmalogens, 125, 125f, 207, 208f
  biosynthesis of, 208f
Plasmids, 390, 391f, 391t, 392f, 404
Plasmin, 588–589, 588f
Plasminogen, 588
  activators of, 58, 588f, 589f, 591t
Plasmodium falciparum, 525
Platelet-activating factor, 205, 604t
  synthesis of, 206f, 207, 208f
Platelets, 593
  activation/aggregation of, 583, 589–591, 590f
    aspirin affecting, 591
Pleckstrin, in platelet activation, 589
PLP. See Pyridoxal phosphate

Pluripotent stem cells, 593
PNMT. See Phenylethanolamine-N-methyltransferase
pOH, in pH calculation, 9
Point mutations, 356
  recombinant DNA technology in detection of, 397, 397f, 398t
Pol II
  phosphorylation of, 344
  in preinitiation complex formation, 345
  in transcription, 343, 345f
    formation, 343–344
Polarity
  of DNA replication/synthesis, 326–327
  of protein synthesis, 359
  of xenobiotics, metabolism and, 609
Polyacrylamide gel electrophoresis, for protein/peptide purification, 23–24, 24f, 25f
Polyadenylation sites, alternative, 386
Polyamines, synthesis of, 264, 265f, 266f
Polycistronic mRNA, 370
Polycythemia, 49
Polydystrophy, pseudo-Hurler, 524, 537t
Polyelectrolytes, peptides as, 19
Polyfunctional acids, nucleotides as, 288
Polyisoprenoids, in cholesterol synthesis, 225, 226f
Polyisoprenol, in N-glycosylation, 513
Polymerase chain reaction (PCR), 60, 395, 395f, 404
  in microsatellite repeat sequence detection, 319
  in primary structure determination, 25–26
Polymerases
  DNA, 322–324, 323f, 324t
    in recombinant DNA technology, 390t
  RNA, DNA-dependent, in RNA synthesis, 336–337, 336f, 339t
Polymorphisms, 396
  acetyltransferase, 613
  cytochrome P450, 611, 613t
  microsatellite, 319, 404
  microsatellite DNA, 400
  plasma protein, 567–568
  restriction fragment length. See Restriction fragment length polymorphisms single nucleotide, 404
Polynucleotide kinase, in recombinant DNA technology, 390t
Polynucleotides, 289–291
  posttranslational modification of, 290–291
Polyol (sorbitol) pathway, 182
Polypeptides
  protein synthesis on, 22f
  receptors for, 427
  sequencing of
    cleavage in, 25
    Sanger's determination of, 24–25
Polyphosphoinositide pathway, platelet activation and, 589–591
Polyprenoids, 128, 128f
Polyribosomes (polysomes), 308, 365, 488
  protein synthesis on, 487, 488f, 489f, 495
    plasma proteins, 567
  signal hypothesis of binding of, 493–495, 494t, 494f
Polysaccharides, 113, 117–119, 118f, 119f. See also specific type
Polysomes. See Polyribosomes
Poly(A) tail, of mRNA, 308, 349
  in initiation of protein synthesis, 361

Polytene chromosomes, 315, 316f
Polyubiquitinated target protein, 498
Polyunsaturated fatty acids, 122, 123t. *See also*
        Fatty acids; Unsaturated fatty acids
    dietary, cholesterol levels affected by, 231
    eicosanoids formed from, 198, 200, 201f, 202f
    essential, 198, 198f
    synthesis of, 199, 199f
POMC. *See* Pro-opiomelanocortin (POMC)
        peptide family
Pompe disease, 160t
Porcine stress syndrome, 552
Porphobilinogen, 272, 273f, 275f
Porphyrias, 271, 273, 275–278, 276f, 277f, 277t
Porphyrinogens, 273
    accumulation of in porphyria, 276–277
Porphyrins, 271–272, 271f, 272f
    absorption spectra of, 274–275, 276f
    heme synthesis and, 272–274, 273f, 274f, 275f,
        276f
    reduced, 273
    spectrophotometry in detection of, 275
Positive nitrogen balance, 465
Positive regulators, of gene expression, 369, 370t,
        373, 375
Posthepatic jaundice, 282f
Posttranslational processing, 31, 40–41, 41f, 366
    of collagen, 529, 529t
    in membrane assembly, 504
Posttranslational translocation, 488
Potassium, 480t, 627
    in extracellular and intracellular fluid, 407, 407t
    permeability coefficient of, 409f
Power stroke, 548
PPI. *See* Peptidyl prolyl isomerase
PPi. *See* Pyrophosphate
PR. *See* Progesterone
Pravastatin, 232

PRE. *See* Progestin response element
Precise metabolic changes that commit a cell to
        dying, 639
Prednisone, 629
Pregnancy
    estriol synthesis in, 432
    fatty liver of, 191
    hypoglycemia during, 172
    iron needs during, 571
Pregnancy toxemia of ewes (twin lamb disease)
    fatty liver and, 218
    ketosis in, 191
Pregnenolone, 430f
    in adrenal steroidogenesis, 430, 430f, 431f
    in testicular steroidogenesis, 432, 433f
Prehepatic jaundice, 282f
Preinitiation complex, 337, 337, 345–346
    assembly of, 345–346
    in protein synthesis, 359, 360f
Prekallikrein, 584f, 585
Premenstrual syndrome, vitamin B6 in manage-
        ment of, sensory neuropathy and,
        476
Prenatal diagnosis, recombinant DNA technology
        in, 398
Preprocollagen, 529
Preprohormone, insulin synthesized as, 438f, 439
Preproparathyroid hormone (preproPTH), 439,
        439f
Preproprotein, albumin synthesized as, 569
Preproteins, 487, 567

Presequence. *See* Signal peptide
Preventive medicine, biochemical research affect-
        ing, 2
Primaquine-sensitive hemolytic anemia, 598
Primary amyloidosis, 576
Primary hypothyroidism, 635
Primary structure, 21–29, 31. *See also* Protein
        sequencing
    amino acid sequence determining, 18
    Edman reaction in determination of, 25, 26
    genomics in analysis of, 28
    molecular biology in determination of, 25–26
    of polynucleotides, 291
    proteomics and, 28–29
    Sanger's technique in determination of, 24–25
Primary transcript (pri-miRNAs), 350, 336
pri-miRNAs. *See* Primary transcripts
Primases, DNA, 323f, 323t, 324
Primosome, 324, 404
Prion diseases (transmissible spongiform enceph-
        alopathies), 39–40
Prion-related protein (PrP), 40
Prions, 39–40
Pro-inflammatory molecules, 519
Pro-opiomelanocortin (POMC) peptide family,
        441, 441f. *See also* specific type
Pro-oxidants, 485, 597. *See also* Free radicals
Proaccelerin (factor V), 585, 585t, 586f
Proalbumin, 502
Proaminopeptidase, 462
Probes, 392–393, 404. *See also* DNA probes
Procarcinogens, 609
Processivity, DNA polymerase, 324
Prochymotrypsin, activation of, 79, 80f
Procollagen, 366, 480, 529
Procollagen aminoproteinase, 529
Procollagen carboxyproteinase, 529
Procollagen N-proteinase, disease caused by defi-
        ciency of, 530t
Proconvertin (factor VII), 584f, 585, 585t
    coumarin drugs affecting, 587–588
Prodrugs, 74, 609
    metabolic transformation of, 74
Products, 51
Proelastase, 462
Proenzymes, 79
    rapid response to physiologic demand and, 79
Profiling, protein, RNA transcript and, 400–403
Progeria, 586
Progesterone, 429f, 430f
    binding of, 442t, 443
    receptors for, 455
    synthesis of, 430, 432, 434f
Progestin response element, 445t
Progestins, binding of, 443
Prohormones, 366
Proinsulin, 438, 438f
Prokaryotic gene expression. *See also* Gene
        expression
    eukaryotic gene expression compared with,
        384t, 385–387
    as model for study, 370
    unique features of, 370–371
Prolactin, 427
    localization of gene for, 396t
    receptor for, 426
Proline, 15t
    accumulation of (hyperprolinemia), 249–250
    catabolism of, 249–250, 249f
    metabolism of, 263f

synthesis of, 236, 237f
Proline-*cis*, *trans*-isomerase, protein folding and,
        39, 39f
Proline dehydrogenase, block of praline catabo-
        lism at, 249
Proline hydroxylase, vitamin C as coenzyme for,
        480
Prolyl and lysyl hydroxylases, deficiencies of,
        530
Prolyl hydroxylase reaction, 238, 238f, 527
Promoter recognition specificity, 337
Promoters, in transcription, 336, 336f
    alternative use of in regulation, 348, 349f, 386
    bacterial, 338–339, 340f
    eukaryotic, 339–343, 340f, 341f, 342f, 377
Promoter site, in operon model, 371, 372f
Proofreading, DNA polymerase, 324
Proparathyroid hormone (proPTH), 439, 438f
Propionate
    blood glucose and, 170
    in gluconeogenesis, 166f, 165
    metabolism of, 165–166, 1267f
Propionic acid, 122t
Propionyl-CoA
    fatty acid oxidation yielding, 185
    methionine in formation of, 257, 258f
Propionyl-CoA carboxylase, 165, 167f
Proproteins, 40, 79, 366
Propyl gallate, as antioxidant/food preservative,
        128
Prostacyclins, 122
    clinical significance of, 202
    clotting/thrombosis affected by, 591, 591t
Prostaglandin E2, 122, 123f
Prostaglandin H synthase, 200
Prostaglandins, 122, 123f, 193, 200
    cyclooxygenase pathway in synthesis of,
        200–201, 201f, 202f
Prostanoids, 122, 129
    clinical significance of, 202
    cyclooxygenase pathway in synthesis of,
        200–201, 201f, 202f
Prosthetic groups, 52
    in catalysis, 52, 53f
Protamine, 587
Protease rennin, 51
Proteases/proteinases, 8, 462, 607t. *See also* spe-
        cific type
    in cartilage, 542
    as catalytically inactive proenzymes, 79
    cleaving synaptobrevin, 502
    mucin resistance to, 510
    of neutrophils, 607, 607t
    in protein degradation, 239, 240f, 462
Proteasome, 495
    degradation in, 498
Proteasomes, misfolded proteins in, 497
    ubiquination in, 498–499, 499f
Protein. *See also* specific proteins
    life cycle of, 22f
    prenylation, 225
    translocation of, 22f
Protein 4.1, in red cell membranes, 600f, 600t,
        601
Protein C, in blood coagulation, 586t, 587
Protein-conducting channel, 495
Protein Database, 86
Protein deficiency, 636
Protein degradation, ubiquitin in, 240, 498–499,
        499f

Protein disulfide isomerase, protein folding and, 39, 497

Protein-DNA interactions, bacteriophage lambda as paradigm for, 373–376, 373f, 374f, 375f, 376f

Protein-energy malnutrition, 636–637

Protein folding, 22f, 38–39, 39f, 626
    after denaturation, 39
    chaperones and, 488, 496–497, 497t
    misfolding in, 497
        endoplasmic reticulum–associated degradation of, 497, 498f
    ubiquination in, 498–499, 499f

Protein Information Resource (PIR), 86

Protein kinase A (PKA), 447, 448f

Protein kinase B (PKB), in insulin signal transmission, 452, 452f

Protein kinase C (PKC)
    in calcium-dependent signal transduction, 450, 451f
    in platelet activation, 589, 590f

Protein kinase D1, in insulin signal transmission, 452–453, 452f

Protein kinase-phosphatase cascade, as second messenger, 427, 428t

Protein kinases, 80, 619
    in cAMP-dependent signal transduction, 447–448, 448f
    in cGMP-dependent signal transduction, 449
    DNA-dependent, in double-strand break repair, 333
    in glycogen metabolism, 160–162, 161f, 163, 163f
    in hormonal regulation, 426–427, 450–454
        of lipolysis, 221–222, 221f
    in initiation of protein synthesis, 359
    in insulin signal transmission, 451–453, 452f
    in Jak/STAT pathway, 453, 453f
    in protein phosphorylation, 80f, 81, 82f

Protein-losing gastroenteropathy, 568

Protein misfolding
    accumulation in endoplasmic reticulum in, 497
    endoplasmic reticulum associated degradation of, 497, 498f
    ubiquination in, 498, 499f

Protein phosphatase-1, 161f, 162, 163f

Protein phosphatases, 80–82, 82f. See also Phosphatases

Protein phosphorylation. See Phosphorylation, protein

Protein profiling, RNA transcript and, 400–403

Protein-RNA complexes, in initiation, 359–362, 360f

Protein S, in blood coagulation, 586t, 587

Protein sequencing
    Edman reaction in, 25, 27f
    genomics and, 28
    mass spectrometry in, 26–28, 27f, 27t
    molecular biology in, 25–26
    peptide purification for, 21–24
    polypeptide cleavage and, 24–25
    proteomics and, 28–29
    purification for, 21–24, 24f, 25f
    Sanger's method of, 24–25

Protein sorting, 487–504
    chaperones and, 496–497, 497t
    cotranslational insertion and, 494f, 495
    disorders, mutations in genes encoding, 504
    Golgi apparatus in, 487, 489f, 496, 487
    importins and exportins in, 491, 492f

KDEL amino acid sequence and, 496, 488t
membrane assembly and, 502–504, 503f, 504t
mitochondria in, 488–490, 490f
peroxisomes/peroxisome disorders and, 493, 493f
retrograde transport and, 496
signal hypothesis of polyribosome binding and, 493–495, 494f, 494t
signal sequences and, 488, 488f
transport vesicles and, 499–502, 500t, 501f
unfolded protein response in, 496

Protein structure, 619
    primary, 21–30. See also Primary structure
    quaternary, 32, 35, 36f
        of hemoglobins, allosteric properties and, 44–48
        stabilizing factors and, 36–37
    secondary, 31–35, 32f, 33f, 34f
        peptide bonds affecting, 32, 32f
        supersecondary, 35
    tertiary, 32, 34f, 35
        stabilizing factors and, 36–37

Protein synthesis
    amino acids in, 133, 133f
    on ribosomes, 22f

Protein turnover, 76, 239
    membranes affecting, 503–504
    rate of enzyme degradation and, 76–77

Proteinases. See Proteases/proteinases

Proteins. See also specific type and Peptides
    acute phase, 568–569, 569t
        negative, vitamin A as, 470
    L-amino acids in, 14–16
    asymmetry of, membrane assembly and, 503, 503f
    binding, 442–443, 442t
    catabolism of, 239–246
    classification of, 31
    configuration of, 31
    conformation of, 31
        peptide bonds affecting, 19
    core, 533, 534
        in glycosaminoglycan synthesis, 534–535
    degradation of, to amino acids, 239, 240f
    denaturation of
        protein refolding and, 39
        temperature and, 66
    dietary
        digestion and absorption of, 462
        metabolism of, in fed state, 137–139
        requirements for, 465
    dimeric, 35
    domains of, 35, 37f
    in extracellular and intracellular fluid, 407, 407t
    fibrous, 31
        collagen as, 40
    function of, bioinformatics in identification of, 29
    fusion, in enzyme study, 60–61, 60f
    globular, 31
    Golgi apparatus in glycosylation and sorting of, 487
    identification, by homology, 87–88
    import of, by mitochondria, 488–491, 491t
    loss of in trauma/infection, 465
    in membranes, 409–410, 410t, 506. See also Glycoproteins; Membrane proteins
        ratio of to lipids, 407–408, 407f
    modular principles in construction of, 31, 39

monomeric, 35
phosphorylation of, 79, 80, 80f, 81t. See also Phosphorylation
posttranslational modification of, 31, 40–41, 41f, 366
purification of, 21–24
receptors as, 421, 426–427
sequences or molecules directing, 488t
soluble, 31
structure of, 32–38
    diseases associated with disorders of, 39–40
    folding and, 38–39, 39f
    higher orders of, 31–41
    molecular modeling and, 38
    nuclear magnetic resonance spectroscopy in analysis of, 38
    primary, 21–29, 31. See also Primary structure
    prion diseases associated with alteration of, 39–40
    quaternary, 32, 35, 36f
    secondary, 31–35, 32f, 33f, 34f, 35f, 36f
    supersecondary, 35
    tertiary, 32, 35, 36f
    x-ray crystallography in analysis of, 37–38
synthesis of, 139, 353–368. See also Protein sorting
    elongation in, 362–364, 362f
    environmental threats affecting, 365
    in fed state, 137–139
    genetic code/RNA and, 306–307, 307t, 354–355. See also Genetic code
    inhibition of by antibiotics, 367–368, 367f
    initiation of, 359–362, 360f, 362f
    by mitochondria, 488–491, 491t
    modular principles in, 31, 39
    polysomes in, 365, 487, 488f
    posttranslational processing and, 366–367
    recognition and attachment (charging) in, 35, 355f
    recombinant DNA techniques for, 396
    reticulocytes in, 596
    in ribosomes, 135, 136f
    termination of, 364–365, 364f
    translocation and, 363
    viruses affecting, 365–366, 367f
targeting, to matrix, 488
transmembrane
    ion channels as, 416, 417f, 417t
    in red cells, 600–601, 600f, 600t
transport, 442–443, 442t
unknown, identification of, 88–89
xenobiotic cell injury and, 613

Proteoglycans, 119, 527, 529, 533–534, 533f, 537–539, 538t. See also Glycosaminoglycans
    in bone, 542t
    carbohydrates in, 533, 533f, 534
    in cartilage, 542–543
    disease associations and, 539
    functions of, 537–538, 538t
    galactose in synthesis of, 179–180, 180f
    link trisaccharide in, 510

Proteolysis, 502
    in covalent modification, 79, 80f
    in prochymotrypsin activation, 79, 80f

Proteolytic cleavage, 22f

Proteolytic enzymes, 51

Proteome/proteomics, 28–29, 404
    plasma, 566
Proteomics, 4
Prothrombin (factor II), 585t, 585, 586f
    activation of, 585
    coumarin drugs affecting, 473, 587–588
    in vitamin K deficiency, 473
Prothrombin to thrombin activation, by factor
        Xa, 585
Proton acceptors, bases as, 10
Proton donors, acids as, 10
Proton motive force, 106
Proton pumps, 106
    respiratory chain as, 103
Proton-translocating transhydrogenase, 110–111
Protons, transport of, by hemoglobin, 47, 48f
Protoporphyrin, 272, 272f
    incorporation of iron into, 272f, 273
Protoporphyrin III, 273, 276f
Protoporphyrinogen III, 273, 276f
Protoporphyrinogen oxidase, 273, 275f, 276f
Provitamin A carotenoids, 468
Proximal histidine (histidine F8)
    in oxygen binding, 43, 44f
    replacement of in hemoglobin M, 49
Proximity, catalysis by, 53
PrP (prion-related protein), 40
PRPP
    in purine synthesis, 293, 294f
    in pyrimidine synthesis, 295, 297, 298f
PRPP glutamyl amidotransferase, 293, 294f
PRPP synthetase, defect in, gout caused by, 297
Pseudo-Hurler polydystrophy, 524, 536, 537t
    as milder condition, 524
Pseudogenes, 321, 404
Pseudomonas aeruginosa infection, in cystic
        fibrosis, 525
Psi angle, 32, 32f
PstI, 389t
PstI site, insertion of DNA at, 391, 392f
Psychiatric disorders, 619
PTA. *See* Plasma thromboplastin antecedent
PTC. *See* Plasma thromboplastin component
Pteroylglutamic acid. *See* Folic acid
PTH. *See* Parathyroid hormone
PTS1, 492
PTS2, 492
PTSs. *See* Peroxisomal-matrix targeting
        sequences
PubMed, 86, 648
"Puffs," polytene chromosome, 315, 316f
Pulmonary infection/heart failure, 625
Pumps, 406, 413
    in active transport, 419, 419f
Purification, protein/peptide, 21–24
Purine nucleoside phosphorylase deficiency, 299
Purines, 285
Purines/purine nucleotides, 285–289, 285f, 288f
    biosynthesis of, 292–293, 293f, 294f, 295f, 296f
        catalysts in, 293, 293f
        pyrimidine synthesis coordinated with,
            297
    dietary nonessential, 292
    metabolism of, 292–301
        disorders of, 297–299
        gout as, 297
        uric acid formation and, 297, 299f
        regulation of, 297f, 299f
        "salvage" reactions in, 293, 294f, 296f
    ultraviolet light absorbed by, 288

Puromycin, 367, 367f
Putrescine, in polyamine synthesis, 265f
Pyranose ring structures, 114, 114f, 115f
Pyridoxal phosphate, 52, 475, 475f
    in heme synthesis, 272
    in urea biosynthesis, 241
Pyridoxine/pyridoxal/pyridoxamine (vitamin B6),
        468t, 475–476, 475f
    deficiency of, 468t, 475
        xanthurenate excretion in, 257, 257f
    excess/toxicity of, 476
Pyrimethamine, 477
Pyrimidine analogs, in pyrimidine nucleotide
        biosynthesis, 297
Pyrimidine nucleotide biosynthesis, 295–296
    regulation of, 296f, 297f
Pyrimidines, 285, 299f
Pyrimidines/pyrimidine nucleotides, 285–289,
        285f, 288f
    dietary nonessential, 292
    metabolism of, 292–301, 300f
        diseases caused by catabolite overproduction
            and, 300–301
        water-soluble metabolites and, 299–300,
            300f
    precursors of, deficiency of, 300
    synthesis of, 285–289, 298f
        catalysts in, 295
        purine synthesis coordinated with, 297
        regulation of, 297, 298f
    ultraviolet light absorbed by, 288
Pyrophosphatase, inorganic
    in fatty acid activation, 96, 184
    in glycogen biosynthesis, 157, 158f
Pyrophosphate
    free energy of hydrolysis of, 94t
    inorganic, 96, 96f
Pyrrole, 43, 44f
Pyruvate, 132
    formation of, in amino acid carbon
            skeleton catabolism, 248–254,
            251f–252f
    in gluconeogenesis, 137
    oxidation of, 146–147, 147f, 153–155, 153f,
            154f, 155t. *See also* Acetyl-CoA;
            Glycolysis
        clinical aspects of, 155
        enzymes in, 168t
        gluconeogenesis and, 165, 166f
Pyruvate carboxylase, 146, 146f, 168t, 630
    in gluconeogenesis regulation, 146, 146f, 165,
            168t
Pyruvate dehydrogenase, 146, 147, 147f, 154, 153f,
            168t
    deficiency of, 155
    regulation of, 154–155, 154f
        acetyl-CoA in, 153–155
        acyl-CoA in, 154f, 201
        thiamin diphosphate as coenzyme for, 473
Pyruvate dehydrogenase complex, 154
Pyruvate kinase, 168t
    deficiency of, 155, 598
    gluconeogenesis regulation and, 167
    in glycolysis, 150, 151f, 168t
        regulation and, 152–153

Q cycle, 103, 106f, 104, 107f
Q-cytochrome *c* oxidoreductase, 103, 104, 105f, 106f
Q10 (temperature coefficient), enzyme-catalyzed
        reactions and, 66

QT interval, congenitally long, 423t
Quarter staggered alignment, 528
Quaternary structure, 32, 35, 36f
    of hemoglobins, allosteric properties and, 44–48
    stabilizing factors and, 36–37

R groups, amino acid properties affected by, 18,
        18t
    pK/pKa, 18
R (relaxed) state, of hemoglobin, oxygenation and,
        47, 46f, 47f
Rab effector proteins, 500, 502
Rab molecules, 500
Rab protein family, 502
Rab proteins, 500
RAC3 coactivator, 457, 457t
Radiation, nucleotide excision-repair of DNA
        damage caused by, 332
Radioimmunoassays (RIA), 58
Ran protein, 491, 492f
Rancidity, peroxidation causing, 128
Rapamycin, mammalian target of (mTOR), in in-
        sulin signal transmission, 452, 452f
RAR. *See* Retinoic acid receptor
RARE. *See* Retinoic acid response element
Rate constant, 65
    Keq as ratio of, 65
Rate-limiting reaction, metabolism regulated by,
        76
Rb protein. *See* Retinoblastoma protein
Reactant concentration, chemical reaction rate
        affected by, 64–65
Reactive oxygen species. *See* Free radicals
Rearrangements, DNA
    in antibody diversity, 322, 385, 580
    recombinant DNA technology in detection of,
            397–398, 398t
recA, 375, 376f
Receptor-associated coactivator 3 (RAC3 coactiva-
        tor), 457, 457t
Receptor-effector coupling, 426
Receptor-mediated endocytosis, 421, 421f
Receptors, 421, 426–427. *See also* specific type
    activation of in signal generation, 444–446, 445f
    nuclear, 427, 455, 456t, 457f
Recognition domains, on hormone receptors, 426
Recognition marker, 523
Recombinant DNA/recombinant DNA
        technology, 388–403
    base pairing and, 393
    blotting techniques in, 393–394, 393f
    chimeric molecules in, 388–395
    cloning in, 390–391, 391f, 392f, 391t
    definition of, 404
    DNA ligase in, 389–390
    DNA sequencing in, 394, 394f
    double helix structure and, 394
    in enzyme study, 60–61, 60f
    gene mapping and, 396, 396t
    in genetic disease diagnosis, 396, 397f, 398,
            398t, 399f
    hybridization techniques in, 393–394
    libraries and, 392
    oligonucleotide synthesis in, 394–395
    polymerase chain reaction in, 394, 395f
    practical applications of, 395–403
    restriction enzymes and, 388–390, 389t, 391f,
            393t, 399f
    terminology used in, 403–4
    transcription and, 396, 397f

Recombinant erythropoietin (epoetin alfa/EPO), 517, 595
Recombinant fusion proteins, in enzyme study, 60–61, 61f
Recombination, chromosomal, 319–320, 320f, 321f
Recruitment hypothesis, of preinitiation complex formation, 346
Recurrent bronchopulmonary infections, 625
Recurrent infections, 523, 576
Recycle, 500
Red blood cells, 593–599. *See also* Erythrocytes
Red thrombus, 583
Red (slow) twitch fibers, 561, 561t
Redox (oxidation-reduction) potential, 98, 99t
Redox state, 187
Reduced porphyrins, 273
Reducing equivalents
    in citric acid cycle, 143–145, 145f
    in mitochondria, 103, 104f
    in pentose phosphate pathway, 177
Reduction, definition of, 98
Reductive activation, of molecular oxygen, 610
Reference values, 616
Refsum disease, 191, 493, 493t
Regional asymmetries, membrane, 410
Regulated secretion, 487
Regulatory domains, 35
Regulatory proteins, binding of to DNA, motifs for, 381–384, 381t, 382f, 383f, 384f
Regurgitation hyperbilirubinemia, 281
Relatively stable components, 586
Relaxation phase
    of skeletal muscle contraction, 549, 551
    of smooth muscle contraction
        calcium in, 557
        nitric oxide in, 559–560, 559f
Relaxed (R) state, of hemoglobin, oxygenation and, 46f, 47, 47f
Releasing factors (RF1/RF3), in protein synthesis termination, 364f, 365
Remnant removal disease, 232t
Renal excretion, 631
Renal glomerulus, laminin in basal lamina of, 532–533
Renal stones (urate calculi), 632
Renal threshold for glucose, 172
Renaturation, DNA, base pair matching and, 304
Renin, 440, 440f
Repeat sequences
    amino acid, 510–511, 511f
    short interspersed (SINEs), 318, 404
Repetitive-sequence DNA, 318
Replacement of glycine, 541
Replication bubbles, 327–328, 327f
Replication fork, 324, 327f
Replication/synthesis. *See* DNA; RNA
Reporter genes, 379–380, 380f
Repression, enzyme
    enzyme synthesis control and, 77
    in gluconeogenesis regulation, 167
Repressor protein/gene, lambda (cI), 373–376, 373f, 374f, 375f, 376f
Repressors, 343
    in gene expression, 369, 371, 372, 379
        tissue-specific expression and, 379
Reproduction, prostaglandins in, 193
Residues, peptide, 19
Respiration, oxygen for, 98
Respiratory and gastrointestinal tracts, 625

Respiratory burst, 464, 606
Respiratory chain, 103–112. *See also* Oxidative phosphorylation
    chemiosmotic theory on respiratory control and uncouplers in, 107f, 109–110
    clinical aspects of, 112
    complex I and II in, 103, 104, 105f, 106f
    complex III (Q cycle) in, 103, 104, 107f
    complex IV in, 103, 104–105, 106f
    dehydrogenases in, 100
    energy captured in catabolism from, 95f, 106–107, 155t
    exchange transporters in, 110–112, 109f–111f
    flavoproteins and iron-sulfur proteins in, 103–104
    in mitochondria, 104f
    mitochondrial protein complexes in, 99t, 103, 105f, 106f
    NADH-Q oxidoreductase as electron acceptor in, 104, 104f, 145f
    oxidation of reducing equivalents in, 103, 104f
    oxidative phosphorylation at, 106, 155t
    poison inhibition of, 108–109, 109f
    protein gradient driving ATP synthesis from electron transport in, 106, 107f, 108f
    as proton pump, 103
    substrates for, citric acid cycle providing, 143, 144f
Respiratory control, 93, 147
    ATP supply from, 108, 108t
    chemiosmotic theory on, 109–110, 109f
Respiratory distress syndrome, surfactant deficiency causing, 124, 209–210
Restriction endonucleases/enzymes, 60, 311, 389–390, 389f, 389t, 390
    in recombinant DNA technology, 389–390, 389t, 389f, 390t, 391f
Restriction enzymes. *See* Restriction endonucleases/enzymes
Restriction fragment length polymorphisms (RFLPs), 60, 398, 399f
    in forensic medicine, 395
Restriction map, 389
Retention hyperbilirubinemia, 281
Reticulocytes, in protein synthesis, 596
Retina
    gyrate atrophy of, 250
    retinaldehyde in, 468, 469f
Retinal. *See* Retinol
Retinaldehyde, 468, 469f
Retinitis pigmentosa, essential fatty acid deficiency and, 200
Retinoblastoma protein, 329
Retinoic acid, 468, 469f. *See also* Retinol
    functions of, 468
    receptors for, 455, 468
Retinoic acid response element, 445t
Retinoid X receptor (RXR), 455, 456, 456f, 469
Retinoids, 468, 469f. *See also* Retinol
Retinol, 468, 468t, 469f. *See also* Vitamin A
    deficiency of, 468t
    functions of, 468t, 469, 469f
Retinol-binding protein, 569t
Retrograde strand, in DNA replication, 324
Retrograde transport, 496, 499
    from Golgi apparatus, 496
    of misfolded proteins, 497
Retroposons/retrotransposons, 318
Retrotranslocation, 497
Retroviruses, reverse transcriptases in, 307, 332

Reverse cholesterol transport, 217, 217f, 225, 230
Reverse engineering approach, 90
Reverse genetics, 625
Reverse transcriptase/reverse transcription, 307, 332, 404
    in recombinant DNA technology, 390t
Reversed-phase high-pressure chromatography, for protein/peptide purification, 23
Reversible covalent modifications, 80–81, 80f, 81t. *See also* Phosphorylation
Reye syndrome, orotic aciduria in, 300
RFLPs. *See* Restriction fragment length polymorphisms
RFs. *See* Releasing factors
Rheumatoid arthritis
    extracellular matrix in, 527
    glycosylation alterations in, 523
Rho-dependent termination signals, 339, 340f
Rhodopsin, 469, 469f
RIA. *See* Radioimmunoassays
Riboflavin (vitamin B2), 98, 468t, 474
    in citric acid cycle, 146
    coenzymes derived from, 52, 474
    deficiency of, 468t, 474
    dehydrogenases dependent on, 99–100
Ribonucleases, 311
Ribonucleic acid. *See* RNA
Ribonucleoprotein particles (RNPs), 365
Ribonucleoside diphosphates (NDPs), reduction of, 295, 297f
Ribonucleosides, 286, 286f
Ribonucleotide reductase complex, 295, 297f
Ribose, 113
    in nucleosides, 286, 286f
    pentose phosphate pathway in production of, 132, 174, 177
D-Ribose, 115f, 116t, 286
Ribose 5-phosphate, in purine synthesis, 292–293, 294f
Ribose 5-phosphate ketoisomerase, 174–177, 176f
Ribose phosphate, pentose phosphate pathway in production of, 174–177, 175f
Ribosomal dissociation, in protein synthesis, 359, 360f
Ribosomal RNA (rRNA), 307, 308–309, 335, 336t. *See also* RNA
    as peptidyltransferase, 362–363, 363t
    processing of, 349
Ribosomes, 308, 310t
    bacterial, 367
    protein synthesis in, 22f, 135, 136f
        dissociation and, 365
Ribozymes, 307, 309, 351
D-Ribulose, 115f, 116t
Ribulose 5-phosphate 3-epimerase, 174, 176f
Richner-Hanart syndrome, 254
Ricin, 367, 510t
Rickets, 468t, 470, 541t
Rieske Fe-S, 104
Right operator, 373–376, 374f, 376f
Rigor mortis, 550, 552
RISC. *See* RNA induced silencing complex
RNA, 302, 305–310, 335–351
    as catalyst, 351
    in chromatin, 312
    classes/species of, 306–307, 307t, 335, 336t
    complementarity of, 306, 307f
    heterogeneous nuclear (hnRNA), 308
        gene regulation and, 348–349

messenger (mRNA), 306, 307–308, 308f, 309f, 335, 336t, 354
 alternative splicing and, 348, 348f, 386
 codon assignments in, 353–354, 354t
 editing of, 350–351
 modification of, 349–350
 nucleotide sequence of, 353–354
 mutations caused by changes in, 356–359, 356f, 357f, 358f
 polycistronic, 370
 recombinant DNA technology and, 388
 relationship of to chromosomal DNA, 318f
 stability of, regulation of gene expression and, 386–387, 386f
 micro (mi) and small interfering (si), 310
modification of, 349–351
processing of, 346–349
 alternative, in regulation of gene expression, 348–349, 349f, 386
 in protein synthesis, 306–307, 307t
ribosomal (rRNA), 307, 308–309, 335, 336t
 as peptidyltransferase, 362–363, 363t
 processing of, 348
silencing, 404
small, 309–310
small nuclear (snRNA), 307, 307t, 310, 335, 336t, 404
splicing, 346–348, 405
 alternative, in regulation of gene expression, 348, 348f, 386
 recombinant DNA technology and, 388
structure of, 305–310, 306f, 307f, 309f, 310f
synthesis of, 303, 335–351
 initiation/elongation/termination in, 336–337, 336f, 338f, 337–338
transfer (tRNA), 307, 308, 310f, 335, 336t, 355–356, 355f
 aminoacyl, in protein synthesis, 362
 anticodon region of, 354
 processing and modification of, 349, 351
 suppressor, 358–359
 xenobiotic cell injury and, 613
RNA editing, 350–351
RNA induced silencing complex (RISC), 350
RNA polymerase III, 337t
RNA polymerases, DNA-dependent, in RNA synthesis, 336–337, 336f, 337t
RNA primer, in DNA synthesis, 324–326, 325f, 326f
RNA probes, 392–393, 404
RNA-RNA duplexes, imperfect, 310
RNA-RNA hybrids, 310
RNA transcript, and protein profiling, 400–403
RNAP. See RNA polymerases
RNase. See Ribonucleases
ROS (reactive oxygen species). See Free radicals
Rossmann fold, 35
Rotor syndrome, 282
Rough endoplasmic reticulum
 glycosylation in, 515f, 516
 in protein sorting, 487, 488f, 489f
 protein synthesis and, 365
 routes of protein insertion into, 495–496, 496f
 signal hypothesis of polyribosome binding to, 493–495, 494f, 494t
Rough ER branch, 493
rRNA. See Ribosomal RNA
RT-PCR, 404
RXR. See Retinoid X receptor

Ryanodine, 551
Ryanodine receptor, 551, 551f
 mutations in gene for, diseases caused by, 552–553, 553f, 613t
RYR. See Ryanodine receptor

S phase of cell cycle, DNA synthesis during, 328–330, 329f, 329t
S_{50}, 70
Saccharopine, in lysine catabolism, 255, 255f
SADDAN phenotype, 543
Salt (electrostatic) bonds (salt bridges/linkages), 8
 oxygen binding rupturing, Bohr effect protons and, 47–48, 48f
"Salvage" reactions
 in purine synthesis, 293, 294f, 297f
 in pyrimidine synthesis, 295
Sanfilippo syndrome, 537t
Sanger's method
 for DNA sequencing, 394, 394f
 for polypeptide sequencing, 24–25
Sanger's reagent (1-fluoro-2,4-dinitro-benzene), for polypeptide sequencing, 25
Sar1, 500
Sarcoglycan complex, 553
Sarcolemma, 545, 628
Sarcomere, 545, 545f
Sarcoplasm, 545
 of cardiac muscle, 553–554
Sarcoplasmic reticulum, calcium level in skeletal muscle and, 551–552, 551f
Sarcosine (N-Methylglycine), 266
SARs. See Structure-activity relationships
Saturated fat, 622
Saturated fatty acids, 121, 122, 122t
 in membranes, 408, 409f
Saturation kinetics, 67f, 69
 sigmoid substrate, Hill equation in evaluation of, 69–70, 69f
Scavenger receptor B1, 217, 217f
Scheie syndrome, 537t
Schiff bases, 519
Schindler disease, 524, 524t
SCID. See Severe combined immunodeficiency disease
Scrapie, 40
Screening, 633
Screening conditions, drug, enzyme kinetics on, 73–74
Scurvy, 234, 468t, 480
 collagen affected in, 41, 480, 530
SDS-PAGE. See Sodium dodecyl sulfatepolyacryl-amide gel electrophoresis
Se gene, 602
Sec12, 500
Sec61p complex, 495
Second messengers, 79, 427, 428t, 446–454, 448t, 450t. See also specific type
 calcium as, 427, 428t, 446
 cAMP as, 159, 427, 428t, 446, 447–449, 447t, 448f
 cGMP as, 289, 427, 428t, 446, 449
 diacylglycerol as, 450, 451f
 inositol trisphosphate as, 450, 451f
 precursors of
  phosphatidylinositol as, 124, 125f
  phospholipids as, 205
Secondary amyloidosis, 576
Secondary hypothyroidism, 635

Secondary structure, 31–35, 32f, 33f, 34f, 35f
 peptide bonds affecting, 32, 32f
 supersecondary, 35
Secreted protein, 494
Secretor (Se) gene, 602
Secretory component, of IgA, 579f
Secretory (exocytotic) pathway, 487
Secretory proteins, 495
Secretory vesicles, 487, 489f
D-Sedoheptulose, 115f
L-Selectin, 521f, 521t
Selectins, 520–522, 521f, 521t
 recruitment of, 521
Selective advantage for growth, 617
Selectivity filter, 417
Selectivity/selective permeability, membrane, 406, 413–416, 414f, 414t, 417f, 417t
Selenium, 480t
 in glutathione peroxidase, 100, 177
Selenocysteine, synthesis of, 238, 238f
Selenophosphate synthetase/synthase, 238, 238f
Self-assembly
 in collagen synthesis, 529
 of lipid bilayer, 408–409
Self-association, hydrophobic interactions and, 7
Self-renewal, 593
Sensory neuropathy, in vitamin B6 excess, 476
Sepharose-lectin column chromatography, in glycoprotein analysis, 507t
Sequential displacement reactions, 72, 73f
Serine, 15t
 catabolism of, pyruvate formation and, 251, 251f
 conserved residues and, 56, 57t
 in cysteine and homoserine synthesis, 236, 237f
 in glycine synthesis, 236, 236f
 phosphorylated, 266
 synthesis of, 236, 236f
 tetrahydrofolate and, 477, 477f
Serine 195, in covalent catalysis, 55, 55f
Serine hydroxymethyltransferase, 251, 251f, 477
Serine protease inhibitor, 574
Serine proteases. See also specific type
 conserved residues and, 56, 57t
 in covalent catalysis, 54–55, 55f
 zymogens of, in blood coagulation, 583, 584, 586t
Serotonin, 264–265, 604t
 biosynthesis and metabolism of, 267f
Serpin, 574
Serum, 627
Serum amyloid A (SAA), 576
Serum complement, 418
Serum enzymes, in clinical diagnosis, 59t
Serum prothrombin conversion accelerator (SPCA/factor VII), 584f, 585t
 coumarin drugs affecting, 587–588
Severe combined immunodeficiency disease (SCID), 616–617, 617f
Sex (gender), xenobiotic-metabolizing enzymes affected by, 613
Sex hormone-binding globulin (testosterone-estrogen-binding globulin), 442t, 443, 569t
SGLT 1transporter protein, 460, 460f
SGOT. See Aspartate aminotransferase
SGPT. See Alanine aminotransferase
SH2 domains. See Src homology 2 (SH2) domains
SHBG. See Sex hormone-binding globulin
Shell, of coat proteins, 500

Short interspersed repeat sequences (SINEs), 318, 404
Short-limbed dwarfism., 543
Short-term memory, loss of, 618
Shoshin beriberi, 473
SI nuclease, in recombinant DNA technology, 390t
si RNAs, 310
SI units (Systeme International d'Unites), 616
Sialic acids, 119, 119f, 126, 180, 181f
    in gangliosides, 181f, 209, 210f
    in glycoproteins, 119t, 508t
Sialidosis, 524, 524t, 536, 537t
Sialoprotein, bone, 538t, 540
Sialyl-Lewis^x, selectins binding, 521–522, 521f
Sialylated oligosaccharides, selectins binding, 521
Sickle cell disease, 357
    pedigree analysis of, 398, 399f
    recombinant DNA technology in detection of, 397
Side chain cleavage enzyme P450 (P450scc), 430, 430f, 432
Side chains, in porphyrins, 271–272, 271f
Side effects, undesirable, 617
Sigmoid substrate saturation kinetics, Hill equation in evaluation of, 69–70, 69f
Signal. See also Signal peptide
    generation of, 444–446, 445f, 445t, 446f
    in recombinant DNA technology, 404
    transmission of, 406, 422. See also Signal transduction across membrane
Signal hypothesis, of polyribosome binding, 493–495, 494f, 494t
Signal peptidase, 494f, 495
Signal peptide, 495, 487, 494
    albumin, 569
    in protein sorting, 488, 488f, 489f, 494, 494f
    in proteins destined for Golgi apparatus membrane, 499
Signal recognition particle, 494
Signal sequence, 500, 502. See also Signal peptide
Signal transducers and activators of transcription (STATs), 453, 453f
Signal transduction, 444–458
    GPI-anchors in, 519
    hormone response to stimulus and, 444, 445f
    intracellular messengers in, 446–454, 448t, 450t. See also specific type
    in platelet activation, 590, 590f
    signal generation and, 444–446, 445f, 445t, 446f
    transcription modulation and, 454–458, 456f, 456t, 457f
Silencers, 343
Silencing mediator for RXR and TR (SMRT), 457, 457t
Silencing RNA, 404
Silent mutations, 356
Silicon, 480t
Simple diffusion, 413, 414f, 414t, 502
Simvastatin, 232
SINEs. See Short interspersed repeat sequences
Single displacement reactions, 72, 73f
Single molecule enzymology, 56–57, 57f
Single nucleotide polymorphism (SNP), 86–87, 404
Single-pass membrane proteins, glycophorins as, 601, 600f, 600t
Single-stranded DNA, replication from, 322. See also DNA
Single-stranded DNA-binding proteins (SSBs), 323f, 323t, 324

siRNA-miRNA complexes, 310
Sister chromatid exchanges, 321–322, 322f
Sister chromatids, 315–316, 316f
Site-directed mutagenesis, in enzyme study, 61
Site-specific DNA methylases, 388
Site specific integration, 320–321
Size exclusion chromatography, for protein/peptide purification, 21–22, 24f
SK. See Streptokinase
Skeletal muscle, 545, 555t. See also Muscle; Muscle contraction
    glycogen stores in, 560–561
    metabolism in, 134–135, 134f
        lactate production and, 150–152
    as protein reserve, 562
    slow (red) and fast (white) twitch fibers in, 561, 561t
Skin
    essential fatty acid deficiency affecting, 201–202
    mutant keratins and, 564
    vitamin D3 synthesis in, 434, 435f, 469f, 470
Sleep, prostaglandins in, 193
Sliding filament cross-bridge model, of muscle contraction, 546, 547f
Slow acetylators, 613
Slow-reacting substance of anaphylaxis, 202
Slow (red) twitch fibers, 561, 561t
Sly syndrome, 537t
Small interfering (si) RNAs, 310
Small intestine
    monosaccharide digestion in, 460, 460f
Small molecules, 625
    developing, 619
Small monomeric GTPases, 491, 502
Small nuclear RNA (snRNA), 307, 307t, 310, 335, 336t, 346, 404
Small peptides, presenting, 499
Small RNA, 309–310
Smoking
    CYP2A6 metabolism of nicotine and, 611
    cytochrome P450 induction and, 611
    on methionine, 574
    nucleotide excision-repair of DNA damage, 332
Smooth endoplasmic reticulum, cytochrome P450 isoforms in, 610
Smooth muscle, 545, 555t
    actin-myosin interactions in, 558t
    contraction of
        calcium in, 557, 558f
        myosin-based regulation of, 557
        myosin light chain phosphorylation in, 557
    relaxation of
        calcium in, 557–558
        nitric oxide in, 559–560, 559f
SMRT, 457, 457t
SNAP 25, 502
α-SNAP, 500
SNAP (soluble NSF attachment factor) proteins, 500, 501f, 502
SNARE proteins, 499–500, 501f, 502
SNAREpins, 500
SNP. See Single nucleotide polymorphism
snRNA. See Small nuclear RNA
Sodium, 480t
    in extracellular and intracellular fluid, 407, 407t
    permeability coefficient of, 409f
Sodium-calcium exchanger, 449
Sodium dodecyl sulfate-polyacrylamide gel electrophoresis

for protein/peptide purification, 24, 25f, 25f
    red cell membrane proteins determined by, 599–600, 600f
Sodium-potassium pump (Na^+-K^+ ATPase), 419, 419f
    in glucose transport, 419–420, 420f
Solid food, 620
Solubility point, of amino acids, 18
Soluble NSF attachment factor (SNAP) proteins, 500, 501f, 502
Soluble proteins, 495, 516
Solutions, aqueous, K_w of, 9
Solvent, water as, 6, 7f
Sorbitol, in diabetic cataract, 182
Sorbitol dehydrogenase, 179, 179f
Sorbitol intolerance, 182
Sorbitol (polyol) pathway, 182
Sore throat, 627
Soret band, 274
Southern blot transfer procedure, 304, 393, 393f, 405
Southwestern blot transfer procedure, 393, 405
SPARC (bone) protein, 538t
Sparteine, CYP2D6 in metabolism of, 611
SPCA. See Serum prothrombin conversion accelerator
Special translocation mechanisms, 491
Specific acid/base catalysis, 53
Specificity, enzyme, 51, 52f
Spectrin, 600, 600f, 601f, 600t, 601
    abnormalities of, 601–602
Spectrometry
    covalent modifications detected by, 26–28, 26f, 26t
    for glycoprotein analysis, 507, 507t
Spectrophotometry
    for NAD(P)^+-dependent dehydrogenases, 57–58, 58f
    for porphyrins, 275
Spectroscopy, nuclear magnetic resonance (NMR)
    for glycoprotein analysis, 507, 507t
    protein structure demonstrated by, 38
Spermidine, synthesis of, 264, 265f
Spermine, synthesis of, 264, 265f
Spherocytosis, hereditary, 423t, 601–602, 602f
Sphingolipidoses, 210–211, 210t
Sphingolipids, 205
    metabolism of, 208–209, 209f, 210f
        clinical aspects of, 209–211, 210t
    in multiple sclerosis, 210
Sphingomyelins, 125, 126f, 208, 209f, 503
    in membranes, 408
    membrane asymmetry and, 410–411
Sphingophospholipids, 121
Sphingosine, 125, 126f
Spina bifida, folic acid supplements in prevention of, 478
Spliceosome, 346, 405
Spongiform encephalopathies, transmissible (prion diseases), 40
Sporadic genetic basis of Alzheimer disease, 618
Squalene
    in cholesterol synthesis, 224, 226f
    synthesis of, 227f
Squalene epoxidase, in cholesterol synthesis, 224, 227f
Squamous cell carcinoma, 644
SR-B1. See Scavenger receptor B1
SRC-1 coactivator, 457, 457t

Src homology 2 (SH2) domains
    in insulin signal transmission, 451–453, 452f
    in Jak/STAT pathway, 453, 453f
SRP. See Signal recognition particle
SRP-R, 494
SRS-A. See Slow-reacting substance of anaphylaxis
ssDNA. See Single-stranded DNA
Stabilization, of unfolded or partially folded inter-
        mediates, 496
STAR. See Steroidogenic acute regulatory protein
Starch, 117, 118f
    glycemic index of, 459
    hydrolysis of, 459
Starling forces, 566
Starvation, 92
    clinical aspects of, 141
    fatty liver and, 218
    ketosis in, 191
    metabolic fuel mobilization in, 137–140, 140f,
        140t
    triacylglycerol redirection and, 215
Statin drugs, 224, 232
STATs (signal transducers and activators of tran-
        scription), 453, 453f
Stearic acid, 122t
Steely hair disease (Menkes disease), 573
Stem cell biology, 4
Stem cell factor, 593
Stem cells, 623
    differentiation of to red blood cells, erythropoi-
        etin in regulation of, 594–595, 595f
Stereochemical (-sn-) numbering system, 124,
        124f
Stereoisomers, 127, 127f. See also Isomerism of
        steroids
Steroid nucleus, 126–127, 126f, 127f
Steroid receptor coactivator 1 (SRC-1 coactivator),
        457, 457t
Steroid sulfates, 209
Steroidogenesis. See Steroids
Steroidogenic acute regulatory protein (StAR), 432
Steroids, 126–128, 126f, 127f, 128f. See also spe-
        cific type
    adrenal. See also Glucocorticoids;
        Mineralocorticoids
        synthesis of, 429–432, 430f, 431f
    receptors for, 427
    stereoisomers of, 127, 127f
    storage/secretion of, 441–442, 442t
    synthesis of, 132, 133f, 429–435, 429f, 430f, 431f
    transport of, 442t, 443
Sterol 27-hydroxylase, 228
Sterols, 127
    in membranes, 408
Stickler syndrome, 543
Sticky end ligation/sticky-ended DNA, 389, 389f,
        391f, 425
"Sticky foot," 518
"Sticky patch," in hemoglobin S, 49, 49f
Stoichiometry, 62
Stokes radius, in size exclusion chromatography,
        21
Stop codon, 364–365, 364f
Stop-transfer signal, 495
Strain, catalysis by, 53–54
Streptokinase, 60, 589, 589f
Streptomycin, 116
Striated muscle, 545, 546f. See also Cardiac
        muscle; Skeletal muscle
    actin-myosin interactions in, 558t

Strong acids, 10
Strong bases, 10
Structural proteins, 527
Structure-activity relationships (SARs), 89
Stuart-Prower factor (factor X), 584f, 585, 585t
    activation of, 584f, 585
    coumarin drugs affecting, 587–588
Substrate analogs, competitive inhibition by,
        70, 70f
Substrate level, phosphorylations at, 107f, 106
Substrate shuttles, 111, 110f, 111f
    coenzymes as, 52
Substrate specificity, of cytochrome P450 isoforms,
        610
Substrates, 51
    competitive inhibitors resembling, 70, 70f
    concentration of, enzyme-catalyzed reaction
        rate affected by, 66–67, 66f, 67f
        Hill model of, 69–70, 69f
        Michaelis-Menten model of, 67–69, 68f
    conformational changes in enzymes caused by,
        54, 54f
    multiple, 72–73
α subunit, 494, 495
Subunits of V cholerae endotoxin, 620
Succinate, 144, 145f
Succinate dehydrogenase, 100, 144, 145f
    inhibition of, 70, 70f
Succinate Q reductase, 103, 104, 105f, 106f
Succinate semialdehyde, 269, 269f
Succinate thiokinase (succinyl-CoA synthetase),
        144, 145f
Succinic acid, pK/pKa value of, 13t
Succinyl-CoA, in heme synthesis, 272–284, 273f,
        274f, 276f
Succinyl-CoA-acetoacetate-CoA transferase (thio-
        phorase), 144, 189, 189f
Succinyl-CoA synthetase (succinate thiokinase),
        144, 145f
Sucrase-isomaltase complex, 460
Sucrose, 117, 117f, 117t
    glycemic index of, 459
Sugarcode of life, 506
Sugars. See also Carbohydrates
    amino (hexosamines), 116, 117f
        glucose as precursor of, 180, 181f
        in glycosaminoglycans, 119, 180, 181f
        in glycosphingolipids, 180, 181f
        interrelationships in metabolism of, 181f
    classification of, 113, 114t
    deoxy, 116, 117f
    "invert," 117
    isomerism of, 113–115, 114f, 115f
    nucleotide, in glycoprotein biosynthesis,
        508–509, 508t
"Suicide enzyme," cyclooxygenase as, 200–201
Sulfatases, 536
Sulfate, 535
    active (adenosine 3′-phosphate-5′-
        phosphosulfate), 288, 288f, 612
    in glycoproteins, 508
    in mucins, 510
Sulfatide, 126
Sulfation, of xenobiotics, 611–612
Sulfo(galacto)-glycerolipids, 209
Sulfogalactosylceramide, 209
    accumulation of, 211
Sulfonamides, hemolytic anemia precipitated by,
        598
Sulfonylurea drugs, 191

Sulfotransferases, in glycosaminoglycan synthesis,
        535
Sunlight. See Ultraviolet light
Supercoils, DNA, 304–305, 327, 328f
Superoxide, 484
Superoxide anion free radical, 102, 597, 597t. See
        also Free radicals
    production of in respiratory burst, 606
Superoxide dismutase, 102, 128, 597, 597t, 606
Supersecondary structures, 35
Suppressor mutations, 358–359
Suppressor tRNA, 358–359
Surfactant, 124, 205
    deficiency of, 124, 209–210
Swainsonine, 518, 518t
Symport systems, 415, 415f
Symptomatic treatment, 629
Syn conformers, 286, 286f
Synaptic function, 619
Synaptic vesicles, 502
Synaptobrevin, 502
Syntaxin, 502
Synthetic biology, 4
Systems biology, 4, 90

T lymphocytes, 576, 617
t-PA. See Tissue plasminogen activator
t-SNARE proteins, 500, 502
T (taut) state, of hemoglobin
    2,3-bisphosphoglycerate stabilizing, 48, 47f
    oxygenation and, 46, 46f, 47f
T tubular system, in cardiac muscle, 553
T-type calcium channel, 554
$t_{1/2}$. See Half-life
T_3. See Triiodothyronine
T_4. See Thyroxine
T_4, decreased, 635
TAFs. See TBP-associated factors
Tag SNPs, 87
Talin, 532, 532f
Tandem, 405
Tandem mass spectrometry, 28
    neonatal blood analysis with, 246
Tangier disease, 232t
TAP SNPs, 87
TaqI, 389t
Target, 519
Target cells, 425–426, 426t
    receptors for, 426, 427f
Targeted gene disruption/knockout, 400
Tarui disease, 160t
TATA binding protein, 339, 343, 344
TATA box, in transcription control, 339, 340, 340f,
        341f, 344t
Taurochenodeoxycholic acid, synthesis of, 231f
Taut (T) state, of hemoglobin
    2,3-bisphosphoglycerate stabilizing, 48, 48f
    oxygenation and, 46, 46f, 47f
Tay-Sachs disease, 210t
TBG. See Thyroxine-binding globulin
tblastn, 88
tblastx, 88
TBP. See TATA binding protein
TBP-associated factors, 339, 343, 344–345
TEBG. See Testosterone-estrogen-binding
        globulin
Telomerase, 315
Telomeres, 315, 316f
Temperature
    chemical reaction rate affected by, 64, 64f

Temperature (*Continued*)
  enzyme-catalyzed reaction rate affected by, 66
  in fluid mosaic model of membrane structure, 412
Temperature coefficient (Q_{10}), enzyme-catalyzed reactions and, 66
Template binding, in transcription, 336, 336f
Template strand DNA, 303, 305, 305f
  transcription of in RNA synthesis, 335–336, 336f
Tenase complex, 585
Tenofovir disoproxil fumarate, 73
Terminal transferase, 390t, 405
Termination
  chain
    in glycosaminoglycan synthesis, 535
    in transcription cycle, 336, 336f
  of protein synthesis, 365, 364f
  of RNA synthesis, 336, 336f, 338, 338f
Termination signals, 354
  or bacterial transcription, 338–339, 339f
  for eukaryotic transcription, 343
Tertiary hypothyroidism, 635
Tertiary structure, 32, 35
  stabilizing factors and, 36–37
Testes, hormones produced by, 427, 432, 433f. *See also* specific type
Testosterone, 429f, 430f
  binding of, 442t, 443
  metabolism of, 432, 433f
  synthesis of, 432, 433f
Testosterone-estrogen-binding globulin (sex hormone-binding globulin), 443, 442t, 569t
Tethering, 499, 500
Tetracycline (tet) resistance genes, 391, 392f
Tetrahedral transition state intermediate, in acid-base catalysis, 54, 55f
Tetrahydrobiopoterin, 237f
Tetrahydrofolate, 477, 477f
Tetraiodothyronine (thyroxine/T_4), 428, 436–437
  storage/secretion of, 441–442, 442t
  synthesis of, 436, 437f
  transport of, 442, 442t
Tetramers
  hemoglobin as, 45
  histone, 312–313
Tetroses, 113, 114t
Tf. *See* Transferrin
TFIIA, 343
TFIIB, 343
TFIID, 339, 343, 344
  in preinitiation complex formation, 346
TFIIE, 343
TFIIF, 343
TFIIH, 343
TFPI. *See* Tissue factor pathway inhibitor
TfR. *See* Transferrin receptor
TGF-β, 531
Thalassemias, 49–50
Thanatophoric dysplasia, 541t, 543
Theca cells, hormones produced by, 432
Theobromine, 288
Theophylline, 288
  hormonal regulation of lipolysis and, 222
Thermodynamics
  biochemical (bioenergetics), 92–96. *See also* ATP
  glycolysis reversal and, 165–167
  laws of, 92–93
    hydrophobic interactions and, 7–8

Thermogenesis, 222f, 223
  diet-induced, 223, 464, 641
Thermogenin, 109, 222f, 223, 641
Thiamin (vitamin B₁), 468t, 474f
  in citric acid cycle, 146
  coenzymes derived from, 52
  deficiency of, 468t, 473–474
  pyruvate metabolism affected by, 154, 155, 473–474
Thiamin diphosphate, 154, 175, 473, 474f
Thiamin pyrophosphate, 52
Thiamin triphosphate, 473
Thick (myosin) filaments, 545–546, 547f
Thickness, of bilayer, 503
Thin (actin) filaments, 545–546, 547f, 548f
Thioesterase, 194
6-Thioguanine, 289, 290f
Thiokinase (acyl-CoA synthetase)
  in fatty acid activation, 184, 185f
  in triacylglycerol synthesis, 207, 221f, 223
Thiol-dependent transglutaminase. *See* Transglutaminase
Thiol ester plasma protein family, 575
Thiolase, 185, 186f, 189
  in mevalonate synthesis, 224, 225f
Thiophorase (succinyl-CoA-acetoacetate-CoA transferase), 144, 188f, 189
Thioredoxin, 295
Thioredoxin reductase, 295, 297f
Threonine, 15t
  catabolism of, 251
  phosphorylated, 266
  requirements for, 465
Thrombin, 585, 586, 587f
  antithrombin III affecting, 587
  circulating levels of, 587
  conserved residues and, 57t
  formation of fibrin and, 585–586, 587f
  in platelet activation, 589, 590f
  from prothrombin, factor Xa activation of, 585
Thrombolysis
  laboratory tests in evaluation of, 591
  t-PA and streptokinase in, 589, 589f
Thrombomodulin, in blood coagulation, 586t, 587, 591, 591t
Thrombopoietin, 593
Thrombosis, 583–591. *See also* Coagulation
  antithrombin III in prevention of, 587
  circulating thrombin levels and, 587
  endothelial cell products in, 591, 591t
  hyperhomocysteinemia and, folic acid supplements in prevention of, 478
  phases of, 583
  in protein C or protein S deficiency, 587
  t-PA and streptokinase in management of, 589, 589f
  types of thrombi and, 583
Thromboxane A2, 123f
  in platelet activation, 590, 590f
Thromboxanes, 122, 123f, 193, 200
  clinical significance of, 2–6
  cyclooxygenase pathway in formation of, 200, 201f
Thrombus, 638
Thymidine, 287t
  base pairing of in DNA, 302, 303, 304f
Thymidine monophosphate (TMP), 287t
Thymidylate, 302
Thymine, 287t
Thymine dimers, 644

Thyroglobulin, 436
Thyroid-binding globulin, 442, 569t
Thyroid hormone receptor-associated proteins (TRAPs), 457, 457t
Thyroid hormone response element, 445t
  storage/secretion of, 441, 442t
Thyroid hormones, 427–428
  in lipolysis, 221, 221f
  receptors for, 427, 455
  synthesis of, 436, 437f
  transport of, 442, 442t
Thyroid-stimulating hormone (TSH), 427, 428, 429f, 437–438
Thyroperoxidase, 438
Thyrotropin-releasing hormone (TRH), 428, 429f, 635
Thyroxine (T_4), 428, 436
  storage/secretion of, 441–442, 442t
  synthesis of, 436–437, 437f
  transport of, 442, 442t
Thyroxine-binding globulin, 442, 442t
TIF2 coactivator, 457, 457t
Tightly bound inhibitors, 71–72
Tiglyl-CoA, catabolism of, 260f
TIM. *See* Translocase-of-the-inner membrane
Time of flight mass spectrometry, 27
Timnodonic acid, 123t
Tin, 480t
Tissue damage in diabetes mellitus, 519
Tissue differentiation, retinoic acid in, 469
Tissue factor (factor III), 584, 584f, 585t
Tissue factor complex, 584
Tissue factor pathway inhibitor, 584
Tissue plasminogen activator (alteplase/t-PA), 60, 588, 589f, 591t
Tissue-specific gene expression, 379
Titin, 553t
Tm. *See* Melting temperature/transition temperature
TMP (thymidine monophosphate), 287f, 287t
Tocopherol, 468t, 471–472, 471f, 472f. *See also* Vitamin E
  as antioxidant, 102, 128, 471, 472f
  deficiency of, 468t
α-Tocopherol, 485
Tocotrienol, 471–472, 471f. *See also* Vitamin E
Tolbutamide, 191
TOM. *See* Translocase-of-the-outer membrane
Tophi, 631
Topogenic sequences, 495
Topoisomerases, DNA, 305, 323t, 327, 328f
Total body iron, 633
Total iron-binding capacity, 571
Total ischemia of myocardium, 638
Totipotent stem cells, 593
Toxemia of pregnancy of ewes, ketosis and, 191
Toxic hyperbilirubinemia, 282
Toxicity, vitamin, 467
Toxicology, 1
Toxins, microbial, 418
Toxopheroxyl free radical, 471
TpC. *See* Troponin C
TpI. *See* Troponin I
TpT. *See* Troponin T
TRAM (translocating chain-associated membrane) protein, 495
TRAM-1 coactivator, 457, 457t
Trans fatty acids, 122, 200
Trans Golgi network (TGN), 487
Transaldolase, 174, 175

Transaminases. *See* Aminotransferases

Transamination, 133, 133f
  in amino acid carbon skeleton catabolism, 248, 248f, 249f
  citric acid cycle in, 146, 146f
  in urea biosynthesis, 241–242, 241f

Transcortin (corticosteroid-binding globulin), 442, 442t

Transcript profiling, 400–402

Transcription, 304, 343–346, 344t, 345t, 405
  activators and coactivators in control of, 344–345, 344t, 345t
  bacterial promoters in, 338–339, 340f
  control of fidelity and frequency of, 338–343
  eukaryotic promoters in, 339–343, 340f, 341f
  in gene expression regulation, 376–381, 384, 384t. *See also* Gene expression
  hormonal regulation of, 454–458, 445, 445f, 456f, 457f, 457t
  initiation of, 336–337, 336f
  nuclear receptor coregulators in, 455–458, 456t
  retinoic acid in regulation of, 469
  reverse, 405
    in retroviruses, 307, 327
  in RNA synthesis, 304, 305f, 335–337, 336f

Transcription complex, eukaryotic, 343–346, 344t, 345t

Transcription control elements, 344–345, 344t

Transcription domains, definition of, 380–381

Transcription factors, 345, 345t, 593
  nuclear receptor superfamily, 455, 456t, 457f

Transcription start sites, alternative, 386

Transcription unit, 336, 340f

Transcriptional control, 518

Transcriptional intermediary factor 2 (TIF2 coactivator), 457, 457t

Transcriptome information, 402, 405

Transcriptomics, 4

Transcytosis, 502

Transfection, identification of enhancers/regulatory elements and, 379

Transfer RNA (tRNA), 307, 308, 335, 336t, 355–356, 355f. *See also* RNA
  aminoacyl, in protein synthesis, 362
  anticodon region of, 354
  processing and modification of, 349, 351
  suppressor, 358–359

Transferases, 52

Transferrin, 463, 569t, 570, 571, 571f, 571t

Transferrin receptor, 572

Transferrin saturation, 633

Transfusion, ABO blood group and, 602

Transgenic animals, 379, 400, 405
  enhancers/regulatory elements identified in, 379

Transglutaminase, in blood coagulation, 584, 586, 586f, 586t

Transhydrogenase, proton-translocating, 110

Transient insertion signal. *See* Signal peptide

Transition mutations, 356, 356f

Transition state analogs, 54

Transition state intermediate,
  formation during simple chemical reaction, 63f
  tetrahedral, in acid-base catalysis, 54, 55f

Transition states, 63–64

Transition temperature/melting temperature (Tm), 303–304, 412

Transketolase, 174, 176f, 181
  erythrocyte, in thiamin nutritional status assessment, 474

thiamin diphosphate in reactions involving, 175, 181, 474

Translation, 353, 405

Translation arrest, 310

Translocase-of-the-inner membrane, 488

Translocase-of-the-outer membrane, 488

Translocating chain-associated membrane (TRAM) protein, 495

Translocation
  into lumen, 493
  out of ER, 516

Translocation, protein, 22f, 488

Translocation complexes, 488

Translocon, 495

Transmembrane proteins, 410
  ion channels as, 416, 417f, 417t
  in red cells, 600–601, 600f, 600t, 601f

Transmembrane signaling, 406, 422, 589
  in platelet activation, 590, 590f

Transmigration, of neutrophils, 521

Transmissible spongiform encephalopathies (prion diseases), 40

Transport proteins, 442t, 449, 569t

Transport systems/transporters. *See also* specific type
  active, 413, 414–415, 414f, 414t, 415f
  ATP-binding cassette, 217, 217f
  in cotranslational insertion, 495, 496f
  disorders associated with mutations in genes encoding, 504, 504t
  facilitated diffusion, 413, 414t, 414f, 415–416, 416f
  glucose. *See* Glucose transporters
  in inner mitochondrial membrane, 109–112, 109f–111f
  membrane, 413–416, 415f
  for nucleotide sugars, 508

Transport vesicles, 487, 495, 499–502, 500t, 501f
  defined, 499
  in intracellular traffic, 499
  in vesicle coating, 500–502

Transporters, 409, 414–416, 495
  comparison with ion channels, 415t

Transposition, 321
  retroposons/retrotransposons and, 318

Transthyretin, 569t, 576

Transverse asymmetry, 503

Transverse movement, of lipids across membrane, 410

Transversion mutations, 356, 356f

TRAPs, 457, 457t

Trauma, protein loss and, 465

TRE. *See* Thyroid hormone response element

Trehalase, 460

Trehalose, 117t

TRH. *See* Thyrotropin-releasing hormone

Triacylglycerols (triglycerides), 124, 124f, 212, 222
  in adipose tissue, 131
  digestion and absorption of, 460, 461f
  excess of. *See* Hypertriacylglycerolemia
  interconvertability of, 137
  in lipoprotein core, 212, 213f
  metabolism of, 132, 133f, 134, 135f
    in adipose tissue, 220–221, 220f
    fatty liver and, 218, 219f
    hepatic, 218, 219f
    high-density lipoproteins in, 216–217, 217f
    hydrolysis in, 205
  reduction of serum levels of, drugs for, 232
  synthesis of, 206f, 207
  transport of, 214–215, 214f, 215f, 216f

Tricarboxylate anions, transporter systems for lipogenesis regulation and, 197

Tricarboxylic acid cycle. *See* Citric acid cycle

Triglycerides. *See* Triacylglycerols

Triiodothyronine (T3), 428, 436
  storage/secretion of, 441–442, 442t
  synthesis of, 436, 437f
  transport of, 442, 442t

Trimethoprim, 477

Trinucleotide repeat expansions, 319

Triokinase, 179, 179f

Triose phosphate isomerase, 34f, 35

Triose phosphates, acylation of, 132

Trioses, 113, 114t

Triphosphates, nucleoside, 286, 286f

Triple helix structure, of collagen, 40–41, 41f, 527–530, 528f

Triplet code, genetic code as, 353, 354t

tRNA. *See* Transfer RNA

Tropocollagen, 40, 41f

Tropoelastin, 530

Tropomyosin, 545–546, 548f, 550
  in red cell membranes, 600t
  as striated muscle inhibitor, 551

Troponin C, 550

Troponin I, 550

Troponin T, 550, 638

Troponin/troponin complex, 545–546, 548f, 550
  as striated muscle inhibitor, 551

Troponin, in diagnosis of myocardial infarction, 59, 60

Trypsin, 60, 462
  conserved residues and, 57t
  in digestion, 462

Trypsinogen, 462

Tryptophan, 16t, 264–265, 474–475
  catabolism of, 257, 256f, 257f
  deficiency of, 475
  niacin synthesized from, 474
  permeability coefficient of, 409f
  requirements for, 465

L-Tryptophan dioxygenase (tryptophan pyrrolase), 101

Tryptophan oxygenase/L-tryptophan oxygenase (tryptophan pyrolase), 101, 257, 256f

TSEs. *See* Transmissible spongiform encephalopathies

TSH, 635. *See also* Thyroid-stimulating hormone

Tumor cells, migration of, hyaluronic acid and, 539

Tumor progression, 623

Tumor promoter, 623

Tumor suppressor genes, p53, 333, 623

Tunicamycin, 518, 518t

Twin lamb disease. *See* Pregnancy toxemia of ewes

Twitch fibers, slow (red) and fast (white), 561, 561t

Two-dimensional electrophoresis, protein expression and, 29

TXs. *See* Thromboxanes

Tyk-2, in Jak-STAT pathway, 453

Type A response, in gene expression, 369, 370f

Type B response, in gene expression, 369–370, 370f

Type C response, in gene expression, 370, 370f

L-type calcium channel, 554

Tyrosine, 15t, 16t, 266, 268f
  catabolism of, 253f, 254
  epinephrine and norepinephrine formed from, 268f
  in hemoglobin M, 49
  in hormone synthesis, 428, 429f
  phosphorylated, 266

Tyrosine (*Continued*)
  requirements for, 465
  synthesis of, 237, 237f
Tyrosine aminotransferase, defect in, in
        tyrosinemia, 254
Tyrosine hydroxylase, catecholamine biosynthesis
        and, 435–436, 436f
Tyrosine kinase
  in insulin signal transmission, 450–454, 452f
  in Jak/STAT pathway, 453, 453f
Tyrosinemia, 254
Tyrosinosis, 254

Ubiquination, 22f, 239–240
  of misfolded proteins, 498, 499f
Ubiquinone (Q/coenzyme Q), 128
  in cholesterol synthesis, 225, 226f
  in respiratory chain, 103, 105f, 106f
Ubiquitin, in protein degradation, 240f, 498–499,
        499f
UDP-glucose. *See* Uridine diphosphate glucose
UDPGal. *See* Uridine diphosphate galactose
UDPGlc. *See* Uridine diphosphate glucose
UFA (unesterified fatty acids). *See* Free fatty acids
Ulcers, 459
Ultraviolet light
  nucleotide absorption of, 288
  nucleotide excision-repair of DNA damage
        caused by, 332
  vitamin D synthesis and, 469f, 470
UMP (uridine monophosphate), 287f, 287t
Uncleaved insertion sequences, 495
Uncouplers/uncoupling proteins
  on respiratory chain, 108–109, 109f
  undernutrition and, 464
Uncoupling protein, 109
Undernutrition, 459, 463–464
Unequal crossover, 320, 321f
Unesterified fatty acids. *See* Free fatty acids
Unfolded protein response, 497
Uniport systems, 414–415, 414f
UniProt, 86
Unique-sequence (nonrepetitive) DNA, 318
Universal donor/universal recipient, 602
Unsaturated fatty acids, 121, 122, 123t. *See also*
        Fatty acids
  cis double bonds in, 122–123, 123f
  dietary, cholesterol levels affected by, 230–231
  eicosanoids formed from, 193, 200, 201f, 202f
  essential, 198, 198f, 200
    abnormal metabolism of, 202
    deficiency of, 200, 201–202
    prostaglandin production and, 193
  in membranes, 408, 409f
  metabolism of, 198–200
  oxidation of, 187
  structures of, 198f
  synthesis of, 199, 199f
Unwinding, DNA, 324
  RNA synthesis and, 337–338
Uracil, 287t
  deoxyribonucleosides of, in pyrimidine synthe-
        sis, 295–297, 298f
Uraciluria-thyminuria, 292, 300f
Urate, as antioxidant, 128
Urea, 627
  amino acid metabolism and, 133, 133f
  nitrogen catabolism producing, 240–241, 243–245
  permeability coefficient of, 409f
  synthesis of, 241–242, 241f, 242f

metabolic disorders associated with, 245–246
  gene therapy for, 246
Urea cycle disorders, 245
Uric acid, 288, 288f
  daily excretion of, 631
  purine catabolism in formation of, 297, 299f
Uricosuric drugs, 632
Uridine, 286f, 287t
Uridine diphosphate galactose (UDPGal), 180,
        508, 508t
Uridine diphosphate galactose (UDPGal)
        4-epimerase, 180, 180f
  inherited defects in, 182
Uridine diphosphate glucose (UDP/UDPGlc), 157,
        158f, 508, 508t
  in glycogen biosynthesis, 157, 158f
Uridine diphosphate glucose dehydrogenase, 177,
        178f
Uridine diphosphate glucose pyrophosphorylase,
        177, 178f
  in glycogen biosynthesis, 157, 158f
Uridine diphosphate-glucuronate/glucuronic acid,
        177, 178f, 289
Uridine diphosphate N-acetylgalactosamine
        (UDP-GalNAc), 508t
Uridine diphosphate N-acetylglucosamine (UDP-
        GlcNAc), 508t
Uridine diphosphate xylose (UDP-Xyl), 508t
Uridine monophosphate (UMP), 287f, 287t
Uridine triphosphate (UTP), in glycogen
        biosynthesis, 157, 158f
Uridyl transferase deficiency, 182
Urine, 627, 631
Urobilinogens
  conjugated bilirubin reduced to, 280
  in jaundice, 283, 283t
  normal values for, 283t
Urocanic aciduria, 250
Urokinase, 589, 589f
Uronic acid pathway, 174, 177, 178f
  disruption of, 181
Uronic acids, 119
  in heparin, 536, 536f
Uroporphyrinogen decarboxylase, 273, 274f, 275f
  in porphyria, 277t
Uroporphyrinogen I, 273, 274f, 275f
Uroporphyrinogen I synthase, in porphyria, 277t
Uroporphyrinogen III, 273, 274f, 275f
Uroporphyrins, 272, 272f
  spectrophotometry in detection of, 275
UTP, in phosphorylation, 96

V region/segment. *See* Variable regions/segments
v-SNARE proteins, 499–500, 502
  pair with cognate t-SNAREs, 500
Valeric acid, 122t
Valine, 15t
  catabolism of, 258, 259f, 260f
  interconversion of, 238
  requirements for, 465
Valinomycin, 110
Vanadium, 480t
Van der Waals forces, 8
Variable numbers of tandemly repeated units
        (VNTRs), in forensic medicine, 400
Variable regions/segments, 578f
  gene for, 579
    DNA rearrangement and, 321, 385, 580
    immunoglobulin heavy chain, 577, 577f, 579,
        578f

immunoglobulin light chain, 322, 385, 577,
        577f, 579, 578f
  of immunoglobulins, 577, 579
Vascular system, nitric oxide affecting, 559–560,
        559f, 560t
Vasodilators, 545
  nitric oxide as, 559–560, 559f, 560t
VDRE. *See* Vitamin D response element
Vector, 405
  cloning, 390–391, 391f, 392f, 391t, 405
  expression, 392
Vegetarian diet, vitamin B12 deficiency and, 476
Velocity
  initial, 66
    inhibitors affecting, 71, 71f
  maximal (*V*max)
    allosteric effects on, 78
    inhibitors affecting, 71, 71f
    Michaelis-Menten equation in determination
        of, 67–69, 68f
    substrate concentration and, 66–67, 67f
Venoms, 598
Very low carbohydrate diets, weight loss from, 173
Very low density lipoprotein receptor, 214, 215
Very low density lipoproteins, 135, 212, 213t, 230
  in fed state, 138
  hepatic secretion of, dietary and hormonal
        status and, 218, 219f
  metabolism of, 135, 135f, 214–216, 216f
  in triacylglycerol transport, 220, 214f, 216f
Vesicles
  coating, 500–502, 501f
    brefeldin A affecting, 502
  genetic approachs to study, in yeast, 499
  secretory, 487, 489f
  targeting, 49–500, 501f
  transport, 487, 499–502, 500t, 501f
  types and functions, 499t
vi. *See* Initial velocity
Vimentins, 563t, 564
Vinblastine, 563
Vinculin, 532, 532f
Viral infection, 498, 627
Viral oncogenes. *See* Oncogenes
Virtual cells, 90
Viruses
  glycan binding of, 524–525
  host cell protein synthesis affected by, 365–366,
        367f
Viscous mucus, 625
Vision, vitamin A in, 468t, 469, 469f
Vitamin A, 468–470, 468t, 469f, 469f
  deficiency of, 468t, 470
  excess/toxicity of, 470
  functions of, 468t, 469
  in vision, 468t, 469
Vitamin B complex. *See also* specific vitamin
  in citric acid cycle, 146
  coenzymes derived from, 52, 53f
Vitamin B1 (thiamin), 468t, 473–474, 474f
  in citric acid cycle, 146
  coenzymes derived from, 52
  deficiency of, 468t, 473–474
    pyruvate metabolism affected by, 154, 155,
        473–474
Vitamin B2 (riboflavin), 98, 468t, 474
  in citric acid cycle, 146
  coenzymes derived from, 52, 474
  deficiency of, 468t, 474
  dehydrogenases dependent on, 99–100

Vitamin B6 (pyridoxine/pyridoxal/pyridoxamine), 468t, 475–476, 475f
  deficiency of, 468t, 475–476
    xanthurenate excretion in, 257, 257f
  excess/toxicity of, 476
Vitamin B12 (cobalamin), 468t, 476–477, 476f
  absorption of, 476
    intrinsic factor in, 462, 476
  deficiency of, 468t, 477
    functional folate deficiency and, 477–478
  in methylmalonicaciduria, 167
Vitamin B12-dependent enzymes, 476–477, 476f
Vitamin C (ascorbic acid), 174, 468t, 479–480, 479f
  as antioxidant, 128
  in collagen synthesis, 41, 480, 527
  deficiency of, 468t, 480
    collagen affected in, 41, 480, 530
  iron absorption and, 463, 480
  supplemental, 480
Vitamin D, 467, 468t, 470–471
  in calcium absorption, 462, 470
  deficiency of, 468t, 470, 471
  ergosterol as precursor for, 127, 127f
  excess/toxicity of, 471
  metabolism of, 470, 471f
  receptor for, 455
Vitamin D-binding protein, 434
Vitamin D receptor-interacting proteins (DRIPs), 457, 457t
Vitamin D response element, 445t
Vitamin D2 (ergocalciferol), 470
Vitamin D3 (cholecalciferol)
  synthesis of in skin, 434, 435f, 470, 469f
  in vitamin D metabolism, 469f, 470, 471f
Vitamin E, 468t, 471–472, 471f, 485, 625
  as antioxidant, 102, 128, 471, 472f
  deficiency of, 468t, 472
Vitamin H. See Biotin
Vitamin K, 468t, 472–473, 472f, 473f, 587–588
  calcium-binding proteins and, 473, 473f
  in coagulation, 472, 472f
    coumarin anticoagulants affecting, 587–588
  deficiency of, 468t
Vitamin K hydroquinone, 473, 473f
Vitamins, 2, 467–480, 468t. See also specific vitamins
  in citric acid cycle, 146
  digestion and absorption of, 462–463
  lipid- (fat-) soluble, 467, 468–473
    absorption of, 460
  metabolic functions of, 467, 468t
  water-soluble, 467, 473–480
VLA-1/VLA-5/VLA-6, 605t
VLDL. See Very low density lipoproteins

Vmax. See Maximal velocity
VNTRs. See Variable numbers of tandemly repeated units
Voltage-gated channels, 416, 417, 556t
von Gierke disease, 160t, 298
Von Hippel-Lindau syndrome, 240
von Willebrand disease, 588
von Willebrand factor, 588
  in platelet activation, 589

Warfarin, 472, 588
  phenobarbital interaction and, cytochrome P450 induction affecting, 611
  vitamin K affected by, 472
Water, 2, 6–9
  as biologic solvent, 6, 7f
  biomolecular structure and, 7–8, 7t
  dissociation of, 9
  in hydrogen bonds, 6, 7f
  as nucleophile, 8–9
  permeability coefficient of, 409f
  structure of, 6, 7f
Water channels, 418
Water solubility, of xenobiotics, metabolism and, 609
Watson-Crick base pairing, 8, 302
Waxes, 121
Weak acids, 10
  buffering capacity of, 11–12, 12f
  dissociation constants for, 10–11, 12
  Henderson-Hasselbalch equation describing behavior of, 11–12, 12f
  physiologic significance of, 10–11
  pK/pKa values of, 10–13, 13t
Weak bases, 10
Wernicke-Korsakoff syndrome, 468t
Wernicke encephalopathy, 473
Western blot transfer procedure, 393, 393f, 405
White blood cells, 593, 605–607. See also specific type
  growth factors regulating production of, 595
  recombinant DNA technology in study of, 607
White thrombus, 583
White (fast) twitch fibers, 561, 561t
Williams-Beuren syndrome, 530
Wilson disease, 423t, 573
  ceruloplasmin levels in, 573
  gene mutations in, 423t, 574
  methylhistidine in, 262
Wobble, 356
Women, iron needs of, 571

X-linked degenerative disease, 628
X-linked disorders, RFLPs in diagnosis of, 399

X-ray crystallography, Laue, 38
X-ray diffraction and crystallography, protein structure demonstrated by, 37–38
Xanthine, 288, 288f
Xanthine oxidase, 99
  deficiency of, hypouricemia and, 299
Xanthurenate, excretion of in vitamin B6 deficiency, 257, 257f
Xenobiotics, metabolism of, 609–615
  conjugation in, 609, 611–613
  cytochrome P450 system/hydroxylation in, 610–611, 612t
  factors affecting, 613
  pharmacogenetics in drug research and, 614–615
  responses to, 613–614, 613t
    toxic, 613–614, 614f
Xeroderma pigmentosum, 332
  case study, 643–645
Xerophthalmia, vitamin A deficiency in, 468t, 470
XP. See Xeroderma pigmentosum
D-Xylose, 115f, 116t
Xylose, in glycoproteins, 508t
D-Xylulose, 115f
L-Xylulose, 116t
  accumulation of in essential pentosuria, 181

Yeast artificial chromosome (YAC) vector, 390, 391t
Yeast cells, mitochondrial protein import studied in, 488

Z line, 545, 546f, 547f
Zellweger (cerebrohepatorenal) syndrome, 191, 492–493, 493t
Zinc, 480t
  α_2-macroglobulin transport of, 575
Zinc finger motif, 381, 381t, 382, 383f
  in DNA-binding domain, 454
Zona fasciculata, steroid synthesis in, 430
Zona glomerulosa, mineralocorticoid synthesis in, 430
Zona pellucida, glycoproteins in, 520
Zona reticularis, steroid synthesis in, 430
Zoology, 1
ZP. See Zona pellucida
ZP1-3 proteins, 520
Zwitterions, 16–17
Zymogens, 79, 462
  in blood coagulation, 583, 584, 586t
  rapid response to physiologic demand and, 79
ZZ genotype, 574